T0406140

Handbook of Nonlocal Continuum Mechanics for Materials and Structures

George Z. Voyiadjis
Editor

Handbook of Nonlocal Continuum Mechanics for Materials and Structures

Volume 1

With 572 Figures and 59 Tables

Editor
George Z. Voyiadjis
Department of Civil and Environmental Engineering
Louisiana State University
Baton Rouge, LA, USA

ISBN 978-3-319-58727-1 ISBN 978-3-319-58729-5 (eBook)
ISBN 978-3-319-58728-8 (print and electronic bundle)
https://doi.org/10.1007/978-3-319-58729-5

Library of Congress Control Number: 2018960886

This Springer imprint is published by the registered company Springer Nature Switzerland AG
The registered company address is: Gewerbestrasse 11, 6330 Cham, Switzerland

Preface

This handbook discusses the integral and gradient formulations of nonlocality, computational aspects, micromechanical considerations, and comparison of approaches and emphasizes recent developments in the bridging of material length and time scales. The contributions in this handbook are on nonlocal continuum plasticity in terms of the experimental, theoretical, and numerical investigations. This handbook presents a comprehensive treatment of the most important areas of nonlocality (integral and gradient) of time-dependent inelastic deformation behavior and heat transfer responses. The following aspects of the advanced material modeling are presented: enhanced (generalized) continuum mechanics, microscopic mechanisms and micro-mechanical aspects responsible for size effect and micro-scale heat transfer, thermodynamic framework, and multiscale computational aspects with detailed nonlocal computational algorithms in the context of finite element. Measures for length scales are introduced together with their appropriate evolution relations.

The work addresses the thermal and mechanical responses of small-scale metallic compounds under fast transient processes based on small and large deformation framework. This handbook presents a comprehensive treatment of the most important areas of nonlocality (integral and gradient) of time-dependent inelastic deformation behavior and heat transfer responses.

From the experimental aspects, a wide spectrum of materials are included in these chapters. For the case of glassy polymers, their nanostructural responses and nanoindentation measurements were discussed in detail by Voyiadjis et al. For metals, the hydrogen embrittlement cracking was discussed by Yonezu and Chen, whereas its size effect of nanoindentation was elucidated by Voyiadjis et al. Composites and their nonlocal behaviors have received particular attention, including the studies of their cracking initiation (Xu et al.), interface stability (Meng et al.), buckling (Chen et al.), and dynamic properties (Tomar et al.). Indentation and its nonlocal characteristics were also emphasized through modeling (Mills et al., Liu et al.) and fatigue testing (Xu et al.).

The micromorphic approach has aroused strong interest from the materials science and computational mechanics communities because of its regularization power in the context of softening plasticity and damage. The micromorphic and Cosserat theories in gradient plasticity are introduced and analyzed in these chapters to address the instabilities in the materials and structures. The micromorphic approach for the gradient continuum plasticity/damage and crystal plasticity is elucidated by Forest et al. The micropolar theory for the crystal plasticity is also introduced (Mayeur et al.). The micromorphic and Cosserat approach are applied to localization in geomaterials (Stefanou et al.) and dispersion of waves in metamaterials (Madeo et al.) in these chapters.

The section on "Mathematical Methods in Nonlocal Continuum Mechanics" combines the original concepts for the description of nonlocal effects, including applications to nonstandard mathematical models. Both quasi-static and dynamic processes are included, together with micro- and macro-level of description, and furthermore multiphysics (in the sense of, e.g., thermomechanical interaction) is covered accordingly. In the chapter by Lazopoulos et al., the fractional calculus is applied to obtain the space-fractional nonlocal continuum mechanics formulation and then applied to the analysis of fractional Zener viscoelastic model. The contribution by Ostoja-Starzewski et al. includes modeling of fractal materials utilizing fractional integrals and application of homogenized continuum mechanics together with the framework of calculus in non-integer dimensional spaces. The work by Sumelka et al. presents implicit time nonlocal modeling of metallic materials including damage anisotropy, and furthermore stress-fractional extensions are suggested. Next, the chapter by Tarasov includes modeling of physical lattices with long-range interactions utilizing exact fractional-order difference operators. Finally, in the chapter by Voyiadjis et al., the strain gradient plasticity is presented with the appropriate flow rules of the grain interior and grain boundary areas within the thermodynamically consistent framework and applied for modeling metallic materials. One can recapitulate that the overall section contents open new fields of investigations of mathematical models for bodies exhibiting strong scale effect.

The section on "Computational Modeling for Gradient Plasticity in Both Temporal and Spatial Scales" introduces a variety of numerical examples for the gradient-enhanced plasticity. The failure mechanisms of metallic materials for high-velocity impact loading are simulated based on the nonlocal approach (Voyiadjis et al.). The gradient plasticity is combined to micro-/mesoscale crystal plasticity (Yalcinkaya et al., Ozdemir et al.) and fracture mechanics (Lancioni et al.). The transverse vibration of microbeams and axial vibration of micro-rods are studied by Civalek et al. using the strain gradient plasticity theory. In addition, in this section, the temperature-driven ductile-to-brittle transition fracture in Ferritic steels is modeled and analyzed (Deliktaş et al.).

Nonlocal peridynamic models for damage and fracture are introduced and analyzed in these chapters. The theory guaranteeing the existence of solution for continuum models of fracture evolution is essential for the development of mesh-independent discretizations. A bond is exhibited based on peridynamic models with damage and softening and shows that they are well-posed evolutions both over the space of Holder continuous functions and Sobolev functions (Jha and Lipton). This feature is used to develop numerical convergence rates in space and time for the associated finite element and finite-difference schemes (Jha and Lipton). These are the first convergence rates for numerical schemes applied to fully nonlinear and nonlocal peridynamic models. A more general state-based peridynamic model is developed for free fracture evolution. In the limit of vanishing nonlocal interaction, this model is shown to converge to the equation of elastic momentum balance away from the crack set (Said et al.). In the final chapter, a state-based and history-dependent dynamic damage model is developed. Several numerical examples are provided illustrating the theory (Said et al.).

This handbook integrates knowledge from the theoretical, numerical, and experimental areas of nonlocal continuum plasticity. This book is focused mainly for graduate students of nonlocal continuum plasticity, researchers in academia and industry who are active or intend to become active in this field, and practicing engineers and scientists who work in this topic and would like to solve problems utilizing the tools offered by nonlocal mechanics. This handbook should serve as an excellent text for a series of graduate courses in mechanical engineering, civil engineering, materials science, engineering mechanics, aerospace engineering, applied mathematics, applied physics, or applied chemistry.

This handbook is basically intended as a textbook for university courses as well as a reference for researchers in this field. It will serve as a timely addition to the literature on nonlocal mechanics and will serve as an invaluable resource to members of the international scientific and industrial communities.

It is hoped that the reader will find this handbook a useful resource as he/she progresses in their study and research in nonlocal mechanics. Each of the individual sections of this handbook could be considered as a compact self-contained mini-book right under its own title. However, these topics are presented in relation to the basic principles of nonlocal mechanics.

What is finally presented in the handbook is the work contributed by celebrated international experts for their best knowledge and practices on specific and related topics in nonlocal mechanics.

The editor would like to thank all the contributors who wrote chapters for this handbook. Finally, the editor would like to acknowledge the help and support of his family members and the editors at Springer who made this handbook possible.

Baton Rouge, USA Dr. George Z. Voyiadjis
December 2018

Contents

Volume 1

Volume 2

About the Editor

George Z. Voyiadjis is the Boyd Professor at the Louisiana State University in the Department of Civil and Environmental Engineering. This is the highest professorial rank awarded by the Louisiana State University System. He is also the holder of the Freeport-McMoRan Endowed Chair in Engineering. He joined the faculty of Louisiana State University in 1980. He is currently the Chair of the Department of Civil and Environmental Engineering. He holds this position since February of 2001. He currently also serves since 2012 as the Director of the Louisiana State University Center for GeoInformatics (LSU C4G; http://c4gnet.lsu.edu/c4g/).

Voyiadjis is a Foreign Member of both the Polish Academy of Sciences, Division IV (Technical Sciences), and the National Academy of Engineering of Korea. He is the recipient of the 2008 Nathan M. Newmark Medal of the American Society of Civil Engineers and the 2012 Khan International Medal for outstanding lifelong contribution to the field of plasticity. He was also the recipient of the Medal for his significant contribution to Continuum Damage Mechanics, presented to him during the Second International Conference on Damage Mechanics (ICDM2), Troyes, France, July, 2015. This is sponsored by the *International Journal of Damage Mechanics* and is held every 3 years.

Voyiadjis was honored in April of 2012 by the International Symposium on "Modeling Material Behavior at Multiple Scales" sponsored by Hanyang University, Seoul, Korea, chaired by T. Park and X. Chen (with a dedicated special issue in the *Journal of Engineering Materials and Technology* of the ASME). He was also honored by an International Mini-Symposium on "Multiscale and Mechanism Oriented Models:

Computations and Experiments" sponsored by the International Symposium on Plasticity and Its Current Applications, chaired by V. Tomar and X. Chen, in January 2013.

He is a Distinguished Member of the American Society of Civil Engineers; Fellow of the American Society of Mechanical Engineers, the Society of Engineering Science, the American Academy of Mechanics and the Engineering Mechanics Institute of ASCE; and Associate Fellow of the American Institute of Aeronautics and Astronautics. He was on the Board of Governors of the Engineering Mechanics Institute of the American Society of Civil Engineers, and Past President of the Board of Directors of the Society of Engineering Science. He was also the Chair of the Executive Committee of the Materials Division (MD) of the American Society of Mechanical Engineers. Dr. Voyiadjis is the Founding Chief Editor of the Journal of Nanomechanics and Micromechanics of the ASCE and is on the editorial board of numerous engineering journals. He was also selected by Korea Science and Engineering Foundation (KOSEF) as one of the only two World-Class University foreign scholars in the area of civil and architectural engineering to work on nanofusion in civil engineering. This is a multimillion research grant.

Voyiadjis' primary research interest is in plasticity and damage mechanics of metals, metal matrix composites, polymers, and ceramics with emphasis on the theoretical modeling, numerical simulation of material behavior, and experimental correlation. Research activities of particular interest encompass macro-mechanical and micro-mechanical constitutive modeling, experimental procedures for quantification of crack densities, inelastic behavior, thermal effects, interfaces, damage, failure, fracture, impact, and numerical modeling.

Dr. Voyiadjis' research has been performed on developing numerical models that aim at simulating the damage and dynamic failure response of advanced engineering materials and structures under high-speed impact loading conditions. This work will guide the development of design criteria and fabrication processes of high-performance materials and structures under severe loading conditions. Emphasis is placed on survivability area that aims to develop and field a contingency armor that is thin and lightweight, but with a very high level of an overpressure protection system that provides low penetration depths. The formation of cracks and voids in the adiabatic shear bands, which are the precursors to fracture, is mainly investigated.

He has 2 patents, over 320 refereed journal articles, and 19 books (11 as editor) to his credit. He gave over 400 presentations as plenary, keynote, and invited speaker as well as other talks. Over 62 graduate students (36 Ph.D.) completed their degrees under his direction. He has also supervised numerous postdoctoral associates. Voyiadjis has been extremely successful in securing more than $25.0 million in research funds as a principal investigator/investigator from the National Science Foundation, the Department of Defense, the Air Force Office of Scientific Research, the Department of Transportation, and major companies such as IBM and Martin Marietta.

He has been invited to give plenary presentations and keynote lectures in many countries around the world. He has also been invited as guest editor in numerous volumes of the *Journal of Computer Methods in Applied Mechanics and Engineering*, *International Journal of Plasticity*, *Journal of Engineering Mechanics of the ASCE*, and *Journal of Mechanics of Materials*. These special issues focus in the areas of damage mechanics, structures, fracture mechanics, localization, and bridging of length scales.

He has extensive international collaborations with universities in France, the Republic of Korea, and Poland.

Associate Editors

Xi Chen
Department of Earth and Environmental Engineering
Columbia Nanomechanics Research Center
Columbia University
New York, NY, USA

Samuel Forest
Centre des Materiaux
Mines ParisTech CNRS
PSL Research University
Paris, Evry Cedex, France

Wojciech Sumelka
Institute of Structural Engineering
Poznan University of Technology
Poznan, Poland

Contributors

Bekir Akgöz Civil Engineering Department, Division of Mechanics, Akdeniz University, Antalya, Turkey

Burak Aksoylu Department of Mathematics, Wayne State University, Detroit, MI, USA

Florin Bobaru Mechanical and Materials Engineering, University of Nebraska–Lincoln, Lincoln, NE, USA

Pavel Bochev Center for Computing Research, Sandia National Laboratories, Albuquerque, NM, USA

Ferhun C. Caner School of Industrial Engineering, Institute of Energy Technologies, Universitat Politècnica de Catalunya, Barcelona, Spain

Department of Materials Science and Metallurgical Engineering, Universitat Politècnica de Catalunya, Barcelona, Spain

Fatih Celiker Department of Mathematics, Wayne State University, Detroit, MI, USA

Xi Chen Department of Earth and Environmental Engineering, Columbia Nanomechanics Research Center, Columbia University, New York, NY, USA

Youlong Chen International Center for Applied Mechanics, State Key Laboratory for Strength and Vibration of Mechanical Structures, School of Aerospace, Xi'an Jiaotong University, Xi'an, China

Ziguang Chen Department of Mechanics, Huazhong University of Science and Technology, Wuhan, Hubei Sheng, China

Hubei Key Laboratory of Engineering, Structural Analysis and Safety Assessment,Wuhan, China

N. Chiba National Defense Academy of Japan, Yokosuka, Japan

Ömer Civalek Civil Engineering Department, Division of Mechanics, Akdeniz University, Antalya, Turkey

Qiang Cui Department of Chemistry and Theoretical Chemistry Institute, University of Wisconsin-Madison, Madison, WI, USA

Marta D'Elia Optimization and Uncertainty Quantification Department Center for Computing Research, Sandia National Laboratories, Albuquerque, NM, USA

Valentín de Carlos Blasco School of Industrial Engineering, Institute of Energy Technologies, Universitat Politècnica de Catalunya, Barcelona, Spain

Babür Deliktaş Faculty of Engineering-Architecture, Department of Civil Engineering, Uludag University, Bursa, Görükle, Turkey

Paul N. Demmie Sandia National Laboratories, Albuquerque, NM, USA

Qiang Du Department of Applied Physics and Applied Mathematics, Columbia University, New York, NY, USA

Samuel Forest Centre des Materiaux, Mines ParisTech CNRS, PSL Research University, Paris, Evry Cedex, France

Mikil D. Foss Department of Mathematics, University of Nebraska-Lincoln, Lincoln, NE, USA

Mercè Ginjaume Egido School of Industrial Engineering, Institute of Energy Technologies, Universitat Politècnica de Catalunya, Barcelona, Spain

Max Gunzburger Department of Scientific Computing, Florida State University, Tallahassee, FL, USA

Prashant K. Jha Department of Mathematics, Louisiana State University, Baton Rouge, LA, USA

Orsan Kilicer Department of Mathematics, Wayne State University, Detroit, MI, USA

Giovanni Lancioni Dipartimento di Ingegneria Civile, Edile e Architettura, Università Politecnica delle Marche, Ancona, Italy

A. K. Lazopoulos Mathematical Sciences Department, Hellenic Army Academy, Vari, Greece

K. A. Lazopoulos National Technical University of Athens, Rafina, Greece

Jun Li Department of Mechanical Engineering, University of Massachusetts, Dartmouth, MA, USA

Robert Lipton Department of Mathematics and Center for Computation and Technology, Louisiana State University, Baton Rouge, LA, USA

David J. Littlewood Center for Computing Research, Sandia National Laboratories, Albuquerque, NM, USA

L. Liu Department of Mechanical and Aerospace Engineering, Utah State University, Logan, UT, USA

Yilun Liu International Center for Applied Mechanics, State Key Laboratory for Strength and Vibration of Mechanical Structures, School of Aerospace, Xi'an Jiaotong University, Xi'an, China

Tomasz Łodygowski Institute of Structural Engineering, Poznan University of Technology, Poznan, Poland

Angela Madeo SMS-ID, INSA-Lyon, Université de Lyon, Villeurbanne cedex, Lyon, France

Institut universitaire de France, Paris Cedex 05, Paris, France

Leila Malekmotiei Department of Civil and Environmental Engineering, Louisiana State University, Baton Rouge, LA, USA

J. R. Mayeur Theoretical Division, Los Alamos National Laboratory, Los Alamos, NM, USA

D. L. McDowell Woodruff School of Mechanical Engineering, School of Materials Science and Engineering, Georgia Institute of Technology, Atlanta, GA, USA

Wen Jin Meng Department of Mechanical and Industrial Engineering, Louisiana State University, Baton Rouge, LA, USA

James A. Mills Department of Civil Engineering and Engineering Mechanics, Columbia University, New York, NY, USA

Patrizio Neff Fakultät für Mathematik, Universität Duisburg-Essen, Essen, Germany

Sina Niazi Mechanical and Materials Engineering, University of Nebraska–Lincoln, Lincoln, NE, USA

N. Ogasawara Department of Mechanical Engineering, National Defense Academy of Japan, Yokosuka, Japan

Martin Ostoja-Starzewski Department of Mechanical Science and Engineering, Institute for Condensed Matter Theory and Beckman Institute, University of Illinois at Urbana–Champaign, Urbana, IL, USA

İzzet Özdemir Department of Civil Engineering, İzmir Institute of Technology, İzmir, Turkey

Mauro Perego Center for Computing Research, Sandia National Laboratories, Albuquerque, NM, USA

Chandra Prakash School of Aeronautics and Astronautics, Purdue University, West Lafayette, IN, USA

Petronela Radu Department of Mathematics, University of Nebraska-Lincoln, Lincoln, NE, USA

Hadrien Rattez Navier (CERMES), UMR 8205, Ecole des Ponts, IFSTTAR, CNRS, Champs-sur-Marne, France

Eyad Said Department of Mathematics, Louisiana State University, Baton Rouge, LA, USA

Aref Samadi-Dooki Computational Solid Mechanics Laboratory, Department of Civil and Environmental Engineering, Louisiana State University, Baton Rouge, LA, USA

Pablo Seleson Computer Science and Mathematics Division, Oak Ridge National Laboratory, Oak Ridge, TN, USA

Shuai Shao Department of Mechanical and Industrial Engineering, Louisiana State University, Baton Rouge, LA, USA

S. A. Silling Sandia National Laboratories, Albuquerque, NM, USA

Yooseob Song Department of Civil and Environmental Engineering, Louisiana State University, Baton Rouge, LA, USA

Ioannis Stefanou Navier (CERMES), UMR 8205, Ecole des Ponts, IFSTTAR, CNRS, Champs-sur-Marne, France

Jean Sulem Navier (CERMES), UMR 8205, Ecole des Ponts, IFSTTAR, CNRS, Champs-sur-Marne, France

Wojciech Sumelka Institute of Structural Engineering, Poznan University of Technology, Poznan, Poland

Vasily E. Tarasov Skobeltsyn Institute of Nuclear Physics, Lomonosov Moscow State University, Moscow, Russia

Vikas Tomar School of Aeronautics and Astronautics, Purdue University, West Lafayette, IN, USA

Ismail Cem Turtuk Mechanical Design Department, Meteksan Defence, Ankara, Turkey

Department of Civil Engineering, Uludag Univeristy, Bursa, Turkey

Devendra Verma School of Aeronautics and Astronautics, Purdue University, West Lafayette, IN, USA

George Z. Voyiadjis Department of Civil and Environmental Engineering, Louisiana State University, Baton Rouge, LA, USA

Hong Wang Department of Mathematics, University of South Carolina, Columbia, SC, USA

Hang Xiao School of Chemical Engineering, Northwest University, Xi'an, China

Baoxing Xu Department of Mechanical and Aerospace Engineering, University of Virginia, Charlottesville, VA, USA

Jun Xu Department of Automotive Engineering, School of Transportation Science and Engineering, Beihang University, Beijing, China

Advanced Vehicle Research Center (AVRC), Beihang University, Beijing, China

Mohammadreza Yaghoobi Department of Civil and Environmental Engineering, Louisiana State University, Baton Rouge, LA, USA

Tuncay Yalçinkaya Aerospace Engineering Program, Middle East Technical University Northern Cyprus Campus, Guzelyurt, Mersin, Turkey

Department of Aerospace Engineering, Middle East Technical University, Ankara, Turkey

Yuan Yan School of Chemical Engineering, Northwest University, Xi'an, China

Akio Yonezu Department of Precision Mechanics, Chuo University, Tokyo, Japan

Zhufeng Yue Department of Engineering Mechanics, Northwestern Polytechnical University, Xi'an, Shaanxi, China

Cheng Zhang Medtronic, Inc., Tempe, AZ, USA

Guanfeng Zhang Mechanical and Materials Engineering, University of Nebraska–Lincoln, Lincoln, NE, USA

Yanting Zheng China Automotive Technology and Research Center, Tianjin, China

Liangliang Zhu Columbia Nanomechanics Research Center, Department of Earth and Environmental Engineering, Columbia University, New York, NY, USA

International Center for Applied Mechanics, State Key Laboratory for Strength and Vibration of Mechanical Structures, School of Aerospace, Xi'an Jiaotong University, Xi'an, China

Part I
Nanoindentation for Length Scales

Size Effects and Material Length Scales in Nanoindentation for Metals

1

George Z. Voyiadjis and Cheng Zhang

Contents

Abstract

In *nanoindentation* experiments at submicron indentation depths, the hardness decreases with the increasing indentation depth. This phenomenon is termed as the indentation size effect. In order to predict the indentation size effect, the classical continuum needs to be enhanced with the strain gradient plasticity theory. The strain gradient plasticity theory provides a nonlocal term in addition to the classical theory. A material length scale parameter is required to be incorporated into the constitutive expression in order to characterize the size effects in different materials. By comparing the model of hardness as a function

G. Z. Voyiadjis (✉)
Department of Civil and Environmental Engineering, Louisiana State University, Baton Rouge, LA, USA
e-mail: voyiadjis@eng.lsu.edu

C. Zhang
Medtronic, Inc., Tempe, AZ, USA
e-mail: cheng.zhang2@medtronic.com

G. Z. Voyiadjis (ed.), *Handbook of Nonlocal Continuum Mechanics for Materials and Structures*, https://doi.org/10.1007/978-3-319-58729-5_27

of the indentation depth with the nanoindentation experimental results, the length scale can be determined. Recent nanoindentation experiments on polycrystalline metals have shown an additional hardening segment in the hardness curves instead of the solely decreasing hardness as a function of the indentation depth. It is believed that the accumulation of dislocations near the grain boundaries during nanoindentation causes the additional increase in hardness. In order to isolate the influence of the grain boundary, bicrystal metals are tested near the grain boundary at different distances. The results show that the hardness increases with the decreasing distance between the indenter and the grain boundary, providing a new type of size effect. The length scales at different distances are determined using the modified model of hardness and the nanoindentation experimental results on bicrystal metals.

Keywords
Nanoindentation · Indentation size effect · Nonlocal theory · Strain gradient plasticity · Grain boundary · Bicrystal · Length scale

Introduction

It has been found in experiments at the microscopic scale that the mechanical response increases as the size of the specimen decreases. The phenomenon is termed as the size effect. Size effects have been reported in tests of metals in small scales such as microbending, microtorsion, and bulge tests (Fleck and Hutchinson 1997; Stolken and Evans 1998; Chen et al. 2007; Xiang and Vlassak 2005). It is believed that the size effects are attributed to nonuniform deformation in the small scale during the tests (Nye 1953). Geometrically necessary dislocations (GNDs) are generated in order to accommodate the deformation. The GNDs act as barriers of the generation of the statistically stored dislocations (SSDs), resulting in the increase of the mechanical response (De Guzman et al. 1993; Stelmashenko et al. 1993; Fleck et al. 1994; Ma and Clarke 1995). The classical continuum theory is only able to predict the macroscopic and therefore not capable in capturing the changes of mechanical responses. A nonlocal term is required in addition to the classical theory. In the study of the plastic deformation, the strain gradient plasticity theory is applied by adding a strain gradient term into the classical expression (Aifantis 1992; Zbib and Aifantis 1998). In order to characterize the influence of the strain gradient, a length scale parameter is required to be incorporated into the expression of the strain gradient plasticity theory. The length scale is an intrinsic parameter and each material has its unique length scale. Therefore, it becomes of great importance to determine the material intrinsic length scale. With the length scale determined, the strain gradient plasticity theory is able to predict the mechanical behaviors in both microscopic and macroscopic scales, bridging the gap between large and small scales.

When using the strain gradient plasticity theory to derive the expression of the length scales, material parameters are needed in order to represent the behavior for different materials. In order to determine the length scale for a specific material, the materials need to be determined from experimental results. Nanoindentation experiments are believed to be the most effective technique to determine the material length scales (Begley and Hutchinson 1998). In nanoindentation at small depth, the hardness increases with the decreasing indentation depth (McElhaney et al. 1998). The size effect encountered in nanoindentation is referred to the indentation size effect. With the use of the conical or Berkovich indenter, the indentation is said to be self-similar. The GND density can be calculated for a self-similar indent from the geometry of the indenter (Nix and Gao 1998). The hardness can be mapped from the stress-strain relation using Tabor's factor in macroscopic scale (Tabor 1951). In microscopic scales, the hardness is related to the dislocation density according to Taylor's hardening law. The hardness expression needs to incorporate the expression of the length scale in order to predict the indentation size effect. By comparing the experiments and the expression of hardness, the material parameters in the expression of the length scale can be determined.

More recent nanoindentation experiments in polycrystalline materials have shown that instead of the solely decreasing hardness with the increasing indentation depth, there is a hardening-softening phenomenon observed (Yang and Vehoff 2007; Voyiadjis and Peters 2010). The additional increase in hardness is believed to be due to the interaction between the dislocations generated by the penetration of the indenter and the grain boundaries. As there is difficulty for dislocations to transfer across the grain boundary, the dislocations accumulate near the grain boundaries. The local dislocation density increases due to the accumulation of dislocations, resulting in the additional increase in hardness. Once the dislocations start to move across the grain boundary at certain point, the dislocation density starts to decrease, which explains the softening effect following the hardening. In order to characterize the grain boundary effect as well the temperature and strain rate dependency of the length scale, a model considering the temperature and strain rate dependency of indentation size effect (TRISE) is developed (Voyiadjis et al. 2011). The length scale is written as a function of the temperature, plastic strain rate, and the grain size. By comparing the TRISE model with nanoindentation experiments at different temperatures, plastic strain rates, and grain sizes, the length scales can be determined.

In order to confirm the contribution of grain boundaries to the hardening-softening phenomenon, the investigation of the grain boundary is isolated through bicrystalline materials, where there is only one grain boundary. Nanoindentation experiments are conducted near the grain boundary at different distances between the indenter and the grain boundary (Voyiadjis and Zhang 2015; Zhang and Voyiadjis 2016). Single crystal behavior is observed for indents made with large distances to the grain boundary, without showing the additional hardening. Only the indents in close proximity of the grain boundary show the hardening-softening effect and there is a stronger hardening effect when the distance is smaller, providing a new type of size effect. The influence of the grain boundary on the hardness during nanoindentation is thus confirmed by experiments. The TRISE model is rewritten by replacing the grain size with the distance between the grain boundary and the indenter. The length scales of bicrystalline materials can be determined through the comparisons between the developed TRISE model and the nanoindentation experiments in order to characterize the size effect regarding the grain boundary.

In this chapter, the dependencies of temperature, plastic strain rate, grain size, and the distance between the indents and the grain boundary are addressed during nanoindentation experiments on single crystal, polycrystalline, and bicrystalline materials. The TRISE model is developed and applied in order to predict the hardness as a function of the indentation depth. The equivalent plastic strain as a function of the indentation depth during nanoindentation is determined through the finite element method. ABAQUS/Explicit software is used in order to simulate the indentation problem. The materials tested are body-centered cubic (BCC) and face-centered cubic (FCC) materials. In order to show the distinct behaviors between FCC and BCC materials during the simulation, user material subroutines VUMAT are incorporated for BCC and FCC metals, respectively. The length scales are determined through the comparison between the nanoindentation experiments and the developed TRISE models.

Nonlocal Theory

According to the nonlocal theory, the material properties at a given material point are not only dependent on their local counterparts but also depend on the state of the neighboring space. While the classical continuum theory only provides the predictions of the local point, the nonlocal theory enhances the classical theory by giving a nonlocal gradient term. The nonlocal term is taken as the characterization of the interactions from the neighboring space. The nonlocal expression was incorporated through an integral form for the elastic models of materials (Kroner 1967; Eringen and Edelen 1972). In the integral format, the nonlocal measure $\overline{A}$ at a given material point x is expressed through the weighted average of its local counterpart A over the surrounding volume V within a small distance d from the point x as follows:

$$\overline{A} = \frac{1}{V}\int_{V} w(d)A\,(x+d)\,dV \tag{1}$$

where $\boldsymbol{w}(d)$ is a weight function that decays gradually with the distance d. It should be noted that there is a limit of the distance d, which is the internal characteristic length that shows the range of the influence of the nonlocal term.

The integration can be solved analytically for the elasticity problems. However, it is not practical to solve the integration for more complicated plasticity problems. Therefore, the integral expression needs to be simplified. The local counterpart $\boldsymbol{A}$ in the integral can be approximated using the Taylor's series expansion at $d = 0$ as follows (Muhlhaus and Aifantis 1991; Vardoulakis and Aifantis 1991):

$$A\,(x+d) = A(x) + \nabla A\,(x)\,d + \frac{1}{2!}\nabla^2 A(x)d^2 + \frac{1}{3!}\nabla^3 A(x)d^3 + \ldots \quad (2)$$

where ∇^{i} denotes the ith order of the gradient operator. The expression can be further simplified considering only the isotropic behaviors. Therefore, the odd terms vanish in Eq. 2 and the nonlocal expression can be rewritten as follows:

$$\overline{A} = \frac{1}{V}\int_V w(d)A(x)dV + \frac{1}{2!V}\int_V w(d)\nabla^2 A(x)d^2 dV \quad (3)$$

Rewriting Eq. 3 as a partial differential equation yields the following expression:

$$\overline{A} = A + \left(\frac{1}{2!V}\int_V w(d)d^2 \mathrm{dV}\right)\nabla^2 p \quad (4)$$

with

$$\frac{1}{V}\int_V w(d)\mathrm{dV} = 1 \quad (5)$$

Therefore, the simplified nonlocal expression can be eventually expressed as follows:

$$\overline{A} = A + l^2\nabla^2 A \quad (6)$$

In Eq. 6, the second-order gradient term is incorporated in addition to the original local counterpart. A length scale parameter l is used in order to weigh the gradient term, reflecting the characteristic length of the influence of the gradient.

Considering the nonuniform deformation of materials in micro- and nanoscales, the strain is usually used in order to characterize the material behaviors. By writing the nonlocal expression using the expression of strain and strain gradient, the strain gradient plasticity (SGP) theory can be written in the format of Eq. 6 as follows:

$$\overline{p} = p + l^2\eta^2 \quad (7a)$$

where $\overline{p}$ is the total accumulated plastic strain with its local counterpart p and the gradient term η, and l is the material length scale that characterizes the influence of the strain gradient.

Another approach was developed through a phenomenological theory of strain gradient plasticity based on gradients of rotation, which fits the framework of the couple stress theory (Fleck and Hutchinson 1993). By assuming that the strain energy density is only dependent on the second von Mises invariant of strain, the strain gradient plasticity theory can be expressed based on that the strain energy density is only dependent on the overall effective strain as follows:

$$\overline{p} = \sqrt{p^2 + l^2\eta^2} \tag{7b}$$

A more general expression (Voyiadjis and Abu Al-Rub 2005) for the strain gradient plasticity theory considering the expressions in Eqs. 6, 7a, and 7b was later proposed as follows:

$$\overline{p} = [p^\gamma + (l\eta)^\gamma]^{1/\gamma} \tag{8}$$

where γ is a fitting parameter.

Physically Based Material Length Scale

As shown in Eq. 8, the total accumulated plastic strain can be determined through its local counterpart and the strain gradient term. It becomes of great importance in the determination of the material length scale parameter, as it captures the intrinsic gradient effects for different materials. From the continuum theory, the flow stress can be written through the plastic strain given in Eq. 8 as follows:

$$\sigma = k[p^\gamma + (l\eta)^\gamma]^{1/\mathrm{m}\gamma} \tag{9}$$

where m and k are material constants. Equation 8 is capable to predict the constitutive relations in both macroscopic scale through the local counterpart and microscopic scale through the coupling between the local and gradient terms. Therefore, by determining the length scale parameter, the strain gradient enhanced classical continuum theory is able to bridge the gap between large and small scales.

In Eq. 9, the local term p is related to the SSDs as it represents the uniform deformation at macroscopic scales. The plastic shear strain γ^p can be defined as a function of SSD density ρ_S as follows (Bammann and Aifantis 1982):

$$\gamma^p = b_S L_S \rho_S \tag{10}$$

where b_S is the magnitude of Burgers vector of SSDs and L_S is the mean space between SSDs. The local plastic strain $\varepsilon^p_{\mathrm{ij}}$ can then be determined from γ^p using Schmid orientation tensor M_{ij} as follows:

$$\varepsilon^p_{\mathrm{ij}} = \gamma^p M_{\mathrm{ij}} \tag{11}$$

where

$$p = \sqrt{2\varepsilon^p_{\mathrm{ij}}\varepsilon^p_{\mathrm{ij}}/3} \tag{12}$$

When relating the plastic strain between macroscopic and microscopic scales, the Schmid orientation factor which is an average form of the Schmid tensor is always used (Bammann and Aifantis 1987; Dorgan and Voyiadjis 2003). With that being said, the local plastic strain can be expresses as follows:

$$p = b_S L_S \rho_S M \tag{13}$$

where M is the Schmid factor which is usually taken to be 0.5 (Bammann and Aifantis 1982).

The gradient term in Eq. 9 is related to the GNDs as it represents the nonuniform deformation that occurred in small scales. The gradient η can be expressed through GND density ρ_G and the Nye factor as follows (Arsenlis and Parks 1999):

$$\eta = \frac{\rho_G b_G}{\overline{r}} \tag{14}$$

where b_G is the magnitude of the Burgers factor of GNDs and $\overline{r}$ is the Nye factor.

In microscopic scales, the flow stress can be related to the dislocation densities generated by the plastic deformation through Taylor's hardening law as follows:

$$\tau_S = \alpha_S \mathrm{Gb}_s \sqrt{\rho_S} \tag{15}$$

$$\tau_G = \alpha_G \mathrm{Gb}_G \sqrt{\rho_G} \tag{16}$$

where τ_S and τ_G are the shear flow stress corresponding to SSD density and GND density, respectively; b_S and b_G are the magnitudes of the Burgers vectors for SSDs and GNDs, respectively; G is the shear modulus; and α_S and α_G are statistical coefficients which account for the deviation from regular spatial arrangements of SSDs and GNDs populations, respectively. The total shear flow stress, τ_f, which is required to initiate a significant plastic deformation, can be obtained by coupling the flow stresses given by Eqs. 15 and 16 as follows:

$$\tau_f = \left[\tau_S^{\beta} + \tau_G^{\beta}\right]^{1/\beta} \tag{17}$$

where β is a constant fitting parameter.

A more general expression for the total shear flow stress can be written in terms of the total dislocation density from the combination of Eqs. 15, 16, and 17 as follows:

$$\tau = \alpha_S \mathrm{Gb}_S \sqrt{\rho_T} \tag{18}$$

with

$$\rho_T = \left[\rho_S^{\beta/2} + \left(\alpha_G^2 b_G^2 \rho_G / \alpha_S^2 b_S^2\right)^{\beta/2}\right]^{2/\beta} \tag{19}$$

The plastic flow stress is related to the shear flow stress through a constant Z as follows:

$$\sigma = \mathrm{ZT} = \alpha_S \mathrm{ZGb}_S \sqrt{\rho_T} \tag{20}$$

Comparing Eqs. 9 and 20 and considering Eqs. 13, 14, and 19, the following expressions can be determined:

$$\gamma = \beta/2 \tag{21}$$

$$m = 2 \tag{22}$$

$$k = \alpha_G \mathrm{GZ} \sqrt{b_S / L_S M} \tag{23}$$

$$l = (\alpha_G/\alpha_S)^2 \, (b_G/b_S) \, L_S M \overline{r} \tag{24}$$

At the beginning, the length scales were determined as a constant value. However, it has been reported that L_S is not a constant but it is equal to the grain size initially and saturates toward values in the order of micrometer in large strains (Gracio 1994). L_S can be written as a function of the grain size and the plastic strain as follows:

$$L_S = \frac{\mathrm{Dd}}{D + \mathrm{dp}^{1/m}} \tag{25}$$

where d is the average grain size, D is the macroscopic characteristic size of the specimen, p is the equivalent plastic strain, and m is the hardening exponent. It can be seen from Eq. 24 that the length scale is proportional to L_S. Therefore, the length scales of crystalline materials are not constant during the deformation. This has also been experimentally and theoretically verified in the work of Voyiadjis and Abu Al-Rub (2005).

Due to the fact that the GND density during deformation is strain rate and temperature dependent, the Nye factor in Eq. 24 can be expressed as a function of the strain rate and temperature as follows (Voyiadjis and Almasri 2009):

$$\overline{r} = \frac{\mathrm{Ae}^{(-E_r/\mathrm{RT})}}{1 + \mathrm{C}\dot{\mathrm{p}}^q} \tag{26}$$

where the temperature factor is incorporated through the Arrhenius equation; $\dot{p}$ is the equivalent plastic strain rate; and A, C, and q are constants. Substituting Eqs. 25 and 26 into Eq. 24 yields a variable length scale that varies with the grain size, temperature, strain rate, and equivalent plastic strain as follows (Voyiadjis et al. 2011):

$$l = \left(\alpha_G/\alpha_S\right)^2 \left(b_G/b_S\right) M \left(\frac{\delta_1 \mathrm{d}e^{(E_r/RT)}}{\left(1 + \delta_2 \mathrm{d}p^{(1/m)}\right)\left(1 + \delta_3 (\dot{p})^q\right)} \right) \tag{27}$$

where δ_1, δ_2, and δ_3 are material parameters that need to be determined through experimental results. For single crystal materials, there is no effect of the grain size during the deformation. Therefore, the length scale for single crystal materials can be expressed without grain size d as follows:

$$l = \left(\alpha_G/\alpha_S\right)^2 \left(b_G/b_S\right) M \left(\frac{\delta_1 e^{(-E_r/\mathrm{RT})}}{\left(1 + \delta_2 \mathrm{p}^{(1/m)}\right)\left(1 + \delta_3 (\dot{p})^q\right)} \right) \tag{28}$$

Determination of the Length Scales

Nanoindentation experiments are believed to be the most effective technique to determine the length scales. The indentations made using a conical or a Berkovich indenter have self-similar shapes, which means once the indent is initially formed, the size of the indent grows and the shape keeps similar. The GND and SSD densities can be calculated during the indentation process. The material hardness is related to the flow stress according to Tabor's factor. Therefore, the hardness can be written as a function of the dislocation densities through Eqs. 18, 19, and 20 as follows:

$$H = \kappa \sigma_f = Z \kappa \alpha_S G b_S \left[{\rho_S}^{\beta/2} + \left(\alpha_G b_G / \alpha_S b_S\right)^{\beta} {\rho_G}^{\beta/2} \right]^{1/\beta} \tag{29}$$

where κ is Tabor's factor relating the hardness and the flow stress. As shown in Eq. 29, the hardness is a function of both GND density and SSD density, which represents the deformation in the micro- and nanoscales where GNDs and SSDs interact with each other. However, in the case of large deformation, the effect of GNDs vanishes and the deformation is mainly attributed to the SSDs. The hardness in macroscopic scales, H_0, is thus only related to the SSD density as follows:

$$H_0 = Z \kappa \tau_S = Z \kappa \alpha_S \mathrm{G} b_S \sqrt{\rho_S} \tag{30}$$

Using Eqs. 29 and 30 and assuming that $\alpha_G = \alpha_S$ and $b_G = b_S$, the ratio $(H/H_0)^{\beta}$ can be derived as follows:

$$\left(\frac{H}{H_0}\right)^{\beta} = 1 + \left(\frac{\rho_G}{\rho_S}\right)^{\beta/2} \tag{31}$$

The nanoindentation experiments provide the material hardness as a function of the indentation depth. If the dislocation densities can be calculated as through the

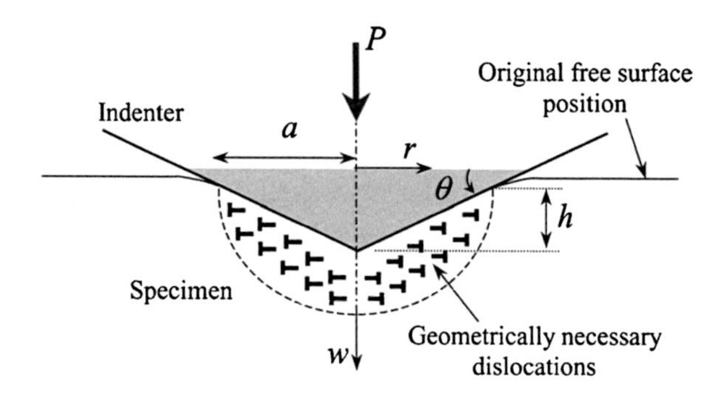

Fig. 1 Sample being indented by a conical indenter (Reprinted from Voyiadjis and Zhang (2015))

indentation depth, the hardness from the model can be compared with the hardness determined by the experiments.

The GND density during nanoindentation can be calculated through the geometry of the indenter. The cross-sectional profile of the indentation process using a conical indenter is shown in Fig. 1. The profile of a Berkovich indenter can be represented by the conical indenter if the surface angle θ is taken to be 0.358 (Radian), since the cross-sectional areas and the volumes are identical between the conical and pyramidal Berkovich indenters.

The dislocation loops are generated by the penetration of the indenter and therefore the plastic deformation volume can be assumed as a semisphere. The GND density can be determined as the total length of dislocations divided by the volume of the semisphere, where the dislocation length can be calculated based on the geometry of the indenter. Without considering the influence of the grain boundaries, the GND density can be written as follows (Nix and Gao 1998):

$$\rho_G = \frac{3\tan^2\theta}{2b_G h} \tag{32}$$

It can be seen from Eq. 32 that the GND density decreases with the increasing indentation depth, proving that the nonuniform deformation decays with the increase of deformation and the nonlocal gradient becomes less significant.

In order to capture the hardening-softening phenomenon, grain boundaries were incorporated into the profile as shown in Fig. 2. When the plastic zone expands and reaches the grain boundaries, the diameter of the semisphere is equal to the average grain size.

However, a more accurate calculation can be approached by removing the volume occupied by the indenter from the semisphere (Yang and Vehoff 2007). The GND density can be written as follows (Voyiadjis and Peters 2010)

$$\rho_G = \frac{\frac{\pi}{b_G}\frac{h^2}{\tan\theta}}{\frac{\pi}{12}d^3 - 8.19h^3} \tag{33}$$

where d represents the average grain size in polycrystalline materials.

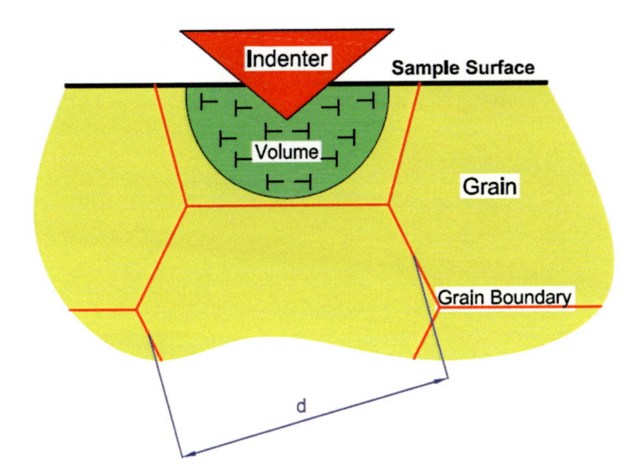

Fig. 2 Nanoindentation cross-sectional profile of a polycrystalline sample (Reprinted from Voyiadjis et al. (2011))

The SSD density can be derived from the Tabor's mapping of hardness-indentation depth from flow stress-plastic strain (Voyiadjis and Peters 2010). The SSD density ρ_S can be written as follows:

$$\rho_S = \frac{c\bar{r}\alpha_G^2 b_G \tan\theta}{\mathrm{l}b_S^2\alpha_S^2} \tag{34}$$

where c is a material constant of order 1 from Tabor's mapping.

Combining Eqs. 31, 32, and 34 and assuming that $\alpha_G=\alpha_S$ and $b_G=b_S$, the hardness of single crystal materials can be derived as a function of the indentation depth as follows:

$$H = H_0\left\{1 + \left[\frac{3M\tan\theta}{2ch} \times \frac{\delta_1 e^{-E_r/RT}}{\left(1+\delta_2 p^{1/m}\right)\left(1+\delta_3 \dot{p}^q\right)}\right]^{\beta/2}\right\}^{1/\beta} \tag{35}$$

For polycrystalline materials, similar derivation can be made through Eqs. 31, 33, and 34 as follows:

$$H = H_0\left\{1 + \left[\frac{M\pi(h-h_1)^2}{\left(\frac{\pi}{12}d^3 - 8.19h^3\right)c\tan^2\theta} \times \frac{\delta_1 \mathrm{d}e^{-E_r/RT}}{\left(1+\delta_2 dp^{1/m}\right)\left(1+\delta_3 \dot{p}^q\right)}\right]^{\beta/2}\right\}^{1/\beta} \tag{36}$$

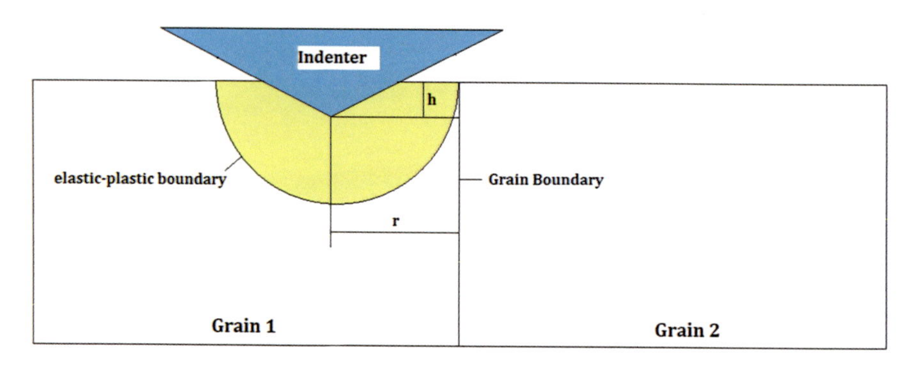

Fig. 3 Nanoindentation cross-sectional profile of a bicrystalline sample (Reprinted with permission from Voyiadjis and Zhang (2015)).

In order to isolate the influence of the grain boundary, bicrystalline materials are used and nanoindentations are made close to the single grain boundary near the grain boundary. The grain size in bicrystalline materials is not a parameter as it is in the size of the macroscopic scale. The distance r between the grain boundary and the indenter tip is used to represent the size of the plastic zone. When the plastic zone reaches the grain boundary, the radius of the semisphere is equal to the distance r as shown in Fig. 3.

By using the distance r instead of the grain size d, the GND density during nanoindentation of bicrystalline materials can be written as follows:

$$\rho_G = \frac{\frac{\pi}{b_G}\frac{h^2}{\tan\theta}}{\frac{2}{3}\pi r^3 - 8.19h^3} \tag{37}$$

Therefore, the hardness expression of a bicrystalline material during nanoindentation can be written as follows:

$$H = H_0\left\{1 + \left[\frac{M\pi(h-h_1)^2}{\left(\frac{2\pi}{3}r^3 + 8.19h^3\right)\mathrm{ctan}^2\theta} \times \frac{\delta_1 \mathrm{r}e^{-E_r/\mathrm{RT}}}{\left(1+\delta_2\mathrm{rp}^{1/m}\right)\left(1+\delta_3\dot{p}^q\right)}\right]^{\beta/2}\right\}^{1/\beta} \tag{38}$$

The length scale expression given by Eqs. 27 and 28 are incorporated into Eqs. 35, 36, and 38. The material parameters δ_1, δ_2, and δ_3 can be determined by comparing Eqs.35, 36, and 38 with nanoindentation experimental results for single crystal, polycrystalline materials, and bicrystalline materials, respectively. Therefore, the length scale parameters can be determined for materials with different microstructures, respectively.

Furthermore, a better approach of the hardness expression can be applied through a cyclic plasticity model (Voyiadjis and Abu Al-Rub 2003) as follows:

$$H_{new} = H_{old} + C(h)H_{old} \tag{39}$$

with

$$C(h) = \sqrt{\frac{1}{h}} \tag{40}$$

where H_{old} is the hardness given by Eqs. 35, 36, and 38 and H_{new} is the corrected value using the cyclic model.

In order to write the hardness H as a function of the indentation depth h in Eqs. 35, 36, and 38, the equivalent plastic strain p is also required to be as a function of h. It is difficult to determine p as a function of h experimentally as the plastic deformation during nanoindentation is complicated. Therefore, finite element simulation is needed to capture this relationship (Voyiadjis and Peters 2010). Commercial finite element analysis software ABAQUS is used throughout an indentation problem. The testing sample is represented by a cube with dimension of 50 μm. The Berkovich indenter is modeled on top of the cubic sample as a blunt three-side pyramid as shown in Fig. 4. The tip of the indenter is set to be an equilateral triangle with 20 nm sides. The ABAQUS interaction module is used in order to simulate the contact between the indenter and the sample. A specific velocity field is assigned in order to drive the indenter to penetrate to the desired indentation depths according to the experiments. In order to show the different behaviors between FCC and BCC materials, a user material subroutine VUMAT is used during the simulation (Voyiadjis et al. 2011). After the indentation process is completed, the equivalent plastic strain p under the indenter can be viewed as shown in Fig. 5. The value of p at each time step can be determined through the contour plots as well as the indentation depth at each step. Therefore, the equivalent plastic strain as a function of the indentation depth can be determined by taking values of p and $\hbar$ at each step.

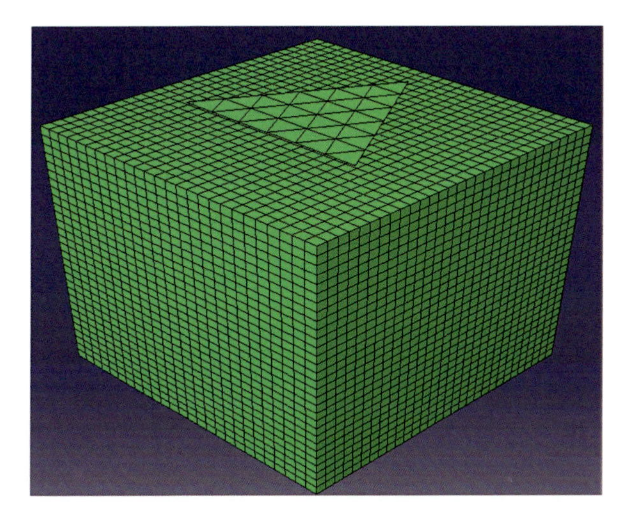

Fig. 4 Indentation model before the simulation (Reprinted from Voyiadjis and Zhang (2015))

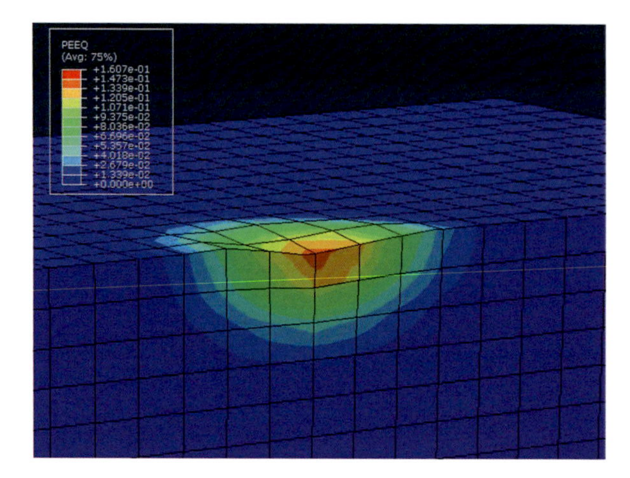

Fig. 5 Contour plot of equivalent plastic strain at the maximum indentation depth (Reprinted from Voyiadjis and Zhang (2015))

Applications on Single Crystal, Polycrystalline, and Bicrystalline Metals

Nanoindentation experiments are performed on metals with different microstructures in order to validate the prediction of the computational models and to determine the material parameters of the expressions of the length scales. Based on the temperature and strain rate indentation size effect (TRISE), nanoindentation experiments are conducted on single crystal and polycrystalline samples at different temperatures and strain rates. On bicrystalline samples, the experiments are carried out near the grain boundary at different distances in order to characterize the influence of the grain boundary on the material hardness. Nanoindentations at different strain rates are also performed at the same distance between the grain boundary and the indenter in order to confirm the rate dependency during nanoindentation on bicrystalline materials. The experimental results are compared with the prediction of hardness of the TRISE model. The length scales of materials with different microstructures are determined through the determinations of the material parameters from the comparisons. All testing samples are polished to acquire the accurate and consistent experimental results.

Sample Preparations

In nanoindentation, the hardness is not a direct measurement from the test. It is determined from the direct measurement of load and displacement through a tip area function (Oliver and Pharr 1992). The tip area function is determined through a model assuming that the indenter penetrates perpendicularly into a flat sample surface. Therefore, in order to obtain accurate experimental results, the surface of the sample needs to be polished in order to approach the flat condition as assumed by the theory. It is required by the nanoindentaion technique that the surface roughness must be smaller than one tenth of the maximum indentation depth. Therefore, if the hardness information at smaller indentation depths is needed, the surface roughness needs to be controlled to be lower.

Several polishing procedures are applied in order to improve the surface roughness. Mechanical polishing is usually applied firstly in order to level the entire surface to be horizontal. Silica carbide polishing papers with polishing particles of different sizes are used depending on the initial surface condition: the rougher the surface, the greater the size of polishing particles. After the use of each polishing paper, the surface is examined using a light microscope to make sure that the scratches on the surface are in the same size. Chemical-mechanical polishing is applied following the mechanical polishing when the polishing paper with the minimum size of polishing particles is used. In the chemical-mechanical polishing process, 50 nm colloidal silica or alumina polishing particles are used depending on the type of materials to be polished. By adjusting the PH values of the polishing slurries, chemical reactions between the polishing particles and the polishing sample can be initiated so that the surface roughness is lowered more effectively compared to the mechanical polishing process. The improvement of the surface quality of an iron sample is shown in Fig. 6. It shows that the surface roughness is improved and the defects such as voids and scratches are also improved.

In addition to the surface roughness, the plastic deformation layer on top of the surface is another concern for nanoindentation experiments. Due to the mechanical abrasion between the polishing particles and the sample, plastic deformation layer is induced no matter how small the polishing particles are used. The hardness varies if there is a plastic deformation that occurred in the indenter area, especially for tests in bicrystalline materials when the indentation depth is as small as a few tens of nanometers where the hardening-softening phenomenon is observed. Electro-polishing is widely used in order to remove the plastic deformation layer. Chemical reactions occur in the electrolyte solution and the peak material on the surface is removed by the electrical current during the electro-polishing process. As

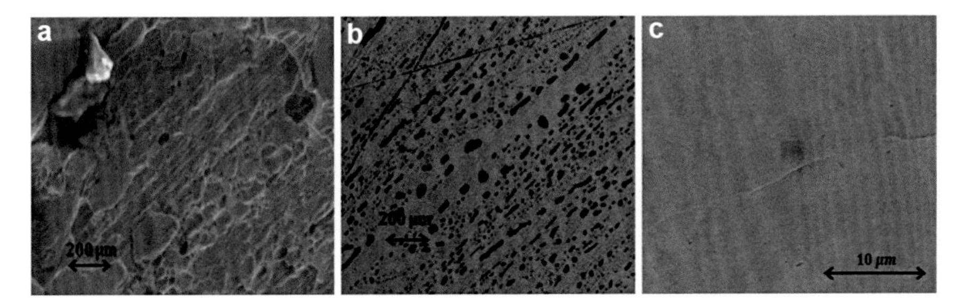

Fig. 6 SEM images of surfaces of testing sample in different polishing conditions: (**a**) without polishing, (**b**) after mechanical polishing, and (**c**) after chemical-mechanical polishing

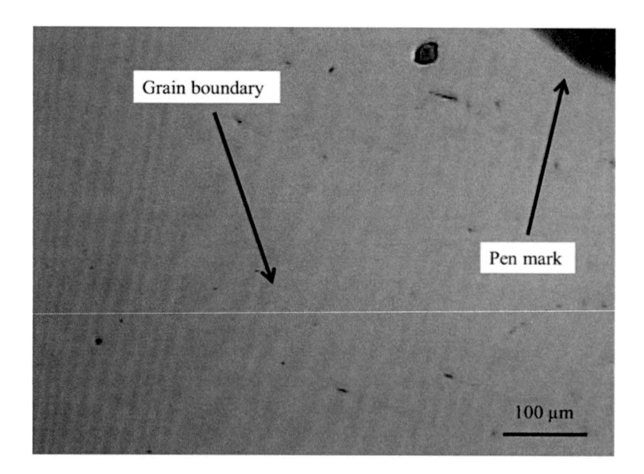

Fig. 7 Visualization of the grain boundary after vibratory on a bicrystalline aluminum (Reprinted from Voyiadjis and Zhang (2015))

no mechanical abrasion is induced, there is no plastic deformation created in the electro-polishing process. Although electro-polishing process is very effective in removing the plastic deformation, it has its disadvantages that the electrical current must be controlled carefully at a constant value; otherwise damages are induced on the surface. Vibratory polishing is another process that can remove the plastic deformation layer. In a vibratory polisher, a vertical vibration is added in addition to the rotation of the polishing pad. The down force pressure during polishing is thus minimized by the vibration so that only the peak material on the surface is removed in a very gentle manner. Due to the minimal down force pressure, the mechanical abrasion between the particles (usually alumina or colloidal silica) and the sample does induce significant plastic deformation. After the vibratory polishing process for bicrystalline samples, the grain boundary is observed under the light microscope as shown in Fig. 7. The pen mark is made on the surface of the bicrystal where the single grain boundary is formed during the growth of the bicrystal.

Temperature and Strain Rate Dependency on Single Crystal and Polycrystalline Metals

The TRISE model incorporates the temperature and strain rate parameters into the expression of the length scale. As the expression of hardness is written using the length scale, the hardness is predicted to be dependent on the temperature and the strain rate. In order to verify the prediction of the TRISE model and material parameters needed for the length scales, nanoindentation experiments are conducted on various metals with single crystal and polycrystalline microstructures at different temperatures and strain rates. The material parameters are determined by comparing the TRISE model and the experiments. The length scales are eventually

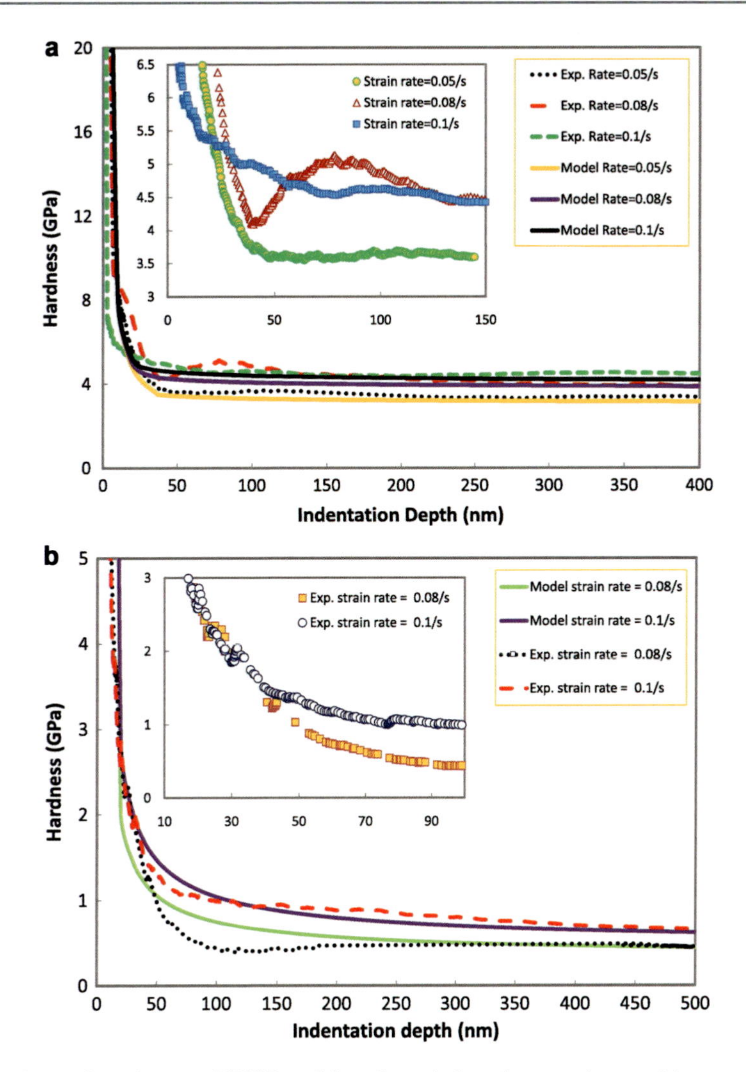

Fig. 8 Comparisons between TRISE models and nanoindentation experiments: (**a**) copper single crystal and (**b**) aluminum single crystal (Reprinted from Voyiadjis et al. (2011))

determined for single crystal and polycrystalline materials at different strain rates and temperatures.

In aluminum and copper single crystals, nanoindentation experiments are performed on a polished surface at different strain rates of 0.05 s^{-1}, 0.08 s^{-1}, and 0.10 s^{-1}. It shows from the experimental results in Fig. 8 that the hardness increases with the increasing strain rate for both metals. The TRISE model for single crystal given by Eq. 35 is applied by giving different values for the parameter equivalent plastic strain rate $\dot{p}$. It shows a good agreement between the experimental results and the predictions from the hardness expression. This proves that the TRISE

Table 1 Materials parameters used in Eq. 35 for single crystals (Reprinted from Voyiadjis et al. 2011)

Parameters	h_1 (nm)	δ_1	δ_2	δ_3	M	c	θ (*Rad*)	m	β	q
Copper	32	7,050	50	0.01	0.5	1	0.358	0.474	2	0.3
Aluminum	50	4,000	48	0.01	0.5	1	0.358	0.474	1.5	0.3

model is able to predict the influence given by the strain rates. Moreover, there is no hardening-softening phenomenon observed which validates the assumption that no dislocation is accumulated with the absence of grain boundaries.

The material parameters shown in Table 1 are obtained through the comparisons and the length scales of copper and aluminum single crystals as shown in Fig. 9.

Although no hardening-softening phenomenon is observed due to the absence of the grain boundaries, the experimental results and the TRISE model both show that the hardness decreases with increasing indentation depth. This is due to the generation of GNDs at small depth where nonuniform deformation is significant during the initial formation of the indent. The GNDs also interact with the SSDs as obstacles of the movement of SSDs, giving rise to the increase of hardness at smaller depths. The material length scales for both single crystals show the length scale decrease with the increase of the equivalent plastic strain, which means the strain gradient is greater when deformation is smaller. The determination of the length scale verifies the strain gradient plasticity theory that the stress mechanical responses are greater when the strain gradients are higher. It also shows in the length scales that a variable length scale expression needs to be used instead of the constant length scale determined in the previous research, which proves that a variable length scale is more realistic and the expression can be applied to different problems.

Nanoindentation experiments are conducted on polycrystalline copper and aluminum in order to capture the hardening-softening phenomenon as well as the strain rate dependency. TRISE model given by Eq. 36 is applied to compare with the experimental results. As shown in Fig. 10, the hardness initially decreases with indentation depth at very small depth less than 50 nm. The hardening-softening phenomenon is observed following the decreasing hardness segment. The experimental results show that at very small indentation depth where the plastic deformation does not reach the grain boundaries, the polycrystalline materials behave similarly with respect to single crystal materials. However, due to the presence of the grain boundaries, the expansion of the plastic zone is constrained and dislocations start to accumulate near the grain boundary. The local increase of the dislocation density gives rise to the increase of the material hardness, causing the hardening-softening phenomenon. Nanoindentation experiments are also performed at different strain rates. It shows in both experiments and TRISE model that the hardness increases with the increasing strain rate, verifying the strain rate dependency on polycrystalline materials.

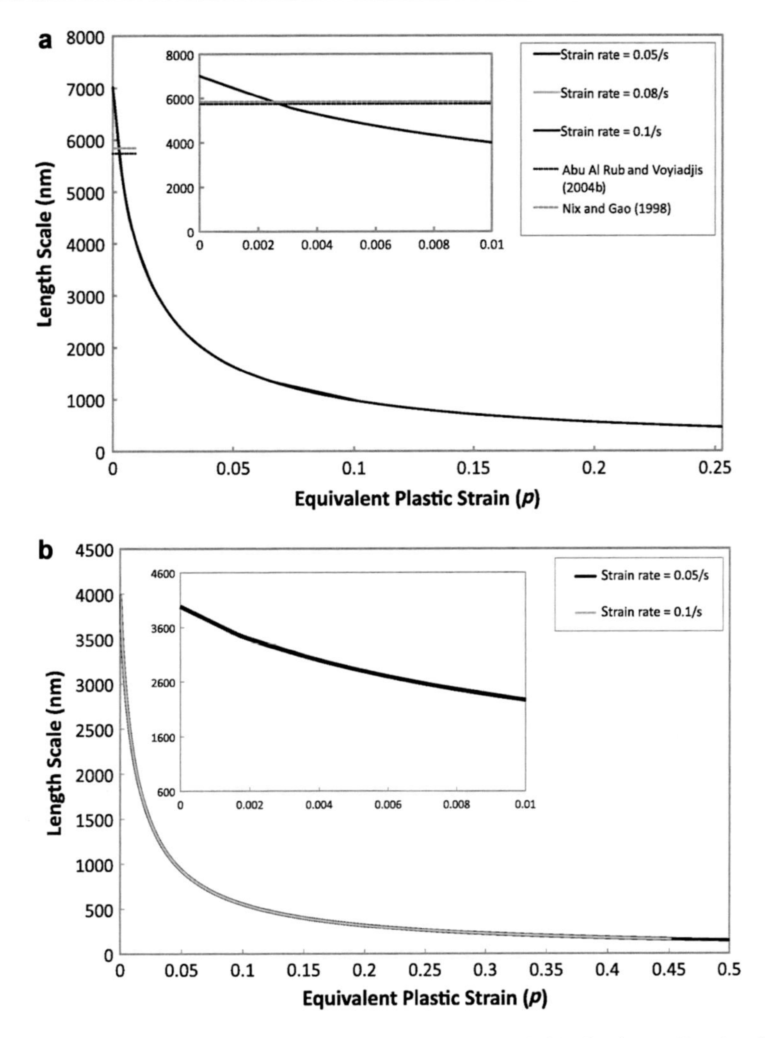

Fig. 9 Material length scales of single crystals: (**a**) copper and (**b**) aluminum (Reprinted from Voyiadjis et al. (2011))

The length scales for polycrystalline copper and aluminum can be determined by comparing the experimental results and TRISE model in determining the material parameters. Using the parameters given in Table 2, the length scales of both materials are obtained as a function of the equivalent plastic strain as shown in Fig. 11.

As shown in Fig. 11, the length scale decreases with the increasing strain rate. When the strain rate is increased, the deformation occurs faster and the dislocations do not have the sufficient time to generate and are trapped by the local nonuniform deformation. This causes the additional increase of the strain gradient, resulting in the increase of the material hardness.

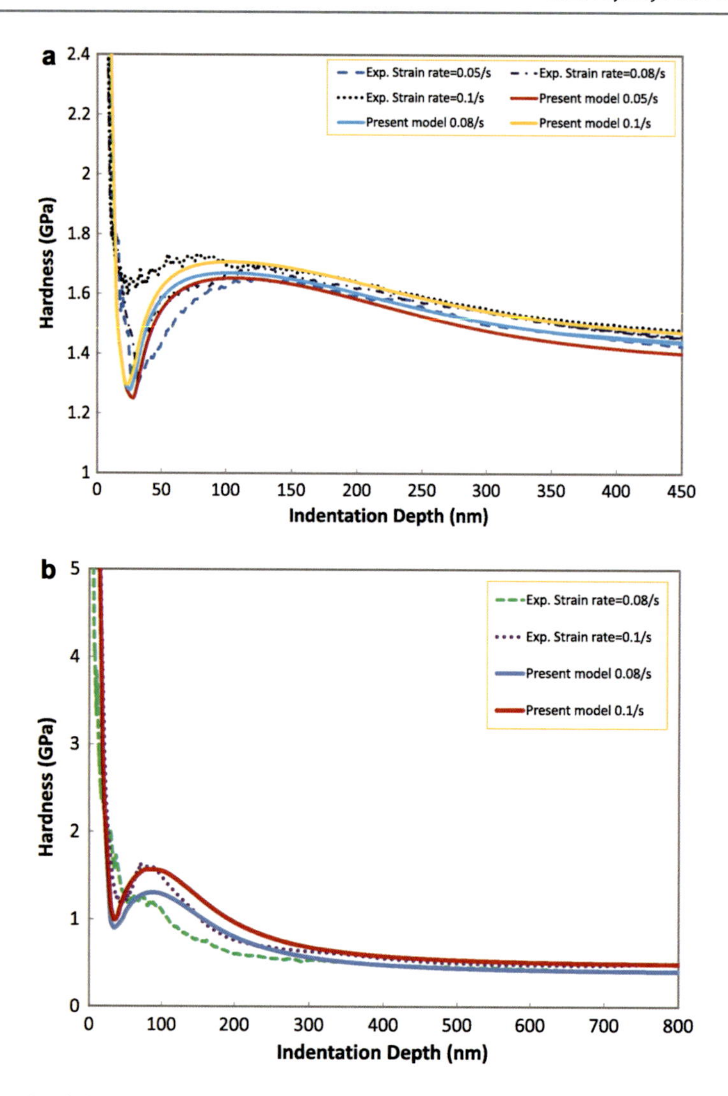

Fig. 10 Nanoindentation experiments and TRISE model of polycrystalline materials: (**a**) copper and (**b**) aluminum at different strain rates (Reprinted from Voyiadjis et al. (2011))

Table 2 Materials parameters used in Eq. 36 for polycrystalline materials (Reprinted from Voyiadjis et al. 2011)

Parameters	h_1 (nm)	δ_1	δ_2	δ_3	d (nm)	M	c	θ (*Rad*)	m	β	q
Copper	26	2.1	1	0.01	1,000	0.5	1	0.358	0.474	2	0.3
Aluminum	41	1.9	1	0.01	1,000	0.5	1	0.358	0.474	1.5	0.3

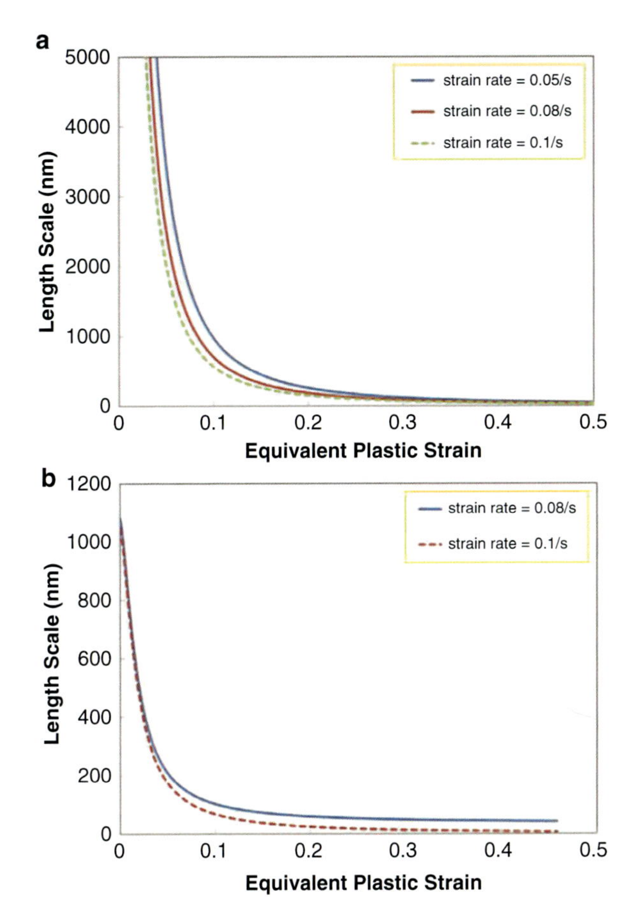

Fig. 11 Length scales as a function of the equivalent plastic strain for polycrystalline materials: (**a**) copper and (**b**) aluminum at different strain rates (Reprinted from Voyiadjis et al. 2011)

Nanoindentation experiments are performed on iron (Bahr et al. 1999) and gold (Volinsky et al. 2004) at different temperatures in order to study the temperature dependency. Different temperatures are given to the TRISE model and the comparisons between the experiments and models are presented in Fig. 12 Voyiadjis and Faghihi (2012). It shows that the hardness decreases with the increase of the temperature. The higher mobility of the dislocations causes the decrease of the dislocation density and thus the hardness decreases.

The length scales of iron and gold are obtained through the comparison and it shows in Fig.13 that the length scales increase with the increasing temperature, causing the decrease in the material hardness. As iron is a BCC metal and gold is an FCC metal, it shows that the TRISE model works for both BCC and FCC metals in predicting the indentation size effect and determining the length scales.

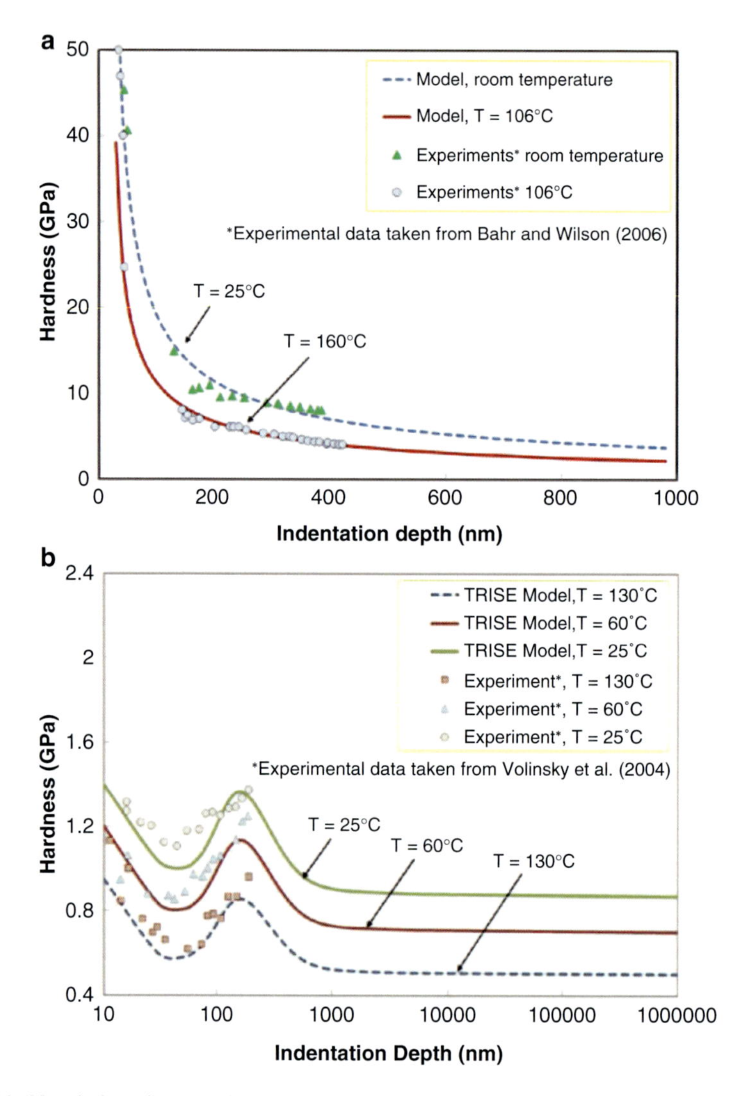

Fig. 12 Nanoindentation experiments at different temperatures on (**a**) iron single crystal and (**b**) gold polycrystalline film (Reprinted from Voyiadjis and Faghihi (2012))

Influence of the Grain Boundaries on Bicrystal Metals

As discussed in the previous section, the distance between the indenter and the grain boundary during nanoindentation has an impact on the hardness behavior of materials with grain boundaries. In the case of polycrystalline materials, the grain size d characterizes the interactions between the dislocations and the grain boundaries. The distance between the indenter and the grain boundary becomes

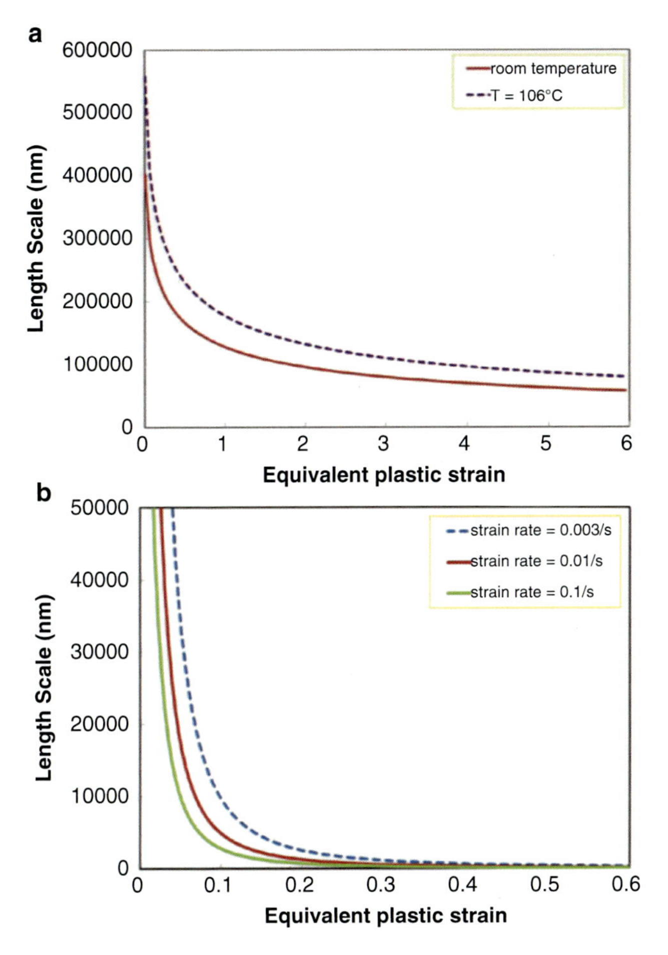

Fig. 13 Length scales at different temperatures: (**a**) iron and (**b**) gold (Reprinted from Voyiadjis and Faghihi (2012))

greater when the grain size increases. Nanoindentation experiments are performed on Nickel samples with different grain sizes (Yang and Vehoff 2007). The TRISE model given by Eq. 36 is applied by assigning different values to the grain size d (Voyiadjis and Faghihi 2012).The comparison between the experiments and the prediction of TRISE model is shown in Fig. 14. It shows in the experiments that as the grain size increases, the hardness during the hardening-softening segment decreases. As the dislocations only accumulate when they move to the grain boundaries, the greater grain size allows the dislocations to move in a greater space comparing to the condition of smaller grain size. The dislocation density thus decreases, resulting in the decrease in hardness. The experimental results show a good agreement with the predictions given by the TRISE model at different grain sizes as shown in Fig. 14.

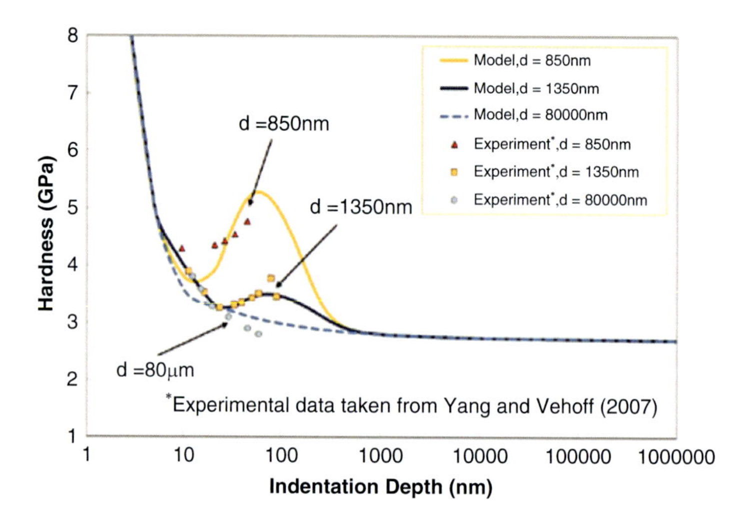

Fig. 14 Comparison between the experiments and the TRISE model at different strain rates (Reprinted from Voyiadjis and Faghihi (2012))

The size of the plastic zone increases with the increasing indentation depth. When the grain size is smaller, the plastic zone reaches the grain boundary at lower depth. As the grain size increases, the plastic zone reaches the grain boundary at greater depth. When the grain size is as much as 80 μm, the size of the plastic zone is smaller than the grain size. There is no accumulation of the grain boundary and no hardening-softening phenomenon is observed. The length scales at different grain sizes are obtained from the comparison. It shows in Fig. 15 that the length scale decreases as the grain size decreases. The grain boundaries act as constrains that block the influence of the strain gradient, which causes the smaller characteristic length scale when the grain size is smaller.

In order to isolate the investigation on the influence of the grain boundary, bicrystalline materials are tested near the single grain boundary at different distances. After the Aluminum sample is polished, the single grain boundary is observed as shown in Fig. 7. However, it is only visible using a 10× microscope in the Nanoindenter. The distance between the indenter and the grain boundary can be only measured using a 40× microscope. In order to identify the position of the grain boundary, the two indents are made on two points on the grain boundary under a 10× microscope. The grain boundary can thus be represented by the straight line connecting the two marking indents under the 40× microscope. A straight line of nanoindentations are performed near the grain boundary with the angle between the line of indents and the grain boundary to be 20° as shown in Fig. 16a (Voyiadjis and Zhang 2015).

It is noted that the depths of the indentations are 500 nm and the spacing is 5 μm. In order to make sure that there is no interaction between the close indents, experiments are made on other locations and the results show that there is no

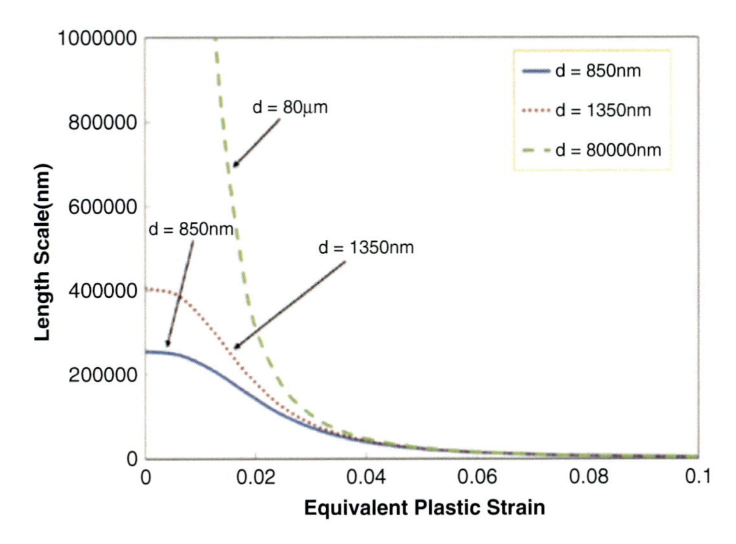

Fig. 15 Length scales as a function of the equivalent plastic strain at different grain sizes (Reprinted from Voyiadjis and Faghihi (2012))

significant interaction observed. The difference in the microstructures from two grains is not considered here. Therefore, the five indents on one side of the grain boundary are selected for the investigation. The distances between the center of the indents and the grain boundary are measured using the scale in the image as shown in Fig. 16b. The experimental results of the five indents are presented in Fig. 17. It shows that for the two indents that are far from the grain boundary, there is no hardening-softening phenomenon observed. The grain boundary does not have the impact on the accumulation of dislocations when the distance is large, showing the single crystal behavior. It also shows in Fig. 17 that hardening-softening phenomenon is observed for the rest three indents and as the distance becomes smaller, there is a greater hardening effect. This observation verifies the assumption of the influence of the grain boundary that more dislocations are accumulated between the indenter and the grain boundary when the distance is smaller. The higher dislocation density at smaller distances causes the increase in hardness.

The developed TRISE model for bicrystalline materials given by Eq. 38 is applied by assigning different values of the distance r measured from the experiments. The comparison between the TRISE model and the experimental results of the three indents with the grain boundary effect is given in Fig. 18. The developed TRISE model shows its capability to predict the hardness with the presence of the single grain boundary. It also reflects the prediction that there is a higher hardening effect as the indents are made closer to the grain boundary. The material parameters in Eq. 38 are determined by the comparison between the TRISE model and the experimental results. The material length scales are obtained as shown in Fig. 19 through the determination of material parameters given in Table 3.

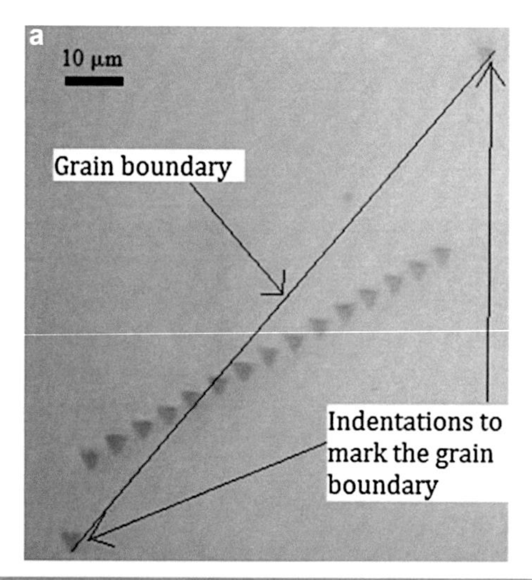

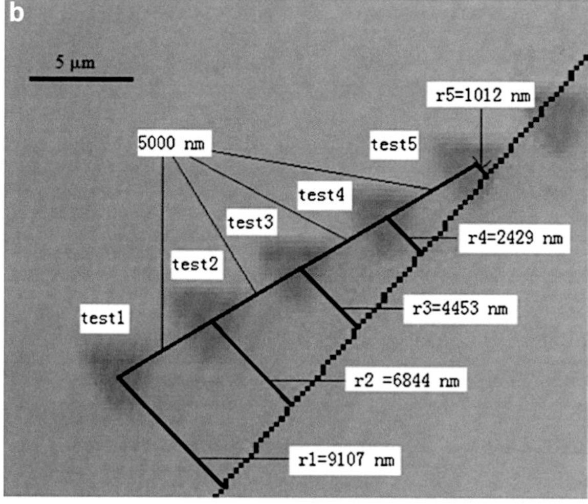

Fig. 16 Nanoindentations near the grain boundary at different distances: (**a**) image under the 40× microscope and (**b**) the magnified image of (**a**) (Reprinted from Voyiadjis and Zhang (2015))

The length scale decreases as the distance between the indenter and the grain boundary decreases. The grain boundary and the indenter both act as obstacles of the influence of the strain field. Therefore, the impact range of the strain gradient is shorter when the distance becomes smaller, resulting in a smaller characteristic length scale. The decrease in the length scale, in return, causes the increase of the strain gradient and thus causes the additional increase on material hardness.

Similar nanoindentation experiments are performed on a bicrystalline copper which is also an FCC metal. The grain boundary is also represented by a straight line connecting the two marking indents made on two points of the straight

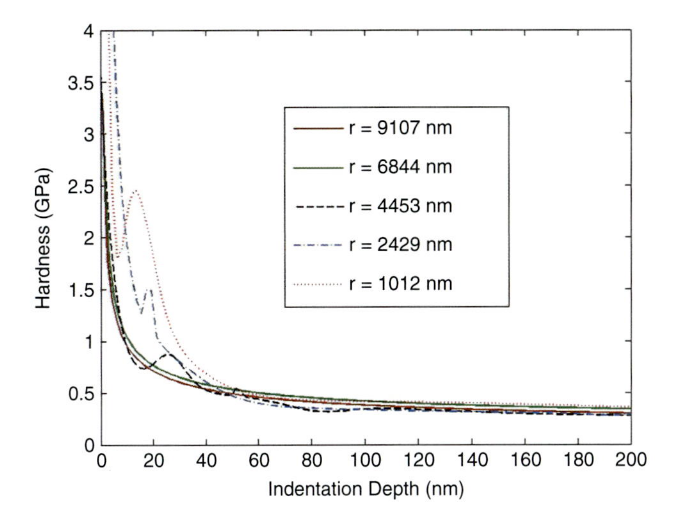

Fig. 17 Hardness versus indentation depth curves for five indents with different distances between the indenter and the grain boundary. The two *solid curves* represent the indents without hardening-softening phenomenon. The three *dashed lines* show the influence of the grain boundary on material hardness (Reprinted from Voyiadjis and Zhang (2015))

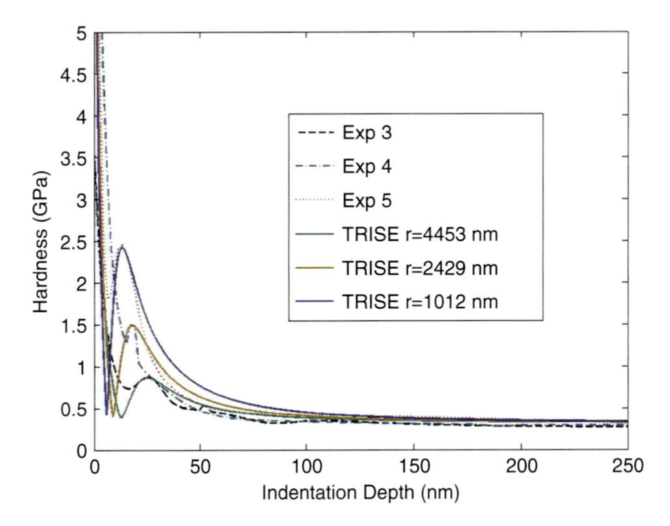

Fig. 18 Comparison of TRISE model with experimental results for the three indents with hardening-softening effect. The *solid lines* are obtained from the TRISE model and the *dashed lines* are the experimental results (Reprinted from Voyiadjis and Zhang (2015))

grain boundary. A line of indents are made across the grain boundary with the angle between indentation line and the grain boundary to be 5° as shown in Fig. 20a (Zhang and Voyiadjis 2016). The image in Fig. 20a is zoomed in order to show the details of the distance between the indent and the grain boundary as

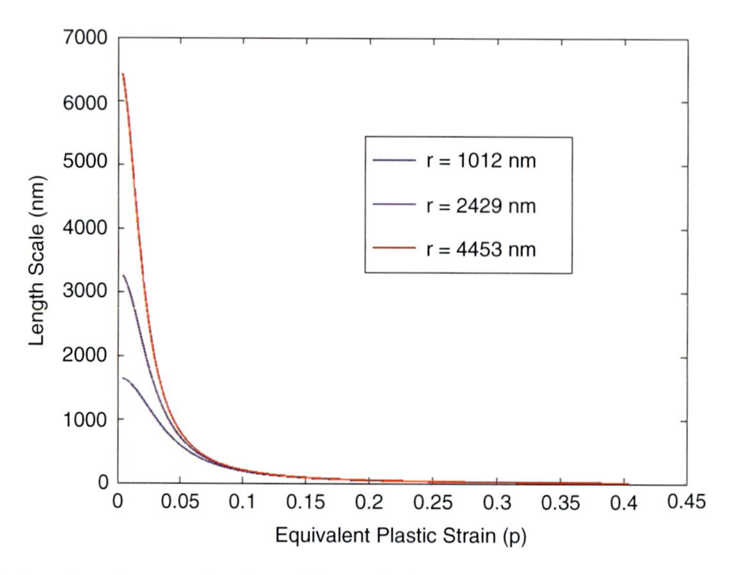

Fig. 19 Length scales as a function of the equivalent plastic strain at different values of r (Reprinted from Voyiadjis and Zhang (2015))

Table 3 Material parameters for bicrystalline Aluminum (Reprinted from Voyiadjis and Zhang 2015)

Material parameters	h_1 (nm)	δ_1	δ_2	δ_3
Aluminum	10	1.9	135	0.01

shown in Fig. 20b. The distances are measured using the scale considering the magnification. Artificial effect is added to Fig. 20b in order to show the clear contrast between the indents and the sample surface.

The experimental results of the three indents shown in Fig. 20b on the same side of the grain boundary as well as the indent made right on the grain boundary are presented in Fig. 21. The indent on the grain boundary solely decreases with the increasing indentation depth. When the indent is made on the grain boundary, the dislocations start to generate into the two grains on both sides of the grain boundary. There is no additional obstacle of the generation of dislocations and therefore no hardening-softening phenomenon is observed. This proves the assumption of the influence of the grain boundary on the other side. Hardening-softening effect is observed for the three indents within a close distance to the grain boundary. It also shows in Fig. 21 that as the distance becomes smaller, there is a greater hardening effect, similarly to the observation made from the experiments of Aluminum bicrystal. The experiments on Copper bicrystal confirm the influence of the grain boundary on the material hardness on FCC metal.

Due to the tip rounding of the indenter, strain hardening may be induced during nanoindentation at very small indentation depths. In order to confirm that the hardening is solely attributed to the accumulation of dislocations, the elastic moduli

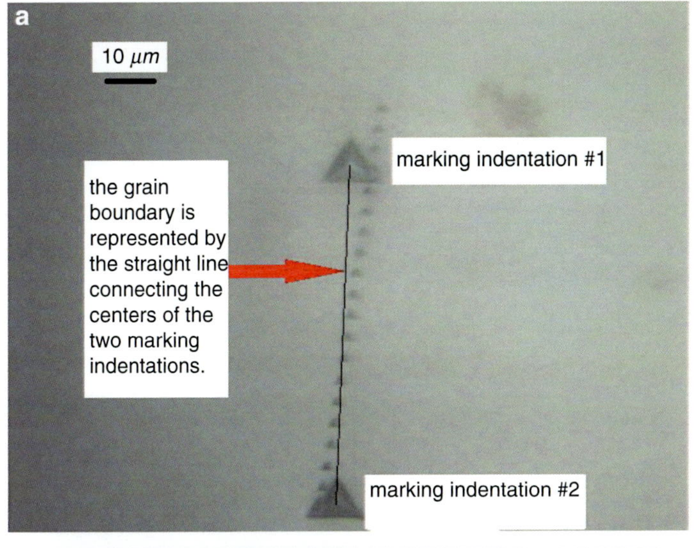

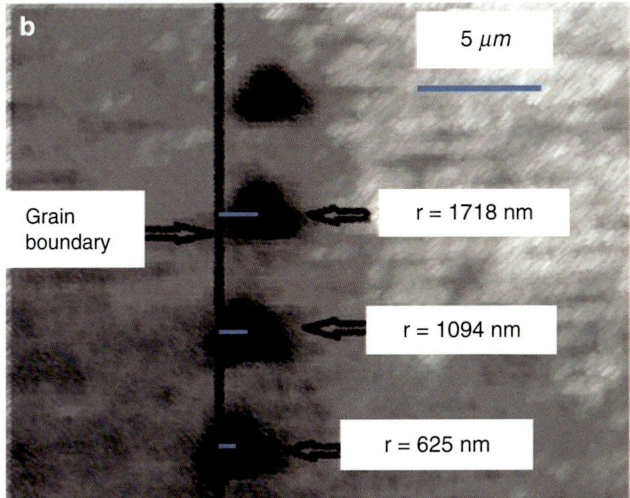

Fig. 20 Nanoindentation experiments with different distances from the grain boundary: (**a**) image under the 40× microscope and (**b**) a zoomed view of (**a**) (Reprinted from Zhang and Voyiadjis (2016))

of the three indents at different distances are presented in Fig. 22. It shows that after the depth of 20 nm, the elastic moduli from the three indents are constants on the average. This means that after the depth of 20 nm where the hardening effect is captured, there is no strain hardening effect induced as the elastic modulus is not dependent on the indentation size effect but only dependent on the strain hardening. The information of elastic moduli confirms that the hardening effect observed for

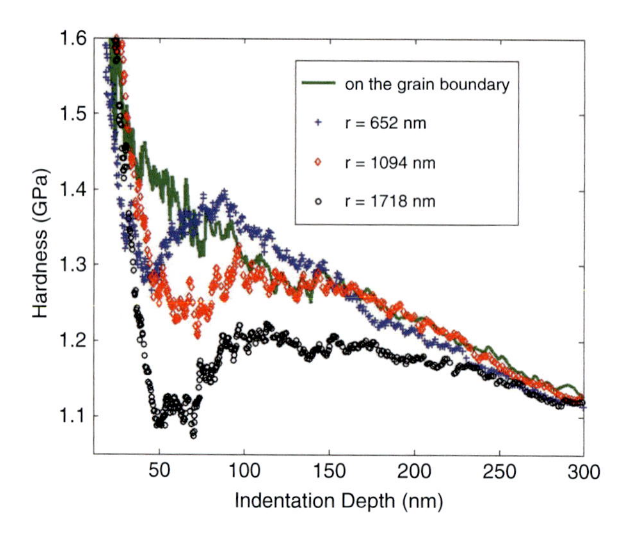

Fig. 21 Hardness versus indentation depth curves from nanoindentation experiments at different distances to the grain boundary and *right* on the grain boundary (Reprinted from Zhang and Voyiadjis (2016))

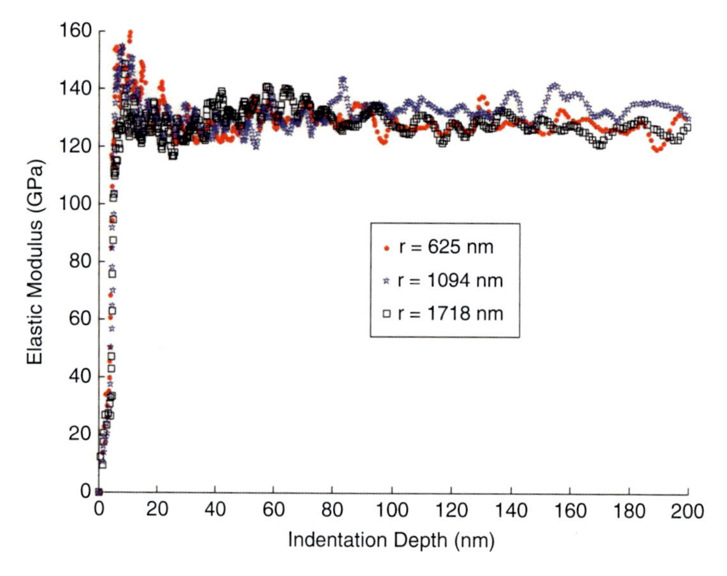

Fig. 22 Elastic modulus as a function of the indentation depth for three indents near the grain boundary (Reprinted from Zhang and Voyiadjis (2016))

the Copper bicrystal is only attributed to the accumulation of dislocations between the indenter and the grain boundary.

The developed TRISE model given by Eq. 38 is applied for the copper bicrystal with different values of distances. The comparison between the TRISE model and

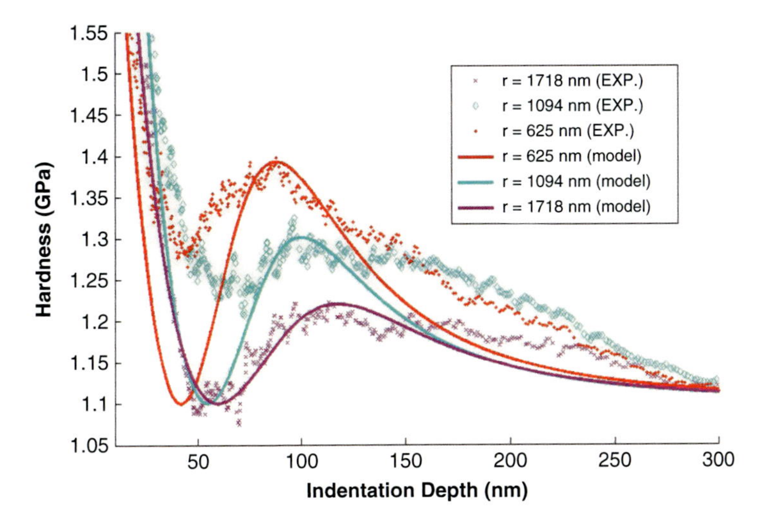

Fig. 23 Comparison between the TRISE model and nanoindentation experiments at different distances between the indenter and the grain boundary (Reprinted from Zhang and Voyiadjis (2016))

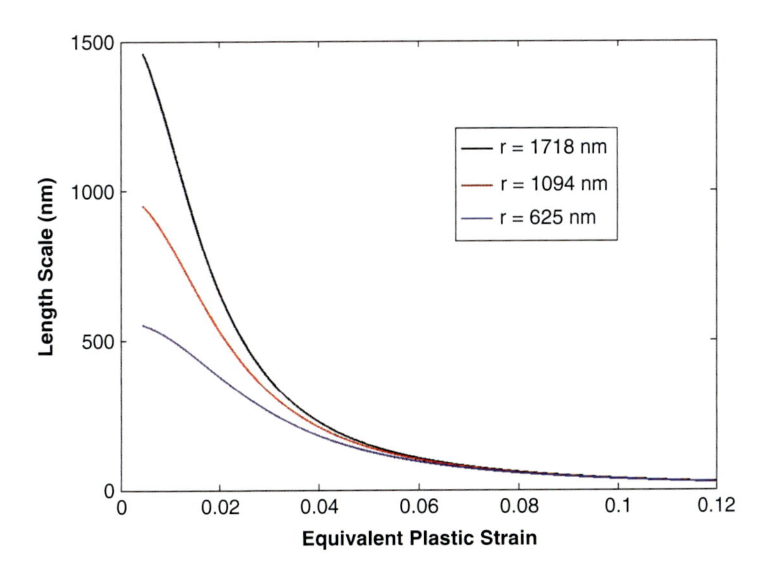

Fig. 24 Material length scales as a function of the equivalent plastic strain for Copper bicrystal at different distances between the indenter and the grain boundary (Reprinted from Zhang and Voyiadjis (2016))

Table 4 Material parameters of Copper bicrystal (Reprinted with permission from Zhang and Voyiadjis 2016)

Material parameters	$h_1(nm)$	δ_1	δ_2	δ_2
Copper	42	2.0	3.0	0.01

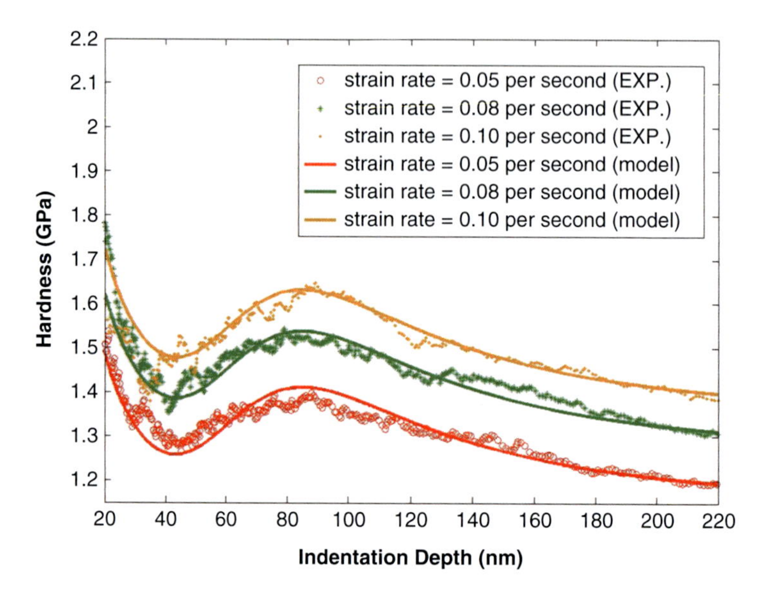

Fig. 25 Hardness vs. indentation depth for nanoindentation experiments at different strain rates with the same distance r (Reprinted from Zhang and Voyiadjis (2016))

experimental results is presented in Fig. 23. The length scales at different distances for the Copper bicrystal are determined as shown in Fig. 24 using the material parameters determined from the comparison as shown in Table 4.

The comparison shows the capability of the developed TRISE model on Copper bicrystal. As shown in Fig. 24, the length scale decreases as the distance r becomes smaller, which confirms the influence of the grain boundary on the material behavior of FCC metals.

Moreover, the strain rate dependency is studied for Copper bicrystal. Nanoindentation experiments are performed at the same distance from the grain boundary at different strain rates of $0.05s^{(-1)}$, $0.08s^{-1}$, and $0.10s^{-1}$. As shown in Fig. 25, the hardness increases with the increasing strain rates. It is worth noting that this increase is not only for the hardening-softening segment but for the entire depth. This is because the strain rate has the influence on material behaviors in both micro- and macroscales.

The TRISE model given by Eq. 38 is applied with different values of strain rates. The material length scales are determined using different values of strain rates and the material parameters given by Table 4. It shows in Fig. 26 that the length scale

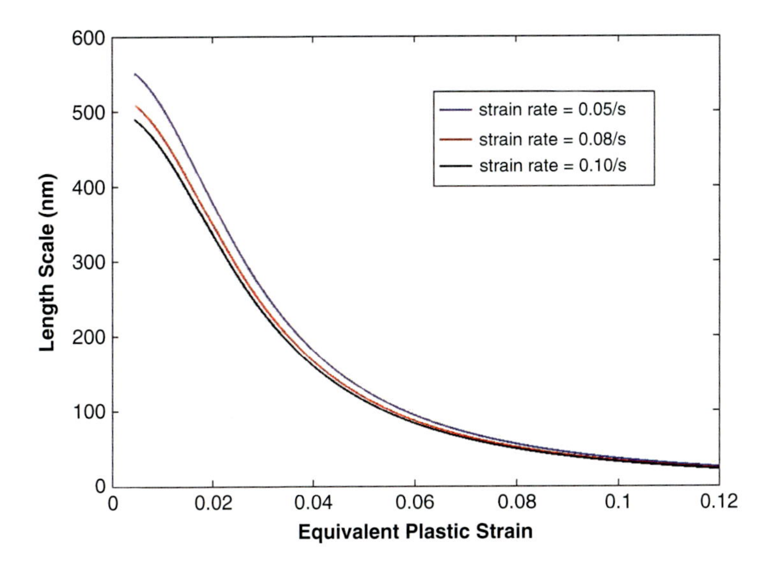

Fig. 26 Length scales as a function of the equivalent plastic strain at different strain rates for Copper bicrystal (Reprinted from Zhang and Voyiadjis (2016))

decreases with the increase of the strain rate. The strain rate dependency for Copper bicrystal is similar with the observation made from the polycrystalline copper.

Summary and Conclusion

The size effect during nanoindentation is addressed in this chapter. The classical continuum theory is not capable in predicting the indentation size effect as it does not incorporate the length scale in the constitutive expression. Strain gradient plasticity theory is applied in order to capture the size effect. The expression of a physically based length scale is determined from the strain gradient plasticity theory. In order to determine the material length scales, nanoindentation experiments are performed on single crystals, polycrystalline materials and bicrystalline materials. The constitutive equation is mapped into hardness versus indentation depth and the material parameters are determined from the expression of hardness and the experimental results.

The material length scale is dependent on the temperature, equivalent plastic strain rate, and grain size in polycrystalline materials. A TRISE model is applied in order to capture the dependencies of temperature, strain rate, and grain size. Nanoindentation experiments are performed on single crystal and polycrystalline materials. From the comparison between the TRISE model and experimental results, the length scales at different temperatures, strain rates, and grain sizes are determined.

The results of nanoindentation experiments on single crystal Copper and Aluminum show that the hardness solely decreases with the increase of the indentation depth. The TRISE model for single crystal is able to predict the hardness obtained from experiments of single crystal materials. The nanoindentation experiments on polycrystalline Copper and Aluminum show the hardening-softening effect due to the presence of grain boundaries. The TRISE model is capable in predicting the hardening-softening phenomenon. The temperature dependency is addressed by the TRISE model at different temperatures and nanoindentation experiments on iron and gold. Both TRISE model and experiments show that the hardness decreases with the increasing temperatures. Nanoindentation experiments are performed at different strain rates on both polycrystalline and single copper and aluminum. It is shown by the TRISE model and experimental results that hardness increases as the strain rate increases. The material length scales are determined by the comparison between the TRISE model and the experiments. The dependencies of length scales on the temperature, strain rate, and grain size verify the theory that higher strain gradient causes greater material hardness.

In order to isolate the influence of the grain boundary, bicrystalline copper and aluminum are used and nanoindentation experiments are performed near the grain boundary at different distances. The TRISE model is developed based on the structure of bicrystals. The length scales are determined by comparing the developed TRISE model and the experimental results on bicrystalline materials.

The experimental results on bicrystalline copper and aluminum show that the material behaves like a single crystal when the indents are made with large distances from the grain boundary. The hardening-softening phenomenon is only observed for the indents made in close proximity to the grain boundary. The experimental observation validates that the increase of hardness is attributed to the presence of the grain boundary. The accumulation of dislocations near the grain boundary causes the increase of dislocation density, resulting in the additional hardening effect. The influence of the grain boundary is further investigated through nanoindentation experiments near the grain boundary at different distances. The TRISE model is developed for bicrystalline materials by replacing the grain size d in polycrystalline models with the distance r between the indents and the grain boundary. The developed TRISE model shows its capability to predict the fact that there is a stronger hardening effect when the distance r becomes smaller as shown by nanoindentation experiments, providing a new type of size effect with respect to the distance r. The length scales of bicrystalline copper and aluminum show that as the indents are made closer to the grain boundary, the length scales decrease. The grain boundary and the indenter act as obstacles that prevent the strain gradient from influencing the space outside the volume constrained by the obstacles. Therefore, the characteristic length scale at smaller distance is lower and in return causes the increase in the strain gradient and dislocation density. The increase of dislocation density causes the additional hardening effect during nanoindentation. The rate dependency is also investigated on bicrystalline copper through nanoindentation experiments at different strain rates with a fixed distance from the grain boundary. Both experimental results and the TRISE model show an increase of hardness with the increasing strain rate, similarly to the rate dependency in single crystal and polycrystalline materials.

For all the length scales determined in this chapter, it shows that the length scale decreases with the increasing equivalent plastic strain and approaches to zero when the deformation is large. This behavior of the length scale proves the strain gradient plasticity theory in addressing the size effect during nanoindentation. When the indentation depth is small, the shape of the indent is initially formed. GNDs are required to accommodate the nonuniform deformation in the formation of the indent. As the indenter penetrates deeper, the indent grows to a greater size with a similar shape because of the use of the self-similar Berkovich indenter. The amount of uniform deformation increases and SSDs are required to generate the uniform deformation. Therefore, the GND density decreases gradually with the increase of the indentation depth and the strain gradient becomes smaller. The influence of the decreasing strain gradient causes the decrease of the characteristic length. When the deformation becomes larger in the macroscopic scale, uniform deformation dominates and there is no influence of the strain gradient, resulting in a zero value of the length scale in large deformations.

References

E.C. Aifantis, Int. J. Eng. Sci. **30**, 10 (1992)
A. Arsenlis, D.M. Parks, Acta Mater. **47**, 5 (1999)
D.F. Bahr, D.E. Wilson, D.A. Crowson, J. Mater. Res. **14**, 6 (1999)
D.J. Bammann, E.C. Aifantis, Acta Mech. **45**, 1–2 (1982)
D.J. Bammann, E.C. Aifantis, Mater. Sci. Eng.: A **309** (1987)
M.R. Begley, J.W. Hutchinson, J. Mech. Phys. Solids **46**, 10 (1998)
X. Chen, N. Ogasawara, M. Zhao, N. Chiba, J. Mech. Phys. Solids **55**, 8 (2007)
M.S. De Guzman, G. Neubauer, P. Flinn, W.D. Nix, MRS Proc **308**, 613 (1993)
R.J. Dorgan, G.Z. Voyiadjis, Mech. Mater. **35**, 8 (2003)
A.C. Eringen, D.G.B. Edelen, Int. J. Eng. Sci. **10**, 3 (1972)
N.A. Fleck, J.W. Hutchinson, J. Mech. Phys. Solids **41**, 12 (1993)
N.A. Fleck, J.W. Hutchinson, Adv. Appl. Mech. **33** (1997)
N.A. Fleck, G.M. Muller, M.F. Ashby, J.W. Hutchinson, Acta Metall. Mater. **42**, 2 (1994)
J.J. Gracio, Scr. Metall. Mater. **31**, 4 (1994)
E. Kroner, Int. J. Appl. Phys. **40**, 9 (1967)
Q. Ma, D.R. Clarke, J. Mater. Res. **10**, 4 (1995)
K.W. McElhaney, J.J. Vlassak, W.D. Nix, J. Mater. Res. **13**, 05 (1998)
H.B. Muhlhaus, E.C. Aifantis, Acta Mech. **89**, 1–4 (1991)
W.D. Nix, H. Gao, J. Mech. Phys. Solids **46**, 3 (1998)
J.F. Nye, Acta Metall. **1**, 2 (1953)
W.C. Oliver, G.M. Pharr, J. Mater. Res. **7**, 6 (1992)
N.A. Stelmashenko, M.G. Walls, L.M. Brown, Y.V. Milman, Acta Metall. Mater. **41**, 10 (1993)
J.S. Stolken, A.G. Evans, Acta Mater. **46**, 14 (1998)
D. Tabor, J. Inst. Met. **79**, 1 (1951)
I. Vardoulakis, E.C. Aifantis, Acta Mech. **87**, 3–4 (1991)
A.A. Volinsky, N.R. Moody, W.W. Gerberich, J. Mater. Res. **19**, 9 (2004)
G.Z. Voyiadjis, R.K. Abu Al-Rub, Int. J. Plast. **19**, 12 (2003)
G.Z. Voyiadjis, R.K. Abu Al-Rub, Int. J. Solids Struct. **42**, 14 (2005)

G.Z. Voyiadjis, A.H. Almasri, J. Eng. Mech. **135**, 3 (2009)
G.Z. Voyiadjis, D. Faghihi, Procedia ITUTAM, 3, pp. 205–227 (2012)
G.Z. Voyiadjis, R. Peters, Acta Mech. **211**, 1–2 (2010)
G.Z. Voyiadjis, C. Zhang, Mater. Sci. Eng. A **621**, 218 (2015)
G.Z. Voyiadjis, D. Faghihi, C. Zhang, J. Nanomech Micromech **1**, 1 (2011)
Y. Xiang, J.J. Vlassak, Scr. Mater. **53**, 2 (2005)
B. Yang, H. Vehoff, Acta Mater. **55**, 3 (2007)
H.M. Zbib, E.C. Aifantis, Res. Mechanica. **23**, 2–3 (1998)
C. Zhang, G.Z. Voyiadjis, Mater. Sci. Eng. A **659**, 55 (2016)

Size Effects During Nanoindentation: Molecular Dynamics Simulation

2

George Z. Voyiadjis and Mohammadreza Yaghoobi

Contents

Abstract

In this chapter, the molecular dynamics (MD) simulation of nanoindentation experiment is revisited. The MD simulation provides valuable insight into the atomistic process occurring during nanoindentation. First, the simulation details and methodology for MD analysis of nanoindentation are presented. The effects of boundary conditions on the nanoindentation response are studied in more detail. The dislocation evolution patterns are then studied using the information provided by atomistic simulation. Different characteristics of metallic sample during nanoindentation experiment, which have been predicted by theoretical models, are investigated. Next, the nature of size effects in samples with small length scales are studied during nanoindentation. The results indicate that the size effects at small indentation depths cannot be modeled using the forest

G. Z. Voyiadjis (✉) · M. Yaghoobi
Department of Civil and Environmental Engineering, Louisiana State University, Baton Rouge, LA, USA
e-mail: voyiadjis@eng.lsu.edu; myagho1@lsu.edu

G. Z. Voyiadjis (ed.), *Handbook of Nonlocal Continuum Mechanics for Materials and Structures*, https://doi.org/10.1007/978-3-319-58729-5_41

hardening model, and the source exhaustion mechanism controls the size effects at the initial stages of nanoindentation. The total dislocation length increases by increasing the dislocation density which reduces the material strength according to the exhaustion hardening mechanisms. The dislocation interactions with each other become important as the dislocation content increases. Finally, the effects of grain boundary (GB) on the controlling mechanisms of size effects are studied using molecular dynamics.

Keywords

Nanoindentation · Molecular dynamics · Size effects · Dislocation · Grain boundary

Introduction

Indentation is a common experiment to investigate the material properties at different length scales. During indentation, a required force to press a hard indenter into the sample is measured. In the case of nanoindentation, unlike the traditional indentation experiment, it has been observed that the hardness is not a constant value and varies during the test (Nix and Gao 1998; Al-Rub and Voyiadjis 2004; Voyiadjis and Al-Rub 2005). Many researchers have tried to study the variation of hardness, which is commonly termed as size effects, during nanoindentation. The variation of geometrically necessary dislocations (GNDs) density has been commonly considered as the mechanism which controls the hardness. Corcoran et al. (1997) investigated the dislocation nucleation and its effects on the response of Au during nanoindentation experiment. Suresh et al. (1999) studied the effects of sample thickness on the mechanical and dislocation nucleation of Cu thin films during nanoindentation. The grain boundary (GB) effects on the defect nucleation and evolution of bicrystal Fe-14 wt. %Si alloy during nanoindentation were investigated by Soer and De Hosson (2005). It was observed that the dislocations pile up against the GB (Soer and De Hosson 2005). Almasri and Voyiadjis (2010) conducted the nanoindentation of polycrystalline thin films and observed that the GB may enhance the sample hardness.

The interaction of dislocations with each other governs the material strength in bulk metallic samples which are usually captured by Taylor-like hardening models (Nix and Gao 1998; Al-Rub and Voyiadjis 2004; Voyiadjis and Al-Rub 2005). The models generally relate the strength to the dislocation density and state that the stress increases by increasing the dislocation density. Recently, researchers have been able to experimentally measure the GNDs content in samples of confined volume (Kysar and Briant 2002; Kysar et al. 2007; Zaafarani et al. 2008; Demir et al. 2009; Dahlberg et al. 2014). However, the experimental observations cannot be fully described by the bulk size models. Demir et al. (2009, 2010) conducted the nanoindentation and microbending experiments and observed that the governing mechanisms of size effects at smaller length scales are not similar to those of the large size samples. Demir et al. (2009) observed that the hardness decreases by increasing the GNDs density during nanoindentation of Cu single crystal thin films which cannot be described using the bulk-sized models. Demir et al. (2010) also showed the breakage of the dislocations mean-field theory at small length scales during the microbending of Cu single crystal thin films.

One approach to investigate the governing atomistic process of size effects during nanoindentation is to simulate the sample with the full atomistic details using MD. Many deformation mechanisms during nanoindentation of metallic thin films have been captured using MD. Incorporating the MD simulation, Kelchner et al. (1998) investigated the defect nucleation and evolution of Au during nanoindentation. The surface step effects on the response of Au during nanoindentation were investigated by Zimmerman et al. (2001) using atomistic simulation. Lee et al. (2005) conducted a comprehensive study on the defect nucleation and evolution patterns during nanoindentation of Al and tried to explain the nanoindentation response using those patterns. Hasnaoui et al. (2004), Jang and Farkas (2007), and Kulkarni et al. (2009) have studied the interaction between the dislocations and GB during nanoindentation experiment using molecular dynamics. Yaghoobi and Voyiadjis (2014) investigated the effects of the MD boundary conditions on the sample response and defect nucleation and evolution patterns during nanoindentation. Voyiadjis and Yaghoobi (2015) investigated the theoretical models developed to capture the size effects during nanoindentation using MD. Yaghoobi and Voyiadjis (2016a) investigated the governing mechanisms of size effects during nanoindentation using MD. Voyiadjis and Yaghoobi (2016) incorporated large-scale MD to study the GB effects on the material strength as the grain size varies.

This chapter is designed as follows. In section "Simulation Methodology," the general details for atomistic simulation of nanoindentation are described. In section "Boundary Conditions Effects," the effects of selected boundary conditions for MD simulation of nanoindentation are elaborated. In section "Comparing MD Results with Theoretical Models," the obtained results from MD simulation are compared to those predicted by the available theoretical models. In section "Size Effects in Small-Length Scales During Nanoindentation," the governing mechanisms of size effects in thin films of confined volumes are presented. In section "Effects of Grain Boundary on the Nanoindentation Response of Thin Films," the effects of grain boundary and grain size on the nanoindentation response of thin films are elaborated using the results obtained from MD simulation.

Simulation Methodology

The Newton's equations of motion for N interacting monoatomic molecules can be described as below:

$$m_i \ddot{\mathbf{r}}_i = -\nabla_i U + \mathbf{f}_i, i = 1, 2, \ldots, N \tag{1}$$

where $\ddot{\mathbf{r}}_i$ is the second time derivative of ith particle trajectory $\mathbf{r}_i$, m_i is the mass of ith particle, $\mathbf{f}_i$ is an external force on the ith particle, and

$$\nabla_i U = \left(\frac{\partial}{\partial x_i} + \frac{\partial}{\partial y_i} + \frac{\partial}{\partial z_i} \right) U, i = 1, 2, \ldots, N \tag{2}$$

where $U(\mathbf{r}_1, \mathbf{r}_2, \ldots, \mathbf{r}_N)$ is the potential energy. A metallic system can be described using Eq. 1 by approximating the atoms as mass points. Equation 1 should be numerically solved for the whole system. In the case of atomistic simulation, N can be a large number, and a very efficient parallel code should be used to solve the equation. The atomic interactions in metallic systems have been modeled using many different potentials such as Lennard–Jones (LJ), Morse, embedded-atom method (EAM), and modified embedded-atom method (MEAM).

The LJ potential E^{LJ} can be described as below:

$$E^{\mathrm{LJ}}\left(r_{\mathrm{ij}}\right) = 4\varepsilon \left[\left(\frac{\sigma}{r_{\mathrm{ij}}} \right)^{12} - \left(\frac{\sigma}{r_{\mathrm{ij}}} \right)^{6} \right] \tag{3}$$

where σ is the distance from the atom at which $E^{\mathrm{LJ}} = 0$ and ε is the potential well depth. A cutoff distance should be chosen for LJ potential.

Morse potential E^{Morse} can be written as follows:

$$E^{\mathrm{Morse}}\left(r_{\mathrm{ij}}\right) = D \left\{ e^{\left[-2\alpha\left(r_{\mathrm{ij}} - r_0\right)\right]} - 2e^{\left[-\alpha\left(r_{\mathrm{ij}} - r_0\right)\right]} \right\} \tag{4}$$

where D is the cohesive energy, α is the elastic parameter, and r_0 is the equilibrium distance.

EAM is a popular potential to model the metallic atoms interactions (Daw and Baskes 1984). The EAM potential E^{EAM} is described as below:

$$E^{\mathrm{EAM}}\left(r_{\mathrm{ij}}\right) = \frac{1}{2} \sum_{i,j} V\left(r_{\mathrm{ij}}\right) + \sum_{i} F\left(\rho_i\right), \quad \rho_i = \sum_{i \neq j} \varphi\left(r_{\mathrm{ij}}\right) \tag{5}$$

where $V(r_{\mathrm{ij}})$ is the pair interaction potential, $F(\rho_i)$ denotes the embedding potentials, and $\varphi(r_{\mathrm{ij}})$ is a function which is defined using the electron charge density.

The MEAM potential is the modification of EAM potential which can be described as follows (Baskes 1992):

$$E^{\mathrm{MEAM}}\left(r_{\mathrm{ij}}\right) = \frac{1}{2} \sum_{i,j} V\left(r_{\mathrm{ij}}\right) + \sum_{i} F\left(\rho_i'\right) \tag{6}$$

where

$$\rho_i' = \sum_{i \neq j} \varphi\left(r_{\mathrm{ij}}\right) + \frac{1}{2} \sum_{k,j \neq i} f_{\mathrm{ij}}\left(r_{\mathrm{ij}}\right) f_{\mathrm{ik}}\left(r_{\mathrm{ik}}\right) g_i\left(\cos\theta_{\mathrm{ijk}}\right) \tag{7}$$

θ_{ijk} is the angle between the ith, j^{th}, and k^{th} atoms. The explicit three-body term is included by introducing the functions $f_{\mathrm{ij}}, f_{\mathrm{ik}}$, and g_i.

The indenter can be modeled as a cluster of atoms. However, in order to manage the computational costs, the indenter is commonly modeled as a repulsive potential. Two types of repulsive models can be used to simulate the interaction between the indenter and thin film atoms:

- First, the spherical indenter is modeled using an indenter repulsive force which is described as below (Yaghoobi and Voyiadjis 2014):

$$\begin{aligned} F^{\mathrm{ind}}(r) &= -K^{\mathrm{ind}}(r - R)^2 \quad \text{for} \quad r < R \\ F^{\mathrm{ind}}(r) &= 0 \qquad\qquad\qquad \text{for} \quad r \geq R \end{aligned} \tag{8}$$

  where K^{ind} is the force constant, r is the atomic distance to the indenter surface, and R is the radius of indenter.
- Second, the indenter, with general geometry can be simplified using a repulsive potential which is described as below (Voyiadjis and Yaghoobi 2015):

$$E^{\mathrm{ind}}(r) = \varepsilon^{\mathrm{ind}}(r - r_c)^2 r < r_c \tag{9}$$

  where $\varepsilon^{\mathrm{ind}}$ is the force constant, r is the distance between the particle and indenter surface, and r_c is the cutoff distance.

Five different indenter geometries of right square prismatic, spherical, cylindrical, blunt conical, and conical with the spherical tip are incorporated in this chapter. During nanoindentation, the precise contact area (A) should be captured to calculate the hardness at each step. A 2D-mesh is produced from the projections of atoms in contact with the indenter. The total contact area is then calculated using the obtained 2-D mesh.

The true indentation depth h is different from the tip displacement d during nanoindentation. A conical indenter of h can be obtained as below:

$$h = \frac{(a_c - a_0)}{\tan\theta} \tag{10}$$

where θ is the cone semi-angle, $a_c = \sqrt{A/\pi}$ is the contact radius, and $a_0 = r_2 + r_c(1/\cos\theta - \tan\theta)$. The indentation depth of a spherical indenter is obtained as below:

$$h = R - \sqrt{R^2 - a_c^2} \tag{11}$$

In the case of a conical indenter with the spherical tip, h can be calculated using Eq. 11 for the spherical part. The indentation depth can be obtained as follows for the conical part:

$$h = \frac{(a_c - a_0)}{\tan(\theta/2)} + h_0 \tag{12}$$

where h_0 is the depth at which the indenter geometry changes from spherical to conical, and a_0 is the contact radius at h_0 (Fig. 1). In the cases of cylindrical and right square prismatic indenters, it is assumed that $h \approx d$ because there is no relation between indentation depth and contact area.

In order to visualize the defects, several methods have been introduced such as energy filtering, bond order, centrosymmetry parameter, adaptive common neighbor analysis, Voronoi analysis, and neighbor distance analysis which have been compared with each other by Stukowski (2012). Also, the Crystal Analysis Tool developed by Stukowski and his coworkers (Stukowski and Albe 2010; Stukowski et al. 2012; Stukowski 2012, 2014) have been incorporated to extract the dislocations from the atomistic data. Here, centrosymmetry parameter (*CSP*) and Crystal Analysis Tool are explained in more detail. *CSP* can be described as below (Kelchner et al. 1998):

$$\text{CSP} = \sum_{i=1}^{N_p} \left|\mathbf{R}_i + \mathbf{R}_{i+N_p}\right|^2 \tag{13}$$

where $\mathbf{R}_i$ and $\mathbf{R}_{i+N_p}$ are vectors from the considered atom to the ith pair of neighbors, and N_p depends on the crystal structure. For example, $N_p = 6$ for fcc

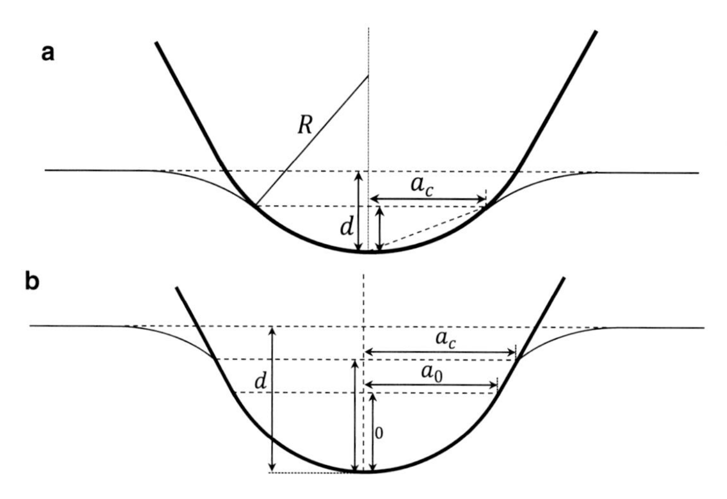

Fig. 1 The true indentation depth h for: **a** spherical part of the indenter **b** conical part of the indenter (Reprinted with permission from Yaghoobi and Voyiadjis 2016a)

materials. *CSP* is equal to zero for perfect crystal structure. However, the atomic vibration introduces a small *CSP* for atoms which are not defects. Accordingly, a cutoff should be introduced in a way that if $CSP_i < CSP_{\text{cutoff}}$, the ith atom is not considered as a defect (Yaghoobi and Voyiadjis 2014). Also, point defect is removed to clearly illustrate stacking faults. Second, the MD outputs can be postprocessed using the Crystal Analysis Tool (Stukowski and Albe 2010; Stukowski 2012, 2014; Stukowski et al. 2012). The common-neighbor analysis method (Faken and Jonsson 1994) is the basic idea of this code. The code is able to calculate the dislocation information such as the Burgers vector and total dislocation length. To extract the required information, the Crystal Analysis Tool constructs a Delaunay mesh which connects all atoms. Next, using the constructed mesh, the elastic deformation gradient tensor is obtained. The code defines the dislocations using the fact that the elastic deformation gradient does not have a unique value when a tessellation element intersects a dislocation.

Boundary Conditions Effects

One of the most important parts of the MD simulation is to select the appropriate boundary conditions which can accurately mimic the considered phenomenon. In the case of nanoindentation, the selected boundary conditions may influence the response of the simulated material. Up to now, four different boundary conditions types have been incorporated in MD to simulate the nanoindentation experiment which can be described as below (Fig. 2):

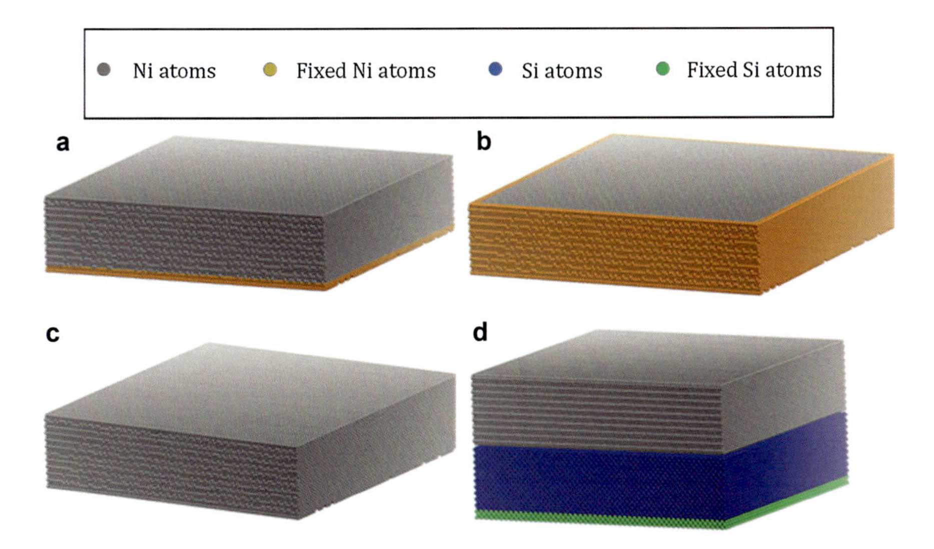

Fig. 2 Boundary conditions of thin films **a** BC1, **b** BC2, **c** BC3, and **d** BC4 (Reprinted with permission from Yaghoobi and Voyiadjis 2014)

- *BC1*: Fixing some atomic layers at the sample bottom to act as a substrate, using free surface for the top, and periodic boundary conditions for the remaining surfaces (see e.g., Nair et al. 2008; Kelchner et al. 1998; Zimmerman et al. 2001)
- *BC2*: Fixing some atomic layer at the surrounding surfaces and using free surfaces for the sample top and bottom (see e.g., Medyanik and Shao 2009; Shao and Medyanik 2010)
- *BC3*: Using free surface for the sample top and bottom, incorporating the periodic boundary conditions for the remaining surfaces, and putting a substrate under the thin film (see e.g., Peng et al. 2010)
- *BC4*: Incorporating the free surfaces for the sample top and bottom, using periodic boundary conditions for the remaining surfaces, and equilibrating the sample by adding some forces (see e.g., Li et al. 2002; Lee et al. 2005)

Yaghoobi and Voyiadjis (2014) studied the different types of boundary conditions and their effects on the dislocation nucleation and evolution patterns using samples with various thicknesses (t_f) indented by spherical indenters with different radii (R). The parallel code LAMMPS (Plimpton 1995), which was developed at Sandia National Laboratories, was selected to conduct the MD simulation. The numerical time integration of Eq. 1 was performed using the velocity Verlet algorithm. Three different types of interatomic interaction were incorporated:

- The interaction of Nickel atoms with each other (Ni-Ni)
- The interaction of Silicon atoms with each other (Si-Si)
- The interaction of Nickel atoms with Silicon ones (Ni-Si)

The three interactions were modeled using the embedded-atom method (EAM) potential for Ni-Ni interaction, Tersoff potential for Si-Si interaction, and Lennard–Jones (LJ) potential for Ni-Si interaction. The Ni-Ni interaction is modeled using the EAM potential parameterized by Mishin et al. (1999). To capture the Si-Si interaction, a three-body Tersoff potential (Tersoff 1988) was chosen. Table 1 presents the potential parameters of Si. The LJ potential E^{LJ} was used to model the Ni-Si interaction and required parameters ($\varepsilon_{\mathrm{Ni-Si}}$ and $\sigma_{\mathrm{Ni-Si}}$) are presented in Table 2. The cutoff distance of 2.5σ was selected for LJ potential. The indenter was spherical modeled using the repulsive force F^{ind} presented in Eq. 8. The centrosymmetry parameter (CSP) was incorporated to visualize the defects with the cutoff equal to 1.5 ($CSP_{\mathrm{cutoff}} = 1.5$).

Table 1 Tersoff potential parameters of Si-Si (Yaghoobi and Voyiadjis 2014)

$A = 3264.7$ eV	$B = 95.373$ eV	
$\lambda_1 = 3.2394$ Å^{-1}	$\lambda_2 = 1.3258$ Å^{-1}	
$\alpha = 0$	$\beta = 0.33675$	$n = 22.956$
$c = 4.8381$	$d = 2.0417$	$h = 0.0000$
$\lambda_3 = \lambda_2$	$R = 3.0$ Å	$D = 0.2$ Å

Table 2 LJ potential parameters of Ni-Ni, Si-Si, and Ni-Si (Yaghoobi and Voyiadjis 2014)

	ε (J)	σ (Å)
Ni-Ni	8.3134e-20	2.2808
Si-Si	2.7904e-21	3.8260
Ni-Si	1.5231e-20	3.0534

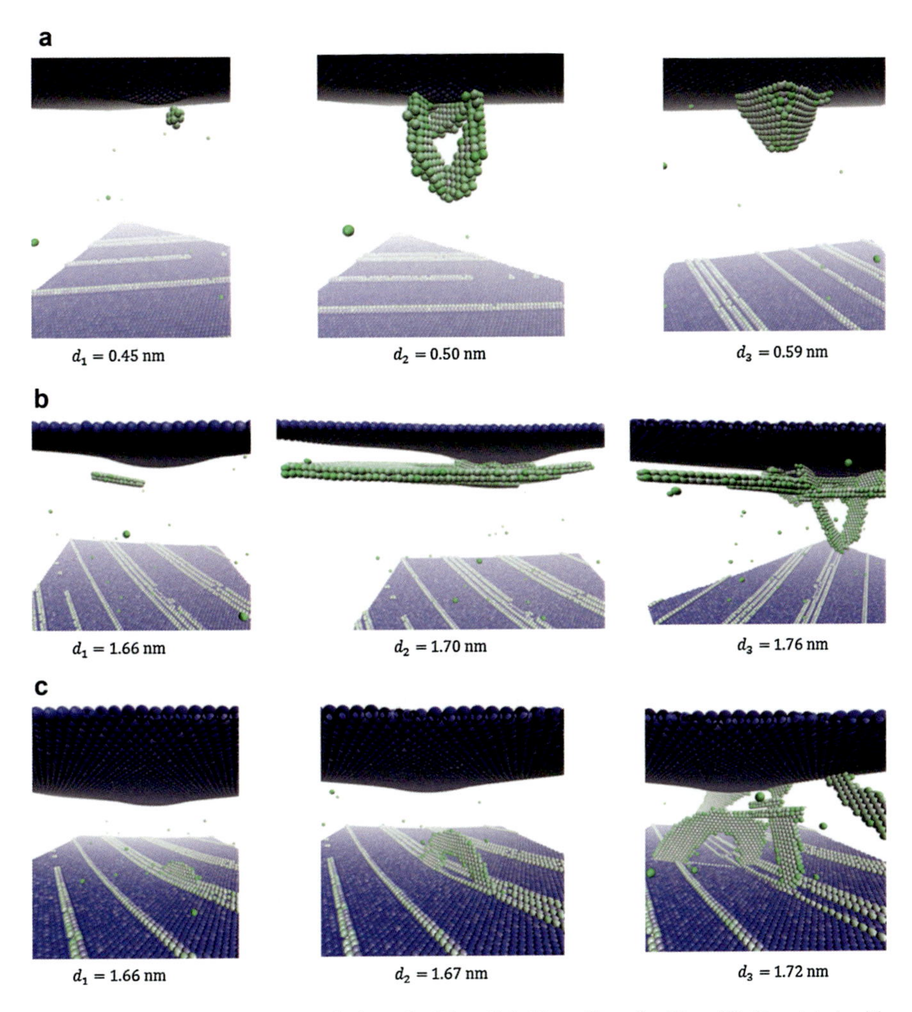

Fig. 3 Defect nucleation and evolution of **a** Type I, **b** Type II, and **c** Type III (Reprinted with permission from Yaghoobi and Voyiadjis 2014)

It was observed that the bending and indentation mechanisms control the initial stages of defect nucleation and evolution for samples with different thicknesses (Yaghoobi and Voyiadjis 2014). Accordingly, three patterns of so-called *Type I*, *Type II*, and *Type III* were observed which can be described as follows (Fig. 3) (Yaghoobi and Voyiadjis 2014):

- *Type I*: The location of initial defect nucleation is beneath the indenter. Two faces of embryonic dislocation loops evolve on $(1\bar{1}1)$ and $(11\bar{1})$ planes. Eventually, a tetrahedral sessile lock is formed. This dislocation pattern is controlled by the indentation mechanism.
- *Type II*: Again, the location of initial nucleation is beneath the indenter. However, the defect evolution occurs on the plane which is parallel to (111), which is the indentation plane. A dislocation pattern similar to *Type I* starts evolving as the indentation depth increases. Both bending and indentation mechanisms are important in this pattern.
- *Type III*: The initial dislocation is nucleated at the sample bottom. The dislocations are evolved on $\{111\}$ planes while they are moving towards the sample top. Bending is the dominant mechanism of deformation.

Figure 4 presents the nanoindentation responses of samples visualized in Fig. 3. In the case of sample with *Type I* defect structure pattern, the first load relaxation occurs due to the initial defect nucleation beneath the indenter (Fig. 4a). Two faces of embryonic dislocation loops evolve on $(1\bar{1}1)$ and $(11\bar{1})$ planes. Another load relation occurred when the tetrahedral sessile lock is shaped. Figure 4b illustrates the effects of defect evolution on the response of the sample with dislocation pattern of *Type II* during nanoindentation. The indentation force is initially relaxed due to the first defect nucleation beneath the indenter. The defect evolution occurs on the plane which is parallel to (111), which is the indentation plane. Due to the complexity of dislocation evolution pattern and activation of both bending and indentation mechanisms, the effects of defects pattern on the nanoindentation response become complicated. Figure 4c shows that indentation load is initially released due to the first dislocation nucleation at the bottom for sample with *Type III* pattern. After the initial load relaxation, the complex pattern of dislocation nucleation and evolution leads to the oscillatory response. However, the general trend is the indentation load increases during indentation.

The governing mechanisms of deformation also depend on the film thickness t_f and indenter radius R (Yaghoobi and Voyiadjis 2014). Samples with BC1 experience no bending independent of the film thickness or indenter radius which leads to the *Type I* pattern controlled by indentation mechanism. Sample with BC2 and BC3 may experience all the patterns of *Type I*, *Type II*, and *Type III* depending on the value of R/t_f. For very small values of R/t_f, the indentation governs the deformation mechanism and *Type I* pattern occurs. Increasing R/t_f, bending mechanism also becomes important and *Type II* pattern occurs. Further increasing R/t_f leads to the *Type III* pattern which is governed by the bending mechanism.

The contact pressure at the onset of plasticity p_m^y is one of the properties which is commonly investigated during nanoindentation. Yaghoobi and Voyiadjis (2014) showed that the p_m^y is also influenced by the choice of MD boundary conditions. The effect can be predicted using the pattern of dislocation nucleation and evolution. In the case of *Type I* pattern, p_m^y is independent of film thickness. However, p_m^y depends on the film thickness for samples with *Type II* and *Type III* in a way that p_m^y increases by increasing the film thickness. Comparing the dislocation nucleation and evolution

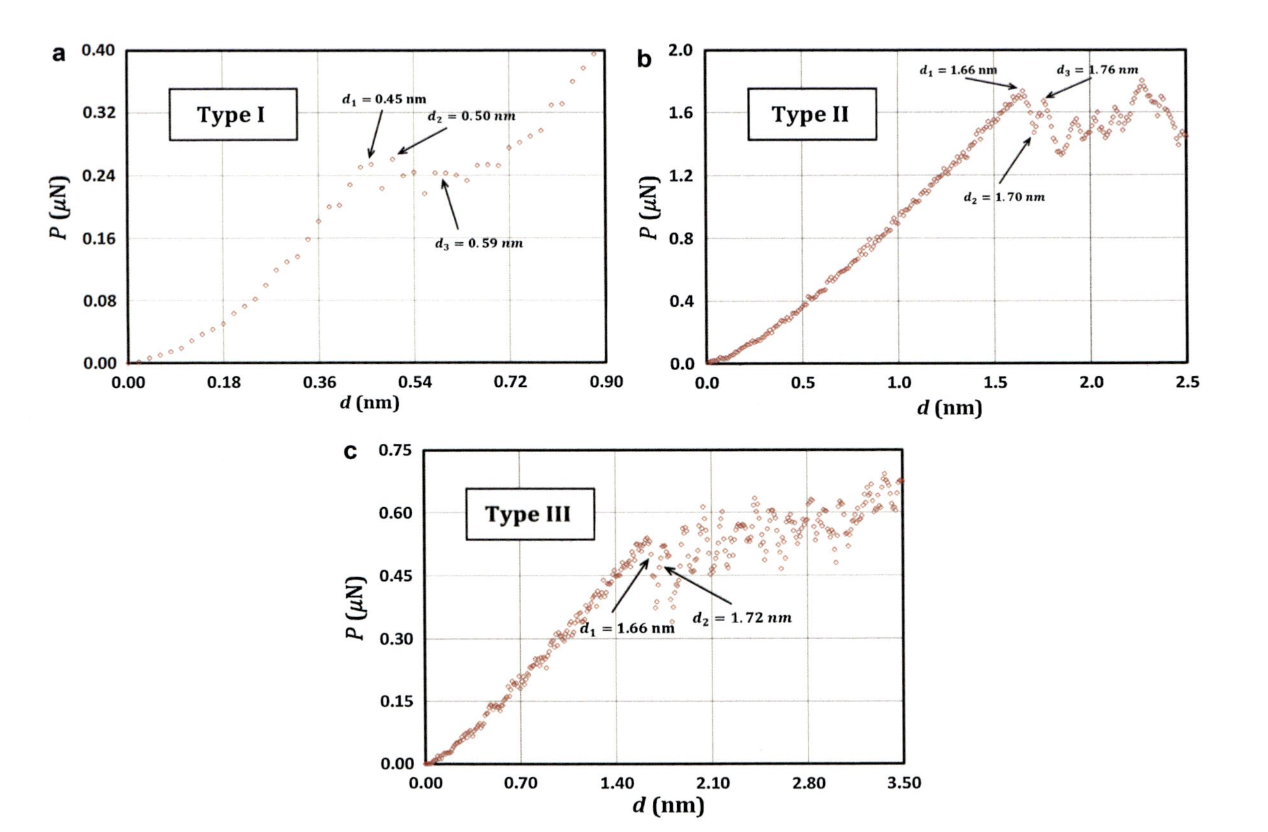

Fig. 4 Nanoindentation responses of thin films with different dislocation nucleation patterns of **a** Type I, **b** Type II, and **c** Type III (Reprinted with permission from Yaghoobi and Voyiadjis 2014)

pattern with each other, *Type I* has the largest and *Type III* has the smallest contact pressure at the time of plasticity initiation.

Comparing MD Results with Theoretical Models

Several theoretical models have been proposed to predict the variation of dislocation length during nanoindentation experiment. Nix and Gao (1998) predicted the variation of geometrically necessary dislocations (GNDs) by integrating the dislocation loops induced during nanoindentation. The dislocation length produced during nanoindentation using a conical indenter can be described as follows (Nix and Gao 1998; Swadener et al. 2002):

$$\lambda_{co} = \frac{\pi a_c h}{b} \tag{14}$$

Pugno (2007) generalized the method by replacing the surface of indentation with the staircase-like surface and presented a generalized equation for dislocation length prediction:

$$\lambda = \frac{S}{b} \tag{15}$$

where $S = \Omega - A$ and Ω is the total contact surface. The variation of dislocation length during nanoindentation using cylindrical and right square prismatic indenters can be obtained using Eq. 15 as below:

$$\lambda_{cy} = \frac{2\pi a_c h}{b} \tag{16}$$

$$\lambda_{pr} = \frac{4\mathrm{ch}}{b} \tag{17}$$

where b is the magnitude of the Burgers vector and $c = \sqrt{A}$.

Voyiadjis and Yaghoobi (2015) conducted MD simulation of Ni thin film during nanoindentation to investigate the proposed theoretical models for dislocation length using different indenter geometries of right square prismatic, conical, and cylindrical. The Ni thin film dimensions were 1,200 nm, 1,200 nm, and 600 nm along $[1\bar{1}0]$, $[11\bar{2}]$, and [111] directions, respectively. The radius of cylindrical indenter was $r_1 = 4.8$ nm. The indentation surface of the right square prismatic indenter was a $7.5 \times 7.5\ \mathrm{nm}^2$. The smaller radius of blunt conical indenter was $r_2 = 0.3$ nm with the cone semi-angle of $\theta = 56.31°$. The parallel code LAMMPS (Plimpton 195) was selected to conduct the MD simulation. The numerical time integration was performed using the velocity Verlet algorithm. *BC4* was incorporated for MD simulation of nanoindentation. The EAM potential is used for Ni-Ni interaction which was parameterized by Mishin et al. (1999). The indenter was modeled

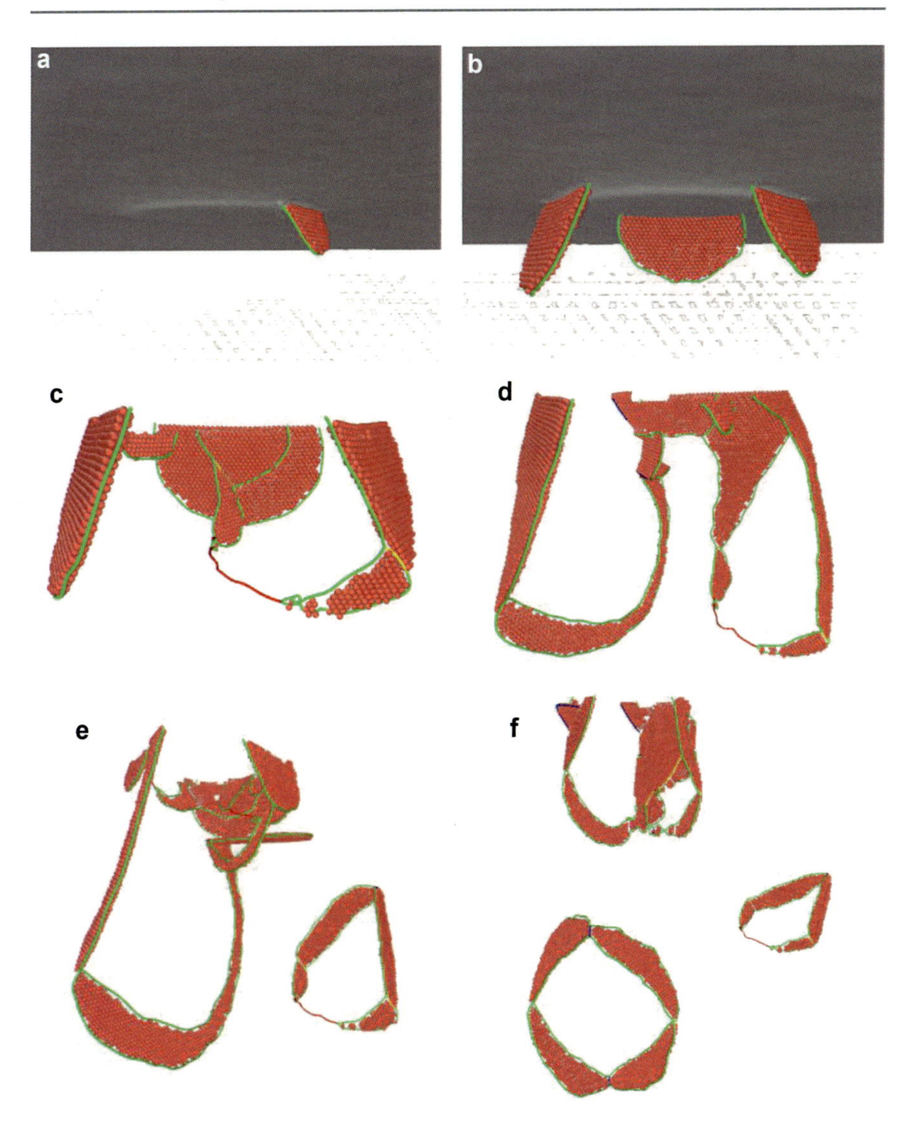

Fig. 5 Defect nucleation and evolution of Ni thin film indented by the cylindrical indenter at **a** $d \approx 0.70$ nm **b** $d \approx 0.86$ nm **c** $d \approx 0.96$ nm **d** $d \approx 1.02$ nm **e** $d \approx 1.05$ nm **f** $d \approx 1.12$ nm (Reprinted with permission from Voyiadjis and Yaghoobi 2015)

using the repulsive potential E^{ind} presented in Eq. 9. The dislocation was extracted from atomistic data using the Crystal Analysis Tool (Stukowski and Albe 2010; Stukowski 2012, 2014; Stukowski et al. 2012).

First, Voyiadjis and Yaghoobi (2015) investigated the dislocation nucleation and evolution pattern during nanoindentation. As an example, Fig. 5 shows the dislocation nucleation and evolution for Ni thin film indented by a cylindrical indenter during nanoindentation. The dislocations and stacking faults are visualized

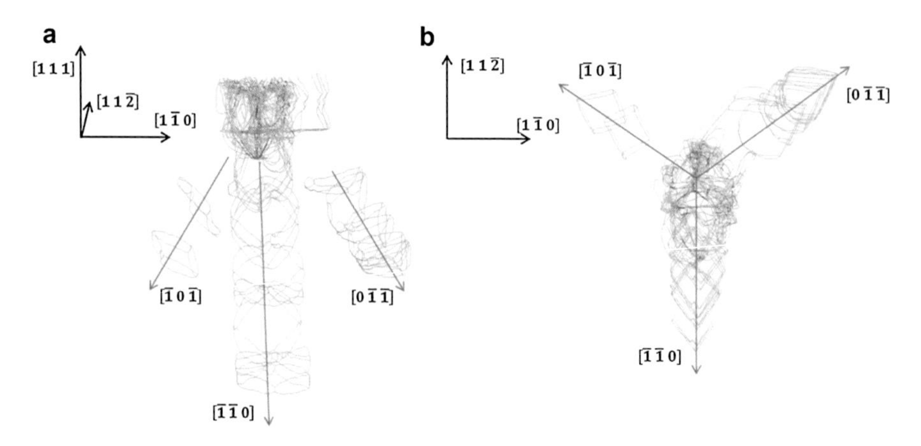

Fig. 6 Prismatic loops forming and movement in Ni thin film indented by the cylindrical indenter during nanoindentation **a** side view **b** top view (Reprinted with permission from Voyiadjis and Yaghoobi 2015)

while the perfect atoms are removed. The color of Shockley, Hirth, and stair-rod partial dislocations and perfect dislocations are green, yellow, blue, and red, respectively. Figure 6 illustrates the dislocation loop formation and movement along three directions of $\left[\bar{1}\,0\bar{1}\right]$, $\left[\bar{1}\bar{1}0\right]$, and $\left[0\bar{1}\bar{1}\right]$. Figure 7 compares the dislocation lengths obtained from atomistic simulation with those predicted by Eqs. 14–17. The results show that the theoretical predictions can accurately capture the dislocation lengths during nanoindentation. However, some discrepancies are observed which can be described as follows:

- Atomistic simulation captures the total dislocation length including both geometrically necessary and statistically stored dislocations, while the theoretical models only calculate geometrically necessary dislocations. This is the reason that the MD simulation dislocation lengths are mostly higher than that of the theoretical ones.
- The theoretical models incorporate the Burgers vector of the Shockley partial dislocations which comprised most of the dislocation content. However, a few stair-rod and Hirth partial and perfect dislocations are nucleated with the Burgers vectors different from the one for Shockley partial dislocations.
- The dislocations which are detached from the main dislocations network as the prismatic loops and leave the plastic zone around the indenter are not considered in the total dislocation length calculation.

Size Effects in Small-Length Scales During Nanoindentation

Demir et al. (2009) observed that the governing mechanisms of size effects at smaller length scales are not similar to those of the large size samples during nanoindentation. They observed that increasing the dislocation density decreases

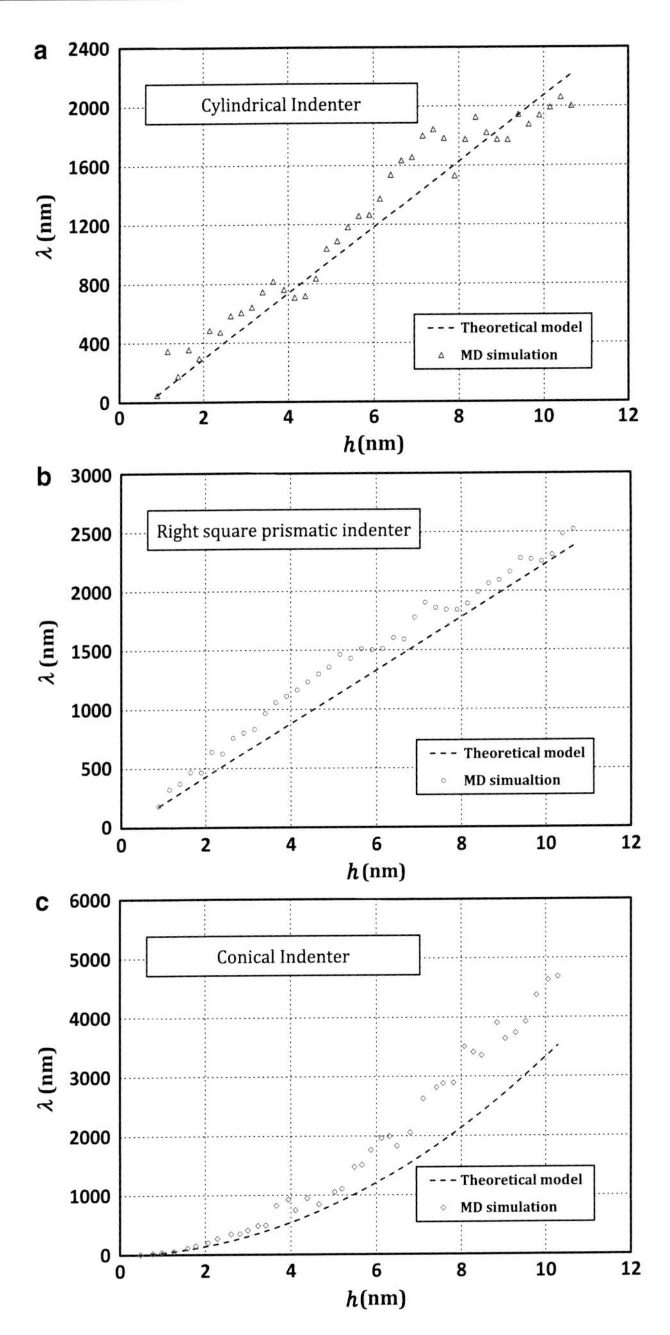

Fig. 7 Total dislocation length obtained from simulation and theoretical models in samples indented by the **a** cylindrical indenter **b** *right square* prismatic indenter **c** conical indenter (Reprinted with permission from Voyiadjis and Yaghoobi 2015)

the strength (Demir et al. 2009). Uchic et al. (2009), Kraft et al. (2010), and Greer (2013) reviewed the different sources of size effects occurring at smaller length scales. Three mechanisms of source exhaustion, source truncation, and weakest link theory have been introduced to capture the size effects. In samples with very small length scales, lacking enough dislocations to sustain the imposed plastic flow leads to the strength enhancement which is commonly termed as source exhaustion hardening (Rao et al. 2008; El-Awady 2015). The dislocation content reduction may happen due to dislocation starvation, i.e., when the dislocations escape from the sample free surfaces, mechanical annealing, or dislocation source shut down. The material strength also depends on the length of dislocation sources in a way that decreasing the length of dislocation source increases the strength. The dislocation source length becomes smaller by decreasing the sample size through a procedure so-called source truncation. In this procedure, the double-ended dislocation sources transform to a single-ended ones due to the surface effects which decreases the length of dislocation sources. Accordingly, decreasing the sample size leads to a smaller single-ended dislocation source which enhances the material strength (Parthasarathy et al. 2007; Rao et al. 2007). The weakest link theory states that the material strength increases by decreasing the sample size, because decreasing the sample length scale will increase the strength of the weakest source available in the sample (Norfleet et al. 2008; El-Awady et al. 2009).

Yaghoobi and Voyiadjis (2016a) incorporated the large scale MD to study the sources of size effects at smaller length scales during nanoindentation. They selected single crystal Ni thin films with the dimensions of 120 nm, 120 nm, and 60 nm along $[1\bar{1}0]$, $[11\bar{2}]$, and [111] directions, respectively. A conical indenter with a spherical tip was incorporated which is similar to the one used by Demir et al. (2009). The remaining simulation methodology is similar to the section "Comparing MD Results with Theoretical Models."

Swadener et al. (2002) approximated the spherical indenter geometry with a parabola and presented the following equation to predict the GNDs length during nanoindentation:

$$\lambda_{sp} \approx \frac{2\pi}{3} \frac{a_c^3}{\mathrm{bR}} \tag{18}$$

The approximation, however, is only applicable for small indentation depths. Yaghoobi and Voyiadjis (2016a) introduced a theoretical equation to predict the dislocation length of sample indented by a spherical tip using the precise geometry of the indenter. The total dislocation length can be described as below (Yaghoobi and Voyiadjis 2016a):

$$\begin{aligned} \lambda_{sp} &= \int_0^{a_c} \frac{2\pi r}{b} \left(\frac{\mathrm{dh}}{\mathrm{dr}}\right) \mathrm{dr} = \frac{2\pi}{b} \int_0^{a_c} \left(\frac{r^2}{\sqrt{R^2 - r^2}}\right) \mathrm{dr} \\ &= \frac{2\pi}{b} \left[\frac{R^2}{2} \sin^{-1}\left(\frac{a_c}{R}\right) - \frac{1}{2}\left(a_c \sqrt{R^2 - a_c{}^2}\right)\right] \end{aligned} \tag{19}$$

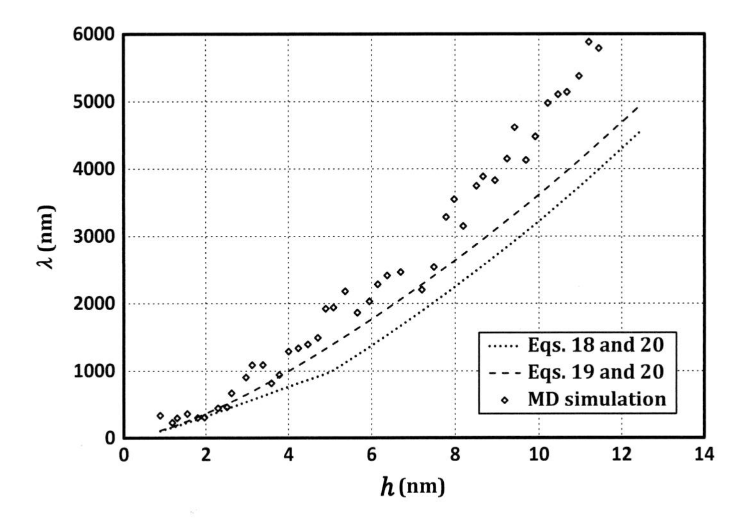

Fig. 8 Comparison between the dislocation lengths obtained from theoretical models and MD simulation during nanoindentation (Reprinted with permission from Yaghoobi and Voyiadjis 2016a)

The GNDs length for the conical part of the indenter can be calculated as below:

$$\lambda_{co} = \lambda_{sp}\big|_{h=h_0} + \int_{a_0}^{a_c} \frac{2\pi r}{b\tan\left(\frac{\theta}{2}\right)}\mathrm{dr} = \lambda_{sp}\big|_{h=h_0} + \frac{\pi\left({a_c}^2 - {a_0}^2\right)}{b\tan\left(\frac{\theta}{2}\right)} \tag{20}$$

Figure 8 compares the dislocation length obtained from atomistic simulation with those calculated from the approximate and precise theoretical models during nanoindentation. The GNDs length calculated from the theoretical model is a lower bound for the total dislocation length obtained from MD which includes all types of dislocations.

A plastic zone should be defined to obtain the dislocation density. Yaghoobi and Voyiadjis (2016a) assumed that the plastic zone is a hemisphere with the radius of $R_{pz} = fa_c$ where f is a constant. The value of $f = 1.9$ was selected by Yaghoobi and Voyiadjis (2016a) for the theoretical dislocation density calculations which is similar to Durst et al. (2005). The density of dislocations can be described as below:

$$\rho = \lambda/V \tag{21}$$

where V is the plastic zone volume. Figure 9 compares the approximate and precise theoretical dislocation densities. Figure 9 shows that approximating a sphere using a parabola leads to a constant dislocation density. However, the density of dislocations increases during nanoindentation by incorporating the precise indenter geometry.

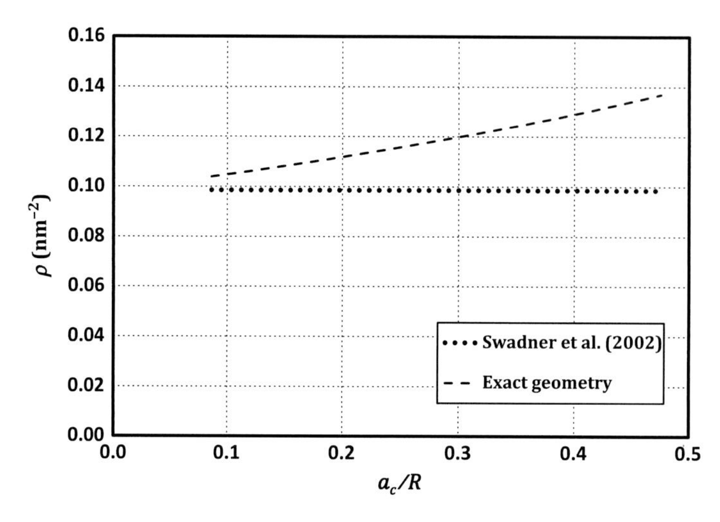

Fig. 9 Theoretical GNDs density obtained from the approximate (Swadener et al. 2002) and exact (see Eq. 19) geometries of the indenter versus the normalized contact radius a_c/R (Reprinted with permission from Yaghoobi and Voyiadjis 2016a)

Yaghoobi and Voyiadjis (2016a) incorporated five different values of $f = 1.5$, 2.0, 2.5, 3.0, and 3.5 and obtained the corresponding dislocation density for each one. The volume of plastic zone can be obtained by removing the volume occupied by the indenter (V_{indenter}) as follows:

$$V = (2/3)\,\pi\,(\text{fa}_c)^3 - V_{\text{indenter}} \tag{22}$$

Figure 10 illustrates the variation of dislocation density ρ during nanoindentation for different sizes of plastic zone. The results show that for all values of f, the dislocation density increases during nanoindentation which is in agreement with the trend predicted by the precise theoretical prediction presented in Fig. 9.

Figure 11 presents the variation of the mean contact pressure ($p_m = P/A$), which is equivalent to the hardness H in the plastic region, during nanoindentation. Figure 11 shows that the mean contact pressure follows the Hertzian theory in the elastic region. After the initial dislocation nucleation, however, the results show that the hardness decreases by increasing the indentation depth.

The forest hardening mechanism governs the material strength in bulk-sized samples which relate the material strength to the interaction of dislocations with each other. The famous Taylor hardening-type models are usually incorporated to describe the shear strength in the case of forest hardening mechanism as follows (Voyiadjis and Al-Rub 2005):

$$\begin{gathered}\tau = \alpha_S \mu b_S \sqrt{\rho} \\ \rho = \left[\rho_S^{\beta/2} + \left(\alpha_G^2 b_G^2 \rho_G / \alpha_S^2 b_S^2\right)^{\beta/2}\right]^{2/\beta}\end{gathered} \tag{23}$$

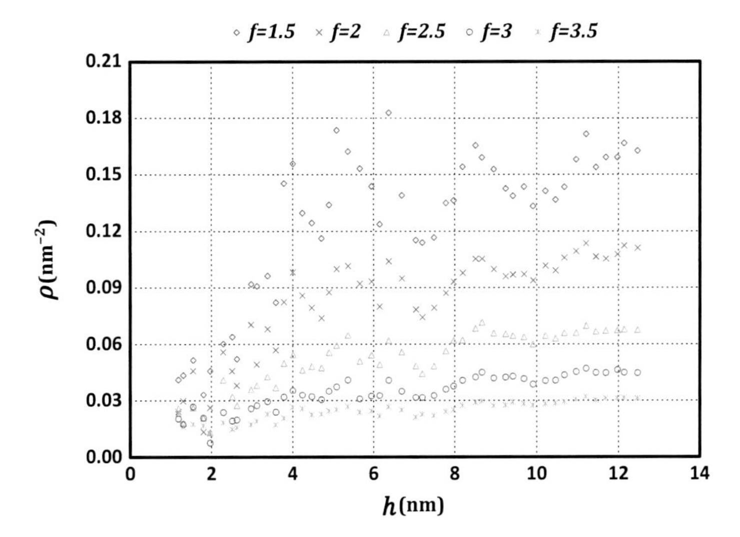

Fig. 10 Dislocation density obtained from MD simulation for different values of f during nanoindentation (Reprinted with permission from Yaghoobi and Voyiadjis 2016a)

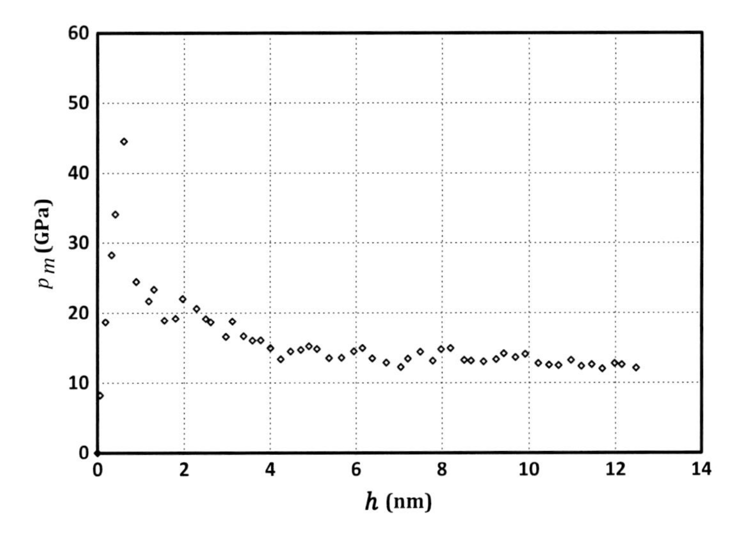

Fig. 11 Variation of mean contact pressure p_m as a function of indentation depth h (Reprinted with permission from Yaghoobi and Voyiadjis 2016a)

where α is a constant, μ is the shear modulus, and the indices G and S designate GNDs and SSDs parameters, respectively. Equation 23 states the material strength increases by increasing the dislocation density. Figure 10 shows that the dislocation density increases by increasing the indentation depth. According to the forest hardening mechanism, the strength should also increase. However, Fig. 11 illustrates that the hardness decreases during nanoindentation, and the material size effects cannot be captured by the forest hardening mechanism.

The dislocation nucleation and evolution should be investigated in addition to the nanoindentation response to unravel the controlling mechanisms of size effects. Figure 12 depicts the initial stages of dislocation evolution. It shows that the cross-slip is the dominant mechanism to produce the dislocation sources. Elongation of dislocations pinned at their ends provides the required dislocation length to

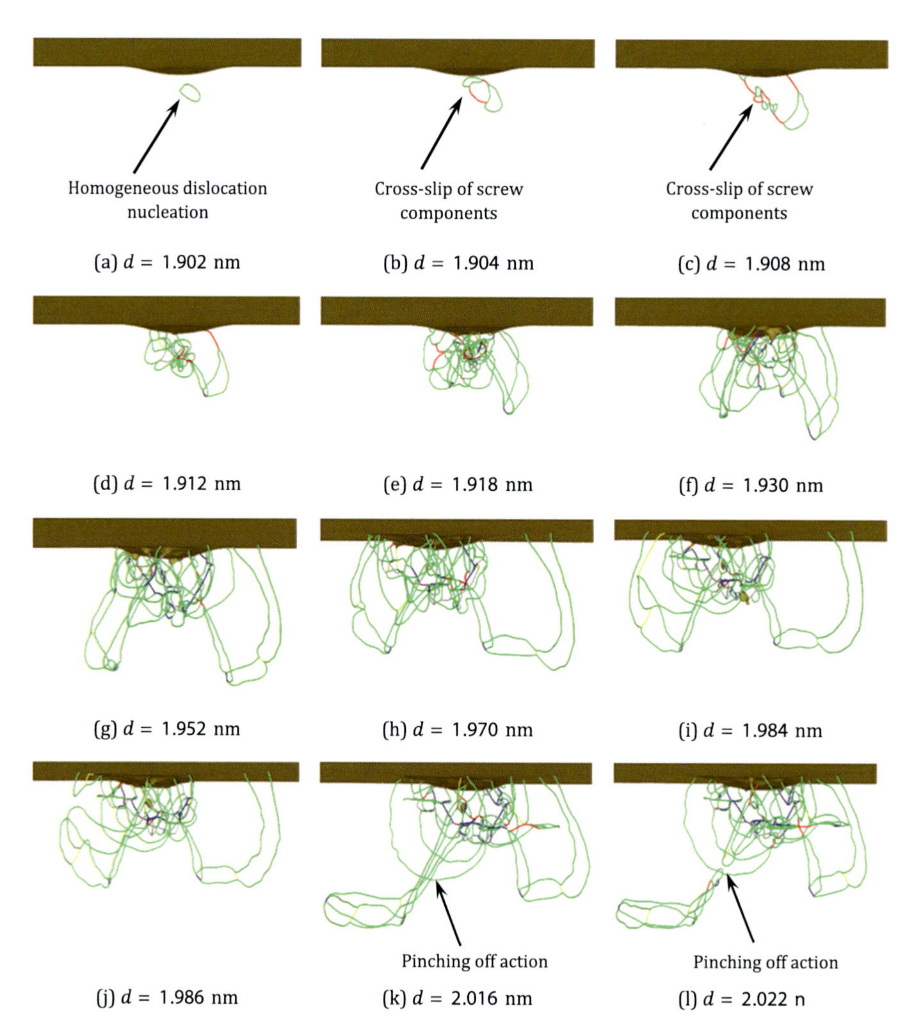

Fig. 12 Dislocation nucleation and evolution at small tip displacements: **a** initial homogeneous dislocation nucleation beneath the indenter which has a Burgers vector of $1/6\left[\bar{2}1\bar{1}\right]$ (Shockley partial dislocation); **b–j** cross-slip of screw components which produces new pinning points; **k–l** first loop is released by pinching off action (Reprinted with permission from Yaghoobi and Voyiadjis 2016a)

sustain the imposed deformation. Dislocation loops are then released by pinching off of screw dislocations and glide along the three directions of $[\bar{1}\bar{1}0]$, $[0\bar{1}\bar{1}]$, and $[\bar{1}\,0\bar{1}]$. After the initial dislocation nucleation, the available dislocation length is insufficient to sustain the imposed deformation, and the source exhaustion controls the size effects. Consequently, the required stress reduces as the dislocation length and density increase during nanoindentation. The dislocation density and length are eventually reaching the value required to sustain the imposed plastic flow, and the hardness tends to a constant value. Also, the interaction of dislocation with each other becomes important by increasing the dislocation length. However, the dislocation density reaches a constant value and forest hardening mechanism does not lead to any size effects.

Effects of Grain Boundary on the Nanoindentation Response of Thin Films

Grain boundary (GB) has a key role in deformation mechanism of crystalline material (Meyers et al. 2006; Koch et al. 2007; Zhu et al. 2008). For crystalline materials with large grains, the Hall-Petch relation describes the effects of grain size which states the material strength increases by decreasing the grain size. Hall-Petch effect is commonly attributed to the dislocation pile-up mechanism (Meyers et al. 2006; Koch et al. 2007; Zhu et al. 2008). The Hall-Petch relation breaks down for grain size smaller than some limits, and other deformation mechanisms control the size effects in crystalline materials such as the grain boundary rotation and sliding (Meyers et al. 2006; Koch et al. 2007; Zhu et al. 2008).

Atomistic simulation is a powerful tool to study the interaction of dislocations with GBs. Several mechanisms of dislocation reflection, transmission, and absorption were investigated by De Koning et al. (2003) by incorporating the atomistic simulation. Hasnaoui et al. (2004) studied the interaction between the dislocations and GB during nanoindentation experiment using molecular dynamics. Jang and Farkas (2007) conducted the atomistic simulation of bicrystal nickel thin film nanoindentation and observed that the GBs can contribute to the nanoindentation hardness. Kulkarni et al. (2009), however, observed that the GBs mainly reduce the hardness of the metallic samples. They showed that the CTB has the least hardness reduction compared to the other types of GBs (Kulkarni et al. 2009). Tsuru et al. (2010) investigated the effect of indenter distance from the GB using MD. Stukowski et al. (2010) conducted MD simulation of nanoindentation for metallic samples with twin boundaries and observed that the effects of twin GBs on the material response depends on the unstable stacking fault and twin boundary migration energies. Sangid et al. (2011) proposed an inverse relation between the GB energy barrier and GB energy based on the MD simulation results.

Effects of GB on the response of thin film during nanoindentation has been studied by many researchers (Hasnaoui et al. 2004; Jang and Farkas 2007; Kulkarni et al. 2009; Tsuru et al. 2010). However, a study which addresses a wide range of grain sizes is not a trivial task due to the MD simulation limitations. Voyiadjis and Yaghoobi (2016) incorporated the large-scale MD to study the effects of grain size and grain boundary geometry on the nanoindentation response. They incorporated Ni thin films with two sizes of $24 \times 24 \times 12$ nm (S1) and $120 \times 120 \times 60$ nm (S2). Four symmetric tilt boundaries of $\sum 3(111)\ [1\bar{1}0]\ \left(\theta = 109.5^{\circ}\right)$, $\sum 11(113)\ [1\bar{1}0]\ \left(\theta = 50.5^{\circ}\right)$, $\sum 3\,(1\,1\,2)\ [1\bar{1}0]\ \left(\theta = 70.5^{\circ}\right)$, and $\sum 11(332)\ [1\,\bar{1}\,0]\ \left(\theta = 129.5^{\circ}\right)$ and three asymmetric tilt boundaries of $\sum 11$ (225)/(441) ($\varphi = 54.74^{\circ}$), $\sum 3\,(112)/\left(\overline{55}2\right)\ \left(\varphi = 19.47^{\circ}\right)$, and $\sum 3$ (114)/(110) ($\varphi = 35.26^{\circ}$) were generated at the two third of the sample from bottom to compare the governing mechanisms of size effects with those of the single crystal thin films. θ and φ are the interface misorientation and inclination angles, respectively. The spherical indenters with two different radii of $R_1 = 10$ nm and $R_2 = 15$ nm were modeled using the repulsive potential E^{ind} presented in Eq. 9. The procedure to generate and equilibrate the GBs was elaborated by Voyiadjis and Yaghoobi (2016). The equilibrium structures of grain boundaries are illustrated in Fig. 13 using the *CSP*. The remaining simulation methodology is similar to the section "Comparing MD Results with Theoretical Models."

Figure 14 depicts the variation of mean contact pressure p_m during nanoindentation for S1 thin films, i.e., the smaller samples. It can be observed that the GB generally reduces the material strength for S1 thin films. However, in the case of coherent twin boundary (CTB), i.e., $\sum 3\ (111)\ \left[1\bar{1}0\right]$, the hardness is slightly enhanced for some indentation depths. Generally, in the cases of smaller thin films, i.e., S1 samples, the CTB has the best performance. Further investigation was conducted by depicting the variation of dislocation length λ during nanoindentation. Figure 15 illustrates the variation of p_m and λ during nanoindentation for CTB. Voyiadjis and Yaghoobi (2016) divided the nanoindentation response to five different regions:

- *Region I*: The bicrystal and single crystal thin films show similar responses during the initial indentation phase which are elastic, and CTB is the only defect that exists in the bicrystal thin film.
- *Region II*: In this region, the dislocation nucleation occurs for the bicrystal thin film beneath the indenter followed by a stress relaxation while the single crystal sample remains elastic. In the case of bicrystal thin film, the size effects is initially governed by the dislocation nucleation and source exhaustion. The dislocation density increases during nanoindentation which decreases the required stress to sustain the imposed plastic flow. Consequently, the hardness decreases by nucleation and evolution of new dislocations.
- *Region III*: The plasticity is initiated in single crystal thin film beneath the indenter followed by a stress relaxation. The thin film strength reduces according to the dislocation nucleation and source exhaustion mechanisms. The dislocation content does not change for bicrystal thin film. Accordingly, the stress should

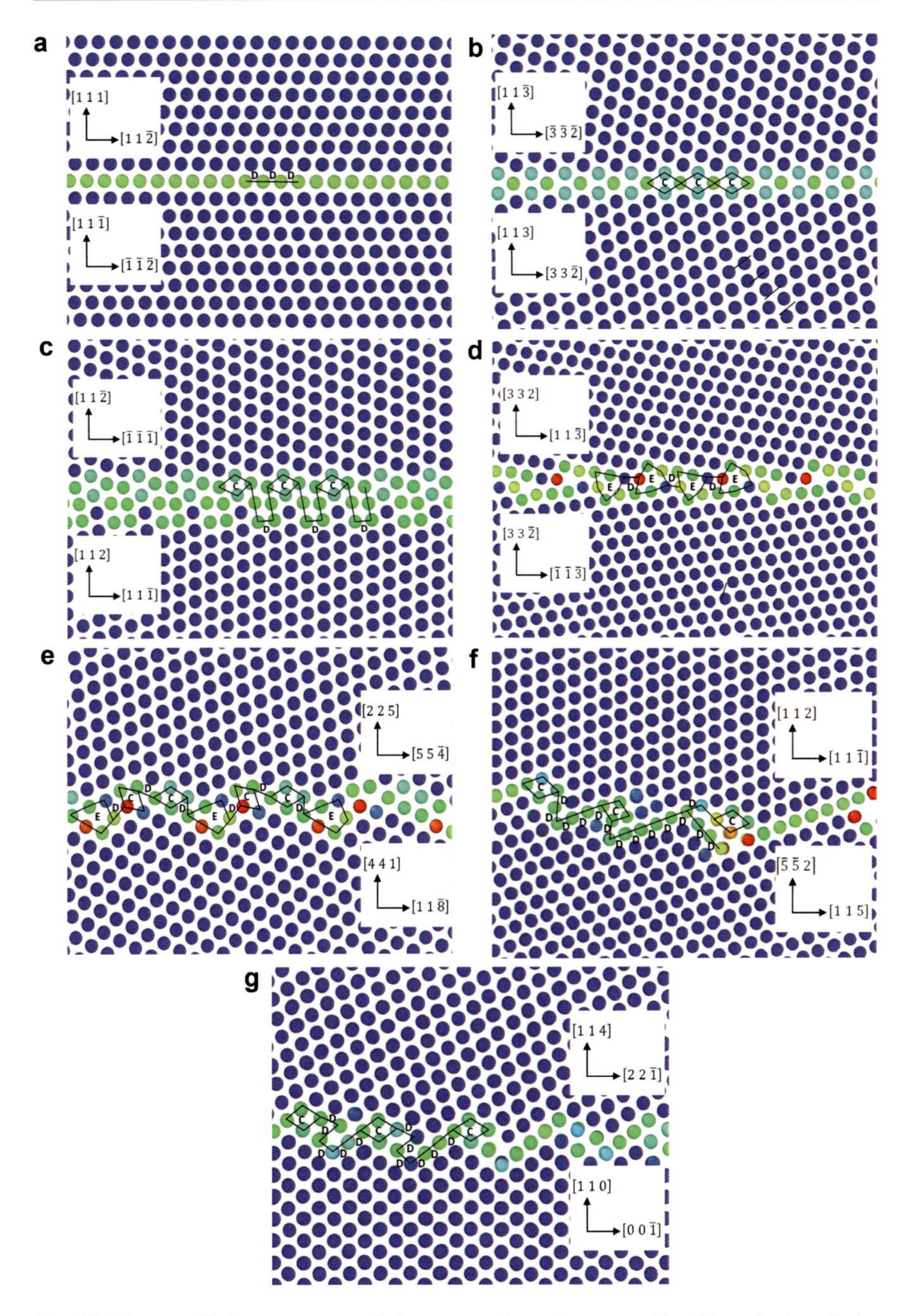

Fig. 13 The equilibrium structure of the symmetric and asymmetric tilt grain boundaries of **a** $\sum 3(111)$ $[1\bar{1}0]$ $(\theta = 109.5^{\circ})$, **b** $\sum 11(113)$ $[1\bar{1}0]$ $(\theta = 50.5^{\circ})$ $\sum 3(1\ 1\ 2)$, **c** $[1\bar{1}0]$ $\left(\theta = 70.5^{\circ}\right)$, **d** $\sum 11(332)$ $[1\ \bar{1}\ 0]$ $\left(\theta = 129.5^{\circ}\right)$, **e** $\sum 11$ (225)/(441) $(\varphi = 54.74^{\circ})$, **f** $\sum 3$ (112)/$(\bar{5}\bar{5}2)$ $\left(\varphi = 19.47^{\circ}\right)$, and **g** $\sum 3$ (114)/(110) $(\varphi = 35.26^{\circ})$, along $[1\bar{1}0]$ axis (Reprinted with permission from Voyiadjis and Yaghoobi 2016)

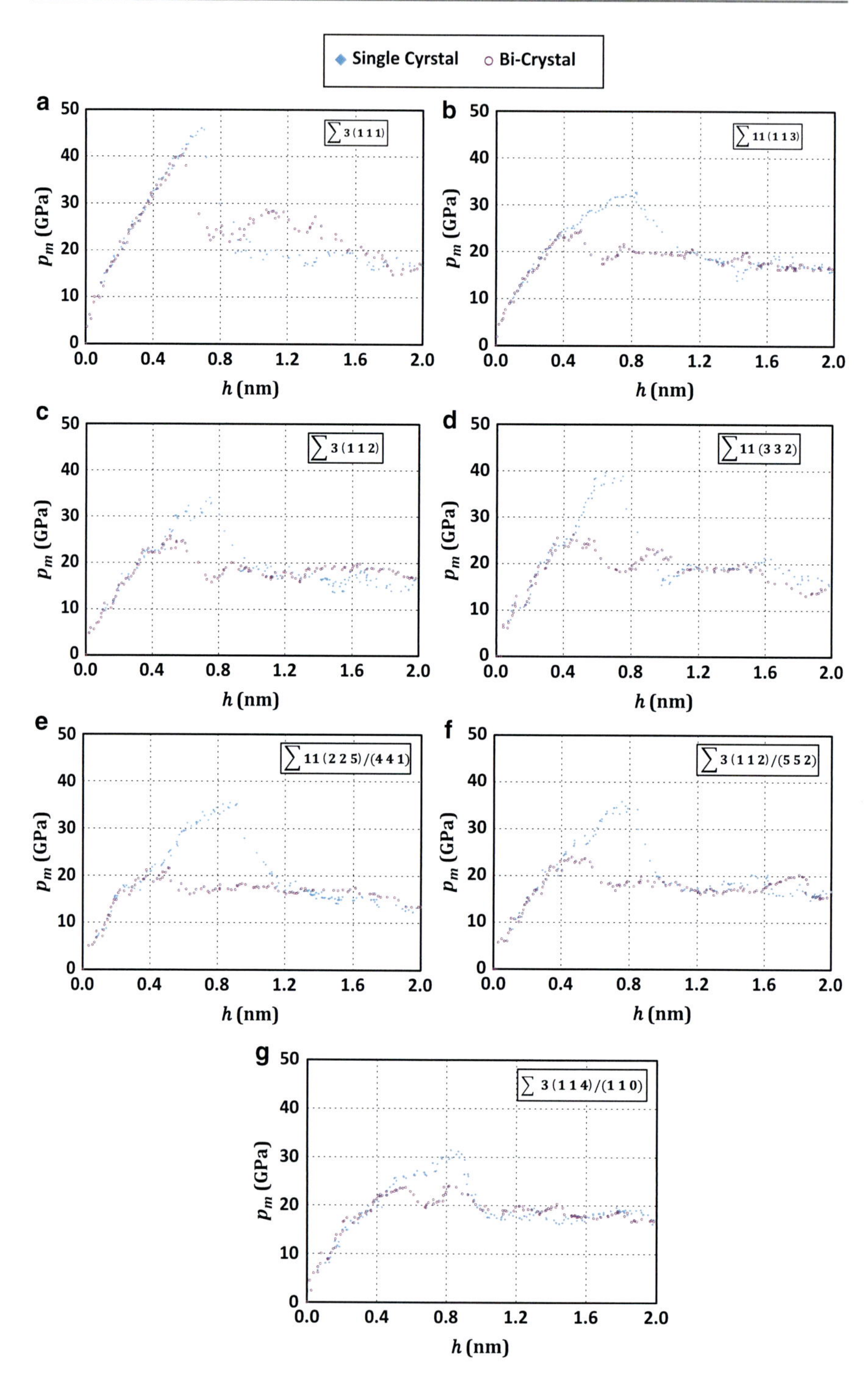

Fig. 14 Variation of mean contact pressure p_m as a function of indentation depth h for S1 single crystal and their related bicrystal samples with the grain boundaries of **a** $\sum 3(111)$, **b** $\sum 11(113)$, **c** $\sum 3(112)$, **d** $\sum 11(332)$, **e** $\sum 11(225)/(441)$, **f** $\sum 3\ (112)/\left(\overline{552}\right)$, and **g** $\sum 3(114)/(110)$ (Reprinted with permission from Voyiadjis and Yaghoobi 2016)

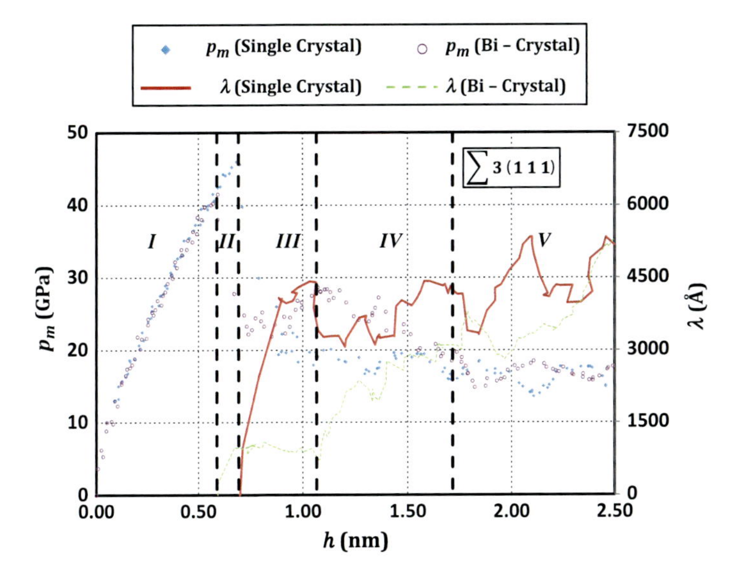

Fig. 15 Variation of mean contact pressure p_m and dislocation length λ as a function of indentation depth h for S1 bicrystal sample with $\sum 3(111)$ GB and its related single crystal sample (Reprinted with permission from Voyiadjis and Yaghoobi 2016)

be increased to sustain the imposed deformation based on the source exhaustion mechanism. The dislocations eventually reach the GB which blocks the dislocations. However, the blocked dislocations do not contribute to the strength.

- *Region IV*: The strength in both single and bicrystal thin films is decreased by increasing the dislocation length which follows the source exhaustion mechanism. However, the influence of the source exhaustion mechanism decreases as the dislocation length increases which decreases the slope of the hardness reduction. Also, the dislocations which are blocked by the GB start dissociating into the next grain.
- *Region V:* In this region, the available dislocation content is sufficient to sustain the imposed plastic flow and no further stress reduction occurs. Also, the single and bicrystal thin films reach a similar hardness which shows that the dislocation blockage by GB does not have any contribution to the size effects.

The structures of dislocations in different regions are illustrated in Fig. 16 for bicrystal thin film with CTB and related single crystal sample. Figure 16a, b shows the dislocation structure in *Region II* at which the single crystal sample is defect free and the nucleation occurs beneath the indenter for bicrystal thin film. The results show that the dominant mechanism of dislocation multiplication is cross-slip. Cross-slip introduces the new pinning points and provides the required dislocation length to sustain the plastic flow. Dislocations are elongated while they are pinned at their ends. Figure 16c, d illustrates the dislocation structure in *Region III* while the cross-slip is still the governing mechanism of deformation for both single and

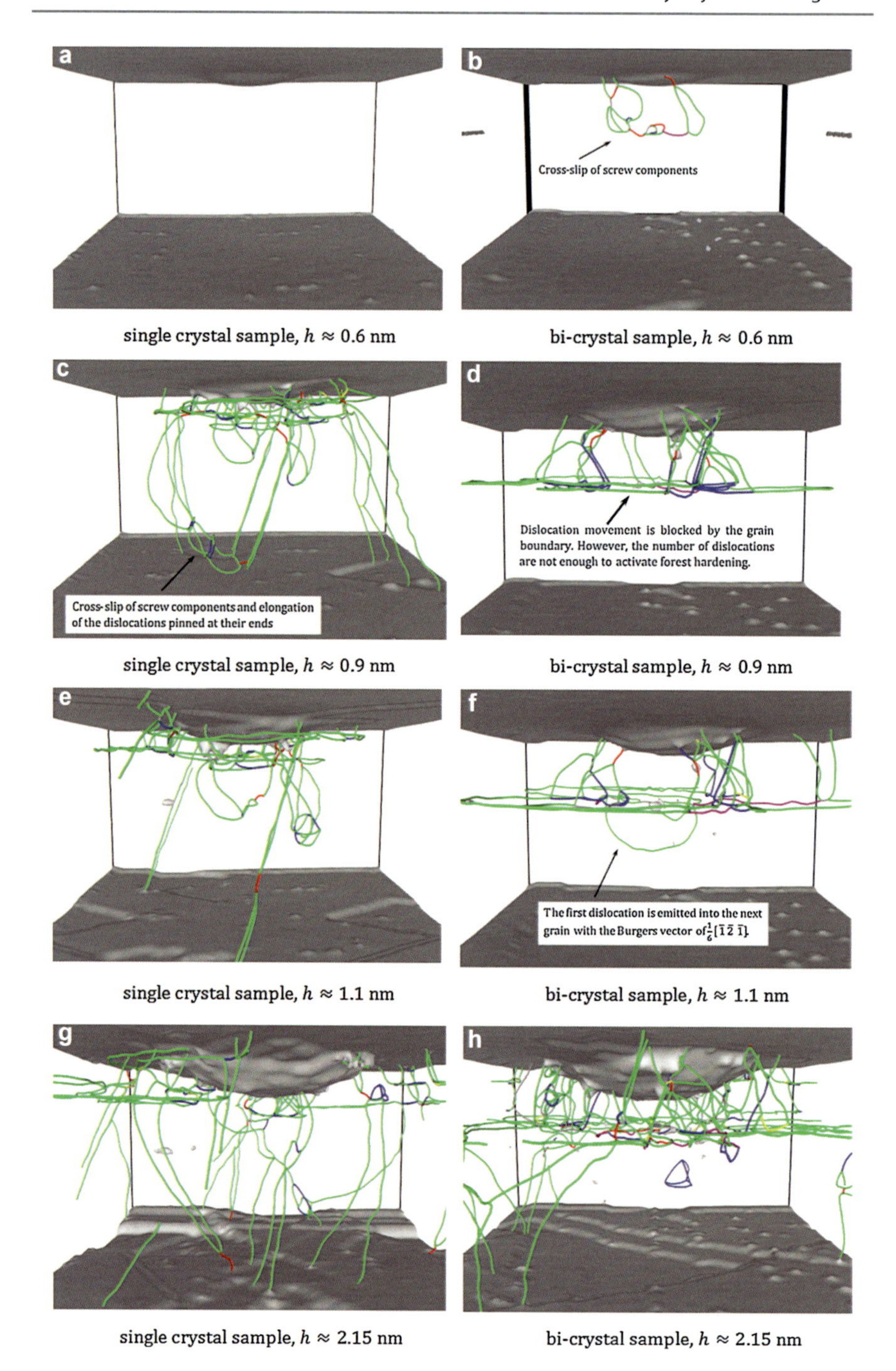

Fig. 16 Dislocation nucleation and evolution: **a** *Region II*, single crystal sample; **b** *Region II*, bicrystal sample; **c** *Region III*, single crystal sample; **d** *Region III*, bicrystal sample; **e** *Region IV*, single crystal sample; **f** *Region IV*, bicrystal sample; **g** *Region V*, single crystal sample; **h** *Region V*, bicrystal sample (Reprinted with permission from Voyiadjis and Yaghoobi 2016)

bicrystal thin films. Figure 16d depicts the dislocation blockage by CTB. Figure 16e, f illustrates the dislocation structure in *Region IV*. Many dislocation multiplications are observed in both single and bicrystal thin films which are induced according to the cross-slip mechanism. Figure 16f shows the initial dislocation dissociation into the next grain in the case of bicrystal sample which is a Shockley partial dislocation with the Burgers vector of $\frac{1}{6}\left[121\right]$. In the case of *Region V*, Fig. 16g, h depicts the dislocation structure which shows enough dislocation length is provided to sustain the imposed deformation. Also, the interaction of dislocations with each other cannot be neglected anymore.

Figure 17 shows the variation of p_m and λ during nanoindentation for the S1 samples with different GBs and their related single crystal thin films. In contrast to

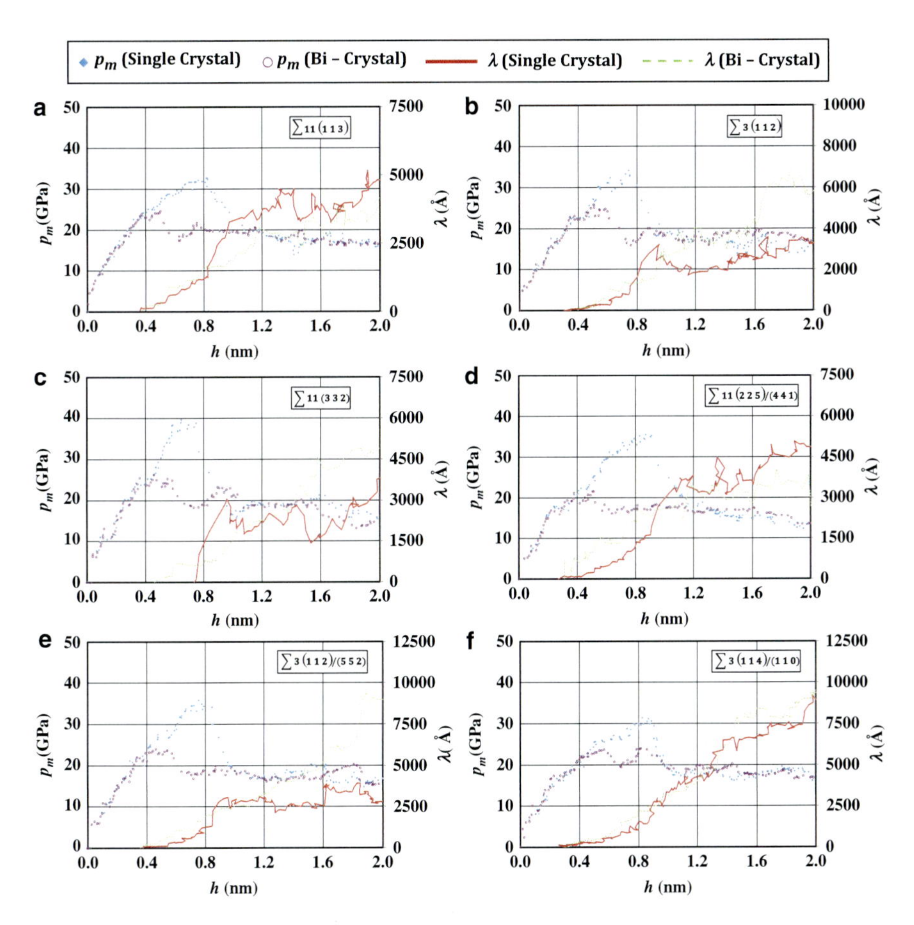

Fig. 17 Variation of mean contact pressure p_m and dislocation length λ as a function of indentation depth h for S1 bicrystal and their related single crystal samples with grain boundaries of: **a** $\sum$11(113), **b** $\sum$3(112), **c** $\sum$11(332), **d** $\sum$11(225)/(441), **e**$\sum$3 (112)/$(\overline{55}2)$, and **f**$\sum$3(114)/(110) (Reprinted with permission from Voyiadjis and Yaghoobi 2016)

CTB, Fig. 17 shows that the first stress relaxation does not occur immediately after the first dislocation nucleation. The first large stress relaxation occurs with the first jump in dislocation density for single crystal thin films. The nature of first apparent strength drop in thin films with GB is more complicated due to the interaction of dislocations with GB. Figure 17 shows that the GB decreases the depth at which the first large stress relaxation occurs, and the bicrystal thin films have larger dislocation length at that depth compared to the single crystal samples. The GB itself can be a source of dislocation nucleation which can be activated at different stages of indentation. Figure 18 shows that the GB is the initial source of dislocation nucleation for $\sum 3\ (112)/(\overline{55}2)$ and $\sum 11(225)/(441)$ GBs, i.e., the initial dislocation nucleation occurs from the GB and not beneath the indenter. The nucleated dislocations are Shockley partial dislocations with the Burgers vectors of $\frac{1}{6}[11\overline{2}]$ and $\frac{1}{6}[112]$ for the GBs of $\sum 3\ (112)/(\overline{55}2)$ and $\sum 11(225)/(441)$, respectively. If the dislocation nucleation from GB occurs at the initial steps of dislocation nucleation and evolution, it will severely decrease the thin film strength which can be noted for $\sum 3\ (112)/(\overline{55}2)$ and $\sum 11(225)/(441)$ GBs in Fig. 17d, e. The size effects during nanoindentation can be described for all GBs incorporating the variation of total dislocation length and dislocation visualization during nanoindentation. The results show that the source exhaustion is the controlling mechanism of size effects for the initial stages of dislocation nucleation and evolution. Increasing the total dislocation length, however, the required dislocation length for sustaining the imposed deformation is provided and the source exhaustion mechanism becomes less dominant. Also, the dislocation interactions with each other become nonnegligible by increasing the dislocation content. Eventually, both bicrystal and their related single crystal thin films reach a similar hardness which indicates that the dislocation pileup does not enhance the hardening in the cases of studied S1 samples.

Figure 19 shows mean contact pressure versus indentation depth in the cases of S2 samples, i.e., larger samples. The initial responses of both single crystal and bicrystal thin films are similar. However, GB enhances the hardness for higher indentation depths. In order to unravel the underlying mechanisms of size effects for S2 samples, the variations of mean contact pressure and total dislocation length should be studied. Figure 20 compares the mean contact pressure and total dislocation density of the single crystal thin film with those of the bicrystal sample with CTB. The nanoindentation response can be divided to three different regions:

- *Region I*: There is no plasticity at this region. GB does not change the nanoindentation response of thin film.
- *Region II*: The plasticity is initiated beneath the indenter for both single and bicrystal thin films followed by a sharp stress relaxation. After the initial nucleation, the source exhaustion governs the size effects, and the required stress to maintain the plastic flow decreases by increasing the total dislocation length. In this region, the GB does not significantly change the total dislocation

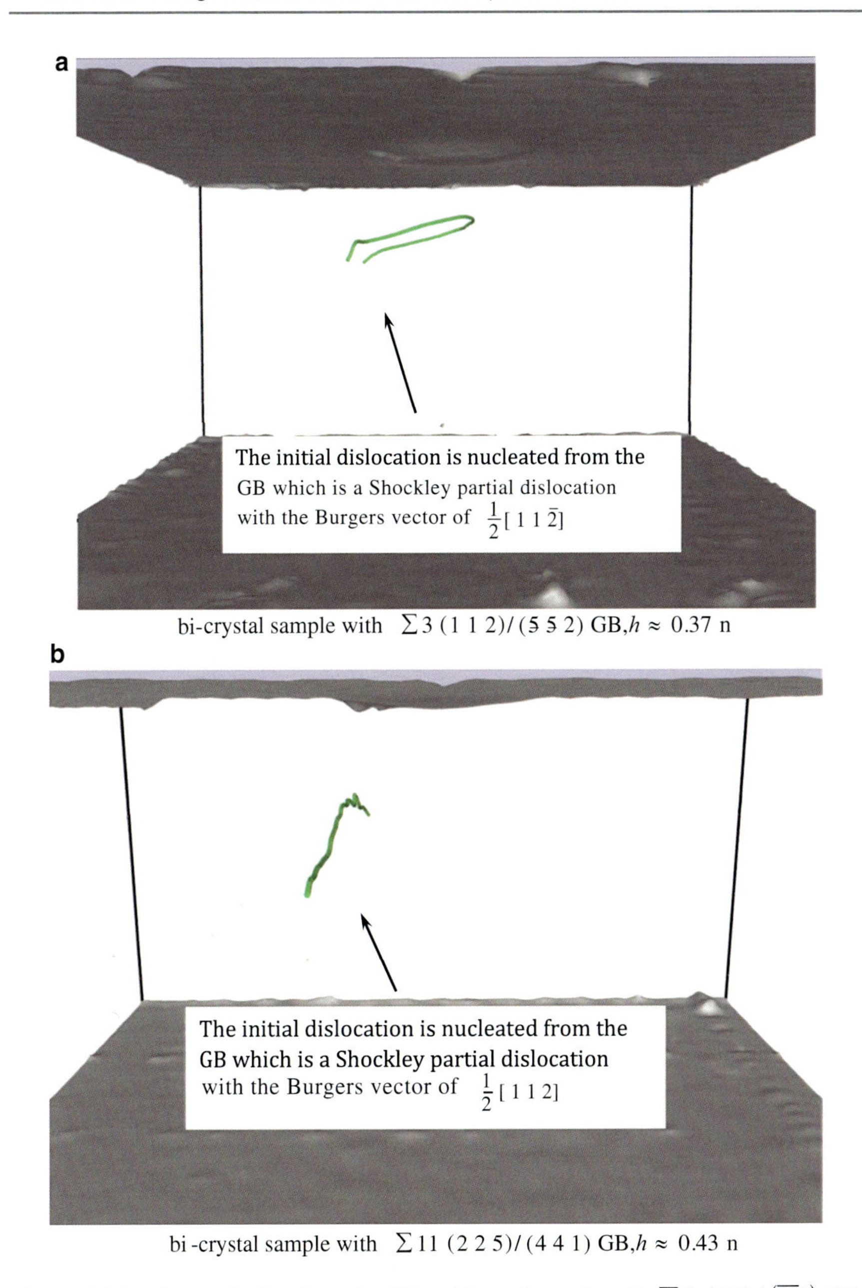

Fig. 18 Dislocation nucleation from the GB: **a** bicrystal sample with $\sum 3\ (112)/(\overline{552})$ GB, $h \approx 0.37$ nm; **b** bicrystal sample with $\sum 11(225)/(441)$ GB, $h \approx 0.43$ nm (Reprinted with permission from Voyiadjis and Yaghoobi 2016)

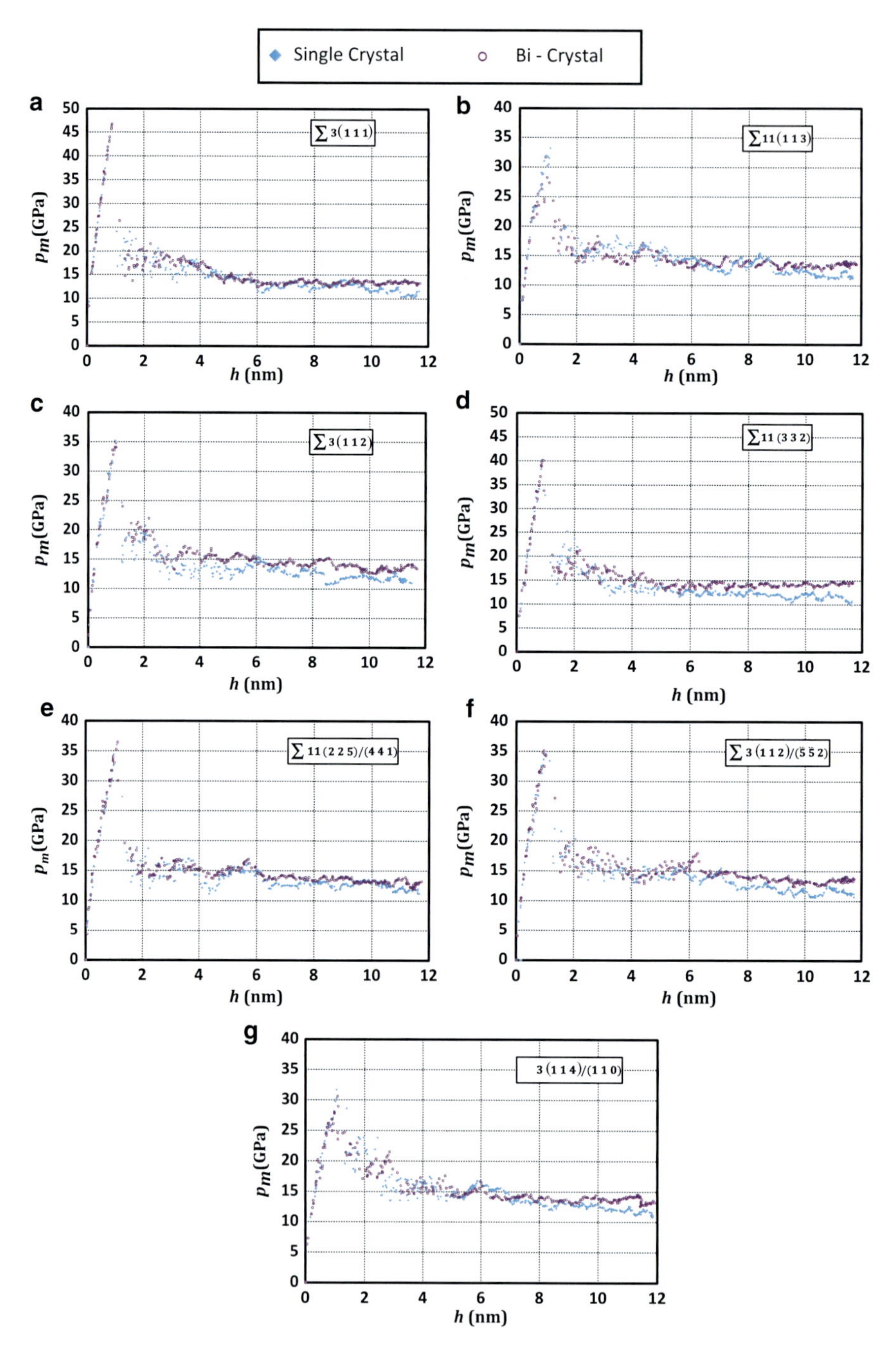

Fig. 19 Variation of mean contact pressure p_m as a function of indentation depth h for S2 single crystal and their related bicrystal samples with the grain boundaries of **a** $\sum 3(111)$, **b** $\sum 11(113)$, **c** $\sum 3(112)$, **d** $\sum 11(332)$, **e** $\sum 11(225)/(441)$, **f** $\sum 3\ (112)/\ (\overline{552})$, and **g** $\sum 3(114)/(110)$ (Reprinted with permission from Voyiadjis and Yaghoobi 2016)

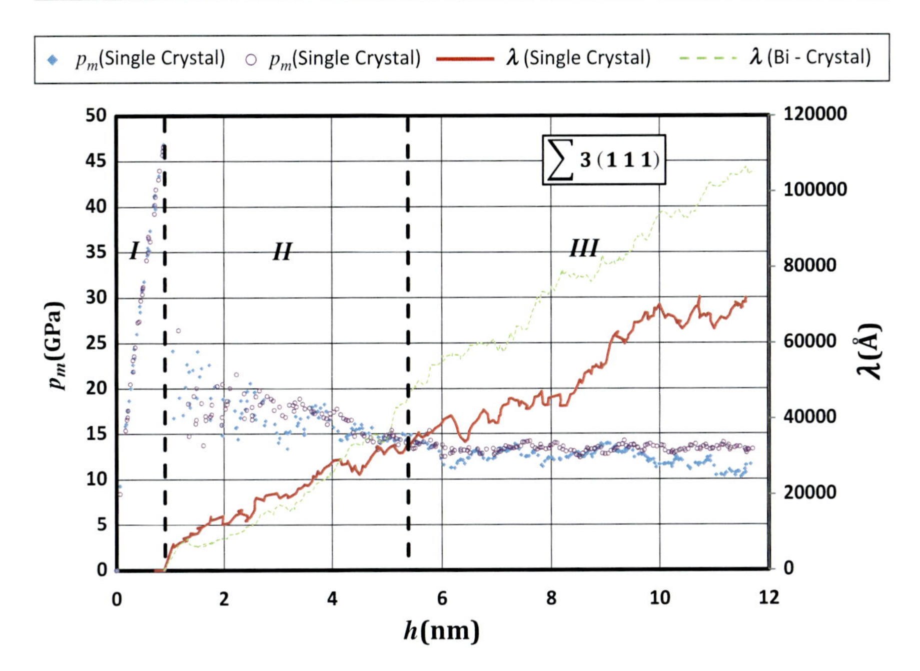

Fig. 20 Variation of mean contact pressure p_m and dislocation length λ as a function of indentation depth h for S2 bicrystal sample with $\sum 3(111)$ GB and its related single crystal sample (Reprinted with permission from Voyiadjis and Yaghoobi 2016)

length and consequently the hardness. The dominancy of source exhaustion decreases during nanoindentation as more dislocations are provided to sustain the imposed deformation. Accordingly, the hardness reduction slope decreases during nanoindentation. Eventually, the dislocations reach the GB which blocks the dislocations.

- *Region III*: Enough dislocation length is provided to sustain the imposed deformation, and the source exhaustion hardening is not active anymore. The interactions of dislocations with each other and GB become important by increasing the dislocation content. Also, the number of dislocations blocked by GB becomes considerable and the produced pile-up enhances the sample strength. Consequently, the GB enhances the nanoindentation response of thin film for S2 sample.

The dislocation visualization of the S2 thin film with and without CTB is illustrated in Fig. 21 during nanoindentation. Figure 21a, b illustrates that the initial dislocation is homogeneously nucleated beneath the indenter which is a Shockley partial dislocation with the Burgers vector of $\frac{1}{6}\left[\bar{2}11\right]$. The results show that the GB does not change the nucleation pattern. After the initial nucleation, Fig. 21c, d shows that the cross-slip is the controlling mechanism of deformation which increases the number of dislocation sources and provides the required dislocation content.

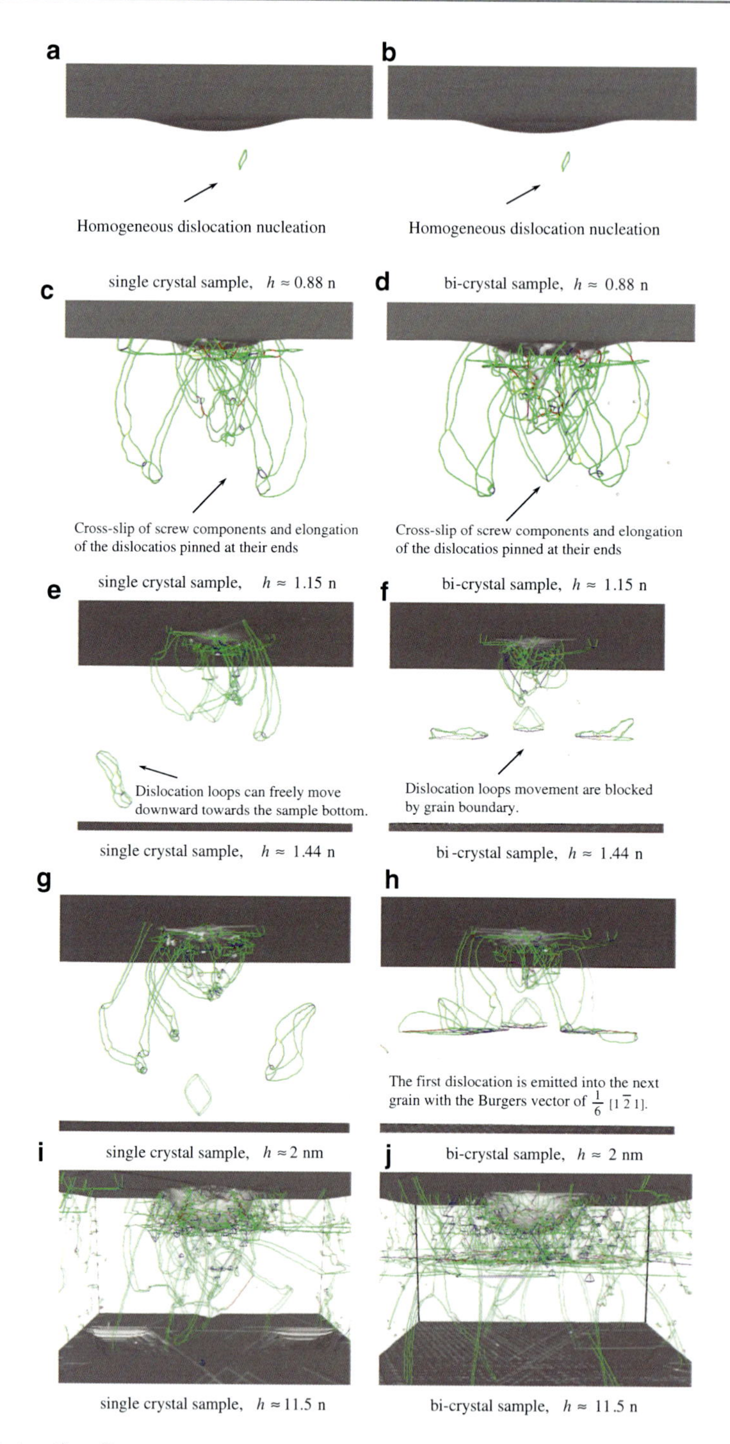

Fig. 21 (continued)

The effects of GB is still negligible on the dislocation pattern. The dislocation loops are induced by cross-slipping and pinching off of screw dislocations as the indentation depth increases. The induced loops are moving downward which are blocked by GB. Consequently, the GB starts to change the pattern of dislocation evolution. Figure 21e, f illustrates the dislocation loops movement in thin films with and without GB, respectively. The dislocation blocked by GB are eventually emitting into the next grain by increasing the indentation depth. Figure 21h illustrates the first dislocation emitting into the next grain which is a Shockley partial dislocation with the Burgers vector of $\frac{1}{6}\left[1\bar{2}1\right]$. Figure 21i, j shows the dislocation visualization of the sample at the higher indentation depths for thin film with and without GB, respectively. Although some dislocations are emitted into the next grain, the visualization results show a considerable pile-up behind the GB, while the dislocations are moving downward freely for single crystal thin film.

Figure 22 shows the variation of p_m and λ during nanoindentation for the S2 samples with different GBs and their related single crystal thin films. The observed microstructural behavior for CTB can be incorporated for all other GBs except $\sum 11(332)$ and $\sum 11(225)/(441)$ GBs. In the cases of two latter GBs, the GB enhances the hardness while the total dislocation length of thin film with GB is very close to the one without GB. The observed discrepancy is due to the fact the total dislocation length is not an appropriate factor to study the forest hardening mechanism. In the case of source exhaustion hardening, the total dislocation length dictates the amount of stress required to sustain the plastic flow. On the other hand, the density of dislocation in the plastic zone should be taken as the representative factor for the forest hardening mechanism. Voyiadjis and Yaghoobi (2016) assumed that the plastic zone is located in the upper grain. Accordingly, the total dislocation length in the upper grain λ_{upper} should be investigated during nanoindentation. The dislocations located in the upper one third is considered for single crystal thin film, and the obtained results are compared with those of bicrystal thin films. Figure 23 compares the variations of mean contact pressure and the total dislocation length located in the plastic zone during nanoindentation for thin films with and without GB. The results show that the GB increases the total dislocation length located in plastic zone and consequently enhances the hardness according to the forest hardening mechanism. The results show that the main role of GB in the cases of large thin films, i.e., S2 samples, is to modify the pattern of dislocation in a way that increases the dislocation density located in the

Fig. 21 Dislocation nucleation and evolution: **a** single crystal sample, $h \approx 0.88$ nm; **b** bicrystal sample, $h \approx 0.88$ nm; **c** single crystal sample, $h \approx 1.15$ nm; **d** bicrystal sample, $h \approx 1.15$ nm; **e** single crystal sample, $h \approx 1.44$ nm; **f** bicrystal sample, $h \approx 1.44$ nm; **g** single crystal sample, $h \approx 2.03$ nm; **h** bicrystal sample, $h \approx 2.03$ nm; **i** single crystal sample, $h \approx 11.5$ nm; **j** bicrystal sample, $h \approx 11.5$ nm (Reprinted with permission from Voyiadjis and Yaghoobi 2016)

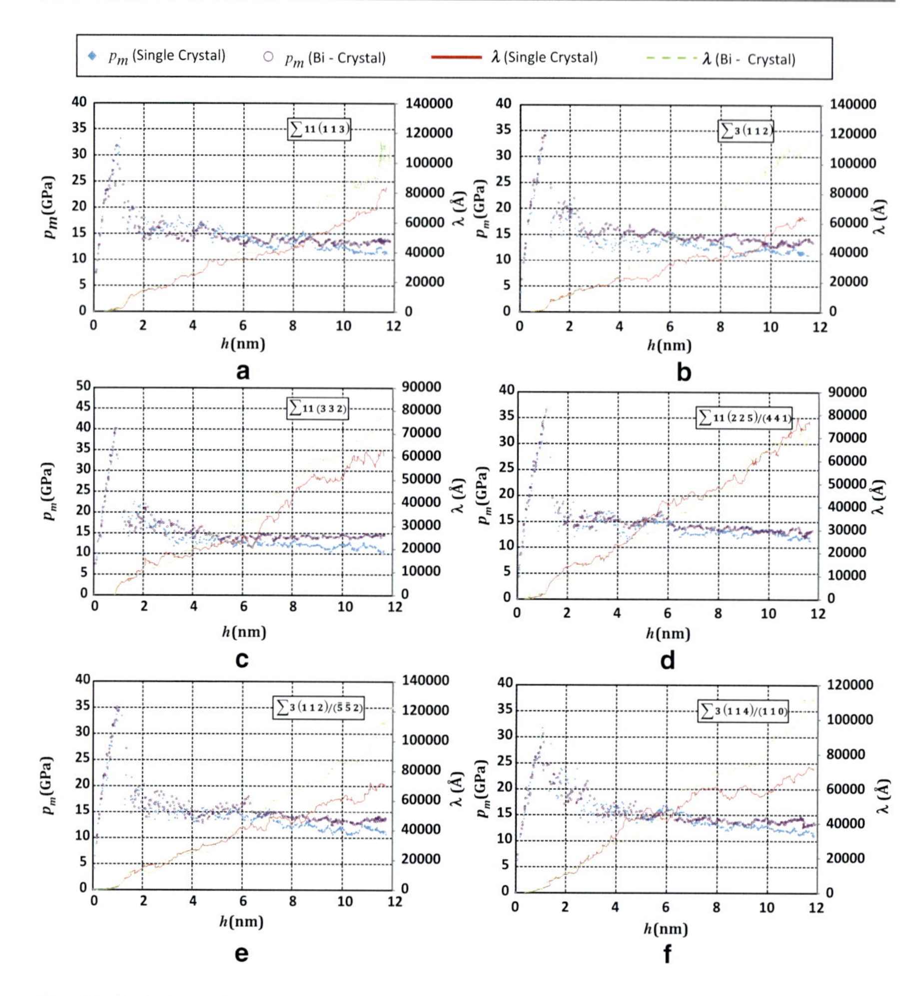

Fig. 22 Variation of mean contact pressure p_m and dislocation length λ as a function of indentation depth h for S2 bicrystal and their related single crystal samples with grain boundaries of: **a** $\sum 11(113)$, **b** $\sum 3(112)$, **c** $\sum 11(332)$, **d** $\sum 11(225)/(441)$, **e** $\sum 3\ (112)/(\overline{5}\overline{5}2)$, and **f** $\sum 3(114)/(110)$ (Reprinted with permission from Voyiadjis and Yaghoobi 2016)

plastic zone and accordingly strengthen the thin films. One should note that the strain rates incorporated in the atomistic simulation are much higher than those selected for experiments. Accordingly, the interpretation of the obtained results should be carefully handled. The applied strain rate can influence both hardening mechanisms and dislocation network properties (see, e.g., Yaghoobi and Voyiadjis, 2016b; Voyiadjis and Yaghoobi, 2017; Yaghoobi and Voyiadjis, 2017). In other words, one should ensure that the observed mechanisms are not artifacts of the high strain rates used in the atomistic simulation.

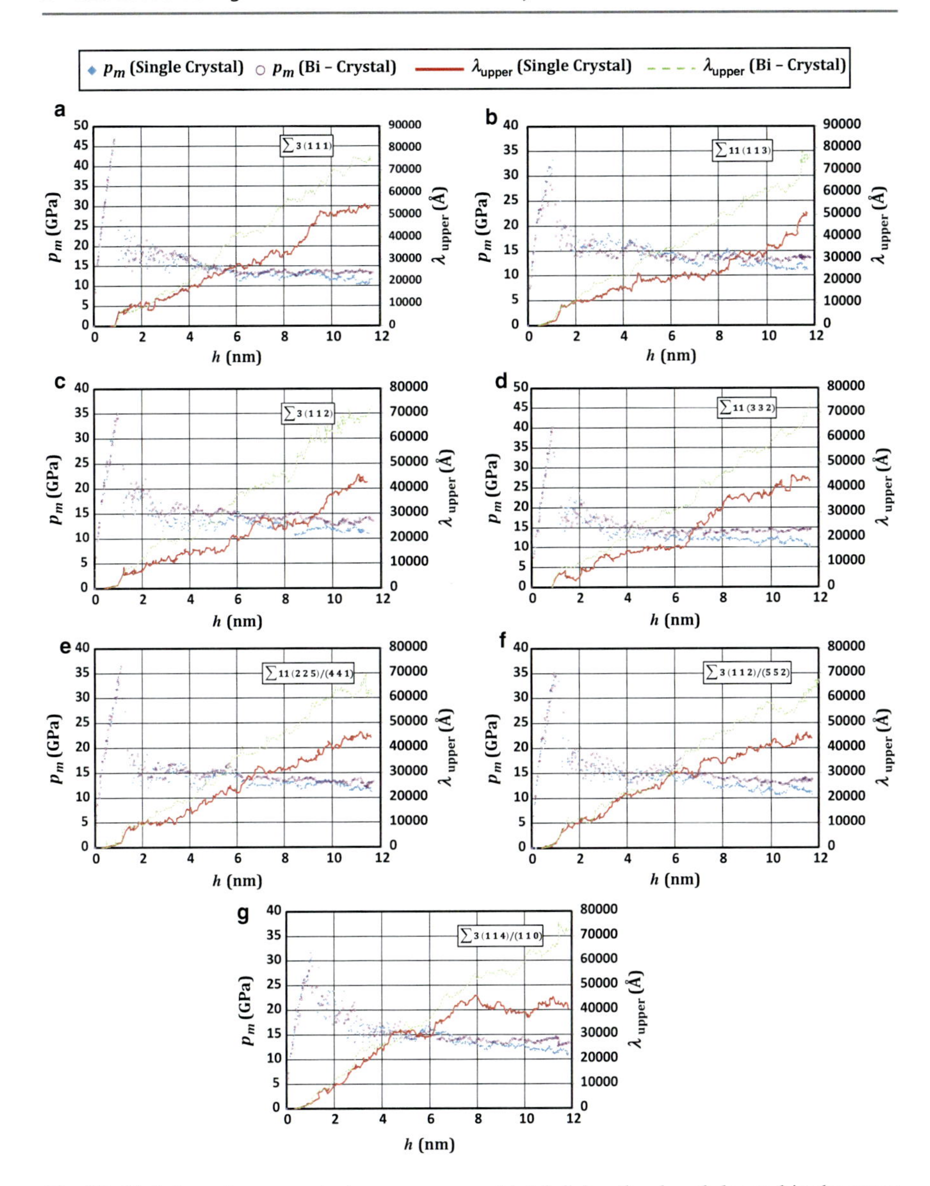

Fig. 23 Variation of mean contact pressure p_m and total dislocation length located in the upper grain λ_{upper} as a function of indentation depth h for S2 bicrystal and their related single crystal samples with grain boundaries of **a** $\sum 3(111)$, **b** $\sum 11(113)$, **c** $\sum 3(112)$, **d** $\sum 11(332)$, **e** $\sum 11(225)/(441)$, **f** $\sum 3\ (112)/\left(\overline{552}\right)$, and **g** $\sum 3(114)/(110)$ (Reprinted with permission from Voyiadjis and Yaghoobi 2016)

References

A.H. Almasri, G.Z. Voyiadjis, Nano-indentation in FCC metals: experimental study. Acta Mech. **209**, 1–9 (2010)

R.K.A. Al-Rub, G.Z. Voyiadjis, Analytical and experimental determination of the material intrinsic length scale of strain gradient plasticity theory from micro-and nano-indentation experiments. Int. J. Plast. **20**, 1139–1182 (2004)

M.I. Baskes, Modified embedded-atom potentials for cubic materials and impurities. Phys. Rev. B **46**, 2727 (1992)

S.G. Corcoran, R.J. Colton, E.T. Lilleodden, W.W. Gerberich, Anomalous plastic deformation at surfaces: nanoindentation of gold single crystals. Phys. Rev. B **55**, 16057–16060 (1997)

C.F.O. Dahlberg, Y. Saito, M.S. Öztop, J.W. Kysar, Geometrically necessary dislocation density measurements associated with different angles of indentations. Int. J. Plast. **54**, 81–95 (2014)

M.S. Daw, M.I. Baskes, Embedded-atom method: derivation and application to impurities, surfaces, and other defects in metals. Phys. Rev. B **29**, 6443–6453 (1984)

E. Demir, D. Raabe, N. Zaafarani, S. Zaefferer, Investigation of the indentation size effect through the measurement of the geometrically necessary dislocations beneath small indents of different depths using EBSD tomography. Acta Mater. **57**, 559–569 (2009)

E. Demir, D. Raabe, F. Roters, The mechanical size effect as a mean-field breakdown phenomenon: example of microscale single crystal beam bending. Acta Mater. **58**, 1876–1886 (2010)

K. Durst, B. Backes, M. Göken, Indentation size effect in metallic materials: correcting for the size of the plastic zone. Scr. Mater. **52**, 1093–1097 (2005)

J.A. El-Awady, Unravelling the physics of size-dependent dislocation-mediated plasticity. Nat. Commun. **6**, 5926 (2015)

J.A. El-Awady, M. Wen, N.M. Ghoniem, The role of the weakest-link mechanism in controlling the plasticity of micropillars. J. Mech. Phys. Solids **57**, 32–50 (2009)

D. Faken, H. Jonsson, Systematic analysis of local atomic structure combined with 3D computer graphics. Comput. Mater. Sci. **2**, 279–286 (1994)

J.R. Greer, Nano and Cell Mechanics: Fundamentals and Frontiers. Wiley, Chichester, pp 163–190 (2013)

A. Hasnaoui, P.M. Derlet, H. Van Swygenhoven, Interaction between dislocations and grain boundaries under an indenter – a molecular dynamics simulation. Acta Mater. **52**, 2251–2258 (2004)

H. Jang, D. Farkas, Interaction of lattice dislocations with a grain boundary during nanoindentation simulation. Mater. Lett. **61**, 868–871 (2007)

C.L. Kelchner, S.J. Plimpton, J.C. Hamilton, Dislocation nucleation and defect structure during surface indentation. Phys. Rev. B **58**, 11085–11088 (1998)

C.C. Koch, I.A. Ovid'ko, S. Seal, S. Veprek, *Structural Nanocrystalline Materials: Fundamentals and Applications* (Cambridge University Press, Cambridge, 2007)

M. de Koning, R.J. Kurtz, V.V. Bulatov, C.H. Henager, R.G. Hoagland, W. Cai, M. Nomura, Modeling of dislocation–grain boundary interactions in FCC metals. J. Nucl. Mater. **323**, 281–289 (2003)

O. Kraft, P. Gruber, R. Mönig, D. Weygand, Plasticity in confined dimensions. Annu. Rev. Mater. Res. **40**, 293–317 (2010)

Y. Kulkarni, R.J. Asaroa, D. Farkas, Are nanotwinned structures in fcc metals optimal for strength, ductility and grain stability? Scr. Mater. **60**, 532–535 (2009)

J.W. Kysar, C.L. Briant, Crack tip deformation fields in ductile single crystals. Acta Mater. **50**, 2367–2380 (2002)

J.W. Kysar, Y.X. Gan, T.L. Morse, X. Chen, M.E. Jones, High strain gradient plasticity associated with wedge indentation into face-centered cubic single crystals: geometrically necessary dislocation densities. J. Mech. Phys. Solids **55**, 1554–1573 (2007)

Y. Lee, J.Y. Park, S.Y. Kim, S. Jun, Atomistic simulations of incipient plasticity under Al (111) nanoindentation. Mech. Mater. **37**, 1035–1048 (2005)

J. Li, K.J. Van Vliet, T. Zhu, S. Yip, S. Suresh, Atomistic mechanisms governing elastic limit and incipient plasticity in crystals. Nature **418**, 307–310 (2002)
S.N. Medyanik, S. Shao, Strengthening effects of coherent interfaces in nanoscale metallic bilayers. Comput. Mater. Sci. **45**, 1129–1133 (2009)
M.A. Meyers, A. Mishra, D.J. Benson, Mechanical properties of nanocrystalline materials. Prog. Mater. Sci. **51**, 427–556 (2006)
Y. Mishin, D. Farkas, M.J. Mehl, D.A. Papaconstantopoulos, Interatomic potentials for monoatomic metals from experimental data and ab initio calculations. Phys. Rev. B **59**, 3393–3407 (1999)
A.K. Nair, E. Parker, P. Gaudreau, D. Farkas, R.D. Kriz, Size effects in indentation response of thin films at the nanoscale: a molecular dynamics study. Int. J. Plast. **24**, 2016–2031 (2008)
W.D. Nix, H.J. Gao, Indentation size effects in crystalline materials: a law for strain gradient plasticity. J. Mech. Phys. Solids **46**, 411–425 (1998)
D.M. Norfleet, D.M. Dimiduk, S.J. Polasik, M.D. Uchic, M.J. Mills, Dislocation structures and their relationship to strength in deformed nickel microcrystals. Acta Mater. **56**, 2988–3001 (2008)
T.A. Parthasarathy, S.I. Rao, D.M. Dimiduk, M.D. Uchic, D.R. Trinkle, Contribution to size effect of yield strength from the stochastics of dislocation source lengths in finite samples. Scr. Mater. **56**, 313–316 (2007)
P. Peng, G. Liao, T. Shi, Z. Tang, Y. Gao, Molecular dynamic simulations of nanoindentation in aluminum thin film on silicon substrate. Appl. Surf. Sci. **256**, 6284–6290 (2010)
S. Plimpton, Fast parallel algorithms for short-range molecular dynamics. J. Comput. Phys. **117**, 1–19 (1995)
N.M. Pugno, A general shape/size-effect law for nanoindentation. Acta Mater. **55**, 1947–1953 (2007)
S.I. Rao, D.M. Dimiduk, M. Tang, T.A. Parthasarathy, M.D. Uchic, C. Woodward, Estimating the strength of single-ended dislocation sources in micron-sized single crystals. Philos. Mag. **87**, 4777–4794 (2007)
S.I. Rao, D.M. Dimiduk, T.A. Parthasarathy, M.D. Uchic, M. Tang, C. Woodward, Athermal mechanisms of size-dependent crystal flow gleaned from three-dimensional discrete dislocation simulations. Acta Mater. **56**, 3245–3259 (2008)
M.D. Sangid, T. Ezaz, H. Sehitoglu, I.M. Robertson, Energy of slip transmission and nucleation at grain boundaries. Acta Mater. **59**, 283–296 (2011)
S. Shao, S.N. Medyanik, Dislocation–interface interaction in nanoscale fcc metallic bilayers. Mech. Res. Commun. **37**, 315–319 (2010)
W.A. Soer, J.T.M. De Hosson, Detection of grain-boundary resistance to slip transfer using nanoindentation. Mater. Lett. **59**, 3192–3195 (2005)
A. Stukowski, Structure identification methods for atomistic simulations of crystalline materials. Model. Simul. Mater. Sci. Eng. **20**, 045021 (2012)
A. Stukowski, Computational analysis methods in atomistic modeling of crystals. JOM **66**, 399–407 (2014)
A. Stukowski, K. Albe, Extracting dislocations and non-dislocation crystal defects from atomistic simulation data. Model. Simul. Mater. Sci. Eng. **18**, 085001 (2010)
A. Stukowski, K. Albe, D. Farkas, Nanotwinned fcc metals: strengthening versus softening mechanisms. Phys. Rev. B **82**, 224103 (2010)
A. Stukowski, V.V. Bulatov, A. Arsenlis, Automated identification and indexing of dislocations in crystal interfaces. Model. Simul. Mater. Sci. Eng. **20**, 085007 (2012)
S. Suresh, T.G. Nieh, B.W. Choi, Nanoindentation of copper thin films on silicon substrates. Scr. Mater. **41**, 951–957 (1999)
J.G. Swadener, E.P. George, G.M. Pharr, The correlation of the indentation size effect measured with indenters of various shapes. J. Mech. Phys. Solids **50**, 681–694 (2002)
J. Tersoff, New empirical approach for the structure and energy of covalent systems. Phys. Rev. B **37**, 6991–7000 (1988)

T. Tsuru, Y. Kaji, D. Matsunaka, Y. Shibutani, Incipient plasticity of twin and stable/unstable grain boundaries during nanoindentation in copper. Phys. Rev. B **82**, 024101 (2010)

M.D. Uchic, P.A. Shade, D.M. Dimiduk, Plasticity of micrometer-scale single crystals in compression. Annu. Rev. Mater. Res. 39, 361–386 (2009)

G.Z. Voyiadjis, R.K.A. Al-Rub, Gradient plasticity theory with a variable length scale parameter. Int. J. Solids Struct. **42**, 3998–4029 (2005)

G.Z. Voyiadjis, M. Yaghoobi, Large scale atomistic simulation of size effects during nanoindentation: dislocation length and hardness. Mater. Sci. Eng. A **634**, 20–31 (2015)

G.Z. Voyiadjis, M. Yaghoobi, Role of grain boundary on the sources of size effects. Comput. Mater. Sci. **117**, 315–329 (2016)

G.Z. Voyiadjis, M. Yaghoobi, Size and strain rate effects in metallic samples of confined volumes: dislocation length distribution. Scr. Mater. **130**, 182–186 (2017)

M. Yaghoobi, G.Z. Voyiadjis, Effect of boundary conditions on the MD simulation of nanoindentation. Comput. Mater. Sci. **95**, 626–636 (2014)

M. Yaghoobi, G.Z. Voyiadjis, Atomistic simulation of size effects in single-crystalline metals of confined volumes during nanoindentation. Comput. Mater. Sci. **111**, 64–73 (2016a)

M. Yaghoobi, G.Z. Voyiadjis, Size effects in fcc crystals during the high rate compression test. Acta Mater. **121**, 190–201 (2016b)

M. Yaghoobi, G.Z. Voyiadjis, Microstructural investigation of the hardening mechanism in fcc crystals during high rate deformations. Comp. Mater. Sci. **138**, 10–15 (2017)

N. Zaafarani, D. Raabe, F. Roters, S. Zaefferer, On the origin of deformation-induced rotation patterns below nanoindents. Acta Mater. **56**, 31–42 (2008)

T.T. Zhu, A.J. Bushby, D.J. Dunstan, Materials mechanical size effects: a review. Mater. Technol. **23**, 193–209 (2008)

J.A. Zimmerman, C.L. Kelchner, P.A. Klein, J.C. Hamilton, S.M. Foiles, Surface step effects on nanoindentation. Phys. Rev. Lett. **87**, 165507 (2001)

Molecular Dynamics-Decorated Finite Element Method (MDeFEM): Application to the Gating Mechanism of Mechanosensitive Channels

3

Liangliang Zhu, Qiang Cui, Yilun Liu, Yuan Yan, Hang Xiao, and Xi Chen

Contents

L. Zhu
Columbia Nanomechanics Research Center, Department of Earth and Environmental Engineering, Columbia University, New York, NY, USA

International Center for Applied Mechanics, State Key Laboratory for Strength and Vibration of Mechanical Structures, School of Aerospace, Xi'an Jiaotong University, Xi'an, China
e-mail: seven.zhu@qq.com; zhu.liangliang@stu.xjtu.edu.cn

Q. Cui
Department of Chemistry and Theoretical Chemistry Institute, University of Wisconsin-Madison, Madison, WI, USA
e-mail: cui@chem.wisc.edu

Y. Liu
International Center for Applied Mechanics, State Key Laboratory for Strength and Vibration of Mechanical Structures, School of Aerospace, Xi'an Jiaotong University, Xi'an, China
e-mail: yilunliu@mail.xjtu.edu.cn

Y. Yan · H. Xiao
School of Chemical Engineering, Northwest University, Xi'an, China
e-mail: deeplake@qq.com; xiaohang007@gmail.com

X. Chen (✉)
Department of Earth and Environmental Engineering, Columbia Nanomechanics Research Center, Columbia University, New York, NY, USA
e-mail: xichen@columbia.edu

G. Z. Voyiadjis (ed.), *Handbook of Nonlocal Continuum Mechanics for Materials and Structures*, https://doi.org/10.1007/978-3-319-58729-5_46

Abstract

Many fundamentally important biological processes rely on the mechanical responses of membrane proteins and their assemblies in the membrane environment, which are multiscale in nature and represent a significant challenge in modeling and simulation. For example, in mechanotransduction, mechanical stimuli can be introduced through macroscopic-scale contacts, which are transduced to mesoscopic-scale (micron) distances and can eventually lead to microscopic-scale (nanometer) conformational changes in membrane-bound protein or protein complexes. This is a fascinating process that spans a large range of length scales and time scales. The involvement of membrane environment and critical issues such as cooperativity calls for the need for an efficient multi-scale computational approach. The goal of the present research is to develop a hierarchical approach to study the mechanical behaviors of membrane proteins with a special emphasis on the gating mechanisms of mechanosensitive (MS) channels. This requires the formulation of modeling and numerical methods that can effectively bridge the disparate length and time scales. A top-down approach is proposed to achieve this by effectively treating biomolecules and their assemblies as integrated structures, in which the most important components of the biomolecule (e.g., MS channel) are modeled as continuum objects, yet their mechanical/physical properties, as well as their interactions, are derived from atomistic simulations. Molecular dynamics (MD) simulations at the nanoscale are used to obtain information on the physical properties and interactions among protein, lipid membrane, and solvent molecules, as well as relevant energetic and temporal characteristics. Effective continuum models are developed to incorporate these atomistic features, and the conformational response of macromolecule(s) to external mechanical perturbations is simulated using finite element (FEM) analyses with in situ mechanochemical coupling. Results from the continuum mechanics analysis provide further insights into the gating transition of MS channels at structural and physical levels, and specific predictions are proposed for further experimental investigations. It is anticipated that the hierarchical framework is uniquely suited for the analysis of many biomolecules and their assemblies under external mechanical stimuli.

Keywords

Mechanotransduction · Multi-scale simulation · Mechanosensitive channels · Gating mechanism · Continuum mechanics · Continuum solvation

Introduction

Occurring over large time and length scales, various biological signal transduction processes rely on the mechanical response of biomolecules and their assemblies to external stimuli. Muscle contraction or stretch, as a prominent example, involves the cooperative mechanical response of a large number of myosin molecules (Geeves and Holmes 1999, 2005), and structural changes from molecular scale to organ scales of muscle contribute greatly to its various and remarkable adaptations under different mechanical stimuli (Wisdom et al. 2014). Another example is mechanosensation (Hamill and Martinac 2001), during which the mechanosensitive (MS) channels play important roles in living cells of diverse phylogenetic origin (Martinac 2004) and have been identified in more than 30 cell types (Sackin 1995). By converting mechanical forces exerted on the cell membrane into biochemical or electrical signals, MS channels are involved in a wide range of cellular processes including cell growth and differentiation (Wang and Thampatty 2006) and blood pressure and cell volume regulation (Hamill and Martinac 2001; Martinac and Kloda 2012; Sun et al. 2009) and are essential to sensations such as touching, balance, and hearing (Hamill and Martinac 2001; Ingber 2006; Martinac 2004). A direct link between the lipid membrane and the structure/function of some MS channels has been revealed (Phillips et al. 2009). And in eukaryotic cells, the cytoskeleton was shown to play a similar role in the activation of MS channels (Hayakawa et al. 2008). With an increasing number of MS channels being identified, their atomic structures, gating characteristics, and functional mechanisms have been studied extensively in the past decades. Among the families of MS channels, a much-studied system is the mechanosensitive channel of small/large conductance (MscS/MscL) in *Escherichia coli* (*E. coli*), which serves as a paradigm for understanding the gating behaviors of the MS family of ion channels. Functioning as the "safety valve" of bacteria that regulates turgor pressure, MscS/MscL is sensitive to tension in the membrane, and the opening of MscS/MscL allows exchange of ions (nonselective between anions and cations) and small molecules (including water) between the cytoplasm and the environment (Berrier et al. 1996; Blount et al. 1997; Martinac et al. 2014; Saimi et al. 1992).

A Brief Review of MscS Studies

The first crystal structure for *E. coli*-MscS was solved at 3.95 Å resolution (Bass et al. 2002) and subsequently refined to a higher resolution of 3.7 Å (Steinbacher et al. 2007). The structure has a pore of less than 5 Å in diameter, and because of its hydrophobic constriction, the pore is thus considered as nonconducting or closed (Steinbacher et al. 2007; Vora et al. 2006). The open form of the A106V mutant of MscS was subsequently crystallized at 3.45 Å resolution (Wang et al. 2008). The crystal structures of *E. coli*-MscS have been constantly challenged, mainly due to the large voids between the TM3 helix (the third transmembrane helix) and the closely packed TM1 and TM2 helices (also referred as TM pockets). Using extrapolated motion dynamics (EMD) (Akitake et al. 2007) and continuous wave electron paramagnetic resonance (cwEPR) (Vasquez et al. 2008a, b), alternative MscS structures of both closed and open forms were independently generated, in which the apparent voids in the crystal structures were absent. The three approaches lead to three sets of mutually incompatible models of the closed and open structures and thus models for the gating transition. A pulsed electron-electron double resonance (PELDOR) approach reevaluated these competing structural models both in detergent (Pliotas et al. 2012) and in bilayer mimics (Ward et al. 2014); the results supported the arrangement of helices seen in the crystal structures. Another study reported the crystal structure of β-dodecylmaltoside-solubilized wild-type *E. coli*-MscS at 4.4 Å resolution and further supported that the A106V structure resembles the open state (Lai et al. 2013). Finally, a higher-resolution structure of the *E. coli*-MscS identified alkyl chains inside the pockets/voids formed by the transmembrane helices (Pliotas et al. 2015), strongly support that the voids in *E. coli*-MscS crystal structures are realistic (Pliotas and Naismith 2016). Based on the above evidence, the present study starts with the assumption that the crystal structure at 3.7 Å resolution (Steinbacher et al. 2007) and the A106V mutant at 3.45 Å resolution (Wang et al. 2008) represent the closed (resting) and open states of MscS, respectively, but noting that the specific functional states of these structures of MscS may be still in debate. Nevertheless, despite the available closed and open structures, little is known about the partially expanded intermediate structures during MscS gating transition.

Numerous studies have explored residues and interactions that are important to the gating characteristics of MscS. Some of the established cases include the Asp62-Arg131/Arg128 salt bridges (Nomura et al. 2008) and the Phe68-Leu111/Leu115 (Belyy et al. 2010) apolar interaction, which affect channel gating and inactivation; Leu105 and Leu109 form a hydrophobic lock at the channel pore (Anishkin and Sukharev 2004; Vora et al. 2006); the interaction between the lower part of TM3 and the cytoplasmic β domain and Gly113 is crucial to inactivation (Edwards et al. 2008; Koprowski et al. 2011; Petrov et al. 2013); and a number of residues were shown to influence force transmission at the protein-lipid interface (Malcolm et al. 2011; Nomura et al. 2006). The physical origins for the importance of these interactions are not always well understood. Another much studied mechanistic issue concerns inactivation of MscS: under prolonged exposure to subthreshold membrane tension, the channel desensitizes into an inactivated and nonconducting state from which it must relax back to the closed state in lower membrane tension before reactivation can be induced (Akitake et al. 2005; Edwards et al. 2008; Koprowski and Kubalski 1998; Levina et al. 1999). Up to now, while multiple residues and interactions are known to be important to inactivation (Vasquez 2013), the structural mechanism underlying inactivation remains elusive.

Furthermore, it has been reported that the open probability of MscS can be significantly increased by membrane depolarization (Cui et al. 1995; Martinac et al. 1987). A later study, however, showed that the activation of *E. coli*-MscS by membrane tension is essentially independent of the transmembrane voltage (Akitake et al. 2005), though depolarizing membrane voltage strongly promotes inactivation

(Akitake et al. 2005). In addition, a recent study reported that the arginine residues at positions of 46, 54, and 74 in TM1 and TM2 helices are not responsible for the voltage dependence of inactivation (Nomura et al. 2016). The structural response and inactivation mechanism under membrane potential thus remain to be better clarified.

Atomistic simulations have made valuable contributions to the understanding of ion channels, including MscS, in recent years (Akitake et al. 2007; Anishkin and Sukharev 2004; Deplazes et al. 2012; Masetti et al. 2016; Pliotas et al. 2015; Sotomayor and Schulten 2004; Sotomayor et al. 2006). Nevertheless, such simulations remain computationally intensive, making it difficult to study gating transitions that occur on the millisecond time scale and explore contributions of specific structural motifs and interactions.

A Brief Review of MscL Studies

A large body of experimental, theoretical, and simulation work has focused on elucidating the molecular mechanism of MscL gating (Booth et al. 2007; Haswell et al. 2011). Experiments directly probing the gating transition of MscL were primarily patch-clamp measurements, which simultaneously monitor the membrane tension and ionic currents through the channel (thus opening probability) (Sukharev et al. 1997, 1999). An important clue from these studies on MscL reconstituted into purified lipid bilayers is that the mechanical property of the membrane plays a principal role in determining the gating behaviors of MscL. A model with five subconducting states was established (Sukharev et al. 1999), in which the tension-dependent conformational transition was primarily attributed to the pore area variation that occurs between the closed state and a low subconductance state. In addition, other experimental studies have been used to probe MscL's conformational transition, and these include electron paramagnetic resonance spectroscopy (EPR) with site-directed spin labeling (SDSL) (Perozo et al. 2002a, b) for monitoring the structural rearrangements, cysteine scanning for identifying residue contacts in the transmembrane helices (Levin and Blount 2004), and numerous mutation studies for probing the importance of residues in different structural motifs (Anishkin et al. 2005; Blount et al. 1997; Levin and Blount 2004; Tsai et al. 2005). Besides, a single-molecule fluorescence resonance energy transfer (FRET) method (Wang et al. 2014) or the combination of data from FRET spectroscopy and simulations (Corry et al. 2010) has enabled a more detailed description of the open form of MscL in the natural lipid environment. Based on geometrical constraints provided by various measurements and the crystal structure of MscL in *Mycobacterium tuberculosis* (*Tb*) (Chang et al. 1998), structural models for the closed-open transition of *E. coli*-MscL were constructed (Sukharev et al. 2001a, b); revised models were proposed subsequently where the conformational changes of the S3 helices are much smaller in scale (Sukharev and Anishkin 2004; Sukharev and Corey 2004). Although highly valuable, these structural models need to be evaluated for validity in a systematic and physical manner (Chen et al. 2008; Tang et al. 2006, 2008).

Meanwhile, analytical models for the gating transition in MscL have been developed by several groups. Markin and Sachs presented a general thermodynamic model for mechanotransduction that relates the probability of channel opening to membrane properties such as thickness, curvature, and stiffness (Markin and Sachs 2004). An analytical continuum model was developed by Wiggins and Phillips to characterize the free energy of the protein-bilayer system (Wiggins and Phillips 2004); the model highlighted that the competition of hydrophobic mismatch could be a physical gating mechanism. As an alternative to dilatational gating, a gating-by-tilt model was proposed (Turner and Sens 2004) in which the gate opening is due to the swinging of the lipids near the channel with respect to a pivot. Although these analytical models are valuable for highlighting the potential contribution of specific physical factors (e.g., hydrophobic mismatch), they lack sufficient structural details to make specific connection with experimental studies.

Numerical simulation is a powerful approach for exploring the fundamental principles of mechanotransduction. To properly assign structural features to important intermediate states along the closed-open transition, it is important to simulate the structural response of the channel to mechanical perturbation consistent with the experimental protocol. Due to the large length scale and time scale involved, however, this is usually beyond the capability of atomistic simulations despite the rapid progresses being made (Deplazes et al. 2012; Dror et al. 2012; Gullingsrud and Schulten 2003; Karplus and Kuriyan 2005; Klepeis et al. 2009; Monticelli et al. 2008; Sawada et al. 2012; Snow et al. 2005; Yefimov et al. 2008). Hence, developing coarse-grained models to access longer time scales has become an important topic in the simulation community (Ingolfsson et al. 2014; Marrink et al. 2007; Marrink and Tieleman 2013; Monticelli et al. 2008; Praprotnik et al. 2008; Saunders and Voth 2012; Shi et al. 2006; Shinoda et al. 2012; Yefimov et al. 2008). Most of these efforts have focused on developing particle-based models in which one bead represents a group of atoms. Specifically for MscL, building upon their success in developing an effective coarse-grained model for lipids, Marrink et al. developed a coarse-grained model (Yefimov et al. 2008) for MscL based on the transfer free energy of amino acids between water and lipids. The gating transition was successfully observed in the simulation although the pore radius in the final state is somewhat smaller than that estimated in the literature (Sukharev et al. 2001b).

Molecular Dynamics-Decorated Finite Element Method (MDeFEM)

In light of the limitations of previous experimental/theoretical/numerical efforts concerning the structural rearrangements of MS channels during gating, it is worthwhile pursuing the alternative approach of continuum mechanical modeling, which has been used in a broad set of mechanics problems (Scarpa et al. 2010; Tang et al. 2006; Tserpes and Papanikos 2009; Zeng et al. 2012). Along this line, the establishment of molecular dynamics-decorated finite element method (MDeFEM) (Chen et al. 2008; Tang et al. 2006, 2008) represents an attractive alternative to atomistic and particle-based coarse-grained simulations, allowing the analysis of bimolecular systems at long time scales while maintaining sufficient molecular details to capture the most essential characteristics of the system under study.

In MDeFEM, biomolecules are modeled as integrated continuum motifs and the finite element simulation framework allows efficient treatment of deformations at large length scales and complex deformation modes inaccessible to conventional all-atom simulations.

In this chapter, MDeFEM (Chen et al. 2008; Tang et al. 2006, 2008) is firstly adapted to study the gating mechanism of MscS. The high computational efficiency of MDeFEM allows us to analyze the contributions of various structural motifs and interactions to the gating process as well as the effect of voltage. The observation of different gating characteristics upon perturbation of material properties of the helical kink region in TM3 at Gly113 also leads to the proposal of a mechanism for inactivation. Overall, the current simulations not only provide new insights into the gating transition of MscS with structural details but also lead to specific predictions that can be tested by future experimental studies.

Secondly, a number of major limitations of previous MDeFEM models of MscL (Chen et al. 2008; Tang et al. 2006, 2008) (or the MDeFEM models of MscS) need to be alleviated for more quantitative analysis. These limitations include (1) not sufficient structural/energetic details of MscL. The helices were represented by rounded sticks, and the inter-component interactions were computed based on surface-to-surface interactions; the nonbonded interactions among loops and those between the loops and the helices were not considered. (2) The lipid bilayer was treated as an elastic solid slab, whereas the realistic membrane should be fluidic and does not sustain large shear stress. (3) No solvation contribution was considered. It has been proposed that solvation plays a major role in stabilizing the open conformation of MscL due to the exposure of hydrophilic residues (Anishkin et al. 2005; Anishkin and Kung 2005). These limitations may have led to the exceedingly high membrane strain required for the full-gating transition of MscL in the previous studies (Chen et al. 2008; Tang et al. 2008). The MDeFEM approach is here improved significantly to address these issues through a coupled mechanical-chemical approach. More realistic models for the MscL molecule and the membrane are developed. To include solvation effects, a force-based protocol is established to integrate a continuum mechanics model for the mechanical properties of the macromolecule with a continuum treatment of solvation. A similar approach has been applied to study the salt concentration dependence of DNA bendability (Ma et al. 2009). It is envisioned that the high computational efficiency and flexibility will make this hierarchical multi-scale framework uniquely applicable to the study of mechanical behaviors of various biological systems, interpreting existing and stimulating new experimental investigations.

Gating and Inactivation of *E. coli*-MscS

Models and Methods

In MDeFEM, the protein structure is described by continuum FEM models, and nonbonded interactions between different components are represented by nonlinear

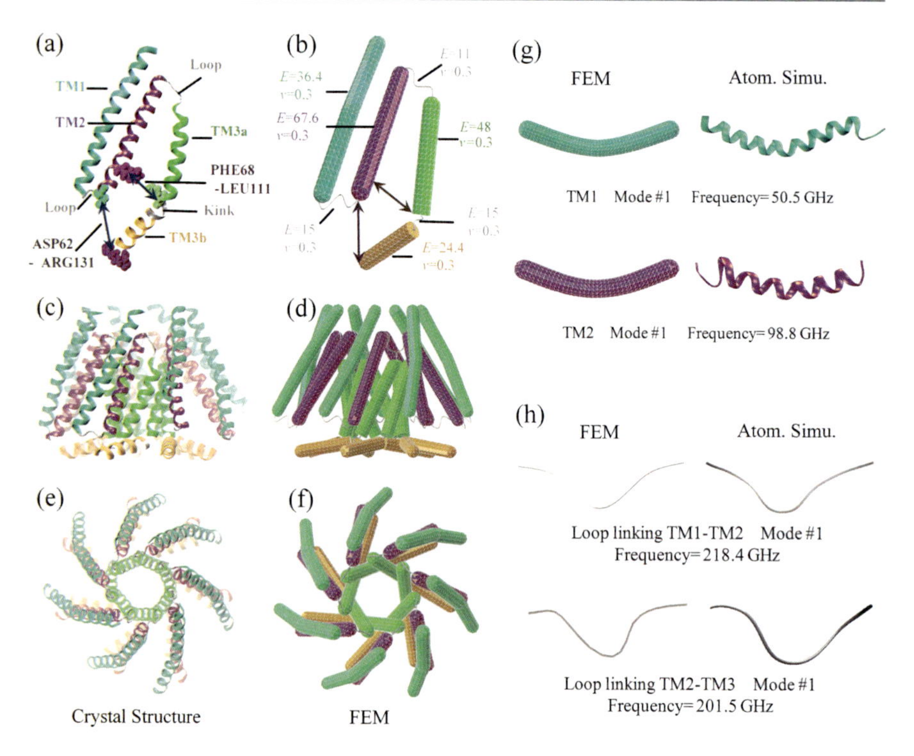

Fig. 1 Continuum mechanics modeling of MscS. (**a**) The backbone structure of one MscS monomer with two key interaction pairs (Asp62-Arg131 and Phe68-Leu111) highlighted by arrows. (**b**) The corresponding FEM model of one monomer. The Young's modulus, E (GPa), and Poisson's ratio, ν, of each component in one monomer are indicated aside. (**c–d**) Side views of the crystal structure and FEM model. (**e–f**) Top views of the crystal structure and FEM model. Examples of several lowest eigenmodes and frequencies of helices and loops are displayed in (**g**) and (**h**), respectively. Results of molecular mechanics are compared to those of the finite element simulations

distance-dependent pressures or point-to-point connectors. In the following, the main continuum construction procedures of *E. coli*-MscS are introduced briefly with the commercial package ABAQUS. More details can be found in our previous works (Chen et al. 2008; Zhu et al. 2016). In the subsequent discussion, unless otherwise specified, MscS refers to *E. coli*-MscS.

A MscS channel (Bass et al. 2002) is formed by seven chains assembled as a heptamer with a large cytoplasmic region that exhibits a mixed α/β structure with several strands and α-helices intertwined together to form a balloon-like osmolyte filter cage (Gamini et al. 2011). In the transmembrane domain (Fig. 1c, e), each monomer (Fig. 1a) of MscS consists of three helices, referred to as TM1, TM2, and TM3, the last of which has a pronounced kink at the Gly113 region and is thus split into TM3a and TM3b. Each TM1 helix and its nearest neighbor, a TM2 helix, are assembled into a pair through a periplasmic loop, forming the outer boundary that interacts extensively with the lipid membrane; the TM3a helices form the inner boundary for the pore, and TM3b helices splay out to be nearly parallel to the membrane plane.

Based on the closed backbone structure of MscS (Bass et al. 2002; Steinbacher et al. 2007), within the continuum mechanics framework, each helix is modeled as a three-dimensional elastic cylinder of 5 Å diameter and the loops as quasi-one-dimensional winding beams with cross-sectional diameter of 2.5 Å (Fig. 1b) (Tang et al. 2006; Zhu et al. 2016). For simplicity, only the transmembrane domain of MscS is considered (Anishkin et al. 2008b). Although there is evidence supporting that the cytoplasmic domain swells up (Machiyama et al. 2009) during gating, it is also suggested to be nonessential but only responsible for increased stability and activity (Schumann et al. 2004); the cytoplasmic domain is not expected to undergo large changes during the gating transition (Pliotas et al. 2012; Wang et al. 2008). The endpoints where the cytoplasmic domain is truncated are softly restrained in the continuum model (1 kcal/mol/Å^2) (Anishkin et al. 2008a, b; Spronk et al. 2006). Furthermore, the first 26 residues (in the N-termini) of the 286 amino acids of each MscS monomer were not resolved in the crystal structure (Bass et al. 2002) and therefore are also excluded in the continuum model; an experimental study showed that MscS can tolerate small deletions at the N-terminus (Miller et al. 2003).

Material properties of each component of the continuum model are calibrated by matching results of normal mode analysis (NMA) at the atomistic and continuum levels (Fig. 1g–h). The key mechanical properties for the helices and loops are shown in Fig. 1b, and these are much larger than the estimated range of the Young's modulus of MscL α-helices (0.2 to 12.5 GPa) in Martinac et al.'s recent work (Bavi et al. 2017) with constant-force steered molecular dynamics (SMD). This is because, in the present study, the helices, for instance, are modeled as much thinner elastic cylinders (with a diameter of 5 Å) by considering only the main chain. The bending stiffness of the helices in the present continuum model and that of MscS helices by SMD are expected to be consistent. As a qualitative comparison, for example, the bending stiffness *EI* (the product of Young's modulus and moment of inertia) of MscS helices in vacuum is in the range of 70~200 (10^{-10} N Å^2) in this work, while based on Martinac et al.'s study (Bavi et al. 2017), the bending stiffness of MscL helices in water can be calculated to be in the range of 1~80 (10^{-10} NÅ^2). For simplicity, the softening effect of the helices due to hydration (Anishkin and Sukharev 2017; Bavi et al. 2017) is not considered here, which may be one of the reasons for the difference in the above comparison of the bending stiffness. In both MscL (Bavi et al. 2017) and MscS, the second helix, TM2, is the stiffest one among the α-helices in each channel.

The MscS continuum model is embedded into an elastic membrane modeled as a sandwich panel that consists of three layers to mimic the lipid head and tail regions. A flat square membrane with a size of 400 × 400 Å is employed. To embed the channel (Fig. 1) into the continuum membrane, a cavity with the shape of a multi-petal flower (Fig. 2) is created in the middle of the membrane with the size and shape of the cavity conforming to those of MscS transmembrane helices in the closed state with an equilibrium distance of ~5.5 Å (Chen et al. 2008) from the surface of the cavity to the surfaces of the helices. The lipids buried in the voids within the transmembrane helices (Pliotas et al. 2015) are neglected; their specific role in gating is not quite clear. The membrane model is parameterized based on MD calculations of the density map of water and lipids and the lateral pressure profiles of the POPE lipid membrane (Chen et al. 2008; Gullingsrud and Schulten 2004). Mechanical properties of the lipid membrane are shown in Fig. 2, and the in-plane shear modulus of the continuum membrane model is reduced (Zhu et al. 2016) to take into account the fluidity of the lipid membrane, i.e., its incapability of sustaining a large shear stress. It would be of interest to investigate the effect of the buried lipids (Pliotas et al. 2015) or more complex lipid properties (such as viscoelasticity (Deseri et al. 2016)) on gating, but we leave it to future studies since more sophisticated continuum models are required.

The assembled protein-membrane continuum model is shown in Fig. 2. Helices of MscS are meshed by four-node tetrahedron finite elements and loops by one-dimensional beam elements (Fig. 1). Without over-resolving the system, appropriate mesh density is ensured through convergence studies, and a typical element size of 1.5 Å is chosen. Four-node tetrahedron finite elements are also used to mesh the lipid membrane model with the mesh gradually more refined toward the boundary of the inner hole where it interacts with the protein directly (Fig. 2). To be consistent with the simple description of the protein-lipid continuum model, solvent molecules are not included, though the hydration effect can be further studied with a more sophisticated continuum mechanics-solvation coupled approach (Zhu et al. 2016) which is to be introduced in section "Coupled Continuum Mechanical-Continuum Solvation Approach with Application to Gating Mechanism of MscL."

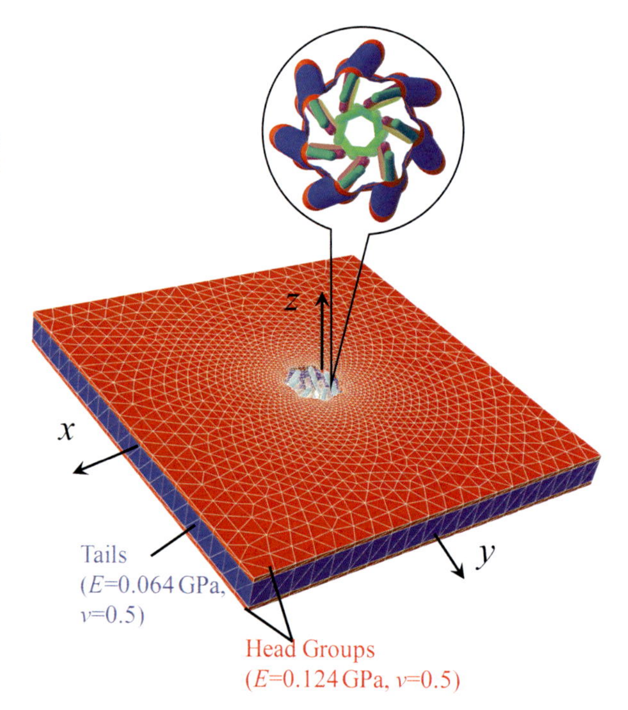

Fig. 2 Schematic of the continuum protein-lipid membrane model. A zoomed-in view is shown to illustrate the multi-petal lipid hole encompassing the MscS protein

In this work, we explicitly consider two key interaction pairs known to be essential to MscS gating, i.e., the Asp62-Arg131 salt bridge and the Phe68-Leu111 apolar interaction (Fig. 1a) mentioned in the section "Introduction." The nonbonded interaction energy (electrostatic and van der Waals) between each pair of residues is calculated using the CHARMM 36 force field (Best et al. 2012; Brooks et al. 1983) as a function of distance along the center-of-mass separation vector (Fig. 3b). An electrostatic screening factor of 6 is used when calculating the electrostatic interaction for the Asp62-Arg131 salt bridge, considering that the gap between Asp62-Arg131 residues in the same monomer or across monomers is spanned mainly by protein atoms or lipid molecules rather than water (Pliotas et al. 2015); varying this factor from 3 to 10 generally has little impact on the computed gating behavior (data not show). The nonlinear distance-dependent interaction force between two residues in each pair is derived accordingly from the energy profile and applied to the closest finite element nodes by invisible nonlinear connector elements (arrows in Fig. 1b) in ABAQUS (Zhu et al. 2016). The connector element representing a pair of key interaction is defined both within the same monomer and across neighbor monomers.

Apart from the above two specific residue pairs, interactions between different helices and between the helix and the lipid membrane are described by a pair-wise effective pressure-distance relationship where the atom-to-atom interaction in the atomic structure are averaged to the surfaces of different components (Tang et al. 2006) (see Fig. 3a). The nonbonded (electrostatic and van der Waals) interactions are calculated without any cutoff. Incorporating an implicit membrane environment through the GBSW model (Im et al. 2003) in CHARMM in the inter-helical calculations has a negligible impact on the energy profiles. Taking the first derivative of the interaction energy with respect to distance, the pressure-distance relationship between two surfaces (adopting the sign convention that repulsive pressure is positive) is obtained and takes the following form:

$$p(\alpha_i) = \frac{Cn}{d_0}\left[(d_0/\alpha_i)^{m+1} - (d_0/\alpha_i)^{n+1}\right] \tag{1}$$

where p and d_0 are the interaction pressure and equilibrium distance between two surfaces, respectively, C is the energy well-depth, and α_i is the instantaneous distance between two deformed surfaces for the i-th element. This nonbonded pressure model implicitly includes both electrostatic and van der Waals (VDW) interactions and has been successfully applied to study the deformation and buckling of carbon nanotubes (Cao and Chen 2011; Chen et al. 2006; Pantano et al. 2003), as well as gating transition of the *E. coli* and *Mycobacterium tuberculosis* MscL (Bavi et al. 2016b; Chen et al. 2008; Tang et al. 2006). Figure 3a shows examples of the interaction energy between different protein components in closed and open crystal structures and the fitted curves used in FEM. The consistency between the interaction energy curves of the closed and open states suggests that the fitted FEM parameters are fairly transferable. Table 1 summarizes the fitted values of C, d_0, and the exponents (m, n) for the closed crystal structure. When calculating

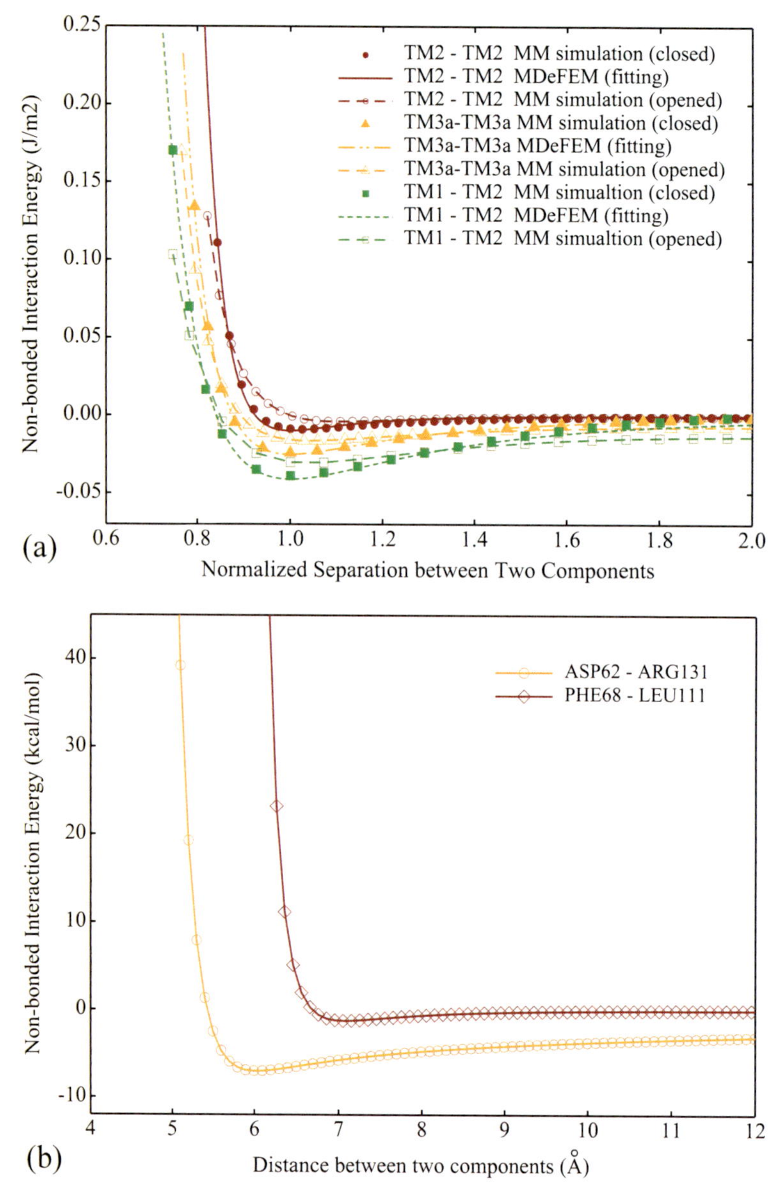

Fig. 3 Examples of the fitting of nonbonded interactions between helices of *E. coli*-MscS (**a**) and the interaction energy for two key interaction pairs (**b**). In (**a**), the distance between two components is normalized by their equilibrium distances. Data points are obtained through molecular mechanics (MM)

the interaction between TM2 and TM3a, the Phe68-Leu111 pair is excluded since it is already considered separately as discussed above. The constitutive interaction relationship is called in each increment through the analysis based on the relative position between two surfaces that are continuously updated. Similar

Table 1 FEM fitting parameters for the nonbonded interactions

Interaction pairs	d_0 (Å)	$\psi = 6Cn/d_0$ (GPa)	m	n
Lipid-TM1	5.5	4.0	8	3
Lipid-TM2	5.5	3.0	6	3
TM1-TM1	11.0	0.35	24	8
TM2-TM2	7.7	1.05	19	8
TM3a-TM3a	3.6	1.5	12	4
TM3b-TM3b	6.6	1.59	6	5
TM1-TM2	4.0	2.25	18	3
TM1-TM3a	10	29.8	24	23
TM2-TM3a	4.5	7.46	12	8
TM2-TM3b	7.7	7.48	24	20
TM3a-TM3b	3.0	2.6	11	2

to the connector elements, the pressure-based nonlinear interaction is defined both within the same monomer and across neighbor monomers as long as the distance between two components is smaller than 16 Å.

The different treatment of the key interaction pairs (described by node-to-node connectors) and other "less important" interactions (described by the averaged surface-to-surface pressure) allows a simple description of the continuum model while providing the opportunity to explore how these key interactions affect gating. A typical tension simulation of the gating transition of MscS takes only about 1 h, on a Thinkpad laptop with four 2.5 GHz CPUs and 8 GB of RAM.

Results and Discussion

In this section, gating transition of MscS in response to membrane stretch (tension) is firstly obtained. The open FEM structure is compared to the crystal open structure, and the intermediate structure is identified in FEM. Similar analysis is then conducted with some key interaction pairs (Asp62-Arg131 or Phe68-Leu111) or loops excluded to explore their role in the gating transition. Next, the kink between TM3a and TM3b is considered as helical (thus having larger cross section and Young's modulus) rather a loop, leading to the discussion of a plausible inactivation mechanism. Finally, the effects of transmembrane voltage are analyzed.

MscS Gating Pathway

Shown in Fig. 4 are several snapshots of MscS during the membrane tension-driven gating process in comparison to the crystal structures for the closed (Bass et al. 2002) and open states (Wang et al. 2008). The results indicate that, during gating, all helices shift radially away from the center of the pore in a manner reminiscent of a mechanical camera iris, similar to the gating transition described for MscL (Betanzos et al. 2002; Tang et al. 2008). The TM1 and TM2 helices in

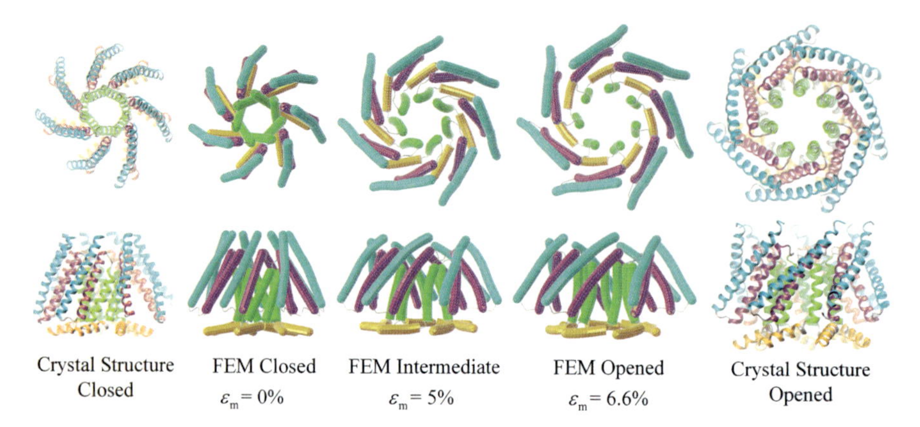

Fig. 4 Comparison between MscS gating pathways under equi-biaxial membrane tension between FEM results and the crystal models in both top (the top panel) and side views (the lower panel)

the same monomer move concurrently almost as a rigid body due to the strong inter-helical interactions. The TM3a helices rotate around and move outward from the central axis and finally become parallel to each other and normal to the membrane plane. Notably, the loop that connects TM1 and TM2 in the same monomer transitions from interacting with the end of TM3b in the neighboring chain (in the counterclockwise direction) to interact with the end of TM3b within the same monomer. Besides, significant rotation of TM3b helices around the central pore axis is observed (Fig. 4), though the average distance from TM3b ends to the pore axis is essentially unchanged. The conformational transitions discussed above are reversible in the FEM simulations once the membrane strain is removed.

On a qualitative level, the current FEM results are in agreement with the crystal structure models, regarding both the orientation and displacement of the transmembrane helices and connecting loops (Fig. 4). Small differences lie in the titling of TM1 helices; from the top view (fist panel in Fig. 4), the TM1 helices form a more compact bundle in the open crystal structure than in the FEM model. In addition to the open structure, intermediate structures during the gating process are obtained in the FEM simulation, and an example is given in Fig. 4.

In the FEM model of MscS in the closed form, the radius of the area lined by TM3a has a radius $R' \approx 7.0$Å (inset in Fig. 5), while the actual pore radius is $R \approx 2.4$Å (Wang et al. 2008) considering the pore's irregular inner surfaces. In the following, the pore radius (R) is estimated by $R = R' - 4.6$Å based on the FEM results of R', and the pore diameter is thus $2R$. Figure 5 depicts the evolution of pore diameter during the gating process. As the membrane strain (ε_m) increases, the pore size of MscS increases slowly until ε_m reaches 5%, after which the pore diameter ($2R$) experiences a rapid increase leading to pore opening ($2R = 13$ Å) (Wang et al. 2008). Since the hydrophobic pore remains nonconducting until approximately $2R > 9$ Å (the threshold diameter for hydration) (Beckstein et al.

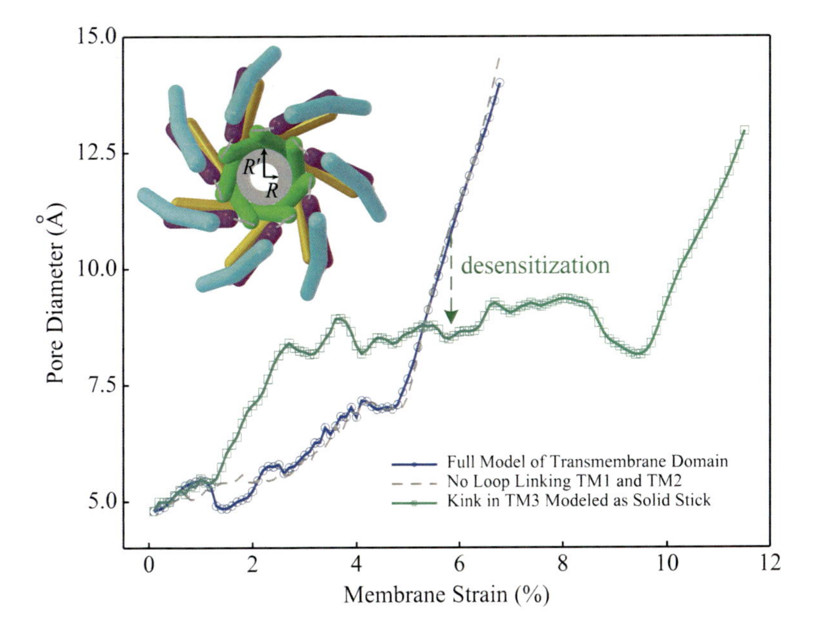

Fig. 5 Pore diameter evolution for different FEM models

2001), this observation is in qualitative agreement with the patch-clamp study of MscS (Martinac and Kloda 2003), in which the gating membrane tension for full opening of MscS was measured to be about half of that of MscL (Martinac and Kloda 2003). The gating membrane strain for McsL was estimated to be ~10% (Zhu et al. 2016), thus for MscS, the gating strain is expected to be 5%; the lipid membrane strain required for full opening of MscS in the present FEM simulation is 6.6% (Fig. 5), which is slightly larger than the 5% estimation.

Effect of the Key Interaction Pairs

Mutation studies have identified two key interaction pairs that greatly affect the gating behavior of MscS, and these involve the Asp62-Arg131 salt-bridge (Nomura et al. 2008) and the apolar interaction between Phe68-Leu111 (Belyy et al. 2010). For example, when the negatively charged Asp62 was replaced with either a neutral (Cys or Asn) or basic (Arg) amino acid, the gating threshold increased significantly (Nomura et al. 2008). Both F68S and L111S substitutions also led to severe loss-of-function phenotypes (Belyy et al. 2010). To provide a structural understanding of how these interactions influence the gating transition of MscS, we conduct FEM simulations with either pair of interactions excluded.

Firstly, the electrostatic interaction between Asp62-Arg131 in the continuum model (Fig. 1) is removed, while the van der Waals interaction between them is preserved (Fig. 6a) so as to mimic the replacement of Asp62 with a charge neutral amino acid. The structural transition upon membrane stretch up to 6.6% is shown in Fig. 6b–c. Without the strong electrostatic interaction, the TM1-TM2 helices tend to

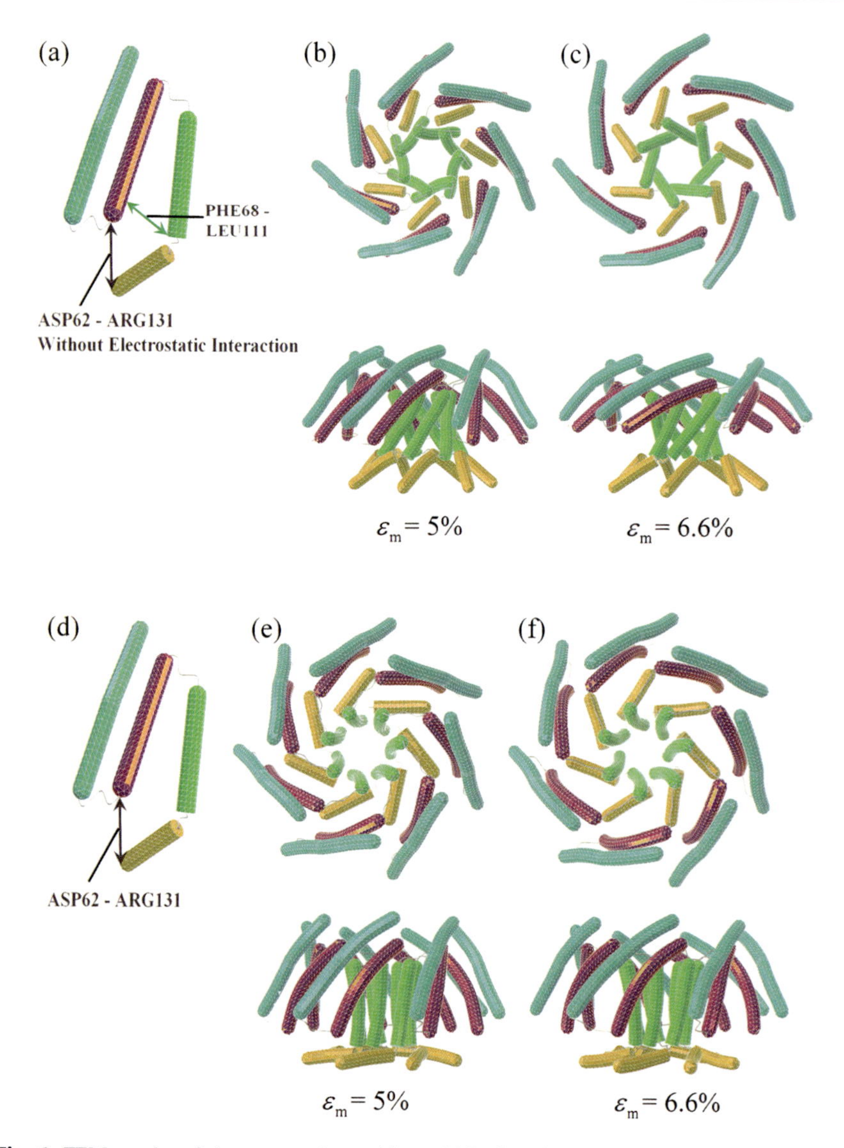

Fig. 6 FEM results of the structural transition of MscS under equi-biaxial membrane tension without electrostatic interaction between Asp62 and Arg131 (**a**–**c**) or without the Phe68-Leu111 traction (**d**–**f**)

detach from the end of TM3b. Under equi-biaxial membrane tension, all helices are lifted up and the kink in TM3 is straightened. As a result, the size of the channel pore surrounded by TM3a helices remains essentially unchanged despite the outward displacements of TM1 and TM2. Therefore, opening of the pore in this "mutant" requires an exceedingly large membrane strain (data not shown).

Similarly, a simulation without the apolar interaction between Phe68-Leu111 is conducted (Fig. 6d–f). During lipid membrane stretching, the TM1 and TM2 helices move outward, but the TM3a helices do not follow and the kink angle in TM3 remains unchanged. With a membrane strain of 6.6%, the pore size actually becomes smaller since the lower ends of TM3a helices are more closely packed because of the loss of TM2-to-TM3a attraction (F68-L111).

In short, the FEM simulations recapitulate the expected behaviors of the relevant mutants in which the key interactions are perturbed (Belyy et al. 2010; Nomura et al. 2008). The observed structural evolution in the FEM simulations also provides a physical understanding of how these interactions contribute to the gating transition. These results support the use of the FEM model to explore other contributions, to which we turn to next.

Effect of Structural Motifs (Loops) in Force Transmission

In this section, we explore the roles that periplasmic loops play in MscS gating by repeating FEM simulations with specific loops excluded from the model. When the loops connecting TM1 and TM2 helices are removed (with the Asp62-Arg131 interaction reserved), the structural response (data not shown) closely resembles that of the full channel model; likewise, the pore diameter evolution is not significantly perturbed (see Fig. 5). When the loops connecting TM2 and TM3 helices are removed, while the rearrangements of TM1 and TM2 helices are similar to those of the full channel model (Fig. 4), the TM3a helices collapse at the upper ends, leading to an essentially closed pore (Fig. 7). This suggests that the loops linking TM2 and TM3 helices are essential to the force transmission from lipid membrane to the pore-lining helices during MscS gating; this is quite different from the situation of MscL (Tang et al. 2008), in which loops that connect the transmembrane helices generally constrain channel opening (Ajouz et al. 2000; Tang et al. 2008). To the best of our knowledge, the importance of the loop between TM2 and TM3 helices to MscS gating has not been pointed in the literature, and this prediction can be further tested by studying loop deletion mutants in patch-clamp experiments as done for MscL (Ajouz et al. 2000; Bavi et al. 2016b).

Effect of the Helical Propensity of the TM3a-TM3b Kink: Inactivation

The Gly113 introduces a kink in TM3, and the region exhibits a weak helical propensity; thus, the kink segment of G1y113 is modeled as a loop (with an arc length of $\sim$8 Å corresponding to the length of the backbone of Gly113) with a small cross-section and a lower Young's modulus (Fig. 1) in models described so far. To explore the significance of the helical kink, FEM simulations are performed in which the Gly113 region in TM3 is treated as helical but remained (enforced) as the bent shape and modeled as a curved solid stick (Fig. 8a) whose material properties take that of the TM3b helix. The structural response of this modified channel model under membrane tension is depicted in Fig. 8a and the pore diameter evolution process in Fig. 5 (square curve). When the membrane strain reaches 6.6%, the pore expands from 4.8 Å to 8 Å in diameter (Fig. 8a). At this point, the expanded pore still remains nonconducting since, for a hydrophobic pore, the threshold diameter

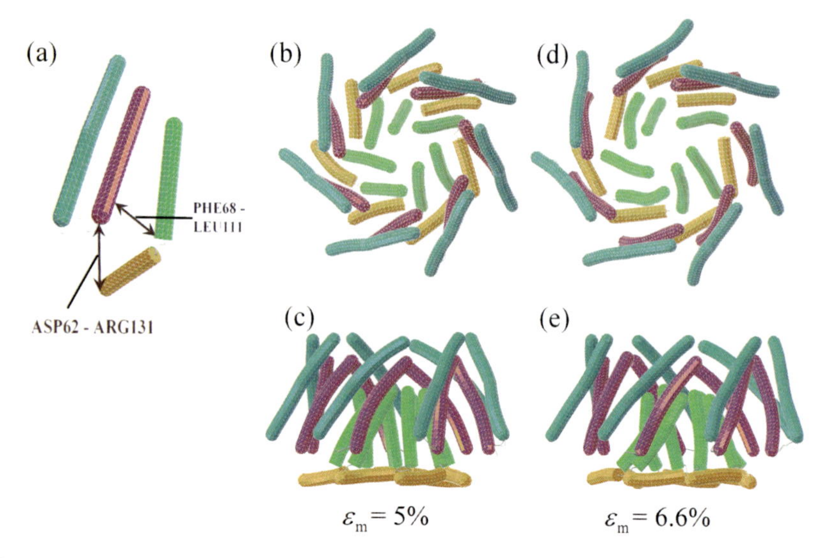

Fig. 7 Structural transition of MscS under equi-biaxial membrane tension with the TM2-TM3 loop removed

for hydration is ~9 Å (Beckstein et al. 2001). To fully open the channel ($2R=13$ Å), a high membrane strain of 11.4% is required (Fig. 8a); this value is comparable to the gating strain of MscL (Zhu et al. 2016). Combining Fig. 4 and Fig. 8a, it seems that a flexible end of TM3a at the kink allows a more effective pulling by the Phe68-Leu111 interaction, thus leading to a lower gating strain. The interaction between TM3b and the cytoplasmic β domain where the TM3b-β interface is quite hydrophobic (Koprowski et al. 2011) promotes the kink around Gly113 (Koprowski et al. 2011; Petrov et al. 2013), which, in the closed state, may prevent backbone hydrogen bonding in the kink area, thus weakening the constraint at the lower ends of TM3a. The TM3b-β interface is indicated in Fig. 8a (ellipse), though noting that the cytoplasmic part is not explicitly included in the present study. Introducing higher polarity (charged residues) into the TM3b-β interface (e.g., the N117 K or G168D mutant) results in a weaker interaction between TM3b and the cytoplasmic β domain (Koprowski et al. 2011), which could be in favor of the backbone hydrogen bonding in the kink area. It has been observed that neither N117 K nor G168D mutant shows visible inactivation behavior, while both of them require a much larger membrane pressure for gating, which is consistent with the present results (Fig. 8a).

With these observations, it is tempting to speculate an inactivation mechanism for MscS: in the resting state, the kink of TM3 behaves like a loop, and upon rapid membrane tension, the channel is easily opened (Fig. 4 and the circle curve in Fig. 5). Under prolonged exposure to subthreshold membrane tension, however, the channel desensitizes (dashed arrow in Fig. 5) into a nonconductive state (the circle curve); this occurs because of the local transition of the kink region into a

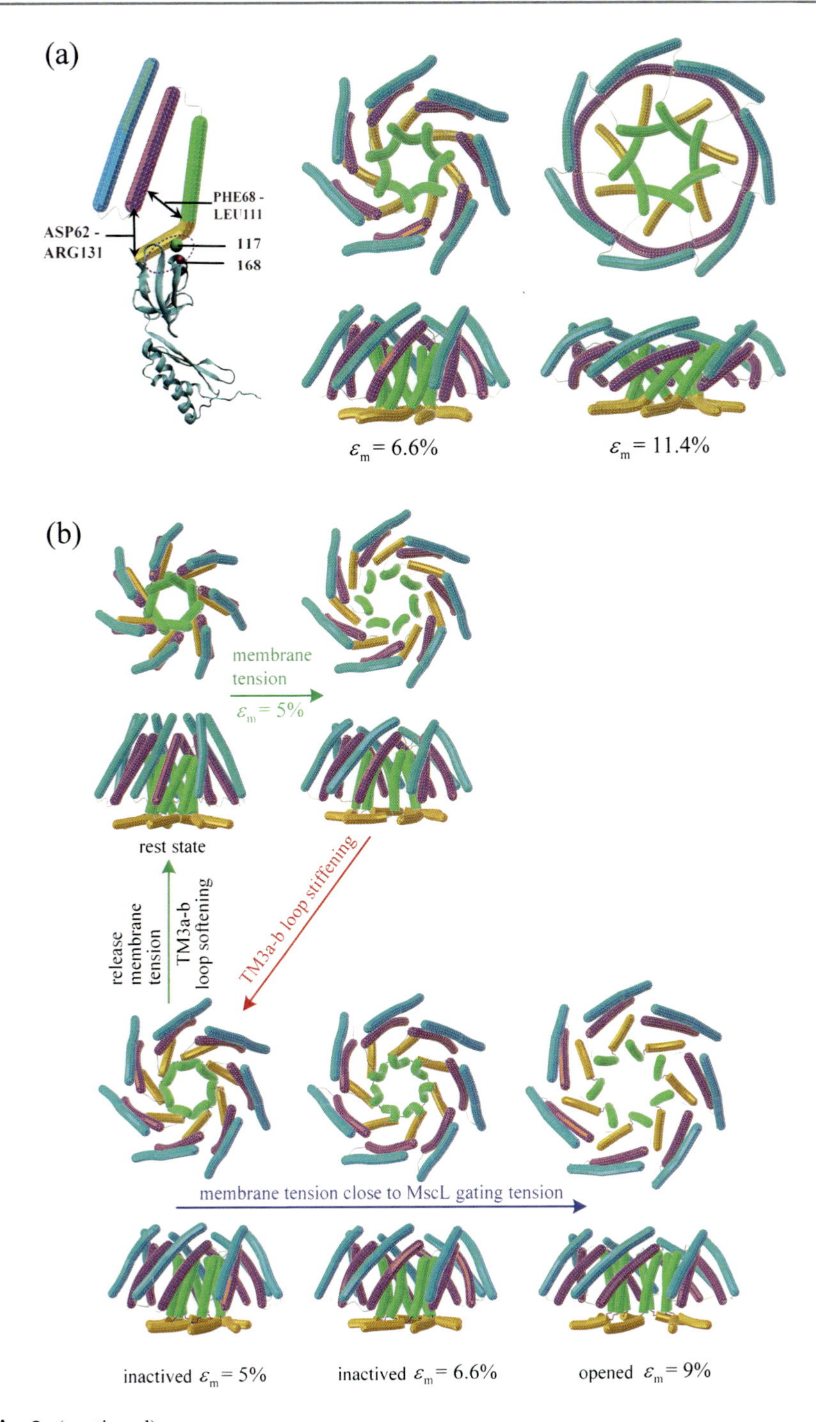

Fig. 8 (continued)

helical conformation that hinders further pore opening. The helical transition of the kink could be induced by the weakening of the TM3b-β interface because of the increased hydration around the interface domain caused by the rotation and outward movement of TM3 helices (Wang et al. 2008) or the reduction of lipids in the TM pockets (Pliotas et al. 2015).

To further test this hypothesis, a FEM simulation is conducted in which, at a subthreshold membrane strain, the Young's modulus of the TM3a-b loop is increased to the extent that its bending stiffness is equal to that of the TM3b helix. In Fig. 8b, after stretching the initial model (model in Fig. 5) by 5% (ε_m=5%), the TM3a-b loop is stiffened, which leads to reduced pore size; further opening the "inactivated structure" requires a very large membrane strain (9%) that is close to the gating strain of MscL. Increasing the stiffness of the TM3a-b loop at other reduced values (<5%) of membrane strain leads to similar results. By contrast, if the stiffening of the TM3a-b loop occurs at a much larger membrane strain (e.g., ε_m=6.6%), the impact on the subsequent pore evolution is minimal. These results are in qualitative agreement with the experimental observation of Sukharev et al. (Kamaraju et al. 2011) that MscS inactivates primarily from the nonconducting state and channel opening prevents inactivation. It should be noted that the present quasi-static approach does not show the dependence of inactivation on (long) time (Kamaraju et al. 2011), which could be a result of desensitization against hydration inside the TM3a pore.

The mechanism for MscS inactivation has remained a mystery for decades. Though some key interactions and residues have been proposed to be involved, how these interactions and residues induce inactivation is poorly understood. The current study suggests that this behavior of MscS could be related to the conformational state of the helical kink at Gly113; a loop conformation allows other interactions to effectively pull TM3a helices outward thus opening the channel, while a more ordered helical conformation hinders the pulling, leading to channel inactivation. Previous experimental studies highlighted the importance of the Gly113 region as well. For example, another MscS-like channel from *Silicibacter pomeroyi* (MscSP) has a conserved glycine residue at the position equivalent to Gly113 in MscS (Petrov et al. 2013). However, the N117 residue in MscS is replaced by a charged residue Glu in MscSP based on the alignments of MscS and MscSP (Petrov et al. 2013), which leads to a weaker interaction between TM3b and the cytoplasmic β domain in MscSP (Koprowski et al. 2011; Petrov et al. 2013) and may facilitate the helical transition of the kink region, resulting in a higher gating threshold (square curve in Fig. 5); indeed, it was reported that the threshold gating strain of MscSP is ~1.5 times of that of MscS (Petrov et al. 2013). In another two MscS-type channels,

Fig. 8 Structural transition of MscS under equi-biaxial membrane tension with the TM3a-TM3b kink considered as helical (**a**) and the proposed mechanism for gating, desensitizing, and inactivation of MscS. The purple ellipse in (**a**) indicates the TM3b-β interface. The increased helical propensity of the kink could be induced by weakening the interaction between TM3b and the cytoplasmic β domain (e.g., the G168D mutant), thus, to an extent, relaxing the kink

MscMJ of *M. jannaschii* (Kloda and Martinac 2001) and MscCG of *C. glutamicum* (Borngen et al. 2010), inactivation was not observed (Petrov et al. 2013); this might be explained by the observation that Gly113 in MscS is replaced in the equivalent position by Asp and Ser in MscMJ and MscCG, respectively (Petrov et al. 2013).

In addition to the connection to these previous studies, the inactivation mechanism proposed here (Fig. 8b) can be tested by replacing Gly113 with amino acids of higher helical propensity or extending the kink area (e.g., multiple Gly insertion) to further increase its structural flexibility; the latter was done by Martinac et al. for MscL (Bavi et al. 2016b). Both mutations are expected to result in the absence of inactivation in MscS.

Interestingly, an earlier work (Akitake et al. 2007) has conducted a series of mutation studies both at the Gly113 kink region and the TM3b-β interface area (Gly121), though that study was aimed to show that the Gly113 kink is a unique feature of the inactivated state and the closed structure favors buckling at Gly121, which was not supported by later studies (Pliotas et al. 2012, 2015; Ward et al. 2014). Thus we here attempt to interpret the results of mutation studies in Akitake et al. (2007) from the perspective of our proposed inactivation mechanism. First, the G113A mutant in Akitake et al. (2007) had increased helical propensity in the kink region and indeed showed no inactivation behavior; this agrees with our inactivation mechanism. The gating tension of G113A mutant was, however, observed to be comparable to that of the wild-type MscS in Akitake et al. (2007); this observation is not consistent with the predicted gating pathway in Fig. 5 (square curve), probably because the Ala substitution interacts with phospholipids inside the TM pockets (Pliotas et al. 2015) (not considered in the present model), thus strengthens the apolar interaction between the lower end of TM2 and that of TM3a (the main traction that pulls TM3a bundle open (Fig. 6)); alternatively, the Ala substitution may lead to structural rearrangements not considered in the present model (e.g., a certain extent of kink straightening (Akitake et al. 2007)) that enhance the interaction between the ends of TM2 and TM3a. To test these possibilities, a simulation is conducted where the kink region is treated as helical and modeled as a curved solid stick (Fig. 8a), while at the same time the TM2-TM3a traction force (green arrow in Fig. 8a) is increased by 50%. The gating strain is found to be 6.0% in this case, close to the gating strain (6.6%) of the model in Fig. 4 where the kink is modeled as a loop.

Second, the Q112G mutation (Akitake et al. 2007) increased the flexibility of the kink area, but instead of resulting in the absence of inactivation as suggested in our proposed inactivation mechanism, the Q112G mutant exhibited faster inactivation (Edwards et al. 2008). We conjecture that the Q112G mutation may weaken the apolar interaction between the lower end of TM2 and that of TM3a and may make the kink area "too flexible" that upon channel opening or even in the initial state, water molecules enter the TM pocket between TM1-TM2 and TM3a thus further causing the detachment between TM2 and TM3a. To test this assumption, a simulation is conducted for a modified Q112G mutant model, where the kink region is treated as a loop and modeled as a flexible quasi-one-dimensional winding beam (Fig. 1b), while at the same time the TM2-TM3a traction force (green arrow in Fig. 1b) is decreased by 10%. The simulation results in a structure close to that in Fig. 6 where the Phe68-Leu111 traction is missing and the channel cannot be opened, which may explain the faster inactivation of Q112G. As replacing one residue could change a set of interactions nearby (Wang et al. 2008), especially for the subtle situation at the Gly113 area which also affects the TM2-TM3a apolar interaction or possibly the hydrophobic lock of the channel pore (Anishkin and Sukharev 2004; Vora et al. 2006), mutation studies in this region require careful analysis and interpretation. The two simulations mentioned above serve as exploratory models by controlling the state of the kink (helix vs. loop) and the traction between the ends of TM2 and TM3a. More detailed experimental and simulation works are required in future to determine the structural and interaction changes around the kink area due to mutations.

At the TM3b-β interface, two mutations were conducted in Akitake et al. (2007), G121A and A120G. The G121A substitution, as discussed above, can reinforce the hydrophobic TM3b-β interface, promoting the kink around Gly113 (Koprowski et al. 2011; Petrov et al. 2013) and resulting in a more flexible kink region at Gly113. As expected from our mechanism, the G121A mutant was observed to open easily and showed no inactivation (Akitake et al. 2007). The A120G mutation, on the other hand, weakened TM3b-β interface, thus promoting the helical propensity at the kink area. Again as expected from our mechanism, the A120G mutant exhibited a high degree of inactivation (Akitake et al. 2007). Therefore, the results for both G121A and A120G mutants are supportive of the inactivation mechanism proposed here.

Effect of Transmembrane Voltage

Transmembrane voltage is applied to the continuum model to investigate its effects on MscS gating (Akitake et al. 2005; Nomura et al. 2016). For simplicity, we only consider the charged residues in the atomic model which are mapped to the finite element nodes in the continuum model (inset of Fig. 9). There are only a few charged residues in the transmembrane domain, which are Arg46, Arg54, Arg59, Lys60, Asp62, Asp67, Arg74, Arg88, and Arg128. Transmembrane voltages from -100 mV to $+100$ mV (Akitake et al. 2005) are applied with the exterior of the cell being positive with respect to its interior. No evident dependence of the activation pathways (Fig. 9) on transmembrane voltages is observed in the FEM simulations. This finding is consistent with recent experimental results (Akitake et al. 2005; Nomura et al. 2016). The rectifying behavior that the conductance is larger at positive voltages observed in some studies (Martinac et al. 1987; Petrov et al. 2013) could be a result of the inside-out nature in patch-clamp experiments and the slight anion preference of MscS (Martinac et al. 1987; Sukharev 2002).

Another observation of depolarizing voltage-dependent inactivation (Akitake et al. 2005), however, remains unsettled. How negative voltages promote inactivation rate must await further studies with more sophisticated models combined with experimental analyses. It is possible that, though the electric field causes only small displacements on the charges in MscS during activation (Akitake et al. 2005), its presence may disrupt some key interactions like the TM3b-β interface (e.g., Asn117). Besides, voltage-dependent inactivation may be related to the

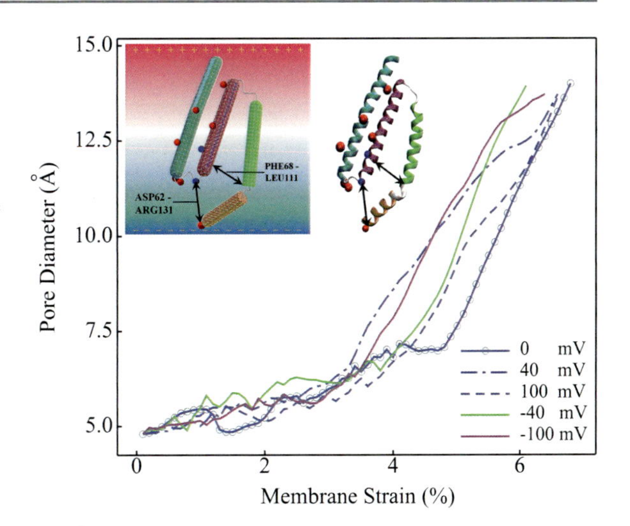

Fig. 9 Pore diameter evolutions of MscS under equi-biaxial membrane tension and different transmembrane voltages. Residue charges and electric field are incorporated into the continuum model based on the charge distribution in the crystal structure (inset). Red particles represent positive residues and blue particles negative residues

conformational changes of the cytoplasmic domain (not included in the present study) as indicated in Rowe et al.'s work (Rowe et al. 2014) where the G168D mutant is insensitive to applied voltages.

Coupled Continuum Mechanical-Continuum Solvation Approach with Application to Gating Mechanism of MscL

Models and Methods

In this section, the construction process of the continuum mechanics model for *E. coli*-MscL is presented in detail. Relevant material and interaction parameterizations are either calibrated by matching results at the atomistic and continuum level or obtained from a previous study (Chen et al. 2008). The computational framework includes essentially two components: continuum mechanics (CM) simulations based on finite element method (FEM), which solve for the deformed structure under specific external loads, and the continuum solvation model (CS), which computes solvation forces based on structural information from the CM model. The commercial software ABAQUS (2011) (for CM calculations) and open-source software APBS (Baker et al. 2001) (for CS model) are used in this fully integrated (CM/CS) simulation framework. Detailed ABAQUS-APBS co-simulation protocols are presented.

Continuum Modeling of *E. coli*-MscL and Interactions Within the Protein

Although *E. coli*-MscL is one of the most studied MS channels, the only available X-ray crystal structure in the literature is for the MscL from bacteria *Mycobacterium tuberculosis* (*Tb*), which was captured in its closed state by Rees Lab (Chang et al. 1998). By retaining the main features of the crystal structure of *Tb*-MscL, the atomic structure of *E. coli*-MscL was developed based on homology modeling along with other experimental constraints (Sukharev and Anishkin 2004; Sukharev et al. 2001a; Sukharev and Corey 2004); the closed state model is shown in Fig. 10g. The crystal structure of *Tb*-MscL was further refined by Stefan Steinbacher et al. (2007), where its N-terminal (termed S1) was an amphipathic helix positioned approximately parallel to the cytoplasmic surface of the membrane. This refined structure for *Tb*-MscL was supported by later studies (Iscla et al. 2008). Based on the new structure for *Tb*-MscL and MD simulations, S1 domain was suggested to interact closely with the lipid membrane, which may facilitate the opening of the channel (Iscla and Blount 2012; Vanegas and Arroyo 2014). While developing the new atomic *E. coli*-MscL structure based on the revived *Tb*-MscL crystal structure (Steinbacher et al. 2007) is beyond the scope of the present study, it should be noted that the effect of S1 domain as a "sliding anchor" to the lipid membrane thus helping channel opening (Iscla and Blount 2012; Vanegas and Arroyo 2014) is not available for studying in this work since we are using Sukharev S. et al.'s model (Sukharev and Anishkin 2004) whose S1 domain assembles as a helix bundle. The S1 domain's interaction to the lipid and its hydration process may be to some extent different from what we present in the following context but is not expected to bias the principal outcomes of the current model.

An *E. coli*-MscL molecule is formed by five chains (from chain-1 to chain-5) assembled as a fivefold structure around its symmetry axis, and each single chain (Fig. 10a) of MscL consists of four helices (referred as TM1, TM2, S1, and S3) and several loops. Within the transmembrane helix bundles, the five TM2 helices form the outer boundary that interact extensively with the lipid membrane, while the longer TM1 helices form the inner boundary for the pore and have limited contact with the lipid. The TM1 and TM2 helix bundles share the same fivefold symmetry axis, denoted as the z-axis here, which is also the direction of the membrane normal. Each TM1 helix and its nearest neighbor, a TM2 helix, are assembled into a pair through a periplasmic loop. A closer view of the TM1 shows that there is a break near the top of the helix due to Pro-43; thus, the segment above Pro-43 is also referred to as the S2 helix. The cytoplasmic region contains two different types of helix bundles, referred to as the S1 helices and S3 helices; each bundle contains five subunits, and each subunit of the S1 or S3 bundles is connected, respectively, to a TM1 or TM2 helix through a loop. The TM1, TM2, S1, and S3 helices correspond to residues Asn-15–Gly-50, Val-77–Glu-107, Ile-3–Met-12, and Lys-117–Arg-135, respectively (Sukharev and Anishkin 2004). In the subsequent discussion, unless otherwise specified, MscL refers to the *E. coli*-MscL.

In the continuum model, we start from the backbone structure with side chains removed because when side chains are presented, some subcomponents are unable to be separated, such as the TM1 and TM2 helices in the same chain which are very close to each other. More dedicated calculation of the surfaces of the protein could be considered in future work to get a more precise model. The molecular surface (Fig. 10c) of each single chain of MscL molecule is calculated and triangularized using the MSMS program (Sanner et al. 1996) with appropriate probe radius to avoid overlapping of subcomponents (such as TM1 and TM2 helices). Among the multiple subcomponents of MscL, helices are modeled to have larger cross sections because

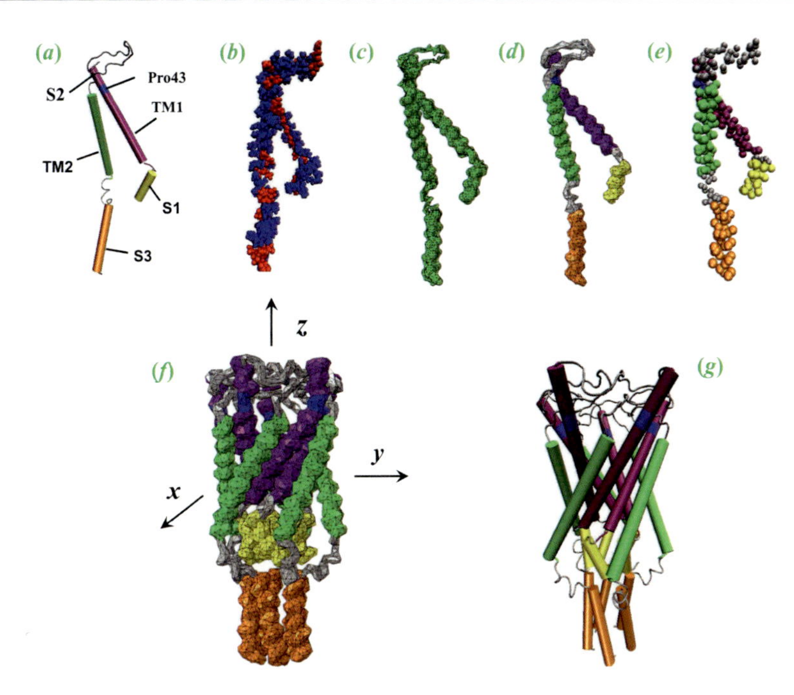

Fig. 10 Illustration of various steps in constructing the FEM Model of *E. coli*-MscL. (**a**, **b**) the atomic structure of one chain of *E. coli*-MscL in cartoon and van der Waals (VDW) representation where blue indicates hydrophobic and red hydrophilic; (**c**) the triangularized molecular surface; (**d**) the simplified triangularized molecular surface and the volume enclosed by this surface are discretized into tetrahedral elements; (**e**) the coarse-grained model of one chain of *E. coli*-MscL; (**f**) the FEM model of E. coli-MscL; (**g**) structural model of *E. coli*-MscL based on homology modeling (Sukharev and Anishkin 2004). Zhu et al. (2016), reprinted with permission of Springer

they are strengthened by the hydrogen bonding between residues, while the loops are modeled much thinner (Fig. 10c). The triangularized surface is then simplified by the QSLIM program (Heckbert and Garland 1999) to reduce the number of the surface triangles to 2000 (Fig. 10d). The volume enclosed by this simplified surface is subsequentially discretized into a 3-D mesh consisting of tetrahedral elements. The final FEM model consists of 2,685 nodes and 9,699 finite elements for each chain of MscL.

Assuming that the mechanical properties of each component vary little with respect to sequence, the properties of each component are assumed to be homogeneous, isotropic, and constant during the gating process. The only exception is for the break between S2 and TM1 (Pro43, illustrated by the blue segment in Fig. 10a), whose properties are determined separately and less stiff than the rest of the helix. The material properties of each component are calibrated by matching results of normal mode analysis (NMA) at the continuum and atomistic levels. The NMA for individual components of the channel is conducted in vacuum using the Gromacs MD simulation package with the Gromos96 vacuum parameter set (Van Der Spoel et al. 2005) so as to be consistent with the continuum calculations which are also

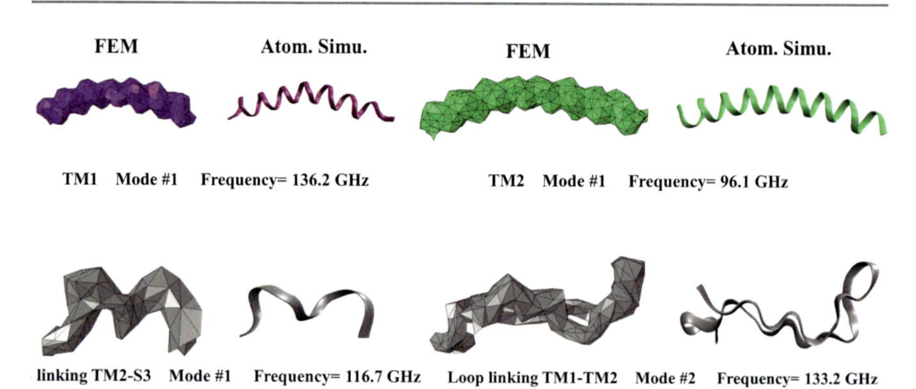

Fig. 11 Examples of several lowest eigenmodes and frequencies of helices and loops. Results of molecular mechanics are compared to those of the finite element simulations. Here the TM1 helix only includes to the segment below Pro43. Zhu et al. (2016), reprinted with permission of Springer

Table 2 Phenomenological material properties of the continuum components

	Helices						Loops			Lipid membrane	
Properties of MscL	TM1	TM2	Pro43	S1	S2	S3	TM1-TM2	TM1-S1	TM2-S3	Head groups	Tails
Young's modulus E (Gpa)	69	54	11	9	14	14	20	10	12	0.124	0.064
Poisson's ratio ν	0.3	0.3	0.3	0.3	0.3	0.3	0.3	0.3	0.3	0.5	0.5

conducted in vacuum. A possible way to get more realistic elastic properties is to conduct NMA considering the water and lipid environment in both continuum and atomistic calculations though this kind of NMA at the continuum level is highly complex. The Young's modulus is then varied at the continuum calculations such that the eigenvalues and eigenvectors for the three lowest-frequency modes best fit the results from the atomistic normal mode calculations. The lowest eigenmode of the TM1 or TM2 helices, for example, is essentially flexural bending as shown with both continuum and atomistic configurations in Fig. 11. The lowest frequencies are 136.2 and 96.1 GHz for TM1 and TM2 helices, respectively, which lead to the fitting of their effective Young's moduli as 69 GPa (TM1) and 54 Gpa (TM2). Summarized in Table 2 are the key mechanical properties for the helices and loops. These are smaller than estimated in the previous work (Chen et al. 2008) where the helices, for instance, were modeled as thinner (with a diameter of 5 Å) elastic cylinders by considering only the main chain and larger than the estimated range of the Young's modulus of MscL α-helices (0.2 to 12.5 GPa) in Martinac et al.'s recent work (Bavi et al. 2017) with constant-force steered molecular dynamics (SMD).

The nonbonded interactions within MscL are treated by adopting a sufficiently detailed yet computationally efficient protocol to replace the oversimplified surface-to-surface contact model used previously (Chen et al. 2008; Tang et al. 2006). To reduce computational cost, the new approach first simplifies the full-atomic structure into a coarse-grained (CG) model consisting of a much smaller number of particles based on the Martini force field v2.1 (Monticelli et al. 2008). This CG process reduces the number of particles from 2166 to 283 for each chain of *E. coli*-MscL with side chains included (Fig. 10e). The coarse-grained particles are then mapped to the nodes of the continuum MscL model so that it consists of two types of nodes: ordinary finite element (FE) nodes and the "chemical nodes." The "chemical nodes" in the continuum MscL model refer to the FE nodes that have the same coordinates to that of the CG particles (Fig. 10d, e). Interactions between the Martini particles are then applied to the corresponding FE nodes in the continuum mechanics simulation. These interactions include the particle-to-particle interactions between different components within the same chain and between components in different chains. For example, within chain-1, one particle on the TM1 may interact with another particle on S3, TM2, S1, or loops of chain-1, and as for different chains, one particle on the TM1 of chain-1 may interact with any particle on chain-2 (from S3 to S1). In the MDeFEM framework, the interactions among atoms within each continuum component (e.g., particles on TM1 of chain-1) are not computed explicitly because the corresponding energy is already included in the elastic representation of the continuum components; this is one reason that the computational cost associated with the continuum framework is substantially lower than all-atom simulations. The nonbonded interactions between CG particles in different continuum components are calculated using pair-wise terms following the standard cutoff schemes commonly used in atomistic simulations. Specifically, we adopt a group of "connector elements" in the mechanical space, each of which characterizes the nonbonded interaction between a pair of interacting CG particles. These connector elements are not actually FEM elements but a special kind of invisible connectors in ABAQUS used to define force-based nonlinear interactions, just like the nonbonded interactions in MD. The connector element behavior is nonlinear elastic including two possible forms:

$$\begin{array}{l} \text{Lennard-Jones form}: \; V_{LJ} = 4\varepsilon\left[(\sigma/l)^{12} - (\sigma/l)^{6}\right] \\ \text{Coulomb's form}: \; V_{Elec} = \frac{1}{4\pi\varepsilon_0\varepsilon_r}\frac{q_1 q_2}{l^2} \end{array} \tag{2}$$

where ε and σ are the energy and distance parameters, respectively, to characterize the Lennard-Jones interactions. ε_0 is the vacuum dielectric constant, $\varepsilon_r = 15$ is the relative dielectric constant for electrostatic screening in the Martini force field v2.1 (Monticelli et al. 2008), and q_i is the partial charge of the i-th particle. l denotes the distance between a pair of interacting particles. We note that with the polarizable MARTINI water model (Yesylevskyy et al. 2010), the relative dielectric constant is substantially smaller (2.5); test calculations are also done with this value, and the results are generally very similar (data not shown) since the behavior of MscL appears to be largely dictated by the nonpolar solvation (see discussions below). For the interactions among the MscL components, all the parameters involved in the Lennard-Jones interaction are obtained from the Martini force field v2.1; all distances are calculated based on the CG particle coordinates that are continuously updated and mapped from the continuum model. Also continuously updated are the connector elements to include newly emerged interactions when two remote particles come closer during the simulation process. The Lennard-Jones interaction cutoff length is set to be 1.0 nm and the electrostatic cutoff length 1.2 nm.

The major advantage of the irregular molecular surface FEM model over the highly simplified cylindrical stick model (Chen et al. 2008; Tang et al. 2006) is that they take into account the molecular nature of MscL, such as its irregular shape. In addition, the mapping between CG particles and corresponding FEM nodes makes it straightforward to define "chemical nodes," which encodes the key chemical characteristics of the molecule (i.e., charge distribution and solute/solvent interface) that are required in the solvation calculations.

Continuum Modeling of Lipid Bilayers and Interactions Between MscL and Lipid

Motivated by the natural difference in the chemical and physical properties of these regions, the lipid membrane bilayer is modeled as a sandwich panel that consists of three layers (Fig. 12): a soft layer in the middle with a thickness of 2.5 nm and two hard layers in two sides with a thickness of 0.5 nm each (Chen et al. 2008). These values for thickness estimation are based on density map of water and lipids from a MD simulation of the POPE lipid system (Gullingsrud and Schulten 2004), and both the derived head group thickness and the tail group thickness are consistent with general thickness estimations for the POPE lipid membrane. Meanwhile, the work of Andrew M. Powl et al. identified the hydrophobic thickness of *E. coli*-MscL associated with the lipid membrane as 25 Å (Powl et al. 2005b). Thus, there is a good match between the hydrophobic thickness of helix and bilayer; and hydrophobic mismatch is not considered in the present model.

For the case of in-plane membrane stretching, which is the major driving force for MscL gating (Tang et al. 2006), a flat square membrane (within the x-y plane) with a size of 400 × 400 Å is employed. The equi-biaxial membrane tension is most likely induced by osmotic pressure: assuming the liposome is spherical with a typical diameter of 1.0 μm, a patch of membrane with the size of 400 × 400 Å corresponds to a center angle of ~0.08, which suggests that the curvature of the patch is negligible. To embed the channel into the continuum membrane, a cavity (hole) with the shape of a 10-petal flower (Fig. 12) is created in the middle of the membrane with the size and shape of the cavity conform to those of MscL transmembrane helices in the closed state with an equilibrium distance of ~5.5 Å, as was measured from the trajectories in previous all-atom simulation (Gullingsrud and Schulten 2003).

The three-layer phenomenological model (with a thickness of ~35 Å (Gullingsrud and Schulten 2003)) for the palmitoyloleoylphosphatidylethanolamine (POPE) lipid bilayer is first assumed to be homogeneous and isotropic, and its mechanical properties are obtained from a previous study (Chen et al. 2008) where the effective elastic properties of the head group layers and the tail group layer were derived from MD simulations (Gullingsrud and Schulten 2004). And these parameters are overall consistent with general simulation or experimental estimations (Binder and Gawrisch 2001; Chacon et al. 2015; Venable et al. 2015), though they may be further optimized through recent developments about the new

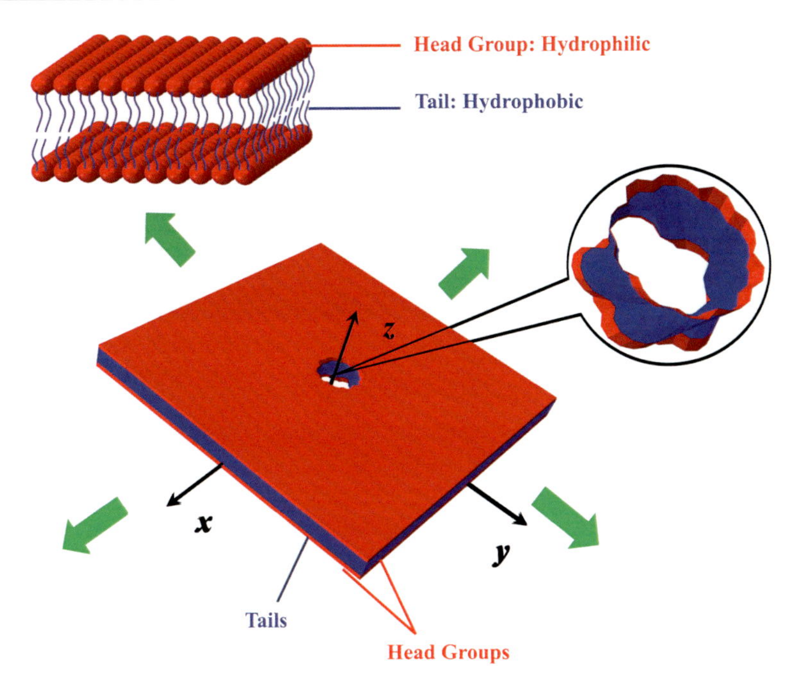

Fig. 12 Schematic of equi-biaxial tension of the lipid membrane model. A zoomed-in view is shown to illustrate the 10-petal lipid hole that encompasses the MscL protein in the CM model. Zhu et al. (2016), reprinted with permission of Springer

force decomposition methods (Torres-Sánchez et al. 2015; Vanegas et al. 2014). For isotropic linear elasticity, the constitutive relation (stress-strain relationship) is given by

$$\begin{Bmatrix} \sigma_{11} \\ \sigma_{22} \\ \sigma_{33} \\ \sigma_{12} \\ \sigma_{23} \\ \sigma_{13} \end{Bmatrix} = \begin{pmatrix} \frac{E(1-\nu)}{(1+\nu)(1-2\nu)} & \frac{E\nu}{(1+\nu)(1-2\nu)} & \frac{E\nu}{(1+\nu)(1-2\nu)} & 0 & 0 & 0 \\ \frac{E\nu}{(1+\nu)(1-2\nu)} & \frac{E(1-\nu)}{(1+\nu)(1-2\nu)} & \frac{E\nu}{(1+\nu)(1-2\nu)} & 0 & 0 & 0 \\ \frac{E\nu}{(1+\nu)(1-2\nu)} & \frac{E\nu}{(1+\nu)(1-2\nu)} & \frac{E(1-\nu)}{(1+\nu)(1-2\nu)} & 0 & 0 & 0 \\ & & & G_{12} & 0 & 0 \\ & \text{sym} & & & G_{23} & 0 \\ & & & & & G_{13} \end{pmatrix} \begin{Bmatrix} \varepsilon_{11} \\ \varepsilon_{22} \\ \varepsilon_{33} \\ \varepsilon_{12} \\ \varepsilon_{23} \\ \varepsilon_{13} \end{Bmatrix} \quad (3)$$

where σ_{ij} and ε_{ij} are the stress and strain tensor components. For isotropic linear elastic materials, the shear moduli $G_{12} = G_{23} = G_{13} = \frac{E}{2(1+\nu)}$; thus, the elastic properties for each layer of the continuum lipid bilayer model are governed by Young's modulus, E, and Poisson's ratio, ν, listed in Table 2. To take into account the prominent characteristics of the fluidic lipid membrane, i.e., its incapability of sustaining a large shear stress, we model each layer of the lipid membrane as orthotropic and reduce the in-plane shear modulus of the continuum lipid model, G_{12} (Eq. 3), to a small value (without losing generality, $\frac{E}{2(1+\nu)}/1000$ is adopted), while all other parameters are inherited from the previous isotropic continuum lipid model. Such a change does not affect the area compressibility under equi-biaxial tension, which is used to estimate the lipid membrane's Young's modulus and Poisson's ratio (Chen et al. 2008) because of the decoupling between tension and shear. It is expected that the in-plane fluidity of the lipid bilayer allows a closer interaction between the surface of the 10-petal lipid hole and the MscL molecule, thus facilitating the gating process since the in-plane lipid tension has been regarded as the major driving force that pulls the channel open (Moe and Blount 2005; Tang et al. 2006). Upon applied tension, the behavior of a real membrane may not be strictly elastic or orthotropic as described, and it was found in a molecular dynamics study that MscL inclusion may increase the rigidity of the membrane (Jeon and Voth 2008). For simplicity, these details are ignored in the present work and are not expected to have a significant influence on the structure or behavior of MscL based on our previous studies (Chen et al. 2008; Tang et al. 2008).

The nonbonded interactions between the lipid hole and the MscL molecule are treated as in Chen et al. (2008) where the interactions between helix and lipid are represented by a pair-wise effective pressure-distance relationship in the following form:

$$p(\alpha_i) = \frac{Dn}{d_0}\left[(d_0/\alpha_i)^{m+1} - (d_0/\alpha_i)^{n+1}\right] \tag{4}$$

where p and d_0 are the interaction pressure and equilibrium distance between two surfaces, respectively, and α_i is the instantaneous distance between two deformed surfaces for the i-th element. This nonbonded interaction model implicitly includes both electrostatic and van der Waals interactions and has been successfully applied to study the radial elastic properties of multiwalled carbon nanotubes (Chen et al. 2006) and the deformation and buckling of double-walled carbon nanotubes (Pantano et al. 2003) as well as nanoindentation of nanotubes (Cao and Chen 2006). To estimate the helix-lipid interactions, the insertion energy profile of a single helix (TM1, TM2, or S1) is calculated with an implicit membrane model; i.e., the helix is gradually transferred from the implicit membrane to the implicit bulk solution. Determined from molecular mechanics calculations (Chen et al. 2008), shown in Table 3 are the well-depth D and the exponents (m, n) which are fairly transferable to the current model since we are using the same atomic structure of *E. coli*-MscL.

It is noted the irregular surface of the lipid membrane is not included in this work, and the particle-to-particle lipid-channel interaction force is averaged to the

Table 3 Parameters for the nonbonded interactions between different helix-lipid pairs (Chen et al. 2008)

Interaction pairs	d_0(Å)	$\psi = 6Dn/d_0$(GPa)	m	n
Lipid-TM1	5.5	2.0	9	3
Lipid-TM2	5.5	2.0	7	3
Lipid-S1	7.0	0.025	4	1

surfaces of them; thus, the heterogeneous binding of the lipid to the channel is lost. Further refinements of the continuum mechanics lipid model and its heterogeneous interaction to the channel could be considered in future work with different levels of sophistication.

Continuum Solvation Modeling

As described above, the new continuum MscL model is meshed in a way so that it consists of two types of nodes: ordinary finite element (FE) nodes and the "chemical nodes" which are also FE nodes but correspond to the CG particles. While all FE nodes contribute to the mechanical deformation of MscL, the "chemical nodes" are subjected to additional force contributions from solvation and inter-component interactions. Since the new continuum representation adopted here retains the information about the irregular shape of the protein and spatial distribution of the amino acid residues, the solvation contribution (electrostatic plus apolar solvation forces) can be readily calculated using a popular continuum solvation model. Shown in Fig. 10e are the CG particles of *E. coli*-MscL that bear solvation forces in the continuum solvation calculations. Following the mapping of these CG particles to the nodes of the continuum MscL model, the solvation forces are also transferred from the CG particles to the corresponding "chemical nodes" and included in the subsequent continuum mechanics simulations.

The total solvation free energy is usually decomposed into apolar and electrostatic contributions. For the electrostatic component, the nonlinear Poisson-Boltzmann (NLPB) (Davis and Mccammon 1990; Honig and Nicholls 1995) model is used. The van der Waals radius ($2^{1/6}\sigma/2$) of each CG particle is used to approximately determine the dielectric boundary between MscL and the solvent, and ions in solution have a finite radius of 2.0 Å. The spline-based (Im et al. 1998) molecular surface, which permits stable solvation force calculations, is used with a 0.3 Å spline window. The channel is surrounded by the nonpolar membrane environment, which is represented crudely by a low dielectric slab following the procedure in APBS (Baker et al. 2001). Accordingly, the MscL protein-lipid system has three dielectric regions: the high dielectric solvent exterior (80.0), the intermediate dielectric protein interior (10.0), and the low dielectric interior of the membrane (2.0). The membrane environment is applied except for the hole in which the MscL is located. Test calculations indicate that for MscL, the results are not sensitive to the value of the protein dielectric used.

As tests of the electrostatic solvation protocol, we compare the computed polar solvation free energy of two systems to values obtained based on the standard solvation protocol in APBS using atomistic structures. One system is a short peptide (with 27 residues and a length of ~4 nm) from APBS, for which the computed polar transferring free energy from bulk water into the membrane is 127.1 and 120.9 KJ/mol for the atomistic protocol and current CG-based protocol, respectively. The second system is the closed state of *E. coli*-MscL, for which the computed polar transferring free energy with the atomistic and CG models are 61.8 and 70.9 KJ/mol, respectively. These two examples suggest that the Martini/NLPB combination appears to provide a reasonable estimate for the polar solvation effects.

The apolar solvation force calculations follow the very generic framework described in Wagoner and Baker (2006) where the solvent-accessible surface area contribution is supplemented with volume and dispersion integral terms. A 0.3 spline window is used for the spline-based molecular surface, and the coefficients for surface tension and the volume term of the apolar calculations are 0.0042 KJ/mol/Å^2 and 0.23 KJ/mol/Å^3, respectively. It is noted that the lipid membrane is not present in apolar solvation force calculations, while in fact most of the residues of TM2 and some residues on TM1 are buried in the membrane. Therefore, the apolar solvation forces on the outer particles of the transmembrane helices are not included in the continuum mechanics calculation (the interactions between these particles and lipids are included explicitly as discussed above). Care is taken to identify these particles based on the coordinates and van der Waals radius of each CG particle in the transmembrane helices.

Besides the outer particles of the transmembrane helices buried in lipids, the inner pore constriction at the closed state for the wild-type (WT) MscL is actually not hydrated at the first stage of gating until the pore gets large enough (> 0.45 nm (Beckstein et al. 2001)). In the above calculations, the channel pore is assumed to be already wetted at the beginning, which is probably more close to the gain-of-function (GOF) mutant MscL (e.g., the V23 T GOF mutant (Anishkin et al. 2005)). The hydration of the channel pore depends on the effective pore radius, and since the pore has different radii along the symmetry axis, at the beginning, only part of the pore surface is hydrated and the rest is not. In the closed state of MscL, about half of the TM1 bundle's inner surface is exposed to solvent and the lower half is not. When considering the hydration process depending on the effective pore radius as a more realistic model for the WT MscL, similar to those outer residues buried in lipids, the apolar solvation forces on the inner hydrophobic constriction region are not included in the continuum mechanics calculations. This is likely a reasonable approximation in the context of continuum modeling, although the hydration of MscL is a complex case that requires thorough MD studies (Beckstein et al. 2001; Beckstein and Sansom 2004).

In the following context, when considering hydrophobic interactions between water and the channel, the effect of gradual hydration is not included first; and the pore radius depended hydration process is considered in the section of "Different Pathways for the GOF (Gain-of-Function) mutant and WT (Wild-Type) MscL."

During the continuum calculations, the positions of the CG model are constantly updated based on the coordinates of the "chemical nodes" of continuum mechanics MscL model, while the particle type, radius, charge, and interaction parameters are retained. And solvation forces of the constantly updated CG model calculated through the above protocols in turn participate in the FEM calculations. Solvation effects on the lipid membrane are neglected.

A Force-Based CM/CS Co-simulation Protocol

Shown in Fig. 13 is the assembled continuum model of *E. coli*-MscL under equi-biaxial tension with zoomed-in views near the protein (the 10-petal lipid hole surface and the *E. coli*-MscL). Four-node tetrahedron finite elements are used

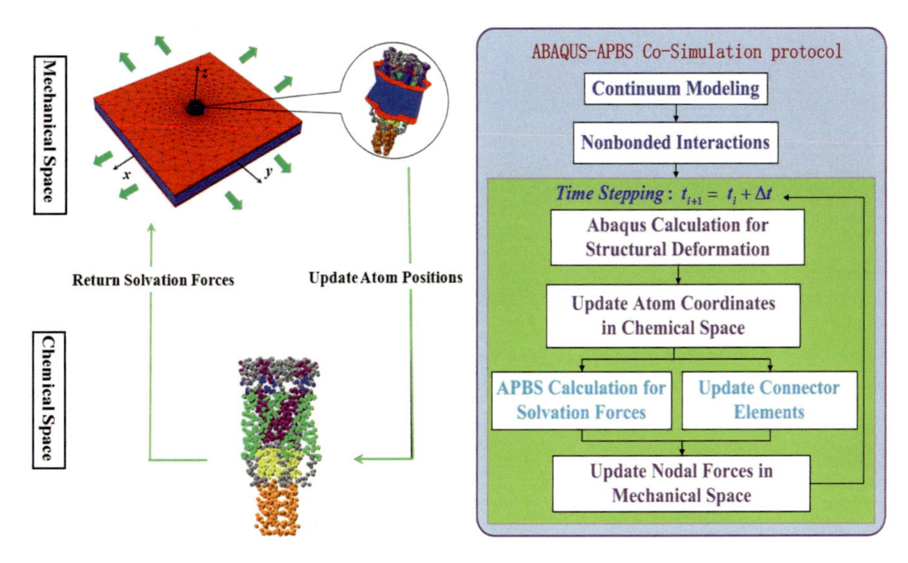

Fig. 13 Force-based ABAQUS-APBS co-simulation protocol: schematic (with a zoomed-in view near the protein) and flowchart representation. Zhu et al. (2016), reprinted with permission of Springer

to mesh both the helices and the lipid membrane with the commercial package ABAQUS (2011). Each chain of the continuum MscL model consists of 2,685 nodes and 9,699 finite elements. The lipid bilayer incorporates 2,412 nodes and 10,680 elements, with the mesh gradually more refined toward the boundary of the inner hole where it interacts with the protein extensively. As a reference calculation, MscL gating is modeled without including the solvation effects. Through ABAQUS simulation, the structural deformation of the lipid-protein system is explicitly calculated in response to an external load, where the lipid membrane is stretched (Fig. 13) by applying equi-biaxial displacement on its outer boundary (relevant to osmotic pressure).

To explicitly explore the solvation effect on MscL gating pattern, continuum solvation forces (electrostatic or apolar) need to be integrated into the continuum model that solves for lipid-protein system deformation under external force. This requires an iterative procedure that alternates between CM simulation and CS calculations (illustrated in Fig. 13):

1. Set up nonbonded interactions within MscL and those between MscL and lipid membrane.
2. A small equi-biaxial displacement on the lipid's outer boundary is applied, and a deformed conformation of the system is obtained by finite element analysis using ABAQUS.
3. The positions of all "chemical nodes" of MscL molecule are extracted from the deformed mesh, and then continuum solvation forces are calculated by APBS.

4. At the same time, the connector elements (interactions within MscL) are updated: new connector elements are added for pairs of two "chemical nodes" that are originally remote but come close during the gating process. The forces exerted by the connector element are directly applied to the corresponding nodes in the continuum model without interrupting the integrity of the element community. Since the gating mechanism of MscL is a gradually expanding process, there are only a few of newly applied connector elements (forces).
5. The solvation forces on the chemical nodes and forces from the newly emerged connector elements are included in the FEM simulation, and a new deformed structure is obtained in the next time step, leading to a new iteration.
6. Dozens of iterations (steps) are carried out until the channel is fully opened, i.e., when the channel pore radius of the continuum model reaches the experimentally estimated value (~19 Å) for the fully opened state (Sukharev et al. 2001b). Note that the solvation contribution includes both electrostatic and apolar components, and their effects are explored separately below. With different salt concentrations, the procedure above can be repeated to validate Sukharev et al.'s experimental observation (Sukharev et al. 1999) that there is no significant change in the gating pattern for salt concentrations between 0.05 and 1.0 M.

For a typical tension simulation of the gating of E. coli-MscL, the computational time is ~8 h on a Dell workstation with 3.2 GHz CPU and 4 Gb RAM, highlighting the efficiency of the MDeFEM framework compared to atomistic MD simulations, which take from days to months for large membrane protein systems.

Results and Discussion

In this section, we compare the gating pathways of three systems: (1) a pure CM model with isotropic membrane properties (i.e., in absence of membrane fluidity and solvation effects), (2) a CM model with membrane fluidity but lacking solvation effects, and (3) a complete model that incorporates the electrostatic/apolar solvation effect through the ABAQUS-APBS co-simulation protocol developed above. Simulation results are compared to structural models, previous all-atom simulations, as well as available experimental results.

Gating Pathway of MscL Without Membrane Fluidity and Solvation Effects

As a reference model, same as our previous MDeFEM study (Chen et al. 2008; Tang et al. 2008), we first treat each layer of the lipid as an isotropic slab without any explicit solvation effects; this is referred to as the preliminary model below. Shown in Fig. 14 are several snapshots of *E. coli*-MscL during the tension-driven gating process in comparison to the structural model of Sukharev and Anishkin (2004) at closed, half-, and fully opened states. Not surprisingly, the transmembrane region (TM1 and TM2 helices) exhibits the most striking conformational changes, in which both TM1 and TM2 helices shift radially away from

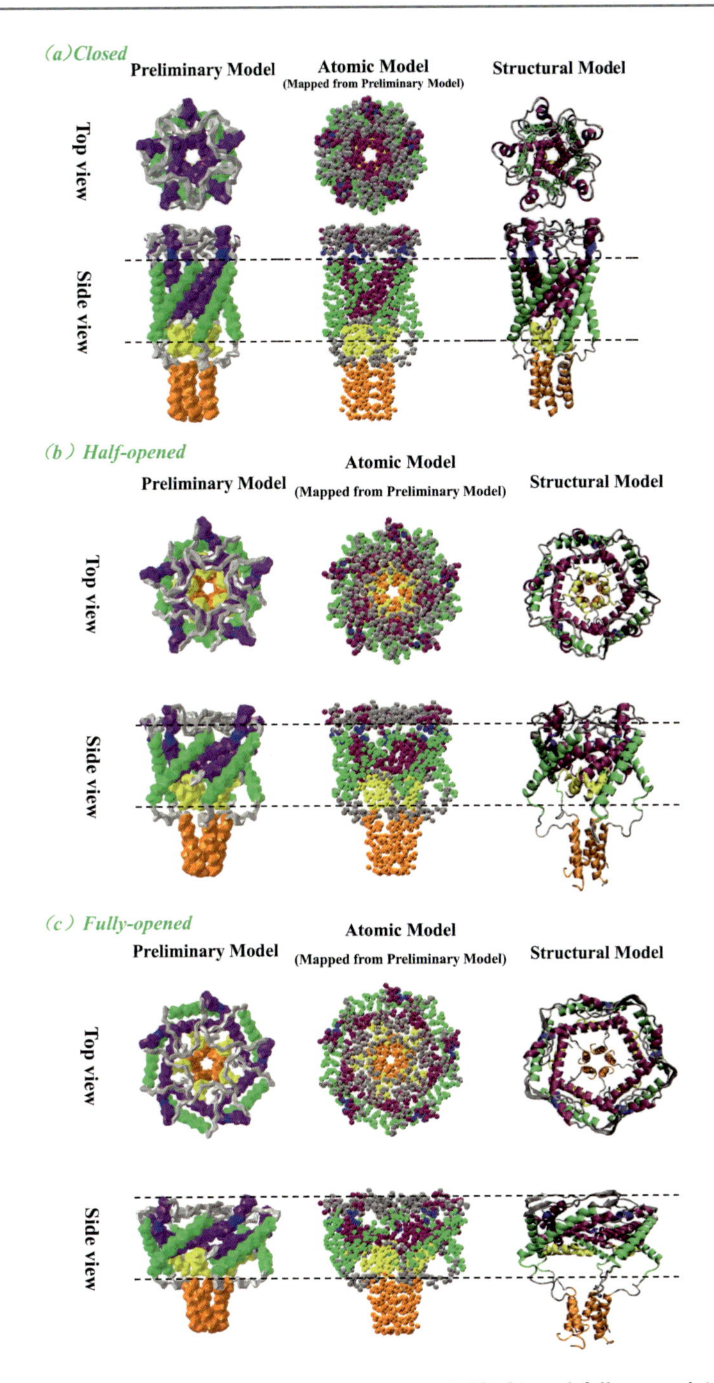

Fig. 14 Gating pathway of *E. coli*-MscL at closed (**a**), half- (**b**), and fully opened (**c**) states. Configurations are illustrated by continuum model in mechanical space and atomic model (VDW) in chemical space with comparison to the structural models (Sukharev and Anishkin 2004). The dashed lines indicate the approximate location of membrane/water interface. Zhu et al. (2016), reprinted with permission of Springer

the fivefold symmetry axis. Gating is primarily realized through the nonbonded interactions between transmembrane helices and the lipid membrane, in which the pore enclosed by TM1s is pulled open. The S1 bundle expands in the radial direction following the path of TM1 though it does not open as large as the TM1 bundle. At the current maximum lipid strain (ε_m=21%), no disassembly of S3 helices is observed; although the top region of the S3 bundle is expanded slightly, the lower ends remain assembled. This observation is in agreement with the revised version of the structural model (Sukharev and Anishkin 2004), as opposed to the earlier one (Sukharev et al. 2001b). Beside radial expansion of the TM helices, visible shrinking in the height of MscL is also observed, which is correlated with the significant titling of the helices and uplifting of the S3 bundle. Similar to previous studies (Tang et al. 2006, 2008), the loops and transmembrane helices are considerably stretched and bent during the gating process to maintain mechanical equilibrium; these features may be verified from experimental studies with sufficient resolution. On a qualitative level, the current MDeFEM results are in good agreement with the structural models (Sukharev and Anishkin 2004), regarding both the orientation and displacement of the helices and loops (Fig. 14). The only exception regards the periplasmic loops that link TM1 and TM2 helices; they remain well packed in our simulations but expand radially in the structural model of Sukharev and Anishkin (2004) along with the tips of the transmembrane helices.

The conductance of MscL is directly correlated with the size of the gate, and it is commonly believed that the TM1 helices in the core of the transmembrane bundle constitute the most important gate of the channel (Sukharev and Anishkin 2004; Tang et al. 2006), which is pentagon-shaped (insert of Fig. 15) when projected onto the *x*-*y* (membrane) plane. To characterize the pore size, we define an effective pore radius (denoted as *r*) as the radius of a circle that has the same area as the pentagon-shaped TM1 pore (see insert of Fig. 15). To be consistent with previous studies (Chen et al. 2008; Tang et al. 2008), the area of the pentagon-shaped TM1 pore is calculated as that surrounded by the five TM1 helical axes. From the closed state to the fully opened state, the effective radius increases from ~6.5 Å to ~19 Å, echoing the structural model (Sukharev and Anishkin 2004). Considering the pore's irregular inner surface, the actual pore's radius ranges from <1 Å to ~14 Å which is consistent with the estimation from fluorescence resonance energy transfer (FRET) experiments (Corry et al. 2010; Wang et al. 2014). For the preliminary model, the percentage of increment of the effective pore radius and the actual pore radius is depicted in Fig. 15 as a function of membrane strain (the inverted triangle curve). Due to the more detailed representation of MscL and the nonbonded interactions, the relationship between the membrane strain and the lipid cavity expansion is much less linear as found in the previous study (the triangle curve (Tang et al. 2008)). As mentioned in the Introduction, the surface-based nonbonded interaction model used in the previous study is oversimplified, and most importantly, it neglects all the nonbonded interactions involving the loops that connect helices. The current work demonstrates that as the membrane strain increases, the pore size of MscL is firstly stabilized at a small value and then experiences a rapid increase leading to pore opening when the membrane strain becomes sufficiently large.

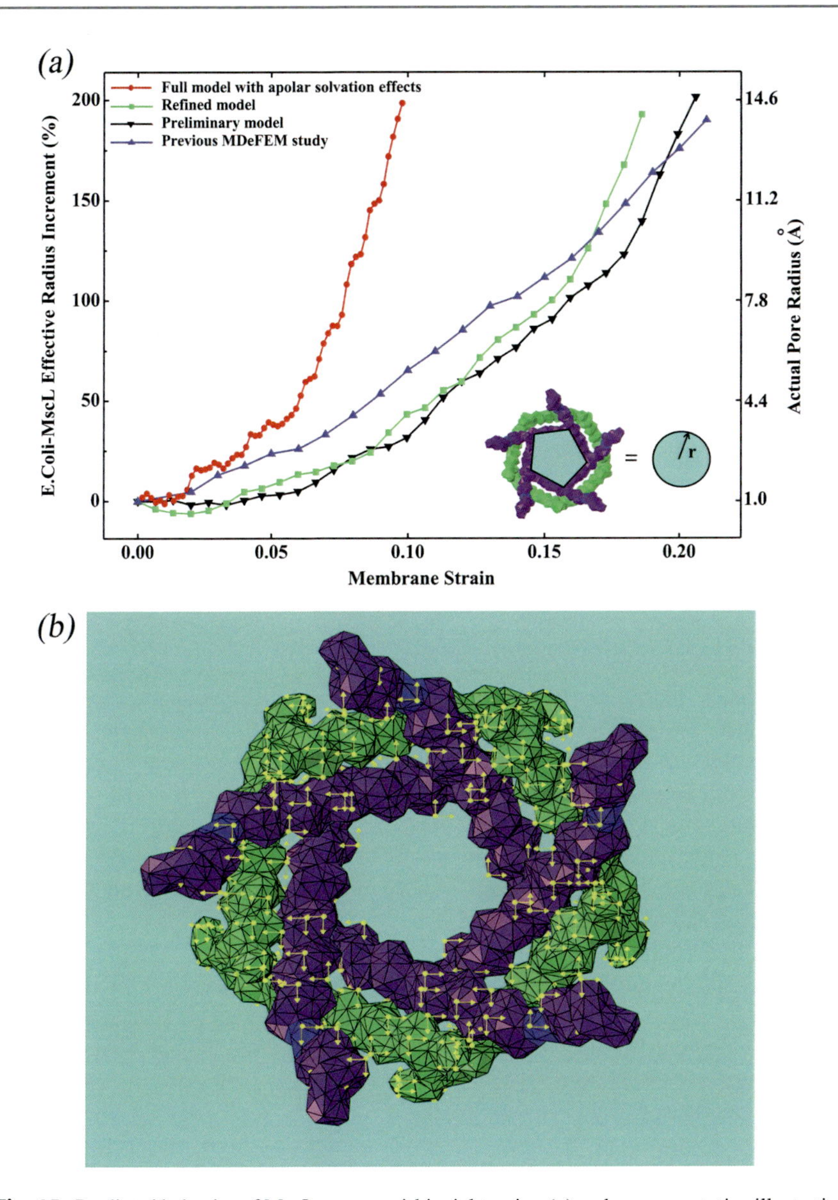

Fig. 15 Predicted behavior of MscL upon equi-biaxial tension (**a**) and a representative illustration of the apolar solvation forces (directions only) during MscL gating (**b**). Evolution of the effective pore radius and approximate actual radius of MscL is shown as a function of membrane strain. The results obtained currently are compared with those of the previous MDeFEM model (Tang et al. 2008). The refined model includes the effect of the lipid's fluidity, and the full model further includes the apolar solvation effects. Zhu et al. (2016), reprinted with permission of Springer

This observation is in qualitative agreement with the patch-clamp experiment of Sukharev et al. (1999), in which the tension-activation data was fitted to a Boltzmann form.

Gating Pathway of MscL with Fluidic Membrane and Apolar Solvation Effects

As emphasized above, two major limitations of our previous work (Chen et al. 2008; Tang et al. 2008) and the preliminary model concern the lack of membrane fluidity and solvation effects. In the following, we first include the fluidic properties of the membrane, leading to the "refined model." Next, we further include solvation effects, leading to the "full model." Since we expect that the interactions between the apolar surface of the MscL pore and the solvent make a dominant contribution to the gating process, in this section we only include effects of apolar solvation; the relative importance of electrostatic and apolar solvation forces will be discussed in the next section.

With the fluidic lipid model and continuum apolar solvation model employed successively, new effective pore radius evolution curves are obtained and shown in Fig. 15 (square curve for the refined model and circle curve for the full model with apolar solvation). The overall trend of the gating pathway remains the same as that of the preliminary model (inverted triangle curve) in that the pore size is firstly stabilized at a small value and then experiences a rapid increase as the lipid strain further increases. Compared to the preliminary model, the most notable difference lies in the lipid membrane strain required for fully opening MscL. With the more realistic (orthotropic) model of the lipid, the membrane strain for full opening of MscL is reduced slightly ($\sim$10%); with the apolar solvation included, the gating strain is further reduced by as much as $\sim$50%. Figure 15b shows the solvation forces (directions only) on the "hydrophobic chemical nodes (residues)" on TM1s and TM2s in one step of the simulation of MscL gating. Since most of the residues on the transmembrane helices are hydrophobic rather than hydrophilic (Fig. 10b), the interactions between water- and solvent-exposed residues are expected to push the channel outward, providing another driving force (in addition to the membrane tension) for the gating transition of MscL.

Take the hydrophobic residues on the constriction area as an example. In the closed state, these residues are very close, and hydrophobic confinement is thermodynamically favorable since exposure of these hydrophobic residues is energetically unfavorable. But when the pore constriction is opened wide enough (mainly by membrane stretch), water molecules will be "driven" to fill in the constriction space despite the fact that exposure of hydrophobic residues on pore constriction area is still energetically unfavorable. This "driving" factor that compel the water to fill into the opened space is probably the system's internal pressure, change of which is usually not considered in MD simulations. Thus the "unfavorable" energy required to expose the hydrophobic residues on the pore constriction is compensated by drop of the internal pressure which may be very small on the scope of the whole system but critical for MscL gating. After that, the repulsion interaction between water and these hydrophobic residues is helping channel gating.

It is known that the gating of MscL can be triggered solely by membrane stretching, although the importance of apolar solvation to gating has only been indirectly probed. The release of content through *Tb*-MscL has been studied by a pioneering coarse-grained MD simulation work (Louhivuori et al. 2010) which, however, did not explicitly elucidate the importance of the hydrophobic effect in MscL gating. A molecular dynamics study on *Tb*-MscL (Jeon and Voth 2008) observed that the water chain formation across the channel pore took place at the same time as the channel pore radius increased to a certain value, and they suggested that the two processes may provide positive feedback to each other. The results we present here provide direct evidence for this point. Though the importance of hydrophobic interactions in channel gating has been demonstrated previously by hydrophilic mutations or molecular dynamics studies (Anishkin et al. 2005; Jeon and Voth 2008), this work, to the best of our knowledge, represents the first explicit evaluation of the hydrophobic contribution.

Another work that concerned hydration process of MscL gating is conducted by Anishkin et al. (2010) who focused on the hydrophilic interaction between water and buried hydrophilic residues instead of hydrophobic interaction as we present in this work. Based on the analysis of the hydration energy of the pore constriction and the fact that hydrophilic mutations in the pore constriction area make the channel opening more easy, a conclusion is drawn that the process of the glycine (hydrophilic residue) exposure with pore expansion and their favorable hydration create disjoining pressure that assists opening (Anishkin et al. 2010).

Based on Anishkin et al.'s hydration energy analysis and the analysis of the results in our present work, detailed effects of the hydrophilic or hydrophobic residues on channel gating are elucidated as follows. In the process of the exposure of initially buried hydrophilic residues in the pore constriction area or buried in other area, due to their favorable hydration, disjoining pressure will be generated that assists conducting as suggested by Anishkin et al. (2010). But once these hydrophilic residues are exposed, their interaction with water molecules will resist opening of the channel based on our analysis primarily due to the hydrogen bonding between the polar residue and water molecules. On the other hand, the effects of the hydrophobic residues also act in a similar but contrary way. Before exposure of the hydrophobic residues in the hydrophobic constriction area, these hydrophobic residues are closely packed and the interaction within helps the channel keep closed. But once these hydrophobic residues are exposed, their interaction with water molecules will facilitate opening of the channel primarily due to the repulsion interaction between water and nonpolar residues. Collectively, that is, the effects of the hydrophobic or hydrophilic residues are reversed before and after exposure to water. These mechanisms may be further verified by future experimental and simulation studies.

Shown in Fig. 16 are the configurations at half-/fully opened states of the refined and full models in comparison to the structural model of Sukharev and Anishkin (2004). As with the preliminary model, the conformational transitions observed with the refined or full model are also in qualitative agreement with the structural model. Closer inspection shows that opening of the S1 bundle with the full model is much more significant and the pore size surrounded (not enclosed) by S1 helices are also comparable with that of the TM1 pore (Fig. 16). Note that the position of the S1 domain in the opened state may be different from the real case for biological MscL molecule since the S1 domain of the initial MscL structure (Sukharev and Anishkin

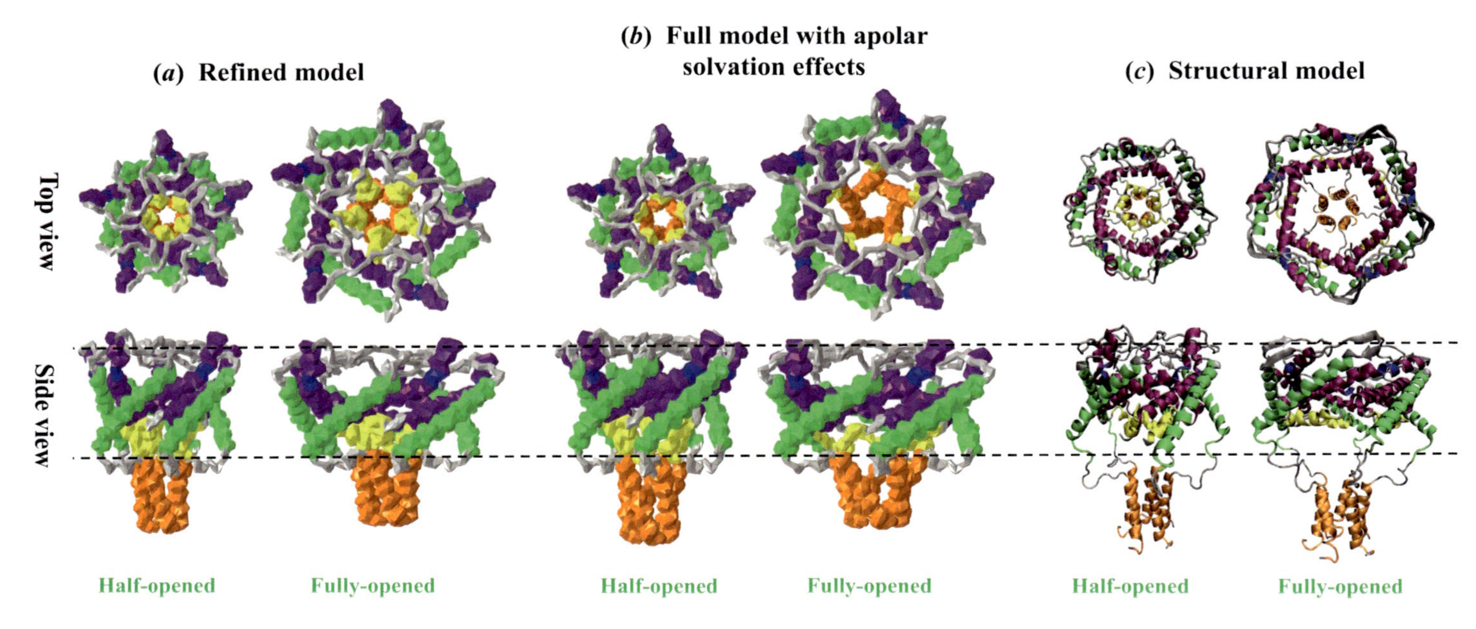

Fig. 16 MscL gating pathway of the refined model (**a**) and the full model (**b**). Results at half- and fully opened states are shown with comparison to the structural model (**c**) (Sukharev and Anishkin 2004). The dashed lines indicate the approximate location of membrane/water interface. Here, the refined model takes into account the fluidity of the lipid membrane, and the full model further includes the apolar solvation effects. Zhu et al. (2016), reprinted with permission of Springer

2004) we use may not be appropriately positioned (Iscla et al. 2008). The S3 bundle also separates more in the full model especially for the upper ends though its lower ends remain assembled together. Similar to the preliminary model, the loops linking TM1-TM2 remain essentially unseparated, and this result is consistent with that of the SMD study (Gullingsrud and Schulten 2003) in which the periplasmic loop region in all simulations remained stable. Experimentally, it was found that the channel remains mechanosensitive even with the external loops cleaved (Ajouz et al. 2000); thus, the precise functions of these loops remain unclear. The stable behavior of the periplasmic loops during the simulated gating process suggests that they may function as a filter screen to prevent large molecules from entering and occluding the channel while at the same time prolonging channel opening.

Respective Effects of Electrostatic and Apolar Solvation Forces

Previous patch-clamp experiments (Sukharev et al. 1999) showed that there was no significant change in the MscL gating pattern in the range of salt concentration between 0.05 and 1 M. This observation suggests that electrostatic interactions are unlikely to dominate in MscL gating. To further explore this hypothesis quantitatively, we simulate MscL gating at several ion concentrations (0.05 M and 2.0 M) in the framework of continuum solvation.

As shown in Fig. 17, there are minor differences between the effective pore radius evolution curves of the refined model, the full model with electrostatic solvation contributions with ion concentration of 0.05 M or 2.0 M, respectively. Configurations from the full model with electrostatic solvation effects (not shown) are also very close to those from the refined model (ion concentration, 0.00 M) shown in Fig. 16. Compared to the refined model where no solvation contributions are included, the effective pore radius evolution does not change much when only the electrostatic solvation forces are considered. The apolar solvation forces, on the other hand, contribute significantly to the gating process (circle curve in Fig. 17 or Fig. 15) as described in the above section.

Different Pathways for the GOF (Gain-of-Function) mutant and WT (Wild-Type) MscL

The hydrophobic core of the TM1 bundle of WT MscL appears to be dehydrated according to the EPR data and molecular dynamics results (Gullingsrud and Schulten 2003; Perozo et al. 2002a). The energy landscape for WT MscL indicates that the major energy cost for MscL opening is between the closed state and the substate $S_{0.13}$ (Anishkin et al. 2005) (0.13 means the relative conductance to the fully opened state) and this substate is believed to be already well hydrated (Anishkin et al. 2005). Experiments showed that hydrophilic substitutions in the hydrophobic restriction of TM1s led to a reduction of the gating tension, while a more hydrophobic substitution resulted in a channel that requires a greater tension to open (Anishkin et al. 2005). Based on these data, one could associate the passage of the main energy barrier with the hydration of the largely hydrophobic pore (Anishkin et al.

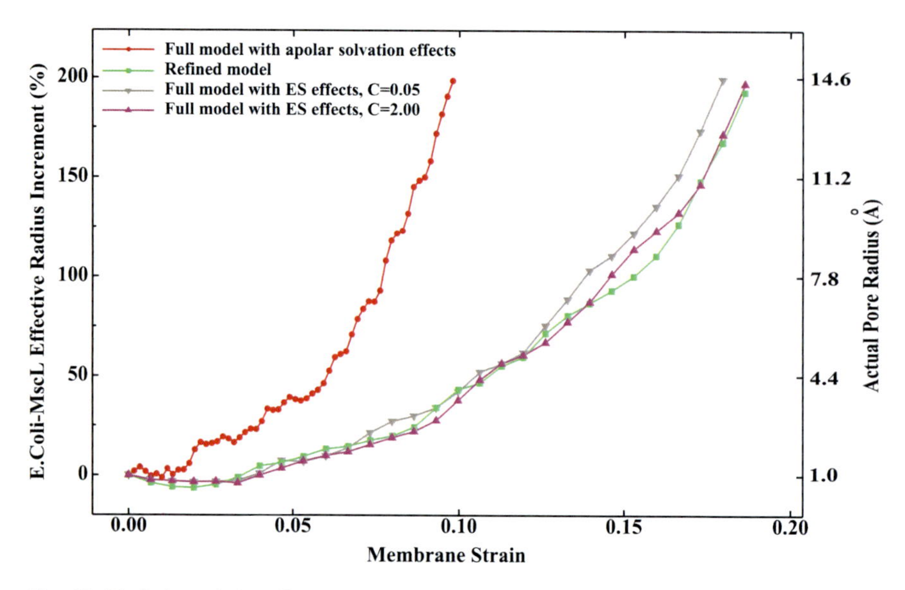

Fig. 17 Evolution of the effective pore radius of MscL versus membrane strain for different models. The different effects of the apolar and electrostatic solvation forces on the gating of *E. coli* MscL are obvious. The full model includes the effects of the electrostatic solvation (ES) (with ion concentration *C*) or apolar solvation (AS) forces. Zhu et al. (2016), reprinted with permission of Springer

2005), which is a prerequisite for conduction. The hydrophilic substitution in the hydrophobic restriction of TM1s may disrupt this initial hydrophobic restriction, leading to facile hydration of the pore and reduction of the first free energy barrier.

In the present simulation, the hydration of the channel pore depends on the effective pore radius as described in the "Models and methods" section. The threshold radius for hydration is taken to be 0.45 nm (Beckstein et al. 2001) as indicated in the right bottom inset of Fig. 18. In the closed state of MscL, about half of the TM1's inner surface is exposed to solvent and the lower half is not. When considering the hydration process here, accordingly, the apolar solvation forces on the inner hydrophobic constriction region are not included in the continuum mechanics calculations.

Figure 18 shows the effective radius evolution curves from full model simulations (including apolar solvation effects) for the WT and the GOF mutant MscL (hypothetical hydrophilic substitution in the hydrophobic constriction of TM1s). A rightward shift of the effective radius evolution curve is observed for the WT MscL as compared to the GOF mutant. This result is in agreement with experimental observation that an initially better hydrated state (GOF mutant MscL caused by hydrophilic substitution) is easier to open (Anishkin et al. 2005). We note that the TM1 helices are not fully hydrated during the simulation until the membrane strain gets to as large as ~2/3 of the strain required for full gating (indicated by the dashed

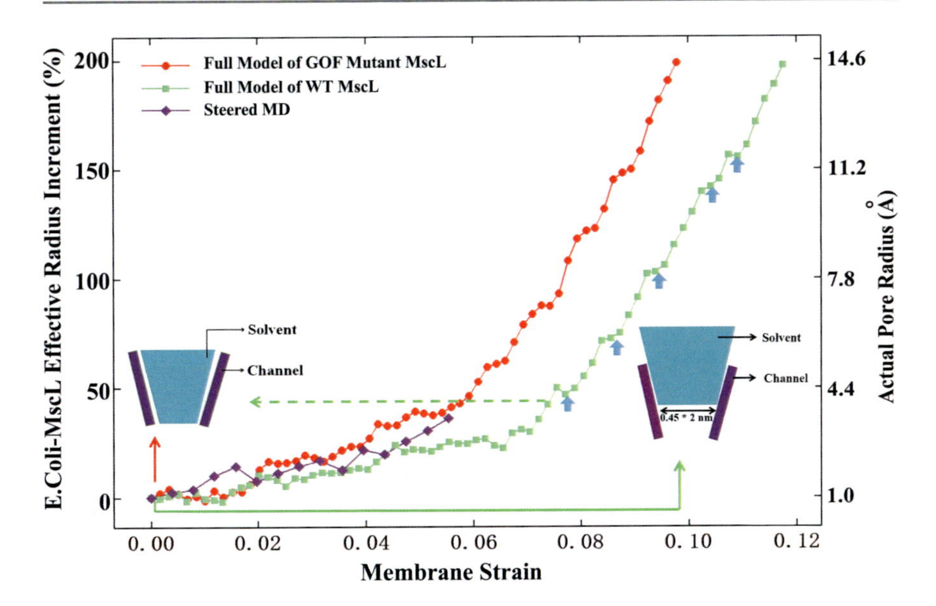

Fig. 18 Effective radius evolution of simulations for the GOF mutant and WT MscL. The GOF mutant MscL has a hydrophilic substitution in its hydrophobic constriction causing full hydration in the initial state, while the initial state of the WT MscL is only half hydrated. The WT channel won't get fully hydrated until the membrane strain reaches ~2/3 of the gating strain (indicated by the dashed line) which agrees well with Sukharev et al.'s experimental study (Sukharev et al. 1999). Zhu et al. (2016), reprinted with permission of Springer

line in Fig. 18), suggesting that MscL is nonconductive until two thirds of its gating tension is reached. This result is in agreement with Sukharev et al.'s experimental observation (Sukharev et al. 1999) that the open probability remains zero until the membrane tension reaches ~10 dyne/m and the gating tension is about 15 dyne/m. Thus, when the membrane strain reaches half of the gating strain for MscL, the mechanosensitive channel of small conductance (MscS) reaches its fully conducting and opened state (Martinac and Kloda 2003), while MscL is not yet conductive, emphasizing MscL's role of a "safety valve" in prokaryotes as a last-ditch effort for survival (Berrier et al. 1996).

The effective radius evolution of these two full models is compared to that in the steered molecular dynamics (SMD) study (Gullingsrud and Schulten 2003) (rhombus curve in Fig. 18). The maximum membrane strain in SMD is estimated to be ~5.4%, and we assume that its increment is proportional to the time step used in SMD. Although the channel is far from fully open in the nanosecond SMD simulation, as shown in Fig. 18, the effective pore radius evolution in SMD is in good agreement with that of the continuum full model at small strain.

The final lipid strain (ε_m) of the full model for the complete opening of MscL is ~11.76% (gating strain). The strain in the direction normal to the membrane is (Tang et al. 2006)

$$\varepsilon_z = \frac{2\overline{\nu_t}}{1 - \overline{\nu_t}} \ \varepsilon_m \tag{5}$$

which leads to a reduction in the membrane thickness of

$$\Delta h = \ \varepsilon_z h \tag{6}$$

where $\overline{\upsilon}_t = 0.5$ is the equivalent Poisson's ratio of the orthotropic lipid membrane model and $h = 35\text{Å}$ is the initial thickness of the membrane. With the membrane strain of 11.76%, the thickness is reduced from 35 Å to 28.77 Å, which is a 23.5% change. This value is in close agreement with the 20% reduction in the thickness of membrane spanning part of MscL measured by experiment (Perozo et al. 2002a) since the flatness of MscL is directly caused by thinning of the lipid membrane during gating.

Another prominent feature of the full model's gating pathway is the stepwise nature of the pore radius evolution curve, as shown in Fig. 18 the circle curve or the square curve. The initial survey of MscL in liposome patches (Sukharev et al. 1999) has recorded three short-lived subconducting states where the pore size stabilizes even in the presence of tension, and a later study of Chiang et al. (2004) identified nine subconducting states. In the present study, when the membrane strain is below ~2/3 of the gating strain, the channel is not fully hydrated and remains nonconducting. Afterward, the channel becomes conductive, and the effective pore radius increases very rapidly with the membrane strain. A closer inspection of the data shows some plateaus (marked with arrows in Fig. 18) of the curve where the pore size stabilizes. Although the current continuum model may not be sufficient enough to capture some delicate molecular phenomena, such as the subconducting states of MscL gating, nevertheless, the results here may imply that the subconducting states may be in part due to the intricate "overall" mechanical interactions of the multiple components of the system, which complement their biophysical functions.

The actual pore radius is about 4.75 Å for the first substate in Fig. 18, and the radius of the inner pore surface in the open state is ~14 Å, which is consistent with the estimation from fluorescence resonance energy transfer (FRET) experiments (Corry et al. 2010; Wang et al. 2014). Assuming that the conductance of the MscL pore is proportional to its size (area enclosed by the inner surfaces of TM1s) (Steinbacher et al. 2007), the relative conductance in the present study can be estimated as the ratio of the pore area to that of the full open state. For example, the relative conductance for the first plateaus in present work would be $(4.75/14)^2 = 0.115$, which is in good agreement of the experimental estimate of 0.13 for the first subconducting state (Anishkin et al. 2005). Their proposal that the main energy costing substate is well hydrated is also consistent with results shown in Fig. 18.

Analytical Effort

An important goal of this section is to establish a closed-form and simple analytical model (Chen et al. 2008), as an alternative approach that can capture the most essential features of MscL gating that might be broadly applicable to MS channels, such as the different contributions of solvation and membrane stretch to gating. An analytical model that couples MscL to both lipid membrane stretch (outer boundary) and apolar solvation effects (inner boundary) is developed as follows.

For the square lipid membrane (Fig. 19a) with a length of 2 *l* (400 Å) and a central cavity radius of *c*, its outer boundary pressure is $-p_{\text{lipid}}$, and the interface pressure between the lipid and MscL is $-p_{\text{lipid-MscL}}$. The whole lipid membrane is treated as an isotropic plate. Since the in-plane fluidity of the lipid membrane contributes only slightly (less than 10%) to the gating process, this treatment is considered a reasonable approximation. During gating, the deformation of the membrane cavity is mainly transferred to the closest TM2 helices via nonbonded interactions in the radial direction. The nonlinear interaction pressure-distance relationship is analogous to a nonlinear elastic medium between the lipid cavity and the TM2 bundle. After the TM2 helices are pulled open, the TM1/TM2 nonbonded interactions (another effective nonlinear medium) may perturb the MS channel radius, which is enclosed by the five TM1 helices. Therefore, a simple analytical model can be established in which the details of protein structures are ignored and the nonbonded interactions are described by effective elastic media. A schematic of such plane stress effective continuum medium model (ECMM) is given in Fig. 19b with *E. coli*-MscL as an example. The inner effective annular medium I accounts for the TM1-TM1 interactions in hoop direction and TM1-TM2 interactions in radial direction, and the outer continuum medium II incorporates TM2-TM2 interactions in hoop direction and TM2-lipid interactions in radial direction. The inner radius, interface radius, and outer radius of the ECMM are denoted by *a*, *b*, and *c*, respectively. Here, *a* is the effective radius of the closed MscL (consistent with the definition in previous sections), which corresponds to the smallest "through" capacity of the TM1 bundle; *b* is defined similarly for the TM2 bundle; and *c* is the interface radius between MscL/lipid cavity. From the closed homology structure of *E. coli*-MscL, *a*, *b*, and *c* are equal to 6.5 Å, 17 Å, and 22 Å, respectively.

The outer boundary of MscL ECMM (Fig. 19b) is coupled to the inner cavity of the lipid model (Fig. 19a), i.e., $\sigma_r^{\text{II}}(c) = -p_{\text{lipid-MscL}}$, $u_r^{\text{II}}(c) = u_r(c)$. Continuity of radial stress and displacement at the interface between medium I and II ($r = b$) requires $\sigma_r^{\text{I}}(b) = \sigma_r^{\text{II}}(b)$, $u_r^{\text{I}}(b) = u_r^{\text{II}}(b)$. The inner boundary pressure of medium I is, $\sigma_r^{\text{I}}(a) = -p_{\text{water}}$, from apolar solvation effects. Following the theoretical analysis in Chen et al. (2008) but considering the coupling between MscL ECMM and the lipid model and the coupling between MscL ECMM and the solvation contribution, a closed-form solution of the MscL pore radius increment can be obtained as

$$\Delta a = 0.085 p_{\text{water}} + 60.205 \varepsilon_m \tag{7}$$

where the units for p_{water} and Δa are MPa and Å, respectively. On the right of Eq. (7), the first term is the contribution of apolar solvation effects to MscL gating and the second the contribution of membrane stretch. The two constants before p_{water} and ε_m depend on the elastic properties of medium I and II (Chen et al. 2008). The value of p_{water} depends on the exposed residues of the TM1 bundle and the state of

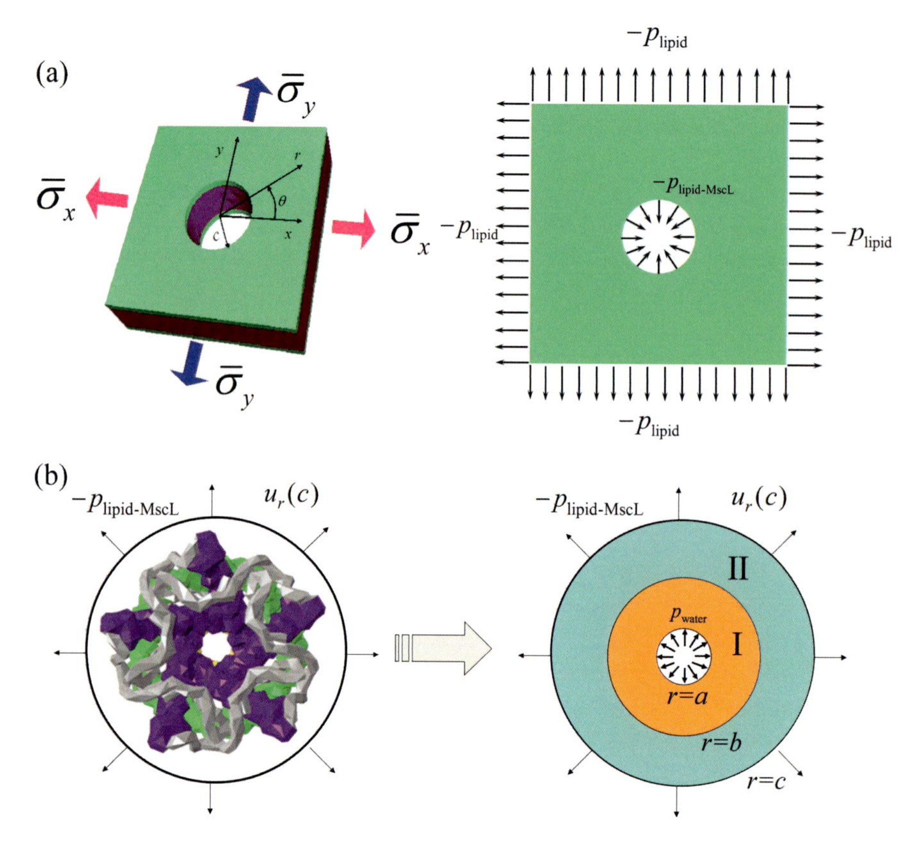

Fig. 19 Schematic of the linear lipid membrane model (**a**) and the effective continuum medium model (ECMM) of MscL (**b**)

the hydrophobic constriction. At the initial state, $p_{\text{water}} = 0$. Averagely, based on the continuum solvation forces calculated in the above sections and the exposed pore surface area, $p_{\text{water}} \approx 68$ MPa. Accordingly, the final membrane strain to fully open the channel is 11.16%, which is in close agreement with that (11.76%) calculated by the coupled CM/CS MDeFEM approach.

Limitations of the Current Implementation and Future Directions

Despite our tremendous endeavor and that the current model has been able to help us gain some useful perspectives for channel gating, there are considerable limitations for the present coupled continuum mechanical-continuum solvation approach, and the room for future improvement is still large.

One of the major limitations concerns the atomic structure of *E. coli*-MscL whose S1 domain may need revision (Iscla et al. 2008) based on the revised crystal structure of *Tb*-MscL (Steinbacher et al. 2007). And the crystal structure of *Tb* MscL could also be subjected to the above developed continuum-based approach to study the effects of the S1 domain on the gating behavior. Through these improvements, the proposed facilitation effect of the S1 domain on gating due to its close interaction to the membrane (Iscla and Blount 2012; Vanegas and Arroyo 2014) could be verified and more specifically quantified.

Another limitation of the current work is the treatment of the lipid membrane and its surface-based interaction to the channel. The particle-to-particle lipid-channel interaction force is averaged to the surfaces of them; thus, the heterogeneous binding of the lipid to the channel is lost. Interfacial lipid-channel interactions are exceedingly complicated due to hydrophobic mismatch, electrostatic effects, dynamic nature of the lipid environment, hydrogen bonding, variation during conformational transition, asymmetry, localization, etc. (Argudo et al. 2016; Powl et al. 2003, 2005a, 2008; Vanegas and Arroyo 2014). While emerging evidence has shown the importance of the interaction of the membrane with membrane channels for many biological processes, it is still difficult to elucidate these interactions in a detailed and comprehensive manner both experimentally and computationally. These characteristics of lipid-channel interactions impose great challenge for continuum modeling of the lipid membrane and its interaction to the channel. For future work, one may consider developing particle-to-particle-based interactions and geometry refinement of the lipid membrane (especially for the lipid closely around the channel), which could be included in a similar way as how we treat the channel in the current work. And solvation forces can then be applied to the lipid membrane model as well to mimic a hydration environment for the lipid model. Thereafter, this well-established lipid model can be exploited to explore the contribution or mechanism of different lipid-channel interactions for channel gating in future work. For example, how and how much does the "anchor" effect between S1 domain and the lipid membrane affects gating (Iscla and Blount 2012)? How and how much does the electrostatic interaction between the lipid head group and the charged transmembrane helix residues affect channel gating? Is the distortion of the lipid membrane around the channel helping channel gating, or is it just a spontaneous adaptation behavior for the system to reach a low energy state? Will the distortion of the membrane around the channel disappear or diminish during gating? Besides, this new lipid membrane is also much more reasonable for a study of the effect of negative or positive hydrophobic mismatch (Perozo et al. 2002b), although the current lipid model can also catch some of the basic principle that, as an example, for the wild-type MscL full model with apolar solvation effects, increasing the hydrophobic (tail group) layer of the current lipid membrane model by 10% will lead to a 2.8% increase of gating tension for channel opening.

Moreover, some other subtle but potentially important molecular details such as inclusion of side chains in the FEM model or initial turgor pressure in living *E. coli* cells (Deng et al. 2011) may also play a part in the gating process and need further refinement in future continuum modeling. We hopefully expect that these refinements will elaborately improve the accuracy of the continuum-based model and greatly expand the application scope and universality of this coupled continuum mechanics-continuum solvation approach.

Concluding Remarks

The analysis of MS channel gating provides us with an effective window to explore how mechanical stimuli induce adaptive cellular behaviors through protein structural transitions across different time and length scales. As shown by the results of the present work, their gating patterns highlight the roles of large-scale helical movements, pore hydration, and protein-lipid interactions during ion channel gating transitions. These processes occur on time and length scales that are too large to be studied directly by regular atomistic simulations. In this chapter, we have modified a molecular dynamics-decorated finite element (MDeFEM) method to incorporate key interaction pairs (e.g., Asp62-Arg131 and Phe68-Leu111) into a continuum mechanics model; this allows us to explore the gating pathway of MscS and how specific interactions and structural motifs impact the gating transition. Besides, a novel simulation protocol is developed that effectively integrates continuum solvation contributions (CS) into continuum mechanics (CM) calculations to study the gating pathway of MscL.

A complete gating transition trajectory of MscS from the closed to the open state along with partially open intermediates is obtained, and the open structure is close to the available structural model from crystallographic studies. It is observed that removing either the Asp62-Arg131 salt bridge or the Phe68-Leu111 nonpolar interaction leads to essentially nonconducting structures. The loop connecting TM2 (the second transmembrane helix) and TM3 is found to be essential for force transmission during gating, while the loop connecting TM1 and TM2 does not make any major contribution. Based on the different structural evolutions observed when the TM3 kink is treated as a loop or a helical segment, we propose that the helical propensity of the kink plays a central role in inactivation. Gating transition of MscS under different transmembrane voltages is also explored.

A novel computational framework is further developed by using MscL as a template. The continuum mechanics is closely coupled with the continuous solvent model. The influence of the solvent and the chemical coupling force is obtained by real-time iteration between the mechanical and chemical spaces. Compared to previous continuum mechanics studies, the present model is capable of capturing the most essential features of the gating process in a much more realistic fashion: due mainly to the apolar solvation contribution, the membrane tension for full opening of MscL is reduced substantially to the experimental measured range. A significant fraction ($\sim$2/3) of the gating membrane strain is required to reach the first subconducting state of our model, which is featured with a relative conductance of 0.115 to the fully opened state. These trends agree well with experimental observations.

We expect that many of the mechanical principles discussed here are also likely to play a role in various membrane-mediated biomechanical processes. The successful application of the MDeFEM approach to MscS/MscL suggests similar studies of the growing families of sensory channels (Chen et al. 2015; Clapham 2003; Dhaka et al. 2006; Krishtal 2003; Pruitt et al. 2014) and their modulations by lipids, lipid-soluble factors, temperature, cell volume, and membrane tension. The approach is particularly powerful in cases that involve large length and long time scales, which are usually not easily accessible to standard particle-based simulations.

References

ABAQUS 6.11 User's Manual. ABAQUS Inc., Providence, RI, (2011)
B. Ajouz, C. Berrier, M. Besnard, B. Martinac, A. Ghazi, J. Biol. Chem. **275**, 1015 (2000)
B. Akitake, A. Anishkin, S. Sukharev, J. Gen. Physiol. **125**, 143 (2005)
B. Akitake, A. Anishkin, N. Liu, S. Sukharev, Nat. Struct. Mol. Biol. **14**, 1141 (2007)
A. Anishkin, C. Kung, Curr. Opin. Neurobiol. **15**, 397 (2005)
A. Anishkin, S. Sukharev, Biophys. J. **86**, 2883 (2004)
A. Anishkin, S. Sukharev, Channels (Austin) **11**, 173 (2017)
A. Anishkin, C.S. Chiang, S. Sukharev, J. Gen. Physiol. **125**, 155 (2005)
A. Anishkin, B. Akitake, S. Sukharev, Biophys. J. **94**, 1252 (2008a)
A. Anishkin, K. Kamaraju, S. Sukharev, J. Gen. Physiol. **132**, 67 (2008b)
A. Anishkin, B. Akitake, K. Kamaraju, C. Chiang, S. Sukharev, J. Phys. Condens. Matter **22**, 454120 (2010)
D. Argudo, N.P. Bethel, F.V. Marcoline, M. Grabe, Biochim. Biophys. Acta Biomembr. **1858**, 1619 (2016)
N.A. Baker, D. Sept, S. Joseph, M.J. Holst, J.A. McCammon, Proc. Natl. Acad. Sci. U. S. A. **98**, 10037 (2001)
R.B. Bass, P. Strop, M. Barclay, D.C. Rees, Science **298**, 1582 (2002)
N. Bavi, O. Bavi, M. Vossoughi, R. Naghdabadi, A.P. Hill, B. Martinac, Y. Jamali, Channels (Austin) **11**, 209, (2017)
N. Bavi et al., Nat. Commun. **7**, 11984 (2016b)
O. Beckstein, M.S. Sansom, Phys. Biol. **1**, 42 (2004)
O. Beckstein, P.C. Biggin, M.S.P. Sansom, J. Phys. Chem. B **105**, 12902 (2001)
V. Belyy, A. Anishkin, K. Kamaraju, N. Liu, S. Sukharev, Nat. Struct. Mol. Biol. **17**, 451 (2010)
C. Berrier, M. Besnard, B. Ajouz, A. Coulombe, A. Ghazi, J Membr. Biol. **151**, 175 (1996)
R.B. Best, X. Zhu, J. Shim, P.E. Lopes, J. Mittal, M. Feig, A.D. Mackerell Jr., J. Chem. Theory Comput. **8**, 3257 (2012)
M. Betanzos, C.S. Chiang, H.R. Guy, S. Sukharev, Nat. Struct. Biol. **9**, 704 (2002)
H. Binder, K. Gawrisch, J. Phys. Chem. B **105**, 12378 (2001)
P. Blount, M.J. Schroeder, C. Kung, J. Biol. Chem. **272**, 32150 (1997)
I.R. Booth, M.D. Edwards, S. Black, U. Schumann, S. Miller, Nat. Rev. Microbiol. **5**, 431 (2007)
K. Borngen, A.R. Battle, N. Moker, S. Morbach, K. Marin, B. Martinac, R. Kramer, Biochim. Biophys. Acta **1798**, 2141 (2010)
B.R. Brooks, R.E. Bruccoleri, B.D. Olafson, D.J. States, S. Swaminathan, M. Karplus, J. Comput. Chem. **4**, 187 (1983)
G.X. Cao, X. Chen, J. Mater. Res. **21**, 1048 (2006)
G. Cao, X. Chen, J. Mater. Res. **21**, 1048 (2011)
E. Chacon, P. Tarazona, F. Bresme, J. Chem. Phys. **143**, 034706 (2015)
G. Chang, R.H. Spencer, A.T. Lee, M.T. Barclay, D.C. Rees, Science **282**, 2220 (1998)
X. Chen, Y. Tang, G. Cao, Proc. Inst. Mech. Eng. Part N **219**, 73 (2006)

X. Chen, Q. Cui, Y. Tang, J. Yoo, A. Yethiraj, Biophys. J. **95**, 563 (2008)
B. Chen, B. Ji, H. Gao, Annu. Rev. Biophys. **44**, 1 (2015)
C.S. Chiang, A. Anishkin, S. Sukharev, Biophys. J. **86**, 2846 (2004)
D.E. Clapham, Nature **426**, 517 (2003)
B. Corry, A.C. Hurst, P. Pal, T. Nomura, P. Rigby, B. Martinac, J. Gen. Physiol. **136**, 483 (2010)
C. Cui, D.O. Smith, J. Adler, J. Membr. Biol. **144**, 31 (1995)
M.E. Davis, J.A. Mccammon, Chem. Rev. **90**, 509 (1990)
Y. Deng, M. Sun, J.W. Shaevitz, Phys. Rev. Lett. **107**, 158101 (2011)
E. Deplazes, M. Louhivuori, D. Jayatilaka, S.J. Marrink, B. Corry, PLoS Comput. Biol. **8**, e1002683 (2012)
L. Deseri, P. Pollaci, M. Zingales, K. Dayal, J. Mech. Behav. Biomed. Mater. **58**, 11 (2016)
A. Dhaka, V. Viswanath, A. Patapoutian, Annu. Rev. Neurosci. **29**, 135 (2006)
R.O. Dror, R.M. Dirks, J.P. Grossman, H. Xu, D.E. Shaw, Annu. Rev. Biophys. **41**, 429 (2012)
M.D. Edwards, W. Bartlett, I.R. Booth, Biophys. J. **94**, 3003 (2008)
R. Gamini, M. Sotomayor, C. Chipot, K. Schulten, Biophys. J. **101**, 80 (2011)
M.A. Geeves, K.C. Holmes, Annu. Rev. Biochem. **68**, 687 (1999)
M.A. Geeves, K.C. Holmes, Adv. Protein Chem. **71**, 161 (2005)
J. Gullingsrud, K. Schulten, Biophys. J. **85**, 2087 (2003)
J. Gullingsrud, K. Schulten, Biophys. J. **86**, 3496 (2004)
O.P. Hamill, B. Martinac, Physiol. Rev. **81**, 685 (2001)
E.S. Haswell, R. Phillips, D.C. Rees, Structure **19**, 1356 (2011)
K. Hayakawa, H. Tatsumi, M. Sokabe, J. Cell Sci. **121**, 496 (2008)
P.S. Heckbert, M. Garland, Comp Geom. Theor. Appl. **14**, 49 (1999)
B. Honig, A. Nicholls, Science **268**, 1144 (1995)
W. Im, D. Beglov, B. Roux, Comput. Phys. Commun. **111**, 59 (1998)
W. Im, M. Feig, C.L. Brooks, Biophys. J. **85**, 2900 (2003)
D.E. Ingber, FASEB J. **20**, 811 (2006)
H.I. Ingolfsson, C.A. Lopez, J.J. Uusitalo, D.H. de Jong, S.M. Gopal, X. Periole, S.J. Marrink, Wiley Interdiscip. Rev. Comput. Mol. Sci. **4**, 225 (2014)
I. Iscla, P. Blount, Biophys. J. **103**, 169 (2012)
I. Iscla, R. Wray, P. Blount, Biophys. J. **95**, 2283 (2008)
J. Jeon, G.A. Voth, Biophys. J. **94**, 3497 (2008)
K. Kamaraju, V. Belyy, I. Rowe, A. Anishkin, S. Sukharev, J. Gen. Physiol. **138**, 49 (2011)
M. Karplus, J. Kuriyan, Proc. Natl. Acad. Sci. U. S. A. **102**, 6679 (2005)
J.L. Klepeis, K. Lindorff-Larsen, R.O. Dror, D.E. Shaw, Curr. Opin. Struct. Biol. **19**, 120 (2009)
A. Kloda, B. Martinac, EMBO J. **20**, 1888 (2001)
P. Koprowski, A. Kubalski, J. Membr. Biol. **164**, 253 (1998)
P. Koprowski, W. Grajkowski, E.Y. Isacoff, A. Kubalski, J. Biol. Chem. **286**, 877 (2011)
O. Krishtal, Trends Neurosci. **26**, 477 (2003)
J.Y. Lai, Y.S. Poon, J.T. Kaiser, D.C. Rees, Protein science: A publication of the protein. Society **22**, 502 (2013)
G. Levin, P. Blount, Biophys. J. **86**, 2862 (2004)
N. Levina, S. Totemeyer, N.R. Stokes, P. Louis, M.A. Jones, I.R. Booth, EMBO J. **18**, 1730 (1999)
M. Louhivuori, H.J. Risselada, E. van der Giessen, S.J. Marrink, Proc. Natl. Acad. Sci. U. S. A. **107**, 19856 (2010)
L. Ma, A. Yethiraj, X. Chen, Q. Cui, Biophys. J. **96**, 3543 (2009)
H. Machiyama, H. Tatsumi, M. Sokabe, Biophys. J. **97**, 1048 (2009)
H.R. Malcolm, Y.Y. Heo, D.E. Elmore, J.A. Maurer, Biophys. J. **101**, 345 (2011)
V.S. Markin, F. Sachs, Phys. Biol. **1**, 110 (2004)
S.J. Marrink, D.P. Tieleman, Chem. Soc. Rev. **42**, 6801 (2013)
S.J. Marrink, H.J. Risselada, S. Yefimov, D.P. Tieleman, A.H. de Vries, J. Phys. Chem. B **111**, 7812 (2007)
B. Martinac, J. Cell Sci. **117**, 2449 (2004)
B. Martinac, A. Kloda, Prog. Biophys. Mol. Biol. **82**, 11 (2003)

B. Martinac, A. Kloda, *Comprehensive Biophysics* (Elsevier, Amsterdam, 2012), p. 108
B. Martinac, M. Buechner, A.H. Delcour, J. Adler, C. Kung, Proc. Natl. Acad. Sci. U. S. A. **84**, 2297 (1987)
B. Martinac et al., Antioxid. Redox Signal. **20**, 952 (2014)
M. Masetti, C. Berti, R. Ocello, G.P. Di Martino, M. Recanatini, C. Fiegna, A. Cavalli, J. Chem. Theory Comput. **12**, 5681 (2016)
S. Miller, W. Bartlett, S. Chandrasekaran, S. Simpson, M. Edwards, I.R. Booth, EMBO J. **22**, 36 (2003)
P. Moe, P. Blount, Biochemistry **44**, 12239 (2005)
L. Monticelli, S.K. Kandasamy, X. Periole, R.G. Larson, D.P. Tieleman, S.J. Marrink, J. Chem. Theory Comput. **4**, 819 (2008)
T. Nomura, M. Sokabe, K. Yoshimura, Biophys. J. **91**, 2874 (2006)
T. Nomura, M. Sokabe, K. Yoshimura, Biophys. J. **94**, 1638 (2008)
T. Nomura, M. Sokabe, K. Yoshimura, BioMed Res. Int. **2016**, 2401657 (2016)
A. Pantano, M.C. Boyce, D.M. Parks, Phys. Rev. Lett. **91**, 145504 (2003)
E. Perozo, D.M. Cortes, P. Sompornpisut, A. Kloda, B. Martinac, Nature **418**, 942 (2002a)
E. Perozo, A. Kloda, D.M. Cortes, B. Martinac, Nat. Struct. Biol. **9**, 696 (2002b)
E. Petrov, D. Palanivelu, M. Constantine, P.R. Rohde, C.D. Cox, T. Nomura, D.L. Minor Jr., B. Martinac, Biophys. J. **104**, 1426 (2013)
R. Phillips, T. Ursell, P. Wiggins, P. Sens, Nature **459**, 379 (2009)
C. Pliotas, J.H. Naismith, Curr. Opin. Struct. Biol. **45**, 59 (2016)
C. Pliotas et al., Proc. Natl. Acad. Sci. U. S. A. **109**, E2675 (2012)
C. Pliotas et al., Nat. Struct. Mol. Biol. **22**, 991 (2015)
A.M. Powl, J.M. East, A.G. Lee, Biochemistry **42**, 14306 (2003)
A.M. Powl, J.M. East, A.G. Lee, Biochemistry **44**, 5873 (2005a)
A.M. Powl, J.N. Wright, J.M. East, A.G. Lee, Biochemistry **44**, 5713 (2005b)
A.M. Powl, J.M. East, A.G. Lee, Biochemistry **47**, 12175 (2008)
M. Praprotnik, L.D. Site, K. Kremer, Annu. Rev. Phys. Chem. **59**, 545 (2008)
B.L. Pruitt, A.R. Dunn, W.I. Weis, W.J. Nelson, PLoS Biol. **12**, e1001996 (2014)
I. Rowe, A. Anishkin, K. Kamaraju, K. Yoshimura, S. Sukharev, J. Gen. Physiol. **143**, 543 (2014)
H. Sackin, Annu. Rev. Physiol. **57**, 333 (1995)
Y. Saimi, B. Martinac, A.H. Delcour, P.V. Minorsky, M.C. Gustin, M.R. Culbertson, J. Adler, C. Kung, Methods Enzymol. **207**, 681 (1992)
M.F. Sanner, A.J. Olson, J.C. Spehner, Biopolymers **38**, 305 (1996)
M.G. Saunders, G.A. Voth, Curr. Opin. Struct. Biol. **22**, 144 (2012)
Y. Sawada, M. Murase, M. Sokabe, Channels (Austin) **6**, 317 (2012)
F. Scarpa, S. Adhikari, A.J. Gil, C. Remillat, Nanotechnology **21**, 125702 (2010)
U. Schumann, M.D. Edwards, C. Li, I.R. Booth, FEBS Lett. **572**, 233 (2004)
Q. Shi, S. Izvekov, G.A. Voth, J. Phys. Chem. B **110**, 15045 (2006)
W. Shinoda, R. DeVane, M.L. Klein, Curr. Opin. Struct. Biol. **22**, 175 (2012)
C.D. Snow, E.J. Sorin, Y.M. Rhee, V.S. Pande, Annu. Rev. Biophys. Biomol. Struct. **34**, 43 (2005)
M. Sotomayor, K. Schulten, Biophys. J. **87**, 3050 (2004)
M. Sotomayor, T.A.v.d. Straaten, U. Ravaioli, K. Schulten, Biophys. J. **90**, 3496 (2006)
S.A. Spronk, D.E. Elmore, D.A. Dougherty, Biophys. J. **90**, 3555 (2006)
S. Steinbacher, R. Bass, P. Strop, D.C. Rees, Mechanosens. Ion Channels Part A **58**, 1 (2007)
S. Sukharev, Biophys. J. **83**, 290 (2002)
S. Sukharev, A. Anishkin, Trends Neurosci. **27**, 345 (2004)
S. Sukharev, D.P. Corey, Sci. Signal. **2004**, re4 (2004)
S.I. Sukharev, P. Blount, B. Martinac, C. Kung, Annu. Rev. Physiol. **59**, 633 (1997)
S.I. Sukharev, W.J. Sigurdson, C. Kung, F. Sachs, J. Gen. Physiol. **113**, 525 (1999)
S. Sukharev, M. Betanzos, C.S. Chiang, H.R. Guy, Nature **409**, 720 (2001a)
S. Sukharev, S.R. Durell, H.R. Guy, Biophys. J. **81**, 917 (2001b)
H. Sun, D.P. Li, S.R. Chen, W.N. Hittelman, H.L. Pan, J. Pharmacol. Exp. Ther. **331**, 851 (2009)
Y. Tang, G. Cao, X. Chen, J. Yoo, A. Yethiraj, Q. Cui, Biophys. J. **91**, 1248 (2006)

Y. Tang, J. Yoo, A. Yethiraj, Q. Cui, X. Chen, Biophys. J. **95**, 581 (2008)
A. Torres-Sánchez, J.M. Vanegas, M. Arroyo, Phys. Rev. Lett. **114**, 258102 (2015)
I.J. Tsai, Z.W. Liu, J. Rayment, C. Norman, A. McKinley, B. Martinac, Eur. Biophys. J. **34**, 403 (2005)
K.I. Tserpes, P. Papanikos, Compos. Struct. **91**, 131 (2009)
M.S. Turner, P. Sens, Phys. Rev. Lett. **93**, 118103 (2004)
D. Van Der Spoel, E. Lindahl, B. Hess, G. Groenhof, A.E. Mark, H.J. Berendsen, J. Comput. Chem. **26**, 1701 (2005)
J.M. Vanegas, M. Arroyo, PLoS One **9**, e113947 (2014)
J.M. Vanegas, A. Torres-Sánchez, M. Arroyo, J. Chem. Theory Comput. **10**, 691 (2014)
V. Vasquez, Biophys. J. **104**, 1391 (2013)
V. Vasquez, M. Sotomayor, J. Cordero-Morales, K. Schulten, E. Perozo, Science **321**, 1210 (2008a)
V. Vasquez, M. Sotomayor, D.M. Cortes, B. Roux, K. Schulten, E. Perozo, J. Mol. Biol. **378**, 55 (2008b)
R.M. Venable, F.L. Brown, R.W. Pastor, Chem. Phys. Lipids **192**, 60 (2015)
T. Vora, B. Corry, S.H. Chung, Biochim. Biophys. Acta **1758**, 730 (2006)
J.A. Wagoner, N.A. Baker, Proc. Natl. Acad. Sci. **103**, 8331 (2006)
H.C. Wang, B.P. Thampatty, Biomech. Model. Mechanobiol. **5**, 1 (2006)
W. Wang, S.S. Black, M.D. Edwards, S. Miller, E.L. Morrison, W. Bartlett, C. Dong, J.H. Naismith, I.R. Booth, Science **321**, 1179 (2008)
Y. Wang, Y. Liu, H.A. Deberg, T. Nomura, M.T. Hoffman, P.R. Rohde, K. Schulten, B. Martinac, P.R. Selvin, elife **3**, e01834 (2014)
R. Ward et al., Biophys. J. **106**, 834 (2014)
P. Wiggins, R. Phillips, Proc. Natl. Acad. Sci. U. S. A. **101**, 4071 (2004)
K.M. Wisdom, S.L. Delp, E. Kuhl, Biomech. Model. Mechanobiol. **14**, 195 (2014)
S. Yefimov, E. van der Giessen, P.R. Onck, S.J. Marrink, Biophys. J. **94**, 2994 (2008)
S.O. Yesylevskyy, L.V. Schafer, D. Sengupta, S.J. Marrink, PLoS Comput. Biol. **6**, e1000810 (2010)
Y. Zeng, A.K. Yip, S.K. Teo, K.H. Chiam, Biomech. Model. Mechanobiol. **11**, 49 (2012)
L. Zhu, J. Wu, L. Liu, Y. Liu, Y. Yan, Q. Cui, X. Chen, Biomech. Model. Mechanobiol. **15**, 1557 (2016)

Spherical Indentation on a Prestressed Elastic Coating/Substrate System

4

James A. Mills and Xi Chen

Contents

Abstract

While there have been many studies on the indentation test of thin film/substrate systems, the primary goal has been determining the film properties. However, there was very little effort to probe the properties of both the film and the substrate (the latter may be as important as the film properties). Moreover, a prestress usually exists in the film, typically resulted from mismatched deformation or material properties. In this study, we establish a spherical indentation framework to examine the material properties of both the film and substrate as well as

J. A. Mills
Department of Civil Engineering and Engineering Mechanics, Columbia University, New York, NY, USA

X. Chen (✉)
Department of Earth and Environmental Engineering, Columbia Nanomechanics Research Center, Columbia University, New York, NY, USA
e-mail: xichen@columbia.edu

G. Z. Voyiadjis (ed.), *Handbook of Nonlocal Continuum Mechanics for Materials and Structures*, https://doi.org/10.1007/978-3-319-58729-5_19

Cheng 2004). On one hand, there are inherent problems associated with performing indentation tests at shallow depths (Saha et al. 2001; Swadener et al. 2002; Chen and Vlassak 2001), that at times makes the test unattractive. For example, the P- δ data of shallow indentation may be affected by localized material surface roughness, indenter tip shape changes due to wear, or possible indenter drag (frictional effects) from surface adhesion. On the other hand, there are instances where the substrate properties are of interest as well. The turbine blade noted earlier is a fine example of where the blade properties (substrate) may be as important as the thermal coating.

The aforementioned gaps may be bridged by performing a moderately deep test, embracing and fully incorporating in the present analysis the coupled effect from both the film and substrate (Chen and Vlassak 2001; Chen et al. 2006; Zhao et al. 2006, 2007), as well as accounting for the film prestress. A relationship accounting for both the film and substrate effects in the P- δ data can be developed based on which an effective reverse analysis algorithm can be established. The moderately deep indentation measurement may help to yield a unique solution for determining the material property through reverse analysis (Chen et al. 2007).

The goal of the present study is to establish the framework of indenting a prestressed film on a semi-infinite substrate, using only one spherical indenter (which prevents penetrating/damaging the film) and without requiring a stress-free reference specimen. The focus is given to the elastic properties of the film and substrate, specifically determining the elastic modulus of each, as well as the film prestress. The next section describes the physical model as well as the parameter groups. Based on extensive finite element simulations of indentation tests over a wide range of material parameters and prestress levels, a set of general functional relationships are developed through the forward analysis. Through the results of two tests at different depths, the reverse analysis can effectively identify two material parameters among the film elastic modulus, substrate elastic modulus, and film prestress (as long as the third one is known a priori). The study is further extended to variable indenter radii or variable film thickness. Finally, a rigorous review of the error sensitivity in the proposed model is discussed.

Model and Computation Method

As shown in Fig. 1, the model is made up of a rigid spherical indenter, and a thin film is fully affixed to a semi-infinite substrate. Both the film and the substrate are modeled as homogeneous, isotropic elastic materials. The film prestress is equibiaxial, σ_{ps}. The indenter acts directly against the film and has a prescribed displacement that is normal to the upper surface of the film. The film has an elastic modulus defined as E_f and Poisson's ratio of ν_f. The substrate has an elastic modulus of E_s and a corresponding Poisson's ratio, ν_s. ν_s and ν_f are fixed as 0.25, and they are minor factors for indentation analysis (Cheng and Cheng 1998).

The study is primarily focused on, through a single indentation test, the determination of two of three variables: σ_{ps}, E_f, and E_s (as long as the third is known a priori). Invoking dimensional analysis, the following functional relationship needs to be determined through forward analysis:

$$\frac{P}{E_f \delta^2} = \prod \left(\frac{E_f}{E_g}, \frac{R}{h}, \frac{\sigma_{ps}}{E_f}, \frac{\delta}{h} \right) \tag{1}$$

To take advantage of the substrate effect (Zhao et al. 2006, 2007; Chen et al. 2006; Zhao et al. 2006), the indentation load can be taken at two indentation depths $\delta_1 = h/2$, and $\delta_2 = h/4$, where h is the film thickness (Mills and Chen 2009), from which two independent relationships are deduced:

$$\frac{P_1}{E_f \delta_1^2} = f_1 \left(\frac{E_f}{E_s}, \frac{R}{h}, \frac{\sigma_{ps}}{E_f} \right) \tag{2}$$

$$\frac{P_2}{E_f \delta_2^2} = f_2 \left(\frac{E_f}{E_s}, \frac{R}{h}, \frac{\sigma_{ps}}{E_f} \right) \tag{3}$$

Subscripts 1 and 2 relate the functional relationships in (2) and (3) to the two prescribed indentation depths: δ_1 and δ_2. Note that in practice, the maximum indentation depth can exceed $h/2$ (and be any value), and only the data taken at $h/2$ and $h/4$ are relevant.

As a first step, it is assumed that the film thickness (h) is known such that one may correspondingly match the indenter radius (R) exactly to this film thickness (R/h will be varied in a later part of this paper). In this simplified case:

$$\frac{P_1}{E_f \delta_1^2} = f_1 \left(\frac{E_f}{E_s}, \frac{\sigma_{ps}}{E_f} \right), \quad \text{with } \frac{R}{h} = 1.0 \tag{4}$$

$$\frac{P_2}{E_f \delta_2^2} f_2 \left(\frac{E_f}{E_s}, \frac{\sigma_{ps}}{E_f} \right), \quad \text{with } \frac{R}{h} = 1.0 \tag{5}$$

After these two equations are established through forward analysis, one will be able to solve for two unknowns E_f, E_s, or σ_{ps} with the third being known a priori.

Numerical simulation of indentation was performed using ABAQUS (Dassault Systèmes (SIMULIA) 2008). The film/substrate was made up of axisymmetric 4-node bilinear, reduced integration elements (CAX4R). Over 180,000 elements (Fig. 2) were incorporated into the model, and the substrate is semi-infinite. The film is equibiaxially prestressed to a desired level prior to indentation. The contact is assumed frictionless which is a relatively minor factor (Bucaille et al. 2003; Cheng and Cheng 2004).

The material properties are varied in moderate ranges during the forward analysis, with the elastic modulus ratio (EMR $= \frac{E_f}{E_s}$) in the range of $0.25 \leq \text{EMR} \leq 15.00$ and the normalized prestress ($K = \sigma\_ps/E\_f$) between $-0.10 \leq K \leq 0.10$. It is verified that film buckling does not take place under such a prestress level.

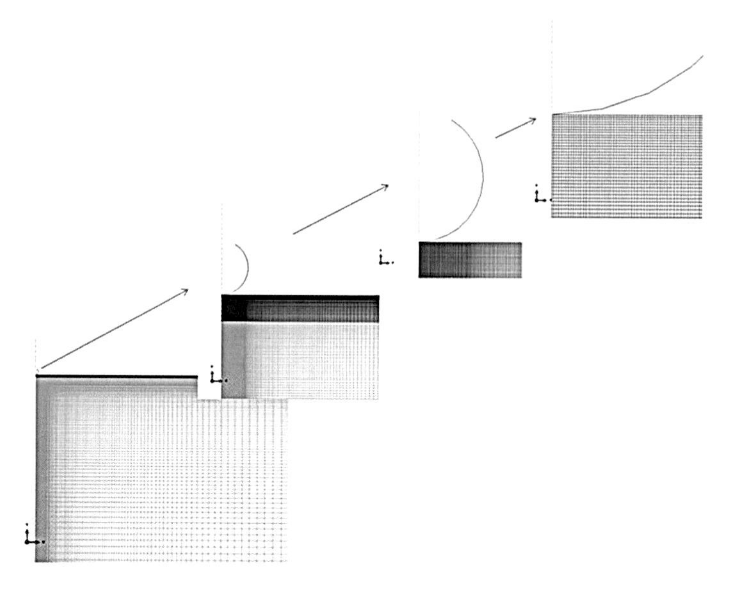

Fig. 2 Finite element model

Formulation for a Fixed Indenter Radius

Forward Analysis

As can be seen in the normalized indentation load-depth relationships in Fig. 3, the substrate effect as well as the spherical geometry indenter plays a significant role in creating a nonlinear force-displacement relationship (Chen et al. 2007). In comparing the curvatures for EMR = 0.25 in Fig. 4a and EMR = 15.0 in Fig. 4b, as the substrate becomes more flexible (less stiff), the plots take on a more linear formulation, which is due to the loss in substrate effect and the indentation behavior becomes dominated by the film-bending effect.

The presence of prestress also affects the indentation force, and the effect is more prominent for the more compliant substrate. As the film prestress increases, the normalized indentation force also increases so as to overcome the tension in film. Through the fitting of extensive numerical simulations, for $\delta_1 = h/2$ and $\delta_2 = h/4$, a set of general functions was formulated according to Eqs. (4) and (5):

$$\frac{P_i}{E_f \delta_i^2} = f_i\left(\frac{E_f}{E_s}, \frac{\sigma_{ps}}{E_f}\right) = A_i\left(\frac{E_f}{E_s} - B_i\right)^{C_i} + D_i\left(\frac{E_f}{E_s}\right)^2 + E_i\left(\frac{E_f}{E_s}\right) + F_i \tag{6}$$

$$i = 1, 2 \text{ for } \delta_1 = \frac{h}{2} \text{ and } \delta_2 = \frac{h}{4}$$

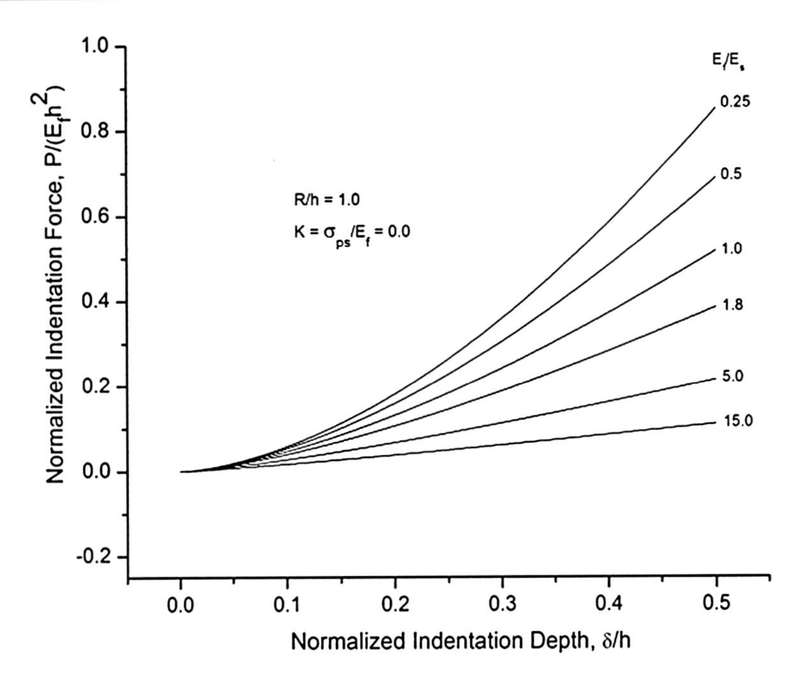

Fig. 3 Normalized indentation load – penetration depth for various elastic modulus ratios (EMR), with $R/h = 1.0$, $K = 0$

with

$$
\begin{aligned}
A_i &= a_{1_i} K + a_{2_i} K^2 + a_{3_i} K^3 + a_{4_i} K^4 \\
B_i &= b_{1_i} K + b_{2_i} K^2 + b_{3_i} K^3 + b_{4_i} K^4 \\
C_i &= C_{1_i} (c_{2_i} - K)^{C_{s_i}} \\
D_i &= d_{1_i} K + d_{2_i} K^2 + d_{3_i} K^3 + d_{4_i} K^4 \\
E_i &= e_{1_i} K + e_{2_i} K^2 + e_{3_i} K^3 + e_{4_i} K^4 \\
F_i &= \text{constant}
\end{aligned}
$$

where coefficients $a_{j_i}, b_{j_i}, c_{k_i}, d_{j_i}, e_{j_i}$ $(i = 1\text{–}2, j = 1\text{–}4, k = 1\text{–}3)$ are fitted constants as shown in Table 1. The representative Eq. (6) is also shown as surface plots in Fig. 5a, b, which correspond to prescribed indenter displacements of δ_1 and δ_2, respectively. The data points generated from the forward analysis (via numerical simulation) are also shown, with excellent agreement with the surface plots.

Reverse Analysis

The two independent equations in (6) allow the solution of any two of the three unknowns E_f, E_s, or σ_{ps}, as long as the third term is known accurately. The reverse

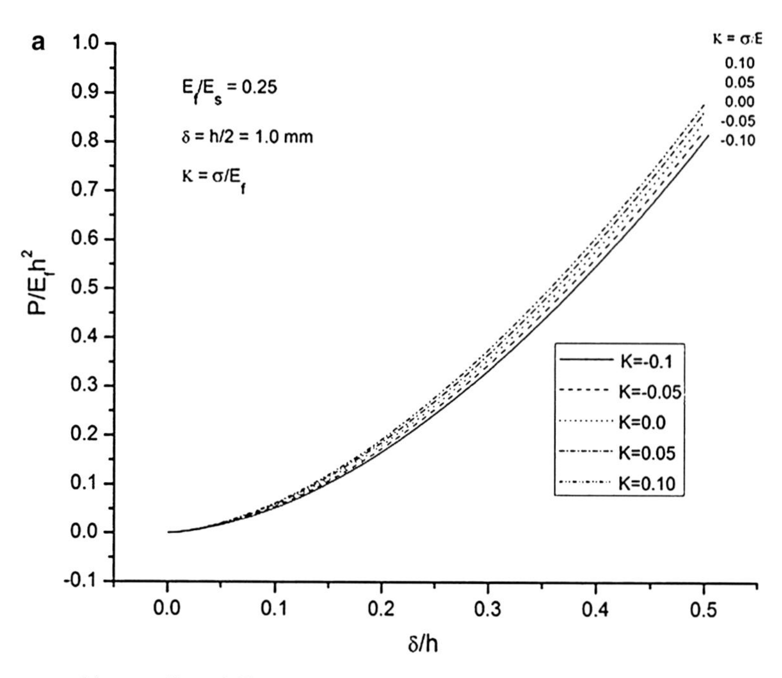

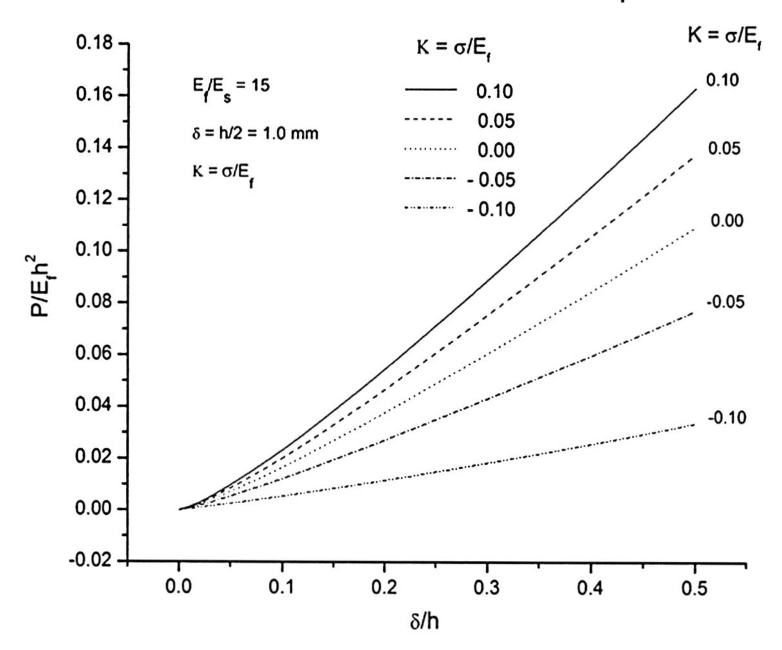

Fig. 4 Normalized load versus normalized indentation depth for various normalized prestress values at (**a**) $E_f/E_s = 0.25$ and (**b**) $E_f/E_s = 15.0$

Table 1 General coefficients for EMR-K fitting function

F_1	−0.138142626	F_2	−0.27345413
a_{01}	2.674281604	a_{02}	4.034252118
a_{11}	0.802831035	a_{12}	1.147974895
a_{21}	1.971999573	a_{22}	5.222286008
a_{31}	−8.544787918	a_{32}	−29.25982734
a_{41}	44.46597133	a_{42}	131.1351472
b_{01}	−0.383265786	b_{02}	−0.568865038
b_{11}	0.549111802	b_{12}	0.763772049
b_{21}	−1.29341713	b_{22}	−2.600832556
b_{31}	3.373917099	b_{32}	10.53364711
b_{41}	−24.85527757	b_{42}	−60.95042138
c_{11}	−0.35865278	c_{12}	−0.29205364
c_{21}	−0.434558478	c_{22}	−0.295299789
c_{31}	−0.63882907	c_{32}	−0.470757762
d_{01}	−0.000225813	d_{02}	−0.000201266
d_{11}	−0.003103271	d_{12}	−0.003870197
d_{21}	0.003580859	d_{22}	−0.00153943
d_{31}	0.102776788	d_{32}	0.277177385
d_{41}	−0.877371942	d_{42}	−2.443837931
e_{01}	0.008209149	e_{02}	0.008712058
e_{11}	0.110259779	e_{12}	0.157132416
e_{21}	−0.24475966	e_{22}	−0.276338029
e_{31}	−0.115013816	e_{32}	−1.035169875
e_{41}	3.664067224	e_{42}	13.69568226

analysis is based on minimizing the total error of the two equations (Mills and Chen 2009), utilizing P-δ data from the corresponding numerical indentation experiment (assuming h is known and $R/h = 1$ is used in the experiment). To verify the reverse analysis procedure, a number of separate numerical indentation tests were performed, using EMR and K parameters that were independent from the parameters used in the forward analysis.

As a first example, assuming E_f is known, the error associated with E_s and σ_{ps} deduced from reverse analysis (with respect to that used in forward analysis) is shown in Fig. 6a, with most errors less than 3% and maximum error about 7%. In the second example (Fig. 6b), with E_s known, the error associated with the reverse analysis of E_f and σ_{ps} is mostly below 2%, and the maximum error is about 6%. Finally, if the prestress is known in advance, the accuracy for determining the film and substrate moduli is very good, with errors well below 1% in most cases (Fig. 6c).

A close examination shows that the maximum errors tend to occur at the edges of the defined problem parameters. That is, the relatively more prominent errors (though only about 7%) are likely to occur at high compressive prestress and at a correspondingly high film moduli (recall that the parameter ranges investigated

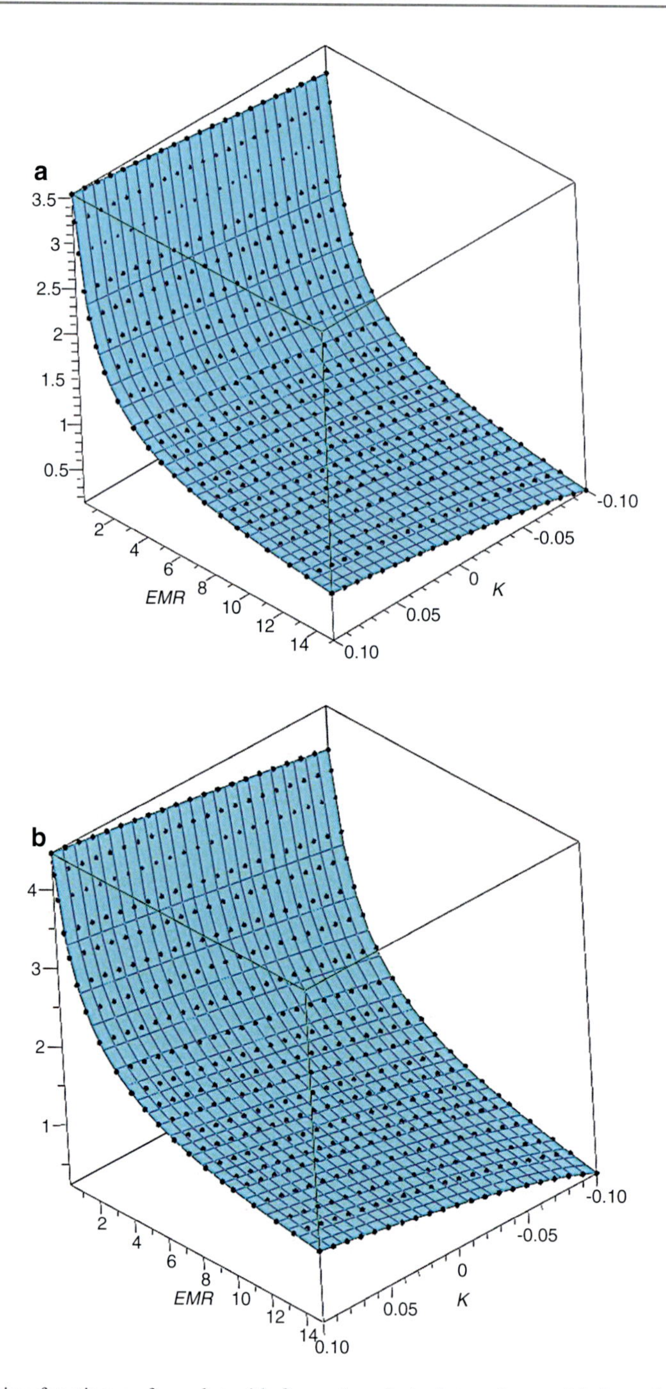

Fig. 5 Fitting function surface plot with forward analysis data points at (**a**) $\delta_1 = h/2$ and (**b**) $\delta_2 = h/4$

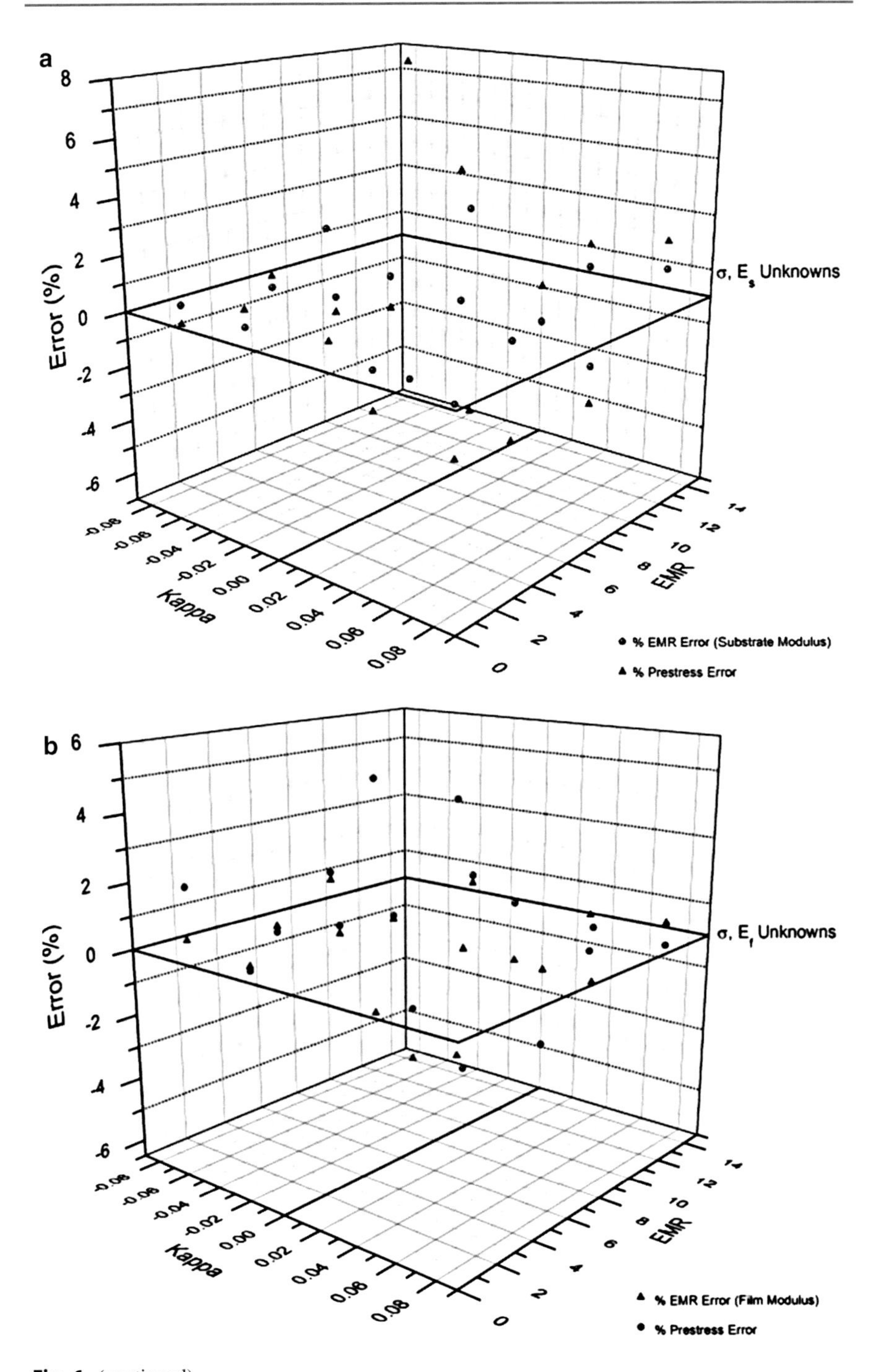

Fig. 6 (continued)

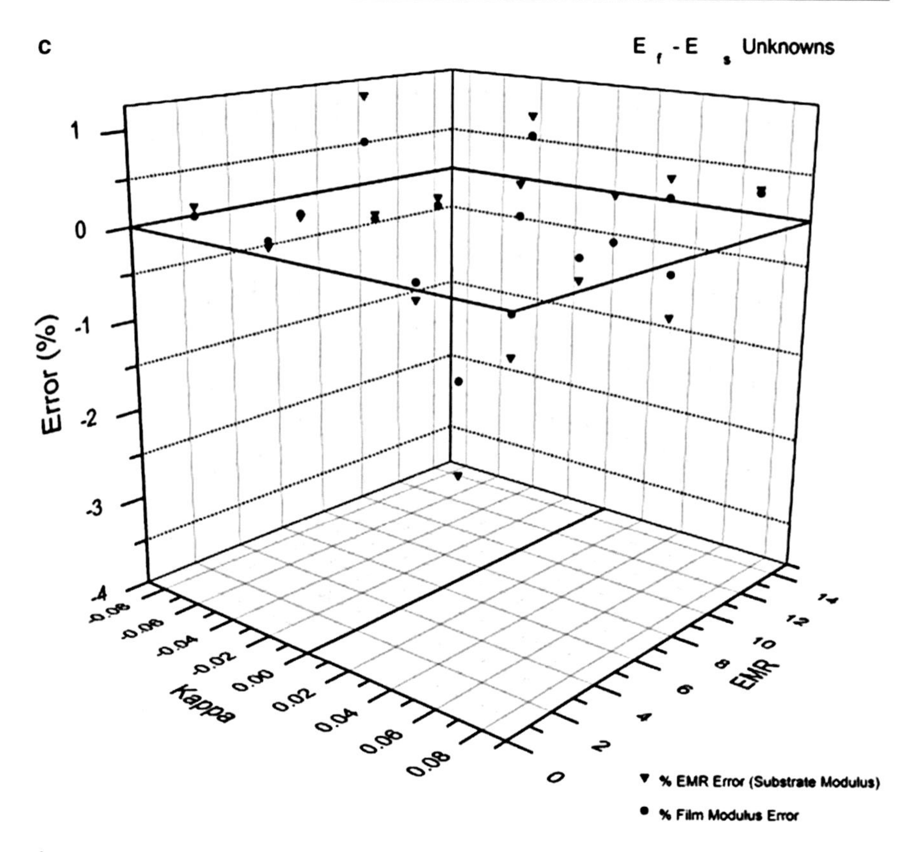

Fig. 6 Reverse analysis error results (**a**) with σ and E_s as unknowns; (**b**) with σ and E_f as unknowns; (**c**) with E_s and E_f as unknowns

in this study are $0.25 \leq \frac{E_f}{E_s} \leq 15.00$ and $-0.10 \leq K \leq 0.10$). This may be due to the inherent instability as the normalized prestress approaches -0.10. Overall, the reverse analysis algorithm is fairly robust and is valid for moderately large ranges of film and substrate moduli and film prestress.

General Formulation with Variable Indenter Radius

Forward Analysis

In many instances, the film thickness may not be known or may not able to be measured. Under certain circumstances, it would be useful to extend the functional relationships developed in section "Formulation for a Fixed Indenter Radius" to the problem of unknown film thickness (or conversely to remove the testing constraint of having to match the indenter radius to the film thickness). In order to show the feasibility without complicating the formulation too much, a proof-of-principle investigation is conducted in this section by choosing a fixed EMR value (EMR = 2.5) whereas allowing K and R/h to vary as $-0.10 \leq K \leq 0.10$ and R/h of 0.5, 0.75, 1.0, 2.0, 5.0, and 10.0, respectively. Plots of the normalized load versus prestress level are given in Fig. 7a, b for the two indentation depths. Each plot shows various sets of curves that were generated at different R/h values, and the separation between them provides confidence for deducing R/h through reverse analysis.

Through extensive forward analyses, the following equations are established:

$$\frac{P_i}{E_f \delta_i^2} = M_i + N_i K + Q_i K^2 \text{ where } i = 1, 2 \text{ and } \delta_1 = h/2 \text{ and } \delta_2 = h/4 \tag{7}$$

$$\begin{aligned} M_i &= m_{1_i} e^{[(R/h)/t_{1_i}]} + m_{2_i} e^{[(R/h)/t_{2_i}]} + y_{0_i} \\ N_i &= n_{1_i} e^{[(R/h)/t_{3_i}]} + n_{2_i} e^{[(R/h)/t_{4_i}]} + y_{1_i} \\ Q_i &= q_{1_i} + q_{2_i} (R/h) + q_{3_i} q_{4_i}^{(R/h)} \end{aligned}$$

Coefficients in M_i, N_i, and Q_i are shown in Table 2.

Reverse Analysis

By performing a single indentation test, the solution of two variables, σ_{ps} and h, can be obtained. In Fig. 8, R/h values were chosen to be 0.6, 4.0, and 9.0 with corresponding σ_{ps}/E_f values of −0.1, 0, and 0.1. Through the minimization of error of Eq. (7), the reverse analysis result errors (deduced parameter vs. the input parameter) are mostly below 1–2%, with the largest error of about 6%. Note that since h is an unknown parameter to be determined, the reverse analysis algorithm, Fig. 9, is an iterative, averaging technique with the only requirement being that the indentation experiment (FEA) be performed to a depth at least as deep as $\delta = h/2$.

Note that (7) is specified for EMR = 2.5; nevertheless, the approach can be readily extended to other EMR values and that would add another dimension in reverse analysis, making the formulation more complicated, yet the strategy remains straightforward.

Error Sensitivity

When performing an indentation study, the accuracy of the P-δ data (P_1, P_2, δ_1, δ_2) is assumed to be exact. In practice, measurement errors can affect the accuracy of the deduced material properties. For these reasons, the error sensitivity

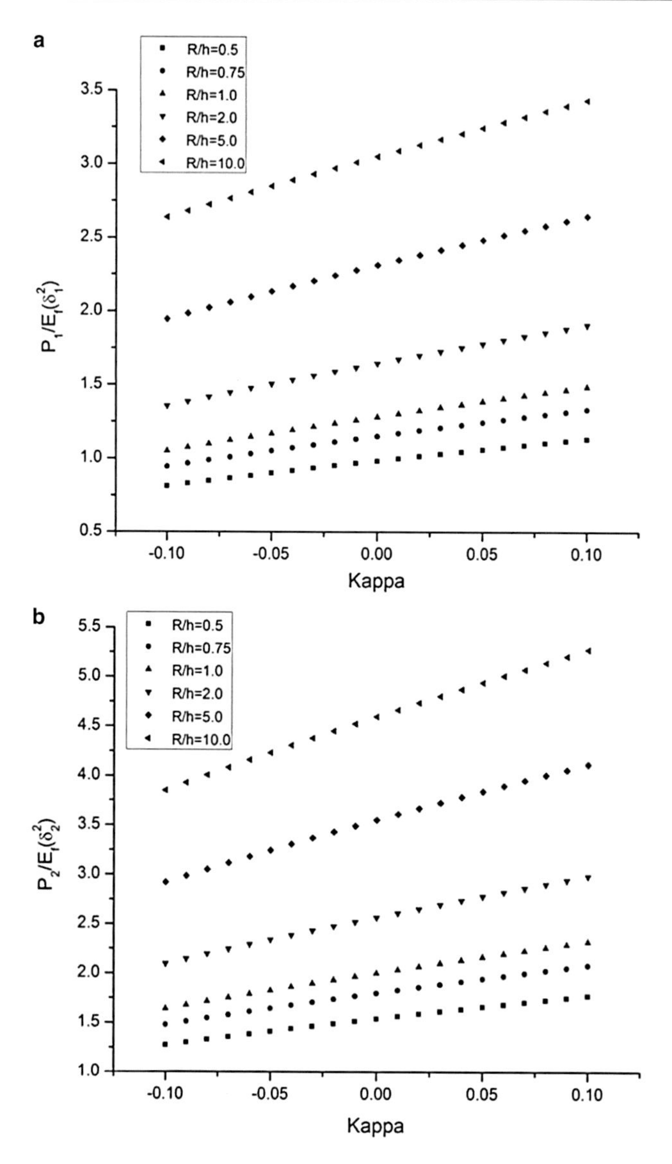

Fig. 7 Normalized indentation load versus $K = \sigma_{ps}/E_f$, at $0.5 \leq R/h \leq 10.0$, (**a**) $\delta_1 = h/2$. (**b**) $\delta_2 = h/4$

and its significance to each variable in the fitting function should be examined. Note that in this analysis the film thickness is assumed to be known, and the indenter radius is matched accordingly.

Differentiating the dimensionless functions from (4) and (5) term by term, we get the following:

Table 2 Coefficients for R/h-K fitting function

M_i			
Y_{01}	4.292937	Y_{02}	6.174527
m_{11}	−3.15663	m_{21}	−4.35566
m_{21}	−0.64938	m_{22}	−1.03513
t_{11}	−10.7228	t_{12}	−9.8578
t_{21}	−0.63728	t_{22}	−0.6755
N_i			
Y_{11}	4.163368	Y_{12}	7.747632
n_{11}	−1.4351	n_{12}	−2.01002
n_{21}	−2.19215	n_{22}	−4.76786
t_{31}	−0.59133	t_{32}	−0.66103
t_{41}	−4.20076	t_{42}	−5.08646
Q_i			
q_{11}	−1.70805	q_{12}	−3.39752
q_{21}	0.018993	q_{22}	−6.92E-04
q_{31}	0.807144	q_{32}	2.102068
q_{41}	0.475559	q_{42}	0.582964

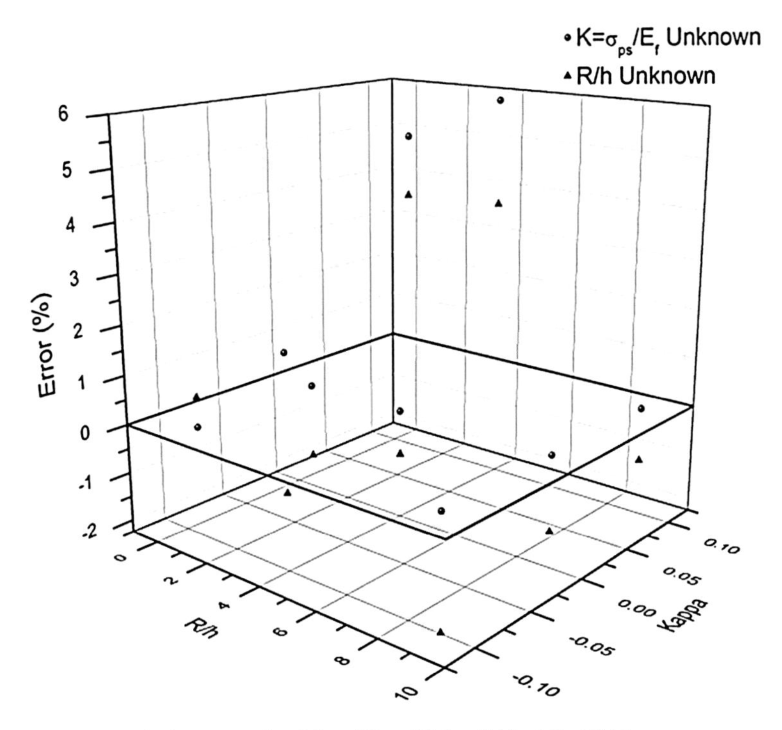

Fig. 8 Reverse analysis – error plot, $0.5 \le R/h \le 10.0$, $-0.10 \le K \le 0.10$

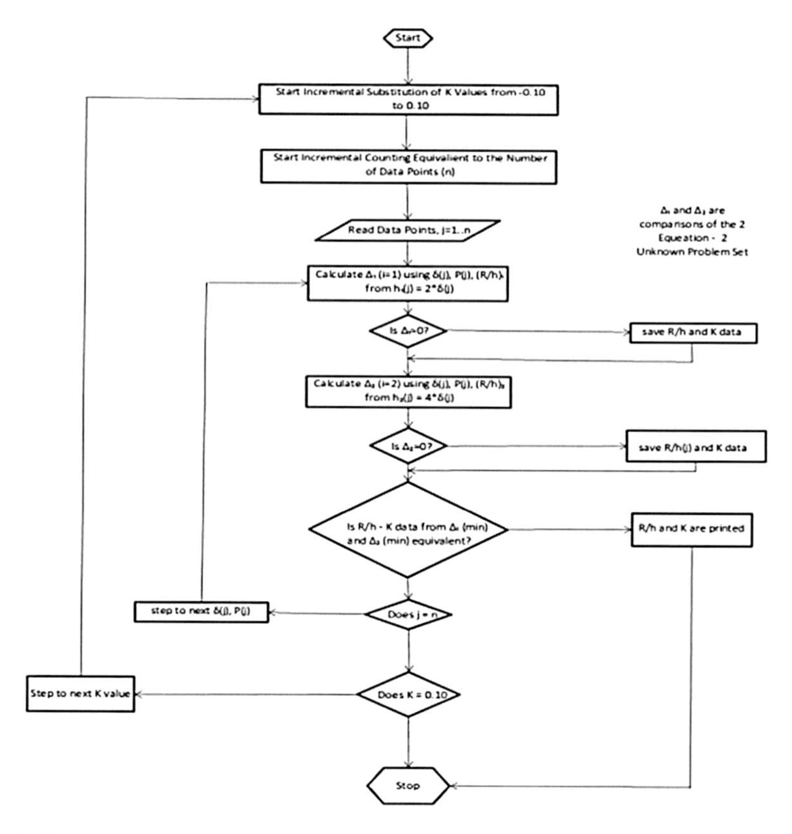

Fig. 9 General algorithm – reverse analysis

$$
\begin{aligned}
\frac{dP_1}{E_f\delta_1^2} - \frac{P_1 dE_f}{E_f^2\delta_1^2} - \frac{2P_1 d\delta_1}{E_f\delta_1^3} &= \frac{\partial\left[f_1\left(\frac{E_f}{E_s},\frac{\sigma_{ps}}{E_f}\right)\right]}{\partial\left(\frac{E_f}{E_s}\right)} d\left(\frac{E_f}{E_s}\right) \\
&\quad + \frac{\partial\left[f_1\left(\frac{E_f}{E_s},\frac{\sigma_{ps}}{E_f}\right)\right]}{\partial\left(\frac{\sigma_{ps}}{E_f}\right)} d\left(\frac{\sigma_{ps}}{E_f}\right)
\end{aligned}
\tag{8}
$$

$$
\begin{aligned}
\frac{dP_2}{E_f\delta_2^2} - \frac{P_2 dE_f}{E_f^2\delta_2^2} - \frac{2P_2 d\delta_2}{E_f\delta_2^3} &= \frac{\partial\left[f_2\left(\frac{E_f}{E_s},\frac{\sigma_{ps}}{E_f}\right)\right]}{\partial\left(\frac{E_f}{E_s}\right)} d\left(\frac{E_f}{E_s}\right) \\
&\quad + \frac{\partial\left[f_2\left(\frac{E_f}{E_s},\frac{\sigma_{ps}}{E_f}\right)\right]}{\partial\left(\frac{\sigma_{ps}}{E_f}\right)} d\left(\frac{\sigma_{ps}}{E_f}\right)
\end{aligned}
\tag{9}
$$

The potential errors due to indenter force and displacement ($\partial P_1/P_1$, $\partial P_2/P_2$, $\partial\delta_1/\delta_1$, $\partial\delta_2/\delta_2$) may arise from measurement errors, noise, edge effects, etc. In rearranging these terms and combining them, they can be related to the perturbation of unknowns, $\partial E_f/E_f$, $\partial E_s/E_s$, $\partial\sigma_{ps}/E_f$.

As an example, if the film modulus is known directly, then $dE_f = 0$. And Eqs. (8) and (9) can be solved directly as two independent equations with two unknowns. In effect the film prestress and substrate modulus are solved in terms of the perturbation forces and displacements:

$$\frac{d\sigma_{ps}}{E_f} = \alpha_1 \frac{dP_1}{P_1} + \alpha_2 \frac{d\delta_1}{p\delta_1} + \alpha_3 \frac{dP_2}{P_2} + \alpha_4 \frac{d\delta_2}{\delta_2} \tag{10}$$

where:

$$\alpha_1 = \frac{\Omega_1 Q_2}{w}, \alpha_2 = \frac{-2\Omega_1 Q_2}{w}, \alpha_3 = \frac{-\Omega_2 Q_1}{w}, \alpha_4 = \frac{2\Omega_2 Q_1}{w}$$

with $w = (Q_2 V_1 - Q_1 V_2)$

$$\frac{dE_s}{E_s} = \beta_1 \frac{dP_1}{P_1} + \beta_2 \frac{d\delta_1}{\delta_1} + \beta_3 \frac{dP_2}{P_2} + \beta_4 \frac{d\delta_2}{\delta_2}$$

$$\beta_1 = \frac{\Omega_1 V_2}{X}, \beta_2 = \frac{-2\Omega_1 V_2}{X}, \beta_3 = \frac{-\Omega_2 V_1}{X}, \beta_4 = \frac{\Omega_2 V_1}{X}$$

with $X = \frac{E_f}{E_s}(Q_2 V_1 - Q_1 V_2)\, f_i\left(\frac{E_f}{E_s}, \frac{\sigma_{ps}}{E_f}\right)$ is as shown in Eq. (6). And Q_i and V_i are:

$$Q_i = \frac{\partial\left[f_i\left(\frac{E_f}{E_s}, \frac{\sigma_{ps}}{E_f}\right)\right]}{\partial\left[\frac{E_f}{E_s}\right]} = A_i C_i \left(\frac{E_s}{E_f} = B_i\right)^{C_i} + 2D_i\left(\frac{E_s}{E_f}\right) + E_i$$

$$V_i = \frac{\partial A_i}{\partial\left(\frac{\sigma_{ps}}{E_f}\right)}\left[\frac{E_f}{E_s} - B_i\right]^{C_i}$$

$$+ A_i\left\{-C_i\left(\frac{E_f}{E_s} - B_i\right)^{C_i - 1}\frac{\partial B_i}{\partial\left(\frac{\sigma_{ps}}{E_f}\right)} + \left(\frac{E_f}{E_s} - B_i\right)^{C_i - 1}\ln\left(\frac{E_f}{E_s} - B_i\right)\frac{\partial C_i}{\partial\left(\frac{\sigma_{ps}}{E_f}\right)}\right\}$$

$$+ \left(\frac{E_f}{E_s}\right)^2 \frac{\partial D_i}{\partial\left(\frac{\sigma_{ps}}{E_f}\right)} + \left(\frac{E_f}{E_s}\right)\frac{\partial E_i}{\partial\left(\frac{\sigma_{ps}}{E_f}\right)} + \frac{\partial F_i}{\partial\left(\frac{\sigma_{ps}}{E_f}\right)}$$

with $A_i, B_i, C_i, D_i, E_i \quad i = 1, 2$ as shown in (6) and

$$\frac{\partial A_i}{\partial\left(\frac{\sigma_{ps}}{E_f}\right)} = a_{1_i} + 2a_{2_i}\left(\frac{\sigma_{ps}}{E_f}\right) + 3a_{3_i}\left(\frac{\sigma_{ps}}{E_f}\right)^2 4a_{4_i}\left(\frac{\sigma_{ps}}{E_f}\right)^3$$

$$\frac{\partial B_i}{\partial\left(\frac{\sigma_{ps}}{E_f}\right)} = b_{1_i} + 2b_{2_i}\left(\frac{\sigma_{ps}}{E_f}\right) + 3b_{3_i}\left(\frac{\sigma_{ps}}{E_f}\right)^2 4b_{4_i}\left(\frac{\sigma_{ps}}{E_f}\right)^3$$

$$\frac{\partial C_i}{\partial\left(\frac{\sigma_{ps}}{E_f}\right)} = c_{1_i} c_{3_i}\left(\frac{\sigma_{ps}}{E_f} - c_{2_i}\right)^{c_{s_i}-1}$$

$$\frac{\partial D_i}{\partial\left(\frac{\sigma_{ps}}{E_f}\right)} = d_{1_i} + 2d_{2_i}\left(\frac{\sigma_{ps}}{E_f}\right) + 3d_{3_i}\left(\frac{\sigma_{ps}}{E_f}\right)^2 4d_{4_i}\left(\frac{\sigma_{ps}}{E_f}\right)^3$$

$$\frac{\partial E_i}{\partial\left(\frac{\sigma_{ps}}{E_f}\right)} = e_{1_i} + 2e_{2_i}\left(\frac{\sigma_{ps}}{E_f}\right) + 3e_{3_i}\left(\frac{\sigma_{ps}}{E_f}\right)^2 4e_{4_i}\left(\frac{\sigma_{ps}}{E_f}\right)^3$$

$$\frac{\partial F_i}{\partial\left(\frac{\sigma_{ps}}{E_f}\right)} = 0$$

$i = 1, 2$ and $a_{r_i}, b_{r_i}, c_{q_i}, d_{r_i}, e_{r_i}$ $(r = 1 - 4, q = 1 - 3)$are listed in Table 1.

The terms α_j and β_j are coefficients for the perturbated errors of indentation load and displacement measurements at δ_1 and δ_2. They can be called error sensitivity coefficients. By examining these coefficients, one may reach an understanding of the scale of error that each physical measurement could contribute to the overall error of our problem solution. As an example, if the normalized error in measuring the indentation force at δ_1, dP_1/P_1, is 1%, then the overall contribution to the error in the solution for the prestress is α_1%. And the contribution to the error in the substrate modulus is β_1%. Therefore the smaller the error sensitivity coefficient, the less sensitive the solution algorithm is to the specific perturbation error. Moreover, the perturbation error due to measuring indentation forces at δ_1 and δ_2 would most likely have the same order of magnitude and occur together in a test. The same should be true for the errors in measuring indentation depth. Therefore error plots are shown with combined error coefficients. For example, this would then be the associated error of the prestress due to the combined perturbated normalized indentation force measurements dP_1 and dP_2: shown as $\alpha_1 + \alpha_3$. In a similar fashion, $\alpha_2 + \alpha_4$, $\beta_1 + \beta_3$, and $\beta_2 + \beta_4$ represent the combined normalized indentation depth measurement error for prestress, combined normalized indentation force measurement error for substrate modulus, and combined normalized indentation depth measurement error for substrate modulus. Representative plots are shown in Fig. 10.

In examining these error plots, it can be seen that the error sensitivity in general is relatively low. However, there are regions where the potential error shown could be quite high. However, if two of the three variables are known, the potential areas of high error sensitivity are no longer present (this is discussed in a later part of this paper).

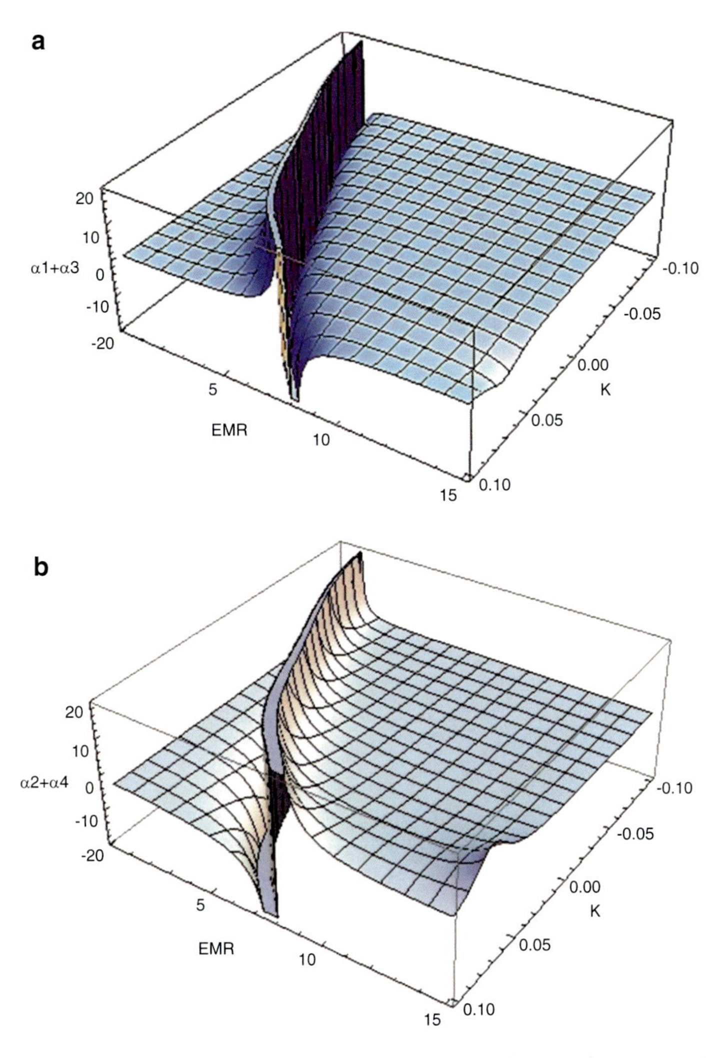

Fig. 10 Plots for (**a**) $\alpha1 + \alpha3$ and (**b**) $\alpha2 + \alpha4$ for the potential error in prestress (σ_{ps}) from measured error in indenter force, with $K = \sigma_{ps}/E_f$, $EMR = E_f/E_s$

Another example with regard to understanding measurement error sensitivity is the case where the substrate modulus is known. In this case, $dE_s = 0$. Therefore Eqs. (8) and (9) can be solved directly for the perturbated normalized film modulus and prestress:

$$\frac{dE_f}{E_f} = \xi_1 \frac{dP_1}{P_1} + \xi_2 \frac{d\delta_1}{\delta_1} + \xi_3 \frac{dP_2}{P_2} + \xi_4 \frac{d\delta_2}{\delta_2} \tag{11}$$

where:

$$\xi_1 = \frac{\Omega_1 V_2}{Y}, \quad \xi_2 = \frac{-2\Omega_1 V_2}{Y}, \quad \xi_3 = \frac{-\Omega_2 V_1}{Y}, \quad \xi_4 = \frac{2\Omega_2 V_1}{Y}$$

$$\text{with } Y = \left[\Omega_1 V_2 - Q_2 V_1 + \frac{E_f}{E_s}\left(V_2 Q_1 - V_1 Q_2\right)\right]$$

$$\frac{d\sigma_{ps}}{E_f} = \varepsilon_1 \frac{dP_1}{P_1} + \varepsilon_2 \frac{d\delta_1}{\delta_1} + \varepsilon_3 \frac{dP_2}{P_2} + \varepsilon_4 \frac{d\delta_2}{\delta_2}$$

where:

$$\varepsilon_1 = \frac{X_2}{U_1}, \quad \varepsilon_2 = \frac{-2X_2}{U_1}, \quad \varepsilon_3 = \frac{-X_1}{U_1}, \quad \varepsilon_4 = \frac{2X_1}{U_1}$$

$$\text{with } U_1 = \frac{X_2 V_1}{\Omega_1} - \frac{X_1 V_2}{\Omega_2}, \quad X_1 = \left(1 + \frac{Q_1 E_f}{\Omega_1 E_s} - \frac{V_1 \sigma_{ps}}{\Omega_1 E_f}\right),$$

$$X_2 = \left(1 + \frac{Q_2 E_f}{\Omega_2 E_s} - \frac{V_2 \sigma_{ps}}{\Omega_2 E_f}\right)$$

As a further example of the potential errors associated with the measurement of indentation data, consider the case where the prestress is known. In this case, $d\sigma_{ps} = 0$. By knowing σ_{ps}, then Eqs. (8) and (9) can be solved directly for the normalized perturbation of both the film and substrate moduli:

$$\frac{dE_f}{E_f} = \tau_1 \frac{dP_1}{P_1} + \tau_2 \frac{d\delta_1}{\delta_1} + \tau_3 \frac{dP_2}{P_2} + \tau_4 \frac{d\delta_2}{\delta_2} \tag{12}$$

where:

$$\tau_1 = \frac{Q_2 \Omega_1}{U_2}, \tau_1 = \frac{-2Q_2 \Omega_1}{U_2}, \tau_1 = \frac{-Q_1 \Omega_2}{U_2}, \tau_1 = \frac{2Q_1 \Omega_2}{U_2}$$

$$\text{with } U_2 = Q_2\left(\Omega_1 - V_1 \frac{\sigma_{ps}}{E_F}\right) - Q_1\left(\Omega_2 - V_2 \frac{\sigma_{ps}}{E_F}\right)$$

and

$$\frac{dE_s}{E_s} = \eta_1 \frac{dP_1}{P_1} + \eta_2 \frac{d\delta_1}{\delta_1} + \eta_3 \frac{dP_2}{P_2} + \eta_4 \frac{d\delta_2}{\delta_2}$$

where:

$$\eta_1 = \frac{R_2}{\psi_1}, \eta_2 = \frac{-2R_2}{\psi_1}, \eta_3 = \frac{-R_1}{\psi_1}, \eta_4 = \frac{2R_1}{\psi_1}$$

$$\text{with } \psi_1 = \frac{E_f}{E_s}\left(\frac{R_1 Q_2}{\Omega_2} - \frac{R_2 Q_1}{\Omega_1}\right), \quad R_1 = \left(1 + \frac{Q_1 E_f}{\Omega_1 E_s} - \frac{V_1 \sigma_{ps}}{\Omega_1 E_f}\right),$$
$$R_2 = \left(1 + \frac{Q_2 E_f}{\Omega_2 E_s} - \frac{V_2 \sigma_{ps}}{\Omega_2 E_f}\right)$$

Again, in solving for either normalized prestress and film modulus or normalized film and substrate modulus, the potential for errors is in general quite low. Nevertheless, there is a small risk for significant error perturbation at two extremes. The first location for potentially significant error is found where the EMR is low combined with high *K*. The second instance is found at a very high compressive

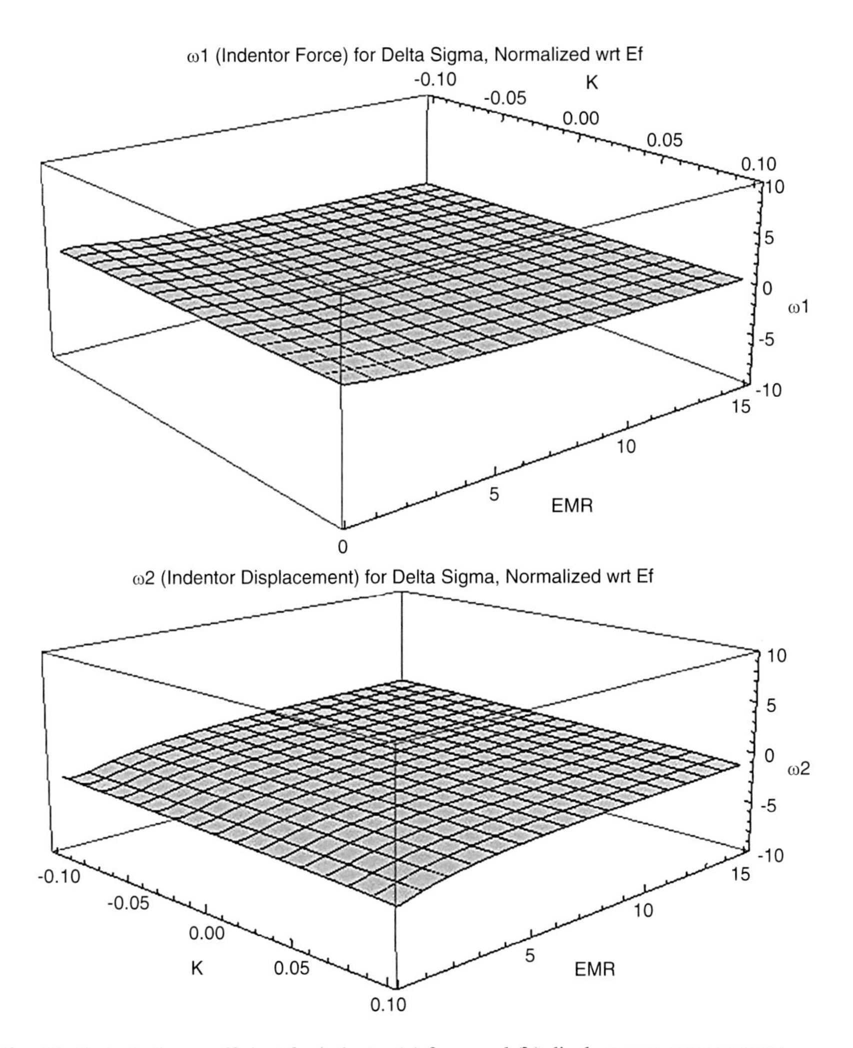

Fig. 11 Perturbation coefficient for indenter (**a**) force and (**b**) displacement measurement

prestress (low K) combined with a very high film (stiff) modulus (high EMR). In the latter case, a practical application may be that the high compressive stress may tend to cause instabilities and in doing so affect the measured indentation force and displacement measurement. In the first instance, the substrate effect is low, and the indentation behavior is dominated by bending of prestretched film (where perhaps better functional forms should be developed instead of the indentation ones).

The above analysis assumes that there are two unknowns that require solution. However there may be instances where there is only one unknown variable. As an example, if the substrate and film moduli are both known, then the potential error in prestress with respect to perturbations for force and displacement can be shown as:

$$\frac{d\sigma_{ps}}{E_f} = \omega 1 \frac{dP_1}{P_1} + \omega 2 \frac{d\delta_1}{\delta_1} \tag{13}$$

where:

$$\omega_1 = \frac{\Omega_1}{V_1}, \quad \omega_2 = -2\frac{\Omega_1}{V_1}$$

with Ω_1 and V_1 as noted above. ω_1 and ω_2 are plotted in Fig. 11 and show that the error sensitivity coefficients are found to be stable and reasonably low, without the potential peaks noted in the two-variable error analysis.

Conclusion

Internal stress of thin films can often lead to premature failure or malfunction of a component. Meanwhile, the substrate properties are sometimes as important as that of the film. It is therefore useful to develop an effective indentation framework to probe both the substrate property and film prestress. Based on a comprehensive numerical analysis, this paper establishes algorithms to effectively deduce film/substrate elastic moduli and equibiaxial prestress. Through a systematic forward analysis, a set of fitting equations were developed for specified indentation depths at one half and one quarter of the film thickness. By incorporating spherical indentation and examining the indentation force at a moderate depth, the effects of both the film and substrate are taken into account. After the force at these specified indentation depths is measured, through the reverse analysis algorithm, one may effectively predict two of the three unknowns, film modulus, substrate modulus, or film prestress, with high accuracy. An extensive examination of error sensitivity was also made which showed that at areas of high instability (high compressive prestress), accurately measuring indentation force and depth is critical. However if two of the three properties (E_f, E_s) are known precisely, the film prestress is shown to be insensitive to minor measurement errors and should provide a robust and consistent solution. The overall strategy can be readily extended to probing elastoplastic properties in future studies.

Acknowledgments The study is supported by the National Natural Science Foundation of China (11172231), DARPA (W91CRB-11-C-0112), the World Class University Program through the National Research Foundation of Korea (R32-2008-000-20042-0), and the National Science Foundation (CMMI-0643726).

References

J.L. Bucaille, S. Stauss, E. Felder, J. Michler, Determination of plastic properties of metals by instrumented indentation using different sharp indenters. Acta Mater. **51**, 1663–1678 (2003)
X. Chen, J.J. Vlassak, Numerical study on the measurement of thin film mechanical properties by means of nanoindentation. J. Mater. Res. **16**, 2974–2982 (2001)
X. Chen, Y. Xiang, J.J. Vlassak, A novel technique to measure mechanical properties of porous materials by nanoindentation. J. Mater. Res. **21**(3), 715–724 (2006)
X. Chen, N. Ogasawara, M. Zhao, N. Chiba, On the uniqueness of measuring Elastoplastic properties from indentation: the indistinguishable mystical materials. J. Mech. Phys. Solids **55**, 1618–1660 (2007)
Y.T. Cheng, C.M. Cheng, Analysis of indentation loading curves obtained using conical indenters. Philos. Mag. Lett. **77**(1), 39–47 (1998)
Y.T. Cheng, C.M. Cheng, Scaling, dimensional analysis, and indentation measurements. Mat. Sci. Eng. **R44**, 91–149 (2004)
Dassault Systèmes (SIMULIA), *ABAQUS 6.8 Documentation* (Dassault Systèmes (Simulia) Corporation, Providence, 2008)
F. Doerner, W.D. Nix, A method for interpreting the data from depth-sensing indentation instruments. J. Mat. Res. **1**(4), 601–609 (1986)
A.G. Evans, J.W. Hutchinson, On the mechanics of delamination and spalling in compressed films. Int. J. Solids Struct. **20**(5), 455–466 (1984)
H. Gao, C. Cheng-Hsin, L. Jin, Elastic contact versus indentation modeling of multi-layered materials. Int. J. Solids Struct. **29**, 2471–2492 (1992)
A. Gouldstone, N. Chollacoop, M. Dao, J. Li, A.M. Minor, Y.L. Shen, Indentation across size scales and disciplines: recent developments in experimentation and modeling. Acta Mater. **55**, 4015–4039 (2007)
X.Z. Hu, B.R. Lawn, A simple indentation stress-strain relation for contacts with spheres on bilayer structures. Thin Solid Films **322**(1–2), 225–232 (1998)
M.S. Hu, M.D. Thouless, A.G. Evans, The decohesion of thin films from brittle substrates. Acta Metall. **36**(5), 1301–1307 (1988)
J.A. Mills, X. Chen, Spherical indentation on an elastic coating/substrate system: determining substrate modulus. J. Eng. Mech. **135**(10), 1189–1197 (2009)
G.M. Pharr, Measurement of mechanical properties by ultra-low load indentation. Mat. Sci. Eng. A **253**, 151–159 (1998)
R. Saha, W.D. Nix, Effects of the substrate on the determination of thin film mechanical properties by nanoindentation. Acta Mater. **50**, 23–38 (2002)
R. Saha, Z. Xue, Y. Huang, W.D. Nix, Indentation of a soft metal on a hard substrate: Strain gradient hardening effects. J. Mech. Phys. Solids **49**, 1997–2014 (2001)
S. Suresh, A.E. Giannakopoulos, A new method for estimating residual stresses by instrumented sharp indentation. Acta Metall. **46**(16), 5755–5767 (1998)
J.G. Swadener, E.P. George, G.M. Pharr, The correlation of the indentation size effect measured with indenters of various shapes. J. Mech. Phys. Solids **50**, 681–694 (2002)
T.Y. Tsui, J.J. Vlassak, W.D. Nix, Indentation plastic displacement field: part I: the case of soft films on hard substrates. J. Mat. Res. **14**, 2196–2203 (1999)
M. Zhao, X. Chen, N. Ogasawara, A.C. Razvan, N. Chiba, D. Lee, Y.X. Gan, A new sharp indentation method of measuring the elastic-plastic properties of soft and compliant materials by using the substrate effect. J. Mat. Res. **21**, 3134–3151 (2006a)

M. Zhao, N. Ogasawara, N. Chiba, X. Chen, A new approach to measure the elastic-plastic properties of bulk materials using spherical indentation. Acta Mater. **54**, 23–32 (2006b)

M. Zhao, X. Chen, Y. Xiang, J.J. Vlassak, D. Lee, N. Ogasawara, N. Chiba, Y.X. Gan, Measuring elastoplastic properties of thin films on an elastic substrate using sharp indentation. Acta Mater. **55**, 6260–6274 (2007)

Experimentation and Modeling of Mechanical Integrity and Instability at Metal/Ceramic Interfaces

5

Wen Jin Meng and Shuai Shao

Contents

Abstract

Controlling the mechanical integrity of metal/ceramic interfaces is important for a wide range of technological applications. Achievement of such control requires a number of key elements, including establishing appropriate experimental protocols for quantifying mechanical response of metal/ceramic interfacial regions under well-defined loading conditions, understanding how interfacial compositional and structural characteristics impact such interfacial mechanical response, and elucidating unit interface physics and predicting interfacial mechanical response via development of multiscale physics-based models. Achieving this combined testing, understanding, and modeling will ultimately lead to effective

W. J. Meng (✉) · S. Shao
Department of Mechanical and Industrial Engineering, Louisiana State University, Baton Rouge, LA, USA
e-mail: wmeng1@lsu.edu; sshao@lsu.edu

G. Z. Voyiadjis (ed.), *Handbook of Nonlocal Continuum Mechanics for Materials and Structures*, https://doi.org/10.1007/978-3-319-58729-5_50

control of mechanical integrity of metal/ceramic interfaces and true interfacial engineering through targeted modification of the interfacial composition and structure.

Major breakthroughs in the improvement of interfacial mechanical integrity can be enabled by understanding and controlling key physical factors, including interfacial architectural and chemical features governing the mechanical response of metal/ceramic interfacial regions (MCIRs), thus leading to unprecedented interfacial mechanical performance that meets/exceeds the demands of future applications. Guided by a multiscale integrated computational materials engineering (ICME) framework, the mechanical integrity of MCIRs can be substantially improved by a variety of architectural and chemical enhancements/refinements. Recent research efforts by the authors aim to provide a fundamental, physics-based understanding of the failure mechanisms of MCIRs by constructing a novel, multiscale, computation-guided, and experiment-validated ICME framework. Interfacial refinements to be explored within this framework include addition of alloying impurities as well as geometrical features such as multilayered and stepped interfacial architectures. The findings can then be consolidated into a high fidelity, experiment-validated, micro- and mesoscale modeling tool to significantly accelerate the discovery-design-implementation cycle of advanced MCIRs. In this chapter, we summarize some preliminary results on shear failure and instability of various metal/ceramic interfacial regions, outline the theoretical background of this research thrust, and identify challenges and opportunities in this area.

Keywords

Metal/ceramic interfaces · Microscale mechanical testing · Integrated computational materials engineering (ICME)

Introduction

Understanding and controlling mechanical integrity of solid/solid interfaces have remained a scientific challenge over the last three decades. The continued interest in this subject stems from the positive and significant impacts to wide ranging technological applications such understanding and control can bring. Notable technological examples include thermal barrier coatings (TBCs) for turbine components (Darolia 2013), hard ceramic coatings for machining tools (Holmberg and Matthews 2009), and hard/lubricious coatings for mechanical components such as gears and bearings (Mercer et al. 2003; Kotzalas and Doll 2010). In TBC systems with a ZrO_2-based ceramic top coating layer, various solid/solid interfaces are present: a metal/metal interface as represented by the bond coat/superalloy substrate interface; a metal/ceramic interface as represented by the thermally grown oxide (TGO)/bond coat interface; and a ceramic/ceramic interface as represented by the TBC top-layer/TGO interface. Failures at these interfaces constitute important mechanisms limiting the durability of TBC systems (Evans et al. 2008).

In coatings for machining tools and mechanical components, the application of thin hard ceramic coatings (CHCs) on metallic substrates likewise requires sufficient mechanical integrity of the interfacial regions between thin ceramic layers and metallic substrates. In the application of thin CHCs onto machining tools and mechanical components, adequate mechanical integrity of the metal/ceramic interfacial region (MCIR) constitutes a "go/no-go" requirement dictating whether a particular application can or cannot be adopted. In the past three decades, CHCs with thickness less than 10 μm have been implemented in many surface engineering applications. In such applications, thin interlayers are often used to promote adhesion between the ceramic coating layers and the substrates, forming thin-ceramic-layer/adhesion-interlayer/substrate sandwich structures (Jiang et al. 2003). A major failure mode for CHC/substrate systems, often with catastrophic consequences, is the mechanical failure of the MCIR, see for example Fig. 1a. Due to the present lack of understanding regarding key physical factors governing mechanical response of MCIRs, "good" or "bad" interfaces can only be distinguished through testing under actual application conditions, a comparative example is shown in Fig. 1a, b. The materials paradigm of synthesis-structure-property-performance circle is thus largely missing at the present time when it comes to MCIRs.

Manipulating interfacial chemistry and architecture may improve its shear/tensile failure stress. For instance, the introduction of low-energy ledges/steps as well as point defects has been demonstrated to control both the shear and dislocation-nucleation resistance of a sharp bi-metal interface (Zhang et al. 2016; Skirlo and Demkowicz 2013). Multiphase nano-adhesion interlayers can perform better than a compositionally sharp MCIR, as shown in Fig. 1a, b (Jiang et al. 2003). Intriguingly, strengthening effects are also observed in metal/ceramic nanolaminates, where the interfaces are arranged in a parallel fashion with nanoscale spacing (Li et al. 2014; Bhattacharyya et al. 2011). Recent work has shown that when the individual layer thickness of a metal/ceramic nano-laminate is very small, e.g., 2 nm, the ceramic layers co-deform plastically with the metal layers (Wang et al. 2017). In this case, under compressive loading normal to the interfaces, the nanolaminated metal/ceramic composites can reach an ultimate compressive stress of 4.7 GPa and a total compressive strain of 13% before the onset of instability, exhibited through the formation of shear bands (Fig. 1c–e). When the individual layer thickness is large, however, plastic co-deformation cannot be achieved which results in cracking in the ceramic layers. While such anecdotal evidence shows promise for engineering the mechanical integrity of MCIRs through interfacial architectural and chemical design, systematic research in this direction has not been carried out and engineering realization has seldomly been demonstrated. Little quantitative data exist at present on the characterization of interfacial mechanical response of coating/metal-adhesion-layer/substrate sandwich structures and on tailoring the chemistry and architecture of the MCIR to influence the interfacial mechanical response. Engineering of a particular coating/interlayer/substrate system has therefore largely proceeded in a trial-and-error manner, necessitating testing under actual application conditions, which is both time consuming and expensive.

Currently, understanding the mechanical response of interfacial regions of thin-ceramic-layer/metal-interlayer/substrate systems is hampered by a lack of

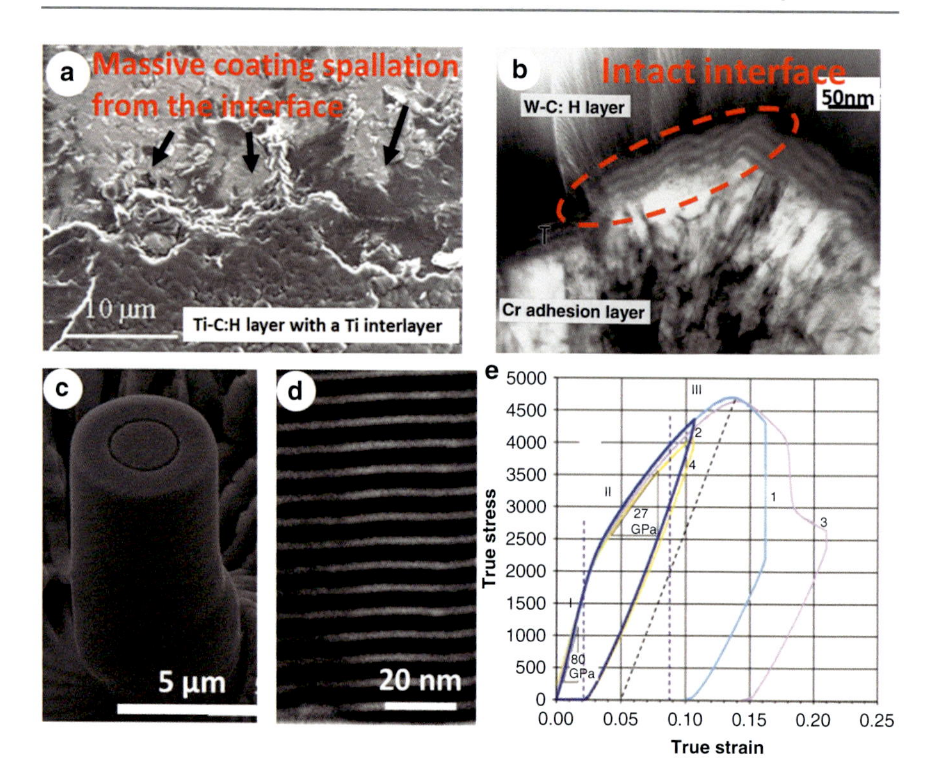

Fig. 1 (**a**) Contact fatigue testing of Ti-containing amorphous hydrogenated carbon (Ti-C:H) coated gears resulting in massive spallation failure at the Ti-C:H/Ti-interlayer interface; (**b**) intact metal/ceramic interfacial region (MCIR) after contact fatigue testing of W-containing amorphous hydrogenated carbon (W-C:H) coated gears. Here, the MCIR has a more complex architecture, with a Cr adhesion layer in contact with the steel substrate, followed by nano-laminated interlayers in contact with the W-C:H top coating layer (Jiang et al. 2003); (**c**–**e**) exceptional mechanical properties of metal/ceramic nano-laminates (Bhattacharyya et al. 2011)

experimental measurement protocols that are capable of generating quantitative data on interfacial mechanical response that can be easily interpreted. This understanding is further limited by a lack of physics-based modeling and simulations on the mechanical response of coating/substrate interfaces. Particularly lacking is an integration of experiments and modeling/simulation efforts, where information obtained from experiments probing the mechanical response of interfacial regions under sufficiently simple loading conditions can be coupled with outputs from multiscale modeling/simulation efforts, so as to develop deeper understandings of the key mechanisms governing mechanical response of interfacial regions under different loading conditions.

Established testing methods for adhesion of thin layers on substrates have mostly been developed to work at the macroscale. For example, adhesion tests under tensile loading conditions (e.g., see ASTM International (2002, 2012, 2014) and shear loading conditions (e.g., see ASTM International (2005, 2007, 2013) both require attachment of the loading device to the thin coating layer by application of a thin epoxy layer. The strength of epoxies is typically insufficient to fail the strong interfacial regions between thin ceramic coatings and metallic substrates typically found in coatings for machining tools and mechanical components, thus resulting in the inapplicability of such macroscale tensile pull-off and shear adhesion tests to engineered thin ceramic coatings. The use of epoxy also limits the range of temperatures at which testing is conducted (Brown 1994, 1995).

While the scratch testing protocol (e.g., ASTM International (2015)) has been widely applied to evaluating adhesion of hard coatings on metallic substrates, the conversion of the so-called critical load data to quantitative measures of coating/substrate interfacial mechanical response has been recognized to be extremely difficult if not impossible. This is due to the many different and complex physical processes that can occur in a scratch test, including energy dissipation within the ceramic coating layer due to coating cracking and energy dissipation within the coating layer or the substrate due to plastic deformation, as well as the complex stress field surrounding the moving scratch indenter (Bull 2001). The critical load measured from scratch testing does not only depend on the adhesion between the coating and the substrate, as it should be in the ideal case (Kutilek and Miksovsky 2011). The complications listed above are responsible for reports in the literature regarding the dependence of the measured critical load on a wide range of parameters of the entire coating/substrate system, including the coating's thickness, hardness, and internal stress state, the hardness of the substrate, and the loading conditions (Lukaszkowicz et al. 2011).

At the present time, the laser spallation test offers, arguably, the best experimental methodology for obtaining a quantitative measure of the interfacial tensile strength between thin films and substrates (Gupta et al. 1992, 2003). In the laser spallation test, an energetic laser beam is incident upon the substrate side of a thin-film/substrate specimen and sets up a compressive stress pulse in the substrate, the reflection of which from the free surface of the film generates tension at the substrate/film interface, and causes film spallation from the substrate when this interfacial tensile stress reaches a critical magnitude. While laser spallation tests have yielded quantitative measures of interfacial tensile strength (Gupta et al. 1994), this testing methodology is associated with limitations due to specimen geometry (typically flat-wafer type specimens) and substrate absorption. Past laser spallation measurements on ceramic-layer/metal-adhesion-layer/metal-substrate systems have also shown an apparent dependence of the interfacial strength on the thickness of metal interlayer (Gupta et al. 1994). Observation of such a dependence of strength value on interlayer thickness indicates that the laser spallation testing protocol also may not yield a value correspondent with an ideal interfacial strength, and suggests the need for an improved understanding of the exact nature of interfacial failure induced by the laser-induced tensile interfacial stress pulse, even though the laser spallation measurement is considered well established and well studied.

The small thickness of the coating layer in ceramic-coating/metal-interlayer/substrate systems, ranging typically from $\sim$0.2 μm to $\sim$10 μm, makes their mechanical coupling to macroscale external loading systems difficult. This difficulty is in part responsible for the lack of progress with respect to the development of reliable and quantitative mechanical testing protocols for the mechanical response of coating/substrate interfacial regions. In this regard, the development of focused ion beam (FIB) micro-/nano-scale machining capabilities (Volkert and Minor 2007) and in situ instrumented micro-/nano-scale mechanical testing capabilities over the last decade opens up new avenues for obtaining quantitative data as well as more detailed observations related to mechanical failure of coating/substrate interfacial regions. Preliminary microscale mechanical testing results presented in this chapter offer some examples for this research direction.

Early simulations related to the mechanical response of solid/solid interfaces, such as ab initio density functional theory (DFT) calculations on metal/ceramic interfaces performed two decades ago, were limited to ideal interfacial configurations without taking into account potential defect configurations at/near interfaces (Smith et al. 1994). The dramatic improvements in computing power occurred over the last two decades have enabled multiscale simulations related to solid/solid interfaces, from DFT to molecular dynamics (MD), to dislocation dynamics (DD), and to grain-level plasticity models, such as crystal-plasticity finite element models (CPFEM) or visco-plastic self-consistent (VPSC) models. Hierarchical multiscale modeling schemes that integrates interfacial deformation physics across a broad spectrum of lengths scales have been developed (Zbib and Diaz de la Rubia 2002; Groh et al. 2009), which have the capabilities of addressing interfacial mechanics problems that are much more realistic and relevant to experimental conditions. We believe that recent advances in microscale mechanical testing capabilities and development of multiscale physics-based modeling/simulation tools offer an opportunity to deepen the current understanding of the mechanical response of solid/solid interfaces. In the particular case of ceramic-coating/substrate interfacial regions, this confluence of micro-/nano-mechanical testing and multiscale modeling/simulations can offer new insights into the critical elements governing the mechanical integrity of interfacial regions between thin ceramic layers and substrates. Preliminary simulation results presented in this chapter again offer some examples.

In this chapter, we first briefly summarize the current understanding in the existing literature regarding the mechanisms of mechanical instabilities in metal/ceramic interfaces (section "Microscopic Mechanistic Understanding of the Failure of MCIRs"). In section "The Computationally-Guided and Experimentally-Validated ICME Framework," we propose a novel integrated computational materials engineering (ICME) framework, currently being implemented by the authors, aimed to improve mechanical integrity of the metal/ceramic interfaces. Details regarding its approach, concerning both computation and experimentation, are discussed. In section "Structure Mechanical Instability of Metal/Ceramic Interfaces – Initial Results," we illustrate a specific effort under such a combined experimentation-simulation framework – the evaluation of shear and compression instabilities of CuN/Cu/Si and CrN/Ti/Si interfacial regions. In this effort, instrumented compression testing was performed, with concurrent scanning electron microscopy (SEM) observations, on microscale cylindrical pillars fabricated from vapor deposited

CrN/Cu/Si(001) and CrN/Ti/Si(001) specimens. Such technique allowed a much more detailed observation of how the interfacial regions fail under load. Concomitant simulations were carried out on Ti/TiN interfaces, employing MD in combination with DFT. The simulation results shed light on the influence of the Ti/TiN interface on the mechanical behavior of the Ti atomic layers nearby, in both shear and tensile strength. The combined DFT/MD simulations illustrate the link between length scales (from the atomic- to nano-) and were effective in probing structural features' influence on mechanical response of realistic semi-coherent interfaces. It is worth noting that this chapter is not meant to be a thorough literature review. Instead, the intention of this chapter is to present a novel ICME-based approach, with a specific example, to evaluate and/or improve the mechanical stability of the MCIRs.

Microscopic Mechanistic Understanding of the Failure of MCIRs

The close proximity of reactive nonmetallic element strongly impacts the electronic structure inside the metal layer near the chemical interface, resulting in weaker theoretical shear and tensile strengths (Zhang et al. 2017; Yadav et al. 2015). Recently, by combining experimental observations made during in situ microscale mechanical testing with MD simulations and DFT calculations, the present authors have provided new insights regarding mechanical failures of thin-ceramic-layer/metal-interlayer/substrate systems (Zhang et al. 2017). DFT calculations performed on the Ti/TiN interface suggest a weakening effect of the metal/ceramic chemical interface to its adjacent metal atomic monolayers in both shear and tension, parallel and normal to the interface. This is evidenced by a reduced generalized stacking fault energy (GSFE) profile as well as work of adhesion (WoA) on the (0001) slip planes in Ti near the chemical interfaces. Similar observation has also been observed for Al/TiN and Al/VN interfaces (Yadav et al. 2015). As a result of the weakened energetic characteristics, the misfit dislocation network (MDN) formed in such planes exhibits increasingly planar dislocation cores (Shao et al. 2013, 2014, 2015), higher dislocation mobility (Hirth and Lothe 1982), as well as vanishing pinning effects of the nodes (dislocation intersections) (Zhang et al. 2017; Shao et al. 2015). On the other hand, MD simulations suggest that the free energy of the MCIR is also at a minimum when the MDN locates away from the chemical interface, coinciding with the MDN position where the shear strength is a minimum (Zhang et al. 2017). This explains the experimental observation that the eventual shear failure occurs not at, but adjacent to, the chemical interface (Zhang et al. 2017).

When the MCIR is subjected to a normal stress, deformation behavior and failure mechanisms are vastly different depending on the length scale of the microstructures. For isolated individual interfaces within the MCIRs (e.g., direct coating without interlayers or laminated interlayers with large layer thickness, e.g., d > ~500 nm), gliding lattice dislocations in metal interact with individual sharp interfaces and form dislocation pile-up (Wang et al. 2017; Hirth and Lothe 1982; Wang and Misra 2011). The significant tensile stress at the tip of a dislocation-pile up may lead to formation of cracks (Meyers and Chawla 2007) for tensile loading normal to the MCIR. Note that due to the reduced theoretical tensile strength, as measured by the WoA, in the metal near the chemical interface in the MCIR, the tensile cracks may preferentially form in the metal layers leading to delamination (Fig. 2a). Under compressive loading, the local normal-to-interface tensile stress component at the tip of the pile up counteracts the applied compressive stress and may be less severe for opening cracks. In such a case, the parallel-to-interface tensile stress component at the tip of the pile-up superimposed with the tensile stress caused by the elasto-plastic deformation incompatibility between ceramic and metal layers may induce cracks perpendicular to the interface (Fig. 2b). When multiple chemical interfaces are adjacent to each other, such as in the MCIRs within nano-laminated interlayers, either of the following situations may apply: (1) the smaller interfacial spacing (d < 50 nm) inhibits the formation of dislocation pile-up, and the propagation of threading dislocation within the metal layer (or confined layer slip (CLS)) is the governing mechanism; (2) the interfacial spacing is so small (d < 10 nm) that CLS is inhibited and nucleation/emission of dislocations from the MDN is preferred. If the thickness of the ceramic is small, e.g., in the range of only several nanometers, the strong dislocation dipoles form on both sides of the ceramic layer may activate slip without cracking (Fig. 2c) (Li et al. 2014; Bhattacharyya et al. 2011; Wang and Misra 2014). For slightly larger ceramic layer thicknesses,

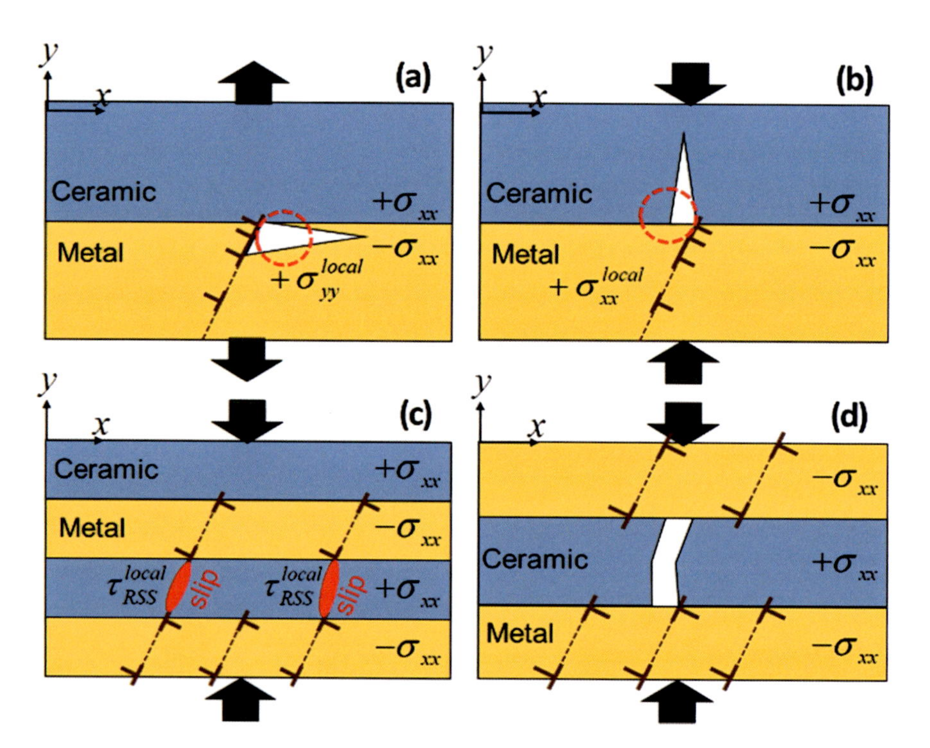

Fig. 2 Schematic illustration of the deformation and failure mechanisms experienced by MCIRs

cracks normal to interfaces may still form, due to the tensile stress in ceramic layer originated from the elasto-plastic deformation incompatibility between ceramic and metal layers (Fig. 2d).

The Computationally Guided and Experimentally Validated ICME Framework

The macroscopic failure of the CHC/substrate system is often governed by the atomic scale defect evolution events at the MCIR (Li et al. 2014; Bhattacharyya et al. 2011; Zhang et al. 2017; Wang and Misra 2014; Yang et al. 2017). For instance, in response to compressive/tensile loading normal to the MCIR, the MDN may nucleate and emit lattice dislocations, which will in turn interact with gliding lattice dislocations. The pile-up of dislocations at an interface may lead to the opening of cracks and hence failure of the CHC. Under shear loading, the MDN may nucleate interface dislocations that propagate parallel to the MCIR. Alternatively, as recently demonstrated by the present authors (Zhang et al. 2017), the MDN may glide collectively in response to an applied shear stress. Both deformation mechanisms lead to shear failure of the MCIR. Therefore, variations in the architecture and/or chemistry at interfaces may hugely impact the defect activities (Wang and Misra 2011; Shao and Wang 2016; Wang et al. 2014; Salehinia et al. 2014, 2015) and consequently improve the mechanical response of MCIRs.

Given the multiscale nature of mechanical failure in MCIRs, we envision a multiscale computationally guided and experimentally validated ICME framework (Fig. 3). At the atomic scale, using density functional theory (DFT) and molecular dynamics (MD) simulations, tensile and shear strength of candidate MCIR with various chemical (metal/ceramic composition as well as interfacial dopants) and architectural (interfacial steps and ledges (Zhang et al. 2016; Hirth and Pond 1996; Henager et al. 2004; Hirth et al. 2006) and arrangement of individual metal/ceramic interfaces in an MCIR) modifications are explored to identify potentially strong interfaces. Selected MCIR designs can then be experimentally implemented through state-of-the-art plasma-assisted vapor phase synthesis, mechanically tested at the microscale under different loading conditions, and characterized through microscopy and spectroscopy techniques. Experimental results, in turn, can validate the computational MCIR design. The so-obtained deformation physics is passed to and implemented in the microscale atomistically informed interface dislocation dynamics (AIDD) model. The output of the validated AIDD model, including the dislocation evolution law and the constitutive relations, is incorporated into a mesoscale visco-plastic self-consistent continuum interface dislocation dynamics (VPSC-CIDD) model (Lyu et al. 2015, 2016, 2017). The output of the VPSC-CIDD model, i.e., the texture evolution and stress-strain response are then validated against experimentation. This VPSC-CIDD model is expected to predict the mechanical/failure strength of MCIRs with arbitrary geometries, such as within metal matrix composites (MMCs), and not limited to the

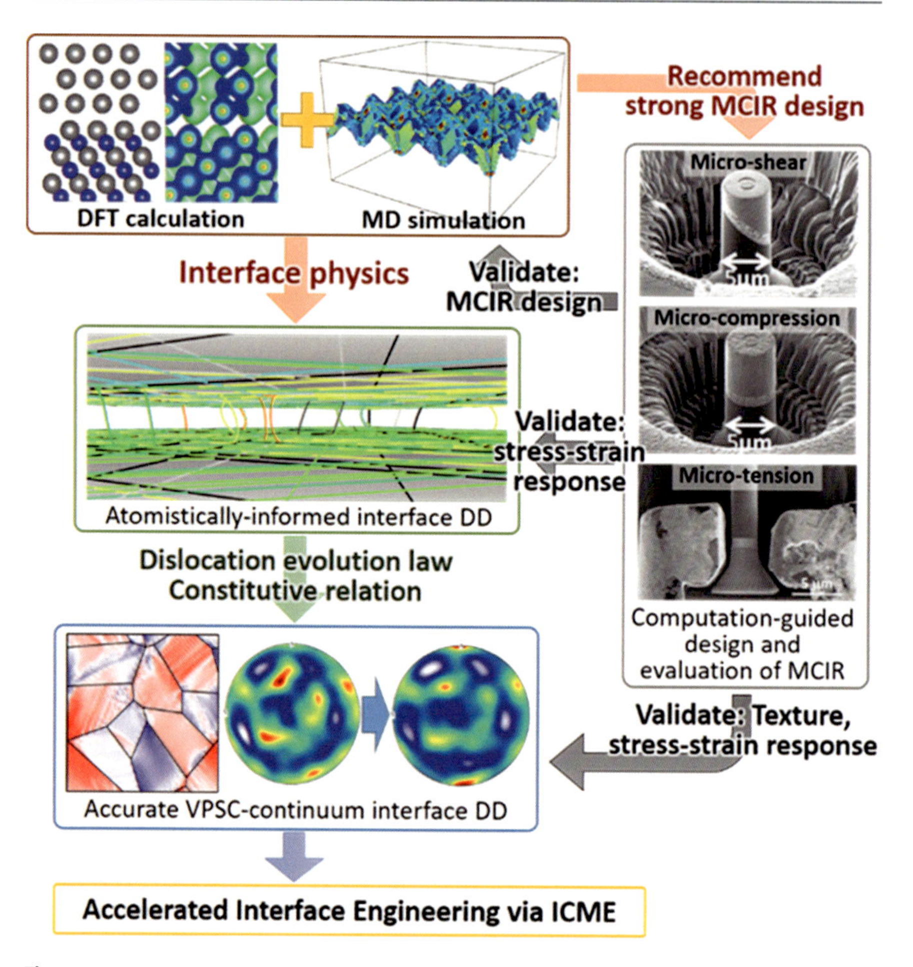

Fig. 3 Illustration of the multiscale, computationally guided, and experimentally validated ICME framework

conventional layered structures (Lyu et al. 2015, 2016, 2017; Lebensohn and Tomé 1993, 1994). Such an ICME framework is expected to advance basic understanding of the physics governing mechanical response of metal/ceramic interfaces, and will have general applicability including, but not limited to, design of CHC systems.

Through regulation of interfacial chemistry and architecture, the structure-property relationship of mechanically strong MCIRs can be explored, established, and validated by combining multiscale modeling/computation with vapor phase synthesis, microscale mechanical testing, and multiscale materials characterization. AIDD and VPSC models can then be advanced to incorporate the structure and deformation physics of the MCIR, which in turn, will enable accelerated design and implementation of strong MCIRs. The multiscale computationally guided and experimentally validated ICME framework will empower a significantly accelerated and more cost-effective design-implementation cycle (encompassing the discovery, validation, modeling, and design) for advanced MCIRs, leading to next generation applications.

The Multiscale AIDD Simulator and VPSC-CIDD Model

The current AIDD model incorporates interface deformation mechanisms and interface structural features in classical dislocation dynamics (DD) models (Zbib and Diaz de la Rubia 2002; Wang et al. 2014; Akasheh et al. 2007; Han and Ghoniem 2005; El-Awady et al. 2008; Ghoniem et al. 2000, 2002; Ghoniem and Han 2005; Zbib et al. 1998). Two distinct types of dislocations are modeled – interface dislocation and lattice dislocation. When lattice dislocations approach within a critical distance to an interface, they are captured by the interface and converted to interface dislocations; their Burgers vectors are conserved. If an interface dislocation with near-parallel line sense is present, the incoming lattice dislocation may react with it and form a new interface dislocation, if the reaction leads to a lower dislocation line energy. On the other hand, when a "weak" interface with relatively low shear strength is considered, in response to the stress field of an incoming lattice dislocation, interface dislocation loops may nucleate, which result in localized shear. The interface is modeled as a special, piecewise flat plane, allowing the glide and climb of interface dislocations, depending on the Burgers vector, i.e., when $\mathbf{b}_{inter} \cdot \widehat{n}_{inter} = 0$, the dislocation is glissile. Otherwise, it can only move by climb. The motion of interface dislocations is governed by a set of equations of motion very similar to lattice dislocations. However, due to the unique characteristics and structure of the interfaces, the parameters for the motion of the interface dislocations, e.g., drag coefficients and vacancy/interstitial migration energies, are different from those for the lattice dislocations and are therefore necessary to be calibrated using atomistic simulations and experimental observations. The reaction between two critically close interface dislocations is also permitted if both reactant dislocations have near-parallel line sense and if the resulting dislocation has lower line energy.

As was discussed previously, the plastic deformation in the nano-laminated interlayers is mainly carried out through the CLS mechanism as well as the nucleation and subsequent propagation of lattice dislocations from the interfaces. MD simulations (Fig. 4a) (Beyerlein et al. 2013a, b; Zhang et al. 2013; Shao et al. 2017) have revealed that a nucleation event is favored if the nucleation source (interface line defect) is parallel to the trace of the slip plane to be activated (Situation 1 in Fig. 4b) (Shao et al. 2017). If no existing interface dislocation is aligned with the trace of slip plane, local reorientation (Situation 2 in Fig. 4b) of line defects and local reaction (Situation 3 in Fig. 4b) between line defects may occur to produce segments of defects aligned with the traces of slip planes (Shao et al. 2017). Accordingly, this interface physics has been implemented in the AIDD model. Further, the nucleation of the lattice dislocation is explicitly treated as a thermally activated process using a Monte Carlo type of approach (Fig. 4c) (Shao et al. 2017).

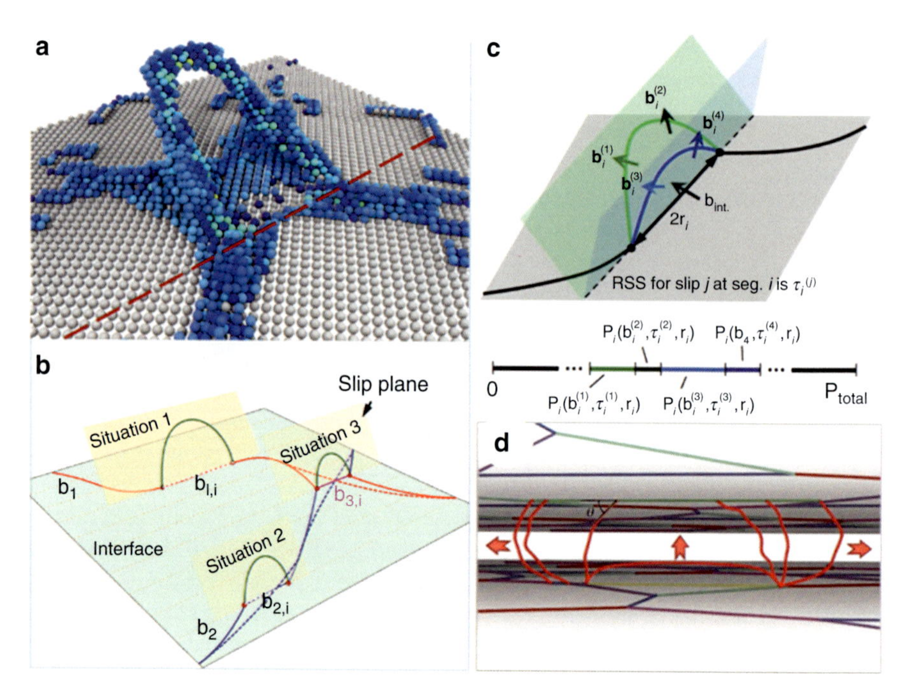

Fig. 4 Nucleation and subsequent propagation of dislocations as modeled in AIDD (Shao et al. 2017)

The newly nucleated dislocation loop bows out, driven by the externally applied stress superimposed by the local stress field of the dislocations (Fig. 4d) (Shao et al. 2017). The nucleated dislocation propagates within the layer. Once it touches the adjacent interface, it is deposited on the interface. If the angle between a dislocation segment and the interface is small, such that $\theta < \theta_c$, it will be continuously deposited on the interface, and the remaining dislocation will "thread" within the layer. Using this nucleation scheme, slip transmission can be captured. Differing from nucleation at an intrinsic interface dislocation, slip transmission occurs through nucleation from an extrinsic interface dislocation. In such an event, the Burgers vector of the residual interface dislocation is significantly reduced, leading to a significantly higher probability.

The AIDD model is limited in both time (micro seconds) and length (below 1 μm) scales and is therefore unable to predict the deformation behavior of materials in the meso- and macroscales. Its advantage is to predict the evolution law of dislocations, including the evolution of dislocation density (both mobile and immobile) as well as rates of dislocations' nucleation and cross-slip events, to name a few. Such rules are relied upon by the higher length scale models, such as crystal plasticity finite element methods (CP-FEM) (Groh et al. 2009; Marin 2006; Casals and Forest 2009; Miller et al. 2004) as well as the VPSC model (Lebensohn and Tomé 1993, 1994; Wang et al. 2010). The VPSC-CIDD model is based upon the conventional VPSC model with expended capabilities in the description of plasticity (Lyu et al. 2015, 2016, 2017). Ignoring elastic deformation, the VPSC-CDD model determines the critical resolved shear stress on slip system α based on a Bailey-Hirsch type of relation (Lyu et al. 2015, 2016, 2017): $\tau_H^\alpha = Cb\mu \sum_m \Omega^{\alpha m} \sqrt{\rho_{TS}^m}$, where C is a numerical constant, b and μ are the Burgers vector and shear modulus of the crystal, $\Omega^{\alpha m}$ is the dislocation interaction matrix between current slip system α and the system m, ρ_{TS}^m is the density of the statically stored dislocations (SSD) on slip system m. ρ_{TS}^m is described based on several different contributions, such as dislocation multiplication, pinning and de-pinning of dislocations, annihilation, cross-slip. The characteristics of all of these processes can be explicitly obtained by AIDD simulations (Lyu et al. 2015, 2016, 2017). The outputs of the VPSC-CIDD model (Fig. 3) include the mesoscale stress-strain relation, the local stress and strain distribution, as well as the texture evolution – all of which can be verified from mechanical testing and material characterization experiments, e.g., electron backscatter diffraction (EBSD) and transmission electron microscopy (TEM).

Strategy for Improving the Mechanical Integrity of Metal-Ceramic Interfaces

Based on the existing literature (Yadav et al. 2015; Lin et al. 2017; Sun et al. 2017) and the recent work by the present authors (Zhang et al. 2017), it is clear that there exists a weak plane in the vicinity of the chemical interface for many MCIRs containing sharp interfaces. Such weak interaction planes give rise to weak shear as well as tensile strengths, which may lead to premature interfacial failure, e.g., failure of interfacial region in CHC/metal substrate systems. On the other hand, the utilizations of the nanolaminate adhesion interlayers, although substantially improves the strength and ductility of the MCIR, may still suffer from interface cracking if the individual layer thickness is not carefully controlled. Therefore, strategies for improving the strength and ductility of MCIRs have to address the failure mechanisms identified above. Three potential paths of structural and chemical modifications to MCIRs are being considered, requiring the understanding of key physical factors controlling the mechanical response in each case, and consequently developing the best approaches and solutions to engineer the mechanical integrity of MCIRs (Fig. 5):

1. **Adding interfacial impurities/solute-atoms** to improve the theoretical shear and tensile strength of a MCIR (Fig. 5a). In doing so, the characteristics of the MDN, i.e., the core structure, mobility, and pinning effects of the dislocations/nodes, may be altered. Such an alteration may result in either a stronger shear resistance of MDN at its minimum energy location or a change in the minimum energy location that corresponds to a higher shear strength;
2. **Utilizing a nanolaminate type adhesion interlayer** that composes of alternating metal and ceramic layers with thickness of only a few nanometers (Fig. 5b) to improve the compressive strength (stress before the onset of instability, such as

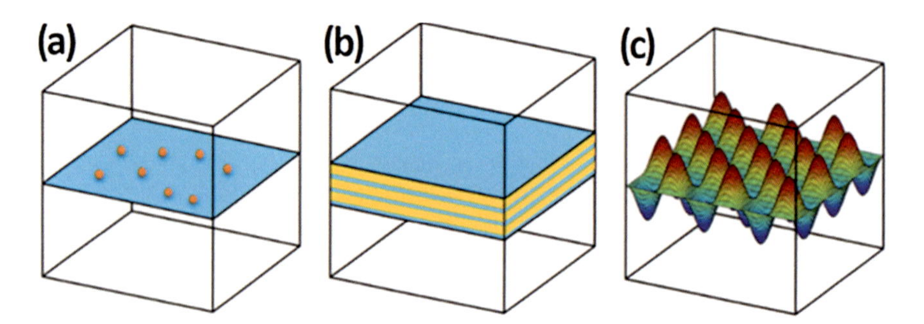

Fig. 5 Three potential paths for improving the strength and ductility of metal/ceramic interface regions. (**a**) Interface with impurity, (**b**) nanolaminate interface, (**c**) 3D structured interface

shear bands or cracks) and ductility of the MCIR. The thickness ratio between the metal and ceramic layers need to be investigated for how it impacts the overall stress response.

3. **Designing an interface with pronounced micro- and/or nanoscale roughness** to improve the shear strength of the MCIR (Fig. 5c). The roughness can be achieved by introducing interface steps/islands with low energy ledges, which will locally transform the parallel-to-interface shear stress into normal-to-interface tensile/compressive stress.

Note that the three potential paths may be combined with each other to achieve an optimum performance.

The extent of possible interfacial strength improvement is expected to vary depending on choice of material/impurity as well as the length scale of the interfacial architectural features. Indeed, implementing such modifications to MCIRs through a random experimental search will be inefficient, if not downright impossible. The presently described multiscale, computationally guided, and experimentally validated ICME framework is expected to generate mechanistic understanding, through multiscale simulation, of how the three paths affect the deformation/failure behavior of MCIRs. This will include (1) using DFT to explore a combination of metal/ceramics that yields the strongest pristine MCIR, (2) using DFT to probe the effect of the various solute atoms on the GSFE and WoA for various metal/ceramic interfaces, (3) parameterizing modified embedded atom method (MEAM) interatomic potentials, and (4) using existing and parameterized MEAM potentials to perform MD studies to investigate the microscopic deformation and failure mechanisms in MCIRs associated to each modification path taken.

The knowledge/insight gained through modeling/simulation can then guide the design of experiments in devising novel interface chemistries/structures. The experimental valuation of the mechanical response of MCIRs in conjunction with detailed multiscale materials characterization will provide feedback to the modeling/simulation efforts. In the meantime, the interface physics obtained through MD simulations can be used to advance the existing AIDD model and VPSC-CIDD model. A closed feedback loop is therefore formed: the in situ and ex situ experimental results are used to validate the VPSC-CIDD model, which in turn will be used to discover, design, and implement novel interfaces.

Structure and Mechanical Instability of Metal/Ceramic Interfaces – Initial Results

In this section, we provide one instance of effort under such a combined experimentation-modeling/simulation framework – the investigation of shear instability of the interfacial regions of two ceramic-coating/metal-adhesion-layer/substrate systems: CrN/Cu/Si and CrN/Ti/Si. The failure of the interfacial regions was measured quantitatively and observed directly through instrumented compression of cylindrical micropillars with concurrent scanning electron microscopy observations. The results indicated, for the first time to our knowledge, that shear failure of the interfacial region occurred in two stages: an initial shear deformation of the entire metal interlayer followed by an unstable dynamic shear-off close to one metal/ceramic interface. The experimentally observed dynamic shear-off was suggested to be concomitant with the metal/ceramic interface going from being "locked," with no relative displacement between materials on the two sides of the interface, to being "unlocked," with significant relative displacements. DFT and MD studies on a related metal/ceramic interface, Ti/TiN, provided further insights into this behavior. It was shown, again for the first time to our knowledge, that a weak interaction plane exists in the metal layer near the chemical interface in a coherent Ti/TiN structure. Consequently, the free energy as well as the theoretical shear strength of the semi-coherent Ti/TiN interface was found to depend on the physical location of the misfit dislocation network (MDN). The minimum energy and strength of the interface occur when the MDN was near, but not at the chemical interface.

Synthesis of Ceramic-Coating/Adhesion-Interlayer/Substrate Systems and Structural/Mechanical Characterization of Coating/Substrate Interfacial Regions

Vapor Phase Deposition of Coating/Substrate Systems

Deposition of ceramic coatings onto the surfaces of machining tools and mechanical components needs to satisfy several general requirements. For surface engineering purposes, thin CHCs are often used, e.g., transition metal carbides and nitrides, with typical physical characteristics of high melting temperatures, high elastic stiffness, high hardness, and high enthalpy of formation (Toth 1971). For satisfactory mechanical and tribological performance, the thin CHC layers deposited onto the substrates need to be fully dense and well adherent to the substrate. Such ceramic layers are typically deposited onto the substrate surface by vapor phase deposition, usually through either chemical vapor deposition (CVD), physical vapor deposition (PVD), or some form of hybrid vapor deposition technique (Vossen and Kern 1991). Deposition of refractory ceramic layers by conventional CVD methods usually involves high substrate temperatures, often exceeding the temperature limitations of the metallic substrates and causing degradation of the substrate mechanical properties. To deposit fully dense refractory ceramic layers at low temperatures typically requires the input of additional energies to the growth surface to counterbalance the lack of adatom surface mobility during growth and is typically achieved through some form of ion beam or plasma assist (Itoh 1989).

Thin transition metal nitride layers are used as examples of CHCs, the deposition of which was carried out at close to room temperature in an ultra-high-vacuum hybrid PVD/CVD deposition tool. This deposition tool housed a 13.56 MHz inductively coupled plasma (ICP) and multiple balanced magnetron sputter sources (Meng et al. 1999). The sputter sources faced the center of the deposition chamber, with a base pressure of $<3 \times 10^{-9}$ Torr. Cleaned silicon and metal substrates were first placed into a load lock, evacuated to $\leq 3 \times 10^{-7}$ Torr, and then transferred to a holder placed at the center of the deposition chamber. The substrates were rotated at $\sim$12 rpm during deposition. Pure metal (99.95%+) targets were operated in the dc current-controlled mode. The entire deposition sequence occurred in $\sim$10 mTorr of Ar (99.999%+). The substrates were first subjected to an Ar ICP etch for $\sim$5 min at a bias voltage of −50 V. Immediately after etching, elemental metal adhesion interlayers were deposited onto the substrate by sputtering the elemental metal targets in Ar with ICP assist. Deposition of metal nitride coating layers, $\sim$5 μm in thickness, occurred immediately after the metal adhesion interlayer deposition in an Ar/N_2 (99.999%+) ICP. To ensure a reasonable deposition rate at close to stoichiometry, the input N_2 flow was kept close to but below the pressure hysteresis point. During metal nitride deposition, a total input ICP power of 1000 W was applied and an electrical bias voltage of −30 V to −100 V was applied to the substrate. Additional details on the methodology of using low pressure high density plasma assisted vapor phase deposition to form fully dense CHC layers have been presented elsewhere (Meng et al. 1999, 2000; Meng and Curtis 1997).

Structural, Compositional, and Mechanical Characterization of Vapor Deposited Thin Film/Coating Specimens

Structures of metal-nitride/metal-adhesion-interlayer/substrate specimens were characterized by combining X-ray diffraction (XRD), scanning electron microscopy (SEM), Ga^+ focused ion beam (FIB) sectioning, and TEM. A PANalytical Empyrean system with Cu Kα radiation was used for XRD measurements. Diffraction patterns were obtained with specimens mounted on a χ-φ-x-y-z stage. Diffraction patterns in the glancing incidence geometry, the symmetric θ-2θ geometry, and the ω-rocking curve geometry are obtained with an incident beam graphite mirror and a Pixel3D detector. X-ray pole figure data were obtained with an incident beam double-cross slit and a scintillation detector.

Scanning imaging with electron- or ion-induced secondary electrons (SE/ISE) and Ga^+ FIB milling were carried out on an FEI Quanta3D Dual-Beam FEG instrument, which housed a 30 kV field-emission electron source, a 30 kV high-current Ga^+ ion source, an OmniProbe for specimen lift-out, and an EDAX X-ray energy-dispersive spectroscopy (EDS) attachment. A JEOL JEM2011 microscope operated at 200 kV was used for TEM examinations. Cross-sectional TEM specimens were made by Ga^+ ion sectioning into rectangular sections, lift-out using an in situ OmniProbe, gluing by Ga^+ catalyzed Pt deposition onto a TEM grid, followed by Ga^+ ion thinning at 30 kV and 5 kV. Final specimen thinning/cleaning was performed on a Gatan PIPS II ion polishing system using an Ar^+ ion beam at 500 eV or less. Further details regarding Ga^+ ion beam sectioning and TEM specimen lift-out have been described elsewhere (Chen et al. 2014a). Additional TEM specimens were made without the final Ar^+ ion thinning/cleaning step. With respect to characterizing the grain structure, no significant difference was found between specimens with or without the final low energy Ar^+ thinning/cleaning step.

Compositional characterization was carried out by X-ray photoelectron spectroscopy (XPS) on a Kratos AXIS165 spectrometer with monochromatic Al Kα excitation. Composition quantification was obtained from raw XPS spectra using factory supplied sensitivity factors. Prior to XPS spectra collection, the specimen surface was sputter cleaned with an Ar^+ ion beam for $\sim$20 min, with the Ar^+ ion beam set at 4 kV and 15 mA.

Instrumented indentation was carried out at room temperature on a Nanoindenter XP system, using a three-sided pyramidal Berkovich diamond indenter. The indenter's Young's modulus and Poisson's ratio are 1170 GPa and 0.07, respectively. The machine compliance and the projected indenter tip area as a function of the indenter contact depth was calibrated using a factory supplied fused silica standard following the Oliver-Pharr method (Oliver and Pharr 1992). The Young's modulus and Poisson's ratio for the fused silica standard were taken, respectively, to be 72 GPa and 0.18, independent of the indenter contact depth. The calibration covered a contact depth range from 40 nm to 2100 nm. Raw indentation loading and unloading curves were obtained in the load-controlled mode using a constant loading and unloading time of 15 s, with a 30 s load hold at the maximum load, L_{max}. Multiple load versus indenter displacement, *L-d*, curves were obtained at one d_{max} value, and d_{max} was varied to obtain a complete set of indentation data.

In what follows, structural, compositional, and mechanical characterization of polycrystalline CrN and Cu thin films deposited onto Si(001) substrates is presented in some detail as an example. The same characterization methodology was applied to all metal-nitride/metal-adhesion-layer/substrate systems. In particular, XRD studies of CrN/Ti/Si specimens showed that the Ti interlayer is hcp in structure and has a strong texture with the Ti hexagonal basal plane parallel to the Si substrate.

A series of single-layer Cu thin films were deposited onto Si(001) substrates. Cu deposition occurred with an Ar ICP assist at a total ICP input power of 1000 W. A substrate bias of −50 V was applied during deposition. The Cu film thickness was controlled through the deposition time. Another series of bilayer CrN/Cu thin films were deposited onto Si(001) substrates. Cu interlayers were first deposited onto Si(001) in pure Ar with ICP assist, followed immediately by deposition of CrN top layers in an Ar/N_2 mixture, also with ICP assist at a total ICP input power of 1000 W. During CrN deposition, a substrate bias of −100 V was applied. To ensure CrN deposition at close to stoichiometry, the input N_2 flow was kept above the pressure hysteresis point.

Typical results of morphological and compositional characterizations from CrN/Cu/Si(001) specimens are illustrated in Fig. 6. Figure 6a shows an SE image of a FIB cross section perpendicular to the original specimen surface, created by Ga^+ ion milling. A Pt protection layer was laid on top of the specimen surface prior to ion milling by Ga^+ catalyzed deposition from an organometallic Pt precursor. Figure 6a shows in sequence the Pt protection layer, the CrN top layer, the Cu interlayer, and the Si substrate. The thicknesses of the CrN and Cu layers were measured, respectively, to be ~5.5 μm and ~810 nm, with typical scatter of ±5% based on repeat measurements. Figure 6b shows a plan-view SE image of the as-deposited CrN top surface, with surface roughness typical of vapor deposited polycrystalline

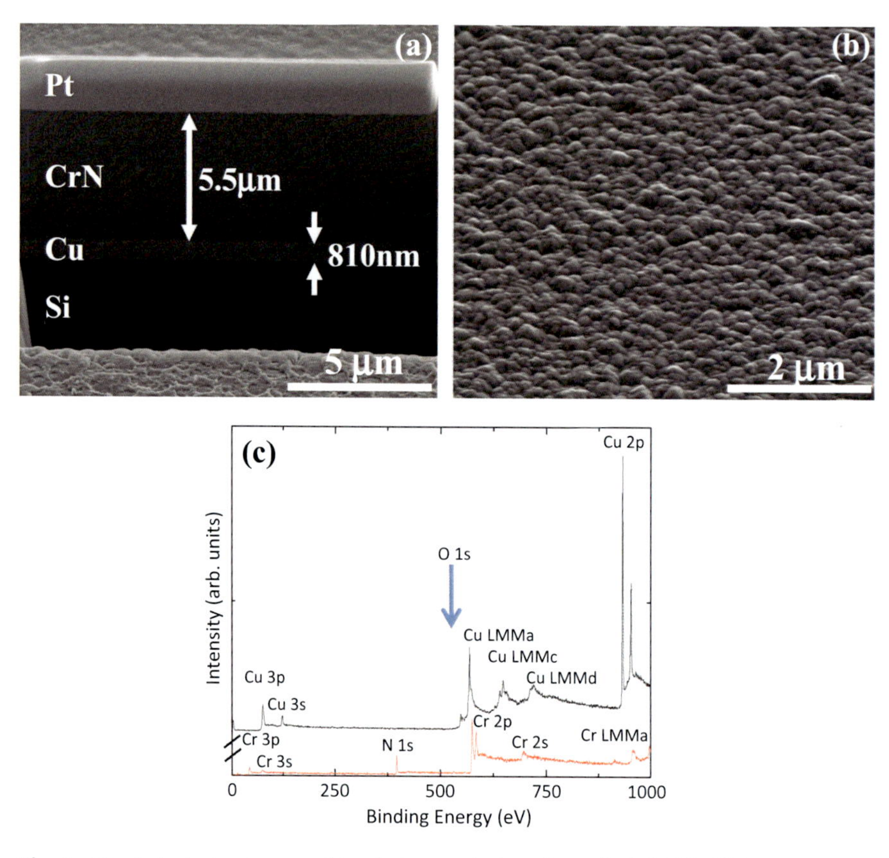

Fig. 6 Morphological and compositional characterization of a typical CrN/Cu/Si(001) specimen: (**a**) a FIB cross sectional SE image, (**b**) a 52° tilted plan-view SE image of the CrN top layer, (**c**) XPS survey spectra collected from top layers of a typical Cu/Si specimen (top trace) and a typical CrN/Cu/Si specimen (bottom trace). In both cases, the oxygen and carbon impurity levels were at or below the XPS detection limit of ~1 at.%

columnar refractory ceramics. Both the cross section and the plan-view images indicate that the CrN top layer is fully dense. Figure 6c shows XPS survey spectra collected from top layers of a typical Cu/Si specimen and a typical CrN/Cu/Si specimen. The Cu spectrum shows peaks consistent with Cu3p, Cu3s, and Cu2p emissions. Signal at the O1s and C1s positions, at binding energies of 532 eV and 284 eV, is not above the noise level. The CrN spectrum shows peaks consistent with Cr3p, Cr3s, Cr2p, Cr2s, and N1 s emissions. Signal at the O1s and C1s positions is also not above the noise level. The Cr:N ratio is determined to be 51:49. XPS data thus show that the Cu layers consist of elemental Cu and that the CrN layers consist of Cr and N at close to the stoichiometric ratio. The O and C impurity levels in both Cu and CrN layers are below the XPS detection limit of ~1 at.%. Similar compositional characterization data showed that all metal-nitride layers deposited were close to stoichiometry and all elemental metal layers deposited were free of O and C contamination above the XPS detection limit.

Results of XRD characterization of the structure of CrN layers in CrN/Cu/Si(001) specimens are illustrated by data shown in Fig. 7. The glancing incidence XRD pattern, shown in Fig. 7a, collected from one typical CrN/Cu/Si(001) specimen at an X-ray beam incidence angle $\omega = 2.5°$, shows diffraction peaks only from the CrN top layer. All diffraction peaks can be indexed to a cubic structure with lattice parameter a = 4.18 Å, close to the bulk B1-CrN lattice parameter of 4.15 Å (Toth 1971). The θ-2θ XRD pattern, shown in Fig. 7b and collected from the same specimen, shows the Si(004) diffraction peak from the Si substrate and one minor Cu(220) diffraction peak from the Cu interlayer in addition to all diffraction peaks indexed to B1-CrN, with a = 4.18 Å. Taken together, glancing incidence and θ-2θ XRD data are consistent with the CrN film having the B1-NaCl structure, being polycrystalline, and with preferential alignment of CrN<111> and CrN<200> along Si[001], the growth direction.

The level of residual stress within the CrN top layer was estimated from analysis of the glancing incidence XRD data shown in Fig. 7a. When the X-ray incidence angle was fixed at $\omega = 2.5°$ and the detector was scanned to obtain diffraction signals in the range of $20° < 2\theta < 120°$, the angle between the scattering vector $\vec{k}$ and the specimen surface normal $\vec{n}$, Ψ, is given by $\Psi = \theta - \omega$ (Perry et al. 1996). The CrN lattice parameter, a_{CrN}, was calculated from the multitude of B1-CrN diffraction peaks shown in Fig. 7a and plotted versus $\sin^2\Psi$ in Fig. 7c. Assuming an equal biaxial residual stress σ_R existing within the CrN layer, the variation of a_{CrN} as the Ψ angle varies relates in the following way to σ_R (Chen et al. 2014a; Cullity and Stock 2001):

$$a_\Psi = \left[\frac{1+\nu}{E}\sigma_R\right] a_o \sin^2\Psi + \left[1 - \frac{2\nu}{E}\sigma_R\right] a_o. \quad (1)$$

In Eq. 1, E and ν are, respectively, the Young's modulus and Poisson's ratio of the CrN layer, and a_o is the bulk lattice parameter of CrN. A linear least squares fit to the data shown in Fig. 7c yielded a slope of −0.022 ± 0.017 Å and an

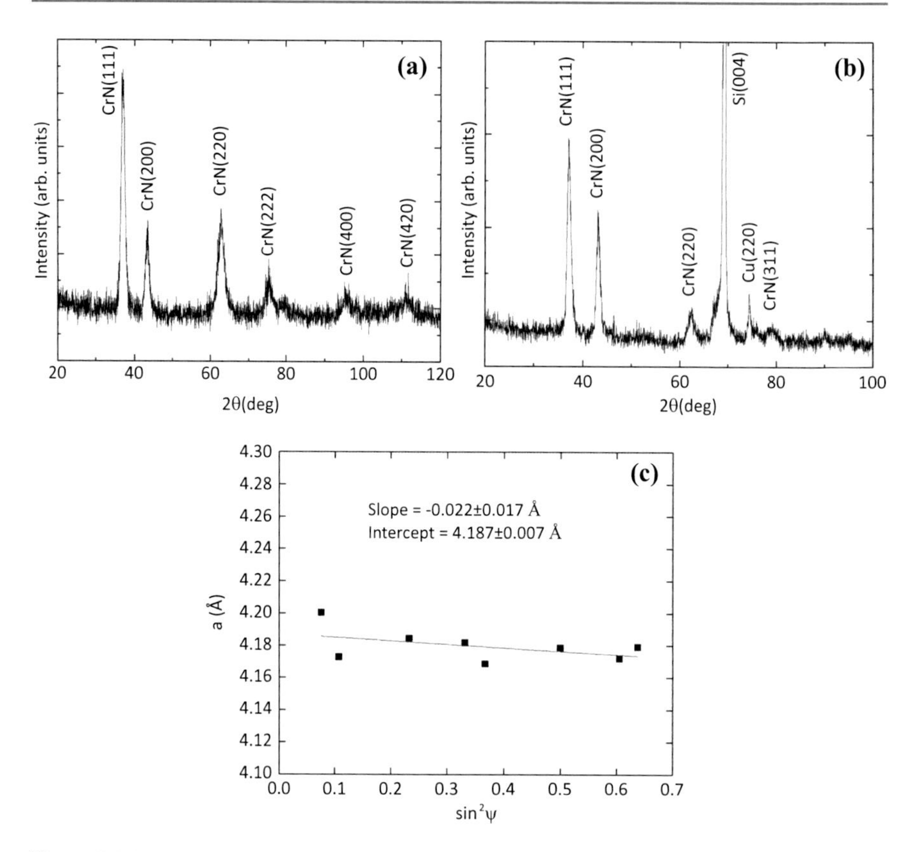

Fig. 7 XRD examination of a typical CrN/Cu/Si(001) specimen: (**a**) glancing incidence ($\omega = 2.5^{\circ}$) diffraction pattern, (**b**) θ-2θ diffraction pattern, (**c**) $\mathrm{Sin}^2\Psi$ analysis of the glancing incidence data shown in (**a**)

intercept of 4.187 ± 0.007 Å. The negative slope of the $\sin^2\Psi$ plot indicates that the residual stress within CrN is compressive. Independent of the values of E and ν, the fact that the slope uncertainty is comparable to the slope itself indicates that the level of residual stress is moderate. Values of E for CrN reported in the literature vary widely from $\sim$ 200 GPa to above 400 GPa (Kral et al. 1998; Chen et al. 2004). Taking a_o to be 4.15 Å, ν to be 0.2 (Sue et al. 1994), and E to be 400GPa (Holleck 1986), the fitted slope of the $\sin^2\Psi$ plot yields $\sigma_R = -1.8 \pm 1.4$ GPa. It is noted that since $E = 400$ GPa is a high value for CrN, this σ_R value represents an upper bond estimate (e.g., taking $E = 300$ GPa would result in a σ_R value of -1.4 ± 1.0 GPa). Both the magnitude and the large uncertainty associated with the σ_R value again indicate that the level of compressive residual stress within the CrN layer is moderate. A moderate residual stress level within CrN makes it easier to deposit thick CrN top layers, e.g., that shown in Fig. 6a. Diffraction data similar to those shown in Fig. 7 were obtained from other CrN/Cu/Si(001) specimens.

Results of XRD characterization of the structure of Cu/Si(001) specimens are illustrated by data shown in Fig. 8. The glancing incidence XRD pattern, shown

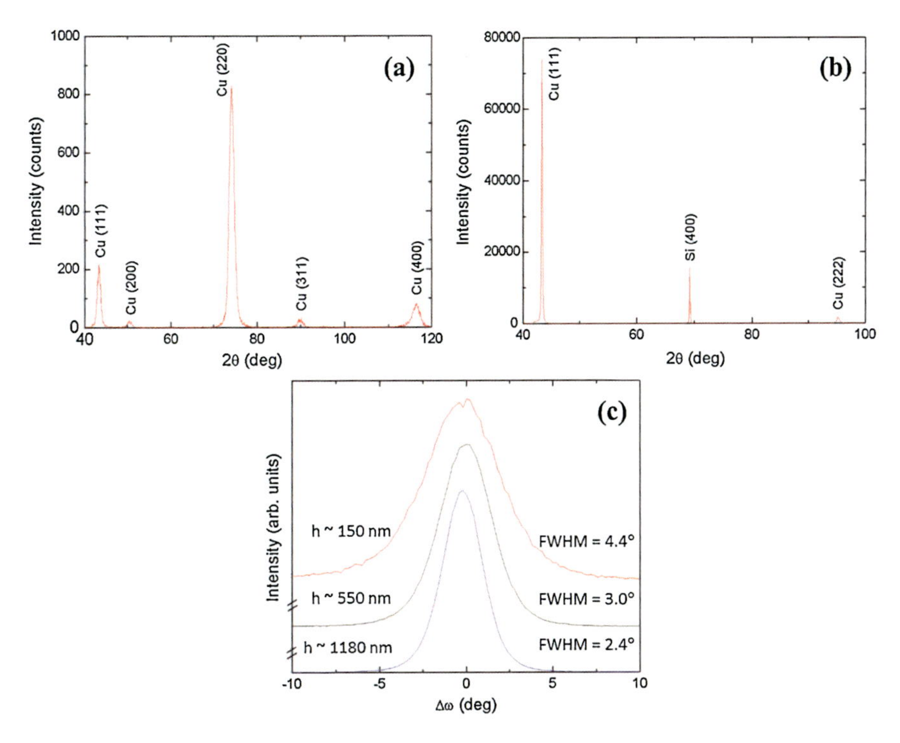

Fig. 8 XRD examination of a typical Cu/Si(001) specimen with a Cu film thickness of ~1180 nm: (**a**) glancing incidence ($\omega = 2.5°$) diffraction pattern, (**b**) θ-2θ diffraction pattern, (**c**) Cu <111> rocking curves from three polycrystalline Cu layers on Si(001). The Cu film thickness values are indicated on top of the rocking curves

in Fig. 8a, collected at an beam incidence angle $\omega = 2.5°$ from one Cu/Si(001) specimen with Cu thickness of ~1180 nm, shows diffraction peaks which can all be indexed to an fcc structure with lattice parameter a of 3.617 Å, close to the bulk fcc-Cu lattice parameter of 3.615 Å (Straumanis and Yu 1969). The θ-2θ XRD pattern, shown in Fig. 8b and collected from the same specimen, shows the Si(004) diffraction peak from the Si substrate and only fcc-Cu (111) and (222) diffraction peaks. The glancing incidence and θ-2θ XRD data are consistent with the Cu film having the fcc structure, being polycrystalline, and with strong preferential alignment of Cu<111> along Si[001]. The Cu(111) ω-rocking curve data are shown in Fig. 8c. Measured rocking curve width decreases monotonically from 4.4° at the Cu film thickness of ~150 nm to 3.0° at ~550 nm and to 2.4° at ~1180 nm. The texture in Cu films was shown in more detail by background corrected Cu(111) pole figure data shown in Fig. 9, collected from three different Cu films deposited on Si(001), with total film thickness values of ~150 nm, ~550 nm, and ~1180 nm. In all cases, the Cu(111) diffraction intensity peaks along the growth direction and drops to $<10\%$ of the peak intensity at directions deviating from the growth direction for $>5°$. Furthermore, the pole figure data clearly show in-plane rotational symmetry

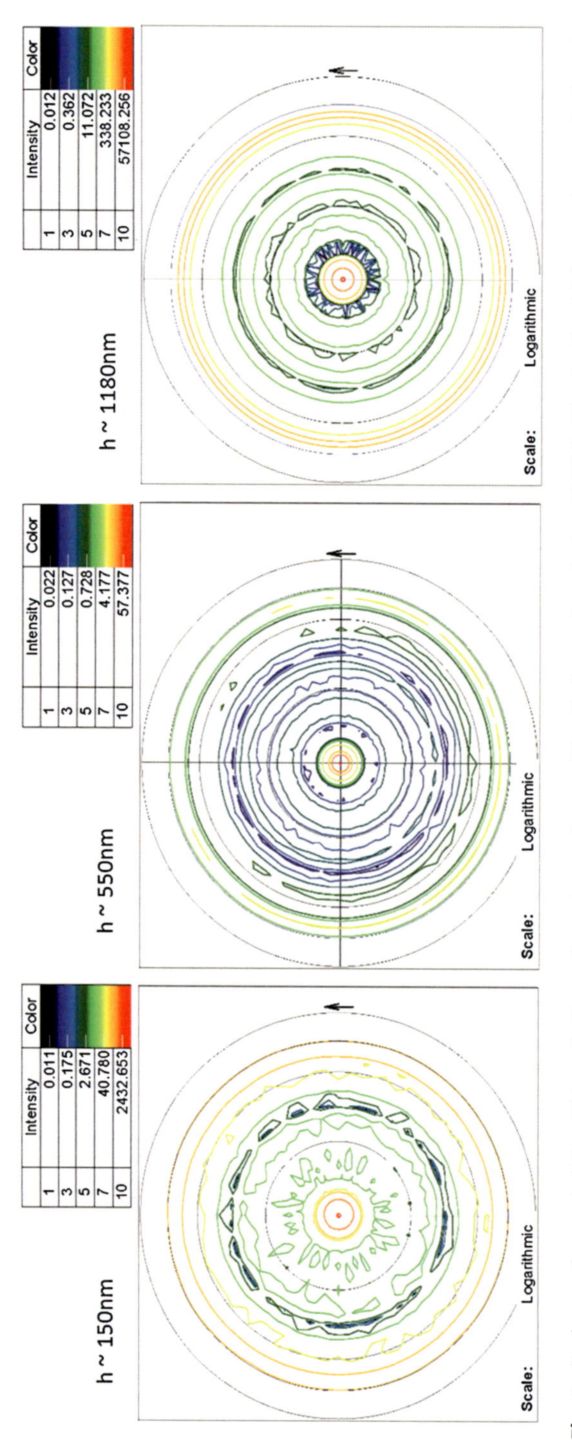

Fig. 9 Background corrected Cu <111> pole figures from three polycrystalline Cu layers deposited on Si(001). The Cu film thickness values are indicated on top of the pole figures

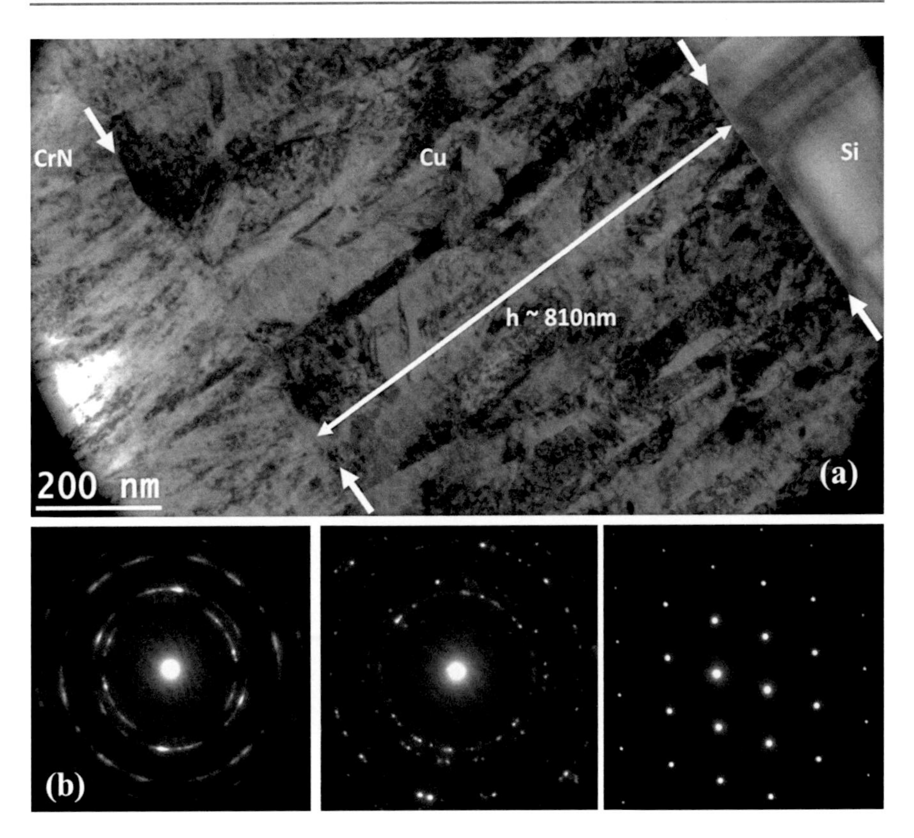

Fig. 10 TEM examination of a typical CrN/Cu/Si(001) specimen: (**a**) a cross sectional BF image across the entire Cu interlayer. The CrN/Cu and Cu/Si interfaces are indicated by white arrows; (**b**) typical SAED patterns collected from the (left to right) CrN, Cu and Si layer

of the Cu(111) diffraction intensity, at all film thicknesses probed. The pole figure data are consistent with the Cu(111) ω-rocking curve data shown in Fig. 8c. Taken together, X-ray diffraction data shown in Figs. 8 and 9 indicate that the Cu films are fcc in structure and have almost perfect fiber texture with Cu<111> along Si[001] to within a few degrees. The Cu films are polycrystalline with random in-plane orientation.

Grain structure of CrN and Cu thin films was further characterized by TEM. Figure 10 shows typical TEM characterization results of CrN/Cu/Si(100) specimens across the entire Cu interlayer. A cross-sectional TEM bright-field (BF) image, shown in Fig. 10a, was obtained from the CrN/Cu/Si(001) specimen whose morphology was shown in Fig. 6.The specimen was oriented with the Si substrate in the [110] zone axis direction. Both CrN and Cu layers are polycrystalline. Both CrN/Cu and Cu/Si interfaces appear clean, without indication of interdiffusion and reaction. For this specimen, the Cu interlayer thickness was measured from Fig. 10a to be $\sim$810 nm, which was consistent with the value measured by FIB

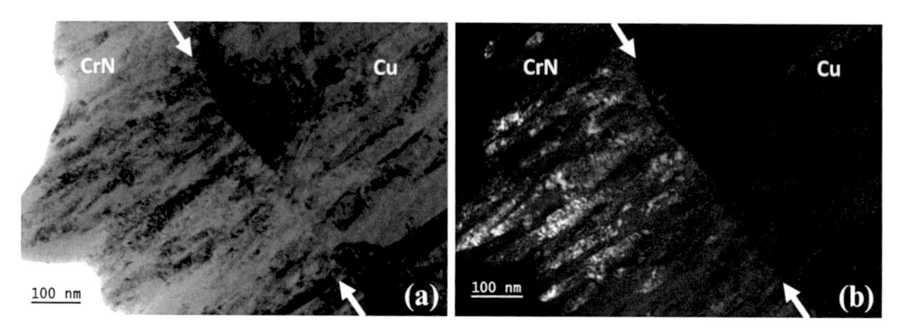

Fig. 11 Cross sectional TEM examination of the CrN top layer: (**a**)/(**b**) a BF/DF image pair of the same area. White arrows indicate the location of the CrN/Cu interface

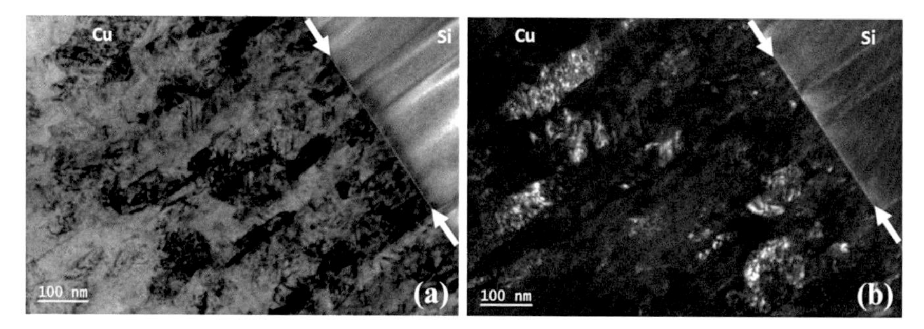

Fig. 12 Cross sectional TEM examination of the Cu interlayer: (**a**)/(**b**) a BF/DF image pair of the same area. White arrows indicate the location of the Cu/Si interface

cross sectioning in Fig. 6a. Figure 10b shows selected area diffraction patterns (SADPs) obtained, respectively, from the CrN top layer, the Cu interlayer, and the Si substrate areas. The SADP from Si shows the Si[110] zone axis diffraction pattern. The SADPs from Cu and CrN are consistent with the structure of Cu being fcc and that of CrN being B1-NaCl. Figure 11 shows a typical TEM bright-field/dark-field (BF/DF) image pair of the CrN top layer near the CrN/Cu interface. The CrN layer has a pronounced columnar structure, with columns with widths ~50 nm and large length/width ratios. The columnar structure of CrN shown in cross sectional TEM is consistent with the rough surface morphology observed in Fig. 6. The CrN layer appears fully dense, without indication of intercolumnar voids. Figure 12 shows a typical TEM BF/DF image pair of the Cu interlayer near the Cu/Si interface. The Cu layer consists of a random mixture of columnar and near equi-axed grains. The presence of nano-twins within the columnar Cu grains is visible in the BF image, with twin spacing <100 nm. Similar image features were observed from Cu films at different thicknesses. Additional DF imaging shows Cu grains ranging from ~20 nm to ~100 nm in widths and with morphologies ranging from near equi-axed grains to columnar grains with large length/width ratios. Similar grain structures were observed from all Cu films, with thicknesses ranging from ~150 nm to ~1180 nm.

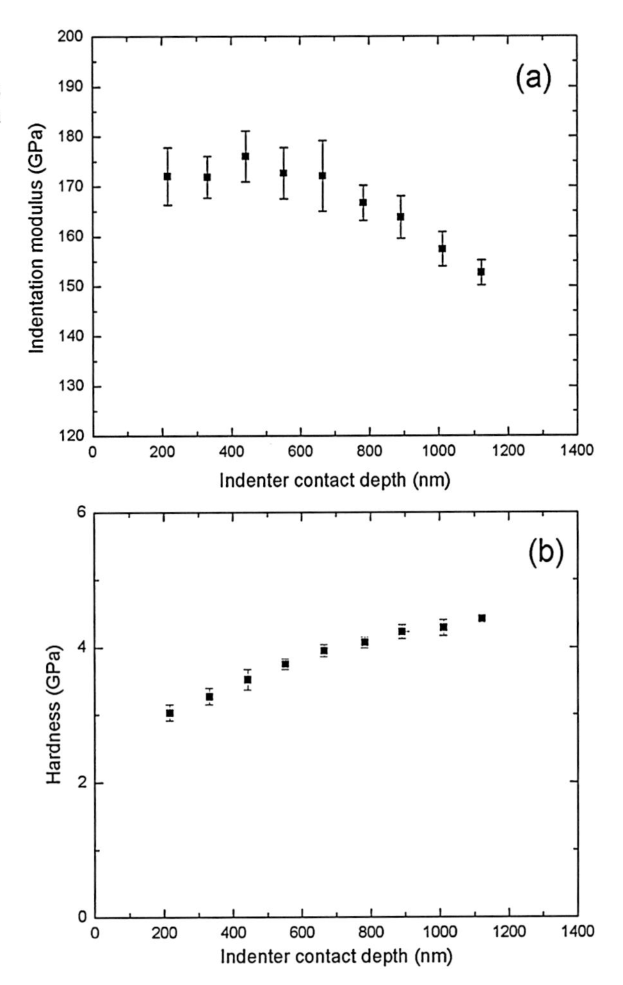

Fig. 13 Instrumented indentation of a Cu/Si(001) specimen with a Cu thickness of ~1180 nm: (**a**) indentation modulus versus contact depth, (**b**) hardness versus contact depth

Figure 13 displays indentation results obtained from a Cu/Si(001) specimen, with a Cu film thickness of ~1180 nm. Figure 13a shows values of measured indentation modulus, $E_{\text{ind}} = E/(1-\nu^2)$, as a function of the indenter contact depth. Error bars for the modulus value are derived from repeat *L-d* curves at the same d_{max} value. In the absence of an independent measurement on the Poisson's ratio, only E_{ind} can be obtained from the elastic unloading portion of the indentation curve. Figure 13a shows that E_{ind} increase slightly from ~150 GPa at large contact depths to ~170 GPa at contact depths between 200 nm and 350 nm.

The elastic stiffness constants for fcc Cu are, respectively, $C_{11} = 168.4$ GPa, $C_{12} = 121.4$ GPa, and $C_{44} = 75.4$ GPa (Kittel 2005), and those for diamond-cubic Si are, respectively, $C_{11} = 165.6$ GPa, $C_{12} = 63.9$ GPa, and $C_{44} = 79.5$ GPa (Hopcroft et al. 2010). The elastic compliance constants for Cu are, respectively, $S_{11} = 14.99$ $(\text{TPa})^{-1}$, $S_{12} = -6.28$ $(\text{TPa})^{-1}$, and $S_{44} = 13.26$ $(\text{TPa})^{-1}$, and those for Si are, respectively, $S_{11} = 7.69\ (\text{TPa})^{-1}$, $S_{12} = -2.14\ (\text{TPa})^{-1}$, and $S_{44} = 12.58\ (\text{TPa})^{-1}$ (Hopcroft et al. 2010). Young's modulus in the <001> and < 111> directions of cubic crystals are obtained from the compliance constants as (Nye 2000),

$$E_{100} = \{s_{11}\}^{-1}, \tag{2}$$

$$E_{111} = \left\{ s_{11} - \frac{2}{3}\left[(s_{11} - s_{12}) - \frac{1}{2}s_{44} \right] \right\}^{-1}. \tag{3}$$

For Cu, E_{100} and E_{111} are, respectively, 67 GPa and 191 GPa. For Si, E_{100} and E_{111} are, respectively, 130 GPa and 188 GPa. Furthermore, Poisson's ratio in the <hkl> direction of cubic crystals are given by (Zhang et al. 2007)

$$\upsilon_{hkl} = \frac{1}{2} - \frac{E_{hkl}}{2\,(C_{11} + 2C_{12})}. \tag{4}$$

For Cu, ν_{100} and ν_{111} are, respectively, 0.42 and 0.27. For Si, ν_{100} and ν_{111} are, respectively, 0.28 and 0.18. In the <001> and < 111> directions, the Cu indentation modulus takes on, respectively, values of 81 GPa and 206 GPa and the Si indentation modulus takes on, respectively, values of 141 GPa and 194 GPa. Based on the XRD results shown in Figs. 8 and 9, the relevant indentation moduli for indentation on the present Cu/Si(001) specimens should, respectively, be $E_{\text{ind}}(111) = 206$ GPa for the Cu film and $E_{\text{ind}}(001) = 141$ GPa for the Si substrate, i.e., the Cu/Si specimen goes from an elastically stiffer film to an elastically more compliant substrate. Thus, the data shown in Fig. 13a are trend-wise consistent with what is expected based on the predominant fiber texture of the Cu films, with Cu < 111>//Si[001]. Measured E_{ind} value is close to E_{ind} for the Si substrate at large contact depths.

Figure 13b shows values of measured hardness as a function of the indenter contact depth. Error bars for the hardness value are derived from repeat *L-d* curves at the same d_{max} value. Measured hardness values decrease from above 4 GPa at large contact depths to ~3 GPa at contact depths between 200 nm and 350 nm. The indentation hardness of Si at room temperature is ~10 GPa, limited by a pressure-induced phase transformation (Vandeperre et al. 2007). The Cu/Si specimen goes from a softer film to a harder substrate. Thus, the observed decrease in hardness with decreasing contact depth is again qualitatively consistent with expectation. Caution should be exercised in taking the measured hardness value of ~3 GPa at small contact depths as an "intrinsic" hardness of the Cu film, for it is influenced by two extra factors: one being the effect of a harder Si substrate and the other being the expected presence of an indentation size effect. Both effects tend to elevate measured hardness values. For these reasons, indentations on Cu/Si specimens with smaller Cu film thicknesses were not performed.

Figure 14 displays indentation results obtained from a 6.2 μm thick polycrystalline CrN film deposited on Si(001), over a range of indenter contact depth from ~1400 nm to ~370 nm. Figure 14a shows that the indentation modulus increases

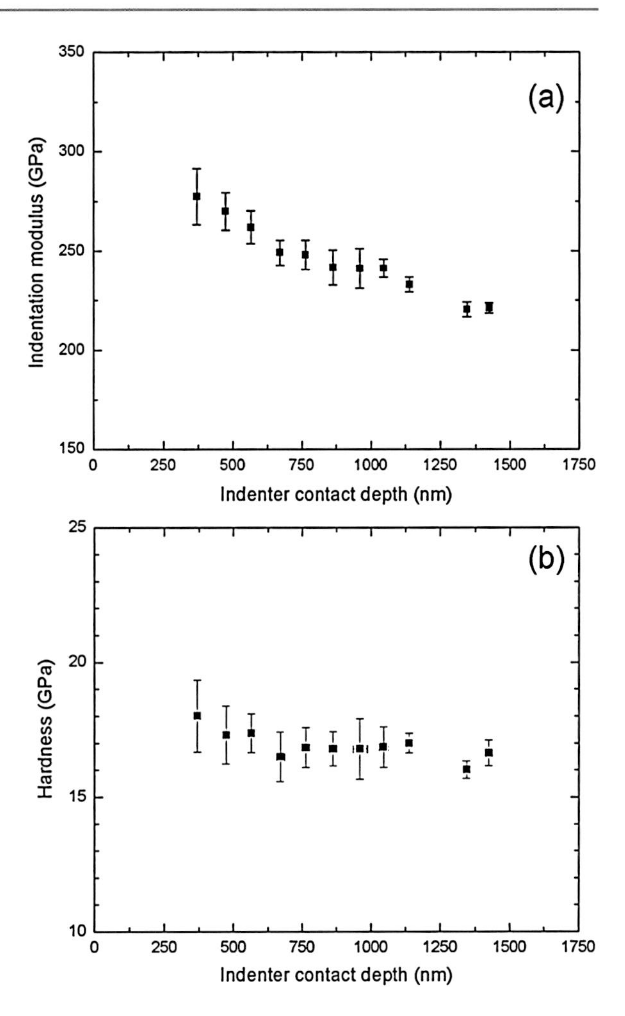

Fig. 14 Instrumented indentation of a CrN/Si(001) specimen with a CrN thickness of ~6.2μm: (**a**) indentation modulus versus contact depth, (**b**) hardness versus contact depth

monotonically with decreasing indenter contact depth, to ~275 GPa at the smallest contact depth of ~370 nm. The observed trend is consistent with the expectation for an elastically stiffer film on an elastically more compliance substrate, where measured indentation modulus decreases monotonically with increasing indenter contact depth (Meng and Eesley 1995). A simple extrapolation of the data shown in Fig. 14a suggests that the indentation modulus of the present CrN film is ~300 GPa. Figure 14b shows that measured hardness values in the same contact depth range exhibit only a modest change, from ~16 GPa to ~18 GPa. Noting that values of indenter contact depth normalized to the CrN film thickness range from 0.06 to 0.23, a simple average of all hardness values shown in Fig. 14b yield a hardness value of ~17 GPa, a value believed to be representative of the presently deposited CrN films.

Values of Young's modulus for CrN as reported in the literature vary widely, from 190 GPa to above 400 GPa (Chen et al. 2004; Holleck 1986). Moduli of sputter deposited CrN films are often reported to be between 200 GPa and 250 GPa, well below 400 GPa (Sue et al. 1994). A CrN indentation modulus of ~300 GPa is consistent with previous reported Young's modulus values of 270–300 GPa on sputter deposited CrN films (Thulasi Raman et al. 2012; Lin et al. 2012). A wide range of CrN hardness values have been reported in the literature, from 10 GPa to above 30 GPa (Zhang et al. 2008). Hardness of CrN was also reported to depend strongly on the level of residual stress (Mayrhofer et al. 2001). The presently measured hardness for CrN of ~17 GPa is consistent with the moderate level of residual compressive stress existing within the CrN layer, obtained from estimates based on glancing incidence XRD measurements data shown in Fig. 7.

Micropillar Fabrication and Axial Compression Measurements on Micropillars Containing Coating/Adhesion-Layer/Substrate Interfacial Regions

Fabrication of monolithic cylindrical micro-/nano-pillars using scripted FIB milling has been well documented (Uchic et al. 2004, 2009). To obtain the mechanical response of the interfacial regions in coating/adhesion-interlayer/substrate systems, two groups of micropillar specimens were fabricated: one with interfaces at a 45° inclination to the pillar axis and the other with interfaces perpendicular to the pillar axis (Chen et al. 2014b; Mu et al. 2014). Figure 15 shows a progressive series of SE images obtained during FIB fabrication of CrN/Cu/Si micropillars, in which the interfacial regions are inclined at 45° with respect to the pillar axis. The vapor-deposited CrN/Cu/Si specimens were first mounted at a 45° angle, mechanically polished to reveal the interfacial region, then FIB milled. Figure 15a–c shows the initial stages of top-down annular FIB milling to define the specimen region containing the interfaces and progressively thin down the pillar region. Figure 15d shows a taper-free micropillar resulting from scripted FIB milling with the pillar shown in Fig. 15c turning in a lathe-like motion and the ion beam incident from the side of the pillar. It is evident that this pillar contains the CrN/Cu/Si interfacial regions at a 45° inclination. Figure 15e shows such an array of taper-free micropillars. Axial compression testing of such micropillar arrays allows repeated and independent testing of the coating/substrate interfacial regions in a combined compression/shear loading. Similar top-down and scripted FIB milling was used to fabricate micropillars with interfacial regions perpendicular to the pillar axis. Figure 16 shows close-up SE images of typical CrN/Cu/Si(001) micropillars with interfaces inclined at 45° and 90° with respect to the pillar axes. Additional information on pillar fabrication was provided elsewhere (Chen et al. 2014b; Mu et al. 2014).

In situ axial compression of micropillars was conducted on a NanoMechanics Inc. NanoFlip® instrumented micro-/nano-mechanical testing device placed within the FEI Quanta3D FIB instrument. A custom-made, flat-ended, ~10 μm × 10μm diamond punch was used for pillar compression. The NanoFlip device, interfaced with the Quanta3D instrument, enabled simultaneous acquisition of the total load force on the indenter, P, the total punch displacement, Δ, and a video of SE images

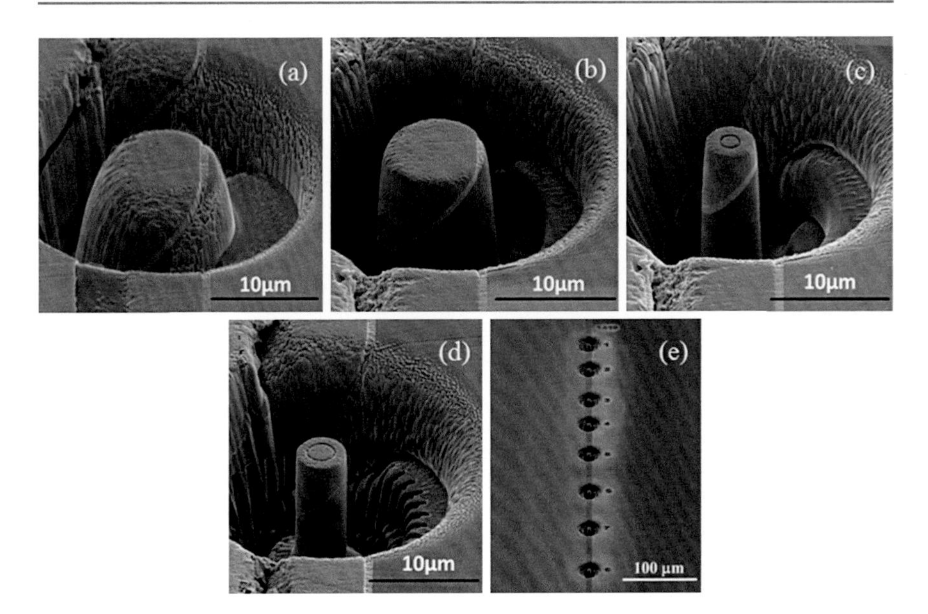

Fig. 15 Fabrication of micropillars containing interfacial regions of coating/adhesion-interlayer/substrate specimens: (**a**) top-down annular FIB milling after mechanical polishing, (**b**) defining specimen region with interfaces at a 45° inclination with respect to the pillar axis, (**c**) progressive thinning down of the pillar region, (**d**) the final pillar after scripted FIB milling with pillar turning and Ga^{+} ion incident from the pillar side, (**e**) A finished pillar array, enabling repeated testing of the same interfacial region in independent experiments

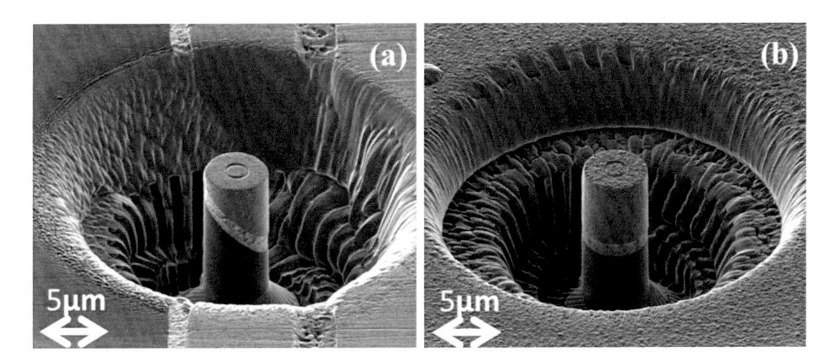

Fig. 16 Typical cylindrical micropillars fabricated by scripted Ga^{+} FIB milling: (**a**) one CrN/Cu/Si(001) micropillar with interfaces inclined at 45° to the pillar axis, (**b**) one CrN/Cu/Si(001) micropillar with interfaces perpendicular to the pillar axis

of the micropillar under testing, thus allowing specific points on the P–Δ curve to be linked to specific images of the pillar morphology at that point. The NanoFlip is fundamentally a load-controlled instrument, but is programmed to enable loading to occur in a "displacement-controlled" mode by providing feedback control of loading force according to deviation of sensed punch displacement rate from the set value.

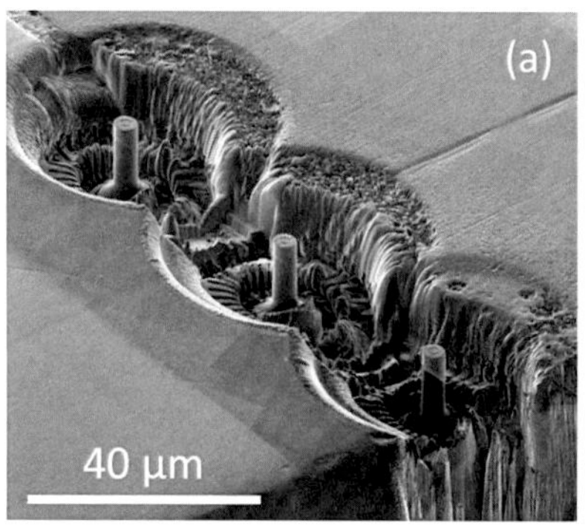

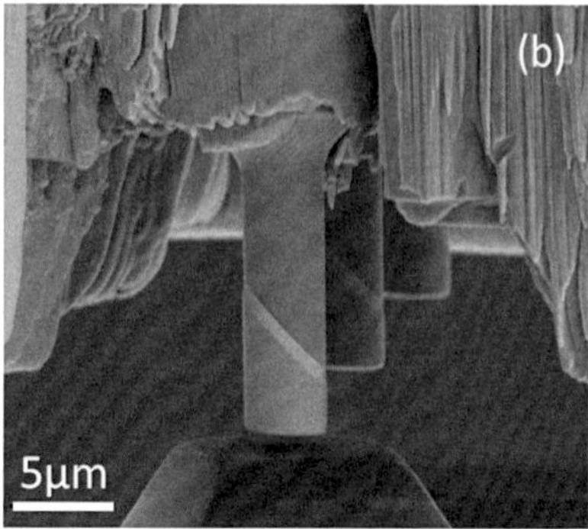

Fig. 17 Instrumented microscale compression testing of micro-pillar specimens in situ an SEM: (**a**) an array of three micropillars with circular surrounding openings to allow for approach of a diamond indenter, (**b**) the experimental configuration prior to contact of the diamond indenter with one micropillar

All in situ pillar compression tests were performed under such a displacement-controlled mode, with the target displacement rate set at 10 nm/s. Additional ex situ axial compression testing of pillars was conducted on a NanoIndenter XP instrument with another custom, flat-ended, ~10 μm × 10 μm diamond punch. These ex situ tests were conducted without concurrent observations, and only post test morphological observations of the tested pillars were made after the compression testing had concluded.

Cylindrical micropillars of CrN/Cu/Si(001) and CrN/Ti/Si(001) specimens, 3–5 μm in diameter, were fabricated with the interfacial regions inclined at 45° and 90° with respect to the pillar axis. From one CrN/interlayer/Si specimen, a number of micropillars were fabricated such that repeat axial compression tests can be performed. The typical experimental situation related to in situ axial compression testing of CrN/interlayer/Si(001) micropillars is illustrated in Fig. 17. The FIB milling process yielded cylindrical micropillars, 3–5 μm in diameter and ~10 μm in length, resting on a larger Si base. Figure 17a shows an array of three micropillars fabricated from one CrN/Cu/Si(001) specimen, with the Cu interlayers inclined at 45° with respect to the pillar axes. The Cu interlayer thickness for this specimen is ~550 nm, measured from SE images of a FIB cross section of the specimen. Finished pillars are taper-free and are subjected to no mechanical contact once the milling process is completed. The circular opening surrounding each pillar, ~30 μm in diameter, is sufficiently large to allow the ~10 μm × 10 μm diamond punch to engage the pillar top without contacting the rim of the circular opening. As shown in Fig. 17a, the right most pillar is located near the specimen edge. The rim material bordering the specimen edge, as well as the rim material separating one pillar from another, has been removed by additional FIB milling to allow SE imaging during axial compression of the pillars. The SE image shown in Fig. 17b illustrates the situation immediately before the flat face of the ~10 μm × 10 μm diamond punch is about to engage the top of the right most pillar shown in Fig. 17a, with the two other pillars visible in the same field of view. The shadow on the flat face of the punch, due to the pillar blocking a portion of secondary electrons emanating from the punch flat face, serves to locate the pillar in the center of the punch face during the engagement process. Similar pillar morphologies and in situ pillar compression test configurations hold for CrN/Ti/Si micropillars with interfaces inclined at either 45° or 90° with respect to the pillar axes.

Figure 18a shows the raw P–Δ curve obtained from in situ axial compression of the right most CrN/Cu/Si(001) micropillar shown in Fig. 17a, with a Cu interlayer thickness of ~550 nm and the interfaces inclined at 45° with respect to the pillar axis. Figure 18a shows an initial rapid increase in P with increasing Δ until a critical load value, P_c, is reached. Once P_c is reached, Δ increases with little further increase in P, exhibiting a stable load plateau. This load plateau, however, does not extend indefinitely. Rather, large and discontinuous punch displacement excursions occur as Δ increases beyond a point. Figure 18b–d shows single frames of the SE imaging video of the same pillar at various stages during the axial compression experiment, corresponding to the points 1, 2, and 3 identified on the P–Δ curve shown in Fig. 18a, with total punch displacements at ~115 nm, ~370 nm, and ~440 nm, respectively. It is apparent from Fig. 18a, b that little or no deformation of the Cu interlayer is observed at point 1 on the rapid load rise portion of the P–Δ curve. Once the critical load P_c is reached and the punch displacement progresses to point 2 on the load plateau, the entire Cu interlayer undergoes a shear deformation. The upward displacement of the punch is accommodated by a rigid upward shift of the top CrN portion of the pillar and is in turn accommodated by a shear deformation of the entire Cu interlayer. The CrN top portion of the pillar shifts rigidly upward and to the left in Fig. 18c. This mode of deformation continues as the punch displacement progresses further on the load plateau to point 3, and the entire Cu interlayers undergoes an increased amount of shear deformation, as evident from Fig. 18d. The large and discontinuous punch displacement excursion observed, to beyond the total punch displacement range of 800 nm displayed in Fig. 18a, is concomitant with a catastrophic shear failure. As shown in Fig. 18e, this dynamic shear-off event causes the diamond punch to quickly move upward in an uncontrolled manner, impacting the bottom Si portion of the pillar and destroying it. The entire dynamic shear-off event occurs suddenly, within the time span for collecting one single SE image frame, thus leaving no image of the pillar during any intermediate stage of this dynamic shear-off. The SE image in Fig. 18f illustrates separate posttest morphological examinations of CrN/Cu/Si pillars subjected to ex situ axial compression, after the large punch displacement excursion had occurred. It is clear that shear loading on the interfacial region, due to the axial compression load, caused a shear failure of the interfacial region, resulting in the top CrN portion of the pillar shifting rigidly with respect to the Si bottom portion. Spot mode EDS spectra collected from the top surface of the exposed bottom pillar portion and the bottom surface of the exposed top pillar portion both show the presence of Cu signal, indicating that the dynamic shear-off occurred within the Cu interlayer. The SE image shown in Fig. 18f and multiple similar ones, not shown, further indicate that the final dynamic shear-off occurred near the Cu/CrN interface, instead of in the

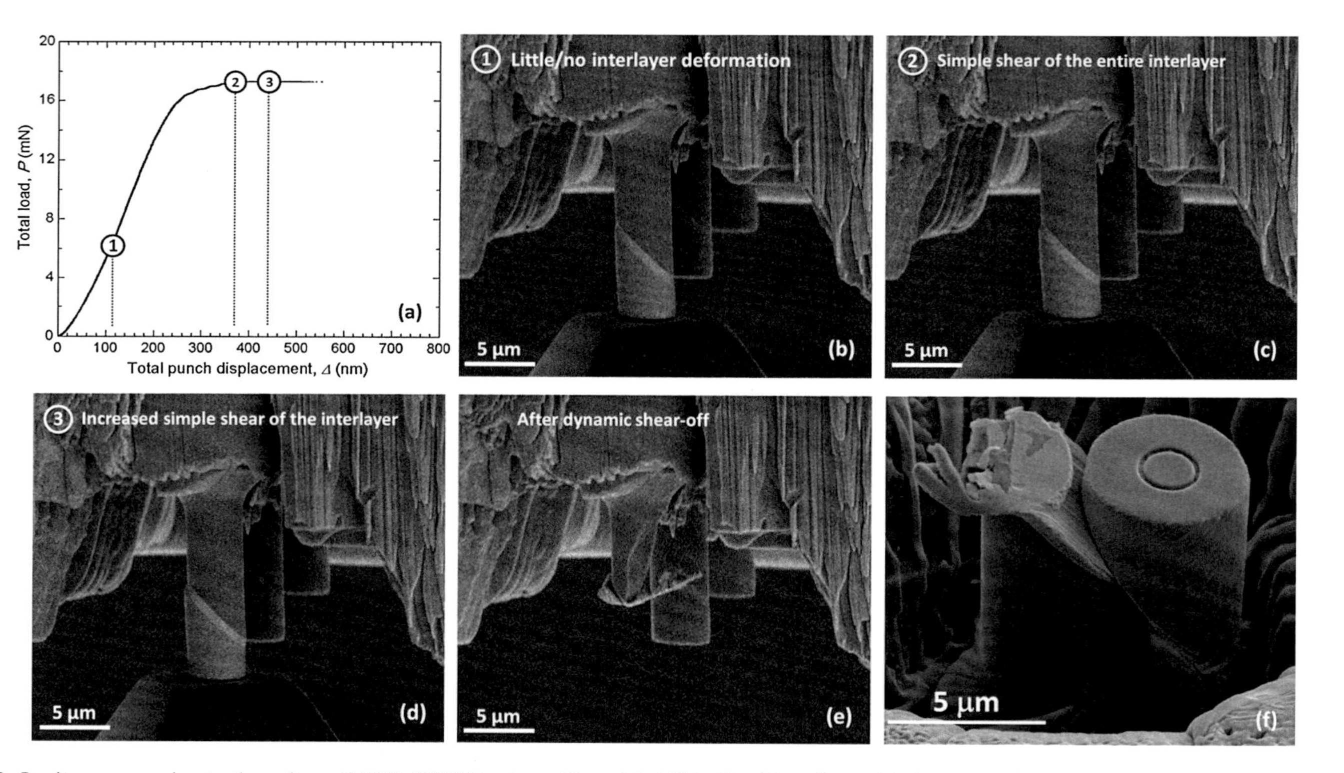

Fig. 18 In situ compression testing of one CrN/Cu/Si(001) micro-pillar with 45° inclined interfaces: (**a**) the raw P–Δ curve, points 1, 2, and 3 correspond, respectively, to Δ values of ~115 nm, ~370 nm, and ~440 nm, (**b**) the SE image of the pillar at Δ ~115 nm (point 1) (**c**) the SE image of the pillar at Δ ~370 nm (point 2), (**d**) the SE image of the pillar at Δ ~440 nm (point 3), (**e**) the SE image of the pillar after occurrence of the large and discontinuous displacement excursion, (**f**) a separate posttest SE images of another CrN/Cu/Si pillar after the dynamic shear-off event occurred, illustrating the consequence of an interfacial shear instability

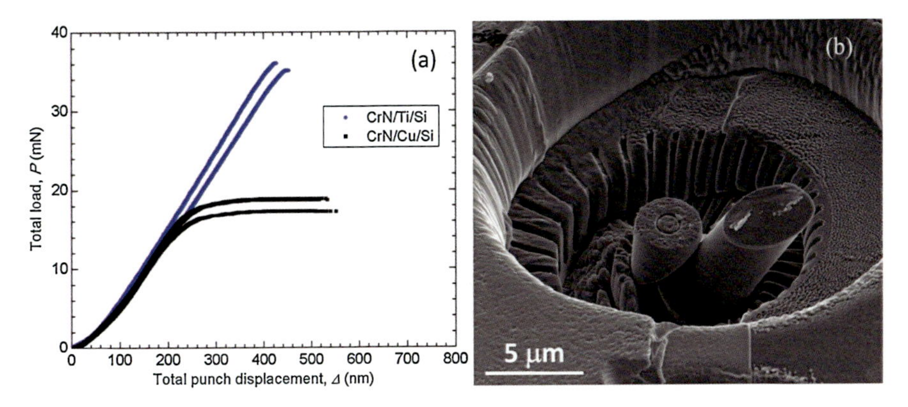

Fig. 19 In situ compression testing of CrN/interlayer/Si(001) micropillars with 45° inclined interfaces: (**a**) raw P–Δ curves obtained from testing pillars with Cu and Ti interlayers, (**b**) a posttest SE image of a pillar with a Ti interlayer, after the occurrence of the unstable dynamic shear-off event

middle of the Cu interlayer. Correlating the P–Δ curve, the concurrent SE imaging during in situ compression testing, and additional posttest SE images obtained from ex situ compression tested pillars, it is concluded that axial compression of CrN/Cu/Si micropillars with 45° inclined interfaces results in shear failure of the interfacial region and that this shear failure occurs in two stages: first by a shear deformation of the entire metal interlayer, then followed by an unstable dynamic shear-off near the Cu/CrN interface.

Figure 19a shows a collection of P–Δ curves obtained from in situ compression testing of CrN/Cu/Si and CrN/Ti/Si micropillars. The three P–Δ curves for CrN/Cu/Si were collected from testing of the three micropillars shown in Fig. 17a. Qualitatively similar P–Δ curves were observed from all three pillars: with an initially rapid rise in P with increasing Δ, followed by a stable and extended load plateau, and finished with large and discontinuous punch displacement excursions concomitant with the final dynamic shear-off. The two P–Δ curves for CrN/Ti/Si were collected from compression testing of two micropillars fabricated from one CrN/Ti/Si(001) specimen. The CrN/Ti/Si pillars are again ~4 μm in diameter and ~10 μm in length, resting on a larger Si base. The Ti interlayer thickness is ~340 nm and the interfaces are 45° inclined with respect to the pillar axis. These P–Δ curves exhibit an initially rapid rise in P with increasing Δ, but show very limited load plateaus when a critical load P_c is reached, followed by large and discontinuous punch displacement excursions concomitant with the final dynamic shear-off.

Comparing P–Δ curves shown in Fig. 19a, the differences in the response of CrN/Ti/Si pillars subjected to axial compression as compared to that of CrN/Cu/Si pillars are apparent. First, the critical load values are significantly different: P_c values are ~18 mN and ~36 mN for CrN/Cu/Si and CrN/Ti/Si pillars, respectively. Second, the extent of the load plateau is much more limited for the CrN/Ti/Si pillars as compared to the extended and stable load plateaus observed for the CrN/Cu/Si pillars. The absence of an extended and stable load plateau during axial compression of CrN/Ti/Si pillars means that very soon after the axial load reaches P_c, dynamic shear-off occurs. The SE image videos during pillar compression testing of CrN/Ti/Si pillars thus contain no frame showing either significant deformation of the Ti interlayer in shear or any intermediate stage of the final dynamic shear-off. Figure 19b shows posttest morphological examination of one CrN/Ti/Si pillar subjected to ex situ axial compression, after the final large punch displacement excursions had occurred. In this case, the punch lurching forward knocked the bottom pillar portion off its base. The top CrN portion of the pillar shifted as a whole with respect to the Si bottom portion along the 45° inclination as a result of the shear loading on the interfacial region. Spot mode EDS analysis showed the presence of Ti signals on both the top surface of the bottom pillar portion and the bottom surface of the top pillar portion. Similar to Fig. 18f, examination of the image shown in Fig. 19b shows that the final dynamic shear-off occurred near the Ti/CrN interface, instead of in the middle of the Ti interlayer.

Data and observations shown in Figs. 18 and 19 indicate that axial compression of CrN/interlayer/Si micropillars with 45° inclined interfaces led to shear failures of the interfacial regions in two stages: an initial shear deformation of the entire metal interlayer, followed by a sudden dynamic shear-off close to the metal/CrN interface. The main difference between the CrN/Cu/Si system and the CrN/Ti/Si system subjected to shear loading on the interfacial region is the existence of an extended and stable load plateau in the former system and a limited load plateau in the latter system. For CrN/Cu/Si, the entire Cu interlayer is subjected to a shear deformation to large strains before the final dynamic shear-off occurs close to the Cu/CrN interface. For CrN/Ti/Si, the final dynamic shear-off close to the Ti/CrN interface occurs soon after the Ti interlayers are subjected to a shear deformation.

Because shear failure of the interfacial regions initiates with a shear deformation of the entire metal interlayer, the critical load P_c measured experimentally from axial compression of CrN/interlayer/Si micropillars with 45° inclined interfaces yields a measure of the average shear stress τ_c necessary to initiate shear failure of the interfacial region,

$$\tau_c = P_c / \left(\pi D^2/4\right) / 2, \tag{5}$$

where D is the pillar diameter. Figure 20 summarizes results of ex situ and in situ axial compression testing on CrN/Cr/Si(001) and CrN/Ti/Si(001) micropillars with 45° inclined interfaces by plotting values of τ_c against the metal interlayer thickness h. Separate data points at the same h value denote repeat measurements on separate pillars. From data presented in Fig. 20, several points should be noted. First, τ_c values measured from ex situ and in situ compression tests, conducted with two separate instruments on the same specimens, are consistent, indicative of the fidelity of the present set of measurements. Second, the value of τ_c increases with decreasing h, significantly for Cu interlayers, somewhat for Ti interlayers although the data scatter is large in this case. An increase in τ_c with decreasing h is

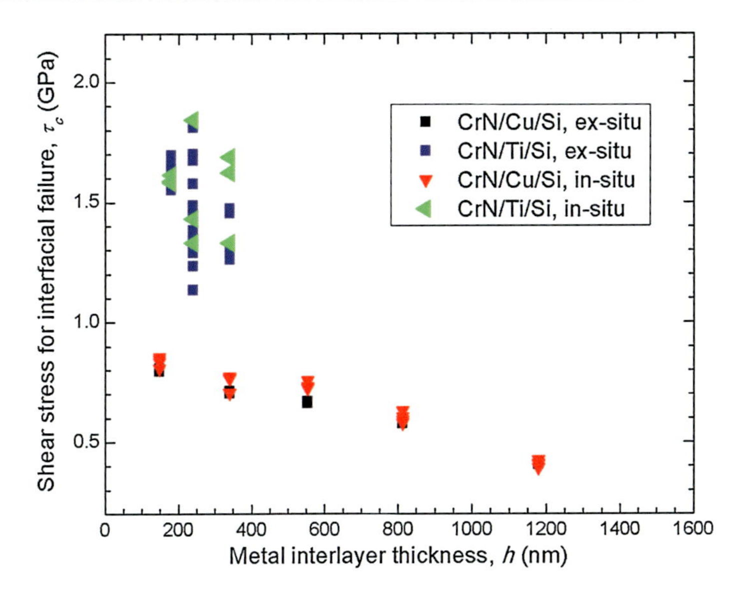

Fig. 20 Average shear stress to initiate shear failure of the interfacial regions of CrN/interlayer/Si specimens plotted versus the thickness of the metal interlayer

qualitatively consistent with strain gradient plasticity model predications, according to which confinement of the metal interlayer in between two elastic-brittle solids offers one reason for the observed increase in interlayer flow stress (Mu et al. 2014). Third, τ_c depends significantly on the metal interlayer: the shear failure stress for Ti interlayers exceeds that for Cu interlayers by about a factor of two, significantly outside the data scatter band even with the large scatter in the Ti case.

Clues to the mechanism through which the final dynamic shear-off occurs in the two CrN/interlayer/Si systems are offered from observations made during in situ axial compression testing of CrN/Cu/Si(001) micropillars with interfaces 90° inclined with respect to the pillar axes. Figure 21a shows a raw P–Δ curve obtained from in situ axial compression of one CrN/Cu/Si(001) micropillar, with a Cu interlayer thickness of ~1180 nm and the interfaces inclined at 90° with respect to the pillar axis. Figure 21a shows an initial rapid increase in P with increasing Δ until a critical load P_c is reached. Once P_c is reached, the P–Δ curve exhibits a plateau region in which P increases only slowly with increasing Δ. Beyond this plateau region, further increases in Δ bring again significant increase in P. Figure 21b–e shows single frames of the SE imaging video of the same pillar at various stages during the axial compression experiment, corresponding to the points 1, 2, 3, and 4 identified on the P–Δ curve shown in Fig. 21a, with total punch displacements at ~225 nm, ~425 nm, ~525 nm, and ~1200 nm, respectively. It is apparent from Fig. 21b that little or no deformation of the Cu interlayer is observed at point 1 on the rapid load rise portion of the P–Δ curve. Once the critical load P_c is reached and the punch displacement progresses to point 2 on the load plateau, Fig. 21c shows

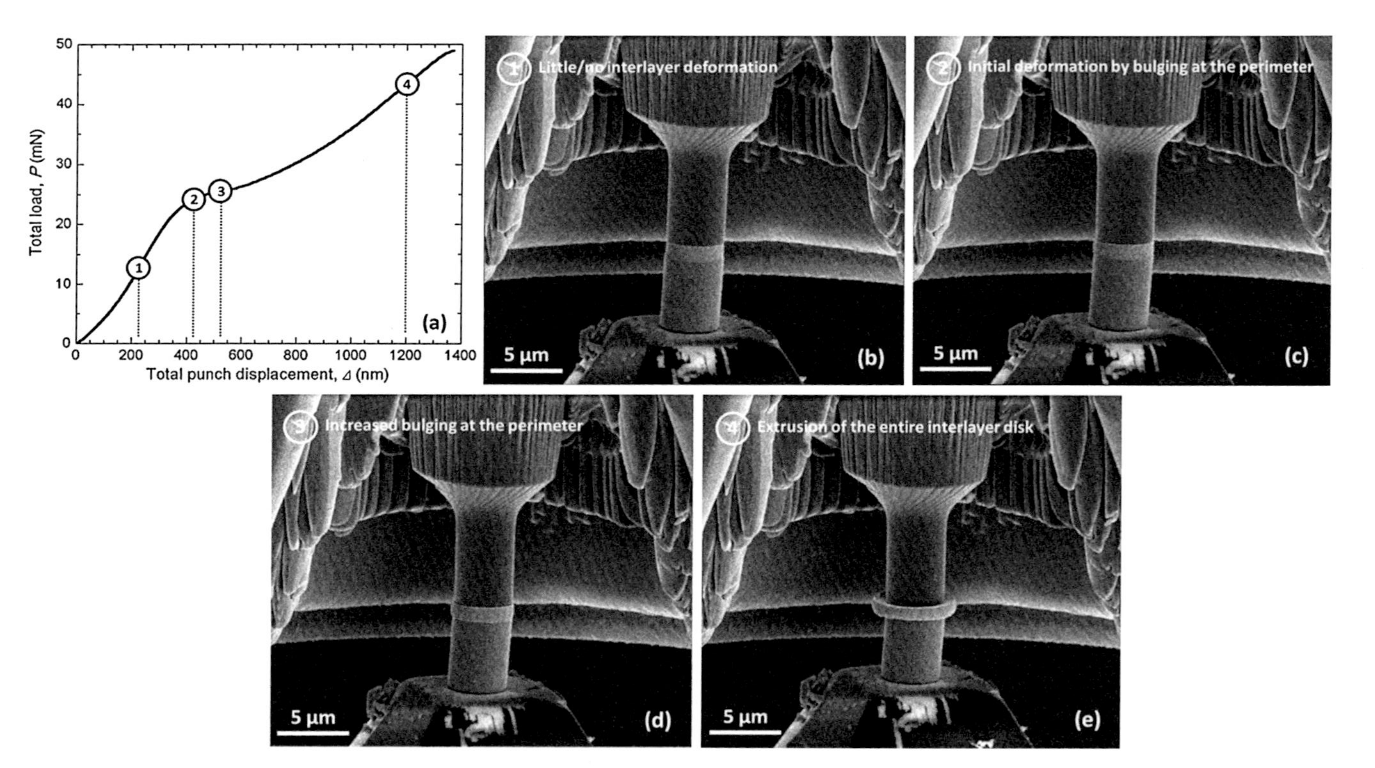

Fig. 21 In situ compression testing of one CrN/Cu/Si(001) micropillar with 90° inclined interfaces: (**a**) the raw P–Δ curve, points 1, 2, 3, and 4 correspond, respectively, to Δ values of ~225 nm, ~425 nm, ~525 nm, and ~1200 nm, (**b**) the SE image of the pillar at Δ ~225 nm (point 1), (**c**) the SE image of the pillar at Δ ~425 nm (point 2), (**d**) the SE image of the pillar at Δ ~525 nm (point 3), (**e**) the SE image of the pillar at Δ ~1200 nm (point 4)

that deformation of the Cu interlayer occurs by bulging out at the perimeter from the center of the interlayer, with the diameters of the Cu interlayer at the Cu side of the Cu/ceramic interfaces appearing not to change. This mode of deformation continues as the punch displacement progresses further on the load plateau to point 3, as shown in Fig. 21d. The entire Cu interlayer undergoes an increased amount of bulging out at the perimeter from the center of the interlayer, still leaving the diameters of the Cu interlayer at the Cu side of the interfaces unchanged. As the punch displacement progresses to beyond the load plateau at point 4 on the P–Δ curve, the entire Cu interlayer is extruded out from the perimeter with its diameter significantly larger than the original pillar diameter, as shown in Fig. 21e.

The in situ compression testing of CrN/Cu/Si micropillars with 90° inclined interfaces offers one insight to how the metal interlayers confined between elastic-rigid materials, CrN and Si in the present case, deform under load. As shown in Fig. 22a, the morphology of the Cu interlayer during an early elasto-plastic stage of deformation is illustrated with an actual SE image of a deformed Cu interlayer. In this posttest examination, Ga^+ FIB milling was used to remove a small section of the pillar after compression loading to reveal better the morphology of the deformed Cu interlayer. At this early deformation stage, plastic deformation of the interlayer occurs by bulging out at the perimeter in the center, while the metal/ceramic interfaces appear to be "locked," with no relative displacement between the metal and ceramic side of the interface. Figure 22b shows two analogous SE images of a deformed Cu interlayer at a later deformation stage. The interlayer deformation leads to the entire interlayer disk being extruded out in an axisymmetric manner. Here the metal/ceramic interfaces appear to be "unlocked," with significant relative displacement between the metal and ceramic side of the interface, thus allowing extrusion of the entire interlayer disk to occur. The sliding marks, visible on the top surface of the extruded Cu interlayer disk, offer clear evidence that relative displacement occurred between the Cu side and the CrN side of the Cu/CrN interface. The FIB milled cut-away view in Fig. 22c shows that the thickness of the Cu interlayer disk is the largest at the outer rim, further indicating that extrusion of the Cu interlayer disk occurred after the interface became "unlocked." Unfortunately, plastic flow stresses of thin Ti layers are too high to allow similar observations to be made in the cases of CrN/Ti/Si pillars: the Si bottom portion of the pillars fail in compression before perceptible deformation of the Ti interlayer occurs in the confined normal compression geometry. Nevertheless, we suggest that a similar transition occurs for both CrN/Cu/Si and CrN/Ti/Si interfacial regions, i.e., the interfaces go from being "locked" to being "unlocked" as the metal interlayers confined between CrN and Si are stressed to increasingly higher levels. We further suggest that this interfacial "unlocking" is physically related to the final dynamic shear-off observed in both systems, as illustrated in Figs. 18 and 19. To our knowledge, both the observation of an interfacial instability under shear loading and the suggestion that this shear instability is linked to a transition from an interfacial "locking" to an interfacial "unlocking" are new and demonstrate the potential of in situ instrumented microscale mechanical testing and its application to studying the mechanical integrity of metal/ceramic interfaces.

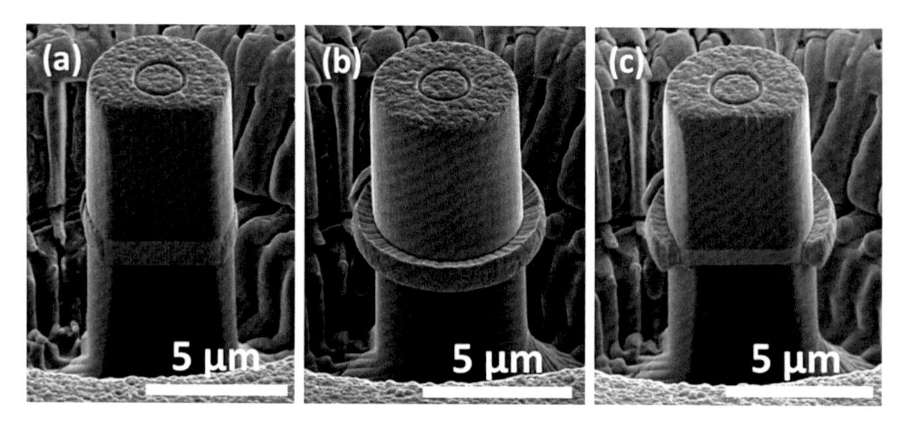

Fig. 22 Posttest images of deformed Cu interlayer: (**a**) in the early elasto-plastic stage, (**b**) in the later elasto-plastic stage

Mechanistic Understanding on the Shear Instability of Metal/Ceramic Interface Provided by Simulations

In this subsection, the structure and mechanical response of the Ti/TiN interfaces are investigated using MD/DFT simulations to shed light on the relevant properties of a general metal/ceramic interface. Part of the reasons for this choice is the availability of the semi-empirical potentials available for metal/ceramic systems (Zhang et al. 2017; Yang et al. 2017; Kim and Lee 2008a). At room temperature, the equilibrium lattice structure of Ti is HCP while that for TiN is of NaCl type. The preferred orientation relation (OR) for the Ti/TiN system is $\langle 0001 \rangle_{Ti} // \langle 111 \rangle_{TiN}$ and $\langle 11\bar{2}0 \rangle_{Ti} // \langle 110 \rangle_{TiN}$ on the interface plane: $(0001)_{\mathrm{Ti}} // (111)_{\mathrm{TiN}}$ (Yang et al. 2017; Sant et al. 2000). Another reason for the choice of the Ti/TiN system is the similarity between TiN and CrN – material used in our experiments (section "Synthesis of Ceramic–Coating/Adhesion–Interlayer/Substrate Systems and Structural/Mechanical Characterization of Coating/Substrate Interfacial Regions"). Both nitrides have identical lattice type and very close lattice parameters ($a_{\mathrm{TiN}} = 4.242$ Å vs. $a_{\mathrm{CrN}} = 4.149$ Å) (Toth 1971; Kim and Lee 2008b). The structures of the Ti/TiN, Ti/CrN, and Cu/CrN interfaces are also expected to be similar, since the hexagonal atomic structure on (0001) Ti planes is also similar to the (111) Cu planes (Shao et al. 2013; Wang et al. 2014). Therefore, understandings gathered from the numerical modeling efforts herein, namely, the structure and properties of Ti/TiN interfaces, should have significant implications on other metal/ceramic interfacial systems, e.g., Ti/CrN, Cu/CrN (Shao et al. 2015; Wang et al. 2008, 2014; Zhang et al. 2013; Chen et al. 2017).

As an outline, in section "The Structure of the Misfit Dislocation Network and the Coherent Regions of the Ti-TiN Interfaces," following the methodology proposed by Hirth and Lothe (1982), the structure of the Ti-TiN interface is analyzed through the "structural relaxation approach." As will be shown, the structural details of the semi-coherent interface are dependent on the intrinsic properties of the coherent Ti-TiN interfacial structures. Hence, in section "The Location of the Misfit Dislocation Network," the energetic characteristics of the coherent structures evaluated by DFT calculations are presented. Finally, in section "Intrinsic Energetic Characteristics of Coherent Ti/TiN Interfacial Structures," the mechanical response of the Ti-TiN interfaces to external shear loading is discussed.

The Structure of the Misfit Dislocation Network and the Coherent Regions of the Ti-TiN Interfaces

Two specific ORs are permitted by the general OR given above, namely:

OR 1: x//$[\bar{1}100]_{Ti}$//$[11\bar{2}]_{TiN}$, y//$[0001]_{Ti}$//$[111]_{TiN}$, and z//$[11\bar{2}0]_{Ti}$//$[1\bar{1}0]_{TiN}$,

OR 2: x//$[1\bar{1}00]_{Ti}$//$[11\bar{2}]_{TiN}$, y//$[0001]_{Ti}$//$[111]_{TiN}$, and z//$[\bar{1}\bar{1}20]_{Ti}$//$[1\bar{1}0]_{TiN}$.

On the interface, the $(111)_{TiN}$ plane and the $(0001)_{Ti}$ plane have the identical hexagonal structure, with lattice match of 1.84%. Due to this mismatch, a number of distinct coherent structures may exist on the unrelaxed Ti/TiN interfaces of the above two ORs, as shown by the plan views of provided in Fig. 23a, b. Such coherent structures include near-FCC, near-HCP, near-Overlap structures and the region separating the former three, based on the relative positions of Ti atoms straddling on both size of the interface, which are shown in the lower portions of Fig. 23. Even though the Ti atoms may have taken the identical relative positions in

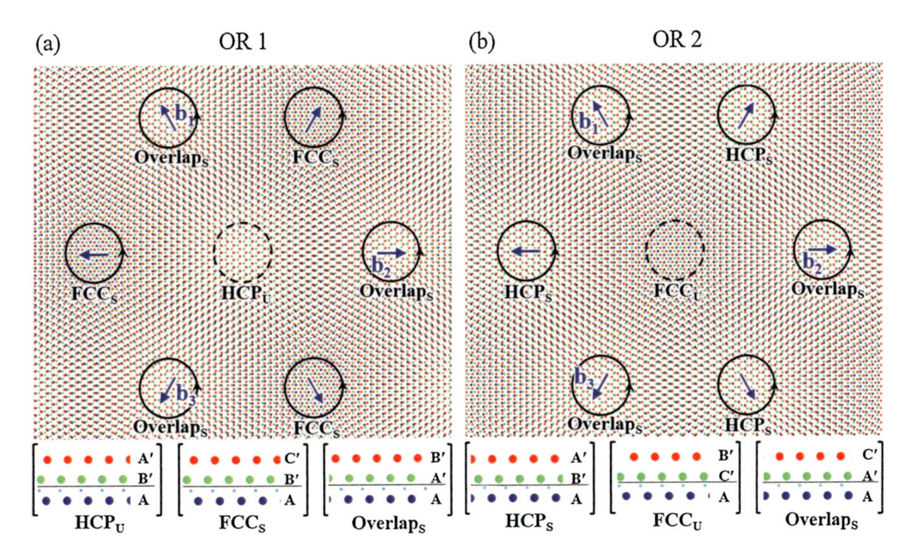

Fig. 23 Structure of the "unrelaxed" Ti/TiN interface prior to formation of the MDN at the chemical interface. The plan views of the interface are provided on the top portions of the figure, and the local atomic stacking is provided on the lower portion. The N atoms are omitted in the plan view for clarity and are shown as smaller spheres in the stacking structures. The atoms are colored according to their vertical positions. The primed notations, i.e., A'B'A', etc., is used for Ti layer to distinguish from stacking in TiN layer. The thin solid lines denote the position of the chemical interface

the respective FCC and HCP regions in the Fig. 23a, b, the structures are clearly not identical, due to the presence of the N atoms. DFT calculations show that (details in the next sections) the FCC and HCP structures in the two ORs have different energy levels. For instance, the FCC structure in OR 1 has the lowest excess energy; therefore, the relative positions of the Ti and N atoms specified in this structure are referred to as the FCC_S structure. On the other hand, the FCC in OR 2, in which the Ti atoms on the Ti side overlap with the N atoms, corresponds to a higher excess energy. According to DFT calculations, this structure is unstable and is thus named FCC_U. Same naming convention and energy characteristics apply for the HCP structures. Unlike the FCC and HCP structures, the Overlap structures are metastable and are identical in both ORs.

Following the approach proposed by Hirth and Lothe (1982), the regions with stable and metastable atomic structure can be regarded as being enclosed by Shockley partial dislocation loops (black solid circles shown in Fig. 23). The loops have Burgers vectors: $b_1, b_2, b_3, -b_1, -b_2$, and $-b_3$ ($b_1 = \frac{a}{3}\left[0\bar{1}10\right], b_2 = \frac{a}{3}\left[10\bar{1}0\right]$, and $b_3 = \frac{a}{3}\left[\bar{1}100\right]$). The arrows on the loops mark the line senses of the loops. Upon relaxation, driven by the minimization of the interface excess energy, the regions of FCC_S, HCP_S, and $Overlap_S$ expand. As a result, larger patches of coherent regions form associated with the in-plane biaxial straining of the Ti and TiN lattices. When the adjacent loops reach close proximity with each other, they react and form the misfit dislocation network (MDN) (schematically shown in Fig. 24, where partial dislocation loops, misfit dislocation lines and nodes are represented in light blue, brown, and green). The nodes (intersections of dislocations) of the MDN originate from the contraction of regions with unstable structures. Dislocation lines in the MDN are also Shockley partials with Burgers vectors b_1, b_2, and b_3. The thermodynamics of the interface relaxation is consistent with the perspective of the classical Peierls-Nabarro dislocation model (Peierls 1940; Nabarro 1947) and is characterized by two competing factors, namely: (1) the reduction of the chemical potential energy (excess interface energy of the coherent structures) and (2) the increase in the core strain energy in the MDN. The former factor is governed by the generalized stacking fault energy (GSFE) of the interface, while the latter is governed by the normal elastic constant of the adjoining crystals. In effect, the core width of the MDN inversely relates to the magnitude of interfaces' GSFE profile (Shao et al. 2014), for instance, higher amplitude in GSFE corresponds to narrower cores.

Two distinct relaxed interface structures exist for the two ORs (Figs. 25a, b). Interface with OR 1 comprises FCC_S (green atoms) and $Overlap_S$ (white atoms) regions, while the interface with OR2 comprises HCP_S (red atoms) and $Overlap_S$ (white atoms) regions. It is interesting to note that the dislocation lines of the MDN (on the edge of the regions in Fig. 25a, b) on the chemical interface are jagged. This is consistent with the observation made by Yang et al. in their recent work (Yang et al. 2017). Accordingly, the blue solid lines in Fig. 25a, b simply mark the neutral positions of the dislocations. Therefore, the net character on dislocation lines is edge, even though the local character of the partial misfit dislocations may be

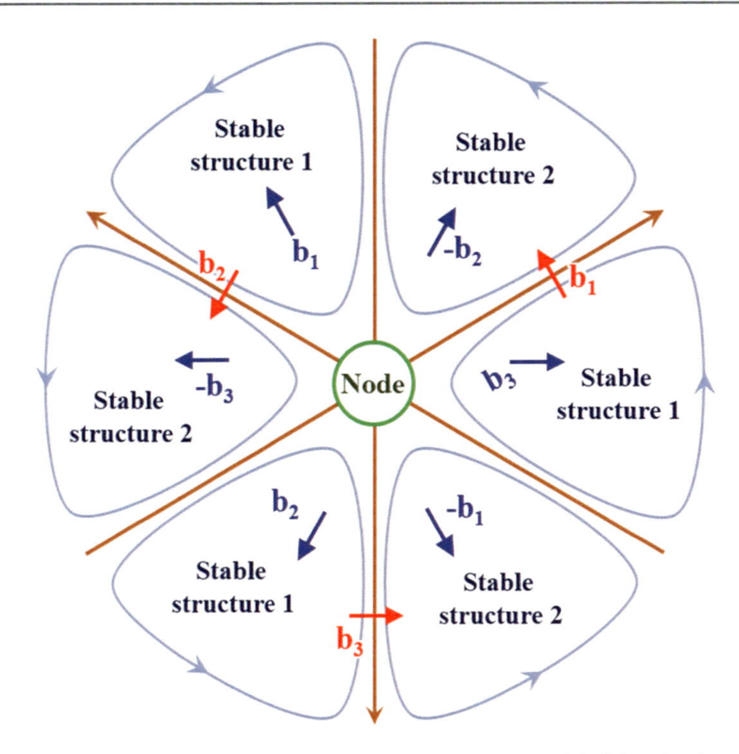

Fig. 24 Schematic of the expansion of and reaction between the partial dislocation loops which encloses the coherent stable structures

mixed. In addition, from Fig. 25a, b, one can see that both FCC_S (green) and HCP_S (red) regions occupy greater area than the $Overlaps_S$ (white) regions, also the area of the HCP_S regions is greater than the area of the FCC_S regions. This implies that the excess specific energies of the coherent structures have the following relation, $E_{\mathrm{Overlaps_S}} > E_{\mathrm{FCC_S}} > E_{\mathrm{HCP_S}}$. Our DFT calculations confirm this prediction, as will be shown in the next sections.

The Location of the Misfit Dislocation Network

The energetically preferred location of the MDN, in general, should not be at the chemical interface. The reasoning is given as follows:

1. The elastic line energy (per unit length) of dislocations is proportional to the shear modulus of the material, i.e., $E/L \propto Gb^2$. A dislocation of the same burgers vector in ceramics would have significantly higher energy than one in metal. Therefore, from the thermodynamics standpoint, the MDN would have significantly higher energy when it is located at the Ti/TiN chemical interface compared to when it is in Ti.
2. The stress field of misfit dislocations decays exponentially as a function of distance from the plane of MDN (Hirth and Lothe 1982; Shao et al. 2018).

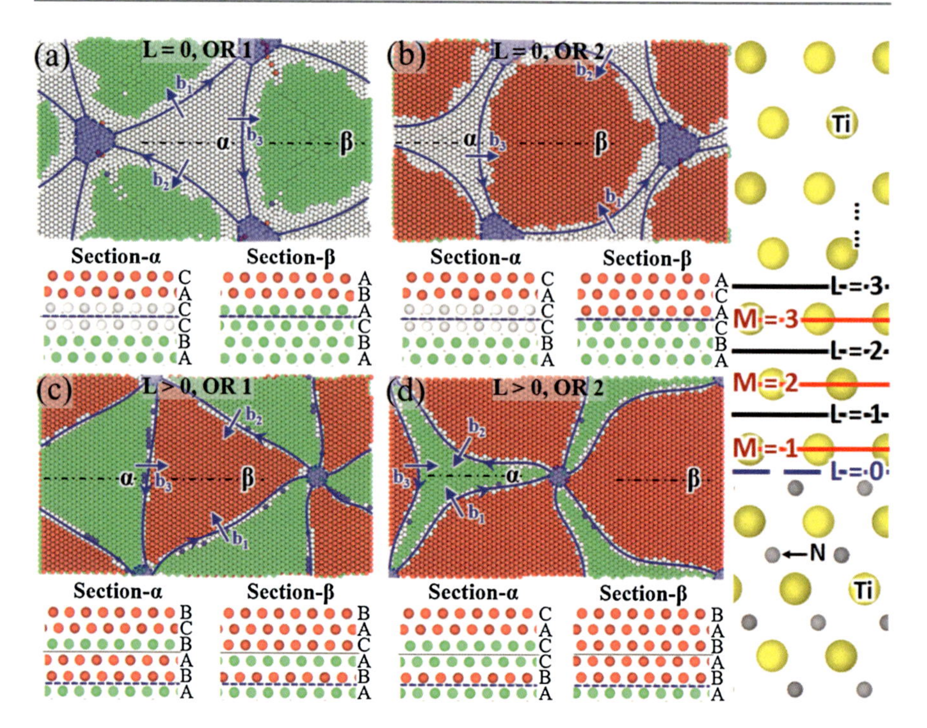

Fig. 25 Typical structures of the MDNs corresponding to different locations, i.e., different L numbers as indicated on the right. (**a**) and (**b**) show Type I and Type II structures for $L = 0$, (**c**) and (**d**) show Type I and Type II structures for $L > 0$. Atoms are colored by common neighbor analysis performed on Ti atoms. Atoms within a local HCP structure are colored red, and those within a local FCC structure are colored green. Atoms that are not in HCP, FCC, BCC, or icosahedral structures are colored white. Misfit dislocation lines and nodes are denoted by blue lines and blue shades

Consequent, the strain energy of the MDN is expected to decrease sharply as it moves away from the chemical interface into Ti.

3. As an MDN is moving away from the chemical interface into Ti, a layer of Ti with in-plane lattice constant coherent to TiN is created. The thickness of this layer is identical to the distance between the MDN to the chemical interface. The total strain energy of this layer therefore increases linearly with its thickness.
4. The true interface configuration depends on the total free energy of the entire interface region (three dimensional entity), which includes the chemical potential energy at the chemical interface, the elastic energy of the misfit dislocation network, and the elastic energy of the coherent Ti atomic layers.

Therefore, it is expected that the MDN prefers to reside inside Ti but very close to the chemical interface. This is confirmed by Yadav et al. (2015) in their recent work, where they found that the interface energy reaches minimum when the MDN is in Al and 2 atomic monolayers away from the chemical interface.

In this section, we present our prior findings, obtained from molecular dynamics simulations, on the variation in the structure as well as the energy of the interface as a function of the position of the MDN in Ti. An L parameter is used here to indicate the position of the MDN. As shown on the right hand side of Fig. 25, the MDN resides either on the chemical interface (blue dashed line), which is in between the atomic monolayers of Ti and N ($L = 0$), or on one of the (0001) slip planes in Ti ($L > 0$, black solid lines). The (0001) slip planes exist in between two adjacent (0001) atomic monolayers of Ti. When the MDN is inside Ti and has n Ti atomic monolayers between the chemical interface, $L = n$. The parameter M indicates the position of (0001) Ti atomic monolayers (red solid lines) near the interface, with $M = 1$ denoting the Ti atomic layer most adjacent to the chemical interface. The structure of the Ti/TiN interface for $L = 0$ has been discussed in detail above. Similar to the case of $L = 0$, the coherent regions are separated by the MDN for $L > 0$. For OR 1, the coherent regions have the structures that comprise single layer of atoms surrounded by local FCC structure (green atoms) and a single layer of atoms surrounded by HCP structure (red atoms). The structures are commonly referred to as the type 1 intrinsic stacking fault (I_1) in HCP (Benoit et al. 2013) and are energetically equivalent. Thus, the area fraction of the coherent regions is the same (see the green and red regions in Fig. 25c). For OR 2, the coherent regions include a higher energy structure (type 2 intrinsic stacking fault of HCP, I_2 (Benoit et al. 2013)), which comprises two layers of atoms surrounded by FCC structures (green atoms) and a low energy HCP structure. Accordingly, the FCC region occupies a significantly less area fraction compared to the HCP region. In addition, a noticeable spiral pattern on the dislocation line around the nodes is clearly visible in Fig. 25c, d, and the same feature was also observed in Cu-Ni interfaces (Shao and Wang 2016).

The energy of the Ti/TiN interfaces, for both ORs, as a function of the L parameter is given in Fig. 26a. The interface energy (γ) is calculated using the following formula:

$$\gamma = \frac{1}{A}\left[E_{\text{interf.}}\left(nTi_c, mTi_m, lN\right) - nE_{Ti}^{c} - mE_{Ti}^{m} - lE_N\right],$$

where A is the area of the interface in the computational cell, $E_{\text{interf.}}$ is the total energy of the structure containing the interface, n is the number of Ti atoms in TiN layer, m is the number of Ti atoms in the Ti layer, and l is the number of N atoms. E_{Ti}^{c} and E_{Ti}^{m} are the cohesive energies of the Ti atoms in TiN (ceramic) and Ti (metal). E_N is the cohesive energy of N. It is apparent from Fig. 26a that the interface energy significantly decreases when the MDN is moved away from the Ti/TiN chemical interface by one monolayer of Ti and gradually increases as the MDN is moved further into Ti. As was discussed above, the dramatic reduction in interface energy is associated with the reduction in the MDN's elastic energy as well as the interface stacking fault energy. The gradual increase in interface energy when $L \geq 1$ is associated with the increase in the number of coherent Ti atomic monolayers situated between the MDN and the Ti/TiN chemical interface.

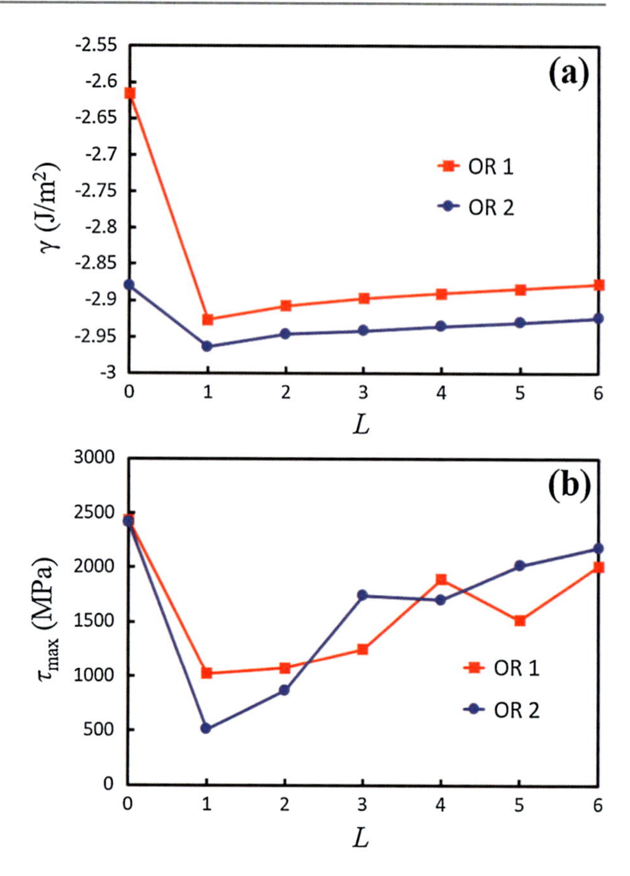

Fig. 26 The variation in interface free energy (**a**) and the interface shear strength (**b**) of the Ti/TiN interface as a function of location of MDN (L)

Another observation from Fig. 26a is that the interface energy for OR 2 is consistently smaller than that for OR 1. This is primarily due to the greater area fraction of low energy structures in the OR 2 interfaces. Next, we discuss this in detail, separately, using the cases of $L = 0$ and $L > 0$. For $L = 0$, the OR 1 interface comprises the higher-energy $\mathrm{Overlap_S}$ structure and the intermediate-energy $\mathrm{FCC_S}$ structure, while the OR 2 interface comprises higher-energy $\mathrm{Overlap_S}$ structure and the low-energy $\mathrm{HCP_S}$ structure. This gives rise to the significantly higher energy of the OR 1 (-2.62 J/m^2) interface compared to OR 2 (-2.88 J/m^2). On the other hand, when $L > 0$, the coherent structures in OR 1 comprise only I_1 stacking fault, while the coherent structures in OR 2 comprise I_2 stacking fault as well as perfect HCP stacking. Energetically, according to the current work and the other studies (Benoit et al. 2013), the stacking fault energy of I_1 is around 1/2 of that of the I_2. In fact, it has been reported, agreeing with the current work, $\gamma_{I_1} = 148.6\ \mathrm{mJ/m^2}$ and $\gamma_{I_2} = 259.1\ \mathrm{mJ/m^2}$ (Benoit et al. 2013). Therefore, the total stacking fault energies for the interfaces are $E_{I_1} = \gamma_{I_1} A$ and $E_{I_2} = \eta\gamma_{I_2} A$, where A is the areal of the unit simulation cell, η is the area fraction of the I2 stacking fault, and $\eta < 0.5$.

It is apparent that $E_{I_1} > E_{I_2}$. Therefore, the interface energy of OR 1 interfaces is always greater than that of the OR 2 interfaces.

It is also interesting to note the correlation between variation of interface free energy (Fig. 26a) and the variation in the theoretical shear strength of the interface (Fig. 26b), as a function of the location of the MDN (L number). As mentioned before, the intrinsic energetic properties of the coherent interface structure (such as the GSFE profile and the work of adhesion) affect the exact characteristics of the MDN (such as core widths of the dislocations and nodes) and, ultimately, the overall mechanical response of the Ti/TiN interface. Therefore, in the following sections, we first discuss the intrinsic properties of the coherent Ti/TiN interfacial structures (section "Intrinsic Energetic Characteristics of Coherent Ti/TiN Interfacial Structures"), then the detailed shear deformation mechanisms of the interfaces are presented (section "The Shear Response of the Ti–TiN Interfaces").

Intrinsic Energetic Characteristics of Coherent Ti/TiN Interfacial Structures

The structural characteristics and, in turn, the mechanical properties of the Ti/TiN interfaces are dictated by the intrinsic properties of the coherent interfacial structures. DFT, although computationally limited to length scales of a few Å in each dimension, has great advantages in probing properties of metal/ceramic interfaces. It explicitly models the valence electrons of atoms, which provide a highly accurate assessment of the fundamental properties of coherent interface structures. As will be discussed in the following sections, it is these properties that govern the structure and mechanical response of "real" interfaces between Ti and TiN layers, i.e., the semi-coherent interfaces. The energy characteristics of the fully coherent Ti/TiN chemical interface as well as the nearby Ti (0001) slip planes obtained using DFT calculations (Zhang et al. 2017) are analyzed and presented below.

Shown in Fig. 27 is the plot of the work of adhesion (W_A) on the Ti/TiN chemical interface ($L = 0$), as well as the various (0001) planes in Ti close to the chemical interface ($L > 0$). Clearly due to the strong ionic bonds formed between the Ti and N atoms, W_A on the chemical interface is very large, $\sim$ 6.11 J/m^2. As the plane of analysis moves away from the chemical interface, W_A decreases dramatically and appears to saturate at 3.74 J/m^2 when $L \geq 2$. Theoretically, the work of adhesion is related to the free surface energy by factor of two, i.e., $W_A = 2\gamma$. The work of adhesion on the (0001) plane of bulk Ti, shown as the red dashed line in Fig. 27, is calculated to be 4.02 J/m^2, which agrees well with the experimental measured (0001) surface energy of 2.1 J/m^2 and 1.92 J/m^2 (Benoit et al. 2013). This indicates that the cohesive strength of the (0001) Ti reaches a minimum due to the presence of the Ti/TiN chemical interface, below that for bulk Ti.

A similar trend is observed in the generalized stacking fault energy (GSFE) profiles of the Ti/TiN chemical interface ($L = 0$), as well as the Ti (0001) slip planes near the chemical interface ($L > 0$). Shown in Fig. 28a is the GSFE surfaces for $L = 0$. The GSFE are calculated using the minimum energy structures as the reference. The minimum energy structure when at the chemical interface ($L = 0$) correspond to point 1 in Fig. 28a. The minimum energy interface structure is such

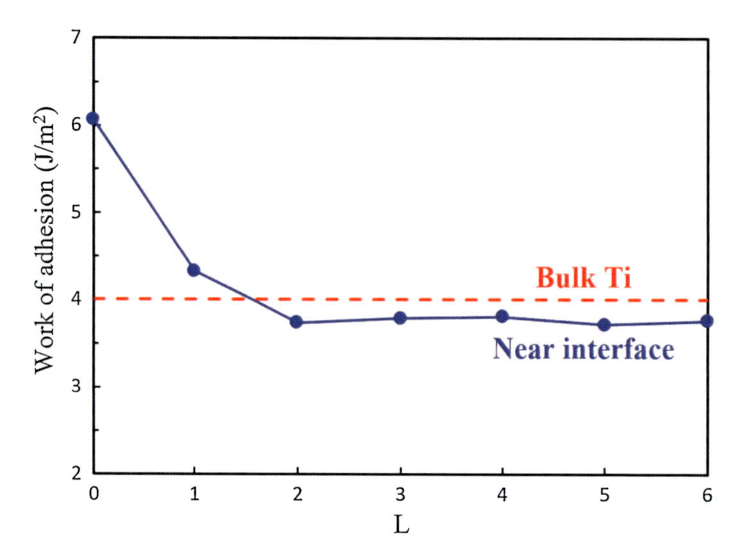

Fig. 27 The variation of the work of adhesion (W_A) on the Ti/TiN chemical interface and various (0001) slip planes of Ti as a function of L number

that the Ti atoms form the stable FCC structure across the interface. The Ti and N atoms immediately adjacent to the interface form the stable NaCl structure. A metastable coherent structure, point 2 on Fig. 28a, also exists on the chemical interface. In this structure, the Ti atoms adjacent to the interface overlap. This metastable structure is named Overlap_S. The energy for Overlap_S structure is 0.89 J/m^2. The maximum of GSFE at the chemical interface (2.71 J/m^2) occurs when, immediately to the interface plane, the Ti atoms are right on top of the N atoms. The maximum GSFE corresponds to point 3 in Fig. 28a and structure 3 in Fig. 28c, where the Ti atoms across the interface form an HCP structure. This structure is unstable and is therefore named HCP_U. Note that maximum stacking fault energy of the chemical interface is so high that it exceeds the (0001) free surface energy of the bulk Ti (2.1 J/m^2) (Benoit et al. 2013).

When $L \geq 1$, the amplitude of the GSFE drastically decreases. For clarity of display, only one typical case ($L = 1$) is shown in Fig. 28b. Note that the energy minimum (Point α in Fig. 28b) corresponds to the stable HCP structure of Ti, which is the reference of the GSFE surface. The metastable structure (I_2 stacking fault of HCP crystals) has GSFE of 0.32 J/m^2 (point β in Fig. 28b). The maximum GSFE for $L = 1$ is 0.63 J/m^2 (Point γ in Fig. 28b). The profile of the GSFE along the $[\bar{1}100]$ direction is shown in Fig. 28d, along with the profile along the same direction on the GSFE surfaces for $L = 2$, 3 and in bulk Ti. The GSFE profile changes slightly from $L = 1$ to $L = 3$. The smallest variation in GSFE occurs when $L = 3$, which is noticeably smaller than in bulk Ti.

The results presented above indicate the existence of a weak interaction plane in the Ti metal near the Ti/TiN chemical interface. As will be discussed in the

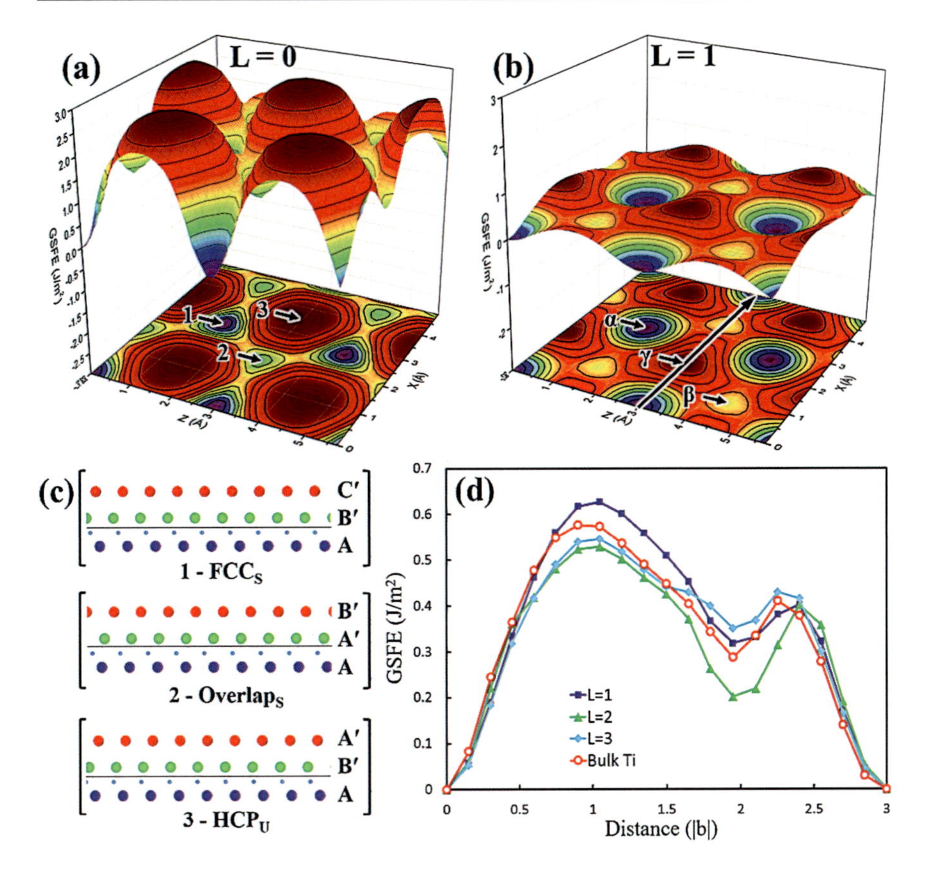

Fig. 28 (**a**) & (**b**) The surface plot of the generalized stacking fault energy of the Ti/TiN chemical interface ($L = 0$) as well as the (0001) slip plane of Ti when $L = 1$. (**c**) The coherent interface structure corresponding to the Points 1, 2, and 3 in (**a**). (**d**) The profile of GSFE along $[\bar{1}100]$ for $L = 1, 2, 3$ and bulk Ti

upcoming section, the weakened interaction on that plane alters the structure therefore the mechanical properties of the MDN which negatively impacts the shear strength of the Ti/TiN interface.

The Shear Response of the Ti–TiN Interfaces

As was shown in section "The Location of the Misfit Dislocation Network," the presence of the chemical interface alters the energy landscape (including the work of adhesion, as well as the generalized stacking fault energy, or the γ surface) of the Ti (0001) slip planes near the Ti/TiN chemical interface. There exists a weak interaction plane inside Ti but near the Ti/TiN chemical interface. Just a small distance away from the chemical interface, the work of adhesion reaches a minimum value of around 3.72~3.76 J/m^2 when $L \geq 2$, which is significantly less than that in the bulk Ti, which is around 4.02 J/m^2 (Fig. 27). In addition, the amplitude of the generalized stacking fault energy profile on Ti (0001) plane also reaches a minimum when $L = 3$, which is noticeably smaller than that of bulk Ti (Fig. 28). One expects that the presence of such a "weak" plane would negatively impact the mechanical response of the interface region.

As was discussed in section "The Structure of the Misfit Dislocation Network and the Coherent Regions of the Ti-TiN Interfaces," the width/diameter of the dislocation/nodal cores is dependent on the amplitude of the GSFE profile. According to GSFE surfaces and profiles obtained from the DFT calculations, the core-size of the dislocations/nodes of the MDN is expected to be the largest when $L = 3$. As was shown by earlier studies, the excessively constricted nodal cores give rise to localization of free volume, which lead to transformation of nodal structures. In fact, the small change in GSFE induced by biaxial tension/compression parallel to the interface is sufficient to produce the concentration of free volume which then lead to transformation of dislocation structure at nodes.

Our MD simulation qualitatively confirmed the predictions above. The dislocation and atomic structures at a node for the cases of $L = 1$ and $L = 5$ are shown in Fig. 29. When $L = 1$ (Fig. 29a), due to the relatively weak interface interaction associated with the GSFE profile with a reduced magnitude, the node is simply an intersection point of dislocations. The size of the node is comparable to the core-width of dislocations. The free volume at the node is smeared, as is evident in the (0001) Ti atomic layer of M = 2 shown in Fig. 29c. The atoms are colored according to centro-symmetry analysis performed only on Ti atoms. Blue color indicates a centro-symmetric FCC structure (a stacking fault in Ti); yellow color indicates a nonsymmetric HCP structure; red color indicates a locally disordered structure. When $L > 1$, including in bulk Ti, the increased GSFE profile (as compared to the case of $L = 1$) constricted the size of the node which gives rise to the concentration of the free volume. Fig. 29b, d shows a typical case of $L = 5$. The highly localized free volume is evident in the (0001) atomic plane with $M = 6$ (Fig. 29d). Once again, there is a qualitative agreement between the results from MD and DFT. According to the MD, the (0001) plane that has the smallest amplitude of GSFE is $L = 1$, while DFT predicted $L = 3$. As was discussed by Shao and others (Shao et al. 2015; Wang and Misra 2014; Henager et al. 2004), the concentration of free volume is the result of the presence of three dislocation jogs (marked by three vertical line segments in green color in Fig. 29b) at the node. Such jogs have Burgers vectors $b_4 = \frac{a}{3}\left[\bar{1}2\bar{1}0\right]$, $b_5 = \frac{a}{3}\left[\bar{1}\bar{1}20\right]$, and $b_6 = \frac{a}{3}\left[2\bar{1}\bar{1}0\right]$ and are glissile on the prismatic planes. However, the aggregation of the three jogs serves as strong pinning point to the motion of MDN due to the following two reasons. First, the global loading condition (simple shear parallel to interface) exerts zero resolved shear stress on the prismatic slip systems ($\left\{10\bar{1}0\right\}\left\langle 1\bar{2}10\right\rangle$). To show this, consider the stress applied to the bilayer system:

$$\boldsymbol{\sigma} = \begin{bmatrix} 0 & \tau & 0 \\ \tau & 0 & 0 \\ 0 & 0 & 0 \end{bmatrix},$$

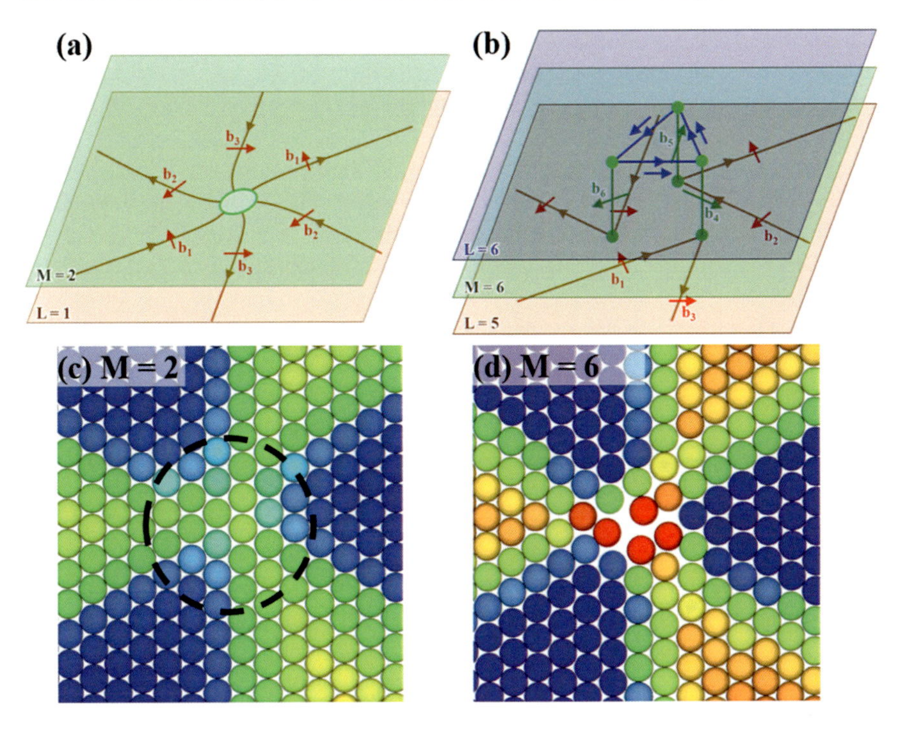

Fig. 29 Dislocation and atomic structures at a node on a Ti (0001) atomic monolayer for $L = 1$ and $L = 5$. (**a**) and (**b**) show the dislocation structures with line senses and Burgers vectors indicated. (**c**) and (**d**) show the corresponding atomic structures. Note that the atomic monolayers are on the tension side of the MDNs. The atoms are colored according to centro-symmetry analysis performed only on Ti atoms. Blue color indicates a centro-symmetric FCC structure (a stacking fault in Ti); yellow color indicates a non-symmetric HCP structure; red color indicates a locally disordered structure

a random prismatic plane with plane normal:

$$\widehat{\boldsymbol{n}} = [n_x \;\; 0 \;\; n_z],$$

and an unit vector parallel to $\frac{a}{3}\langle\bar{1}2\bar{1}0\rangle$ type Burgers vector in the prismatic plane:

$$\widehat{\boldsymbol{v}} = \left[n_z \; 0 \; -n_x \right].$$

Obviously, since $\widehat{\boldsymbol{v}}$ lies within the prismatic plane, $\widehat{\boldsymbol{n}} \cdot \widehat{\boldsymbol{v}} = 0$. The resolved shear stress of this slip system is $\widehat{\boldsymbol{n}} \cdot \boldsymbol{\sigma} \cdot \widehat{\boldsymbol{v}} \equiv 0$. Therefore, the jogs do not move in response to the loading. Second, the jogs are connected by Shockley partial dislocation lines (blue dislocation lines forming a triangle in Fig. 29b) residing in the (0001) slip plane ($L = 6$) immediately above the MDN, which also encloses a stacking fault. As was shown in section "Mechanistic Understanding on the Shear Instability of Metal/Ceramic Interface Provided by Simulations," the stable stacking fault energy of I_2 structure in Ti is very high, ranging between 0.2 and 0.35 J/m^2. Driven by the reduction in energy associated with the extra dislocation lines and the stacking fault, the jogs shown in Fig. 29d are tightly bound. Evidently, any motion of the tightly bound jogs leads to increase in the stacking fault energy and dislocation line energy in the blue triangle shown in Fig. 29d. Therefore, the jogs, as an aggregation, are sessile, and its motion is nonconservative and requires climb.

In bulk Ti, the amplitude of the GSFE profile is large enough to induce the formation of a jogged nodal dislocation structure, as predicted by our MD simulations. Qualitatively agreeing with the DFT calculation, the MD simulation predicts that, at a (0001) glide plane close to the chemical interface (e.g., $L = 1$), the GSFE profile is noticeably diminished which transforms the nodes from the jogged structure (Fig. 29b, d) to a smeared structure (Fig. 29a, c). The smeared node structure does not offer pinning effect to the gliding motion of the MDN. The pinning effect of the jogs shown in Fig. 29b, d is evident in Fig. 26b. As is shown, due to the strong interaction at the chemical interface ($L = 0$), the strength of the interface is extremely high. In fact, in this case, interactions at the chemical interface is so strong, that no shear on the $L = 0$ plane is observed. Instead, slip on $L = 1$ plane is observed in the form of nucleation and propagation of dislocations. When the MDN is located at $L > 1$, the pinning effect of the nodes becomes increasingly stronger, which increases the shear strength of the interface. When $L = 6$, the interface shear strength of both ORs approached 2 GPa, approaching that in bulk Ti.

The shear response of the Ti/TiN interface, in terms of the motion/transformation of the MDN, is shown in Fig. 30. As shown, the MDN, which contains edge dislocation lines, moves towards the positive x direction driven by the Peach-Koehler force induced by the applied shear stress. For $L = 1$ (Fig. 30a–c, g–i), the nodes of the MDN offer minimal pinning, and the entire MDN translates approximately in a rigid manner. In response to the loading, the curvature of the dislocation lines seems to increase associated with the more pronounced spiral pattern at nodes. Upon unloading, the spiral pattern does not vanish. This suggests that the more pronounced spiral pattern is energetically more stable, the mechanical loading simply provided additional perturbation to help the system achieve the global energy minimum. When $L > 1$, the formation of the jogs significantly increased the pinning effect of the nodes for both ORs. Shown in Fig. 30d–f, j–l is the shear response of the MDN when $L = 5$. The strong pinning effect of the nodes is reflected by the bowing of partial dislocation lines. Continued loading with increased stress gives rise to "stress assisted climb" of the nodes (Hirth and Lothe 1982), which leaves vacancies on their trace (Fig. 30f, l). This is a further proof that it is indeed the presence of the jogs that is responsible for the pinning effect of the nodes. The pinning effect of the nodes is reflected by the increased shear strength of the interfaces of both ORs as shown in Fig. 26b, i.e., when the

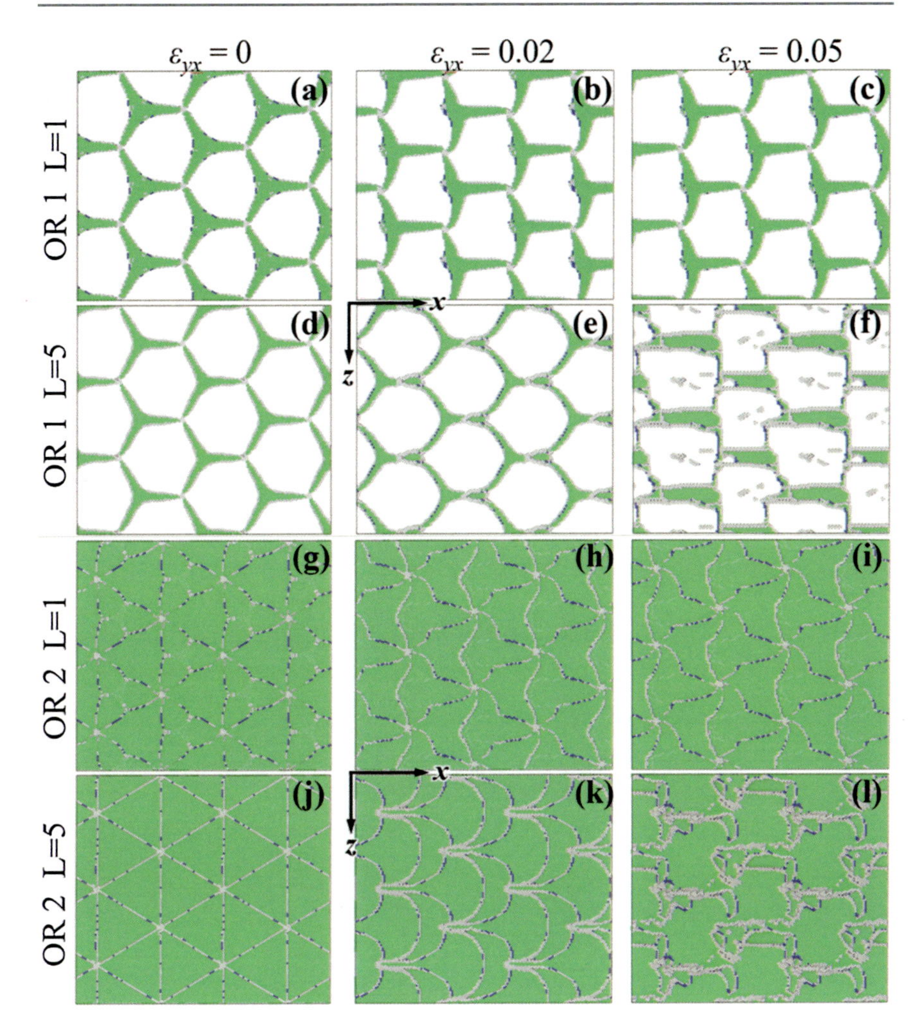

Fig. 30 The sequential snapshots of the MDN at three different effective shear strain levels (0, 0.02 and 0.05) for the two ORs and $L = 1, 5$

location of the MDN changes from $L = 1$ to $L = 6$, the strength increase from 500 to 1000 MPa range to around 2000 MPa. Comparing Fig. 26a, b, it is also worth noting that location of the MDN ($L = 1$) that corresponds to the minimum interface shear strength also corresponds to the minimum interface energy. Our results suggest that the minimum energy Ti/TiN interface has significantly reduced shear strength due to the presence of the MDN. Such metal/ceramic interfaces may exhibit weaker mechanical response (e.g., maximum shear stress) compared to the inside of the metal layer.

Summary

Mechanical integrity of metal/ceramic interfaces is critical for wide ranging applications. The application of ceramic hard coatings on mechanical components, including gears and bearings, can increase their service life by several folds (Jiang et al. 2003). Despite the industrial significance of engineering surfaces of mechanical components and manufacturing tools (Holmberg and Matthews 2009; Mercer et al. 2003; Kotzalas and Doll 2010), engineering of MCIRs still proceeds largely in a trial-and-error manner at the present time. This results in slow development, high cost, inconsistent outcomes, and sometimes failure to deploy. The challenge lies in the incomplete physical understanding and an under-developed defect theory for the mechanical response of MCIRs. The necessary understanding is hampered by (1) a lack of quantitative experimental measurements to assess the mechanical response of the MCIRs; (2) a lack of synergetic, physics-based modeling and simulation efforts to investigate the deformation physics of the MCIRs; (3) a lack of physics-based, experimentally validated, efficient, micro- to mesoscale predictive tools to enable the accelerated discovery-design-implementation cycle for metal/ceramic systems.

To tackle the challenge, research scheme outlined in this chapter centers on understanding the interface failure mechanisms and improving the strength of MCIRs through architectural and chemical refinements following a multiscale ICME approach. The mechanical response and deformation physics of MCIRs with various types of modifications can be investigated by combining first-principles DFT and MD calculations, suggesting experimental designs for strong MCIRs. State-of-the-art vapor phase synthesis in combination with quantitative microscale shear, compression, and tension testing and multiscale materials characterization can be employed to validate model outputs and test MCIR designs. Micro- and mesoscale materials models can then be developed to foster true physics-based MCIR design and accelerate the implementation of next generation applications, e.g., ceramics hard coatings in various surface engineering applications.

In this chapter, we provided an overview on the research on mechanical instability of metal/ceramic interfaces. Challenges and opportunities in the improvements of mechanical integrity of the interfaces have been identified. Theoretical background of this research thrust has been laid out. We concluded by illustrating some preliminary results by the authors on shear instability of various metal/ceramic interfacial regions. In order to achieve the goal of mechanical stability improvement and implementation of next generation MCIRs, true materials-based design, discovery, and understanding of the microscopic deformation physics of MCIRs is crucial. The research scheme described in this chapter outlines an approach to such a problem: (1) developing a multiscale computationally guided, experimentally validated ICME framework that is both time- and cost- efficient; (2) providing fundamental, physics-based understanding that will substantially advance the presently missing defect theories for MCIRs; (3) delivering accurate, experimentally validated, multiscale predictive tools for accelerated MCIR design and implementation.

References

F. Akasheh, H.M. Zbib, J.P. Hirth, R.G. Hoagland, A. Misra, Dislocation dynamics analysis of dislocation intersections in nanoscale metallic multilayered composites. J. Appl. Phys. **101**, 084314 (2007). https://doi.org/10.1063/1.2721093

ASTM International, *ASTM D2095 Standard Test Method for Tensile Strength of Adhesives by Means of Bar and Rod, Annu. B. ASTM Stand. 15* (2002), pp. 1–3. https://doi.org/10.1520/D2095-96R15.2

ASTM International, *ASTM D1002-05: Standard Test Method for Apparent Shear Strength of Single-Lap-Joint Adhesively Bonded Metal Specimens by Tension Loading (Metal-To-Metal), Standards* (2005), pp. 1–5. https://doi.org/10.1520/D1002-10.on

ASTM International, *ASTM D3165-07 Standard Test Method for Strength Properties of Adhesives in Shear by Tension Loading of Single-Lap-Joint Laminated Assemblies* (2007), p. 4. https://doi.org/10.1520/D3165-07

ASTM International, *ASTM D897 Standard Test Method for Tensile Properties of Adhesive Bonds* (ASTM International, West Conshohocken, 2012), pp. 24–26. https://doi.org/10.1520/D0897-08R16

ASTM International, *Standard Test Method for Strength Properties of Double Lap Shear Adhesive Joints by* (ASTM International, 2013). https://doi.org/10.1520/D3528-96R08.2

ASTM International, *ASTM D4541-17: Standard Test Method for Pull-off Strength of Coatings Using Portable Adhesion Testers* (ASTM international, 2014), pp. 1–16. https://doi.org/10.1520/D4541-09E01

ASTM International, *ASTM C1624-05: Standard Test Method for Adhesion Strength and Mechanical Failure Modes of Ceramic Coatings by Quantitative Single Point Scratch Testing* (ASTM international, 2015)

M. Benoit, N. Tarrat, J. Morillo, Density functional theory investigations of titanium γ -surfaces and stacking faults. Model. Simul. Mater. Sci. Eng. **21**, 015009 (2013). https://doi.org/10.1088/0965-0393/21/1/015009

I.J. Beyerlein, J. Wang, R. Zhang, Interface-dependent nucleation in nanostructured layered composites. APL Mater. **1** (2013a). https://doi.org/10.1063/1.4820424

I.J. Beyerlein, J. Wang, R. Zhang, Mapping dislocation nucleation behavior from bimetal interfaces. Acta Mater. **61**, 7488–7499 (2013b). https://doi.org/10.1016/j.actamat.2013.08.061

D. Bhattacharyya, N.A.A. Mara, P. Dickerson, R.G.G. Hoagland, A. Misra, Compressive flow behavior of Al–TiN multilayers at nanometer scale layer thickness. Acta Mater. **59**, 3804–3816 (2011). https://doi.org/10.1016/j.actamat.2011.02.036

S.D. Brown, Adherence failure and measurement: Some troubling questions. J. Adhes. Sci. Technol. **8**, 687–711 (1994). https://doi.org/10.1163/156856194X00438

S.D. Brown, Adherence failure and measurement: Some troubling questions, in *Adhesion Measurement of Films and Coatings*, ed. by K. L. Mittal, 1st edn. (VSP, Utrecht, 1995), pp. 15–39

S.J. Bull, Can the scratch adhesion test ever be quantitative? Adhes. Meas. Film. Coatings **2**, 107 (2001)

O. Casals, S. Forest, Finite element crystal plasticity analysis of spherical indentation in bulk single crystals and coatings. Comput. Mater. Sci. **45**, 774–782 (2009). https://doi.org/10.1016/j.commatsci.2008.09.030

H.Y. Chen, C.J. Tsai, F.H. Lu, The Young's modulus of chromium nitride films. Surf. Coatings Technol. **184**, 69–73 (2004). https://doi.org/10.1016/j.surfcoat.2003.10.064

K. Chen, W.J. Meng, J.A. Eastman, Interface development in cu-based structures transient liquid phase (TLP) bonded with thin Al foil intermediate layers. Metall. Mater. Trans. A. **45**, 3892–3906 (2014a). https://doi.org/10.1007/s11661-014-2339-5

K. Chen, Y. Mu, W.J. Meng, A new experimental approach for evaluating the mechanical integrity of interfaces between hard coatings and substrates. MRS Commun. **4**, 19–23 (2014b). https://doi.org/10.1557/mrc.2014.3

Y. Chen, S. Shao, X.-Y. Liu, S.K. Yadav, N. Li, N. Mara, J. Wang, Misfit dislocation patterns of mg-Nb interfaces. Acta Mater. **126**, 552–563 (2017). https://doi.org/10.1016/j.actamat.2016.12.041

B.D. Cullity, S.R. Stock, *Elements of X-Ray Diffraction*, 3rd edn. (Prentice-Hall, New York, 2001)

R. Darolia, Thermal barrier coatings technology: critical review, progress update, remaining challenges and prospects. Int. Mater. Rev. **58**, 315–348 (2013)

J.A. El-Awady, S. Bulent Biner, N.M. Ghoniem, A self-consistent boundary element, parametric dislocation dynamics formulation of plastic flow in finite volumes. J. Mech. Phys. Solids **56**, 2019–2035 (2008). https://doi.org/10.1016/j.jmps.2007.11.002

A.G. Evans, D.R. Clarke, C.G. Levi, The influence of oxides on the performance of advanced gas turbines. J. Eur. Ceram. Soc. **28**, 1405–1419 (2008). https://doi.org/10.1016/j.jeurceramsoc.2007.12.023

N.M. Ghoniem, X. Han, Dislocation motion in anisotropic multilayer materials. Philos. Mag. **85**, 2809–2830 (2005). https://doi.org/10.1080/14786430500155338

N. Ghoniem, S.-H. Tong, L. Sun, Parametric dislocation dynamics: a thermodynamics-based approach to investigations of mesoscopic plastic deformation. Phys. Rev. B **61**, 913–927 (2000). https://doi.org/10.1103/PhysRevB.61.913

N.M. Ghoniem, J. Huang, Z. Wang, Affine covariant-contravariant vector forms for the elastic field of parametric dislocations in isotropic crystals. Philos. Mag. Lett. **82**, 55–63 (2002). https://doi.org/10.1080/09500830110103216

S. Groh, E.B. Marin, M.F. Horstemeyer, H.M. Zbib, Multiscale modeling of the plasticity in an aluminum single crystal. Int. J. Plast. **25**, 1456–1473 (2009). https://doi.org/10.1016/j.ijplas.2008.11.003

V. Gupta, A.S. Argon, D.M. Parks, J.A. Cornie, Measurement of interface strength by a laser spallation technique. J. Mech. Phys. Solids **40**, 141–180 (1992). https://doi.org/10.1016/0022-5096(92)90296-E

V. Gupta, J. Yuan, A. Pronin, Recent developments in the laser spallation technique to measure the interface strength and its relationship to interface toughness with applications to metal/ceramic, ceramic/ceramic and ceramic/polymer interfaces. J. Adhes. Sci. Technol. **8**, 713–747 (1994). https://doi.org/10.1163/156856194X00447

V. Gupta, V. Kireev, J. Tian, H. Yoshida, H. Akahoshi, Glass-modified stress waves for adhesion measurement of ultra thin films for device applications. J. Mech. Phys. Solids **51**, 1395–1412 (2003). https://doi.org/10.1016/S0022-5096(03)00057-7

X. Han, N.M. Ghoniem, Stress field and interaction forces of dislocations in anisotropic multilayer thin films. Philos. Mag. **85**, 1205–1225 (2005). https://doi.org/10.1080/14786430412331331907

C.H. Henager, R.J. Kurtz, R.G. Hoagland, Interactions of dislocations with disconnections in fcc metallic nanolayered materials. Philos. Mag. **84**, 2277–2303 (2004). https://doi.org/10.1080/14786430410001678235

J.P. Hirth, J. Lothe, *Theory of Dislocations*, 2nd edn. (Krieger Publishing Company, Malabar, 1982)

J.P. Hirth, R. Pond, Steps, dislocations and disconnections as interface defects relating to structure and phase transformations. Acta Mater. **44**, 4749–4763 (1996). http://www.sciencedirect.com/science/article/pii/S1359645496001322 . Accessed 29 Nov 2012

J.P.P. Hirth, R.C.C. Pond, J. Lothe, Disconnections in tilt walls. Acta Mater. **54**, 4237–4245 (2006). https://doi.org/10.1016/j.actamat.2006.05.017

H. Holleck, Material selection for hard coatings. J. Vac. Sci. Technol. A Vac. Surf. Film **4**, 2661–2669 (1986). https://doi.org/10.1116/1.573700

K. Holmberg, A. Matthews, *Coatings Tribology – Properties, Mechanisms, Techniques and Applications in Surface Engineering*, 2nd edn. (Elsevier Science, Amsterdam, 2009)

M.A. Hopcroft, W.D. Nix, T.W. Kenny, What is the Young's modulus of silicon? J. Microelectromech. Syst. **19**, 229–238 (2010). https://doi.org/10.1109/JMEMS.2009.2039697

T. Itoh (ed.), *Ion Beam Assisted Film Growth* (Elsevier, Amsterdam, 1989)

J.C. Jiang, W.J. Meng, A.G. Evans, C.V. Cooper, Structure and mechanics of W-DLC coated spur gears. Surf. Coatings Technol. **176**, 50–56 (2003). https://doi.org/10.1016/S0257-8972(03)00445-6

Y.M. Kim, B.J. Lee, Modified embedded-atom method interatomic potentials for the Ti-C and Ti-N binary systems. Acta Mater. **56**, 3481–3489 (2008a). https://doi.org/10.1016/j.actamat.2008.03.027

Y.-M. Kim, B.-J. Lee, Modified embedded-atom method interatomic potentials for the Ti–C and Ti–N binary systems. Acta Mater. **56**, 3481–3489 (2008b). https://doi.org/10.1016/j.actamat.2008.03.027

C. Kittel, Introduction to solid state physics. Solid State Phys., 703 (2005). https://doi.org/10.1119/1.1974177

M.N. Kotzalas, G.L. Doll, Tribological advancements for reliable wind turbine performance. Philos. Trans. R. Soc. A Math. Phys. Eng. Sci. **368**, 4829–4850 (2010). https://doi.org/10.1098/rsta.2010.0194

C. Kral, W. Lengauer, D. Rafaja, P. Ettmayer, Critical review on the elastic properties of transition metal carbides, nitrides and carbonitrides. J. Alloys Compd. **265**, 215–233 (1998). https://doi.org/10.1016/S0925-8388(97)00297-1

P. Kutilek, J. Miksovsky, The procedure of evaluating the practical adhesion strength of new biocompatible nano-and micro-thin films in accordance with international standards. Acta Bioeng. Biomech. **13**, 87–94 (2011)

R.A. Lebensohn, C.N. Tomé, A self-consistent anisotropic approach for the simulation of plastic deformation and texture development of polycrystals: Application to zirconium alloys. Acta Metall. Mater. **41**, 2611–2624 (1993). https://doi.org/10.1016/0956-7151(93)90130-K

R.A. Lebensohn, C.N. Tomé, A self-consistent viscoplastic model: Prediction of rolling textures of anisotropic polycrystals. Mater. Sci. Eng. A **175**, 71–82 (1994). https://doi.org/10.1016/0921-5093(94)91047-2

N. Li, H. Wang, A. Misra, J. Wang, In situ nanoindentation study of plastic co-deformation in Al-TiN nanocomposites. Sci. Rep. **4**, 6633 (2014). https://doi.org/10.1038/srep06633

J. Lin, W.D. Sproul, J.J. Moore, Microstructure and properties of nanostructured thick CrN coatings. Mater. Lett. **89**, 55–58 (2012). https://doi.org/10.1016/j.matlet.2012.08.060

Z. Lin, X. Peng, T. Fu, Y. Zhao, C. Feng, C. Huang, Z. Wang, Atomic structures and electronic properties of interfaces between aluminum and carbides/nitrides: A first-principles study. Phys. E Low-Dimensional Syst. Nanostruct. **89**, 15–20 (2017). https://doi.org/10.1016/j.physe.2017.01.025

K. Lukaszkowicz, A. Kriz, J. Sondor, Structure and adhesion of thin coatings deposited by PVD technology on the X6CrNiMoTi17-12-2 and X40CrMoV5-1 steel substrates. Arch. Mater. Sci. Eng. **51**, 40–47 (2011)

H. Lyu, A. Ruimi, H.M. Zbib, A dislocation-based model for deformation and size effect in multi-phase steels. Int. J. Plast. **72**, 44–59 (2015). https://doi.org/10.1016/j.ijplas.2015.05.005

H. Lyu, N. Taheri-Nassaj, H.M. Zbib, A multiscale gradient-dependent plasticity model for size effects. Philos. Mag. **96**, 1883–1908 (2016). https://doi.org/10.1080/14786435.2016.1180437

H. Lyu, M. Hamid, A. Ruimi, H.M. Zbib, Stress/strain gradient plasticity model for size effects in heterogeneous nano-microstructures. Int. J. Plast. (2017). https://doi.org/10.1016/j.ijplas.2017.05.009

E.B. Marin, *On the Formulation of a Crystal Plasticity Model* (Livermore, 2006). http://prod.sandia.gov/techlib/access-control.cgi/2006/064170.pdf

P.H. Mayrhofer, G. Tischler, C. Mitterer, Microstructure and mechanical/thermal properties of Cr-N coatings deposited by reactive unbalanced magnetron sputtering. Surf. Coatings Technol. **142–144**, 78–84 (2001). https://doi.org/10.1016/S0257-8972(01)01090-8

W.J. Meng, T.J. Curtis, Inductively coupled plasma assisted physical vapor deposition of titanium nitride coatings. J. Electron. Mater. **26**, 1297–1302 (1997)

W.J. Meng, G.L. Eesley, Growth and mechanical anisotropy of TiN thin films. Thin Solid Films **271**, 108–116 (1995). https://doi.org/10.1016/0040-6090(95)06875-9

W.J. Meng, T.J. Curtis, L.E. Rehn, P.M. Baldo, Temperature dependence of inductively coupled plasma assisted growth of TiN thin films. Surf. Coatings Technol. **120–121**, 206–212 (1999). https://doi.org/10.1016/S0257-8972(99)00457-0

W.J. Meng, E.I. Meletis, L.E. Rehn, P.M. Baldo, Inductively coupled plasma assisted deposition and mechanical properties of metal-free and Ti-containing hydrocarbon coatings. J. Appl. Phys. **87**, 2840–2848 (2000). https://doi.org/10.1063/1.372266

C. Mercer, A.G. Evans, N. Yao, S. Allameh, C.V. Cooper, Material removal on lubricated steel gears with W-DLC-coated surfaces. Surf. Coatings Technol. **173**, 122–129 (2003). https://doi.org/10.1016/S0257-8972(03)00467-5

M.A. Meyers, K.K. Chawla, *Mechanical Behavior of Materials* (Cambridge University Press, 2007)

R.E. Miller, L. Shilkrot, W.A. Curtin, A coupled atomistics and discrete dislocation plasticity simulation of nanoindentation into single crystal thin films. Acta Mater. **52**, 271–284 (2004). https://doi.org/10.1016/j.actamat.2003.09.011

Y. Mu, J.W. Hutchinson, W.J. Meng, Micro-pillar measurements of plasticity in confined cu thin films. Extrem. Mech. Lett. **1**, 62–69 (2014). https://doi.org/10.1016/j.eml.2014.12.001

F.R.N. Nabarro, Dislocations in a simple cubic lattice. Proc. Phys. Soc. **59**, 256–272 (1947). https://doi.org/10.1088/0959-5309/59/2/309

J.F. Nye, *Physical Properties of Crystals: Their Representation by Tensors and Matrices* (Oxford University Press, New York, 2000)

W. Oliver, G. Pharr, An improved technique for determining hardness and elastic modulus using load and displacement-sensing indentation systems. J. Mater. Res. **7**, 1564–1583 (1992). https://doi.org/10.1557/JMR.1992.1564

R. Peierls, The size of a dislocation. Proc. Phys. Soc. **52**, 34–37 (1940). https://doi.org/10.1088/0959-5309/52/1/305

A.J. Perry, J.A. Sue, P.J. Martin, Practical measurement of the residual stress in coatings. Surf. Coatings Technol. **81**, 17–28 (1996). https://doi.org/10.1016/0257-8972(95)02531-6

I. Salehinia, S. Shao, J. Wang, H.M.M. Zbib, Plastic deformation of metal/ceramic Nanolayered composites. JOM **66**, 2078–2085 (2014). https://doi.org/10.1007/s11837-014-1132-7

I. Salehinia, S. Shao, J. Wang, H.M. Zbib, Interface structure and the inception of plasticity in Nb/NbC nanolayered composites. Acta Mater. **86** (2015). https://doi.org/10.1016/j.actamat.2014.12.026

C. Sant, M. Ben Daia, P. Aubert, S. Labdi, P. Houdy, Interface effect on tribological properties of titanium–titanium nitride nanolaminated structures. Surf. Coatings Technol. **127**, 167–173 (2000). https://doi.org/10.1016/S0257-8972(00)00663-0

S. Shao, J. Wang, Relaxation, structure, and properties of semicoherent interfaces. JOM **68**, 242–252 (2016). https://doi.org/10.1007/s11837-015-1691-2

S. Shao, J. Wang, A. Misra, R.G. Hoagland, Spiral patterns of dislocations at nodes in (111) semi-coherent FCC interfaces. Sci. Rep. **3** (2013). https://doi.org/10.1038/srep02448

S. Shao, J. Wang, A. Misra, Energy minimization mechanisms of semi-coherent interfaces. J. Appl. Phys. **116**, 023508 (2014). https://doi.org/10.1063/1.4889927

S. Shao, J. Wang, I.J. Beyerlein, A. Misra, Glide dislocation nucleation from dislocation nodes at semi-coherent {111} cu–Ni interfaces. Acta Mater. **98**, 206–220 (2015). https://doi.org/10.1016/j.actamat.2015.07.044

S. Shao, A. Misra, H. Huang, J. Wang, Micro-scale modeling of interface-dominated mechanical behavior. J. Mater. Sci. (2017). https://doi.org/10.1007/s10853-017-1662-9

S. Shao, F. Akasheh, J. Wang, Y. Liu, Alternative misfit dislocations pattern in semi-coherent FCC {100} interfaces. Acta Mater. **144**, 177–186 (2018). https://doi.org/10.1016/j.actamat.2017.10.052

S.A. Skirlo, M.J. Demkowicz, Viscoelasticity of stepped interfaces. Appl. Phys. Lett. **103**, 171908 (2013). https://doi.org/10.1063/1.4827103

J.R. Smith, T. Hong, D.J. Srolovitz, Metal-ceramic adhesion and the Harris functional. Phys. Rev. Lett. **72**, 4021–4024 (1994). https://doi.org/10.1103/PhysRevLett.72.4021

M.E. Straumanis, L.S. Yu, Lattice parameters, densities, expansion coefficients and perfection of structure of cu and of cu–in α phase. Acta Crystallogr. Sect. A. **25**, 676–682 (1969). https://doi.org/10.1107/S0567739469001549

J.A. Sue, A.J. Perry, J. Vetter, Young's modulus and stress of CrN deposited by cathodic vacuum arc evaporation. Surf. Coatings Technol. **68–69**, 126–130 (1994). https://doi.org/10.1016/0257-8972(94)90149-X

T. Sun, X. Wu, R. Wang, W. Li, Q. Liu, First-principles study on the adhesive properties of Al/TiC interfaces: Revisited. Comput. Mater. Sci. **126**, 108–120 (2017). https://doi.org/10.1016/j.commatsci.2016.09.024

K.H. Thulasi Raman, M.S.R.N. Kiran, U. Ramamurty, G. Mohan Rao, Structural and mechanical properties of room temperature sputter deposited CrN coatings. Mater. Res. Bull. **47**, 4463–4466 (2012). https://doi.org/10.1016/j.materresbull.2012.09.051

L. Toth, *Transition Metal Carbides and Nitrides* (Elsevier Science, 1971)

M.D. Uchic, D.M. Dimiduk, J.N. Florando, W.D. Nix, Sample dimensions influence strength and crystal plasticity. Science **305**, 986–989 (2004). https://doi.org/10.1126/science.1098993

M.D. Uchic, P.A. Shade, D.M. Dimiduk, Plasticity of micrometer-scale single crystals in compression. Annu. Rev. Mater. Res. **39**, 361–386 (2009). https://doi.org/10.1146/annurev-matsci-082908-145422

L.J. Vandeperre, F. Giuliani, S.J. Lloyd, W.J. Clegg, The hardness of silicon and germanium. Acta Mater. **55**, 6307–6315 (2007). https://doi.org/10.1016/j.actamat.2007.07.036

C.A. Volkert, A.M. Minor, Focused ion beam microscopy and micromachining. MRS Bull. **32**, 389–399 (2007). https://doi.org/10.1557/mrs2007.62

J. L. Vossen, W. Kern (eds.), *Thin Film Processes II* (Academic Press, Boston, 1991)

J. Wang, A. Misra, An overview of interface-dominated deformation mechanisms in metallic multilayers. Curr. Opin. Solid State Mater. Sci. **15**, 20–28 (2011). https://doi.org/10.1016/j.cossms.2010.09.002

J. Wang, A. Misra, Strain hardening in nanolayered thin films. Curr. Opin. Solid State Mater. Sci. **18**, 19–28 (2014). https://doi.org/10.1016/j.cossms.2013.10.003

J. Wang, R.G. Hoagland, J.P. Hirth, A. Misra, Atomistic simulations of the shear strength and sliding mechanisms of copper–niobium interfaces. Acta Mater. **56**, 3109–3119 (2008). https://doi.org/10.1016/j.actamat.2008.03.003

H. Wang, P.D. Wu, C.N. Tomé, Y. Huang, A finite strain elastic-viscoplastic self-consistent model for polycrystalline materials. J. Mech. Phys. Solids **58**, 594–612 (2010). https://doi.org/10.1016/j.jmps.2010.01.004

J. Wang, C. Zhou, I.J.I.J. Beyerlein, S. Shao, Modeling interface-dominated mechanical behavior of nanolayered crystalline composites. JOM **66**, 102–113 (2014). https://doi.org/10.1007/s11837-013-0808-8

J. Wang, Q. Zhou, S. Shao, A. Misra, Strength and plasticity of nanolaminated materials. Mater. Res. Lett. **5** (2017). https://doi.org/10.1080/21663831.2016.1225321

S.K.K. Yadav, S. Shao, J. Wang, X.-Y.X.-Y. Liu, Structural modifications due to interface chemistry at metal-nitride interfaces. Sci. Rep. **5**, 17380 (2015). https://doi.org/10.1038/srep17380

W. Yang, G. Ayoub, I. Salehinia, B. Mansoor, H. Zbib, Deformation mechanisms in Ti/TiN multilayer under compressive loading. Acta Mater. **122**, 99–108 (2017). https://doi.org/10.1016/j.actamat.2016.09.039

H.M. Zbib, T. Diaz de la Rubia, A multiscale model of plasticity. Int. J. Plast. **18**, 1133–1163 (2002). https://doi.org/10.1016/S0749-6419(01)00044-4

H.M. Zbib, M. Rhee, J.P. Hirth, On plastic deformation and the dynamics of 3D dislocations. Int. J. Mech. Sci. **40**, 113–127 (1998). https://doi.org/10.1016/S0020-7403(97)00043-X

J.M. Zhang, Y. Zhang, K.W. Xu, V. Ji, Representation surfaces of Young's modulus and Poisson's ratio for BCC transition metals. Phys. B Condens. Matter **390**, 106–111 (2007). https://doi.org/10.1016/j.physb.2006.08.008

Z.G. Zhang, O. Rapaud, N. Bonasso, D. Mercs, C. Dong, C. Coddet, Control of microstructures and properties of dc magnetron sputtering deposited chromium nitride films. Vacuum **82**, 501–509 (2008). https://doi.org/10.1016/j.vacuum.2007.08.009

R.F. Zhang, T.C. Germann, J. Wang, X.-Y. Liu, I.J. Beyerlein, Role of interface structure on the plastic response of cu/Nb nanolaminates under shock compression: Non-equilibrium molecular dynamics simulations. Scr. Mater. **68**, 114–117 (2013). https://doi.org/10.1016/j.scriptamat.2012.09.022

R.F. Zhang, I.J. Beyerlein, S.J. Zheng, S.H. Zhang, A. Stukowski, T.C. Germann, Manipulating dislocation nucleation and shear resistance of bimetal interfaces by atomic steps. Acta Mater. **113**, 194–205 (2016). https://doi.org/10.1016/j.actamat.2016.05.015

X. Zhang, B. Zhang, Y. Mu, S. Shao, C.D.C.D. Wick, B.R. Ramachandran, W.J. Meng, Mechanical failure of metal/ceramic interfacial regions under shear loading. Acta Mater. **138**, 224–236 (2017). https://doi.org/10.1016/j.actamat.2017.07.053

Uniqueness of Elastoplastic Properties Measured by Instrumented Indentation

6

L. Liu, Xi Chen, N. Ogasawara, and N. Chiba

Contents

L. Liu (✉)
Department of Mechanical and Aerospace Engineering, Utah State University, Logan, UT, USA
e-mail: ling.liu@usu.edu

X. Chen (✉)
Department of Earth and Environmental Engineering, Columbia Nanomechanics Research Center, Columbia University, New York, NY, USA
e-mail: xichen@columbia.edu

N. Ogasawara (✉)
Department of Mechanical Engineering, National Defense Academy of Japan, Yokosuka, Japan
e-mail: oga@nda.ac.jp

N. Chiba
National Defense Academy of Japan, Yokosuka, Japan
e-mail: chiba@nda.ac.jp

G. Z. Voyiadjis (ed.), *Handbook of Nonlocal Continuum Mechanics for Materials and Structures*, https://doi.org/10.1007/978-3-319-58729-5_22

Abstract

Indentation is widely used to extract material elastoplastic properties from the measured load-displacement curves. One of the most well-established indentation technique utilizes dual (or plural) sharp indenters (which have different apex angles) to deduce key parameters such as the elastic modulus, yield stress, and work-hardening exponent for materials that obey the power-law constitutive relationship. Here we show the existence of "mystical materials," which have distinct elastoplastic properties, yet they yield almost identical indentation behaviors, even when the indenter angle is varied in a large range. These mystical materials are, therefore, indistinguishable by many existing indentation analyses unless extreme (and often impractical) indenter angles are used. Explicit procedures of deriving these mystical materials are established, and the general characteristics of the mystical materials are discussed. In many cases, for a given indenter angle range, a material would have infinite numbers of mystical siblings, and the existence maps of the mystical materials are also obtained. Furthermore, we propose two alternative techniques to effectively distinguish these mystical materials. In addition, a critical strain is identified as the upper bound of the detectable range of indentation, and moderate tailoring of the constitutive behavior beyond this range cannot be effectively detected by the reverse analysis of the load-displacement curve. The topics in this chapter address the important question of the uniqueness of indentation test, as well as providing useful guidelines to properly use the indentation technique to measure material elastoplastic properties.

Keywords

Indentation · Elastoplastic properties · Unique solution · Numerical study · Indistinguishable load-displacement curve · Reverse analysis · Detectable strain range · Critical strain · Loading curvature · Indenter angle

Introduction

Instrumented indentation is widely used to probe the constitutive relationships of engineering materials. Without losing generality, the uniaxial true stress-strain curve of a stress-free elastoplastic solid can be expressed in a power-law form, which is a good approximation for most metals and alloys (Cheng and Cheng 2004).

$$\sigma = E\varepsilon \text{ for } \varepsilon \leq \frac{\sigma_y}{E} \text{ and } \sigma = \sigma_y\left(\frac{E}{\sigma_y}\right)^n \varepsilon^n = R\varepsilon^n, \tag{1}$$

where E is the Young's modulus, σ_y is the yield stress, and n is the work-hardening exponent. For most metals and alloys, n is between 0.0 and 0.5, E is between 10 and 600 GPa, σ_y is between 10 and 2000 MPa, and E/σ_y is between 100 and 5000 (Ashby 1999) – this is a technical range of the engineering materials suitable for the conventional indentation analysis where finite strains are involved.

Four independent parameters (E, ν, σ_y, n) are needed to completely characterize the elastoplastic properties of a power-law stress-free material. Probing these material parameters by indentation has become a focal point of interest in the indentation literature, and various techniques were proposed; see the review by Cheng and Cheng (2004). However, even for some of the existing techniques that are considered as "well-established," the fundamental question of whether the elastoplastic properties of a specimen can be uniquely determined is still open.

For any indentation technique, the existence of a unique solution requires that the indentation response must be unique for a given material, i.e., one-to-one correspondence between the shape factors of the measured indentation load-displacement curves and material elastoplastic properties. For example, when the apex angle of a sharp indenter is fixed, several research groups have shown that a set of special materials with distinct elastoplastic properties may yield almost the same indentation load-displacement curves (Cheng and Cheng 1999; Capehart and Cheng 2003; Tho et al. 2004; Alkorta et al. 2005). Therefore, the mechanical properties of these specimens cannot be uniquely determined by using one sharp indenter.

In this chapter based on Chen et al. (2007) and Liu et al. (2009), we carry out a systematic numerical study to correlate the indentation responses with a wide range of material properties and a variety of indenter geometries and present an explicit formulation to determine the special sets of materials with distinct elastoplastic properties yet exhibit indistinguishable indentation behaviors even when different indenters are employed. We call such sets of special materials as *mystical materials* – i.e., they are beyond the previous (and conventional) understanding of this topic. For many power-law materials, they have infinite mystical siblings that have indistinguishable loading and unloading behaviors during the indentation test, for a wide range of indenter geometries. Due to the lack of unique solutions, theoretically, these mystical materials are unable to be distinguished by many previously established techniques, including the dual (or plural) sharp indentation method and the conventional spherical indentation analysis with small penetration. The existence map of and the common characteristics of the mystical materials are also established. We then illustrate that the properties of these mystical materials may still be distinguishable by alternative indentation techniques, for example, a film indentation analysis (Zhao et al. 2006a) and an improved spherical indentation technique (Zhao et al. 2006a), which are suitable for specimens with finite thickness and bulk materials, respectively.

Challenging the Uniqueness of Indentation Load-Displacement Curve vs. Material Property

Most indentation techniques, the sharp indentation analysis, the spherical indentation analysis, the film indentation analysis, etc., in essence, require numerical analyses to correlate various shape factors of the P-δ curve with the specimen elastoplastic properties. In doing so, a critical theoretical question emerges: is there a one-to-one correspondence between indentation load-displacement curves (for loading and/or unloading) and material properties (E, ν, σ_y, n)? The uniqueness of the solution of the reverse analysis is the key verification for all indentation analyses – although some of those methods are now widely used in practice and cited in literature, unfortunately, the uniqueness of their solutions has been rarely challenged.

The fundamental question is, does a set of mystical materials which will yield indistinguishable indentation load-displacement curves for not only one particular indenter angle, but also another indenter angle exist? If this is true, then the dual (or plural) indenter method, regardless of the detail of the theory, cannot be used to distinguish these mystical materials. Such a fundamental question is not only the basis for the dual (or plural) sharp indenter method but also the foundation of the spherical indentation and film indentation techniques, as well as most indentation analyses which rely on the load-displacement curves.

In what follows, we will first present a simple and explicit technique to derive special sets of materials (with different elastoplastic properties) that lead to the same loading curves during sharp indentation when different indenter angles are used. Next, we show another technique to derive special sets of materials that yield almost same loading and unloading curves for a given indenter angle. We then extend our analysis to predict the mystical materials that have indistinguishable loading and unloading curves when different sharp indenters are used. Consequently, many of the previously established indentation analyses would fail to distinguish these materials. The existence range, trend, and special features of the mystical materials are discussed. We also show that these mystical materials may still be distinguished by using the improved spherical indentation and film indentation techniques.

Computation Method

In this chapter, the relationships between indentation responses, material properties, and indenter geometries are established from extensive finite element analyses. FEM calculations are performed using the commercial code. The rigid contact surface option is used to simulate the rigid indenter, and the option for finite deformation and strain is employed. A typical mesh for the axisymmetric indentation model comprises about 10,000 4-node elements with reduced integration. The Coulomb's friction law is used between contact surfaces, and the friction coefficient is taken to be 0.15 (Bowden and Tabor 1950), which is a minor factor for indentation (Mesarovic and Fleck 1999; Cheng and Cheng 2004) as long as this value is relatively small. The strain gradient effect is ignored by assuming that the indentation

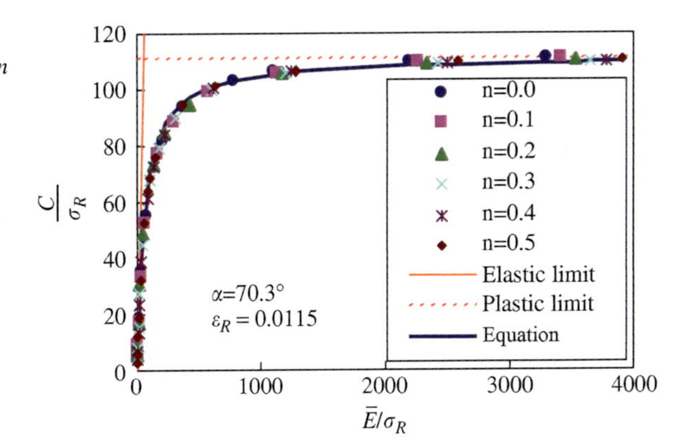

Fig. 1 The relationship between C/σ_R and $\overline{E}/\sigma_R$ as n is varied, for $\alpha = 70.3$. The numerical results (symbols) are shown with both elastic and rigid plastic limits, and the empirical fitting function incorporating these limits is given in Eq. 6 (which can be extended to other angles) (Chen et al. 2007)

depth is sufficiently deep so that the continuum mechanics still applies to the bulk specimen. In addition, the strain rate effect is also ignored. In order to obtain both complete and robust numerical results, the material parameters are varied over a large range to cover essentially all engineering materials with $\overline{E}/\sigma_R = 3 - 3900$ and $n = 0$–0.5, where $\overline{E}$ is the plane-strain modulus and σ_R is the representative stress, see (Ogasawara et al. 2005, 2007a) for details. And, for the same reason, a large indenter angle range is also used, from 60° to 80°, which covers most of the angle range used in literature. To be consistent with most literature, Poisson's ratio is fixed at 0.3.

Determine Special Materials with Same Loading Curves for Dual Sharp Indenters

A Simple Relationship Between the Loading Curvature and Material Properties

When a sharp indenter is penetrating a bulk specimen, the loading P-δ relationship is always quadratic, i.e., $P{=}C\delta^2$. Therefore, two materials must have the same loading curvature C in order to have the same loading P-δ curve. A simple relationship, not a high-order fitting polynomial function, between C and $\overline{E}/\sigma_R$ must be established on a physical basis.

Note that the specimen is essentially elastic when the variable $\overline{E}/\sigma_R \Rightarrow 0$, whereas the material approaches to rigid plastic when $\overline{E}/\sigma_R \Rightarrow \infty$ – the functional form $\prod\left(\overline{E}/\sigma_R\right)$ should incorporate the limits of both mechanisms, such that it remains valid for all materials regardless of the range of data used for fitting. In the representative case of $\alpha = 70.3$ (Fig. 1), both of the elastic and rigid plastic limits can be well defined:

(a) When $\overline{E}/\sigma_R \Rightarrow 0$(elastic), $C/\sigma_R = \prod\left(\overline{E}/\sigma_R\right)$ varies linearly with $\overline{E}/\sigma_R$:

$$\frac{C}{\sigma_R} = m_e \frac{\overline{E}}{\sigma_R} \text{ as } \overline{E}/\sigma_R \Rightarrow 0, \tag{2}$$

where m_e can be derived from the classic solution of indentation on elastic materials (Sneddon 1965), and it agrees well with FEM calculations (Ogasawara et al. 2006):

$$m_e = \frac{2\gamma \tan \alpha}{\pi}. \tag{3}$$

(b) When $\overline{E}/\sigma_R => \infty$(rigid plastic), C/σ_R approaches a constant:

$$\frac{C}{\sigma_R} = m_p \text{ as } \overline{E}/\sigma_R => \infty, \tag{4}$$

where m_p is the rigid plastic limit of conical indentation into a material that obeys the Mises yield criterion.

$$m_p = 13.2\tan^2\alpha + 6.18 \tan \alpha - 8.54 \tag{5}$$

for $50^o \leq \alpha \leq 80^o$. m_p is equal to 112.1 for the Berkovich indenter. In view of the importance of these two limits, we have proposed a very simple empirical form of $\prod \left(\overline{E}/\sigma_R\right)$ to incorporate both limits (Ogasawara et al. 2006, 2007a):

$$\prod = \frac{C}{\sigma_R} = \left(\frac{1}{m_e \frac{\overline{E}}{\sigma_R}} + \frac{1}{m_p}\right)^{-1}, \tag{6}$$

from which the representative stress can be obtained as $\sigma_R = m_e C \overline{E}/m_p \left(m_e \overline{E} - C\right)$ without iteration. The above equation not only incorporates the elastic and plastic limits (thus having physical meaning and wider range of application), but it also involves no fitting parameter if m_p could be solved analytically.

Special Materials with Same Loading Curvature for Dual Sharp Indenters

With the simple Eq. 6 relating C and material properties, it is now possible to explicitly derive material combinations that have the same loading curvature. First consider the case with one indenter (#A) whose half-apex angle is fixed: once α^A is specified, its related elastic limit m_e^A, rigid plastic limit m_p^A, and representative strain $\varepsilon_R^A = 0.0319 \cot \alpha^A$ can be fixed. For two materials, #1 with elastoplastic property (E_1, σ_{y1}, n_1) and #2 with elastoplastic property (E_2, σ_{y2}, n_2), to have the same loading P-δ curve they must satisfy

$$C_1^A = \frac{m_e^A m_p^A \overline{E}_1}{m_e^A \frac{\overline{E}_1}{\sigma_{R1}^A} + m_p^A} = C_2^A = \frac{m_e^A m_p^A \overline{E}_2}{m_e^A \frac{\overline{E}_2}{\sigma_{R2}^A} + m_p^A}, \tag{7}$$

where the representative stresses are

$$\sigma_{R1}^{A} = R_1\left(2\frac{\sigma_{R1}^{A}}{E_1} + 2\varepsilon_R^{A}\right)^{n_1} \tag{8}$$

and

$$\sigma_{R2}^{A} = R_2\left(2\frac{\sigma_{R2}^{A}}{E_2} + 2\varepsilon_R^{A}\right)^{n_2} \tag{9}$$

respectively, and they need to satisfy

$$\frac{1}{\sigma_{R2}^{A}} = \frac{1}{\sigma_{R1}^{A}} + \frac{m_p^{A}}{m_e^{A}}\left(\frac{1}{\overline{E}_1} - \frac{1}{\overline{E}_2}\right). \tag{10}$$

Similarly, for another sharp indenter (#B) with a different angle α^{B}, its elastic limit is m_e^{B}, rigid plastic limit is m_p^{B}, and representative strain is ε_R^{B}. If the two materials will again have the same loading curvature, their representative stresses need to satisfy

$$\frac{1}{\sigma_{R2}^{B}} = \frac{1}{\sigma_{R1}^{B}} + \frac{m_p^{B}}{m_e^{B}}\left(\frac{1}{\overline{E}_1} - \frac{1}{\overline{E}_2}\right) \tag{11}$$

with

$$\sigma_{R1}^{B} = R_1\left(2\frac{\sigma_{R1}^{B}}{E_1} + 2\varepsilon_R^{B}\right)^{n_1} \tag{12}$$

and

$$\sigma_{R2}^{B} = R_2\left(2\frac{\sigma_{R2}^{B}}{E_2} + 2\varepsilon_R^{B}\right)^{n_2}. \tag{13}$$

The procedure of deriving two materials with different elastoplastic properties yet with the same loading curvature (for both indenters #A and #B) can be concluded as

(a) Choose any E_1 and E_2 that are different (with fixed $\nu_1 = \nu_2 = 0.3$).
(b) Choose any value of σ_{y1} and n_1, and derive $R_1 = \sigma_{y1}(E_1/\sigma_{y1})^{n_1}$.
(c) Calculate σ_{R1}^{A} from Eq. 8 and solve for σ_{R2}^{A} from Eq. 10.
(d) Obtain one flow stress-total strain pair of the uniaxial stress-strain curve for material #2 as $\sigma_2^{A} = \sigma_{R2}^{A}$ and $\varepsilon_2^{A} = 2\varepsilon_R^{A} + 2\sigma_{R2}^{A}/E_2$.
(e) Calculate σ_{R1}^{B} from Eq. 12 and solve for σ_{R2}^{B} from Eq. 13.
(f) Obtain another flow stress-total strain pair for material #2 as $\sigma_2^{B} = \sigma_{R2}^{B}$ and $\varepsilon_2^{B} = 2\varepsilon_R^{B} + 2\sigma_{R2}^{B}/E_2$.

(g) From both flow stress-total strain pairs, solve $n_2 = \ln\left(\sigma_2^A/\sigma_2^B\right)/\ln\left(\varepsilon_2^A/\varepsilon_2^B\right)$, $R_2 = \sigma_2^A/\left(\varepsilon_2^A\right)^{n_2}$, and $\sigma_{y2} = \left((E_2)^{n_2/(n_2-1)}/(R_2)^{1/(n_2-1)}\right)$. Finally, from the numerical indentation test, confirm that $C_1^A = C_2^A$ and $C_1^B = C_2^B$.

Therefore, for any given material #1 with elastoplastic property (E_1, σ_{y1}, n_1), we can explicitly derive a special material #2 with elastoplastic property (E_2, σ_{y2}, n_2) such that they not only yield the same loading curvature when indenter #A is used but also have the same loading curvature when indenter #B is used. There are infinite numbers of such special siblings. The procedure outlined above can be readily extended to identify materials with indistinguishable indentation loading P-δ curves for three different sharp indenters (#A, #B, #C).

An example of the set of special materials is given in Fig. 2: for the five materials (mat1–mat5) that have distinct elastoplastic properties, their uniaxial stress-strain relationships are given in the inset of Fig. 2. These materials not only have the same loading curvature for the Berkovich indenter but are also the same when α=63.14°, 75.79°, and 80.0° are used; moreover, for any indenter angle between 63.14° and 80°, their loading curvatures are also the same. From Eqs. 10 and 11, if $\overline{E}_1 > \overline{E}_2$ then $\sigma_{R1}^A < \sigma_{R2}^A$ and $\sigma_{R1}^B < \sigma_{R2}^B$ – this can be verified from the inset where for the special sets of materials, the ones with larger moduli have smaller representative stresses. Moreover, when $\alpha^B > \alpha^A$, we have $\varepsilon_R^B < \varepsilon_R^A$ and $m_p^B/m_e^B < m_p^A/m_e^A$. Therefore, for a pair of such special materials, if $\overline{E}_1 > \overline{E}_2$, the difference between σ_1^B and σ_2^B is larger than that of σ_1^A and σ_2^A, and thus $n_1 > n_2$ and $\sigma_{y1} < \sigma_{y2}$, all can be verified from Fig. 2. In this case, the uniaxial stress-strain curves of these special two materials must intersect outside ε_2^A.

From Fig. 2, during unloading the contact stiffness (and thus unloading work) of these special materials are different, which means that if their Young's moduli are known (with a fixed ν), their plastic properties (σ_y, n) can still be uniquely determined from the loading curvature by using the dual (or plural) indenter method, under the important premise that the two indenter angles are distinct enough such that the two determined total strains ε_2^A and ε_2^B are separated sufficiently apart (to ensure numerical accuracy) (Ogasawara et al. 2006).

Determine Special Materials with Same Loading and Unloading Curves for one Conical Indenter

Relating the Unloading Work with Material Properties

While the normalized C (or equivalently, loading work W_t) is the only shape factor during loading, in principle, there are three shape factors for unloading: the normalized δ_f, contact stiffness S, and unloading work W_e. However, only one of them is completely independent; this is because the curvature of the unloading curve of conical indentation (c.f. Fig. 2) is usually very small (except by the end of the unloading process). Therefore, if either one of the variables (δ_f, S, W_e) is

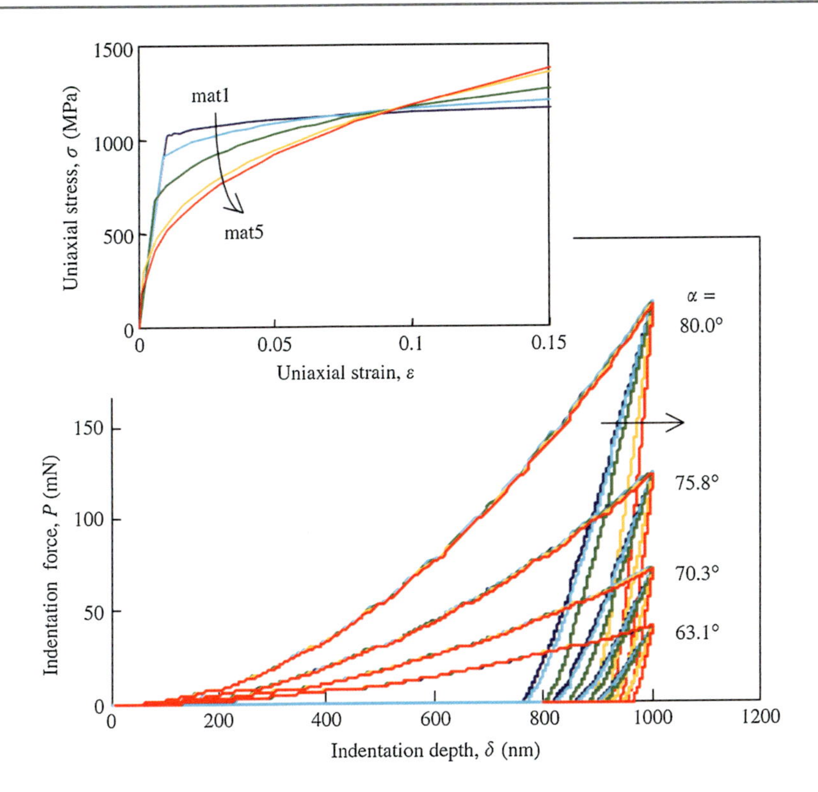

Fig. 2 A set of special materials with same loading curvatures when different indenter angles (63.14°, 70.3°, 75.79°, 80.0°) are used. The five materials have elastoplastic properties (E, σ_y, n) as mat1 (100 GPa, 1022 MPa, 0.05), mat2 (105.0GPa, 911.4 MPa, 0.10), mat3 (120.0 GPa, 678.3 MPa, 0.19), mat4 (200.0 GPa, 299.8 MPa, 0.33), and mat5 (300.0 GPa, 185.0 MPa, 0.37), respectively. The uniaxial stress-strain curves of these special materials are given in the inset on the top-left corner (Chen et al. 2007)

known, from the unloading triangle (plus the knowledge of C), the other two shape factors can be approximately derived. In order to make the best overall matching of the unloading curves, we take the normalized unloading work W_e as the governing unloading shape factor.

Since the unloading work depends on both the contact stiffness (which is related with $\overline{E}$) and the maximum load (which is related with C), a new representative stress for unloading, σ_r, is sought such that the normalized unloading work is related with both $\overline{E}$ and C, but is essentially independent of n. A representative example is given in Fig. 3 for the Berkovich indenter where the unloading work is fitted by

$$\frac{\delta^3 \sigma_r}{W_e} = \lambda_1 \frac{\overline{E}}{\sigma_r} + \lambda_0 \tag{14}$$

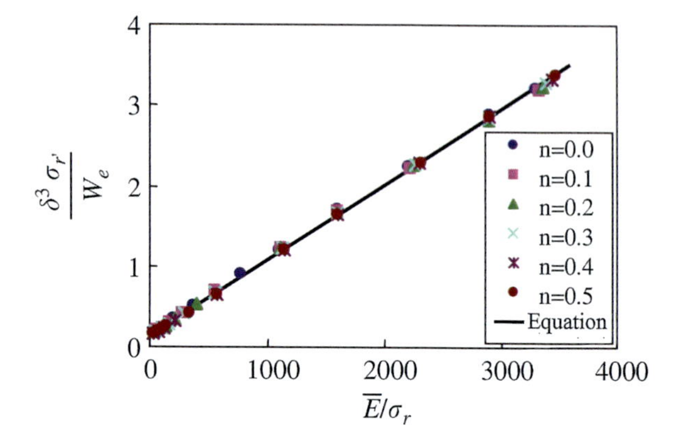

Fig. 3 The relationship between $\delta^3\sigma_r/W_e$ and $\overline{E}/\sigma_r$ for plastic materials, with $\overline{E}/\sigma_r > 30$ or so (Chen et al. 2007)

for plastic material with $\overline{E}/\sigma_r > 30$. $\lambda_1 = 0.0009322$ and $\lambda_0 = 0.1402$ for α=70.3°, and they take different values for other α. The representative stress σ_r is given by

$$\sigma_r = R\left(1.3\frac{\sigma_r}{E} + 2.6\varepsilon_R\right)^n, \tag{15}$$

which is valid for α between 60° and 80°. The simple functional forms derived in this section permit an explicit derivation of special materials with almost same unloading work, elaborated below.

Special Materials with Same Loading and Unloading Curves for a Sharp Indenter

Although Alkorta et al. (2005) have determined special materials with the same loading and unloading curves for one particular indenter, here we report an improved procedure to explicitly derive such materials, which also sets a part of the basis for finding the mystical materials. Based on Eq. 14, for a given indenter #A with half-apex angle α^A and two materials with elastoplastic properties (E_1, σ_{y1}, n_1) and (E_2, σ_{y2}, n_2), if they are to have the same unloading work, they need to satisfy

$$\frac{\delta^3}{W_{e1}^A} = \lambda_1\frac{\overline{E}_1}{\left(\sigma_{r1}^A\right)^2} + \lambda_0\frac{1}{\sigma_{r1}^A} = \frac{\delta^3}{W_{e2}^A} = \lambda_1\frac{\overline{E}_2}{\left(\sigma_{r2}^A\right)^2} + \lambda_0\frac{1}{\sigma_{r2}^A}, \tag{16}$$

where $\varepsilon_R^A = 0.0319\cot\alpha^A$, and $\sigma_{r1}^A = R_1\left(1.3\sigma_{r1}/E_1 + 2.6\varepsilon_R^A\right)^{n_1}$ and $\sigma_{r2}^A = R_2\left(1.3\sigma_{r2}/E_2 + 2.6\varepsilon_R^A\right)^{n_2}$ are the unloading representative stresses for materials #1 and #2 of indenter #A, respectively. Thus,

$$\sigma_{r2}^{A} = \frac{\lambda_0 + \sqrt{\lambda_0^2 + 4\left(\lambda_1 \frac{\overline{E}_1}{\left(\sigma_{r1}^{A}\right)^2} + \lambda_0 \frac{1}{\sigma_{r1}^{A}}\right)\left(\lambda_1 \overline{E}_2\right)}}{2\left(\lambda_1 \frac{\overline{E}_1}{\left(\sigma_{r1}^{A}\right)^2} + \lambda_0 \frac{1}{\sigma_{r1}^{A}}\right)}. \tag{17}$$

Therefore, the procedure of deriving two materials with different elastoplastic properties yet with almost the same loading and unloading curves (for indenter #A) can be concluded as

(a) Choose any E_1 and E_2 that are different (with fixed $\nu_1 = \nu_2 = 0.3$).
(b) Choose any value of σ_{y1} and n_1 (as long as material #1 remains sufficiently plastic), and derive $R_1 = \sigma_{y1}\left(E_1/\sigma_{y1}\right)^{n_1}$.
(c) Calculate σ_{R1}^{A} from Eq. 8 and solve for σ_{R2}^{A} from Eq. 10.
(d) Obtain a flow stress-total strain pair of the uniaxial stress-strain curve for material #2 as $\sigma_2^A = \sigma_{R2}^{A}$ and $\varepsilon_2^A = 2\varepsilon_R^A + 2\sigma_{R2}^{A}/E_2$.
(e) Calculate σ_{R1}^{A} from Eq. 15 and solve for σ_{R2}^{A} from Eq. 17.
(f) Obtain another flow stress-total strain pair for material #2 as $\sigma_2^a = \sigma_{r2}^{A}$ and $\varepsilon_2^a = 2.6\varepsilon_R^A + 1.3\sigma_{r2}^{A}/E_2$.
(g) From the stress-strain pairs, solve $n_2 = \ln\left(\sigma_2^A/\sigma_2^a\right)/\ln\left(\varepsilon_2^A/\varepsilon_2^a\right)$, $R_2 = \sigma_2^A/\left(\varepsilon_2^A\right)^{n_2}$, and $\sigma_{y2} = \left(\left(E_2\right)^{n_2/(n_2-1)}/\left(R_2\right)^{1/(n_2-1)}\right)$. Finally, carry out a numerical test to verify that indeed $C_1^A = C_2^A$ and $W_{e1}^{A} = W_{e2}^{A}$.

Therefore, for any material #1 with (E_1, σ_{y1}, n_1), a special material #2 with (E_2, σ_{y2}, n_2) can be explicitly derived such that they yield indistinguishable loading and unloading curves for a given conical indenter. There are infinite sets of such special materials, and identification of these is no longer based on "trial and error." An example of such is given in Fig. 4, where the effectiveness of the proposed approach is validated (with the difference of C and W_e less than 0.5% – such very small difference is due to the error of fitting functions and numerical solutions which is inevitable). This pair of material cannot be distinguished by only using the Berkovich indenter, yet their P-δ curves may become separable with other distinct indenter angles.

According to Eq. 16, if $\overline{E}_1 > \overline{E}_2$ it can be shown that $\sigma_{r1}^{A} > \sigma_{r2}^{A}$, whereas from Eq. 10, $\sigma_{R1}^{A} > \sigma_{R2}^{A}$. For most plastic materials, the representative strain is much larger than yield strain; thus, the two identified total strains satisfy $\varepsilon_2^a > \varepsilon_2^A$ – this implies that for a pair of special materials, if $\overline{E}_1 > \overline{E}_2$, then $n_1 > n_2$ and $\sigma_{y1} < \sigma_{y2}$; in addition, the stress-strain curves of material #1 and #2 must intersect between ε_2^a and ε_2^A. Thus, if two special materials also have the same loading and unloading curves for indenter #B, their intersection point must also be placed between ε_2^b and ε_2^B; this is only possible when the difference between α^A and α^B is not too extreme (since both ε_R^A and ε_R^B vary as indenter angle changes, and such variation is more prominent for sharper angles). That is, the mystical materials, if they exist, should be valid within a specified range of indenter angles – more discussions are given in the next subsection.

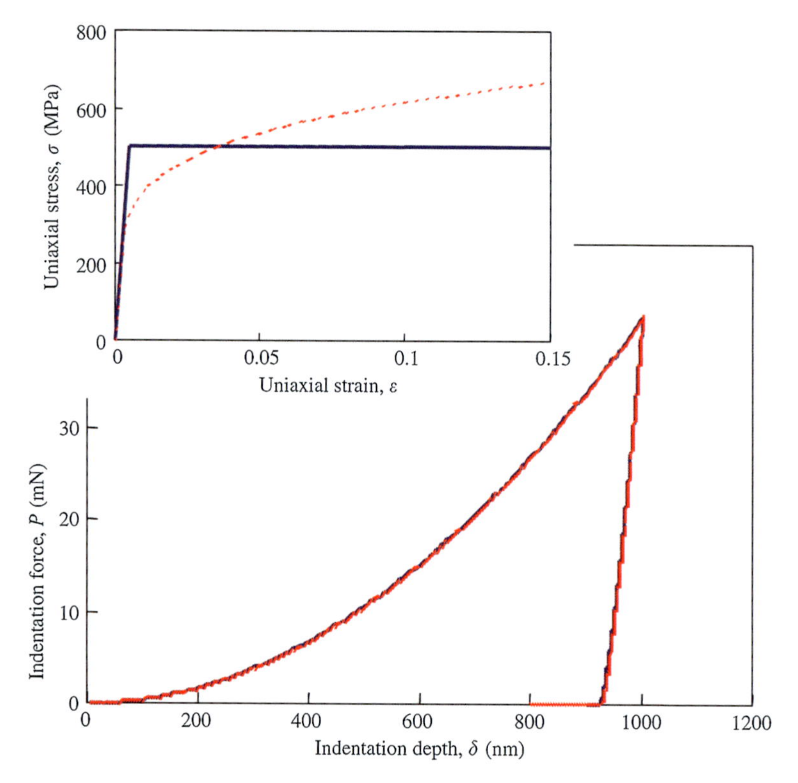

Fig. 4 A pair of special materials with indistinguishable loading and unloading curves when α=70.3. The black solid curve represents the material with (E, σ_y, n) = (100.0GPa, 500.0 MPa, 0.0), and the red dash curve represents the material with (E, σ_y, n) = (110.0GPa, 295.0 MPa, 0.2). The uniaxial stress-strain curves of these special materials are given in the inset on the top-left corner (Chen et al. 2007)

Determine Mystical Materials with Same Loading and Unloading Curve for Dual Indenters

Weak-Form Mystical Materials and Their Possible Existence

In most previous dual (or plural) indentation approaches with Poisson's ratio fixed, the material elastic modulus is first obtained from the unloading curve of one particular indenter (e.g., α=70.3°). Next, the representative stress-based approach is used to determine two (or more) flow stress-total strain points on the uniaxial stress-strain curve, by utilizing the loading curvatures obtained from dual (or plural) sharp indenters. Therefore, if we could identify a pair of special materials with elastoplastic properties (E_1, σ_{y1}, n_1) and (E_2, σ_{y2}, n_2) (their Poisson's ratio $\nu_1 = \nu_2 = 0.3$), such that they yield indistinguishable loading and unloading P-δ curves for indenter #A (e.g., α=70.3°), and also almost identical loading curvature for another indenter #B, then the conventional dual (or plural) indentation analysis would fail since it cannot promise unique solution. This pair of special materials, which does not require their unloading works to match for indenter #B, may be termed as the weak-form mystical materials.

Meanwhile note that in a displacement-controlled experiment where the maximum penetration is fixed, if $C_1^A = C_2^A$ and $C_1^B = C_2^B$, then the maximum indentation load for these two materials are also the same. Thus, to make their unloading works to match for indenter #A $\left(W_{e1}^A = W_{e2}^A\right)$, their contact stiffness must be fairly close $\left(S_1^A \approx S_2^A\right)$, the two materials must have close Young's moduli $\left(\overline{E}_1 \approx \overline{E}_2\right)$, and this also implies their unloading works for indenter #B must also be very close $\left(W_{e1}^B \approx W_{e2}^B\right)$. Thus, the weak-form mystical materials are very close to the mystical materials we are looking for.

What general properties must the weak-form mystical materials satisfy (assuming they are sufficiently plastic such that Eq. 14 applies)? Since these materials must be a subset of the special materials derived from the above procedures, therefore, with ($\alpha^B > \alpha^A$), if $\left(\overline{E}_1 < \overline{E}_2\right)$, the two mystical materials must satisfy $n_1 < n_2$ and $\sigma_{y1} > \sigma_{y2}$; moreover, the uniaxial stress-strain curves of these two materials must intersect outside $2\varepsilon_R^A$ but inside $2.6\varepsilon_R^A$. A schematic showing of the relative status of the two mystical materials is given in Fig. 5a. Since these two candidates need to intersect within a relatively small region, this implies that the mystical materials would only exist for a specified range of indenter angle (with $\alpha^B > \alpha^A$) and material properties. For the current case, the difference between α^B and α^A cannot be too extreme, and the materials need to be sufficiently plastic (with large E_1/σ_{y1} and E_2/σ_{y2}) so as to leave enough possible space for materials #1 and #2 to intersect within the desired region.

Next, we consider the case with $\alpha^B < \alpha^A$, but the difference is not so much such that $2\varepsilon_R^A < 2\varepsilon_R^B < 2.6\varepsilon_R^A$. If $\overline{E}_1 < \overline{E}_2$, then from the above procedures, $\sigma_{R1}^A = \sigma_1^A > \sigma_{R2}^A = \sigma_2^A$ and $\sigma_{R1}^B = \sigma_1^B > \sigma_{R2}^B = \sigma_2^B$; since $m_p^A/m_e^A > m_p^B/m_e^B$, so the difference between σ_1^A and σ_2^A is larger than that of σ_1^B and σ_2^B; moreover, the unloading representative stresses satisfy $\sigma_{r1}^A = \sigma_1^a > \sigma_{r2}^A = \sigma_2^a$. All these features lead to $n_1 < n_2$ and $\sigma_{y1} > \sigma_{y2}$, and a possible solution for the mystical pair #1 and #2 is sketched in Fig. 5b. In this case, the uniaxial stress-strain curves of these two materials must intersect between $2\varepsilon_R^B$ and $2.6\varepsilon_R^A$, and such range is even narrower than that in Fig. 5a. That is, the mystical materials may exist in a small space of material properties and indenter angles; nevertheless, such solution is possible.

When α^B is much smaller than α^A such that $2\varepsilon_R^B > 2.6\varepsilon_R^B$, all the relative magnitudes of the representative stresses discussed in the last paragraph still hold, except that now those related with $2\varepsilon_R^B$ are moved to the right side of those related with $2.6\varepsilon_R^A$ – according to the new schematic in Fig. 5c, it is impossible to find a solution for the mystical material. This implies that if the indenter #A is the Berkovich tip, the indenter #B must be larger than about 64.88° such that the mystical materials may exist according to Fig. 5a, b. Similarly, if #B is taken to be the Berkovich tip (since the weak-form mystical material is very close to the desired mystical material), then #A must be smaller than 74.50°. Therefore, rigorously

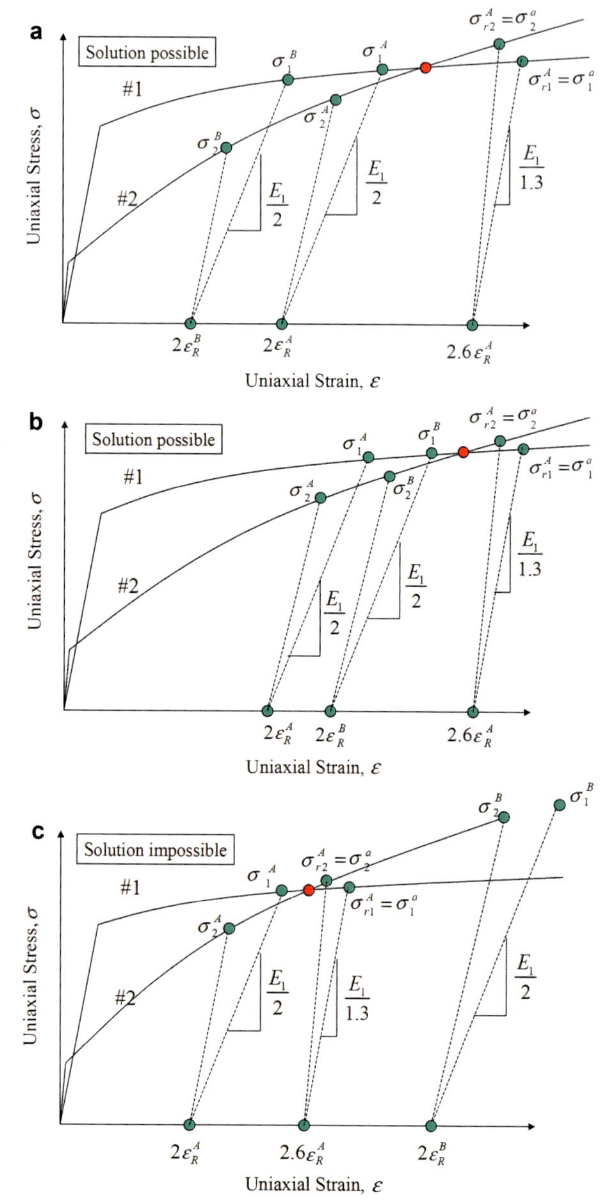

Fig. 5 Schematics of possible solutions of a pair of weak-form plastic mystical materials; assuming $\overline{E}_1 < \overline{E}_2$, then $n_1 < n_2$ and $\sigma_{y1} > \sigma_{y2}$. The indenter angle $A(\alpha^A)$ is fixed (which equals to 70.3° in many existing dual indenter techniques). (**a**) When $\alpha^B > \alpha^A$ solution is possible. (**b**) When $\alpha^B < \alpha^A$ but $2\varepsilon_R^B < 2.6\varepsilon_R^A$, solution is possible. (**c**) When $\alpha^B < \alpha^A$ and the difference between these two angles is large such that $2\varepsilon_R^B > 2.6\varepsilon_R^A$, rigorous solution is not possible (Chen et al. 2007)

speaking, under the premise that Poisson's ratio is always fixed at 0.3, if a Berkovich tip is used in the dual indenter method, the mystical materials would only be possible if the other indenter angle is between 64.88° and 74.50°. Moreover, such mystical materials need to be relatively plastic.

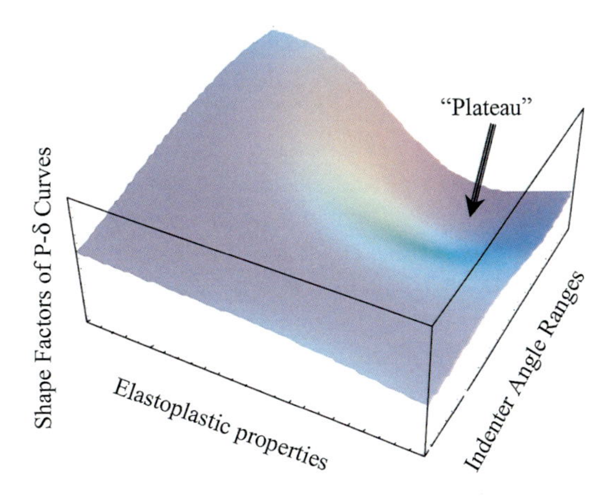

Fig. 6 The schematic of the possible existence of the mystical materials and relevant indenter angle ranges – they correspond to the plateaus on the multidimensional surface of the shape factors of the P-δ curves (Chen et al. 2007)

In a short summary, there is no pair of mystical materials that would be applicable to arbitrary large range of indenter angles, and only for limited material space-indenter angle combinations can the mystical siblings be found – this is part of the reason the mystical materials were not discovered in the past. The current problem of finding the mystical materials is analogous to a multivariable problem (Fig. 6), where the indentation shape factors (the general z-axis) are related with the material elastoplastic properties (the general x-axis) and indenter angle ranges (the general y-axis) through a multidimensional surface. On such surface, the slope may not be large and distinct everywhere, and small regions that are relatively flat may exist, as sketched in Fig. 6. The mystical materials and the relevant indenter angle ranges would correspond to such plateaus (with zero or almost zero slope) of the multidimensional surface – although these plateaus are small compared with the entire parameter space, they may still exist and thus have important theoretical value for probing the uniqueness of indentation analysis. The search of such plateaus is elaborated next.

Search for Mystical Materials with Fixed Poisson's Ratio

In most literature of indentation analysis, Poisson's ratio was fixed by neglecting its influence. In order to challenge the basis of the uniqueness solution and be consistent with the literature, a fixed $\nu = 0.3$ is used in most part of this chapter.

For a given pair of indenters #A and #B, and a given material #1 which is sufficiently plastic, in order to identify the weak-form mystical siblings, Eqs. 10, 11, and 17 need to be satisfied rigorously, from which σ_{R2}^{A}, σ_{R2}^{B}, and σ_{r2}^{A} can be derived (with any specified $\overline{E}_2$), respectively. Unfortunately, the only solution that rigorously satisfies all these equations is exactly material #1. On the other hand, note that there are no perfect finite element solution and numerical fitting, and all numerical results are subject to small error. It is then acceptable if small errors are added to fitting functions Eqs. 10, 11, and 17, such that the resulting loading/unloading P-δ curves of material #2 would be indistinguishable to that of material #1, with the difference between their shape factors below several percent – such small perturbation is also inevitable in the data measured from any real experiment. Another advantage is that, when small perturbation to the shape function is allowed, the existence range of the mystical material is significantly enlarged, in terms of both the material property and indenter angle space. In essence, although it is difficult to search for perfect plateaus (with exactly zero slope) on the multidimensional surface describing the indentation shape factors, it is possible to search for plateaus with slopes that are very close to zero through a numerical algorithm – the results are still the indistinguishable mystical materials given the inevitable small perturbations in numerical and experimental indentation tests. The search process is also relatively straightforward since the simple and explicit formulations of the primary shape factors of loading/unloading P-δ curves are established earlier. Of course, with Poisson's ratio fixed at 0.3, such identification procedure is no longer explicit.

The numerical search process is the following (with fixed$\nu_1 = \nu_2 = 0.3$):

(a) Choose any E_1 and α^{A} (e.g., the Berkovich tip).
(b) Choose any initial value of α^{B} and then iterate such that its difference with respect to α^{A} is increased.
(c) Choose initial values of n_1 and σ_{y1} and then iterate, preferably in the plastic material range.
(d) Choose an initial value of E_2 which is at least 5% different than E_1 and then iterate, such that the difference could become bigger. This ensures that the initial guesses of material #1 and #2 are sufficiently different.
(e) From Eqs. 10, 11, and 17, solve for σ_{R2}^{A}, σ_{R2}^{B}, and σ_{r2}^{A}.
(f) Give small errors (few percent) to Eqs. 10, 11, and 17. For example, if we wish to increase σ_{R2}^{A} by ω times (e.g., ω=1.01 for a 1% error), then from Eq. 6 the new representative stress of material #2 that is related with loading for indenter #A becomes

$$\sigma_{R2-error}^{A} = \left(\frac{1}{\omega \sigma_{R2}^{A}} + \frac{m_p^A}{m_e^A} \frac{1}{\overline{E}_2} \left(\frac{1}{\omega} - 1 \right) \right)^{-1} \tag{18}$$

Similarly, when error is permitted, the two other representative stresses are

$$\sigma_{R2-error}^{B} = \left(\frac{1}{\omega \sigma_{R2}^{B}} + \frac{m_p^B}{m_e^B} \frac{1}{\overline{E}_2} \left(\frac{1}{\omega} - 1 \right) \right)^{-1} \tag{19}$$

and

$$\sigma_{R2-error}^{A} = \left(\lambda_0 + \sqrt{\lambda_0^2 + 4\left(\frac{\lambda_1}{\omega}\frac{\overline{E}_2}{\left(\sigma_{r2}^{A}\right)^2} + \frac{\lambda_0}{\omega}\frac{1}{\sigma_{r2}^{A}}\right)\left(\lambda_1\overline{E}_2\right)}\right) / \left(2\frac{\lambda_1}{\omega}\frac{\overline{E}_2}{\left(\sigma_{r2}^{A}\right)^2} + 2\frac{\lambda_0}{\omega}\frac{1}{\sigma_{r2}^{A}}\right), \tag{20}$$

respectively. Therefore, by plus or minus several percent of error, the upper and lower bounds of the error bars of σ_{R2}^{A}, σ_{R2}^{B}, and σ_{r2}^{A} can be derived – when the three error bars are combined, an error band can be formed.

(g) For all possible combinations of σ_{R2}^{A}, σ_{R2}^{B}, and σ_{r2}^{A} within the error band, the admissible solutions of material #2 are sought; since σ_{R2}^{A}, σ_{R2}^{B}, and σ_{r2}^{A} and E_2 must satisfy certain compatibility, only a small portion of the combinations within the error band could become candidate materials. For any admissible solution, their loading curvatures for indenters #A and #B can be estimated from Eq. 6, and their unloading work for indenter #A is obtained from Eq. 14; the results are then compared with that of material #1, and the more promising pairs with smaller errors are recorded along with the current indenter angle range. A candidate pair of mystical materials is found if the computed error is smaller than 2% for all shape factors.
(h) Iterate E_2.
(i) Iterate n_1 and σ_{y1}.
(j) Iterate α^{B}.
(k) Lastly, numerical indentation analyses are performed on the most promising candidate mystical material pairs using finite element simulations, to confirm that their loading and unloading curves are visually indistinguishable (i.e., the shape factors of their indentation curves are sufficiently close).

The existence range of the mystical materials is first explored through a series of maps in terms of materials space (E/σ_y, n). In Fig. 7a, b, the pairs of identified mystical materials are shown in line segments – both ends of each segment represent two mystical materials with different elastoplastic properties, yet they yield almost indistinguishable loading/unloading P-δ curves. For a given line segment, any sets of materials along the length of the segment are also mystical materials, and the most distinct mystical materials can be found at the ends of the longest segment. The area where the density of the segments is large indicates a possible gold mine of mystical materials.

For any given indenter angle range, the mystical materials can only exist within a certain region. As the difference between the dual indenters becomes larger, the existence range of the mystical material becomes smaller; moreover, the segments are shorter which also indicates the differences between their elastoplastic properties are smaller. Therefore, if a pair of extreme indenter angles (very sharp and very blunt) is used, the mystical materials do not exist (however, in experiments, the

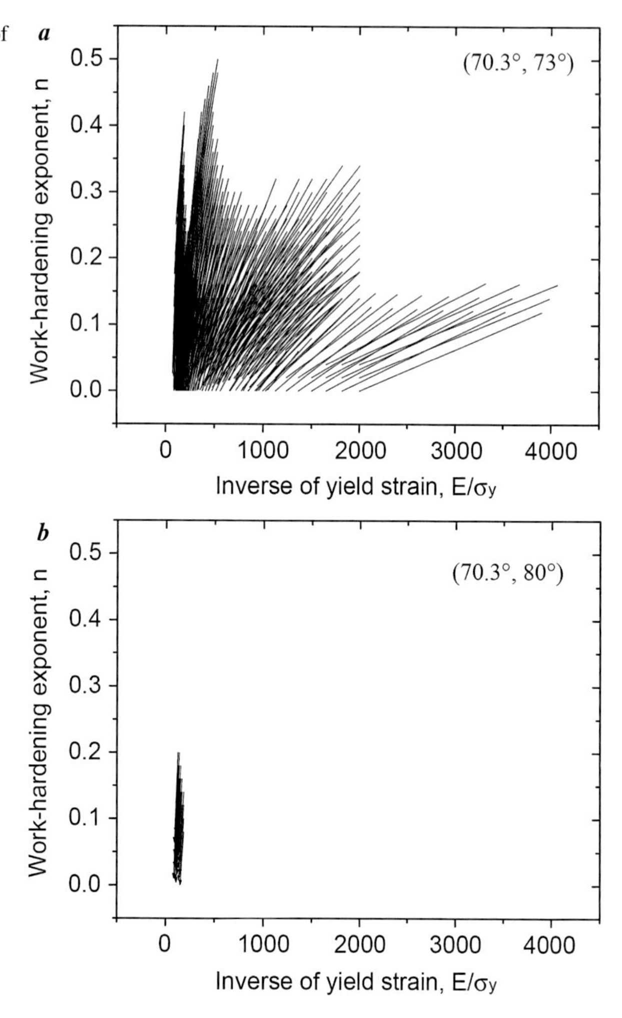

Fig. 7 The existence maps of mystical materials for given indenter angle ranges, (α^A, α^B) = (**a**) (70.3°, 73°), (**b**) (70.3°, 80°). Each segment links a pair of mystical materials within the material space (E/σ_y, n). A gold mine of mystical materials is discovered when E/σ_y is about 100 and when n is small (Chen et al. 2007)

use of a pair of extreme sharp indenters is often impractical). In the examples illustrated in Fig. 7, $\alpha^A = 70.3°$ is always used because many previous studies rely on measuring the elastic modulus from a Berkovich indentation.

When the indenter angle range is relatively small, e.g., (α^A, α^B) = (70.3°, 73°) (Fig. 7a), the mystical materials can be found in a large range, but notably for materials with smaller n. Quite a few of the more plastic mystical materials can exist with large differences of their elastoplastic properties. No mystical materials are available with both large n and large E/σ_y.

By contrary, when the indenter angle difference becomes larger, e.g., (α^A, α^B) = (70.3°, 80°) (Fig. 7b), then the survival range of mystical materials are confined to the lower-left corner of the materials space map. Mystical materials become possible only for materials with n = 0–0.2 and $E/\sigma_y \approx 100$. Note that if a pair of mystical materials is identified for a larger angle range, they are still mystical siblings for any subrange of indenter angles. Moreover, once a map for $(\alpha^{A}, \alpha^{B}) = (70.3°, 60°)$ is made and combined with Fig. 7b, their common elements are the extreme mystical materials that are effective when $(\alpha^{A}, \alpha^{B}) = (60°, 80°)$. Therefore, the materials with small n and with E/σ_y around 100 represent the gold mine around which many mystical materials can be identified. In fact, there are quite a few important engineering metals and alloys near this area, for example, Ti alloys, Ni alloys, Mg alloys, and high strength steel, in addition to a few ceramics and polymers (Ashby 1999) – extra care is needed for the measurement of their elastoplastic properties.

For the mystical materials identified in this section, with $\nu = 0.3$ but without knowing other information in advance, within the specified dual indenter angle range, their elastoplastic properties cannot be uniquely determined from the indentation analysis since their loading and unloading curves are indistinguishable.

Numerical indentation tests are carried out on these materials, and their P-δ curves are compared, which also reveal the characteristics of the mystical materials.

In Fig. 8a, for the indenter angle range $(\alpha^{A}, \alpha^{B}) = (70.3°, 74°)$, a pair of plastic mystical materials is chosen, with $E_1 = 100$ GPa, $\sigma_{y1} = 50$ MPa, $n_1 = 0.06$, and $E_2 = 110$ GPa, $\sigma_{y2} = 29.336$ MPa, and $n_2 = 0.17277$, respectively. Their uniaxial stress-strain curves are shown in the inset, and their corresponding indentation load-displacement curves are given. It is apparent that their indentation behaviors are almost identical (for both loading and unloading curves); specifically, the difference between C_1^A and C_2^A is about 2%, the difference between C_1^B and C_2^B is about 1%, and the difference between W_{e1}^A and W_{e2}^A is about 2%. The difference between W_{e1}^B and W_{e2}^B is about 4%; nevertheless, during the search of weak-form mystical materials, the matching criterion is not applied to the unloading curves with indenter #B. Note that when a larger indenter angle range is used, such as 63.14° and 75.79°, their P-δ curves become quite separable, and therefore it is still possible to use the established dual indenter method to measure the elastoplastic properties of these two materials with the wider indenter angle ranges.

In fact, a wider indenter angle separation also means that the two identified total strains are further apart, which also gives better numerical accuracy – a rule of thumb is that the two identified total strain should be separated by at least 30%, which is qualified for most plastic materials with indenter angles 63.14° and 75.79° (Ogasawara et al. 2007b), including the example in Fig. 8a. On the other hand, for the more elastic materials, the separation between the identified total strain points becomes smaller for a given indenter angle range. To ensure accuracy, a large indenter angle range $\Delta\alpha = |\alpha^{A} - \alpha^{B}|$ is recommended for more elastic materials with $E/\sigma_y < 100$ or so.

Figure 8b gives an intriguing example, which corresponds to $E_1 = 100$ GPa, $\sigma_{y1} = 872.47$ MPa, $n_1 = 0.0$, and $E_2 = 103.75$ GPa, $\sigma_{y2} = 715.61$ MPa, and $n_2 = 0.10663$. This pair of extreme mystical material leads to almost the same indentation loading/unloading behaviors when the indenter angle changes from 60° to 80°, which has reached the limit of the sharp indenter angles used in this study. Apparently, many existing dual (or plural) indentation methods in the literature

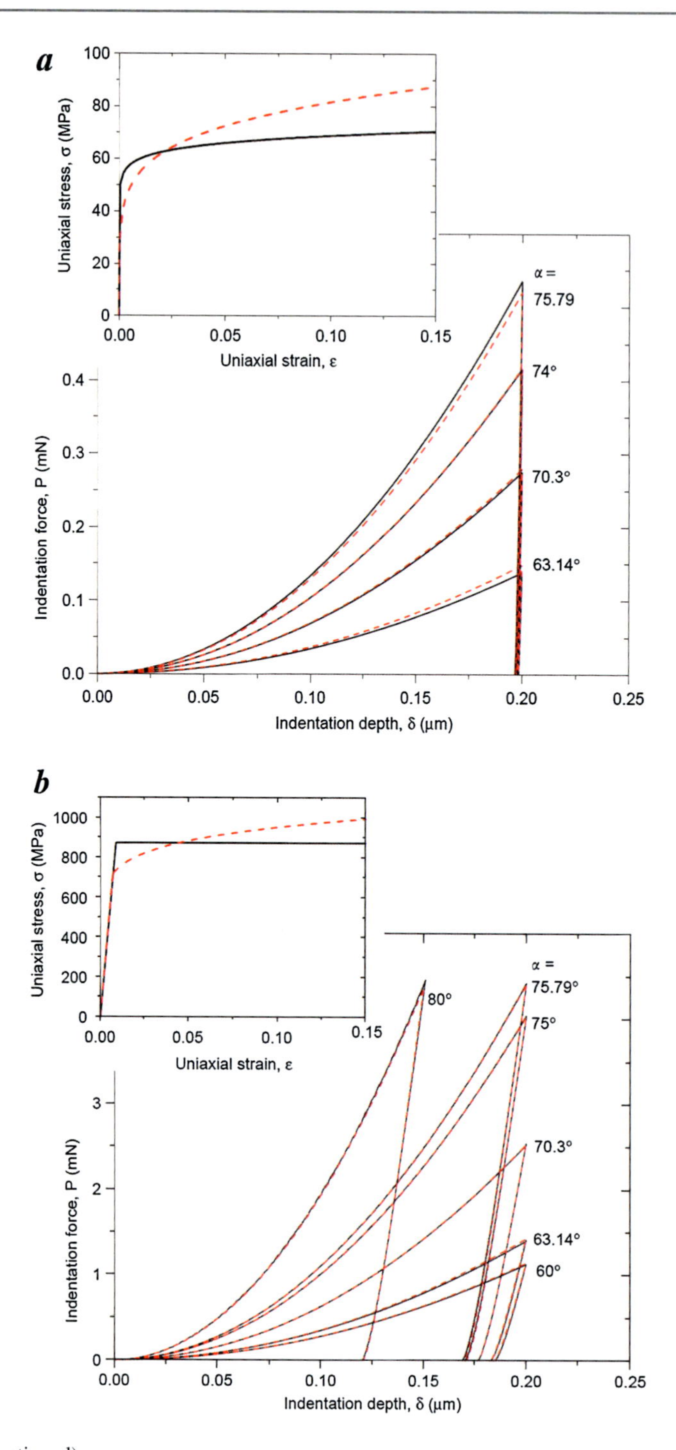

Fig. 8 (continued)

would fail to distinguish them, and more extreme indenters are needed. However, this is often not practical because without advanced knowledge, one could not predetermine what kind of $\Delta\alpha$ needs to be used in an experiment. Moreover, when extreme indenter angles are used, the measured hardness would differ by orders of magnitude, and new problems such as those associated with the resolution of the instrument, size effect, indenter tip alignment, and indentation cracking will emerge. In the next section, we will introduce how to use alternative methods to distinguish such extreme mystical pair.

For a pair of mystical materials near or inside the gold mine, their stress-strain curves tend to intersect around a total strain of 0.05. Moreover, if one material has a larger plane-strain modulus, then it always has larger work-hardening exponent and smaller yield stress.

Alternative Methods to Distinguish Mystical Materials

Improved Spherical Indentation

Spherical indenter has the unique advantage that with one penetration, the loading curvature is reduced as if the indenter angel becomes smaller. Since the mystical materials may be eventually distinguishable by a large $\Delta\alpha$, we only need to control the δ_{max} during spherical indentation. For that matter, the penetration depth has to be sufficiently deep and a few previous studies (e.g., Cao and Lu 2004) where $\delta_{max}/r = 0.1$) do not qualify; indeed, if δ_{max}/r is small (such as 0.1), the spherical indenter method still cannot distinguish the mystical materials because the effective $\Delta\alpha$ is small. Alternatively, in one of our works (Zhao et al. 2006b), an improved spherical indentation technique was proposed, which seems promising since $\delta_{max}/r = 0.3$, which mimics a very sharp indenter angle.

Figure 9a shows the spherical indentation result on the extreme mystical material pair derived from Fig. 8b. Initially, when the penetration is shallow, which is analogous to the blunter indenter angles, the two P-δ curves cannot be separated. However, when δ/r is larger than about 0.15, these two materials become distinguishable. Finally, at the maximum penetration, there is about 8% difference between their C. Although these two materials have very close contact stiffness and also C measured at $\delta/r = 0.13$, the relatively large difference at

Fig. 8 Representative case studies of mystical materials: the P- δ curves of mystical materials and the uniaxial σ- ε curves of mystical materials are given in inset on the top-left corner. (**a**) A pair of plastic mystical materials for $(\alpha^A, \alpha^B) = (70.3°, 74°)$, with $E_1 = 100$ GPa, $\sigma_{y1} = 50$ MPa, $n_1 = 0.06$ (solid curve), and $E_2 = 110$GPa, $\sigma_{y2} = 29.336$ MPa, $n_2 = 0.17277$ (dash curve). They can be distinguished when more different indenter angles are used. (**b**) A pair of extreme mystical materials with $E_1 = 100$ GPa, $\sigma_{y1} = 872.47$ MPa, $n_1 = 0.0$ (solid curve), and $E_2 = 103.75$ GPa, $\sigma_{y2} = 715.61$ MPa, $n_2 = 0.10663$ (dash curve). They cannot be effectively distinguished by indenter angles from 60° and 80° (Chen et al. 2007)

$\delta_{max}/r = 0.3$ is sufficient to make the spherical indentation technique work well. By following the reverse analysis procedure in Zhao et al. (2006b), the determined values are $E_1 = 102.5$ GPa, $\sigma_{y1} = 828.85$ MPa, $n_1 = 0.024$, and $E_2 = 106.34$ GPa, $\sigma_{y2} = 701.30$ MPa, and $n_2 = 0.108$, respectively. All errors (except that for n_1 cannot be counted since its true value is 0.0) are smaller than about 2% (except σ_{y1} which is about 5% and still within reasonable range). Finally, the identified uniaxial stress-strain curve from the reverse analysis of the improved spherical indentation method (Zhao et al. 2006b) is given in the inset, and its excellent capability of distinguishing the mystical material is justified. The improved spherical indentation technique proposed by Zhao et al. (2006b) only requires one simple indentation test, and thus it is convenient and reliable. This technique is therefore recommended for materials inside or near the gold mine of mystical materials.

Film Indentation

When an elastoplastic film with finite thickness is bonded to a rigid substrate, the increased conical penetration dramatically increases the loading curvature as if the sharp indenter angle is increased. Thus, the substrate effect provides an alternative way of obtaining extreme indenter angles to distinguish the mystical materials. We have proposed a theory, where the loading curvatures at the penetration of 1/3 and 2/3 of film thickness (h), along with the unloading work, are used to obtain the material elastoplastic properties from one fil indentation test (Zhao et al. 2006a). When this technique is applied to the extreme mystical materials (found in Fig. 8b, in Fig. 9b), it can be readily seen that their loading curvatures become quite different at $\delta_{max}/h = 2/3$, which is again sufficient to distinguish the extreme mystical pair although their loading curvatures at $\delta/r = 1/3$ are close. In addition, their unloading works are also different.

By following the reverse analysis described in Zhao et al. (2006a), finally the determined properties of the extreme pair of mystical materials are $E_1 = 97.5$ GPa, $\sigma_{y1} = 872.47$ MPa, $n_1 = 0.0$ and $E_2 = 103.75$ GPa, $\sigma_{y2} = 719.19$ MPa, $n_2 = 0.105$, respectively. All errors (except that for n_1 cannot be counted since its true value is 0.0) are smaller than 1.5% except E_1 which is only -2.5%. The identified uniaxial stress-strain curve from the film indentation technique (Zhao et al. 2006a) is given in the inset of Fig. 10, and it is demonstrated to be able to distinguish the mystical materials with high accuracy. Note that the advantage of the film indentation technique is that it may be applied to specimens with finite thickness; however, it also requires the testing platform (i.e., the substrate) to be sufficiently stiff and hard, which may not be practical in some cases (see Zhao et al. (2006a) for discussions). Nevertheless, it could be used as an alternative method to distinguish the mystical materials and is proven to work well.

In fact, from the error analysis of both the improved spherical indentation and film indentation methods (Zhao et al. 2006a, b), both techniques have the best accuracy for materials inside or near the gold mine of mystical materials, which make them complementary to the dual (or plural) sharp indentation analysis.

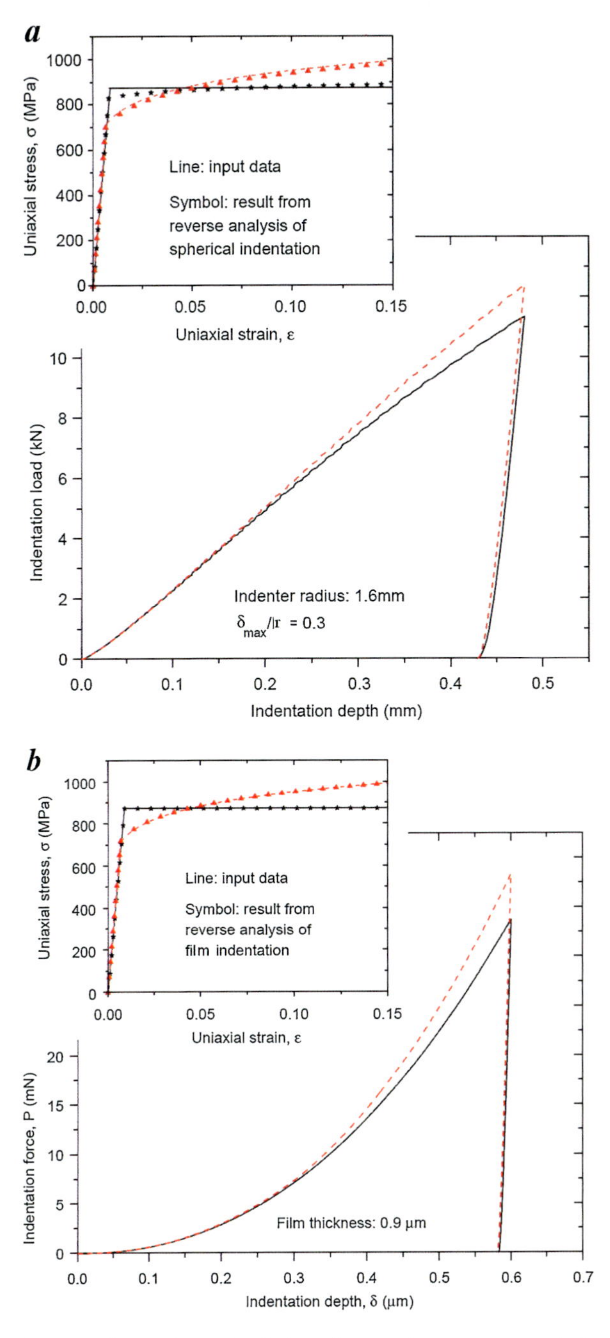

Fig. 9 The P-δ curves of two alternative indentation on the extreme mystical materials found in Fig. 8b. (**a**) Improved spherical indentation method (Zhao et al. 2006b). (**b**) Film indentation method (Zhao et al. 2006a). These methods can distinguish the extreme mystical materials – in the inset on top-left corner, the uniaxial stress-strain curves obtained from reverse analysis (symbols) are compared with the input (true) data and show excellent agreement (Chen et al. 2007)

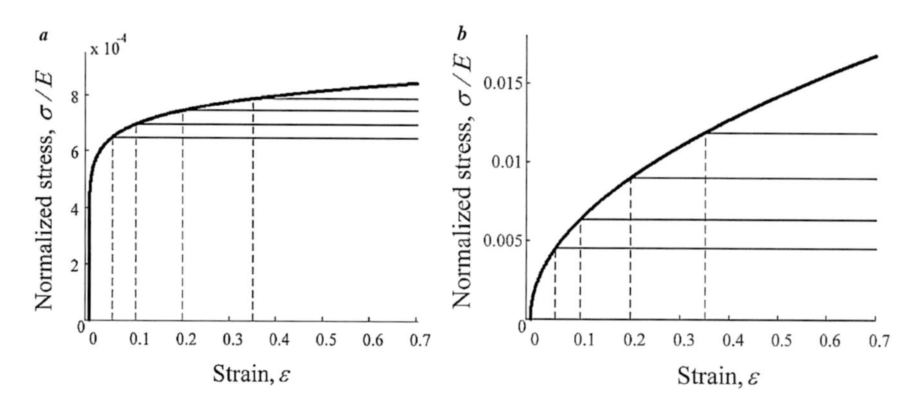

Fig. 10 Materials tailoring of two power-law materials (**a**) $E/\sigma_y = 2500$ and $n = 0.1$ and (**b**) $E/\sigma_y = 2500$ and $n = 0.5$. Thick lines are the uniaxial true stress-strain curves of untailored materials, and each thin line presents a modified material. Every modified material is defined as initially having the same constitutive curve with the original power-law material but becomes perfectly plastic beyond a bifurcation strain, ε_b. In the examples in both (**a**) and (**b**), ε_b equals 0.05, 0.1, 0.2, and 0.35 for the four tailored materials (Liu et al. 2009)

Detectable Strain Range of Indentation Test

The Critical Strain

When a sharp indenter penetrates a bulk specimen, the loading curvature C is only a function of the material elastoplastic; theoretically, any variation of the material constitutive relationship (the stress-strain curve) can, to a certain degree, cause C to deviate from its original value. However, there exists a critical strain beyond which tailoring material properties will no longer induce prominent variation to the indentation response (e.g., less than 1% deviation in the measured C). [The 1% threshold is set because it is a typical order-of-magnitude intrinsic error associated with numerical indentation analyses: below this critical level, one can hardly tell whether the difference is caused by material tailoring or by numerical error. Of course it could set a different threshold (e.g., 0.5%), but the critical strains derived from the new threshold are not going to be much different from the ones identified in this study, and the relevant conclusions still hold.] Therefore, indentation is limited to probing material elastoplastic properties within a particular strain range below critical, and the tailoring of the stress-strain curve beyond this range cannot be detected by indentation reverse analysis, leading to a non-unique solution.

To verify the existence of the critical strain and identify its dependence on material properties and indenter angles, without losing generality, we consider six representative power-law materials with $E/\sigma_y = 2500$, 1000, and 100 and $n = 0.1$ and 0.5. Two example stress-strain curves are given as thick lines in Fig. 10a, b. To tailor the constitutive relationship, one could choose any bifurcation strain $\varepsilon_b > \varepsilon_y$

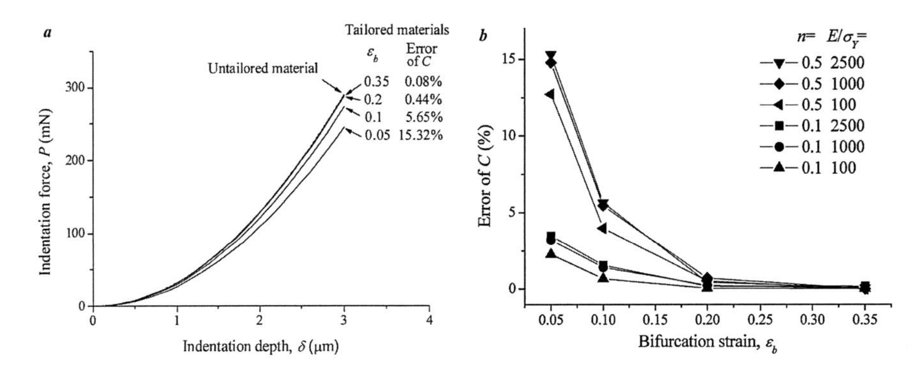

Fig. 11 (**a**) Indentation loading curves (with α=70.3°) for a representative power-law material and four artificial materials tailored at different ε_b. The error of C is calculated as the difference of loading curvatures between each tailored materials and the untailored material. (**b**) The evolution of the error of C (with α=70.3°) as a function of ε_b induced by material tailoring. A critical strain ε_c can be defined when the error of C falls below 1% (Liu et al. 2009)

and assign perfect plastic behavior after this point (the flat thin line in Fig. 10). In other words, the tailored stress-strain relationship becomes

$$\begin{aligned} \sigma &= E\varepsilon, \text{for } \varepsilon \leq \varepsilon_y \\ \sigma &= R\varepsilon^n, \text{for } \varepsilon_y \leq \varepsilon \leq \varepsilon_b \\ \sigma &= R\varepsilon_b^n = \text{constant, for } \varepsilon \geq \varepsilon_b. \end{aligned} \tag{21}$$

Such a tailoring strategy provides a rough upper bound of the moderate modifications of the hardening function. For the selected cases shown in Fig. 10, ε_b equals 0.05, 0.1, 0.2, and 0.35 for the four modified materials. Upon indentation, the modified material will in principle show a different C from that of the original (unmodified) material, but S remains essentially unchanged since the elastic modulus is unaffected; thus, we only focus on the perturbation of C during loading.

Numerical indentation tests are carried out on the original power-law materials as well as their modified counterparts (with ε_b varying over a large range and example load-displacement curves given in Fig. 11a). The indenter angle α is also varied. The percentage error of C (between the original and modified materials) can be plotted as a function of ε_b, given in Fig. 11b for a representative indenter angle $\alpha = 70.3°$. Each line in this Figure represents modifications of one of the six original materials, and each symbol in that line denotes the percentage error of C for a tailored material characterized by a particular ε_b. The error decreases quickly with ε_b, since a higher ε_b means that the constitutive relationship of the modified material is closer to the original one (Fig. 10). In addition, the error of C is smaller when n is smaller or when E/σ_y is smaller (and the effect of n is more prominent), and this is also related to the fact that the difference between the stress-strain curves of the original and tailored materials are smaller when n and/or E/σ_y is smaller (Fig. 10 and Eq. 21).

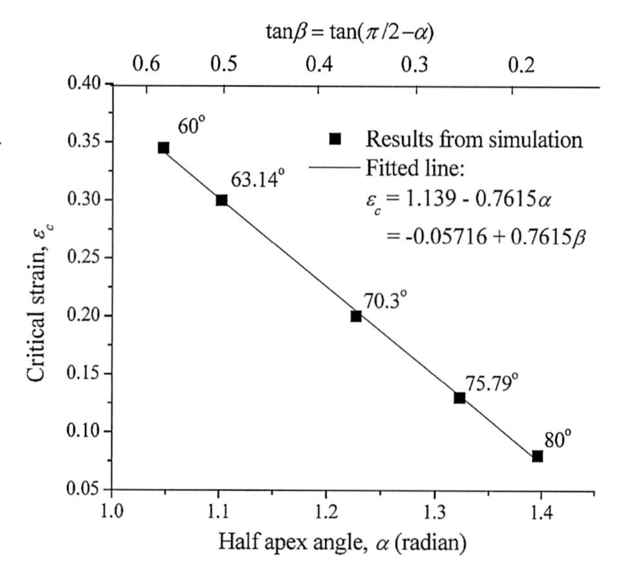

Fig. 12 The dependency of the critical strainε_c on the half-apex angle, α, of sharp indenters. Also shown is the linear fitting Eq. 22 (Liu et al. 2009)

The most important finding is, regardless of the original material, the percentage error of C tends to be lower than the 1% threshold when ε_b exceeds a critical value, ε_c. For $\alpha = 70.3°$, the critical strain ε_c is identified as 0.20 – any modification of the plastic behavior beyond this point cannot be effectively reflected on the P-δ curve, and the modified material would exhibit an indistinguishable indentation response with respect to the original material as long as $\varepsilon_b > \varepsilon_c$; under this circumstance, both the original and modified materials are possible solutions of the reverse analysis of the same P-δ curve. Note that the difference between the constitutive relationships of the original material and modified material can be substantial especially when n is large and/or when the critical strain is small.

Using the sharp indentation technique, the stress-strain curve may only be probed when the strain is between 0 and ε_c (the detectable strain range or sometimes the detectable range for short). When α is fixed, ε_c is essentially material-independent, and this is verified by numerical analyses using a wide range of materials with diverse elastoplastic properties (as long as ε_y is not too large, which is satisfied for most metals and alloys). (Although we derived the critical strain by modifying the power-law material model in this section, the approach can be extended to other material models, and we have verified that the value of critical strain is not sensitive to the material model used.) In Fig. 12, the critical strain ε_c is presented as a function of the indenter angle, where the relationship is almost linear and can be fitted as

$$\varepsilon_c = 1.139 - 0.7615\alpha \tag{22}$$

within the indenter angle range in this chapter (where α is in radians in Eq. 22). It is interesting that the detectable range of sharp indenters can be small for blunt indenters, and one of the popular indenters with $\alpha = 80°$ could only prove strain up to about 8% and thus is sensitive to the non-uniqueness issue; the use of sharper indenters could partially reduce this concern. However, sharp indenters may cause cracking, and the results may be sensitive to friction in practice. In addition, even the sharpest indenter used in this study ($\alpha = 60°$) could only detect up to 35% of strain, and this performance falls well below common expectations. We therefore conclude that it is impossible to measure the entire stress-strain curve uniquely via a sharp indentation test. (Although the plural indenter technique, especially those with sharper indenters (e.g., using $\alpha = 60°$ and 70.3°), could alleviate the non-uniqueness problem than those with blunter indenters (e.g., using $\alpha = 80°$ and $\alpha = 70.3°$), none of them would work well outside the critical strain range.)

Besides challenging the uniqueness of indentation test, the discovery of the critical strain has several other impacts. First, during the verification of an established indentation method, often the stress-strain curve of a real engineering material needs to be fitted into power-law form and serve as a benchmark for examination (Guelorget et al. 2007); however, a different fitting range of the stress-strain curve would lead to different fitted results of the plastic parameters, and the fitting range should be consistent with the detectable range of indentation test, otherwise a large bias would occur (Ogasawara et al. 2008). Second, the critical strain could also guide numerical indentation analyses. From either the stress-strain curve measured in a lab experiment or with respect to a specific material model, a data set of the uniaxial stress and strain is needed as the input for material properties to be used in an FEM program/simulation. Sometimes there is a concern as to how many data points are needed and how refined they need to be. Here we show that regardless of the details, only the input stress-strain data within the detectable strain range is relevant, and outside this range, the data is essentially unimportant (in terms of the resulting P-δ curves).

Variation of Critical Strain: A Qualitative Explanation

A qualitative explanation of the critical strain, along with its dependency on the indenter geometry, lies in the nonuniform plastic deformation below the indenter. Since the work done by the indenter during the penetration of a sharp indenter is $W_t = C\delta^3/3$, the perturbation of C may be understood from that of W_t, which equals the total deformation energy. The material deformation energy includes two parts: the recoverable strain energy and the plastic dissipation; the later part equals (W_t-W_e).

The influence of tailoring the material constitutive behavior is nonuniform in the indented solid. Within the field of equivalent plastic strain (ε_e) produced by indentation, only the regions with $\varepsilon_e > \varepsilon_b$ are more sensitive to the modification of plastic properties beyond ε_b. Figure 13 shows the contour plots of ε_e in the deformed unmodified power-law solid (with $E/\sigma_y = 2500$, and $n = 0.5$), and the contour lines of $\varepsilon_e = 0.1$ and 0.2 are given when $\alpha = 70.3°$ and 75.79°, respectively. For fair comparison, the contours shown in Fig. 13 are taken at the instants when the indenters have done the same amount of work. The material enclosed roughly by

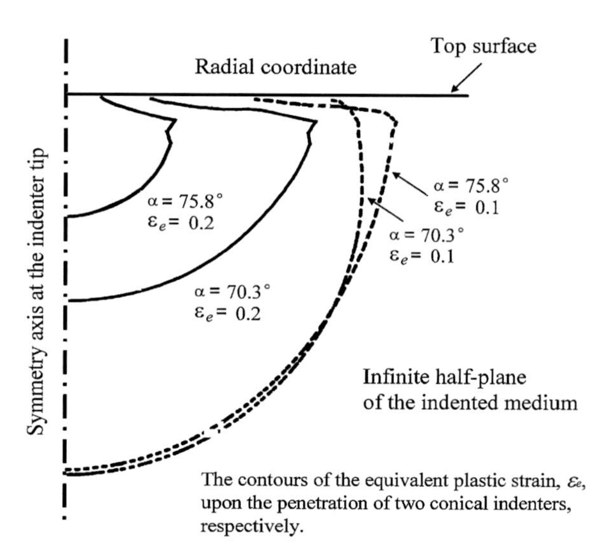

Fig. 13 The contour plot of the equivalent plastic strain, ε_e, in a semi-infinite solid indented by two sharp indenters. The superimposed contours are taken from independent tests at the instants when the indenters have done the same amount of work (Liu et al. 2009)

the solid contour lines (e.g., $\varepsilon_e = 0.2$) may be perturbed if the stress-strain curve is modified beyond the corresponding value, for example, $\varepsilon_b = 0.2$. Within the area where $\varepsilon_e < \varepsilon_b$, if the deformation energy is relatively small, material tailoring would not yield prominent variation of the overall indentation response.

For $\alpha = 70.3°$ for example, the deformation energy in the region enclosed by the contour of $\varepsilon_e = 0.2$ is only about 22% of the energy enclosed by the contour of $\varepsilon_e = 0.1$. Therefore, tailoring the material beyond $\varepsilon_b = 0.1$ will include a larger variation to the overall P-δ curve than tailoring above $\varepsilon_b = 0.2$. This is qualitatively consistent with the descending trend of the error of C due to tailoring Fig. 11b. When the indenter angle is varied, for $\alpha = 75.79°$, the fraction of deformation energy of the region enclosed by the contour of $\varepsilon_e = 0.2$ is only 3% of its counterpart for $\alpha = 70.3°$, and the fraction of deformation energy of the region surrounded by the contour of $\varepsilon_e = 0.1$ is about half that for $\alpha = 70.3°$. This suggests that material tailoring beyond the same strain ($\varepsilon_b = 0.1$ or 0.2) could influence C more significantly for the sharper indenter, and thus qualitatively speaking, ε_c is larger for a sharper indenter and smaller for a blunter indenter.

Conclusion

Although indentation tests have been long used to measure the elastoplastic properties of engineering materials, a systematic study on the uniqueness of indentation analysis, i.e., on the possible existence of the one-to-one correspondence between the indentation load-displacement curves, material parameters, and indenter geometries, is still lacking. Among the available indentation techniques, the dual (or plural) sharp indenter method is often considered as well established, and it is also the foundation of many other similar indentation analyses and has been widely used in practice. In this chapter, through a comprehensive numerical study, the primary shape factors of the indentation load-displacement curves are related with the material properties and indenter angles through simple functional forms. Both explicit and numerical procedures are established to search for mystical materials with distinct elastoplastic properties yet yield indistinguishable load-displacement curves, even when the indenter angles are varied. Consequently, these mystical materials cannot be distinguished by many of the existing dual (or plural) indenter methods or spherical indenter methods (if the indentation depth is shallow). The properties and the existence of such mystical materials are discussed.

In addition, a critical strain is identified as the upper bound of the detectable range of indentation, and moderate tailoring of the constitutive behavior beyond this range cannot be effectively detected by the reverse analysis of the load-displacement curve. That is, for a given indenter geometry, beyond the critical strain, there is no unique solution of the material plastic behavior from the reverse analysis of the load-displacement curve. The critical strain is identified as a function of the sharp indenter angle, through which the analysis of sharp indentation may be qualitatively correlated to some extent – this link also enriches the indentation theory and applications.

The topics in this chapter address the important question of the uniqueness of indentation test, as well as providing useful guidelines to properly use the indentation technique to measure material elastoplastic properties.

Acknowledgments The work is supported in part by National Science Foundation CMS-0407743 and CMMI-CAREER-0643726 and in part by the Department of Civil Engineering and Engineering Mechanics, Columbia University.

References

J. Alkorta, J.M. Martinez-Esnaola, J.G. Sevillano, Absence of one-to-one correspondence between elastoplastic properties and sharp-indentation load–penetration data. J. Mater. Res. **20**, 432–437 (2005)

M.F. Ashby, *Materials Selection in Mechanical Design*, 2nd edn. (Elsevier, Amsterdam, 1999)

F.P. Bowden, D. Tabor, *The Friction and Lubrications of Solids* (Oxford University Press, Oxford, 1950)

Y.P. Cao, J. Lu, A new method to extract the plastic properties of metal materials from an instrumented spherical indentation loading curve. Acta Mater. **52**, 4023–4032 (2004)

T.W. Capehart, Y.T. Cheng, Determining constitutive models from conical indentation: sensitivity analysis. J. Mater. Res. **18**, 827–832 (2003)

X. Chen, N. Ogasawara, M. Zhao, N. Chiba, On the uniqueness of measuring elastoplastic properties from indentation: the indistinguishable mystical materials. J. Mech. Phys. Solids **55**, 1618–1660 (2007)

Y.T. Cheng, C.M. Cheng, Can stress–strain relationships be obtained from indentation curves using conical and pyramidal indenters? J. Mater. Res. **14**, 3493–3496 (1999)

Y.T. Cheng, C.M. Cheng, Scaling, dimensional analysis, and indentation measurements. Mater. Sci. Eng. **R44**, 91–149 (2004)

B. Guelorget, M. Francois, C. Liu, J. Lu, Extracting the plastic properties of metal materials from microindentation tests: experimental comparison of recently published methods. J. Mater. Res. **22**, 1512–1519 (2007)

L. Liu, N. Ogasawara, N. Chiba, X. Chen, Can indentation technique measure unique elastoplastic properties? J. Mater. Res. **24**, 784–800 (2009)

S.D. Mesarovic, N.A. Fleck, Spherical indentation of elastic-plastic solids. Proc. R. Soc. Lond. A **455**, 2707–2728 (1999)

N. Ogasawara, N. Chiba, X. Chen, Representative strain of indentation analysis. J. Mater. Res. **20**, 2225–2234 (2005)

N. Ogasawara, N. Chiba, X. Chen, Limit analysis-based approach to determine the material plastic properties with conical indentation. J. Mater. Res. **21**, 947–958 (2006)

N. Ogasawara, N. Chiba, M. Zhao, X. Chen, Measuring material plastic properties with optimized representative strain-based indentation technique. J. Solid Mech. Mater. Eng. **1**, 895–906 (2007a)

N. Ogasawara, N. Chiba, M. Zhao, X. Chen, Comments on "Further investigation on the definition of the representative strain in conical indentation" by Y. Cao and N. Huber [J. Mater. Res. 21, 1810 (2006)]: A systematic study on applying the representative strains to extract plastic properties through one conical indentation test. J. Mater. Res. **22**, 858–868 (2007b)

N. Ogasawara, M. Zhao, N. Chiba, X. Chen, Comments on "Extracting the plastic properties of metal materials from microindentation tests: experimental comparison of recently published methods" by B. Guelorget, et al. [J. Mater. Soc. 22, 1512 (2007)]: The correct methods of analyzing experimental data and reverse analysis of indentation tests. J. Mater. Res. **23**, 598–608 (2008)

I.N. Sneddon, The relationship between load and penetration in the axisymmetric Boussinesq problem for a punch of arbitrary profile. Int. J. Eng. Sci. **3**, 47–57 (1965)

K.K. Tho, S. Swaddiwudhipong, Z.S. Liu, K. Zeng, J. Hua, Uniqueness of reverse analysis from conical indentation tests. J. Mater. Res. **19**, 2498–2502 (2004)

M. Zhao, X. Chen, N. Ogasawara, A.C. Razvan, N. Chiba, D. Lee, Y.X. Gan, A new sharp indentation method of measuring the elastic-plastic properties of soft and compliant materials by using the substrate effect. J. Mater. Res. **21**, 3134–3151 (2006a)

M. Zhao, N. Ogasawara, N. Chiba, X. Chen, A new approach of measuring the elastic-plastic properties of bulk materials with spherical indentation. Acta Mater. **54**, 23–32 (2006b)

Helical Buckling Behaviors of the Nanowire/Substrate System

7

Youlong Chen, Yilun Liu, and Xi Chen

Contents

Y. Chen (✉)
International Center for Applied Mechanics, State Key Laboratory for Strength and Vibration of Mechanical Structures, School of Aerospace, Xi'an Jiaotong University, Xi'an, China
e-mail: cyl900125@126.com

Y. Liu
International Center for Applied Mechanics, State Key Laboratory for Strength and Vibration of Mechanical Structures, School of Aerospace, Xi'an Jiaotong University, Xi'an, China
e-mail: yilunliu@mail.xjtu.edu.cn

X. Chen
Department of Earth and Environmental Engineering, Columbia Nanomechanics Research Center, Columbia University, New York, NY, USA
e-mail: xichen@columbia.edu

G. Z. Voyiadjis (ed.), *Handbook of Nonlocal Continuum Mechanics for Materials and Structures*, https://doi.org/10.1007/978-3-319-58729-5_47

Abstract

When a nanowire is deposited on a compliant soft substrate or embedded in matrix, it may buckle into a helical coil form when the system is compressed. Using theoretical and finite element method (FEM) analyses, the detailed three-dimensional coil buckling mechanism for a silicon nanowire (SiNW) on a poly-dimethylsiloxne (PDMS) substrate is discussed. A continuum mechanics approach based on the minimization of the strain energy in the SiNW and elastomeric substrate is developed, and the helical buckling spacing and amplitude are deduced, taking into account the influences of the elastic properties and dimensions of SiNWs. These features are verified by systematic FEM simulations and parallel experiments. When the debonding of SiNW from the surface of the substrate is considered, the buckling profile of the nanowire can be divided into three regimes, i.e., the in-plane buckling, the disordered buckling in the out-of-plane direction, and the helical buckling, depending on the debonding density. For a nanowire embedded in matrix, the buckled profile is almost perfectly circular in the axial direction; with increasing compression, the buckling spacing decreases almost linearly, while the amplitude scales with the 1/2 power of the compressive strain; the transition strain from 2D mode to 3D helical mode decreases with the Young's modulus of the wire and approaches to 1.25% when the modulus is high enough, which is much smaller than nanowires on the surface of substrates. The study may shed useful insights on the design and optimization of high-performance stretchable electronics and 3D complex nanostructures.

Keywords

Buckling mode · Nanowire · Soft substrate · Helical mode · In-plane mode · Transition · Embedded wire · Continuum mechanics · FEM · Post-buckling behaviors · Buckling wavelength · Buckling amplitude

Introduction

Recently, stretchable electronics has attracted wide research interests and holds great potential applications, such as precision metrology (Stafford et al. 2004; Wilder et al. 2006), electronic eye cameras (Ko et al. 2008; Rogers et al. 2010), flexible displays (Chen et al. 2002; Crawford 2005), stretchable electronic circuits (Kim et al. 2008; Song et al. 2009), and conformable skin sensors (Lacour et al. 2005; Someya et al. 2004), to name a few. In stretchable electronics, the fragile and stiff elements (e.g., silicon, metal films, or wires) are usually placed on the elastomeric substrates and precompressed to some fundamental buckling modes (Audoly and Boudaoud 2008; Charlot et al. 2008; Chen and Yin 2010). The buckling conformation of the brittle and stiff elements can provide large deformability and tolerance for stretching, compression, bending, twisting, and even combined loading modes of the stretchable electronics (Rogers et al. 2010).

In essence, through buckling the compressive strain is released and replaced by relative small bending strain of the slender compressed structure. In this way, the brittle component of electronics can be fabricated into complex buckled form by precisely adjusting the geometrical structures of the stiff components and the elastomeric substrates, the mechanical properties of every constitutive component, and the adhesion between the stiff components and the substrates (Ko et al. 2008; Xu et al. 2015). Further reduction of the thickness or radius of the buckling members, although beneficial for accommodating more compressibility (Song et al. 2009; Xiao et al. 2010), is unfortunately limited by the fabrication processes and functionality of the flexible electronics. Besides, the bending strain is localized at the crest and valley of the buckling configurations for sinusoidal buckling such as out-of-plane buckling of nanowires and films. Hence, the failure of the brittle components usually initiates at the stress concentration region. Furthermore, the electronic properties of silicon components are closely dependent on the strain applied in the components, and the localized strain may cause nonuniform electronic properties in the components (Peng et al. 2009; Sajjad and Alam 2009). An alternative solution to further enhancing the deformability of the buckling mode is to make the buckles go three-dimensional, such as the helical buckling model of nanowire that has been successfully applied to silicon nanowire (SiNW) which can sustain very large stretchability up to the failure strain of poly-dimethylsiloxne (PDMS) (Xu et al. 2011). This helical buckling mode can easily handle diverse loading models, including multi-axial stretching, compression, bending, and twisting, and extend the usefulness of stretchable electronic elements. Furthermore, systems at micro- or even nanoscales were fabricated (Chen and Yin 2010), and coaxial electrospinning with the helical configuration and other spring-like structures were obtained based on buckling of nanofibers and films on curved substrates (Chen and Yin 2010; Yin and Chen 2010).

The buckling of nanowires on elastomeric substrate is a very common phenomenon and has important applications in stretchable electronics (Durham and Zhu 2013; Kim et al. 2008; Ryu et al. 2009; Wang et al. 2013), and different buckling structures have been reported. For example, the buckled nanowire could lie within the plane of the substrate (in-plane buckling), or perpendicular to the substrate (out-of-plane buckling), or of special interest here is the helical configuration (combination of the in-plane and out-of-plane modes) (Xiao et al. 2008, 2010; Xu et al. 2011). The adhesion strength between SiNW and elastomeric substrate may have an important role in regulating the buckling mode of SiNW on PDMS substrate. It has been reported that the ultraviolet/ozone (UVO) treatment may strengthen the interaction between SiNW and PDMS by forming strong covalent bonds, and without the treatment, there only exist much weaker Van der Waals forces (Efimenko et al. 2002; Qin and Zhu 2011). Consequently, no debonding between SiNW and PDMS is observed after proper UVO treatment and SiNW buckles in the helical mode. While sliding at the interface is distinctly detected without UVO (with low adhesion strength) and hence SiNW shows the in-plane buckling mode with lower strain energy to the out-of-plane buckling.

Buckling behaviors of slender wires under compression have been studied for centuries, and buckling modes are strongly affected by the boundary conditions of the wire. For a beam without continuous support, such as a cantilever or a simply supported beam, the Euler buckling theory is fairly effective to describe the buckling configuration and critical loads (Goriely et al. 2008; Oldfather et al. 1933; Timoshenko and Gere 2009; Zeeman 1976). Although the initial curved nanoribbons with selecting adhesion to the elastomeric substrate can also generate complex three-dimensional buckle modes, the strain distribution along the nanowires are also not uniform and the loose contact between the nanoribbons and the substrate may initiate failure (Xu et al. 2015), and the selecting adhesion is more challenging than perfect (uniform) adhesion.

Meanwhile, when embedded in a soft matrix with low Young's modulus, both 2D sinusoidal (planar) and 3D helical (nonplanar) buckling modes of compressed wires were observed, while a stiffer matrix would drive wires to buckle only in the 3D mode (Su et al. 2014). Based on the Winkler foundation model (Timoshenko and Gere 2009), they developed the 2D buckling theory of a wire embedded in a soft matrix and concentrated on the critical strain and initial buckling wavelength (or wavevector) of sinusoidal buckled wires (Slesarenko and Rudykh 2016; Su et al. 2014), but the 3D helical buckling mechanism and post-buckling behaviors of embedded wires have not been explicitly explained.

In this chapter, we first focus on studying the intrinsic helical buckling mechanism of a straight SiNW on the surface of PDMS substrate with perfect interface adhesion via theoretical analysis and comprehensive finite element method (FEM) simulations, and further verifying by parallel experiments. A continuum mechanics approach is established which is extendable to three-dimensional helical coil buckling on elastomeric substrates. Then, the compressive buckling of a nanowire partially bonded to an elastomeric substrate is studied via finite element method (FEM) simulations and experiments, to explore the mechanism of the transition from in-plane buckling to helical buckling. Thereafter, the helical buckling mechanics is extended to buckling behaviors of embedded wires in an infinite matrix, and the post-buckling of post-buckling evolution of embedded wires is studied.

Helical Buckling Mechanism for a Stiff Nanowire on the Surface of an Elastomeric Substrate

Experimental

The experimental method previously developed (Xu et al. 2011) was employed in this work. The experimental results are shown in section "Helical Buckling of a Nanowire Embedded in a Soft Matrix Under Axial Compression" for comparison with analytical and FEM simulation results. Below, a brief summary of the experimental method is provided. SiNWs were synthesized on the silicon wafer by chemical vapor deposition using gold nanoclusters as catalysts and silane (SiH_4) as vapor-phase reactant. A poly(dimethylsiloxane) (PDMS) substrate with a thickness

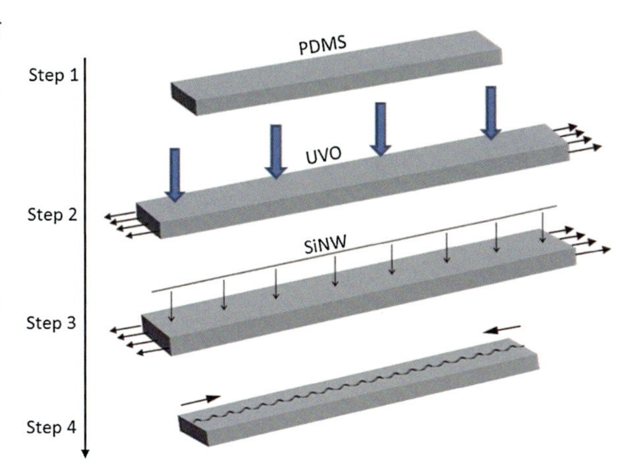

Fig. 1 Schematic diagram of the helical buckling of SiNW on PDMS substrate (Xu et al. 2011). First, the PDMS is prestretched and radiated by ultraviolet/ozone (Step 2), which improves the adhesion strength between PDMS and SiNW. Then, the SiNW is transferred to the surface of PDMS (Step 3) using contact printing. After releasing the prestrain in PDMS, helical buckling occurs in the SiNW (Step 4)

of 2 mm was prepared using Sylgard 184 (Dow Corning) by mixing the "base" and the "curing agent" with a ratio of 10:1. The mixture was first placed in a vacuum oven to remove air bubbles and then thermally cured at 65 °C for 12 h. Rectangular slabs of suitable sizes were cut from the cured piece.

Figure 1 schematically shows the process for fabricating buckled SiNWs. A miniaturized tensile testing stage (Ernest F. Fullam) was used to mechanically stretch the PDMS slab to the desired levels of prestrain, with both ends of the slab clamped. The prestrained substrate was radiated under a UV lamp (low-pressure mercury lamp, BHK) (Fig. 1, Step 2). A contact printing method was used to transfer the SiNWs on the silicon wafer to the PDMS substrate. The silicon wafer was slid along the prestrained direction to align the SiNWs (Fig. 1, Step 3). Releasing the prestrain in PDMS resulted in buckling of the SiNWs (Fig. 1, Step 4). The releasing step was carried out in-situ under an atomic force microscope (AFM). At a number of intermittent strain levels, the releasing was paused and AFM images were taken, based on which the buckling spacing of the SiNWs is deduced. Between such intermittent strain levels, the releasing (unloading) strain rate was between 10^{-4}/s and 10^{-3}/s to eliminate the effect of loading rate on the buckling behaviors. The entire experiment was strain controlled.

Theoretical Analysis

Continuum Model for Helical Buckling Mode

Subjected to an effective compressive strain $\varepsilon_{\mathrm{com}}$ imposed by contraction of the PDMS substrate (from Steps 3 to 4 in Fig. 1), the SiNW buckles into a helical buckling configuration, which is schematically shown in Fig. 2. Here, the x axis is along the axial direction of the SiNW, y axis is perpendicular to the surface of PDMS, and z axis is along lateral direction in PDMS surface. Based on experimental observation (Xu et al. 2011), the helical mode of the SiNW is uniform in the middle.

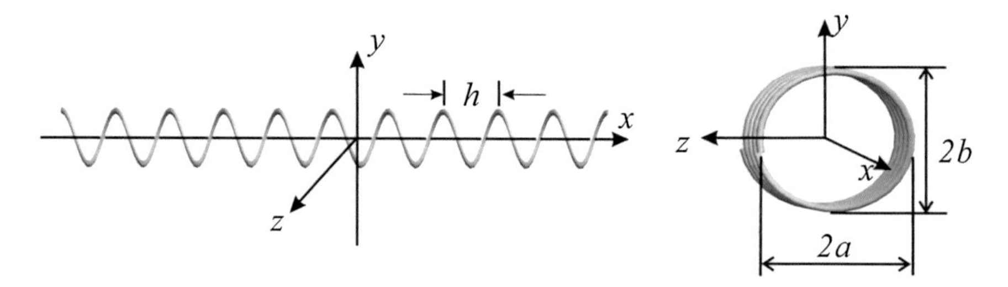

Fig. 2 Schematic diagram of the three-dimensional helical buckling of the SiNW with y direction perpendicular to the surface of PDMS substrate and the profile of the buckled SiNW

Therefore, in theoretical model the deflection of the SiNW is approximated by a helical curve as $w = a\cos(kx)$ in y direction and $v = b\sin(kx)$ in z direction, where a and b denote the deflection amplitudes in y and z directions, respectively, and k is the buckling wavevector, a variable related to the screw pitch (or buckling spacing) h by $k = 2\pi/h$. For the simplification of the theoretical analysis, the constitutive relations of the SiNW and PDMS substrate are assumed as linear elasticity, which is also consistent with the previous literatures (Mei et al. 2011; Song et al. 2009; Su et al. 2014).

Let u denote the axial displacement of the SiNW, and the membrane strain ε_{m} in the SiNW (caused by axial tensile or compressive force in the SiNW) is expressed as

$$\varepsilon_{\mathrm{m}} = \frac{\mathrm{d}u}{\mathrm{d}x} + \frac{1}{2}\left[\left(\frac{\mathrm{d}v}{\mathrm{d}x}\right)^2 + \left(\frac{\mathrm{d}w}{\mathrm{d}x}\right)^2\right]. \tag{1}$$

As the Young's modulus of the SiNW (e.g., 200 GPa) is several orders of magnitude larger than the elastic modulus of the PDMS substrate (e.g., 6 MPa), the membrane stress in SiNW is assumed as a constant and the shear stress along the axial direction of SiNW is ignored in the theoretical model. Indeed, based on the FEM simulation results (see Fig. 16) the membrane strain in SiNW is almost constant and is much smaller than the bending strain. The shear stress along the axial direction of SiNW is about one order of magnitude smaller than the traction stresses perpendicular to the SiNW. However, if the Young's modulus of the wire is comparable to that of the substrate, the shear stress may play an important role in the buckling behaviors of a wire on an elastomeric substrate. The shear stress effect will be systematically explored in our future works. By assuming constant membrane strain ε_{m} in SiNW, Eq. 1 can be rewritten as

$$\frac{\mathrm{d}\varepsilon_{\mathrm{m}}}{\mathrm{d}x} = \frac{\mathrm{d}^2u}{\mathrm{d}x^2} + \frac{\mathrm{d}v}{\mathrm{d}x}\frac{\mathrm{d}^2v}{\mathrm{d}x^2} + \frac{\mathrm{d}w}{\mathrm{d}x}\frac{\mathrm{d}^2w}{\mathrm{d}x^2} = 0. \tag{2}$$

Substituting the helix deflection of the SiNW into Eq. 2, the governing equation of the axial displacement u is

$$\frac{d^2u}{dx^2} + \left(a^2 - b^2\right) k^3 \sin(kx)\cos(kx) = 0. \tag{3}$$

Solving Eq. 3 to obtain the axial displacement

$$u = \frac{\left(a^2 - b^2\right) k}{8} \sin\left(2kx\right) + C_1 x + C_2, \tag{4}$$

where the two parameters C_1 and C_2 can be determined by the boundary conditions. By ignoring the rigid body displacement of the SiNW, we obtain $C_2 = 0$. As the SiNW is perfectly boned to the PDMS substrate, by considering the displacement compatibility between the SiNW and PDMS substrate we obtain $C_2 = -\varepsilon_{\text{com}}$, where ε_{com} is the effective compressive strain applied to the PDMS substrate. Then, the membrane strain ε_{m} can be obtained by substituting Eq. 4 into Eq. 1 as

$$\varepsilon_{\text{m}} = \frac{k^2\left(a^2 + b^2\right)}{4} - \varepsilon_{\text{com}}, \tag{5}$$

and the membrane energy (per unit length) in the SiNW is

$$U_{\text{m}} = \frac{k}{2\pi}\int_0^{2\pi/k} \frac{1}{2} EA{\varepsilon_{\text{m}}}^2 dx = \frac{EA}{2}\left[\frac{k^2\left(a^2 + b^2\right)}{4} - \varepsilon_{\text{com}}\right]^2, \tag{6}$$

where E and A denote the Young's modulus and cross-sectional area of the SiNW, respectively.

The curvature of the SiNW under helical mode is

$$\kappa = \frac{\left[a^2k^4\cos^2(kx) + b^2k^4\sin^2(kx) + a^2b^2k^6\right]^{\frac{1}{2}}}{\left[a^2k^2\sin^2(kx) + b^2k^2\cos^2(kx) + 1\right]^{\frac{3}{2}}}. \tag{7}$$

Then, the bending energy (per unit length) in the SiNW is

$$U_{\text{b}} = \frac{k}{2\pi}\int_0^{2\pi/k} \frac{EI}{2}\kappa^2 dx, \tag{8}$$

where I is the moment of inertia of the SiNW in the corresponding bending direction of the helical curve. In this work, the cross section of SiNW is circular, so that the moment of inertia $I = \pi R^4/4$ is equal in all directions, where R is the radius of SiNW. In fact, the theoretical analysis presented herein is not restricted to the circular cross section of the SiNW. Nevertheless, in practice helical buckling is most likely to appear in nanowires with equal moment of inertial in all directions, e.g., circular cross section. Otherwise, the buckling would occur with respect to the direction with the smallest moment of inertia. By substituting Eq. 7 in to Eq. 8 and integrating, the bending energy per length is

$$U_{\mathrm{b}} = \frac{EIk^4\left(3a^4k^2 + 2a^2b^2k^2 + 3b^4k^2 + 4a^2 + 4b^2\right)}{16\left(a^2k^2 + a^2k^4b^2 + b^2k^2 + 1\right)^{3/2}}. \tag{9}$$

For the helical buckled SiNW, additional torsion energy must be taken into account. As an approximation, the SiNW are regarded as a spring loaded by an axial force $F = EA\varepsilon_{\mathrm{m}}$

$$F = EA\left[\frac{k^2\left(a^2 + b^2\right)}{4} - \varepsilon_{\mathrm{com}}\right]. \tag{10}$$

Then, the shear force F_{S} and torsional moment T acting on the cross section of the SiNW are

$$F_{\mathrm{S}} = F\sqrt{\frac{(\mathrm{d}v/\mathrm{d}x)^2 + (\mathrm{d}w/\mathrm{d}x)^2}{1 + (\mathrm{d}v/\mathrm{d}x)^2 + (\mathrm{d}w/\mathrm{d}x)^2}} \tag{11}$$

and

$$T = \frac{F\left(w\mathrm{d}v/\mathrm{d}x - v\mathrm{d}w/\mathrm{d}x\right)}{\sqrt{1 + (\mathrm{d}v/\mathrm{d}x)^2 + (\mathrm{d}w/\mathrm{d}x)^2}}. \tag{12}$$

The torsion and shear energy (per unit length) in the SiNW can be given as

$$U_{\mathrm{t}} = \frac{k}{2\pi}\int_0^{2\pi/k}\left(\frac{T^2}{2GI_{\mathrm{p}}} + \frac{{F_{\mathrm{S}}}^2}{2GA}\right)\mathrm{d}x, \tag{13}$$

employing the helix deflection of the SiNW and substituting Eqs. 10–12 into Eq. 13, the torsional energy per unit length is

$$U_{\mathrm{t}} = \frac{AE^2\left(a^2k^2 + b^2k^2 - 4\varepsilon_{\mathrm{com}}\right)^2\left(Aa^2b^2k^2 + I_{\mathrm{p}}\left(\sqrt{(1+b^2k^2)(1+a^2k^2)} - 1\right)\right)}{32GI_{\mathrm{p}}\sqrt{(1+b^2k^2)(1+a^2k^2)}}, \tag{14}$$

where G is the shear modulus and $I_{\mathrm{p}} = \pi R^4/2$ the polar moment of inertia.

In order to derive the deformation of the PDMS substrate caused by the helical buckling of SiNW, the helical buckling is decomposed into two buckling modes, i.e., the in-plane buckling mode and out-of-plane buckling mode. Then, the deformation of the PDMS substrate is the superposition of displacement fields caused by the two buckling modes, respectively. As the thickness of PDMS substrate is much larger than the deflection of SiNW, the substrate is regarded as a semi-infinite solid. Based on the beam theory, the lateral distributed load (per unit length) of a beam can be derived by the deflection v and axial force $F = EA\varepsilon_m$ of the beam as $P = EI\mathrm{d}^4v/\mathrm{d}x^4 - EA\varepsilon_m\mathrm{d}^2v/\mathrm{d}x^2$. For the helical buckling mode, the deflection can be decomposed into out-of-plane deflection $w = a\cos(kx)$ and in-plane deflections $v = b\sin(kx)$, respectively. Then, the distributed load applied on the SiNW in y direction (out-of-plane direction) is $T_y = EI\mathrm{d}^4w/\mathrm{d}x^4 - EA\varepsilon_m\mathrm{d}^2w/\mathrm{d}x^2 = -P_y\cos(kx)$ where $P_y = -EAak^2(a^2k^2/4 - \varepsilon_{com}) - EIak^4$, while the distributed load in z direction is $T_z = EI\mathrm{d}^4v/\mathrm{d}x^4 - EA\varepsilon_m\mathrm{d}^2v/\mathrm{d}x^2 = -P_z\sin(kx)$ where $P_z = -EAbk^2(b^2k^2/4 - \varepsilon_{com}) - EIbk^4$.

Using the Green's function method, for unit normal force acting at point $(x_1, 0, z_1)$ on the surface of an incompressible semi-infinite solid, the normal displacement at point $(x, 0, z)$ can be given as $[(x - x_1)^2 + (z - z_1)^2]^{-1/2}/(\pi\overline{E}_S)$, where $\overline{E}_S = E_S/(1 - \nu_S{}^2)$ is the plane-strain modulus of the elastomeric substrate with ν_S the Poisson's ratio of the substrate. For the average normal force $P_y\cos(kx)/(2R)$ over the width $2R$, the normal displacement on the surface of the PDMS substrate can be integrated as $w_{sub} = \int_{-R}^{R}\frac{P_y\cos(kx)}{\pi\overline{E}_S R}K_0(k|z - z_1|)\,\mathrm{d}z_1$, where $K_0(k|z - z_1|)$ is the modified Bessel function of the second kind (Abramowitz and Stegun 1972; Timoshenko et al. 1970). As the buckling spacing of SiNW is much larger than the radius of SiNW, i.e., $kR \ll 1$, the dominant term in the Taylor series expansion of the normal displacement is

$$w_{sub} = \frac{P_y\cos(kx)}{\pi\overline{E}_S R}\left[2R(1-\gamma+\ln 2) - (R+z)\ln(k|R+z|) - (R-z)\ln(k|R-z|)\right], \tag{15}$$

where $\gamma = 0.577$ is the Euler's constant. Following the same procedure, for unit lateral force (z direction) acting at the point $(x_1, 0, z_1)$ on the surface of an incompressible semi-infinite solid, the lateral displacement at point $(x, 0, z)$ is $\frac{(1-\nu_S)(x-x_1)^2+(z-z_1)^2}{\pi(1-\nu_S)\overline{E}_S\left[(x-x_1)^2+(z-z_1)^2\right]^{3/2}}$. Then, for the average lateral force $P_z\sin(kx)/(2R)$ over the width $2R$, the lateral displacement (z direction) on the surface of the PDMS substrate can be integrated as $\upsilon_{sub} = \int_{-R}^{R}\frac{P_z\sin(kx)}{\pi\overline{E}_S R(1-\nu_S)}\left[(1-\nu_S)K_0(k|z-z_1|) + \nu_S k|z-z_1|K_1(k|z-z_1|)\right]\mathrm{d}z_1$, where $K_1(k|z - z_1|)$ is the modified Bessel function of the second kind. For $kR \ll 1$, the dominant term in the Taylor series expansion of the lateral displacement is

$$\upsilon_{sub} = \frac{P_z\sin(kx)}{\pi R\overline{E}_S}\left[2R\left(\frac{1}{1-\nu_S} - \gamma + \ln 2\right) - (R+z)\ln(k|R+z|) - (R-z)\ln(k|R-z|)\right] \tag{16}$$

Therefore, the strain energy of the PDMS substrates (unit length) caused by the out-of-plane and in-plane buckling of the SiNW is

$$U_{\mathrm{S}} = U_{\text{out-of-plane}} + U_{\text{in-plane}} = \frac{k}{2\pi}\int_{0}^{2\pi/k}\int_{-R}^{R}\frac{1}{2}\left[\frac{P_y\cos(kx)}{2R}w_{\mathrm{sub}} + \frac{P_z\sin(kx)}{2R}v_{\mathrm{sub}}\right]\mathrm{d}z\mathrm{d}x$$
$$= \frac{P_y{}^2}{4\pi\overline{E}_{\mathrm{S}}}\left[3 - 2\gamma - 2\ln(kR)\right] + \frac{P_z{}^2}{4\pi\overline{E}_{\mathrm{S}}}\left[5 - 2\gamma - 2\ln(kR)\right], \tag{17}$$

and the total potential energy (per unit length) of the whole system including the SiNW and PDMS substrate is

$$\begin{aligned} U_{\text{total}} &= U_{\mathrm{m}} + U_{\mathrm{b}} + U_{\mathrm{t}} + U_{\mathrm{S}} \\ &- \frac{k}{2\pi}\int_{0}^{\frac{2\pi}{k}}\int_{-R}^{R}\frac{P_y\cos(kx)}{2R}\left[w_{\mathrm{sub}} - a\cos(kx)\right]\mathrm{d}z\mathrm{d}x, \\ &- \frac{k}{2\pi}\int_{0}^{\frac{2\pi}{k}}\int_{-R}^{R}\frac{P_z\sin(kx)}{2R}\left[v_{\mathrm{sub}} - b\sin(kx)\right]\mathrm{d}z\mathrm{d}x, \end{aligned} \tag{18}$$

where the two integral terms in Eq. 18 represent the deformation compatibility between the SiNW and the PDMS substrate. The admissible solution should make the two integral terms equal zero. Substituting Eqs. 6, 9, 14, and 17 into Eq. 18, the total potential energy of the whole system can be written as

$$\begin{aligned} U_{\text{total}} =& \frac{EA}{2}\left[\frac{k^2\left(a^2+b^2\right)}{4} - \varepsilon_{\text{com}}\right]^2 - \frac{\left[EAak^2\left(\frac{a^2k^2}{4} - \varepsilon_{\text{com}}\right) + EIak^4\right]a}{2} \\ &- \frac{\left[EAbk^2\left(\frac{b^2k^2}{4} - \varepsilon_{\text{com}}\right) + EIbk^4\right]b}{2} + \frac{EIk^4\left(3a^4k^2 + 2a^2b^2k^2 + 3b^4k^2 + 4a^2 + 4b^2\right)}{16\left(a^2k^2 + a^2k^4b^2 + b^2k^2 + 1\right)^{\frac{3}{2}}} \\ &+ \frac{AE^2\left(a^2k^2 + b^2k^2 - 4\varepsilon_{\text{com}}\right)^2\left(Aa^2b^2k^2 + I_{\mathrm{p}}\left(\sqrt{\left(1+b^2k^2\right)\left(1+a^2k^2\right)} - 1\right)\right)}{32GI_{\mathrm{p}}\sqrt{\left(1+b^2k^2\right)\left(1+a^2k^2\right)}} \\ &- \frac{\left[EAak^2\left(\frac{a^2k^2}{4} - \varepsilon_{\text{com}}\right) + EIak^4\right]^2}{4\pi\overline{E}_S}\left[3 - 2\gamma - 2\ln(kR)\right] \\ &- \frac{\left[EAbk^2\left(\frac{b^2k^2}{4} - \varepsilon_{\text{com}}\right) + EIbk^4\right]^2}{4\pi\overline{E}_S}\left[5 - 2\gamma - 2\ln(kR)\right]. \end{aligned} \tag{19}$$

As the PDMS substrate is almost volume incompressible, the Poisson's ratio ν_{S} is set as 0.5. Then, the theoretical solution of the helical buckling spacing and critical buckling strain can be obtained by minimizing the total potential energy U_{total} with respective to a, b, and k, that is

$$k = \left[\frac{\overline{E}_s}{EI} \frac{2\pi \left(1+\eta^2\right)\left[1-\gamma-\ln(kR)+\eta^2(2-\gamma-\ln(kR))\right]}{\left[3-2\gamma-2\ln(kR)+\eta^2(5-2\gamma-2\ln(kR))\right]^2} \right]^{1/4}, \tag{20}$$

and

$$\varepsilon_{cr} = \frac{I}{A}k^2 + \frac{\pi \overline{E}_s \left(1+\eta^2\right)}{Ak^2E\left[3-2\gamma-2\ln(kR)+\eta^2\left(5-2\gamma-2\ln(kR)\right)\right]}, \tag{21}$$

respectively, where $\eta = b/a$ is the ratio between the in-plane displacement and out-of-plane displacement amplitude for the helical buckling of SiNW. If the helical buckling mode degenerates to the out-of-plane ($\eta = 0$) or in-plane ($\eta \longrightarrow \infty$) buckling mode, Eq. 20 can degenerates to the buckling wavevector of the out-of-plane or in-plane buckling mode reported in previous work, respectively (Xiao et al. 2008, 2010), which verifies the validity of the theoretical model presented herein. Furthermore, as the right-hand side of Eq. 20 changes very slowly with kR, Eq. 21 can be approximately represented as

$$k = \left(\frac{\overline{E}_s}{EI}\right)^{1/4} \left(\frac{0.4793^{\eta}}{\eta^{2.3}+23.31} + 0.7399\right). \tag{22}$$

For the SiNW with Young's modulus 200 GPa and radius of 15 nm and the PDMS substrate with elastic modulus 6 MPa, the relations between the helical buckling wavevector k and η described by Eqs. 20 and 22 are shown in Fig. 3,

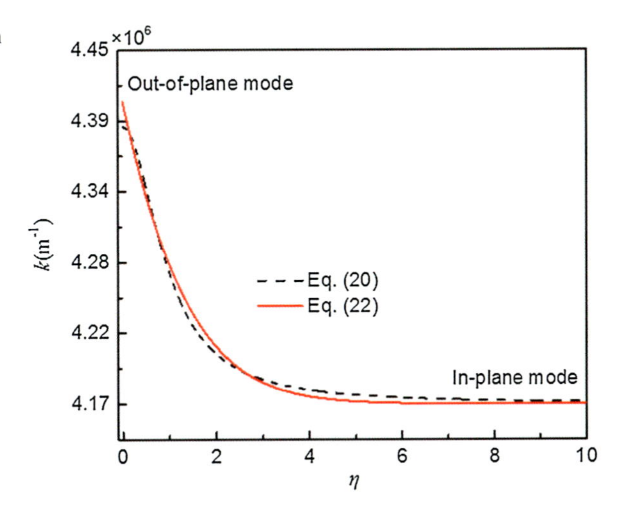

Fig. 3 The relations between the wavevector k and η described by Eqs. 20 and 22, respectively.

respectively. It is suggested the relation between k and η can be well described by Eq. 22. The wavevector decreases with the increasing of η and for the prefect helical buckling mode with $\eta = 1$ the wavevector can be approximately given by $k = \frac{3}{4}\left(\frac{\overline{E}_S}{EI}\right)^{1/4}$, which gives the initial helical buckling spacing as

$$h = \frac{2\pi}{k} = \frac{8\pi}{3}\left(EI/\overline{E}_S\right)^{1/4}. \tag{23}$$

This relation will be further verified by FEM simulation results in section "FEM Simulations."

Substituting Eq. 20 into Eq. 21, the critical buckling strain can be rewritten as

$$\varepsilon_{cr} = \sqrt{\frac{\overline{E}_S}{E}} f(\eta), \tag{24}$$

where $f(\eta)$ is

$$f(\eta) = \sqrt{\frac{(1+\eta^2)}{8\left[(1-\gamma-\ln(kR)) + \eta^2(2-\gamma-\ln(kR))\right]}} \frac{(5-4\gamma-4\ln(kR)) + \eta^2(9-4\gamma-4\ln(kR))}{(3-2\gamma-2\ln(kR)) + \eta^2(5-2\gamma-2\ln(kR))}.$$

Interestingly, if we ignore the very slow change of $f(\eta)$ with kR, the critical buckling strain is independent on the radius of the SiNW, R. The relation between the critical buckling strain ε_{cr} and the displacement amplitude ratio η is shown in Fig. 4. Similar to the relation between the buckling wavevector and η, the critical buckling strain

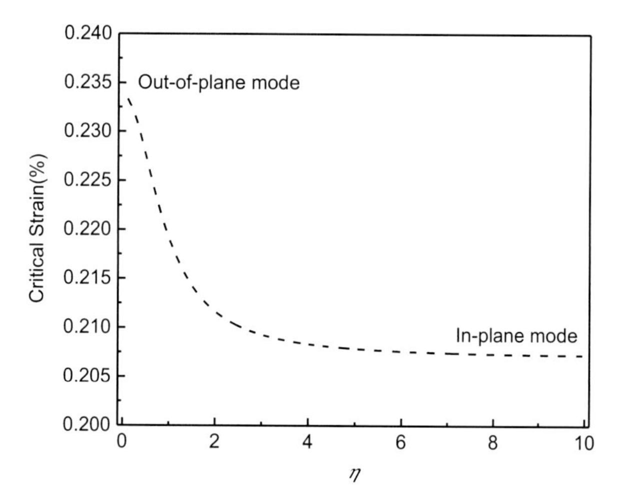

Fig. 4 The relation between the critical buckling strain and the displacement amplitude ratio η described by Eq. 24

decreases with the increasing of η. Besides, the same scaling law $\sqrt{\overline{E}_S/E}$ of the critical bucking strain has also been found in previous literatures for the in-plane or out-of-plane buckling mode of a nanowire on an elastomeric substrate (Xiao et al. 2008, 2010).

Both experiments and FEM simulations (below) indicate that the number of helical buckling coil in SiNW keeps constant during compression of the PDMS substrate (see Fig. 8). Assuming the initial length of SiNW is l_0 and the number of helical bucking coil is n, the initial helical buckling spacing is $h = l_0(1 - \varepsilon_{cr})/n$ and the post-buckling spacing is $h_p = l_0(1 - \varepsilon_{com})/n$. Thus, the relation between the post-buckling spacing h_p and compressive strain ε_{com} is

$$h_p = h\frac{1-\varepsilon_{com}}{1-\varepsilon_{cr}} = \frac{8\pi}{3}\left(\frac{EI}{\overline{E}_S}\right)^{1/4}\frac{1-\varepsilon_{com}}{1-\varepsilon_{cr}}. \tag{25}$$

The post-buckling spacing is determined by the cross section dimension of the nanowire, the Young's modulus of the nanowire and substrate, and the compressive strain, similar to that reported in previous theoretical works (Jiang et al. 2007; Kalita and Somani 2010).

In order to obtain the helical buckling displacement, the total potential energy of the SiNW and PDMS system (Eq. 19) is minimized by conjugate gradient method with respect to a and b. For the effective compressive strain 34%, the out-of-plane and in-plane displacement amplitudes for different diameters of SiNW from 10 nm to 100 nm and different Young's modulus of nanowire from 10 GPa to 250 GPa are shown in Tables 1 and 2, respectively. Here, the elastic modulus of the PDMS substrate is set as 6 MPa. Generally, both of the out-of-plane and in-plane displacement amplitudes a and b increase with the diameter and Young's modulus of the nanowire. Furthermore, the out-of-plane and in-plane displacement amplitudes can be well described by

$$a = \frac{1.259}{k}\sqrt{\varepsilon_{com} - \varepsilon_{cr}} \tag{26}$$

Table 1 The out-of-plane (a) and in-plane (b) displacement amplitudes for different diameters of SiNW

D/nm	10	30	50	70	100
a/nm	59.8	179	299	418	598
b/nm	58.0	174	290	407	580

Table 2 The out-of-plane (a) and in-plane (b) displacement amplitudes for different Young's modulus of nanowire

E/GPa	10	50	100	150	200	250
a/nm	86.7	128	151	167	179	189
b/nm	83	124	147	162	174	184

and

$$b = \frac{1.225}{k}\sqrt{\varepsilon_{\text{com}} - \varepsilon_{\text{cr}}}. \tag{27}$$

The largest discrepancy between the predictions from Eqs. 26 and 27 and the values of a and b obtained by numerically solving Eq. 19 is smaller than 3.5%.

Comparison of the Bending Strain for Different Buckling Modes

The helical buckling can significantly release the bending strain in SiNW, so that SiNW can sustain larger compression in helical mode than that of the in-plane or out-of-plane modes, enhancing the stretchability (Xu et al. 2011). In this section, the maximum bending strain in SiNW for different buckle modes is analyzed.

As studied in section "Model and Method," the deflection of SiNW is represented by a general helical form, that is $w = a\cos(kx)$ in y direction and $v = b\sin(kx)$ in z direction. If $a = 0$, it is the in-plane mode; while for $b = 0$, it is the out-of-plane mode. Based on the general helical form, the maximum bending strain in SiNW can be given as (Xu et al. 2011)

$$\varepsilon_{\max} = \begin{cases} aR/\left(b^2 + (h_{\text{p}}/2\pi)^2\right), & a \geq b \\ bR/\left(a^2 + (h_{\text{p}}/2\pi)^2\right), & a < b \end{cases}. \tag{28}$$

For the diameter of SiNW $D = 30$ nm at a given post-buckling spacing $h_{\text{p}} = 1.2$ μm (corresponding to an effective compressive strain of $\varepsilon_{\text{com}} = 19.36\%$ for helical buckling). The contour map of $\varepsilon_{\max}$ for different values of a and b is shown in Fig. 5. The contour map is symmetric along the line of $a = b$ and has the lowest values when $a = b$. Beyond the line $a = b$, $\varepsilon_{\max}$ increases with the increasing of b (the in-plane displacement amplitude) and decreases with the increasing of a (the out-of-plane displacement amplitude), whereas below the line $a = b$, the trend reverses. Taking the failure strain of SiNWs as 6.5% (Xu et al. 2011; Zhu et al.

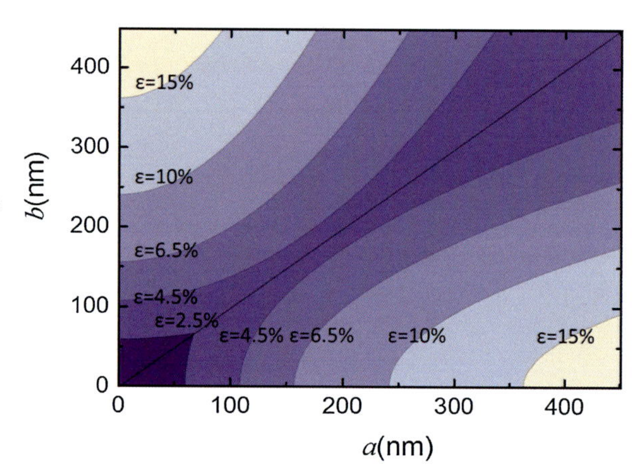

Fig. 5 Maximum local strain along the axial direction of SiNW for the SiNW with Young's modulus 200 GPa, diameter 30 nm, and PDMS with elastic modulus 6 MPa. The line represents the perfect circular buckling mode

2009), from the contour map we can find that the fracture strain is always larger than ε_{max} around the line $a = b$, which represents the perfect circular buckling mode, even if some displacement amplitudes (a or b) become large. This means that the stretchability of helical buckled SiNW can be enhanced, which echoes the fact that during experiments the helical buckled SiNW did not fail even upon the failure of PDMS substrate (Xu et al. 2011). The stretchability of helical coil can further benefit by increasing the failure strain of PDMS and bonding strength of the interface between SiNW and PDMS. For $a = b$, another benefit is that the distribution of the bending strain along the axis of the nanowire is uniform which will be discussed in Sect. 4.5.

Energetically Favorable Helical Buckling Mode of SiNWs

Based on the total potential energy of the SiNW and PDMS substrate (Eq. 19), a typical energy landscape for different buckling profiles (i.e., different displacement amplitudes in the in-plane and out-of-plane direction) is given in Fig. 6, in which the Young's modulus of SiNW is set as 200 GPa and the elastic modulus of PDMS

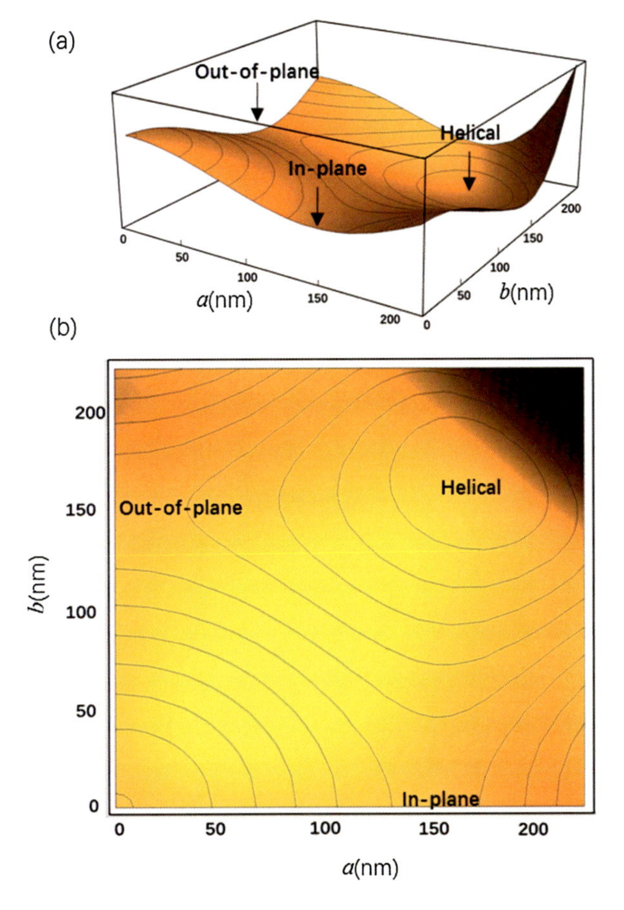

Fig. 6 Energy landscape for different buckling profiles of SiNW on PDMS substrate. (**a**) Three-dimensional view of the energy landscape and (**b**) plane view of the energy landscape

is set as 6 MPa, respectively, and the radius of SiNW is 15 nm and the effective compressive strain is 30%. It is shown that both of the in-plane and out-of-plane buckling modes are at the local minimum points of the potential energy landscape. This can explain why in some of the previous experiments the in-plane buckling for a nanowire on an elastomeric substrate has been observed, while in others the out-of-plane buckling has been observed. However, as shown in Fig. 6 the helical buckling has the global minimum potential energy which means the helical buckling is the most favorable buckling mode. Through properly adjusting the loading rate and interface adhesion strength between the nanowire and elastomeric substrate, the helical buckling can be observed in experiments.

FEM Simulations

To verify the theoretical model of helical buckling, FEM simulations using ABAQUS is conducted. For simplicity, we only consider Step 4 in Fig. 1 which is equivalent to applying an effective compressive strain e_{com} to the PDMS substrate. The in-plane dimensions of PDMS substrate is 40×40 μm with thickness of 5 μm, which are much larger than the initial helical buckle spacing and amplitudes of SiNW. SiNW is taken as beams with circular cross section. Besides, the binding between PDMS substrate and SiNW is assumed strong enough without debonding in all simulations. SiNW is consisted of beam elements (B31) with linear elasticity in ABAQUS and for PDMS substrate it is described by linear elasticity with three-dimensional continuum elements (C3D8R), which is consistent with the theoretical model presented in section "Mechanism of the Transition from In-plane Buckling to Helical Buckling for a Stiff Nanowire on the Surface of an Elastomeric Substrate." Indeed, the linear elasticity has been widely applied to the PDMS substrate and can give reasonable buckling behaviors of a stiff element on an elastomeric substrate in previous literatures (Mei et al. 2011; Song et al. 2009; Su et al. 2014). We have also conducted the FEM simulation by assuming hyperelastic constitutive relation (Neo-Hooke model) for the PDMS substrate. The helical buckling of the SiNW is also observed with small difference of the buckling spacing and amplitude to that of the PDMS substrate described by linear elasticity. Therefore, in order to directly compare with theoretical results, the linear elasticity has been employed to describe the PDMS substrate. Although the Young's modulus of SiNWs may be dependent on the diameter of SiNWs (Zhu et al. 2009), such size dependency is ignored in the current study. Different diameter of the SiNW ranging from 10 nm to 100 nm and Young's modulus ranging from 10 GPa to 250 GPa are also considered in the comprehensive FEM simulations.

Helical Buckling of SiNW on PDMS Substrate

We compare the helical buckle profiles between FEM simulations and experiments to highlight the main geometrical features. The Young's modulus and diameter of the SiNW used in simulation are 187 GPa for a typical diameter of SiNW 28 nm (Xu et al. 2011; Zhu et al. 2009). As the elastic modulus of PDMS is varied after being radiated by ultraviolet/ozone (Step 2), in order to fit the initial-buckle spacing measured in experiments (1.65 μm) its elastic modulus is set as 3.76 MPa, which is in the range of the elastic modulus of PDMS reported in previous literatures (Ryu et al. 2009; Song et al. 2009; Zhou et al. 2015). The typical simulated buckling configuration of SiNW on PDMS substrate after compression is shown in Fig. 7a, in which the helical coil morphology is easily recognized, and it is noticed that the deformation of the middle of SiNW is uniform. The buckling configuration in parallel experiments is demonstrated through planar and 3D atomic force microscopy (AFM) images of the helical SiNW, as indicated in Fig. 7b.

The evolution of the buckling profile of SiNW on PDMS substrate obtained from FEM simulation is shown in Fig. 8a. As compression proceeds, the displacement amplitudes of the SiNW gradually increase, while the post-buckling spacing gradually decreases, which agrees well with the experimental observations, see Fig. 8b. The two arrows in Fig. 8a are fixed reference points on the SiNW and interestingly; they always correspond to the peaks of the buckling waves during compression. This implies that the peaks always locate at the same points of the SiNW, and the number of coil waves is constant during compression. Further analysis indicates that the relative distribution of the buckling displacement on the SiNW does not change.

The detailed relation between the post-buckling spacing and effective compressive strain for the helical buckling is given in Fig. 8. During helical buckling the number of coils in SiNW is constant, for example, there are always seven coils between the two reference arrows when the compressive strain varies from 5.07% to 26.5%. The relation between the post-buckling spacing and effective compressive strain is given in Fig. 9 for the experiments (red squares), simulation results (black

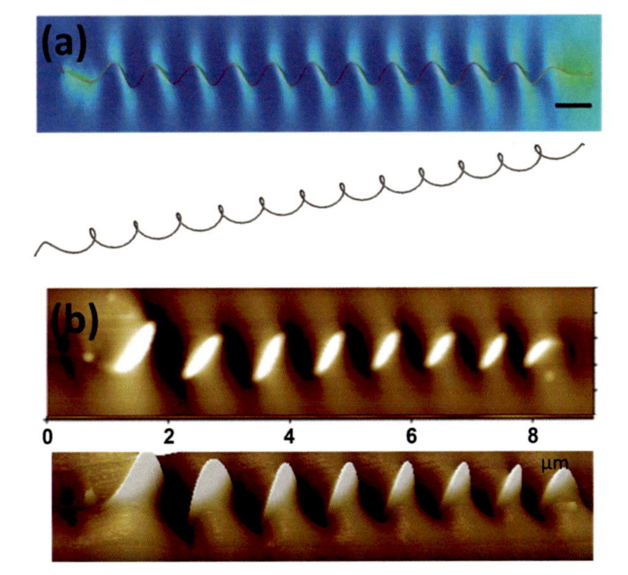

Fig. 7 The typical configurations of the helical buckling of a SiNW on PDMS substrate. (**a**) FEM simulations (Top figure: the color contour represents the distribution of the normal strain in x direction ε_{xx} with figure legend presented underneath and the scale bar is 1 μm. Bottom figure: the three-dimensional configuration of SiNW) and (**b**) experiments (top figure: planar AFM image. Bottom figure: three-dimensional AFM image. The gray level represents the height of the surface of the PDMS substrate in z direction.)

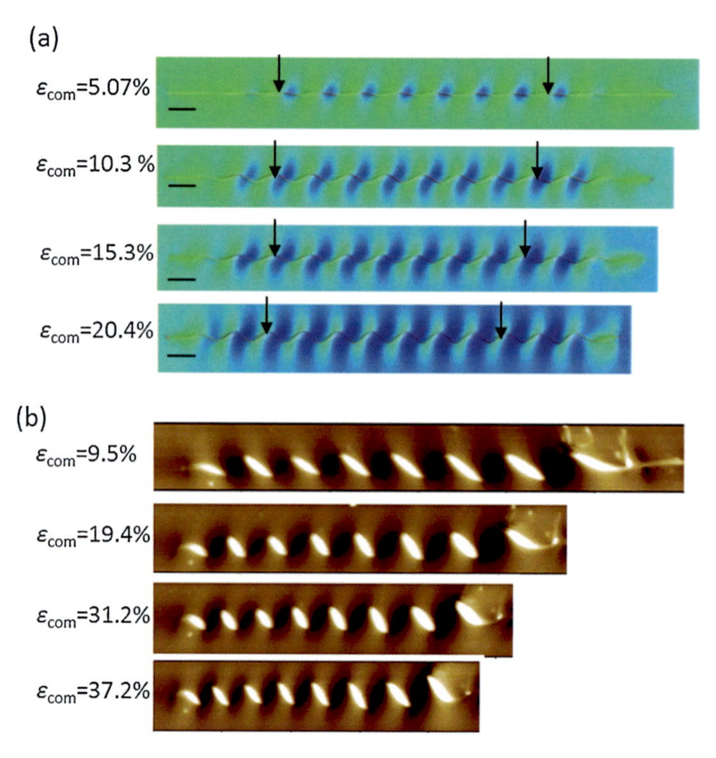

Fig. 8 The snapshots of the typical configuration (top view) for the helical buckling of SiNW on PDMS substrate at different compressive strains. (**a**) FEM simulations (the color contour represents the distribution of ε_{xx}) and (**b**) experiments. The scale bar is 1 μm

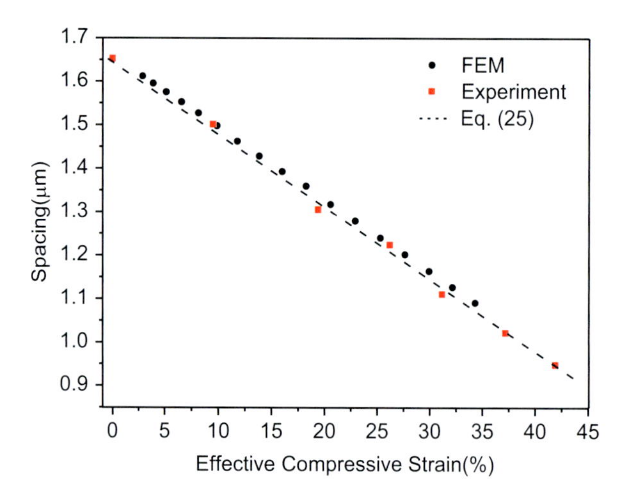

Fig. 9 The post-buckling spacing for helical bucking for different effective compressive strain. The experiments are red squares, simulation results are black circles, and theoretical predictions of Eq. 25 are black dashed line

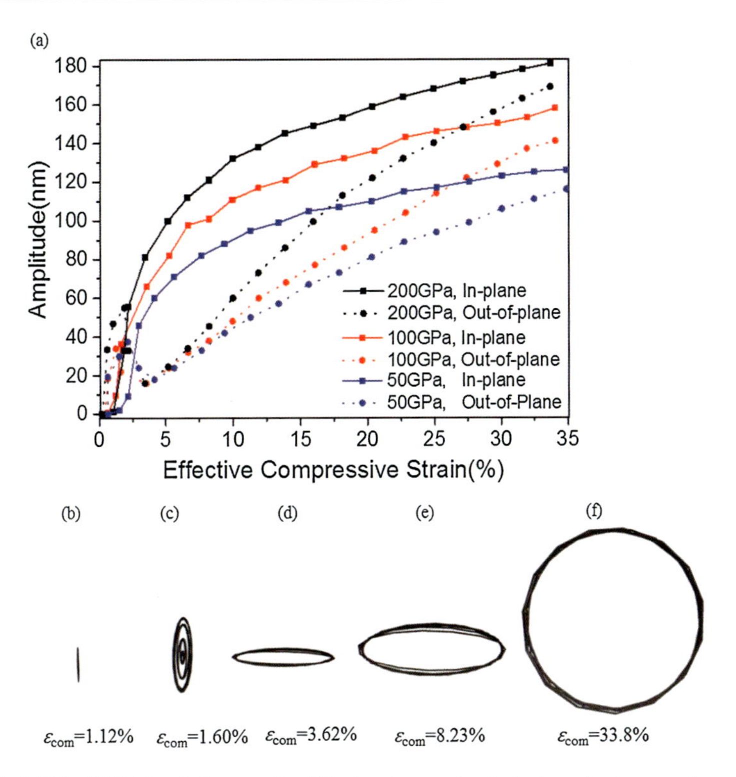

Fig. 10 (**a**) The FEM simulation results of the in-plane and out-of-plane displacement amplitudes for different effective compressive strain. (**b–f**) Side views (along the axial direction of SiNW) of the typical buckling profiles of SiNW for several effective compressive strains. The elastic modulus of PDMS is set as 6 MPa, the diameter and Young's modulus of SiNW are 30 nm and 200 GPa, respectively

circles) and theoretical predictions of Eq. 25 (black dotted line). Nearly perfect accordance with experiments can also verify the effectiveness of both the theoretical model and the FEM results.

Evolution of the Helical Buckling Profile During Compression

FEM simulation results show that the helical buckling profile of SiNW actually changes from a vertical ellipse to a lateral ellipse, and then approaches to a circle when the PDMS substrate is being gradually compressed. Figure 10a shows the amplitude evolution and Fig. 10b–f shows the changes of the SiNW side-view profiles (along the axial direction) with the effective compressive strain. As the effective compressive strain increases exceeding the initial buckling strain, first the out-of-plane displacement increases quickly and the in-plane displacement is almost zero, see Fig. 10a, b. Then, the in-plane displacement becomes more prominent in contrast to slightly decreasing of the out-of-plane displacement as the effective strain gets larger. With further compression, both the in-plane and

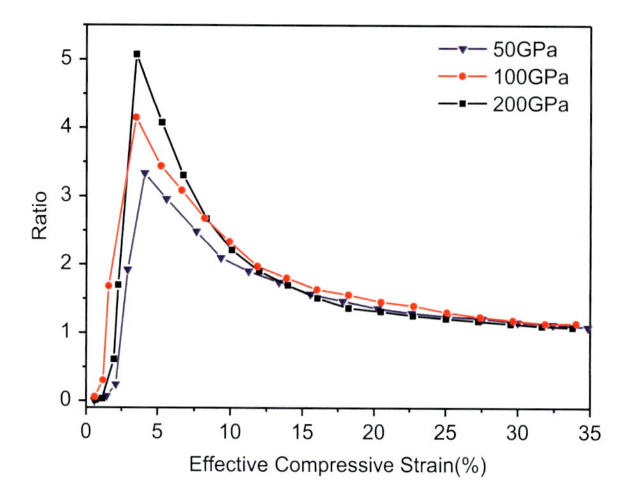

Fig. 11 The FEM simulation results of the ratio of in-plane displacement amplitude to out-of-plane displacement amplitude for the helical buckling of SiNW with different Young's modulus

out-of-plane amplitudes increase, and eventually the helical coil profile of SiNW becomes circular to accommodate the large local strain, with the ratio of in-plane amplitudes to out-of-plane amplitudes shown in Fig. 11. In our current FEM simulations, due to the extensively distorted elements in the buckling area, the maximum compressive strain applied to the PDMS substrate is 35% smaller than the experimental value 42%. However, as the buckling profile is gradually approaching to a circle with compression, which is similar to the experiments, we can predict the helical buckling profile is a circle to release the bending strain in SiNW with further increasing the compressive strain.

As shown in Fig. 11, the ratio of in-plane displacement amplitude to out-of-plane displacement amplitude quickly increases with the compressive strain when the compressive strain is larger than the initial buckling strain, and reaches the local acmes at the compressive strain about 4%. Generally, the nanowires with higher Young's modulus have larger peak values of the displacement amplitude ratio. Thereafter, the ratio conspicuously decreases with the magnitude approaching 1. The same trend in the displacement amplitude ratio was observed in experiments (Xu et al. 2011) (note that the minimum effective compressive strain in the experiments was larger than 5%, so only the decreasing trend was able to observe in experiments). Since the displacement amplitude ratio approaching 1 can accommodate larger effective compressive strain (e.g., the circular profile can take the largest strain), the buckled NWs appear to be self-adaptive or "smart" to search for an optimum displacement amplitude ratio.

Helical Buckling Spacing

In order to obtain the initial buckling spacing in FEM simulations, we should first clarify the onset of buckling. Here, the strain energy of SiNW is given in Fig. 12a

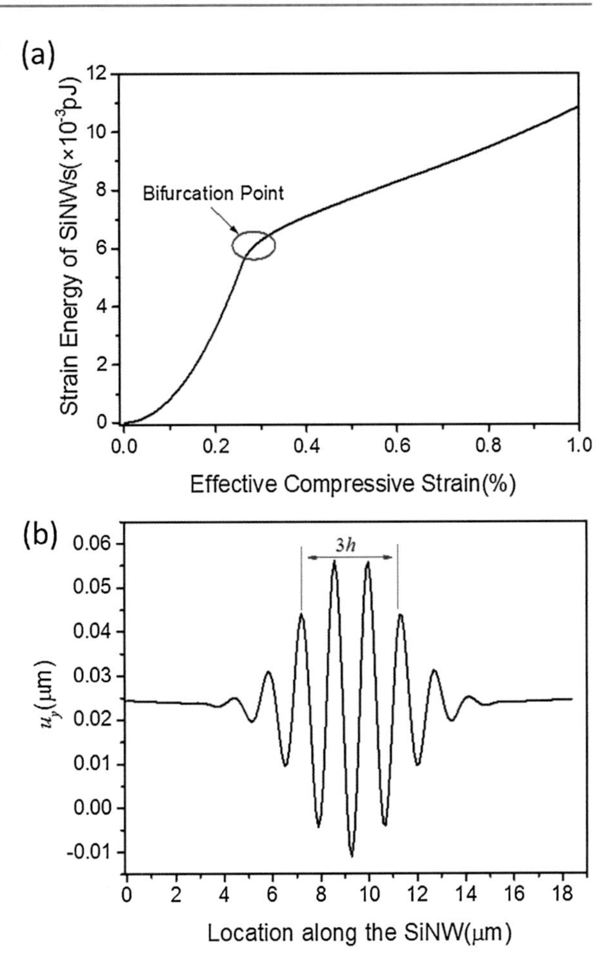

Fig. 12 (**a**) The strain energy of a SiNW on PDMS substrate during compression in FEM simulation and (**b**) the out-of-plane displacement of SiNW at the onset of buckling, where the modulus of SiNW and PDMS is 200 GPa and 6 MPa, respectively, and diameter of SiNW is 30 nm

and a clear bifurcation point is found in the strain energy-compressive strain curve. Therefore, the onset of the buckling is defined as the bifurcation point. As the major displacement of SiNW is perpendicular to the surface of PDMS when the compressive strain is smaller than 2%, as shown in Figs. 10 and 11, the out-of-plane displacement of SiNW at the bifurcation point is used to determine the initial buckling spacing, as shown Fig. 12b. For the case shown in Fig. 12b, the out-of-plane displacement v of SiNW is about 50 nm much smaller than the initial buckling spacing, which echoes this point is very close to the initial buckling point. Because of the computational accuracy, it is impossible to determine the exact initial buckling point in FEM simulation. As the influence of the boundary, the buckling amplitude is not uniform. However, the buckle spacing is quite uniform, and the initial buckling spacing h is 1.47 μm, see Fig. 11b.

The contour of axial normal strain (ε_{xx}) on the surface of PDMS corresponding to the bifurcation point in Fig. 12 is given in Fig. 13a. The initial buckling spacing

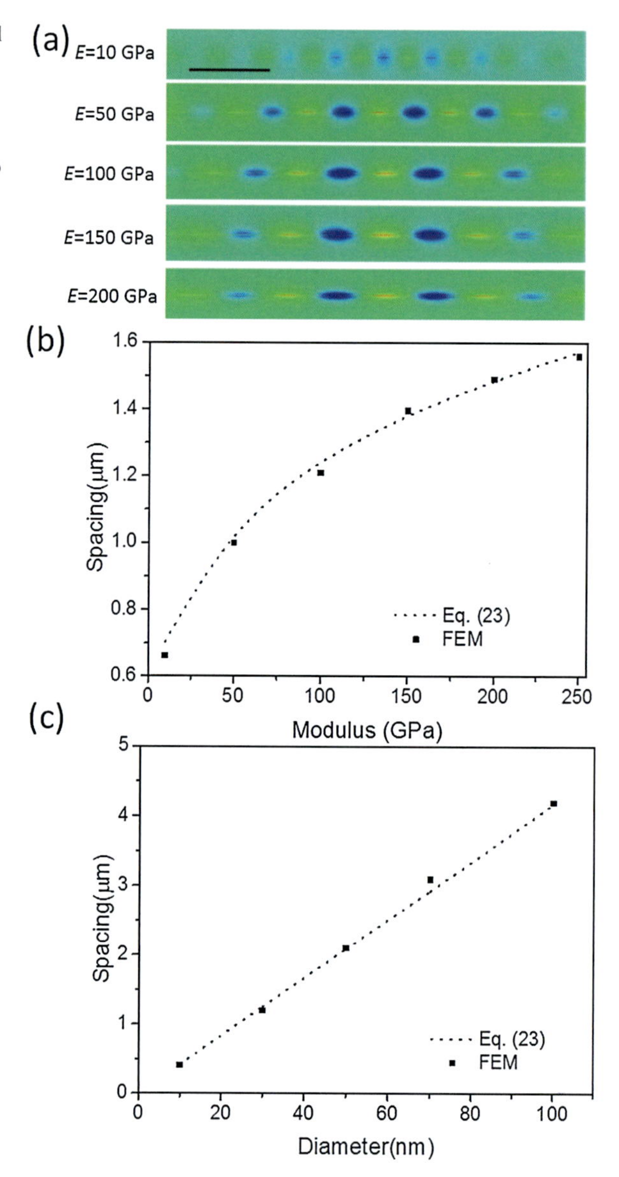

Fig. 13 (**a**) The axial normal strain ε_{xx} contour (top view) at the onset of buckling shown in Fig. 12. The scale bar is 1 μm. The relations of the initial buckling spacing to the Young's modulus of SiNW (**b**) and diameter of SiNW (**c**). The dotted line is the theoretical prediction from Eq. 23

increases as the Young's modulus of SiNW. FEM simulation shows that the initial buckling spacing is almost proportional to the 1/4 power of the Young's modulus of SiNW and is linear to the diameter of SiNW (see Fig. 13b, c), which agrees well with Eq. 23.

After the initial buckling, the deflection of SiNW increases accompanied by the decreasing of the post-buckling spacing as the compression proceeds. Figure 14a shows the axial normal strain contour (ε_{xx}) on the surface of PDMS at the

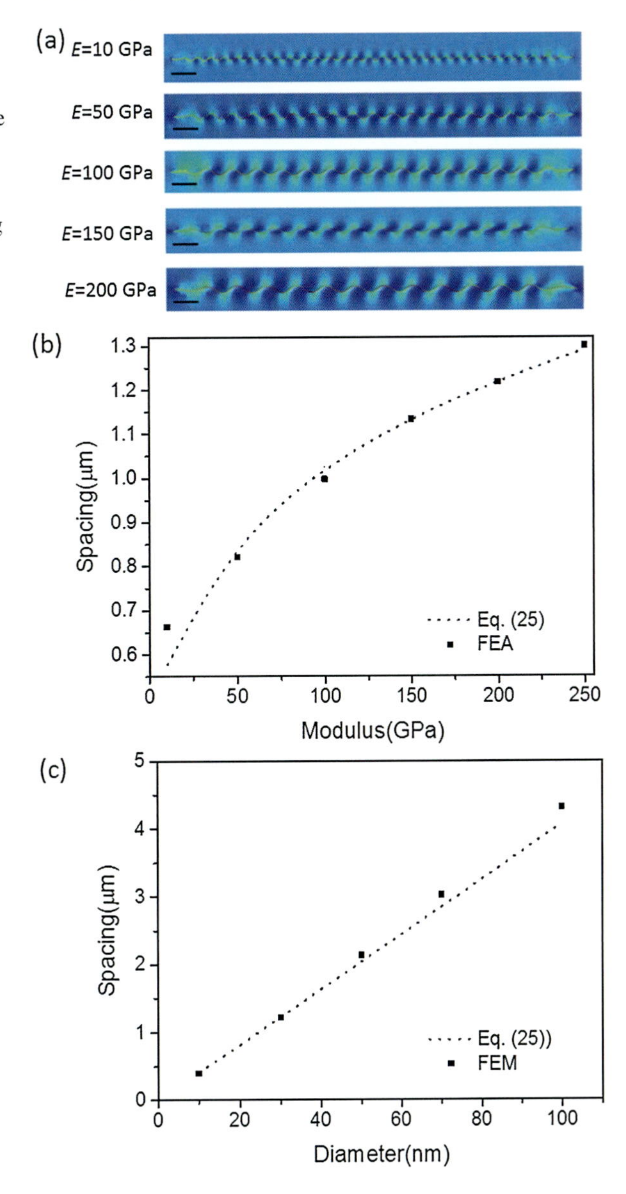

Fig. 14 (**a**) The FEM simulation results of the helical buckling configurations at the effective compressive strain of 18% (ε_{xx} contour and top view). The scale bar is 1 μm. The relations of the post-buckling spacing to Young's modulus of NW (**b**) and diameter of SiNW (**c**). The dotted line is the theoretical prediction from Eq. 25

effective compressive strain of 18%. The post-buckling spacing increases with the increasing of SiNW modulus for a given compressive strain, as shown in Fig. 14b, c. Both the theoretical (see Eq. 25) and FEM simulation results indicate that the post-buckle spacing scales with the 1/4 power of the SiNW modulus and increases linearly with SiNW diameter, the same trends as the initial buckling spacing.

Displacement Amplitudes of Helical Buckling

In Fig. 15, the comparisons of the in-plane and out-of-plane displacement amplitudes between the theoretical predictions (Eqs. 26 and 27) and FEM simulations are presented. The FEM simulation results show that the in-plane and out-of-plane displacement amplitudes increase linearly with SiNW diameter and are almost proportional to the 1/4 power of the SiNW modulus, which fit very well to the theoretical predictions. Besides, the displacement amplitudes in y and z directions are almost equal for the effective compressive strain larger than 34%, which means the SiNW buckles into a circular coil form, helping to accommodate larger bending deformation with less maximum local strain. Previous studies indicate that even if the effective compression strain reaches as high as 50%, the maximum local strain in SiNW is less than 3%, considerably lower than the fracture strain of SiNWs (6.5%) (Xu et al. 2011).

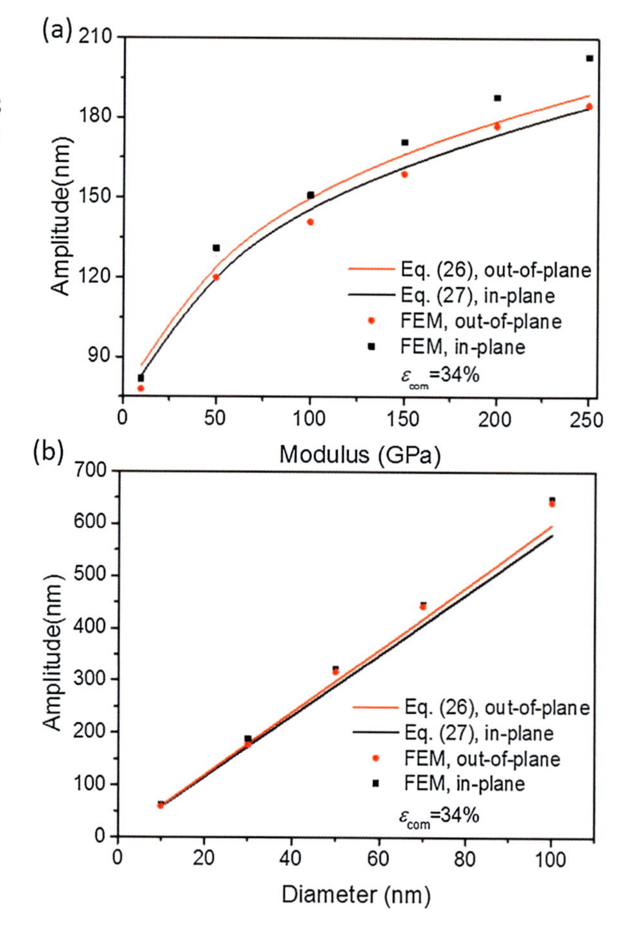

Fig. 15 The in-plane and out-of-plane displacement amplitudes for the helical buckling of SiNW on PDMS substrate versus (**a**) modulus and (**b**) diameter of SiNW. The dashed lines are the theoretical predictions from Eqs. 26 and 27

Strain Distribution in SiNW

The advantage of the helical buckling of SiNW can be revealed from the FEM simulation results in Fig. 16a. The membrane strain in SiNW, compared to the total strain (mainly the bending strain), is considerably small and can be regarded as a constant, verifying the assumption in section "Model and Method." The inset of Fig. 16a shows the distribution of the maximum local strain along the helical buckling configuration. The maximum local strain at the cross section of SiNW is quite uniform and much smaller than the externally imposed compressive strain in the PDMS substrate, which is benefical to ensure the uniform electronic properties along the SiNW. However, there is still a small fluctuation of the maximum local strain with the fluctuation period equal to the post-buckling spacing, see Fig. 16a. This is because the helical buckling profile of SiNW is not a perfect circle and the

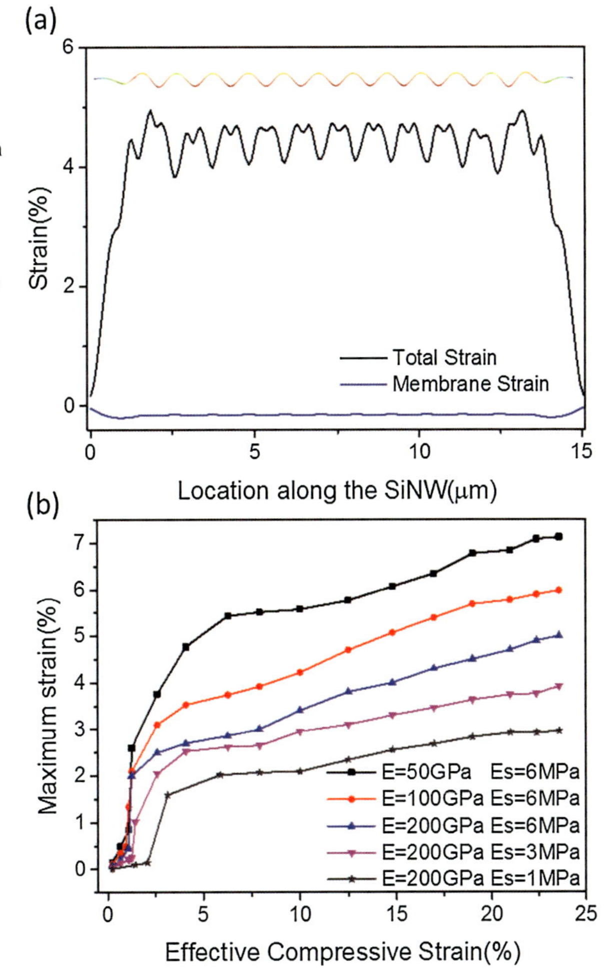

Fig. 16 (**a**) FEM results of the membrane strain and the maximum local strain at the cross section of SiNW along the axial direction of SiNW for a given compressive strain of 25%. The inset shows the distribution of the maximum local strain in the helical buckling configuration of SiNW. The Young's modulus and diameter of SiNW are 200 GPa and 30 nm, respectively. (**b**) The maximum local strain of SiNW versus the effective compressive strain

maximum local strain is the largest at the points of the largest deflection point. The maximum local strain versus the effective compressive strain for different SiNW and PDMS modulus (with constant diameter of SiNW, i.e., 30 nm) is shown in Fig. 16b. At a small effective compressive strain the maximum local strain increases fast, and as the effective compressive strain gets larger, the maximum local strain becomes more stable, which agrees well with the previous studies (Xu et al. 2011).

The nanowire/substrate modulus ratio has a profound impact on the maximum local strain. In Fig. 16b, the maximum local strain for several modulus ratios (E/E_s) are given, and as the modulus ratio decreases, the maximum local strain decreases remarkably. This is because the post-buckling spacing h_p is proportional to the 1/4 power of the nanowire/substrate modulus ratio E/E_s and based on Eq. 28 the larger value of h_p has smaller bending strain. In particular, when the modulus ratio is small, e.g., 1/200,000 (when $Es = 1$ MPa and $E = 200$ GPa), the maximum local strain is smaller than 3% even at a large effective compressive strain 25%, and such a helical buckled coil profile may accomodate large deformation. This phenomenon is attributed to the deformation homogeneity along the SiNW for helical coil buckling. Note that the fracture strain of SiNW with a diameter of 28 nm is 6.5%, and thus one does not expect fracture to occur even for very large effective compressive strain (such as 50% in previous experiments (Xu et al. 2011)).

Mechanism of the Transition from In-plane Buckling to Helical Buckling for a Stiff Nanowire on the Surface of an Elastomeric Substrate

In most previous works about the buckling behaviors of the nanowire/substrate system, the nanowires were assumed perfectly bonded to the elastomeric substrates throughout deformation, and the effect of the interface debonding was ignored. However, for the buckling of the SiNWs on the PDMS substrate, the UVO treatment time plays an important role in the buckling mode of the SiNWs. The SiNWs may partially debond from the PDMS substrate during compressive buckling, especially with less UVO treatment time. Though the competition between the in-plane buckling mode and out-of-plane buckling mode of SiNW has been studied in recent literature, the selection of the buckling mode is determined by minimum moment of inertia due to the noncircular cross section of the SiNW (Duan et al. 2015). In this section, the buckling behaviors of the SiNWs partially bonded to the PDMS substrate are studied to explore the mechanism of the transition from in-plane buckling to helical buckling.

Model and Method

As shown in Fig. 17, a silicon nanowire (SiNW) is placed on the PDMS substrate and a uniaxial compression is applied on the PDMS substrate in x direction (the axial direction of SiNW); this is equivalent to releasing the prestrain in PDMS substrate.

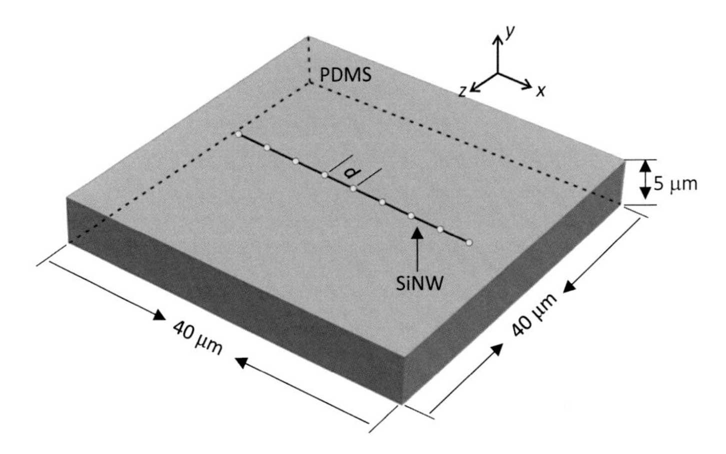

Fig. 17 Model of a SiNW partially bonded to the PDMS substrate. The yellow dots represent the constraint between the SiNW and the PDMS substrate and d is the distance between the adjacent constraint points. A uniaxial compression is applied to the PDMS substrate in x direction

Only the yellow dots in the SiNW are constrained to the PDMS substrate, so as to represent the partial bonding between the SiNW and the PDMS substrate. The constraint between the SiNW and the PDMS is the node-type-constraint, that means the corresponding nodes in the SiNW and the PDMS have the same displacement. Indeed, we have checked different type constraints, e.g., node-type constraint, element-type-constraint, and the influence to the critical buckling strain of SiNW is negligible.

The distance between adjacent constraint points is d, as shown in Fig. 17. As shown in our previous study, the compression buckling mode of a perfectly bonded SiNW ($d/h = 0$) is helical and the buckling spacing is $h = 7.89\ R\left(E_{NW}/\overline{E}_S\right)^{1/4}$ determined by the radius R, the elastic modulus E_{NW} of the SiNW, and the effective elastic modulus $\overline{E}_S$ of the PDMS substrate. Here, the effective elastic modulus of the substrate is $\overline{E}_S = E_S\left(1 - \nu_S^2\right)$, where E_S and ν_S are the elastic modulus and Poisson's ratio of the substrate. Such a helical model pertains when d/h is close to 0, whereas for large d/h, the buckling of the SiNW transfers to the in-plane mode.

Finite element method (FEM) simulations are carried out to study the buckling behaviors of the SiNW on PDMS substrate via the commercial software ABAQUS. The PDMS substrate is simplified as an approximately incompressible isotropic material with Young's modulus $E_S = 3.76$ MPa and Poisson's ratio $\nu_{NW} = 0.475$, and the SiNW is simplified as a beam with circular cross section. The elastic modulus and Poisson's ratio of the SiNW are $E_{NW} = 187$ GPa, $\nu_{NW} = 0.3$ (Zhu et al. 2009). The radius of SiNW is $R = 15$ nm in consistent with parallel experiments. The PDMS substrate and the SiNW are discreted by C3D8R and B31 elements, respectively. In all FEM simulations, the length of the SiNW is 20 μm, which is much longer than the buckling wavelength of the SiNW studied in this work, and the SiNW lies in the middle of the PDMS surface and far away from the surface edges (which is also consistent with the parallel experiments). Therefore, the boundary effect is regarded small. Although finite boundary is utilized in our FEM simulation, in the middle part of SiNW the buckling configuration exhibits excellent periodicity (see Fig. 23). Mesh convergence is carried out to ensure the reliability of numerical results. Different constraint densities (different d/h) are studied to explore the mechanism of the buckling mode transition.

Results and Discussion

Buckling Modes

The buckling modes of the SiNW for different d are shown in Fig. 18, which can be divided into three distinct regimes as the constraint density increases. When the spacing between adjacent constraint points is large enough, examples such as $d = 2.5$ μm and $d = 1$ μm, the buckling of the SiNW is in-plane of the PDMS substrate. The in-plane mode is sinusoidal, and the wavelength of the in-plane buckling is $2d$. The maximum strain in the SiNW locates at the maximum curvature point (middle between two constraint points), whereas the strain (as well as curvature) at the constraint point is almost 0. As the constraint density increases, e.g., $d = 0.8$ μm and $d = 0.5$ μm, the out-of-plane displacement of the SiNW becomes significant. The in-plane displacement is still sinusoid like, but the out-of-plane displacement is disordered. We term this mode as the disordered buckling in the out-of-plane direction. As the constraint density increases further, the configuration of the buckled SiNW becomes the helical coil, as shown for $d = 0.3$ μm and $d = 0.1$ μm in Fig. 18. Besides, for the helical buckling the

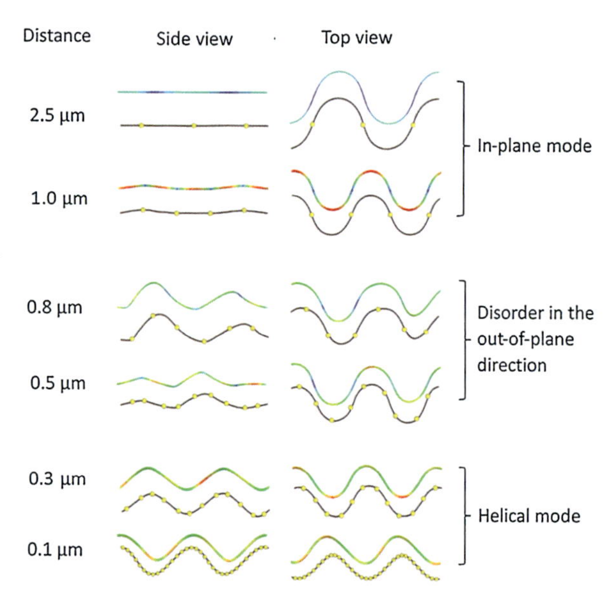

Fig. 18 The buckling modes of the SiNW partially bonded to a PDMS substrate. The yellow dots represent the constraints (bonded sites) between the SiNW and the PDMS substrate. Color contour of the maximum principal strain in the buckled configuration is given

strain distribution in the SiNW is much more uniform than that for the in-plane and disordered modes.

The corresponding in-plane and out-of-plane displacements are shown in Fig. 19 for $d = 1$ μm, 0.5 μm, and 0.1 μm, respectively. For $d = 1$ μm, the in-plane displacement is periodic and much larger than the out-of-plane displacement. While

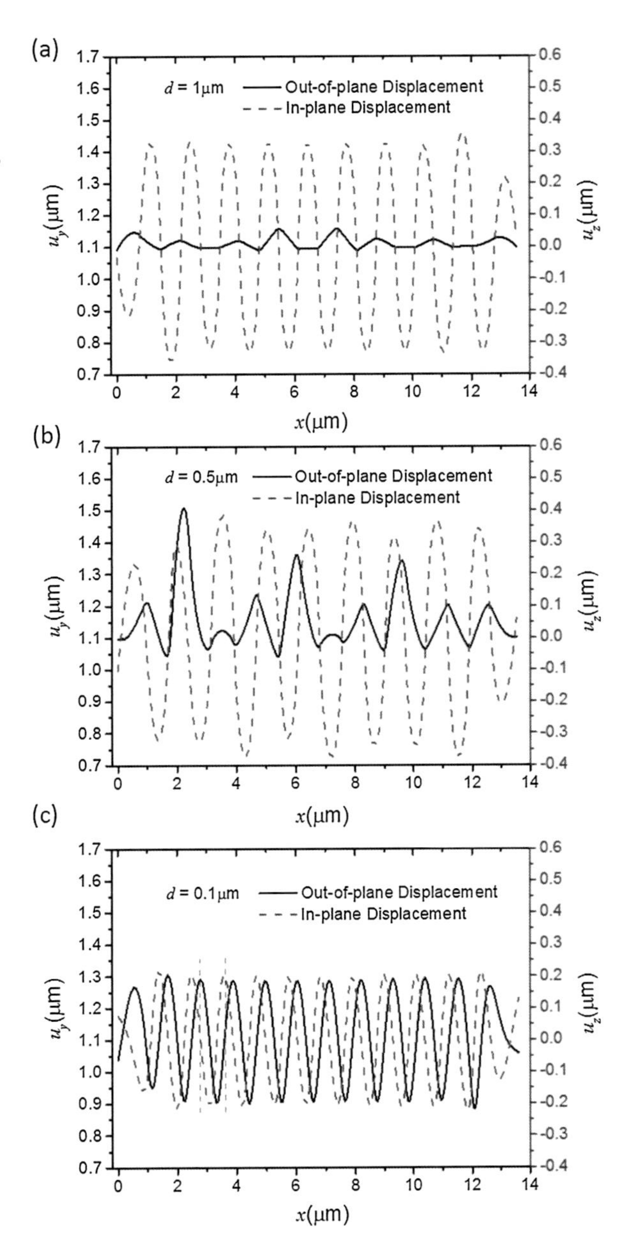

Fig. 19 The in-plane (u_z) and out-of-plane (u_y) displacements of SiNW corresponding to different modes in Fig. 18 ($d = 1$ μm, 0.5 μm, and 0.1 μm), respectively

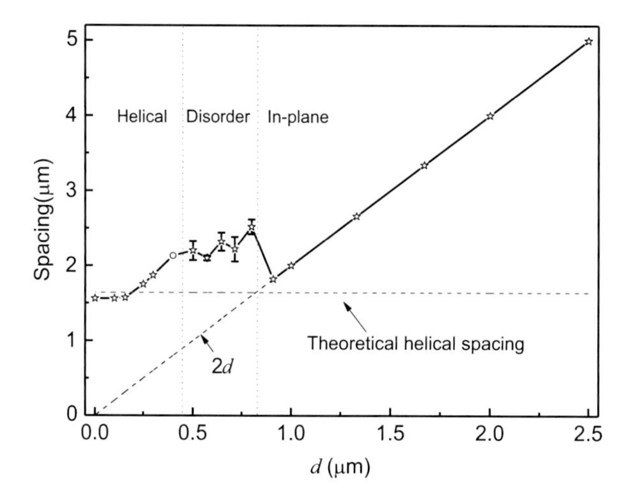

Fig. 20 The buckling wavelength of SiNW for different constraint densities. The red dashed line indicates the theoretical buckling spacing h of the perfectly bonded SiNW and the black dashed line is $2d$. Error bars are given for the out-of-plane disordered buckling mode

for $d = 0.5$ μm, the amplitude of the out-of-plane displacement is of the same order as the in-plane displacement; however, it is fairly irregular (while the in-plane component remains approximately periodic which determines the buckling wavelength). For the helical buckling ($d = 0.1$ μm), the in-plane and out-of-plane displacements are similar. Nevertheless, the phase angle difference of the in-plane and out-of-plane displacements is about 90°, which generates the helical coil form.

The buckle spacing (or buckling wavelength for in-plane buckling) of the SiNW for different constraint densities are shown in Fig. 20, from which the three regimes are obvious. For the in-plane mode, the buckling wavelength is exactly $2d$. While for the disordered mode, the relation between the buckling spacing and the distance d is not monotonous. As the distance d approaches to 0, the profile of the SiNW transfers to the helical buckling, whose wavelength approaches to the theoretical prediction of a perfectly bonded SiNW, $h = 7.89R\left(E_{\mathrm{NW}}/\overline{E}_{\mathrm{S}}\right)^{1/4}$. Substituting the parameters used in the present FEM simulation, the theoretical helical buckling spacing is $h = 1.64$ μm. Further parametric studies show that the buckling mode of the SiNW is mainly governed by the ratio d/h. For $d/h > 0.50$, it is the in-plane buckling and for $0.27 < d/h < 0.50$, it is the disordered mode. While, for roughly $d/h < 0.27$, the buckling of the SiNW is predominantly helical.

The critical buckling strain, namely, the compressive strain applied to the PDMS substrate when the SiNW buckling initiates, is shown in Fig. 21. For the in-plane mode, the buckling behavior between adjacent constraint points is analogous to that of a simple supported beam with length d. Based on the Euler buckling theory, the initial buckling strain for a circular beam with radius R is $(\pi R/2d)^2$, which reasonably fits the simulated buckling strain of the in-plane modes. As the constraint density increases, the critical buckling strain gradually diverges from the Euler beam

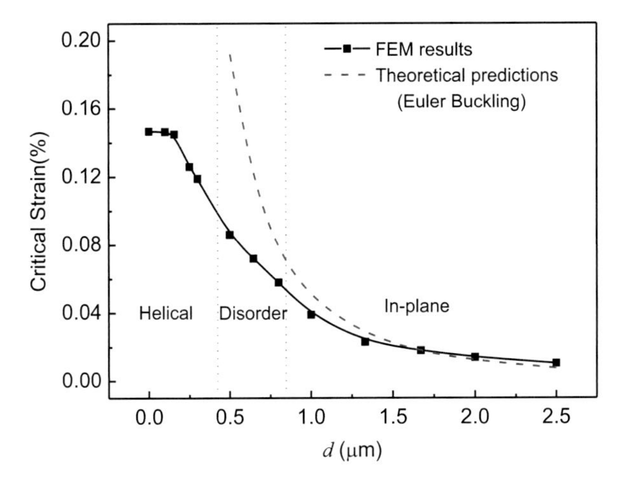

Fig. 21 The critical buckling strain of the partially bonded SiNW with different constraint densities. The red dashed line indicates the Euler beam buckling strain

theory, so as to keep a lower strain energy upon buckling. With further increasing the constraint density, the behavior approaches to that of a perfectly bonded SiNW and the initial buckling strain is independent on the distance *d*.

Strain Distribution in SiNW

The strain distribution in the SiNW is another important factor that influences the performance of the stretchable electronics. For example, fracture may initiate at the largest strain point and the electrical properties of the SiNW are also affected by the strain in it (Peng et al. 2009; Sajjad and Alam 2009). The strain distribution along the axis of the SiNW is shown in Fig. 22a for different constraint densities $d = 2.5$ μm, 0.5 μm, and 0 μm. Here, the strain represents the maximum principal in the cross section of the SiNW. The strain distribution for the in-plane mode ($d = 2.5$ μm for example) fluctuates with the maximum value attained at the middle of two adjacent constraint points, and the minimum value at the constraints. For the disordered mode (e.g., $d = 0.5$ μm), the strain fluctuation is more severe with the maximum exceeding 5% and minimum below 2%. Relatively speaking, the strain distribution in the helical mode is more uniform and oscillates, whose maximum 4.76% is slightly smaller than the disordered mode.

A map of the amplitudes of the maximum strain and the maximum in-plane displacement is given in Fig. 22b; in each case *d* is denoted. Generally, the strain amplitude increases as the constraint density increases. In contrast, the displacement amplitude decreases. For the in-plane buckling mode, it has smaller strain, but has larger displacement, so that larger space is required and the density of the SiNW is limited. While, for the helical buckling mode it has the smallest displacement and moderate strain amplitude, which might be beneficial for stretchable electronics.

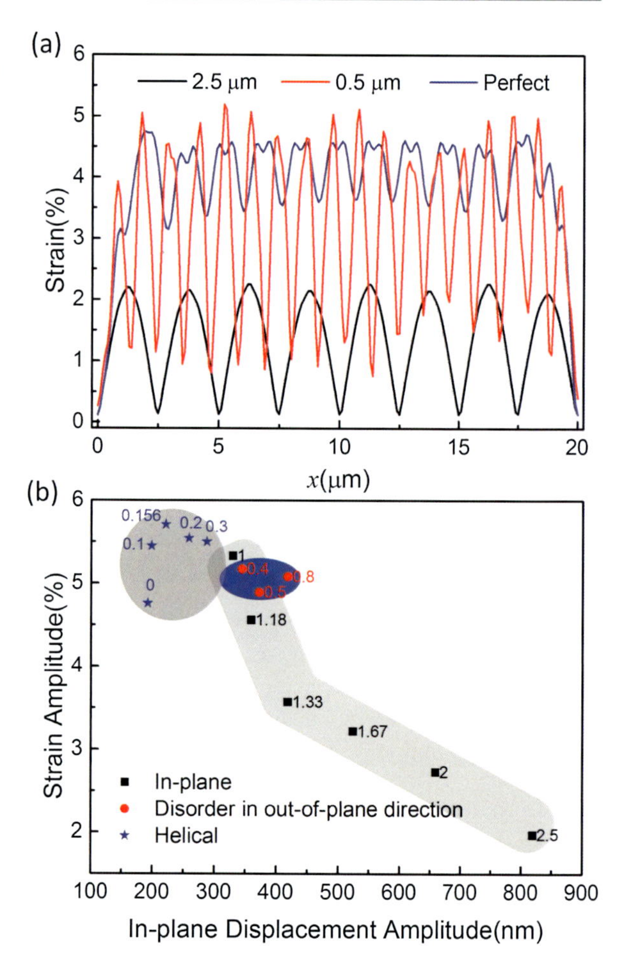

Fig. 22 (**a**) The strain distributions along SiNW for the perfectly bonded SiNW and partially bonded SiNW ($d = 2.5$ μm, 0.5 μm) and (**b**) the maximum strain–maximum displacement relation for different constraint densities

Comparison to Experiments

Parallel experiment is carried out with details outlined in previous literature (Xu et al. 2011). For the SiNW–PDMS system without UVO treatment or less UVO treatment time (<3 min), the buckling mode of the SiNW is the in-plane which fits the prediction from FEM for a partially bonded SiNW ($d = 1$ μm), as shown in Fig. 23c. When the UVO treatment time is between 5 min and 8 min (when strong adhesion is likely to present), the buckling of the SiNW is the helical mode which is also consistent with the FEM simulation for a perfectly bonded SiNW, as shown in Fig. 23a, b. However, when the UVO treatment time is longer than 8 min, the mode changes to in-plane again, because the UVO treatment weakens the compliance of the PDMS near the surface which may debond.

For the in-plane mode, the surface of the PDMS substrate remains very smooth, which indicates the sliding between the SiNW and the PDMS substrate. The relations of the in-plane displacement and out-of-plane displacement amplitudes to the effective compressive strain are given in Fig. 24. Both displacement amplitudes

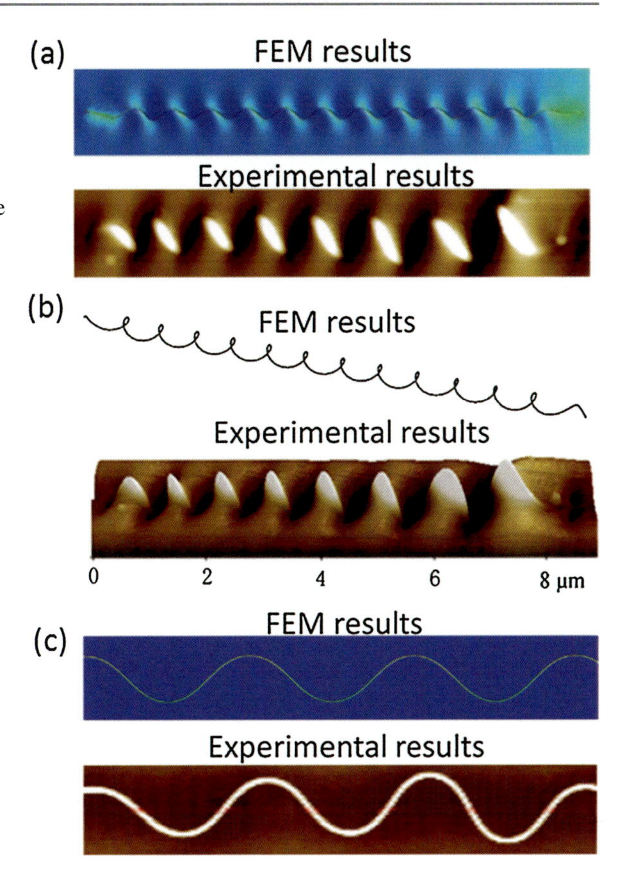

Fig. 23 The typical planar (**a**) and 3D (**b**) configuration of the SiNW for the helical buckling and (**c**) the in-plane buckling mode. The brightness represents the height of the substrate surface

observed in the experiments are within the ranges of the FEM results for the helical buckling ($d = 0.25$ μm, 0.1 μm and perfectly bonded SiNW, $d = 0$), as shown in Fig. 24b. This indicates that the partial debonding of the SiNW is the major reason for the transition from the in-plane buckling to the helical buckling. Due to the absent van der Walls force in our model, the present study cannot yet precisely reproduce every feature in experiment; nevertheless, the mechanism of transition of distinctive buckling modes is elucidated.

In order to explore the underlying mechanism of the buckling mode transition for different constraint density, the strain energy of the three different buckling modes (in-plane, disorder, and helical) for three different constraint distances, 2.5 μm, 0.5 μm, and 0.1 μm, is shown in Fig. 25. Here, in order to obtain the in-plane buckling mode for the constraint distance 0.5 μm and 0.1 μm, the out-of-plane displacement of the SiNW is constrained. As shown in Fig. 25, for the constraint distance 0.5 μm, the strain energy of the in-plane buckling mode is larger than that of the out-of-plane buckling mode. Besides, the strain energy of the out-of-plane disordered buckling mode for the constraint distance 0.5 μm is smaller than that of the helical buckling mode for the constrain distance 0.1 μm. While, for the constraint distance 0.1 μm, the strain energy of the helical buckling mode is smaller

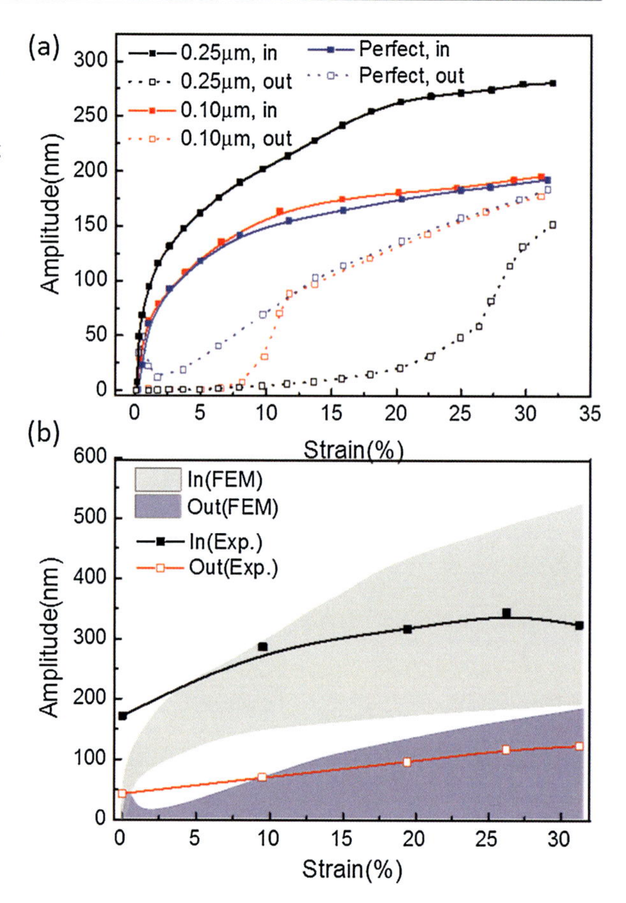

Fig. 24 (**a**) The relations of the in-plane and out-of-plane displacement amplitudes to the effective compressive strain for the helical buckling obtained from FEM simulations. (**b**) The comparison of the in-plane and the out-of-plane displacement amplitudes between experiments (up various UVO treatment time) and FEM simulations. The shadow areas indicate the ranges obtained from FEM simulations

than that of the in-plane buckling mode. Note that we cannot generate the helical buckling mode for the constraint distance 0.5 μm and the out-of-plane disordered bucking mode for the constraint distance 0.1 μm. Here, we cannot directly compare the strain energy between the out-of-plane disordered mode and the helical buckling mode for the constraint distance 0.5 μm and 0.1 μm. Nevertheless, it can be inferred that the out-of-plane disordered buckling mode is energetically more favorable than the in-plane buckling mode for $0.27 < d/h < 0.5$ and the helical buckling mode energy is more favorable for $d/h < 0.27$. In our future work, the underlying mechanism of the buckling mode transition for a nanowire on the elastomeric substrate will be systematically studied.

It should to be noted that our FEM model is a simplification to the buckling process. In the experiments, the debonding of the SiNW is accompanied by the buckling of the SiNW, while in our FEM simulation we assume the debonding occurs prior to the buckling. Moreover, the constraints in experiments may be small segments instead of singular points. In the future work, we will systematically study the debonding characteristics of the SiNW and its influence to the buckling

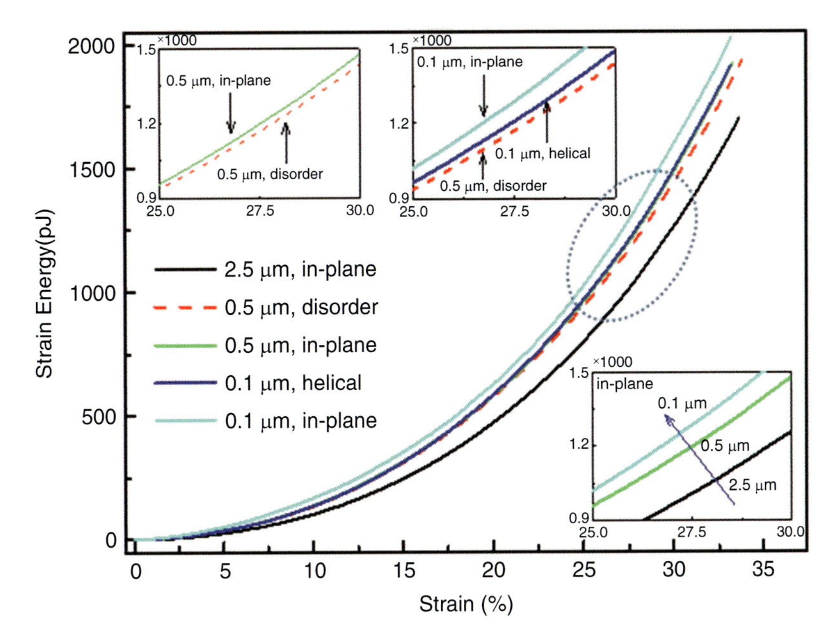

Fig. 25 The strain energy of three buckling modes for the constraint distance 2.5 μm, 0.5 μm, and 0.1 μm, respectively. For the constraint distance 0.5 μm, the strain energy of the in-plane bucking mode (green line) is almost overwritten by the strain energy of the out-of-plane disordered buckling mode (blue line)

behaviors of the SiNW. Besides, the mechanical properties of the PDMS, especially near the surface of the substrate, are influenced by the UVO treatment, for example, increasing the stiffness and decreasing the deformability of the PDMS. In the future work, the gradient of the mechanical properties near the surface of the PDMS will be factored into approach.

Helical Buckling of a Nanowire Embedded in a Soft Matrix Under Axial Compression

In previous sections, we have focused on the 3D helical buckling behaviors and the mechanism of the transition from 2D in-plane buckling to 3D helical buckling of a nanowire on the surface of an elastomeric substrate. Actually, similar 2D and 3D buckling modes were also observed in wires embedded in matrix (Brangwynne et al. 2006; Jiang and Zhang 2008; Li 2008; O'Keeffe et al. 2013; Slesarenko and Rudykh 2016; Su et al. 2014; Zhao et al. 2016). Based on the Winkler foundation model Timoshenko and Gere (2009), and Su et al. (2014) developed the 2D buckling theory of a wire embedded in a soft matrix, and concentrated on the critical strain and initial buckling wavelength (or wavevector) of 2D sinusoidal buckled wires (Slesarenko and Rudykh 2016; Su et al. 2014). However, the 3D helical buckling mechanism and post-buckling behaviors of embedded wires have not been explicitly explained.

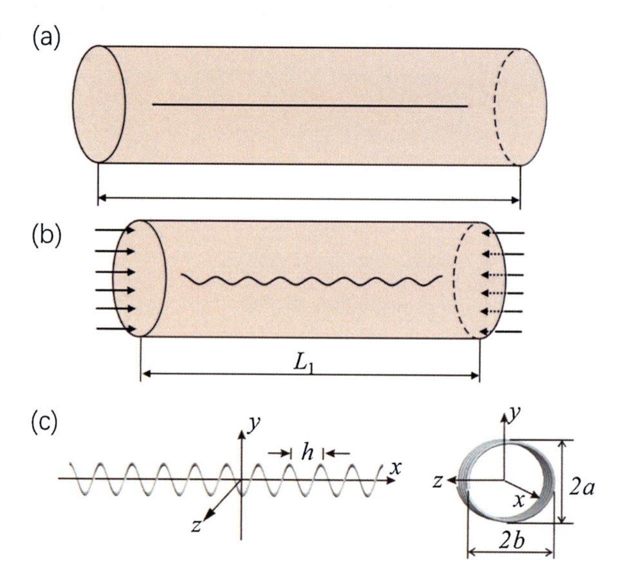

Fig. 26 Schematic of helical buckling of embedded wire under axial compression

In this section, a 3D helical buckling theory for compressed wires embedded in a soft matrix is established, and the initial and post-buckling response of wires in association with parallel FEM simulations is discussed systematically.

Theoretical Analysis

Consider a wire with radius R embedded in an infinite matrix (see Fig. 26). According to the experiments of Su et al. (2014), the matrix is so soft that delamination between the wire and the matrix is prevented, and thus the delamination and friction between them is not considered in present research; such simplification is also utilized in their work. When the matrix is compressed and the effective compressive strain (ε_{com}) exceeds the critical strain (ε_{cr}), the wire buckles into a helical configuration.

For a wire buckling in helical mode, the deformation can be decomposed into two sinusoidal deformations in two perpendicular directions (y and z direction) with x direction denoting the axial direction of the wire. The lateral displacements in y and z direction are assumed as $v = a\cos(kx)$ and $w = b\sin(kx)$, respectively, where a and b are the buckling displacement amplitudes, and k is the buckling wavevector, namely, $2\pi/h$ (h being the initial buckling spacing). The strain energy (per unit length) of the wire can be expressed (Eq. 19) as

$$U_{\text{wire}} = \frac{EA}{2}\left[\frac{k^2\left(a^2+b^2\right)}{4} - \varepsilon_{\text{com}}\right]^2 + \frac{EIk^4\left(3a^4k^2+2a^2b^2k^2+3b^4k^2+4a^2+4b^2\right)}{16\left(a^2k^2+a^2k^4b^2+a^2k^2+1\right)^{\frac{3}{2}}}$$
$$+\frac{(1+\nu)\,AE\left(a^2k^2+b^2k^2-4\varepsilon_{\text{com}}\right)^2\left(Aa^2b^2k^2+2I\left(\sqrt{(1+b^2k^2)(1+a^2k^2)}-1\right)\right)}{32I\sqrt{(1+b^2k^2)(1+a^2k^2)}}, \tag{29}$$

where E and ν are the Young's modulus and Poisson's ratio of the wire, respectively. A and I are the cross section area and the moment of inertia of the wire, and ε_{com} is the applied effective compressive strain. The three terms in Eq. 29 represent the membrane strain energy, bending strain energy, and torsion energy of the wire, respectively.

The lateral force experienced by the wire (per unit length) can be given as $T = EI\mathrm{d}^4v/\mathrm{d}x^4 - EA\varepsilon_{\text{m}}\,\mathrm{d}^2v/\mathrm{d}x^2$, where the v denotes the lateral deflection and ε_{m} is the membrane strain. ε_{m} derived from y- and z-direction sinusoidal deformations can be given as $a^2k^2/4 - \varepsilon_{\text{com}}$ and $b^2k^2/4 - \varepsilon_{\text{com}}$, respectively. Thus, the lateral force in y and z directions can be given as $T_y = -P_y\cos(kx)$ with $P_y = -EAak^2(a^2k^2/4 - \varepsilon_{\text{com}}) - EIak^4$ and $T_z = -P_z\sin(kx)$ with $P_z = -EAbk^2(b^2k^2/4 - \varepsilon_{\text{com}}) - EIbk^4$. Inside an infinite matrix, a unit point force in y direction at $(x_0, y_0, 0)$ induces a y-directional displacement at $(x, y, 0)$, given by $\frac{1+\nu_s}{4\pi E_s\sqrt{(x-x_0)^2+(y-y_0)^2}}$, where E_s and ν_s are the Young's modulus and Poisson's ratio of the matrix. Thus, the y-directional displacement at $(x, y, 0)$ due to the deformation of the wire can be approximately expressed as (Abramowitz and Stegun 1972; Jiang and Zhang 2008; Li 2008; Timoshenko et al. 1970)

$$v_{\text{sub}} = \int_{-R}^{R}\int_{-\infty}^{\infty} \frac{P_y\cos(kx_0)}{2R} \times \frac{1+\nu_s}{4\pi E_s\sqrt{(x-x_0)^2+(y-y_0)^2}}\mathrm{d}x_0\mathrm{d}y_0. \tag{30}$$

Considering R is much smaller than the buckle spacing h, the dominant term of Eq. 30 can be rewritten into

$$\begin{aligned} v_{\text{sub}} =& \frac{(1+\nu_s)P_y\cos(kx)}{4\pi RE_s}\{2R(1-\gamma)+2R\ln 2 \\ & -(R+y)\ln[k(R+y)]-(R-y)\ln[k(R-y)]\}, \end{aligned} \tag{31}$$

where $\gamma \approx 0.577$ is the Euler constant.

Then, the strain energy (per unit length) of the matrix due to the sinusoidal deformation of the wire in y direction is

$$\begin{aligned} U_{sy} &= \frac{k}{2\pi}\int_{-R}^{R}\int_{0}^{2\pi/k} \frac{P_y\cos(kx)}{2R}\cdot v_{\text{sub}}\mathrm{d}x\mathrm{d}y \\ &= \frac{(1+\nu_s)P_y{}^2}{16\pi E_s}\left[\frac{2-\nu_s}{1-\nu_s} - 2\gamma - 2\ln(kR)\right]. \end{aligned} \tag{32}$$

Similarly, the strain energy (per unit length) of the matrix due to the sinusoidal deformation of the wire in z direction can be approximately given as

$$U_{sz} = \frac{(1+\nu_s)P_z^2}{16\pi E_s}\left[\frac{2-\nu_s}{1-\nu_s} - 2\gamma - 2\ln(kR)\right]. \tag{33}$$

Thus, the total potential energy (per unit length) of the wire and matrix can be expressed by

$$U_{\text{total}} = U_{\text{wire}} + U_{sy} + U_{sz} - \frac{k}{2\pi}\int_{-R}^{R}\int_{0}^{2\pi/k} \frac{P_y \cos(kx)}{2R}\left[v_{\text{sub}} - a\cos(kx)\right] dx dy - \frac{k}{2\pi}\int_{-R}^{R}\int_{0}^{2\pi/k} \frac{P_z \sin(kx)}{2R}\left[w_{\text{sub}} - b\sin(kx)\right] dx dy. \tag{34}$$

Here, the two integration terms represent the deformation compatibility between the wire and matrix.

By substituting Eqs. 29, 32, and 33 into Eq. 34, and integrating the last equation, we obtain

$$\begin{aligned} U_{\text{total}} =& \frac{EA}{2}\left[\frac{k^2\left(a^2+b^2\right)}{4} - \varepsilon_{\text{com}}\right]^2 - \frac{\left[EAak^2\left(\frac{a^2k^2}{4} - \varepsilon_{\text{com}}\right) + EIak^4\right]a}{2} \\ &- \frac{\left[EAbk^2\left(\frac{b^2k^2}{4} - \varepsilon_{\text{com}}\right) + EIbk^4\right]b}{2} \\ &+ \frac{(1+\nu)AE\left(a^2k^2+b^2k^2-4\varepsilon_{\text{com}}\right)^2\left(Aa^2b^2k^2 + 2I\left(\sqrt{(1+b^2k^2)(1+a^2k^2)} - 1\right)\right)}{32I\sqrt{(1+b^2k^2)(1+a^2k^2)}} \\ &- \frac{(1+\nu)\left[EAak^2\left(\frac{a^2k^2}{4} - \varepsilon_{\text{com}}\right) + EIak^4\right]^2}{16\pi E_s}\left[3 - 2\gamma - 2\ln(kR)\right] \\ &- \frac{(1+\nu)\left[EAbk^2\left(\frac{b^2k^2}{4} - \varepsilon_{\text{com}}\right) + EIbk^4\right]^2}{16\pi E_s}\left[3 - 2\gamma - 2\ln(kR)\right] \\ &+ \frac{EIk^4\left(3a^4k^2 + 2a^2b^2k^2 + 3b^4k^2 + 4a^2 + 4b^2\right)}{16(a^2k^2 + a^2k^4b^2 + a^2k^2 + 1)^{\frac{3}{2}}} \end{aligned} \tag{35}$$

Minimizing the potential energy gives the buckling vector as

$$k = \left(\frac{E_s}{EI}\right)^{1/4}\left[\frac{16\pi\left(1 - \gamma - \ln(kR)\right)}{3(3 - 2\gamma - 2\ln(kR))^2}\right]^{1/4}. \tag{36}$$

Generally, kR is much smaller than 1, and thus Eq. 36 can be approximated into $k = 1.02\left(\frac{E_s}{EI}\right)^{1/4}$, and thus the initial buckle spacing is $h = 6.160\left(\frac{EI}{E_s}\right)^{1/4}$. In particular, for a wire with circular cross section, the initial buckle spacing can also

Table 3 Buckling amplitudes for wires with different Young's modulus ($R = 15$ nm)

E/GPa	10	50	100	150	200	250
a/nm	67.9	101	119	131	141	149
b/nm	67.9	101	119	131	141	149

Table 4 Buckling amplitudes for wires with radius ($E = 200$ GPa)

R/nm	5	10	15	20	25
a/nm	47.0	94.0	141	188	235
b/nm	47.0	94.0	141	188	235

be written into $h = 5.80R\left(\frac{E}{E_s}\right)^{1/4}$. And the post-buckling spacing can be expressed as (Chen et al. 2016; Jiang et al. 2007; Xu et al. 2011):

$$h = 5.80R\left(\frac{E}{E_s}\right)^{1/4}(1 - \varepsilon_{com}), \tag{37}$$

which shares the similar scaling law $R\left(\frac{E}{E_s}\right)^{1/4}$ as the helical buckling on the surface of substrates, with the prefactor a quarter smaller than the latter.

Substituting Eq. 36 back into Eq. 35, and minimizing with respect to the displacement amplitude a and b, we can numerically obtain these two undetermined profile parameters. The typical amplitudes of wires with different Young's modulus and radius are shown in Tables 3 and 4 (where the effective compressive strain is 35%). The buckling amplitude a and b are equal due to the isotropy of the matrix. Thereafter, we do not differentiate the amplitudes in y and z direction unless specified. By fitting of the theoretical results, the amplitude can be expressed as $a = \frac{1.276}{k}\sqrt{\varepsilon_{com} - \varepsilon_{cr}}$ where ε_{cr} is the critical buckling strain. In the present research, ε_{cr} is very small (0.34% for a 15-nm-radius wire of modulus 200 GPa embedded in a matrix of 6 MPa) and thus negligible compared with the effective compressive strain ε_{com} in the post-buckling phase. Therefore, the buckling amplitude can also be simplified as

$$a = \frac{1.276}{k}\sqrt{\varepsilon_{com}}. \tag{38}$$

Results and Discussion

To verify the theoretical solution, FEM simulations based on commercial software ABAQUS are conducted. The matrix is simplified as a cylinder, which is much larger than the buckling amplitude and spacing. The wire is regarded as a beam with the radius 15 nm along the axial of the cylindrical matrix. Both ends of the cylindrical matrix are slowly compressed toward each other. The Young's modulus

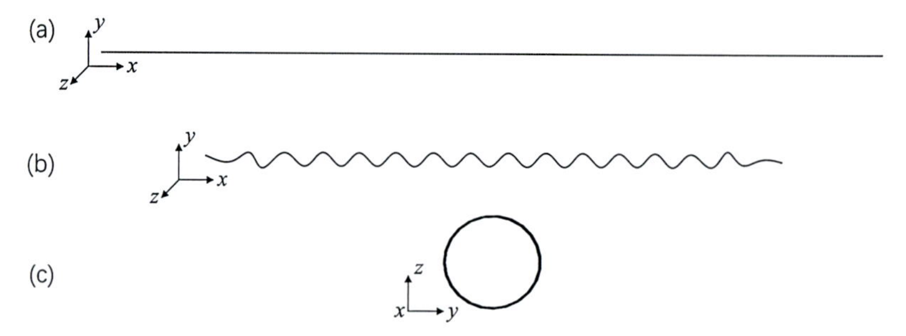

Fig. 27 Typical configurations of the wire before (**a**) and after (**b** and **c**) buckling obtained from FEM simulation

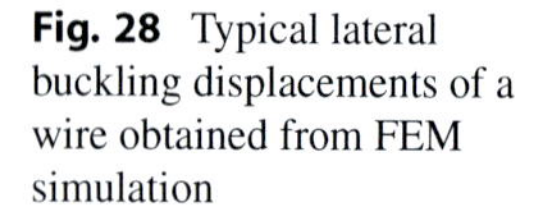

Fig. 28 Typical lateral buckling displacements of a wire obtained from FEM simulation

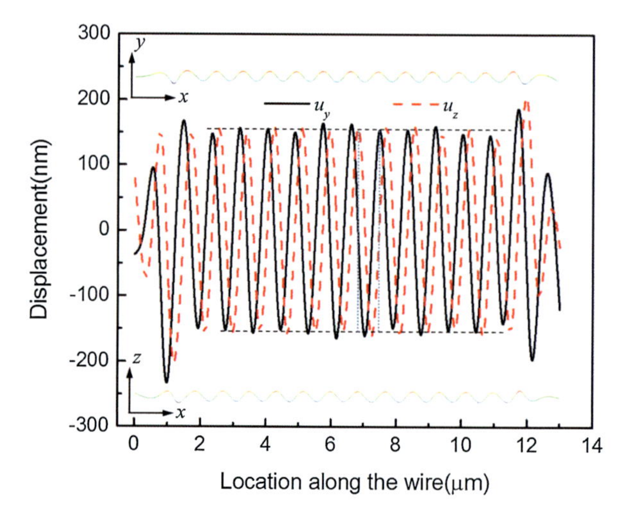

of the wire is 200 GPa, and 6 MPa for the matrix. Mesh convergence is carried out to ensure the reliability of numerical results. Typical buckling configurations of buckled wires are shown in Fig. 27.

Shown in Fig. 28 are the FEM results of typical buckling displacements of the wire in y and z directions. In both x–y and x–z planes, the projections of buckled wires exhibit the sinusoidal curves. View from the axial direction reveals that the profile of the wire is almost perfect circular, since the buckling amplitudes in x–y and x–z planes are almost equal (also see Fig. 27c). Besides, it should also be noticed that the angle phases in the two planes have a difference of $\pi/2$, which further substantiates the helical configuration of the wire.

The consistency between FEM simulation and theoretical result is shown in Fig. 29, in which the effective compressive strain is 35%, and the theoretical results are Eqs. 37 and 38 for the buckling spacing and amplitude, respectively. The spacing is proportional to wire radius, and the amplitude scales with 1/4 power of the Young's modulus of the wire, which agrees with previous studies (Chen et al. 2016; Jiang and Zhang 2008; Xiao et al. 2008, 2010; Zhao et al. 2016).

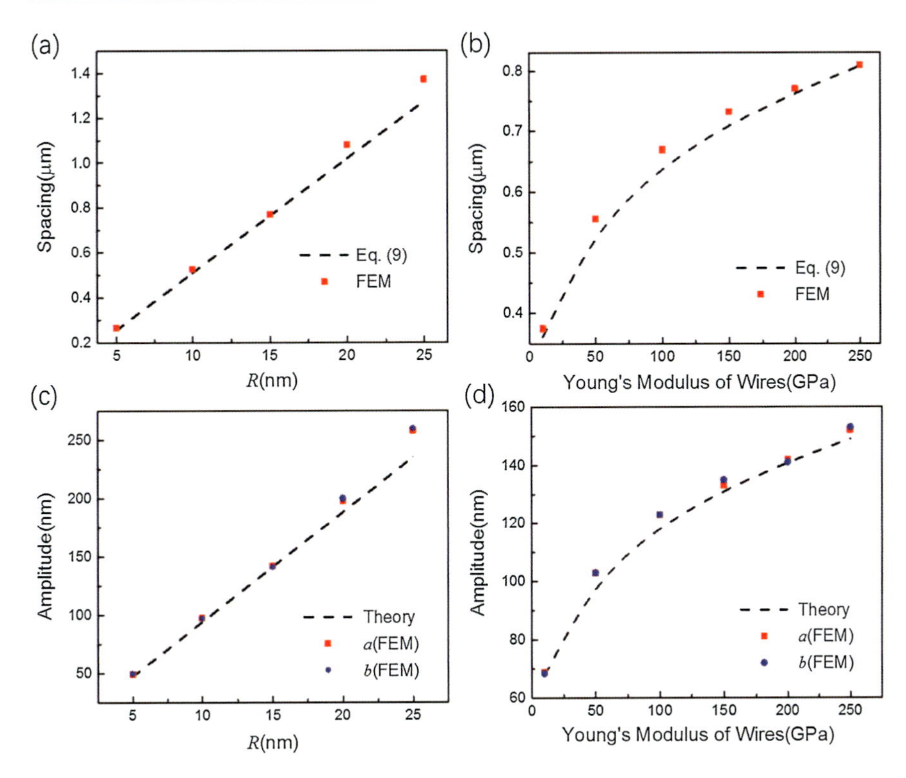

Fig. 29 Post-buckling spacing (**a** and **b**) and amplitude (**c** and **d**) versus the radius (**a** and **c**) and Young's modulus (**b** and **d**) of wires. In (**a**) and (**c**), the Young's modulus of the wire is 200 GPa, and in (**b**) and (**d**) the radius of the wire is 15 nm

The post-buckling behaviors, mainly the buckling spacing and amplitudes, are shown in Fig. 29, which again validates the theoretical model with respect to FEM simulation. The buckling spacing decreases almost linearly with the compressive strain, while the evolution of buckling amplitude is slightly more complex. Once the compressive strain exceeds the critical strain, the sinusoidal buckles initiate mainly in one direction (i.e., almost in a 2D profile); however, the deflection in the other direction quickly catches up and transits the configuration from 2D sinusoidal to 3D helical mode. As shown in Fig. 30, the modal analysis conducted via FEM indicates that the 2D buckling is the first order mode while the 3D helical form is the second order. Thus, the 2D sinusoidal appears first and exhibits lower critical strain, which is also confirmed by the present theory (see Fig. 30b); thereafter, the buckled wire turns into the helical mode with less strain energy. Besides, it should be noted that the critical strain is almost linear with respect to the root of the ratio $(E_s/E)^{1/2}$, which agrees with previous studies (Chen et al. 2016; Su et al. 2014; Xiao et al. 2008, 2010).

As suggested by the FEM results, the amplitude increases with the compressive strain, in accordance with the theoretical prediction. It is noteworthy that even at

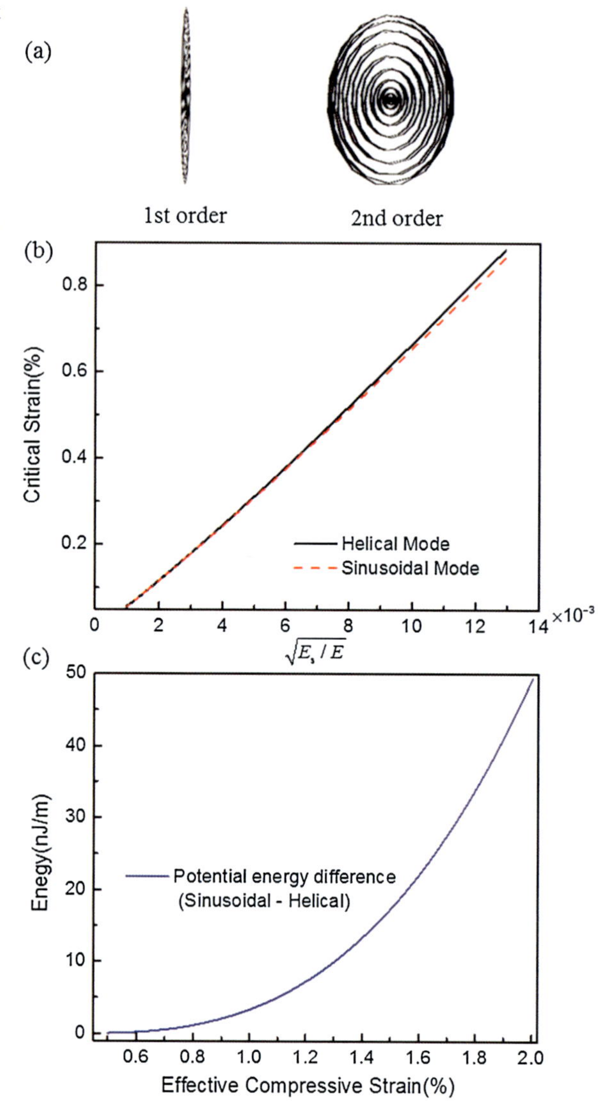

Fig. 30 (**a**) Modal of the first two orders of the wire embedded in matrix (FEM). (**b**) Critical strains for the helical buckling mode and sinusoidal mode ($R = 15$ nm). (**c**) Difference of potential energy (per length) of matrix and wires in the helical mode and sinusoidal mode ($E = 200$ GPa, $E_s = 6$ MPa, $R = 15$ nm)

very small strain (less than 1.5% in the inset of Fig. 31b) the axial view of the buckling configuration turns into almost perfectly circular. By comparison, if the wire sits on top of the surface of a semi-infinite substrate, the axial profile change from an oval toward a circle (Xu et al. 2011) could require a much higher effective compressive strain (over 30%).

Transition from in-plane buckling mode to helical buckling mode can also be represented by the ratio of in-plane displacement amplitude to out-of-plane

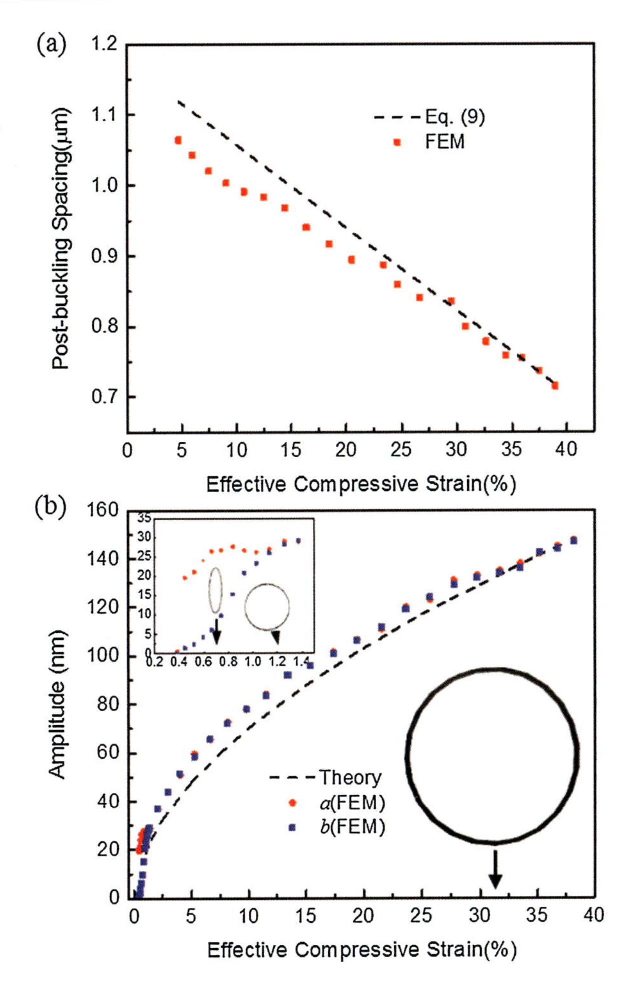

Fig. 31 Theoretical and FEM evolutions of post-buckling behaviors ($E = 200$ GPa, $E_s = 6$ MPa, $R = 15$ nm)

displacement amplitude of wire, as shown in Fig. 32a. Here, the in-plane direction is referred as the initial deformation direction of the wire and the out-of-plane direction is perpendicular to the in-plane direction and the axis of the wire. It is suggested that, when the effective compressive strain reaches a critical value, the buckling of the wire initiates and the ratio is maximum at the beginning (i.e., the in-plane mode); with further compression, the ratio decreases dramatically and approaches 1 (the blue dashed line, i.e., the helical mode). More interestingly, for wires with lower Young's modulus, the effective compressive strain necessary for the transition from 2D to 3D helical mode is higher (see Fig. 32b). Nevertheless, when the Young's modulus is high enough, the transition strain tends to be 1.25%, namely, as the effective compressive strain exceeds 1.25%, wires will buckle into the 3D helical mode.

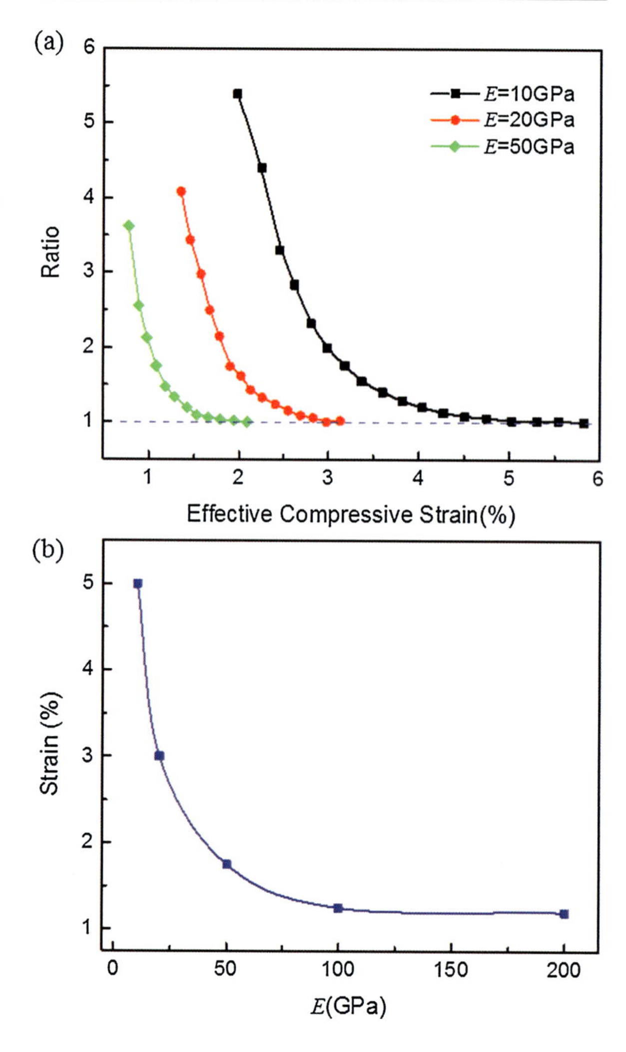

Fig. 32 (**a**) The FEM evolution of the ratio of in-plane to out-of-plane displacement amplitude of wire with different Young's modulus and (**b**) the strain necessary for transition from in-plane mode to helical mode ($E_s = 6$ MPa, $R = 15$ nm)

Conclusion

In summary, we have introduced systematic studies of a special buckling mode – three-dimensional helical coil buckling of SiNW on the UVO-treated PDMS substrate. A continuum mechanics theory based on the minimization of the total potential energy of the SiNW and PDMS substrate is established and the impact of SiNW properties and the geometric dimension on the buckling characteristics, such as the buckling spacing, the in-plane and out-of-plane displacement amplitudes, etc., has been revealed, which agrees quite well with the FEM simulation results. Both the theory and FEM simulation suggest that the buckling spacing and displacement amplitudes increase linearly with the SiNW diameter and are almost proportional to the 1/4 power of the ratio between the SiNW modulus and the substrate modulus.

In addition, the FEM simulation results indicate that the amplitude ratio changes with the effective compressive strain, and when the effective compressive strain is large enough, the buckling shape is almost circular, namely, the amplitude ratio approaching 1.

The partial detachment of SiNW from the PDMS substrate is mainly responsible for the transition of the buckling modes, which can be dictated by the constraint ratio d/h. For $d/h > 0.50$, the buckling of the SiNW is the in-plane which can be described by the Euler beam buckling theory with the buckling wavelength $\lambda = 2d$. For $0.27 < d/h < 0.50$, the buckles are disordered in the out-of-plane direction and the critical buckling strain gradually diverges from the Euler beam theory. Whereas, for $d/h < 0.27$, the buckling is helical and the buckling spacing gradually approaches to the theoretical value of a perfectly bonded SiNW.

Meanwhile a theoretical analysis of three-dimensional helical buckling of wires embedded in matrix is also provided and the buckling spacing and amplitudes are deduced, which are further verified by parallel FEM simulations. It is suggested that the buckled profile is almost perfectly circular in the axial direction; with increasing compression, the buckling spacing decreases almost linearly, while the amplitude scales with the 1/2 power of the compressive strain. Besides the transition strain from 2D mode to 3D helical mode decreases with the Young's modulus of the wire and approaches to 1.25% when the modulus is high enough. This study may shed some lights on the buckling behaviors of wires embedded in matrix and provide some useful instructions of manufacturing complex structures.

References

M. Abramowitz, I.A. Stegun, *Handbook of Mathematical Functions with Formulas, Graphs, and Mathematical Tables* (National Bureau of Standards Applied Mathematics Series 55. Tenth Printing. Engineering: 1076, Washington, DC, 1972)

B. Audoly, A. Boudaoud, Buckling of a stiff film bound to a compliant substrate – part I. J. Mech. Phys. Solids **56**(7), 2401–2421 (2008)

C.P. Brangwynne, F.C. MacKintosh, S. Kumar, N.A. Geisse, J. Talbot, L. Mahadevan, K.K. Parker, D.E. Ingber, D.A. Weitz, Microtubules can bear enhanced compressive loads in living cells because of lateral reinforcement. J. Cell Biol. **173**(5), 733–741 (2006)

B. Charlot, W. Sun, K. Yamashita, H. Fujita, H. Toshiyoshi, In-plane bistable nanowire for memory devices, in *Symposium on Design, Test, Integration and Packaging of MEMS/MOEMS, 2008. MEMS/MOEMS 2008* (IEEE, 2008). http://ieeexplore.ieee.org/document/4752995/

X. Chen, J. Yin, Buckling patterns of thin films on curved compliant substrates with applications to morphogenesis and three-dimensional micro-fabrication. Soft Matter **6**(22), 5667–5680 (2010).

Z. Chen, B. Cotterell, W. Wang, The fracture of brittle thin films on compliant substrates in flexible displays. Eng. Fract. Mech. **69**(5), 597–603 (2002)

Y. Chen, Y. Liu, Y. Yan, Y. Zhu, X. Chen, Helical coil buckling mechanism for a stiff nanowire on an elastomeric substrate. J. Mech. Phys. Solids **95**, 25–43 (2016)

G. Crawford, *Flexible Flat Panel Displays* (Wiley, Chichester, 2005)

Y. Duan, Y. Huang, Z. Yin, Competing buckling of micro/nanowires on compliant substrates. J. Phys. D. Appl. Phys. **48**(4), 045302 (2015)

J.W. Durham 3rd, Y. Zhu, Fabrication of functional nanowire devices on unconventional substrates using strain-release assembly. ACS Appl. Mater. Interfaces **5**(2), 256–261 (2013)

K. Efimenko, W.E. Wallace, J. Genzer, Surface modification of Sylgard-184 poly(dimethyl siloxane) networks by ultraviolet and ultraviolet/ozone treatment. J. Colloid Interface Sci. **254**(2), 306–315 (2002)

A. Goriely, R. Vandiver, M. Destrade, Nonlinear Euler buckling. Proc. R. Soc. Lond. A Math. Phys. Eng. Sci. **464**(2099), 3003–3019 (2008)

H. Jiang, J. Zhang, Mechanics of microtubule buckling supported by cytoplasm. J. Appl. Mech. **75**(6), 061019 (2008)

H. Jiang, D.Y. Khang, J. Song, Y. Sun, Y. Huang, J.A. Rogers, Finite deformation mechanics in buckled thin films on compliant supports. Proc. Natl. Acad. Sci. U. S. A. **104**(40), 15607–15612 (2007)

S.J. Kalita, V. Somani, Al_2TiO_5–Al_2O_3–TiO_2 nanocomposite: structure, mechanical property and bioactivity studies. Mater. Res. Bull. **45**(12), 1803–1810 (2010)

D.H. Kim, J. Song, W.M. Choi, H.S. Kim, R.H. Kim, Z. Liu, Y.Y. Huang, K.C. Hwang, Y.W. Zhang, J.A. Rogers, Materials and noncoplanar mesh designs for integrated circuits with linear elastic responses to extreme mechanical deformations. Proc. Natl. Acad. Sci. U. S. A. **105**(48), 18675–18680 (2008)

H.C. Ko, M.P. Stoykovich, J. Song, V. Malyarchuk, W.M. Choi, C.J. Yu, J.B. Geddes 3rd, J. Xiao, S. Wang, Y. Huang, J.A. Rogers, A hemispherical electronic eye camera based on compressible silicon optoelectronics. Nature **454**(7205), 748–753 (2008)

S.P. Lacour, J. Jones, S. Wagner, T. Li, Z. Suo, Stretchable interconnects for elastic electronic surfaces. Proc. IEEE **93**(8), 1459–1467 (2005)

T. Li, A mechanics model of microtubule buckling in living cells. J. Biomech. **41**(8), 1722–1729 (2008)

H. Mei, C.M. Landis, R. Huang, Concomitant wrinkling and buckle-delamination of elastic thin films on compliant substrates. Mech. Mater. **43**(11), 627–642 (2011)

S.G. O'Keeffe, D.E. Moulton, S.L. Waters, A. Goriely, Growth-induced axial buckling of a slender elastic filament embedded in an isotropic elastic matrix. Int. J. Non Linear Mech. **56**, 94–104 (2013)

W.A. Oldfather, C.A. Ellis, D.M. Brown, Leonhard Euler's elastic curves. Isis **20**(1), 72–160 (1933)

X.H. Peng, A. Alizadeh, S.K. Kumar, S.K. Nayak, Ab initio study of size and strain effects on the electronic properties of Si nanowires. Int. J. Appl. Mech. **1**(3), 483–499 (2009)

Q. Qin, Y. Zhu, Static friction between silicon nanowires and elastomeric substrates. ACS Nano **5**(9), 7404–7410 (2011)

J.A. Rogers, T. Someya, Y. Huang, Materials and mechanics for stretchable electronics. Science **327**(5973), 1603–1607 (2010)

S.Y. Ryu, J. Xiao, W.I. Park, K.S. Son, Y.Y. Huang, U. Paik, J.A. Rogers, Lateral buckling mechanics in silicon nanowires on elastomeric substrates. Nano Lett. **9**(9), 3214–3219 (2009)

R.N. Sajjad, K. Alam, Electronic properties of a strained 100 silicon nanowire. J. Appl. Phys. **105**(4), 044307 (2009)

V. Slesarenko, S. Rudykh, Microscopic and macroscopic instabilities in hyperelastic fiber composites. J. Mech. Phys. Solids **99**, 471–482 (2016)

T. Someya, T. Sekitani, S. Iba, Y. Kato, H. Kawaguchi, T. Sakurai, A large-area, flexible pressure sensor matrix with organic field-effect transistors for artificial skin applications. Proc. Natl. Acad. Sci. U. S. A. **101**(27), 9966–9970 (2004)

J. Song, Y. Huang, J. Xiao, S. Wang, K.C. Hwang, H.C. Ko, D.H. Kim, M.P. Stoykovich, J.A. Rogers, Mechanics of noncoplanar mesh design for stretchable electronic circuits. J. Appl. Phys. **105**(12), 123516 (2009)

C.M. Stafford, C. Harrison, K.L. Beers, A. Karim, E.J. Amis, M.R. VanLandingham, H.C. Kim, W. Volksen, R.D. Miller, E.E. Simonyi, A buckling-based metrology for measuring the elastic moduli of polymeric thin films. Nat. Mater. **3**(8), 545–550 (2004)

T. Su, J. Liu, D. Terwagne, P.M. Reis, K. Bertoldi, Buckling of an elastic rod embedded on an elastomeric matrix: planar vs. non-planar configurations. Soft Matter **10**(33), 6294–6302 (2014)

S.P. Timoshenko, J.M. Gere, *Theory of Elastic Stability* (Courier Corporation, North Chelmsford, 2009)
S.P. Timoshenko, J.N. Goodier, H.N. Abramson, Theory of elasticity (3rd ed.) J. Appl. Mech. **37**(3), 888 (1970)
Y. Wang, J. Song, J. Xiao, Surface effects on in-plane buckling of nanowires on elastomeric substrates. J. Phys. D. Appl. Phys. **46**(12), 125309 (2013)
E.A. Wilder, S. Guo, S. Lin-Gibson, M.J. Fasolka, C.M. Stafford, Measuring the modulus of soft polymer networks via a buckling-based metrology. Macromolecules **39**(12), 4138–4143 (2006)
J. Xiao, H. Jiang, D.Y. Khang, J. Wu, Y. Huang, J.A. Rogers, Mechanics of buckled carbon nanotubes on elastomeric substrates. J. Appl. Phys. **104**(3), 033543 (2008)
J. Xiao, S.Y. Ryu, Y. Huang, K.C. Hwang, U. Paik, J.A. Rogers, Mechanics of nanowire/nanotube in-surface buckling on elastomeric substrates. Nanotechnology **21**(8), 85708 (2010)
F. Xu, W. Lu, Y. Zhu, Controlled 3D buckling of silicon nanowires for stretchable electronics. ACS Nano **5**(1), 672–678 (2011)
S. Xu, Z. Yan, K.I. Jang, W. Huang, H.R. Fu, J. Kim, Z. Wei, M. Flavin, J. McCracken, R. Wang, A. Badea, Y. Liu, D.Q. Xiao, G.Y. Zhou, J. Lee, H.U. Chung, H.Y. Cheng, W. Ren, A. Banks, X.L. Li, U. Paik, R.G. Nuzzo, Y.G. Huang, Y.H. Zhang, J.A. Rogers, Assembly of micro/nanomaterials into complex, three-dimensional architectures by compressive buckling. Science **347**(6218), 154–159 (2015)
J. Yin, X. Chen, Buckling of anisotropic films on cylindrical substrates: insights for self-assembly fabrication of 3D helical gears. J. Phys. D. Appl. Phys. **43**(11), 115402 (2010)
E. Zeeman, Euler buckling, in *Structural Stability, the Theory of Catastrophes, and Applications in the Sciences* (Springer, 1976), pp. 373–395. http://www.springer.com/la/book/9783540077916
Y. Zhao, J. Li, Y.P. Cao, X.Q. Feng, Buckling of an elastic fiber with finite length in a soft matrix. Soft Matter **12**(7), 2086–2094 (2016)
C. Zhou, S. Bette, U. Schnakenberg, Flexible and stretchable gold microstructures on extra soft poly(dimethylsiloxane) substrates. Adv. Mater. **27**(42), 6664–6669 (2015)
Y. Zhu, F. Xu, Q. Qin, W.Y. Fung, W. Lu, Mechanical properties of vapor–liquid–solid synthesized silicon nanowires. Nano Lett. **9**(11), 3934–3939 (2009)

Hydrogen Embrittlement Cracking Produced by Indentation Test

8

Akio Yonezu and Xi Chen

Contents

Abstract

Indentation is a convenient method to evaluate mechanical properties of materials as well as to simulate contact fracture with locally plastic deformation. Indentation experiment has been widely used for brittle solids, including ceramics and glass, for evaluating the fracture properties. With the aid of computational framework, simulation of crack propagation (for quasi-static and dynamic impact) is conducted to characterize "brittleness" of materials. In this review, we explore the applicability of indentation method for hydrogen embrittlement

A. Yonezu (✉)
Department of Precision Mechanics, Chuo University, Tokyo, Japan
e-mail: yonezu@mech.chuo-u.ac.jp

X. Chen
Department of Earth and Environmental Engineering, Columbia Nanomechanics Research Center, Columbia University, New York, NY, USA
e-mail: xichen@columbia.edu

G. Z. Voyiadjis (ed.), *Handbook of Nonlocal Continuum Mechanics for Materials and Structures*, https://doi.org/10.1007/978-3-319-58729-5_23

cracking (HEC). HEC is an important issue in the development of hydrogen-based energy systems. Especially high-strength steels tend to suffer from HE cracking, which leads to a significant decrease in the mechanical properties of the steels, including the critical stress for crack initiation and resistance to crack propagation. For such materials integrity for HEC, convenient material testing is necessary. In this review, the first part describes new indentation methodology to evaluate threshold stress intensity factor K_{ISCC}, and the latter one is investigation into HEC morphology due to residual stress produced by indentation impression. Our findings will be useful for predicting K_{ISCC} for HE instead of conventional long-term test with fracture mechanics testing. It will also indicate the stress criterion of HE cracking from an indentation impression crater, when the formed crater (for instance due to shot peening or foreign object contact) is exposed to a hydrogen environment.

Keywords
Indentation · Hydrogen embrittlement cracking · Fracture strength · Finite element method · Cohesive zone model · Residual stress · High-strength steel

Introduction

Hydrogen embrittlement (HE) cracking is an important issue in the development of hydrogen-based energy systems. High-strength steels tend to suffer from HE cracking, which leads to a significant decrease in the mechanical properties of the steels, including the critical stress for crack initiation and resistance to crack propagation. It is well known that Mode-I tensile stress is responsible for HE cracking (Gangloff 2003). In contrast, degradation of the mechanical properties upon compression loading (e.g., macroscopic hardness) never occurs in a hydrogen environment (Reddy et al. 1992). In order to evaluate the structural integrity of the material with respect to hydrogen embrittlement, tensile or bending loading is usually employed for mechanical testing, producing tensile stress for HE cracking. In contrast, we found that Vickers indentation caused HE cracking in high-strength steel (Yonezu et al. 2010). This is due to the fact that the tensile stress field that develops around the indentation impression reaches the critical value for crack nucleation. Since indentation testing is a convenient method to probe the mechanical properties of materials compared to other mechanical testing methods, such as fracture mechanics testing under tensile or bending loading, the ability to evaluate HE cracking upon indentation is of importance.

In addition, indentation test induces permanent impression crater, which is similar with shot peening. Peening techniques make the material harder but in addition also introduce compressive residual stress. It is well known that compressive stress is effective in preventing HE cracking. With multiple shot impacts on the surface, the surface hardened layer expands plastically and compressive residual stress develops (Kobayashi et al. 1998). However, the boundary of the peening area (outside the crater) sometimes shows tensile stress (Klemenza et al. 2009). This could lead to mechanical degradation such as HE, stress corrosion cracking (SCC), and fatigue fracture. In fact, a single crater (permanent impression due to indentation) forms tensile stress around the impression, balanced by local compressive strain. Such a tensile stress may induce HE cracking, experienced in high-strength steels (Yonezu et al. 2010, 2012).

From the above engineering background, indentation method may bring the possibility to characterize HE cracking. In this review, the first topic is to explore the use of Vickers indentation testing for evaluating the susceptibility of steel to HE cracking. For this purpose, indentation test is applied to hydrogen-charged steel. The resistance to crack propagation, in particular, the threshold stress intensity factor (K_{ISCC}), is an important parameter for understanding the susceptibility to HE cracking. The second topic is to clarify HE cracking produced by residual stress of indentation impression. For this experiment, spherical impression crater is exposed in hydrogen environment, and HE cracking morphology and the mechanics are systematically investigated. This may become a potential fracture mode, when the steel forms a permanent crater on the surface due to surface treatment (e.g., shot peening) or other types of contact loading in a hydrogen environment. It is expected that the present knowledge is of importance and will significantly contribute to material/mechanical design in a hydrogen environment. In addition, the findings (of the first topic) allowed us to propose an alternative method to the fracture mechanics approach for HE cracking evaluation.

Evaluation of Threshold Stress Intensity Factor for HE Cracking

Materials and Experimental Methods

The material used in the present study was 18Ni maraging steel (350 ksi), which is a low-carbon martensitic steel. In order to achieve high strength, the steel was solution heat treated (820 °C for 3 h) followed by cooling in air to room temperature. Subsequently, the steel was aged at 500 °C for 4 h and air-cooled to room temperature. The mechanical properties of the steel are shown in Table 1, indicating that the yield stress and tensile strength were 2.40 and 2.45 GPa, respectively (Boyer and Gall 1985).

This steel has superior tensile strength, but is very susceptible to hydrogen embrittlement (Gangloff and Wei 1974; Pao and Wei 1977; Antlovich et al. 1980;

Table 1 Mechanical properties of 18Ni maraging steel (350ksi) employed in this study

Young's modulus E (GPa)	Poisson's ratioν	Yield strength σ_y (GPa)	Tensile strength σ_B (GPa)	Work hardening coefficient n	Strength for HE cracking σ_{scc} (GPa)
210	0.3	2.4[a]	2.45[a]	0.025	0.65[b]

[a]Boyer and Gall (1985) Metals Handbook
[b]Yonezu et al. (2010)

Reddy et al. 1992; Gangloff 2003; Tsay et al. 2005, 2008). The specimens in the study were disk-shaped plates with a diameter of 20 mm and thickness of 1 mm. After mechanically and electrochemically polishing the specimen surface, cathodic charging was carried out in order to introduce hydrogen into the material in a buffer solution of sodium acetate (0.20 mol/L CH_3COOH + 0.17 mol/L CH_3COONa, pH = 4.7) with a current density of 3 A/m^2. In order to vary the hydrogen content in the steel, the duration of hydrogen charging was varied from 48 to 72 h.

After cathodic charging, the total diffusible hydrogen content in the material was measured by thermal desorption spectroscopy (TDS-KU, ULVAC). Just after the cathodic charging, TDS measurements indicated C_H to be 33.7 ppm for 48-h charging and 49.1 ppm for 72-h charging. When the specimen with absorbed hydrogen is conditioned in air at room temperature, the hydrogen in the material diffuses out into the atmosphere as a function of the exposure time (degassing time). This suggests that C_H can be controlled by changing the degassing time. The discharged C_H as a function of conditioning time was also measured by TDS, indicating that the C_H for both specimens decreased with increasing degassing time (C_H was varied from 49.1 to almost 0 ppm). By referring to this hydrogen degassing behavior, the timing of the indentation test can be determined at a desired C_H in order to examine the relationship between the HE susceptibility and the C_H.

Indentation testing was carried out using a hydraulic servo-controlled fatigue testing machine equipped with a diamond Vickers indenter. The indenter impresses the specimen surface with up to a specific maximum indentation force at a rate of 1 N/s. Subsequently, the indentation force is reduced at a rate of 1 N/s until $F = 0$ N. The maximum force was varied from 100 to 300 N. The detailed conditions used will be described in each section below. After the indentation test, the impression morphology was observed using an optical microscope. These observations were conducted at least 1 h after the indentation test.

Experimental Results

Vickers indentation was applied to the specimen that had absorbed hydrogen as a result of cathodic charging for 72 h (C_H = 49.1 ppm). Figure 1 shows an optical micrograph of the impression and its surrounding showing that four cracks propagated from the corners of the impression. These are similar to surface cracks (radial and halfpenny-shaped cracks) observed in brittle materials subjected to Vickers indentation (Lawn et al. 1980; Niihara 1983; Cook and Pharr 1990). We previously reported that the cracks grew in a zigzag manner, indicating that they propagate along grain boundaries (Yonezu et al. 2010). Such intergranular cracks are often observed in maraging steel when subjected to hydrogen embrittlement (Gangloff 2003; Tsay et al. 2005, 2008). Furthermore, from observations of cross-sectional views, each crack was identified as a type of radial crack, which propagates along the radial direction from one corner of the indentation impression (Yonezu et al. 2010). In contrast, steel with no hydrogen absorption (as-received specimen: C_H less than 0.05 ppm) did not exhibit any cracking around the impression. We thus

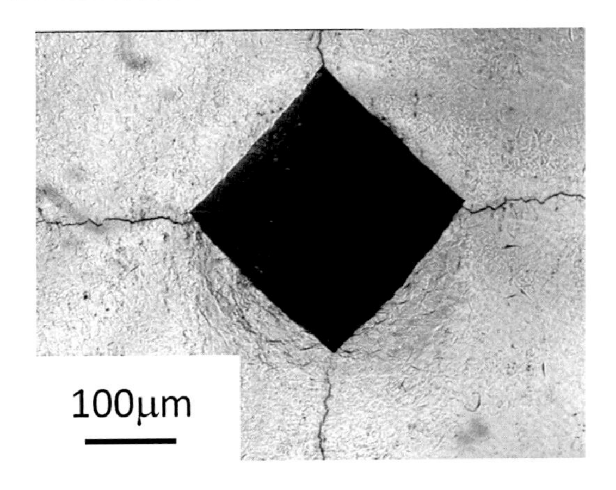

Fig. 1 Micrographs of the indent impression produced in a hydrogen-charged specimen tested with $F_{max} = 300$ N (hydrogen content is 49.1 ppm)

concluded that the crack formation in the present case was caused by hydrogen embrittlement (Yonezu et al. 2010).

Figure 2 shows the relationship between the surface crack length and the maximum indentation force. Here, the tested specimens were charged for 72 h, and all specimens had C_H of 49.1 ppm. The crack length (designed as c) is defined as the distance from the impression corner to the crack tip. The number of tests with $F_{max} = 100$, 200 and 300 N were three, five and nine, respectively. Their standard deviations are also plotted in the figure. It should be noted that for one indentation impression, the number of cracks formed was usually four (such as seen in Fig. 1). However, in some impressions (especially those produced with a smaller $F_{max} = 100$ N), the number of cracks tended to be two or three. Thus, the representative crack length for one test was assumed to be the averaged value, that is, the sum of the crack lengths divided by the number of cracks formed. Figure 2 reveals that the crack length c increases when F_{max} is larger. Figure 2 also shows the results for half of the diagonal length in the impression (designed as a) as a function of F_{max}, also indicating that the impression size is larger with increasing F_{max}.

In order to investigate the effect of hydrogen content on the crack size, indentation tests were conducted at different times after cathodic charging for 72 h. The value of F_{max} was set at 300 N for all tests. Figure 3 shows three micrographs of representative surface cracks in specimens with different hydrogen content which propagated from the impression corners; the tests were conducted at different discharge times (0 h (a), 43 h (b), and 119 h (c)) for hydrogen contents of 49.1, 26.9, and 19.5 ppm. Although F_{max} was set at the same value (300 N) for all the tests, the crack length was strongly dependent on C_H, showing that the length became longer when C_H was larger.

Figure 4 shows the crack length as a function of C_H for all specimens. Here, the specimens with $C_H > 20$ ppm were obtained from 72-h cathodic charging, while the results for C_H around 5 ppm were obtained from the 48-h charging specimen. It can be seen that the crack length became larger with increasing C_H. Figure 4 also shows the Vickers hardness values measured from the impression size (marked

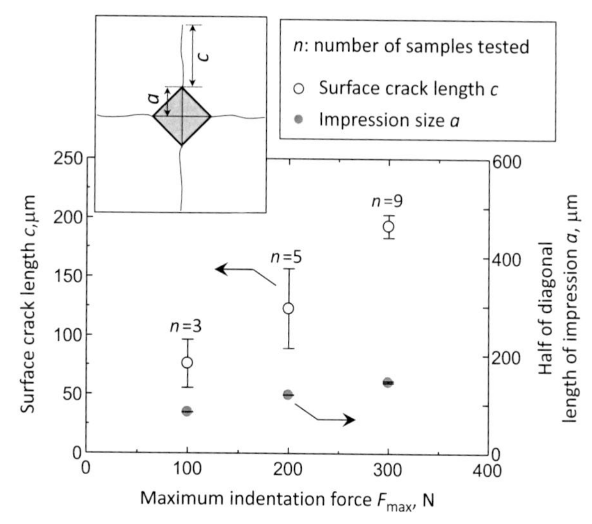

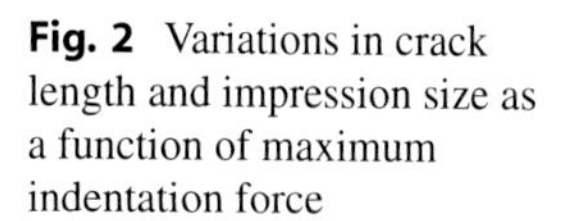
Fig. 2 Variations in crack length and impression size as a function of maximum indentation force

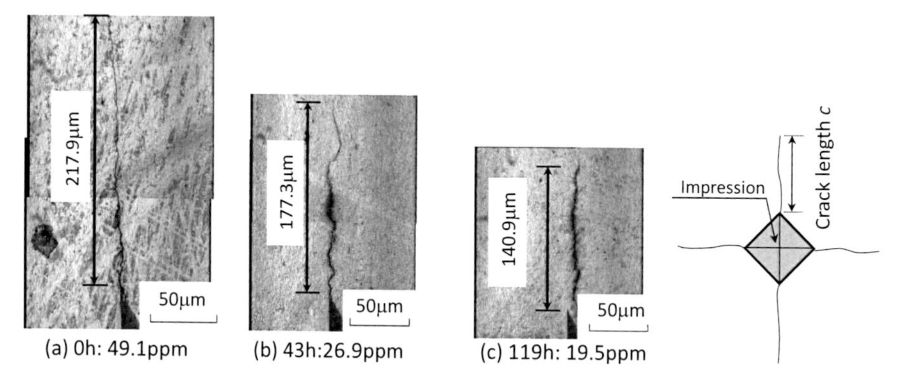

Fig. 3 Micrographs of the crack produced from an indent impression corner tested with $F_{max} = 300$ N. Three tests were conducted at different times after the hydrogen charging: (**a**) 0 h, (**b**) 43 h, and (**c**) 119 h. The hydrogen content was estimated to be (**a**) 49.1, (**b**) 26.9, and (**c**) 19.5 ppm

by a gray circle). The hardness did not depend on the C_H and showed an almost constant value equal to the hardness of the as-received steel with no hydrogen absorption ($HV = 675$) as indicated by the dotted line. Therefore, it is concluded that hydrogen in the steel influences the propagation of the Vickers indentation-induced crack, while the hardness related to the macroscopic elastoplastic properties in compression is not affected by the presence of hydrogen in the steel. Thus, hydrogen in the steel causes a decrease in the critical tensile strength, resulting in HE crack initiation and propagation. Propagation of the crack tip is expected to be suspended upon full unloading under the stress state, where the driving force of the crack propagation (due to the residual stress field from the indent formation) is equilibrated with the resistance to crack growth in the steel having hydrogen.

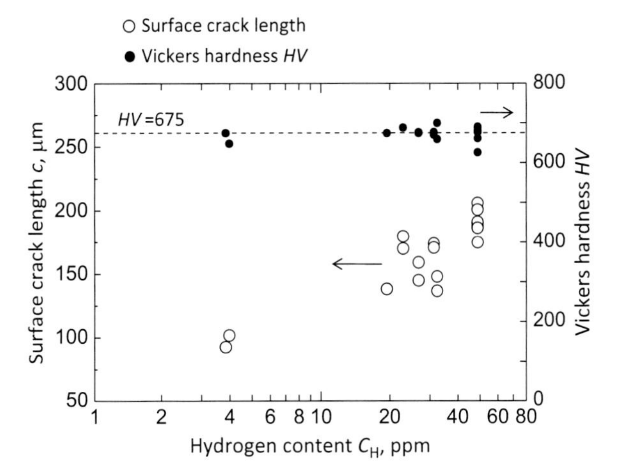

Fig. 4 Variations in radial surface crack length and Vickers hardness as a function of hydrogen content

In other words, the "final" crack length is dependent on resistance to HE crack propagation in the material.

Here, several questions regarding crack propagation arise, such as how the crack grows, what the growth rate is, and when crack propagation stops. In fact, K_{ISCC} tests based on fracture mechanics show that an HE crack starts to propagate when the stress intensity exceeds K_{ISCC} (the threshold resistance to HE crack growth), the growth rate rapidly increases, and the crack propagates stably at the constant crack growth rate (so-called plateau region). Finally, unstable fracture occurs at around K_{IC}. This implies that K_{ISCC} is much lower than K_{IC}. In other words, an HE crack cannot propagate when the stress intensity factor is lower than K_{ISCC}.

In our indentation tests, the total testing time (including both loading and unloading) was about 600 s, for the test condition of $F_{max} = 300$ N and loading rate of $dF/dt = 1$ N/s. If the crack initiates and propagates up to a length of 200 μm (see Fig. 3) during the indentation test, the average rate of crack growth may be about 3.33×10^{-7} m/s. In contrast, fracture mechanics tests yielded HE crack growth rates in the plateau region for 18Ni-300 (300 ksi grade) and 18Ni-250 (250 ksi grade) maraging steels of 5.0×10^{-5} (Sumitomo Precision Products CO. L.1992) and 1.11 $\times 10^{-7}$ m/s (Gangloff 2003), respectively. These values are of the same order as the average rate in the present study. This implies that the HE crack propagation due to indentation corresponds to the plateau region, indicating that the crack grows stably in the present tests.

We will next discuss when the crack stops propagating. Note that the measurement of crack length (in Fig. 3) in the present tests was conducted more than 1 h after the test. Considering the maximum crack length of about 200 μm for $F_{max} = 300$ N (see Fig. 3), assuming the crack continuously propagates until when the microscopic observations were conducted, the average crack growth rate is less than 6.7×10^{-8} m/s. This value is smaller than the stable rate in the plateau region (5.0×10^{-5} and 1.11×10^{-7} m/s for 18Ni-300 and 18Ni-250 maraging steels as mentioned above), suggesting that the crack in the present study underwent stable propagation. In addition, when we continually observed the crack length up to 100 h after the test, the length did not change. In this case, the crack growth rate was less than about 1.1×10^{-12} m/s, since the measurement resolution of the optical microscope was less than 0.4 μm. This rate is almost the same order as the threshold rate (2.8×10^{-12} m/s), when the value of K_{ISCC} was determined in the fracture mechanics test (Yamaguchi et al. 1997). Therefore, after full unloading, the driving force of the crack tip (stress intensity K) is considered to be less than the value of K_{ISCC}. Such a condition of $K \leq K_{ISCC}$ indicates that the HE crack stops propagating, resulting in the final crack length. In the next section, the stress field at the crack is computed to evaluate the threshold stress intensity of the final crack.

Numerical Analysis

Cohesive Zone Model

The stress field at the crack tip is a key issue, as expected from the above results, since the crack length (at full unloading) is dependent on the HE crack growth resistance of steel, K_{ISCC}. Note that the process of indentation loading/unloading remarkably changes the stress field, resulting in HE crack initiation and propagation (Yonezu et al. 2010). Therefore, such crack growth is required to be incorporated with stress analysis. We employed the finite element method (FEM) to compute the indentation stress field in conjunction with crack propagation.

One of the approaches to simulate crack propagation is the cohesive zone model (CZM). The CZM is theoretically well established and is proven to be applicable to both ductile and brittle materials (Barenblatt 1962; Tvergaard and Hutchinson 1992; Chandra et al. 2002; Xia et al. 2004; Hal et al. 2007; Olden et al. 2008). For instance, Mode-I fracture of a brittle coating on a silicon substrate (Xia et al. 2004), hydrogen embrittlement in a duplex stainless steel (Olden et al. 2008), and the interfacial fracture of an IC-interconnect (e.g., copper and low-K materials) (Hal et al. 2007) have been successfully investigated using the CZM.

Figure 5 shows a schematic of crack propagation incorporating the cohesive zone. The CZM essentially models the fracture process zone in a plane ahead of the crack tip. The zone is assumed to be subjected to cohesive traction. The model usually describes the gradual degradation of the adhesion between two regions along the crack propagation plane. The mechanical response of the cohesive zone obeys a traction-separation law that yields the relationship between the separation distance v of the two material faces at an interface and the traction stress σ acting between them. Figure 5 graphically explains the lumping of the nonlinear material response in the cohesive zone. By the action of the external force (i.e., remote stress/strain), the crack tip opens. The opening response at the actual crack tip (designated as v_B) obeyed the cohesive zone element with the traction-separation law, such as the

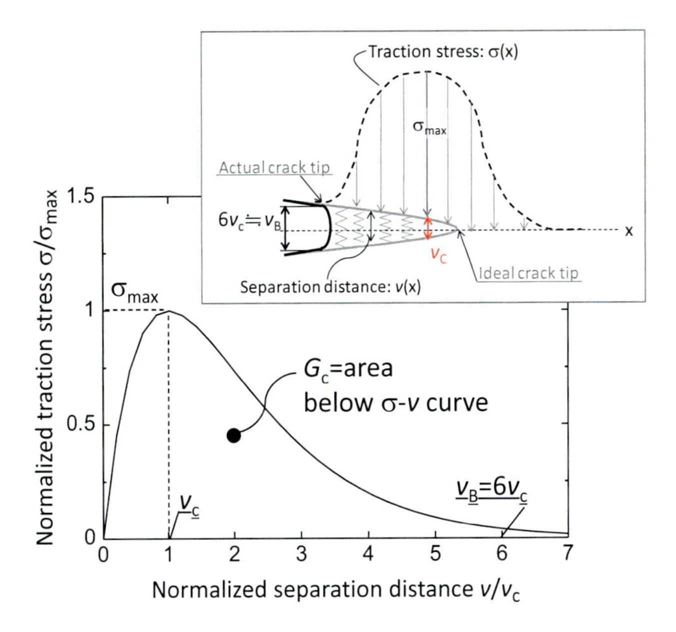

Fig. 5 Schematic of crack advancement with the cohesive zone model. The cohesive zone element having the traction-separation law exists ahead of the crack tip

bottom figure in Fig. 5. In the cohesive zone, after the acting stress σ exceeds the maximum cohesive stress σ_{max} (corresponding to the characteristic separation v_c), the resistance stress (traction force) dramatically decreases to almost zero with larger separation. This indicates that the cohesive zone ahead of the material crack tip opens, and crack propagation then occurs, resulting in the nucleation of new surfaces. Such a simulated crack is permitted to propagate along the specific direction along the cohesive zone element. As shown in Figs. 1 and 3, the surface crack propagates along the radial direction from the indent impression corner. The cohesive element is thus to be inserted in this plane.

Although numerous traction-separation laws for the cohesive zone element have been proposed (e.g., see reference (Chandra et al. 2002) for an overview), an exponential law, called the Smith-Ferrante type (Ortiz and Pandolfi 1999; Hal et al. 2007), is employed due to its simplicity. In fact, the exponential type has been used for various cases, including brittle Mode-I fracture, interfacial delamination, and hydrogen embrittlement cracking (Xia et al. 2004; Hal et al. 2007; Olden et al. 2008) as mentioned above. The constitutive behavior is expressed by the relationship between the normal traction stress σ and corresponding separation distance v across the cohesive element. This is given by

$$\sigma = \sigma_{max} \frac{v}{v_C} \exp\left(1 - \frac{v}{v_C}\right), \tag{1}$$

where σ_{max} is the maximum cohesive stress and v_c is the characteristic opening displacement corresponding to σ_{max} (see Fig. 5). Note that the tangential traction across this surface (i.e., shear component) is ignored in this study, since HE is caused by Mode-I loading and the driving force of the crack initiation and propagation is tensile stress (i.e., principal stress) (Yonezu et al. 2010). For the cohesive zone law, the normal work of fracture G_c corresponds to the area under the traction-separation law, as shown in this figure. It can be described as

$$G_C = \int_0^{\infty} \sigma(v)\, dv = \exp(1)\sigma_{max} v_C. \tag{2}$$

Therefore, Eqs. (1) and (2) lead to

$$\sigma = \frac{G_C}{{v_C}^2} \exp\left(-\frac{v}{v_C}\right) v. \tag{3}$$

This indicates that the traction-separation curve depends on two variables, the critical energy release rate G_c and v_c (or σ_{max}). When these two variables are known, the traction-separation curve can be drawn.

The critical energy release rate G_c is related to the critical stress intensity factor for crack growth of the material. When considering unstable crack growth, the fracture toughness K_{IC} can be obtained in the plane strain condition as shown below.

$$G_c = \frac{K_{IC}^2}{E}\left(1 - \upsilon^2\right) \tag{4}$$

Here, E and ν are the Young's modulus and Poisson's ratio, respectively. As mentioned above, the present study focuses on indentation HE cracking which is expected to be suspended under the condition of the driving force for crack propagation, $K \leq$ threshold stress intensity factor, K_{ISCC}. Thus, in Eq. (4), K_{IC} is hereafter replaced by K_{ISCC}.

Finite Element Method

A three-dimensional model of one-quarter of the specimen was created as shown in Fig. 6. The model contains more than 30,000 nodes and eight-node elements, and the part of the indenter contact and crack propagation (in the cohesive zone element) has a fine mesh. A mesh convergence study was carried out. The calculation was performed using commercially available FEM code (Marc and Mentat 2010.2). A rigid contact surface was used to simulate the rigid Vickers diamond indenter. Coulomb's law of friction was assumed with a friction coefficient of 0.15 (Bowden and Tabor 1950). The indenter penetrated to a maximum force of 300 N and was then withdrawn to zero. The mechanical properties of the as-received maraging steel, listed in Table 1, were used in the FEM computation because the macroscopic

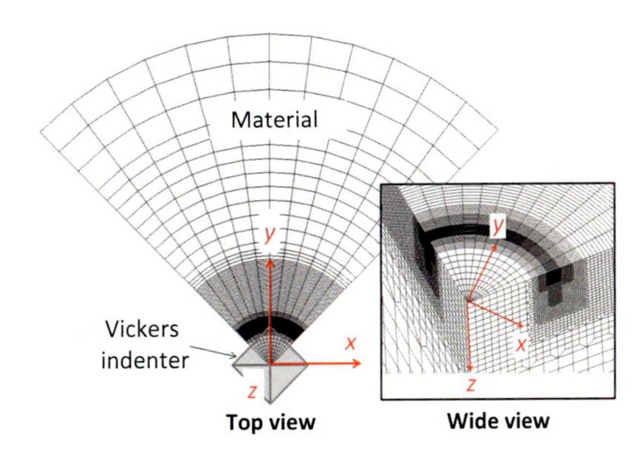

Fig. 6 Stress computation of Vickers indentation with the three-dimensional FEM model

hardness, which is related to the macroscopic plastic properties in compression, does not change with hydrogen content (see Fig. 4).

In order to simulate crack propagation, a cohesive zone element was inserted in the *y*-*z* plane, where the crack extends from the corner of the impression. Thus, during indentation loading/unloading, the simulated crack propagates along the *y*-z plane, and such crack propagation obeys the traction-separation law. We next show computational results for propagation of various cracks. In this study, the actual crack tip is taken as the point where the normal displacement v equals $6v_c$, at which point the normal stress σ on the separated surface is essentially zero (Xia et al. 2004) (see Fig. 5). Since the tensile stress σ_{xx} is responsible for the HE crack propagation, we focused on the distribution of σ_{xx} in this study.

As an example case, the model with crack growth resistance G_c of 108.3 J/m^2 was computed. The maximum stress σ_{max} in Eq. (1) was set to be 0.65 GPa, which was the critical stress for nucleating the present HE crack measured by Vickers indentation (Yonezu et al. 2010). Here, the value of G_c (108.3 J/m^2) for FEM can be converted to a K_{ISCC} value of 5.0 MPa m$^{1/2}$ from Eq. (4). Figure 7 shows a snapshot of the normal stress σ_{xx} distribution at maximum indentation force (a) and upon full unloading (b). For comparison, the model with *no* cohesive zone element was also computed, as shown in Fig. 7c, d. This model simulates the stress field due to indenter contact and does not induce any crack formation: the tensile stress σ_{xx} develops in the elastic field outside the finite plastic deformation region (local impression). Note that the magnitude of σ_{xx} in the fully unloaded condition (Fig. 7d) is higher than that at the maximum indentation force (Fig. 7c). This is because unloading reduces the elastic compressive stress due to the indenter contact, and therefore, σ_{xx} reaches a maximum value at full unloading.

In contrast, in Figs. 7a, b, the model with the cohesive zone element showed the release of tensile stress σ_{xx} (indicated by the dotted line in A) outside the impression (compressive plastic strain field). This is due to crack initiation and propagation. The FEM analysis showed the crack initiated at loading (Fig. 7a) and then propagated significantly under unloading (Fig. 7b). Finally, the crack length reaches a maximum upon full unloading.

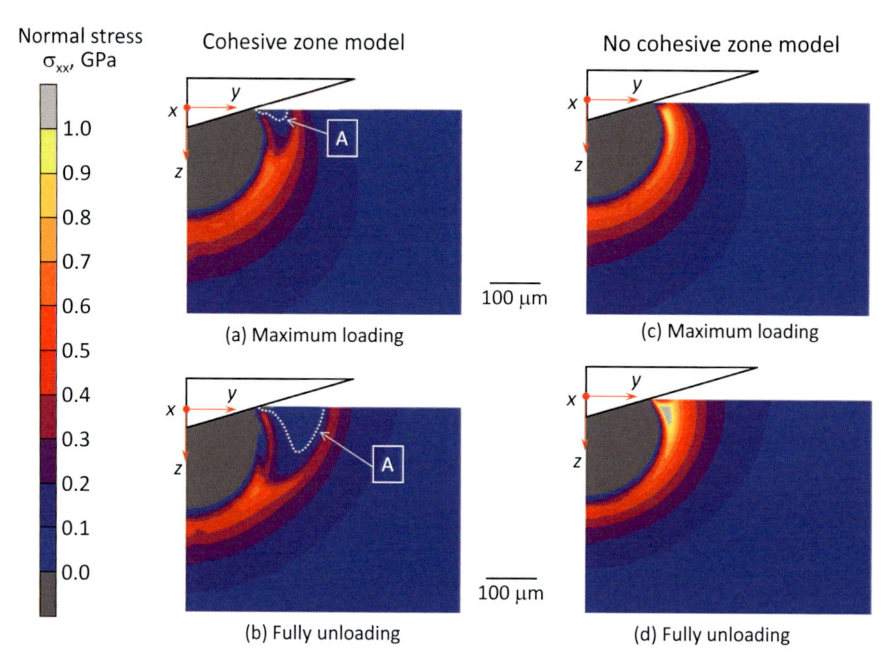

Fig. 7 Contour maps of σ_{xx} for the radial crack with the cohesive zone model (**a**) and (**b**) and without the cohesive zone element (**c**) and (**d**). Note Figs. (**a**) and (**c**) are shown at the maximum indentation force and (**b**) and (**d**) at the end of unloading

Crack Growth Resistance

To evaluate the crack growth resistance of the specimen with C_H of 49.1 ppm, an FEM study was carried out by changing K_{ISCC} from 3 to 8 MPa m$^{1/2}$, in order to analyze the relationship between K_{ISCC} and surface crack length. Here, the FEM model and its constitutive equation for the material were the same as the model described in Fig. 6, and the maximum indentation force was 300 N. Figure 8 shows the relationship between K_{ISCC} and the simulated surface crack length, c, indicating that the crack length is longer with a decrease in K_{ISCC}. The value of $K_{ISCC} \geq 8$ MPa m$^{1/2}$ produced no crack. In this figure, the actual crack length obtained from the indentation experiment (192.7 ± 9.7 μm from Fig. 2) is also shown by the broken line. Here, the thin lines indicate the range obtained from the standard deviation of the experimental results. Compared with the experiments, K_{ISCC} can be estimated to be 4.41 MPa m$^{1/2}$, in the range from 4.28 to 4.55 MPa m$^{1/2}$. It should be emphasized that the variation in the estimated values is not very large, compared with those obtained by the fracture mechanics K_{ISCC} test (Floreen 1978).

To verify the estimated K_{ISCC} obtained from the test with F_{max} = 300 N, the other test results with indentation forces of 100 and 200 N were analyzed. Similar to Fig. 9, the experimental crack length (from Fig. 2) was used to estimate the values of K_{ISCC}. The estimated K_{ISCC} gives almost the same value (about 4 MPa m$^{1/2}$ in Fig. 8). As shown in Fig. 3, the impression size as well as the crack length varied

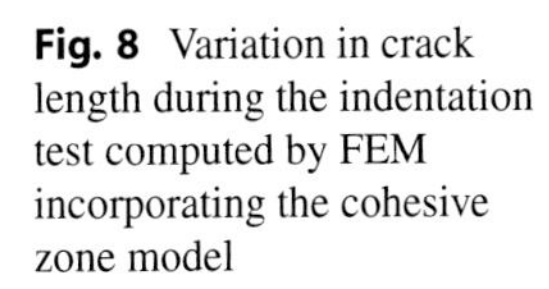

Fig. 8 Variation in crack length during the indentation test computed by FEM incorporating the cohesive zone model

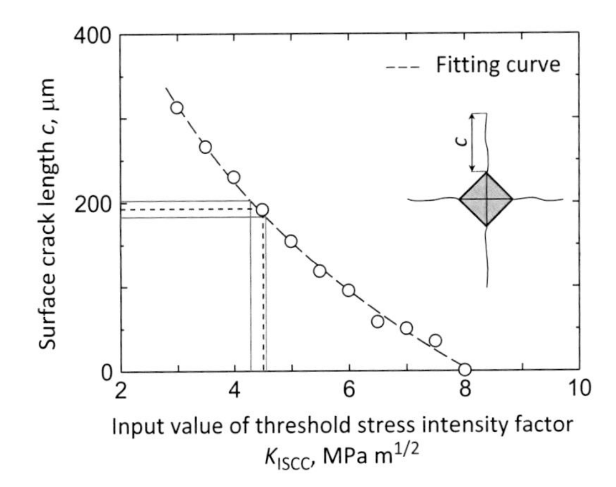

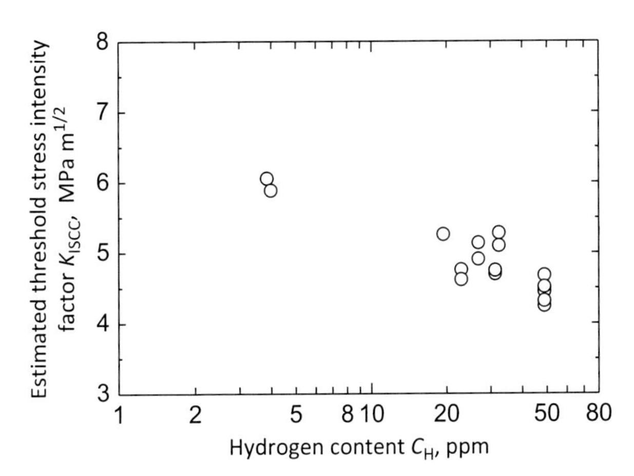

Fig. 9 Estimated threshold stress intensity factor K_{ISCC} as a function of the hydrogen content

depending on F_{max}. This indicates that the stress distribution (which is responsible for crack formation) is different. However, the estimated K_{ISCC} values were almost the same, suggesting that the present estimation is valid.

Figure 9 shows the K_{ISCC} value estimated by the present indentation method as a function of hydrogen content C_H. The K_{ISCC} values were calculated from the data as shown in Fig. 4. This suggests that the estimated values are a function of C_H, indicating that K_{ISCC} is lower with larger C_H. Such a trend was observed in previous studies on HE cracking evaluation for high-strength steels based on the fracture mechanics K_{ISCC} test (Yamaguchi et al. 1997; Gangloff 2003).

Table 2 summarizes the mechanical properties and threshold value of HE crack growth (i.e., threshold stress intensity factor for HE cracking, K_{ISCC}) obtained in the present study together with those for two different grades of maraging steels. The values of K_{ISCC} were obtained by the fracture mechanics test (K_{ISCC} test) in an H_2 environment for 18Ni-250 (250 ksi grade) and in 3.5% NaCl solution for 18Ni-300

Table 2 Comparisons of mechanical properties and threshold stress intensity factor K_{ISCC} among various 18Ni maraging steels

	18Ni maragingsteel grade	Yield stress (GPa)	Tensile strength (GPa)	Fracture toughness (MPa m$^{1/2}$)	K_{ISCC} (MPa m$^{1/2}$)
This study	350ksi	2.4	2.45	40	4.24–6.05[c]
Ref[a]	300ksi	2.07	2.17	75	5.5–11
Ref[b]	250ksi	1.65	1.72	120	20

[a]Sumitomo Precision Products CO, L. 1992, personal communication in 3.5% NaCl
[b]Gangloff (2003), in pressure H_2 ($P_{H2} = 0.17$ MPa)
cK_{ISCC} estimated by the present study

(300 ksi grade) steel. The detailed conditions are described elsewhere (Sumitomo Precision Products CO, L. 1992, personal communication; Gangloff 2003). Since higher strength (higher yield/tensile strength) usually makes the HE susceptibility higher, the value of K_{ISCC} in 300 ksi grade steel is lower than for the 250 ksi grade steel. Our estimation order (4.41–6.05 MPa m$^{1/2}$) is almost the same as the K_{ISCC} values for the other steels (from 5.5 to 11 MPa m$^{1/2}$ for the 300 ksi grade, and 20 MPa m$^{1/2}$ for the 250 ksi grade), although the steel grade and hydrogen content are different. Furthermore, other maraging steels with a yield stress of about 2 GPa were reported to have K_{ISCC} of 10 MPa m$^{1/2}$ (McEvily 1990) and 7.6 MPa m$^{1/2}$ (Floreen 1978), indicating that the values are also close to our estimated values. Thus, this suggests that our method based on indentation can indeed be used to evaluate K_{ISCC}.

Compared with the indentation tests, the fracture mechanics K_{ISCC} test requires much more effort for sample preparation and setting up of the testing method, etc. For instance, the value of K_{ISCC} is usually obtained when crack growth has stopped, with decreasing driving force for crack growth. To recognize the termination of crack propagation, the experiment needs to last a minimum of 1000 h. Thus, the K_{ISCC} test usually requires a long time. In contrast, our method based on indentation saves time and potentially is an alternative technique.

Mechanism of HE Cracking from Indentation Impression

Materials and Experimental Methods

The above section is to investigate whether indentation cracking occurs or not, when the indentation loading is applied to hydrogen-charged steel. On the contrary, indentation cracking from permanent impression due to hydrogen absorption (i.e., when the impression crater is exposed in hydrogen environment) is clarified in this section. We used the same steel, i.e., 18Ni maraging steel (called 350 ksi (Boyer and Gall 1985)), which is a low-carbon martensitic steel. This section used a spherical diamond indenter with a 400 μm radius, which is quite a bit larger than the grain size of the maraging steel (about 8 μm observed in Yonezu et al. 2010) used in this work. The typical shot peening crater has a diameter of several hundred μm, and in this case, we can ignore the effect of individual crystal plasticity on the stress analysis (based on continuum mechanics). Thus, a relatively large impression crater is required. The maximum indentation forces are 100, 200, and 300 N. The loading rate is set to 1 N/s for all tests, thereby simulating quasi-static contact loading.

After the indentation test to create a permanent impression, cathodic charging was carried out to introduce hydrogen into the material using a phosphate buffer solution (2.6 w/v% K_2HPO_4, 0.2 w/v% NaOH, pH = 6.5) with a current density of 5 A/m^2. In order to vary the hydrogen content in the steel, the duration of hydrogen charging was varied, and two durations of 6 and 48 h were used. The impression morphology was observed using an optical microscope (BXM-N33 M, Olympus Corp.). After the hydrogen charging, the sample was cleaned in ethanol in an ultrasonic cleaner (UT-206, SHARP Corp) for more than 1 h. Some specimens were mechanically cleaned along with fine polishing of buffing in order to remove the oxide film (formed during hydrogen charging) for clear observations. We were able to confirm that the length of the crack did not change before and after the fine polishing to remove the oxide film. Observations of the impression were conducted at least 1 h after the hydrogen charging. We confirmed that the hydrogen crack completely stops propagating after hydrogen charging.

Experimental Results

Indentation tests with different maximum indentation forces of F_{max} = 100, 200 and 300 N were carried out. Cathodic charging of the samples was then carried out for 6 and 48 h. For each test, about 30 impressions were performed. Figure 10 shows optical micrographs of representative indentation impressions and the surrounding area for hydrogen charging of 48 h. These figures separately show the results as a function of hydrogen charging time and F_{max} value. In Fig.10a of F_{max} = 100 N test, the impressions show short cracks (less than about 50 μm) from the rim. Indeed, it is difficult to observe these, since the crack width is very narrow. In Fig.10b of F_{max} = 200 N tests, however, long cracks are clearly observed from the rim. The cracks propagate radially from the impression, indicating a radial crack. Short cracks (similar to Fig.10a) are also observed in these samples. Figure 10c (of F_{max} = 300 N test) show the same trend, indicating two types of cracks. It should be noted that the steel with no hydrogen absorption (before hydrogen charging) did not exhibit any cracking around the indentation impression (Yonezu et al. 2010) (as explained before). The present results indicate that permanent indentation impressions in the steel used absorbed hydrogen-induced short cracks and sometimes long cracks. Although the pictures of 6-h charging tests are omitted, it is found that the crack morphology also changes depending on the hydrogen content and applied indentation force (Niwa et al. 2015).

Figure 11 shows the frequencies of different crack lengths for each test condition. Similar to the series, Figs. 10 and 12 show results separately for different maximum

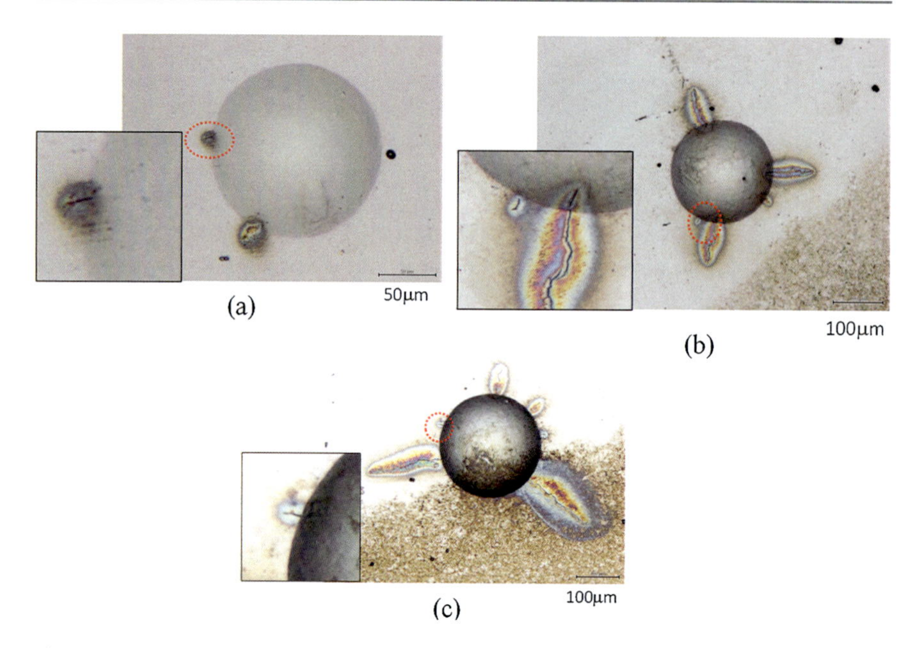

Fig. 10 Micrographs of an impression crater and its surrounding area. The hydrogen charging time is 48 h. The maximum indentation force $F_{max} = 100$ N for (**a**), 200 N for (**b**), and 300 N for (**c**)

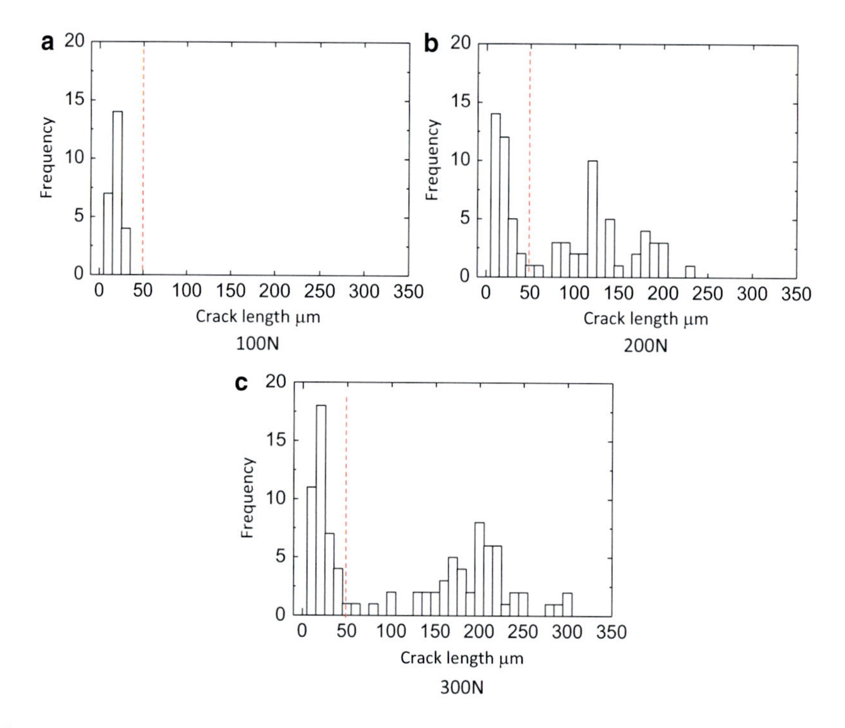

Fig. 11 Frequency of crack lengths from the indentation impressions for $F_{max} = 100$ N (**a**), 200 N (**b**), and 300 N (**c**) (H_2 charging time, 48 h)

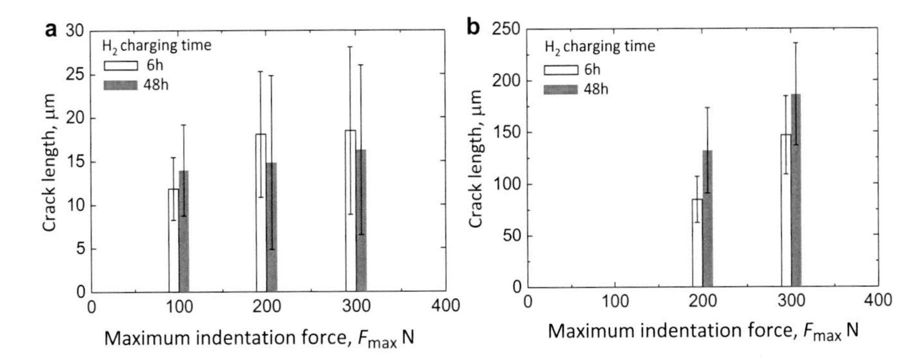

Fig. 12 Changes in crack length (for short cracks (**a**) and long cracks (**b**)) as a function of maximum indentation force

indentation force F_{max} (100, 200, and 300 N). For the $F_{max} = 100$ N test (Fig. 11a), short cracks with length less than 50 μm were observed for both charging times. On the other hand, the tests for $F_{max} = 200$ and 300 N showed two peaks in the distributions, where both short and long cracks exist. As also seen in Fig. 10, the two types of cracks (short and long cracks) can be distinguished, and the boundary between them is set to 50 μm, as shown by the dashed vertical line in Fig. 11.

Subsequently, we measured the crack length for both crack types, as shown in Fig. 12. The short crack length (Fig. 12a) seems to be independent of F_{max} and hydrogen charging time; however, the data shows large scatter. In contrast, Fig. 12b shows that the long crack length has strong dependency on F_{max}. In addition, the crack length is dependent on the hydrogen charging time, and the crack length is longer with larger charging time. The stress distributions and criteria were clarified in the study using finite element simulations, as described in the next section.

Discussion

Mechanism of Crack Propagation

In order to investigate crack formations of Fig. 10, the cohesive zone model (CZM) was employed in FEM to compute the stress field in conjunction with radial crack propagation. As mentioned above (section "Cohesive Zone Model"), this study employed an exponential law (called the Smith-Ferrante law) due to its simplicity. This law requires two independent material parameters, i.e., the maximum stress σ_{max} and the crack growth resistance K_C. The σ_{max} roughly corresponds to the critical stress for crack nucleation. As discussed before, the critical stress (σ_{max}) is estimated to be 0.65 GPa in the present steels which suffer from hydrogen embrittlement (Yonezu et al. 2010, 2012, 2015). In contrast, it is well known that Kc (which is similar to K_{ISCC} as discussed earlier) is strongly dependent hydrogen accumulation (Yamaguchi et al. 1997; Yonezu et al. 2012), and thus K_{ISCC} is not known. In fact, this study explores the actual K_{ISCC} with comparison between the

FEM computation and experiment (see later, section "Crack Length and Stress Intensity Factor").

We created a three-dimensional FEM model of one-quarter of the specimen (similar with Fig. 6). One CZM element for crack propagation is inserted along the center line (at a 45-degree angle from the side surface). It is noted that the situation of crack propagation is different from Fig. 6, because HE crack propagates from permanent impression crater (i.e., crack propagation starts "post" indentation). A permanent impression is first made by spherical indentation with $F_{max} = 100$, 200 and 300 N. After full unloading, the cohesive zone element operates, such that the crack starts propagating. We simulated crack propagation by the residual stress developed around the impression. To make the indenter impression, a rigid spherical indenter was used in a similar manner to that shown in Fig. 6. The mechanical properties used for the FEM are listed in Table 1.

Figure 13 shows representative contour maps of the crack nucleation area upon full unloading. The figure shows cross-sectional views along the crack propagation. For all models, the input value of the stress intensity factor was set to 3.5 MPa $m^{1/2}$, as a representative case. The maximum indentation force was changed to $F_{max} = 300$ N, 200 N, and 100 N, and the results are shown separately in Fig. 13a, b, c, respectively. In the figures, the yellow area indicates the crack nucleation area (i.e., the crack surface). Here, the yellow region defines the crack nucleation area where the CZM interface completely opens, and there is no traction force between the CZM interfaces. It is found that the crack length strongly depends on F_{max}. In particular, the test with $F_{max} = 100$ N produces very small cracks, whose length is about 29 μm (categorized as a "small crack" in this study). On the other hand, the larger F_{max} tests ($F_{max} = 200$ and 300 N) produced long cracks, which were very deep. This computational result corresponds to the experimental results in Figs. 10 and 11.

However, the larger F_{max} tests ($F_{max} = 200$ and 300 N) not only formed long cracks but also small cracks (even if the circumferential stress surrounding the indenter impression was uniform). The question arises as to why two different crack types exist (i.e., the larger F_{max} produces both short and long cracks). The mechanism of such a "multiple crack" scenario was investigated as discussed in the next section.

Mechanism of Multiple Crack Formation

Multiple crack formation with two types of cracks with different lengths was observed, as discussed above. We created a simple FEM model for simulating such a multiple crack propagation. Figure 14 shows a three-dimensional model with half size. Two CZM elements are inserted along the radial direction, so that it can simulate radial crack propagation. Similar to Fig. 13, when the indenter is completely withdrawn, the CZM element starts operating, such that we can simulate crack propagation from the residual stress around the indenter impression. To make the indenter impression, a rigid spherical indenter was used, and the maximum force F_{max} is 300 N. The three-dimensional half model in Fig. 14 comprises 53,200 eight-node elements. The mechanical properties used for the FEM are listed in Table 1.

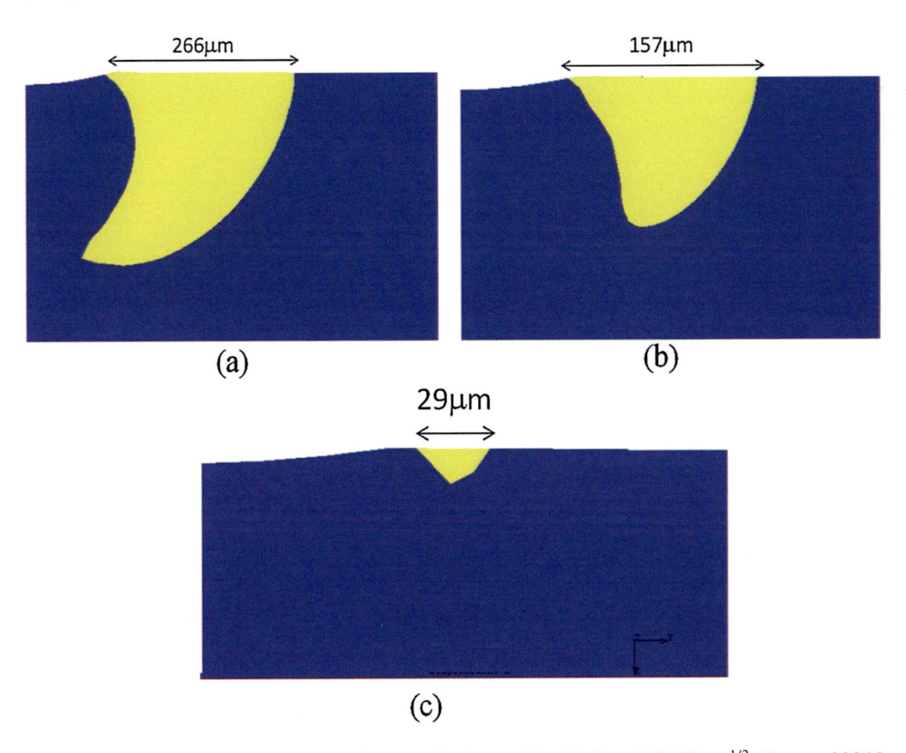

Fig. 13 Contour map of crack nucleation area in the model with $K = 3.5$ MPa m$^{1/2}$. $F_{max} = 300$ N (**a**), 200 N (**b**), and 100 N (**c**)

As shown in Fig. 14, two CZM elements (designated as CZM① and CZM②) are inserted at a certain angle θ. We investigated the relationship between the crack propagation behavior and the crack angle θ, which can be influenced by the interaction of multiple cracks. Indeed, when a crack propagates, the stress around the crack is released, and it may be difficult for a new crack to propagate around that area. However, with increasing distance from the propagated crack, the stress recovers to the original state, resulting in new crack propagation. It should be noted that the distance in this case is the angle in the circumferential direction, and in the present study, this crack angle was set from 30 to 90 degrees, as shown in Fig. 14.

As mentioned above, we experimentally observed the occurrence of both long and short cracks together. In the present case, it seemed that a long crack was generated first, resulting in the surrounding stress field becoming weak. Then, the second crack near the long crack becomes short, while the crack far from the first long one becomes long (i.e., it recovers to the length of the first long crack). The crack length might be dependent on the angle from the first long crack. Based on this assumption, after full unloading, CZM① operates first, followed by the operation of CZM②. Such a "delay" in the computation process is controlled by

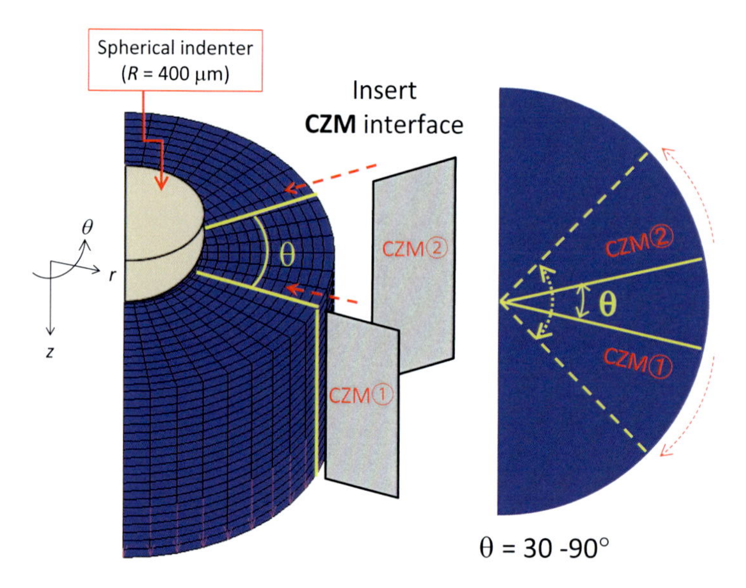

Fig. 14 Three-dimensional FEM model (half model) with cohesive zone element (CZM) for simulating radial crack propagation. This model simulates multiple crack propagation using CZM and CZM whose angle is changed from 30 to 90 degrees

the step increment in the FEM analysis. Of course, if CZM① and CZM② are generated at the same time, their lengths must be the same. In reality, however, it is unlikely to guess that all cracks are generated at the same time during the hydrogen charging, since crack initiation is strongly dependent on both inherent defects (which are present inhomogenously) and the hydrogen accumulation state. Thus, we hypothesize that multiple cracks initiate separately at different times, i.e., CZM① is the "primary crack" and CZM② is the "secondary crack" (in situ monitoring of hydrogen cracking cannot yet be conducted in this study). Note that the material parameters of CZM① and CZM② are identical.

Using the model shown in Fig. 14, we investigated changes in the crack length of CZM① and CZM② as a function of the angle between the two cracks. As a representative case, the threshold stress intensity factor of the crack growth K_{ISCC} was set to 4.5 MPa m$^{1/2}$ for the CZM parameter. Figure 15 shows the results for the computed crack length for two CZM elements. The figure shows top views of the contour map, with the crack propagation area shown as yellow. Figure 15a shows the result for the 30-degree angle, while the result for the 90-degree angle is shown in Fig. 15b. When the angle is 90 degrees (larger angle), the crack lengths of CZM① and CZM② are identical. In contrast, when the angle is 30 degrees, the secondary crack (CZM②) becomes significantly shorter.

We next investigated how the crack length (especially of CZM②) is changed by the angle between the cracks. The results are shown in Fig. 16. It can be seen that the primary crack length of CZM① has little dependence on the angle, while

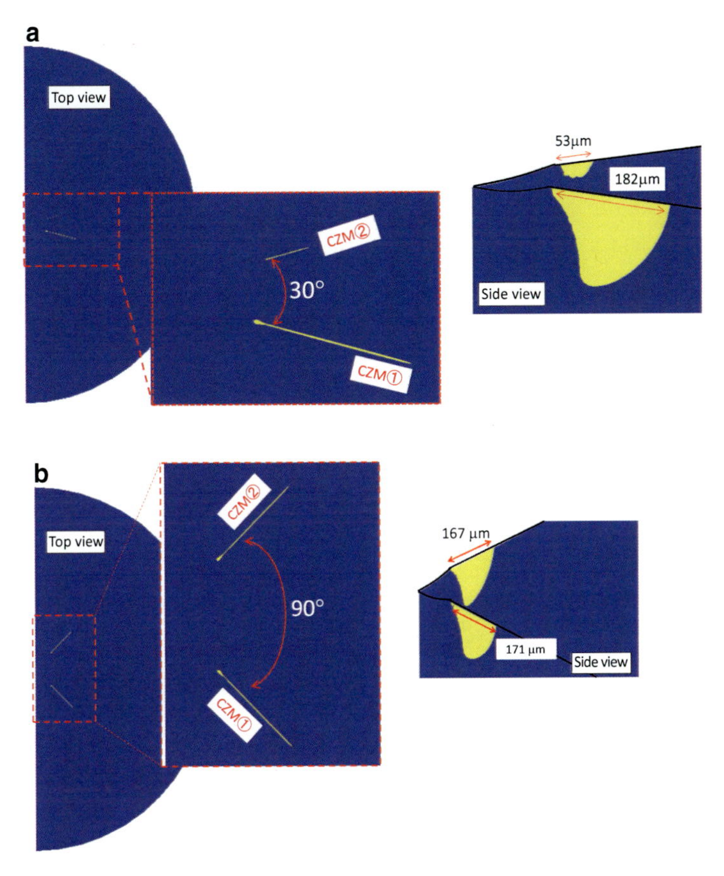

Fig. 15 Contour map of crack nucleation area, showing the multiple crack propagation (crack angle of 30 degrees (**a**) and 90 degrees (**b**))

the secondary crack (CZM②) is strongly dependent on the angle. At an angle of 90 degrees, the length of CZM② agrees well with that of CZM①. This suggests that at 90 degrees, multiple cracks do not interact with each other. However, at a smaller angle (when the position of CZM② is closer to CZM①), the length of CZM becomes significantly shorter. Finally, the length of CZM at about 30 degrees becomes less than 50 μm, and angles smaller than 30 degrees do not produce any cracks in this model. This might be due to the fact that the driving force for the CZM crack becomes small owing to propagation of the primary crack (CZM). Thus, the effect of primary crack generation becomes significant when the crack angle is small. It can therefore be concluded that the above discussion is the mechanism for the appearance of a short crack together with a long crack. It should be noted that this trend (crack length vs. crack angle) is dependent on the crack growth resistance, K_{ISCC}. It is expected that the case of smaller K_{ISCC} leads to a smaller crack angle, while a larger K_{ISCC} leads to a larger angle.

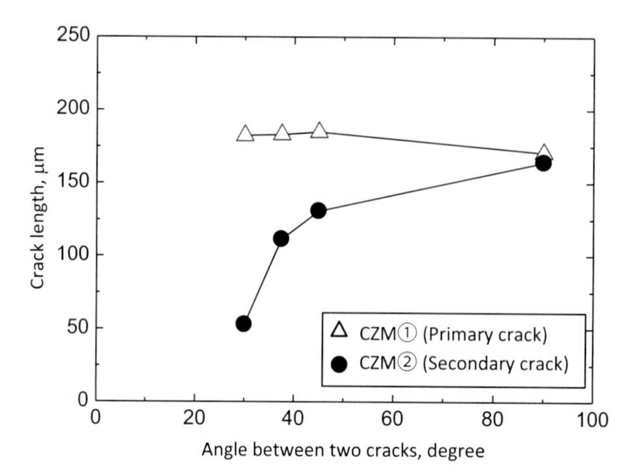

Fig. 16 Simulated crack length with respect to the angle between the two cracks (CZM and CZM)

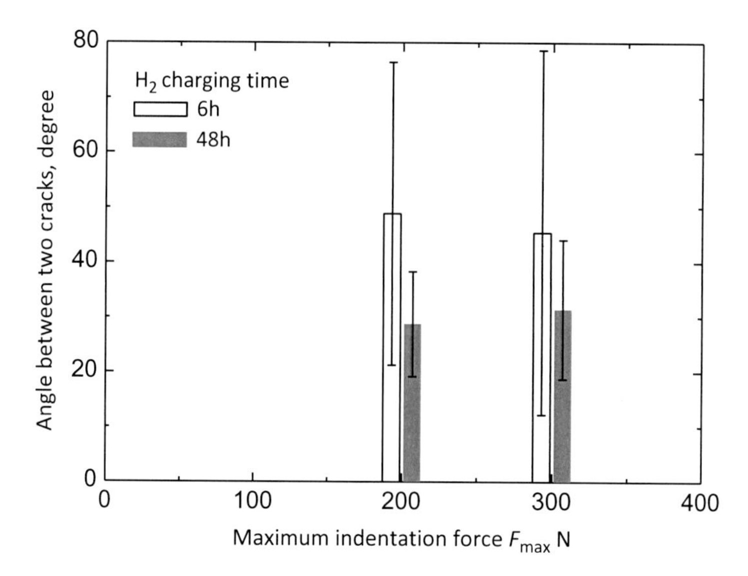

Fig. 17 Experimental result of the angle between the two cracks (short and long cracks)

Experimentally, we measured the angle between the long and short cracks for all indenter impressions. In this study, we only measured the minimum angle between the long and short cracks as the crack angle θ. The results are shown in Fig. 17 as a function of the applied F_{max} and the hydrogen charging time. Although the data show large scatter, the average crack angle θ was about 45 degrees for hydrogen charging for 6 h and 30 degrees for 48-h hydrogen charging. Although θ is dependent on K_{ISCC}, the experimental data are similar to the computational results in Fig. 17, especially for the 48-h charging time (the angle is about 30 degrees).

Crack Length and Stress Intensity Factor

We finally discuss the actual crack length from experimental HE cracking based on stress analysis, especially the stress intensity factor around the crack tip. Here we focus on the "long crack" (see Fig. 16), since it is not influenced by other crack generation. Similar with Fig. 8 of Vickers indentation, we investigated the effect of the input value of the stress intensity factor (crack growth resistance) K_{ISCC} on the crack length. We conducted several computations in this study with various values of the crack growth resistance (stress intensity factor K was changed from 3.0 to 5.5 MPa $m^{1/2}$ in steps of 0.5 MPa $m^{1/2}$) for both the $F_{max} = 200$ N and 300 N tests. As expected, the crack length is strongly dependent on the input K_{ISCC}, indicating that the crack length is longer when K_{ISCC} is smaller (like Fig. 8). In addition, a strong dependence of F_{max} on the crack length was observed, i.e., with larger F_{max}, the radial crack becomes longer. This relationship can yield the actual K_{ISCC} when the crack length is known. Therefore, the experimental crack length is substituted in this relationship in order to estimate the actual K_{ISCC}.

The estimated values are investigated as a function of the maximum indentation force, F_{max}. This indicates that the estimated K_{ISCC} depends on the hydrogen charging time (i.e., hydrogen content), while it is not dependent on F_{max}. For the hydrogen charging time of 6 h, K_{ISCC} is estimated to be 4.9 MPa $m^{1/2}$, while for the charging time of 48 h, K_{ISCC} is 4.3 MPa $m^{1/2}$. As mentioned in section "Crack Growth Resistance," the K_{ISCC} via Vickers indentation is estimated from 4 to 6 MPa $m^{1/2}$ (depending on the hydrogen content). As shown in Table 2, furthermore, fracture mechanics tests for HE cracking reported the values of K_{ISCC} to be in the range of 5.5 to 11 MPa $m^{1/2}$ (although the steel grade and hydrogen content in these tests are different from those in the present study). These values are in reasonable agreement with our estimation, suggesting that the present crack length (long crack) is governed by the threshold stress intensity factor K_{ISCC} for HE. Thus, the present findings may be useful for prediction of HE crack morphology from indentation impression.

Conclusion

This study investigated indentation-induced cracking in a maraging steel due to hydrogen embrittlement (HE). This review addressed two topics. The first one is that Vickers indentation was applied to the hydrogen-charged steel, such that we explored new evaluation method for HE susceptibility. The second topic is to investigate HE crack morphology from the residual stress that develops around a spherical indentation impression crater.

In the first topic, it is discovered that Vickers indentation induces hydrogen embrittlement cracking. The surface crack length is dependent on the hydrogen content in the steel. Since the stress intensity factor is equilibrated with the threshold stress intensity factor, K_{ISCC}, the stress field upon indentation in conjunction with crack growth was computed by the finite element method (FEM) incorporating a cohesive zone model. It was found that the simulated crack starts to grow under loading and then propagates under unloading. The crack length finally reaches a maximum value at full unloading. The FEM with different values of the crack growth resistance (K_{ISCC}) was used to obtain the relationship between K_{ISCC} and surface crack length. By substituting the actual crack length in the relationship, K_{ISCC} could be estimated. The estimated values were found to agree with the K_{ISCC} values measured by fracture mechanics testing (K_{ISCC} test). Furthermore, our method showed a dependence of hydrogen content on K_{ISCC}. This trend is observed when fracture mechanics K_{ISCC} tests are applied to high-strength steel in a hydrogen environment. Such a HE evaluation is usually conducted by K_{ISCC} tests that require significant effort for specimen preparation and setting up of testing, etc. However, the indentation-based evaluation method proposed in this study is very simple and convenient. The present approach is an alternative for evaluation of HE cracking.

The second topic clarified the stress criterion of hydrogen embrittlement cracking (i.e., HE crack initiation and propagation) from the residual stress of indentation impression. Namely, when spherical impression crater is exposed in hydrogen environment, how HE cracking propagates from the crater was investigated. The HE crack propagates radially, and the crack length is dependent on maximum indentation force and critical crack growth resistance K_{ISCC} (which is changed by the hydrogen content). In addition, a larger indentation force results in a complicated HE fracture, showing both long and short cracks. In this case, the appearance of the short cracks is affected by stress release of a neighboring long crack. These findings in our systematic investigation may be useful to predict damage of HE cracks, produced by residual stress in the permanent impression. Especially, this will be useful to determine how the size of an impinging foreign object (indenter size and shape) and the loading magnitude (maximum indentation force) induce HE damage, including crack length. This may become an indicator for HE upon permanent impression crater/dimple formation, such as by shot peening and localized contact loading.

Acknowledgment We would like to thank Professor Kohji Minoshima (Osaka University) for his guidance. This work is supported by JSPS KAKENHI (Grant No. 22760077 and 26420025) from the Japan Society for the Promotion of Science (JSPS) and Research Grant of Suga Weathering Technology Foundation (SWTF).

References

S.D. Antlovich, T.R. Risbeck, et al., The effect of microstructure on the fracture toughness of 300 and 350 grade maraging steels. Eng. Frac. Mech. **13**, 717–739 (1980)

G.I. Barenblatt, Mathematical theory of equilibrium cracks. Adv. Appl. Mech. **7**, 56–129 (1962)

F.P. Bowden, D. Tabor, *The Friction and Lubrications of Solids* (Oxford University Press, Oxford, 1950)

H.E. Boyer, T.L. Gall, *Metals Handbook Desk Edition Metals Park* (American Society for Metals, Ohio, 1985)

N. Chandra, H. Li, et al., Some issues in the application of cohesive zone models for metallic–ceramic interfaces. Int. J. Solids Struct. **39**, 2827–2855 (2002)

R.F. Cook, G.M. Pharr, Direct observation and analysis of indentation cracking in glasses and ceramics. J. Am. Ceram. Soc. **73**, 787–817 (1990)

S. Floreen, Maraging steels/metals handbook 9th edn. vol 1, in *Properties and Selection: Irons and Steels Metals Park*, (American Society for Metals, Ohio, 1978)

R.P. Gangloff, *Hydrogen Assisted Cracking of High Strength Alloys* (Elsevier Science, New York, 2003)

R.P. Gangloff, R.P. Wei, Gaseous hydrogen assisted crack growth in 18 nickel maraging steels. Scri. Metall. **8**, 661–668 (1974)

B.A.E. Hal, R.H.J. Peerlings, et al., Cohesive zone modeling for structural integrity analysis of IC interconnects. Microelectron. Reliab. **47**, 1251–1261 (2007)

M. Klemenza, V. Schulzea, et al., Application of the FEM for the prediction of the surface layer characteristics after shot peening. J. Mater. Process. Technol. **209**, 4093–4102 (2009)

M. Kobayashi, T. Matsui, et al., Mechanism of creation of compressive residual stress by shot peening. Int. J. Fatigue **20**(5), 351–357 (1998)

B.R. Lawn, A.G. Evans, et al., Elastic/plastic indentation damage in ceramics: the median/radial crack system. J. Am. Ceram. Soc. **63**, 574–580 (1980)

A.J. McEvily, *Atlas of Stress Corrosion and Corrosion Fatigue Curves* (ASM International, Ohio, 1990)

K. Niihara, A fracture mechanics analysis of indentation-induced Palmqvist cracks in ceramics. J. Mater. Sci. Lett. **2**, 221–223 (1983)

M. Niwa, T. Shikama, et al., Mechanism of hydrogen embrittlement cracking produced by residual stress from indentation impression. Mater. Sci. Eng. A **624**, 52–61 (2015)

V. Olden, C. Thaulow, et al., Application of hydrogen influenced cohesive laws in the prediction of hydrogen induced stress cracking in 25% Cr duplex stainless steel. Eng. Fract. Mech. **75**, 2333–2351 (2008)

M. Ortiz, A. Pandolfi, Finite-deformation irreversible cohesive elements for three-dimensional crack-propagation analysis. Int. J. Num. Methods Eng. **44**, 1267–1282 (1999)

P.S. Pao, R.P. Wei, Hydrogen assisted crack growth in 18Ni (300) maraging steel. Scr. Metall. **11**(6), 515–520 (1977)

K.G. Reddy, S. Arumugam, et al., Hydrogen embrittlement of maraging steel. J. Mater. Sci. **27**, 5159–5162 (1992)

L.W. Tsay, Y.F. Hu, et al., Embrittlement of T-200 maraging steel in a hydrogen sulfide solution. Corros. Sci. **47**, 965–976 (2005)

L.W. Tsay, H.L. Lu, et al., The effect of grain size and aging on hydrogen embrittlement of a maraging steel. Corros. Sci. **50**, 2506–2511 (2008)

V. Tvergaard, J.W. Hutchinson, The relation between crack-growth resistance and fracture process parameters in elastic plastic solids. J. Mech. Phys. Solids **40**, 1377–1397 (1992)

Z. Xia, W.A. Curtin, et al., A new method to evaluate the fracture toughness of thin films. Acta Mater. **52**, 3507–3517 (2004)

Y. Yamaguchi, H. Nonaka, et al., Effect of hydrogen content on threshold stress intensity factor in carbon steel in hydrogen-assisted cracking environments. Corrosion-NACE **53**(2), 147–155 (1997)

A. Yonezu, M. Arino, et al., On hydrogen-induced Vickers indentation cracking in high-strength steel. Mech. Res. Comm. **37**, 230–234 (2010)

A. Yonezu, T. Hara, et al., Evaluation of threshold stress intensity factor of hydrogen embrittlement cracking by indentation testing. Mater. Sci. Eng. A **531**(1), 147–154 (2012)

A. Yonezu, M. Niwa, et al., Characterization of hydrogen-induced contact fracture in high strength steel. J. Eng. Mater. Technol. **137**, 021007 (2015)

Continuous Stiffness Measurement Nanoindentation Experiments on Polymeric Glasses: Strain Rate Alteration

9

George Z. Voyiadjis, Leila Malekmotiei, and Aref Samadi-Dooki

Contents

Abstract

In many studies using continuous stiffness measurement (CSM) nanoindentation technique, it is assumed that the strain rate remains constant during the whole experiment since the loading rate divided by the load ($\dot{P}/P$) is considered as a constant input parameter. Using the CSM method, the soundness of this assumption in nanoindentation of polymeric glasses is investigated by conducting a series of experiments on annealed poly(methyl methacrylate) (PMMA) and polycarbonate (PC) at different set $\dot{P}/P$ values. Evaluating the variation of the

G. Z. Voyiadjis (✉) · L. Malekmotiei
Department of Civil and Environmental Engineering, Louisiana State University, Baton Rouge, LA, USA
e-mail: voyiadjis@eng.lsu.edu; lmalek1@lsu.edu

A. Samadi-Dooki
Computational Solid Mechanics Laboratory, Department of Civil and Environmental Engineering, Louisiana State University, Baton Rouge, LA, USA
e-mail: asamad3@lsu.edu

G. Z. Voyiadjis (ed.), *Handbook of Nonlocal Continuum Mechanics for Materials and Structures*, https://doi.org/10.1007/978-3-319-58729-5_26

actual $\dot{P}/P$ value during the course of a single test shows that this parameter varies intensely at shallow indentation depths, and it reaches a stabilized value after a significant depth which is not material dependent. In addition, the strain rate variation is examined through two methods: first, using the definition of the strain rate as the descent rate of the indenter divided by its instantaneous depth $(\dot{h}/h)$ and second, considering the relationship between the strain rate and the load and hardness variations during the test. Based on the findings, the strain rate is greatly larger at shallow indentations, and the depth beyond which it attains the constant value depends on the material and the set $\dot{P}/P$ ratio. Lastly, incorporating the relationship between the hardness and strain rate, it is revealed that although the strain rate variation changes the material hardness, its effect does not give a justification for the observed indentation size effect (ISE); therefore, other contributing parameters are discussed for their possible effects on this phenomenon.

Keywords
Glassy polymers · Amorphous · Nanoindentation · Hardness · Indentation strain rate · Continuous stiffness measurement · Loading rate · Poly(methyl methacrylate) · Polycarbonate · Elastic modulus · Polymeric glasses

Introduction

The interest of many researchers has been recently directed to study the mechanical properties of polymers in small size scales due to their extensive use in nano- and microscale elements over the past decades. A large series of constitutive models and experimental methods by which the properties of this class of materials can be thoroughly captured in millimeter or larger size scales has been presented (Hasan et al. 1993; Hoy and Robbins 2006; Van Breemen et al. 2012; Mulliken and Boyce 2006; Anand and Gurtin 2003; Voyiadjis and Samadi-Dooki 2016); however, there are still many questions about the behavior and deformation mechanism of polymers in submicron size scales. To acquire the precise and reliable results for mechanical properties, including elastic modulus and hardness, of very small volumes of materials, instrumented-indentation testing (IIT) can be employed (Al-Haik et al. 2004; Boersma et al. 2004; Lee et al. 2004; Zeng et al. 2012). In this technique, an indenter induces a localized deformation by applying a specified load on the material surface. Basically, there are two different indentation methods: (1) basic mode in which, with monotonic loading and unloading, the mechanical properties are only measured at the predefined maximum load from the unloading curve, and (2) continuous stiffness measurement (CSM) mode in which a small oscillation force is superimposed on the primary loading signal, and the resulting response of the system is analyzed through a frequency-specific amplifier. Employing the second procedure, the material mechanical properties can be continuously measured from zero to the maximum indentation depth during the loading segment. In addition, while in the former, the contact stiffness is measured just at the initial point of the unloading, the measurement of the contact stiffness at any point along the loading segment is possible in the latter with a smaller time constant (Li and Bhushan 2002; Hay et al. 2010; Pethica and Oliver 1988). Therefore, the small time constant of the CSM method makes it more useful for measuring the properties of materials especially those which are strongly time dependent like polymers. The loading (or strain) rate is controlled in a different way in these two modes; during the basic mode, the load is applied with a constant rate on the sample surface by the indenter until it reaches a determined maximum value; however, in the CSM mode, the indenter travels up to a predefined maximum depth and the load is controlled so that the loading rate divided by the load ($\dot{P}/P$) remains constant over the course of a single indentation.

The loading and strain rates are adjustable parameters in nanoindentation experiments, and their variations have shown profound effects on the mechanical response of time-dependent materials like polymers (Odegard et al. 2005; Mazeran et al. 2012; Samadi-Dooki et al. 2016; Malekmotiei et al. 2015; Kraft et al. 2001; White et al. 2005; Zhang et al. 2009; Shen et al. 2004). That being the case, a closer look at the strain (loading) rate variation during the indentation is required since the generated strain and stress fields in the material due to the loading by a self-similar tip is inhomogeneous. In the basic mode nanoindentation, although the test is conducted with a constant loading rate $\dot{P}$, the strain rate is considerably decreasing at shallow depths, and it eventually approaches almost stable value after a long distance travel of the tip into the material. For this reason, an average value of the strain rate over the deep part of the indentation can be considered as the representative strain rate of the test (Schuh and Nieh 2003). On the other hand, in the CSM nanoindentation experiments, the $\dot{P}/P$ ratio is set as a constant value at the beginning of the test. It has been shown that the indentation strain rate can also be assumed to remain constant during the constant $\dot{P}/P$ experiment where the material hardness has the steady-state value, i.e., $\dot{H} = 0$ (Lucas and Oliver 1999). However, the indentation size effect (ISE), which is the increment of hardness as the indentation depth decreases, has been observed during the nanoindentation experiments on many materials including crystalline and amorphous solids (Briscoe et al. 1998; Voyiadjis and Zhang 2015). In a study on Al-based foams, it has been observed that the strain rate varies about three orders of magnitude during the first 200 nm of the indentation before reaching a steady-state value (Kraft et al. 2001). As a result, in the case of the CSM mode, the indentation strain rate can be considered constant in that part of the test where the ISE is negligible.

Conducting the CSM nanoindentation experiments on PMMA and PC as polymeric glasses, the variation of the strain rate during the course of a single test is investigated as a main goal in this chapter. Examining the variation of the $\dot{P}/P$ ratio during the test shows that although the $\dot{P}/P$ ratio is set to remain invariant during the loading segment, it takes a considerable tip travel distance until it stabilizes and reaches the set value. Furthermore, the indentation strain rate, which has been incorrectly considered as the $\dot{P}/P$ ratio in some studies (Shen et al. 2004, 2006; Vachhani et al. 2013), is also found to change at shallow depths of indentation. The obtained results show a good correlation between the instantaneous indentation strain rate, which is evaluated directly from the indentation depth-time data recorded during the loading segment, and the strain rate relation proposed by Lucas and Oliver (1999) based on the variation of the load and hardness. As another purpose, the possible relation between the variation of the strain rate during the nanoindentation and the observed ISE in polymers is also scrutinized in this chapter. While the high values of strain rate in shallow depths can cause the increment of material hardness, it is discussed here that it cannot be the reason for the observed ISE since the obtained high values of hardness could be the result of the indentation strain rates which are orders of magnitude higher than the actual recorded strain rate values.

Materials and Methods

Sample Preparation

The commercially manufactured (Goodfellow, Cambridge, UK) polymeric glasses including PMMA and PC, 2.0 and 5.0 mm-thick sheets, respectively, are considered for this investigation. The sheets are first cut into 20 × 20 mm squares, and then washed with 30% isopropyl alcohol (IPA) to eliminate the remainders of the protective film, and at the end rinsed with distilled water. Using a TA Instruments 2920 differential scanning calorimetry (DSC) machine, the glass transition temperature (T_g) of the specimens is measured to be about 110 °C and 148 °C for PMMA and PC, respectively. The samples are annealed at 120 °C for 4 h to remove any thermal history, and then cooled down to ambient temperature with the rate of 10 °C/h in a vacuum oven. The roughness of sample surface is one of the factors which affects the nanoindentation results since high values of roughness can make inaccuracy in the hardness of material measurements; therefore, to capture the surface topography of the samples, an Agilent 5500 atomic force microscope (AFM) is utilized. Since the average surface roughness, R_a, of the PC and PMMA specimens are 0.411 ± 0.033 and 0.372 ± 0.013 nm, respectively, one can assume the flat surface for samples (Kim et al.2007), and there is no need to modify the obtained results for materials' hardness (Voyiadjis and Malekmotiei 2016).

Nanoindentation Analysis

To address the goal of this chapter which is scrutinizing the strain rate variation during the course of a single nanoindentation experiment and its effect on the observed ISE, an MTS Nanoindenter® XP equipped with a three-sided pyramidal Berkovich diamond tip is employed (Voyiadjis and Malekmotiei 2016). The mechanical properties of the specimens are measured through the CSM mode indentation in which the load-hold-unload sequences are carried out with the constant $\dot{P}/P$ during the loading stage. According to the formulations developed by Oliver and Pharr

(2004), the material hardness is defined as the mean contact pressure under the indenter as follows:

$$H = \frac{P}{A_c} \tag{1}$$

where P is the applied load on the sample surface and A_c is the projected contact area of the hardness impression at that load. Based on this discerption, a precise measurement of the contact area between the sample surface and the indenter tip is required to calculate the hardness. The contact area is a function of the contact depth, h_c, and equal to $A_c = 24.56{h_c}^2$ for a perfect Berkovich indenter tip. The one used for these experiments is not ideally sharp; so, the contact area function is obtained through calibrating the tip which improves the accuracy of the contact area measurements by accounting for the tip imperfections and leads to introducing some additional terms to the above relation as:

$$A_c = 24.56{h_c}^2 + C_1{h_c}^1 + C_2{h_c}^{1/2} + C_3{h_c}^{1/4} + \cdots + C_8{h_c}^{1/128} \tag{2}$$

in which C_1 through C_8 are constant coefficients which are obtained based on the results of the nanoindentation on fused silica as a standard sample. Another important parameter that needs to be accurately determined is the depth over which the material is in contact with the tip (h_c). The contact depth is estimated using

$$h_c = h - \varepsilon\frac{P}{S}. \tag{3}$$

where h is the total penetration depth, S is the elastic contact stiffness, and ε is a constant that depends on the indenter geometry (for a Berkovich indenter $\varepsilon = 0.75$ (Oliver and Pharr 1992)). As already stated, the CSM technique makes the continuous measurement of the contact stiffness as a function of depth possible during the loading segment of the indentation. Considering the imposed driving force as $P = P_0\, e^{i\omega t}$ and the indenter displacement response as $h(\omega) = h_0\, e^{(i\omega t + \alpha)}$, the elastic contact stiffness is calculated as follows:

$$S = \left[\frac{1}{\frac{P_0}{h(\omega)}\cos(\alpha) - (K_s - m\omega^2)} - \frac{1}{K_f}\right]^{-1} \tag{4}$$

in which P_0 is the force oscillation magnitude, ω is the oscillation frequency, h_0 is the resulting displacement oscillation magnitude, and α is the phase angle between the displacement and force signals. The other contributing parameters are the leaf spring constant, K_s, that supports the indenter, the indenter mass, m, and the indenter frame stiffness, K_f (Li and Bhushan 2002).

Another mechanical property measured in the nanoindentation experiments is the elastic modulus of the sample, E, which is calculated by the following relation:

$$\frac{1}{E_r} = \frac{1-\nu^2}{E} + \frac{1-{\nu_i}^2}{E_i}. \tag{5}$$

where E_r is the reduced elastic modulus which attributes to the elastic deformation in both the sample and indenter, ν is the sample Poisson's ratio, and E_i and ν_i are the indenter elastic modulus and Poisson's ratio, respectively. Sneddon (1965) has developed a relation for the reduced elastic modulus as follows:

$$E_r = \frac{S}{2\beta}\sqrt{\frac{\pi}{A_c}}. \tag{6}$$

where β as a constant depends on the indenter geometry and is about 1.034 for the Berkovich tip. Subsequently, as a main feature of the CSM method, the material hardness and elastic modulus are measured as continuous functions of depth with the course of an individual loading-unloading cycle.

The Indentation Strain Rate

Basically in the nanoindentation experiments, the strain rate affects the material in a direction perpendicular to the sample surface and is correlated with the displacement/loading rate of the indentation. For a pyramidal indenter, the indentation strain rate is defined as the penetration rate of the indenter into the material divided by its instantaneous depth as follows (Mayo and Nix 1988):

$$\dot{\epsilon}_i = \left(\frac{1}{h}\right)\left(\frac{dh}{dt}\right). \tag{7}$$

where t is time. In a study, Lucas and Oliver (1999) investigated that by keeping the loading rate divided by the load ($\dot{P}/P$) constant during the CSM nanoindentation, the indentation strain rate can also remain constant. It has been shown that incorporating the loading and hardness data, the indentation strain rate can be obtained as (Lucas and Oliver 1999):

$$\dot{\epsilon}_i = \frac{\dot{h}}{h} = \frac{1}{2}\left(\frac{\dot{P}}{P} - \frac{\dot{H}}{H}\right). \tag{8}$$

in which $\dot{H}$ is the hardness variation rate and other parameters are defined before. According to Eq. 8, the indentation strain rate reaches a constant value $\left(\frac{1}{2}\frac{\dot{P}}{P}\right)$at large indentation depths where the material hardness is almost unvaried, i.e., $\dot{H} = 0$.

It is noteworthy to mention that two main simplifying assumptions have made to get this relation: (a) the projected contact area relation is considered as $A = 24.56h^2$ which is used for an ideal Berkovich indenter tip, and (b) instead of the contact or plastic depth, the total depth is used in the contact area function.

Experimental Procedure

As mentioned in preceding sections, as a first step, the specimens should be thoroughly prepared and the tip should be carefully calibrated; the tests then are triggered by running the loading-hold-unloading cycles as follows: before any measurement, the tip drift should be controlled at a rate below 0.05 nms^{-1}, so it is held on the top of the sample surface until it gains the stabilized rate. The tip then moves downward to reach the material surface. As soon as the tip touches the sample surface, the loading stage begins with a constant $\dot{P}/P$ ratio and it continues until a specified maximum depth of 10 μm. To account for the creep behavior of the polymer, the load is then held at this stage for 10 s, and eventually, the unloading part is carried out with a constant unloading rate until 10% of the maximum load. Since the goals are investigating the variation of $\dot{P}/P$ ratio during the whole nanoindentation experiment from zero to the maximum depth, and also its contribution on the indentation strain rate, a series of tests are performed on annealed PMMA and PC samples with three different set values of $\dot{P}/P$ (0.005, 0.05, and 0.11 s^{-1}). For each $\dot{P}/P$ ratio, 25 indents are accomplished to get the accurate results and to prevent from interaction of the indents, 150 μm distance is considered between them.

Results and Discussion

Variation of the $\dot{P}/P$ in the Course of an Indentation

The applied load on the sample and the tip travel distance are recorded as unbroken curves in the CSM nanoindentation experiments with *nN* and sub-*nm* exactness, respectively. Since the loading rate divided by the load is constant during the loading stage, i.e., $\dot{P}/P = \eta$, the load is expected to be an exponential function of time as follows:

$$P = \beta e^{\eta t}. \tag{9}$$

where β is the constant obtained by solving the ordinary differential equation (ODE). The load variation with time during the loading section of the nanoindentation on PC sample is presented in Fig. 1 (Voyiadjis and Malekmotiei 2016); the figure shows the results for three different set $\dot{P}/P$ ratios and their exponential interpolations (lines) for comparison. The result curves depart from the exponential behavior at shallow indentation depths while they behave in accordance with the exponential variation within the long tip travel distance, especially for higher $\dot{P}/P$ values. This discrepancy at the early stages of loading is due to the fact that based on Eq. 9, the initial loading condition is $P(0) = \beta$; however, the set initial condition for the experiment process is $P(0) = 0$. Thus, the indenter $\dot{P}/P$ ratio can be adjusted to the set $\dot{P}/P$ value after several nanometers of indentation displacement (or several seconds).

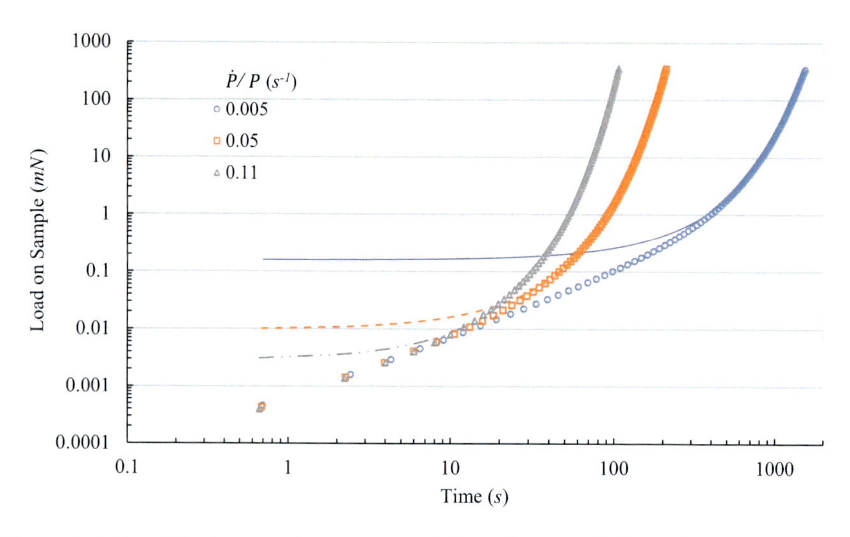

Fig. 1 Variation of load versus time response of the indenter for PC sample measured on the loading segment of the nanoindentation experiment at three different set $\dot{P}/P$ values. The *lines* represent the exponential interpolations (Reprinted from Voyiadjis and Malekmotiei 2016)

The actual variation of $\dot{P}/P$ ratio as a function of the indentation depth is presented in Fig. 2 for PC and PMMA (Voyiadjis and Malekmotiei 2016) at three different set values of this ratio which are shown as horizontal dashed lines. As mentioned before, the actual $\dot{P}/P$ values do not approach their set values right after the indenter tip comes into the contact with the sample surface. Furthermore, the depth at which the $\dot{P}/P$ ratio stabilizes does not depend on the material; however, it is extremely dependent on the set $\dot{P}/P$ value. As depicted in Fig. 2 (Voyiadjis and Malekmotiei 2016), the depth beyond which the actual $\dot{P}/P$ approaches the set value and stabilizes is smaller for the bigger set value of $\dot{P}/P$: it is almost 1000 nm for the set $\dot{P}/P$ value of 0.005 s^{-1} and reduces to 200 and 100 nm for the set $\dot{P}/P$ values of 0.05 and 0.11 s^{-1}, respectively. Interestingly, the starting point of actual value of $\dot{P}/P$ is not dependent on the material and set $\dot{P}/P$ ratio, and it approximately equals 0.3 s^{-1} for all experiments.

Variation of the Strain Rate in the Course of an Indentation

In the CSM nanoindentation method, since the tip displacement is recorded continuously with time, the indentation strain rate can be directly calculated by using Eq. 7 and simple numerical differentiation as a continuous function of the indentation depth. In addition, indirect evaluation of the indentation strain rate during the loading segment of the test is possible by incorporating Eq. 7 and using the recorded load on the sample and the measured material hardness as functions of the depth. Since to employ Eq. 8 the variation of the instantaneous hardness rate divided by hardness ($\dot{H}/H$) is required, this parameter is represented in Fig. 3 (Voyiadjis and Malekmotiei 2016) at three different set $\dot{P}/P$ ratios for PC and PMMA. As depicted

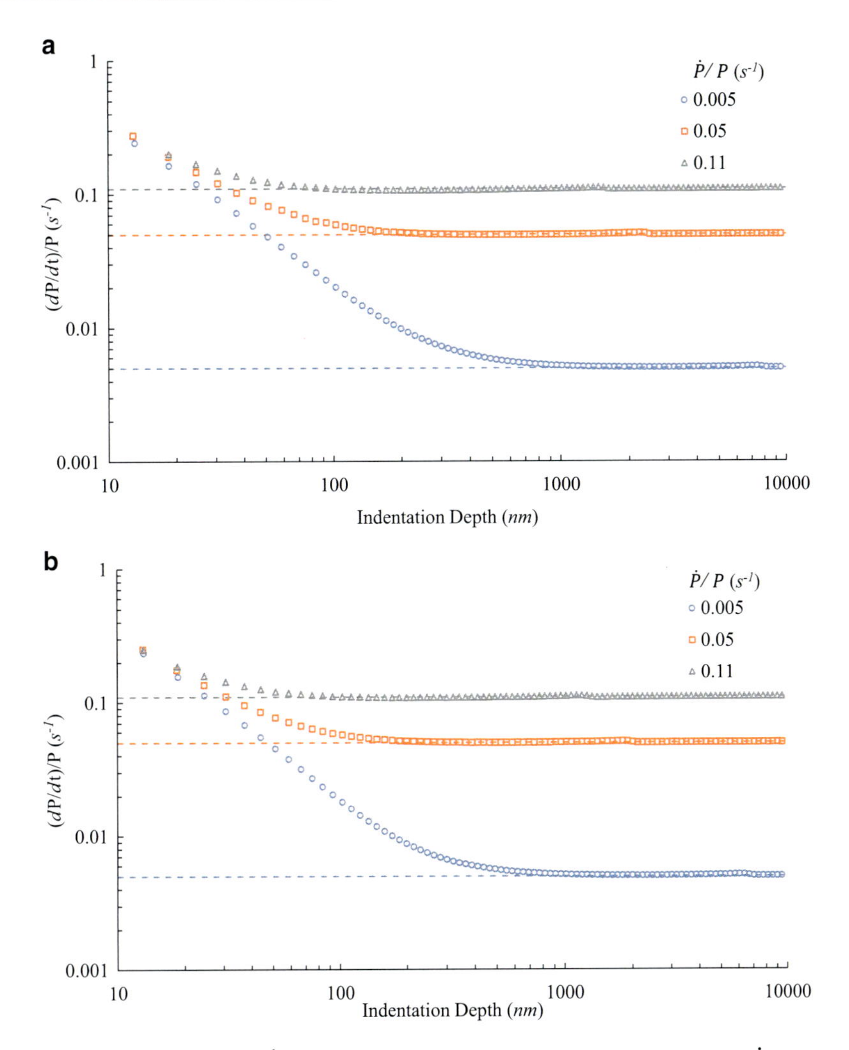

Fig. 2 Variation of the actual $\dot{P}/P$ values with indentation depth at three different set $\dot{P}/P$ values for (**a**) PC and (**b**) PMMA (Reprinted from Voyiadjis and Malekmotiei (2016))

in Fig. 5, the material hardness is higher at shallower indentation depths (indentation size effect) and then reaches a plateau at the certain depth which is the representative of the macroscopic hardness. Therefore, this trend results in the negative values of the $\dot{H}/H$ ratio at the initial stages of the loading section and finally zero values of $\dot{H}/H$ at deep part of the indentation (see Fig. 3). As another result obtained from Fig. 3, the depth beyond which $\dot{H}$ can be assumed zero depends on the material and the set $\dot{P}/P$ value.

Figure 4 (Voyiadjis and Malekmotiei 2016) displays the variation of the indentation strain rate with the indentation depth calculated based on the two different

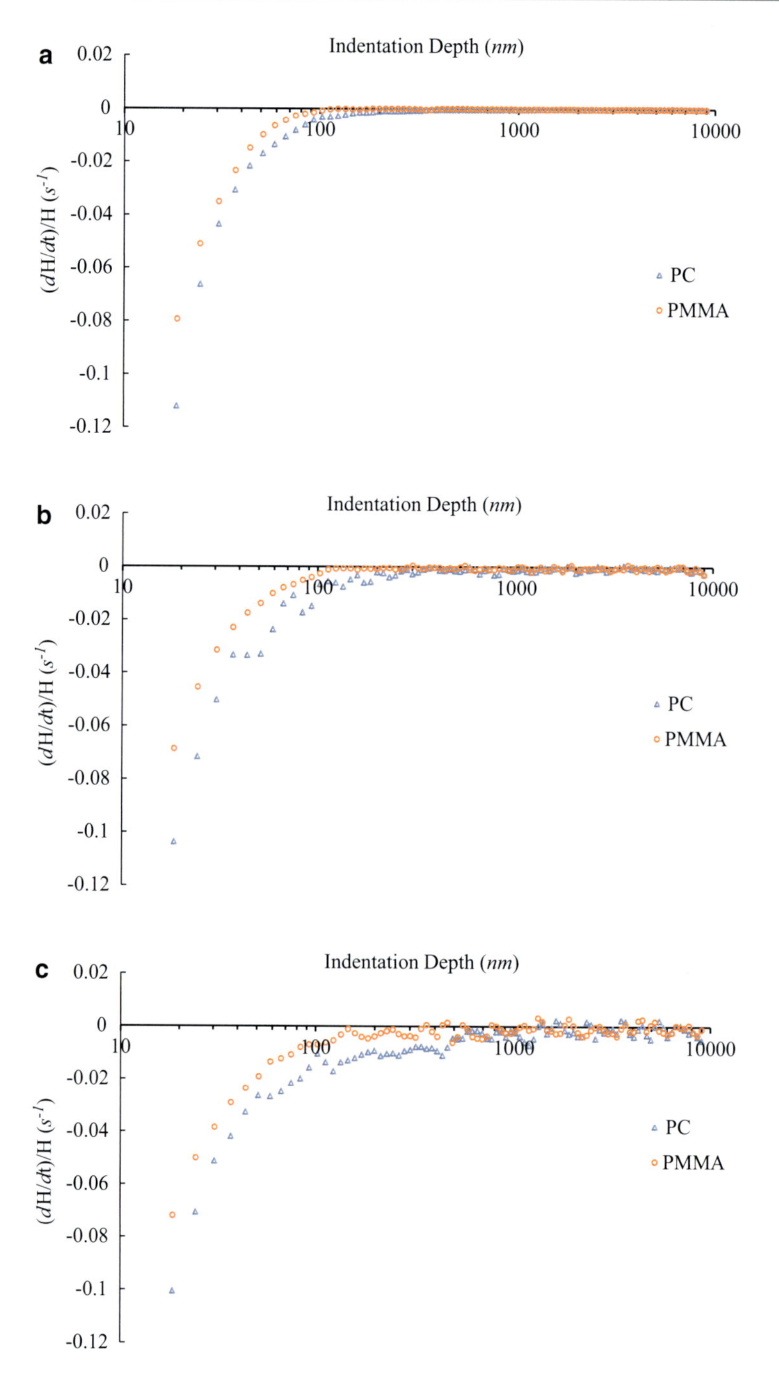

Fig. 3 Variation of the instantaneous hardness rate divided by hardness $\left(\frac{\dot{H}}{H}\right)$ with indentation depth for PC and PMMA at set $\dot{P}/P$ equal to (**a**) 0.005, (**b**) 0.05, and (**c**) 0.11 s^{-1} (Reprinted from Voyiadjis and Malekmotiei (2016))

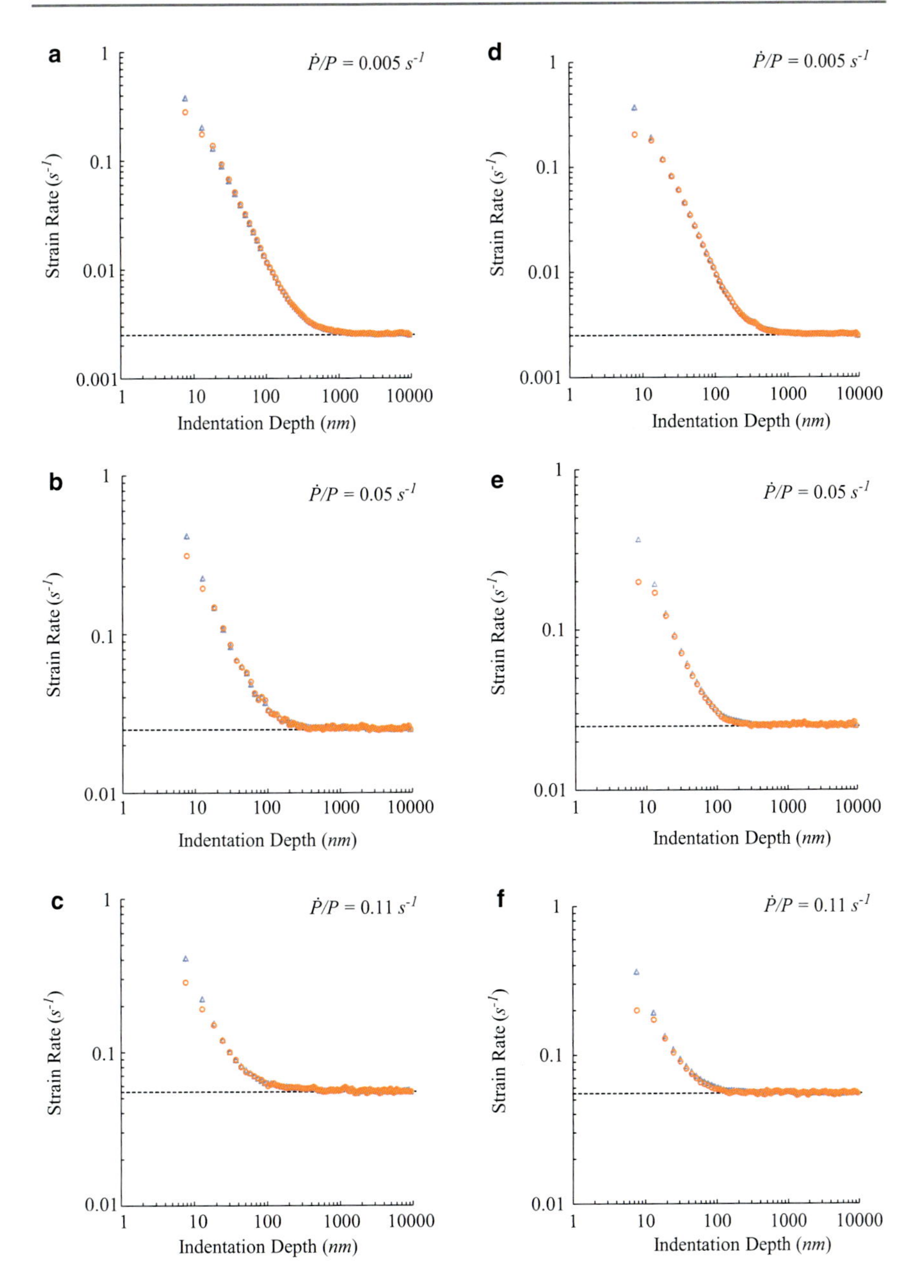

Fig. 4 Variation of the indentation strain rate based on Eq. 7 (△) and Eq. 8 (○) versus indentation depth for experiments on PC (**a–c**) and PMMA (**d–f**) at three different set $\dot{P}/P$ values. The *dashed lines* represent the expected values of strain rate $\left(\frac{1}{2}\frac{\dot{P}}{P}\right)$ (Reprinted from Voyiadjis and Malekmotiei (2016))

approaches. Since both methods show almost the same results, it reveals the validity of the assumptions made by Lucas and Oliver (1999) to obtain Eq. 8 not only for deep but also shallow indentations of polymeric glasses. The most important result is that the indentation strain rate is not constant during the loading segment of the CSM nanoindentation of glassy polymers, and its variation is material and rate dependent. However, it can be assumed to be constant and equal to $\frac{1}{2}\frac{\dot{P}}{P}$for deep enough indentation experiments.

Indentation Size Effect

In rate-dependent materials including polymers, the flow stress extremely depends on the applied loading (strain) rate: the higher the experiment strain rate, the higher the yield stress (Voyiadjis and Samadi-Dooki 2016; Samadi-Dooki et al. 2016; Malekmotiei et al. 2015; Richeton et al. 2006; Rottler and Robbins 2003). Therefore, since there is a relation between the flow stress and hardness of the material through Tabor's relation (Prasad et al. 2009), the higher value of hardness is expected from nanoindentation with the higher strain rate. The variation of hardness versus the tip displacement is presented in Fig. 5 (Voyiadjis and Malekmotiei 2016) for experiments on PC and PMMA samples at three different set $\dot{P}/P$ values. It is observed that the obtained hardness values are higher as the strain rate increases; especially, the macroscopic hardness which is the hardness at the deep part of the nanoindentation and is the plateau for each curve depends on both material and strain rate. It is clear in these figures that the strain rate dependency of PMMA is more considerable which is, physically, demonstrated as smaller shear activation volumes in this material (Malekmotiei et al. 2015). Another observation in Fig. 5 is the profound increment of the hardness as the depth decreases during each indentation which is known as the ISE at nanoscales (Shen et al. 2006; Lam and Chong 1999; Zhang and Xu 2002). Due to the above-mentioned reason, the increased values of the material hardness at shallow indentation depths might be correlated with the higher values of the strain rate at these depths. However, an exact quantitative analysis is needed to understand and evaluate this possible relationship.

Many studies show that there is a linear relationship between the flow stress (or hardness) of the polymeric glasses and the logarithm of the strain rate. The explicit relationships have been previously obtained for PC and PMMA by nanoindentation evaluations (Samadi-Dooki et al.2016; Malekmotiei et al.2015). Using the obtained formulations and the strain rate variation during indentation, the fictitious hardness can be calculated for each test. It should be mentioned that to calculate the fictitious hardness, it is assumed that the hardness variation is just the result of the strain rate variation during the loading stage. Figure 6 shows the actual measured hardness as well as the calculated hardness versus the indentation depth for PC and PMMA at three different set $\dot{P}/P$ values for comparison. The curves of PC sample show that the calculated hardness is almost constant and there is no considerable change

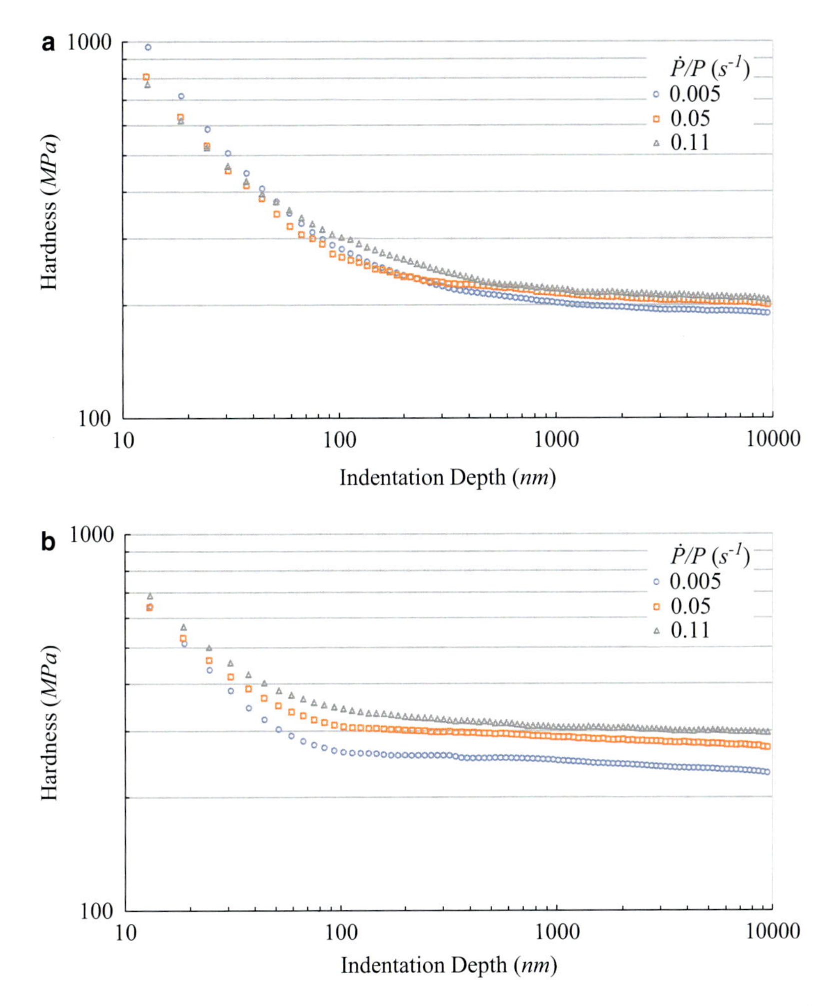

Fig. 5 Variation of the hardness versus indentation depth during the loading segment of the experiments on (**a**) PC and (**b**) PMMA at three different set $\dot{P}/P$ values (Reprinted from Voyiadjis and Malekmotiei (2016))

during a test; however, its variation during each indentation on PMMA sample is notably large in amount and follows almost the same hardening pattern at shallow depths as the actual hardness variation trend. In addition, for PMMA, the calculated hardness is the same (about 336 MPa) for all $\dot{P}/P$ values when the loading stage is triggered, which is acceptable for the reason that the strain rate at the beginning of the indentation is also the same for different $\dot{P}/P$ ratios (Fig. 4), while the actual hardness at this point (maximum hardness in each curve) is different for different $\dot{P}/P$ values in both PC and PMMA. More importantly, as Fig. 4 represents the strain rate variation in the course of an indentation is approximately material

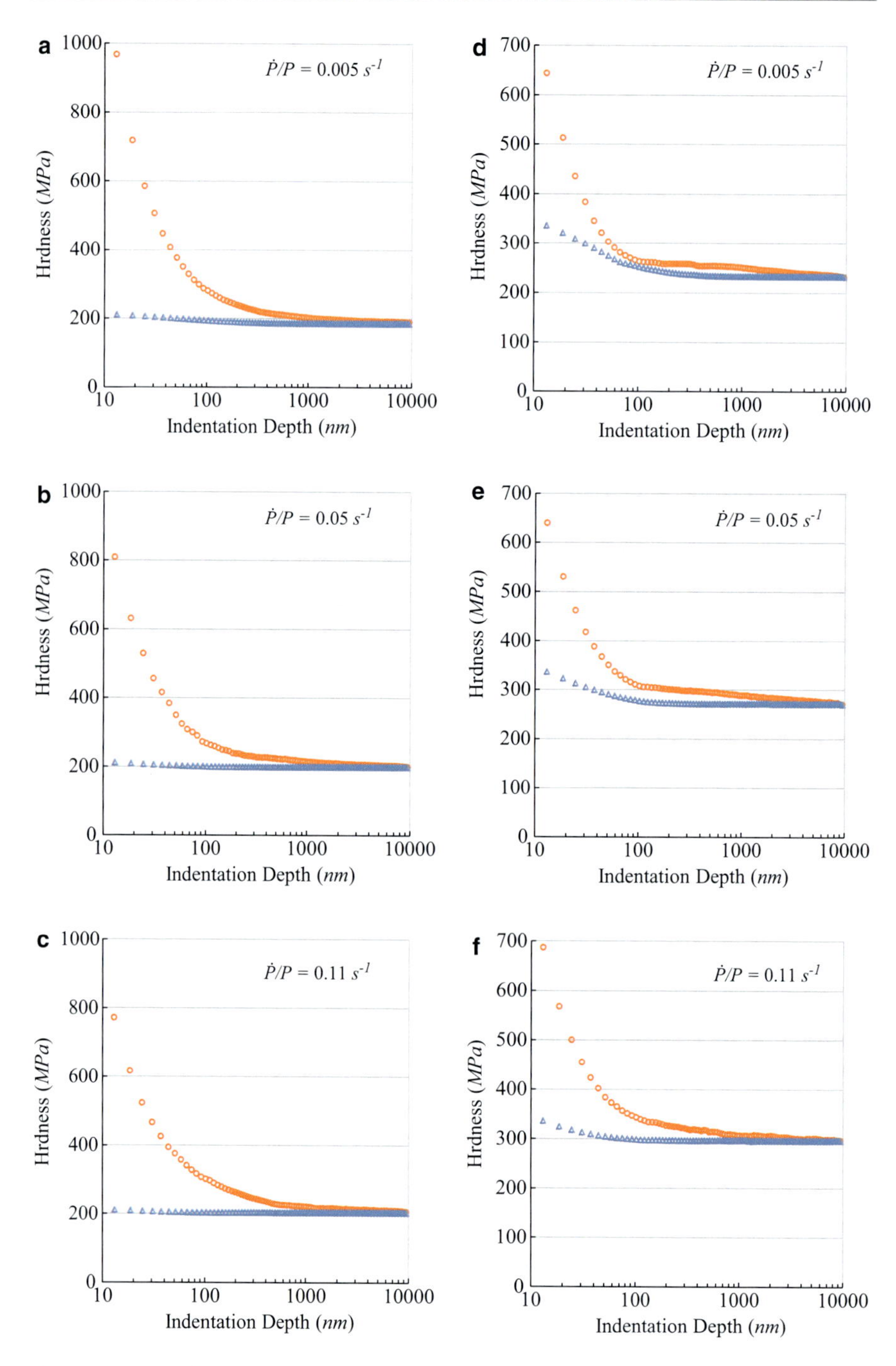

Fig. 6 Variation of the calculated (△) and actual (○) hardness versus indentation depth during the loading segment of the experiments on PC (**a**–**c**) and PMMA (**d**–**f**) at three different set $\dot{P}/P$ values (Reprinted from Voyiadjis and Malekmotiei (2016))

independent, however, PMMA reveals a more intense strain rate dependency of hardness response than PC (Fig. 5); therefore, a more profound indentation size effect is expected to be observed in the nanoindentation of PMMA. Nevertheless, the real situation is different since the hardness variation at shallow depths is more noticeable for PC in comparison with PMMA (see Figs. 5 and 6) (Voyiadjis and Malekmotiei 2016). The main result from these observations is that although the strain rate variation during the loading segment of the CSM nanoindentation on PC and PMMA is notable, it cannot be the major cause of the observed ISE in amorphous polymers. As a matter of fact, the indentation strain rate variation during the loading has *no* contribution to the observed ISE of PC and its contribution to the ISE phenomenon in PMMA is *negligible*. Additionally, assuming the constant strain rate during the CSM nanoindentation of polymeric glasses for the size effect studies seems to be reasonable and there should exist other mechanisms behind this phenomenon which are correlated to the localization and or free surface effects (Alisafaei and Han 2015; Han et al. 2016).

It is noteworthy to mention that another important factor which can affect the contact area and, subsequently, the measured hardness, especially at shallow depths of the indentation, is the material pile-up around the indenter tip. In Fig. 6, since the calculated hardness is obtained from the direct measurement of the indentation strain rate $\left(\dot{h}/h\right)$ (Samadi-Dooki et al. 2016; Malekmotiei et al. 2015), it is not affected by the material pile-up, while the actual measured hardness in this figure could be affected by the pile-up. For this reason, the material pile-up around the tip could be another factor that causes the difference between the calculated hardness of material and the actual one.

Another important phenomenon which is usually observed during the CSM nanoindentation experiments is a small size effect on the recorded elastic modulus of the material. As shown in Fig. 7 (Voyiadjis and Malekmotiei 2016), it is an increased Young's modulus at shallower indentation depths. This phenomenon is in contrast to the earlier observations from particle embedment experiments (Teichroeb and Forrest 2003; Karim and McKenna 2011, 2012, 2013; Hutcheson and McKenna 2007). As discussed comprehensively in the literature (Parry and Tabor 1973, 1974) the applied hydrostatic pressure on the polymer samples can hamper the chain movements which are required for relaxation processes and can subsequently result in a considerable increment of the glass transition temperatures of the material. Based on that, it has been proposed that, in nanoindentation experiments, the contact loading at the indenter tip-polymer interface induces hydrostatic pressure under the tip which increases the glass transition temperature of the sample near the surface, and correspondingly, the increased stiffness of the material at low indentation depths has been related to the increment of T_g (Gacoin et al. 2006; Tweedie et al. 2007). Therefore, the observed considerable material stiffening at shallow indentation depths (for depths of <50 nm in Fig. 7) (Voyiadjis and Malekmotiei 2016) could also attribute to the elevated values of T_g at the surface layer within this tip travel distance compared to the bulk. Moreover, incorporating the shear transformation theory, Voyiadjis and Samadi-Dooki (2016) have proposed a model for yielding

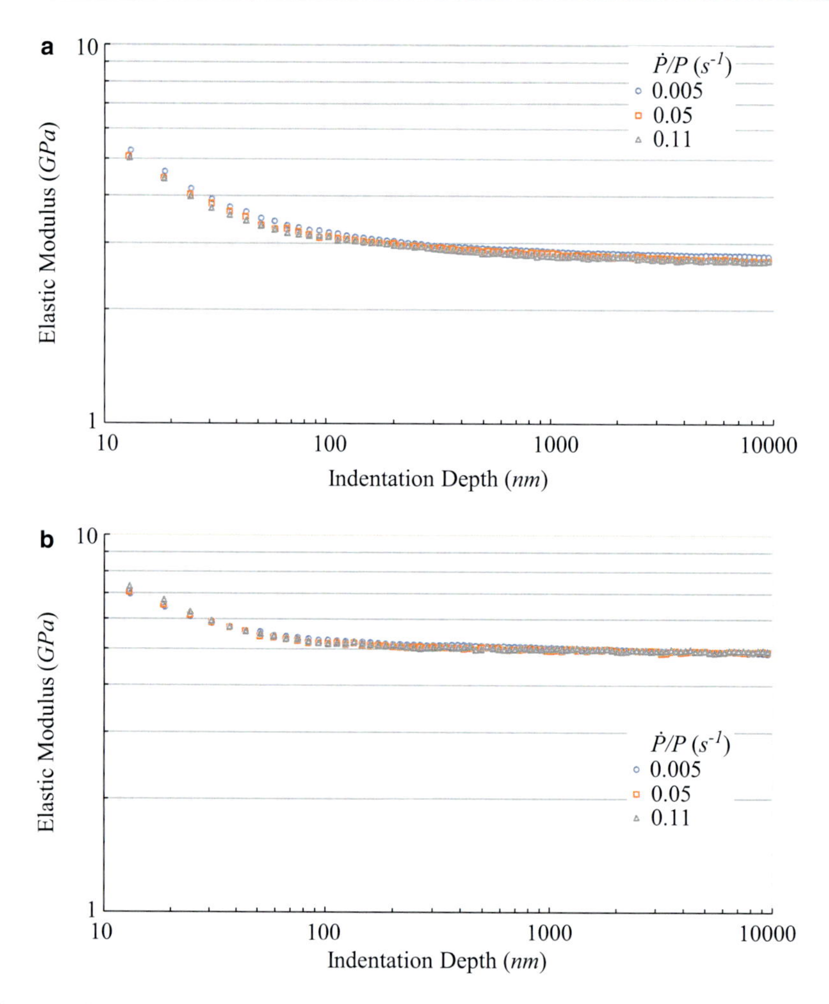

Fig. 7 Variation of the elastic modulus versus indentation depth during the loading segment of the experiments on (**a**) PC, and (**b**) PMMA at three different set $\dot{P}/P$ values (Reprinted from Voyiadjis and Malekmotiei (2016))

and plasticity of amorphous polymers which shows the relationship between the yield stress (which is proportional to hardness through Tabor's relation) and an activation energy which itself is a function of the elastic modulus of the material. Thus, as another confirmation, there exists a possibility that the increased hardness observed at the shallow indentation depths to be interrelated to the increased elastic modulus at these depths. This hypothesis, however, should be viewed as a qualitative observation and treated cautiously since there is a noticeable difference between the length scales during which the elastic modulus and hardness increments are observed (see Figs. 5 and 7).

Conclusions

The mechanical behavior of viscoelastic-viscoplastic materials, including polymers, strongly depends on the rate at which they are loaded. In the case of the nanoindentation, where these behaviors are evaluated at nanoscales, there are different methods in which the loading rate can be controlled in a way that the strain rate changes or remains constant during the experiment. For many reasons, for example, to study the effect of the temperature, thermal history, composition of alloys, etc., it is desirable to conduct a test during which the strain rate remains constant. The continuous stiffness measurement (CSM) nanoindentation is a technique which offers testing at constant loading rate to load ratio ($\dot{P}/P$) ratios; this has been manifested as constant strain rates during the test. In this chapter, the assumption of the constant strain rate during this nanoindentation technique is studied for glassy polymers. Investigating the instantaneous variations of $\dot{P}/P$ and strain rate during the CSM nanoindentation on poly(methyl methacrylate) (PMMA) and polycarbonate (PC) samples showed that this assumption is not valid during the whole test and the strain rate changes in the early stages of the indentation before acceptable stabilization; the depth beyond that this parameter stabilizes depends on the material and the set $\dot{P}/P$ value. It has been shown that although by assuming the constant value of $\dot{P}/P$, an exponential load-time response of the indenter is expected, the load-time curves do not obey the exponential variation in early stages of the loading. The reason is the incompatibility of the actual initial load with the initial condition of the exponential loading which is required to assure a constant $\dot{P}/P$. To overcome this problem, one can apply a very small load prior to the main loading segment of the experiment; this adjusts the aforementioned incompatibility of the initial condition. By this adjustment, it is expected that the $\dot{P}/P$ value during the whole test remains constant and equals the initial set value. However, it may not lead to a constant strain rate since the hardness variation rate also contributes to the strain rate variation as proposed by Lucas and Oliver (1999). The strain rate can be assumed constant only at deep enough indentations where the load-displacement curve obeys the Hertzian relation (Malekmotiei et al. 2015). In this chapter, the possible effect of the variation of the strain rate during the indentation on the observed indentation size effect (ISE) in amorphous polymers has also been discussed (Voyiadjis and Malekmotiei 2016). While it is concluded that the increased strain rate within the shallow indentation depths cannot be the sole reason for the observed profound ISE, contribution of the other factors, such as material pile-up around the tip and stiffening due to the glass transition temperature (T_g) shift induced by the hydrostatic component of the stress, have been qualitatively discussed.

References

M. Al-Haik, H. Garmestani, D. Li, M. Hussaini, S. Sablin, R. Tannenbaum, K. Dahmen, J. Polym. Sci. B Polym. Phys. **42**, 1586 (2004)

F. Alisafaei, C.-S. Han, Adv. Condens. Matter Phys. **2015**, 391579 (2015)

L. Anand, M.E. Gurtin, Int. J. Solids Struct. **40**, 1465 (2003)
A. Boersma, V. Soloukhin, J. Brokken-Zijp, G. De With, J. Polym. Sci. B Polym. Phys. **42**, 1628 (2004)
B. Briscoe, L. Fiori, E. Pelillo, J. Phys. D. Appl. Phys. **31**, 2395 (1998)
E. Gacoin, C. Fretigny, A. Chateauminois, A. Perriot, E. Barthel, Tribol. Lett. **21**, 245 (2006)
C.-S. Han, S.H. Sanei, F. Alisafaei, J. Polym. Eng. **36**, 103 (2016)
O. Hasan, M. Boyce, X. Li, S. Berko, J. Polym. Sci. B Polym. Phys. **31**, 185 (1993)
J. Hay, P. Agee, E. Herbert, Exp. Tech. **34**, 86 (2010)
R.S. Hoy, M.O. Robbins, J. Polym. Sci. B Polym. Phys. **44**, 3487 (2006)
S. Hutcheson, G. McKenna, Eur. Phys. J. E. **22**, 281 (2007)
T.B. Karim, G.B. McKenna, Polymer **52**, 6134 (2011)
T.B. Karim, G.B. McKenna, Macromolecules **45**, 9697 (2012)
T.B. Karim, G.B. McKenna, Polymer **54**, 5928 (2013)
J.Y. Kim, S.-K. Kang, J.-J. Lee, J.-i. Jang, Y.-H. Lee, D. Kwon, Acta Mater. **55**, 3555 (2007)
O. Kraft, D. Saxa, M. Haag, A. Wanner, Z. Metallkd. **92**, 1068 (2001)
D.C. Lam, A.C. Chong, J. Mater. Res. **14**, 3784 (1999)
C. Lee, J. Kwon, S. Park, S. Sundar, B. Min, H. Han, J. Polym. Sci. B Polym. Phys. **42**, 861 (2004)
X. Li, B. Bhushan, Mater. Charact. **48**, 11 (2002)
B. Lucas, W. Oliver, Metall. Mater. Trans. A **30**, 601 (1999)
L. Malekmotiei, A. Samadi-Dooki, G.Z. Voyiadjis, Macromolecules **48**, 5348 (2015)
M. Mayo, W. Nix, Acta Metall. **36**, 2183 (1988)
P.E. Mazeran, M. Beyaoui, M. Bigerelle, M. Guigon, Int. J. Mater. Res. **103**, 715 (2012)
A. Mulliken, M. Boyce, Int. J. Solids Struct. **43**, 1331 (2006)
G. Odegard, T. Gates, H. Herring, Exp. Mech. **45**, 130 (2005)
W.C. Oliver, G.M. Pharr, J. Mater. Res. **7**, 1564 (1992)
W.C. Oliver, G.M. Pharr, J. Mater. Res. **19**, 3 (2004)
E.J. Parry, D. Tabor, J. Mater. Sci. **8**, 1510 (1973)
E.J. Parry, D. Tabor, J. Mater. Sci. **9**, 289 (1974)
J.B. Pethica, W.C. Oliver, MRS Online Proc. Lib. Arch. **130**, 13 (1988)
K.E. Prasad, V. Keryvin, U. Ramamurty, J. Mater. Res. **24**, 890 (2009)
J. Richeton, S. Ahzi, K. Vecchio, F. Jiang, R. Adharapurapu, Int. J. Solids Struct. **43**, 2318 (2006)
J. Rottler, M.O. Robbins, Phys. Rev. E **68**, 011507 (2003)
A. Samadi-Dooki, L. Malekmotiei, G.Z. Voyiadjis, Polymer **82**, 238 (2016)
C.A. Schuh, T. Nieh, Acta Mater. **51**, 87 (2003)
L. Shen, I.Y. Phang, T. Liu, K. Zeng, Polymer **45**, 8221 (2004)
L. Shen, I.Y. Phang, T. Liu, Polym. Test. **25**, 249 (2006)
I.N. Sneddon, Int. J. Eng. Sci. **3**, 47 (1965)
J. Teichroeb, J. Forrest, Phys. Rev. Lett. **91**, 016104 (2003)
C.A. Tweedie, G. Constantinides, K.E. Lehman, D.J. Brill, G.S. Blackman, K.J. Van Vliet, Adv. Mater. **19**, 2540 (2007)
S. Vachhani, R. Doherty, S. Kalidindi, Acta Mater. **61**, 3744 (2013)
L.C. Van Breemen, T.A. Engels, E.T. Klompen, D.J. Senden, L.E. Govaert, J. Polym. Sci. B Polym. Phys. **50**, 1757 (2012)
G.Z. Voyiadjis, L. Malekmotiei, J. Polym. Sci. Part B: Polym. Phys. **54**, 2179 (2016)
G.Z. Voyiadjis, A. Samadi-Dooki, J. Appl. Phys. **119**, 225104 (2016)
G.Z. Voyiadjis, C. Zhang, Mater. Sci. Eng. A **621**, 218 (2015)
C. White, M. Vanlandingham, P. Drzal, N.K. Chang, S.H. Chang, J. Polym. Sci. B Polym. Phys. **43**, 1812 (2005)
F. Zeng, Y. Liu, Y. Sun, E. Hu, Y. Zhou, J. Polym. Sci. B Polym. Phys. **50**, 1597 (2012)
T.-Y. Zhang, W.-H. Xu, J. Mater. Res. **17**, 1715 (2002)
Y.F. Zhang, S.L. Bai, X.K. Li, Z. Zhang, J. Polym. Sci. B Polym. Phys. **47**, 1030 (2009)

Shear Transformation Zones in Amorphous Polymers: Geometrical and Micromechanical Properties

10

George Z. Voyiadjis, Leila Malekmotiei, and Aref Samadi-Dooki

Contents

Abstract

Glassy polymers are extensively used as high impact resistant, low density, and clear materials in industries. Due to the lack of the long-range order in the microstructures of glassy solids, plastic deformation is different from that in crystalline solids. Shear transformation zones (STZs) are believed to be the plasticity

G. Z. Voyiadjis (✉) · L. Malekmotiei
Department of Civil and Environmental Engineering, Louisiana State University, Baton Rouge, LA, USA
e-mail: voyiadjis@eng.lsu.edu; lmalek1@lsu.edu

A. Samadi-Dooki
Computational Solid Mechanics Laboratory, Department of Civil and Environmental Engineering, Louisiana State University, Baton Rouge, LA, USA
e-mail: asamad3@lsu.edu

G. Z. Voyiadjis (ed.), *Handbook of Nonlocal Continuum Mechanics for Materials and Structures*, https://doi.org/10.1007/978-3-319-58729-5_28

carriers in amorphous solids and defined as the localized atomic or molecular deformation patches induced by shear. Despite a great effort in characterizing these local disturbance regions in metallic glasses (MGs), there are still many unknowns relating to the microstructural and micromechanical characteristics of STZs in glassy polymers. This chapter is aimed at investigating the flow phenomenon in polycarbonate (PC) and poly(methyl methacrylate) (PMMA) as glassy polymers and obtaining the mechanical and geometrical characteristics of their STZs. To achieve this goal, the nanoindentation experiments are performed on samples with two different thermal histories: as-cast and annealed, and temperature and strain rate dependency of the yield stress of PC and PMMA are studied. Based on the experimental results, it is showed that the flow in PC and PMMA is a homogeneous phenomenon at tested temperatures and strain rates. The homogeneous flow theory is then applied to analyze the STZs quantitatively. The achieved results are discussed for their possible uniqueness or applicability to all glassy polymers in the context of amorphous plasticity.

Keywords
Glassy polymers · Shear transformation zone · Nucleation energy · Shear activation volume · Homogeneous flow · Amorphous · Transformation shear strain · β-transition · Nanoindentation · Hardness · Plasticity

Introduction

Over the past several years, many researches have been conducted on investigating the mechanical behavior of polymers, due to the extensive use of them for design and development of a variety of components and structures. In general, polymers are divided in two different categories: semicrystalline and amorphous (glassy) polymers. Amorphous polymers are composed of entangled and disordered long molecular chains, and there is no significant chain alignment in their intra- and intermolecular structures. Since the molecular structure of the polymeric glasses (PGs) is totally different from that of the crystalline solids, the plastic deformation process does not obey the crystal plasticity rules. Moreover, in contrast to many crystalline solids, the postyield behavior begins with a softening at the onset of the yielding in the stress-strain characteristic behavior of PGs, and then continues by a hardening which starts at the specific strain and ends up at the break point (Boyce et al. 1988; Stoclet et al. 2010). There is a large body of literature dealing with the process of yielding and characterizing the mechanism of nonlinear elastoplastic deformation in glassy polymers, which resulted in different physical and phenomenological models (Ree and Eyring 1955; Robertson 1966; Argon 1973; Boyce et al. 1988; Arruda et al. 1995; Anand and Gurtin 2003; Mulliken and Boyce 2006; Chen and Schweizer 2011; Voyiadjis and Samadi-Dooki 2016).

In the context of crystal plasticity, crystal dislocations are the principal carriers of plasticity, and their slips result in plastic deformation (Argon 2008). But, since there is no long-range coherence in atomic or molecular structure of glassy solids, there are no analogous mobile defects. Consequently, the flow and the mechanism of plastic response in the microstructural level are different from the crystalline solids. The ongoing widely recognized mechanism for the plastic response of all types of disordered solids, including metallic glasses, glassy polymers, and covalent glasses, is the cooperative localized rearrangement of molecular or atomic patches in small distinct regions which are called shear transformation zones (STZs) (Spaepen 1977; Falk 2007; Schuh et al. 2007; Pauly et al. 2010; Argon 2013). Especially, the presence of seperate plastic deformation units has been experimentally recognized in glassy polymers and attributed to the formation of STZs (Oleinik et al. 2007).

The STZs are isolated irreversible stress relaxation events which form around free volume sites with the thermal fluctuations assistance under the action of an applied shear stress (Argon 1979; Falk et al. 2005; Argon 2013). The absence of long-range coherence in glassy materials results in the sessile transformations which, once formed, do not expand by translational movements of their interfaces. Therefore, the plasticity mediated by the shear transformations is nucleation controlled (Argon 1993; Spathis and Kontou 2001). Since PGs are considered as homogeneous and isotropic materials, a localized disturbance in their bulk can be considered as an Eshelby inclusion problem (Eshelby 1957). The Eshelby leading-edge homogenization method, which has been extensively applied for solving a broad area of problems in inhomogeneous media (Tandon and Weng 1984; Shodja et al. 2003; Malekmotiei et al. 2013), is based on the strain compatibility of a medium containing no-elastic strains (Mura 1987). Using this method for amorphous solids, the STZ's nucleation energy, which is necessary to relate the shear flow stress to the shear flow strain rate through an Arrhenius function, can be obtained (Argon 2013). Based on the Eshelby solution, the microgeometrical and micromechanical properties of the embedded STZ, i.e., size, shape, and transformation shear strain, are determining parameters for evaluating the STZ's nucleation energy in glassy solids. Therefore, these characteristics have been extensively studied, especially for metallic glasses (Yang et al. 2007; Pan et al. 2008; Pan et al. 2009; Ju et al. 2011). It has been found that a single STZ in MGs possesses an avearge volume of less than 10 nm^3 including $\leq$500 atoms (Pan et al. 2008), with the average nucleation energy of about 1.5 eV (Yu et al. 2010) and transformation shear strain of 0.07 (Argon 2013).

In contrast, the number of studies devoted to evaluate the STZs quantitatively in glassy polymers is limited. The plastic deformation units pertaining parameters of some glassy polymers have been obtaind by Argon and Bessonov (1977) based on double kink theory (Argon 1973). Later, in a molecular dynamics simulations, Mott et al. (1993) scrutinized the plastic deformation kinematics in amorphous atactic polypropylene, and found the transformation shear strain of about 0.015 in the spherical plastic flow units with the average 10 nm diameter. Moreover, Ho et al. (2003) also inquired the correlation between the STZ size scale and entanglement density by performing different compressive tests on mixable polystyrene-poly(2,6-dimethyl-1,4-phenylene oxide) (PS-PPO) blends at different mix ratios.

In this chapter, the nanoindentation technique is used to probe the flow nature of poly(methyl methacrylate) (PMMA) and polycarbonate (PC) as glassy polymers, and, consequently, their STZs' micromechanical and geometrical characteristics are obtained. This chapter is organized as follows: in the first section, the experimental procedure is explained in two parts: sample preparation and the nanoindentation technique. In the second section, it is first shown that at tested temperatures and strain rates, the flow is homogeneous in these polymers. The homogeneous flow theory is then elaborated based on the Eshelby solution for nonelastic strain in an embedded inclusion in the representative volume element (RVE) and the Arrhenius function relating the shear flow stress to the shear flow strain rate. In the third section, the obtained results from the nanoindentation experiments on samples with two different thermal histories, as-cast and annealed, are presented. The characteristic properties of STZs including the nucleation energy barrier, size and shape of an STZ, and the shear activation volume are then obtained by utilizing the flow theory. Furthermore, the observed jump in the activation energy of the STZ is ascribed to the β-transition, and the energy barrier for this transition is found to be about 10% of the STZ nucleation energy. At the end, the concluding remarks are summarized in the last section.

Experimental Procedure

Sample Preparation

Commercially available 2.0- and 5.0-mm-thick sheets of amorphous poly (methyl methacrylate) (PMMA) and polycarbonate (PC), Goodfellow® catalogue #ME303020 and #L5433027, Cambridge, UK, are selected for this study, respectively. There is no preexisting molecular chain orientation in the sheets since they have been produced through the traditional method of cell cast. All the sheets are cut into 20×20 mm^2 samples small enough for handling the nanoindentation experiments. The residues of the protective film covering the sheets are then removed by washing them with 30% isopropyl alcohol (IPA), and thoroughly rinsing by distilled water. To eliminate any moisture caused by the washing process, all the specimens are stored in a desiccator for at least 10 days before any experiment. The glass transition temperature (T_g) of the samples are measured by means of a TA Instruments 2920 differential scanning calorimetry (DSC) device, which is operating under the nitrogen flow and by using standard aluminum pans. The DSC cycles are conducted at 10 °C min^{-1} from ambient temperature to 200 and 250 °C for PMMA and PC, respectively. The calorimetric measurements reveal that the glass transition temperature of the PMMA and PC specimens are about 110 and 148 °C, respectively.

Half of the samples, prepared as described above, are subjected to the following thermal treatment to study the effect of thermal history on their STZs' micromechanical and microstructural characteristics. Initially, they are annealed at 120 °C for 4 h in a vacuum oven. They are then cooled down at 10 °C h^{-1} to room temperature. This thermal process is done in a vacuum oven to prevent the specimens' surfaces oxidation (Hirata et al. 1985). An Agilent 5500 atomic force microscope (AFM) and a Wyko Optical Profiler are used to measure the surface roughness of the

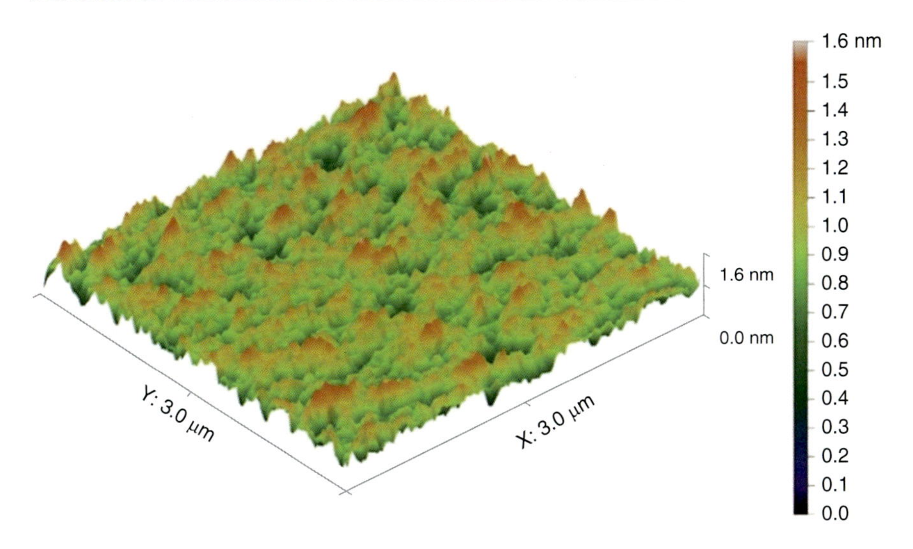

Fig. 1 Sample of AFM scanning of the PMMA sample surface (Reprinted from Malekmotiei et al. 2015)

samples, which markedly has effects on the nanoindentation experiment results (Kim et al. 2007; Nagy et al. 2013). The average surface roughness (R_a) of the PMMA and PC samples are measured to be 0.372 ± 0.013 and 0.305 ± 0.021 nm, respectively. Accordingly, since the samples' surfaces can be assumed as almost flat (Kim et al. 2007), no modification is needed for the obtained experimental results. Figure 1 represents a sample AFM scanning of the specimen surface. To indent the samples, they are mounted on aluminum stubs applying thermoresistant epoxy putty (Drummond™ Nu-Doh Epoxy Repair Compound Titanium Reinforced) for indentations at high temperature on hot-stage apparatus and super glue for indentations at ambient temperature.

Nanoindentation Technique

The nanoindentation technique has been utilized to measure the mechanical properties of the materials including the elastic modulus, *E*, and the hardness, *H*. The indentation load-hold-unload cycles have been performed using an MTS Nanoindenter® XP equipped with a three-sided pyramidal Berkovich diamond tip. The analysis of the applied load-indentation depth curves is mainly based on the original formalism developed by Oliver and Pharr (1992, 2004) as described in the following.

The hardness is described as the mean contact pressure under the indenter as follows:

$$H = \frac{P_{\max}}{A_c} \tag{1}$$

in which P_{max} is the peak indentation load and A_c is the projected area of the tip-sample contact at the maximum load. According to Eq. 1, one needs an accurate measurement of the projected contact area under the load to evaluate the hardness from indentation load-displacement data. For a perfectly sharp Berkovich indenter, the contact area which is a function of the contact depth, h_c, can be calculated as

$$A_c = 24.56{h_c}^2 \tag{2}$$

For a practical indenter, which is not ideally sharp, the contact area function is required to be obtained by tip calibration with introducing additional terms to the aforementioned second-order relation as follows:

$$A_c = 24.56{h_c}^2 + C_1{h_c}^1 + C_2{h_c}^{1/2} + C_3{h_c}^{1/4} + \cdots + C_8{h_c}^{1/128} \tag{3}$$

where C_1 through C_8 are constant coefficients which account for deviations from ideal geometry due to the blunting of the tip, and determined by using the indentation results on a standard fused silica sample and curve fitting performed on the Analyst® software. To obtain the exact contact area, an accurate determination of the depth over which the test material makes contact with the indenter, h_c, is required. The contact depth is generally different from the total penetration depth, and is estimated using

$$h_c = h - \varepsilon \frac{P}{S} \tag{4}$$

in which S is the measured elastic contact stiffness and ε is a constant which depends on the indenter geometry (for a Berkovich indenter $\varepsilon = 0.75$) (Oliver and Pharr 1992).

The elastic modulus of the sample, E, is calculated from Eq. 5 as follows:

$$\frac{1}{E_r} = \frac{1 - \nu^2}{E} + \frac{1 - {\nu_i}^2}{E_i} \tag{5}$$

In this equation, E_i and ν_i are the indenter elastic modulus and Poisson's ratio, respectively, ν is the sample Poisson's ratio, and E_r is the reduced elastic modulus. The reduced modulus which accounts for elastic deformation in both the indenter and the sample can be obtained from the following relation developed by Sneddon (1965)

$$E_r = \frac{S}{2\beta}\sqrt{\frac{\pi}{A_c}} \tag{6}$$

where β is a constant that depends on the indenter geometry and equals to 1.034 for the Berkovich tip.

Room-Temperature CSM Nanoindentation

The room temperature nanoindentation experiments are conducted by employing the continuous stiffness measurement (CSM) technique. The CSM method makes the continuous measurement of the mechanical properties of materials possible during the indentation loading segment from zero to the maximum indentation depth during a single test; this includes the measurement of the elastic contact stiffness at any point along the loading curve, and not just at the point of initial unload as in the basic mode (Pethica and Oliver 1988; Lucas et al. 1998; Li and Bhushan 2002; Hay et al. 2010). In this technique, a small sinusoidally varying load is superimposed on top of the primary loading signal that drives the motion of the indenter, and the resulting response of the system is analyzed by means of a frequency-specific amplifier. The displacement amplitude and the frequency of the superimposed oscillating force are set as 2 nm and 45 Hz, respectively, which are optimum values for the MTS nanoindentation® XP.

Utilizing the so-called CSM technique, the indentation load-hold-unload cycles are performed by keeping the loading rate divided by the load ratio $\left(\dot{P}/P\right)$ constant during the loading segment over the course of a single indentation test. In a deep indentation test, where the indentation size effect (ISE) is negligible (i.e., the hardness value is almost constant), the loading path follows a Hertzian contact relation as $P=\alpha h^{\beta}$, where P represents the indentation load, h is the indentation depth, α is a material dependent parameter, and β is a curve fitting parameter close to 2 for the Berkovich tip (Johnson 1987; Zhang et al. 2005). Accordingly, for a pyramidal indenter, the indentation strain rate $\dot{\varepsilon}_i$, which is defined as the descent rate of the indenter divided by its instantaneous depth, can be obtained by

$$\dot{\varepsilon}_i = \frac{\dot{h}}{h} = \frac{1}{\beta}\frac{\dot{P}}{P} \tag{7}$$

Therefore, since the $\dot{P}/P$ ratio is remained constant during the CSM test, at deep enough indentations, the indentation strain rate approaches a constant value equal to $\frac{1}{2}\frac{\dot{P}}{P}$ (Lucas and Oliver 1999; Voyiadjis and Malekmotiei 2016). Moreover, the effective shear strain rate induced by the indentation is then related to the indentation strain rate as follows:

$$\dot{\gamma} = \sqrt{3}C\dot{\varepsilon}_i = \frac{\sqrt{3}C}{\beta}\frac{\dot{P}}{P} \tag{8}$$

where C is a constant equal to 0.09 (Poisl et al. 1995; Schuh and Nieh 2003; Schuh et al. 2004).

After samples preparation and calibration of the indenter tip function, the tests are carried out by typical loading-hold-unloading sequences as follows: prior to any measurement, the tip is held on the top of the sample surface until its drift rate is stabilized at a rate below 0.05 nms^{-1}. The tip then starts to travel downward until it reaches the surface. Once the indenter tip comes into the contact with the sample surface, the loading segment begins with the constant value of $\dot{P}/P$ until a predefined maximum depth of about 10 μm; the load is then held at the maximum value for 10 s to account for the material creep behavior of the polymer surface; and finally, the unloading stage is carried out with the constant unloading rate until 10% of the maximum load is attained. Since one of the main purposes of this study is investigation of the strain rate dependency of the flow in PMMA and PC, a series of experiments are conducted at room temperature with the values of $\dot{P}/P$ varying from 0.001–0.2 and 0.002–0.4 s^{-1} on annealed and as-cast PMMA and PC samples, respectively. A total of 25 indents are performed for each $\dot{P}/P$ value with a minimum distance of 150 μm between neighboring indents to prevent from interaction.

High-Temperature Nanoindentation Procedure

The MTS Nanoindenter® XP is equipped with temperature control system for elevated temperature nanoindentation tests. The system includes a hot stage, a coolant apparatus to transfer the extra heat to the outside of the instrument, and a heat shield to keep the indenter transducer apart from the heat source. The load control experiments are performed by using the basic method which is employed for high-temperature indentations. To make sure that there is no indentation size effect, and making the obtained date comparable to those of the CSM tests as well, the experiments are carried out with a maximum load of 300 mN. Loading rates of 4, 10, 50, 100, and 300 mNs^{-1} for PMMA and 10, 100, and 300 mNs^{-1} for PC, which are constant during the tests, are applied for temperatures varying from room temperature to 100 and 140 °C (slightly lower than the samples glass transition temperature) for PMMA and PC, respectively.

While the loading rate $\dot{P}$ is kept constant during the loading and unloading segments of the basic mode, evaluating the $\dot{P}/P$ ratio shows that it changes with $1/h^{\beta}$ during the test since the load-depth (P-h) curve follows a Hertzian relation according to the statements in the preceding section. Therefore, the indentation strain rate $\dot{\varepsilon}_i$ varies as $1/h^{\beta}$ during a test as well. Using Eqs. 7 and 8 to calculate the effective shear strain rate, an average value of $\dot{P}/P$ over the deep part of the indentation (5 μm in this study) is considered in each test. Accordingly, the corresponding effective shear strain rates for the load rates given above are 0.0014, 0.0035, 0.0175, 0.035, and 0.105 s^{-1} for PMMA and 0.0035, 0.035, and 0.105 s^{-1} for PC, respectively.

Applying a thermoresistant epoxy putty (Drummond™ Nu-Doh Epoxy Repair Compound Titanium Reinforced), the samples are mounted on the hot stage. Since polymer samples and the adhering thermoresistant have low thermal conductivity, the temperature of the sample surface can be considerably different from the set temperature. For precise evaluation, at the end of the tests, an Omega® SA1-K-SRTC thermocouple is attached to a sample surface to measure its temperature. The temperature is recorded by means of an Omega® HH74K handheld monitoring device coupled with the thermocouple. The discrepancy between the measured and set temperatures is significant for high temperatures. The number of indentation points for each set temperature and loading rate is reduced to 12 for the elevated temperature tests to minimize the risk of the contamination of the tip by the softened polymer at high temperatures; the minimum distance of the adjacent indents is also increased to 200 μm. Prior to performing the loading cycle, each sample is heated to the set temperature inside the indenter; and then it is left to equilibrate for 2 –3 h. On the onset of the test cycle, the tip is held at the distance of about 1 μm from the sample surface for about 10 min to adjust allowable thermal drift. This delayed contact is believed to help the tip to reach a thermal equilibrium with the sample.

Theory of Homogeneous Flow for Glassy Polymers

The inhomogeneous flow in amorphous solids is mediated by the formation of shear bands, whereas their homogeneous flow is triggered by the nucleation of STZs (Spaepen 1977). There are some important differences between inhomogeneous and homogenous flows of glassy solids. While the inhomogeneous flow is strain rate independent in a way that the flow stress does not change significantly with the strain rate, in a homogenous flow, the higher strain rate applied to the sample results in a considerable higher flow stress (Schuh and Nieh 2003; Schuh et al. 2007; Yang et al. 2007). Another significant distinction is the generation of multiple pop-ins in the load-displacement curves during the inhomogeneous flow which appears only in the nanoindentation experiments (Golovin et al. 2001; Zhang et al. 2005; Yang et al. 2007).

It has been shown that the flow behavior of metallic glasses is temperature dependent in a sense that there exists a transition from inhomogeneous to homogeneous flow at a certain temperature. The temperature at which the transition happens is also strain rate sensitive: the higher the applied strain rate, the higher the transition temperature (Yang et al. 2007). To investigate the flow nature in PMMA and PC as glassy polymers, the nanoindentation tests are performed on both as-cast and annealed samples. Figures 2 and 3 represent the variation of the hardness with temperature for as-cast and annealed PMMA and PC samples at different loading rates, respectively. Considering the direct relation between the hardness H and the flow stress σ_y as $H=\kappa\sigma_y$ with κ is Tabor's factor, these figures show that the hardness (or flow stress) is greatly strain rate sensitive in a way that a higher hardness is obtained at a higher strain rate (loading rate) at a given temperature. This observed rate-dependent softening, which is attributed to the thermally activated nature of the flow, indicates that the flow of PMMA and PC at tested temperatures, which are below their glass transition temperatures T_g, is homogeneous. In addition, Figs. 4 and 5 show the load-displacement (*P-h*) curves of the samples at the loading rates of 10 and 300 mNs^{-1} and different temperatures. All the curves are represented with the origin offset of 2 μm except the first one at room temperature. Since the curves are smooth with no pop-in events, Figs. 4 and 5 further confirm the

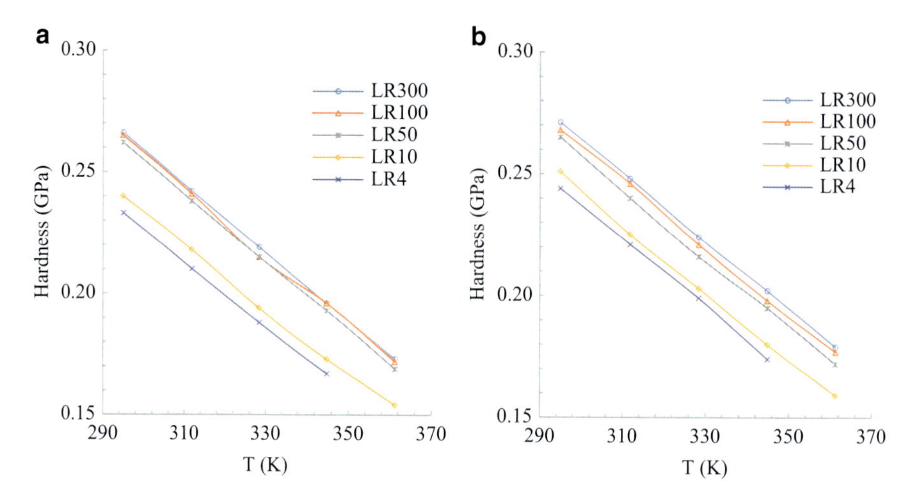

Fig. 2 Variation of the PMMA hardness with temperature at different loading rates for (**a**) as-cast and (**b**) annealed samples. The data points at the load rate of 4 mNs^{-1} and 361 K are not shown due to their high standard deviation values (Reprinted from Malekmotiei et al. 2015)

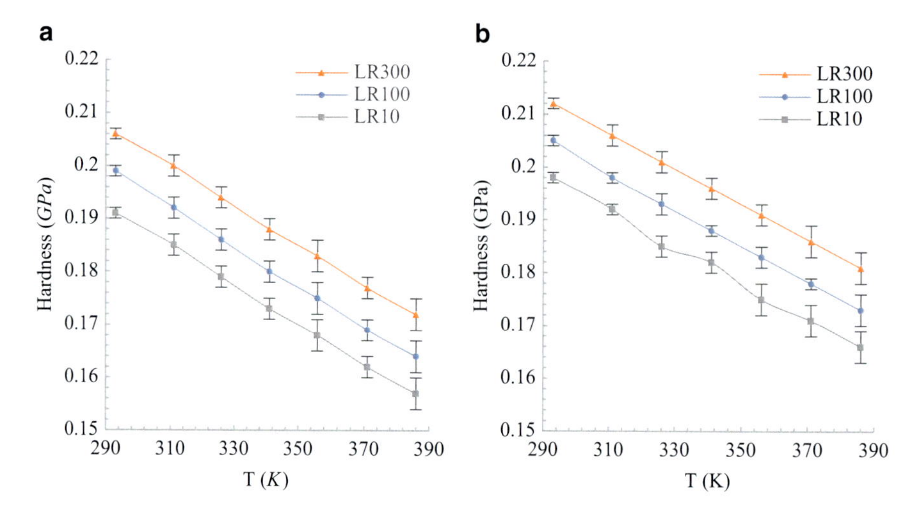

Fig. 3 Variation of the PC hardness with temperature at different loading rates for (**a**) as-cast and (**b**) annealed samples (Reprinted from Samadi-Dooki et al. 2016)

homogenous nature of the flow in PMMA and PC at tested temperatures and strain rates.

The flow in glassy solids is mediated by the irreversible local disturbances which form rearranged atomic (in MGs) or molecular (in PGs) clusters. These cooperative rearrangements result in isolated unit increments of shear, and are known as shear transformation zones (STZs). While in the homogeneous flow regime each volume element has contribution to the total plastic strain, the strain is localized in distinct

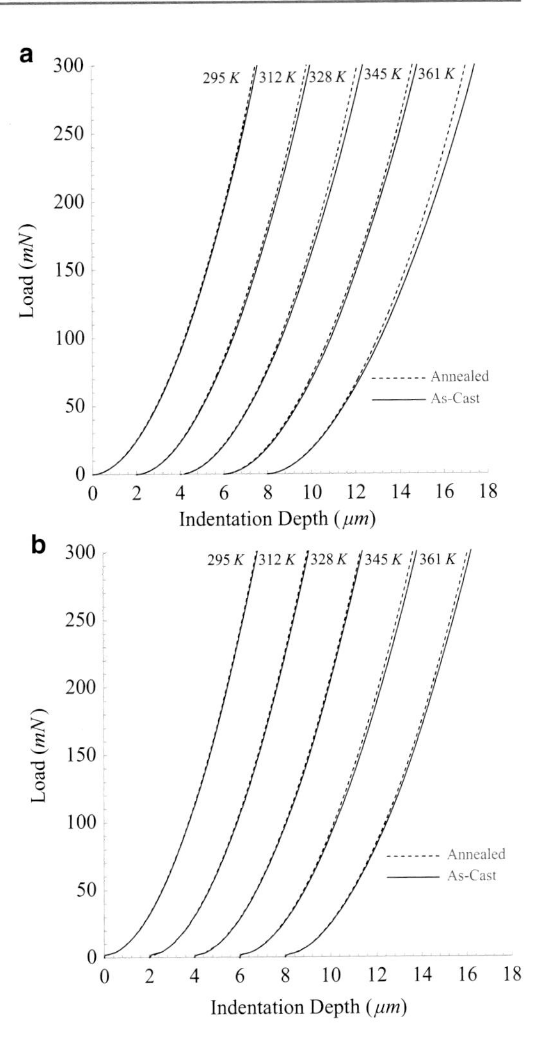

Fig. 4 Load-indentation depth curves for as-cast and annealed PMMA samples at different temperatures and the loading rates of (**a**) 10 mNs^{-1}, and (**b**) 300 mNs^{-1} (Reprinted from Malekmotiei et al. 2015)

shear bands in the inhomogeneous flow regime (Spaepen 1977). The thermally activated homogeneous flow in glassy polymers can be described based on the flow mechanism developed by Spaepen (1977) and Argon (1979). For the STZs-mediated homogeneous flow in glassy solids, the kinetics relation for the shear strain rate $\dot{\gamma}$ due to the applied shear stress τ is well expressed by an Arrhenius relation as follows (Spaepen 1977; Argon 1979):

$$\dot{\gamma} = \dot{\gamma}_0 \exp\left(-\frac{\Delta F_0}{k_B T}\right) \sinh\left(\frac{\gamma^T \Omega \tau}{2k_B T}\right) \tag{9}$$

where $\dot{\gamma}_0$ is the pre-exponential factor proportional to the attempt frequency, k_B is the Boltzmann constant, T is the absolute temperature, and ΔF_0 is the nucleation

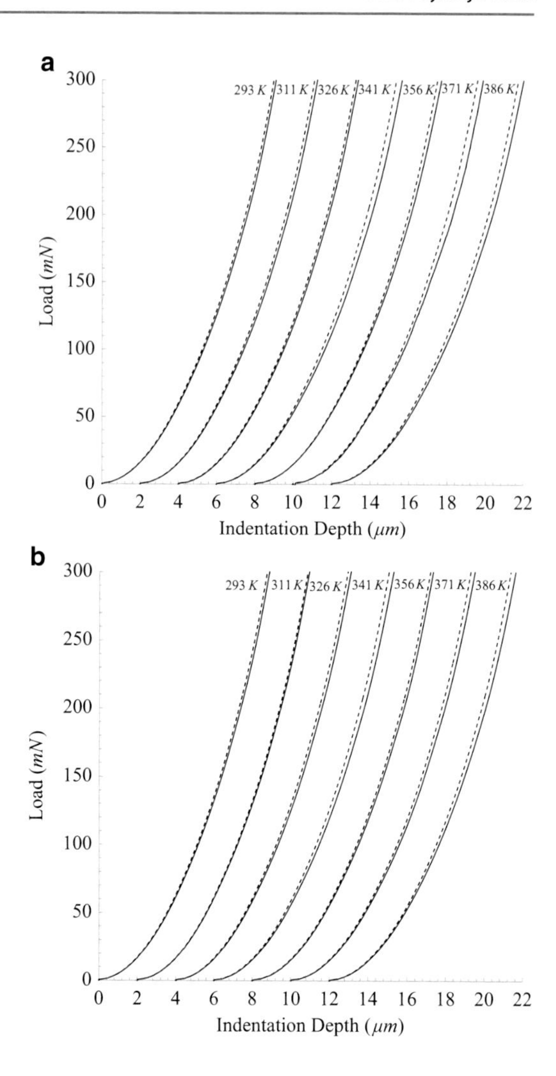

Fig. 5 The load-indentation depth curves for as-cast (*solid line*) and annealed (*dashed line*) PC samples at different temperatures and loading rates of (**a**) 10 mNs^{-1} and (**b**) 300 mNs^{-1} (Reprinted from Samadi-Dooki et al. 2016)

energy of an STZ with the shear strain y^T occurring in a region of volume Ω. The factor 2 in the denominator of the argument of the hyperbolic function is due to the reverse transformation probability (Spaepen 1977). To evaluate the nucleation energy of an STZ, this locally transformed region has been treated as an embedded volume with nonelastic strain the micromechanical field of which can be obtained by using the Eshelby inclusion model (Eshelby 1957). For the RVE shown in Fig. 6, the free energy of the nucleation of a single STZ is given as follows:

$$\Delta F_0 = \left[\Xi\left(\nu\right) + \Psi\left(\nu\right)\beta^2\right]\mu\left(\gamma^T\right)^2\Omega \tag{10}$$

in which μ is the shear modulus, $\Xi(\nu)$ and $\Psi(\nu)$ are functions of the Poisson's ratio ν, and pertain to the shear and dilatational components of the transformation strain

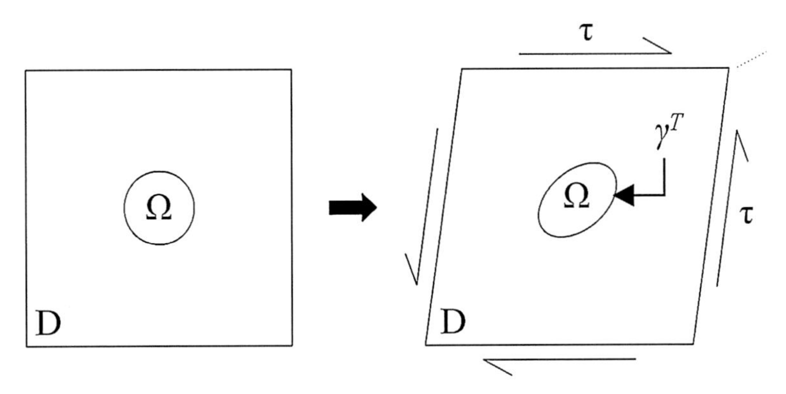

Fig. 6 Representative volume element (RVE) of polymer matrix containing a single shear transformation zone (STZ) (Reprinted from Malekmotiei et al. 2015)

tensor, respectively. The coefficient β is the dilatancy parameter and can be obtained from the pressure sensitivity of the flow. While $\Psi(\nu) = \frac{2(1+\nu)}{9(1-\nu)}$ is independent of the aspect ratio of the ellipsoid, $\Xi(\nu)$ is shape dependent (Mura 1987). The STZ's characteristics in Eq. 9, such as shear strain, size, and activation energy, have been experimentally obtained for different metallic glasses (see Chap. 7 of Argon (2013) for details), but their numerical evaluation for PGs have been limited to some theoretical modeling and simulations (Mott et al. 1993; Argon 2013), and a few experimental studies (Argon and Bessonov 1977; Ho et al. 2003).

Since $\gamma^T \Omega \tau \gg 2k_B T$ for conventional PGs at temperatures below their glass transition, Eq. 9 can be rearranged as:

$$\ln \dot{\gamma} = \frac{\gamma^T \Omega}{2k_B T}\tau + C_1 \tag{11}$$

where $C_1 = \ln \frac{\dot{\gamma}_0}{2} - \frac{\Delta F_0}{k_B T}$ represents a temperature-dependent parameter. In Eq. 11, the material constant $\gamma^T \Omega$, which is proportional to the important characteristic parameter of glassy polymers known as shear activation volume V^* (Ward 1971), can be determined from the derivative of the natural logarithm of the strain rate ($\ln \dot{\gamma}$) with respect to the shear flow stress (τ). Accordingly, considering the effect of hydrostatic pressure on the shear yield stress of polymers, the modified shear activation volume is obtained from Eq. 12 as follows (Ward 1971; Ho et al. 2003):

$$V^* = (1 - \alpha\beta)\,\gamma^T \Omega = 2k_B T\,(1 - \alpha\beta)\,\frac{\partial\, ln\, \dot{\gamma}}{\partial \tau} \tag{12}$$

where β is the yield stress sensitivity to the pressure as defined before, and α is the loading condition constant which is between 0.6 and 0.7 for different compressive loading conditions (Ward 1971; Tervoort 1996). In fact, one can conclude from Eq. 12 that the shear activation volume is a modified STZ volume with considering the dilatation effect. Furthermore, since γ^T is assumed to be a universal constant for all glassy polymers and equals about 0.02 (Mott et al. 1993; Ho et al. 2003), the size of the single shear transformation zone Ω can be obtained. Rearranging Eq. 9 for a constant strain rate, the shear flow stress can be expressed as a linear function of temperature as follows:

$$\tau = \Theta C_2 + \frac{2\Delta F_0}{\gamma^T \Omega} \tag{13}$$

where $C_2 = \frac{2k_B}{\gamma^T \Omega}\left(\ln \dot{\gamma} - \ln \frac{\dot{\gamma}_0}{2}\right)$ is representing a strain rate-dependent constant. As a result, the activation energy of an individual STZ, ΔF_0, can be calculated from the linear interpolation of the variation of the flow shear stress with temperature by incorporating the obtained $\gamma^T \Omega$ values. Knowing all the STZ's parameters in Eq. 10, one can obtain the numerical value of $\Xi(\nu)$ and, consequently, an approximation of the STZ's shape in glassy polymers.

Results and Discussion

Calibrating the Nanoindentation Results

The instrumented-indentation testing (IIT) can be employed for the purpose of probing the mechanical properties including hardness and elastic modulus of very small volumes of materials including polymers, which both of them depend on the applied load on the sample surface and the contact area. In the nanoindentation experiments, the load is recorded with an nN scale precision while the contact area is calculated as a function of the tip geometry and contact depth. Since the tip in a practical indenter is not ideally sharp, the contact area function is required to be obtained by calibrating the tip with introducing a polynomial approximation (Oliver and Pharr 1992; Voyiadjis and Zhang 2015). Using the results of the indentation on standard fused silica sample, the tip areal function is calibrated to obtain reliable results.

It is also worth noting that the material pile-up around the penetrating tip can significantly alter the contact area and affect the measured mechanical properties of the material especially at shallow depths of the indentation. Using the optical profiler, the pile-up values are precisely measured for the indentation on PMMA and PC samples. As can be seen in Fig. 7, which shows a sample of pile-up measurement on PC specimen, the pile-up is highly unsymmetrical with the maximum values around the pyramidal tip faces. The maximum pile-up is measured to be about 400 nm, which in comparison with the maximum indentation depth of 10 μm is very small. For that reason, the pile-up effect is neglected in the nanoindentation measurements.

Another important observation during the nanoindentation experiments is the increment of the hardness values at shallow indentation depths which is known as

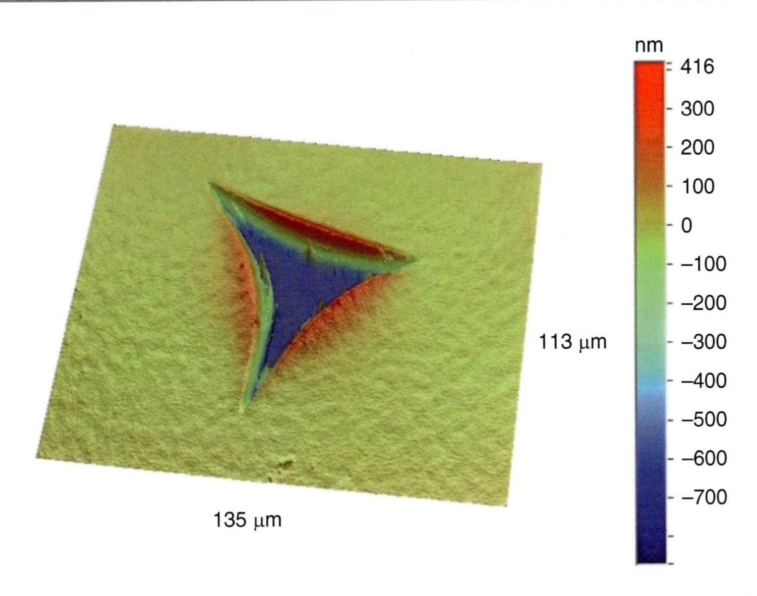

Fig. 7 A sample of the indentation pile-up measurement on PC specimen using Wyko Optical Profiler (Reprinted from Samadi-Dooki et al. 2016)

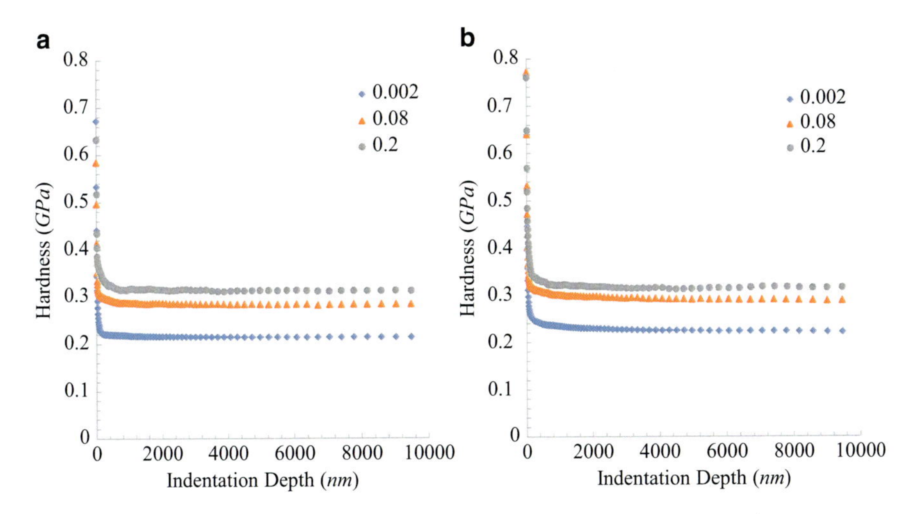

Fig. 8 Variation of the hardness versus the indentation depth for three different $\dot{P}/P$ values for (**a**) as-cast and (**b**) annealed PMMA samples. The legend numbers represent the $\dot{P}/P$ values (Reprinted from Malekmotiei et al. 2015)

indentation size effect (ISE) (Briscoe et al. 1998; Voyiadjis and Zhang 2015).To prevent the ISE in the current study, the CSM indentation results are first evaluated to find the indentation depth beyond which the obtained hardness reaches the stable value. The variation of hardness versus the indentation depth for both annealed and

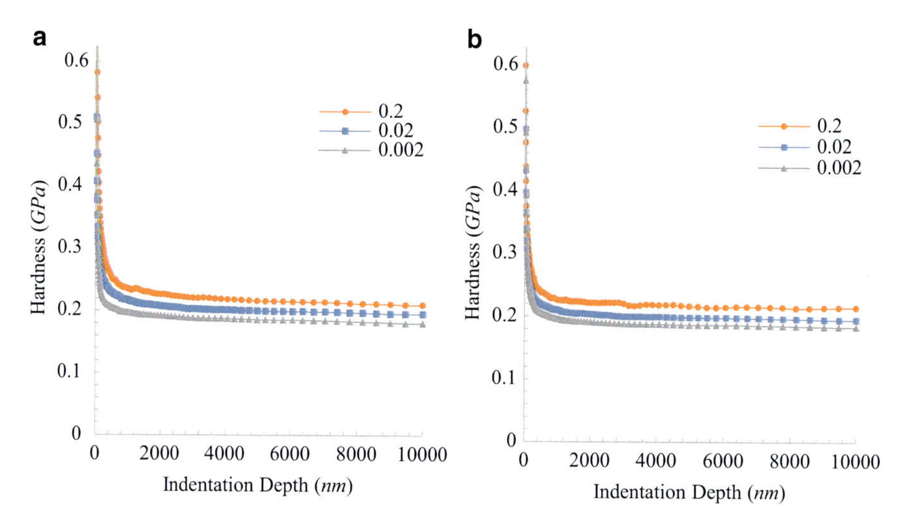

Fig. 9 Variation of the hardness versus the indentation depth for three different $\dot{P}/P$ values for (**a**) as-cast and (**b**) annealed PC samples. The legend numbers represent the $\dot{P}/P$ values (Reprinted from Samadi-Dooki et al. 2016)

as-cast PMMA and PC samples are presented in Figs. 8 and 9, respectively. As these figures show, the profound ISE is completely eliminated for the indentation depth beyond 2 and 4 μm for PMMA and PC samples, respectively. Accordingly, the hardness values are averaged over the indentation depth beyond 5 μm where the indentation size effect is absent. The elevated temperature tests are performed by using the basic mode in which the hardness value is only reported at the maximum indentation depth. While the maximum indentation depth is the input for the CSM technique, the maximum load is the input in the basic mode. As mentioned before, to avoid the indentation size effect, and also make the obtained date comparable to those of the CSM tests, the high temperature experiments are carried out with a maximum load of 300 mN, which corresponds to the maximum indentation depth of about 10 μm. Furthermore, since the ISE reduces at elevated temperatures (Voyiadjis et al. 2011), it is assured that the obtained hardness results are within the stable region.

Shear Activation Volume

In addition to the hardness, the elastic modulus is another mechanical property of the material which is continuously recorded during the loading segment of the CSM nanoindentation as a function of displacement. Figures 10 and 11 represent the variation of the material elastic modulus versus the indentation depth for some selected $\dot{P}/P$ values for PMMA and PC samples, respectively. In comparison to the hardness, the elastic modulus is almost constant for the indentation depth beyond about 100 and 150 nm for both as-cast and annealed PMMA and PC samples,

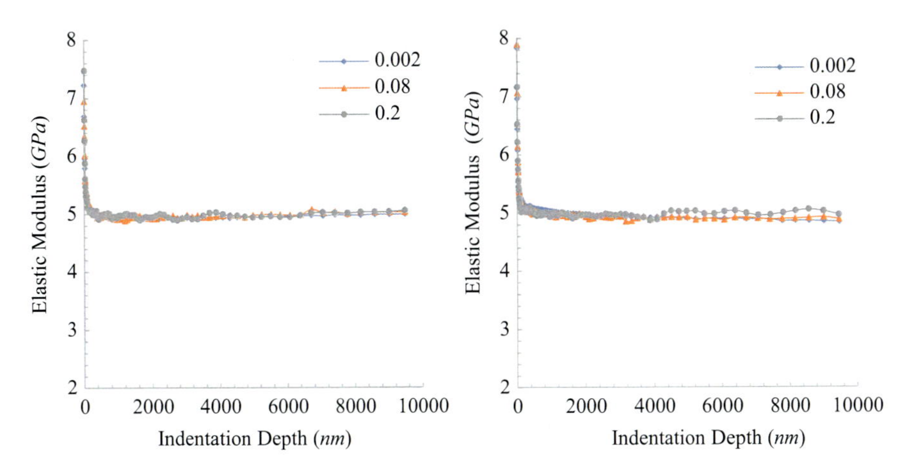

Fig. 10 Variation of the elastic modulus versus the indentation depth for three different $\dot{P}/P$ values for (**a**) as-cast and (**b**) annealed PMMA samples. The legend numbers represent the $\dot{P}/P$ values (Reprinted from Malekmotiei et al. 2015)

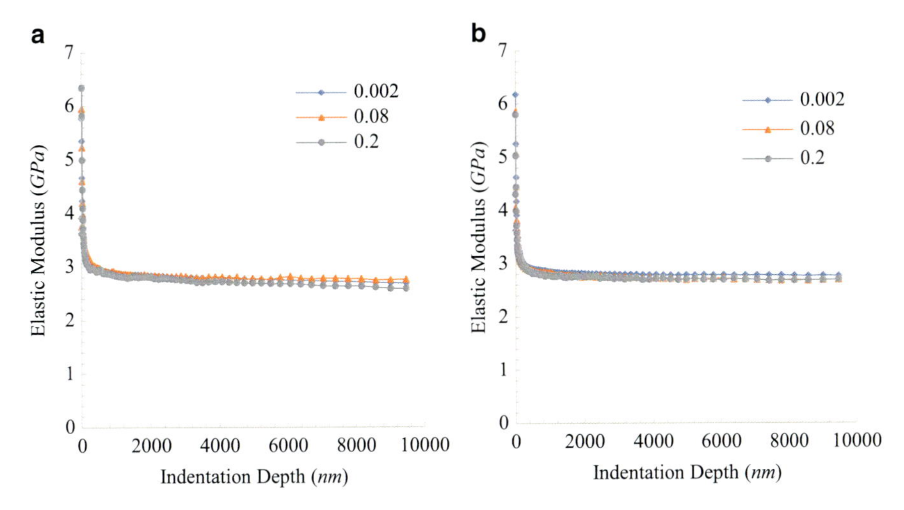

Fig. 11 Variation of the elastic modulus versus the indentation depth for three different $\dot{P}/P$ values for (**a**) as-cast and (**b**) annealed PC samples. The legend numbers represent the $\dot{P}/P$ values (Reprinted from Samadi-Dooki et al. 2016)

respectively. More importantly, while the hardness values significantly change with the strain rate (see Figs. 8 and 9), the elastic modulus does not show any significant strain rate sensitivity.

Using the Tabor's factor, the hardness values H obtained from the nanoindentation experiments can be converted to the yield stress σ_y of the material as follows:

$$H = \kappa \sigma_y \tag{14}$$

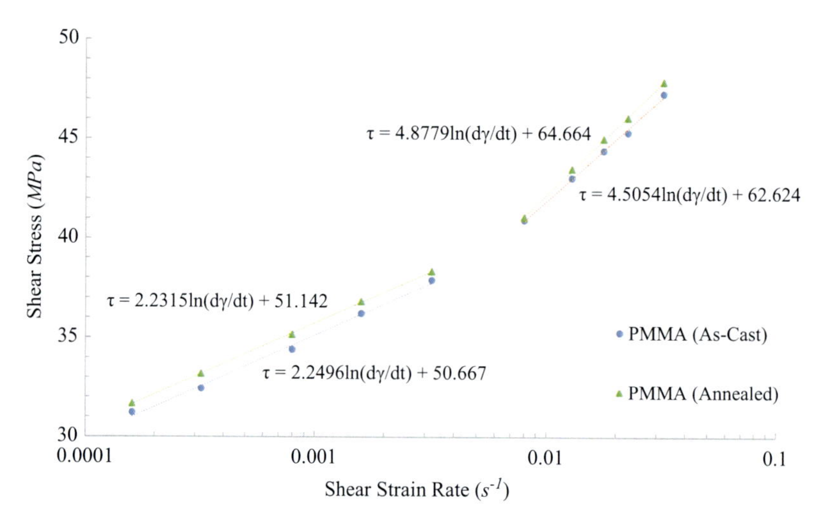

Fig. 12 Variation of the shear flow stress with the shear strain rate for both as-cast and annealed PMMA samples (Reprinted from Malekmotiei et al. 2015)

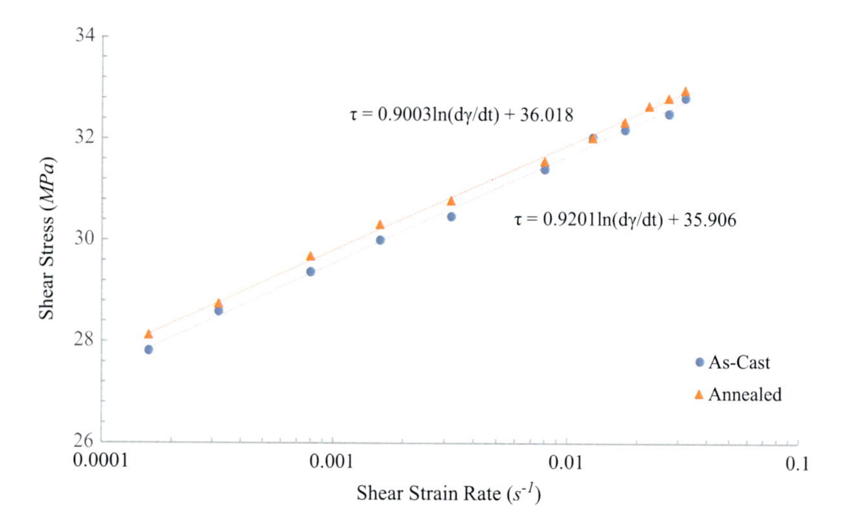

Fig. 13 Variation of the shear flow stress with the shear strain rate for both as-cast and annealed PC samples (Reprinted from Samadi-Dooki et al. 2016)

in which κ is Tabor's factor which is approximately 3.3 for amorphous polymers at high indentation strains (Prasad et al. 2009). Additionally, since the shear flow stress is about half of the yield stress in plane stress condition for monotonic loading (Yang et al. 2007), the ratio of the hardness to the shear flow stress is about 6.6. Figures 12 and 13 illustrate the shear flow stress-shear strain rate data points for the as-cast and annealed PMMA and PC samples measured by CSM method at room temperature, respectively. The obtained results for PC are in good agreement with the ones obtained by Bauwens-Crowet et al. (1969) at 21.5 °C. As it is expected, the annealed samples have slightly bigger shear flow stresses compared to the as-cast ones at the same shear strain rate (Jancar et al. 2013). The outstanding feature of Fig. 12 is the existence of a significant transition at a certain value of the strain rate beyond which the strain rate sensitivity of the shear flow stress increases. This phenomenon is believed to be a result of strain rate shift of the β-relaxation process in the storage modulus of the PMMA, which is related to the restriction of the ester side group rotations at high strain rates, besides the intermolecular and local back bone motion restrictions (Calleja et al. 1994; Mulliken and Boyce 2006; Argon 2013). As obviously shown in Fig. 12, the transition shear strain rate is approximately 0.005 s^{-1} for both as-cast and annealed PMMA samples. Following the descriptions presented in Mulliken and Boyce (2006), the flow stress regimes below and above the transition shear strain rate might be referred to as α and β regimes, respectively. However, it is noteworthy to mention that since the room temperature β-transition strain rate of PC has been previously detected to be about 10^2 s^{-1} (Mulliken and Boyce 2006), which is beyond the strain rates that can be applied in nanoindentation experiments, no considerable jump is observed in Fig. 13 for the range of strain rates in this study.

Based on Eqs. 11 and 12, the shear activation volume V^* for an amorphous polymer can be obtained by linear interpolation of the $\tau - \ln \dot{\gamma}$ curve. In these figures, the slopes of the semilogarithmic stress-strain rate plots are almost the same for samples with different thermal history, which suggest that the shear activation volume and, therefore, the size of a single STZ are almost independent of the thermal history of the samples. Incorporating Eq. 11 and the data represented on Figs. 12 and 13, the factor $\gamma^T \Omega$ for as-cast and annealed samples is obtained about 3.66 and 3.69 nm^3 for PMMA, and 8.94 and 9.14 nm^3 for PC, respectively. Accordingly, by assuming $\alpha = 0.65$, $\beta = 0.204$ for PMMA (Ward 1971) and 0.27 for PC (Rittel and Dorogoy 2008), V^* is found to be 3.17 and 3.20 nm^3 in α regime for PMMA, and 7.37 and 7.54 nm^3 for PC, for as-cast and annealed samples, respectively. The obtained values of shear activation volume are in consonance with the molecular dynamics simulation results (Argon 2013). As results show, the shear activation volume for samples with different thermal histories is almost the same, and this small discrepancy might be due to the short time of annealing in this study (4 h). Since it has been shown that the flow stress of glassy polymers increases logarithmically with the annealing time at temperatures below their glass transitions (Hutchinson et al. 1999), a more profound difference might be expected between the shear flow stress results of the as-cast and annealed samples in Figs. 12 and 13 for longer annealing time. However, the increased difference may or may not result in a considerable difference in the shear activation volume since it depends on the slope of the $\tau - \ln \dot{\gamma}$ plots and not the shear flow stress solely.

STZ's Activation Energy

In light of Eq. 13, linear interpolation of the flow stress as a function of the temperature $\tau - T$ can be used to obtain the STZ's activation energy. Figures 14 and 15 represent the variance of the shear flow stress of PMMA and PC with temperature, respectively, for different strain rates which is well interpolated with linear functions at each loading rate. Assuming the parameter $\gamma^T \Omega$ does not vary with temperature, the STZ's activation energy for both as-cast and annealed samples can be calculated at different shear strain rates as shown in Fig. 16 for PMMA. One of the most important features of this figure is the existence of jump in the activation energy at the strain rate range of 0.0035–0.0175 s^{-1} which is consistent with the β-transition strain rate obtained from room-temperature CSM nanoindentation. Therefore, this jump might be referred to as the β-transition activation energy. Although the discrepancy in the activation energy for the annealed and as-cast samples is small for strain rates above the β-transition strain rate, the difference is profound for strain rates below this transition. Since the annealed PMMA sample is expected to have more ordered chains in comparison to the as-cast one, the slip and rotation of these chains are more restricted in this sample; consequently, the STZ's activation energy increases. In contrast, beyond the β-transition strain rate, the rapid loading does not allow the chains to rotate or slip smoothly which puts the annealed and as-cast samples in the same deformation condition, and as a result, the activation energy of STZs for high strain rates is approximately identical for samples with different thermal histories.

Another important feature of Fig. 16 is that the β-transition activation energy is much bigger for the as-cast PMMA than the annealed one (almost three times).

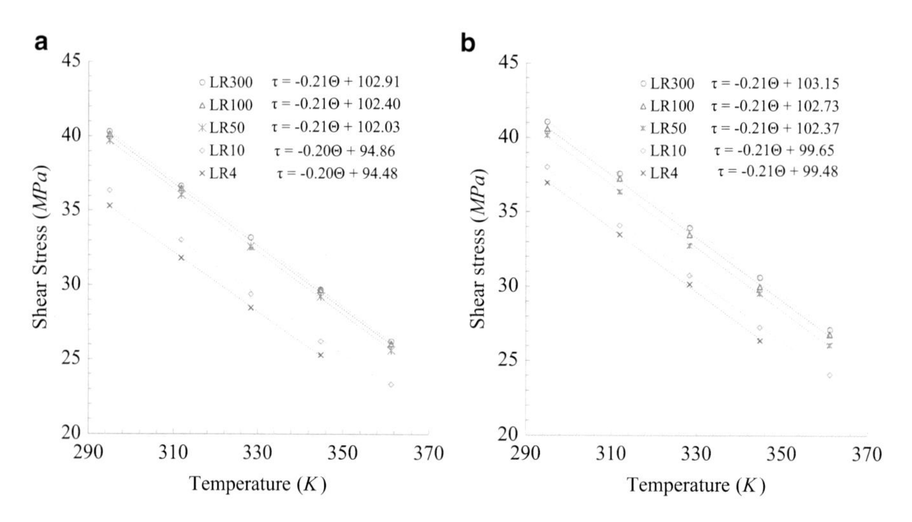

Fig. 14 Variation of the shear flow stress with temperature at different loading rates for (**a**) as-cast and (**b**) annealed PMMA samples (Reprinted from Malekmotiei et al. 2015)

Furthermore, the β-transition energy is about one order of magnitude smaller than the thermal activation energy of an STZ for PMMA, which is in agreement with the findings of Barral et al. (1994) who found almost the same ratio for a system containing a diglycidyl ether of bisphenol A (DGEBA) and 1,3-bisaminomethylcyclohexane (1,3-BAC). In comparison, the β-transition activation energy has been obtained to be almost equal to the STZ's activation energy in the metallic glasses (Yu et al. 2010).

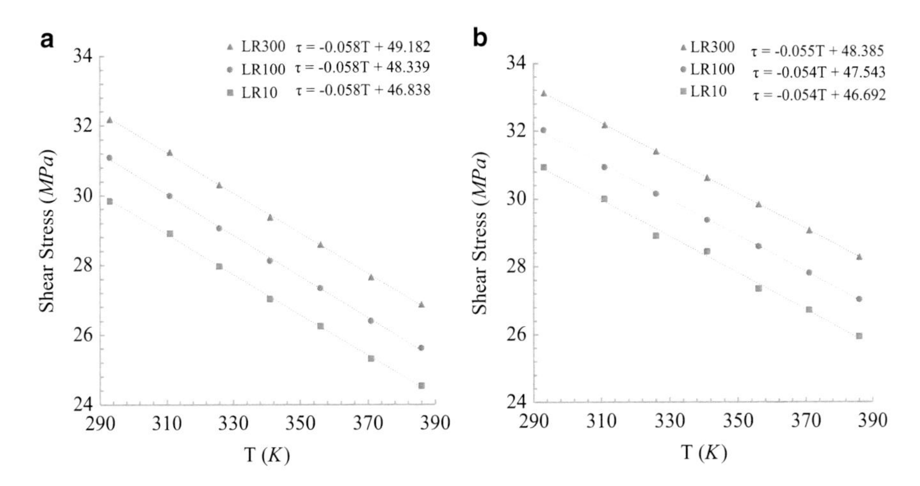

Fig. 15 Variation of the shear flow stress with temperature at different loading rates for (**a**) as-cast and (**b**) annealed PC samples (Reprinted from Samadi-Dooki et al. 2016)

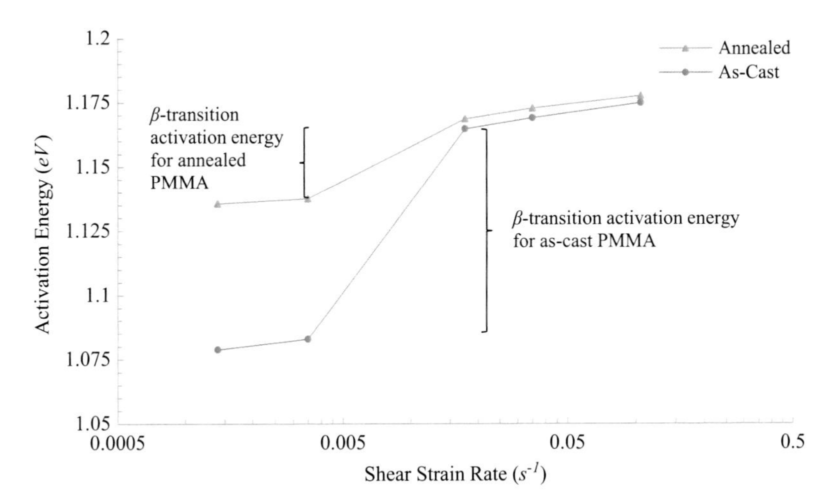

Fig. 16 STZ's activation energy for both as-cast and annealed PMMA samples at different shear strain rates (Reprinted from Malekmotiei et al. 2015)

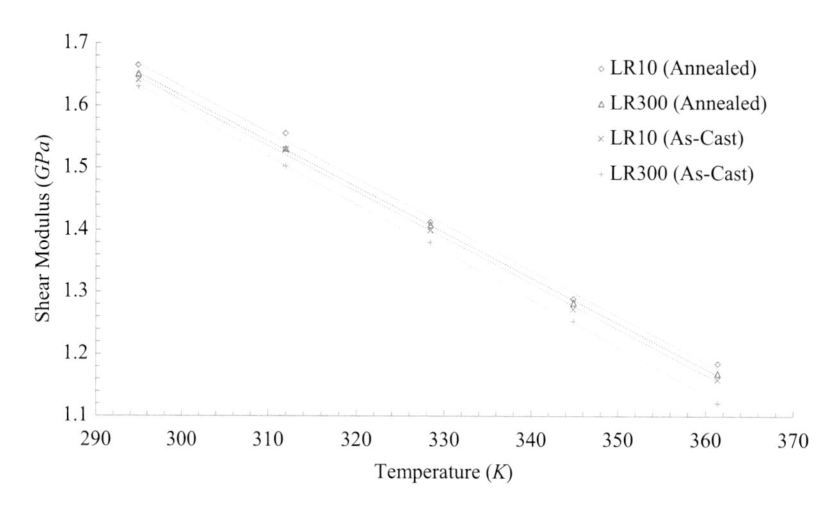

Fig. 17 Variation of the shear modulus with temperature for as-cast and annealed PMMA samples at two different loading rates (Reprinted from Malekmotiei et al. 2015)

As defined in Eq. 10, the Helmholtz free energy also depends on the shear modulus μ of the material which itself is temperature dependent. Since the shear modulus of the PMMA does not change considerably for temperatures below about 100 *K* (Gall and McCrum 1961), Eq. 13 is still valid, and ΔF_0 might be referred to as activation energy at 0–100 *K*. Using the elastic modulus data for different temperatures, obtained by basic method nanoindentation, and the relation between the elastic modulus and shear modulus as: $\mu = \frac{E}{2(1+\nu)}$, the variation of the shear modulus with temperature for both as-cast and annealed PMMA samples at two different strain rates are represented in Fig. 17. Assuming the variation to remain linear for temperatures down to 100 *K*, and constant for temperatures below 100 *K* which is a valid assumption based on the experiments of Gall and McCrum (1961), the variation of the STZ's activation energy with temperature can be obtained as shown in Fig. 18. It is also noticeable that the STZ's activation energy is around 0.6 eV at room temperature which is about one-third of that for metallic glasses (Yu et al. 2010).

Doing the same calculations on the obtained results for PC samples, the activation energy of a single STZ is presented in Table 1. As expected, the activation energy is slightly bigger for the annealed samples, and is almost strain rate insensitive which is in agreement with continuum mechanics principles. Since the obtained activation energies are calculated using the extrapolation of the $\tau - T$ values to the 0 K, they should be referred to as the zero Kelvin STZ's activation energies, and are shown by ΔF_0 hereafter. It should be mentioned that since the room temperature β-transition strain rate of PC is beyond the strain rates in this study, the β-transition activation energy barrier cannot be captured for PC samples.

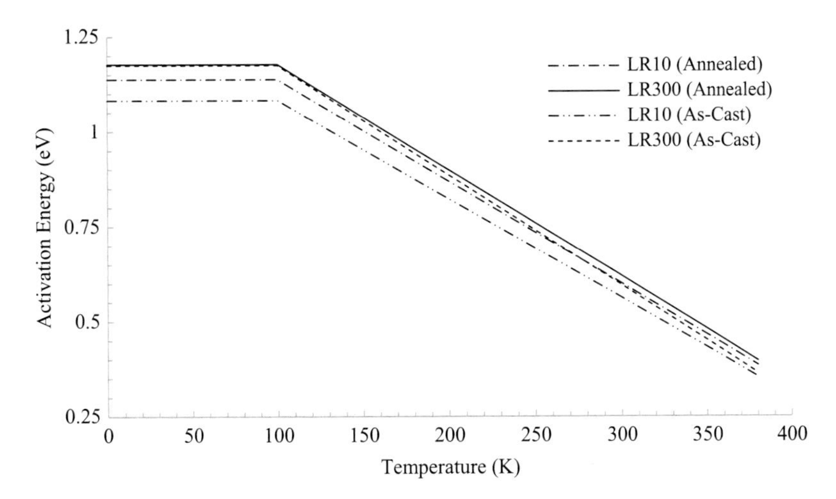

Fig. 18 Variation of the STZ's activation energy with temperature for as-cast and annealed PMMA samples at two different loading rates (Reprinted from Malekmotiei et al. 2015)

Table 1 Characteristic properties of the STZs in PC for samples with different thermal histories (Reprinted from Samadi-Dooki et al. 2016)

Thermal history	$\gamma^T \Omega (nm^3)$	V* (nm^3)	Activation energy (eV) at the shear strain rate of			γ^T	$\Omega (nm^3)$	Number of monomers
			0.0035 (s^{-1})	0.035 (s^{-1})	0.105 (s^{-1})			
As-cast	8.94	7.37	1.32	1.35	1.37	0.019	470	5,596
Annealed	9.14	7.54	1.33	1.36	1.38	0.0187	488	5,810

STZ's Geometry

Incorporating Eqs. 10 and 13, the intercept of $\tau - T$ linear interpolation with τ axis is equal to $\frac{2\Delta F_0}{\gamma^T \Omega} = 2\left[\Xi(\nu) + \Psi(\nu)\beta^2\right]\mu\gamma^T$. As it is mentioned before, $\Psi(\nu) = \frac{2(1+\nu)}{9(1-\nu)}$ does not vary with the STZ shape, and is about 0.5 and 0.48 for PMMA and PC, respectively (considering the Poisson's ratio of 0.38 for PMMA and 0.37 for PC). However, $\Xi(\nu)$ is determined by the shape of the STZ which varies between $\frac{7-5\nu}{30(1-\nu)} = 0.27$ and 0.5 pertaining to the spherical and flat ellipsoidal shape of STZ, respectively (Mura 1987). Taking $\mu_0 = 3.1$ GPa from Fig. 17 and $\beta = 0.204$, the value of γ^T varies between 0.03 and 0.05 for PMMA, with the lower and upper bonds pertaining to the flat ellipsoidal and spherical transformation zones, respectively. Based on the molecular dynamics simulation (Mott et al. 1993), the shear strain (γ^T) of 0.05 is a large value for polymeric glasses; therefore, the shape of the transformation zone is more likely to be a flat ellipsoid for PMMA instead of a sphere. Using the previously obtained factor $\gamma^T \Omega$ and assuming $\gamma^T = 0.03$, the volume of an individual plastic deformation unit Ω, which is almost the same for as-cast and annealed PMMA samples, is obtained about 123 nm^3 which is at least one order of magnitude bigger than that of the metallic glasses (Pan et al. 2008; Yu et al. 2010). Assuming the PMMA monomers as cylinders with the length of 1.55 Å and radius of 2.85 Å (Argon and Bessonov 1977), the single STZ is found to contain about 3000 monomers.

In a same way, using $\mu_0 = 2.4$ GPa (Argon 2013) and $\beta = 0.27$ (Rittel and Dorogoy 2008), γ^T is obtained equal to 0.035 for the spherical and about 0.02 for the flat ellipsoidal shapes of the STZ for both as-cast and annealed PC samples. Due to the aforementioned reason, it can be inferred that the STZs in PC are regions with the shapes close to flat ellipsoids rather than spheres. In addition, the volume of an STZ, Ω, is calculated to be about 470 and 488 nm^3 which has in average 5,600 and 5,800 idealized cylindrical shape monomers with 2.8 Å length and 3.09 Å radius (Argon and Bessonov 1977) for the as-cast and annealed PC samples, respectively. All the obtained characteristic properties of a single STZ in PC samples are presented in Table 1.

Concluding Remarks

In summary, nanoindentation experiments conducted on as-cast and annealed specimens are studied to explore the temperature and strain rate sensitivity of the flow in poly(methyl methacrylate) (PMMA) and polycarbonate (PC) as amorphous polymers. Showing that the flow is homogeneous in these polymers at temperatures below their glass transitions and incorporating a homogeneous flow theory, geometrical and micromechanical characteristics of the shear transformation zones (STZs), as main carriers of plasticity in amorphous polymers, have been investigated in the molecular level. Since the experimental studies on the STZs in glassy polymers is less addressed in the literature in comparison to the metallic glasses, the nanoindentation technique is employed as an accurate, repeatable, and nondestructive method to answer some of the current open questions in this area. The findings suggest that the STZs are flat ellipsoidal regions with the volume of about 123 and 480 nm^3 and the transformation shear strain of about 0.03 and 0.02 in PMMA and PC samples, respectively. In addition, the nucleation energy of the shear transportation zones for both samples as well as the β-transition energy barrier for PMMA samples has been obtained. The procedure used for obtaining the β-transition energy based on the nanoindentation technique is an innovative approach which can be used for other glassy solids. The experimentally evaluated parameters produce unequivocal values as inputs for further theoretical and numerical investigations of yielding and plasticity in polymeric glasses.

In light of the obtained results for PMMA and PC, the following remarks are made:

1. The STZ's nucleation energy in amorphous polymers is about 1 eV which is not considerably smaller than that of the metallic glasses. Since the nucleation energy of an STZ directly depends on the material shear modulus, this energy is expected to be much higher for MGs due to their considerably bigger shear modulus. However, the nucleation energy also depends on the STZ size. Therefore, the bigger size of the STZs in PGs compared to MGs compensates for their shear modulus discrepancy, and levels the activation energy in two materials.
2. The transformation shear strain γ^T is slightly bigger than what is believed to be the universal value for PGs. The transformation shear strain has been considered to be about 0.015 in all types of glassy polymers (Argon 2013); however, the current study suggests that this parameter is unique to a particular polymer, and is about 0.02 in PC and 0.03 in PMMA.
3. While this work suggests that the shear transformation zones are formed in the regions with the shape close to the flat ellipsoid in PMMA and PC, the STZ shape is assumed to be spherical in all types of glassy solids (Ho et al. 2003; Argon 2013; Li et al. 2013).
4. Since all experiments are performed at temperatures beyond 0.6 Tg, the observed homogeneous flow is in accordance with the amorphous flow theory at elevated temperatures (Argon 1979, 2013). With the current nanoindentation technology, it seems impossible to perform experiments at temperatures below 0.6 Tg for available PGs (−43 and −20 °C for PMMA and PC, respectively). Undoubtedly, experiments at temperatures below 0.6 Tg would result in a better understanding of the flow nature in glassy polymers at a wider temperature range.
5. The obtained results for the STZs shape are based on the acceptable values of transformation shear strain in polymers which is considerably smaller than that in metallic glasses. A precise evaluation requires the STZs direct observation, which is not possible and convenient since STZs are local transition events rather than being actual defects like dislocations, or their indirect observation via localized stress field monitoring, which to the best of the authors' knowledge has not been reported yet. The only indirect experimental measurement of the STZ's size scale is one by Liu et al. (2011) in which the 2.5 nm size of the viscoelastic heterogeneities observed by transmission electron microscopy (TEM) has been related to the size of the STZ in a metallic glass.

References

L. Anand, M.E. Gurtin, Int. J. Solids Struct. **40**, 1465 (2003)
A. Argon, Philos. Mag. **28**, 839 (1973)
A. Argon, Acta Metall. **27**, 47 (1979)
A.S. Argon, Mater. Sci. Technol. (1993)
A. Argon, *Strengthening Mechanisms in Crystal Plasticity* (Oxford University Press on Demand, Oxford, 2008)
A.S. Argon, *The Physics of Deformation and Fracture of Polymers* (Cambridge University Press, Cambridge, 2013)
A.S. Argon, M. Bessonov, Philos. Mag. **35**, 917 (1977)

E.M. Arruda, M.C. Boyce, R. Jayachandran, Mech. Mater. **19**, 193 (1995)
L. Barral, J. Cano, A. López, P. Nogueira, C. Ramírez, J. Therm. Anal. Calorim. **41**, 1463 (1994)
C. Bauwens-Crowet, J. Bauwens, G. Homes, J. Polym, Sci. Part A-2: Polym. Phys. **7**, 735 (1969)
M.C. Boyce, D.M. Parks, A.S. Argon, Mech. Mater. **7**, 15 (1988)
B. Briscoe, L. Fiori, E. Pelillo, J. Phys. D. Appl. Phys. **31**, 2395 (1998)
R.D. Calleja, I. Devine, L. Gargallo, D. Radić, Polymer **35**, 151 (1994)
K. Chen, K.S. Schweizer, Macromolecules **44**, 3988 (2011)
J.D. Eshelby, The determination of the elastic field of an ellipsoidal inclusion, and related problems. Proc. R. Soc. Lond., Ser. A **241**, 376 (1957)
M.L. Falk, Science **318**, 1880 (2007)
M.L. Falk, J.S. Langer, L. Pechenik, Toward a Shear-Transformation-Zone Theory of Amorphous Plasticity. In: Yip S. (eds) Handbook of Materials Modeling (Springer, Dordrecht, 2005), pp 1281–1312
W. Gall, N. McCrum, J. Polym. Sci. **50**, 489 (1961)
Y.I. Golovin, V. Ivolgin, V. Khonik, K. Kitagawa, A. Tyurin, Scr. Mater. **45**, 947 (2001)
J. Hay, P. Agee, E. Herbert, Exp. Tech. **34**, 86 (2010)
T. Hirata, T. Kashiwagi, J.E. Brown, Macromolecules **18**, 1410 (1985)
J. Ho, L. Govaert, M. Utz, Macromolecules **36**, 7398 (2003)
J. Hutchinson, S. Smith, B. Horne, G. Gourlay, Macromolecules **32**, 5046 (1999)
J. Jancar, R.S. Hoy, A.J. Lesser, E. Jancarova, J. Zidek, Macromolecules **46**, 9409 (2013)
K.L. Johnson, *Contact Mechanics* (Cambridge University Press, Cambridge, 1987)
J. Ju, D. Jang, A. Nwankpa, M. Atzmon, J. Appl. Phys. **109**, 053522 (2011)
J.-Y. Kim, S.-K. Kang, J.-J. Lee, J.-i. Jang, Y.-H. Lee, D. Kwon, Acta Mater. **55**, 3555 (2007)
X. Li, B. Bhushan, Mater. Charact. **48**, 11 (2002)
L. Li, E. Homer, C. Schuh, Acta Mater. **61**, 3347 (2013)
Y. Liu, D. Wang, K. Nakajima, W. Zhang, A. Hirata, T. Nishi, A. Inoue, M. Chen, Phys. Rev. Lett. **106**, 125504 (2011)
B. Lucas, W. Oliver, Metall. Mater. Trans. A **30**, 601 (1999)
B. Lucas, W. Oliver, J. Swindeman, The dynamics of frequency-specific, depth-sensing indentation testing, MRS Online Proceedings Library Archive 522 (1998)
L. Malekmotiei, F. Farahmand, H.M. Shodja, A. Samadi-Dooki, J. Biomech. Eng. **135**, 041004 (2013)
L. Malekmotiei, A. Samadi-Dooki, G.Z. Voyiadjis, Macromolecules **48**, 5348 (2015)
P. Mott, A. Argon, U. Suter, Philos. Mag. A **67**, 931 (1993)
A. Mulliken, M. Boyce, Int. J. Solids Struct. **43**, 1331 (2006)
T. Mura, Micromechanics of defects in solids (Springer Science & Business Media, 2013)
P. Nagy, I.I. Kükemezey, S. Kassavetis, P. Berke, M.-P. Delplancke-Ogletree, S. Logothetidis, Nanosci. Nanotechnol. Lett. **5**, 480 (2013)
E. Oleinik, S. Rudnev, O. Salamatina, Polym. Sci. Ser. A **49**, 1302 (2007)
W.C. Oliver, G.M. Pharr, J. Mater. Res. **7**, 1564 (1992)
W.C. Oliver, G.M. Pharr, J. Mater. Res. **19**, 3 (2004)
D. Pan, A. Inoue, T. Sakurai, M. Chen, Proc. Natl. Acad. Sci. **105**, 14769 (2008)
D. Pan, Y. Yokoyama, T. Fujita, Y. Liu, S. Kohara, A. Inoue, M. Chen, Appl. Phys. Lett. **95**, 141909 (2009)
S. Pauly, S. Gorantla, G. Wang, U. Kühn, J. Eckert, Nat. Mater. **9**, 473 (2010)
J. Pethica, W. Oliver, Mechanical properties of nanometre volumes of material: use of the elastic response of small area indentations, in MRS Proceedings (Cambridge University Press, 1988)
W. Poisl, W. Oliver, B. Fabes, J. Mater. Res. **10**, 2024 (1995)
K.E. Prasad, V. Keryvin, U. Ramamurty, J. Mater. Res. **24**, 890 (2009)
T. Ree, H. Eyring, J. Appl. Phys. **26**, 800 (1955)
D. Rittel, A. Dorogoy, J. Mech. Phys. Solids **56**, 3191 (2008)
R.E. Robertson, J. Chem. Phys. **44**, 3950 (1966)
A. Samadi-Dooki, L. Malekmotiei, G.Z. Voyiadjis, Polymer **82**, 238 (2016)
C.A. Schuh, T. Nieh, Acta Mater. **51**, 87 (2003)

C.A. Schuh, A.C. Lund, T. Nieh, Acta Mater. **52**, 5879 (2004)
C.A. Schuh, T.C. Hufnagel, U. Ramamurty, Acta Mater. **55**, 4067 (2007)
H. Shodja, I. Rad, R. Soheilifard, J. Mech. Phys. Solids **51**, 945 (2003)
I.N. Sneddon, Int. J. Eng. Sci. **3**, 47 (1965)
F. Spaepen, Acta Metall. **25**, 407 (1977)
G. Spathis, E. Kontou, J. Appl. Polym. Sci. **79**, 2534 (2001)
G. Stoclet, R. Seguela, J. Lefebvre, S. Elkoun, C. Vanmansart, Macromolecules **43**, 1488 (2010)
G. Tandon, G. Weng, Polym. Compos. **5**, 327 (1984)
T. A. Tervoort, Constitutive modelling of polymer glasses: finite, nonlinear viscoelastic behaviour of polycarbonate (Technische Universiteit Eindhoven, (1996)
G.Z. Voyiadjis, L. Malekmotiei, J. Polym. Sci. Part B: Polym. Phys. **54**, 2179 (2016)
G.Z. Voyiadjis, A. Samadi-Dooki, J. Appl. Phys. **119**, 225104 (2016)
G.Z. Voyiadjis, C. Zhang, Mater. Sci. Eng. A **621**, 218 (2015)
G.Z. Voyiadjis, D. Faghihi, C. Zhang, J. Nanomech. Micromech. **1**, 24 (2011)
I.M. Ward, J. Mater. Sci. **6**, 1397 (1971)
B. Yang, J. Wadsworth, T.-G. Nieh, Appl. Phys. Lett. **90**, 061911 (2007)
H. Yu, W. Wang, H. Bai, Y. Wu, M. Chen, Phys. Rev. B **81**, 220201 (2010)
G. Zhang, W. Wang, B. Zhang, J. Tan, C. Liu, Scr. Mater. **52**, 1147 (2005)

Properties of Material Interfaces: Dynamic Local Versus Nonlocal

11

Devendra Verma, Chandra Prakash, and Vikas Tomar

Contents

Abstract

Interfaces in the materials are known entities since last century described as early as in the interfacial excess energy formulations by Gibbs (Boßelmann et al. 2007). The interface effect (or surface effect) is also widely referred to as the interface stress (or surface stress) that consists of two parts, both arise from the distorted atomic structure near the interface (or surface): the first part is the interface (or surface) residual stress which is independent of the deformation of solids, and the second part is the interface (or surface) elasticity which contributes to the stress field related to the deformation. Plastic deformation, in particular, the initial yielding point (i.e., the yield surface), is sensitive to the local stress (or local strain) of a heterogeneous material, which includes both the local (surface/interface) residual stress and local stress–strain relationship. The plastic

D. Verma · C. Prakash · V. Tomar (✉)
School of Aeronautics and Astronautics, Purdue University, West Lafayette, IN, USA
e-mail: tomar@purdue.edu

G. Z. Voyiadjis (ed.), *Handbook of Nonlocal Continuum Mechanics for Materials and Structures*, https://doi.org/10.1007/978-3-319-58729-5_21

deformation at the interfaces also considers the tension and compression along the interface and stress mismatch because of the material property differences. In the nanomaterials, the surface and interface stresses become even more important owing to the nanoscale size of the particles and interface areas.

Keywords
Nanomechanical Raman Spectroscopy · Interface · GB · CZM

Introduction

Interfaces in the materials are known entities since last century described as early as in the interfacial excess energy formulations by Gibbs (Boßelmann et al. 2007). The interface effect (or surface effect) is also widely referred to as the interface stress (or surface stress) that consists of two parts, both arise from the distorted atomic structure near the interface (or surface): the first part is the interface (or surface) residual stress which is independent of the deformation of solids, and the second part is the interface (or surface) elasticity which contributes to the stress field related to the deformation. Plastic deformation, in particular, the initial yielding point (i.e., the yield surface), is sensitive to the local stress (or local strain) of a heterogeneous material, which includes both the local (surface/interface) residual stress and local stress–strain relationship. The plastic deformation at the interfaces also considers the tension and compression along the interface and stress mismatch because of the material property differences. In the nanomaterials, the surface and interface stresses become even more important owing to the nanoscale size of the particles and interface areas.

The naturally occurring materials have been of keen interest in the materials community with the aim of understanding and reproducing the exceptional strength and toughening mechanisms present in exoskeletons of shrimps and lobsters (Boßelmann et al. 2007; Raabe et al. 2006), crabs (Mayer 2011; Chen et al. 2008), nacre (Flores-Johnson et al. 2014), etc. All these materials share some common traits such as a strong hierarchical structure, layered structure, composition of material with both minerals and fibers, gradient in the thickness of layers, etc. These naturally occurring materials have been able to manipulate the characteristics mentioned above to customize the design of their exoskeletons. The one important parameter along with the material composition in these designs is the role of the interfaces in the multilayered structure. In our articles, we highlighted the difference between the mechanical properties of two similar species of shrimp *Pandalus platyceros* and *Rimicaris exoculata* with interlayer structure as shown in Fig. 1a that are found at sea level and at 2300 m depth in the sea as a function of habitat, wet versus dry (Verma and Tomar 2014a), as a function of temperature (Verma and Tomar 2014b), and as a function of mineral composition (Verma and Tomar 2015a, b).

One unique feature that determines the properties of materials is the interfacial interaction between organic and inorganic phases in the form of protein (e.g., chitin (CHI) or tropocollagen (TC))-mineral (e.g., calcite (CAL) or hydroxyapatite

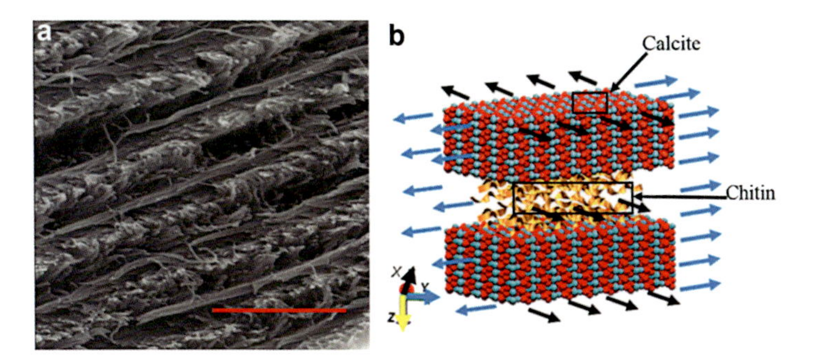

Fig. 1 (**a**) Interfacial structure in shrimp exoskeleton and (**b**) interfacial molecular structure made of Calcite and Chitin

(HAP)) interfaces. The size of the protein-mineral interfaces can be enormous as the mineral bits have nanoscale eliminations. These interfaces control biological reactions, and provide unique organic microenvironments that can enhance specific affinities, as well as self-assembly in the interface plane that can be used to orient and space molecules with precision. Interfaces also play a significant role in determining structural integrity and mechanical creep and strength properties of biomaterials. The length scale and complexity of microstructure of hybrid interfaces in biological materials make it difficult to study them and to understand the underlying mechanical principles, which are responsible for their extraordinary mechanical performance. One of the most important aspects of understanding the influence of interfaces on natural material properties is the knowledge of how stress transfer occurs across the organic-inorganic interfaces. Molecular modeling provides a way to study these phenomena at the length scale of the individual components. The effect of different kinds of interfaces was modeled by our research in the case of chitin layers. The interfaces with different thickness and different phases were compared to study the effectiveness of the stress transfer of the layers as shown in Fig. 1b. A comprehensive study on these is given in the articles by Tao et al. (Qu et al. 2015a, b).

In this article, a discussion is given on the interface mechanics. Nanomechanical Raman spectroscopy is used in this paper to measure the interface stresses. A comparison between the numerically and experimentally measured interface elastic constants is presented in addition to the stress-stress response of interface and its effect on material properties of the interface and its adjacent phases.

Interfaces

Interface Mechanical Properties

Interfaces in composite materials can be considered as a material phase confined between two separate grains or phases. Single interface samples of glass and epoxy were prepared with an epoxy interface sandwiched between two glass phases. The samples were prepared using two-part industrial epoxy procured from Composite Polymer Design (South St. Paul, MN, USA).

The resin, CPD4505A, and hardener, CPD 4507B, were thoroughly mixed in recommended proportions of 100A: 28B by weight. The epoxy layer thickness was controlled by putting tabs of appropriate thickness in between the glass slides. The samples were cured at a prescribed temperature of 250 °F for 1 day.

The thickness of the interface in samples was measured with a microscope to make sure that it was in the error margin of 10 ± 0.5 μm. The sample surfaces as shown in Fig. 2a were polished to remove scratches that could interfere with the data measurement during experiments.

The Raman spectroscopy is based on "Raman Effect," which provides a unique "fingerprint" of every individual substance as a characteristic for its identification. It is an inelastic process in which energy is exchanged between the incident photon and molecule. We have used the Raman spectroscopy to measure the stresses in the interface at different applied loads during indentation to compare the stress distribution. The Mechanical load was applied using the nanoindentation platform manufactured by Micro Materials Inc., UK, with load range from 0.1 to 500 mN, with the accuracy of better than 0.1 mN. The experimental setup to measure the Raman signal is shown in Fig. 2c.

The first step to measure the stress across the interface is to establish a calibration curve of Raman shift with the applied stress. A uniaxial load was applied on a block

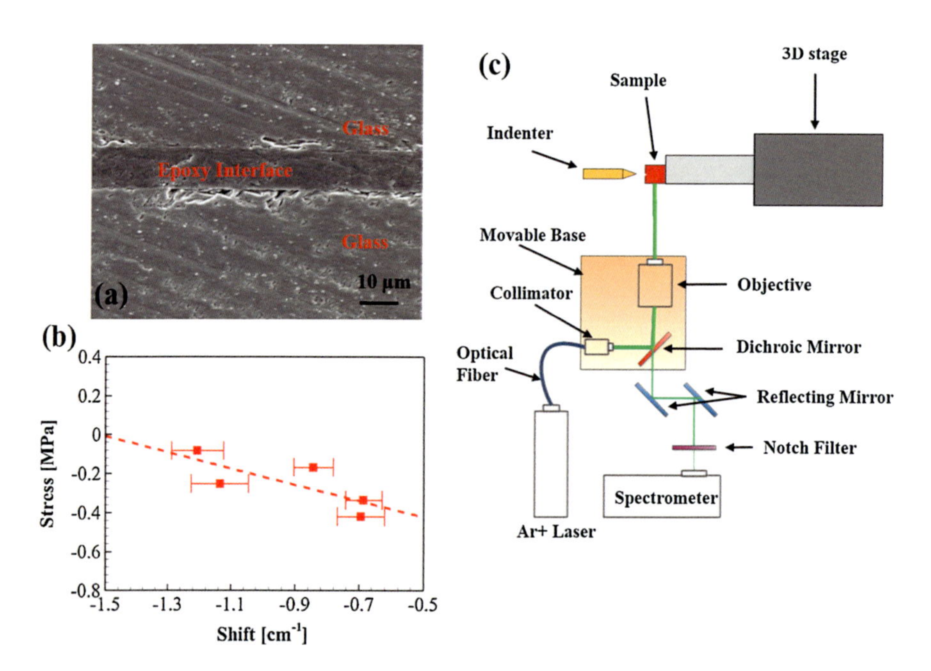

Fig. 2 (**a**) Image of the epoxy interface sample, (**b**) Raman shift-stress calibration *curve*, and (**c**) schematic of Raman measurements setup

of epoxy and the Raman shift was measured at applied loads of 100, 200, 300, 400, and 500 mN. The stresses were obtained by dividing the load by the area of the calibration sample. The measured Raman peak data was converted into shift by the equation

$$\Delta\omega = \left(\frac{1}{\lambda_{\text{laser}}} - \frac{1}{\lambda_{\text{measured}}}\right)^{*} 10\hat{}7\ \text{cm}^{-1}. \tag{1}$$

The change in shift was obtained by subtracting the shift at the applied load for the shift at zero load. The calibration curve for shift versus load for epoxy is given in Fig. 2b. The measurements were performed on the epoxy interface while holding the load constant. The stress distribution across interfaces is shown in Fig. 3c. The load direction is from the top of the picture. The figure shows the average values of

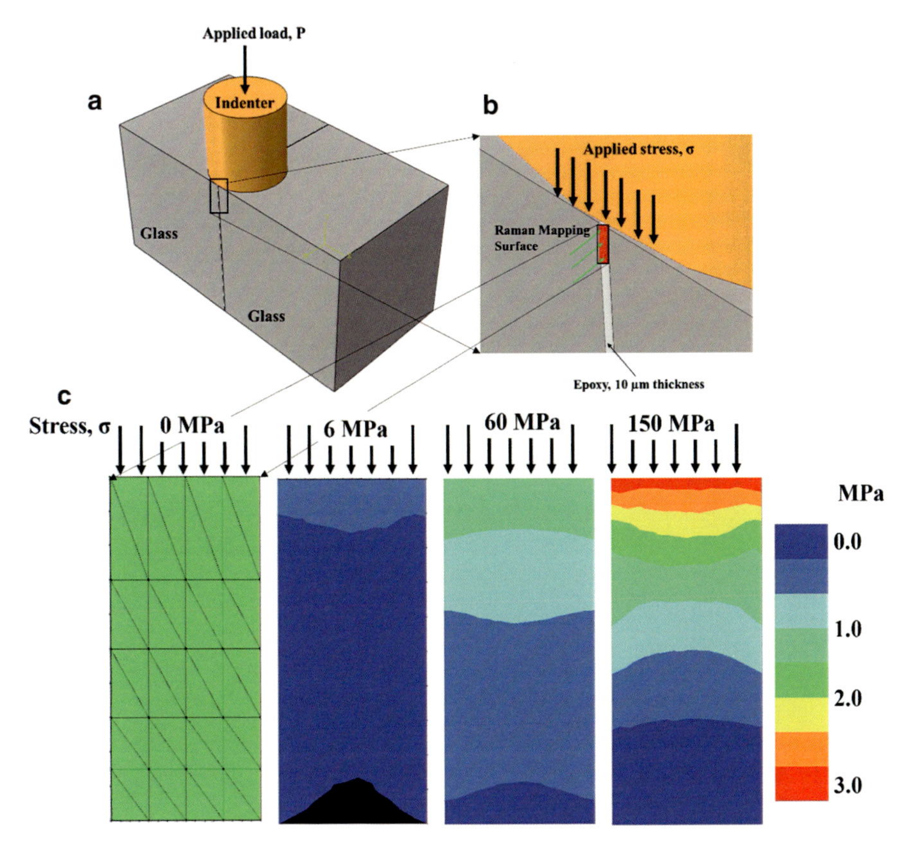

Fig. 3 (**a**) Schematic of epoxy interface between glasses with the indenter direction, (**b**) the magnified region showing the area for Raman map collection, and (**c**) Raman stress map on the interface with applied load of 0, 20, 200, and 500 mN. The height in each map is 80 μm and width is 25 μm. Figure with 0 mN load shows the Raman data collection points

stresses as measured on the interface. This technique has been used in the literature to measure stresses tensors in the crystalline materials such as Silicon (Gan and Tomar 2014) but for the case polymers such as epoxy the average stresses can be measured (Colomban et al. 2006).

The Raman spectroscopy only provides the average stress at the interface but stress tensors in different directions are needed to fully understand the behavior of interfaces. Even in the experiments, it is difficult to measure the lateral stresses. An analytical solution is therefore developed to calculate the lateral stresses during indentation of interfaces. A schematic of the contact problem of interface is illustrated in Fig. 3. The details on the analytical solution are given in authors' previous articles.

Interface Elastic Constants

The interface elastic constants were calculated using equations in the article by Ustinov et al. (2013) for different ratios of interface modulus as compared to the main phase. The results are given in Fig. 4 for the range of rations from 0.001 to 100. The absolute values of the interface elastic constants are also plotted to observe

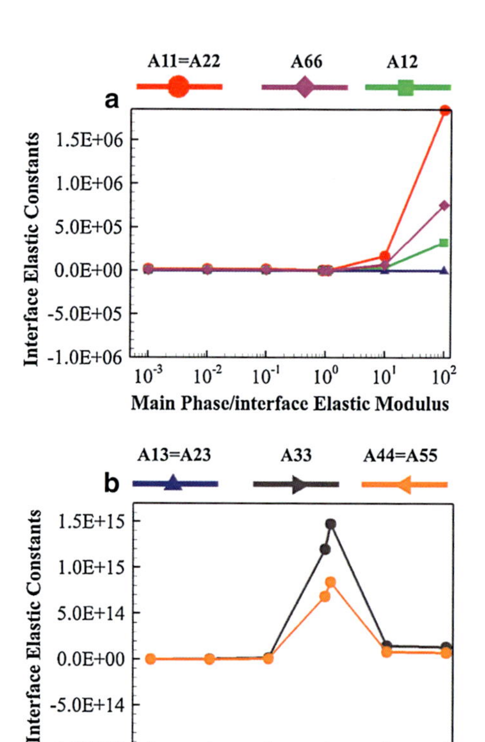

Fig. 4 Absolute values of interface elastic constants calculated for different ratios of elastic modulus of main phase to interface with (**a**) plotting the elastic constants A11, A66, A12, A13, and (**b**) plotting A33 and A44

Table 1 Comparison of interface elastic constants by both methods

	Experimental solutions	Theoretical formula Ustinov et al. (2013)	Theoretical approximations Ustinov et al. (2013)
A11 (Pa-m)	34029.1	−1253260.483	53571.42857
A22 (Pa-m)	6821.673	−1253260.483	53571.42857
A33 (Pa/m)	2.03E + 14	5.74205E + 14	5.35714E + 14
A13 (Pa)	2.8E + 09	2208480565	3571428571

the variation with the change in the ratio as shown in Fig. 4. It can be seen that in the present isotropic material A11, A66, A12, A13 keeps decreasing as the modulus of both phases matches and then starts increasing again with the minimum at the point of same materials, but the opposite is true for A33 and A44 with the maximum values at the point when the interface and bulk phase has the same properties.

The interface elastic constants by both Dingreville and Qu (2008), Dingreville et al. (2014), and Ustinov et al. (2013) are calculated based on the strain energy. Per Dingreville, the interface contribution to the thermodynamic properties is defined as the excess over the values that would obtain if the bulk phases retained their properties constant up to an imaginary surface (of zero thickness) separating the two phases. The surface free (excess) energy of a near-surface atom is defined by the difference between its total energy and that of an atom deep in the interior of a large crystal. Surface free energy corresponds to the work of creating a unit area of surface, whereas surface stress is involved in computing the work in deforming a surface.

The interface elastic constants given in Table 1 are converted into surface constants based on equation from Ustinov et al. for a 10 μ thickness interface. The reduced modulus measured from the indentation experiments after conversion to surface constant is also listed in the same table for comparison. The interface constants after the conversion fall under the same order as the ones calculated from the analytical relations given for interfaces.

These results show the comparison between different formulations given in the literature and a first comparison with the experimental results. The interface stress contribution is well recognized but still there is more work needed to be able to address and implement the constitutive behavior of interfaces in the engineering problems. It has a very high potential of application in areas such as interface between glass/fiber in composites, interfaces at grain boundaries in metals to mention a few.

Interface Dynamic Properties

To understand the interface mechanical properties, an idealized system with epoxy interface between two glass slides was examined under static and dynamic loading. The sample surface is shown in Fig. 2a. The impact experiments were performed on the epoxy interface to extract the stress-strain effect of the interface at different strain rates (Verma et al. 2015). It was found that interfaces are affected by both strain rate and confinement effects during deformation. A new constitutive model is developed that couples the effect of both strain rate and lateral stresses given as

$$\sigma = (A + B\varepsilon^n)\left(1 + C\ln\dot{\varepsilon}^*\right)\left(1 + k\sigma_1{}^*\right). \quad (2)$$

Here, σ is the equivalent stress, ε is the equivalent plastic strain, A is the yield stress, B is the strain hardening constant, n is the strain hardening coefficient, C is the strain rate strengthening coefficient, and k is the confinement factor. $\dot{\varepsilon}^* = \frac{\dot{\varepsilon}}{\dot{\varepsilon}_{\text{ref}}}$ is the dimensionless strain rate normalized with reference strain rate, $\sigma_1{}^* = \frac{\sigma_1}{\sigma_{\text{compressive strength}}}$ is the dimensionless lateral stress normalized with the compressive strength of the material. The reference strain rate is taken as 1 s^{-1} and $\sigma_{\text{compressive strength}}$ is taken as 100 MPa. In the current experiments, the temperature was constant so the temperature effects were neglected.

The stress-strain response for the epoxy interface was analyzed at high and low strain rates as shown in Fig. 5 with the confined interface behavior labeled with closed symbols and unconfined epoxy behavior labeled with open symbols. At lower strain rates, the difference in stresses is mostly because of the confinement

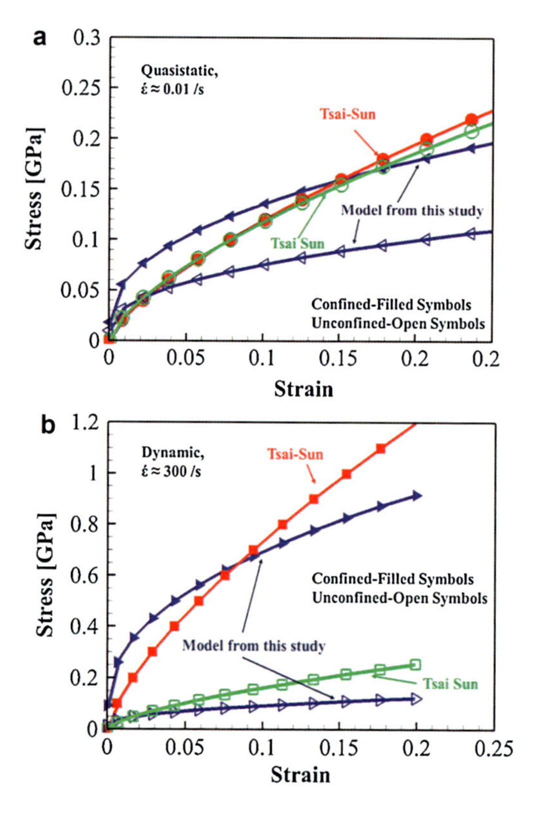

Fig. 5 Fit of the current model with stress strain for (**a**) quasistatic loading and (**b**) dynamic loading

effect while in the dynamic case the strain rate effect also plays a major role. The difference in the stresses is higher in the dynamic case as evident in Fig. 5b compared to Fig. 5a. It is also compared to the stress strain behavior fitted by Tsai-Sun model on the same plot. Thus, the current model is better suited to model the behavior of materials under confined spaces such as interfaces in the composite materials, metals, ceramics, etc. A conventional way is to consider interface as a zero thickness and to not consider interface effect on the material deformation. The model in Eq. 2 takes into account the effect of interface mechanical properties on the mechanical deformation and should be considered for cases that have interface dominant geometries such as the biological materials analyzed in this work. The size effect of the interfaces is being analyzed by the authors and will be published in the future papers.

Fracture Properties of Interfaces

Cohesive Zone Model

Fracture or crack formation is the creation of new surfaces in the domain of the body. This surface creation invariably leads to loss in the global stiffness and load-bearing ability of the material, often leading to failure. Traditionally, either energy-based or stress-intensity based approaches have been employed to predict this mode of failure. The energy-based theory of failure introduced by Griffith (1921) was motivated by the inadequacy of the elastic solution that renders singular stresses at the mathematically sharp crack tip. The key idea behind the stress-intensity based theory (Irvin 1957) is the observation that the near tip crack field in isotropic linear elastic materials is similar for all specimen geometry and loading conditions, to within a constant.

To address the inconsistency of infinite stresses at the crack-tip, a theory involving a process zone, equivalent to a plastic zone in elastic-plastic fracture mechanics, was presented by Barenblatt (1962) and Dugdale (1960), for crack propagation in homogeneous isotropic materials. The main assumption was that in the vicinity of crack tip, opposing faces of the crack are bounded by the molecular cohesive forces. As the body is loaded, the two faces undergo significant deformation, which eventually leads to loss of interatomic cohesion and traction free surface creation. The cohesive forces are concentrated near a small, but finite region of the continuum crack tip and drop to zero within few atomic distances from the crack tip. This idea of the cohesive traction applied in a small cohesive zone removes the difficulty of crack tip singularity in LEFM and is called Cohesive Zone Model (CZM). A cohesive zone model assumes a relation between the normal (and shear) traction and the opening (and sliding) displacement, and is capable of capturing the debonding process of interfaces.

There are several numerical studies of failure based on the cohesive zone models (Barua et al. 2012a, b, 2013a, b; Barua and Zhou 2011; Kim et al. 2014; Tvergaard 2003; Tvergaard and Hutchinson 1992, 1993; Camacho and Ortiz 1996; Zhong

1999; Zhong and Knauss 2000, 1997; Zhong and Meguid 1997; Needleman and Xu 1993; Kubair et al. 2002a, b; Samudrala et al. 2002; Samudrala and Rosakis 2003). Using Rose et al.'s (1981) cohesive law that is primarily for bimetallic interfaces, Levy (1994) studied the debonding of a circular fiber in an infinite matrix subject to equibiaxial load. There are also analytical studies of interface debonding based on cohesive zone models, and these studies focus on the effect of interface debonding on the macroscopic behavior of composite materials. Almost all analytic studies are limited to the linear cohesive law, i.e., a linear relation between the normal (and shear) traction(s) and the opening (and sliding) displacement(s) of interfaces ((!!! INVALID CITATION !!!) n.d.; Hashin 1991a, b, 2002; Qu 1993; Wu et al. 1999). There are very few analytic studies on the effect of nonlinear cohesive law (Levy 1996, 2000).

Experimental observations show distinctive characteristics in micromechanical failure mechanisms in peel and shear fracture, thus the cohesive behavior is expected to be mode dependent (Chai 2004, 2003; Chai and Chiang 1996; Roy et al. 1999). A frequently used coupled cohesive law is developed by Tvergaard and Hutchinson (1993), using a dimensionless separation parameter. A drawback with this model is that the fracture energy is the same in all mode mixities (Högberg 2006). Another theoretically motivated exponential cohesive law is developed by Xu and Needleman (Needleman and Xu 1993). Van den Bosch et al. (2006) showed that this exponential formulation only realistically describes mixed mode behavior in a special case: The fracture energy is equal in normal and tangential direction. Park and Paulino (2013) have extensively reviewed a number of CZMs and concluded that the available models should be used with great caution for mix-mode cohesive fracture.

There are several CZMs that have been suggested in literature (Park and Paulino 2013; Needleman 2014; Shet and Chandra 2004). The constitutive behavior of the cohesive model is formulated as a traction-separation law (TSL), which relates the traction, T, to the separation, Δ, the latter representing the displacement jump across the interface. These cohesive parameters can be extracted from load-displacement curve obtained from a simple tension or shear test (Zhu et al. 2009). One approach is to use a J-integral versus end opening curve to get the cohesive strength and the critical separation. Molecular dynamics (MD) simulations have also been used previously to obtained cohesive zone parameters of the interfaces (Dandekar and Shin 2011; Zhou et al. 2008, 2009; Yamakov et al. 2006). Stress-displacement curve obtained from MD simulation is used as a traction-separation curve and the maximum stress is used as cohesive strength of the interface. The total area under stress-separation curve is taken as the cohesive fracture energy.

Interface Thickness Effect

In this section we will show the effect of interface thickness on the fracture properties by studying the effect of grain boundary thickness in a polycrystalline material. In a typical polycrystalline metal, GBs are significantly thinner ($\sim$10 nm) than the grains (∼μm). The accurate prediction of crack propagation through GBs and interfaces while simultaneously predicting crack propagation though grains can be difficult due to the factors that include: significant difference in the length scales of GBs and grains, respectively; unknown GB strength properties; embrittlement effects of GBs owing to GB chemistry, etc. Considering the fact that Ni additions are likely to be segregated along GBs in polycrystalline W, (Gupta et al. 2007), change in GB mechanical property plays a significant role in enhancing or degrading of the materials failure resistance.

The GB simulation model is constructed based on the image of a realistic Ni doped W GB from a HRTEM image (Gupta et al. 2007). Fully saturated Ni-doped W GB has a thickness of about 0.6 nm (Gupta et al. 2007). However, when calculating the GB properties one must insert a few atoms on either side of GB, Fig. 1, resulting in an interface structure.

A 3-D GB interface model of Ni-doped W GB, Fig. 6d, is developed for (Prakash et al. 2016) fracture simulations using a combination of XFEM with the cohesive finite element method (CFEM) in order to characterize the influence of GB thickness on fracture toughness. For the constitutive description of W grains and Ni doped W GBs in a polycrystalline W microstructure, a general elastic-plastic material model is used with parameters derived based on the data presented in the ab-initio modeling work of Lee and Tomar (2013, 2014). Cohesive parameters were evaluated from the stress-displacement curve obtained from the ab-initio calculation.

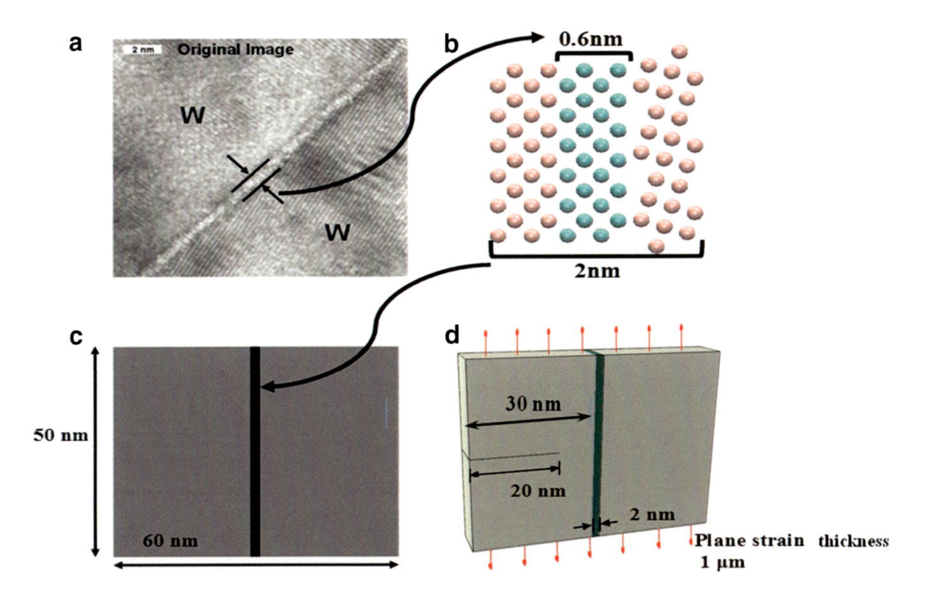

Fig. 6 A schematic showing (**a**) HRTEM image of polycrystalline Ni doped W GB (Gupta et al. 2007), (**b**) atomic structure of W GB analyzed by Lee and Tomar (2013, 2014), (**c**) 2-D, and (**d**) 3-D image of the continuum GB element analyzed for crack propagation

Although the total GB volume is relatively small, the thickness of GBs can greatly affect crack propagation. Lee and Tomar (2013, 2014) found that the minimum thickness of such an interphase that contains GB and few atomic layers around GB should be at least 2 nm in order to predict thickness independent properties of GB in ab initio simulations. In order to understand the influence of GB thickness on simulation results, crack propagation through interfaces with GB thickness of 2 nm, 4 nm, and 8 nm was analyzed. As shown in Fig. 7, change in thickness of GB does not significantly influence either the crack tip position change or crack tip elastic strain energy release (calculated based on elastic strains surrounding crack tip) indicating that GB thickness change at the length scale of

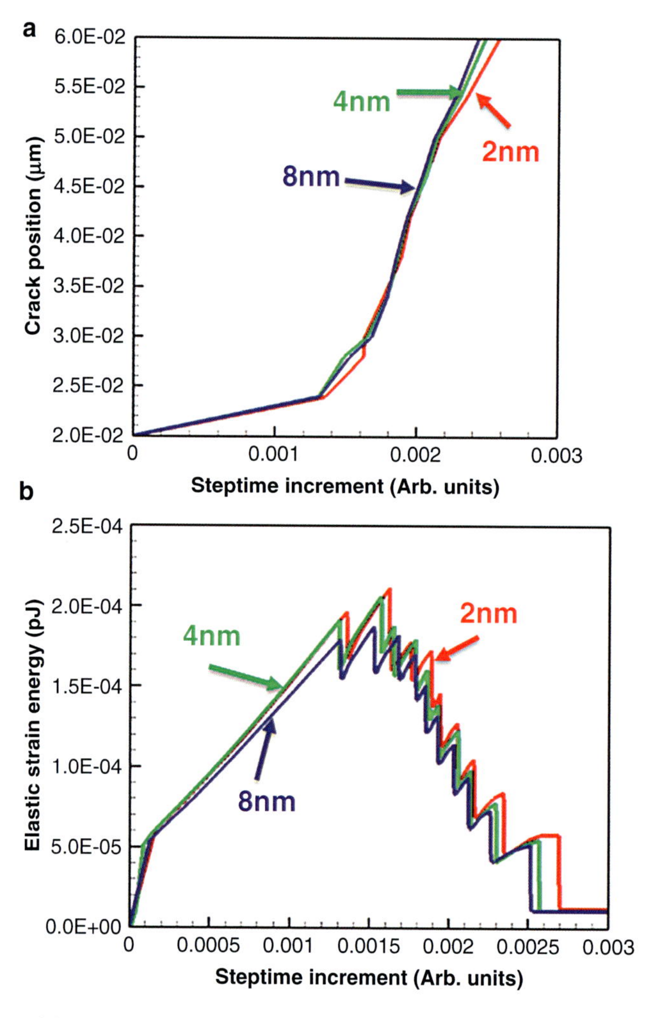

Fig. 7 Effect of GB *thickness* on (**a**) crack tip position and (**b**) strain energy release with respect to the time-step increment

materials model that corresponds to atomic scale does not affect interface crack propagation characteristics.

In order to understand the effect of length scale, as discussed earlier, scaled up interface structured with different GB thickness are analyzed for crack propagation. Figure 7 shows the total plastic energy dissipation per unit volume and crack tip strain energy change per unit volume change as interface specimen is loaded and crack propagation undergoes. The plastic energy dissipation is calculated based on plastic strain in the region surrounding the crack tip. As shown, the effect of length scale on fundamental energetic quantities related to crack propagation is insignificant (Fig. 8).

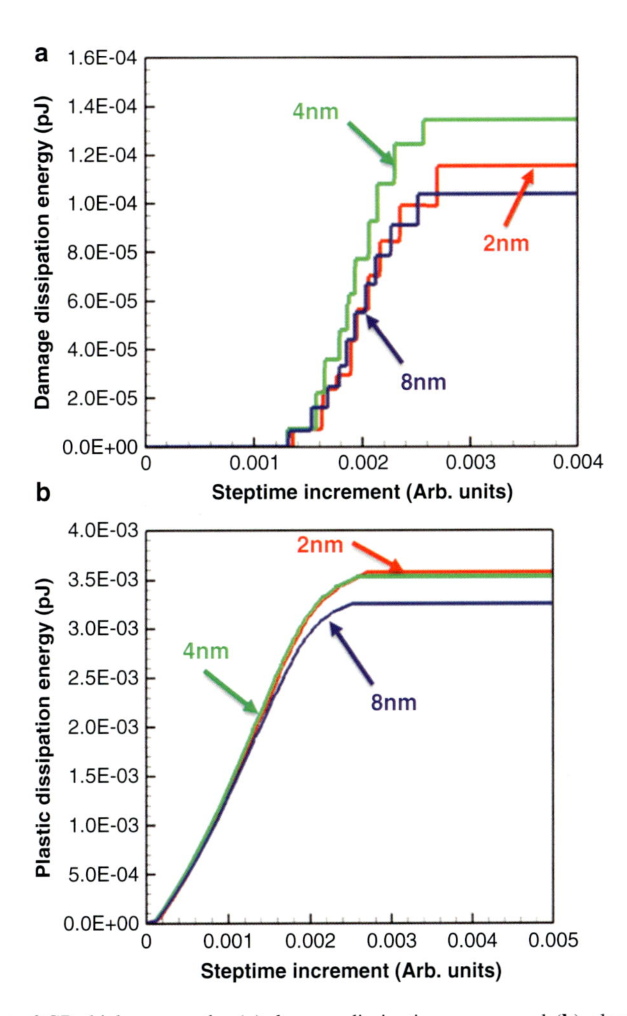

Fig. 8 Effect of GB thickness on the (**a**) damage dissipation energy and (**b**) plastic dissipation energy of the 3-D model

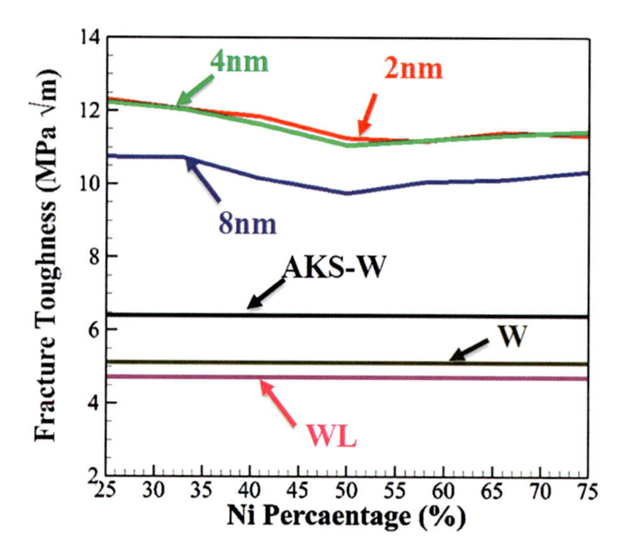

Fig. 9 Fracture toughness with respect to Ni percentage in GB and a comparison of fracture toughness of W interface specimens at the three scales examined and experimentally available values for polycrystalline W (Gludovatz et al. 2010)

As shown in Fig. 9, a clear scale dependence of fracture toughness is seen with the value corresponding to 100 nm thick interface length scale approaching experimental values. Earlier, Mai and Lawn (1987) have used theoretical fracture mechanics-based approach to calculate scale-dependent fracture toughness (Mai and Lawn 1987).

Theoretical calculations showed that the fracture toughness has an intrinsic length-scale dependence, which matches with the result in the presented study. In the work of Mai and Lawn (1987), the range of fracture toughness was found to be 2–7 MPa for Al, and it has been observed that the fracture toughness tends to decrease significantly as length-scale of system becomes smaller, similar to what is observed here. This result establishes the length scale dependence of fracture toughness in examined interfaces and does not refute the use of continuum elements at the small length scale of 2 nm.

Conclusions

A new technique based on nanomechanical Raman spectroscopy is presented to calculate the interface stresses and elastic constants using an analytical model. The elastic constants are then compared with the strain energy frameworks provided in the literature. The comparison between both methods shows the dependence of the interface elastic constants on the thickness of the interface. The elastic constants calculated from the stress-strain data matches the theoretical values after the thickness effect correction.

The interfaces are believed to play an important role in the observed behavior. The role of interfaces is identified by performing experiments on an idealized system of glass-epoxy interfaces. The confinement effect on the interfaces along with the effect of strain-rate was found to play a major role in the deformation of the examined interfaces. A new model capturing both strain rate and confinement effects is developed for strain rates up to 10^3 s^{-1} in this paper to account for confinement effect and strain rate effect coupling.

Interface thickness effects were studied for the Ni-doped Tungsten polycrystalline material. For small length scale, as in the case of GB, it has been shown that the fracture toughness at the interface is affected by the GB thickness. However, the crack tip stress and the strain energy at the interface are not affected. Further investigation on the effect of interface thickness at large length scale is being studied by the authors.

References

G.I. Barenblatt, *The Mathematical Theory of Equilibrium Cracks Formed in Brittle Fracture* (Armed Services Technical Information Agency, Arlington, 1962)
A. Barua, Y. Horie, M. Zhou, J. Appl. Phys. **111**, 054902 (2012a)
A. Barua, Y. Horie, M. Zhou, Proceedings of the royal society a: mathematical. Phys. Eng. Sci. **468**, 3725 (2012b)
A. Barua, S. Kim, Y. Horie, M. Zhou, J. Appl. Phys. **113**, 184907 (2013a)
A. Barua, S.P. Kim, Y. Horie, M. Zhou, Mater. Sci. Forum **767**, 13 (2013b)
A. Barua, M. Zhou, Model. Simul. Mater. Sci. Eng. **19**, 055001 (2011)
M.J. van den Bosch, P.J.G. Schreurs, M.G.D. Geers, Eng. Fract. Mech. **73**, 1220 (2006)
F. Boßelmann, P. Romano, H. Fabritius, D. Raabe, M. Epple, Thermochim. Acta **463**, 65 (2007)
G.T. Camacho, M. Ortiz, Int. J. Solids Struct. **33**, 2899 (1996)
H. Chai, Int. J. Solids Struct. **40**, 6023 (2003)
H. Chai, Int. J. Fract. **130**, 497 (2004)
H. Chai, M.Y.M. Chiang, J. Mech. Phys. Solids **44**, 1669 (1996)
P.-Y. Chen, A.Y.-M. Lin, J. McKittrick, M.A. Meyers, Acta Biomater. **4**, 587 (2008)
P. Colomban, G. Gouadec, J. Mathez, J. Tschiember, P. Pérès, Compos. A: Appl. Sci. Manuf. **37**, 646 (2006)
C.R. Dandekar, Y.C. Shin, Compos. A: Appl. Sci. Manuf. **42**, 355 (2011)
R. Dingreville, A. Hallil, S. Berbenni, J. Mech. Phys. Solids **72**, 40 (2014)
R. Dingreville, J. Qu, J. Mech. Phys. Solids **56**, 1944 (2008)
D.S. Dugadale, J. Mech. Phys. Solids **8**, 100 (1960)
E. Flores-Johnson, L. Shen, I. Guiamatsia, G.D. Nguyen, Compos. Sci. Technol. **96**, 13 (2014)
M. Gan, V. Tomar, Rev. Sci. Instrum. **85**, 013902 (2014)
B. Gludovatz, S. Wurster, A. Hoffmann, R. Pippan, Int. J. Refract. Met. Hard Mater. **28**, 674 (2010)
A.A. Griffith, Philosophical transactions of the royal society a: mathematical. Phys. Eng. Sci. **221**, 163 (1921)
V.K. Gupta, D.-H. Yoon, H.M. Meyer Iii, J. Luo, Acta Mater. **55**, 3131 (2007)
Z. Hashin, J. Mech. Phys. Solids **39**, 745 (1991a)
Z. Hashin, Trans. ASME **58**, 444 (1991b)
Z. Hashin, J. Mech. Phys. Solids **50**, 2509 (2002)
J.L. Högberg, Int. J. Fract. **141**, 549 (2006)
N. Pagano and G.P. Tandon: Mechanics of Materials, **9**, 49 (1990)
G.R. Irwin, J. Appl. Mech. **24**, 361 (1957)
S. Kim, A. Barua, Y. Horie, M. Zhou, J. Appl. Phys. **115**, 174902 (2014)
D.V. Kubair, P.H. Geubelle, Y.Y. Huang, Eng. Fract. Mech. **70**, 685 (2002a)
D.V. Kubair, P.H. Geubelle, Y.Y. Huang, J. Mech. Phys. Solids **50**, 1547 (2002b)
H. Lee, V. Tomar, Comput. Mater. Sci. **77**, 131 (2013)

H. Lee, V. Tomar, Int. J. Plast. **53**, 135 (2014)
A. Levy, J. Mech. Phys. Solids **42**, 1087 (1994)
A. Levy, J. Appl. Mech. **63**, 357 (1996)
A. Levy, J. Appl. Mech. **67**, 727 (2000)
Y.-W. Mai, B.R. Lawn, J. Am. Ceram. Soc. **70**, 289 (1987)
G. Mayer, J. Mech. Behav. Biomed. Mater. **4**, 670 (2011)
A. Needleman, Procedia IUTAM **10**, 221 (2014)
A. Needleman, X.P. Xu, Model. Simul. Mater. Sci. Eng. **1**, 111 (1993)
K. Park, G.H. Paulino, Appl. Mech. Rev. **64**, 060802 (2013)
C. Prakash, H. Lee, M. Alucozai, V. Tomar, Int. J. Fract. **199**, 1 (2016)
J. Qu, Mech. Mater. **14**, 269 (1993)
T. Qu, D. Verma, M. Alucozai, V. Tomar, Acta Biomater. **25**, 325 (2015)
T. Qu, D. Verma, M. Shahidi, B. Pichler, C. Hellmich, V. Tomar, MRS Bull. **40**, 349 (2015)
D. Raabe, P. Romano, C. Sachs, H. Fabritius, A. Al-Sawalmih, S.-B. Yi, G. Servos, H. Hartwig, Mater. Sci. Eng. A **421**, 143 (2006)
J.H. Rose, J. Ferrante, J.R. Smith, Am. Phys. Soc. **47**, 675 (1981)
A.Y. Roy, R. Narashimhan, P.R. Arora, Acta Mater. **47**, 1587 (1999)
O. Samudrala, Y. Huang, A.J. Rosakis, J. Mech. Phys. Solids **50**, 1231 (2002)
O. Samudrala, A.J. Rosakis, Eng. Fract. Mech. **70**, 309 (2003)
C. Shet, N. Chandra, Mech. Adv. Mater. Struct. **11**, 249 (2004)
V. Tvergaard, Eng. Fract. Mech. **70**, 1859 (2003)
V. Tvergaard, J.W. Hutchinson, J. Mech. Phys. Solids **40**, 1377 (1992)
V. Tvergaard, J.W. Hutchinson, J. Mech. Phys. Solids **41**, 1119 (1993)
K.B. Ustinov, R.V. Goldstein, V.A. Gorodtsov, On the modeling of surface and interface elastic effects in case of eigenstrains, in *Surface Effects in Solid Mechanics: Models, Simulations and Applications*, ed. by H. Altenbach, F. N. Morozov (Springer Berlin Heidelberg, Berlin, 2013), p. 167
D. Verma, J. Singh, A.H. Varma, V. Tomar, JOM **67**, 1694 (2015)
D. Verma, V. Tomar, Mater Sci Eng C Mater Biol Appl **44**, 371 (2014a)
D. Verma, V. Tomar, J Bionic Eng **11**, 360 (2014b)
D. Verma, V. Tomar, J. Mater. Res. **30**, 1110 (2015a)
D. Verma, V. Tomar, Mater Sci Eng C Mater Biol Appl **49**, 243 (2015b)
Y. Wu, Z. Ling, Z. Dong, Int. J. Solids Struct. **37**, 1275 (1999)
V. Yamakov, E. Saether, D.R. Phillips, E.H. Glaessgen, J. Mech. Phys. Solids **54**, 1899 (2006)
X.A. Zhong, Mech. Adv. Mater. Struct. **6**, 1 (1999)
X.A. Zhong, W.G. Knauss, Trans. ASME **119**, 198 (1997)
X.A. Zhong, W. Knauss, Mech. Adv. Mater. Struct. **7**, 35 (2000)
Z. Zhong, S.A. Meguid, J. Elast. **46**, 91 (1997)
X.W. Zhou, N.R. Moody, R.E. Jones, J.A. Zimmerman, E.D. Reedy, Acta Mater. **57**, 4671 (2009)
X.W. Zhou, J.A. Zimmerman, E.D. Reedy, N.R. Moody, Mech. Mater. **40**, 832 (2008)
Y. Zhu, K.M. Liechti, K. Ravi-Chandar, Int. J. Solids Struct. **46**, 31 (2009)

Nanostructural Response to Plastic Deformation in Glassy Polymers

12

George Z. Voyiadjis and Aref Samadi-Dooki

Contents

Abstract

A closed form stress-strain relation is proposed for modeling the postyield behavior of amorphous polymers based on the shear transformation zones (STZs) dynamics and free volume evolution. Use is made of the classical free volume theory by Cohn and Turnbull (J Chem Phys 31:1164, 1959), and also STZ-mediated plasticity model for amorphous metals by Spaepen (Acta Metall 25:407, 1977) and Argon (Acta Metall 27:47, 1979) for developing a new homogenous plasticity framework for glassy polymers. The variations

G. Z. Voyiadjis (✉)
Department of Civil and Environmental Engineering, Louisiana State University, Baton Rouge, LA, USA
e-mail: voyiadjis@eng.lsu.edu

A. Samadi-Dooki
Computational Solid Mechanics Laboratory, Department of Civil and Environmental Engineering, Louisiana State University, Baton Rouge, LA, USA
e-mail: asamad3@lsu.edu

G. Z. Voyiadjis (ed.), *Handbook of Nonlocal Continuum Mechanics for Materials and Structures*, https://doi.org/10.1007/978-3-319-58729-5_42

of free volume content and STZs activation energy during large deformation are parametrized considering the previous experimental measurements using positron annihilation lifetime spectroscopy (PALS) and thermal analysis with differential scanning calorimetry (DSC), respectively. The proposed model captures the softening-hardening behavior of glassy polymers at large strains with a single formula. This study shows that the postyield softening of the glassy polymers is a result of the reduction of the STZs nucleation energy as a consequence of increased free volume content during the plastic straining up to a steady-state point. Beyond the steady-state strain where the STZ nucleation energy reaches a plateau, the increased number density of STZs, which is required for finite strain, brings about the secondary hardening continuing up to the fracture point. This model also accurately predicts the effect of strain rate, temperature, and thermal history of the sample on its postyield behavior which is in consonance with experimental observations. Implication of the model for interpreting the localization and indentation size effect of polymers is also discussed.

Keywords

Plasticity · Polymer · Amorphous · Free volume · Shear transformation · Stress · Strain · Energy · Deformation · Microstructure · Chain · Rate · Glass

Introduction

Many industries have turned their attention to polymers as the materials that can be used as key structural elements with excellent compatibilities with the efficiency-centered design criterion. Polymers are light-weight, inexpensive, transparent, and extremely formable, and with some minor enhancements (like reinforcing with fibers in polymer matrix composites) can be utilized as load-bearing elements under severe loading conditions. Polymeric glasses (PGs), which consist the majority of polymers, are amorphous materials with no arrangement in their microstructure. From the mechanical point of view, while the preyield behavior of this type of materials can be considered as viscoelastic, they show a very distinct deformation response when loaded beyond their yield point (Argon 2013; Boyce et al. 1988). Of the characteristic behavior of glassy polymers within the postyield region is the considerable primary softening which takes place right after yielding and continues up to a certain steady state strain, and a secondary hardening which extends up to the fracture point as demonstrated in Fig. 1.

Many efforts have been made to unfold and quantitatively describe the aforementioned unique postyield behavior of PGs during the past one-half century. The earliest significant model for the plastic deformation of PGs is the one proposed by Eyring (1936). With considering the molecular rearrangement as the key factor mediating the plastic deformation, Eyring (and later Ree and Eyring 1955) suggested that the restriction or activation of the spatial movement of the chains with altering the loading rate and/or the temperature is responsible for elevation or decreasing of the PGs' resistance to deformation. With considering

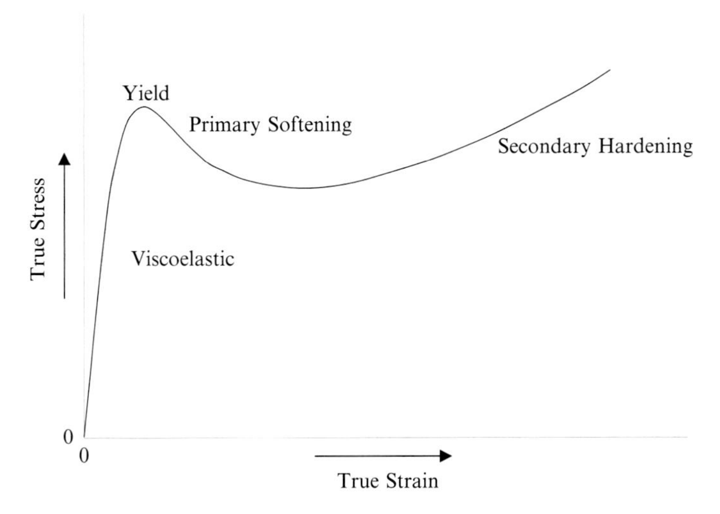

Fig. 1 A schematic representation of the generic uniaxial stress-strain behavior of glassy polymers (Reproduced from Voyiadjis and Samadi-Dooki 2016 with permission from AIP)

the chain deformation as the main contributing factor to the strain resistance of the macromolecular systems, Robertson (1966) proposed a model for the plastic deformation of these materials which was different from the rubber-like deformation mechanism. Robertson's model, however, is able to correctly predict the large deformation behavior of PGs at temperatures close to the glass transition temperature (Tg) only. Later, Argon (1973) came up with a model which considered the localized antisymmetric twists along the macromolecular chains as the mechanism for the permanent deformation. This model was formulated by considering the energy of the twisting mechanism to be calculated similar to that of a pair of adjacent disclinations. Known as "double kink" model, Argon's formalism successfully predicts the rate, temperature, and pressure dependence of the yielding behavior of amorphous polymers. Hence, it was extended by others for modeling the postyield plastic deformation (Arruda and Boyce 1991; Arruda et al. 1993, 1995; Boyce et al. 1988) and strain gradient plasticity (Lam and Chong 1999; Voyiadjis et al. 2014) of PGs. Among the most noticeable developments to Argon's model is the one proposed by Boyce et al. (1988) who used this theory to express the resistance of the macromolecules to segmental rotation. Accordingly, this model suggests that once this resistance is overcome, the chain entanglement opening causes a considerable softening in stress-strain behavior of the PGs. At higher strains, the chain alignment is considered to be the dominant factor which results in secondary hardening of the material behavior.

For crystalline materials with defined microstructural order, the primary carriers of plastic deformation are nanostructural defects like dislocations and point defects (Argon 2008; Shodja et al. 2013; Voyiadjis and Yaghoobi 2015). In contrast, the plastic deformation of amorphous polymers has traditionally been thought as the conformational changes in macromolecular systems, like chain rearrangements.

Since no preferred long-range order exists in the intra- and intermolecular space these amorphous solids, considering the chain rearrangement as the sole mechanism responsible for the permanent deformation of this type of solids, seem to be reasonable. Accordingly, it is expected that the stored energy in the permanently deformed polymers to be small which can be justified by conformational rearrangement. However, the calorimetric analyses by Oleinik and coworkers (Oleinik 1991; Oleinik et al. 2007) for a number of amorphous polymers revealed a large amount of stored energy during the course of plastic deformation which can never be attributed to the chain rearrangement mechanism. Instead, Oleinik (1991) suggested that this large amount of observed stored energy is due to the elastic field around some defect sites which form during the plastic deformation, like Eshelby's inclusion model (Eshelby 1957; Mura 1987). These defect sites are, in fact, the localized permanent deformation inclusions which encompass slip, glide, and shear rotation of macromolecular chains. The formation of these inclusions is mostly enforced by the shear component of the applied stress (Oleinik 1991), hence, they are termed shear transformation zones (STZs). It is worth noting that STZs were originally proposed to be responsible for the plastic deformation of metallic glasses (MGs) (Argon 1979; Spaepen 1977), but they are now believed to be the main mechanism for the plastic deformation of all types of disordered solids (Argon 2013). Unfortunately, the direct observation of STZs is not practical at this time, because they are localized transition zones rather than being actual defects (Wang et al. 2013); nevertheless, physical evidences exist for proving the nucleation of these sites during the inelastic deformation of solid state glasses (Falk 2007; Oleinik 1991; Pauly et al. 2010).

From the shear transformation plasticity point of view, the excessive applied stress triggers the nucleation of STZs, and the kinetics of inelastic deformation is determined by the rate of the formation of these sites (Argon 1979; Spaepen 1977). Hence, the dynamics of the STZs seems to be a determining factor for analyzing the plastic deformation of glassy state solids. For the case of MGs, this hypothesis has become the principal approach to study their yielding and postyield behavior. Most noticeably, Langer and Pechenik (2003) have proposed that the rate of the STZs' annihilation and creation needs to be proportional to the rate of the energy dissipation during plastic deformation in order to satisfy the thermodynamic consistency with considering the first and second laws. Since this model was successful in qualitatively capturing the inelastic deformation behavior of MGs, it was extended by Falk et al. (2004) with considering the role of thermal fluctuations on the atomic rearrangement during large deformation of MGs.

From the microscopic point of view, the localized deformation sites tend to form around the "weak" areas in the bulk of the solids. Amorphous solids possess a considerable amount of "voids" in their structure, the presence of which is dictated by the very nature of this type of materials as disordered solids (Flores et al. 2002; Hofmann et al. 2002; Jean et al. 2013; Utracki and Jamieson 2011; Yavari et al. 2005). While these voids can serve as weak points for STZs formation, existence of a correlation between the evolution of STZs and that of the free volume sites which is observed during plastic deformation of amorphous materials like MGs (de Hey et al. 1997) seems to be reasonable. Accordingly, for the case of MGs, some researchers developed models which consider the adjustment of STZs nucleation energy with the variation of free volume content as a state variable, similar to the one proposed by Li et al. (2013).

Although the shear transformation-mediated plasticity is the accepted inelastic deformation mechanism for all of the amorphous solids (Argon 2013), the number of studies using this approach for PGs is limited. Among these studies, the works by Hasan and coworkers notably consider the evolution of the number density of STZs during the large deformation as a determining factor for plastic deformation (Hasan and Boyce 1995; Hasan et al. 1993). This model is capable of capturing the effect of the rate, temperature, and thermal history of the sample on the postyield softening behavior of PGs. Argon (2013) also proposed a relation for shear transformation-mediated plasticity of PGs based on his molecular dynamics simulations of amorphous silicon (Argon and Demkowicz 2006, 2008) to capture the effect of the thermal history of the sample on PGs' overall postyield softening behavior. With defining the liquid-like sites as the fertile sites for shear transformation, Argon's model describes the evolution of these liquid-like densities during the plastic deformation.

In this chapter, a new model is presented for the postyield deformation of PGs which formulates the competition between STZs' nucleation energy variation due to the evolution of the free volume content during large deformation and the increase of STZs' number density to bring about the permanent plastic deformation. The variation of the number density of the STZs is directly related to the plastic strain, and the variation of the deformation energy and free volume are quantified based on the previous experimental studies by Oleinik (1991) and Hasan et al. (1993), respectively. Accordingly, effects of the temperature, strain rate, and thermal history of the polymer on its plastic deformation are investigated. With incorporating the microgeometrical and micromechanical properties of the STZs of Poly(methyl methacrylate) (PMMA) obtained from nanoindentation experiments and presented in the authors previous study (Malekmotiei et al. 2015), the results based on the proposed model are obtained and compared with those of previous experimental observations for uniaxial compressive deformation of this PG.

Theory

Free Volume Evolution

In developing the current theory, the plastic deformation is assumed to be a homogeneous flow which is mainly mediated by the nucleation and accumulation of shear events rather than a sudden development of shear bands. Accordingly, the kinetics relation which presents the relation between the plastic shear strain rate

$\dot{\gamma}^p$ and the shear stress in the plastic state τ^p can be used which was originally developed by Spaepen (1977) and Argon (1979) as:

$$\dot{\gamma}^p = \dot{\gamma}_0^p c_f \exp\left(\frac{-\Delta F}{k_B T}\right) \sinh\left(\frac{\gamma^T \Omega \tau^p}{2k_B T}\right), \tag{1}$$

in which $\dot{\gamma}_0^p$ is a preexponential constant, c_f represents the fraction of the potential jump sites, ΔF is the nucleation energy barrier, γ^T is the transformation strain, Ω is the volume of an STZ, T is the absolute temperature, and k_B is the Boltzmann constant.

For the case of MGs, potential jump sites are the interatomic spaces with excessive free volume which can act as sources for atomic jumps that result in STZs formation. The fraction of potential jump sites in an amorphous material can be expressed by the relation developed by Cohen and Turnbull (1959) based on their probability analysis as:

$$c_f = \exp\left(\frac{-\lambda \nu^*}{\nu_f}\right), \tag{2}$$

where λ is a geometrical factor which is in the range of 1/2 to 1, ν^* is the minimum size of an excessive free volume which can serve as a jump site, and ν_f is the average free volume per atom. Assuming an even distribution of free volume in the bulk of the solid, the average free volume per atom may be defined as:

$$\nu_f = \frac{V_f}{N}, \tag{3}$$

in which V_f is the total free volume content of the sample and N is its total number of atoms.

Despite MGs which have atomic microstructures, PGs are mainly composed of huge covalently bonded molecules. Hence, the "atomic jump" concept seems to require a revisit and redefinition for STZ-mediated plasticity of polymers. Glassy polymers possess an average free volume in the order of 5% with the average size of about a couple of hundred cubic Angstroms (Jean 1990). This size is about the size of a handful of monomers of common glassy polymers. On the other hand, each STZ in PGs is calculated to encompass a few thousands of monomers (Argon 2013; Malekmotiei et al. 2015; Samadi-Dooki et al. 2016). Although the size of the free volume voids seems to be small compared to that of the STZs, these voids can be assumed to provide weak spots for nucleation of STZs. The fact that free volume content increases with increasing the STZs number density in PGs (Hasan et al. 1993) suggests a reciprocal effect between the evolution of these two parameters in polymers. Hence, it could be reasonably assumed that free volume is redistributed at the STZs' sites upon their formation. Accordingly, N in Eq. 3 might be referred to as the number of STZs in PG, and the total free volume can be assumed to be a fraction of the total plastic volume V_p as:

$$V_f = f V_p, \tag{4}$$

where f stands for the fraction of the free volume.

Several experimental techniques have been used for evaluating the free volume content of polymers. The positron annihilation lifetime spectroscopy (PALS) is one of them which offers a reliable method for measuring the concentration and average size of the free volume holes in the bulk of the polymeric glass (Jean et al. 2013). In this method, positrons are injected into the bulk of the polymer from a radioactive source. Upon injection, part of the positrons interact with the electrons and form a particle called positronium (Ps), which can exist in form of para-positronium (p-Ps) with antiparallel spins and ortho-positronium (o-Ps) with parallel spins. Among these two particles, the transitional properties of o-Ps have important implications on the free volume of polymers. In particular, the intensity of o-Ps particles represents the concentration of the free volume sites, and its lifetime correlates with the average radius, R, of these voids as (Jean 1990):

$$\frac{1}{\tau_{o-Ps}} = 2\left[1 - \frac{R}{R + \Delta R} + \frac{1}{2\pi}\sin\left(\frac{2\pi R}{R + \Delta R}\right)\right], \tag{5}$$

in which $\Delta R = 1.656$ Å is an empirical constant and τ_{o-Ps} is the o-Ps lifetime in nanoseconds. The free volume content of the polymer can be related to the average volume and concentration of these sites as (Jean 1990; Wang et al. 1990):

$$f = \frac{4}{3}\pi R^3 \left(A I_{o-Ps} + B\right), \tag{6}$$

with I_{o-Ps} representing the o-Ps concentration and $A = 0.018$ and $B = 0.39$ are the experimentally found constants. The two o-Ps pertaining parameters, τ_{o-Ps} and I_{o-Ps}, are both temperature dependent; however, only the o-Ps annihilation lifetime τ_{o-Ps} is dependent on the plastic strain and thermal history of the polymer (Hasan et al. 1993). Accordingly, the free volume content of the polymer can be taken as a function of temperature T, shear plastic strain γ^p, and thermal history of the sample θ as:

$$f \equiv f\left(\gamma^p, T, \theta\right). \tag{7}$$

To come up with a relation for the fraction of the jump sites of Eq. 2 which is fully defined based on the measurable parameters, one needs to present alternate definitions for the number of STZs, N, and the factor $\lambda \nu^*$. Since it presents a factor for the minimum size of the free volume that can accommodate an STZ, the factor $\lambda \nu^*$ can be assumed to be proportional to the shear activation volume V^* (Li et al. 2013), which itself is defined as:

$$V^* = \alpha \gamma^T \Omega, \tag{8}$$

in which α is a factor for considering the dilatation effect and is close to unity for PGs (Ho et al. 2003; Malekmotiei et al. 2015; Samadi-Dooki et al. 2016). On the other hand, the number of the shear transformation zones can be assumed to be proportional to the total shear plastic strain (Argon 2013) and may be calculated as:

$$N \propto \frac{\gamma^p V_p}{\gamma^T \Omega}, \tag{9}$$

the parameters in which have already been defined. Finally, by incorporating Eqs. 3, 4, 7, 8, and 9, the fraction of the potential jump sites of Eq. 2 can be rearranged as:

$$c_f = \exp\left(-\frac{k\gamma^p}{f\left(\gamma^p, T, \theta\right)}\right), \tag{10}$$

in which k is a single constant that accounts for the various proportionality factors. Apparently, this equation presents the fractional free volume as a function of measurable variables such as temperature, plastic strain, and thermal history of the sample. While plastic strain is a dependent variable which is analytically present in Eq. 10, the variation of f with temperature, plastic strain, and thermal history of the sample is extracted from appropriate experimental studies using PALS (Hasan et al. 1993; Hristov et al. 1996).

STZ Nucleation Energy Evolution

As it was mentioned before, the STZs formed within the bulk of the glassy materials can be treated as nonelastic inclusions. Eshelby (1957) proposed a rigorous mathematical model for finding the elastic fields inside and outside the inclusions which has been used by many researchers as a homogenization technique in various types of problems (Bedayat and Taleghani 2014; Khoshgoftar et al. 2007; Malekmotiei et al. 2013). Accordingly, the energy of an inclusion ΔF can be found by a simple integration scheme as (Mura 1987):

$$\Delta F = \left[\Xi\left(\nu\right) + \beta^2 \Psi\left(\nu\right)\right]\mu\left(\gamma^T\right)^2 \Omega, \tag{11}$$

in which γ^T and Ω are the predefined transformation shear strain and volume of an STZ, respectively, μ is the shear modulus of the sample, β is the dilatency parameter which can be found from the pressure sensitivity of the flow, $\Psi(\nu)$ is a function of the Poisson's ratio as $\Psi\left(\nu\right) = \frac{2(1+\nu)}{9(1-\nu)}$, and $\Xi(\nu)$ is a parameter which depends on both the Poisson's ratio and shape of the inclusion and varies between $\frac{7-5\nu}{30(1-\nu)}$ and 0.5 for spherical and flat ellipsoidal shapes, respectively. For the sake of simplicity, the STZs are assumed to be noninteracting inclusions (Argon 2013); hence, the total energy of the deformation can be obtained by superposition of the nucleation energy of the individual STZs as inclusions (Bedayat and Taleghani 2014; Mura 1987; Shodja and Khorshidi 2013).

The use of the aforementioned form of energy for an inclusion which was proposed by Eshelby gives a constant value for the STZ's nucleation energy. However, as investigated by some researchers, this quantity evolves during the plastic deformation of amorphous solids (Argon 1979; Li et al. 2013). For a number of amorphous polymers, Oleinik (1991) obtained the total energy of deformation during their elastoplastic compression. According to his investigation, the total energy of deformation constantly increases with increasing the strain; however, the rate of the increment of the energy decreases beyond a certain strain (see Fig. 5 of Oleinik 1991). If one considers the aforementioned relation between the number of STZs and the plastic strain presented in Eq. 9, and the fact that the total plastic volume is equal to the total volume of the sample in uniaxial compression, the variation of the energy per STZ during the deformation can be obtained based on Oleinik's work, as presented in Fig. 2a for PMMA. Interestingly, the value of the energy of nucleation of a single STZ at yield which is found to be about 0.6 eV based on Oleinik's experiments is very close to that based on the authors' nanoindentation study on PMMA at room temperature (Malekmotiei et al. 2015). However, the latter method is not able to find the variation of this energy with increasing the strain.

Since the free volume void sites serve as the preferred spots for STZs formation, some investigators tried to link the variation of the STZs nucleation Helmholtz free energy to the evolution of the free volume during plastic strain to theoretically justify this phenomenon. The variation of the free volume holes' size and concentration with the strain were experimentally investigated by Hasan et al. (1993). Figure 2b shows the variation of the total free volume with the true uniaxial strain based on the work mentioned above and using Eqs. 5 and 6. Noticeably, the evolution of the free volume content in this figure is similar to that of the STZ nucleation energy as shown in Fig. 2a, but in the opposite direction. Both of these quantities change sharply upon yielding, and reach a stable value at a specific strain beyond which they show no significant change. This observation further supports the idea of existence of a relation between the free volume and shear transformation energy evolutions. Accordingly, some hypothetical models have been proposed to formulize this phenomenon. Notably for the case of MGs, Argon (1979) proposed a model which considered the dilatational component of the shear transformation to be stored as an excess free volume, the rate of the variation of which is calculated from creation-annihilation balance. In this way, the free volume creation takes place during the shear transformation events and its annihilation is mediated by the diffusive rearrangement of atoms. The free volume variation based on this model was recently used by Li et al. (2013) as a state variable for adjusting the STZ nucleation energy during the deformation. While the free volume content based on Argon's model alters with the dilatation component of the shear transformation, its effect needs to be considered on the dilatational part of the STZ nucleation energy as what was presented in Li et al. (2013). This, basically, means that the parameter β in Eq. 11 has to be considered as a function of free volume content. However, for the case of PGs, the parameter β is in the range of 0.2 (Malekmotiei et al. 2015;

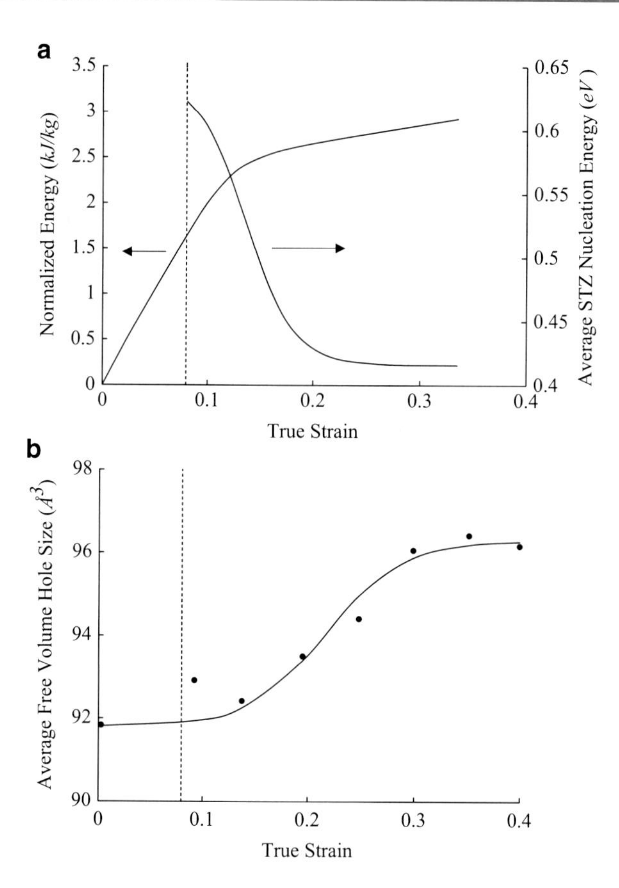

Fig. 2 (**a**) Variation of the stored internal energy in PMMA with true strain at room temperature for the compressive strain rate of 8.3×10^{-4} (s^{-1}) based on Oleinik (1991) (*left axis*) and the average nucleation energy per STZ calculated using Eq. 8 (*right axis*). (**b**) Data points show the variation of the average free volume size with true strain based on the compressive experiments of Hasan et al. (1993) on PMMA at room temperature and strain rate below β-transition, and the *solid line* is the *tanh* interpolation as discussed later in this chapter. The *dashed line* in both figures is the projected yield strain (Reproduced from Voyiadjis and Samadi-Dooki (2016) with permission from AIP)

Samadi-Dooki et al. 2016) and its variation (which is squared in this equation) cannot bring about the variation of STZ nucleation energy required to capture the behavior presented in Fig. 2a. Hence, in the current study, the quantitative evolution of STZ nucleation energy with plastic strain is obtained directly from the analysis of Oleinik's experiments as shown in Fig. 2a.

Other parameters which can affect the nucleation energy of the STZ are the temperature and the thermal history of the PG sample. The nucleation Helmholtz free energy is directly dependent on the shear modulus of the material as understood from Eq. 11. Hence, it is reasonable that one assumes the alteration of the STZ

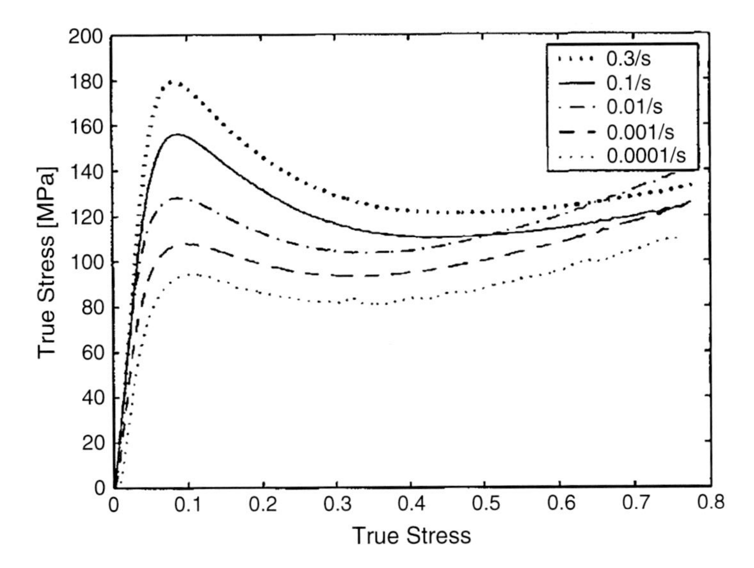

Fig. 3 Uniaxial compressive stress-strain curves of PMMA at different strain rates (Reproduced from Mulliken (2004) with permission from MIT)

nucleation energy to be a function of temperature in the same fashion of that for the shear modulus. This assumption is valid because other parameters in Eq. 11 can be assumed to be independent of temperature (see the discussion in Voyiadjis and Samadi-Dooki (2016) and the referencing to Holt (1968) and Roetling (1965)). The authors' previous investigation on PMMA shows that the STZ nucleation energy ΔF decreases with increasing temperature and almost vanishes at glass transition temperature T_g. This variation scheme is used in this study for investigating the temperature dependence of the flow. In addition, the authors' previous studies demonstrate a slight increase of the STZ nucleation energy with a short annealing time. Since the free volume content of the PG decreases with thermal aging, and an inverse relation between the free volume and shear transformation energy evolution seems to exist, the increase of the STZ nucleation energy could be qualitatively justified. The quantitative investigation of this phenomenon, however, is deferred to a forthcoming study by the authors.

Concluding from these remarks, the STZ nucleation energy in a PG can be taken as a function of temperature T, shear plastic strain γ^p, and thermal history of the sample θ as:

$$\Delta F = \Delta F\left(\gamma^p, T, \theta\right). \tag{12}$$

Finally, in light of Eqs. 10 and 12, and assuming $\gamma^T \Omega \tau^p \gg 2k_B T$ for PGs at temperatures below their T_g (Argon 2013), Eq. 1 can be reorganized at a constant strain rate as:

$$\tau^p\left(\gamma^p, \dot{\gamma}^p, T, \theta\right) = \frac{2k_BT}{\gamma^T\Omega}\left[\ln\left(\frac{2\dot{\gamma}^p}{\dot{\gamma}_0^p}\right) + \frac{\Delta F\left(\gamma^p, T, \theta\right)}{k_BT} + \frac{k\gamma^p}{f\left(\gamma^p, T, \theta\right)}\right]. \quad (13)$$

Equation 13 represents the explicit plastic shear stress-strain relation of a PG. To compare the results based on the current theory and those based on experiments, the shear stress and strain quantities can be simply converted to uniaxial parameters according to the von Mises' equivalency concept as:

$$\sigma^p = \sqrt{3}\tau^p \quad (14a)$$

and

$$\gamma^p = \sqrt{3}\epsilon^p, \quad (14b)$$

with σ^p and ϵ^p representing the uniaxial plastic stress and strain, respectively.

To complete the model, the viscoelastic preyield behavior is considered as a simple Maxwell viscoelastic material, the governing differential equation of which reads as:

$$\dot{\epsilon}^{ve}(t) = \frac{\dot{\sigma}^{ve}(t)}{E} + \frac{\sigma^{ve}(t)}{\eta}, \quad (15)$$

in which $\sigma^{ve}(t)$ and $\epsilon^{ve}(t)$ are the preyield viscoelastic stress and strain, respectively, $\dot{\epsilon}^{ve}(t)$ is the strain rate, E is the Young's modulus which is temperature dependent, η is the viscosity which is strain rate and temperature dependent (Kobayashi et al. 2001; Sun 2007), and t is time. The solution of Eq. 15 for constant strain rates, where $\dot{\epsilon}^{ve}(t) = \dot{\epsilon}^{ve}$, is readily found as:

$$\sigma^{ve}(t) = \dot{\epsilon}^{ve}\eta + A_1 e^{-\frac{Et}{\eta}}, \quad (16)$$

where A_1 is the constant of the solution which is easily obtained from the boundary condition of $\sigma^{ve}(0) = 0$. Assuming that the contribution of the elastic component of the strain rate during the postyield deformation is negligible (Hasan et al. 1993), Eqs. 16 and 13 can be considered to independently represent the stress-strain behavior in the preyield and postyield regions.

Results and Discussion

To evaluate the accuracy of the current model, the large deformation of PMMA as a glassy polymer is investigated using the proposed formalism which is compared with the available experimental data for this material. PMMA is a transparent polymer which is of both industrial and medical importance. Accordingly, it has attracted many researchers to study its manufacturing process and characteristic properties, including its mechanical behavior. Notably, Ghadipasha et al. (2016) developed a framework for investigating the polymerization of this polymer using integrated simulation, optimization, and online feedback control; Gunel and Basaran (2009, 2010, 2011a, b, 2013) developed comprehensive experimental, analytical, and numerical approaches for scrutinizing the deformation of PMMA and its composites with taking into account the effect of temperature and damage evolution; and Nasraoui et al. (2012) studied its mechanical behavior dependence on the strain rate and temperature both experimentally and analytically.

In this study, effects of strain rate, temperature, and thermal history of the sample on the postyield stress-strain behavior of PMMA are obtained based on the proposed model and compared with the experimental observations. The input parameters of the model are adopted from relevant experimental studies. In particular, the numerical values for the characteristic parameters of the STZ in PMMA, like its strain, volume, nucleation energy at yield, and the variation of this energy with temperature and thermal history of the sample, are extracted from the experimental evaluation of Malekmotiei et al. (2015). In addition, the evolution of the STZ nucleation energy during inelastic deformation is taken into account with considering the calculated values based on the measurements by Oleinik (1991) as presented in Figs. 1 and 2a. The variation of the free volume content of PMMA with strain, temperature, and thermal history of the sample is also calculated using the results of the work by Hasan et al. (1993) and Hristov et al. (1996).

Effect of the Rate

When tested at a wide range of strain rates, the variation of the yield stress of glassy polymers shows two distinct regions, each of which is almost linear with respect to the logarithm of the strain rate. The separation point might be called the β-transition point according to the study by Mulliken and Boyce (2006), and the trend of the variation of yield point is sharper in post-β-transition area compared to the pre-β-transition region. While the β-transition strain rate for PMMA in uniaxial loading condition was shown to be temperature dependent (Roetling 1965), it is generally considered to be about 0.01 s^{-1} at room temperature (Malekmotiei et al. 2015; Mulliken and Boyce 2006). The distinct pre- and post-β-transition behaviors are not just limited to the yield point of PMMA. The overall plastic deformation of this polymer shows rate dependence (as seen in Fig. 3) such that the postyield behavior might be divided into two distinguishable parts. Obviously, in each region, only the yield stress is strain rate sensitive, and the curves for the postyield behavior are parallel. In addition, while the steady-state is reached at the true strain of about 0.3 for pre-β-transition strain rates, it is shifted to the strain of about 0.45 for the strain rates of above the β-transition. In this study, we limit the investigation of the effect of the rate on the PMMA postyield behavior to the sub-β-transition strain rates. The reason for this adjustment is that the numerical values pertaining to the microstructural properties of the STZs, free volume evolution, and stored deformation energy variation of PMMA are given or calculated for the strain rates below the β-transition (in Malekmotiei et al. 2015; Hasan et al. 1993; Oleinik 1991, respectively).

During the plastic deformation at a constant strain rate, the free volume content of the PG increases as previously shown in Fig. 2b. The trend of the increment is such that the variation is insignificant prior to the yield, then it increases sharply until it gets to a stabilized value; this variation is acceptably interpolated by hyperbolic tangent (*tanh*) function. Interestingly, this function is compatible with the solution of the differential equation for the variation of c_f with the plastic strain at a constant strain rate which is presented in De Hey et al. (1998) as:

$$\frac{dc_f}{d\gamma} = -k_r c_f \left(c_f - c_{f.eq}\right) + a_f, \tag{17}$$

with k_r standing for a rate constant which is also dependent on the activation energy, a_f for a temperature-dependent parameter, and $c_{f.eq}$ for the equilibrium value of the fraction of the potential jump sites. In addition, since the evolution of the STZ nucleation energy with the plastic strain follows the same trend as that for the free volume (see Fig. 2), the same *tanh* interpolation format with required adjustments is used to represent its variation.

In Table 1 the numerical values for model parameters pertaining to PMMA at three sub-β-transition strain rates and at room temperature are shown. Accordingly, the true stress versus true strain curves based on the proposed model are obtained and compared to the numerical values based on the experiments of Mulliken (2004) as shown in Fig. 4. A very important fact which is indicated in this figure is that all curves are almost parallel in the postyield region, and the yield stress is only affected by the strain rate. Luckily, the structure of Eq. 13 readily reveals the reason. In this equation ΔF and f are independent of the strain rate, and the only factor which varies with this parameter is the first term in the right-hand side; while this term is independent of the plastic strain, it causes a constant jump in the curves proportional to the logarithm of the strain rate. In addition, Fig. 4 demonstrates that the proposed model accurately captures the primary softening and secondary hardening of PMMA. When the results are scrutinized in detail, it is observed that the secondary hardening is close to a linear variation. Since the free volume and the STZ nucleation energy converge to their respective plateaus beyond the steady state

Table 1 Numerical values for model parameters pertaining to PMMA. In this table, η is the viscosity, E is the Young's modulus, ΔF_0 is the STZ nucleation energy at yield, and $\gamma^T\Omega$ is the STZ size scale

Parameter Strain rate (s^{-1})	η (*Pa.s*)	E (*GPa*)	ΔF_0 (*eV*)	$\gamma^T\Omega$ (nm^3)
0.0001	10^{12}			
0.001	10^{11}	4.8	0.62	3.7
0.01	1.2×10^{10}			
Reference(s)	Sun (2007) and Kobayashi et al. (2001)	Gall and McCrum (1961)	Malekmotiei et al. (2015)	Malekmotiei et al. (2015)

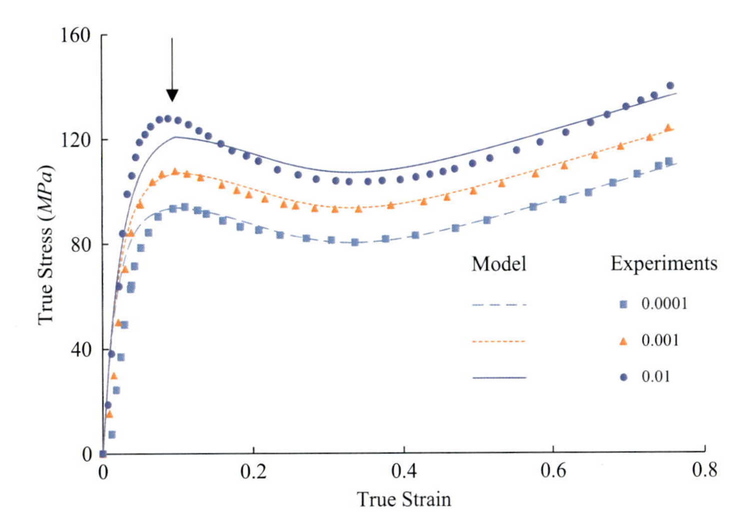

Fig. 4 True stress versus true strain curves of PMMA deformation at three different strain rates based on the proposed model as compared to the experiments of Mulliken (2004). The *vertical arrow* shows the yield point (Reproduced from Voyiadjis and Samadi-Dooki (2016) with permission from AIP)

strain, according to the last term of Eq. 13 the stress should increase with the number of STZs which has a linear variation with strain. While this behavior is very close to the experimental observations, some studies suggest that at higher plastic strains, PMMA might show a higher order hardening rather than a linear one (Arruda et al. 1995). This latter could be a result of chain alignment inside or on the border of the STZs which can increase the overall material resistance to further deformation. As another significant observation, at the strain rate of 0.01 s^{-1} which is close to the β-transition strain rate, the behavior of the material starts to deviate from that at sub-β-transition strain rates and also the viscoelastic-viscoplastic formulation developed in the study. The proper treatment of this phenomenon requires the consideration of the adiabatic heat generation in the PG at higher deformation rates, and is deferred to a forthcoming study by the authors.

Effect of the Temperature

The thermoplastics response to mechanical loading is undeniably affected by the temperature at which it is loaded. The formalism presented in this study considers the effect of the temperature on pre- and postyield behavior of PGs as reflected in Eqs. 13 and 16, respectively. For the preyield response, the temperature is a determining factor such that both viscosity and elastic modulus in Eq. 16 are temperature dependent, and hence temperature variation alters the viscoelastic response of the polymer. On the other hand, in the constitutive relation of Eq. 13

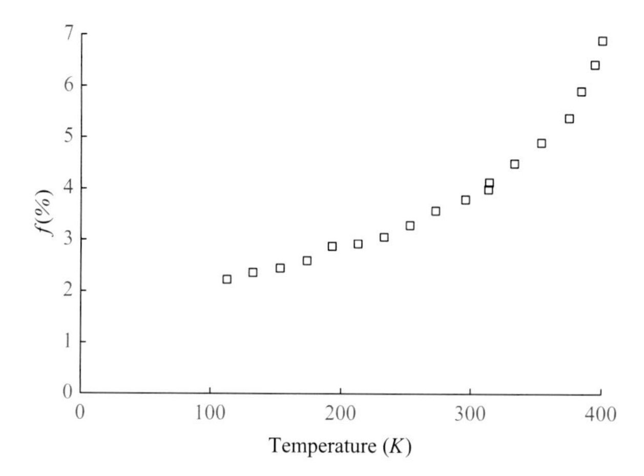

Fig. 5 Variation of the fraction of the free volume of PMMA with temperature (Reproduced from Hristov et al. (1996) with permission from ACS)

Table 2 Numerical values for temperature-dependent model parameters. The free volume content at yield is f_0. Other parameters hold the same definition as presented in the caption of Table 1

Parameter Temp. (K)	η ($Pa.s$)	E (GPa)	ΔF_0 (eV)	f_0 (%)
300	10^{11}	4.8	0.62	3.8
325	7×10^{10}	3.6	0.47	4.5
375	3.5×10^{10}	1.7	0.33	5.4
Reference(s)	Sun (2007) and Kobayashi et al. (2001)	Gall and McCrum (1961)	Malekmotiei et al. (2015)	Hristov et al. (1996)

for the postyield stage, the STZ nucleation energy ΔF and free volume fraction f are both functions of temperature (see Eqs. 12 and 7, respectively). Since ΔF is a function of shear modulus μ, it can be assumed to vary with temperature in the same fashion as the variation of shear modulus with temperature. In addition, since both the free volume hole size and concentration have been shown to vary almost linearly with temperature (Hasan et al. 1993), the free volume fraction f is expected to vary with a rate faster than a linear variation with temperature according to Eq. 6. This higher order variation was beheld by Hristov et al. (1996) for a number of glassy polymers including PMMA as shown in Fig. 5.

The numerical values of the temperature-dependent parameters involved in Eqs. 13 and 16 are presented in Table 2. Using these values, the true stress versus true strain curves of PMMA at the compressive strain rate of 0.001 s^{-1} are obtained for three different temperatures as depicted in Fig. 6 which shows a good correlation with the experimental observations of Arruda et al. (1995). The results of Fig. 6 demonstrate two notable alterations of the PGs' behavior with variation of the temperature. Firstly, with increasing the temperature the yield stress decreases; this

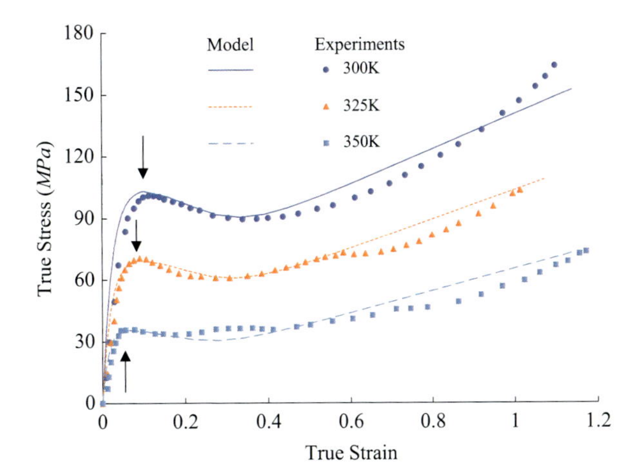

Fig. 6 True stress versus true strain curves of PMMA at the compressive strain rate of 0.001 s^{-1} for three different temperatures based on the proposed model and experiments of Arruda et al. (1995). Both yield strain and stress reduce with increasing the temperature (Reproduced from Voyiadjis and Samadi-Dooki (2016) with permission from AIP)

is a direct result of the STZs nucleation energy reduction at higher temperature as extensively discussed in Malekmotiei et al. (2015). Secondly, the variation of the temperature also changes the postyield behavior such that higher temperature PMMA undergoes primary softening and secondary hardening which are dampened compared to lower temperature behavior; this phenomenon is also well captured by the proposed model. As a justification, it could be mentioned that the higher order variation of f with temperature in the denominator of the third term in the right-hand side of Eq. 13 is faster than the linear variation of the numerator. As a result, the variation of the stress with plastic strain is damped by the effect of the increased temperature.

Effect of the Thermal History

Heat treatment is a process for enhancing the mechanical resistance of many solid state materials by relaxing the microstructural disturbances which might exist in these solids as residual stress or strain, or extra cavities and defects. For glassy polymers, such extra defects might exist if the cooling rate from molten state to solid state is relatively high (quenching). Since instabilities which are the result of unrelaxed microstructure can negatively affect the mechanical properties of the PGs, they could be physically aged at temperatures close to their glassy temperature for several hours in order to regain their nominal strength. This process is called annealing at it is shown to increase the yield stress of PGs (Chen and Schweizer 2011; Hutchinson et al. 1999; Nanzai et al. 1999). Nevertheless, both quenched

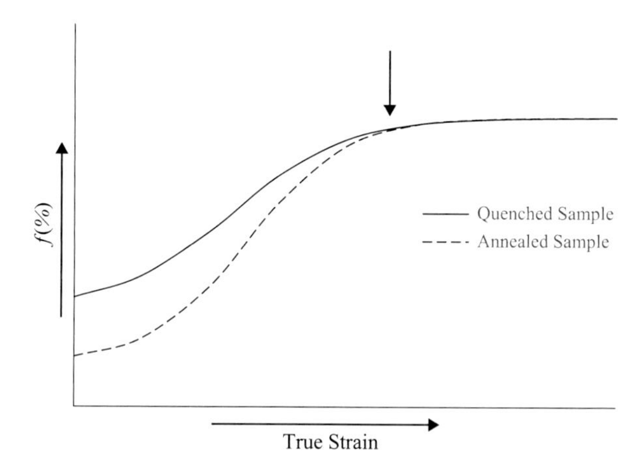

Fig. 7 Schematic representation of the variation of the free volume fraction with strain for annealed and quenched samples based on Hasan et al. (1993). *Vertical arrow* shows the steady-state strain (Reproduced from Voyiadjis and Samadi-Dooki (2016) with permission from AIP)

and annealed samples of the same polymer are shown to reach the same postyield steady-state stress which takes place at the same strain, and afterwards they both exhibit identical secondary hardening (Hasan et al. 1993; Klompen et al. 2005; Xiao and Nguyen 2015). This phenomenon seems to be correlated with the free volume evolution of the polymeric samples with different thermal histories. For example, the investigation by Hasan et al. (1993) shows that while the annealed PMMA sample possesses a lower free volume content compared to the quenched one, the free volume fraction for both of the samples reaches the same value when they are deformed up to the steady-state plastic strain. This is schematically shown in Fig. 7. On the other hand, since it is believed that there exists an inverse correlation between the free volume content and the STZ nucleation energy, one might expect a higher STZ activation energy at yield for annealed sample compared to the quenched one. The experimental evaluation of the STZ nucleation energy at yield for PMMA and Polycarbonate (PC) with different thermal histories confirms this hypothesis with revealing a slightly bigger STZ nucleation energy for annealed samples (Malekmotiei et al. 2015; Samadi-Dooki et al. 2016).

According to the aforementioned correlation between the free volume content and the STZ nucleation energy, it seems to be reasonable to assume that the STZ nucleation energy of quenched and annealed PMMA samples also reaches the same value at the steady-state strain and remains identical after that. Hence, with considering the STZ activation energy variation with the strain as presented in Fig. 2a, the *tanh* interpolation function for this quantity could be utilized with the required adjustments for quenched and annealed samples. Accordingly, and with using the numerical values for model parameters as presented in Table 3, the true stress versus true strain curves for samples with different thermal histories could be obtained as presented in Fig. 8. The prediction of the model for considering

Table 3 Numerical values for model parameters for PMMA with different thermal histories (the value of η for quenched sample is assumed for curve fitting purpose)

Parameter Thermal History	η (*Pa.s*)	*E* (*GPa*)	ΔF_0 (*eV*)	$\gamma^T \Omega$ (nm^3)	f_0 (%)
Annealed	10^{12}	4.8	0.62	3.7	3.8
Quenched	0.86×10^{12}		0.58	3.6	3.9
Reference(s)	Sun (2007), Kobayashi et al. (2001)	Gall and McCrum (1961)	Malekmotiei et al. (2015)	Malekmotiei et al. (2015)	Hristov et al. (1996) and Hasan et al. (1993)

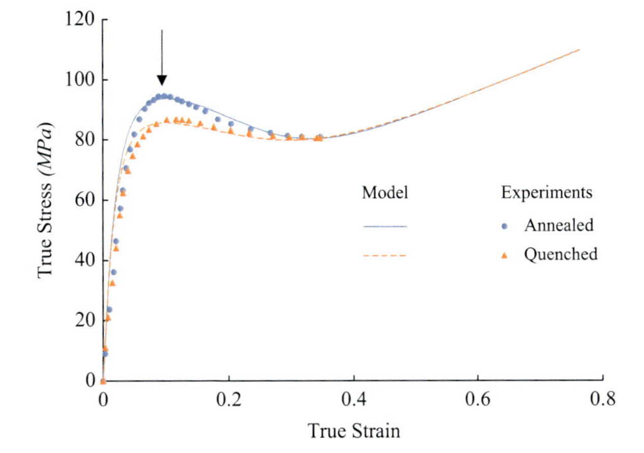

Fig. 8 True stress versus true strain of PMMA for annealed and quenched samples based on the proposed model and experiments of Hasan and Boyce (1995) at strain rate of 0.0001 s^{-1}. The *vertical arrow* shows the yield (Reproduced from Voyiadjis and Samadi-Dooki (2016) with permission from AIP)

the thermal history of PMMA on its stress-strain behavior is acceptably close to the experimental observation by Hasan and Boyce (1995) for the softening region. The secondary hardening is predicted to be identical for the samples with different thermal histories which is in agreement with the experimental observation for other glassy polymers like PC (see Fig. 2 of Hasan et al. 1993).

Implication of the Model for Shear Banding and Indentation Size Effect in PGs

Since the model parameters in this study are obtained from compressive experiments, the authors were careful to make the comparisons with compressive behavior of PMMA where no localization effect is expected to exist. However,

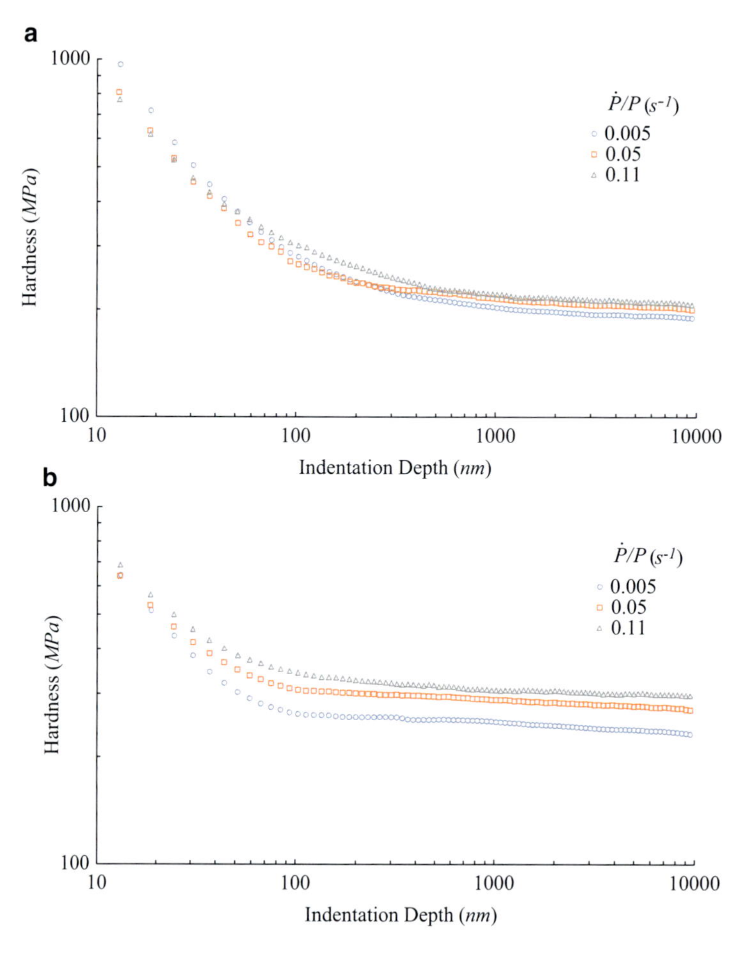

Fig. 9 Variation of the hardness with indentation depth at three different rates for (**a**) PC and (**b**) PMMA (Reproduced from Voyiadjis and Malekmotiei (2016) with permission from Wiley Online Library)

the localization can happen in polymeric samples under more complex loading conditions with formation of shear bands (Anand and Spitzig 1982; Li and Buckley 2009; Voyiadjis et al. 2014) or where the strain gradient can generate an additional hardening which, for example, emerges as indentation size effect (ISE) in nanoindentation of glassy polymers (Malekmotiei et al. 2015; Samadi-Dooki et al. 2016; Voyiadjis and Malekmotiei 2016; Voyiadjis et al. 2014). The latter is the increased hardness at shallow indentation in nanoindentation experiments with self-similar tips like the Berkovich tip (see Fig. 9). Since the proposed model takes into account the free volume and STZs evolutions, it can be used to evaluate these phenomena. For example, the formation of shear bands in PGs can be thought of as coalescence of STZ within a small region where extra defects (free volume) exist. In addition, the ISE in polymers might also be formulated with considering the probability of presence of a free volume site in the small anticipated plastic region which can accommodate the required number of STZs to bring about the plastic deformation. While modeling these effects is an ongoing topic of study by the authors, they differ further discussions to their forthcoming publications.

Concluding Remarks

In this study, the plastic deformation of glassy polymers is formulated based on the evolutions of the free volume and the shear transformation nucleation energy during the course of large inelastic deformation. In order to be used as input parameters, these variations were quantified based on previous experimental observations. The reason for this strategy is that the authors have found that the direct use of the formulations developed for the evolution of these two parameters in MGs cannot account for their variations in polymeric glasses. To test the accuracy of the model, effects of the strain rate, temperature, and thermal history of the samples on the postyield behavior of PMMA, as a widely used PG, have been analyzed and compared with the previous experimental observations. The results show acceptable consonance with experiments for both the primary postyield softening and the secondary hardening behavior of PMMA. Although the authors have made some simplifying assumptions, the proposed model is promising in a sense that it captures the softening-hardening deformation behavior of PGs with a single formula. This suggests that the overall plastic deformation behavior of PGs is a unified mechanism. This hypothesis, which is well proved in this study, is fundamentally different from previous studies which suggest different mechanisms for the postyield softening-hardening behavior of PGs. With additional developments, the proposed model can stand as a conclusive theory for plastic deformation of amorphous polymers.

References

L. Anand, W. Spitzig, Acta Metall. **30**, 553 (1982)
A. Argon, Philos. Mag. **28**, 839 (1973)
A. Argon, Acta Metall. **27**, 47 (1979)
A.S. Argon, *Strengthening Mechanisms in Crystal Plasticity* (Oxford University Press, Oxford, 2008)
A.S. Argon, *The Physics of Deformation and Fracture of Polymers* (Cambridge University Press, New York, 2013)
A. Argon, M. Demkowicz, Philos. Mag. **86**, 4153 (2006)
A. Argon, M. Demkowicz, Metall. Mater. Trans. A **39**, 1762 (2008)
E.M. Arruda, M.C. Boyce, in *Anisotropy and Localization of Plastic Deformation*, ed. by J-P. Boehler, A.S. Khan (Elsevier Applied Science, London and New York, 1991), p. 483
E.M. Arruda, M.C. Boyce, H. Quintus-Bosz, Int. J. Plast. **9**, 783 (1993)
E.M. Arruda, M.C. Boyce, R. Jayachandran, Mech. Mater. **19**, 193 (1995)
H. Bedayat, A.D. Taleghani, Mech. Mater. **69**, 204 (2014)
M.C. Boyce, D.M. Parks, A.S. Argon, Mech. Mater. **7**, 15 (1988)

K. Chen, K.S. Schweizer, Macromolecules **44**, 3988 (2011)
M.H. Cohen, D. Turnbull, J. Chem. Phys. **31**, 1164 (1959)
P. De Hey, J. Sietsma, A. Van Den Beukel, Acta Mater. **46**, 5873 (1998)
J.D. Eshelby, Proc. Roy. Soc. London Ser. A 241, 376 (1957)
H. Eyring, J. Chem. Phys. **4**, 283 (1936)
M.L. Falk, Science **318**, 1880 (2007)
M. Falk, J. Langer, L. Pechenik, Phys. Rev. E **70**, 011507 (2004)
K. Flores, D. Suh, R. Dauskardt, P. Asoka-Kumar, P. Sterne, R. Howell, J. Mater. Res. **17**, 1153 (2002)
W. Gall, N. McCrum, J. Polym. Sci. **50**, 489 (1961)
N. Ghadipasha, A. Geraili, J.A. Romagnoli, C.A. Castor, M.F. Drenski, W.F. Reed, Processes **4**, 5 (2016)
E. Gunel, C. Basaran, Mater. Sci. Eng. A **523**, 160 (2009)
E. Gunel, C. Basaran, J. Eng. Mater. Technol. **132**, 031002 (2010)
E. Gunel, C. Basaran, Mech. Mater. **43**, 979 (2011a)
E. Gunel, C. Basaran, Mech. Mater. **43**, 992 (2011b)
E. Gunel, C. Basaran, Mech. Mater. **57**, 134 (2013)
O. Hasan, M. Boyce, Polym. Eng. Sci. **35**, 331 (1995)
O. Hasan, M. Boyce, X. Li, S. Berko, J. Polym. Sci. B Polym. Phys. **31**, 185 (1993)
P. de Hey, J. Sietsma, A. Van Den Beukel, Mater. Sci. Eng. A **226**, 336 (1997)
J. Ho, L. Govaert, M. Utz, Macromolecules **36**, 7398 (2003)
D. Hofmann, M. Heuchel, Y. Yampolskii, V. Khotimskii, V. Shantarovich, Macromolecules **35**, 2129 (2002)
D.L. Holt, J. Appl. Polym. Sci. **12**, 1653 (1968)
H. Hristov, B. Bolan, A. Yee, L. Xie, D. Gidley, Macromolecules **29**, 8507 (1996)
J. Hutchinson, S. Smith, B. Horne, G. Gourlay, Macromolecules **32**, 5046 (1999)
Y. Jean, Microchem. J. **42**, 72 (1990)
Y. Jean, J.D. Van Horn, W.-S. Hung, K.-R. Lee, Macromolecules **46**, 7133 (2013)
M. Khoshgoftar, S. Najarian, F. Farmanzad, B. Vahidi, F. Ghomshe, Am. J. Appl. Sci. **4**, 918 (2007)
E. Klompen, T. Engels, L. Govaert, H. Meijer, Macromolecules **38**, 6997 (2005)
H. Kobayashi, H. Takahashi, Y. Hiki, J. Non-Cryst. Solids **290**, 32 (2001)
D.C. Lam, A.C. Chong, J. Mater. Res. **14**, 3784 (1999)
J. Langer, L. Pechenik, Phys. Rev. E **68**, 061507 (2003)
H. Li, C. Buckley, Int. J. Solids Struct. **46**, 1607 (2009)
L. Li, E. Homer, C. Schuh, Acta Mater. **61**, 3347 (2013)
L. Malekmotiei, F. Farahmand, H.M. Shodja, A. Samadi-Dooki, J. Biomech. Eng. **135**, 041004 (2013)
L. Malekmotiei, A. Samadi-Dooki, G.Z. Voyiadjis, Macromolecules **48**, 5348 (2015)
A.D. Mulliken, *Low to high strain rate deformation of amorphous polymers: experiments and modeling* (Massachusetts Institute of Technology, Massachusetts, 2004)
A. Mulliken, M. Boyce, Int. J. Solids Struct. **43**, 1331 (2006)
T. Mura, in *Micromechanics of Defects in Solids* (Martinus Nijhoff Publishers, The Netherlands, 1987)
Y. Nanzai, A. Miwa, S.Z. Cui, JSME Int. J. Ser. A Solid Mech. Mater. Eng. **42**, 479 (1999)
M. Nasraoui, P. Forquin, L. Siad, A. Rusinek, Mater. Des. **37**, 500 (2012)
E. Oleinik, in *High Performance Polymers*, ed. by E. Baer, A. Moet (Hanser, New York, 1991), p. 79
E. Oleinik, S. Rudnev, O. Salamatina, Polym. Sci. Ser. A **49**, 1302 (2007)
S. Pauly, S. Gorantla, G. Wang, U. Kühn, J. Eckert, Nat. Mater. **9**, 473 (2010)
T. Ree, H. Eyring, J. Appl. Phys. **26**, 793 (1955)
R.E. Robertson, J. Chem. Phys. **44**, 3950 (1966)
J. Roetling, Polymer **6**, 311 (1965)
A. Samadi-Dooki, L. Malekmotiei, G.Z. Voyiadjis, Polymer **82**, 238 (2016)
H. Shodja, A. Khorshidi, J. Mech. Phys. Solids **61**, 1124 (2013)

H. Shodja, M. Tabatabaei, A. Ostadhossein, L. Pahlevani, Open Eng. **3**, 707 (2013)
F. Spaepen, Acta Metall. **25**, 407 (1977)
Y. Sun, J. Cent. S. Univ. Technol. **14**, 342 (2007)
L.A. Utracki, A.M. Jamieson, *Polymer Physics: From Suspensions to Nanocomposites and Beyond* (Wiley, New York, 2011)
G.Z. Voyiadjis, L. Malekmotiei, J. Polym. Sci. Part B: Polym. Phys. **54**, 2179 (2016)
G.Z. Voyiadjis, A. Samadi-Dooki, J. Appl. Phys. **119**, 225104 (2016)
G.Z. Voyiadjis, M. Yaghoobi, Mater. Sci. Eng. A **634**, 20 (2015)
G.Z. Voyiadjis, A. Shojaei, N. Mozaffari, Polymer **55**, 4182 (2014)
Y. Wang, H. Nakanishi, Y. Jean, T. Sandreczki, J. Polym. Sci. B Polym. Phys. **28**, 1431 (1990)
D. Wang, Z. Zhu, R. Xue, D. Ding, H. Bai, W. Wang, J. Appl. Phys. **114**, 173505 (2013)
R. Xiao, T.D. Nguyen, J. Mech. Phys. Solids **82**, 62 (2015)
A.R. Yavari, A. Le Moulec, A. Inoue, N. Nishiyama, N. Lupu, E. Matsubara, W.J. Botta, G. Vaughan, M. Di Michiel, Å. Kvick, Acta Mater. **53**, 1611 (2005)

Indentation Fatigue Mechanics

13

Baoxing Xu, Xi Chen, and Zhufeng Yue

Contents

B. Xu (✉)
Department of Mechanical and Aerospace Engineering, University of Virginia, Charlottesville, VA, USA
e-mail: bx4c@virginia.edu

X. Chen
Department of Earth and Environmental Engineering, Columbia Nanomechanics Research Center, Columbia University, New York, NY, USA
e-mail: xichen@columbia.edu

Z. Yue
Department of Engineering Mechanics, Northwestern Polytechnical University, Xi'an, Shaanxi, China
e-mail: zfyue@nwpu.edu.cn

G. Z. Voyiadjis (ed.), *Handbook of Nonlocal Continuum Mechanics for Materials and Structures*, https://doi.org/10.1007/978-3-319-58729-5_25

Abstract

Instrumented indentation has been widely used in the determination of mechanical properties of materials due to its fast, simple, precise, and nondestructive merits over the past few years. In this chapter, we will present an emerging indentation technique, referred to as indentation fatigue, where a fatigue load is applied on a sample via a flat punch indenter, and establish the framework of mechanics of indentation fatigue to extract fatigue properties of materials. Through extensive experimental, theoretical, and computational investigations, we demonstrate a similarity between the indentation fatigue depth propagation and the fatigue crack growth, and propose an indentation fatigue depth propagation law and indentation fatigue strength law to describe indentation fatigue-induced deformation and failure of materials, respectively. This study provides an alternative approach for determining fatigue properties, as well as for studying the fatigue mechanisms of materials, especially for materials that are not available or feasible for conventional fatigue tests.

Keywords

Indentation · Fatigue loading · Indentation depth propagation · Crack propagation · Strength

Introduction

Fatigue is a process of accumulations of material deformation and damage due to a repeated loading and unloading and is considered as one of major threats to the mechanical integrity of materials and structures. As a technical problem, the study of fatigue can be traced to as early as the mid-nineteenth century with the data collection of failure of railroad system by Germany technologist A Wohler (Suresh 1998). Since then, it has attracted tremendous attention over the last century, and great achievements have been made from macroscaled phenomena to micro-/nanoscaled deformation mechanisms (Estrin and Vinogradov 2010; Connolley et al. 2005), from qualitative descriptions to quantitative predictions (Rao and Farris 2008; Fleck and Smith 1984), and from engineering metals to biological and soft materials (Teoh 2000; Dirks et al. 2013; Tang et al.). Meanwhile, with the ever-growing applications of small structures, including nano-/micro-electromechanical systems (N/MEMS) and thin films over the last decades and emerging low-dimensional nanomaterials such as nanowires, nanofibers, and graphene, their properties usually exhibit differences than their materials counterpart at the macroscale (Alsem et al. 2007, 2008; Höppel et al. 2009; Li et al. 2003, 2014; Luo et al. 2015; Lee et al. 2008; Zhu et al. 2007), and the characterization of their mechanical properties in particular fatigue properties is challenging the traditional uniaxial mechanical testing technique that is mainly based on standard "dog-bone" shaped specimens. The development of an alternative testing technique that could help to address these challenges is highly desired.

Indentation, which origins from the hardness measurement, provides a compelling solution to measure mechanical properties of materials down to the nanosize (Cheng and Cheng 2004; Fischer-Cripps 2000). The merits of indentation include minor requirements of testing materials, easy to conduct, and high resolutions of measurement. During the experiment, the indentation load and displacement can be recorded continuously with the perpetration of the indenter to samples, and even deformation of materials can also be observed when the testing platform is integrated with advanced observational facilities such as transmission electron microscopy (TEM) and scanning electron microscope (SEM), usually referred to as in situ TEM-indentation (Warren et al. 2007) and SEM-indentation (Nowak et al. 2010), respectively. Significant efforts have been made to extract mechanical properties of materials from indentation data including elastic properties (Oliver and Pharr 2004; Yang et al. 2016; Li and Bhushan 2002; Dao et al. 2001; Lan and Venkatesh 2007; Jiang et al. 2009), plastic properties (Cheng and Cheng 1998; Chen et al. 2007; Xu and Chen 2010; Lee et al. 2010; Bucaille et al. 2003), fracture properties (Lawn and Wilshaw 1975; Tang et al. 2008; Xia et al. 2004; Sakaguchi et al. 1992; Miranzo and Moya 1984; Quinn and Bradt 2007), creep properties (Xu et al. 2008a; Cheng and Cheng 2001; Li et al. 1991; Yang and Li 1995; Chen et al. 2010; Stone et al. 2010), and relaxation properties (Xu et al. 2008b; Baoxing et al. 2010; Chu and Li 1977, 1980a; Hu et al. 2011; Chan et al. 2012) with a broad range of materials from hard metals, to composites, to biological materials, to soft matters. Usually, these properties can be determined from the measured indentation load and displacement curves and deformation profiles of indented materials at monotonic loading conditions. However, fatigue properties such as fatigue strength and fatigue-induced crack growth that rely on fatigue/cyclic loading conditions usually cannot be measured.

Indentation that is conducted under a cyclic loading with cyclic numbers >1000 is referred to indentation fatigue here with an emphasis on measurement of fatigue properties of materials. An early study by Li and Chu found that the flat punch indenter could continue to penetrate into the β-tin single crystal with applied cycles, and the recorded curves of indentation displacement-cyclic numbers were similar to that of fatigue crack propagation with the absence of third stage (Li and Chu 1979). With the same β-tin single crystal, later on, Chu and Li changed the fatigue conditions to fatigue spectrum with interruption by a peak loading and further found the similarity with that of fatigue crack propagation with a delayed retardation of indentation depth propagation caused by a peak loading (Chu and Li 1980b). By employing a Vickers diamond indenter, Kaszynski et al. investigated the indentation fatigue response of 316 L stainless steel and found that the indentation depth showed an approximately linear increase with the logarithm of the number of cycles (Kaszynski et al. 1998). This difference is believed to be led by the geometric shape of indenters and associated change of contact areas between indenter and materials in experiments (Li 2002; Yang and Li 2013). The similarities between indentation fatigue and fatigue crack propagation provide a critical hint to probe the fatigue properties of materials through instrumented indentation technique

(Yang and Li 2013). *However, why do such similarities exist? What is the deformation mechanism of the evolution of indentation depth propagation under a fatigue loading? What fatigue properties can be determined and how? How to quantitatively describe the depth propagation of indentation fatigue? What is the quantitative relationship between the depth propagation of indentation fatigue and fatigue crack propagation of conventional fatigue test?* Over the last 10 years, we have been actively working on the study of indentation fatigue by integrating tools of experiments, theories, and computations to address these questions (Xu et al. 2010, 2007a, b, 2009; Xu and Yue 2006, 2007).

In this book chapter, we will summarize several important results based on our progresses and present quantitative mechanics descriptions and validations of indentation fatigue from indentation fatigue depth propagation to indentation fatigue strength of materials. In section "Mechanics Theory of Indentation Fatigue," an elastic indentation fatigue mechanics model will be first developed, and its reduction to and comparison with indentation mechanical model under a monotonic loading will be discussed. And then we will show theoretical analysis on a similarity of stress field between the rim of contact area between flat punch indenter and surface of a sample and crack tips. In section "Indentation Fatigue Deformation," extensive experimental results on polycrystalline copper and finite element analysis will be performed to demonstrate similarities between indentation fatigue depth propagation and fatigue crack growth, and an indentation fatigue depth propagation law will be developed. In section "Indentation Fatigue Damage," we will extend the indentation fatigue mechanics to indentation fatigue damage with a focus on indentation fatigue-induced failure of materials and predict the fatigue strength of materials from indentation fatigue. Concluding remarks will be given in section "Concluding Remarks."

Mechanics Theory of Indentation Fatigue

Indentation Load-Depth Curve

Consider a semi-infinite homogeneous elastic solid indented by a rigid frictionless flat punch, as schematized in Fig. 1, when the indentation load P is applied to the indenter, the penetration of the indenter (i.e., indentation depth) to the solid is h, and based on Hertz contact theory, it can be written as

$$h = \frac{\left(1 - \nu^2\right)}{Ed} P \tag{1}$$

where d is the diameter of the indenter, and E and ν are Young's modulus and Poisson's ratio of the solid, respectively. Equation 1 is also referred to as Sneddon's solution and serves the fundamental of determining Young's modulus of materials from a flat punch indentation (Sneddon 1965). Besides, because of the constant

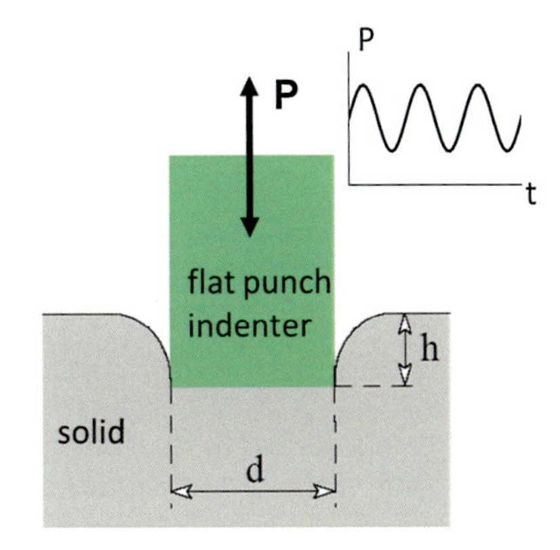

Fig. 1 Schematic illustration of indentation fatigue on a semi-infinite solid via a flat punch indenter subjected to a fatigue loading P. d is the diameter of flat punch indenter, h is the indentation depth. The inset illustrates the cyclic fatigue loading

diameter of the punch indenter, Eq. 1 suggests a linear variation of indentation depth with indentation load at a small deformation, indicating one of the benefits of flat punch indenter in indentation technique in comparison with other indenter shapes such as sharp indenter, Vicker indenter, or spherical indenter that will result in increased contact area with the increase of the indentation depth. Obviously, when the indentation load P increases or decreases, Eq. 1 holds instantaneously for a semi-infinite homogeneous elastic solid, and thus it can be used for both loading and unloading conditions. Assume the indentation load P is applied with a sinusoidal loading spectrum, we will have

$$P(t) = P_m + \Delta P \sin(\omega t) \tag{2}$$

where ΔP, P_m, and ω are the indentation load range, mean, and frequency, respectively. With Eqs. 1 and 2, when a sinusoidal loading is applied to a semi-infinite homogeneous elastic solid via a flat punch indenter, the indentation load-depth relationship will be

$$h(t) = \frac{\left(1 - \nu^2\right)}{Ed} \left(P_m + \Delta P \sin(\omega t)\right) \tag{3}$$

That is

$$h(t) = \frac{\left(1 - \nu^2\right)}{Ed} P_m \left(1 + \frac{\Delta P}{P_m} \sin(\omega t)\right) \tag{4}$$

Apparently, the variation of indentation depth with experimental time is sinusoidal. Assume

$$h(t) = h_m + \Delta h \sin(\omega t) \tag{5}$$

where h_m and Δh are the mean and magnitude of the indentation depth. Thus, Eq. 4 can be rewritten as

$$h_m = \frac{(1 - \nu^2)}{Ed} P_m \tag{6}$$

$$\Delta h = \frac{(1 - \nu^2)}{Ed} \Delta P \tag{7}$$

Equation 6 is similar to Eq. 1 and indicates the indentation depth is induced by the mean load of the sinusoidal loading. Meanwhile, Eq. 7 indicates the dynamic response of indentation under the sinusoidal loading and can be written as

$$\frac{E}{1 - \nu^2} = \frac{1}{d} \frac{\Delta P}{\Delta h} \tag{8}$$

$E/(1 - \nu^2)$is usually referred to as the stiffness of materials. Apparently one can easily probe the variation of stiffness of materials by applying a dynamic loading. Equation 8 has served the theoretical model in the measurement of dynamic contact stiffness of materials and has been employed in the continuous stiffness measurement (CSM) technique in the instrumented indentation technique (Li and Bhushan 2002; Asif et al. 1999; Loubet et al. 2000).

Indentation Stress Intensity Factor

Consider the solid in Fig. 2a, the boundary condition of stress field beneath the flat punch indenter is (Sneddon 1965; Johnson 1985)

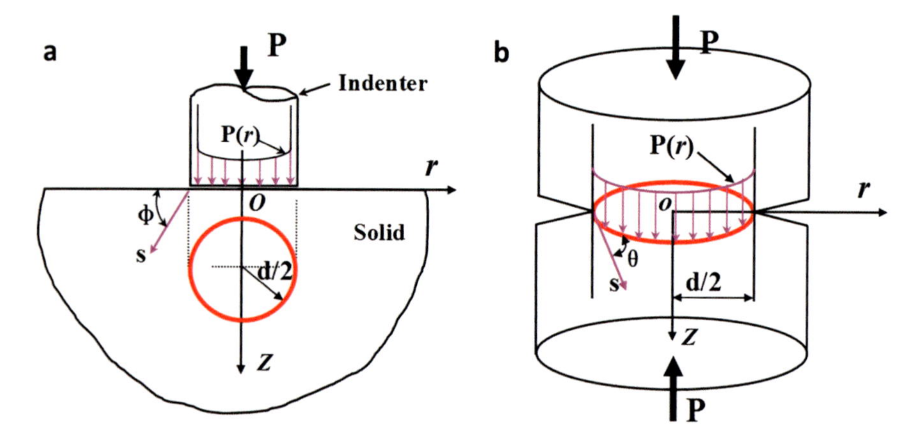

Fig. 2 (**a**) Schematic representation of contact between the flat punch indenter and a semi-infinite solid. (**b**) Circumferentially cracked notch specimen subjected to far-field compressive load Xu et al. 2009 (reprinted with the permission from Elsevier)

$$\sigma_{zz}(x,0) = 0, \quad | x | > d/2 \tag{9a}$$

$$\tau_{xz}(x,0) = 0, \quad 0 \leq | x | \leq d/2 \tag{9b}$$

In the cylindrical coordinate systemr, θ, z, the stress field can be written as (Sneddon 1965; Johnson 1985)

$$\begin{cases} \sigma_{zz} = \frac{Eh}{2(1-\nu^2)} \int\limits_0^\infty (1+mz)\, e^{-mz} J_o(mr) \sin(md/2)\, dm \\ \sigma_{rr} = \frac{Eh}{2(1-\nu^2)} \left[\int\limits_0^\infty (1-mz)\, e^{-mz} J_o(mr) \sin(md/2)\, dm \right. \\ \left. \quad -\frac{1}{r} \int\limits_0^\infty \frac{1}{m} (1-2\nu-mz)\, e^{-mz} J_1(mr) \sin(md/2)\, dm \right] \\ \sigma_{\theta\theta} = \frac{Eh}{2(1-\nu^2)} \left[\int\limits_0^\infty 2\nu m^2 e^{-mz} J_o(mr) \sin(md/2)\, dm \right. \\ \left. \quad +\frac{1}{r} \int\limits_0^\infty \frac{1}{m} (1-2\nu-mz)\, e^{-mz} J_1(mr) \sin(md/2)\, dm \right] \\ \sigma_{zr} = \frac{Eh}{2(1-\nu^2)} \int\limits_0^\infty mze^{-mz} J_1(mr) \sin(mz) dm \end{cases} \tag{10}$$

where m is the variable function. J_α is the Bessel function of the first kind and is

$$J_\alpha(x) = \sum_{m=0}^{\infty} \frac{(-1)^m}{m!\Gamma(m+\alpha+1)} \left(\frac{x}{2}\right)^{2m+\alpha}$$

Within the contact range at $z = 0$, one can have

$$\sigma_{zz}|_{z=0} = \frac{Et_0}{\pi(1-\nu^2)} \int\limits_0^\infty J_o(mr) \sin(md/2)\, dm \tag{11}$$

Because $\int\limits_0^\infty J_o(mr) \sin(md/2)\, dm = \frac{1}{\sqrt{(d/2)^2 - r^2}}$, $r \leq d/2$, one will have

$$\begin{aligned} \sigma_{zz}|_{z=0} &= \frac{Et_0}{\pi(1-\nu^2)} \int\limits_0^\infty J_o(mr) \sin(md/2)\, dm \\ &= \frac{Et_0}{\pi(1-\nu^2)} \frac{1}{\sqrt{(d/2)^2 - r^2}} \qquad r \leq d/2 \end{aligned} \tag{12}$$

That is

$$\sigma_{zz}(r,0) = \frac{Eh}{\pi(1-\nu^2)} \frac{1}{\sqrt{(d/2)^2 - r^2}}, \quad r \leq d/2 \tag{13}$$

At the rim of indentation, $|x|$ is approaching to $d/2$, one will have

$$\sigma_{zz}(r,0) = \frac{Eh}{\pi\left(1-\nu^2\right)}\left(\frac{1}{\sqrt{dr}}+\cdots\right), \quad r \to d/2 \tag{14}$$

Equation 14 indicates there is a stress singularity at the edge of indentation with respect to $\sqrt{r}$.

Similarly, for a semi-infinite model-I crack subjected to a far-field compressive load P, as schematized in Fig. 2b, given the distance$d/2$from the origin of the coordinate system to the crack tip, the stress field at the crack tip is (Hertzberg 1995)

$$\begin{pmatrix} \sigma_{zz} \\ \sigma_{rz} \\ \sigma_{rr} \\ \sigma_{\varphi\varphi} \end{pmatrix} \to -\frac{2P}{\pi d^2\sqrt{s}}\cos\frac{\varphi}{2}\begin{pmatrix} 1+\sin\frac{\varphi}{2}\sin\frac{3\varphi}{2} \\ \sin\frac{\varphi}{2}\cos\frac{3\varphi}{2} \\ 1-\sin\frac{\varphi}{2}\sin\frac{3\varphi}{2} \\ 2\nu \end{pmatrix} \tag{15}$$

Comparison between Eqs. 14 and 15 implies that σ_{zz} in Eq. 15 is an asymptote of Eq. 13 at $r \to d/2$, leading to an equivalence between Eqs. 14 and 15, and thus suggesting the same stress singularity with respect to $\sqrt{}$ at the rim of flat punch indentation and crack tip.

Given Eq. 1, Eq. 13 can be rewritten as

$$\sigma_{zz}(r,0) = -\frac{4P}{\pi d^2}, \quad r \le d/2 \tag{16}$$

According to the definition of stress intensity factor, we will have

$$\sigma_{zz} = -\frac{K_I}{\sqrt{2\pi\left(d/2-r\right)}} \tag{17}$$

And thus the stress intensity factor near the contact rim of flat punch and solid can be expressed as

$$K_I = \frac{F}{d\sqrt{\pi d/2}} \tag{18}$$

And Eq. 18 indicates the same stress intensity factor near the contact rim of flat punch and solid with that at the crack tip.

Under a fatigue loading condition, the stress intensity factor can be obtained similar to that of Eq. 2, and is

$$K_{\max} = \frac{P_{\max}}{d\sqrt{\pi d/2}}, \quad \Delta K = \frac{\Delta P}{d\sqrt{\pi d/2}} \tag{19}$$

where $P_{\max} = \Delta P/2 + P_{\mathrm{m}}$ is the maximum indentation load.

Indentation Fatigue Deformation

Indentation Fatigue Depth Propagation Law

When a solid (engineering metals otherwise specified) is subjected to a fatigue loading, crack will initiate and propagate with the increase of cyclic number. The steady state (i.e., the 2nd stage) of the fatigue crack growth $(\mathrm{d}a/\mathrm{d}N)_s$ can be described by the well-known Paris equation (Ritchie 1977)

$$(\mathrm{d}a/\mathrm{d}N)_s = C(\Delta K)^q \tag{20}$$

where $\Delta K = K_{\max} - K_{\min}$ is the nominal stress intensity factor range, and $K_{\max}$ and $K_{\min}$ are the maximum and minimum stress intensity factors, respectively. To highlight the competition between the intrinsic mechanism of crack tip growth and extrinsic mechanism of crack-tip shielding behind the tip, Eq. 20 can be modified to (Liu and Chen 1991; Dauskardt et al. 1992)

$$(\mathrm{d}a/\mathrm{d}N)_s = C(K_{\max})^n(\Delta K)^m \tag{21}$$

where ΔK and $K_{\max}$ describe the intrinsic and extrinsic mechanisms of fatigue crack growth, respectively, and their dominance is reflected by the power indices m and n. m, n, and C are empirical constants and depend on material and microstructure, fatigue frequency, loading mode and environment, etc.

When an elastoplastic solid is subjected to a fatigue loading via a flat punch indenter, given the similarity of stress singularity in theory in section "Indentation Stress Intensity Factor" between the crack tip and the rim of indentation, in particular, the same stress intensity factor near the rim of indentation and crack tip, and inspired by Eq. 21, the indentation fatigue depth propagation at the steady state can be described quantitatively by using a power law

$$(\mathrm{d}h/\mathrm{d}N)_s = C_i {K_{\max}}^{n_i} \Delta K^{m_i} \tag{22}$$

where ΔK and $K_{\max}$ are obtained from Eq. 19. C_i, n_i, and m_i are constants of indentation fatigue and depend on material/microstructure and testing environments. Similar to the fatigue crack growth, the continuous sinking of the indenter into the solid suggests the accumulation of plastic deformation beneath the indenter and is driven by the stress concentration near the rim of contact between the indenter and solid surface. It reflects the intrinsic mechanism of indentation depth propagation with the cyclic number and is determined byΔK. In contrast, $K_{\max}$ indicates the elastic recovery during the indentation depth propagation and represents the extrinsic mechanism. We will validate Eq. 22 through both computations and experiments in the following.

Computational Validation

Computational Method and Modeling

Modeling plastic behavior of materials under a fatigue loading has received considerable attention, and it requires the including of the Bauschinger effect, elastic shakedown, cyclic hardening or softening, ratcheting, and mean stress relaxation. A significant number of models have been developed over the last decades such as developed by Chaboche and Nouailhas (1989a, b), Ohno and Wang (1995), and Delobelle et al. (1995), and most of these models are based on the kinematic hardening model of Armstrong and Frederick (A-F model) (Jiang and Kurath 1996). A-F model has also been employed in the study of cyclic spherical indentation behavior by Huber and Tsakmakis (Moosbrugger and Morrison 1997; Abdel-Karim and Ohno 2000), but limited to a few number of cycles. In our simulations, two different polycrystalline copper alloys whose elastoplastic property proves to obey the classical kinematic hardening rule (namely the A-F model) will also be employed. The mechanical properties for a brittle copper alloy are 122.5 GPa for Young's modulus, 0.35 for Poisson's ratio, 33.32 MPa for initial yield stress, and 1.607 GPa for a linear hardening rate; the mechanical properties for a ductile copper alloy are 119.9 GPa for Young's modulus, 0.35 for Poisson's ratio, 73.50 MPa for initial yield stress, and 0.369 GPa for a linear hardening rate (Mclean 1965). We note that these two copper alloys can be distinguished based on measurements from the conventional uniaxial fatigue cracking test. For example, for fatigue cracking behavior of the brittle copper alloy, the index n is higher than m in Eq. 21, and the index n is less than m in Eq. 21 for fatigue cracking behavior of the ductile copper alloy. The plat punch indenter with radius $d/2 = 1.0$ mm is assumed to be rigid and frictionless. Figure 3a illustrates the history of fatigue loading with a sinusoidal manner. The application of a preloading to the mean fatigue load will help to minimize the effect of initial loadings. The simulations were conducted by the finite element software ABAQUS.

Results and Discussion

For the brittle copper as a representative, under $\Delta P = 100$ MPa and $P_m = 100$ MPa, the indentation load (P)-depth (h) curve is shown in Fig. 3b, and three stages are observed in response to the loading history in Fig. 3a. Stage II corresponds to the cyclic response of indentation to the fatigue loading, and is our focus. At the beginning of this stage, the indentation depth propagation rate (per cycle) is quite high, and then it decays with the increase of number of cyclic loading till to approximately reaching a constant. Overall, the indentation depth keeps increasing upon cyclic loading and shows a clear difference from the static elastoplastic

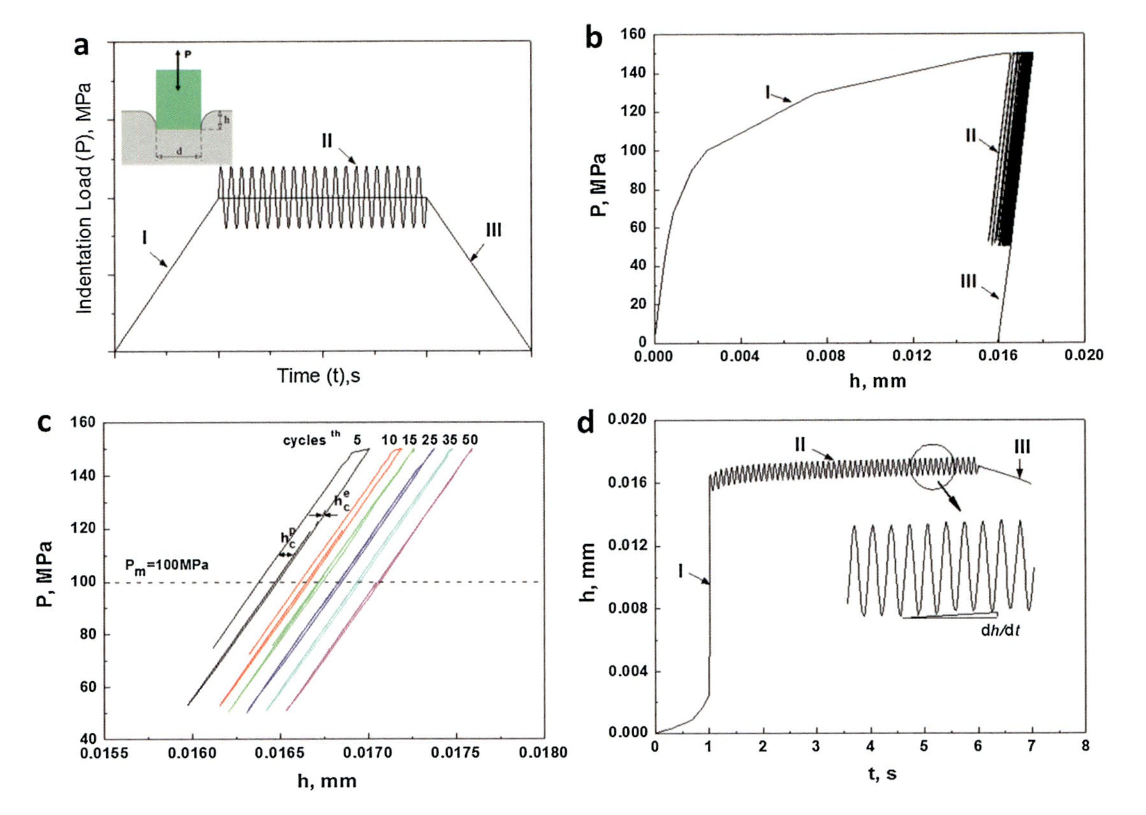

Fig. 3 (**a**) Schematic of the load history used in the numerical simulation of indentation fatigue. (**b**) The typical indentation load (P)-depth (h) curve. (**c**) The hysteresis loop of indentation load (P)-depth (h) curves under fatigue loading cycles 5th, 10th, 15th, 25th, 35th, and 50th. (**d**) Indentation depth (h)-time (t) curves. $\Delta P = 100$ MPa and $P_m = 100$ MPa Xu et al. 2009 (reprinted with the permission from Cambridge University Press)

indentation response. Besides, the propagation of indentation depth shows the same sinusoidal frequency as that of the applied indentation load. More importantly, there is a hysteresis loop in the indentation fatigue *P-h* curve at each cyclic loading, and several of them are magnified in Fig. 3c. As the number of cycles increases, the hysteresis loop tends to be more "closed" in comparison with the early ones, leading to a smaller enclosed area. Further, each hysteresis loop can be decomposed into two parts: the cyclic elastic indentation depth range h_c^e and the cyclic plastic indentation depth range h_c^p. The former suggests the recovery of indentation depth during indentation fatigue, and the latter is associated with the continuous increase of indentation depth due to the propagation of plastic zone beneath the indenter. Both h_c^e and h_c^p decrease with the increase of indentation depth toward the overall closed hysteresis loop, and the decreasing rate for h_c^p seems to be larger. Essentially, the area of hysteresis loop reflects the dissipated energy that is required to advance the indentation depth, and the dissipation rate will slow down as the indentation depth propagation increases (Xu et al. 2007b).

Figure 3d presents the variation of indentation depth with the history of cyclic loadings. Similar to the indentation fatigue *P-h* curve in Fig. 3b, three stages that correspond to the loading history are also obtained, and the second stage under a cyclic loading will be investigated. It further confirms that an indentation depth rate ($\mathrm{d}h/\mathrm{d}N$) arrives in a stable state with the increase of cyclic numbers at the second loading stage, and this stable indentation depth propagation ratio $(\mathrm{d}h/\mathrm{d}N)_s$ depends on the indentation load range and the maximum load, and radius of indenter, i.e., ΔK and $K_{\max}$ and $d/2$. With $d/2 = 1.0$ m, Fig. 4a gives the variation of the steady-state depth rate $(\mathrm{d}h/\mathrm{d}N)_s$ with ΔK and $K_{\max}$. A power-law function (i.e., Eq. 22) of $(\mathrm{d}h/\mathrm{d}N)_s$as both the stress intensity range ΔK and the maximum stress intensity $K_{\max}$ is obtained. When the indenter radius $d/2 = 3.0$ mm, the functional form of Eq. 22 holds. Further, the exponents n_i, m_i can be obtained, and n_i=5.2>m_i=1.5, suggesting that $(\mathrm{d}h/\mathrm{d}N)_s$ is dominated by $K_{\max}$ in comparison with ΔK. Similar to the fatigue crack propagation, ΔK can be considered the driving force for propagation of plastic zone due to the stress concentration at the rim of contact (associated with the cyclic plasticity), which promotes indentation depth propagation and represents the intrinsic mechanism. Meanwhile, $K_{\max}$ is associated with the elastic recovery of indentation depth, which recovers elastic deformation during indentation fatigue and represents the extrinsic mechanism. The dominance of $K_{\max}$in the indentation fatigue depth propagation for the brittle copper alloy is qualitatively similar to the conventional fatigue cracking behavior of brittle materials (Liu and Chen 1991; Dauskardt et al. 1992; Ritchie 1999), where their crack growth rate is dominated by $K_{\max}$ in Eq. 21.

Following the similar computational procedures, we performed the finite element analysis of indentation fatigue on the ductile copper alloys. As ΔK and $K_{\max}$ change with different indenter radii, Fig. 4b shows the corresponding obtained$(\mathrm{d}h/\mathrm{d}N)_s$, which confirms that the indentation fatigue depth propagation law (Eq. 22) holds. The exponents are n_i=4.2<m_i=5.6, which indicates that the dominance of ΔK over that of $K_{\max}$ for ductile copper alloys, also consistent well with the fatigue cracking behaviors of ductile materials (Liu and Chen 1991; Dauskardt et al. 1992).

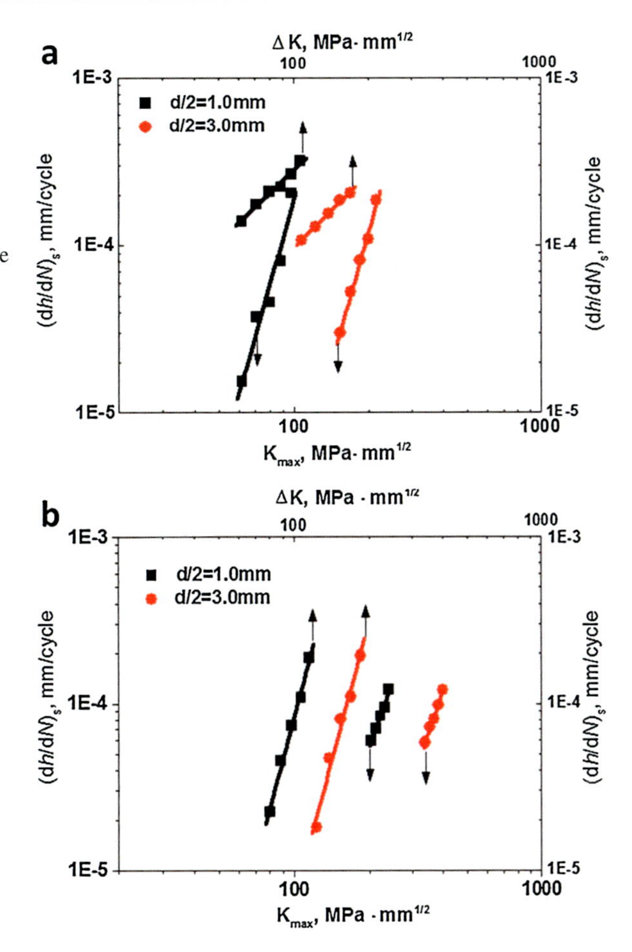

Fig. 4 Variation of the steady-state rate of indentation fatigue depth $(dh/dN)_s$ with the stress intensity range ΔK and the maximum stress intensity K_{max} from the numerical simulation on (**a**) a brittle copper alloy and (**b**) a ductile copper alloy Xu et al. 2009 (reprinted with the permission from Elsevier)

In summary, these computational results validate the proposed power-law function, i.e., Eq. 22, and the steady-state indentation fatigue depth propagation rate is a power-law function of K_{max} and ΔK for both brittle and ductile metals, similar to that fatigue crack growth. Besides, the dominance of K_{max} and ΔK can be determined by comparing their exponents. In addition, note that the coefficients in Eq. 22 C_i, n_i, and m_i depend on materials and testing conditions but are insensitive to indenter radii.

Experimental Validation

Material Choice and Testing Platform

99.9% polycrystalline copper (with 0.0262% Zn, 0.0145% P, 0.003% Pb, and 0.1266% Fe) with an average grain size 32μm was chosen, and the surface of each specimen (15 mm in diameter, by 15 mm in length) was polished to minimize the

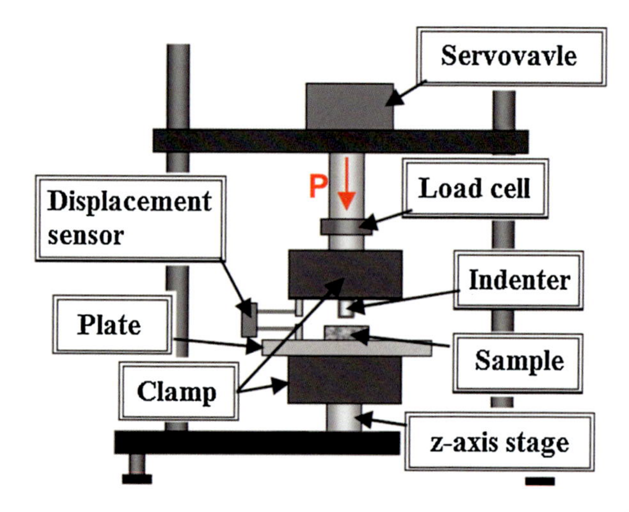

Fig. 5 Schematic of indentation fatigue testing system

effect of surface roughness before each experiment. Besides, annealing procedures for specimens were made at temperatures 150 °C for 35 minutes to relieve the residual stress. The material of the flat punch indenter was made by high temperature alloys and its deformation was neglected in comparison with the indentation depth. The diameter of the indenter was 0.993 mm.

A sinusoidal wave was employed to apply the indenter and was used to mimic a fatigue loading condition and the loading frequency ($f=\omega/2\pi$, Eq. 2) was chosen to be 1 Hz. To keep contact between the indenter and surface of the specimens during the experiment, the minimum indentation load (P_{min}) was set to *20 N*. The conventional tensile machine INSTRON8871 that can easily realize the fatigue loading has been designed to conduct the indentation fatigue experiment with a customized load cell, as schematized in Fig. 5. All experiments were performed at room temperature, and the indentation load-depth data were recorded automatically for each 10 loading cycles during an indentation fatigue testing.

Experimental Results

For representative fatigue loadings, $P_{max} = 200$ N and $P_{min} = 20$ N were employed and the corresponding $K_{\max} = 485.26\text{MPa} \cdot \text{mm}^{1/2}$ and $\Delta K = 469.12\text{MPa} \cdot \text{mm}^{1/2}$ can be calculated by utilizing Eq. 22. Figure 6a shows the recorded indentation load-depth curve. The indentation depth increases with the cyclic number. Besides, there is a quick penetration of the indenter at the initial stage, and then a relative steady state of indentation penetration rate arrives. More importantly, a hysteresis loop is observed in each loading-unloading cycle and seems to remain unclosed even after 3000 cycles. Figure 6b presents the curve of the indentation depth-loading cycle, where the width of the bands is expected to associate with the elastic recovery in each loading cycle. Similar to that observation in simulations in Fig. 3d, the indentation depth propagation shows a quick increase at the beginning and then followed by a steady-state stage. Essentially, these features can be understood from

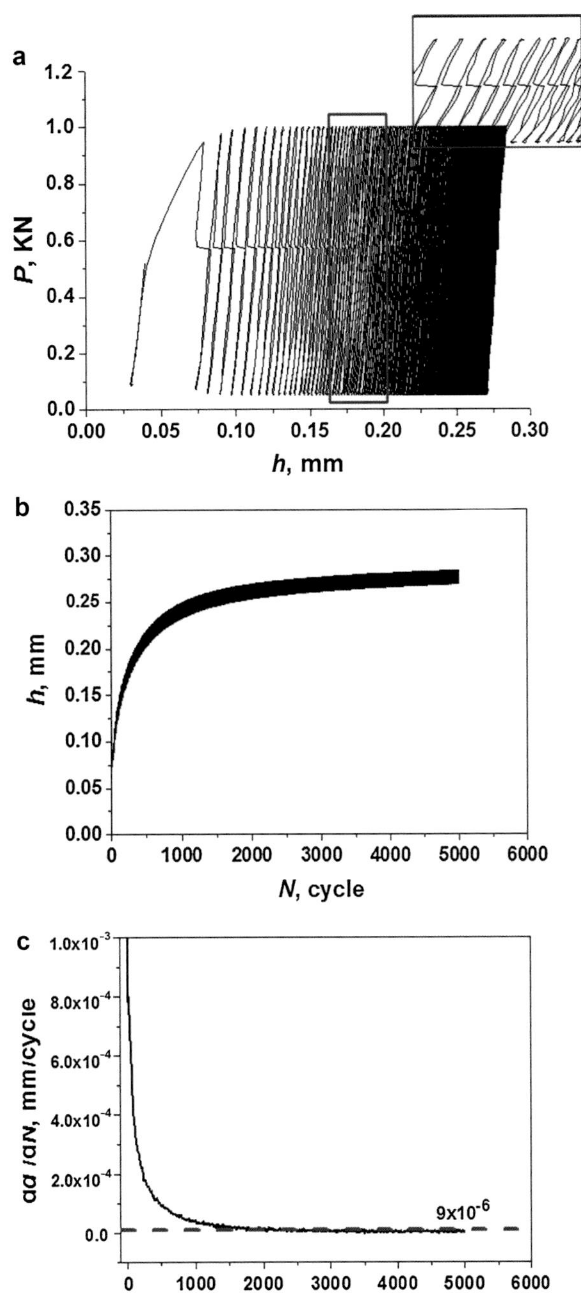

Fig. 6 Indentation fatigue experiment on polycrystalline copper. (**a**) Indentation load P - indentation depth h. (**b**) Indentation depth h and (**c**) indentation depth per cycle dh/dN with the applied fatigue cycle number.$K_{\max} = 485.26 \text{MPa} \cdot \text{mm}^{1/2}$ and $\Delta K = 469.12 \text{MPa} \cdot \text{mm}^{1/2}$ Xu and Yue 2006 (reprinted with the permission from Cambridge University Press)

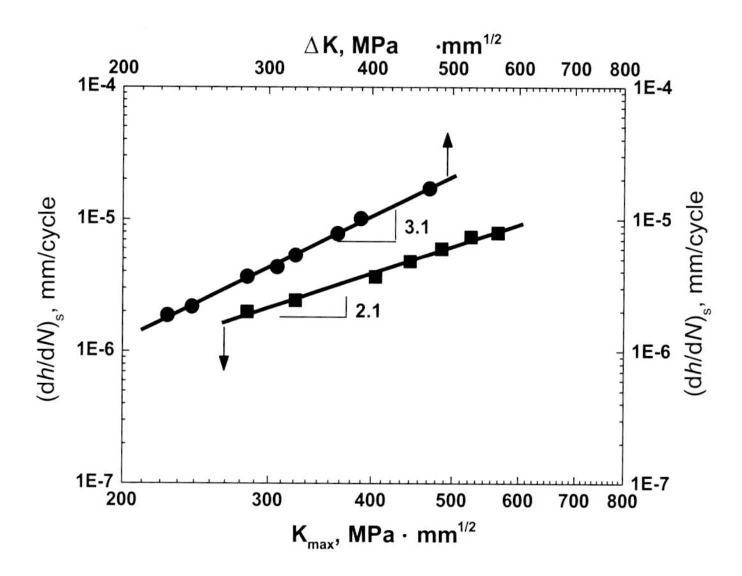

Fig. 7 Variation of the steady-state rate of indentation fatigue depth $(\mathrm{d}d/\mathrm{d}N)_s$ with the stress intensity range ΔK and maximum stress intensity $K_{\max}$, measured from experiment on a polycrystalline copper Xu et al. 2009 (reprinted with the permission from Elsevier)

the dislocation activation and retraction, and more details see section "Discussion on Deformation Mechanism." In fact, these features are quite analogous to the first two stages of uniaxial tensile fatigue crack growth. The approximate steady-state propagation of indentation fatigue is further confirmed when the indentation fatigue depth propagation rate $\mathrm{d}h/\mathrm{d}N$ is plotted as a function of number of cycles in Fig. 6c, and the steady-state propagation rate is$(\mathrm{d}h/\mathrm{d}N)_s$as denoted in Eq. 22.

Given $(\mathrm{d}h/\mathrm{d}N)_s$, we can examine its variation with ΔK and $K_{\max}$ and validate the indentation fatigue depth propagation law (i.e., Eq. 22). Figure 7 shows there is a linear variation of $(\mathrm{d}h/\mathrm{d}N)_s$ with both ΔK and $K_{\max}$ on log-scale plots, which agrees with Eq. 22. Besides, the power exponents are $m_i = 3.1$ and $n_i = 2.1$, respectively. The higher m_i implies that the steady-state propagation of indentation fatigue is dominated by ΔK, which is also consistent with fatigue crack growth in ductile metals (Ritchie 1999). We should note the power exponents n_i and m_i may not exactly equal to the n and m obtained from uniaxial tensile fatigue crack propagation because of influences of local microstructures in indentation experiments, and this difference is expected to increase with the decrease of indentation scales.

Extend to Overloading and Underloading

In previous sections, we have revealed the similarity between indentation fatigue depth propagation and fatigue crack growth under a normal fatigue loading spectrum. It is known that the fatigue crack growth is greatly affected by the applied fatigue loading interactions such as sudden increase of peak load followed by a normal fatigue loading conditions (referred to as overloading here) and sudden decrease of peak load followed by a normal fatigue loading conditions (referred to as underloading here). For example, the normal fatigue crack growth can be delayed and boosted by the overloading and underloading, respectively. In this section, we will continue to employ polycrystalline copper to investigate whether the overloading/underloading will lead to delay/acceleration of indentation fatigue depth propagation so as to further study its similarity to fatigue crack growth.

Figure 8a shows the indentation fatigue *P*-h curve under an overloading condition, and the inset illustrates the overloading spectrum, where two loading blocks with a low-high loading sequence are considered. P_{max0} and P_{min} are the maximum load and the minimum load of the first load block, respectively. P_{max1} is the maximum load of the second load block. N_0 and N_1 are the number of loading cycles of the two load blocks, respectively. The indentation depth increases with the number of cycles, and its propagation arrives at a relatively steady stage with a constant rate after an initial sharp increase in each loading block. Besides, hysteresis loops are also found in each block and seem not be closed, even after 7000 cycles in the second loading block.

The indentation depth-loading cycle is plotted in Fig. 8b, and it further indicates the approximately constant evolution of indentation depth propagation after initial rapid adjustment in each loading block, similar to that under a single normal fatigue loading block. Figure 8b further shows the rate of indentation depth propagation dh/dN with the number of fatigue loading. In each block, a clear steady stage with a constant dh/dN is observed after a quick decrease. More importantly, when the maximum load increases from 500 N in the first loading block to 700 N in the second loading block, a dh/dN shows an obvious increase first and then gradually decreases until to a new higher steady state. Experiments further indicate that the new steady state is nearly independent of the maximum loads in the second load block, except that it needs more cycles to reach the new steady state for the larger maximum load, as shown in Fig. 8b, where the maximum load changes from 700 N to 800 N in the second load block. Generally, a larger increase of the maximum load will require more number of cycles to eliminate the effect of load interaction. The enhancement of the indentation depth propagation by the overloading further shows similarity with that in conventional fatigue crack propagation (Tvergaard 2005; Sadananda and Vasudevan 2003; Huang and Ho 2000, 2003; Borrego et al. 2003; Kumar et al. 1996).

Similar to overloading conditions, the underloading condition also consists of two fatigue loading blocks with a decrease of the maximum load, i.e., a high-low loading sequence in the two load blocks, as illustrated in the inset in Fig. 8c. Consider a decrease of the maximum load from $P_{max0} = 600$ N to $P_{max1} = 550$ N with $N_0 = 2000$ cycles, and $N_1 = 7000$ cycles, Fig. 8c shows the indentation depth - number of cycles. The indentation depth keeps increasing instead of ceasing when the maximum load decreases. Besides, similar to that under overloading conditions, dh/dN arrive at a new steady state after lowering the maximum load, yet within a very short cyclic numbers, as shown in Fig. 8d. The steady state dh/dN for the new

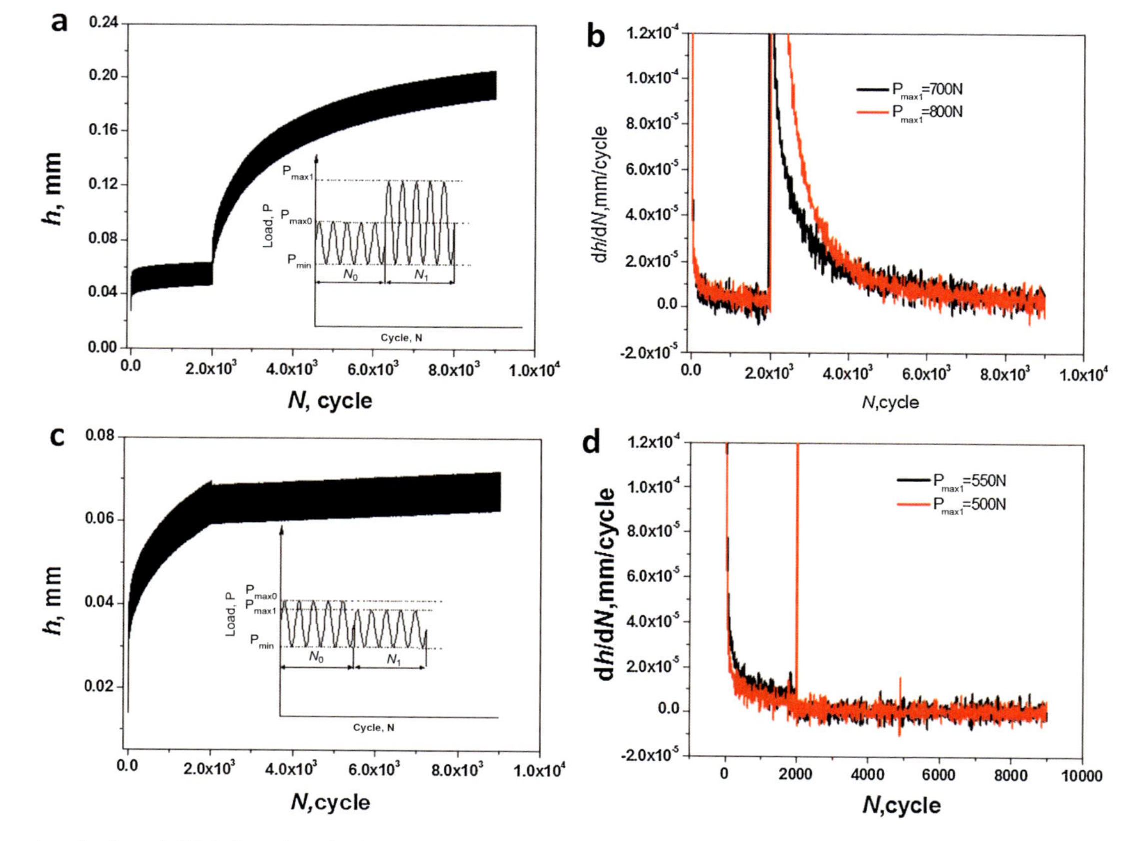

Fig. 8 (**a**) Indentation depth and (**b**) indentation depth per cycle - number of cycles curves under a low-high overloading (inset in **a**). $P_{max0} = 500\ N$, $P_{max1} = 700\ N$, $N_0 = 2000\ cycles$, $N_1 = 7000\ cycles$. (**c**) Indentation depth and (**b**) indentation depth per cycle - number of cycles curves under a high-low underloading (inset in **c**). $P_{max0} = 600\ N$, $P_{max1} = 550\ N$, $N_0 = 2000\ cycles$, $N_1 = 7000\ cycles$ Xu et al. 2007a (reprinted with the permission from Elsevier)

maximum load is obtained in the second cyclic loading block and smaller than that for the first cyclic loading block, indicating a delayed propagation of indentation depth. When the decrease of the maximum load is much larger (from $P_{max0} = 600$ N to $P_{max1} = 500$ N), Fig. 8d shows that the indentation depth propagation becomes slower. The delay of indentation fatigue depth propagation due to underloading is also similar to those findings in conventional fatigue crack propagation.

Microstructural Observation

Figure 9 presents the optical image of indentation on polycrystalline copper after cyclic loadings (Fig. 8c for indentation fatigue load-depth curve), where $P_{max0} = 600$ N, $P_{max1} = 550$ N, $N_0 = 2000$ cycles, $N_1 = 7000$ cycles. Pile-up, gap, and wrinkles (i.e., overlapping layers) are observed, similar to those under monotonous loading conditions. Generally, the pile-up or sink-in is caused by material hardening and when the hardening exponent of materials, n, is less than 0.33 (Storakers and Larsson 1994; Taljat and Pharr 2004), the pile-up will appear near the flat punch indentation (Storakers and Larsson 1994). For the polycrystalline copper used in our experiment, $n \approx 0.3 < 0.33$was measured from our uniaxial tensile testing (Xu et al. 2006), and its cyclic hardening exponent $n \approx 0.1$is also less than 0.33, which agrees well with the appearance of pile-up. Given the constant contact area between the flat punch indenter and surface of materials, deformation of materials will deviate from the indenter in accommodation with the pile-up,

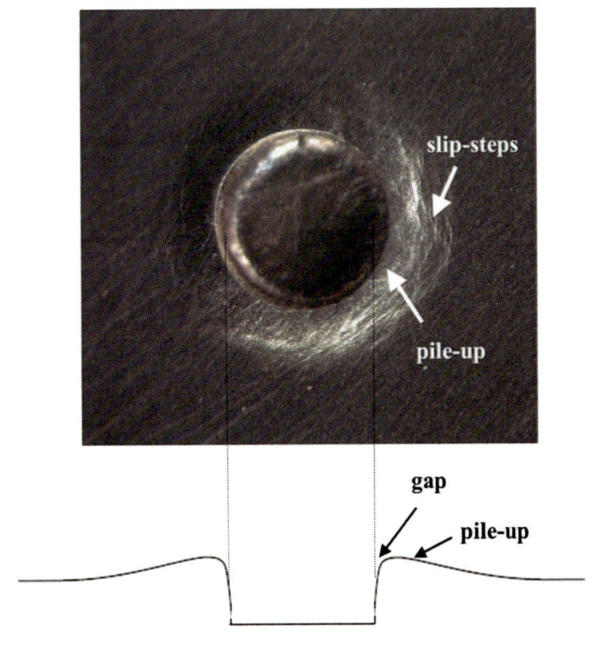

Fig. 9 Optical microscope (OM) indentation image of polycrystalline copper (top view) and its sketch map (slide view) Xu et al. 2007a (reprinted with the permission from Elsevier)

leaving a gap between pile-up and indentation. This gap will increase with the continuous sinking of indenter into the materials, which also agrees with the finite element analysis (Xu and Yue 2007). Wrinkles reflect the slip-steps of shear bands surrounding the indentation, and the formation and propagation of these shear bands can be deemed the major plastic deformation mechanism in the present indentation fatigue on polycrystalline copper. The closer to the indentation, the more wrinkles due to the more severe plastic deformation. From crystal plasticity theory point of view, the wrinkles result from activation of multiple slip systems, and these wrinkles are also referred to multislip-steps in quasi-static indentation (Lloyd et al. 2005; Zaafarani and Raabe 2008; Nibur and Bahr 2003) .

Figure 10 shows the SEM image of cross section of indentation. Pile-up and gap are observed more clearly. Besides, some light and dark cyclic lines with radii that decrease with increasing distance from the indenter are also observed. Those lines are expected to be dependent of the slip lines beneath the flat punch indenter, which is considered major deformation mechanism (Hill 1998). Figure 10 (i–iii) shows the local deformation near the indentation, highlighted in the boxes. Figure 10i shows the SEM images of pile-up around the samples. The pile-up is flat at first and then

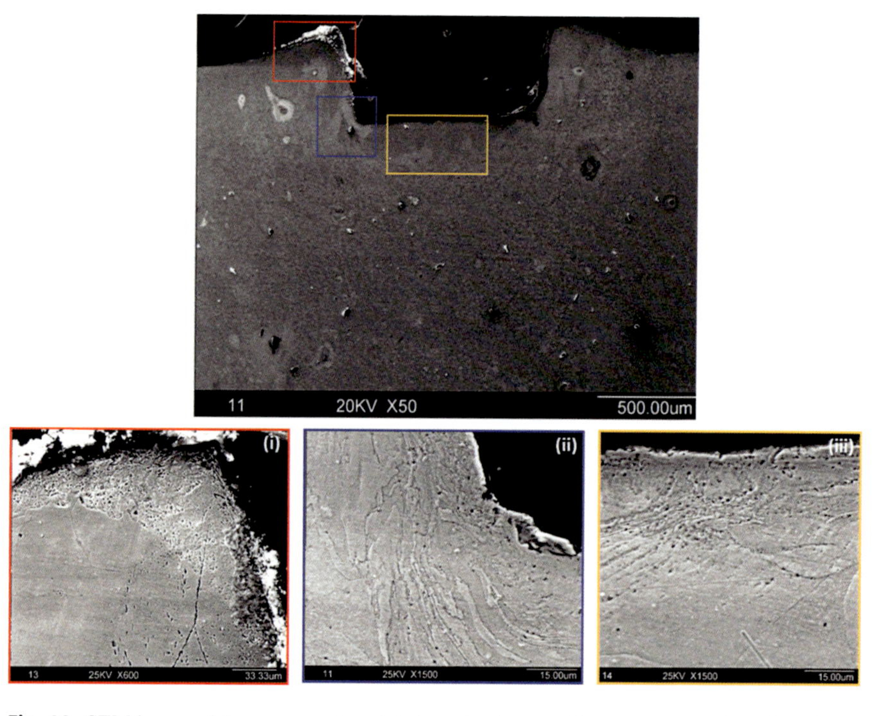

Fig. 10 SEM image of the cross section of indented polycrystalline copper sample under a high-low underloading (inset in Fig. 8c) and higher magnification (*i*) around the pile-up on the surface, (*ii*) near the corner of indentation, (*iii*) below the indentation; $P_{max0} = 600\ N$, $P_{max1} = 550\ N$, $N_0 = 2000\ cycles$, $N_1 = 7000\ cycles$ Xu et al. 2007a (reprinted with the permission from Elsevier)

vanishes with the distance far from the indentation. The flat transition is expected to be related with the amount of plastic deformation of materials around the indenter, and a cavity layer or cavity strip is similar to the form of a pile-up in geometry. From Fig. 10i, many cavities appear along the cyclic lines, and even some cracks in higher density cyclic lines near the rim of indentation. These cavities are expected to be nucleated at an early stage of indentation fatigue and accumulate to cracks, and then further spread with the increase of the number of loading cycles. Once the cracks interact, peel-offs of materials will happen, especially near the rim of indentation with higher cyclic lines associated with a high stress concentration. Figure 10iii gives the SEM images in the regions below the indentation. More cavities and cracks are found at a distance of about ~*15um* far from the free end of indentation. The orientation of these cracks is in line with the dark and light cyclic lines, which are approximately level near the end of the indentation and steeper further from the end of indentation.

Discussion on Deformation Mechanism

In the viewpoint of indentation load–indentation depth curves, the increase of indentation depth results from deformation ratcheting behavior of materials during cyclic indentation due to a nonzero mean fatigue stress. The ratcheting behavior reflects the accumulation of plastic deformation preceding in one direction. Besides, the nucleation and development of the cracks further facilitate the development of plastic zone, leading to a continuous increase of indentation depth.

When a new loading block is applied following a normal fatigue loading history, the retardation or acceleration of indentation fatigue depth propagation is caused by interactions of loading blocks and can be understood by considering activation of dislocations. For example, the effective applied stress σ_{eff} acting on mobile dislocation is

$$\sigma_{eff} = \sigma_{app} - \sigma_r \tag{23}$$

Where σ_{app} is the maximum fatigue stress and σ_r is the residual stress. For example, when the maximum load increases (i.e., overloading), the effective applied stress σ_{eff} is very large, meanwhile the residual stress σ_r is small, leading to a high indentation depth per cycle. At the same time, this difference will require numbers of loading cycles to increase the residual stress $\sigma_r$until reaching a new balance with a new steady state of indentation depth propagation. Hence, the greater the overloading amplitude, the more number of cycles are needed to remove the effect of overloading, consistent with experimental results in Fig. 8b. Besides, with the increase of the effective applied stress σ_{eff}, the dislocation density becomes intensive, the initial indentation depth per cycle will be very large and then gradually reduces and approaches to a new steady-state value with the arrival of new balance. The variation of the residual stress σ_r will lead to rearrangement of the dislocations beneath the indenter. The larger increase of the maximum load, the more cyclic numbers of loading to rearrange dislocation, and the larger steady indentation depth propagation ratio. These interpretations further show the similarity with the generation and annihilation of dislocations around the crack tip under the fatigue loadings.

Similarly, the decrease in the maximum cyclic load after reaching steady indentation depth ratio will reduce the residual stress σ_r. Consequently, underloading with a smaller maximum load produces a smaller effective stress σ_{eff} for plastic zone propagation, leading to a decreased indentation depth propagation after initial adjustments, as shown in Fig. 8d, which is also similar to that of fatigue crack growth under fatigue loadings.

Indentation Fatigue Damage

Indentation Fatigue Strength Law

With the continuous increase in indentation load cycles, damage of indented materials will nucleate and propagate, and eventually leads to failure of materials (Lawn and Wilshaw 1975; Xu et al. 2007a; Guiberteau et al. 1993; Bhowmick et al. 2007). Consider the number of indentation cycles to failure as N_f and the amplitude of the cyclic indentation load as F_a, in this section, we will build the relationship between N_f and F_a.

For a uniaxial fatigue testing, the relationship between the amplitude of fatigue load, σ_a, and the number of cycles to failure of materials, N_f, is the well-known S-N curve and can be expressed as a power-law relationship (Suresh 1998; Basquin 1910)

$$\sigma_a = \sigma_f \left(N_f\right)^n \tag{24}$$

where σ_f and n are the fatigue strength coefficient and exponent, respectively, and are related to the material/microstructure and testing environment/condition. In general, σ_f equals the uniaxial fracture strength of the material under quasi-static loading.

Inspired by this equation (Eq. 24), we propose that a similar power-law relationship exists for indentation fatigue damage

$$F_a = F_f \left(N_f\right)^m \tag{25}$$

where F_f and m are the indentation fatigue strength coefficient and exponent, respectively, and both depend on the material/microstructure and testing environment/condition. In the following, we will validate this power law equation in experiments.

Experimental Validation

Material Choice and Testing Platform

Different from the employment of copper alloy, to highlight the damage, relatively brittle materials were used in this section, and they were PVC bulk material, and TiN, NiP thin films on steel (SUS304) substrate. Besides, in order to avoid the early damage by stress concentration at contact, a spherical indenter was employed (instead of a flat punch indenter or a pyramidal/conical indenter). The cyclic indentation load applied on the indenter varies in a sinusoidal manner; the load ratio, i.e., the ratio between the minimum and maximum indentation loads, was fixed at 0.1 for the thin-film specimens and fixed at 0.05 for the PVC bulk specimen. Similar to the study of indentation fatigue depth propagation on polycrystalline copper, the minimum load was above zero to keep the "compressive" contact all the time. The frequency of load is 20 Hz for thin films and 5 Hz for the bulk material. The maximum indentation force is constrained so as not to induce cracking under static/monotonic loading conditions.

The failure was monitored in situ by acoustic emission (AE) during the indentation fatigue, as schematized in Fig. 11. This technique was used to diagnose the nucleation and propagation of damage and cracks in the static and cyclic indentation tests and showed good agreement with the observations of microstructures (Yonezu et al. 2009, 2010). Since AE may also be sensitive to contact friction/noise due to cyclic indentation, the initiation of major crack, or material failure, is quantitatively defined if more than 20 new AE counts per second are detected. This ensured the detection of the initiation of major cracks (macrocracks), which is regarded as failure in this study. From the variation of the AE count, the critical number of fatigue cycles (upon failure) can be readily determined, which is regarded as the fatigue strength of the specimen under a particular testing condition.

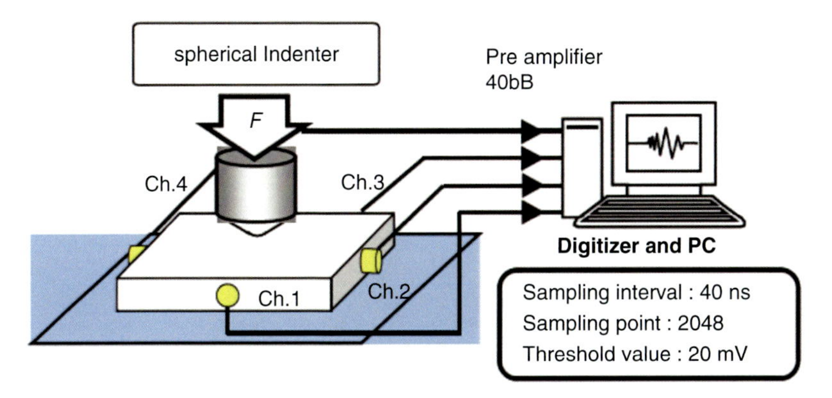

Fig. 11 Experimental setup of spherical indentation with AE and corrosion potential fluctuation monitoring system

Results and Discussion

Indentation fatigue tests on the PVC bulk material were first performed. Figure 12a shows a typical curve of indentation fatigue depth versus the number of cycles with a maximum indentation force of 880 N. Both the maximum and minimum depth histories are shown and the difference between them is due to the elastic recovery. The indentation depth increases with the number of cycles, and after an initial stage, a steady-state indentation depth propagation rate is achieved. These features are similar to those of polycrystalline copper in section "Experimental Validation." The accompanied AE count shows that at $N = 109{,}245$ cycles there is a significant increase of new AE counts. This implies that a macrocrack nucleates around the indentation, which is identified as the failure in the PVC, as shown in the inset in Fig. 12a. For the same specimen, more indentation fatigue tests with different load amplitudes were carried out, and the resulting number of cycles to failure obtained as a function of the fatigue load amplitude. This is shown in Fig. 12b. A linear

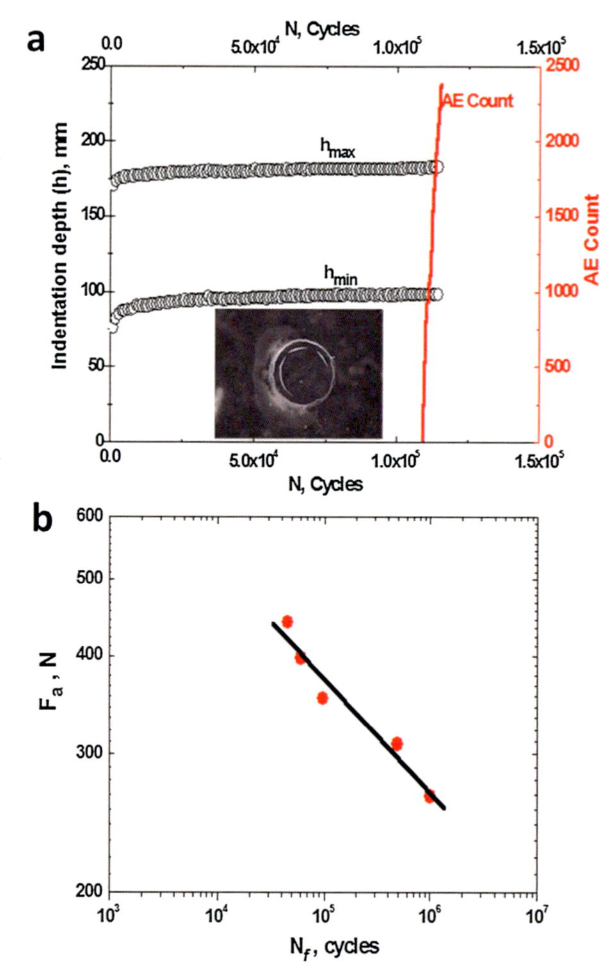

Fig. 12 (**a**) Experimental evolution of indentation fatigue depth and AE count with number of cycles on the PVC bulk material. The inset shows the surface topography near the indentation (after failure). (**b**) The variation of the indentation fatigue load amplitude with number of cycles to failure for the PVC bulk material. The maximum indentation fatigue load is 880 N and the load ratio is 0.05. The frequency is 5 Hz Xu et al. 2010 (reprinted with the permission from Taylor & Francis)

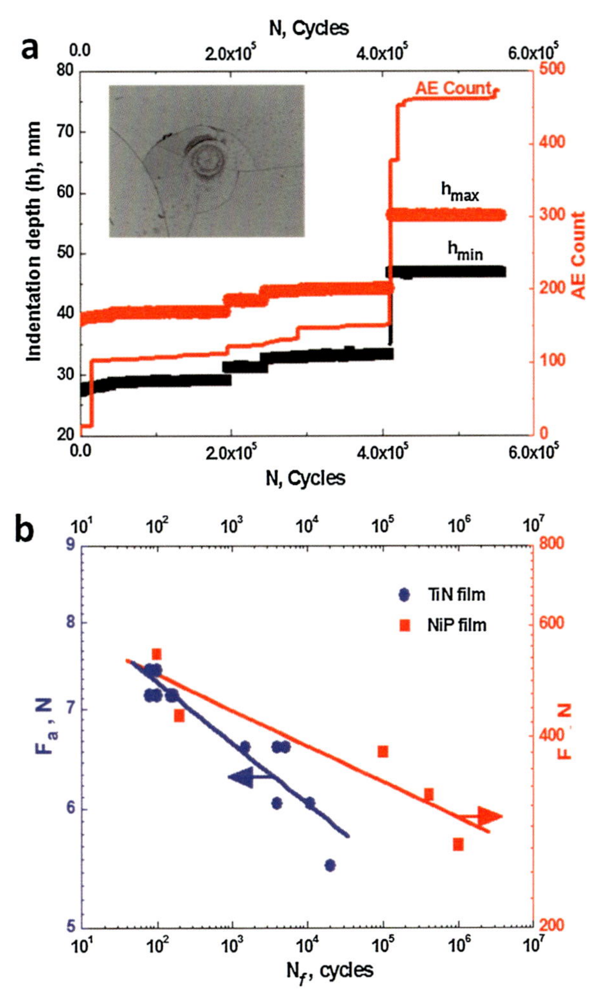

Fig. 13 (**a**) Experimental evolution of indentation depth and AE count with number of cycle on the NiP thin film/SUS304 steel substrate system. The inset shows the surface topography near the indentation (after failure). (**b**) Variation of indentation fatigue load amplitude with number of cycles to failure for the Ni-P film/SUS 304 steel substrate system, as well as that of the TiN film/SUS304 substrate system. The maximum indentation fatigue load is 600 N and the load ratio is 0.1. The frequency is 20 Hz Xu et al. 2010 (reprinted with the permission from Taylor & Francis)

relationship is obtained in the log-log plot, which validates the feasibility of using Eq. 25 to describe the indentation fatigue strength of the bulk material.

Next, indentation fatigue on film/substrate systems was also carried out. Figure 13a shows the evolution of indentation depth with number of cycles for the NiP film on a steel substrate, where the maximum indentation force was 600 N. A similar characteristic of the evolution of the indentation depth with that of PVC bulk material is observed: the indentation depth rate remains steady for a number of cycles, and then undergoes a sudden change when the cycle number is about 409,860 cycles. Accordingly, a significant increase of new AE counts (well over 20 per second) was also detected, indicating macrocracking failure in the system. These findings are consistent with the post-observation (inset in Fig. 13a) where the NiP thin film develops many cracks.

Table 1 Comparisons between the fatigue strength exponents obtained from the indentation fatigue test (shaded rows) and uniaxial fatigue test in literature, for several thin film/substrate systems Xu et al. 2010 (reprinted with the permission from Taylor & Francis)

Thin film/substrate	Film thickness (μm)	Fatigue strength exponent n	Error of indentation (present) and tensile (literatures) results
TiN/SUS304	4	−0.041 ± 0.005	–
TiN/AISI 316 L (Puchi-Cabrera et al. 2004)	1.4	−0.037	2.7∼24.3%
$TiN_{0.75}$/AISI 316 L (Berr'ıos et al. 2001)	3	−0.035	2.7∼31.4%
NiP/SUS304	200	−0.056 ± 0.013	–
NiP/SAE4340 (D'ıaz et al. 2002)	18	−0.084	17.9∼48.8%
NiP/AISI 1045 (Contreras et al. 1999)	17	−0.065	6.2∼33.8%

With the similar procedures, a series of indentation fatigue tests were also performed with different load amplitudes on the NiP and TiN film/steel substrate systems. Figure 13b shows the relationship between the indentation fatigue load amplitude and the number of cycles to their failure. In both cases, a power-law relationship (i.e., Eq. 25) is validated. Moreover, the indentation fatigue strength exponent, m, can be determined by fitting Fig. 13b, and it equals to −0.041 ± 0.005 and −0.056 ± 0.013 for TiN and NiP thin films on steel substrate, respectively. At the same time, the fatigue strength coefficient F_f can be obtained from the extrapolation and is 8.9 ± 1N and 660 ± 120N for TiN and NiP films on steel substrate, respectively, which agrees well the measurements (14.4 N and 1000 N counterparts (Yonezu et al. 2009, 2010)) at the static indentation tests on the same TiN and NiP films on steel substrate, thus suggesting an underlying bridge between the indentation fatigue strength and uniaxial fatigue strength.

Table 1 lists the further comparison of the fatigue stress exponents between the current measurements with those obtained from uniaxial tests. Although the fatigue stress exponents vary with respect to the thickness of films, they are on the same order of magnitude. Therefore, we may conclude that the indentation fatigue strength law, Eq. 25, is suitable for describing the failure behavior with measurement fatigue parameter close to those in conventional uniaxial fatigue test. Note that although the indentation fatigue failure mechanisms of PVC bulk materials, and TiN/SUS304 and NiP/SUS304 film/substrate systems are different, the general potential applicability of Eq. 25 seems to not be affected. This qualitatively echoes the fact that in conventional uniaxial fatigue tests, different failure mechanisms may be involved for different materials, e.g., for brittle and ductile materials, yet Eq. 24 holds for all common mechanisms. Thus, the direct comparison of fatigue strength exponents indicates a quantitative connection between the indentation fatigue test and uniaxial fatigue test.

Concluding Remarks

Indentation fatigue is emerging as an alternative approach to measure the fatigue properties and to probe fatigue-induced deformation and failure mechanism of materials and structures in particular materials and structures with small volumes. This chapter presents, in theory, experiment and computation, extensive evidences of similarity between indentation fatigue depth propagation and fatigue crack growth, and builds a quantitative foundation in mechanics. The recent indentation fatigue experiments performed at the nanoscale on nanomaterials (Cavaliere 2010; Wei et al. 2008) and nanofilms (Bhat 2012) further confirmed the proposed similarity and indentation fatigue laws/equations. Further development of this indentation fatigue technique is envisioned to be integrated with multiple different measurements (e.g., AE) and/or observatory (e.g., TEM, SEM) tools to monitor in situ deformation of materials under cyclic fatigue loadings. At the same time, advanced computational models that can reproduce the experimental results of indentation fatigue will be critically important, and they may require to span several length scales to capture both indentation fatigue phenomena at the macroscale and deformation mechanism at the nanoscale. Moreover, the application of indentation fatigue to measure the cyclic/fatigue properties of broader materials such as soft matter and biological materials and structures will help to extend theories of indentation fatigue mechanics beyond the current focus of engineering metals, and may also offer new opportunities to probe mechanics of soft materials and structures.

References

M. Abdel-Karim, N. Ohno, Kinematic hardening model suitable for ratchetting with steady-state. Int. J. Plast. **16**, 225–240 (2000)

D.H. Alsem, O.N. Pierron, E.A. Stach, C.L. Muhlstein, R.O. Ritchie, Mechanisms for fatigue of Micron-scale silicon structural films. Adv. Eng. Mater. **9**, 15–30 (2007)

D.H. Alsem, C.L. Muhlstein, E.A. Stach, R.O. Ritchie, Further considerations on the high-cycle fatigue of micron-scale polycrystalline silicon. Scr. Mater. **59**, 931–935 (2008)

S.A.S. Asif, K.J. Wahl, R.J. Colton, Nanoindentation and contact stiffness measurement using force modulation with a capacitive load-displacement transducer. Rev. Sci. Instrum. **70**, 2408–2413 (1999)

X. Baoxing, Y. Zhufeng, C. Xi, Characterization of strain rate sensitivity and activation volume using the indentation relaxation test. J. Phys. D. Appl. Phys. **43**, 245401 (2010)

O.H. Basquin, The exponential law of endurance tests. Proc. Am. Soc. Test. Mater. **10**, 625–630 (1910)

J.A. Berr'ıos, D.G. Teer, E.S. Puchi-Cabrera, Fatigue properties of a 316L stainless steel coated with different TiNx deposits. Surf. Coat. Technol. **148**, 179–190 (2001)

T. S., Bhat, *Indentation Analysis of Transversely Isotropic Materials,* PhD thesis, Stony Brook University, Stony Brook, 2012

S. Bhowmick, J.J. Meléndez-Martínez, B.R. Lawn, Bulk silicon is susceptible to fatigue. Appl. Phys. Lett. **91**, 201902 (2007)

L.P. Borrego, J.M. Ferreira, J.M. Pinho da Cruz, J.M. Costa, Evaluation of overload effects on fatigue crack growth and closure. Eng. Fract. Mech. **70**, 1379–1397 (2003)

J.L. Bucaille, S. Stauss, E. Felder, J. Michler, Determination of plastic properties of metals by instrumented indentation using different sharp indenters. Acta Mater. **51**, 1663–1678 (2003)
P. Cavaliere, Cyclic deformation of ultra-fine and nanocrystalline metals through nanoindentation: Similarities with crack propagation. Procedia Eng. **2**, 213–222 (2010)
J.L. Chaboche, D. Nouailhas, Constitutive modeling of ratchetting effects—part I: experimental facts and properties of the classical models. J. Eng. Mater. Technol. **111**, 384–392 (1989a)
J.L. Chaboche, D. Nouailhas, Constitutive modeling of ratchetting effects—part II: possibilities of some additional kinematic rules. J. Eng. Mater. Technol. **111**, 409–416 (1989b)
E.P. Chan, Y. Hu, P.M. Johnson, Z. Suo, C.M. Stafford, Spherical indentation testing of poroelastic relaxations in thin hydrogel layers. Soft Matter **8**, 1492–1498 (2012)
X. Chen, N. Ogasawara, M. Zhao, N. Chiba, On the uniqueness of measuring elastoplastic properties from indentation: The indistinguishable mystical materials. J. Mech. Phys. Solids **55**, 1618–1660 (2007)
R. Chen, Y.C. Lu, F. Yang, G.P. Tandon, G.A. Schoeppner, Impression creep of PMR-15 resin at elevated temperatures. Polym. Eng. Sci. **50**, 209–213 (2010)
Y.T. Cheng, C.M. Cheng, Scaling approach to conical indentation in elastic-plastic solids with work hardening. J. Appl. Phys. **84**, 1284–1291 (1998)
Y.T. Cheng, C.M. Cheng, Scaling relationships in indentation of power-law creep solids using self-similar indenters. Philos. Mag. Lett. **811**, 9–16 (2001)
Y.T. Cheng, C.M. Cheng, Scaling, dimensional analysis, and indentation measurements. Mater. Sci. Eng. R **44**, 91–149 (2004)
S.N.G. Chu, J.C.M. Li, Impression creep; a new creep test. J. Mater. Sci. **12**, 2200–2208 (1977)
S.N.G. Chu, J.C.M. Li, Localized stress relaxation by impression testing. Mater. Sci. Eng. **45**, 167–171 (1980a)
S.N.G. Chu, J.C.M. Li, Delayed retardation of overloading effects in impression fatigue. J. Eng. Mater. Technol. **102**, 337–340 (1980b)
T. Connolley, P.E. McHugh, M. Bruzzi, A review of deformation and fatigue of metals at small size scales. Fatigue Fract. Eng. Mater. Struct. **28**, 1119–1152 (2005)
G. Contreras, C. Fajardo, J.A. BerrõÂos, A. Pertuz, J. Chitty, H. Hintermann, E.S. Puchi, Fatigue properties of an AISI 1045 steel coated with an electroless Ni-P deposit. Thin Solid Films **355–356**, 480–486 (1999)
J.A. D'ıaz, M. Passarelli, J.A. Berr'ıos, E.S. Puchi-Cabrera, Fatigue behavior of a 4340 steel coated with an electroless Ni-P deposit. Surf. Coat. Technol. **149**, 45–56 (2002)
M. Dao, N. Chollacoop, K.J. VanVliet, T.A. Venkatesh, S. Suresh, Computational modeling of the forward and reverse problems in instrumented sharp indentation. Acta Mater. **49**, 3899–3918 (2001)
R.H. Dauskardt, M.R. James, J.R. Porter, R.O. Ritchie, Cyclic fatigue-crack growth in SiC-whiskerreinforced alumina ceramic composite: Long and small-crack behavior. J. Am. Ceram. Soc. **75**, 759–771 (1992)
P. Delobelle, P. Robinet, L. Bocher, Experimental study and phenomenological modelization of ratchet under uniaxial and biaxial loading on an austenitic stainless steel. Int. J. Plast. **11**, 295–330 (1995)
J.-H. Dirks, E. Parle, D. Taylor, Fatigue of insect cuticle. J. Exp. Biol. **216**, 1924–1927 (2013)
Y. Estrin, A. Vinogradov, Fatigue behaviour of light alloys with ultrafine grain structure produced by severe plastic deformation: an overview. Int. J. Fatigue **32**, 898–907 (2010)
A.C. Fischer-Cripps, *Nanoindentation* (Spring-Verlag, New York, 2000)
N.A. Fleck, R.A. Smith, Fatigue life prediction of a structural steel under service loading. Int. J. Fatigue **6**, 203–210 (1984)
F. Guiberteau, N.P. Padture, H. Cai, B.R. Lawn, Indentation fatigue: a simple cyclic hertzian test for measuring damage accumulation in polycrystalline ceramics. Philos. Mag. A **68**, 1003–1016 (1993)
R.W. Hertzberg, *Deformation and Fracture Mechanics of Engineering Materials* (Wiley, Oxford, 1995)
R. Hill, *The Mathematical Theory of Plasticity* (Oxford University Press, 1998)

H.-W. Höppel, H. Mughrabi, A. Vinogradov, *Bulk Nanostructured Materials* (Wiley-VCH Verlag GmbH & Co. KGaA, Weinheim, 2009), pp. 481–500
Y. Hu, E.P. Chan, J.J. Vlassak, Z. Suo, Poroelastic relaxation indentation of thin layers of gels. J. Appl. Phys. **110**, 086103 (2011)
H.L. Huang, N.J. Ho, The model of crack propagation in polycrystalline copper at various propagating rates. Mater. Sci. Eng. A **279**, 254–260 (2000)
H.L. Huang, N.J. Ho, The observation of dislocation reversal in front of crack tips of polycrystalline copper after reducing maximum load. Mater. Sci. Eng. A **345**, 215–222 (2003)
Y. Jiang, P. Kurath, Characteristics of the Armstrong-Frederick type plasticity models. Int. J. Plast. **12**, 387–415 (1996)
P. Jiang, T. Zhang, Y. Feng, R. Yang, N. Liang, Determination of plastic properties by instrumented spherical indentation: expanding cavity model and similarity solution approach. J. Mater. Res. **24**, 1045–1053 (2009)
K.L. Johnson, *Contact Mechanics* (Cambridge University Press, Cambridge, 1985)
P. Kaszynski, E. Ghorbel, D. Marquis, An experimental study of ratchetting during indentation of 316L stainless steel. J. Eng. Mater. Technol. **120**, 218–223 (1998)
R. Kumar, A. Kumar, S. Kumar, Delay effects in fatigue crack propagation. Int. J. Press. Vessel. Pip. **67**, 1–5 (1996)
H. Lan, T.A. Venkatesh, Determination of the elastic and plastic properties of materials through instrumented indentation with reduced sensitivity. Acta Mater. **55**, 2025–2041 (2007)
B. Lawn, R. Wilshaw, Indentation fracture: principles and applications. J. Mater. Sci. **10**, 1049–1081 (1975)
C. Lee, X. Wei, J.W. Kysar, J. Hone, Measurement of the elastic properties and intrinsic strength of monolayer graphene. Science **321**, 385–388 (2008)
J.H. Lee, T. Kim, H. Lee, A study on robust indentation techniques to evaluate elastic–plastic properties of metals. Int. J. Solids Struct. **47**, 647–664 (2010)
J.C.M. Li, Impression creep and other localized tests. Mater. Sci. Eng. A **322**, 23–42 (2002)
X.D. Li, B. Bhushan, A review of nanoindentation continuous stiffness measurement technique and its applications. Mater. Charact. **48**, 11–36 (2002)
J.C.M. Li, S.N.G. Chu, Impression fatigue. Scr. Metall. **13**, 1021–1026 (1979)
W.B. Li, J.L. Henshall, R.M. Hooper, K.E. Easterling, The mechanisms of indentation creep. Acta Metall. Mater. **39**, 3099–3110 (1991)
X. Li, H. Gao, C.J. Murphy, K.K. Caswell, Nanoindentation of silver nanowires. Nano Lett. **3**, 1495–1498 (2003)
P. Li, Q. Liao, S. Yang, X. Bai, Y. Huang, X. Yan, Z. Zhang, S. Liu, P. Lin, Z. Kang, Y. Zhang, In situ transmission electron microscopy investigation on fatigue behavior of single ZnO wires under high-cycle strain. Nano Lett. **14**, 480–485 (2014)
S.Y. Liu, I.W. Chen, Fatigue of yttria-stabilized zirconia - I. Fatigue damage, fracture origins and lifetime prediction. J. Am. Ceram. Soc. **74**, 1197–1205 (1991)
S.J. Lloyd, A. Castellero, F. Giuliani, Y. Long, K.K. McLaughlin, J.M. Molina-Aldareguia, N.A. Stelmashenko, L.J. Vandeperre, W.J. Clegg, Observations of nanoindents via cross-sectional transmission electron microscopy: a survey of deformation mechanisms. Proc. R. Soc. A **461**, 2521–2543 (2005)
J.L. Loubet, W.C. Oliver, B.N. Lucas, Measurement of the loss tangent of low-density polyethylene with a nanoindentation technique. J. Mater. Res. **15**, 1195–1198 (2000)
J. Luo, K. Dahmen, P.K. Liaw, Y. Shi, Low-cycle fatigue of metallic glass nanowires. Acta Mater. **87**, 225–232 (2015)
D. Mclean, *Mechanical Properties of Metals* (Wiley Press, New Jersey, 1965)
P. Miranzo, J.S. Moya, Elastic/plastic indentation in ceramics: a fracture toughness determination method. Ceram. Int. **10**, 147–152 (1984)
J.C. Moosbrugger, D.J. Morrison, Nonlinear kinematic hardening rule parameters – direct determination from completely reversed proportional cycling. Int. J. Plast. **13**, 633–668 (1997)
K.A. Nibur, D.F. Bahr, Identifying slip systems around indentations in FCC metals. Scr. Mater. **49**, 1055–1060 (2003)

J.D. Nowak, K.A. Rzepiejewska-Malyska, R.C. Major, O.L. Warren, J. Michler, In-situ nanoindentation in the SEM. Mater. Today **12**(Supplement 1), 44–45 (2010)
N. Ohno, J. Wang, On modelling of kinematic hardening for ratcheting behaviour. Nucl. Eng. Des. **153**, 205–212 (1995)
W.C. Oliver, G.M. Pharr, Measurement of hardness and elastic modulus by instrumented indentation: advances in understanding and refinements to methodology. J. Mater. Res. **19**, 3–20 (2004)
E.S. Puchi-Cabrera, F. Mat'ınez, I. Herrera, J.A. Berr'ios, S. Dixit, D. Bhat, On the fatigue behavior of an AISI 316L stainless steel coated with a PVD TiN deposit. Surf. Coat. Technol. **182**, 276–286 (2004)
G.D. Quinn, R.C. Bradt, On the Vickers indentation fracture toughness test. J. Am. Ceram. Soc. **90**, 673–680 (2007)
Y. Rao, R.J. Farris, Fatigue and creep of high-performance fibers: deformation mechanics and failure criteria. Int. J. Fatigue **30**, 793–799 (2008)
R.O. Ritchie, Influence of microstructure on near-threshold fatigue crack propagation in ultra-high strength steel. Met. Sci. **11**, 368–381 (1977)
R.O. Ritchie, Mechanisms of fatigue-crack propagation in ductile and brittle solids. Int. J. Fract. **100**, 55–83 (1999)
K. Sadananda, A.K. Vasudevan, Fatigue crack growth mechanisms in steels. Int. J. Fatigue **25**, 899–914 (2003)
S. Sakaguchi, N. Murayama, Y. Kodama, & F. Wakai, in *Fracture Mechanics of Ceramics: Fracture Fundamentals, High-Temperature Deformation, Damage, and Design*, eds. by R. C. Bradt, D. P. H. Hasselman, D. Munz, M. Sakai, V. Ya Shevchenko (Springer US, 1992), pp. 509–521
I.N. Sneddon, The relationship between load and penetration in the axisymmetric Boussinesq problem for a punch of arbitrary profile. Int. J. Eng. Sci. **3**, 47–57 (1965)
D.S. Stone, J.E. Jakes, J. Puthoff, A.A. Elmustafa, Analysis of indentation creep. J. Mater. Res. **25**, 611–621 (2010)
B. Storakers, P.L. Larsson, On Brinell and Boussinesq indentation of creeping solids. J. Mech. Phys. Solids **42**, 307–332 (1994)
S. Suresh, *Fatigue of materials* (Cambridge University Press, Cambridge, 1998)
B. Taljat, G.M. Pharr, Development of pile-up during spherical indentation of elastic-plastic solids. Int. J. Soilds Struct. **41**, 3891–3904 (2004)
Y. Tang, A. Yonezu, N. Ogasawara, N. Chiba, X. Chen, On radial crack and half-penny crack induced by Vickers indentation. Proc. R. Soc. A: Math. Phys. Eng. Sci. **464**, 2967–2984 (2008)
J. Tang, J. Li, J.J. Vlassak, Z. Suo, Fatigue fracture of hydrogels. Extreme Mech. Lett. **10**, 24–31 (2017)
S.H. Teoh, Fatigue of biomaterials: a review. Int. J. Fatigue **22**, 825–837 (2000)
V. Tvergaard, Overload effects in fatigue crack growth by crack-tip blunting. Int. J. Fatigue **27**, 1389–1397 (2005)
O.L. Warren, Z. Shan, S.A.S. Asif, E.A. Stach, J.W. Morris Jr., A.M. Minor, In situ nanoindentation in the TEM. Mater. Today **10**, 59–60 (2007)
P.J. Wei, Y.C. Wang, J.F. Lin, Retardation of cyclic indentation creep exhibited in metal alloys. J. Mater. Res. **23**, 2650–2656 (2008)
Z. Xia, W.A. Curtin, B.W. Sheldon, A new method to evaluate the fracture toughness of thin films. Acta Mater. **52**, 3507–3517 (2004)
B. Xu, X. Chen, Determining engineering stress–strain curve directly from the load–depth curve of spherical indentation test. J. Mater. Res. **25**, 2297–2307 (2010)
B. Xu, Z. Yue, Study of the ratcheting by the indentation fatigue method with a flat cylindrical indenter: part I. Experimental study. J. Mater. Res. **21**(7), 1793–1797 (2006)
B. Xu, Z. Yue, Study of the ratcheting by the indentation fatigue method with a flat cylindrical indenter: part II. Finite element simulation. J. Mater. Res. **22**, 186–192 (2007)
B. Xu, B. Zhao, Z. Yue, Investigation of residual stress by the indentation method with the flat cylindrical indenter. J. Mater. Eng. Perform. **15**, 299–305 (2006)
B. Xu, Z. Yue, J. Wang, Indentation fatigue behaviour of polycrystalline copper. Mech. Mater. **39**(12), 1066–1080 (2007a)

B. Xu, X. Wang, Z. Yue, Indentation behavior of polycrystalline copper under fatigue peak overloading. J. Mater. Res. **22**, 1585–1592 (2007b)

B.X. Xu, X.M. Wang, B. Zhao, Z.F. Yue, Study of crystallographic creep parameters of nickel-based single crystal superalloys by indentation method. Mater. Sci. Eng. A **478**, 187–194 (2008a)

B.X. Xu, X.M. Wang, Z.F. Yue, Determination of the internal stress and dislocation velocity stress exponent with indentation stress relaxation test. J. Mater. Res. **23**, 2486–2490 (2008b)

B. Xu, Z. Yue, X. Chen, An indentation fatigue depth propagation law. Scr. Mater. **60**(10), 854–857 (2009)

B. Xu, Z. Yue, X. Chen, Numerical investigation of indentation fatigue on copper polycrystalline. J. Mater. Res. **24**(3), 1007–1015 (2009)

B.X. Xu, A. Yonezu, X. Chen, An indentation fatigue strength law. Philos. Mag. Lett. **90**(5), 313–322 (2010)

F. Yang, J.C.M. Li, Impression creep by an annular punch. Mech. Mater. **21**, 89–97 (1995)

F. Yang, J.C.M. Li, Impression test – a review. Mater. Sci. Eng. R. Rep **74**, 233–253 (2013)

W. Yang, S. Mao, J. Yang, T. Shang, H. Song, J. Mabon, W. Swiech, J.R. Vance, Z. Yue, S.J. Dillon, H. Xu, B. Xu, Large-deformation and high-strength amorphous porous carbon nanospheres. Sci. Rep. **6**, 24187 (2016)

A. Yonezu, B. Xu, X. Chen, Indentation induced lateral crack in ceramics with surface hardening. Mater. Sci. Eng. A **507**, 226–235 (2009)

A. Yonezu, B. Xu, X. Chen, An experimental methodology for characterizing fracture of hard thin films under cyclic contact loading. Thin Solid Films **8**, 2082–2089 (2010)

N. Zaafarani, D. Raabe, F. Roters, S. Zaefferer, On the origin of deformation-induced rotation patterns below nanoindents. Acta Mater. **56**, 31–42 (2008)

Y. Zhu, C. Ke, H.D. Espinosa, Experimental techniques for the mechanical characterization of one-dimensional nanostructures. Exp. Mech. **47**, 7 (2007)

Crack Initiation and Propagation in Laminated Composite Materials

14

Jun Xu and Yanting Zheng

Contents

Abstract

PVB laminated glass is a kind of typical laminated composite material and its crack characteristics are of great interest to vehicle manufacturers, safety engineers, and accident investigators. Because crack morphology on laminated

J. Xu (✉)
Department of Automotive Engineering, School of Transportation Science and Engineering, Beihang University, Beijing, China

Advanced Vehicle Research Center (AVRC), Beihang University, Beijing, China
e-mail: junxu@buaa.edu.cn

Y. Zheng
China Automotive Technology and Research Center, Tianjin, China
e-mail: yantingzheng@buaa.edu.cn

G. Z. Voyiadjis (ed.), *Handbook of Nonlocal Continuum Mechanics for Materials and Structures*, https://doi.org/10.1007/978-3-319-58729-5_24

windshield contains important information on energy mitigation, pedestrian protection, and accident reconstruction. In this chapter, we investigated the propagation characteristics for both radial and circular cracks in PVB laminated glasses by theoretical constitutive equations analysis, numerical simulation, experiments, and tests of impact. A damage-modified nonlinear viscoelastic constitutive relations model of PVB laminated glass were developed and implemented into FEA software to simulate the pedestrian head impact with vehicle windshield. Results showed that shear stress, compressive stress, and tensile stress were main causes of plastic deformation, radial cracks, and circumferential cracks for the laminated glass subject to impactor. In addition, the extended finite element method (XFEM) was adopted to study the multiple crack propagation in brittle plates. The effects of various impact conditions and sensitivity to initial flaw were discussed. For experiment analysis, crack branching was investigated and an explicit expression describing the crack velocity and number of crack branching is proposed under quasi-static Split Hopkinson Pressure Bar (SHPB) compression experiments. And the radial crack propagation behavior of PVB laminated glass subjected to dynamic out of - plane loading was investigated. The steady-state cracking speed of PVB laminated glass was lower pure glass, and it increased with higher impactor speed and mass. The supported glass layer would always initiate before the loaded layer and the final morphologies of radial cracks on both sides are completely overlapped. Two different mechanisms of crack propagation on different glass layers explained the phenomenon above. Then further parametric dynamic experiments study on two dominant factors, i.e., impact velocity and PVB thickness are investigated: Firstly, a semiphysical model describing the relationship between the maximum cracking velocity and influential factors was established; Then the Weibull statistical model was suggested considering various factors to describe the macroscopic crack pattern in this chapter; Finally, the relation between radial crack velocity and crack numbers on the backing glass layer and the relation between the crack length and the capability of energy absorption on the impacted glass layer were proposed.

Keywords

PVB Laminated windshield · Crack Propagation and Initiation · Theoretical Constitutive Relations · Numerical Simulation of Crack Propagation · Extended Finite Element Method · Experiments Analysis of Crack Propagation · Quasi-Static Loading · Dynamic Out-of-Plane Loading

Introduction

As an typical laminated composite material, the standard PVB laminated glass comprises of a PVB interlayer sandwiched by two mono soda-lime and it has extensive applications in architecture, automobile industry, as well as the structural parts during the past few decades where they usually play an important role in human protection and structural integrity due to its excellent energy-absorbing and fragment-holding ability. However, mechanical behavior especially for the cracks of PVB laminated glass subject to loading has not been fully explored yet due to the complicity caused by the composite material, i.e., polymer material sandwiched by two pieces of brittle material. As a result, it becomes a pressing need to investigate the mechanical loading information contained in the macroscopic crack initiation and propagation which enables improved design or prevention for laminated PVB glass as a structural part.

In this chapter, there will be three methods mainly applied to investigate PVB laminated glass damage mechanism and crack growth: theoretical constitutive equations describing the failure criteria of PVB laminated glass under load, numerical simulations of crack initiation and growing on PVB laminated glass, experiments and tests of impact between impactor and laminated glass plate. For theoretical investigation, a damage-modified nonlinear viscoelastic constitutive equation is developed based on the updated Lagrangian approach. For numerical simulation investigation, a FEA model with proper material parameters verified by a classical example is established to describing the impact between pedestrian head and PVB laminated windshield. In addition, the extended finite element method (XFEM) is employed to study the low-speed impact-induced cracking in brittle plates in the present chapter. Upon head impact, the interaction between the stress field and the initiation and propagation of the radial and circumferential cracks are computed using XFEM. The effects of various impact conditions and sensitivity to initial flaw are discussed. The experiment analysis of PVB laminated glass is mainly based on quasi-static loading and dynamic impact. We carried out quasi-static Split Hopkinson Bar (SHPB) experiments to study the crack velocity and number of crack branching. For the dynamic out-of-plane impact tests carried out to study the crack propagation behavior of PVB laminated glass subjected to dynamic impact by using high-speed photography, the time histories of the averaged radial crack tip position, propagation velocity, and acceleration are recorded and the parametric study of impactor speed and mass, impact velocity, PVB thickness for PVB laminated glass are investigated. Then two different mechanisms of crack propagation on different glass layers are explained. Besides, a semiphysical model describing the relationship between the maximum cracking velocity and influential factors is established. In addition, the Weibull statistical model is suggested considering various factors to describe the macroscopic crack pattern thus providing a theoretical evidence for engineering practice. Finally, the relation between radial crack velocity and crack numbers on the backing glass layer and the relation between the crack length and the capability of energy absorption on the impacted glass layer are investigated.

Constitutive Relations

The constitutive relation of laminated composite material can well reflect the mechanical behavior of that. In order to study the crack in laminated composite material, the constitutive relation should be established firstly.

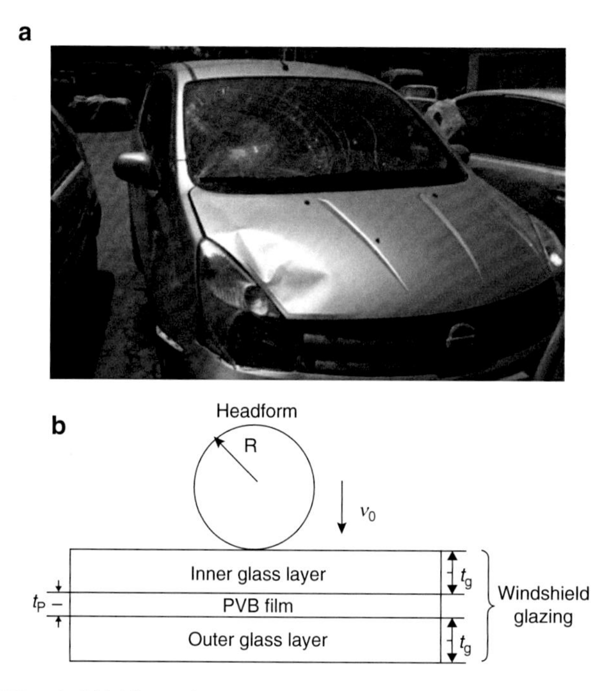

Fig. 1 (**a**) PVB windshield crack morphology in real-world pedestrian-vehicle accident. (**b**) Schematic illustration of a sphere pedestrian headform impacting on a windshield glazing

PVB windshield used in automotive industry are a kind of laminated composite material which comprise of a PVB interlayer sandwiched by two momo soda-lime glass sheet. Thus, pedestrian head impacts on the vehicle PVB windshield can be regarded as a scenario where a composite plate is subjected to a concentrated force by a spherical indenter on its top surface shown in Fig. 1. The composite plate consists of $n = 2$ individual brittle soda-lime glass layers of thickness t_g and modulus E_g, sandwiched with a PVB interlayer of thickness t_p and modulus E_p.

By studying the constitutive relations of the PVB windshield, the crack in PVB windshield can be analyzed thoroughly. Lili et al. (Wang et al. 2003; Lili et al. 1991) study the mechanical behavior of polymer with performing dynamic impact and quasi-static tensile experiments. According to the experimental results, the behavior of the polymer can be described as the following constitutive relation:

$$\sigma = \sigma_e + \sigma_{t_1} + \sigma_{t_2} \tag{1}$$

σ_t stand for the tensile stress.

$$\sigma_e = \sigma_m \left[1 - \exp\left(-\sum\nolimits_{t=1}^{n} \frac{(m\varepsilon)^t}{t}\right)\right] \tag{2}$$

$$\sigma_{t_1} = E_1 \int_0^t \dot{\varepsilon}(\tau) \exp\left(-\frac{t-\tau}{\theta_1}\right) d\tau \tag{3}$$

$$\sigma_{t_2} = E_2 \int_0^t \dot{\varepsilon}(\tau) \exp\left(-\frac{t-\tau}{\theta_2}\right) d\tau \tag{4}$$

where m, τ, E_1, E_2 are a nondimensional relative initial elastic modulus, relaxation time as defined in Maxwell body, elastic constants of first Maxwell body and elastic constants of second Maxwell body, respectively. θ_1, θ_2, are constants as well, standing for the relaxation times of first and second Maxwell body. Where σ_e stands for the stress with the damage material. σ_{t1}, σ_{t2} standing for tensile stress of first and second Maxwell body. D is a “damage” variable.

Nevertheless, the above constitutive relations are only valid when strain is less than 7% (Lili et al. 1991), and it cannot express the relation completely. To further utilize the above relations, Fenghua et al. (Zhou et al. 1992) suggested the following equation to describe the constitutive relations:

$$\sigma = (1 - D)\,\sigma_e + \sigma_{t_1} + \sigma_{t_2} \tag{5}$$

where D is a “damage” variable and is defined as:

$$D = \begin{cases} 0 & \varepsilon \le \varepsilon_{th} \\ \dot{D}_0 \dot{\varepsilon}^{\delta-1} (\varepsilon - \varepsilon_{th}) & \varepsilon > \varepsilon_{th} \end{cases} \tag{6}$$

And $\varepsilon_{\mathrm{th}}$ here regards as strain threshold of damage evolution. $\dot{D}_0$, δ both are material parameters determined by experiments. Then the dynamic damage criterion is put forward:

$$D_f = \dot{D}_0 \dot{\varepsilon}^{\delta-1} (\varepsilon - \varepsilon_{th}) \tag{7}$$

Incremental Form of Constitutive Relations When the Material Is Not Damaged

When employing the constitutive relations, we should consider some conditions like whether the strain is larger than the threshold strain. Thus we need an equivalent strain to determine whether it is or not. As following is the equivalent:

$$\varepsilon_{eq} = \frac{\sqrt{2}}{3} \sqrt{(\varepsilon_1 - \varepsilon_2)^2 + (\varepsilon_2 - \varepsilon_3)^2 + (\varepsilon_3 - \varepsilon_1)^2 + \frac{3}{2}\left(\gamma_{12}^2 + \gamma_{23}^2 + \gamma_{31}^2\right)^2} \tag{8}$$

where subscripts “1,” “2,” and “3” denote the directions of principal stresses.

On the basis of updated Lagrangian virtual work method, stress-strain relations of material that undergoes large deformation with the ignorance of physical nonlinear properties can be regarded as (Fischer and Washizu 1982):

$$\{S_{ij}\} = \int_0^t E(t-\tau)[A]\frac{\partial\{e_{kl}\}}{\partial\tau}d\tau \tag{9}$$

$\{S_{ij}\}$, $\{e_{kl}\}$ are Kirchhoff stress tensors and Green strain tensors, respectively.

$$[A] = \frac{1-v}{(1+v)(1-2v)}\cdot \begin{bmatrix} 1 & \frac{v}{(1-v)} & \frac{v}{(1-v)} & 0 & 0 & 0 \\ \frac{v}{(1-v)} & 1 & \frac{v}{(1-v)} & 0 & 0 & 0 \\ \frac{v}{(1-v)} & \frac{v}{(1-v)} & 1 & 0 & 0 & 0 \\ 0 & 0 & 0 & \frac{1-2v}{2(1-v)} & 0 & 0 \\ 0 & 0 & 0 & 0 & \frac{1-2v}{2(1-v)} & 0 \\ 0 & 0 & 0 & 0 & 0 & \frac{1-2v}{2(1-v)} \end{bmatrix} \tag{10}$$

Similarly, as PVB windshield is a three-dimensional structure, Equations from 1 to 6 can be rewritten as (Swanson and Christensen 2015; Wang et al. 2007; Shen et al. 1987):

$$S_{ij} = S_{ij,e} + S_{ij,t_1} + S_{ij,t_2} \tag{11}$$

where

$$S_{ij,e} = \sigma_m\left\{1 - \exp\left(-mAE_{kl} - \sum\nolimits_{t=1}^{n}\frac{(mE_{ij})^t}{t}\right)\right\} \tag{12}$$

$$S_{ij,t_1} = E_1\int_0^t A\frac{\partial E_{kl}}{\partial\tau}\exp\left(-\frac{t-\tau}{\theta_1}\right)d\tau \tag{13}$$

$$S_{ij,t_2} = E_2\int_0^t A\frac{\partial E_{kl}}{\partial\tau}\exp\left(-\frac{t-\tau}{\theta_2}\right)d\tau \tag{14}$$

On the basis of Boltzmann superposition (Feng 1995), incremental form of constitutive relations can be obtained. Also, implementation in commercial FEA software needs the incremental form of the stress tensor of Eqs. 12, 13, and 14 to be rewritten. Thus, it is a must for us to reach the incremental form first. According to proofs of Chain Rule, Eq. 12 can be rewritten as follows:

$$\begin{aligned} dS_{ij,e} = {} & \sigma_m\exp\left(-mAE_{kl} - \sum\nolimits_{t=2}^{n}\frac{(mE_{ij})^t}{t}\right) \\ & \cdot\left[mA\frac{\partial E_{kl}}{\partial\tau} + \sum\nolimits_{t=2}^{n}\frac{(mE_{ij})^{t-1}}{t}\frac{\partial E_{kl}}{\partial\tau}\right]\cdot\Delta t \end{aligned} \tag{15}$$

Considering Δt is short enough yields (Feng 1995):

$$\frac{\partial E_{kl}}{\partial \tau} = \frac{\partial E_{kl}}{\partial t} = \frac{\partial E_{kl}}{\Delta \tau} \tag{16}$$

Thus Eq. 13 becomes:

$$\Delta S_{ij,e} = \sigma_m \exp\left(-mAE_{kl} - \sum_{t=2}^{n} \frac{(mE_{ij})^t}{t}\right) \cdot \left[mA\Delta E_{kl} + \sum_{t=2}^{n} \frac{(mE_{ij})^{t-1}}{t} \Delta E_{kl}\right] \tag{17}$$

Similarly, viscoelastic stress tensors for both low strain rate and high strain rate at t are:

$$\Delta S_{ij,t_1} = E_1\theta_1 A \frac{1-\exp\left(-\frac{\Delta t}{\theta_1}\right)}{\Delta t} \Delta E_{kl} - \left(1-\exp\left(-\frac{\Delta t}{\theta_1}\right)^t\right) S_{ij,t_1} \tag{18}$$

$$\Delta S_{ij,t_2} = E_2\theta_2 A \frac{1-\exp\left(-\frac{\Delta t}{\theta_2}\right)}{\Delta t} \Delta E_{kl} - \left(1-\exp\left(-\frac{\Delta t}{\theta_2}\right)^t\right) S_{ij,t_2} \tag{19}$$

Therefore, the incremental form of stress tensor during one unit time step Δt can be obtained:

$$\begin{aligned}\Delta S_{ij,n+1} &= \Delta S_{ij,e} + \Delta S_{ij,t_1} + \Delta S_{ij,t_2} = \sigma_m \exp\left(-mAE_{kl} - \sum_{t=2}^{n} \frac{(mE_{ij})^t}{t}\right) \\ &\cdot \left[mA\Delta E_{kl} + \sum_{t=2}^{n} \frac{(mE_{ij})^{t-1}}{t} \Delta E_{kl}\right] + E_1\theta_1 A \frac{1-\exp\left(-\frac{\Delta t}{\theta_1}\right)}{\Delta t} \Delta E_{kl} \\ &- \left(1-\exp\left(-\frac{\Delta t}{\theta_1}\right)^t\right) S_{ij,t_1} + E_2\theta_2 A \frac{1-\exp\left(-\frac{\Delta t}{\theta_2}\right)}{\Delta t} \Delta E_{kl} \\ &- \left(1-\exp\left(-\frac{\Delta t}{\theta_2}\right)^t\right) S_{ij,t_2}\end{aligned} \tag{20}$$

In the following way, the stress tensor can be calculated:

$$S_{ij,n+1} = S_{ij,n} + \Delta S_{ij,n+1} \quad \varepsilon < \varepsilon_{th} \tag{21}$$

where $t_{n+1} = t_n + \Delta t$.

Incremental Form of Constitutive Relations When the Material Is Damaged

When the material is not damaged, the incremental form of constitutive relations can be described using function relation above. However, when the strain is larger than the threshold strain, material turns into "damage" phase. As mentioned above, stress tensor in "damage" phase at time t_{n+1} can be concluded as:

$$\{S_{ij}\}_{n+1} = \{S_{ij}\}_n + \{\Delta S_{ij}\}_{n+1} \tag{22}$$

and Eq. 21 can be rewritten as:

$$S_{ij,n+1} = S_{ij,n} + \Delta S_{ij,n+1} \tag{23}$$

Therefore,

$$\begin{aligned} S_{ij,n+1} &= S_{ij,n} + \Delta S_{ij,n+1} \\ &= \frac{1}{1-D_{n+1}} S^d_{ij,n+1} + 1^d - \frac{1}{1-D_n} S^d_{ij,n} \\ &= \frac{1}{1-D_{n+1}} \Delta S^d_{ij,n+1} + \frac{\Delta D_{n+1} S^d_{ij,n}}{(1-D_{n+1})(1-D_n)} \end{aligned} \tag{24}$$

where $\Delta D_{n+1} = D_{n+1} - D_n$ is the increase of damage variable.

Rewriting Eq. 23 yields:

$$\begin{aligned} \Delta S^d_{ij,n+1} &= (1 - D_{n+1})\, \Delta S_{ij,n+1} - \frac{\Delta D_{n+1} S^d_{ij,n}}{(1 - D_n)} \\ &= (1 - D_{n+1})\, \Delta S_{ij,n+1} - \Delta D_{n+1} \Delta S_{ij,n} \end{aligned} \tag{25}$$

Similarly, the stress tensor under damage can be calculated as:

$$S^d_{ij,\mathrm{n}+1} = S^d_{ij,\mathrm{n}} + \Delta S^d_{ij,\mathrm{n}+1} \quad \varepsilon \geq \varepsilon_{th} \tag{26}$$

Verification of Constitutive Relations

The constitutive relation of PVB windshield are gotten preliminary and the accuracy of that should be studied next. In order to verify the above constitutive relations, a classical example is chosen. A thin plate with infinite length and 2b = 12 mm in width is under the uniform load$q(t) = 1\ \mathrm{kN} \cdot m$. There is a small crack in center with length 2a = 1 mm, see Fig. 2. Material properties are listed in Table 1.

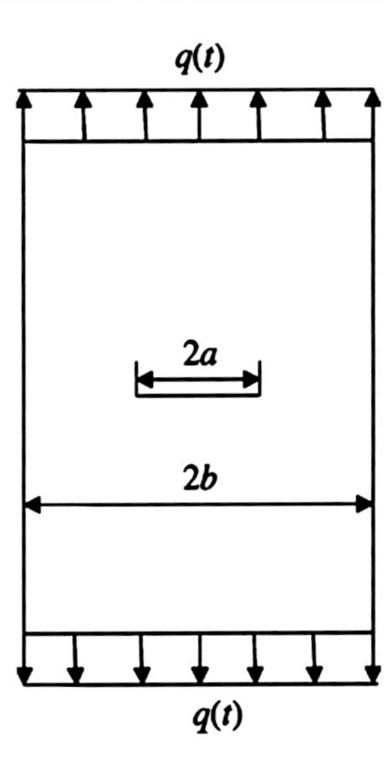

Fig. 2 Thin plate with center crack under uniform load

Table 1 Material parameters for plate

σ_m/MPa	n	m	E_1/MPa	θ_1/s	E_2/MPa	$\theta_2/\mu s$	ε_{th}	δ	D_0	D_f
1,100	4	19.9	949	13.8	3,981	67.4	0.055	1.22	0.863	0.0797

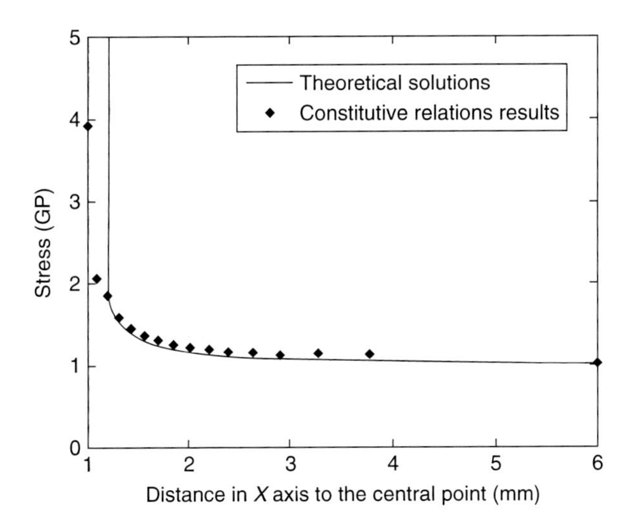

Fig. 3 A comparison between theoretical solutions and FEA results

By implementing the above constitutive relations into FEA software numerical and adding the command of failure of material, results can be obtained. Comparing numerical results with theoretical ones, we can see that the constitutive relations results can coincide with the theoretical solutions well, shown in Fig. 3.

Numerical Simulation

PVB Windshield FEA Modeling Based on Constitutive Relations of PVB Laminated Glass

By verifying the constitutive relations above with a classical example, the relations have a sufficient accuracy. Then the FEA modeling of PVB windshield are introduced in detail following.

Contact Analysis

A nonconforming two-body contact model of a PVB windshield loaded by a headform (sphere) is illustrated in Fig. 4. The thickness of PVB windshield is 8.76 mm, the length of it is 350 mm, and the thickness of interlayer PVB is 0.76. Shell elements were chosen to mesh the two bodies in contact. Hourglass control and automatic surface- to-surface contact with dynamic friction coefficient 0.1 were employed in LS-Dyna. Initial nodal gaps between the two contact surfaces were prescribed. The contact elements were not activated until the penetration of the sphere into the composite glass occurred for the sake of saving computation time.

An extra command that allows the failure of material is added in the input file of LS-Dyna. Then the above-mentioned constitutive relations with damage variable could be fully employed in FEA. The accuracy of the contact FEA model is firstly examined by classical Hertzian pressure calculation method. Illustration of classical Hertzian theory is shown in Fig. 5 (Daphalapurkar et al. 2007). The δ and a_0 stand for mutual approach between two bodies and their contact radius.

Deviations between FEA and theoretical results are all within 1.5%. Comparison shown in Fig. 6 demonstrates the sufficient accuracy of the FEA model to stimulate the impact between pedestrian head and PVB windshield.

Material Model

The PVB laminated glass is modeled according to the following principles: if the glass does not fail, the composite is treated as a shell; otherwise, the interlayer PVB film acts as a membrane. A Belytschko–Tsay shell element for the glass material and a membrane element for the interlayer are employed (Zhou et al. 1992; Fischer

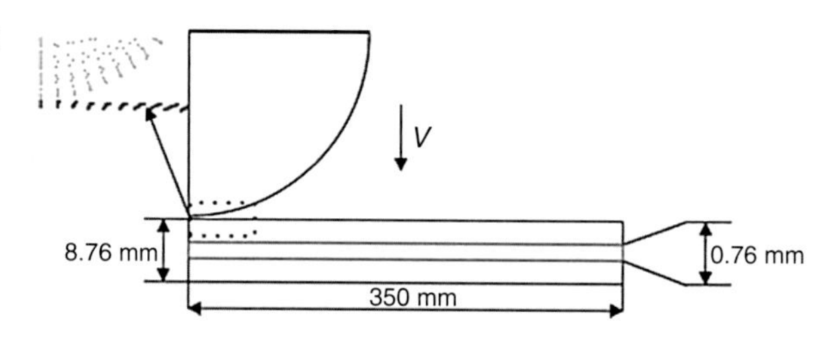

Fig. 4 A simplified physical model of impact between head and windshield

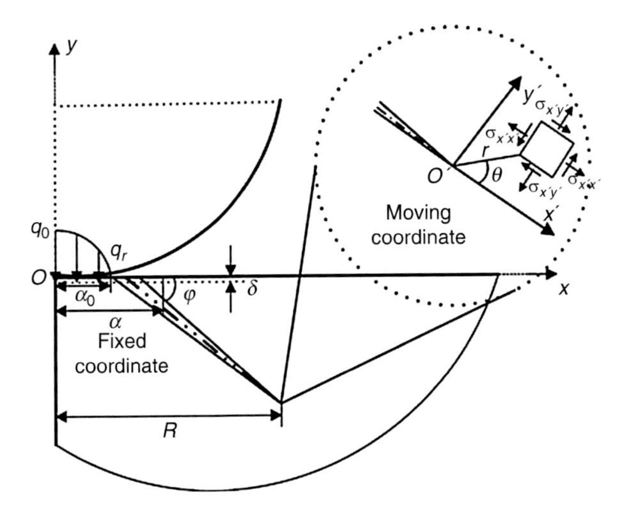

Fig. 5 A simplified crack-model based on Hertzian pressure with moving coordinate at the crack tip is shown

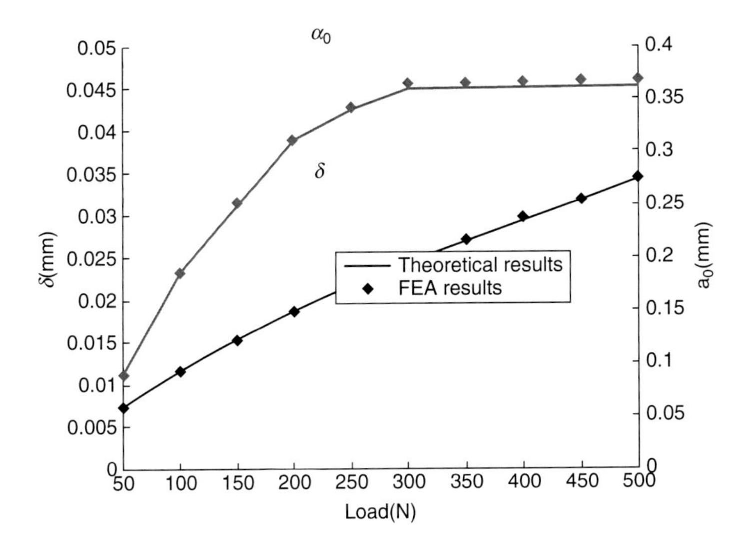

Fig. 6 Comparison between FEA results and theoretical results

and Washizu 1982). Both types of elements are fully integrated in simulations so that no hourglass modes should be expected (Sharon and Fineberg 1999).

Glass and PVB film both can be treated as isotropic materials according to their physical and material properties. Therefore, we assume that each layer of composite windshield is isotropic with respect to its material symmetry lines and obeys Hooke's law. It is assumed that, due to the influence of dynamic loading, there is no sliding between two contact layers. It then becomes reasonable to consider the

Table 2 Parameters used in both constitutive relations and finite element analysis

Components	Parameters and values
Headform	$E = 65\,\text{GPa}, \rho = 1412\,\text{kg/m}^3, \nu = 0.22$
Glass	$E = 74\,\text{GPa}, \rho = 2500\,\text{kg/m}^3, \nu = 0.25, t_g = 2\,\text{mm}$
PVB film	$k = 20\,\text{GPa}, \rho = 1100\,\text{kg/m}^3, t_p = 0.76\,\text{mm}$
Windshield dimension	Panel dimensions (a × b):1,320 × 630 mm, with a small curvature in x and z direction in coordinates
Other configuration	Initial velocity: $v_0 = 10$ m/s with no gravity field

composite windshield as integrity. As a result, an equivalent elastic modulus and Poisson's ratio should be introduced as follows (Dwivedi and Espinosa 2003).

$$\overline{E} = \frac{2t_g E_g + t_p E_p}{2t_g + t_p} \tag{27}$$

$$\overline{v} = \frac{2t_g v_g + t_p v_p}{2t_g + t_p} \tag{28}$$

where $\overline{E}$ and $\overline{v}$ are equivalent Young's modulus and equivalent Poisson's ratio of windshield glazing.

Material parameters are listed in Table 2.

Results and Discussions

Internal Stress Analysis

The cracks in PVB windshield are demonstrations of internal and external stresses. Thus, the FEA model above is used to analyze the internal stresses. Only through investigating the internal stresses can we know the cracks on windshields better.

During the dynamic crack evolution shown in FEA analysis, stages of damage evolution could be sketchily drawn as Fig. 7. Three kinds of cracks are: circumferential cracks, web crack, and plastic crack. The circumferential cracks are mainly caused by pressure stress while the web cracks are mainly caused by tensor stress. Plastic cracks are the last stage that the windshield deforms plastically. In Fig. 8, three kinds of internal stresses, tensile stress σ_t, compressive stress σ_c, and shear stress σ_s, are plotted versus impact time. We can find out that the value of stresses sorted from large to small is σ_s, σ_c and σ_t. Lawn and Wilshaw (1975) stated that in addition to the Hertzian cone crack and the median vent crack, there is also an inelastic deformation zone where shear and hydrostatic compression is greatest. Quite similarly, it is obvious to see in Fig. 8 that shear stress is the greatest among three stresses, causing the windshield to deform plastically.

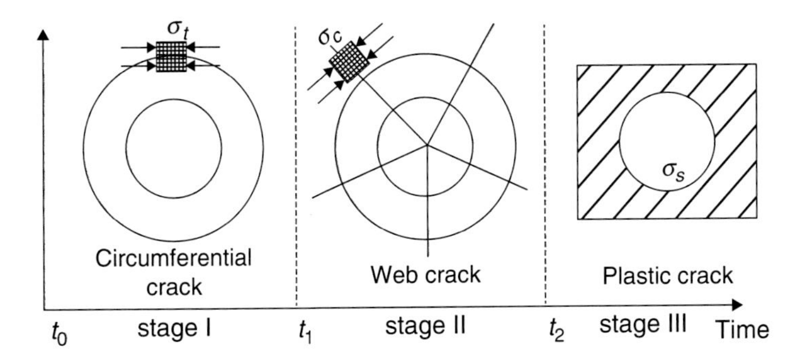

Fig. 7 Diagram of different crack-growing stage

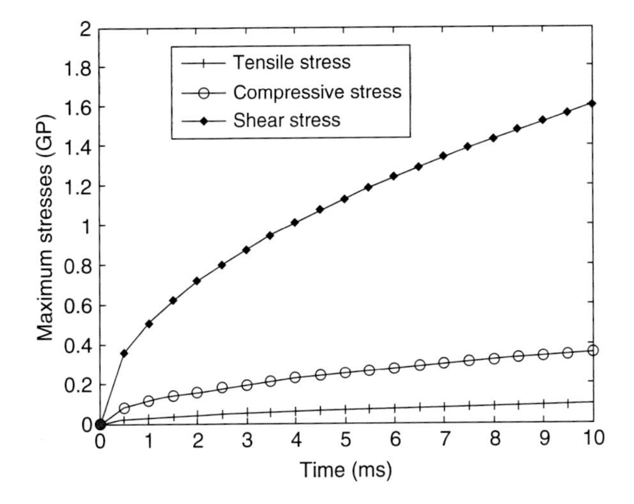

Fig. 8 Three kinds of maximum internal stress in PVB windshield, tensile stress, compressive stress, and shear stress curves versus impact time

Internal Stress Variation Among Different Point on Windshield Plate

Internal stresses of different points on windshield plate are far different and crack information can prove it. The cracks become weaker as the radius grows larger for circumferential cracks, as do the radial cracks. In Fig. 9, maximum stresses appear at the very point that impacted by pedestrian head. Stresses decrease as the distances from the central point increase. The decreasing rates of three stresses are nearly the same according to the computational results.

Effects of Poisson's Ratio on Crack Angle

Poisson's ratio plays an important role of cone crack angle. A parametric study is performed to manifest the effects on crack angle. Poisson's ratio is ranged from 0.2 to 0.4, with a step-width of 0.05. Table 3 shows the results of FEA analysis

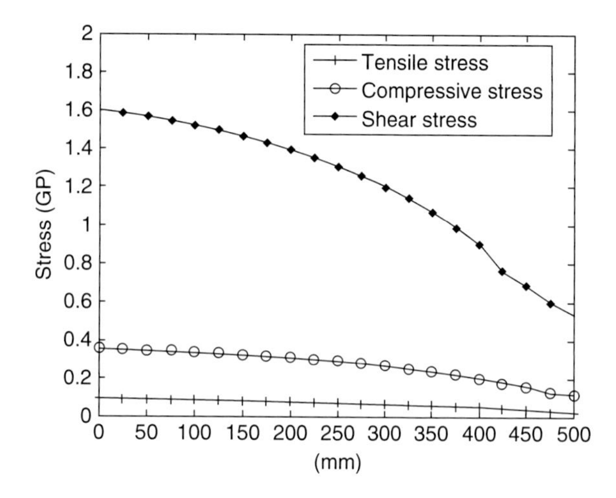

Fig. 9 Three kinds of maximum internal stress in PVB windshield, tensile stress, compressive stress, and shear stress curves versus the displacement from the central point

Table 3 Comparison results of two models on Poisson's ratio effect to crack angle

Poisson's ratio	0.20	0.25	0.30	0.35	0.40
SIF analysis	12.4	15.3	19.6	24.5	27.7
FEA analysis	11.8	15.1	18.9	23.1	26.2
Differences (%)	5.20	1.32	3.70	6.10	5.80

compared to that of stress intensity factor analysis (SIF). Maximum difference between the results is below 7%. With the increase in Poisson's ratio, the conical crack angle also increases though the increase pattern is still unknown.

Effects of Impact Velocity

Vehicle speed is a critical factor in pedestrian-vehicle traffic accident. Extends and degrees of injuries largely depend on the impact velocity between pedestrian and vehicle. Therefore, studies of cracks on windshields under different head impact velocity would provide the essential foundations in pedestrian protection. According to the census data in NTADTU, in most cases, vehicles impact pedestrians at the speed of about 40–70 km/h. Thus, we vary the impact velocity of the head from 0 to 20 m/s and make impact velocity V0 as an independent variable. On the other hand, we choose the maximum radius of circumferential cracks R_c, the maximum length of radial cracks L_r, and the maximum radius of plastically deformed area R_p as dependent variables. Three dependent variables are illustrated and defined in Fig. 10.

In Fig. 11, the impact velocity has very great influence to three chosen variables. At the impact velocity of 20 m/s, R_c can reach up to about 642 mm, much larger than $R_c \approx 89$ mm under the impact velocity of 5 m/s. Vehicle velocity plays an extremely important role in both pedestrian head injury and damage of windshield. In addition, it is reasonable to infer the extend of injuries of pedestrian head and impacting velocity of vehicle roughly through R_c, L_r, and R_p.

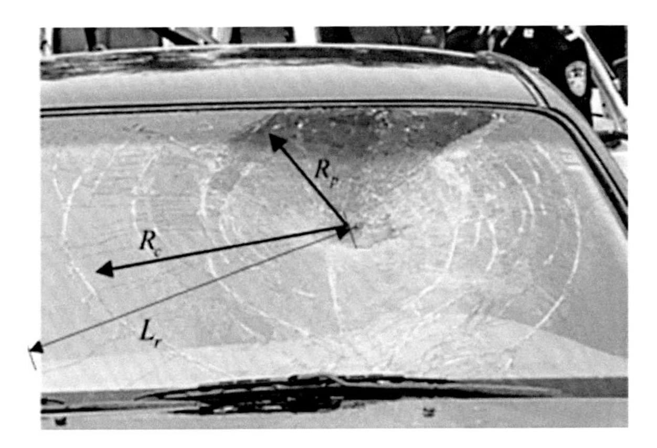

Fig. 10 Effect of initial impact velocity of the stress

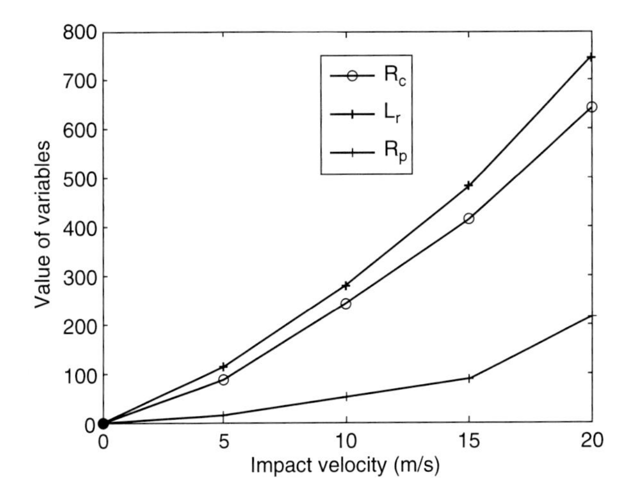

Fig. 11 Effect of initial impact velocity to cracks

Numerical Simulation Based on the Extended Finite Element Method

After the head impacts on the windshield material, cracks including radial crack and circumferential crack appear in the glass material. The fracture characteristics of the windshield are widely recognized as one of the most important factors in automotive crashworthiness. The crack profiles (e.g., length, pattern, etc.) on the windshield material contain critical information for impact speed (which is extremely useful for accident reconstruction), vehicle crashworthiness, and insights for improving pedestrian and passenger protection. This motivates us to study the characteristics of low-speed head impact-induced crack propagation on the windshield glass material. In the past studies of impact on windshield glass, due to the intrinsic complexity of numerical analyses of cracking, researchers in automotive engineering often avoid explicit simulation of crack propagation. The following study aims to study the crack propagation characteristics (including both radial and circumferential cracks) when a model windshield undergoes low-speed impact and the numerical investigation of multiple crack propagation is based on the extended finite element method (XFEM) in this study.

Fundamentals of XFEM

XFEM incorporates a discontinuous displacement field across the crack facing away from the crack tip, in the form of:

$$u^h = \sum_{i=1}^{n} N_i(x) \left(u_i + b_i H(x) + \sum_{l=1}^{4} \sum_{j=1}^{2} c_{il} F_l^i(x) \right) \tag{29}$$

where n is the number of nodes in the mesh, $N_i(x)$ is the shape function of node i, u_i are the classical DOFs of node i. b_i and c_{il} are the DOFs associated with the Heaviside "jump" function $H(x)$, with value 1 above the crack and below the crack. The crack-tip function $F_l^i(x)$ is:

$$F_l(r, \theta) = \left(\sqrt{r}\sin\left(\frac{\theta}{2}\right), \sqrt{r}\cos\left(\frac{\theta}{2}\right), \sqrt{r}\sin\left(\frac{\theta}{2}\right)\sin\theta, \sqrt{r}\cos\left(\frac{\theta}{2}\right)\sin\theta \right) \tag{30}$$

where (r, θ) are the local polar coordinates at the crack tip.

Figure 12 illustrates the enriched nodes near a crack tip, where $H(x)$ is discontinuous across the crack surface. Under general mixed mode-loadings, the asymptotic near-tip hoop and shear stress components are

$$\begin{cases} \sigma_{\theta\theta} = \frac{1}{4\sqrt{2\pi r}} \left(k_1 \left(3\cos(\theta/2) + \cos(3\theta/2) + k_{II}\left(-3\sin(\theta/2) - 3\sin(3\theta/2)\right)\right)\right. \\ \sigma_{\theta\theta} = \frac{1}{4\sqrt{2\pi r}} \left(k_1 \left(\sin(\theta/2) + \sin(3\theta/2) + k_{II}\left(\cos(\theta/2) - 3\cos(3\theta/2)\right)\right)\right. \\ \sigma_{\theta\theta} = \frac{1}{4\sqrt{2\pi r}} \left(k_1 \left(5\cos(\theta/2) - \cos(3\theta/2) + k_{II}\left(5\sin(\theta/2) - 3\sin(3\theta/2)\right)\right)\right. \end{cases} \tag{31}$$

For fracture propagation, we use the popular criterion of maximum local hoop stress, where the crack propagation direction is determined from Eq. 31 as

$$\theta_c = arc\cos\left(\frac{3k_{II}^2 + \sqrt{K_I^4 + 8k_I^2 k_{II}^2}}{K_I^2 + 9k_{II}^2} \right) \tag{32}$$

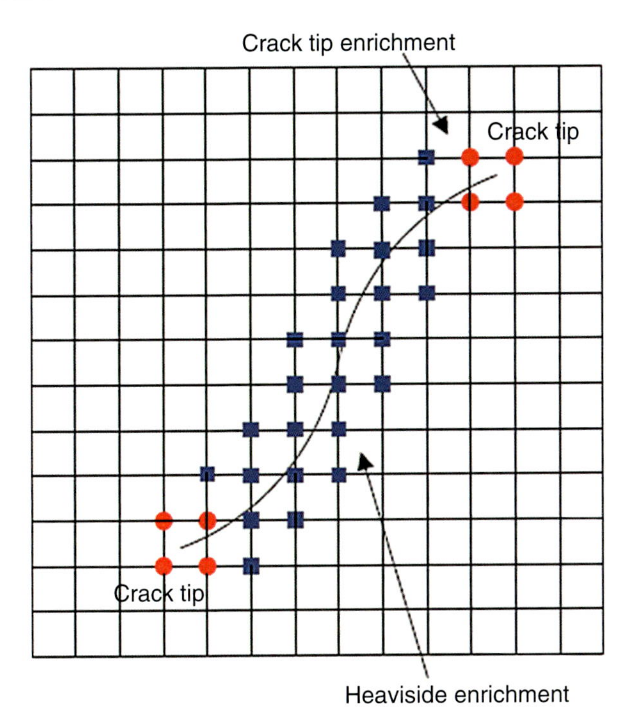

Fig. 12 Schematic of uniform mesh surrounding a crack used in XFEM simulation. *Blue square nodes* stand for the Heaviside function enrichment and *red circular nodes* stand for the crack tip enrichment. (For interpretation of the references to colour in this figure legend, the reader is referred to the web version of this article)

Here, θ_c is measured with respect to a local polar coordinate system with its origin at the crack tip and aligned with the direction of the existing crack. Numerical simulations based on the above XFEM algorithm is carried out using commercial code ABAQUS.

Model and Methods

Model Setup

A featureless spherical headform is used to model the human head; The model head has a radius $R = 90$ mm and mass 4.5 kg. Note that effective head weights are changeable in different accident cases because human body may also get involved due to inertia, causing the effective impact mass larger than the mass of the pure head, especially during higher speed impact. Thus, the effective head mass M is also varied from 4.5 kg up to 90 kg by varying the density (from 1,500 to 30,000 kg/m^3) and keeping volume the same in this study. According to 151 pedestrian–vehicle accident cases from the National Traffic Accident Database of Tsinghua University (NTADTU), most of the impact speeds of the vehicle in accident fall into the range of 40–80 km/h, i.e. 11.1–22.2 m/s, and that is the range of low-speed impact we

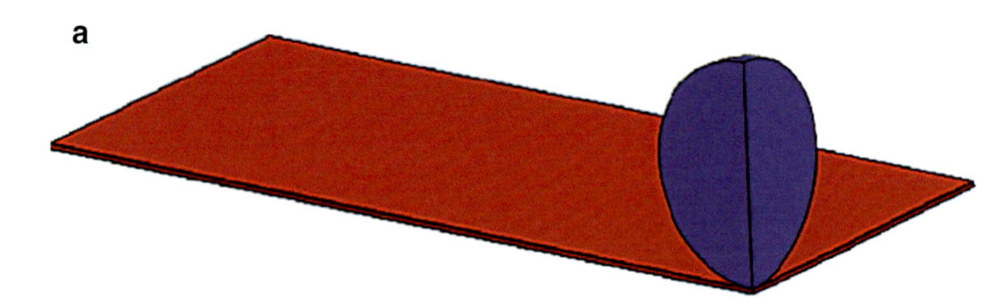

Global view of the quarter model of head impact on a model windshield plate

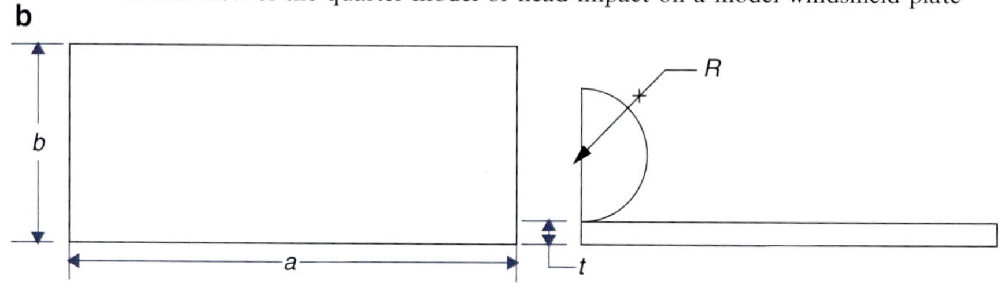

Dimensions of the featureless headform and windshield plate in a quarter model

Fig. 13 Computational model setup. (**a**) Global view of the quarter model of head impact on a model windshield plate. (**b**) Dimensions of the featureless headform and windshield plate in a quarter model

consider. We focus on normal impact and assume that impact occurs at the center of the plate, we neglect the small windshield curvature and focus on a flat rectangular plate with dimensions a × b = 700 × 300 mm and thickness $t = 4.76$ mm as Fig. 13 shows.

In terms of the material properties, typical material parameters of glass are assigned to the plate, with density $\rho = 2{,}500$ kg/m^3, Young's modulus to the plate, Possion's ratio $v = 0.22$, mode I energy release rate $G_{\mathrm{I}} = 10$ J/m^2, mode II and III energy release rate $G_{\mathrm{II}} = G_{\mathrm{III}} = 50$ J/m^2. The properties are assumed to be rate independent. To take advantage of symmetry, only one quarter of the model is needed for simulation. The boundary condition of the windshield is assumed to be fixed. The plate and head are modeled with eight-node linear brick elements. The contact between the head and plate obeys the Coulomb's friction law with coefficient $f = 0.1$ (since according to our accident scene investigation, a small slip exists between head and windshield).

Computational Method

Suppose there is no crack or damage, we first compute the deformation response with response to impact. For a given effective head mass M and impact speed v, the maximum displacement D is computed and presented in a dimensionless form in Fig. 14. For the range of impact speed and head mass considered in this study, D/t can be fitted as a dimensionless function of v/v_0 and M/M_0, as

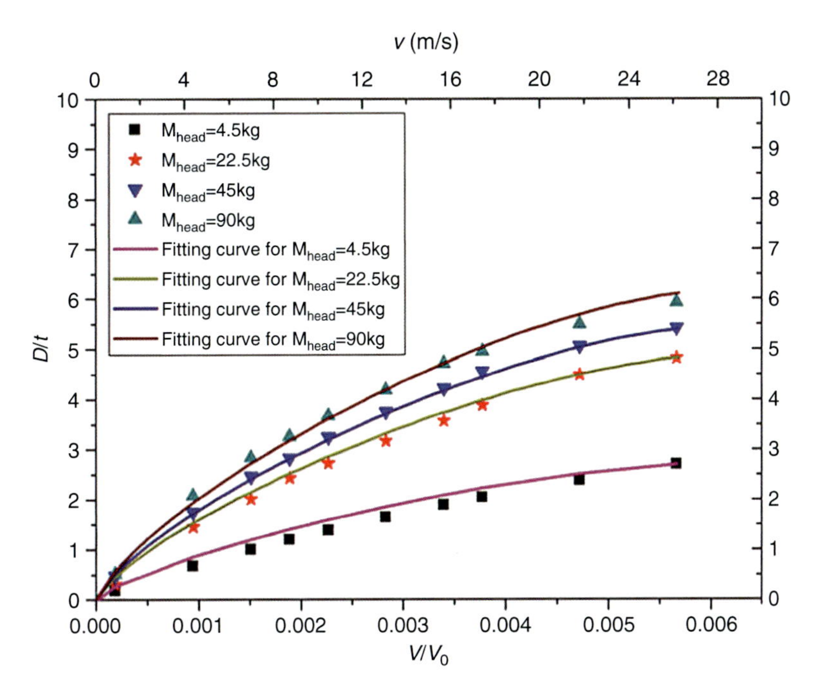

Fig. 14 Relationship between the initial impact velocity and displacement under different head masses; *lines* denote curve fitting, Eq. 33

$$D/t = \left(135.3(v/v_0)^{0.7115} - 0.0305^{518.5(v/v_0)}\right)(M/M_0)^{0.2669} \tag{33}$$

where $v_0 = \sqrt{E/\rho}$ and $M_0 = \text{abt}\,\rho$.

Since the impact speed is relatively low (much smaller than the wave speed), our simulation shows that the dynamic effect is relatively minor for the range of parameters investigated in this chapter. That is, for a given head mass, the computed stress field from dynamic impact (with incident speed v) is very close to that of quasi-static indentation (with maximum indentation depth D/t) as long as they induce the same deformation curvature of the plate (i.e., when v and D/t satisfy Eq. 33); see an example in Fig. 15 for the stress field. This enables us to simplify the dynamic impact problem into a quasi-static indentation one, where the normalized indentation depth D/t becomes the governing variable to indicate the effective "impact condition."

Our previous study showed that for windshield fracture, circumferential cracks always grow after the radial crack has developed, except near the vicinity of impact where extensive damage can be found (Xu et al. 2009). This is consistent with Fig. 15 that the hoop stress field is more prominent than radial stress. Thus, the crack simulation can be decoupled into two steps. First, we embed a number of

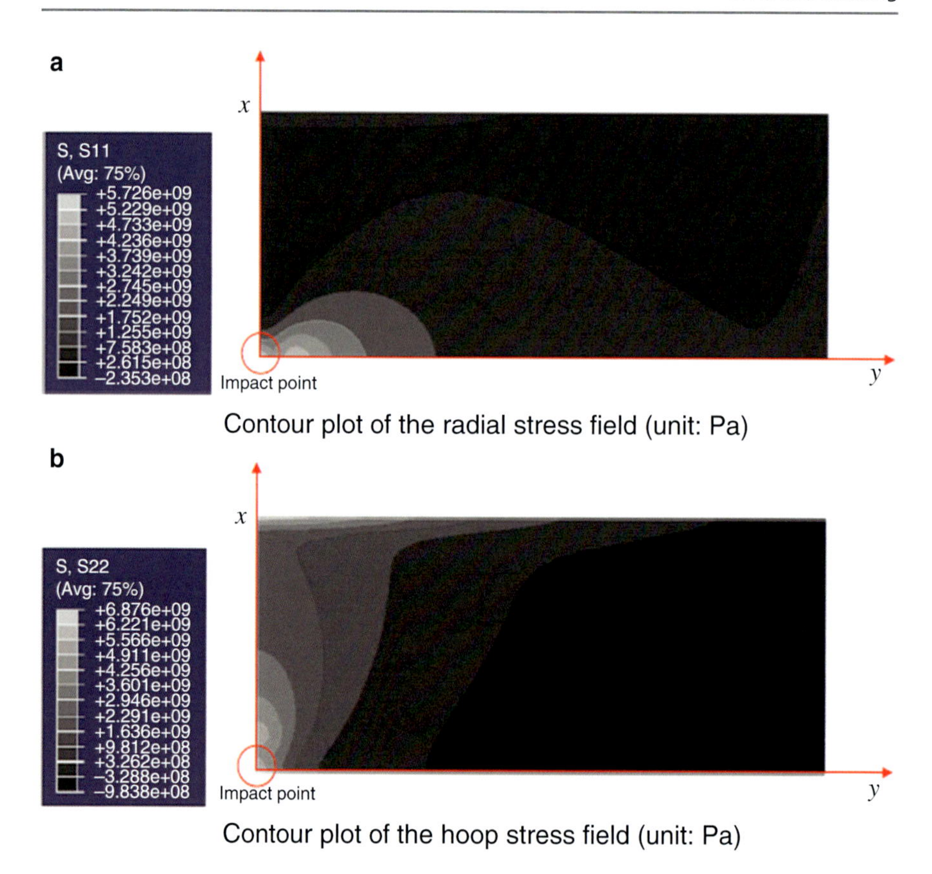

Fig. 15 Comparison between radial and hoop stress field resulted purely from impact deformation; to relieve the prominent stress, cracks need to form and radial crack is preferred to propagate before the circumferential crack owing to the more significant hoop stress field. (**a**) Contour plot of the radial stress field (unit: Pa). (**b**) Contour plot of the hoop stress field (unit: Pa)

initial flaws at the center of the plate (impact location); this is consistent with the fact that extensive damage is initiated in the contact zone and that facilitates subsequent crack initiation and propagation. The size of the initial flaw is $a_0 = 3$ mm with variable directions. When these initial flaws are very small, only the radial flaws will grow to become radial cracks upon impact, so as to release strain energy. For a given indentation depth D, the radial crack grows gradually until a certain length; with the increase of D, the final length of radial crack also increases. In the second step, based on the radially cracked specimen, new circumferential flaws with size $a_0 = 3$ mm are introduced (the sensitivity of the flaw geometry and distribution is discussed below), and without further increasing the load, with the residual stress and strain fields inherited from the first step, the development of circumferential cracks is simulated by XFEM.

Rack Propagation Characteristics

Radial Crack Propagation Characteristics

Overall Crack Feature

In Fig. 16, for a typical displacement of $D/t = 0.62$, the evolved radial crack is shown along with the hoop stress field. Note that with the fully developed crack, this stress field is already much relaxed compared to its counterpart without cracking (see for example the difference between Figs. 15 and 16); moreover, the current hoop stress field is specified for the global coordinate, whereas during crack propagation, it is the local hoop stress that determines the crack trajectory. When the local maximum hoop stress exceeds a critical threshold, referred to as the damage stress σ_d, the crack grows incrementally toward such a local principal direction.

In the beginning stage of impact, since the growth of radial crack can release more energy than the circumferential crack (due to the more prominent hoop stress field, Fig. 15), radial crack grows with the increase of the indentation depth D/t. It is noticeable that the crack growth direction keeps changing slightly with growth direction 22° approximately. This is mainly caused by the rectangular shape of the windshield plate, which makes the overall strain in the y-direction larger than that in the x-direction. When the crack grows further, the boundary constraint changes and the crack direction approaches about 45°.

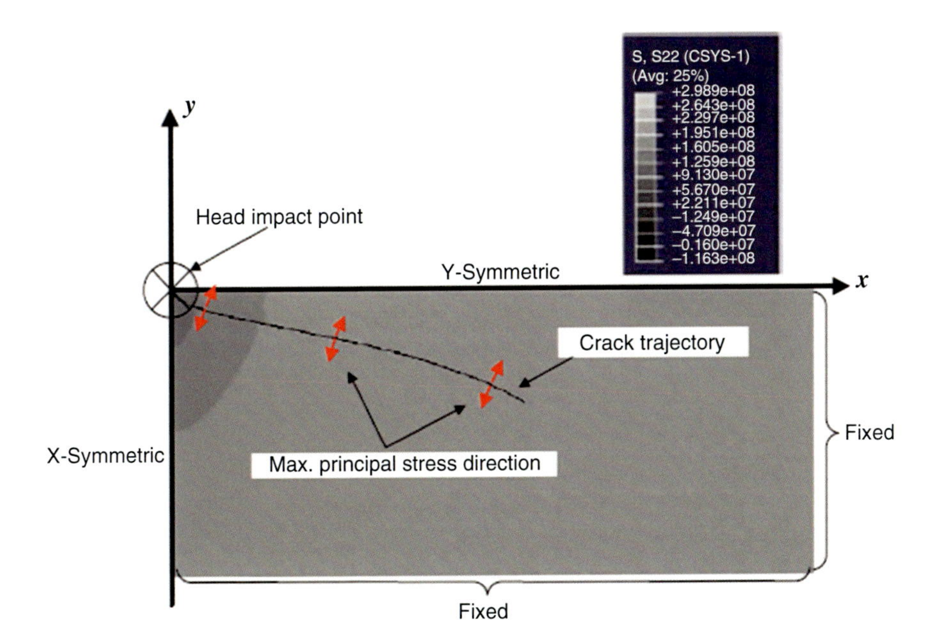

Fig. 16 View of crack trajectory and the hoop stress field after the radial crack is fully developed at $D/t = 0.62$

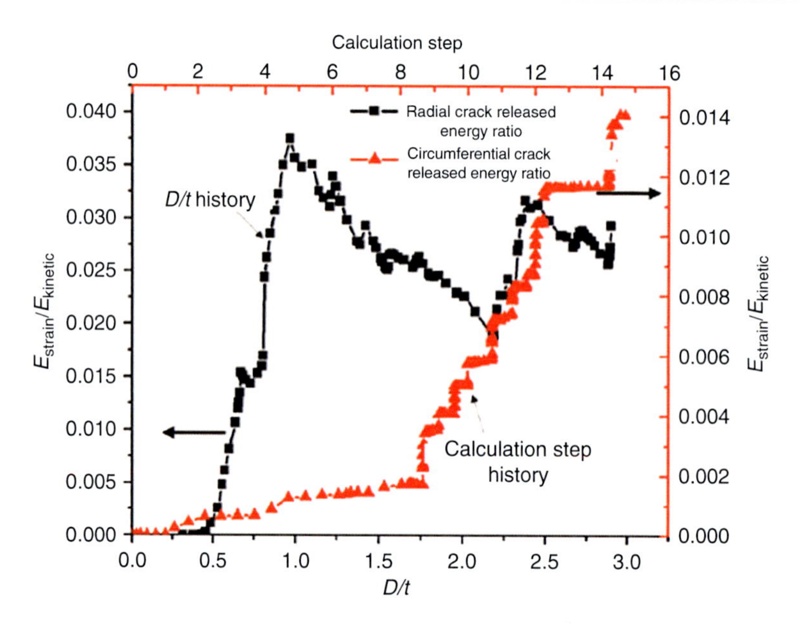

Fig. 17 The normalized released strain energy during the propagation of radial and circumferential cracks. For radial crack, the released strain energy is a function of D/t (or equivalently the impact speed through Eq. 33); it is normalized by the incident kinetic energy with $M = 4.5$ kg. The circumferential crack propagates at $D/t = 2.89$ and it is normalized by the corresponding incident kinetic energy

During its growing process, the crack seeks to release as much energy as possible through mode I fracture. The released strain energy E_{strain} is computed as the difference between the strain energies of two specimens, one cracked and one without crack (at the same D/t). In Fig. 17, E_{strain} is normalized by the initial impact kinetic energy of the head E_{kinetic} (with $M = 4.5$ kg, the impact speed can be deduced from Eq. 33). It is also illustrated that the released strain energy is relatively small compared with the overall system energy, and it increases as D/t enlarges (although when it is normalized by the incident kinetic energy, which also increases as D/t gets larger, the dimensionless number $E_{\text{strain}}/E_{\text{kinetic}}$ fluctuates somewhat).

Sensitivity of Initial Flaw and Fracture Criterion

The damage stresses σ_d of different soda-lime glass specimens are different due to its intrinsic stochastic flaws, usually from 10 to 60 MPa. Under quasi-static indentation, the effect of σ_d on crack evolution is shown in Fig. 18. The results indicate that as the critical stress for crack initiation σ_d becomes larger, the crack angle is smaller (the boundary constraint is more prominent) and the crack is shorter. In other words, if the impact condition is known, one can roughly estimate the damage stress threshold value according to the crack path and length. In what follows, $\sigma_d = 10$ MPa is employed.

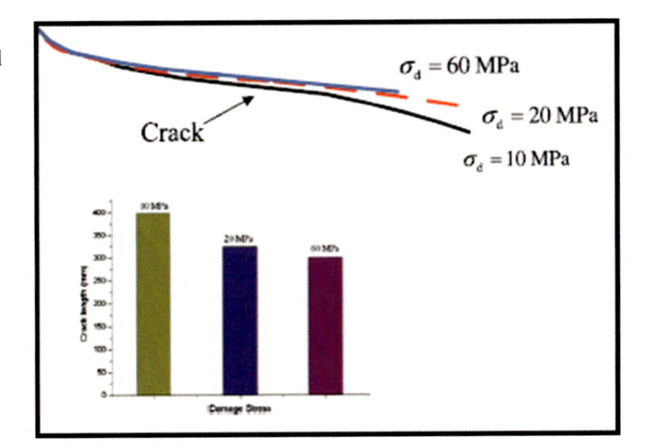

Fig. 18 Effect of different damage stresses on the radial crack trajectory, for$D/t = 0.62$

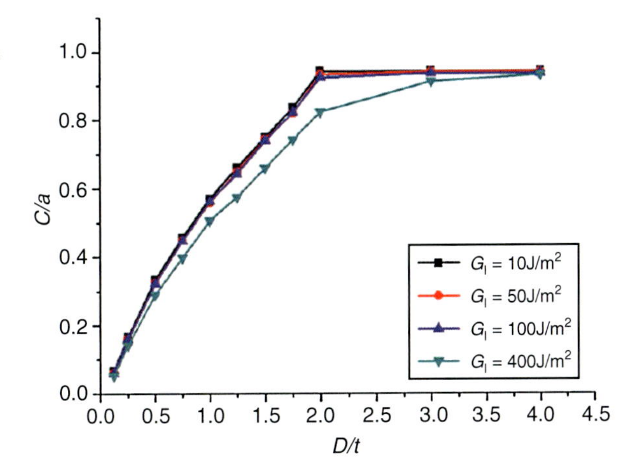

Fig. 19 The effect of model I energy release rate effect on the radial crack length

The model I energy release rate is another very important parameter governing the brittleness of the material. Consider typical material candidates for windshield such as soda-lime glass and PMMA, the former is a brittle material whereas the later has certain degree of ductility, their mode I energy release rates are about $G_I = 10$ J/m^2 and $G_I = 400$ J/m^2, respectively. Without loss of generality, we normalized the crack length C with plate size a. In Fig. 19, the effect on the resulting length of crack is examined by varying G_I from 10 to 400 J/m^2. Numerical study demonstrates that if the G_I value is within the brittle material domain, there is little effect on the crack length; otherwise, if the windshield is made by the more ductile material, the crack length is shorter under the same plate deformation. In what follows, $G_I = 10$ J/m^2 is employed.

The characteristics of initial flaw have some minor effects on crack propagation. We first keep the initial flaw length to be 3 mm and let the flaw angle θ_i to be 30°, 45°, and 60° with respect to the x-axis, and the final crack pattern is shown in Fig. 20. It is observed that θ_i does have some influence on the initial crack trajectory;

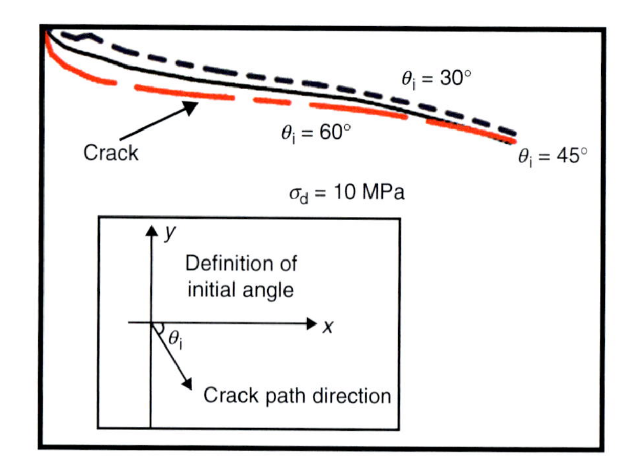

Fig. 20 The effect of the initial flaw angle hi on the radial crack pattern, for$D/t = 0.62$

however, eventually, all crack paths converge with similar lengths, indicating both the robustness of the current model. The length of the initial flaw has negligible effect on both the crack propagation pattern and crack length (as long as the initial flaw is short). When the initial flaw length is taken to be 0.5 mm, 1 mm, 3 mm, and 5 mm, the resulting radial crack lengths in these four scenarios are exactly the same; if the initial flaw is 10 mm, the crack propagation pattern is slightly different. Therefore, the 3 mm initial flaw used in the present study is validated and it is a relatively insensitive parameter in the present XFEM simulation.

Radial Crack Length

The length of the radial crack can be represented as a function of the quasi-indentation depth D/t, which is in turn related to the impact speed v for a given effective head mass (see Eq. 34). Figure 19 also shows the normalized crack lengths, i.e., C/a versus the normalized head displacement D/t. Initially, the crack length is almost proportional to head displacement; however, after a certain length owing to the fixed boundary, the crack length almost remains a constant. The relationship between C/a and D/t is nonlinear and may be fitted as

$$C/a = -0.4046 + 0.9231\,(D/t) - 0.05074(D/t)^3 + 0.03803\exp(D/t) - 0.4243\sqrt{(D/t)}\ln(D/t) \tag{34}$$

By combining Eqs. 33 and 34, a new method can be derived to estimate the impact speed directly based on the knowledge of radial crack length (assuming the effective head mass is known) using Newton–Raphson method, thus useful for traffic accident reconstruction. This is illustrated in the next section.

Equation 34 shows that if the head mass is fixed, the crack length increases nonlinearly with impact speed. In order to estimate the influence of the effective head mass, we can combine Figs. 14 and 19, along with Eqs. 33 and 34. For instance, from Fig. 14 (or Eq. 33), if the impact speed is fixed at 5 m/s, for different effective head masses, the variation of crack length (as a function of the corresponding D/t) can be obtained from either Fig. 19 or Eq. 34. Compared with the crack length of $M = 22.5$ kg, $M = 45$ kg, and $M = 90$ kg its 193.27%, 241.35%, and 294.23% for $M = 4.5$ kg, respectively. That is to say, the effective head mass has a strong influence over the crack length, indicating that during a low-speed impact accident, the shape of vehicle front end and pedestrian–vehicle contact point is critical in impact speed estimation since these factors will largely determine the pedestrian revolution angle. In addition, the influence of M becomes relatively smaller when M is large, implying that the influence of the effective mass starts to decay if the effective head mass is much larger than the windshield mass.

Circumferential Crack Propagation Characteristics

On the basis of the radially cracked geometry, in the second step, we introduce circumferential flaws of length $a_0 = 3$ mm along the radial crack, with a spacing of 50 mm (the spacing is found to be a relatively minor factor on the final circumferential crack pattern). The stress field calculated from the end of first step is transferred into the second step. Since the first step has relaxed most of the hoop stresses, the remaining excessive radial stress needs to be relaxed via the development of circumferential crack, and the most efficient way of releasing the radial stress is to develop the circumferential crack with maximum radius, i.e., on the same order of C. This is verified from Fig. 21: Cracks also grow near the contact zone because of the high stress. This is consistent with the observation that intensive cracks are identified within the contact area while the crack density is lower outside, and the largest circumferential crack almost bounds the radial cracks. The strain energy released by circumferential crack is given in Fig. 17, where the total released energy is less than 50% than that of radial crack. This evidence strongly supports the argument that crack is prone to grow in a radial way rather than a circumferential way, since more energy can be released. From Fig. 17, we can see that less than about 5% of the total kinetic impact energy is consumed by both radial cracks and circumferential cracks and thus there is enough room for improving the automotive windshield for enhanced energy mitigation efficiency and pedestrian protection (for example, using transparent energy absorption materials).

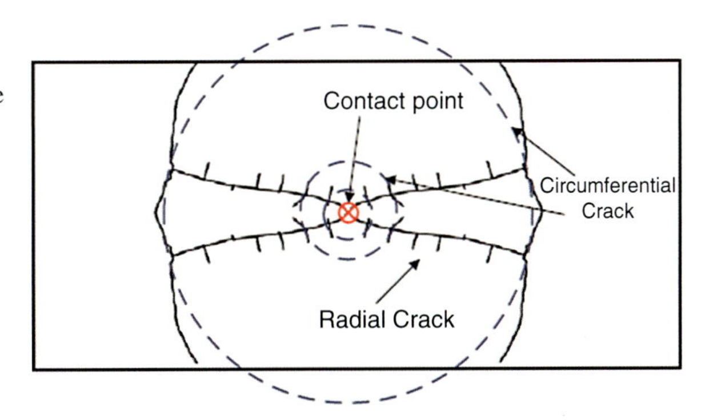

Fig. 21 Final crack pattern (including both radial and circumferential cracks) on the model windshield plate subject to head impact with $D/t = 0.62$

Experiment Analysis

The mechanical behavior of PVB laminated glass is quite complex, especially under dynamic loadings due to the combination of rate-dependent effect of PVB and brittle-polymer composite structure. Therefore, experimental data are very important for the proper suggestion of constitution model and FE model to further investigation mechanical behavior of PVB laminated glass under both static and dynamic loadings.

Typically, one can categorize experiments for studying mechanical response of such a composite material into three regimes in terms of different strain rates: quasi-static, low impact speed, and high impact speed (Stout et al. 1999).

Experiment of Quasi-Static Loading

Because of the difficulties in conducting a quasi-static tension experiment with a slippery surface and the windshield undergoes compression subject to human head impact, we adopted a compression experiment rather than the bending test which the shear effect will influence the mechanical property of the specimen to a large extent.

Experimental Setup

PVB laminated glass circular flake is introduced to be the testing specimen, with the dimension of $u = 10$ mm, where 0.76 mm-thick PVB interlayer is sandwiched by two pieces of 2 mm-thick glass, shown in Fig. 22. Note that the glass, as a typical rate-independent brittle material, has a density of $\rho = 2{,}500$ kg/m^3, Young's modulus of $E = 70$ GPa, and Poisson's ratio of $m = 0.22$ (Forquin and Hild 2010). It is worthy to point out that PVB laminated glass is made by putting two glass sheets into an autoclave and compressed with 0.76 mm-thick of PVB interlayer under specific cycle of pressure (10 bar) and heat (120 °C). Thus, we can regard there is no slide between three layers (Wang et al. 2007). To make the experiment more engineering applicable, the thickness of specimen used here has the same thickness as that in most passenger cars. In experiment, specimen is put in a cylinder container which has a steel pin with a diameter $u = 38$ mm inside for compression and the experiments are carried out on Material Testing System (MTS). The load is applied perpendicularly through the pin to the sample. A clip-gage type extensometer is installed between the gaps to measure the displacement of the pin as a function of load during the experiment, shown in Fig. 23. 1 mm/min, 0.1 mm/min, and 0.01 mm/min are three loading rates, corresponding to the strain rate on the upper and lower sides of the specimen of $\dot{\varepsilon} \approx 1 \times 10^{-3} s^{-1}$, $\dot{\varepsilon} \approx 1 \times 10^{-4} s^{-1}$, $\dot{\varepsilon} \approx 1 \times 10^{-5} s^{-1}$ chosen to measure the quasi-static mechanical property of PVB laminated glass. During tests, unloading threshold value is set to be 80% of the maximum loading value and thus complete load–displacement history is recorded. The temperature and relative humidity during the experiments remain in the range of 25–26 °C and 40–45%, respectively.

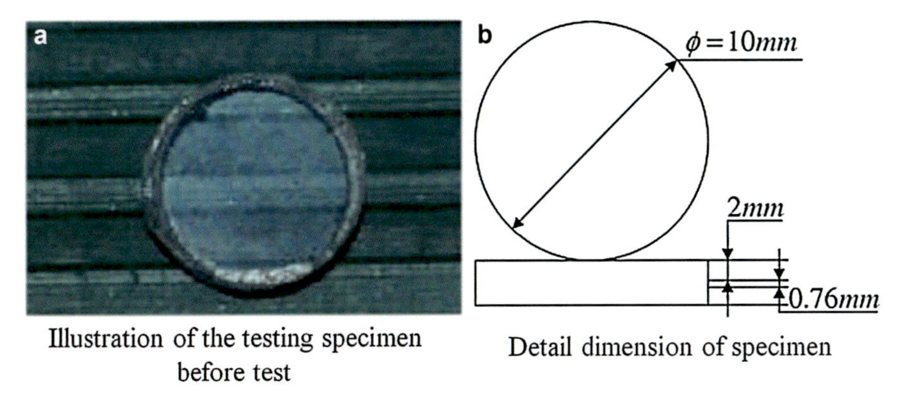

Fig. 22 Illustrations and dimensions of specimen used in both quasi-static and dynamic impact experiments. (**a**) Illustration of the testing specimen before test. (**b**) Detail dimension of specimen

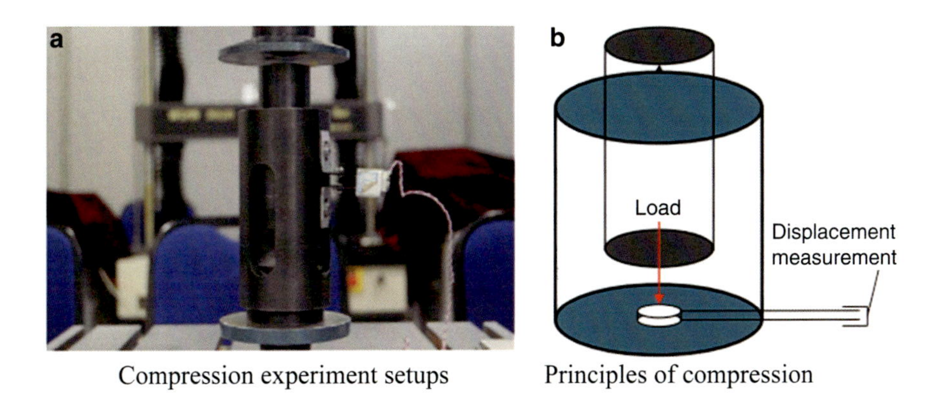

Fig. 23 Quasi-static compression experimental setup on MTS. (**a**) Compression experiment setups. (**b**) Principles of compression

Experiment Results

It is commonly accepted that the failure process of composite subject to quasi-static compression loading usually involves a sequential damage accumulation process (Xu et al. 2009). As aforementioned, we set the unloading threshold value to be 80% of the maximum load value due to the major load decrease occurred during the experiment when major crack could be observed clearly. Thus, we define the stress and strain at this moment as major failure on-set (MFO) stress and strain. Figure 24 shows us the "load decrease" phenomenon and the morphology of the cracked samples.

Figure 25 illustrates three different stress–strain relation curves under three loading rates. The stress–strain curves show "nonlinearity" characteristic in mechanical behavior of PVB laminated glass and this "nonlinearity" phenomenon is probably caused by both nonlinear mechanical response of PVB interlayer and progressive

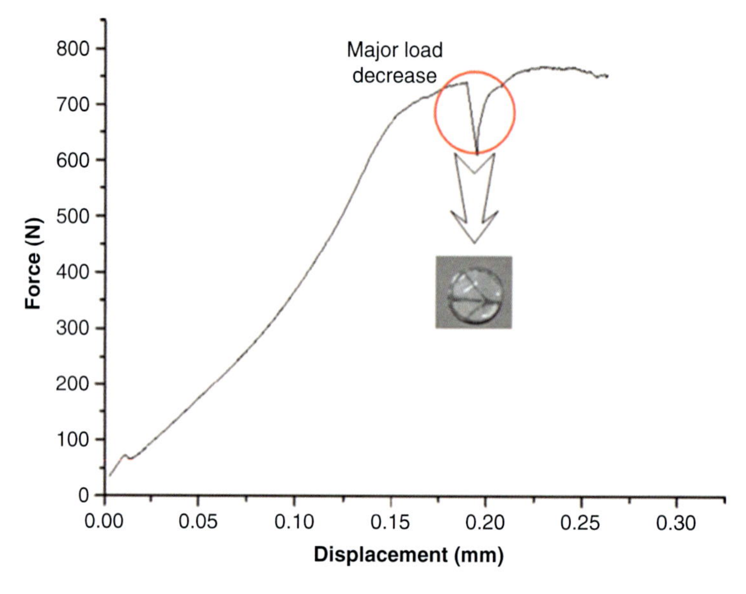

Fig. 24 Major load decrease occurrence with the onset of major crack

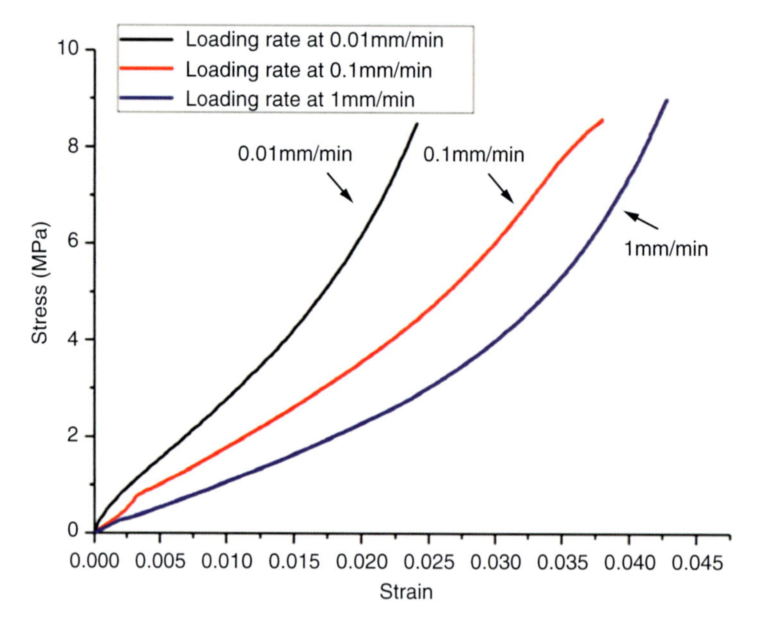

Fig. 25 Stress–strain curves obtained from quasi-static compression experiments under three different loading rates, i.e., 1 mm/min, 0.1 mm/min, and 0.01 mm/min

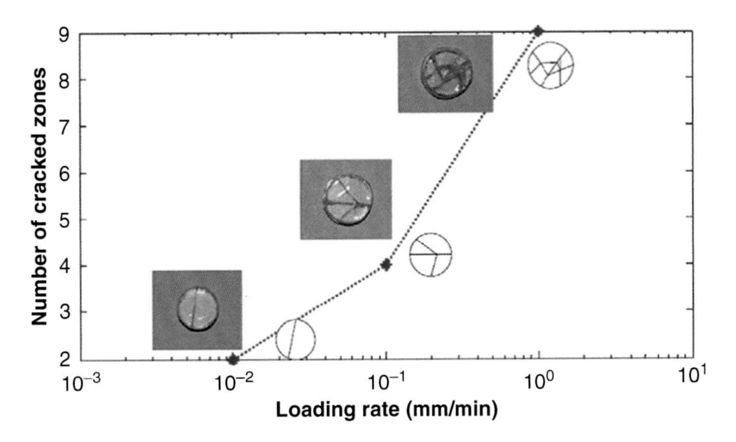

Fig. 26 Number of cracked branching at different loading rates

microcrack growth in sample. As the loading rate increasing, the MFO strain increases while the MFO stress remains nearly the same. The major responsible reasons are: (i) in extremely low strain rate (quasi-static) situation, the outer glass panel plays a critical role in mechanical response and (ii) glass is a rate-independent material whereas PVB is a rate-dependent one.

Number of Crack Branching

The numbers of cracks in testing samples according to different quasi-static loading rates are illustrated in Fig. 26. The number of complete cracks (or major cracked branching) increases with the loading rate. As aforementioned, in quasi-static situation, the stress remains the same whereas the strain rises with the increasing loading rate. Therefore, the number of cracked branching, namely the "damage degree," depends on the value of strain, not the stress, in quasi-static circumstance that is in accordance with Wang's (Wang et al. 2006) study. Further, if a nonlinear fitting is tried on the number of crack branching versus loading rate, we will harvest the following explicit formula with fitting coefficient $R = 0.999$.

$$N = \left[8.975W^{0.339}\right] \tag{35}$$

where $[X]$ is Round function, e.g. $[5.2] = 5$. N is the number of crack branching and W is loading rate with unit of mm/min. From Eq. 35, we can infer the numbers of crack zones under different loading rates in the quasi-static domain.

Mandelbrot (1977) pointed out that crack branching contained critical information of crack propagation speed and complexity. Bouchaud et al. (2012) further investigated the relationship between crack length and crack branching mode and concluded

$$D_f = \frac{\lg b}{\lg k} \tag{36}$$

where $D_f = \frac{\lg b}{\lg k}$ is degree of branching, b is the number of basic branching pattern, and $K \approx 0.4$ in brittle material.

Assuming that the initial crack branching length is a_0 and the length becomes $a_1 = a_0 + \mathrm{k}a_0$ (Bouchaud et al. 2012), thus after the initial crack branches for i times, the length finally comes to

$$a_i = \left(1 + \cdots + K^i\right) a_0 \tag{37}$$

Since the number of branching can be written as N ¼ biþ1, we are able to know that

$$i = \log_b^N - 1 \tag{38}$$

According to Freund's (Michel and Freund 1990) theory, the dynamic crack propagation velocity in a finite body obeys the following relation:

$$v = C_r \left(1 - \frac{a_0}{a_i}\right) \tag{39}$$

where C_r is the Rayleigh wave velocity and v is the crack propagation velocity.

By combination of Eqs. 36, 37, 38, and 39, the relation between crack velocity and crack branching number is

$$v = C_r \frac{1 + b^{-1/D_f} - 2N^{-1/D_f}}{2\left(1 - N^{-1/D_f}\right)} \tag{40}$$

Further, we substitute Eq. 35 into Eq. 40 and get

$$v = C_r \frac{1 + b^{-1/D_f} - 2\left[8.975W^{0.339}\right]^{-1/D_f}}{2\left(1 - \left[8.975W^{0.339}\right]^{-1/D_f}\right)} \tag{41}$$

Equation 40 may help us to clarify the crack propagation speeds in different loading rates. We may also conclude that crack propagation speeds are different during the three loading rates in quasi-static experiments based on Eq. 41.

Experiments of Dynamic Out-of-Plane Loading

By starting from a fundamental point, quasi-static mechanical behavior of PVB laminated glass has been widely investigated experimentally above (Xu et al. 2011a). Unlike quasi-static cracking problem, dynamic fracture is more complex and challenging but there are few reported investigations that focused on the dynamic fracture of PVB laminated glass except for some numerical simulations those based on probabilistic damage mathematical model (Forquin and Hild 2010), continuum damage mechanics (Zhao et al. 2005, 2006a, b; Sun et al. 2005), explicit finite element method (Bois et al. 2003; Timmel et al. 2007; Sun et al. 2009), and extended finite element method (Xu et al. 2010) that often neglected the important PVB layer and most of them did not involve explicit crack growth and pattern analysis. The dynamic experimental investigation is pressing need.

System Setup of Dynamic Out-of-Plane Loading

The impact system setup is sketched in Fig. 27, where a weight block can slide freely along the two standing poles. At the end of the drop-weight, there is a tip which is used to hit the force direction converter on the testing sample and it can be referred in Fig. 28. It is a necessity to convert the vertical impact direction into a horizontal one, which would facilitate the capture of the film.

Upon impact high speed photography is employed for capturing the dynamic crack growth in the plate in situ. The entire system consists of a high-voltage charging controller, multispark box, two field lenses, and an array camera with films (see in Fig. 28). In addition, a multispark high-speed camera is adopted to provide enough spot light source. The multispark box provides 16 independently spot lights triggered by a high-voltage charging controller by a preset adjustable interval time from 1 to 9,999 μs whose charging voltage is able to reach up to 30 kV. Meanwhile, a 4×4 array camera with 16 films is set correspondingly on the other side to receive the light and therefore record the crack image information. In order to make sure that the images recorded on the films are in the time series of the ignition of the spot lights, two field lenses are used to make up the optical path (see Fig. 28b). Moreover, the delay controller is introduced to control the time interval between each spot light. In addition, to minute the exact ignition time for each spot light, the electric-light receiver along with the oscillograph monitor are used together where the electric-light receiver is employed to collect the electric-light signals and send them to the oscillograph monitor. Reference control experiments of impact-induced fracture are carried out using pure glass sheets, with results in good agreement with the literature (Nielsen et al. 2009; Sharon and Fineberg 1999) (see in Fig. 29), thus validating the current test system.

The Preliminarily Experiment to Investigate the Radial Crack Propagation

Experiment Condition

In the preliminarily impact fracture experiment to investigate the radial crack propagation behavior of PVB laminated glass, the minimum mass of the drop-weight is 0.5 kg and the maximum height of the drop-weight is 1.84 m (corresponds to the maximum impact velocity of 6 m/s); both the mass and height of the drop-weight are also varied in this experimental investigation. Prior to impact, the drop-weight is suspended and fixed through electromagnet. The temperature and relative humidity during the experiments remain in the range of 20–21 and 20–25% respectively.

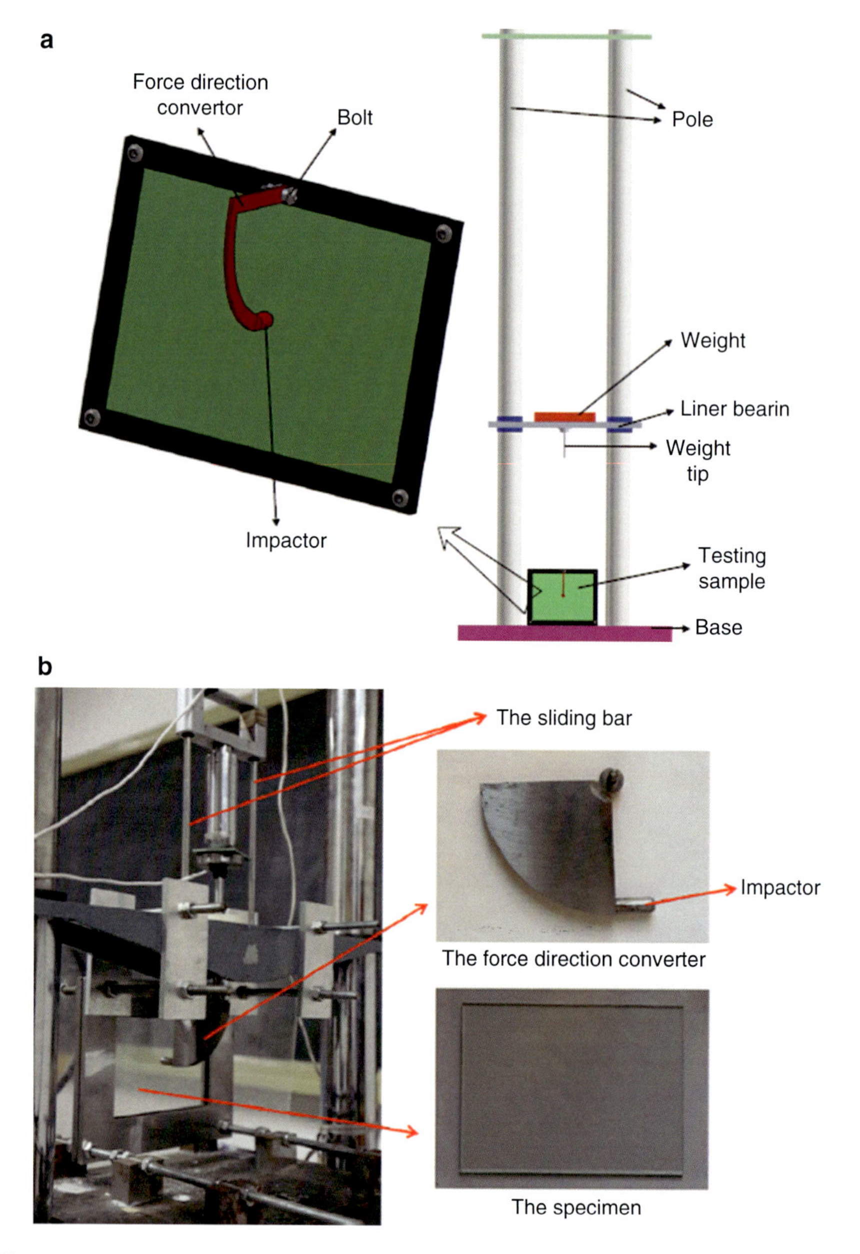

Fig. 27 (**a**, **b**) Schematic of the drop-weight tower experiment setup. The testing sample is illustrated along with the force direction convertor

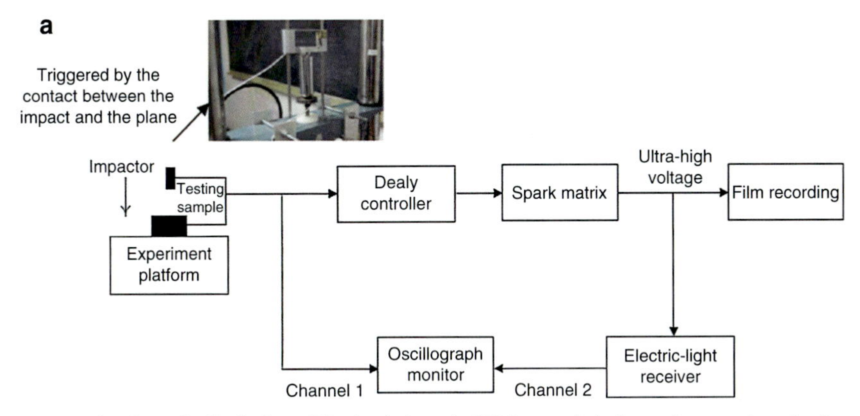

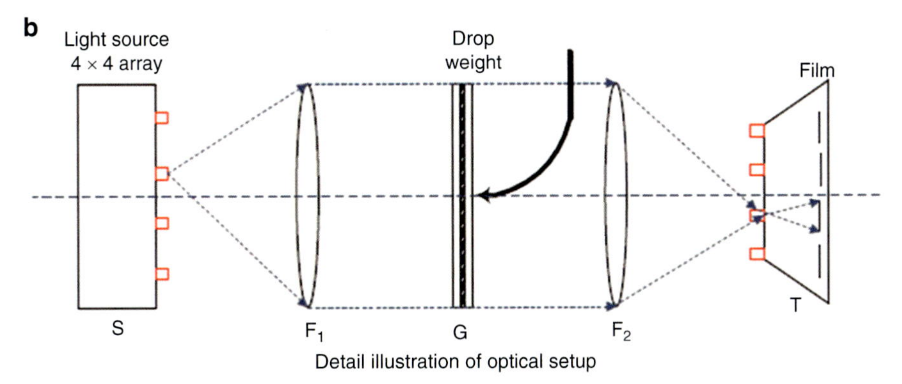

Fig. 28 The schematic illustration of the entire kit of experiment setup. (**a**) A schematic illustration of the basic layout of high-speed photography experiment setup. (**b**) Detail illustration of optical setup

Specimen Preparation

The standard PVB layer has a Young's modulus $E_{\mathrm{PVB}} = 100$ MPa, Poisson's ratio $\upsilon_{\mathrm{PVB}} = 0.48$, and mass density $\rho_{\mathrm{PVB}} = 870$ kg/m^3 (Xu and Li 2009). To make close connection to the automotive industry, the PVB thickness is $t_{\mathrm{PVB}} = 0.76$ mm for passenger cars. In addition to the middle PVB layer, for the upper and lower soda-lime glass sheets in laminated glass, the thickness $t_{\mathrm{glass}} = 2$ mm, with $\rho_{\mathrm{glass}} = 2{,}500$ kg/m^3, $E_{\mathrm{glass}} = 70$ GPa, and $\upsilon_{\mathrm{glass}} = 0.22$. They are bonded together at 10 bar and 120 °C following standard procedures, and the effective in-plane dimension is 200 × 150 mm.

Experimental Results and Discussion

Crack Morphology

A series of experiments under different loading conditions, i.e., with different height and mass of the drop-weight are recorded. Figure 30 shows the typical dynamic

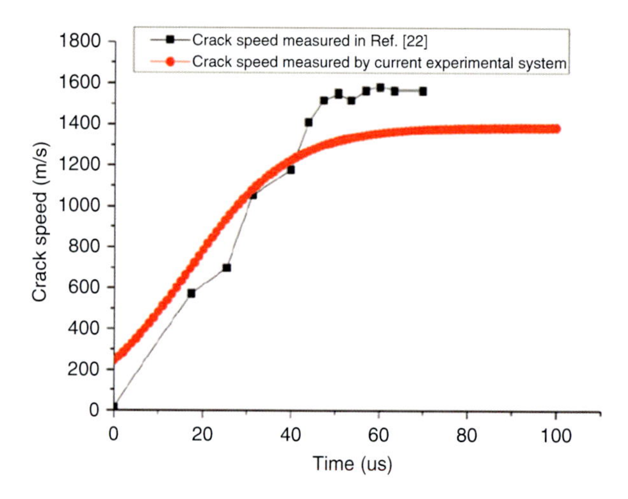

Fig. 29 Comparison of the crack velocity propagation speed of pure glass between experiments results extracted from the current one and the one in Nielsen et al. (2009). Note that the external impact energy is 1 J in the current experiment while it is 0.2 J in the experiment in Nielsen et al. (2009)

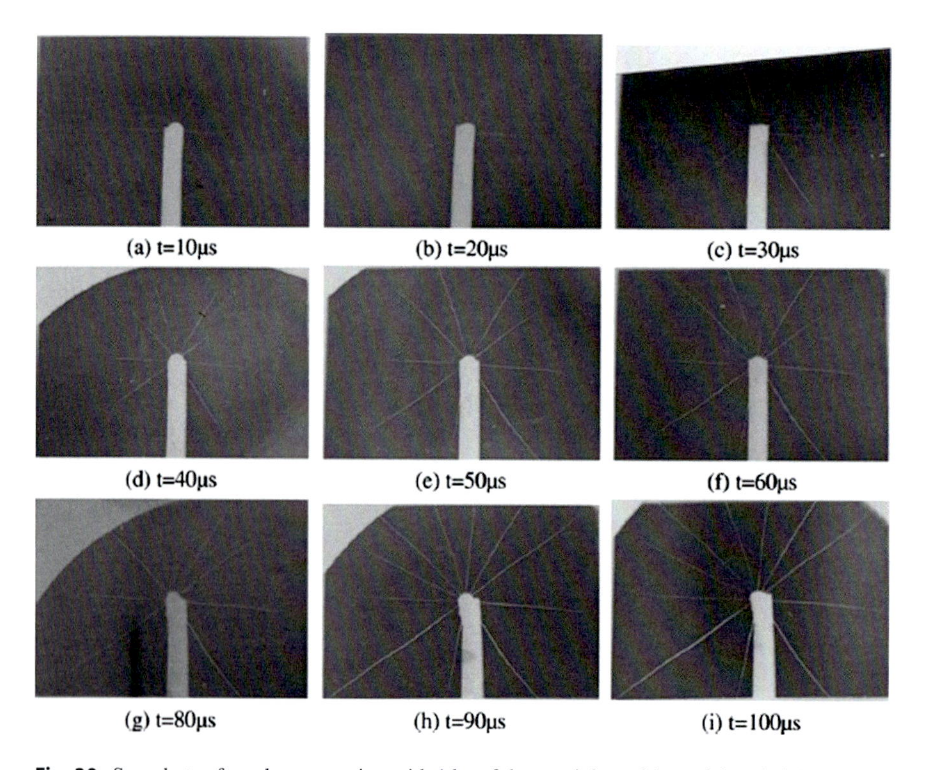

Fig. 30 Snapshots of crack propagation with 1 kg of drop weight and 1 m of drop height

crack propagation snapshots of the PVB laminated glass sheet, in response to a 1 kg weight dropped from 1 m height. Extensive radial fracture is observed, with no appearance of circumferential crack, which is consistent with numerical simulation that when a foreign object impacts a large windshield, the radial crack always appears before the circumferential crack (Xu et al. 2010). It is seen that there are about ten radial cracks on the composite sheet, and the development and evolution of each crack are quite different due to the inevitable stochastic flaw in glass. Here, the average crack length is employed to discuss the behavior of radial crack propagation. By using nonlinear fitting method to the experimental points obtained from the experiments, we finally got the crack tip position in time history. Then the averaged time evolution history of the radial crack tip velocity and acceleration can be obtained from the time history of the radial position of the crack tip as showed in Fig. 31.

The measured crack velocity first increases and then fluctuates slightly, which goes accordance with the experiment results of cracking speeds of pure soda-lime glass at various cracking modes in Nielsen et al. (2009) and Sharon and Fineberg (1999), as well as being consistent with previous numerical simulation results of several brittle materials (Song et al. 2008; Grégoire et al. 2009) and composites (Daphalapurkar et al. 2007; Dwivedi and Espinosa 2003).

As proposed by Ravichandar and Knauss (1984), near the vicinity of crack tip, there is a microcrack zone which impedes the further propagation of crack. Within the impact area, the spacing between the initial cracks is so small and such an intensively damaged zone may reduce the cracking speed substantially. With the

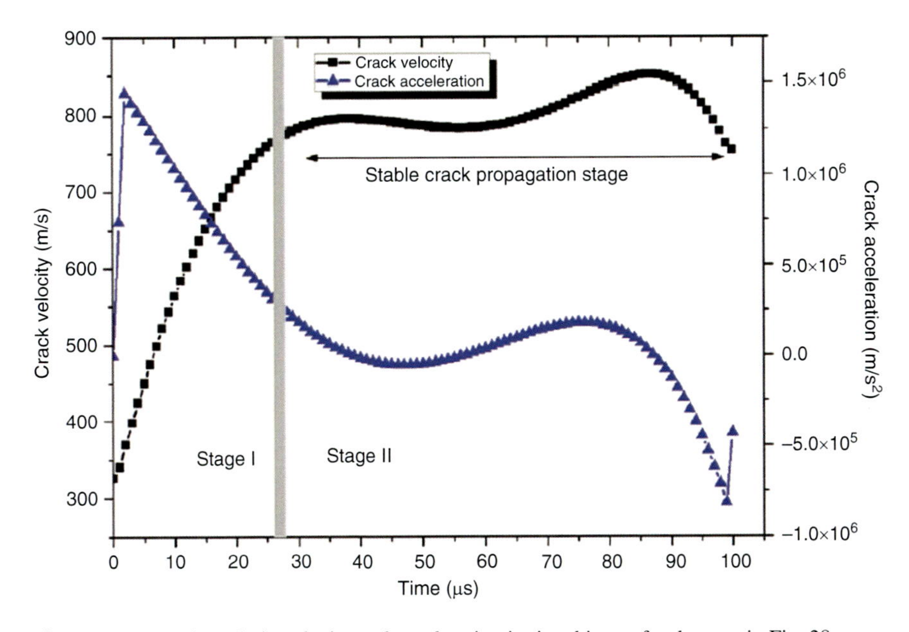

Fig. 31 Averaged crack tip velocity and acceleration in time history for the case in Fig. 28

advancement of cracks, the retardment effect will be greatly lessened until it reaches an almost constant influence, when the crack speed becomes steadier in the later stage. Thus, the crack evolution may be qualitatively divided into two stages: in stage I, the cracking speed keeps rising, and in stage II the crack speed becomes more stable but with some fluctuations. The transition occurs roughly at about time (further discussion will be made later). During stage I, the crack propagation acceleration is decreasing until stage II begins; that is, as the stress in the crack tip accumulates and develops, finally the crack tip will be able to break through the energy barrier surrounding material to reach the equilibrium.

In this example, the averaged steady-state cracking speed during stage II is about $\overline{v}_{PVB} \approx 811\ m/s$. Comparing with the stable cracking speed of Polymethyl Methacrylate (PMMA) $\overline{v}_{PMMA} \approx 600\ m/s$ (Sharon and Fineberg 1999) and tempered soda-lime glass $\overline{v}_{GLASS} \approx 1470\ m/s$ (Nielsen et al. 2009) (One may refer to Fig. 32 for detail comparison. Note that although the impact energy is a little bit different in (Xu and Li 2009) and this experiment, the comparison is still valid), the PVB laminated glass may be regarded a composite material whose overall effective "fracture toughness" is between that of PMMA and pure glass. In Fig. 32, we can also see that the general variation trend of crack speeds in two different materials are almost the same. One may use the Rayleigh wave speed v_R of the glass sheet to normalize the crack speed (since cracking occurs in the glass sheet), i.e., $v_R = 3{,}098$ m/s. Thus, the stable cracking speed of PVB laminated glass is about 0.26 v_R in this particular case, which is lower than the typically measured

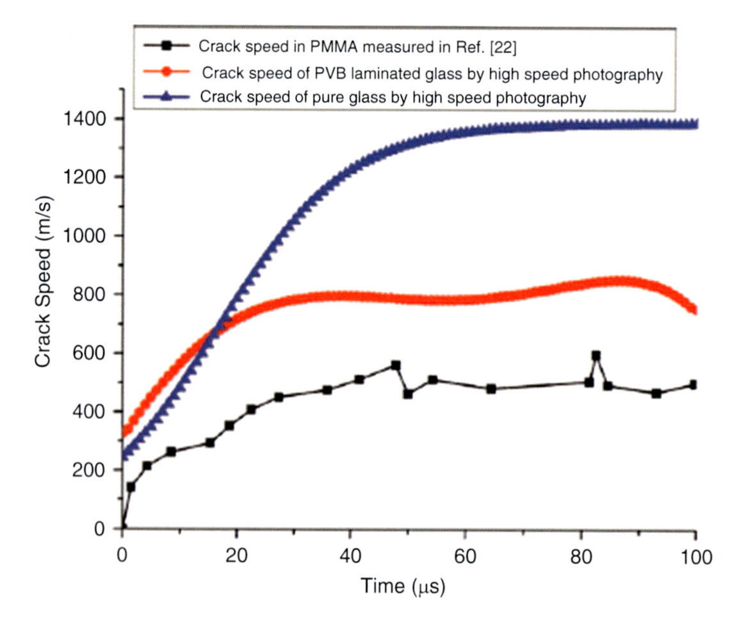

Fig. 32 Comparison of the crack velocity propagation speed among pure glass, PVB laminated glass, and PMMA. Note that data for PMMA is extracted from

maximum speed of about 0.4–0.5 v_R in brittle materials and we can clearly see that the PVB interlayer serves to slow down the crack speed, which would also make the material as well as the brittle-polymer sandwiched structure a candidate for energy absorption/shielding.

Different Driving Mechanisms of In-Plane Cracking on Two Glass Layer

Figure 33 shows a typical crack propagation comparison on supported and loaded glass layers at the loading speed with PVB thickness $h = 0.76$ m. Here, time $t = 0$ μs is used to denote crack tip initiation for the radial cracks on supported glass layer. It is clear that the radial cracks (Fig. 33a–d) initiated firstly on the "supported glass layer," then the radial cracks on the "loaded glass layer" generated after about $t = 600$ μs as shown in Fig. 33f. On the other hand, it is also interesting to observe that the cracks on the two glass layers are completely overlapped even though they propagate at totally different times.

In fact, the stress wave dominated inertial effects during impact loading on the PVB laminated glass play an important role in crack propagation (Singh and Parameswaran 2003; Clements et al. 1996). Due to the high modulus of glass material and the fixed mode of the specimen, the plate is small enough to ignore the bending deformation during the impact. For the convenience of analysis, a schematic diagram in a layer system consisting of two glass plates bonded by a PVB interlayer is shown in Fig. 34. Impact occurs on the surface of the "loaded glass layer" of the PVB laminates, the compressive waves caused by the impact travel

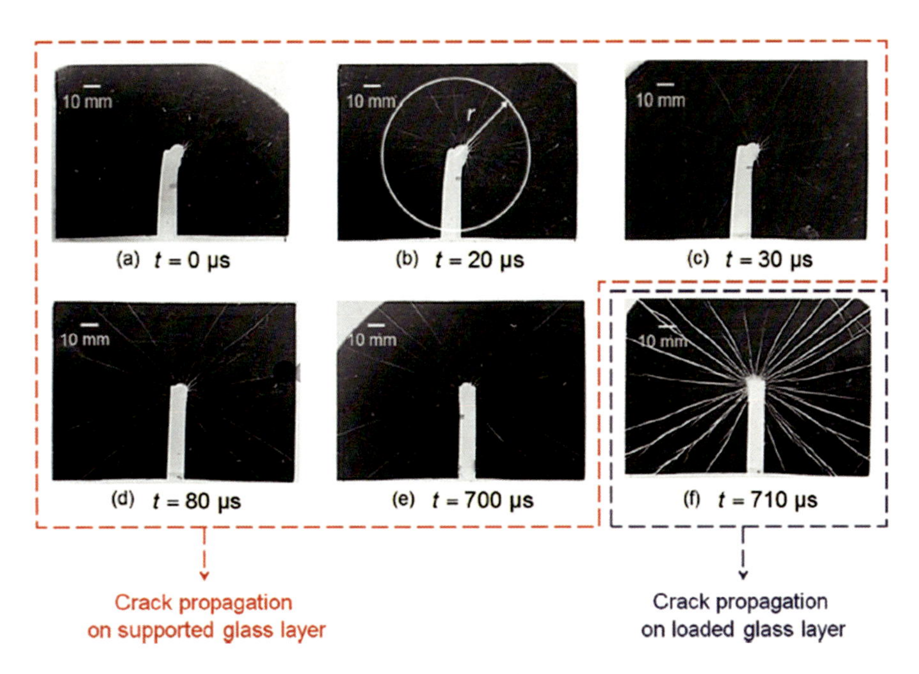

Fig. 33 Selected sequence of images depicting the radial crack growth on both glass layers

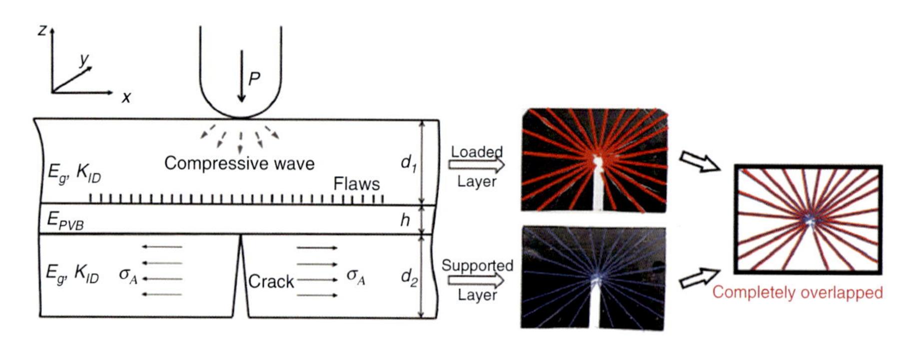

Fig. 34 Schematic of the impact process in a layer system consisting of two glass plates bonded by a PVB interlayer. The cracks propagate into the loaded glass layer by reinitiation from a surface flaw

down the depth direction (Fig. 34). Due to its viscoelasticity, the PVB layer plays an important role in reducing the amplitude and the speed of waves that eventually pass through by internal friction (Thom 2005), compared with the glass layer. Finally, the compressive waves reflect from the boundaries as tensile waves, which continuously arrive at the supported plate thus the tensile dominated cracks initiate while reaching the limit of dynamic fracture toughness KID of the material (Lambros and Rosakis 1997). Considering that glass plate is more sustainable in compressive stress wave but prone to fail in tensile stress wave, the radial cracks initiate first on supported glass layer.

Effect of Drop Weight and Height

Next, we investigate the influence of drop weight and height on the crack propagation. The drop mass ranges from 0.5 to 2 kg, and the height varies from 0.6 to 1.4 m. Table 1 shows the detail combination of the parametric experiments. Figures 7 and 8 show that with the increase of drop weight and height, the basic trend of crack would remain almost the same while the value would increase, i.e., larger mass and higher impact speed will cause faster crack propagation. Zhang et al. (2010) investigated the effect of loading rate on crack velocities in concrete and found with the increase of loading rate, the crack velocity increase proportionally; in quasi-static indentation, it was also found that with larger indentation load, the radial and lateral cracks in ceramic also became longer. All these conclusions are qualitatively consistent with our experimental findings. Particularly, $\overline{v}_{PVB}$ increases 17.70% and 12.21% when the impact speed changes from 3.43 to 4.43 m/s and 5.24 m/s, respectively, and $\overline{v}_{PVB}$ increases 26.72% and 14.92% when the drop weight is increased from 0.5 to 1 kg and 2 kg, respectively. Furthermore, if we combine the effect of impact mass and velocity together into the term of impact energy E, and normalize it by a reference potential energy $E_n = 9.81$ J, then the normalized steady-state cracking speed, i.e., $\overline{v}_{PVB}/v_R$ can be fitted as

$$\overline{v}_{PVB}/v_R = 0.167 \pm 0.137/\left(1 \pm 10\exp\left(1.50\left(0.739 - E/E_0\right)\right)\right) \quad (42)$$

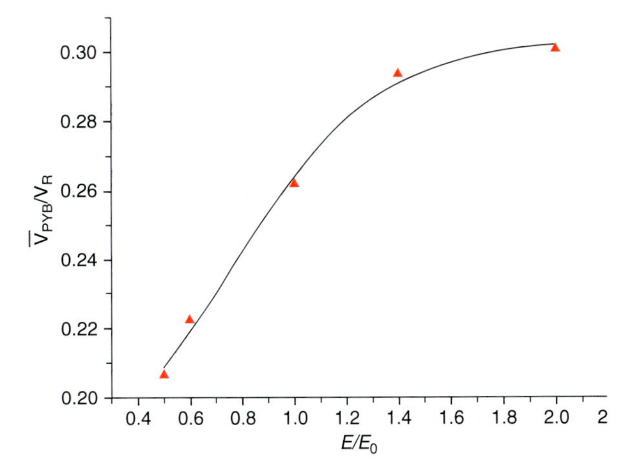

Fig. 35 The relationship between the steady-state crack propagation velocity and the external impact energy

for the range of parameters in this study. From Fig. 35, it can also be found that the stable crack velocity will not likely to increase indefinitely as the external impact energy gets higher, and it has a theoretical limit, i.e., Rayleigh wave speed v_R, but not reaching it.

Regarding the transition time between stages I and II, the general trend is that the smaller the external energy, the faster the transition time (although the difference is relatively small and still on the same order of magnitude). That is to say, if the impact energy is higher, it is easier for the crack speed to achieve steady-state. The detailed physical mechanism of the variation of such a transition time will be explored in future.

The Further Experiment Investigating the Radial and Circular Crack

Experiment Condition

In the further experiment to investigate the radial and circular crack propagation of PVB laminated glass, a weight block with mass of 2 kg is first evaluated to a certain height and then released to create impact loading. The height of the drop-weight adopted in our work varies from 300 to 900 mm. An impactor is fixed at the tip of the force direction converter to control the contact shape and area. Different shapes may be designed for the impactor, and in this study, the impactor shape is confined to be cylindrical and the top is hemispherical. The diameter of the cylindrical is 10 mm and the total length of the impactor is 25 mm. The mass of the impactor (including the force direction converter) is 0.545 kg. In addition, a force sensor with sufficient accuracy and frequency is attached right below the drop-weight to record force–time history and the signals recorded by the force sensor are sent to the oscillograph monitor to record force–time history.

Specimen Preparation

The plate is consisted of a PVB interlayer sandwiched by two brittle glass sheets (For PVB interlayer: Young's modulus $E_p = 0.1$ GPa, short time shear modulus $G_0 = 0.33$ GPa, long time shear modulus $G_1 = 0.69$ MPa, bulk modulus $K = 20$ GPa, Poison's ratio $v_p = 0.49$, density $\rho_p = 1{,}100$ kg/m^3, and decay factor $\beta = 12.6$ s^{-1}; For glass: Young's modulus, Poison's ratio and thickness of the glass. $E_g = 70$ GPa, Poison's ratio $v_g = 0.22$ and thickness of the glass is 2 mm. The thickness of the PVB interlayer varies from 0.38 to 3.04 mm during our parametric experimental study (where thickness of 0.76 mm is for common passenger cars). The specimen is prepared using the same manufacturing process (with compression at 10 bar and 120 °C) as that used in automotive windshield. Glass is a typical linear elastic brittle material where its facture strain is about 0.1% while the PVB is a rubber-like material which may sustain a much larger deformation during tension. The unique design enables the possible protection against impact with small amount of energy and dissipates larger impact energy through the large deformation of interlayer.

The specimen is clamped within two metal cover sheets with thin layers of rubber pad inside to avoid possible scratches and stress concentration on the sample surface and distribute the boundary force more uniformly. The tightening torque for the screw bolt is set to be 4 Nm which properly mimics the clamping boundary condition. Six screws penetrating the metal cover sheets are fixed to make sure the metal cover sheets are placed in right horizontal position and reduce the rebound under impact. A force direction conversion part (indicated in Fig. 27) is employed to transfer the vertical impact force into a horizontal one, providing an out-of-plane dynamic loading to the sample.

Experimental Results and Discussion

Crack Morphology

Two kinds of macroscopic crack patterns, radial crack and circular crack, are observed and all the pieces of glass plates after impact are stay connected to the PVB layer. Hoop stress is larger than radial stress such that radial crack appears before circular crack, therefore, it gives us an opportunity to capture and study the evolution history of radial and circular cracks separately. Figure 36 is typical crack propagation process which contains a series of selected sequence of images depicting the radial and circular crack growth in PVB laminated glass specimen respectively. Time $t = 0$ μs is used to denote crack tip initiation for both radial and circular cracks. The crack propagation velocity v is determined as the crack length increase Δl over the time increase Δt. For radial crack, at time t_i, multiple major crack tips conform approximately a circle with radius r_i on average, shown in Fig. 36a. Therefore, radial crack propagation velocity $v_r = \Delta l/\Delta t = (r_{i+1} - r_i)/(t_{i+1} - t_i)$. For circular crack, the total length of n main cracks at t_i is added and divided by the crack number as $\overline{l}_i$ and such that circular crack propagation velocity $v_c = \Delta l/\Delta t = \left(\overline{l}_{i+1} - \overline{l}_i\right) / (t_{i+1} - t_i)$. In addition,

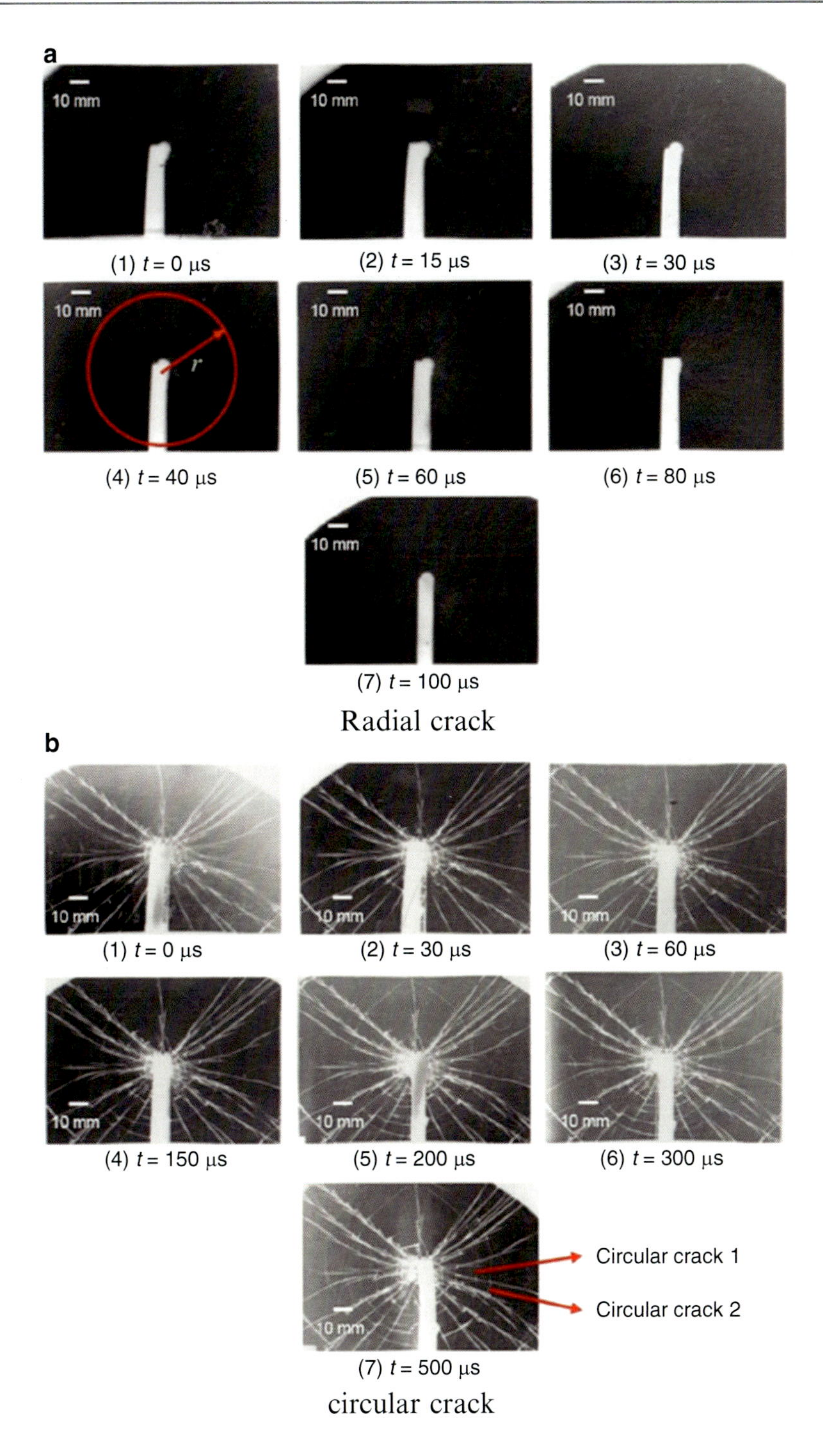

Fig. 36 Selected sequence of images depicting the crack growth at the impact velocity of 3.13 m/s, with PVB thickness of 0.76 mm. (**a**) Radial crack. (**b**) Circular crack

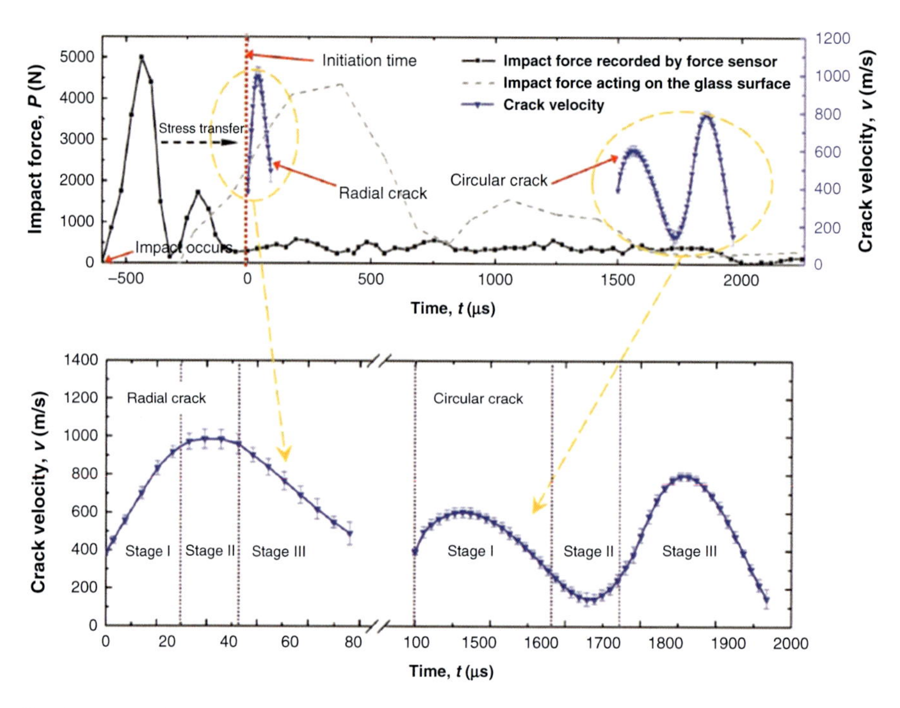

Fig. 37 Load–time curve and crack growth velocity history of the radial crack and the circular crack, in response to 500 mm drop height (impact velocity:3.13 m/s) and 0.76 mm PVB thickness

the B-spline interpolation method is used to smooth the experimental points in this study.

Crack Propagation

To enhance the credibility of the experimental results, the dynamic strain indicator is introduced. Locations near the force sensor and the top of the impactor are fitted with strain gages respectively so as to monitor in situ stains (and stresses) transmit during the impact process. Thus, we could estimate the delay of the impact force from the force sensor to the glass surface and depict the accurate force–time curve acting directly on the glass surface (see dashed line in Fig. 37). Figure 37 shows a typical crack propagation velocity time history curves for radial and circular crack propagation velocities respectively in response to 2 kg weight from 500 mm drop height (impact velocity: 3.13 m/s) to a specimen with 0.76 mm PVB interlayer. Time $t = 0$ μs is employed to denote crack tip initiation for radial crack. Therefore, the negative time correspond to preinitiation conditions (see Fig. 37). Note that the crack only exists in the glass sheets such that the stress component normal to the crack front σ_{11} is the major driving force for crack initiation and propagation. The dynamic crack propagation velocity v may be expressed as

$$v^2 = \frac{\sigma_{11} + \sum_{11}}{\rho_g \left(1 + \varepsilon_{11}\right)} \tag{43}$$

where the contact stress at the crack tip which can be calculated based on existing dynamic contact model by attaching a strain gage near the impactor. $\sum_{11}$ is the strain component. By supposing the proportionality coefficient D between σ_{11} within the framework of linear elasticity (Zhang et al. 2010), the above equation may be rewritten as

$$v^2 = \frac{1 + D}{\rho_g \left(1/\sigma_{11} + 1/E_g\right)} \tag{44}$$

In this impact loading condition, the maximum crack propagation velocity is 990 m/s and 790 m/s for radial and circular cracks, respectively. As Xu et al. (2010) indicated that σ_{11} is larger in hoop stress than circumferential stress so radial crack propagates faster than circular crack. The radial crack propagation velocity is lower than that in pure soda-lime glass (Nielsen et al. 2009) where no obvious circular crack occurs. The PVB interlayer acts as an energy-absorbing material thus to reduce the crack propagation velocity. Meanwhile, PVB interlayer provides a connection media for cracked glass panel such that circular crack could continue to dissipate impact energy after the full growth of radial cracking.

Impact occurs on the outer surface of the glass plate at around −250 μs (see Fig. 37). A compressive loading wave front travels down at around 5,300 m/s (Zhao et al. 2006b) in the glass layer. The PVB layer presents the dual advantages of delaying the passage of the elastic wave into the backing glass layer and reducing the amplitude and the wavelength of wave that eventually passes through by internal friction caused by its viscoelasticity. Both of these effects are beneficial in utilizing the energy absorbing capabilities of the PVB interlayer to the fullest. The compressive waves arrive at the free surface of the backing layer and reflect as tensile waves. Thus, the delay of the crack initiation probably results from the transmitting and reflection of the stress wave and the increasing process of the stress intensity factor caused by increasing elastic energy stored near the crack tip. Similar phenomena have also been observed in Lambros and Rosakis (1997). More and more tensile waves continually arrive at the backing plate and thus the radial cracks initiate while reaching the limit of dynamic crack initiation toughness K_{IC}^d. For the rate-sensitive material, the PVB interlayer is rate dependent while the fracture occurs in the glass plate which is rate independent. Considering the fact that cracks grow in the glass layer, therefore, the impact velocity has little effect on the K_{IC}^d value, which means the effect of rate sensitivity can be ignored. The propagation of the radial crack can be divided into three stages (see Fig. 37). After initiation, the crack tip accelerates from about 400 m/s to around 990 m/s within 30 μs. We can see that the impact force remain increasing after the initiation of the radial crack, which means the elastic energy stored near the crack tip keeps increasing thus causing the acceleration sequences. Then the crack velocity remains constant at about 990 m/s during the next 25 μs. However, it decelerates to less than 600 m/s in under 100 μs after initiation. The entry of crack tip into the region of compressive stress as it approaches the boundaries and the arrival of less stress wave at the crack tip are two responsible reasons for the deceleration phenomenon (Lambros and Rosakis 1997).

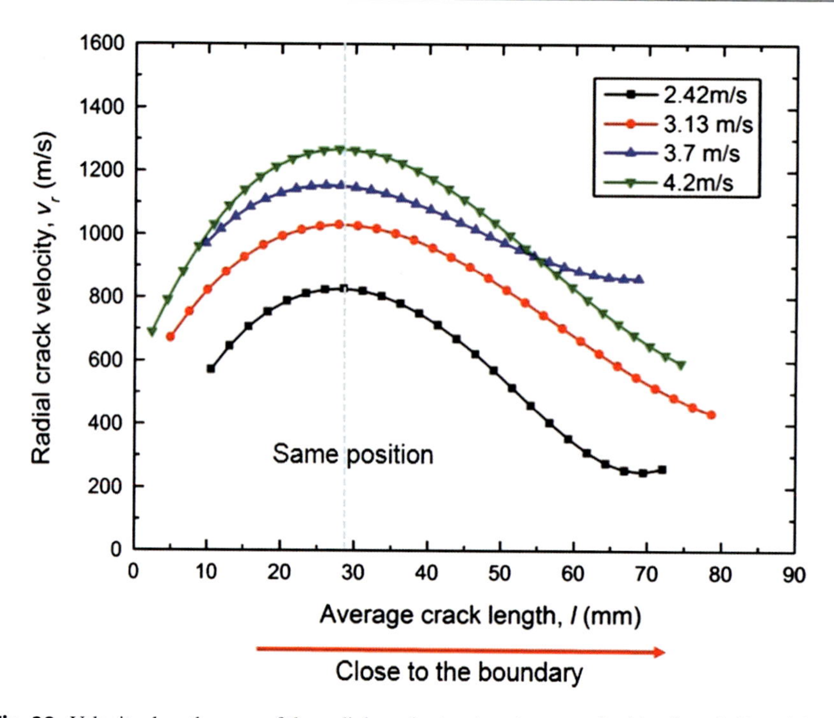

Fig. 38 Velocity–length curve of the radial crack at various impact velocities from 2.42 to 4.2 m/s with the PVB thickness of 0.76 mm

Through further investigation, the reason for the deceleration and crack arrestment phenomenon is relevant to the boundaries. The history of the radial crack velocity at various impact velocities from 2.42 to 4.2 m/s with the PVB thickness of 0.76 mm is plotted as a function of crack length in Fig. 38 and it is clearly shown that the crack velocity start decreasing at the same position.

For the circular crack, the crack initiation occurs 1,750 μs after the impact on the glass surface (see Fig. 37). Combined with the images depicting the radial and circular crack growth, it is believed that circular cracks initiate long after the radial crack growth as we stated before. An observation of the specimen, at the end of the test, showed that circular cracks merely existed on the loaded layer, according to which we can conclude that the circular cracks nuclear under the influence of surface acoustic waves (namely "Rayleigh waves"). The propagation of this kind of elastic waves is limited to the loaded surface of the solids. The Rayleigh waves spread at a much lower speed compared with the longitudinal compression waves. Besides, the lateral dimensions of the sample are much larger than those in longitudinal direction. Therefore, the circular crack appears generally later than the radial crack which is confirmed by the previous finding (Xu et al. 2010). The circular crack propagates under the tensile stress caused by the reflected tensile Rayleigh waves, which results in that the circular cracks always emerge first close to the boundaries. The propagation of the circular crack can be divided into three stages as well (see

Fig. 37). Both the first and the third stage contain the acceleration, the maintenance, and the deceleration process similar as those in radial cracks. There is more space for circular crack to grow than radial crack such that the total crack lengths are longer in circular cracks. Besides, the circular crack has the lower peak value. However, between the two stages, there exists a special stage which could be called "slowing-down stage." In this stage, the crack propagation velocity decelerates to almost 0. This is a direct consequence of the fracture at the glass interface caused by the radial crack. We can clearly see that the circular crack propagation always delays at the relatively large fracture where the stress waves have to propagate through the interlayer, without which the circular crack cannot transfer effectively across the interface (Park and Chen 2011).

Effect of Impact Velocity

To investigate the effect of impact velocity v_d, the both radial and circular crack propagation characteristics, respectively, series of parametric experiments are conducted. Figure 7 shows the crack propagation velocity time history of both radial and circular cracks under various impact velocities from 2.42 to 4.2 m/s (drop height from 300 to 900 mm). Two kilogram drop weight and 0.76 mm PVB thickness are employed in this study. Three repeated experiments under each condition are carried out to ensure the results accuracy (see error bars in Fig. 39). Time $t = 0$ μs is

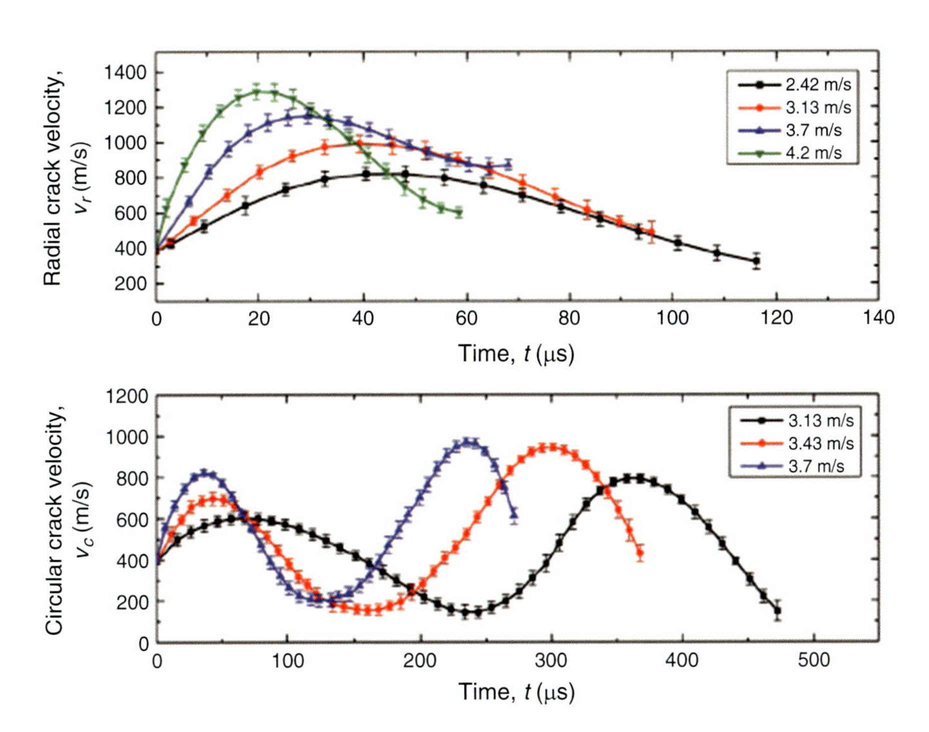

Fig. 39 Crack velocity history of the radial and circular crack at various impact velocities from 2.42 to 4.2 m/s with the PVB thickness of 0.76 mm

employed to denote crack tip initiation. For the radial crack, in the case with higher impact velocity, a shorter acceleration time is needed to reach the top propagation velocity where the top value is also higher. This is consistent with the fact that at higher impact velocity with larger wave amplitude, the load applied to the specimen increases at a faster rate and thus crack acceleration can be larger during the dynamic loading process. Besides, additional experiments also indicate that the higher the impact velocity, the shorter delay time is needed before initiation (see Fig. 40). However, the case with higher impact velocity would also decelerate faster and earlier due to the limited available cracking space in the panel plane. Similar to the radial crack, higher impact velocity results in the shorter initiation time, the shorter acceleration time, and the larger peak value for the circular crack. To establish the model combining the crack velocity with the impact conditions, we chose the variable "the maximum velocity" to quantify our study, which apparently depends on both the impact conditions and the material property. We can suppose that the maximum velocity v_{max} at the crack tip during the whole propagation process has an implicit relationship with the impact force, the duration of the loading time and the dynamic initiation fracture toughness of the material in the present investigation.

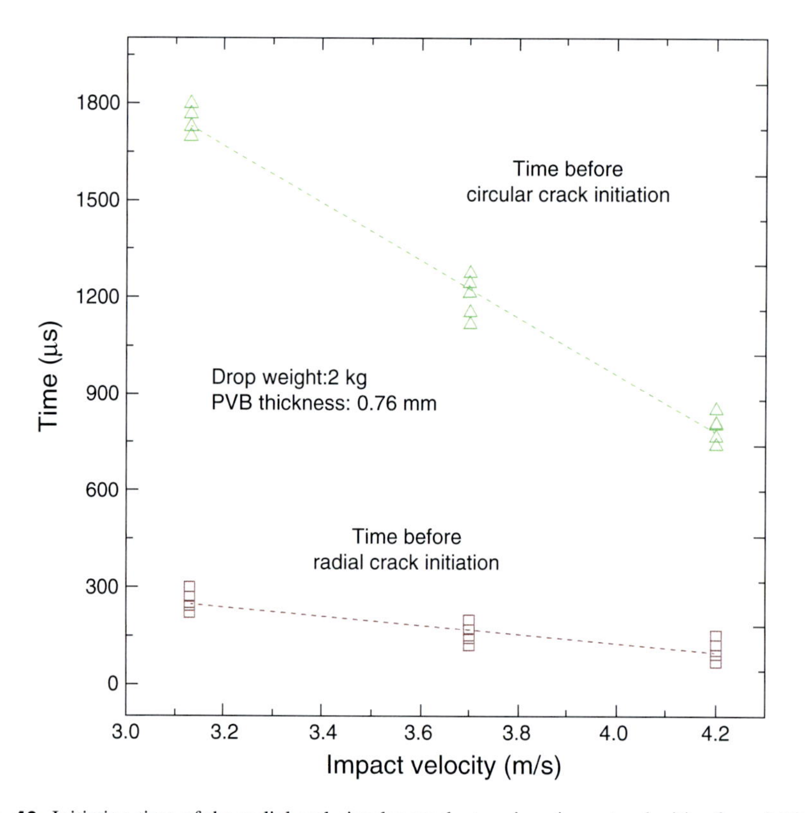

Fig. 40 Initiation time of the radial and circular crack at various impact velocities from 3.13 to 4.2 m/s with the PVB thickness of 0.76 mm

According to the principle of dimensionless unifying, the equation of $v_{\max}$ could have the following form as:

$$v_{\max} = \frac{AP_d^{2/3}}{\left(k_{IC}^d\right)^{2/3} t_d} \tag{45}$$

where p_d is the average contact force over the contact period and t_d is the duration of the loading, respectively, A is a macroscopic crack pattern dependent coefficient. According to the momentum theorem, we have:

$$p_d t_d = m_d v_d \tag{46}$$

This means that the maximum velocity at the crack tip can be expressed in terms of the impact velocity v_d and the drop weight m_d

$$v_{\max} = \frac{Am_d^{2/3} v_d^{2/3}}{\left(k_{IC}^d\right)^{2/3} t_d^{5/3}} \tag{47}$$

Thus, the maximum velocity contains two model parameters: the dynamic initiation fracture toughness k_{IC}^d, which can be geometry dependent, and the macroscopic crack pattern dependent parameter $A > 0$. Both parameters are subject to the experimental data via best-fitting.

Effect of PVB Thickness

Series of parametric experiments are conducted to study the effect of PVB thickness t_p on the crack propagation characteristics as well. Three different PVB thicknesses, i.e., 0.38 mm, 0.76 mm, and 1.52 mm are chosen for the parametric experiments. Two kilogram drop weight is employed. The crack propagation velocity time history of both radial and circular cracks under three different PVB thicknesses can be seen in Fig. 41. The impact velocity used to investigate the radial crack is 2.42 m/s (i.e., drop height is set to be 300 mm) while the impact velocity is 3.13 m/s (i.e., drop height is set to be 500 mm) for circular crack. Martinez et al. (1998) reported that the capability for transmitting the impact energy depends on the thickness and type of the adhesive used. Therefore, the variation in PVB thickness will have a great effect on delaying the passage of the elastic wave into the backing glass layer and reducing the amplitude and the wavelength of wave through internal friction (Xu et al. 2011b), which results in reducing the maximum radial crack speed as show in Fig. 41. Considering the wave speed in the through thickness direction of around 5,320 m/s in the glass layer and 300 m/s in the PVB interlayer (which means the PVB interlayer plays an important role in slowing down the stress wave), the transit time is around 3 μs, 6 μs, and 11 μs, respectively with the PVB thickness ranged from 0.76, 1.52, to 3.04 mm. Compared with the delay

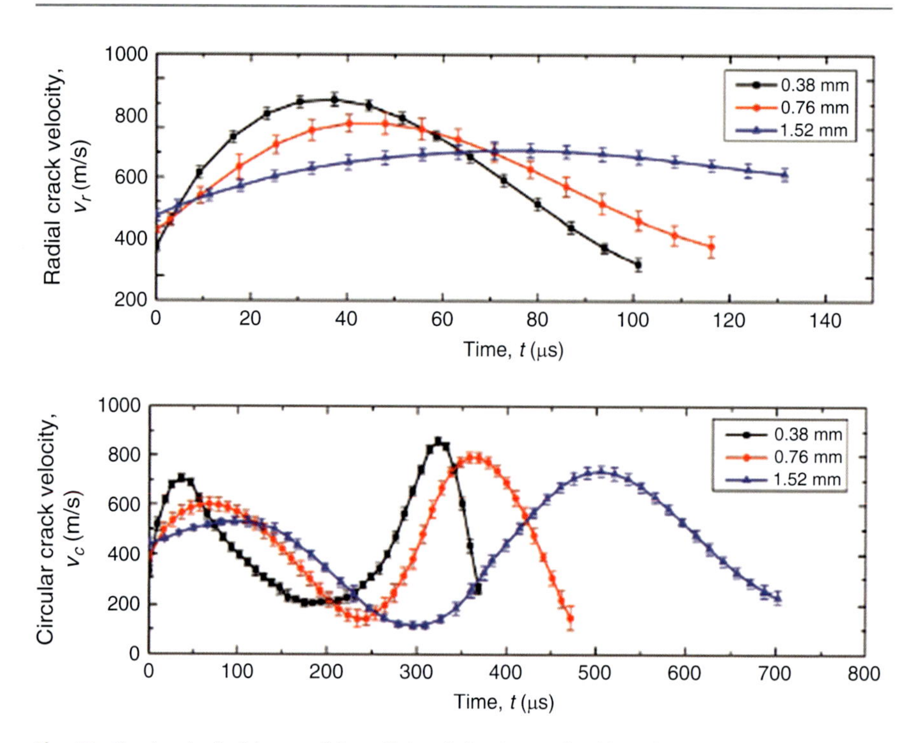

Fig. 41 Crack velocity history of the radial and circular crack with various PVB thicknesses from 0.38 to 1.52 mm at the impact velocity of 2.42 m/s and 3.13 m/s, respectively

time in the order of 102 μs for radial crack initiation in Fig. 10, we can conclude that the delay of the radial crack initiation mainly results from the increasing process of the stress-intensity factor. The thicker PVB thickness would reduce the amplitude of elastic wave crossing the PVB interlayer by internal friction thus to weaken the increasing process of the stress-intensity factor. Therefore, in the case with thicker PVB layer, a longer delay time is needed to initiate. For the circular crack caused by Rayleigh wave, similar phenomenon is observed. Considering the Rayleigh surface wave speed in the in-plane direction of around 3,370 m/s in the glass layer (Sharon et al. 2002), the transit time is around 37 μs, which is much longer than the transmit of the elastic wave in the medium. Similarly, the delay of the circular crack initiation mainly also results from the increasing process of the stress-intensity factor. In addition, from the force–time curve, we can see that the compressive waves stay around the loaded side for a relatively long time during the impact, which also have an effect on the increasing process of the stress-intensity factor.

It is clearly seen that the glass sheet with thicker PVB layer starts at a higher initiation velocity v_0 for both crack patterns. Since the exact expression of v_0 is not clarified yet, we may implicitly represent the expression to the initiation velocity v_0 as $v_0(t_p)$. Due to the fact that $k_{IC}^{d} \propto \sigma_c \propto v_0$ (see Eq. 44), where σ_c is the fracture stress, we have the following expression for the initiation velocity (Berezovski and Maugin 2007).

$$v_0^2\left(t_p\right) = c_R^2\left(1+D\right)\left(1-\left(1+\frac{\left(k_{IC}^d\right)^2}{M}\right)^{-1}\right) \tag{48}$$

where M is a constant coefficient and C_R is the Rayleigh wave velocity. In this case, we arrive at an expression for the dynamic initiation fracture toughness in Eq. 47 in the form of

$$k_{IC}^d = \sqrt{\frac{M v_0^2\left(t_p\right)}{C_R^2\left(1+D\right) - v_0^2\left(t_p\right)}} \tag{49}$$

Therefore, we may come to the expression for the maximum velocity at the crack tip as follows:

$$v_{\max} = \left[\frac{B m_d^2 v_d^2 t_d^{-5}\left(c_R^2\left(1+D\right) - v_0^2\left(t_p\right)\right)}{v_0^2\left(t_p\right)}\right] \tag{50}$$

where $B > 0$ is another constant coefficient depending on the macroscopic crack pattern. According to Eq. 50, the thicker the PVB layer, the lower of the maximum velocity for both radial and circular crack, which is consistent with the experiment result (see Fig. 41). This is expected since the glass sheet with thicker PVB layer would absorb more impact energy via molecule vibration and friction by the interlayer as polymer-like material, verified by Xu et al. (2011c). As the Rayleigh wave velocity $C_R \approx 3370$ m/s (Hauch and Marder 2010), which is sufficiently high for the initiation velocity v_o. The expression may be simplified as

$$v_{\max} = \left[\frac{B m_d^2 v_d^2 t_d^{-5}\left(c_R^2\left(1+D\right)\right)}{v_0^2\left(t_p\right)}\right]^{1/3} = c\left[\frac{m_d^2 v_d^2 t_d^{-5}}{v_0^2\left(t_p\right)}\right]^{1/3} \tag{51}$$

Then the agreement between theory and experiment may be achieved by an appropriate choice of the values of the constant coefficient c (c_1 and c_2 for radial and circular crack, respectively). The fitted models for the maximum velocity of radial crack $v_{\max}^r$ and circular crack $v_{\max}^c$ are in the following form of:

$$\begin{cases} v_{\max}^r \approx c_1\left[\dfrac{m_d^2 v_d^2 t_d^{-5}}{v_0^2\left(t_p\right)}\right]^{1/3} = 0.0657\left[\dfrac{m_d^2 v_d^2 t_d^{-5}}{v_0^2\left(t_p\right)}\right]^{1/3} \\ v_{\max}^c \approx c_2\left[\dfrac{m_d^2 v_d^2 t_d^{-5}}{v_0^2\left(t_p\right)}\right]^{1/3} = 0.0545\left[\dfrac{m_d^2 v_d^2 t_d^{-5}}{v_0^2\left(t_p\right)}\right]^{1/3} \end{cases} \tag{52}$$

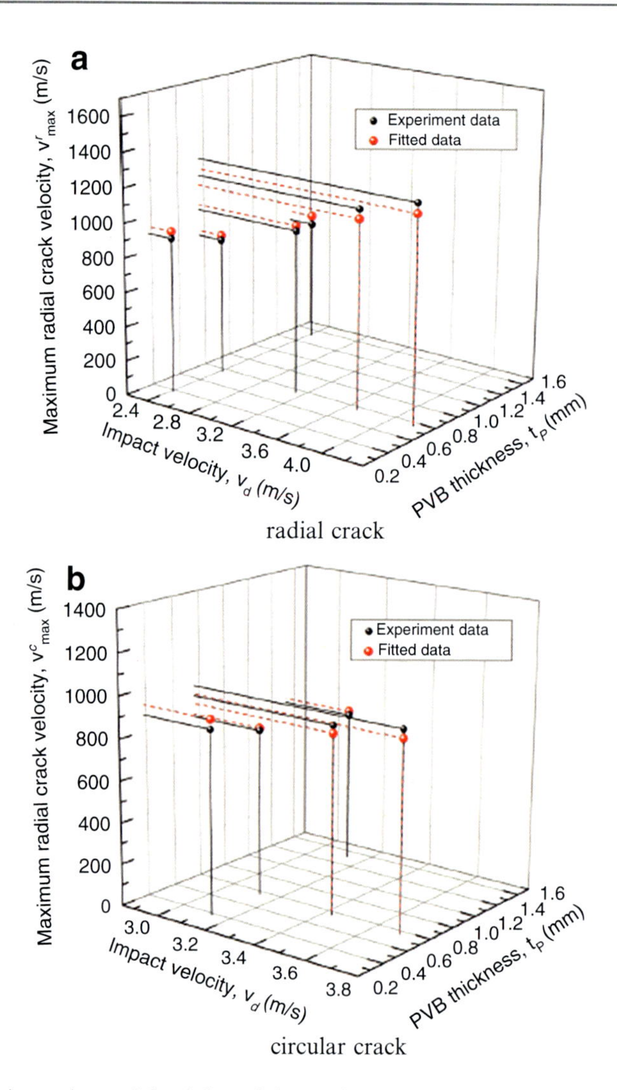

Fig. 42 Experiment data and fitted data of the maximum crack velocity from the semiphysical with various impact velocities from 2.42 to 4.2 m/s with PVB thickness from 0.38 to 1.52 mm. (**a**) Radial crack. (**b**) Circular crack

The fitted data and experimental data are shown together in Fig. 42 (the experimental data in the figure is the average maximum crack velocity). The good agreement between the two proves the model to be dependable.

Statistical Analysis on Macroscopic Cracking Morphology

To confirm the crack propagation mechanism analysis before and understand the crack propagation mechanism from another perspective, the relationship between

Table 4 Parametric statistical experiments: linear regression estimates of the Weibull parameters under different experiment conditions

Crack type	PVB thickness (mm)	Impact velocity (m/s)	Mean crack number	Weibull char. number (N_0)	Weibull modulus (m)
Radial crack	0.38	2.42	87	96	4.4
	0.76	2.42	81	89	4.0
	1.52	2.42	70	79	3.1
	3.04	2.42	66	75	2.7
	0.76	3.13	90	99	4.4
	0.76	3.7	99	107	5.1
	0.76	4.2	118	128	6.0
Circular crack	0.38	2.42	2.5	2.7	2.8
	0.76	2.42	2	2.2	2.5
	1.52	2.42	1.7	2.0	2.2
	3.04	2.42	1.5	1.6	1.7
	0.76	3.13	2.7	3.1	2.7
	0.76	3.7	3.9	4.2	5.1
	0.76	4.2	7.0	7.4	9.3

the number of cracks and the impact velocity or the PVB thickness is established. Since all cracks locate in the surface of glass sheet and glass is a material with internal stochastic flaws which may serve as crack nuclei, a statistical model may be needed to describe the two macroscopic crack patterns. Table 4 shows the detail combination of the statistical experiments. One hundred repeated experimental data for each specified condition have been used for each experiment condition to obtain a statistical result. The two-parameter Weibull model is employed to study the distribution of both radial and circular crack number under each experiment condition. Weibull statistics are commonly used in the engineering community, especially in the fracture statistics. In the specific case of Weibull statistics analysis in this study, the cumulative probability function is written as follows:

$$p_f = 1 - \exp\left(-\left(\frac{N}{N_0}\right)^m\right) \tag{53}$$

where P_f is the cumulative probability, N is the number of radial or circular crack (N_r or N_c), m and N_0 is the distribution shape parameter and characteristic crack number, respectively. To improve the statistical accuracy, the number of circular crack is counted with decimal during the statistical process (that is the number of circular crack with incomplete circle is determined by the length ratio of incomplete circle and the complete one). For the ease of accessing information, the double logarithm of the Weibull Eq. 55 is used in number analysis, yielding the Weibull parameters in a simple graphical representation of the data. Figure 43 shows the

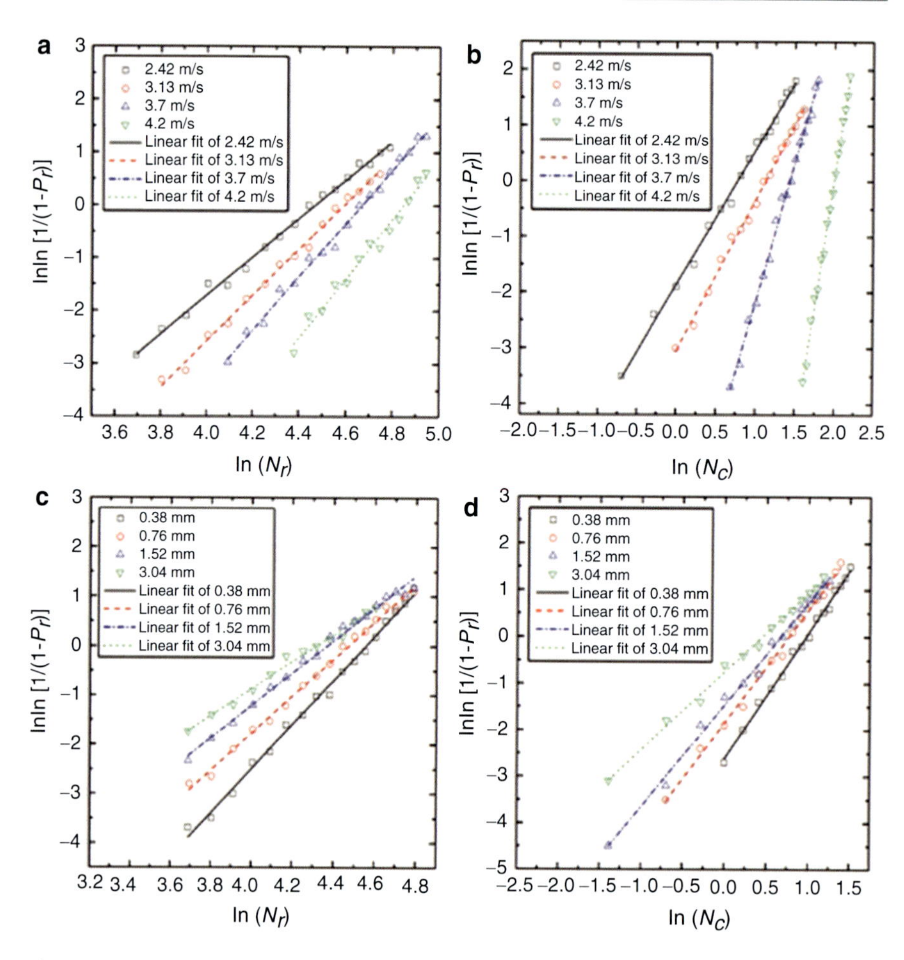

Fig. 43 Fitted Weibull model plots for macroscopic cracking morphology: (**a**) radial crack at various impact velocities from 2.42 to 4.2 m/s; (**b**) circular crack at various impact velocities 2.42–4.2 m/s; (**c**) radial crack with different PVB thicknesses from 0.38 to 3.04 mm; and (**d**) circular crack with different PVB thicknesses from 0.38 to 3.04 mm

graphs with $X = 1$ nN plotted on the horizontal axis and $Y = \ln[\ln(1/(1-p_f))]$ on the vertical axis. One can perform a simple linear regression analysis to get the Weibull parameters under different experiment conditions. The slope of the line is the Weibull modulus, m. The characteristic crack number, N_0 is the value of crack number for which $\ln[\ln(1/(1-p_f))]$ is zero. In this study, the linear regression analysis is used for the Weibull parameter estimates under each experiment condition (see Table 4).

$$\ln\left(1 - p_f\right) = -\left(\frac{N}{N_0}\right)^m \tag{54}$$

$$\ln\left[\ln\left(\frac{1}{1-p_f}\right)\right]=m\ln\ N-m\ln\ N_0 \tag{55}$$

Firstly, the effect of different impact velocities on the distribution of both radial and circular crack number is investigated. Figure 44a shows the probability density curves of radial and circular crack number under different impact velocities. The characteristic crack number indicates the distribution location along the X axis and is expected to vary according to the impact velocity or the PVB thickness. The distributions for both radial and circular crack number are proved to move to the right (higher value) as impact velocity increases. This is reasonable because higher impact velocity produces more elastic energy, and thus it is necessary to create larger new surface to release the excess energy. In addition, the probability density curves considering different PVB thicknesses to describe the radial and circular crack number are shown in Fig. 44b. Contrary to the effect of impact velocity, the distributions for both radial and circular crack number move to the left (lower value) as PVB thickness increases since thicker PVB layer absorbs more elastic energy as stated before. This statistical method could also be applied to the engineering applications and play an important role in the laminated glass design.

The Relation Between Radial Crack Velocity and Crack Numbers on the Backing Glass Layer

Here, the effect of the radial crack number on their own crack propagation velocity as well as the crack generation on the other glass layer is thoroughly investigated. Firstly, the radial crack velocity on the backing glass layer with different crack numbers is calculated. A theoretical model from the perspective of energy conversion is established to depict the cracking process of backing glass layer and elucidate the fundamental reason that causes the variation of the crack velocity.

The specimen in response to impact speed $V = 3.7$ m/s is chosen for 100 repeated experiments by considering the intrinsic stochastic flaws in glass laminated samples. A series of images depicting the crack growth on both glass plates are recorded at the set time intervals. Typical crack propagation process is shown in Fig. 45. The glass plate directly contacting with the impactor is defined as "impacted glass layer," while the glass plate on the other side is referred as "backing glass layer." $t = 0$ μs represents the time when impact is triggered. Figure 45a–d demonstrate the whole cracking process on the backing glass layer where cracks always initiate first, and only radial crack pattern is observed while both radial and circular cracks appear on the impacted glass layer (under the effect of Rayleigh waves, whose propagation is limited to the loaded surface of the solids) long after the cracking of the backing one (Fig. 45e–f). However, the radial crack number on the backing glass plate varies from 18 to 112 within 100 repeated experiments, which is quite a large deviation. Further, it is also qualitatively observed that the number of radial

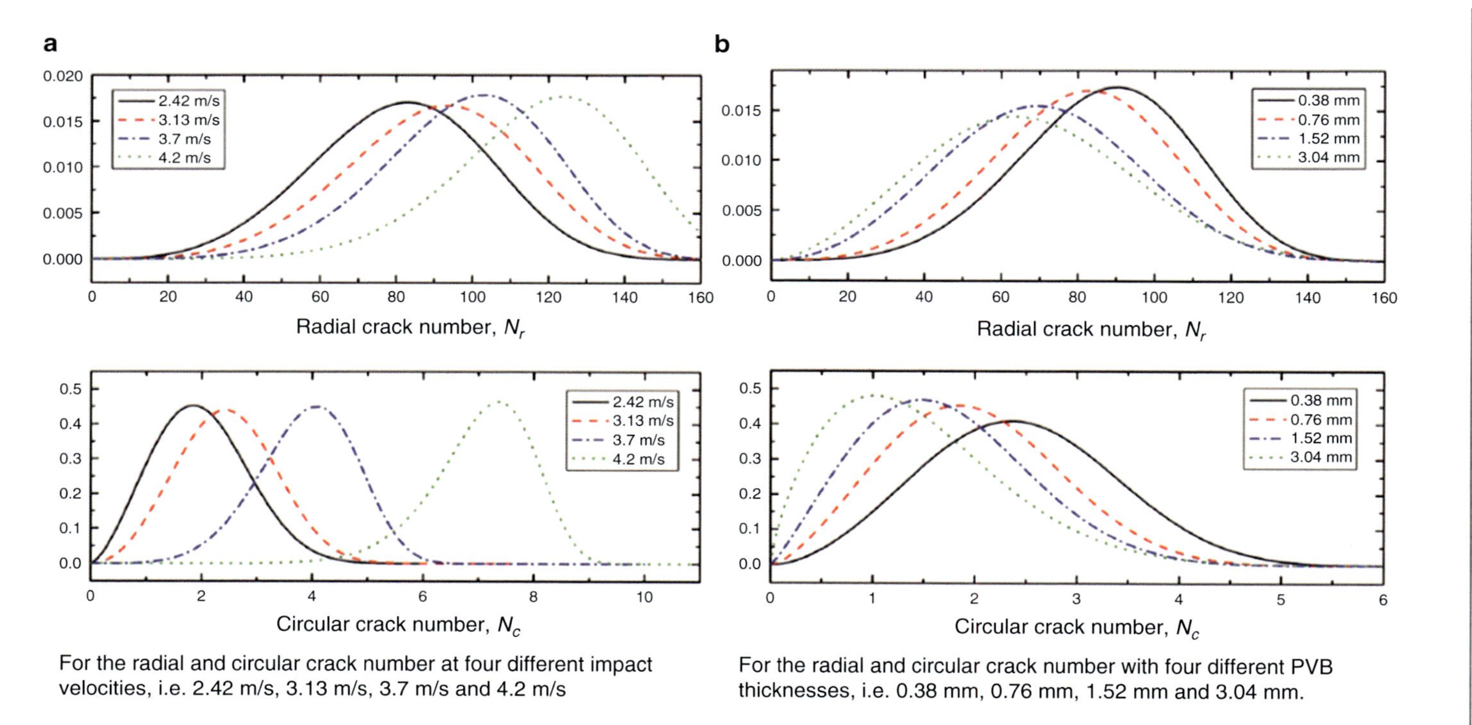

Fig. 44 Probability density curves of the radial and circular crack number under various boundary conditions. (**a**) For the radial and circular crack number at four different impact velocities, i.e., 2.42 m/s, 3.13 m/s, 3.7 m/s, and 4.2 m/s. (**b**) For the radial and circular crack number with four different PVB thicknesses, i.e., 0.38 mm, 0.76 mm, 1.52 mm, and 3.04 mm

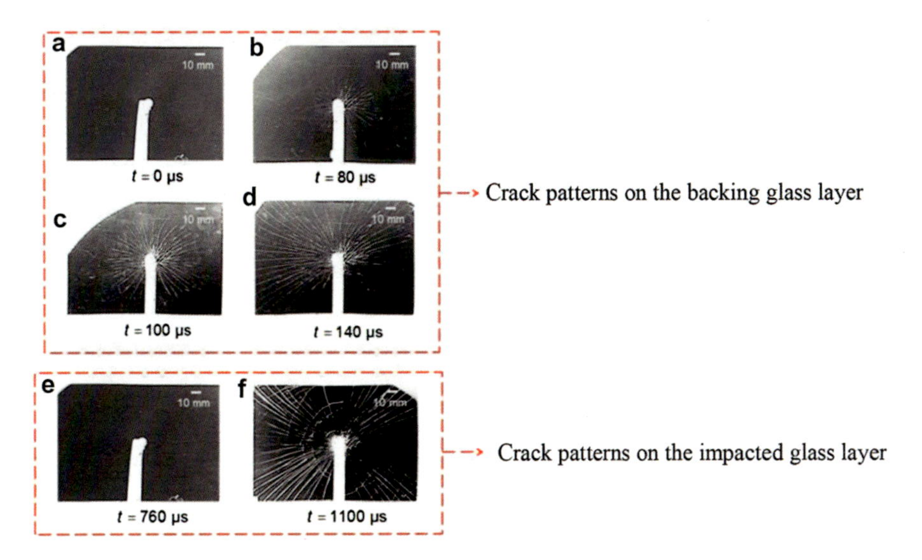

Fig. 45 Images depicting the crack growth on the two glass layers ($V = 3.7$ m/s)

cracks have great effect on the later crack generation on the impacted glass layer as well as the propagation characteristics of radial crack on the backing glass layer itself. In the following section, the effect of the different radial crack numbers of backing glass layer would be thoroughly discussed.

The crack velocity history on the backing glass layer with different radial crack number is calculated separately. All the radial cracks present approximately the same length during the propagation, conforming a circle overall (Fig. 45). Therefore, the crack velocity v_r is calculated as the radius increase Δr over the time increase Δt, i.e., $\Delta r/\Delta t$. Figure 46 shows the crack velocity variation with the crack length l_r under three selected radial crack numbers ($N_r = 18$, 64, and 108). Crack propagations far from the boundaries and the center of the glass plate are focused to eliminate the possible boundary effect (i.e., the crack length between 20 and 40 mm). Results indicate that the number of radial cracks on the backing glass layer regularly negatively influences cracking propagation velocity themselves, i.e., the crack velocity v_r always decreases with the radial crack number increasing.

During the propagation of the cracks on the backing glass layer, the impacted glass layer remains intact as stated before (Fig. 46). However, due to the adhesion by PVB interlayer, the two glass layers have the same radius of deformed region r_f (Fig. 46). The bending energy of the impacted glass plate caused by the impact can be estimated by assumption that the backing glass layer is still intact, in which case the two glass plates can be regarded as "single one with $2h_g$ thickness." Therefore, the bending energy of impacted glass layer is equal to half the energy in the single glass plate with $2h_g$ thickness. After the contact between the impactor and the glass plate, kinetic and bending energies balance $\rho_g \Omega V^2 \sim E_g (2h_g \kappa)^2 \Omega$ (Vandenberghe et al. 2013), where $\Omega = 2h_g \pi r_f^2$ is the volume of deformed region and V is the impact

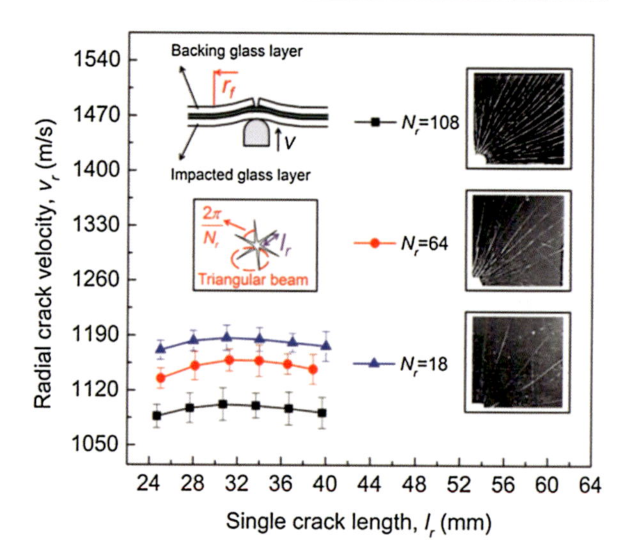

Fig. 46 Radial crack velocity versus crack length with different radial crack number ($V = 3.7$ m/s)

speed whose variation can be ignored during such a short period. Thus, the curvature $\kappa \sim w_0/r_f^2 \sim V/2h_g c_g$, yielding $r_f \sim (2h_g c_g t)^{1/2}$, where $w_0 = V_t$ is the indentation and $c_g = \sqrt{E_g/\rho_g}$ is the sound speed in the glass material. The bending energy of impacted glass plate U_{b1} can be expressed as

$$U_{b1} = k_1 \pi E_g \left(\frac{V}{c_g}\right) r_f^2 h_g \tag{56}$$

where k_1 is a coefficient.

Considering the low Young's modulus of PVB material (i.e., $E_p = 0.1$ GPa), the bending energy of PVB U_{b3} is about 0.001 of that in the glass plate, which has been proved by the finite element simulation with a three-dimensional laminated plate FE model using 1×1 mm surface size element. The boundary condition of the glass plate model is set to be clamped, which is consistent with the real experimental condition. Thus, U_{b3} is small enough to be ignored compared to that of U_{b1}.

While for the backing glass layer in the cracking process with crack length l_r, the number of radial cracks N_r on the backing layer has been set and would not change as the cracks extend. Thus, the bending energy of the backing glass plate U_{b2} can be estimated by the energy of N_r triangular beams of length l_r with the neglect of transverse bending (Vandenberghe et al. 2013):

$$\begin{aligned} U_{b2} &= E_g h_g^3 w_0^2 N_r \tan\left(\frac{\pi}{N_r}\right)/3l_r^2 \\ &\approx k_2 \pi E_g h_g r_f^2 \left(\frac{V}{c_g}\right)^2 \left(\frac{1}{2} + \frac{\pi^2}{6N_r^2}\right) \end{aligned} \tag{57}$$

where k_2 is another coefficient. Therefore, the bending energy of the whole laminated plates U_b can be estimated as

$$\begin{aligned} U_b &= (U_{b1} + U_{b2}) \\ &\approx \pi E_g h_g r_f^2 \left(\frac{V}{c_g}\right)^2 \left[k_1 + k_2 \left(\frac{1}{2} + \frac{\pi^2}{6N_r^2}\right)\right] \end{aligned} \tag{58}$$

From the perspective of energy transferring, during the cracking process, the elastic energy of the laminated plates (i.e., the bending energy) caused by the impact is converted into the glass surface energy through the fracture behavior. Here we define the increasing rate of the plates elastic energy as "η_b," and the energy release rate is "η_c." Therefore, we have

$$\eta_b = \frac{dU_b}{dt} \approx \pi E_g h_g^2 \frac{V^2}{c_g} \left[k_1' + k_2' \left(\frac{1}{2} + \frac{\pi^2}{6N_r^2}\right)\right] \tag{59}$$

$$\eta_c = \frac{d\left(2N_r \Gamma h_g l_r\right)}{dt} = 2N_r \Gamma h_g v_r \tag{60}$$

where k_1' and k_2' are both coefficients. Γ is the glass material fracture (surface) energy. There should be a balance between the "input energy" (i.e., the increasing elastic energy) and the "output energy" (i.e., the fracture energy). Suppose that the proportion between the two parameters η_b/η_c, which is named as "energy conversion factor," indicating the stability of the system or the ability for further crack growth, then the larger proportion would obviously lead to a greater degree of instability for the whole system. Thus for the system of PVB laminated plates in our study, the "energy conversion factor" η_b/η_c can be expressed as

$$\frac{\eta_b}{\eta_c} = \frac{\pi E_g h_g}{2\Gamma c_g} \cdot V^2 \cdot \frac{1}{v_r} \left\{ \frac{1}{N_r} \left[k_1' + k_2' \left(\frac{1}{2} + \frac{\pi^2}{6N_r^2}\right)\right]\right\} \tag{61}$$

One may clearly see that the expression of the "energy conversion factor" contains three parts: (1) $\pi E_g h_g / 2\Gamma c_g$ is determined by the material properties (here refers to the glass material), (2) V^2 refers to the impact condition, and (3) rest of the expression $f(v_r, N_r)$ refers to the impact responses of the laminates. Thus the stability of the system is determined together by the material properties, impact condition, and the impact response of the material according to Eq. 61. Previous studies (Xu et al. 2011d; Chen et al. 2013) investigating the effect of the impact speed on the crack velocity have demonstrated that, with higher impact speed V that increasing the system instability η_b/η_c, both of the crack velocity v_r and the radial crack number N_r tend to increase as well thus to lower the value of η_b/η_c such to maintain the balanced energy input and output in a large extent. Therefore, as the crack number N_r on the backing glass layer decreases under the same impact condition, the value of η_b/η_c would correspondingly increase due to the result of Eq. 61, which means the instability of the system is rising. Thus the system tends to response in a way to reduce this instability (i.e., the value of η_b/η_c) through propagating the cracks at higher speed, improving the value of v_r in Eq. 61, which is consistent with our experiment results (Fig. 46).

The Relation Between the Crack Length and the Capability of Energy Absorption on the Impacted Glass Layer

Further, the total crack length for both two crack patterns on the impacted glass layer is calculated to study the capability of energy absorption of laminated glass influenced by the radial crack number on the backing glass layer.

Figure 47 shows the crack morphology on the impacted glass layer where both of the two crack patterns (i.e., radial crack and circular crack) appear long after the cracking of the backing glass plate. For the radial cracks on the impacted layer that are indirectly caused by the stress concentration brought by each generated crack on the backing glass layer as initial flaw (Lee et al. 2007) (which is also proved by our experiment investigation), they are completely overlapped with the radial cracks on the backing layer (Fig. 47). Thus, it is obvious that the morphology as well as the number of radial cracks in the impacted glass layer depends completely on that of the backing layer. The radial cracks on the impacted glass initiates from those on the backing glass plate through the depth direction (Lee et al. 2007).

Specimen in response to lower impact speed $V = 2.42$ m/s is chosen for 100 repeated experiments. As expected, the radial crack number on the backing glass plate varies from 15 to 65 in these experiments. The total crack length for both two patterns on the impacted glass layer is calculated with different crack numbers on the

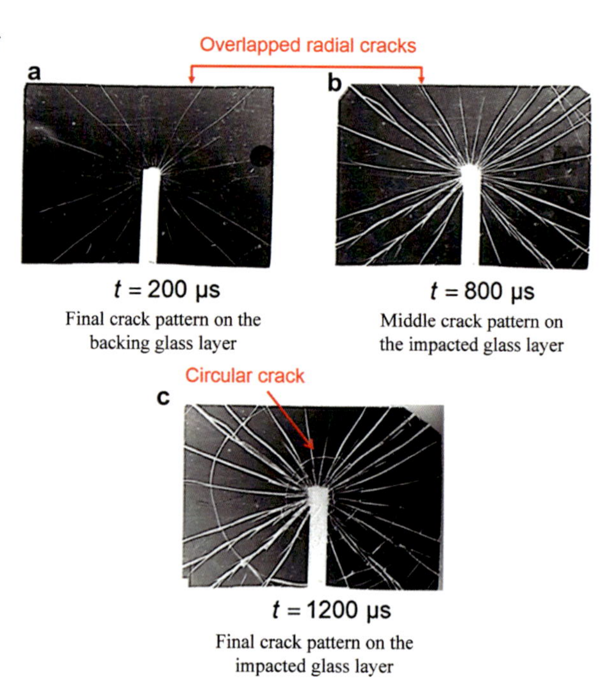

Fig. 47 Selected sequence of images depicting the crack patterns on both glass layers at the loading speed $V_d = 3.7$ m/s, with PVB thickness $h = 0.76$ mm

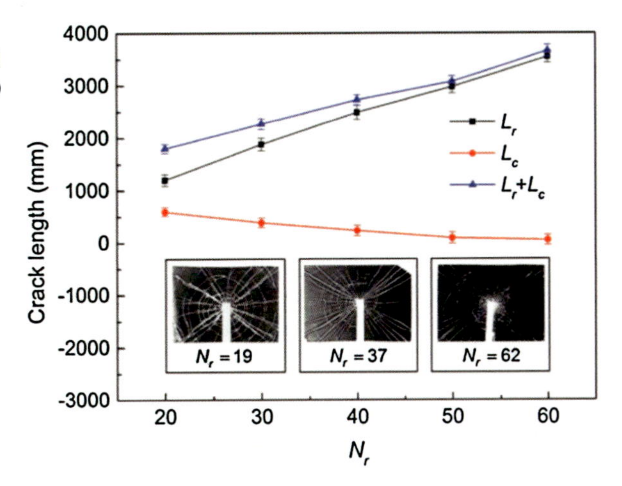

Fig. 48 Crack length on the inner glass layer versus radial crack number ($V = 2.42$ m/s)

backing glass layer (Fig. 48). L_r and L_c are defined as the total radial and circular crack length on the impacted glass layer, respectively. Thus, ($L_r + L_c$) refers to the total crack length on the impacted glass plate. The radial crack length linearly increases with crack number on the backing glass layer due to the uniform growth of all cracks while the variation trend of circular crack length is decreasing (Fig. 48). Circular cracks always initiate long after the completed propagation of radial cracks on the impacted glass (Xu et al. 2010) (Fig. 47). Thus, before the initiation of circular cracks, the laminated plates can be regarded as N_r triangular beams with thickness $2h_g$ and length l_0 (which can be estimated as the side length of specimen) since the radial cracks on the two glass plates are completely overlapped (Fig. 47). The bending energy U_{b0} of the plates is

$$U_{b0} = 4\pi E_g h_g^3 w_0^2 \left(1 + \pi^2/3N_r^2\right) / \left(3l_0^2\right) \tag{62}$$

In the case with less radial crack number N_r, the bending energy would increase, resulting in stronger stress concentration in the radial direction, which makes the later generation of circular cracks easier. In addition, taking the law of energy conservation into account, the total fracture energy on the whole laminated plates can be estimated as $2\Gamma \mathrm{L} \mathrm{h}_g$, where L is the total crack length on the two glass plates. Thus, the fracture energy released on two glass plates can be estimated by calculating the crack length on each plate. For the plates with shorter radial crack length, leaving more residual energy, it will benefit the generation of circular cracks as well.

Although the increase of circular crack number contributes positively to the energy absorption of the laminated plates, the effect is much less compared with the radial cracks (Fig. 48). As the radial crack morphology on the impacted glass is directly determined by that on the backing glass layer, it is safe to conclude that the backing glass plate is the key that decides the fracture morphology and further the capability of energy absorbing of the whole laminated structure such that the material selection and structure design for the backing layer is extremely important for impact protection.

Summary

In this chapter, we have investigated the propagation characteristics for both radial and circular cracks in PVB laminated glasses by theoretical constitutive equations analysis, numerical simulation, experiments, and tests of impact.

Firstly, a damage-modified nonlinear viscoelastic constitutive relations model of PVB laminated windshield is suggested based on updated Lagrangian method to study the crack evolutions on windshield plate subject to pedestrian head.

Then, the constitutive relations are implemented into FEA software to simulate the pedestrian head impact with vehicle windshield after verified by a classical example and classical Hertzian pressure calculation model. Based on the FE model above, the internal stress of PVB laminated glass is analyzed and found that shear stress, compressive stress, and tensile stress are main causes of plastic deformation, radial cracks, and circumferential cracks, respectively for the laminated glass subject to impactor. By a parametric study based on the FE model above, the increase of Poisson's ratio will lead the increase of crack angle. Impact velocity has a great influence of crack formations as well as the injuries of pedestrian head. Comparing with other numerical techniques, XFEM is arguably more efficient since it does not require remeshing during crack propagation. So the extended finite element method (XFEM) is adopted to numerically investigate the multiple crack propagation of PVB laminated glass subject to quasi-static indentation or low-speed impact. It is found that the critical accident information, such as the impact speed or damage stress, can be deduced from the crack pattern characteristics. A qualitative bridge can be established between numerical simulation result and real-world accident via the crack growth mechanism.

Lastly, for the experimental analysis of PVB laminated glass, crack branching based on crack fractal theory is investigated and an explicit expression describing the crack velocity and number of crack branching is proposed under quasi-static Split Hopkinson Pressure Bar (SHPB) compression experiments. For dynamic out-of-plane loading, we carried out preliminarily impact fracture experiments firstly to investigate the radial crack propagation behavior of PVB laminated glass subjected to light-weight impact by using high-speed photography. The time histories of the averaged radial crack tip position, propagation velocity, and acceleration are obtained. It is found that the steady-state cracking speed of PVB laminated glass is lower than that of pure glass, and it also increases with higher impactor speed and mass. It is also found that the supported glass layer would always initiate before the loaded layer and the final morphologies of radial cracks on both sides are completely overlapped even if they propagate at different time. Here, two different mechanisms of crack propagation on different glass layers are explained: the cracks on supported glass layer are dominantly motivated by the in-plane crack-tip stress such that a relatively longer time is needed for its propagation while for the cracks appearing much later on loaded layer, which are caused by the stress concentrated in depth direction for each generated crack on supported layer, the propagation time needed is largely reduced. Then, the further experiment investigating the radial and circular crack propagation of PVB laminated glass thoroughly is carried out. The parametric study on two dominant factors, i.e., impact velocity and PVB thickness is discussed aiming at the crack propagation of PVB laminated glass. Results show that cracking speeds of both radial crack and circular crack increase with higher impact velocity. Radial crack appears earlier than circular one and it propagates at higher speeds. Based on the mechanism of crack initiation and propagation, a semiphysical model describing the relationship between the maximum cracking velocity and influential factors is established. In addition, the Weibull statistical model is suggested considering various factors to describe the macroscopic crack pattern thus providing a theoretical evidence for engineering practice. Finally, the relation between radial crack velocity and crack numbers on the backing glass layer and the relation between the crack length and the capability of energy absorption on the impacted glass layer are investigated. Results show that the radial crack velocity on the backing glass layer decreases with the crack number under the same impact conditions during large quantities of repeated experiments. Thus, the "energy conversion factor" is suggested to elucidate the physical relation between the cracking number and the crack propagation speed. However, the fracture energy of the whole laminated plates is mainly determined by the radial crack length, which means that the radial crack number generated first on the backing glass plate is the key that decides the capability of energy absorbing of the whole laminated structure.

References

A. Berezovski, G.A. Maugin, On the propagation velocity of a straight brittle crack. Int. J. Fract. **143**, 135–142 (2007)

P.A.D. Bois, S. Kolling, W. Fassnacht, Modelling of safety glass for crash simulation. Comput. Mater. Sci. **28**, 675–683 (2003)

E. Bouchaud, J.P. Bouchaud, J.P.S.G. Lapasset, The statistics of crack branching during fast crack propagation. Fractals-Compl. Geom. Patterns Scaling Nat. Soc. **1**, 1051–1058 (2012)

J. Chen, J. Xu, X. Yao, B. Liu, X. Xu, Y. Zhang, Y. Li, Experimental investigation on the radial and circular crack propagation of PVB laminated glass subject to dynamic out-of-plane loading. Eng. Fract. Mech. **112–113**, 26–40 (2013)

B.E. Clements, J.N. Johnson, R.S. Hixson, Stress waves in composite materials. Phys. Rev. E Stat. Phys. Plasmas Fluids Relat. Interdisc. Top. **54**, 6876–6888 (1996)

N.P. Daphalapurkar, H. Lu, D. Coker, R. Komanduri, Simulation of dynamic crack growth using the generalized interpolation material point (GIMP) method. Int. J. Fract. **143**, 79–102 (2007)

S.K. Dwivedi, H.D. Espinosa, Modeling dynamic crack propagation in fiber reinforced composites including frictional effects ☆. Mech. Mater. **35**, 481–509 (2003)

F. Zhigang and Z. Jianping, On the viscoelastic finite element method, Shanghai J. Mech. **16**(1), 20–26 (1995)

U. Fischer, K. Washizu, *Variational Methods in Elasticity & Plasticity*, 3rd edn. (Pergamon Press, Oxford/New York, 1982.) XV, 630 S., £ 40.00. US $ 100.00. ISBN 0 08 026723 8, Zamm

Journal of Applied Mathematics & Mechanics Zeitschrift Für Angewandte Mathematik Und Mechanik, 64 (1984) 70–71

P. Forquin, F. Hild, A probabilistic damage model of the dynamic fragmentation process in brittle materials. Adv. Appl. Mech. **44**, 1–72 (2010)

D. Grégoire, H. Maigre, A. Combescure, New experimental and numerical techniques to study the arrest and the restart of a crack under impact in transparent materials. Int. J. Solids Struct. **46**, 3480–3491 (2009)

J.A. Hauch, M.P. Marder, Energy balance in dynamic fracture, investigated by a potential drop technique. Int. J. Fract. **90**, 133–151 (2010)

J. Lambros, A.J. Rosakis, Dynamic crack initiation and growth in thick unidirectional graphite/epoxy plates. Compos. Sci. Technol. **57**, 55–65 (1997)

B. Lawn, R. Wilshaw, Review: indentation fracture: principles and applications. J. Mater. Sci. **10**, 1049–1081 (1975)

J.W. Lee, I.K. Lloyd, H. Chai, Y.G. Jung, B.R. Lawn, Arrest, deflection, penetration and reinitiation of cracks in brittle layers across adhesive interlayers. Acta Mater. **55**, 5859–5866 (2007)

W. Lili, Z. Xixiong, S. Shaoqiu, G. Su, and B. Hesheng, An impact dynamics investigation on some problems in bird strike on windshields of high speed aircrafts. Acta Aeromauticaet Astronautica Sinica, **12**(2), B27–B33 (1991)

B.B. Mandelbrot, *Fractal geometry of nature*. WH Freeman and Company (1983)

M.A. Martínez, I.S. Chocron, J. Rodríguez, V.S. Gálvez, L.A. Sastre, Confined compression of elastic adhesives at high rates of strain. Int. J. Adhes. Adhes. **18**, 375–383 (1998)

B. Michel, L.B. Freund, *Dynamic Fracture Mechanics* (Cambridge University Press, Cambridge, 1990.) XVII, 563 pp., L 40.00 H/b. ISBN 0-521-30330-3 (Cambridge Monographs on Mechanics and Applied Mathematics), Zeitschrift Angewandte Mathematik Und Mechanik, 72 (1992) 383–384

J.H. Nielsen, J.F. Olesen, H. Stang, The fracture process of tempered soda-lime-silica glass. Exp. Mech. **49**, 855–870 (2009)

H. Park, W.W. Chen, Experimental investigation on dynamic crack propagating perpendicularly through interface in glass. J Appl. Mech-t ASME **78**(5) (2011)

K. Ravi-Chandar, W.G. Knauss, An experimental investigation into dynamic fracture: IV. On the interaction of stress waves with propagating cracks. Int. J. Fract. **26**, 189–200 (1984)

E. Sharon, J. Fineberg, Confirming the continuum theory of dynamic brittle fracture for fast cracks. Nature **397**, 333–335 (1999)

E. Sharon, G. Cohen, J. Fineberg, Crack front waves and the dynamics of a rapidly moving crack. Phys. Rev. Lett. **88**, 47–103 (2002)

Y.P. Shen, Y.H. Chen, Y.F. Pen, The finite element method of viscoelastic large deformation plane problem with Kirchhoff stress tensors-green strain tensors constitutive relation. Acta Mech. Solida Sin. 87–91 (1987)

R.P. Singh, V. Parameswaran, An experimental investigation of dynamic crack propagation in a brittle material reinforced with a ductile layer. Opt. Lasers Eng. **40**, 289–306 (2003)

J.H. Song, H. Wang, T. Belytschko, A comparative study on finite element methods for dynamic fracture. Comput. Mech. **42**, 239–250 (2008)

M.G. Stout, D.A. Koss, C. Liu, J. Idasetima, Damage development in carbon/epoxy laminates under quasi-static and dynamic loading. Compos. Sci. Technol. **59**, 2339–2350 (1999)

X. Sun, M.A. Khaleel, X. Sun, M.A. Khaleel, Effects of different design parameters on stone-impact resistance of automotive windshields. Proc. Inst. Mech. Eng. Part D J. Automob. Eng. **219**, 1059–1067 (2005)

X. Sun, W. Liu, W. Chen, D. Templeton, Modeling and characterization of dynamic failure of borosilicate glass under compression/shear loading ☆. Int. J. Impact Eng. **36**, 226–234 (2009)

S.R. Swanson, L.W. Christensen, A constitutive formulation for high-elongation propellants. J. Spacecr. Rocket. **20**, 559–566 (2015)

A. Thom, Experimental and modeling studies of stress wave propagation in multilayer composite materials: low modulus interlayer effects. J. Compos. Mater. **39**, 981–1005 (2005)

M. Timmel, S. Kolling, P. Osterrieder, P.A.D. Bois, A finite element model for impact simulation with laminated glass. Int. J. Impact Eng. **34**, 1465–1478 (2007)

N. Vandenberghe, R. Vermorel, E. Villermaux, Star-shaped crack pattern of broken windows. Phys. Rev. Lett. **110**, 285–291 (2013)
H.W. Wang, Z.J. Lai, X.U. Sun, D.J. Ming-Qiao, S.Q. Huang, Shi, Dynamic deformation and fracture of polymers taking account of damage evolution. J Ningbo University, **16**(4), 373–381 (2003)
L.L. Wang, X.L. Dong, Z.J. Sun, Dynamic constitutive behavior of materials at high strain rate taking account of damage evolution. Explosion Shock Waves **26**, 193–198 (2006)
X. Wang, Z. Feng, F. Wang, Z. Yue, Dynamic response analysis of bird strike on aircraft windshield based on damage-modified nonlinear viscoelastic constitutive relation. Chinese J. Aeronaut. **20**, 511–517 (2007)
J. Xu, Y. Li, Model of vehicle velocity calculation in vehicle-pedestrian accident based on deflection of windshield. J. Mech. Eng. **45**, 210–215 (2009)
J. Xu, Y. Li, G. Lu, W. Zhou, Reconstruction model of vehicle impact speed in pedestrian–vehicle accident. Int. J. Impact Eng. **36**, 783–788 (2009)
J. Xu, Y. Li, X. Chen, Y. Yan, D. Ge, M. Zhu, B. Liu, Characteristics of windshield cracking upon low-speed impact: numerical simulation based on the extended finite element method. Comput. Mater. Sci. **48**, 582–588 (2010)
J. Xu, Y. Li, B. Liu, M. Zhu, D. Ge, Experimental study on mechanical behavior of PVB laminated glass under quasi-static and dynamic loadings. Compos. Part B **42**, 302–308 (2011a)
J. Xu, Y.B. Li, X. Chen, D.Y. Ge, B.H. Liu, M.Y. Zhu, T.H. Park, Automotive windshield – pedestrian head impact: energy absorption capability of interlayer material. Int. J. Automot. Technol. **12**, 687–695 (2011b)
J. Xu, Y. Sun, B. Liu, M. Zhu, X. Yao, Y. Yan, Y. Li, X. Chen, Experimental and macroscopic investigation of dynamic crack patterns in PVB laminated glass sheets subject to light-weight impact. Eng. Fail. Anal. **18**, 1605–1612 (2011c)
J. Xu, Y. Li, X. Chen, D. Ge, B. Liu, M. Zhu, T.H. Park, Automotive windshield – pedestrian head impact: energy absorption capability of interlayer material. Int. J. Automot. Technol. **12**, 687–695 (2011d)
X.X. Zhang, R.C. Yu, G. Ruiz, M. Tarifa, M.A. Camara, Effect of loading rate on crack velocities in HSC. Int. J. Impact Eng. **37**, 359–370 (2010)
S. Zhao, L.R. Dharani, X. Liang, L. Chai, S.D. Barbat, Crack initiation in laminated automotive glazing subjected to simulated head impact. Int. J. Crashworthiness **10**, 229–236 (2005)
S.M. Zhao, L.R. Dharani, L. Chai, S.D. Barbat, Analysis of damage in laminated automotive glazing subjected to simulated head impact. Eng. Fail. Anal. **13**, 582–597 (2006a)
S. Zhao, L.R. Dharani, L. Chai, S.D. Barbat, Dynamic response of laminated automotive glazing impacted by spherical featureless headform. Int. J. Crashworthiness **11**, 105–114 (2006b)
F. Zhou, L. Wang, S. Hu, A damage-modified nonlinear VISCO-elastic constitutive relation and failure criterion of PMMA at high strain-rates. Combust. Explosion Shock Waves **12**, 333–342 (1992)

Part II
Micromorphic and Cosserat in Gradient Plasticity for Instabilities in Materials and Structures

Micromorphic Approach to Gradient Plasticity and Damage

15

Samuel Forest

Contents

S. Forest (✉)
Centre des Materiaux, Mines ParisTech CNRS, PSL Research University, Paris, Evry Cedex, France
e-mail: samuel.forest@mines-paristech.fr

G. Z. Voyiadjis (ed.), *Handbook of Nonlocal Continuum Mechanics for Materials and Structures*, https://doi.org/10.1007/978-3-319-58729-5_9

Abstract

Eringen and Mindlin's original micromorphic continuum model is presented and extended towards finite elastic-plastic deformations. The framework is generalized to any additional kinematic degrees of freedom related to plasticity and/or damage mechanisms. It provides a systematic method to develop size–dependent plasticity and damage models, closely related to phase field approaches, that can be applied to hardening and/or softening material behavior. The regularization power of the method is illustrated in the case of damage in single crystals. Special attention is given to the various possible finite deformation formulations enhancing existing frameworks for finite elastoplasticity and damage.

Keywords

Gradient plasticity · Gradient damage · Micromorphic media · Regularization · Finite deformations · Generalized continua · Microstrain · Microstretch · Strain localization · Elasto-plasticity · Cleavage

Introduction

Micromorphic media are examples of three-dimensional generalized continua including additional degrees of freedom complementing the usual displacement vector. A classification of generalised mechanical continuum theories is proposed in Fig. 1 in order to locate more precisely the class of micromorphic media. The present chapter is limited to continuum media fulfilling the principle of local action, meaning that the mechanical state at a material point $\underline{X}$ depends on variables defined at this point only (Truesdell and Toupin 1960; Truesdell and Noll 1965). The classical Cauchy continuum is called *simple material* because its response at

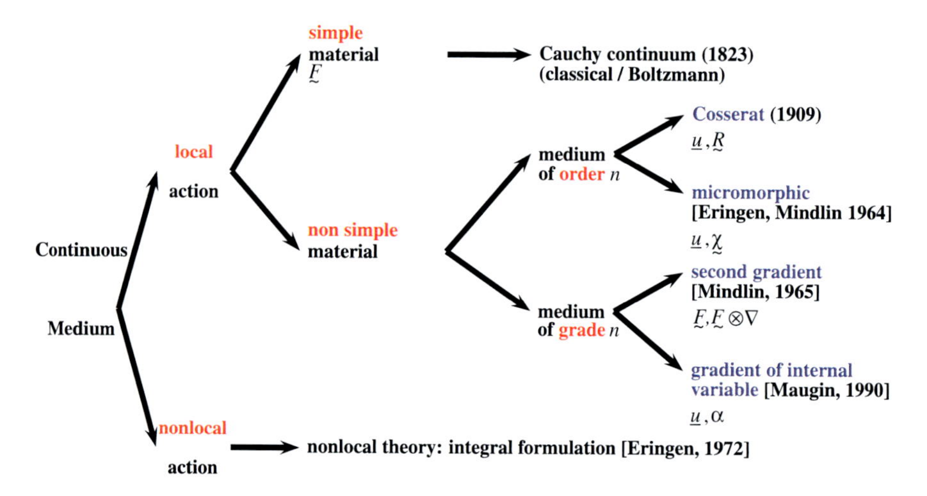

Fig. 1 A classification of the mechanics of generalised continua.

material point $\underline{X}$ to deformations homogeneous in a neighborhood of $\underline{X}$ determines uniquely its response to every deformation at $\underline{X}$. In *higher grade* materials, homogeneous deformations are not sufficient to characterise the material behaviour because they are sensitive to higher gradients of the displacement field. Mindlin formulated for instance the theories that include the second and third gradients of the displacement field (Mindlin 1965). The gradient effect may be limited to the plastic part of deformation which leads to strain gradient plasticity models (Aifantis 1984; Forest and Bertram 2011) or, more generally, theories that include the gradient of some internal variables (Maugin 1990). *Higher order* materials are characterised by additional degrees of freedom of the material points (Eringen 1999). Directors can be attached to each material point that evolve in a different way from the material lines. Cosserat directors can rotate. In the micromorphic continuum designed by Eringen and Suhubi (1964) and Mindlin (1964), the directors can also be distorted, so that a second order tensor is attributed to each material point. Tensors of higher order can even be introduced as proposed in Germain's general micromorphic theory (Germain 1973b; Forest and Sab 2017).

Higher order media are sometimes called continua with *microstructure*. This name is misleading because Cauchy material models can also integrate some aspects of the underlying microstructure as illustrated by classical homogenisation methods used to derive the effective properties of composites. However generalised continua incorporate a feature of the microstructure which is not accounted for by standard homogenisation methods, namely their size–dependent material response. They involve intrinsic lengths directly stemming from the microstructure of the material. The mechanics of generalized continua represents a way of introducing, in the continuum description of materials, some characteristic length scales associated with their microstructure (Mühlhaus 1995). Such intrinsic lengths and generalized constitutive equations can be identified in two ways. Direct identification is possible from experimental curves exhibiting clear size effects in plasticity or fracture or from full–field strain measurements of strongly heterogeneous fields (Geers et al. 1998). The effective properties of such generalized continua can also be derived from scale transition and homogenization techniques by prescribing appropriate boundary conditions on a representative volume of material with microstructure (Cailletaud et al. 2003).

The multiplication of generalized continuum model formulations from Cosserat to strain gradient plasticity in literature may leave an impression of disorder and inconsistency. Recent accounts have shown, on the contrary, that unifying presentations of several classes of generalized continuum theories are possible (Hirschberger and Steinmann 2009; Forest 2009). One of them, called the micromorphic approach, encompasses most theories incorporating additional degrees of freedom from the well–established Cosserat, microstretch and micromorphic continua (Eringen 1999) up to Aifantis and Gurtin strain gradient plasticity theories. Gradient theories are obtained from the micromormphic approach by imposing some internal constraints linking the additional degrees of freedom and other model variables.

The micromorphic theory now arouses strong interest from the materials science and computational mechanics communities because of its regularisation power in the context of softening plasticity and damage and of its rather simple implementation in a finite element program. The number of degrees of freedom is not an obstacle any more with constantly increasing computer power and parallel solvers.

The objective of this chapter is first to present the elastoviscoplasticity theory of micromorphic media at finite deformation. This presentation is based on the fundamental work of Eringen and on recent developments in the context of plasticity. It represents an update of the corresponding chapter in the CISM book (Forest 2012). The second objective of this chapter is to present an extension of the micromorphic theory to other kinds of additional degrees of freedom like plasticity and damage related variables. This micromorphic approach to gradient effects in materials' behavior is a systematic method for incorporating intrinsic lengths in non–linear continuum mechanical models. It is illustrated here in the case of an anisotropic plasticity and damage model. The so–called *microdamage* model takes into account the crystallography of plasticity and fracture in metal single crystals.

The nonlinear theory of micromorphic media is presented in section "The Micromorphic Theory After Eringen and Mindlin and Its Extension to Plasticity." The micromorphic approach is exposed in section "The Micromorphic Approach to Various Gradient Field Theories" together with the closely related phase field approach. Differences and similarities between the micromorphic framework and the phase field approach are pointed out following the general framework provided in Gurtin (1996). A single crystal plasticity and damage model is explored in section "Application to a Continuum Damage Model for Single Crystals and Its Regularization" up to crack propagation simuation. This presentation of the micromorphic approach and the corresponding example are taken from the formulation presented in Aslan and Forest (2011). A first account of the method is given at small strains for the sake of simplicity. It is followed by extensions to finite deformations, following the guidelines given in Forest (2016). The presentation is limited to the static case, the reader being referred to Eringen's original dynamical formulations and to Forest and Sab (2017) for the consideration of inertial terms.

Intrinsic notations are used throughout the chapter. In particular, scalars, vectors, tensors of second, third and fourth ranks are denoted by $a, \underline{\boldsymbol{a}}, \underset{\sim}{\boldsymbol{a}}, \underset{\simeq}{\boldsymbol{a}}, \underset{\approx}{\boldsymbol{a}}$, respectively. Contractions are written as:

$$\underset{\sim}{\boldsymbol{a}} : \underset{\sim}{\boldsymbol{b}} = a_{ij}b_{ij}, \; \underset{\simeq}{\boldsymbol{a}} \vdots \underset{\simeq}{\boldsymbol{b}} = a_{ijk}b_{ijk}, \; \underset{\approx}{\boldsymbol{a}} :: \underset{\approx}{\boldsymbol{b}} = a_{ijkl}b_{ijkl} \tag{1}$$

using the Einstein summation rule for repeated indices. The tensor product is denoted by $\otimes$. For example, the component $\left(\underset{\sim}{\boldsymbol{a}} \otimes \underset{\sim}{\boldsymbol{b}}\right)_{ijkl}$ is $a_{ij}b_{kl}$. A modified tensor product $\boxtimes$ is also used: the component $\left(\underset{\sim}{\boldsymbol{a}} \boxtimes \underset{\sim}{\boldsymbol{b}}\right)_{ijkl}$ is $a_{ik}b_{jl}$.

The gradient operators ∇_x or ∇_X are introduced when the functions depend on microscopic coordinates $\underline{\boldsymbol{x}}$ or macroscopic coordinates $\underline{\boldsymbol{X}}$. The following notation is used:

$$\underline{U} \otimes \nabla_X = U_{i,j} \underline{e}_i \otimes \underline{e}_j, \quad \text{with} \quad U_{i,j} = \frac{\partial U_i}{\partial X_j} \tag{2}$$

$$\underline{u} \otimes \nabla_x = u_{i,j} \underline{e}_i \otimes \underline{e}_j, \quad \text{with} \quad u_{i,j} = \frac{\partial u_i}{\partial x_j} \tag{3}$$

where $(\underline{e}_i)_{i=1,2,3}$ is a Cartesian orthonormal basis.

The Micromorphic Theory After Eringen and Mindlin and Its Extension to Plasticity

Kinematics of Micromorphic Media

The degrees of freedom of the theory are the displacement vector $\underline{u}$ and the microdeformation tensor $\underset{\sim}{\chi}$:

$$DOF := \left\{ \underset{\sim}{u}, \underset{\sim}{\chi} \right\}$$

The current position of the material point is given by the transformation Φ according to $\underline{x} = \Phi(\underline{X}, t) = \underline{X} + \underline{u}(\underline{X}, t)$. The microdeformation describes the deformation of a triad of directors, $\underline{\Xi}^i$ attached to the material point.

$$\underline{\xi}^i(\underline{X}, t) = \underset{\sim}{\chi}(\underline{X}, t) \cdot \underline{\Xi}^i \tag{4}$$

As such, its determinant is taken as strictly positive. The polar decomposition of the generally incompatible microdeformation field $\underset{\sim}{\chi}(\underline{X})$ is introduced

$$\underset{\sim}{\chi} = \underset{\sim}{R}^\sharp \cdot \underset{\sim}{U}^\sharp \tag{5}$$

Internal constraints can be prescribed to the microdeformation. The micromorphic medium reduces to the Cosserat medium when the microdeformation is constrained to be a pure rotation: $\underset{\sim}{\chi} \equiv \underset{\sim}{R}^\sharp$. The microstrain medium is obtained when $\underset{\sim}{\chi} \equiv \underset{\sim}{U}^\sharp$ (Forest and Sievert 2006). Finally, the second gradient theory is retrieved when the microdeformation coincides with the deformation gradient, $\underset{\sim}{\chi} \equiv \underset{\sim}{F}$. A hierarchy of higher order continua can be established by specialising the micromorphic theory and depending on the targeted material class, see Table 1.

The following kinematical quantities are then introduced:

- the velocity field $\underline{v}(\underline{x}, t) := \underline{\dot{u}}\left(\Phi^{-1}(\underline{x}, t)\right)$
- the deformation gradient $\underset{\sim}{F} = \underset{\sim}{1} + \underline{u} \otimes \nabla_X$

Table 1 A hierarchy of higher order continua

Name	Number of DOF	DOF	References
Cauchy	3	$\underline{u}$	Cauchy (1822)
Microdilatation	4	$\underline{u},\ \chi$	Goodman and Cowin (1972) and Steeb and Diebels (2003)
Cosserat	6	$\underline{u},\ \underset{\sim}{R}$	Kafadar and Eringen (1971)
Microstretch	7	$\underline{u},\ \chi,\ \underset{\sim}{R}$	Eringen (1990)
Microstrain	9	$\underline{u},\ \underset{\sim}{C}^{\sharp}$	Forest and Sievert (2006)
Micromorphic	12	$\underline{u},\ \underset{\sim}{\chi}$	Eringen and Suhubi (1964) and Mindlin (1964)

- the velocity gradient $\underline{v} \otimes \nabla_x = \underset{\sim}{\dot{F}} \cdot \underset{\sim}{F}^{-1}$
- the microdeformation rate $\underset{\sim}{\dot{\chi}} \cdot \underset{\sim}{\chi}^{-1}$
- the third rank Lagrangean microdeformation gradient $\underset{\approx}{\underline{K}} := \underset{\sim}{\chi}^{-1} \cdot \underset{\sim}{\chi} \otimes \nabla_X$
- the gradient of the microdeformation rate tensor

$$\left(\underset{\sim}{\dot{\chi}} \cdot \underset{\sim}{\chi}^{-1}\right) \otimes \nabla_x = \underset{\sim}{\chi} \cdot \underset{\approx}{\underline{\dot{K}}} : \left(\underset{\sim}{\chi}^{-1} \boxtimes \underset{\sim}{F}^{-1}\right) \tag{6}$$

and the corresponding index notation:

$$\left(\dot{\chi}_{il}\chi_{lj}^{-1}\right)_{,k} = \chi_{ip}\dot{K}_{pqr}\chi_{qj}^{-1}F_{rk}^{-1}$$

Principle of Virtual Power

The method of virtual power is used to introduce the generalised stress tensors and the field and boundary equations they must satisfy (Germain 1973b).

The modelling variables are introduced according to a first gradient theory:

$$MODEL = \left\{\underline{v},\ \underline{v} \otimes \nabla_x,\ \underset{\sim}{\dot{\chi}} \cdot \underset{\sim}{\chi}^{-1},\ \left(\underset{\sim}{\dot{\chi}} \cdot \underset{\sim}{\chi}^{-1}\right) \otimes \nabla_x\right\}$$

The virtual power of internal forces of a subdomain $\mathcal{D} \subset \mathcal{B}$ of the body is

$$\mathcal{P}^{(i)}\left(\underline{v}^*, \underset{\sim}{\dot{\chi}}^* \cdot \underset{\sim}{\chi}*^{-1}\right) = \int_{\mathcal{D}} p^{(i)}\left(\underline{v}^*, \underset{\sim}{\dot{\chi}}^* \cdot \underset{\sim}{\chi}*^{-1}\right)\, dV$$

The virtual power density of internal forces is a linear form on the fields of virtual modeling variables:

$$
\begin{aligned}
p^{(i)} &= \underset{\sim}{\boldsymbol{\sigma}} : \left(\dot{\underset{\sim}{\boldsymbol{F}}} \cdot \underset{\sim}{\boldsymbol{F}}^{-1}\right) + \underset{\sim}{\boldsymbol{s}} : \left(\dot{\underset{\sim}{\boldsymbol{F}}} \cdot \underset{\sim}{\boldsymbol{F}}^{-1} - \dot{\underset{\sim}{\boldsymbol{\chi}}} \cdot \underset{\sim}{\boldsymbol{\chi}}^{-1}\right) + \underset{\approx}{\boldsymbol{M}} \,\vdots\, \left(\left(\dot{\underset{\sim}{\boldsymbol{\chi}}} \cdot \underset{\sim}{\boldsymbol{\chi}}^{-1}\right) \otimes \nabla_x\right) \\
&= \underset{\sim}{\boldsymbol{\sigma}} : \left(\dot{\underset{\sim}{\boldsymbol{F}}} \cdot \underset{\sim}{\boldsymbol{F}}^{-1}\right) + \underset{\sim}{\boldsymbol{s}} : \left(\underset{\sim}{\boldsymbol{\chi}} \cdot \left(\underset{\sim}{\boldsymbol{\chi}}^{-1} \cdot \underset{\sim}{\boldsymbol{F}}\right)^{\cdot} \cdot \underset{\sim}{\boldsymbol{F}}^{-1}\right) \\
&\quad + \underset{\approx}{\boldsymbol{M}} \,\vdots\, \left(\underset{\sim}{\boldsymbol{\chi}} \cdot \dot{\underset{\simeq}{\boldsymbol{K}}} : \left(\underset{\sim}{\boldsymbol{\chi}}^{-1} \boxtimes \underset{\sim}{\boldsymbol{F}}^{-1}\right)\right)
\end{aligned}
\tag{7}
$$

where the relative deformation rate $\dot{\underset{\sim}{\boldsymbol{F}}} \cdot \underset{\sim}{\boldsymbol{F}}^{-1} - \dot{\underset{\sim}{\boldsymbol{\chi}}} \cdot \underset{\sim}{\boldsymbol{\chi}}^{-1}$ is introduced and expressed in terms of the rate of the relative deformation $\underset{\sim}{\boldsymbol{\chi}}^{-1} \cdot \underset{\sim}{\boldsymbol{F}}$. The virtual power density of internal forces is invariant with respect to virtual rigid body motions so that $\underset{\sim}{\boldsymbol{\sigma}}$ must be symmetric. The generalised stress tensors conjugate to the velocity gradient, the relative deformation rate and the gradient of the microdeformation rate are the simple stress tensor $\underset{\sim}{\boldsymbol{\sigma}}$, the relative stress tensor $\underset{\sim}{\boldsymbol{s}}$ and the double stress tensor $\underset{\approx}{\boldsymbol{M}}$ of third rank.

The Gauss theorem is then applied to the power of internal forces.

$$
\begin{aligned}
\int_{\mathcal{D}} p^{(i)} dV &= \int_{\partial\mathcal{D}} \underline{\boldsymbol{v}}^* \cdot \left(\underset{\sim}{\boldsymbol{\sigma}} + \underset{\sim}{\boldsymbol{s}}\right) \cdot \underline{\boldsymbol{n}}\, dS + \int_{\partial\mathcal{D}} \left(\dot{\underset{\sim}{\boldsymbol{\chi}}}^* \cdot \underset{\sim}{\boldsymbol{\chi}}^{*-1}\right) : \underset{\approx}{\boldsymbol{M}} \cdot \underline{n}\, dS \\
&\quad - \int_{\mathcal{D}} \underline{\boldsymbol{v}}^* \cdot \left(\underset{\sim}{\boldsymbol{\sigma}} + \underset{\sim}{\boldsymbol{s}}\right) \cdot \nabla_x dV \\
&\quad - \int_{\mathcal{D}} \left(\dot{\underset{\sim}{\boldsymbol{\chi}}}^* \cdot \underset{\sim}{\boldsymbol{\chi}}^{*-1}\right) : \left(\underset{\approx}{\boldsymbol{M}} \cdot \nabla_x + \underset{\sim}{\boldsymbol{s}}\right) \cdot dV
\end{aligned}
$$

The form of the previous boundary integral dictates the possible form of the power of contact forces acting on the boundary $\partial\mathcal{D}$ of the subdomain $\mathcal{D} \subset \mathcal{B}$.

$$
\begin{aligned}
\mathcal{P}^{(c)}\left(\underline{\boldsymbol{v}}^*, \dot{\underset{\sim}{\boldsymbol{\chi}}}^* \cdot \underset{\sim}{\boldsymbol{\chi}}^{*-1}\right) &= \textstyle\int_{\partial\mathcal{D}} p^{(c)}\left(\underline{\boldsymbol{v}}^*, \dot{\underset{\sim}{\boldsymbol{\chi}}}^* \cdot \underset{\sim}{\boldsymbol{\chi}}^{*-1}\right) dV \\
p^{(c)}\left(\underline{\boldsymbol{v}}^*, \dot{\underset{\sim}{\boldsymbol{\chi}}}^* \cdot \underset{\sim}{\boldsymbol{\chi}}^{*-1}\right) &= \underline{\boldsymbol{t}} \cdot \underline{\boldsymbol{v}}^* + \underset{\sim}{\boldsymbol{m}} : \left(\dot{\underset{\sim}{\boldsymbol{\chi}}}^* \cdot \underset{\sim}{\boldsymbol{\chi}}^{*-1}\right)
\end{aligned}
$$

where the simple traction $\underline{\boldsymbol{t}}$ and double traction $\underset{\sim}{\boldsymbol{m}}$ are introduced.

The power of forces acting at a distance is defined as

$$
\begin{aligned}
\mathcal{P}^{(e)}\left(\underline{\boldsymbol{v}}^*, \dot{\underset{\sim}{\boldsymbol{\chi}}}^* \cdot \underset{\sim}{\boldsymbol{\chi}}^{*-1}\right) &= \textstyle\int_{\mathcal{D}} p^{(e)}\left(\underline{\boldsymbol{v}}^*, \dot{\underset{\sim}{\boldsymbol{\chi}}}^* \cdot \underset{\sim}{\boldsymbol{\chi}}^{*-1}\right) dV \\
p^{(e)}\left(\underline{\boldsymbol{v}}^*, \dot{\underset{\sim}{\boldsymbol{\chi}}}^* \cdot \underset{\sim}{\boldsymbol{\chi}}^{*-1}\right) &= \underline{\boldsymbol{f}} \cdot \underline{\boldsymbol{v}}^* + \underset{\sim}{\boldsymbol{p}} : \left(\dot{\underset{\sim}{\boldsymbol{\chi}}}^* \cdot \underset{\sim}{\boldsymbol{\chi}}^{*-1}\right)
\end{aligned}
$$

including simple body forces $\underline{f}$ and double body forces $\underset{\sim}{p}$. More general double and triple volume forces could also be incorporated according to Germain (1973b).

The principle of virtual power is now stated in the static case,

$$\forall \underline{v}^*, \forall \underset{\sim}{\chi}^*, \forall \mathcal{D} \subset \mathcal{B}, \mathcal{P}^{(i)}\left(\underline{v}^*, \dot{\underset{\sim}{\chi}}^* \cdot \underset{\sim}{\chi}^{*-1}\right) = \mathcal{P}^{(c)}\left(\underline{v}^*, \dot{\underset{\sim}{\chi}}^* \cdot \underset{\sim}{\chi}^{*-1}\right) + \mathcal{P}^{(e)}\left(\underline{v}^*, \dot{\underset{\sim}{\chi}}^* \cdot \underset{\sim}{\chi}^{*-1}\right)$$

This variational formulation leads to.

$$\begin{aligned}
&\int_{\partial\mathcal{D}} \underline{v}^* \cdot \left(\underset{\sim}{\sigma} + \underset{\sim}{s}\right) \cdot \underline{n}\, ds + \int_{\partial\mathcal{D}} \left(\dot{\underset{\sim}{\chi}}^* \cdot \underset{\sim}{\chi}^{*-1}\right) : \underset{\approx}{M} \cdot \underline{n}\, dS \\
&- \int_{\mathcal{D}} \underline{v}^* \cdot \left(\left(\underset{\sim}{\sigma} + \underset{\sim}{s}\right) \cdot \nabla_x + \underline{f}\right) dV \\
&- \int_{\mathcal{D}} \left(\dot{\underset{\sim}{\chi}}^* \cdot \underset{\sim}{\chi}^{*-1}\right) : \left(\underset{\approx}{M} \cdot \nabla_x + \underset{\sim}{s} + \underset{\sim}{p}\right) dV = 0
\end{aligned}$$

which delivers the field equations of the problem (Kirchner and Steinmann 2005; Lazar and Maugin 2007; Hirschberger et al. 2007):

- balance of momentum equation (static case)

$$\left(\underset{\sim}{\sigma} + \underset{\sim}{s}\right) \cdot \nabla_x + \underline{f} = 0, \quad \forall \underline{x} \in \mathcal{B} \tag{8}$$

- balance of generalized moment of momentum equation (static case)

$$\underset{\approx}{M} \cdot \nabla_x + \underset{\sim}{s} + \underset{\sim}{p} = 0, \; \forall \underline{x} \in \mathcal{B} \tag{9}$$

- boundary conditions

$$\left(\underset{\sim}{\sigma} + \underset{\sim}{s}\right) \cdot \underline{n} = \underline{t}, \quad \forall \underline{x} \in \partial\mathcal{B} \tag{10}$$

$$\underset{\approx}{M} \cdot \underline{n} = \underset{\sim}{m}, \quad \forall \underline{x} \in \partial\mathcal{B} \tag{11}$$

Elastoviscoplasticity of Micromorphic Media

This section is dedicated to the formulation of constitutive equations for micromorphic media. The general case of hyperelastic-viscoplastic materials is considered.

Elastic–Plastic Decomposition of the Generalised Strain Measures

According to Eringen (1999), the following Lagrangean strain measures are adopted:

$$STRAIN = \left\{ \underset{\sim}{C} := \underset{\sim}{F}^T \cdot \underset{\sim}{F},\ \underset{\sim}{\Upsilon} := \underset{\sim}{\chi}^{-1} . \underset{\sim}{F},\ \underset{\approx}{K} := \underset{\sim}{\chi}^{-1} . \left(\underset{\sim}{\chi} \otimes \nabla_X \right) \right\}$$

i.e. the Cauchy–Green strain tensor, the relative deformation and the microdeformation gradient.

In the presence of plastic deformation, the question arises of splitting the previous Lagrangean strain measures into elastic and plastic contributions. Following Mandel (1973), a multiplicative decomposition of the deformation gradient is postulated:

$$\underset{\sim}{F} = \underset{\sim}{F}^e \cdot \underset{\sim}{F}^p = \underset{\sim}{R}^e \cdot \underset{\sim}{U}^e \cdot \underset{\sim}{F}^p \tag{12}$$

which defines an intermediate local configuration at each material point, see Fig. 2. Uniqueness of the decomposition requires the suitable definition of directors. Such directors are available in any micromorphic theory.

A multiplicative decomposition of the microdeformation is also considered:

$$\underset{\sim}{\chi} = \underset{\sim}{\chi}^e \cdot \underset{\sim}{\chi}^p = \underset{\sim}{R}^{e\sharp} \cdot \underset{\sim}{U}^{e\sharp} \cdot \underset{\sim}{\chi}^p \tag{13}$$

according to Forest and Sievert (2003, 2006). The uniqueness of the decomposition also requires the suitable definition of directors. As an example, lattice directions in a single crystal are physically relevant directors for an elastoviscoplasticity micromorphic theory, see Aslan et al. (2011). Finally, a partition rule must also be proposed for the third strain measure, namely the microdeformation gradient. Sansour (1998a, b) introduced an additive decomposition of curvature:

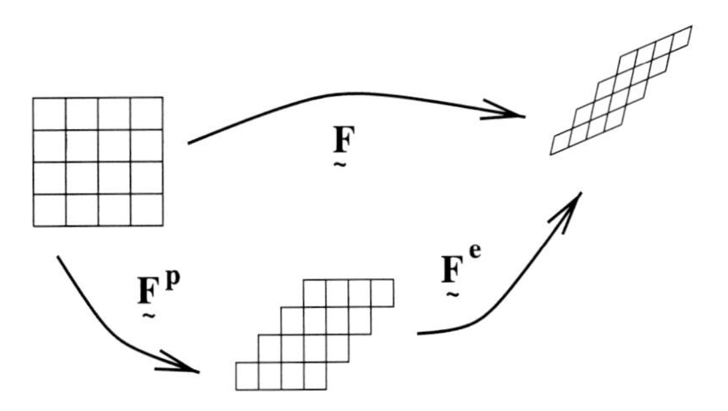

Fig. 2 Multiplicative decomposition of the deformation gradient

$$\underset{\approx}{\boldsymbol{K}} = \underset{\approx}{\boldsymbol{K}}^e + \underset{\approx}{\boldsymbol{K}}^p \tag{14}$$

A quasi–additive decomposition was proposed by Forest and Sievert (2003) with the objective of defining an intermediate local configuration for which all generalised stress tensor are simultaneously released, as it will become apparent in the next section:

$$\underset{\approx}{\boldsymbol{K}} = \underset{\sim}{\chi}^{p-1} \cdot \underset{\cong}{\boldsymbol{K}}^e : \left(\underset{\sim}{\chi}^p \boxtimes \underset{\sim}{\boldsymbol{F}}^p \right) + \underset{\approx}{\boldsymbol{K}}^p \tag{15}$$

Constitutive Equations

The continuum thermodynamic formulation is essentially unchanged in the presence of additional degrees of freedom provided that all functionals are properly extended to the new sets of variables. The local equation of energy balance is written in its usual form:

$$\rho \dot{\epsilon} = p^{(i)} - \underset{-}{\boldsymbol{q}} \cdot \nabla + r \tag{16}$$

where ε is the specific internal energy density, and $p^{(i)}$ is the power density of internal forces according to Eq. (7). The heat flux vector is $\underset{-}{\boldsymbol{q}}$ and r is a heat source term (Fig. 3). The local form of the second principle of thermodynamics is written as.

$$\rho \dot{\eta} + \left(\frac{\underset{-}{\boldsymbol{q}}}{T} \right) . \nabla - \frac{r}{T} \geq 0$$

where η is the specific entropy density. Introducing the Helmholtz free energy function the ψ, second law becomes

$$p^{(i)} - \rho \dot{\Psi} - \eta \dot{T} - \frac{\underset{-}{\boldsymbol{q}}}{T} \cdot (\nabla T) \geq 0$$

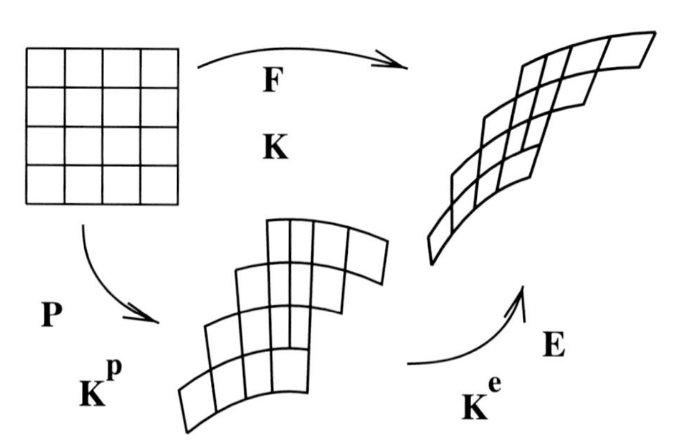

Fig. 3 Definition of an intermediate local configuration for micromorphic elastoplasticity

The state variables of the elastoviscoplastic micromorphic material are all the elastic strain measures and a set of internal variables q. The free energy density is a function of the state variables:

$$\Psi\left(\underset{\sim}{C}^e := \underset{\sim}{F}^{eT}.\underset{\sim}{F}^e,\ \underset{\sim}{\Upsilon}^e := \underset{\sim}{\chi}^{e-1}\cdot\underset{\sim}{F}^e,\ \underset{\approx}{K}^e, q\right)$$

The exploitation of the entropy inequality leads to the definition of the hyperelastic state laws in the form:

$$\underset{\sim}{\sigma} = 2\underset{\sim}{F}^e\cdot\rho\frac{\partial\Psi}{\partial\underset{\sim}{C}^e}\cdot\underset{\sim}{F}^{eT},\ \underset{\sim}{s} = \underset{\sim}{R}^{e\sharp}\cdot\underset{\sim}{U}^{e\sharp-1}\cdot\rho\frac{\partial\Psi}{\partial\underset{\sim}{\Upsilon}^e}\cdot\underset{\sim}{F}^{eT}$$

$$\underset{\approx}{M} = \underset{\sim}{\chi}^{-T}\cdot\rho\frac{\partial\Psi}{\partial\underset{\approx}{K}^e} : \left(\underset{\sim}{\chi}^T\boxtimes\underset{\sim}{F}^T\right) \tag{17}$$

while the entropy density is still given by $\eta = -\frac{\partial\Psi}{\partial T}$. The thermodynamic force associated with the internal variable q is

$$R = -\rho\frac{\partial\Psi}{\partial q}$$

The hyperelasticity law (17) for the double stress tensor was derived for the additive decomposition (14). The quasi–additive decomposition (15) leads to an hyperelastic constitutive equation for the conjugate stress $\underset{\approx}{M}$ in the current configuration, that has also the same form as for pure hyperelastic behaviour. One finds:

$$\underset{\approx}{M} = \underset{\sim}{\chi}^{e-T}\cdot\rho\frac{\partial\Psi}{\partial\underset{\approx}{K}^e} : \left(\underset{\sim}{\chi}^{eT}\boxtimes\underset{\sim}{F}^{eT}\right) \tag{18}$$

The residual intrinsic dissipation is

$$D = \underset{\sim}{\Sigma} : \left(\underset{\sim}{\dot{F}}^p.\underset{\sim}{F}^{p-1}\right) + \underset{\sim}{\mathcal{S}} : \left(\underset{\sim}{\dot{\chi}}^p.\underset{\sim}{\chi}^{p-1}\right) + \underset{\approx}{\mathcal{M}}\vdots\underset{\approx}{\dot{K}}^p + R\dot{q} \geq 0$$

where generalised Mandel stress tensors have been defined.

$$\underset{\sim}{\Sigma} = \underset{\sim}{F}^{eT}\cdot\left(\underset{\sim}{\sigma}+\underset{\sim}{s}\right)\cdot\underset{\sim}{F}^{e-T},\ \underset{\sim}{\mathcal{S}} = -\underset{\sim}{U}^{e\sharp}\cdot\underset{\sim}{R}^{e\sharp T}\cdot\underset{\sim}{s}\cdot\underset{\sim}{R}^{e\sharp}\cdot\underset{\sim}{U}^{e\sharp-1} \tag{19}$$

$$\underset{\approx}{\mathcal{M}} = \underset{\sim}{\chi}^T\cdot\underset{\simeq}{S} : \left(\underset{\sim}{\chi}^{-T}\boxtimes\underset{\sim}{F}^{-T}\right) \tag{20}$$

At this stage, one may define a dissipation potential, function of the Mandel stress tensors, from which the viscoplastic flow rule and the evolution equations for the internal variables are derived.

$$\Omega\left(\underset{\sim}{\boldsymbol{\Sigma}}, \underset{\sim}{\boldsymbol{\mathcal{S}}}, \underset{\approx}{\boldsymbol{\mathcal{M}}}, R\right)$$

$$\dot{\underset{\sim}{\boldsymbol{F}}}^{p}.\underset{\sim}{\boldsymbol{F}}^{p-1} = \frac{\partial\Omega}{\partial\underset{\sim}{\boldsymbol{\Sigma}}}, \dot{\underset{\sim}{\boldsymbol{\chi}}}^{p}.\underset{\sim}{\boldsymbol{\chi}}^{p-1} = \frac{\partial\Omega}{\partial\underset{\sim}{\boldsymbol{\mathcal{S}}}}, \dot{\underset{\approx}{\boldsymbol{K}}}^{p} = \frac{\partial\Omega}{\partial\underset{\approx}{\boldsymbol{\mathcal{M}}}}, \dot{q} = \frac{\partial\Omega}{\partial R}$$

The convexity of the dissipation potential with respect to its arguments ensures the positivity of the dissipation rate at each instant.

Explicit constitutive equations can be found in Forest and Sievert (2003), Grammenoudis and Tsakmakis (2009), Grammenoudis et al. (2009), Regueiro (2010), and Sansour et al. (2010). Examples of application of elastoplastic micromorphic media can be found in Dillard et al. (2006) for plasticity and failure of metallic foams.

The Micromorphic Approach to Various Gradient Field Theories

The previous micromorphic model can be extended to other types of additional degrees of freedom. This leads to a systematic approach for the construction of generalized continuum models with enriched kinematics. The method is presented first within the small deformation framework. The general formulation is discussed in the next section.

Thermomechanics with Additional Degrees of Freedom

One starts from an elastoviscoplasticity model formulation within the framework of the classical Cauchy continuum and classical continuum thermodynamics according to Germain et al. (1983) and Maugin (1999). The material behaviour is characterized by the reference sets of degrees of freedom and state variables.

$$DOF0 = \{\underline{\boldsymbol{u}}\}, \; STATE0 = \left\{\underset{\sim}{\varepsilon}, T, q\right\} \tag{21}$$

which the free energy density function ψ may depend on. The small strain tensor is denoted by $\underset{\sim}{\varepsilon}$ whereas q represents the whole set of internal variables of arbitrary tensorial order accounting for nonlinear processes at work inside the material volume element, like isotropic and kinematic hardening variables. The absolute temperature is T.

Additional degrees of freedom ϕ_χ are then introduced in the previous original model. They may be of any tensorial order and of different physical nature (deformation, plasticity or damage variable). The notation χ indicates that these variables eventually represent some microstructural features of the material so that we will call them micromorphic variables or microvariables (*microdeformation, microdamage...*). The *DOF* and *STATE* spaces are extended as follows:

$$DOF = \{\underline{u},\ \phi_\chi\},\ STATE = \left\{\underset{\sim}{\varepsilon}, T, q,\ \phi_\chi, \nabla\phi_\chi\right\} \tag{22}$$

Depending on the physical nature of ϕ_χ, it may or may not be a state variable. For instance, if the microvariable is a microrotation as in the Cosserat model, it is not a state variable for objectivity reasons and will appear in *STATE* only in combination with the macrorotation. In contrast, if the microvariable is a microplastic equivalent strain, as in Aifantis model, it then explicitly appears in the state space.

The virtual power of internal forces is then extended to the power done by the micromorphic variable and its first gradient:

$$\begin{aligned} \mathcal{P}^{(i)}\left(\underline{v}^*, \dot{\phi}_\chi^*\right) &= -\int_{\mathcal{D}} p^{(i)}\left(\underline{v}^*, \dot{\phi}_\chi^*\right) dV \\ p^{(i)}\left(\underline{v}^*, \dot{\phi}_\chi^*\right) &= \underset{\sim}{\sigma} : \nabla\underline{v}^* + a\dot{\phi}_\chi^* + \underline{b}\cdot\nabla\dot{\phi}_\chi^* \end{aligned} \tag{23}$$

where $\mathcal{D}$ is a subdomain of the current configuration of the body. The Cauchy stress is $\underset{\sim}{\sigma}$ and a and $\underline{b}$ are generalized stresses associated with the micromorphic variable and its first gradient. Similarly, the power of contact forces must be extended as follows:

$$\mathcal{P}^{(c)}\left(\underline{v}^*, \dot{\phi}_\chi^*\right) = \int_{\mathcal{D}} p^{(c)}\left(\underline{v}^*, \dot{\phi}_\chi^*\right) dV,\quad p^{(c)}\left(\underline{v}^*, \dot{\phi}_\chi^*\right) = \underline{t}.\underline{v}^* + a^c\dot{\phi}_\chi^* \tag{24}$$

where $\underline{t}$ is the traction vector and a^c a generalized traction. For conciseness, we do not extend the power of forces acting at a distance and keep the classical form:

$$\mathcal{P}^{(e)}\left(\underline{v}^*, \dot{\phi}_\chi^*\right) = \int_{\mathcal{D}} p^{(e)}\left(\underline{v}^*, \dot{\phi}_\chi^*\right) dV,\quad p^{(e)}\left(\underline{v}^*, \dot{\phi}_\chi^*\right) = \rho\underline{f}.\underline{v}^* \tag{25}$$

where $\rho\underline{f}$ accounts for given simple body forces. Following Germain (1973a), given body couples and double forces working with the gradient of the velocity field could also be introduced in the theory. The generalized principle of virtual power with respect to the velocity and micromorphic variable fields, is presented here in the static case only:

$$\mathcal{P}^{(i)}\left(\underline{v}^*, \dot{\phi}_\chi^*\right) + \mathcal{P}^{(e)}\left(\underline{v}^*, \dot{\phi}_\chi^*\right) + \mathcal{P}^{(c)}\left(\underline{v}^*, \dot{\phi}_\chi^*\right) = 0,\ \forall\mathcal{D}\subset\Omega,\ \forall\underline{v}^*, \dot{\phi}_\chi^* \tag{26}$$

The method of virtual power according to Maugin (1980) is used then to derive the standard local balance of momentum equation:

$$\operatorname{div}\underset{\sim}{\sigma} + \rho\underline{f} = 0,\ \forall\underline{x}\in\Omega \tag{27}$$

and the generalized balance of micromorphic momentum equation:

$$\operatorname{div}\underline{b} - a = 0,\ \forall\underline{x}\in\Omega \tag{28}$$

The method also delivers the associated boundary conditions for the simple and generalized tractions:

$$\underline{t} = \underset{\sim}{\sigma}.\underline{n},\ a^c = \underline{b}.\underline{n},\ \forall \underline{x} \in \partial\mathcal{D} \tag{29}$$

The local balance of energy is also enhanced by the generalized micromorphic power already included in the power of internal forces (23):

$$\rho\dot{\epsilon} = p^{(i)} - \text{div}\ \underline{q} + \rho r \tag{30}$$

where ϵ is the specific internal energy, $\underline{q}$ the heat flux vector and r denotes external heat sources. The entropy principle takes the usual local form:

$$-\rho\left(\dot{\psi} + \eta\dot{T}\right) + p^{(i)} - \frac{\underline{q}}{T}.\nabla T \geq 0 \tag{31}$$

where it is assumed that the entropy production vector is still equal to the heat vector divided by temperature, as in classical thermomechanics according to Coleman and Noll (1963). Again, the enhancement of the theory goes through the enriched power density of internal forces (7). The entropy principle is exploited according to classical continuum thermodynamics to derive the state laws. At this stage it is necessary to be more specific on the dependence of the state functions ψ, η, $\underset{\sim}{\sigma}$,a, $\underline{b}$ on state variables and to distinguish between dissipative and non–dissipative mechanisms. The introduction of dissipative mechanisms may require an increase in the number of state variables. These different situations are considered in the following subsections.

Non–dissipative Contribution of Generalized Stresses

Dissipative events are assumed here to enter the model only via the classical mechanical part. Total strain is split into elastic and plastic parts:

$$\underset{\sim}{\varepsilon} = \underset{\sim}{\varepsilon}^e + \underset{\sim}{\varepsilon}^p \tag{32}$$

The following constitutive functional dependencies are then introduced

$$\begin{aligned} &\psi = \widehat{\psi}\left(\underset{\sim}{\varepsilon}^e, T, q, \phi_\chi, \nabla\phi_\chi\right),\ \underset{\sim}{\sigma} = \underset{\sim}{\widehat{\sigma}}\left(\underset{\sim}{\varepsilon}^e, T, q, \phi_\chi, \nabla\phi_\chi\right),\\ &\eta = \widehat{\eta}\left(\underset{\sim}{\varepsilon}^e, T, q, \phi_\chi, \nabla\phi_\chi\right)\ \ a = \widehat{a}\left(\underset{\sim}{\varepsilon}^e, T, q, \phi_\chi, \nabla\phi_\chi\right),\\ &\underline{b} = \underline{\widehat{b}}\left(\underset{\sim}{\varepsilon}^e, T, q, \phi_\chi, \nabla\phi_\chi\right) \end{aligned} \tag{33}$$

The entropy inequality (31) can be expanded as:

$$\left(\underset{\sim}{\boldsymbol{\sigma}} - \rho\frac{\partial\widehat{\psi}}{\partial\underset{\sim}{\boldsymbol{\varepsilon}}^e}\right) : \dot{\underset{\sim}{\boldsymbol{\varepsilon}}}^e + \rho\left(\eta + \frac{\partial\widehat{\psi}}{\partial T}\right)\dot{T} + \left(a - \rho\frac{\partial\widehat{\psi}}{\partial\phi_\chi}\right)\dot{\phi}_\chi$$
$$+ \left(\underline{\boldsymbol{b}} - \rho\frac{\partial\widehat{\psi}}{\partial\nabla\phi_\chi}\right).\nabla\dot{\phi}_\chi + \underset{\sim}{\boldsymbol{\sigma}} : \dot{\underset{\sim}{\boldsymbol{\varepsilon}}}^p - \rho\frac{\partial\widehat{\psi}}{\partial q}\dot{q} - \frac{\underline{\boldsymbol{q}}}{T}.\nabla T \geq 0 \tag{34}$$

Assuming that no dissipation is associated with the four first terms of the previous inequality, the following state laws are found.

$$\underset{\sim}{\boldsymbol{\sigma}} = \rho\frac{\partial\widehat{\psi}}{\partial\underset{\sim}{\boldsymbol{\varepsilon}}^e},\ \eta = -\frac{\partial\widehat{\psi}}{\partial T},\ R = -\rho\frac{\partial\widehat{\psi}}{\partial q} \tag{35}$$

$$a = \rho\frac{\partial\widehat{\psi}}{\partial\phi_\chi}, \quad \underline{\boldsymbol{b}} = \rho\frac{\partial\widehat{\psi}}{\partial\nabla\phi_\chi} \tag{36}$$

and the residual dissipation is.

$$D^{res} = W^p + R\dot{q} - \frac{\underline{q}}{T}\cdot\nabla T \geq 0 \tag{37}$$

where W^p represents the (visco–)plastic power and R the thermodynamic force associated with the internal variable q. The existence of a convex dissipation potential, $\Omega\left(\underset{\sim}{\boldsymbol{\sigma}}, R\right)$ depending on the thermodynamic forces can then be assumed from which the evolution rules for internal variables are derived, that identically fulfill the entropy inequality, as usually done in classical continuum thermomechanics (Germain et al. 1983):

$$\dot{\underset{\sim}{\boldsymbol{\varepsilon}}}^p = \frac{\partial\Omega}{\partial\underset{\sim}{\boldsymbol{\sigma}}}, \quad \dot{q} = \frac{\partial\Omega}{\partial R} \tag{38}$$

Micromorphic Model

After presenting the general approach, we readily give the most simple example which provides a direct connection to several existing generalized continuum models. An element ϕ is selected in the STATE0 set, see Eq. (21) or among other variables present in the original model. Cases are first considered where ϕ and ϕ_χ are observer invariant quantities. The free energy density function ψ is chosen as a function of the generalized relative strain variable e defined as:

$$e = \phi - \phi_\chi \tag{39}$$

thus introducing a coupling between macro and micromorphic variables. Assuming isotropic material behavior for brevity, the additional contributions to the free energy can be taken as quadratic functions of e and $\nabla\phi_\chi$:

$$\psi\left(\underset{\sim}{\boldsymbol{\varepsilon}}, T, q, \phi_\chi, \nabla\phi\chi\right) = \psi^{(1)}\left(\underset{\sim}{\boldsymbol{\varepsilon}}, T, q\right) + \psi^{(2)}\left(e = \phi - \phi_\chi, \nabla\phi_\chi, T\right), \quad \text{with} \tag{40}$$

$$\rho\psi^{(2)}\left(e, \nabla\phi_\chi, T\right) = \frac{1}{2}H_\chi(\phi - \phi_\chi)^2 + \frac{1}{2}A\nabla\phi_\chi \cdot \nabla\phi_\chi \tag{41}$$

where H_χ and A are the additional moduli introduced by the micrmorphic model. After inserting the state laws (36)

$$a = \rho\frac{\partial\psi}{\partial\phi_\chi} = -H_\chi\left(\phi - \phi_\chi\right), \underline{\boldsymbol{b}} = \rho\frac{\partial\psi}{\partial\nabla\phi_\chi} = A\nabla\phi_\chi \tag{42}$$

into the additional balance equation (28), the following partial differential equation is obtained, at least for a homogeneous material under isothermal conditions:

$$\phi = \phi_\chi - \frac{A}{H_\chi}\Delta\phi_\chi \tag{43}$$

where Δ is the Laplace operator. This type of equation is encountered at several places in the mechanics of generalized continua especially in the linear micromorphic theory (Mindlin 1964; Eringen 1999; Dillard et al. 2006) and in the so–called implicit gradient theory of plasticity and damage (Peerlings et al. 2001, 2004; Engelen et al. 2003). Note however that this equation corresponds to a special quadratic potential and represents the simplest micromorphic extension of the classical theory. It involves a characteristic length scale defined by:

$$l_c^2 = \frac{A}{H_\chi} \tag{44}$$

This length is real for positive values of the ratio A/H_χ. The additional material parameters H_χ and A are assumed to be positive in this work. This does not exclude a softening material behaviour that can be induced by the proper evolution of the internal variables (including $\phi = q$ itself).

Viscous Generalized Stress and Phase Field Model

Generalized stresses can also be associated with dissipation by introducing the viscous part a^v of a:

$$\underset{\sim}{\boldsymbol{\varepsilon}} = \underset{\sim}{\boldsymbol{\varepsilon}}^e + \underset{\sim}{\boldsymbol{\varepsilon}}^p, \; a = a^e + a^v \tag{45}$$

The entropy inequality (31) now becomes

$$\left(\underset{\sim}{\boldsymbol{\sigma}} - \rho\frac{\partial\widehat{\psi}}{\partial\underset{\sim}{\boldsymbol{\varepsilon}}^e}\right) : \underset{\sim}{\dot{\boldsymbol{\varepsilon}}}^e + \rho\left(\eta + \frac{\partial\widehat{\psi}}{\partial T}\right)\dot{T} + \left(a^e - \rho\frac{\partial\widehat{\psi}}{\partial\phi_\chi}\right)\dot{\phi}_\chi + \left(\underline{\boldsymbol{b}} - \rho\frac{\partial\widehat{\psi}}{\partial\nabla\phi_\chi}\right).\nabla\dot{\phi}_\chi$$
$$+\underset{\sim}{\boldsymbol{\sigma}} : \underset{\sim}{\dot{\boldsymbol{\varepsilon}}}^p - \rho\frac{\partial\widehat{\psi}}{\partial q}\dot{q} + a^\upsilon\dot{\phi}_\chi - \frac{\underline{\boldsymbol{q}}}{T}.\nabla T \geq 0 \tag{46}$$

Assuming that no dissipation is associated with the four first terms of the previous inequality, the following state laws are found

$$\underset{\sim}{\boldsymbol{\sigma}} = \rho\frac{\partial\widehat{\psi}}{\partial\underset{\sim}{\boldsymbol{\varepsilon}}^e},\ \eta = -\frac{\partial\widehat{\psi}}{\partial T},\ R = -\rho\frac{\partial\widehat{\psi}}{\partial q} \tag{47}$$

$$a^e = \rho\frac{\partial\widehat{\psi}}{\partial\phi_\chi},\ \underline{\boldsymbol{b}} = \rho\frac{\partial\widehat{\psi}}{\partial\nabla\phi_\chi} \tag{48}$$

and the residual dissipation is

$$D^{res} = \underset{\sim}{\boldsymbol{\sigma}} : \underset{\sim}{\dot{\boldsymbol{\varepsilon}}}^p + R\dot{q} + a^\upsilon\dot{\phi}_\chi - \frac{\underline{\boldsymbol{q}}}{T}.\nabla T \geq 0 \tag{49}$$

Evolution rules for viscoplastic strain, internal variables, and the additional degrees of freedom can be derived from a dissipation potential $\Omega\left(\underset{\sim}{\boldsymbol{\sigma}}, R, a^\upsilon\right)$:

$$\underset{\sim}{\dot{\boldsymbol{\varepsilon}}}^p = \frac{\partial\Omega}{\partial\underset{\sim}{\boldsymbol{\sigma}}},\ \dot{q} = \frac{\partial\Omega}{\partial R},\ \dot{\phi}_\chi = \frac{\partial\Omega}{\partial a^\upsilon} \tag{50}$$

Convexity of the dissipation potential then ensures positivity of dissipation rate for any process.

Note that no dissipative part has been assigned to the generalized stress $\underline{\boldsymbol{b}}$ since then exploitation of second principle does not seem to be straightforward. Instead, the total gradient $\nabla\phi_\chi$ can be split into elastic and plastic parts, as it will be done in section "Elastic-Plastic Decomposition of Generalized Strains."

Phase Field Model

The dissipation potential can be decomposed into the various contributions due to all thermodynamic forces. Let us assume for instance that the contribution of the viscous generalized stress a^υ is quadratic:

$$\Omega\left(\underset{\sim}{\boldsymbol{\sigma}}, R, a^\upsilon\right) = \Omega_1\left(\underset{\sim}{\boldsymbol{\sigma}}, R\right) + \Omega_2\left(a^\upsilon\right),\ \Omega_2\left(a^\upsilon\right) = \frac{1}{2\beta}a^{\upsilon 2} \tag{51}$$

The use of the flow rule (50) and of the additional balance equation (28) then leads to

$$\beta \dot{\phi} = a^{v} = a - a^{e} = a - \rho \frac{\partial \widehat{\psi}}{\partial \phi_{\chi}} = \operatorname{div}\left(\rho \frac{\partial \widehat{\psi}}{\partial \nabla \phi_{\chi}}\right) - \rho \frac{\partial \widehat{\psi}}{\partial \phi_{\chi}} \tag{52}$$

One recognizes the Landau–Ginzburg equation that arises in phase field theories. The previous derivation of Landau–Ginzburg equation is due to Gurtin (1996), see also Ammar et al. (2009). The coupling with mechanics is straightforward according to this procedure and more general dissipative mechanisms can be put forward, see Rancourt et al. (2016).

Elastic–Plastic Decomposition of Generalized Strains

Instead of the previous decomposition of generalized stresses, the introduction of additional dissipative mechanisms can rely on the split of all strain measures into elastic and plastic parts:

$$\underset{\sim}{\boldsymbol{\varepsilon}} = \underset{\sim}{\boldsymbol{\varepsilon}}^{e} + \underset{\sim}{\boldsymbol{\varepsilon}}^{p},\ \phi_{\chi} = \phi_{\chi}^{e} + \phi_{\chi}^{p},\ \underline{\boldsymbol{\kappa}} = \nabla \phi_{\chi} = \underline{\boldsymbol{\kappa}}^{e} + \underline{\boldsymbol{\kappa}}^{p} \tag{53}$$

The objectivity of ϕ_{χ} is required for such a unique decomposition to be defined. We do not require here that

$$\underline{\boldsymbol{\kappa}}^{e} = \nabla \phi_{\chi}^{e},\ \underline{\boldsymbol{\kappa}}^{p} = \nabla \phi_{\chi}^{p} \tag{54}$$

although such a model also is possible, as illustrated by the *gradient of strain theory* put forward in Forest and Sievert (2003). The Clausius–Duhem inequality then writes

$$\begin{aligned}&\left(\underset{\sim}{\boldsymbol{\sigma}} - \rho \frac{\partial \widehat{\psi}}{\partial \underset{\sim}{\boldsymbol{\varepsilon}}^{e}}\right) : \underset{\sim}{\dot{\boldsymbol{\varepsilon}}}^{e} + \rho\left(\eta + \frac{\partial \widehat{\psi}}{\partial T}\right) \dot{T} + \left(a - \rho \frac{\partial \widehat{\psi}}{\partial \phi_{\chi}}\right) \dot{\phi}_{\chi}^{e} + \left(\underline{\boldsymbol{b}} - \rho \frac{\partial \widehat{\psi}}{\partial \nabla \phi_{\chi}}\right) . \underline{\dot{\boldsymbol{\kappa}}}^{e}\\ &+ \underset{\sim}{\boldsymbol{\sigma}} : \underset{\sim}{\dot{\boldsymbol{\varepsilon}}}^{p} - \rho \frac{\partial \widehat{\psi}}{\partial q} \dot{q} + a \dot{\phi}_{\chi}^{p} + \underline{\boldsymbol{b}} . \underline{\dot{\boldsymbol{\kappa}}}^{p} - \frac{\underline{\boldsymbol{q}}}{T} . \nabla T \geq 0\end{aligned} \tag{55}$$

Assuming that no dissipation is associated with the four first terms of the previous inequality, the following state laws are found

$$\underset{\sim}{\boldsymbol{\sigma}} = \rho \frac{\partial \widehat{\psi}}{\partial \underset{\sim}{\boldsymbol{\varepsilon}}^{e}},\ \eta = -\frac{\partial \widehat{\psi}}{\partial T},\ R = -\rho \frac{\partial \widehat{\psi}}{\partial q} \tag{56}$$

$$a = \rho \frac{\partial \widehat{\psi}}{\partial \phi_{\chi}^{e}},\ \underline{\boldsymbol{b}} = \rho \frac{\partial \widehat{\psi}}{\partial \underline{\boldsymbol{\kappa}}^{e}} \tag{57}$$

and the residual dissipation is

$$D^{res} = \underset{\sim}{\sigma} : \underset{\sim}{\dot{\varepsilon}}^p + R\dot{q} + a\dot{\phi}_\chi^p + \underline{b}.\underline{\dot{\kappa}}^p - \frac{\boldsymbol{q}}{T}.\nabla T \geq 0 \tag{58}$$

Evolution rules for viscoplastic strain, internal variables, and the additional degrees of freedom can be derived from a dissipation potential $\Omega\left(\underset{\sim}{\sigma}, R, a, \underline{b}\right)$:

$$\underset{\sim}{\dot{\varepsilon}}^p = \frac{\partial \Omega}{\partial \underset{\sim}{\sigma}}, \quad \dot{q} = \frac{\partial \Omega}{\partial R}, \quad \dot{\phi}_\chi^p = \frac{\partial \Omega}{\partial a}, \quad \underline{\dot{\kappa}}^p = \frac{\partial \Omega}{\partial \underline{b}} \tag{59}$$

As a result of the additional balance equation (28) combined with the previous state laws, the type of derived partial differential equation can be made more specific by adopting a quadratic free energy potential for $\underline{b}$ (modulus A) and a quadratic dissipation potential with respect to a (parameter β). We obtain:

$$\beta\dot{\phi}_\chi = a + \beta\dot{\phi}_\chi^e = \text{div } A\underline{\kappa} - \text{div } A\underline{\kappa}^p + \beta\dot{\phi}_\chi^e \tag{60}$$

Decompositions of stresses and strains can also be mixed, for instance in the following way:

$$\underset{\sim}{\varepsilon} = \underset{\sim}{\varepsilon}^e + \underset{\sim}{\varepsilon}^p, \quad a = a^e + a^\upsilon, \quad \underline{\kappa} = \nabla\phi_\chi = \underline{\kappa}^e + \underline{\kappa}^p \tag{61}$$

based on which a constitutive theory can be built.

Application to a Continuum Damage Model for Single Crystals and Its Regularization

The micromorphic approach is now illustrated in the case of a constitutive model for damaging viscoplastic single crystals. The objective is to simulate crack initiation and propagation. The micromorphic model is used in order to obtain a regularized continuum damage formulation with a view to simulating mesh-independent crack propagation in single crystals.

Constitutive Equations

In the proposed crystal plasticity model taken from Marchal, et al. (2006a), viscoplasticity and damage are coupled by introducing an additional damage strain variable $\underset{\sim}{\varepsilon}^d$, into the strain rate partition equation:

$$\underset{\sim}{\dot{\varepsilon}} = \underset{\sim}{\dot{\varepsilon}}^e + \underset{\sim}{\dot{\varepsilon}}^p + \underset{\sim}{\dot{\varepsilon}}^d \tag{62}$$

where $\underset{\sim}{\dot{\boldsymbol{\varepsilon}}}^{e}$ and $\underset{\sim}{\dot{\boldsymbol{\varepsilon}}}^{p}$ are the elastic and the plastic strain rates, respectively. The flow rule for plastic part is written at the slip system level by means of the orientation tensor $\underset{\sim}{\boldsymbol{m}}^{s}$:

$$\underset{\sim}{\boldsymbol{m}}^{s} = \frac{1}{2}\left(\underline{\boldsymbol{n}}^{s} \otimes \underline{\boldsymbol{l}}^{s} + \underline{\boldsymbol{l}}^{s} \otimes \underline{\boldsymbol{n}}^{s}\right) \tag{63}$$

where $\underline{\boldsymbol{n}}^{s}$ is the normal to the plane of slip system s and $\underline{\boldsymbol{l}}^{s}$ stands for the corresponding slip direction. Then, plastic strain rate reads:

$$\underset{\sim}{\dot{\boldsymbol{\varepsilon}}}^{p} = \sum_{s=1}^{N_{slip}} \dot{\gamma}^{s} \underset{\sim}{\boldsymbol{m}}^{s} \tag{64}$$

where N_{slip} is the total number of slip systems. The flow rule on slip system s is a classical Norton rule with threshold:

$$\dot{\gamma}^{s} = \left\langle \frac{|\tau^{s} - x^{s}| - r^{s}}{K} \right\rangle^{n} sign\left(\tau^{s} - x^{s}\right) \tag{65}$$

where r^{s} and x^{s} are the variables for isotropic and kinematic hardening respectively and K and n are viscosity material parameters to be identified from experimental curves.

Material separation is assumed to take place w.r.t. specific crystallographic planes, like cleavage planes in single crystals. The word *cleavage* is written in a more general sense that its original meaning in physical metallurgy associated with brittle fracture of non–f.c.c. crystals. In the continuum mechanical model, *cleavage* means cracking along a specific crystallographic plane as it is often observed in low cycle fatigue of f.c.c. crystals like single crystal nickel–base superalloys. The damage strain $\underset{\sim}{\dot{\boldsymbol{\varepsilon}}}^{d}$ is decomposed in the following crystallographic contributions:

$$\underset{\sim}{\dot{\boldsymbol{\varepsilon}}}^{d} = \sum_{s=1}^{N_{damage}} \dot{\delta}_{c}^{s} \underline{\boldsymbol{n}}_{d}^{s} \otimes \underline{\boldsymbol{n}}_{d}^{s} + \dot{\delta}_{1}^{s} \underline{\boldsymbol{n}}_{d}^{s} \overset{sym}{\otimes} \underline{\boldsymbol{l}}_{d_1}^{s} + \dot{\delta}_{2}^{s} \underline{\boldsymbol{n}}_{d}^{s} \overset{sym}{\otimes} \underline{\boldsymbol{l}}_{d_2}^{s} \tag{66}$$

where $\dot{\delta}^{s}$, $\dot{\delta}_{1}^{s}$ and $\dot{\delta}_{2}^{s}$ are the strain rates for mode I, mode II and mode III crack growth, respectively and N_{damage}^{d} stands for the number of damage planes which are fixed crystallographic planes depending on the crystal structure. Cleavage damage is represented by the opening $\dot{\delta}_{c}^{s}$ of crystallographic cleavage planes with the normal vector $\underline{\boldsymbol{n}}^{s}$. Additional damage systems must be introduced for the in-plane accommodation along orthogonal directions $\underline{\boldsymbol{l}}_{1}^{s}$ and $\underline{\boldsymbol{l}}_{2}^{s}$, once cleavage has started (Fig. 4). Three damage criteria are associated to one cleavage and two accommodation systems:

$$f_{c}^{s} = \left| \underline{\boldsymbol{n}}_{d}^{s} \cdot \underset{\sim}{\boldsymbol{\sigma}} \cdot \underline{\boldsymbol{n}}_{d}^{s} \right| - Y_{c}^{s} \tag{67}$$

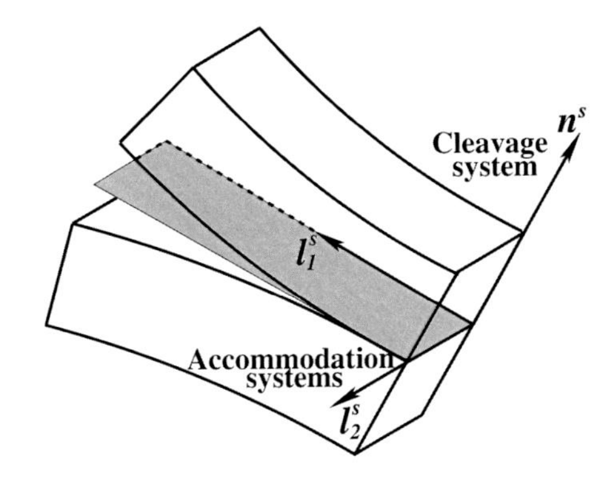

Fig. 4 Illustration of the cleavage and two accommodation systems to be associated to the crystallographic planes

$$f_i^s = \left| \underline{n}_d^s \cdot \underset{\sim}{\sigma} \cdot \underline{l}_{d_i}^s \right| - Y_i^s \ (i = 1,\ 2) \tag{68}$$

The critical normal stress Y^s for damage decreases as δ increases:

$$Y_c^s = Y_0^s + H\delta_c^s,\ Y_i^s = Y_0^s + H\delta_i^s \tag{69}$$

where Y_0^s is the initial damage stress (usually coupled to plasticity) and H is a negative modulus which controls material softening due to damage. Finally, evolution of damage is given by the following equations;

$$\dot{\delta}_c^s = \left\langle \frac{f_c^s}{K_d} \right\rangle^{n_d} \text{sign}\left(\underline{n}_d^s \cdot \underset{\sim}{\sigma} \cdot \underline{n}_d^s \right) \tag{70}$$

$$\dot{\delta}_i^s = \left\langle \frac{f_i^s}{K_d} \right\rangle^{n_d} \text{sign}\left(\underline{n}_d^s \cdot \underset{\sim}{\sigma} \cdot \underline{l}_{d_i}^s \right) \tag{71}$$

where K_d and n_d are material parameters.

These equations hold for all conditions except when the crack is closed $\left(\delta_c^s < 0\right)$and compressive forces are applied $\left(\underline{n}_d^s \cdot \underset{\sim}{\sigma} \cdot \underline{n}_d^s < 0\right)$. In this case, damage evolution stops $\left(\dot{\delta}_c^s = \dot{\delta}_i^s = 0\right)$, corresponding to the unilateral damage conditions.

Note that the damage variables δ introduced in the model differ from the usual corresponding variables of standard continuum damage mechanics that vary from 0 to 1. In contrast, the δs are strain–like quantities that can ever increase.

Coupling between plasticity and damage is generated through initial damage stress Y_0 in (69) which is controlled by cumulative slip variable γ_{cum}:

$$\dot{\gamma}_{cum} = \sum_{s=1}^{N_{slips}} |\dot{\gamma}^s| \tag{72}$$

Then, Y_0 takes the form:

$$Y_0^s = \sigma_n^c\, e^{-d\gamma_{cum}} + \sigma_{ult} \tag{73}$$

This formulation suggests an exponential decaying regime from a preferably high initial cleavage stress value σ_n^c, to an ultimate stress, σ_{ult} which is close but not equal to zero for numerical reasons and d is a material constant.

This model, complemented by the suitable constitutive equations for viscoplastic strain, has been used for the simulation of crack growth under complex cyclic loading at high temperature (Marchal et al. 2006a). Significant mesh dependency of results was found Marchal et al. (2006b).

In the present work, the model is further developed by switching from classical to microdamage continuum in order to assess the regularization capabilities of a higher order theory. The coupling of the model with microdamage theory is achieved by introducing a cumulative damage variable calculated from the damage systems and a new threshold function $Y_0(\delta, \gamma_{cum})$:

$$\dot{\delta}_{cum} = \sum_{s=1}^{N_{planes}} \dot{\delta}^s, \quad \text{where } \dot{\delta}^s = \left|\dot{\delta}_c^s\right| + \left|\dot{\delta}_1^s\right| + \left|\dot{\delta}_2^s\right| \tag{74}$$

$$Y_0 = \sigma_n^c\, e^{-d\gamma_{cum} - H\delta_{cum}} + \sigma_{ult} \tag{75}$$

where the modulus H accounts for damage induced softening and σ_{ult} is a ultimate stress.

Microdamage Continuum

In microdamage theory, the introduced microvariable is a scalar microdamage field δ_χ:

$$DOF = \{\underline{\boldsymbol{u}}, \delta_\chi\} \; STRAIN = \left\{\underset{\sim}{\boldsymbol{\varepsilon}}, \delta_\chi, \nabla\delta_\chi\right\} \tag{76}$$

The power of internal forces is extended as

$$p^{(i)} = \underset{\sim}{\boldsymbol{\sigma}} : \dot{\underset{\sim}{\boldsymbol{\varepsilon}}} + a\dot{\delta}_\chi + \underline{\boldsymbol{b}} \cdot \nabla\dot{\delta}_\chi \tag{77}$$

where generalized stresses $a, \underline{\boldsymbol{b}}$ have been introduced. The generalized balance equations are:

$$\text{div}\ \underset{\sim}{\boldsymbol{\sigma}} = 0,\ a = \text{div}\ \underline{\boldsymbol{b}} \tag{78}$$

The free energy density is taken as a quadratic potential in the elastic strain, damage δ, relative damage $\delta - \delta_\chi$ and microdamage gradient $\nabla\delta_\chi$:

$$\rho\psi = \frac{1}{2}\underset{\sim}{\boldsymbol{\varepsilon}}^e : \underset{\approx}{\boldsymbol{c}} : \underset{\sim}{\boldsymbol{\varepsilon}}^e + \frac{1}{2}\sum_{s=1}^{N_{damage}} H\delta_s^2 + \frac{1}{2}H_\chi\left(\delta - \delta_\chi\right)^2 + \frac{1}{2}A\nabla\delta_\chi \cdot \nabla\delta_\chi \tag{79}$$

where H, H_χ and A are scalar material constants. The tensor of elastic moduli is called $\underset{\approx}{\boldsymbol{c}}$. Then, the elastic response of the material becomes:

$$\underset{\sim}{\boldsymbol{\sigma}} = \rho\frac{\partial\psi}{\partial\underset{\sim}{\boldsymbol{\varepsilon}}^e} = \underset{\approx}{\boldsymbol{c}} : \underset{\sim}{\boldsymbol{\varepsilon}}^e \tag{80}$$

The generalized stresses read:

$$a = \rho\frac{\partial\psi}{\partial\delta_\chi} = -H_\chi\left(\delta - \delta_\chi\right), \ \underline{\boldsymbol{b}} = A\nabla\delta_\chi \tag{81}$$

and the driving force for damage can be derived as:

$$Y^s = \rho\frac{\partial\psi}{\partial\delta^s} = H\delta^s + H_\chi\left(\delta^s - \delta_\chi\right) \tag{82}$$

The damage criterion now is:

$$f^s = \left|\underline{\boldsymbol{n}}^s \cdot \underset{\sim}{\boldsymbol{\sigma}} \cdot \underline{\boldsymbol{n}}^s\right| - Y_0 - Y^s = 0 \tag{83}$$

Substituting the linear constitutive equations for generalized stresses into the additional balance equation (78), assuming homogeneous material properties, leads to the following partial differential equation for the microdamage

$$\delta_\chi - \frac{A}{H_\chi}\boldsymbol{\Delta}\delta_\chi = \delta \tag{84}$$

where the macrodamage δ acts as a source term. Exactly this type of Helmholtz equation has been postulated in the so–called implicit gradient theory of plasticity and damage (Peerlings et al. 2001, 2004; Engelen et al. 2003; Germain et al. 2007), where the microvariables are called non local variables and where the generalized stresses a and $\underline{\boldsymbol{b}}$ are not explicitly introduced (see Forest (2009) and Dillard et al. (2006) for the analogy between this latter approach and the micromorphic theory).

The solution of the problem of failure of a 1D bar in tension/compression was treated by Aslan and Forest (2009). The characteristic size of the damage zone was shown to be controlled by the characteristic length corresponding to the inverse of

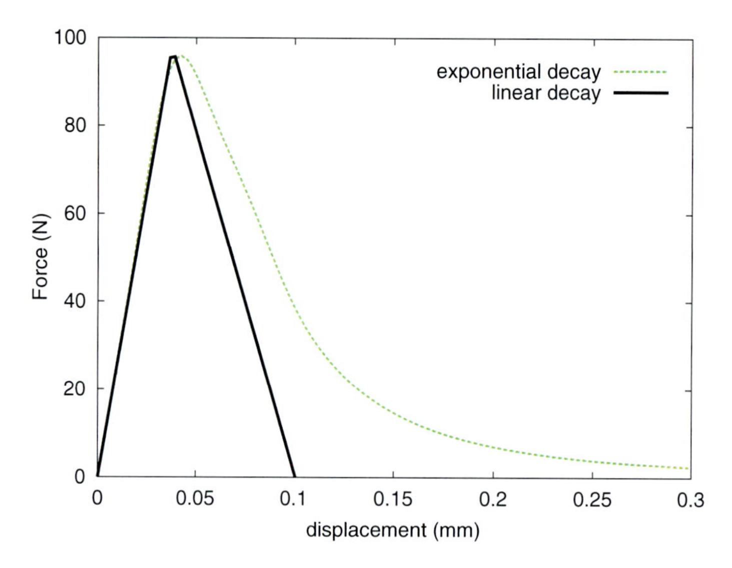

Fig. 5 Comparison between force vs. displacement diagram of a 1D softening rod for linear and exponential decay

$$\omega = \sqrt{\left| H\ H_\chi / A \left(H + H_\chi \right) \right|} \tag{85}$$

In comparison with the standard strain gradient approaches (Peerlings et al. 2001; Germain et al. 2007), microdamage theory eliminates the final fracture problem without any modification to the damage function, since there exists no direct coupling between the force stress $\underset{\sim}{\boldsymbol{\sigma}}$ and the generalized stresses, a and $\underline{\boldsymbol{b}}$. For a better representation of a cracked element, an exponential drop is used for both the damage threshold Y_0 and the modulus A, since the element should become unable to store energy due to the generalized stresses when broken (see Fig. 5):

$$Y_0 = \sigma_n^c\ e^{-H\delta} + \sigma_{ult},\ \underline{\boldsymbol{b}} = Ae^{-H\delta}\nabla^\chi\delta \tag{86}$$

Numerical Examples

As a 2D example, a single crystal CT-like specimen under tension is analysed. The corresponding FE mesh is given in Fig. 6. Analyses are performed for two different crack widths, obtained by furnishing different material parameters which control the characteristic length (Fig. 7). The propagation of a crack, corresponding

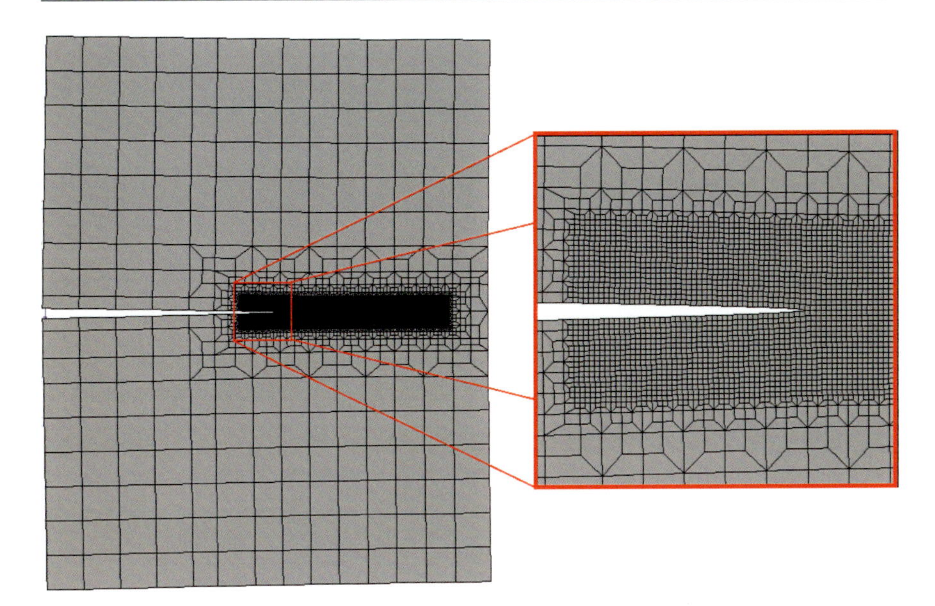

Fig. 6 FE mesh of a CT-like specimen

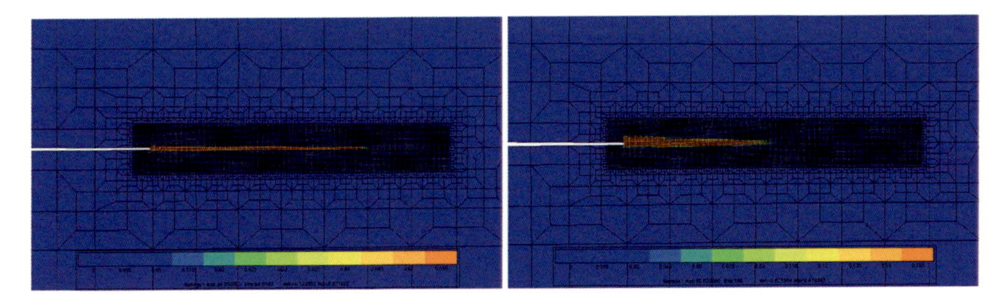

Fig. 7 Crack growth in a 2D single crystal CT-like specimen with a single cleavage plane aligned through the horizontal axis under vertical tension. Field variable δ (Left) A = 100 MPa.mm^2, H = −20000 MPa, H_χ = 30,000 MPa, (Right) A = 150 MPa.mm^2, H = -10000 MPa, H_χ = 30000 MPa

stress fields and the comparison with classical elastic solutions are given in Fig. 8. This comparison shows that the microdamage model is able to reproduce the stress concentration at the crack tip except very close to the crack tip where finite stress values are predicted.

The Micromorphic Approach at Finite Deformations

Extensions of the previous systematic micromorphic approach are presented for finite deformations along the lines of Forest (2016). They generalize some aspects

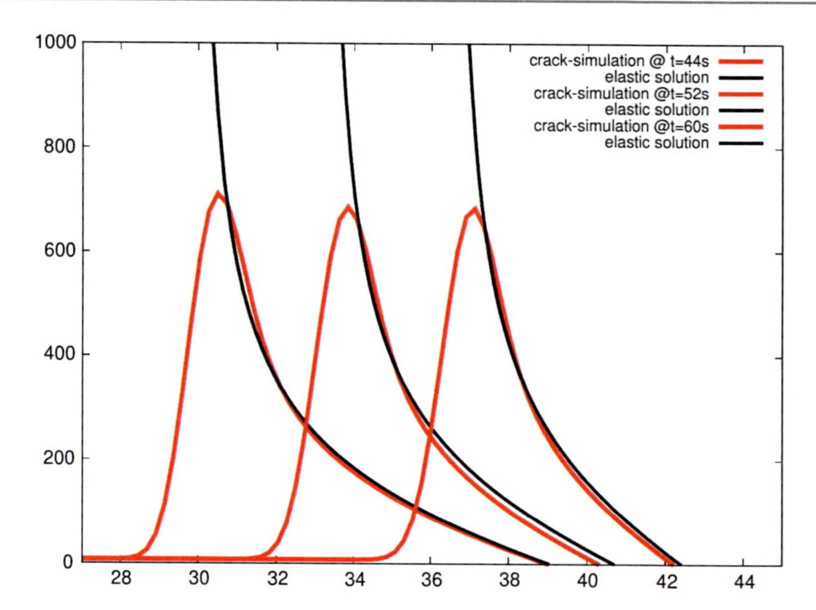

Fig. 8 Mode I stress profile (in MPa) vs. position along the ligament (in mm) at three distinct crack propagation steps: micromorphic solution compared to the linear elastic one

of the finite deformation micromorphic theory presented in section "The Micromorphic Theory After Eringen and Mindlin and Its Extension to Plasticity".

Micromorphic *and Gradient Hyperelasticity*

Nonlinearity arises not only from nonlinear material response but also from the consideration of finite deformations. The impact of finite strains on regularisation operators is largely unexplored. It is first illustrated in the pure hyperelastic case, leaving aside for a moment the inclusion of plastic effects. It is recalled that the Lagrangian coordinates of the material points are denoted by $\underline{X}$ on the reference configuration $_0$, whereas their positions in the current configuration are called $\underline{\boldsymbol{x}}$. The gradient operators with respect to Lagrangian (reference) and Eulerian (current) coordinates are denoted by ∇_X and ∇_x, respectively. The deformation gradient is $\underset{\sim}{\boldsymbol{F}} = \underset{\sim}{\mathbf{1}} + \underline{\boldsymbol{u}} \otimes \nabla_X$ where the displacement vector is the function $\underline{\boldsymbol{u}}\,(\underline{X}, t)$. The initial and current mass density functions are $\rho_0\,(\underline{X}, t)$ and $p\,(\underline{\boldsymbol{x}}, t)$, respectively.

It is appropriate here to recall the Eulerian balance equation and boundary conditions in the case of a scalar micromorphic degree of freedom, in the absence of body forces:

$$\text{div}\ \underset{\sim}{\boldsymbol{\sigma}} = 0,\ a = \text{div}\ \underline{\boldsymbol{b}},\ \forall \underline{\boldsymbol{x}} \in \Omega,\ \underline{\boldsymbol{t}} = \underset{\sim}{\boldsymbol{\sigma}} \cdot \underset{\sim}{\boldsymbol{n}},\ a_c = \underline{\boldsymbol{b}} \cdot \underline{\boldsymbol{n}},\ \forall \underline{\boldsymbol{x}} \in \partial\Omega \tag{87}$$

Finite Microstrain Tensor Model

The microstrain model was introduced in Table 1 as a reduced micromorphic model where Eringen's microdeformation tensor is taken as symmetric, as proposed in Forest and Sievert (2006). The additional degrees of freedom are the six components of a microstrain tensor $\underset{\sim}{\chi}$, a second order symmetric tensor associated with the right Cauchy-Green strain measure of the micromorphic deformation. Accordingly, the microstrain is regarded here as a Lagrangian variable. The Lagrangian power density of internal forces takes the form:

$$p_0^{(i)} = \frac{1}{2}\underset{\sim}{\Pi} : \dot{\underset{\sim}{C}} + \underset{\sim}{a}_0 : \dot{\underset{\sim}{\chi}} + \underset{\approx}{b}_0 \vdots \left(\dot{\underset{\sim}{\chi}} \otimes \nabla_X\right) = Jp^{(i)} \tag{88}$$

$p^{(i)}$ being the Eulerian internal power density and the Jacobian $J = \det \underset{\sim}{F}$. The right Cauchy-Green tensor is $\underset{\sim}{C} = \underset{\sim}{F}^T \cdot \underset{\sim}{F}$ and $\underset{\sim}{\Pi}$ is the Piola stress tensor. The method of virtual power can be used to show that the generalized Lagrangian stress tensors, a second and a third rank tensor, fulfill the following balance equation:

$$\text{Div}\ \underset{\approx}{b}_0 = \underset{\sim}{a}_0 \tag{89}$$

in addition to the balance of momentum equation

$$\text{Div}\ \underset{\sim}{F} \cdot \underset{\sim}{\Pi} = 0 \tag{90}$$

in the absence of body and inertial forces, the divergence operator Div being computed with respect to Lagrangian coordinates. The corresponding Eulerian forms of the balance equations are

$$\text{div}\ \underset{\sim}{\sigma} = 0, \quad \text{div}\ \underset{\approx}{b} = \underset{\sim}{a} \tag{91}$$

with the Neumann boundary conditions for tractions and double tractions: $\underline{t} = \underset{\sim}{\sigma} \cdot \underline{n}, \underset{\sim}{b} = \underset{\approx}{b} \cdot \underline{n}$, involving the normal vector of the current surface element. The relations between the Lagrangian and Eulerian generalized stress tensors

$$\underset{\sim}{a}_0 = J\underset{\sim}{a}, \quad \underset{\approx}{b}_0 = J\underset{\approx}{b} \cdot \underset{\sim}{F}^{-T} = J\underset{\sim}{F}^{-1} \cdot \underline{b} \tag{92}$$

The hyperelastic free energy density function is $\psi_0\left(\underset{\sim}{C}, \underset{\sim}{\chi}, \underset{\sim}{\chi} \otimes \nabla_X\right)$ and the stress–strain relations read:

$$\underset{\sim}{\Pi} = 2\rho_0 \frac{\partial \psi_0}{\partial \underset{\sim}{C}}, \quad \underset{\sim}{a}_0 = \rho_0 \frac{\partial \psi_0}{\partial \underset{\sim}{\chi}}, \quad \underset{\approx}{b}_0 = \rho_0 \frac{\partial \psi_0}{\partial \underset{\sim}{\chi} \otimes \nabla_X} \tag{93}$$

The relative strain tensor $\underset{\sim}{e} = \underset{\sim}{C} - \underset{\sim}{\chi}$ was defined in Forest and Sievert (2006) and measures the difference between macro and micro–strain. As an example, the following potential is proposed:

$$\rho_0 \psi_0 = \rho_0 \psi_{ref} \left(\underset{\sim}{C} \right) + \frac{1}{2} H_\chi \left(\underset{\sim}{C} - \underset{\sim}{\chi} \right)^2 + \frac{1}{2} \underset{\sim}{\chi} \otimes \nabla_X \vdots \underset{\approx}{A} \vdots \underset{\sim}{\chi} \otimes \nabla_X \tag{94}$$

where a penalty modulus H_χ is introduced and where ψ_{ref} is a standard hyperelastic strain energy potential (neo–Hookean, etc.). The higher order term involves a sixth–rank tensor of elasticity moduli which is symmetric and assumed definite positive, see Auffray et al. (2013). The stress–strain relations (93) become:

$$\underset{\sim}{\Pi} = 2\rho_0 \frac{\partial \psi_{ref}}{\partial \underset{\sim}{C}} + H_\chi \left(\underset{\sim}{C} - \underset{\sim}{\chi} \right), \underset{\sim 0}{a} = -H_\chi \left(\underset{\sim}{C} - \underset{\sim}{\chi} \right), \underset{\approx 0}{b} = \underset{\approx}{A} \vdots \left(\underset{\sim}{\chi} \otimes \nabla_X \right) \tag{95}$$

Note that the classical hyperelastic relation is complemented by a coupling term involving the microstrain tensor. Taking the balance equation (90) into account, the set of p.d.e. for the microstrain components is found to be:

$$\underset{\sim}{C} = \underset{\sim}{\chi} - \frac{1}{H_\chi} \mathrm{Div} \left(\underset{\approx}{A} \vdots \left(\underset{\sim}{\chi} \otimes \nabla_X \right) \right) \tag{96}$$

The associated regularisation operator is now given in the simplified case where the sixth rank tensor of higher order moduli is assumed to be the identity multiplied by the single modulus A:

$$\mathrm{Op} = \underset{\sim}{1} - \frac{A}{H_\chi} \Delta_X \tag{97}$$

It involves the Lagrangian Laplace operator $\Delta_X(\bullet) = (\bullet)_{,KK}$ in a Cartesian frame where capital indices refer to Lagrange coordinates and the comma to partial derivation. It is linear w.r.t. Lagrangian coordinates but the full problem is of course highly non linear. The associated Eulerian partial differential operator is nonlinear in the form: $\chi_{IJ,KK} = \chi_{IJ,kl} F_{kK} F_{lK}$, where small index letters refer to the current coordinate system.

An Eulerian formulation of the proposed constitutive equations is possible. It will be illustrated in the next section in the case of a scalar microstrain variable for the sake of brevity.

Equivalent Microstrain Model at Finite Deformation

According to the micromorphic approach, it is not necessary to consider the full microstrain tensor. Instead, the set of degrees of freedom of a reduced micromorphic model can be the usual displacement vector and a scalar microstrain variable $\chi(\underline{X}, t)$. The latter variable is assumed to be a Lagrangian quantity, invariant w.r.t. change of observer. The free energy density is a function of the following argument $\psi_0\left(\underset{\sim}{C}, \chi, \nabla_x \chi\right)$. The corresponding hyperelastic state laws fulfilling the Clausius–Duhem inequality take the form:

$$\underset{\sim}{\Pi} = 2\rho_0 \frac{\partial \psi_0}{\partial \underset{\sim}{C}}, \quad a_0 = \rho_0 \frac{\partial \psi_0}{\partial \chi}, \quad \underline{b}_0 = \rho_0 \frac{\partial \psi_0}{\partial \nabla_x \chi} \tag{98}$$

As an example, the following Lagrangian potential is proposed:

$$\rho_0 \psi_0 = \rho_0 \psi_{ref}\left(\underset{\sim}{C}\right) + \frac{1}{2} H_\chi (C_{eq} - \chi)^2 + \frac{1}{2} \nabla_x \chi \cdot \underset{\sim}{A} \cdot \nabla_X \chi \tag{99}$$

The microstrain is compared to some equivalent strain measure, C_{eq}, function of the invariants of $\underset{\sim}{C}$. The stress–strain relations (98) become:

$$\underset{\sim}{\Pi} = 2\rho_0 \frac{\partial \psi_{ref}}{\partial \underset{\sim}{C}} + H_\chi \left(C_{eq} - \chi\right) \frac{\partial C_{eq}}{\partial \underset{\sim}{C}}, \quad a_0 = -H_\chi \left(C_{eq} - \chi\right), \quad \underline{b}_0 = \underset{\sim}{A} \cdot \nabla_X \chi \tag{100}$$

Taking the balance equation $a_0 = \nabla_X \cdot \underline{b}_0$ into account, the p.d.e. governing χ is:

$$C_{eq} = \chi - \frac{1}{H_\chi} \mathrm{Div}\left(\underset{\sim}{A} \cdot \nabla_X \chi\right) \tag{101}$$

Let us mention the corresponding isotropic form, when $\underset{\sim}{A} = A\underset{\sim}{1}$:

$$C_{eq} = \chi - \frac{A}{H_\chi} \Delta_X \chi \tag{102}$$

where Δ_X is the Laplace operator with respect to Lagrangian coordinates. An example of equivalent strain measure which the microstrain is compared with is

$$C_{eq} = \sqrt{\underset{\sim}{C} : \underset{\sim}{C}}, \quad \frac{\partial C_{eq}}{\partial \underset{\sim}{C}} = \underset{\sim}{C} / C_{eq} \tag{103}$$

The penalty modulus H_χ constrains the microstrain degree of freedom to remain close to the equivalent strain measure C_{eq}. In the limit of large values of H_χ, the generalized stress a_0 can be regarded as a Lagrange multiplier enforcing the constraint. As a result, the microstrain gradient becomes equal to the equivalent strain gradient and the micromorphic model degenerates into a gradient model, see Forest and Sab (2017) for a detailed description of internal constraints in micromorphic theories.

The formulation of a constitutive law based on Eulerian strain measures, $\underset{\sim}{\boldsymbol{B}} = \underset{\sim}{\boldsymbol{F}} \cdot \underset{\sim}{\boldsymbol{F}}^T$ and $\nabla_x \chi$, is now envisaged. It relies on the choice of a free energy potential $\rho\psi\left(\underset{\sim}{\boldsymbol{B}}, \chi, \nabla_x \chi\right)$. Galilean invariance of the constitutive law requires that this function fulfills the following conditions:

$$\psi\left(\underset{\sim}{\boldsymbol{Q}} \cdot \underset{\sim}{\boldsymbol{B}} \cdot \underset{\sim}{\boldsymbol{Q}}^T, \chi, \underset{\sim}{\boldsymbol{Q}} \cdot \nabla_x \chi\right) = \psi\left(\underset{\sim}{\boldsymbol{B}}, \chi, \cdot\nabla_x \chi\right) \tag{104}$$

for all constant orthogonal transformations $\underset{\sim}{\boldsymbol{Q}}$. This amounts to stating isotropy of the function ψ. Representation theorems are available for such functions, $\psi\left(\underset{\sim}{\boldsymbol{B}}, \chi, \nabla_x \chi\right) = \psi\left(B_1, B_2, B_3, \chi, \|\nabla_x \, \chi\|\right)$, where the B_i are the eigenvalues of $\underset{\sim}{\boldsymbol{B}}$. The Cauchy stress tensor, $\underset{\sim}{\boldsymbol{\sigma}}$, is known then to commute with $\underset{\sim}{\boldsymbol{B}}$ such that:

$$\underset{\sim}{\boldsymbol{\sigma}} : \underset{\sim}{\boldsymbol{D}} = \left(\underset{\sim}{\boldsymbol{\sigma}} \cdot \underset{\sim}{\boldsymbol{B}}^{-1}\right) : \left(\underset{\sim}{\dot{\boldsymbol{F}}} \cdot \underset{\sim}{\boldsymbol{F}}^T\right) = \left(\underset{\sim}{\boldsymbol{B}}^{-1} \cdot \underset{\sim}{\boldsymbol{\sigma}}\right) : \left(\underset{\sim}{\boldsymbol{F}} \cdot \underset{\sim}{\dot{\boldsymbol{F}}}^T\right) = \frac{1}{2}\left(\underset{\sim}{\boldsymbol{B}}^{-1} \cdot \underset{\sim}{\boldsymbol{\sigma}}\right) : \underset{\sim}{\dot{\boldsymbol{B}}} \tag{105}$$

where the strain rate tensor, $\underset{\sim}{\boldsymbol{D}}$, is the symmetric part of the velocity gradient, $\underset{\sim}{\dot{\boldsymbol{F}}} \cdot \underset{\sim}{\boldsymbol{F}}^{-1}$. The hyperelastic state laws then take the form:

$$\underset{\sim}{\boldsymbol{\sigma}} = 2\underset{\sim}{\boldsymbol{B}} \cdot \rho\frac{\partial\psi}{\partial\underset{\sim}{\boldsymbol{B}}}, \quad a = \rho\frac{\partial\psi}{\partial\chi}, \quad \underline{\boldsymbol{b}} = \rho\frac{\partial\psi}{\partial\nabla_X\chi} \tag{106}$$

As an example, the following Eulerian potential is proposed:

$$\rho\psi = \rho\psi_{ref}\left(\underset{\sim}{\boldsymbol{B}}\right) + \frac{1}{2}H_\chi(B_{eq} - \chi)^2 + \frac{1}{2}A\nabla_x\chi : \nabla_x\chi, \tag{107}$$

where ψ_{ref} refers to a standard isotropic elasticity potential from the classical finite elasticity theory. Note that $B_{eq} = C_{eq}$ since $\underset{\sim}{\boldsymbol{B}}$ and $\underset{\sim}{\boldsymbol{C}}$ share the same eigenvalues. The state laws (106) become:

$$\underset{\sim}{\boldsymbol{\sigma}} = 2\rho\frac{\partial\psi_{ref}}{\partial\underset{\sim}{\boldsymbol{B}}} + H_\chi\left(B_{eq} - \chi\right)\frac{\partial B_{eq}}{\partial\underset{\sim}{\boldsymbol{B}}}, \quad a = -H_\chi\left(B_{eq} - \chi\right), \quad \underline{\boldsymbol{b}} = A\nabla_x\chi \tag{108}$$

The Eulerian regularisation operator follows from (91):

$$B_{eq} = \chi - \frac{A}{H_\chi} \Delta_x \chi \tag{109}$$

where Δ_x is the Laplace operator with respect to the Eulerian coordinates.

It is essential to notice that the isotropic regularisation operators (102) and (109) are distinct. For, if the Lagrangian higher order elastic law is linear with respect to the constitutive quantities involved, the deduced Eulerian law is NOT linear:

$$\underline{b}_0 = A \nabla_X \chi \quad \Rightarrow \quad \underline{b} = J^{-1} \underset{\sim}{F} \cdot \underline{b}_0 = A J^{-1} \underset{\sim}{B} \cdot \nabla_x \chi \tag{110}$$

so that the Eulerian regularisation operator will not be linear, i.e. different from (109).

Finite Deformation Micromorphic Elastoviscoplasticity Using an Additive Decomposition of a Lagrangian Strain

The most straightforward extension of the previous framework to viscoplasticity is to introduce a finite plastic strain measure in the decomposition of a Lagrangian total strain tensor. Such Lagrangian formulations of elastoviscoplasticity involve the additive decomposition of some Lagrangian total strain measure into elastic and viscoplastic parts:

$$\underset{\sim}{E}_h = h\left(\underset{\sim}{C}\right) = \underset{\sim}{E}_h^e + \underset{\sim}{E}_h^p \tag{111}$$

Many choices are possible for the invertible function h with restrictions ensuring that $\underset{\sim}{E}_h$ is a strain measure (symmetric, vanishing for rigid body motions, differentiable at 0 so that the tangent is the usual small strain tensor $\underset{\sim}{\varepsilon}$, used before, see Besson et al. 2009). Seth-Hill's strain measures are obtained for power law functions such that: $\underset{\sim}{E}_m = \frac{1}{m}\left(\underset{\sim}{C}^{\frac{m}{2}} - \underset{\sim}{1}\right)$, for $m > 0$, $\underset{\sim}{E}_0 = \log \underset{\sim}{C}^{\frac{1}{2}}$, the latter being the Lagrangian logarithmic strain. The case $m = 2$ corresponds to the Green–Lagrange strain measure for which this finite deformation theory was first formulated by Green and Naghdi (see Lee and Germain 1972; Bertram 2012) for the pros and the cons of various such formulations. This Lagrangian formulation is preferable to Eulerian ones based on corresponding Eulerian strain measures in order not to limit the approach to isotropic material behaviour (Simo and Miehe 1992). The additive decomposition of the Lagrangian logarithmic strain is put forward in the computational plasticity strategies developed in Miehe et al. (2002) and Helfer (2015). However, there is generally no physical motivation for the selection of one or another Lagrangian strain measure within this framework. In addition, this approach favours one particular reference configuration for which the corresponding strain is decomposed into elastic and plastic parts, again without clear physical argument. Changes of reference configuration lead to complex hardly interpretable transformation rules for the plastic strain variables, see Shotov and Ihlemann (2014) for a comparison of finite deformation constitutive laws with respect to this issue. Note also that limitations arise from using a symmetric plastic strain variable $\underset{\sim}{\boldsymbol{E}}{}_h^p$ especially when plastic spin relations are needed for anisotropic materials (Bertram 2012). This framework was applied to micromorphic and gradient plasticity and damage theories in Geers (2004), Horak and Jirasek (2013), and Miehe (2014).

A Lagrangian conjugate stress tensor $\underset{\sim}{\prod}{}_h$ is defined for each strain measure $\underset{\sim}{\boldsymbol{E}}{}_h$ such that

$$\frac{1}{2}\underset{\sim}{\boldsymbol{\Pi}} : \dot{\underset{\sim}{\boldsymbol{C}}} = \underset{\sim}{\boldsymbol{\Pi}}{}_h : \dot{\underset{\sim}{\boldsymbol{E}}}{}_h \text{ with } \underset{\sim}{\boldsymbol{\Pi}}{}_h = \underset{\sim}{\boldsymbol{\Pi}} : \left(\frac{\partial h}{\partial \underset{\sim}{\boldsymbol{C}}}\right)^{-1} \tag{112}$$

The power density of internal forces is:

$$p_0^{(i)} = \underset{\sim}{\boldsymbol{\Pi}}{}_h : \dot{\underset{\sim}{\boldsymbol{E}}}{}_h + a_0 \dot{\chi} + \underline{\boldsymbol{b}}_0 \cdot \nabla_X \dot{\chi}$$

and the free energy density function has the following arguments: $\psi_0 \left(\underset{\sim}{\boldsymbol{E}}{}_h^e, q, \chi, \underline{\boldsymbol{K}} := \nabla X \chi\right)$. The dissipation inequality then reads:

$$\begin{aligned} \left(\underset{\sim}{\boldsymbol{\Pi}}{}_h - \rho_0 \frac{\partial \psi_0}{\partial \underset{\sim}{\boldsymbol{E}}{}_h^e}\right) : \dot{\underset{\sim}{\boldsymbol{E}}}{}_h^e + \left(a - \rho_0 \frac{\partial \psi_0}{\partial \chi}\right) \dot{\chi} + \left(\underline{\boldsymbol{b}}_0 - \rho_0 \frac{\partial \psi_0}{\partial \nabla_X \chi}\right) \cdot \dot{\underline{\boldsymbol{K}}} \\ + \underset{\sim}{\boldsymbol{\Pi}}{}_h : \dot{\underset{\sim}{\boldsymbol{E}}}{}_h^p - \rho_0 \frac{\partial \psi_0}{\partial q} \dot{q} \geq 0 \end{aligned} \tag{113}$$

from which the following state laws are selected:

$$\underset{\sim}{\boldsymbol{\Pi}}{}_h = \rho_0 \frac{\partial \psi_0}{\partial \underset{\sim}{\boldsymbol{E}}{}_h^e}, \quad a = \rho_0 \frac{\partial \psi_0}{\partial \text{ffl}}, \; \underline{\boldsymbol{b}}_0 = \rho_0 \frac{\partial \psi_0}{\partial \nabla_X \chi}, \quad R = \rho_0 \frac{\partial \psi_0}{\partial q} \tag{114}$$

The flow and hardening rules can be determined from the suitable choice of dissipation potential $\Omega \left(\underset{\sim}{\boldsymbol{\Pi}}{}_h, R\right)$:

$$\dot{\underset{\sim}{\boldsymbol{E}}}{}_h^p = \frac{\partial \Omega}{\partial \underset{\sim}{\boldsymbol{\Pi}}{}_h}, \quad \dot{q} = -\frac{\partial \Omega}{\partial R} \tag{115}$$

The existence of such a dissipation potential is not necessary but assumed in the whole chapter for convenience. Alternative methods of exploitation of the dissipation principle exist for micromorphic continua, for instance based on the extended Liu procedure (Ván et al. 2014; Berezovski et al. 2014).

Two straightforward extensions of the micromorphic approach to finite strain viscoplasticity based on an additive decomposition of a Lagrangian strain measure are presented:

$$\rho_0\psi_0 = \frac{1}{2}\underset{\sim}{\boldsymbol{E}}^e_h : \underset{\approx}{\boldsymbol{c}} : \underset{\sim}{\boldsymbol{E}}^e_h + \rho_0\psi_q(q) + H_\chi\left(E_{heq} - \chi\right)^2 + \rho_0\psi_\nabla\left(\underline{\boldsymbol{K}}\right) \tag{116}$$

where E_{heq} is an equivalent total strain measure, or, alternatively,

$$\rho_0\psi_0 = \frac{1}{2}\underset{\sim}{\boldsymbol{E}}^e_h : \underset{\approx}{\boldsymbol{c}} : \underset{\sim}{\boldsymbol{E}}^e_h + \rho_0\psi_q(q) + H_\chi(p - \chi)^2 + \rho_0\psi_\nabla\left(\underline{\boldsymbol{K}}\right) \tag{117}$$

where $\dot{p} = \sqrt{2/3\dot{\underset{\sim}{\boldsymbol{E}}}^p_h : \dot{\underset{\sim}{\boldsymbol{E}}}^p_h}$ is the cumulative plastic strain in the present context. These choices respectively provide the following regularisation partial differential equations:

$$E_{heq} = \chi - \mathrm{Div}\,\rho_0\frac{\partial\psi_0}{\partial\underline{\boldsymbol{K}}}, \quad \text{or} \quad p = \chi - \mathrm{Div}\,\rho_0\frac{\partial\psi_0}{\partial\underline{\boldsymbol{K}}} \tag{118}$$

If, $\underline{\boldsymbol{b}}_0 = A\underline{\boldsymbol{K}}$, then the regularisation operator involves the Lagrangian Laplace operator Δ_X in the same way as in Eq. (97).

Finite Deformation Micromorphic Viscoplasticity Using Local Objective Frames

An alternative and frequently used method to formulate anisotropic elastoviscoplastic constitutive equations at finite deformations that identically fulfill the condition of Euclidean invariance (also called material frame indifference, see Besson et al. 2009), is to resort to local objective rotating frames, as initially proposed by Dogui and Sidoroff (1985) and Sidoroff and Dogui (2001). A local objective rotating frame is defined by the rotation field $\underset{\sim}{\boldsymbol{Q}}\left(\underline{x}, t\right)$, objective w.r.t. to further change of observer, and taking different values at different material points and different times:

$$\underline{\boldsymbol{x}}^\dagger = \underset{\sim}{\boldsymbol{Q}}^T\left(\underline{x}, t\right) \cdot \underline{\boldsymbol{x}} \tag{119}$$

It is based on the idea that there exists for each material point a privileged observer w.r.t. which the constitutive law takes a simple form. The method is described in details in Besson et al. (2009) and is used in many commercial finite element codes with the standard choices: corotational frame, such that $\underset{\sim}{\boldsymbol{W}}^\dagger = \dot{\underset{\sim}{\boldsymbol{Q}}} \cdot \underset{\sim}{\boldsymbol{Q}}^T = \underset{\sim}{\boldsymbol{W}}$, $\underset{\sim}{\boldsymbol{W}}$ being the skew–symmetric part of the velocity gradient, and polar frame, such that $\underset{\sim}{\boldsymbol{Q}}(\underline{\boldsymbol{x}}, t) = \underset{\sim}{\boldsymbol{R}}(\underline{\boldsymbol{x}}, t)$, $\underset{\sim}{\boldsymbol{R}}$ being the rotation part in the polar decomposition of the deformation gradient $\underset{\sim}{\boldsymbol{F}}$. The main drawback of this method is the absence of thermodynamic background since, depending on specific constitutive choices within this framework, a free energy function of the strain may not exist.

This method is now applied to a micromorphic model including a scalar additional d.o.f. χ. Extension to higher order tensor–valued additional degrees of freedom is straightforward. The field equations are still given by (87). The stresses w.r.t. the local objective frames are:

$$\underset{\sim}{\boldsymbol{\sigma}}^{\dagger} := \underset{\sim}{\boldsymbol{Q}}^{T} \cdot \underset{\sim}{\boldsymbol{\sigma}} \cdot \underset{\sim}{\boldsymbol{Q}}, \quad a^{\dagger} = a, \quad \underline{\boldsymbol{b}}^{\dagger} = \underset{\sim}{\boldsymbol{Q}}^{T} \cdot \underline{\boldsymbol{b}} \tag{120}$$

Time–derivation of these relations shows that the rotated stress derivatives are given by

$$\underset{\sim}{\boldsymbol{Q}} \cdot \dot{\underset{\sim}{\boldsymbol{\sigma}}}^{\dagger} \cdot \underset{\sim}{\boldsymbol{Q}}^{T} = \dot{\underset{\sim}{\boldsymbol{\sigma}}} + \underset{\sim}{\boldsymbol{\sigma}} \cdot \underset{\sim}{\boldsymbol{W}}^{\dagger} - \underset{\sim}{\boldsymbol{W}}^{\dagger} \cdot \underset{\sim}{\boldsymbol{\sigma}}, \quad \underset{\sim}{\boldsymbol{Q}} \cdot \dot{\underline{\boldsymbol{b}}}^{\dagger} = \dot{\underline{\boldsymbol{b}}} - \underset{\sim}{\boldsymbol{W}}^{\dagger} \cdot \underline{\boldsymbol{b}} \tag{121}$$

i.e. objective derivatives of the corresponding Eulerian stress tensors. If the corotational frame is used, the corresponding time derivative is the Jaumann rate, whereas it is the Green–Naghdi stress rate when the polar rotation is used. The same procedure is applied to the strain rates:

$$\underset{\sim}{\boldsymbol{D}}^{\dagger} := \underset{\sim}{\boldsymbol{Q}}^{T} \cdot \underset{\sim}{\boldsymbol{D}} \cdot \underset{\sim}{\boldsymbol{Q}}, \quad \dot{\underline{\boldsymbol{k}}}^{\dagger} = \underset{\sim}{\boldsymbol{Q}}^{T} \cdot \nabla_x \dot{\chi} \tag{122}$$

The time integration of the second equation in the rotating frame provides the variable $\underline{\boldsymbol{k}}^{\dagger}$. It must be underlined that $\underline{\boldsymbol{k}}^{\dagger}$ is NOT equal to $\underset{\sim}{\boldsymbol{Q}}^{T} \cdot \nabla_x \chi$. It is NOT the exact material time derivative of a constitutive variable, in general. The standard procedure then consists in postulating an additive decomposition of the rotated strain rate into elastic and plastic parts as

$$\underset{\sim}{\boldsymbol{D}}^{\dagger} = \dot{\underset{\sim}{\boldsymbol{e}}}^{e} + \dot{\underset{\sim}{\boldsymbol{e}}}^{p} \tag{123}$$

where the elastic and plastic strain $\underset{\sim}{\boldsymbol{e}}^{e}$ and $\underset{\sim}{\boldsymbol{e}}^{p}$ are solely defined in the rotated frame. Anisotropic elastic laws are assumed to take the form:

$$\underset{\sim}{\boldsymbol{\sigma}}^{\dagger} = \underset{\approx}{\boldsymbol{c}} : \underset{\sim}{\boldsymbol{e}}^{e}, \quad a = -H_{\chi}(p - \chi), \quad \underline{\boldsymbol{b}}^{\dagger} = \underset{\sim}{\boldsymbol{A}} \cdot \underline{\boldsymbol{b}}^{\dagger} \tag{124}$$

Time–derivation of these equations and consideration of Eq. (120) show that the elasticity laws are in fact hypoelastic and that, generally, there does not exist a free energy density function from which they can be derived (Toll 2011).

The yield function and the flow rule are formulated within the rotated frame:

$$f\left(\underset{\sim}{\boldsymbol{\sigma}}^{\dagger}, R\right) = \sigma_{eq}^{\dagger} - R, \quad \underset{\sim}{\dot{\boldsymbol{e}}}^{p} = \dot{p}\frac{\partial f}{\partial \underset{\sim}{\boldsymbol{\sigma}}^{\dagger}} \tag{125}$$

where normality is assumed for convenience and the viscoplastic multiplier $\dot{p}$ is given by some viscoplastic law. The evolution of internal variables is of the form $\dot{q} = H\left(q, \underset{\sim}{\dot{\boldsymbol{e}}}^{p}\right)$ for suitable functions H. The yield radius is chosen as the following expression inspired by the previous thermodynamically based models: $R = R_{ref}(p) - a = R_{ref} - \text{div } \underline{\boldsymbol{b}}$. This extension of the micromorphic approach to finite deformations using rotating frames has been proposed first by Saanouni and Hamed (2013) and used by these authors for metal forming simulations involving regularised damage laws. As a result, the regularisation operator can be written as:

$$p = \chi - \frac{1}{H_\chi} \text{div } \underline{\boldsymbol{b}} = \chi - \frac{1}{H_\chi} \text{div}\left(\underset{\sim}{\boldsymbol{Q}} \cdot \underline{\boldsymbol{b}}^{\dagger}\right) = \chi - \frac{1}{H_\chi} \text{div}\left(\underset{\sim}{\boldsymbol{Q}} \cdot \underset{\sim}{\boldsymbol{A}} \cdot \underline{\boldsymbol{k}}^{\dagger}\right) \tag{126}$$

In the isotropic case, $\underset{\sim}{\boldsymbol{A}} = A\underset{\sim}{\boldsymbol{1}}$, the regularisation operator reduces to

$$p = \chi - \frac{A}{H_\chi} \text{div}\left(\underset{\sim}{\boldsymbol{Q}} \cdot \underline{\boldsymbol{k}}^{\dagger}\right) \tag{127}$$

It is worth insisting on the fact that, in general, $\underset{\sim}{\boldsymbol{Q}} \cdot \underline{\boldsymbol{k}}^{\dagger} \neq \nabla_x \chi$. Accordingly, the previous equation does not involve the Laplace operator and the regularisation is therefore nonlinear even with respect to rotated quantities.

Among all choices of rotating frames, the one associated with the logarithmic spin rate tensor (Xiao et al. 1999) was claimed to be the only one such that, when $\underset{\sim}{\dot{\boldsymbol{e}}}^{p} = 0$, the isotropic hypoelastic strain-stress relation turns out to be hyperelastic. However, this property does not pertain to the general case $\underset{\sim}{\dot{\boldsymbol{e}}}^{p} \neq 0$, so that this specific choice does not in general provide any explicit form of the regularisation operator.

Alternative constitutive choices are possible for the higher order stresses if Laplacian operators are preferred. They amount to restricting the use of the rotating frame only to the classical elastoviscoplasticity equations and to writing independently, $\underline{\boldsymbol{b}} = A\ \nabla_x \chi$, so that the regularisation operator is expressed in terms of the Eulerian Laplace operator Δ_x, see Eq. (148) in the next section, or $\underline{\boldsymbol{b}}_0 = A\ \nabla_X \chi$ which leads to the Lagrangian Laplace operator Δ_X, see Eq. (143) in the next section.

Note that limitations in the formulation of anisotropic plasticity arise from using symmetric plastic strain variable $\underset{\sim}{\boldsymbol{e}}^{p}$ and that generalisations are needed in order to introduce necessary plastic spins for materials with microtructures, which is possible within the rotating frame approach (Forest and Pilvin 1999).

Finite Deformation Micromorphic Elastoviscoplasticity Based on the Multiplicative Decomposition

The most appropriate thermodynamically based framework for the formulation of finite deformation anisotropic elastoviscoplasticity relies on the multiplicative decomposition of the deformation gradient, as settled by Mandel (1971). This method is applied here to a generalised continuum model again limited to one scalar degree of freedom, χ, in addition to the displacement vector, $\underline{u}$. The gradients of the degrees of freedom can be computed with respect to the reference or current coordinates:

$$\underset{\sim}{F} = \underset{\sim}{1} + \text{Grad}\,\underline{u} = \underset{\sim}{1} + \underline{u} \otimes \nabla_X \tag{128}$$

$$\underline{K} = \text{Grad}\;\chi = \nabla_X \chi, \quad \underline{k} = \text{grad}\;\chi = \nabla_X \chi = \underset{\sim}{F}^{-T} \cdot \underline{K} = \underline{K} \cdot \underset{\sim}{F}^{-1} \tag{129}$$

The consideration of microdeformation degrees of freedom of higher order is possible without fundamental modification of the approach below, see Forest and Sievert (2006).

A multiplicative decomposition is envisaged in this section, see Eq. (12), in the form:

$$\underset{\sim}{F} = \underset{\sim}{F}^e \cdot \underset{\sim}{F}^p \tag{130}$$

which assumes the existence of a triad of directors attached to the material point in order to unambiguously define the isoclinic intermediate local configuration, labelled (#) in the sequel, see Mandel (1971, 1973). The directors are usually related to non–material microstructure directions like lattice directions in single crystals or fibre directions in composites. The existence of such directors is required for the formulation of objective anisotropic constitutive equations (Besson et al. 2009). The Jacobians of all contributions in Eq. (130) are denoted by

$$J = \det \underset{\sim}{F},\; J_e = \det \underset{\sim}{F}^e,\; J_p = \det \underset{\sim}{F}^p,\; J = J_e J_p,\; \rho_0 = \rho_\sharp J_p = \rho J \tag{131}$$

They are used to relate the mass densities with respect to the three local configurations: In the present section, the microdeformation gradient $\underline{K}$ is not split into elastic and plastic contributions, although it is possible as done in section "Elastic-Plastic Decomposition of the Generalized Strain Measures," at the expense of additional evolution laws to be determined and of drastically different regularisation operators.

The power density of internal and contact forces are

$$p^{(i)} = \underset{\sim}{\sigma} : \dot{\underset{\sim}{F}}\,\underset{\sim}{F}^{-1} + a\dot{\chi} + \underline{k} \cdot \nabla_x \dot{\chi},\; \forall \underline{x} \in \Omega,\; p^{(c)} = \underline{t} \cdot \underline{v} + a_c \dot{\chi},\; \forall \underline{x} \in \partial\Omega \tag{132}$$

The invariance of $p^{(i)}$ with respect to any change of observer requires the Cauchy stress $\underset{\sim}{\boldsymbol{\sigma}}$ to be symmetric. The scalar microstrain is assumed to be invariant. The corresponding balance and boundary conditions are still given by Eq. (87).

Lagrangian Formulation

The Lagrangian free energy density is a function $\psi_0\left(\underset{\sim}{\boldsymbol{C}}^e, q, \chi, \underline{\boldsymbol{K}}\right)$, where $\underset{\sim}{\boldsymbol{C}}^e = \underset{\sim}{\boldsymbol{F}}^{eT} \cdot \underset{\sim}{\boldsymbol{F}}^e$ is the elastic strain and q a set of internal variables accounting for material hardening. Note that the usual elastic strain tensor $\underset{\sim}{\boldsymbol{C}}^e$ is defined with respect to the intermediate configuration to comply with standard anisotropic plasticity, whereas $\underline{\boldsymbol{K}}$ is Lagrangian. The presented formulation is therefore not purely Lagrangian but rather mixed. The local Lagrangian form of the entropy inequality is: $\mathcal{D}_0 = p_0^{(i)} - \rho_0 \dot{\psi}_0 \geq 0, \quad p_0^{(i)} = Jp^{(i)}$. Accounting for the multiplicative decomposition (130), the power of internal forces is expanded as:

$$
\begin{aligned}
p_0^{(i)} &= J\underset{\sim}{\boldsymbol{\sigma}} : \underset{\sim}{\dot{\boldsymbol{F}}} \cdot \underset{\sim}{\boldsymbol{F}}^{-1} + Ja\dot{\chi} + J\underline{b} \cdot \nabla_x \dot{\chi} = \frac{J_p}{2} \underset{\sim}{\boldsymbol{\Pi}}^e : \underset{\sim}{\dot{\boldsymbol{C}}}^e \\
&+ J_p \underset{\sim}{\boldsymbol{\Pi}}^M : \underset{\sim}{\dot{\boldsymbol{F}}}^p \cdot \underset{\sim}{\boldsymbol{F}}^{p-1} + a_0 \dot{\chi} + \underline{\boldsymbol{b}}_0 \cdot \underline{\dot{\boldsymbol{K}}}
\end{aligned}
\tag{133}
$$

where the Piola stress tensor w.r.t. the intermediate configuration and the Mandel stress tensor according to (Haupt 2000) are respectively defined as:

$$
\underset{\sim}{\boldsymbol{\Pi}}^e = J_e \underset{\sim}{\boldsymbol{F}}^{e-1} \cdot \boldsymbol{\sigma} \cdot \underset{\sim}{\boldsymbol{F}}^{e-T}, \quad \underset{\sim}{\boldsymbol{\Pi}}^M = \underset{\sim}{\boldsymbol{C}}^e \cdot \underset{\sim}{\boldsymbol{\Pi}}^e = J_e \underset{\sim}{\boldsymbol{F}}^{eT} \cdot \underset{\sim}{\boldsymbol{\sigma}} \cdot \underset{\sim}{\boldsymbol{F}}^{e-T} \tag{134}
$$

The Lagrangian generalized stresses in (133) are $a_0 = Ja$ and $\underline{\boldsymbol{b}}_0 = J\underset{\sim}{\boldsymbol{F}}^{-1}\underline{\boldsymbol{b}}$. As a result the dissipation rate becomes:

$$
\begin{aligned}
&\left(\frac{J_p}{2}\underset{\sim}{\boldsymbol{\Pi}}^e - \rho_0 \frac{\partial \psi_0}{\partial \underset{\sim}{\boldsymbol{C}}^e}\right) : \underset{\sim}{\dot{\boldsymbol{C}}}^e + \left(a_0 - \rho_0 \frac{\partial \psi_0}{\partial \chi}\right)\dot{\chi} + \left(\underline{\boldsymbol{b}}_0 - \rho_0 \frac{\partial \psi_0}{\partial \underline{\boldsymbol{K}}}\right) \cdot \underline{\dot{\boldsymbol{K}}} \\
&+ J_p \underset{\sim}{\boldsymbol{\Pi}}^M : \underset{\sim}{\dot{\boldsymbol{F}}}^p \cdot \underset{\sim}{\boldsymbol{F}}^{p-1} - \rho_0 \frac{\partial \psi_0}{\partial q}\dot{q} \geq 0
\end{aligned}
\tag{135}
$$

The following state and evolution laws ensure the positivity of $\mathcal{D}_0$:

$$
\underset{\sim}{\boldsymbol{\Pi}}^e = 2\rho_\sharp \frac{\partial \psi_0}{\partial \underset{\sim}{\boldsymbol{C}}^e}, \; a_0 = \rho_0 \frac{\partial \psi_0}{\partial \chi}, \; \underline{\boldsymbol{b}}_0 = \rho_0 \frac{\partial \psi_0}{\partial \underline{\boldsymbol{K}}}, \; R = \rho_0 \frac{\partial \psi_0}{\partial q} \tag{136}
$$

$$
\underset{\sim}{\dot{\boldsymbol{F}}}^p \cdot \underset{\sim}{\dot{\boldsymbol{F}}}^{p-1} = \frac{\partial \Omega}{\partial \underset{\sim}{\boldsymbol{\Pi}}^M}\left(\underset{\sim}{\boldsymbol{\Pi}}^M, R\right), \; \dot{q} = -\frac{\partial \Omega}{\partial R}\left(\underset{\sim}{\boldsymbol{\Pi}}^M, R\right), \tag{137}
$$

Following Mandel (1971), a dissipation potential $\Omega\left(\underset{\sim}{\boldsymbol{\Pi}}^{M}, R\right)$ function of the driving forces for plasticity, is introduced to formulate the flow and hardening variable evolution rule. If the dissipation function is convex w.r.t. $\underset{\sim}{\boldsymbol{\Pi}}^{M}$ and concave w.r.t. R, the positivity of dissipation is ensured. Specific expressions for within the context of viscoplasticity can be found in Besson et al. (2009).

As an example, the following free energy potential is proposed:

$$\rho_0\psi_0 = \frac{1}{2}J_p\underset{\sim}{\boldsymbol{E}}^e : \underset{\approx}{\boldsymbol{c}} : \underset{\sim}{\boldsymbol{E}}^e + \rho_0\psi_q(q) + \frac{1}{2}H_\chi(p-\chi)^2 + \rho_0\psi_{\nabla X}\left(\underline{\boldsymbol{K}}\right) \tag{138}$$

where ψ_q is the appropriate free energy contribution associated with usual work-hardening (not specified here) and $2\underset{\sim}{\boldsymbol{E}}^e = \underset{\sim}{\boldsymbol{C}}^e - \underset{\sim}{\boldsymbol{1}}$ is the Green-Lagrange elastic strain measure. The microstrain variable is compared to the cumulative plastic strain variable p defined as:

$$\dot{p} = \sqrt{\frac{2}{3}\left(\underset{\sim}{\dot{\boldsymbol{P}}}\cdot\underset{\sim}{\boldsymbol{P}}^{-1}\right):\left(\underset{\sim}{\dot{\boldsymbol{P}}}\cdot\underset{\sim}{\boldsymbol{P}}^{-1}\right)} \tag{139}$$

According to the state laws (136), we obtain

$$\underset{\sim}{\boldsymbol{\Pi}}^e = \rho_\sharp\frac{\partial\psi_0}{\partial\underset{\sim}{\boldsymbol{E}}^e} = \underset{\approx}{\boldsymbol{c}} : \underset{\sim}{\boldsymbol{E}}^e,\ a_0 = -H_\chi\left(p-\chi\right),\ \underline{\boldsymbol{b}}_0 = \rho_0\frac{\partial\psi_{\nabla X}}{\partial\underline{\boldsymbol{K}}} \tag{140}$$

The regularisation operator then follows from the combination of the previous constitutive equations with the balance equation (89):

$$p = \chi - \frac{1}{H_\chi}\mathrm{Div}\left(\rho_0\frac{\partial\psi_{\nabla X}}{\partial\underline{\boldsymbol{K}}}\right) \tag{141}$$

The specific choice $\rho_0\psi_{\nabla X}\left(\underline{\boldsymbol{K}}\right) = \frac{1}{2}\underline{\boldsymbol{K}}\cdot\underset{\sim}{\boldsymbol{A}}\cdot\underline{\boldsymbol{K}}$ leads to a regularisation operator that is linear with respect to Lagrangian coordinates:

$$p = \chi - \frac{1}{H_\chi}\mathrm{Div}\left(\underset{\sim}{\boldsymbol{A}}\cdot\mathrm{Grad}\ \chi\right) \tag{142}$$

which involves the Laplacian operator Δ_X in the isotropic case, i.e. $\underset{\sim}{\boldsymbol{A}} = A\underset{\sim}{\boldsymbol{1}}$,

$$\mathrm{Op} = 1 - \frac{A}{H_\chi}\Delta_X \tag{143}$$

The impact on hardening can be seen by choosing, as an example, $q = p$, according to (139), as an internal variable in (138). The dissipation potential can be chosen in such a way that the residual dissipation takes the form

$$\mathcal{D}_0 = J_p \underset{\sim}{\boldsymbol{\Pi}}^M : \left(\dot{\underset{\sim}{\boldsymbol{F}}}^p \cdot \underset{\sim}{\boldsymbol{F}}^{p-1} \right) - \rho_0 \frac{\partial \Psi_q}{\partial p} \dot{p} = J_p \Pi^M_{eq} \dot{p} - R\dot{p} \geq 0 \tag{144}$$

with

$$\dot{p} = \frac{\partial \Omega}{\partial f}, \; f\left(\underset{\sim}{\boldsymbol{\Pi}}^M, R \right) = J_p \Pi^M_{eq} - R$$

where f is the yield function. As a result, the yield stress R is given by the following enhanced hardening law:

$$R = \rho_0 \frac{\partial \psi_0}{\partial p} = \rho_0 \frac{\partial \psi_q}{\partial p} + H_\chi (p - \chi) = \rho_0 \frac{\partial \psi_q}{\partial p} - \text{Div} \left(\rho_0 \frac{\partial \psi_{\nabla X}}{\partial \underline{\boldsymbol{K}}} \right) \tag{145}$$

Eulerian Formulation

In the Eulerian formulation, the free energy density is taken as a function of $\underline{\boldsymbol{k}}$ instead of $\underline{\boldsymbol{K}}$, according to (129): $\psi\left(\underset{\sim}{\boldsymbol{C}}^e, \chi, \underline{\boldsymbol{k}} \right)$, so that the state laws for generalised stresses become:

$$a = \rho \frac{\partial \psi}{\partial \chi}, \; \underline{\boldsymbol{b}} = \rho \frac{\partial \psi}{\partial \underline{\boldsymbol{k}}} \tag{146}$$

The arguments of the free energy mix the invariant quantities $\underset{\sim}{\boldsymbol{C}}^e, \chi$ and the observer–dependent variable $\underline{\boldsymbol{k}}$. Galilean invariance then requires ψ to be isotropic with respect to $\underline{\boldsymbol{k}}$.

The constitutive choice (138) is now replaced by

$$\rho\psi = \frac{1}{2} J \underset{\sim}{\boldsymbol{E}}^e : \underset{\approx}{\boldsymbol{c}} : \underset{\sim}{\boldsymbol{E}}^e + \rho\psi_q(q) + \frac{1}{2} H_\chi (p - \chi)^2 + \rho\psi_\nabla (\underline{\boldsymbol{k}}) \tag{147}$$

A quadratic potential ψ_∇ is necessarily of the form $A\|\underline{\boldsymbol{k}}\|^2/2$, for objectivity reasons, so that the regularisation operator involves the Laplace operator Δ_x w.r.t. Eulerian coordinates:

$$\text{Op} = 1 - \frac{A}{H_\chi} \Delta_x \tag{148}$$

If the same viscoplastic yield function $f\left(\underset{\sim}{\boldsymbol{\Pi}}^M, R \right)$ as in the previous subsection is adopted, the hardening rule is enhanced as follows:

$$R = \rho \frac{\partial \psi}{\partial p} = \rho \frac{\partial \psi_q}{\partial p} + H_\chi (p - \chi) = \rho \frac{\partial \psi_q}{\partial p} - \text{div} \; \underline{\boldsymbol{b}} = \rho \frac{\partial \psi_q}{\partial p} - A \Delta_x \chi \tag{149}$$

This therefore yields a finite strain generalisation of Aifantis strain gradient plasticity model (Aifantis 1987; Forest and Aifantis 2010).

Formulation Using the Local Intermediate Configuration Only

In the two previous formulations, Lagrangian or Eulerian generalized strain variables were mixed with the elastic strain variable $\underset{\sim}{\boldsymbol{C}}^e$ attached to the intermediate local configuration, as the arguments of the free energy function. It is possible to assign the free energy function with a consistent set of arguments solely attached to the intermediate configuration. For that purpose, a generalised strain $\underline{\boldsymbol{K}}^\sharp$ and generalised stresses $a_\sharp, \underline{\boldsymbol{b}}^\sharp$ are now defined on the intermediate local configuration:

$$\underline{\boldsymbol{K}}^\sharp = \underline{\boldsymbol{k}} \cdot \underset{\sim}{\boldsymbol{F}}^e = \underset{\sim}{\boldsymbol{F}}^{eT} \cdot \underline{\boldsymbol{k}} = \underline{\boldsymbol{K}} \cdot \underset{\sim}{\boldsymbol{F}}^{p-1} = \underset{\sim}{\boldsymbol{F}}^{p-T} \cdot \underline{\boldsymbol{K}} \tag{150}$$

$$a_\sharp = J_e a = J_p^{-1} a_0, \quad \underline{\boldsymbol{b}}^\sharp = J_p^{-1} \underset{\sim}{\boldsymbol{F}}^p \cdot \underline{\boldsymbol{b}}_0 = J_e \underset{\sim}{\boldsymbol{F}}^{e-1} \cdot \underline{\boldsymbol{b}} \tag{151}$$

The power density of internal forces expressed w.r.t. the intermediate local configuration then takes the form:

$$p_\sharp^{(i)} = J_p^{-1} p_0^{(i)} = \frac{1}{2} \underset{\sim}{\boldsymbol{\Pi}}^e : \dot{\underset{\sim}{\boldsymbol{C}}}^e + \left(\underset{\sim}{\boldsymbol{\Pi}}^M + \underline{\boldsymbol{K}}^\sharp \otimes \underline{\boldsymbol{b}}^\sharp \right) : \dot{\underset{\sim}{\boldsymbol{F}}}^p \cdot \underset{\sim}{\boldsymbol{F}}^{p-1} + a_\sharp \dot{\chi} + \underline{\boldsymbol{b}}_\sharp \cdot \dot{\underline{\boldsymbol{K}}}^\sharp \tag{152}$$

where $p_0^{(i)}$ is still given by Eq. (133). To establish this expression, the following relation was used:

$$\dot{\underline{\boldsymbol{K}}} = \underset{\sim}{\boldsymbol{F}}^{pT} \cdot \dot{\underline{\boldsymbol{K}}}^\sharp + \dot{\underset{\sim}{\boldsymbol{F}}}^{pT} \cdot \underline{\boldsymbol{K}}^\sharp \tag{153}$$

The dissipation rate density measured w.r.t. the intermediate local configuration is then:

$$\mathcal{D}_\sharp = p_\sharp^{(i)} - \rho_\sharp \dot{\psi}_\sharp \geq 0 \tag{154}$$

The free energy density function is chosen as $\psi_\sharp \left(\underset{\sim}{\boldsymbol{C}}^e, q, \chi, \underline{\boldsymbol{K}}^\sharp \right)$. As such, it is invariant w.r.t. change of observer. The Clausius–Duhem inequality is now derived as

$$\begin{aligned} &\left(\tfrac{1}{2} \underset{\sim}{\boldsymbol{\Pi}}^e - \rho_\sharp \tfrac{\partial \psi_\sharp}{\partial \underset{\sim}{\boldsymbol{C}}^e} \right) : \dot{\underset{\sim}{\boldsymbol{C}}}^e + \left(a_\sharp - \rho_\sharp \tfrac{\partial \psi_\sharp}{\partial \chi} \right) \dot{\chi} + \left(\underline{\boldsymbol{b}}^\sharp - \rho_\sharp \tfrac{\partial \psi_\sharp}{\partial \underline{\boldsymbol{K}}^\sharp} \right) \cdot \dot{\underline{\boldsymbol{K}}}^\sharp \\ &+ \left(\underset{\sim}{\boldsymbol{\Pi}}^M + \underline{\boldsymbol{K}}^\sharp \otimes \underline{\boldsymbol{b}}^\sharp \right) : \dot{\underset{\sim}{\boldsymbol{F}}}^p \cdot \underset{\sim}{\boldsymbol{F}}^{p-1} - \rho_\sharp \tfrac{\partial \psi_\sharp}{\partial q} \dot{q} \geq 0 \end{aligned} \tag{155}$$

This expression reveals the existence of a generalised Mandel tensor, $\underset{\sim}{\mathbf{\Pi}}^M + \underline{\mathbf{K}}^\sharp \otimes \underline{\mathbf{b}}^\sharp$, conjugate to the plastic deformation rate, that is a function of the classical Mandel stress and of microdeformation related stress and strain. Positivity of dissipation is ensured by the choice of the following state laws and plastic flow and hardening rules:

$$\underset{\sim}{\mathbf{\Pi}}^e = 2\rho_\sharp \frac{\partial \psi_\sharp}{\partial \underset{\sim}{\mathbf{C}}^e},\ a_\sharp = \rho_\sharp \frac{\partial \psi_\sharp}{\partial \chi},\ \underline{\mathbf{b}}^\sharp = \rho_\sharp \frac{\partial \psi_\sharp}{\partial \underline{\mathbf{K}}^\sharp},\ R = \rho_\sharp \frac{\partial \psi_\sharp}{\partial q} \tag{156}$$

$$\dot{\underset{\sim}{\mathbf{F}}}^p \cdot \underset{\sim}{\mathbf{F}}^{p-1} = \frac{\partial \Omega}{\partial \left(\underset{\sim}{\mathbf{\Pi}}^M + \underline{\mathbf{K}}^\sharp \otimes \underline{\mathbf{b}}^\sharp \right)},\ \dot{q} = -\frac{\partial \Omega}{\partial R} \tag{157}$$

provided that the dissipation potential $\Omega \left(\underset{\sim}{\mathbf{\Pi}}^M + \underline{\mathbf{K}}^\sharp \otimes \underline{\mathbf{b}}^\sharp, R \right)$ displays suitable convexity properties with respect to both arguments.

The yield criterion is taken as a function $f\left(\underset{\sim}{\mathbf{\Pi}}^M + \underline{\mathbf{K}}^\sharp \otimes \underline{\mathbf{b}}^\sharp, R \right) = \Pi^M_{eq} - R$ where the Π^M_{eq} is an equivalent stress measure based on the generalized Mandel stress tensor. Choosing $q = p$, where p is still given by Eq. (139), the residual dissipation takes the form:

$$\mathcal{D}_\sharp = \left(\underset{\sim}{\mathbf{\Pi}}^M + \underline{\mathbf{K}}^\sharp \otimes \underline{\mathbf{b}}^\sharp \right) : \dot{\underset{\sim}{\mathbf{F}}}^p \cdot \underset{\sim}{\mathbf{F}}^{p-1} - R\dot{p} = \Pi^M_{eq}\dot{p} - R\dot{p} \tag{158}$$

Note that the contribution $\underline{\mathbf{K}}^\sharp \otimes \underline{\mathbf{b}}^\sharp$in the generalised Mandel stress acts as a size–dependent kinematic hardening component which comes in addition to isotropic hardening represented by R. This is a specific feature of the model formulation w.r.t. the intermediate configuration.

As an example, a typical form of the free energy density function based on constitutive variables defined on the intermediate configuration, and hyperelastic laws are:

$$\rho_\sharp \psi_\sharp = \frac{1}{2} \underset{\sim}{\mathbf{E}}^e : \underset{\approx}{c} : \underset{\sim}{\mathbf{E}}^e + \rho_\sharp \psi_\sharp(q) + \frac{1}{2} H_\chi (p - \chi)^2 + \frac{1}{2} \rho_\sharp \psi_{\sharp\nabla} \left(\underline{\mathbf{K}}^\sharp \right) \tag{159}$$

$$\underset{\sim}{\mathbf{\Pi}}^e = \underset{\approx}{c} : \underset{\sim}{\mathbf{E}}^e,\ a_\sharp = -H_\chi (p - \chi),\ \underline{\mathbf{b}}^\sharp = \rho_\sharp \frac{\partial \psi_\sharp}{\partial \underline{\mathbf{K}}^\sharp} \tag{160}$$

These generalized stresses can be inserted into the balance equation

$$a_0 = \mathrm{Div}\ \underline{\mathbf{b}}_0 \Rightarrow J_p a_\sharp = \mathrm{Div} \left(J_p \underset{\sim}{\mathbf{F}}^{p-1} \cdot \underline{\mathbf{b}}^\sharp \right) \tag{161}$$

This provides the form of the regularisation operator:

$$p = \chi - \frac{1}{J_p H_\chi} \text{Div} \left(J_p \underset{\sim}{\boldsymbol{F}}^{p-1} \rho_\sharp \frac{\partial \psi_\sharp}{\partial \underline{\boldsymbol{K}}^\sharp} \right) \tag{162}$$

A quadratic dependence of the contribution $\rho_\sharp \psi_{\sharp \nabla} = 1/2 \underline{\boldsymbol{K}}^\sharp \cdot \underset{\sim}{\boldsymbol{A}} \cdot \underline{\boldsymbol{K}}^\sharp$ leads to the following linear relationship between $\underline{\boldsymbol{b}}^\sharp$ and $\underline{\boldsymbol{K}}^\sharp$:

$$\underline{\boldsymbol{b}}^\sharp = A \underline{\boldsymbol{K}}^\sharp \quad \Rightarrow \underline{\boldsymbol{b}} = J_e^{-1} A \underset{\sim}{\boldsymbol{B}}^e \cdot \underline{\boldsymbol{k}} \quad \text{and} \quad \underline{\boldsymbol{b}}^\sharp = J_p A \underset{\sim}{\boldsymbol{C}}^{p-1} \cdot \underline{\boldsymbol{k}} \tag{163}$$

with $\underset{\sim}{\boldsymbol{B}}^e = \underset{\sim}{\boldsymbol{F}}^e \cdot \underset{\sim}{\boldsymbol{F}}^{eT}$ and $\underset{\sim}{\boldsymbol{C}}^p = \underset{\sim}{\boldsymbol{F}}^{pT} \cdot \underset{\sim}{\boldsymbol{F}}^p$. However, in that case, the regularisation operator (161) is nonlinear and does not involve a Laplace operator, even in the isotropic case $\underset{\sim}{\boldsymbol{A}} = A \underset{\sim}{\boldsymbol{1}}$. As a result, the hyperelastic relationships for the higher order stresses are not linear w.r.t. to the associated strain gradient measures.

Conclusion

Eringen and Mindlin's micromorphic theory offers real opportunities for the modeling of size effects in the mechanics of materials. Elastic-viscoplastic constitutive laws have been formulated at finite deformations. They remain to be further specialized and calibrated with respect to size effects observed in metal and polymer plasticity. Successful applications deal for example with the ductile fracture of metallic alloys (Enakoutsa and Leblond 2009; Hutter 2017b) and porous metals (Dillard et al. 2006). Micromorphic elasticity has been recently revisited and further developed to account for the dispersion of elastic waves in architectured and metamaterials (Neff et al. 2014; Rosi and Auffray 2016; Madeo et al. 2016). Intensive work is needed to establish connections between the micromorphic continuum theories and the actual underlying microstructure (Forest and Trinh 2011; Hutter 2017a; Biswas and Poh 2017).

The proposed systematic treatment of the thermomechanics of continua with additional degrees of freedom leads to model formulations ranging from micromorphic to phase field models. In particular, a general framework for the introduction of dissipative processes associated with the additional degrees of freedom has been proposed. If internal constraints are enforced on the relation between macro and microvariables in the micromorphic approach, standard second gradient and strain gradient plasticity models can be retrieved.

As a variant of micromorphic continuum, microdamage continuum and its regularization capabilities for the modelling of crack propagation in single crystals have been studied. First, a crystallographic constitutive model which accounts for continuum damage with respect to fracture planes has been presented. Then, the theory has been extended from classical continuum to microdamage continuum. It has been shown that the approach can be a good candidate for solving mesh dependency and the prediction of final fracture in anistropic media. Analytical fits and numerical results showed that the theory is well suited for FEA and possesses a great potential for the future modelling aspects. Comparison with available data on crack growth especially cyclic loading in nickel-based superalloys, will be decisive to conclude on the ability of the approach to reach realistic prediction of component failure.

The presented extensions to finite deformations show that the regularisation operator cannot be postulated in an intuitive way. It is rather the result of a constitutive choice regarding the dependence of the free energy function on the gradient term. Purely Lagrangian and Eulerian formulations are straightforward and lead to Helmholtz–like operators w.r.t. Lagrangian of Eulerian coordinates. Two alternative standard procedures of extension of classical constitutive laws to large strains, widely used in commercial finite element codes, have been combined with the micromorphic approach. In particular, the choice of local objective rotating frames leads to new nonlinear regularisation operators that are not of the Helmholtz type. Three distinct operators were proposed within the context of the multiplicative decomposition of the deformation gradient. A new feature is that a free energy density function depending on variables solely defined with respect to the intermediate isoclinic configuration leads to the existence of additional kinematic hardening induced by the gradient of a scalar micromorphic degree of freedom.

Note that the results obtained for the micromorphic theory with additional degrees of freedom are also valid for gradient theories (gradient plasticity or gradient damage) once an internal constraint is imposed linking the additional degrees of freedom to strain or internal variables. This amounts to selecting high values of parameter H_χ or introducing corresponding Lagrange multipliers. The analysis was essentially limited to scalar micromorphic degrees of freedom for the sake of simplicity, even though tensorial examples were also given. Scalar plastic microstrain approaches suffer from limitations like indeterminacy of flow direction at cusps of the cumulative plastic strain in bending for instance, see Peerlings (2007), Poh et al. (2011), and Wulfinghoff et al. (2014). Those limitations can be removed by the use of tensorial micromorphic variable (microstrain or microdeformation tensors). The micromorphic approach is not limited to the gradient of strain–like, damage or phase field variables. It can also be applied to other internal variables, as demonstrated for hardening variables in Dorgan and Voyiadjis (2003) and Saanouni and Hamed (2013).

It remains that the regularisation properties of the derived nonlinear operators are essentially unknown, except through examples existing in the mentioned literature. For instance, the Eulerian and Lagrangian variants of the Helmholtz-type equation for scalar micromorphic strain variables have been assessed by Wcislo et al. (2013) giving the advantage to the latter, based on finite element simulations of specific situations. The regularising properties of more general operators should be explored in the future from the mathematical and computational perspectives in order to select the most relevant constitutive choices that may depend on the type of material classes.

It may be surprising that the constitutive theory underlying the construction of regularisation operators for plasticity and damage, mainly relies on the enhancement of the free energy density function instead of the dissipative laws. It is in fact widely recognised that plastic strain gradients, e.g. associated with the multiplication of geometrically necessary dislocations, lead to energy storage that can be released by further deformation or heat treatments. However, the limitation to the enhancement of free energy potential is mainly due to the simplicity of the theoretical treatment and to the computational efficiency of the operators derived in that way. Dissipative higher order contributions remain to be explored from the viewpoint of regularisation, as started in Forest (2009). Constitutive models of that kind are already available for plasticity, damage and fracture (Amor et al. 2009; Pham et al. 2011; Vignollet et al. 2014; Miehe et al. 2016).

Acknowledgments The first author thanks Prof. O. Aslan for his contribution to the presented micro-morphic damage theory. These contributions are duly cited in the references quoted in the text and listed below.

References

E. Aifantis, On the microstructural origin of certain inelastic models. J. Eng. Mater. Technol. **106**, 326–330 (1984)

E. Aifantis, The physics of plastic deformation. Int. J. Plast. **3**, 211–248 (1987)

K. Ammar, B. Appolaire, G. Cailletaud, F. Feyel, F. Forest, Finite element formulation of a phase field model based on the concept of generalized stresses. Comput. Mater. Sci. **45**, 800–805 (2009)

H. Amor, J.J. Marigo, C. Maurini, Regularized formulation of the variational brittle fracture with unilateral contact: Numerical experiments. J. Mech. Phys. Solids **57**, 1209–1229 (2009)

O. Aslan, S. Forest, Crack growth modelling in single crystals based on higher order continua. Comput. Mater. Sci. **45**, 756–761 (2009)

O. Aslan, S. Forest, The micromorphic versus phase field approach to gradient plasticity and damage with application to cracking in metal single crystals, in *Multiscale Methods in Computational Mechanics*, ed. by R. de Borst, E. Ramm. Lecture Notes in Applied and Computational Mechanics, vol. 55, (Springer, New York, 2011), pp. 135–154

O. Aslan, N.M. Cordero, A. Gaubert, S. Forest, Micromorphic approach to single crystal plasticity and damage. Int. J. Eng. Sci. **49**, 1311–1325 (2011)

N. Auffray, H. Le Quang, Q. He, Matrix representations for 3D strain-gradient elasticity. J. Mech. Phys. Solids **61**, 1202–1223 (2013)

A. Berezovski, J. Engelbrecht, P. Van, Weakly nonlocal thermoelasticity for microstructured solids: Microdeformation and microtemperature. Arch. Appl. Mech. **84**, 1249–1261 (2014)

A. Bertram, *Elasticity and Plasticity of Large Deformations* (Springer, Heidelberg, 2012)

Besson, J., Cailletaud, G., Chaboche, J.-L., Forest, S., Bletry, M., *Non-linear Mechanics of Materials*. Solid Mechanics and Its Applications, vol. 167 (Springer, Dordrecht, 2009), 433 p. ISBN:978-90-481-3355-0

R. Biswas, L. Poh, A micromorphic computational homogenization framework for heterogeneous materials. J. Mech. Phys. Solids **102**, 187–208 (2017)

G. Cailletaud, S. Forest, D. Jeulin, F. Feyel, I. Galliet, V. Mounoury, S. Quilici, Some elements of microstructural mechanics. Comput. Mater. Sci. **27**, 351–374 (2003)

A. L. Cauchy, Recherches sur l'équilibre et le mouvement intérieur des corps solides ou fluides, élastiques ou non élastiques. Bulletin de la Société Philomatique, 9–13 (1822)

B. Coleman, W. Noll, The thermodynamics of elastic materials with heat conduction and viscosity. Arch. Ration. Mech. Anal. **13**, 167–178 (1963)

T. Dillard, S. Forest, P. Ienny, Micromorphic continuum modelling of the deformation and fracture behaviour of nickel foams. Eur. J. Mech. A/Solids **25**, 526–549 (2006)

A. Dogui, F. Sidoroff, Kinematic hardening in large elastoplastic strain. Eng. Fract. Mech. **21**, 685–695 (1985)

J. Dorgan, G. Voyiadjis, Nonlocal dislocation based plasticity incorporating gradients of hardening. Mech. Mater. **35**, 721–732 (2003)

K. Enakoutsa, J. Leblond, Numerical implementation and assessment of the glpd micromorphic model of ductile rupture. Eur. J. Mech. A/Solids **28**, 445–460 (2009)

R. Engelen, M. Geers, F. Baaijens, Nonlocal implicit gradient-enhanced elasto-plasticity for the modelling of softening behaviour. Int. J. Plast. **19**, 403–433 (2003)

A. Eringen, Theory of thermo-microstretch elastic solids. Int. J. Eng. Sci. **28**, 1291–1301 (1990)

A. Eringen, *Microcontinuum Field Theories* (Springer, New York, 1999)

A. Eringen, E. Suhubi, Nonlinear theory of simple microelastic solids. Int. J. Eng. Sci. **2**(189–203), 389–404 (1964)

S. Forest, The micromorphic approach for gradient elasticity, viscoplasticity and damage. ASCE J. Eng. Mech. **135**, 117–131 (2009)

S. Forest, Micromorphic media, in *Generalized Continua – From the Theory to Engineering Applications. CISM International Centre for Mechanical Sciences*, ed. by H. Altenbach, V. Eremeyev. Courses and Lectures No. 541 (Springer, New York, 2012), pp. 249–300

S. Forest, Nonlinear regularisation operators as derived from the micromorphic approach to gradient elasticity, viscoplasticity and damage. Proc. R. Soc. A **472**, 20150755 (2016)

S. Forest, E.C. Aifantis, Some links between recent gradient thermo-elasto-plasticity theories and the thermomechanics of generalized continua. Int. J. Solids Struct. **47**, 3367–3376 (2010)

S. Forest, A. Bertram, Formulations of strain gradient plasticity, in *Mechanics of Generalized Continua*, ed. by H. Altenbach, G.A. Maugin, V. Erofeev. Advanced Structured Materials, vol. 7 (Springer, Berlin/Heidelberg, 2011), pp. 137–150

S. Forest, P. Pilvin, Modelling finite deformation of polycrystals using local objective frames. Z. Angew. Math. Mech. **79**, S199–S202 (1999)

S. Forest, K. Sab, Finite deformation second order micromorphic theory and its relations to strain and stress gradient models. Math. Mech. Solids (2017). https://doi.org/10.1177/1081286517720844

S. Forest, R. Sievert, Elastoviscoplastic constitutive frameworks for generalized continua. Acta Mech. **160**, 71–111 (2003)

S. Forest, R. Sievert, Nonlinear microstrain theories. Int. J. Solids Struct. **43**, 7224–7245 (2006)

S. Forest, D.K. Trinh, Generalized continua and non–homogeneous boundary conditions in homogenization methods. ZAMM **91**, 90–109 (2011)

M.G.D. Geers, Finite strain logarithmic hyperelasto-plasticity with softening: A strongly non-local implicit gradient framework. Comput. Methods Appl. Mech. Eng. **193**, 3377–3401 (2004)

M. Geers, R.D. Borst, W. Brekelmans, R. Peerlings, On the use of local strain fields for the determination of the intrinsic length scale. Journal de Physique IV **8**(Pr8), 167–174 (1998)

P. Germain, La méthode des puissances virtuelles en mécanique des milieux continus, premiere partie : théorie du second gradient. J. de Mécanique **12**, 235–274 (1973a)

P. Germain, The method of virtual power in continuum mechanics. Part 2: microstructure. SIAM J. Appl. Math. **25**, 556–575 (1973b)

P. Germain, Q. Nguyen, P. Suquet, Continuum thermodynamics. J. Appl. Mech. **50**, 1010–1020 (1983)

N. Germain, J. Besson, F. Feyel, Simulation of laminate composites degradation using mesoscopic non–local damage model and non-local layered shell element. Model. Simul. Mater. Sci. Eng. **15**, S425–S434 (2007)

M. Goodman, S. Cowin, A continuum theory for granular materials. Arch. Ration. Mech. Anal. **44**, 249–266 (1972)

P. Grammenoudis, C. Tsakmakis, Micromorphic continuum part I: Strain and stress tensors and their associated rates. Int. J. Non–Linear Mech. **44**, 943–956 (2009)
P. Grammenoudis, C. Tsakmakis, D. Hofer, Micromorphic continuum part II: Finite deformation plasticity coupled with damage. Int. J. Non–Linear Mech. **44**, 957–974 (2009)
M. Gurtin, Generalized Ginzburg–landau and Cahn–Hilliard equations based on a microforce balance. Physica D **92**, 178–192 (1996)
P. Haupt, *Continuum Mechanics and Theory of Materials* (Springer, Berlin, 2000)
T. Helfer, Extension of monodimensional fuel performance codes to finite strain analysis using a Lagrangian logarithmic strain framework. Nucl. Eng. Des. **288**, 75–81 (2015)
C. Hirschberger, P. Steinmann, Classification of concepts in thermodynamically consistent generalized plasticity. ASCE J. Eng. Mech. **135**, 156–170 (2009)
C. Hirschberger, E. Kuhl, P. Steinmann, On deformational and configurational mechanics of micromorphic hyperelasticity - theory and computation. Comput. Methods Appl. Mech. Eng. **196**, 4027–4044 (2007)
M. Horak, M. Jirasek, An extension of small-strain models to the large-strain range based on an additive decomposition of a logarithmic strain. Programs Algorithms Numer. Math. **16**, 88–93 (2013)
G. Hütter, Homogenization of a cauchy continuum towards a micromorphic continuum. J. Mech. Phys. Solids **99**, 394–408 (2017a)
G. Hütter, A micromechanical gradient extension of Gurson's model of ductile damage within the theory of microdilatational media. Int. J. Solids Struct. **110-111**, 15–23 (2017b)
C. Kafadar, A. Eringen, Micropolar media: I the classical theory. Int. J. Eng. Sci. **9**, 271–305 (1971)
N. Kirchner, P. Steinmann, A unifying treatise on variational principles for gradient and micromorphic continua. Philos. Mag. **85**, 3875–3895 (2005)
M. Lazar, G. Maugin, On microcontinuum field theories: The eshelby stress tensor and incompatibility conditions. Philos. Mag. **87**, 3853–3870 (2007)
E.H. Lee, P. Germain, Elastic–plastic theory at finite strain, in *Problems of Plasticity*, ed. by A. Sawczuk (Ed), (Noordhoff International Publishing, 1972), pp. 117–133
A. Madeo, G. Barbagallo, M.V. d'Agostino, L. Placidi, P. Neff, First evidence of non-locality in real band-gap metamaterials: Determining parameters in the relaxed micromorphic model. Proc. R. Soc. Lond. A **472**, 20160169 (2016)
J. Mandel, *Plasticité Classique et Viscoplasticité*, CISM Courses and Lectures No. 97, Udine (Springer, Berlin, 1971)
J. Mandel, Equations constitutives et directeurs dans les milieux plastiques et viscoplastiques. Int. J. Solids Struct. **9**, 725–740 (1973)
N. Marchal, S. Flouriot, S. Forest, L. Remy, Crack–tip stress–strain fields in single crystal nickel–base superalloys at high temperature under cyclic loading. Comput. Mater. Sci. **37**, 42–50 (2006a)
N. Marchal, S. Forest, L. Rémy, S. Duvinage, Simulation of fatigue crack growth in single crystal superalloys using local approach to fracture, in *Local Approach to Fracture, 9th European Mechanics of Materials Conference, Euromech–Mecamat*. ed. by J. Besson, D. Moinereau, D. Steglich (Presses de l'Ecole des Mines de Paris, Moret–sur–Loing, 2006b), pp. 353–358
G. Maugin, The method of virtual power in continuum mechanics: Application to coupled fields. Acta Mech. **35**, 1–70 (1980)
G. Maugin, Internal variables and dissipative structures. J. Non–Equilib. Thermo-dyn. **15**, 173–192 (1990)
G. Maugin, *Thermomechanics of Nonlinear Irreversible Behaviors* (World Scientific, Singapore/River Edge, 1999)
C. Miehe, Variational gradient plasticity at finite strains. Part I: Mixed potentials for the evolution and update problems of gradient-extended dissipative solids. Comput. Methods Appl. Mech. Eng. **268**, 677–703 (2014)
C. Miehe, N. Apel, M. Lambrecht, Anisotropic additive plasticity in the logarithmic strain space: Modular kinematic formulation and implementation based on incremental minimization principles for standard materials. Comp. Methods Appl. Mech. Eng **191**, 5383–5425 (2002)

C. Miehe, S. Teichtmeister, F. Aldakheel, Phase-field modelling of ductile fracture: a variational gradient-extended plasticity-damage theory and its micromorphic regularization. Philos. Trans. R. Soc. Lond. A **374**(2066) (2016)
R. Mindlin, Micro–structure in linear elasticity. Arch. Ration. Mech. Anal. **16**, 51–78 (1964)
R. Mindlin, Second gradient of strain and surface–tension in linear elasticity. Int. J. Solids Struct. **1**, 417–438 (1965)
H. Mühlhaus, *Continuum Models for Materials with Microstructure* (Wiley, Chichester, 1995)
P. Neff, I. Ghiba, A. Madeo, L. Placidi, G. Rosi, A unifying perspective: The relaxed linear micromorphic continuum. Contin. Mech. Thermodyn. **26**, 639–681 (2014)
R. Peerlings, On the role of moving elastic-plastic boundaries in strain gradient plasticity. Model. Simul. Mater. Sci. Eng. **15**, S109–S120 (2007)
R. Peerlings, M. Geers, R. de Borst, W. Brekelmans, A critical comparison of nonlocal and gradient–enhanced softening continua. Int. J. Solids Struct. **38**, 7723–7746 (2001)
R. Peerlings, T. Massart, M. Geers, A thermodynamically motivated implicit gradient damage framework and its application to brick masonry cracking. Comput. Methods Appl. Mech. Eng. **193**, 3403–3417 (2004)
K. Pham, H. Amor, J.J. Marigo, C. Maurini, Gradient damage models and their use to approximate brittle fracture. Int. J. Damage Mech. **20**, 618–652 (2011)
L. Poh, R. Peerlings, M. Geers, S. Swaddiwudhipong, An implicit tensorial gradient plasticity model - formulation and comparison with a scalar gradient model. Int. J. Solids Struct. **48**, 2595–2604 (2011)
V.D. Rancourt, B. Appolaire, S. Forest, K. Ammar, Homogenization of viscoplastic constitutive laws within a phase field approach. J. Mech. Phys. Solids **88**, 35–48 (2016)
R. Regueiro, On finite strain micromorphic elastoplasticity. Int. J. Solids Struct. **47**, 786–800 (2010)
G. Rosi, N. Auffray, Anisotropic and dispersive wave propagation within strain-gradient framework. Wave Motion **63**, 120–134 (2016)
K. Saanouni, M. Hamed, Micromorphic approach for finite gradient-elastoplasticity fully coupled with ductile damage: Formulation and computational aspects. Int. J. Solids Struct. **50**, 2289–2309 (2013)
C. Sansour, A theory of the elastic–viscoplastic cosserat continuum. Arch. Mech. **50**, 577–597 (1998a)
C. Sansour, A unified concept of elastic–viscoplastic Cosserat and micromorphic continua. Journal de Physique IV **8**(Pr8), 341–348 (1998b)
C. Sansour, S. Skatulla, H. Zbib, A formulation for the micromorphic continuum at finite inelastic strains. Int. J. Solids Struct. **47**, 1546–1554 (2010)
A.V. Shotov, J. Ihlemann, Analysis of some basic approaches to finite strain elasto-plasticity in view of reference change. Int. J. Plast. **63**, 183–197 (2014)
F. Sidoroff, A. Dogui, Some issues about anisotropic elastic-plastic models at finite strain. Int. J. Solids Struct. **38**, 9569–9578 (2001)
J.C. Simo, C. Miehe, Associative coupled thermoplasticity at finite strains: Formulation, numerical analysis and implementation. Comput. Methods Appl. Mech. Eng. **98**, 41–104 (1992)
H. Steeb, S. Diebels, A thermodynamic–consistent model describing growth and remodeling phenomena. Comput. Mater. Sci. **28**, 597–607 (2003)
S. Toll, The dissipation inequality in hypoplasticity. Acta Mech. **221**, 39–47 (2011)
C. Truesdell, W. Noll, The non-linear field theories of mechanics, in *Handbuch der Physik*, ed. by S. Flügge, reedition (Springer, Berlin/Heidelberg, 1965)
C. Truesdell, R. Toupin, The classical field theories, in *Handbuch der Physik*, ed. by S. Flügge, vol. 3 (Springer, Berlin, 1960), pp. 226–793
P. Ván, A. Berezovski, C. Papenfuss, Thermodynamic approach to generalized continua. Contin. Mech. Thermodyn. **26**, 403–420 (2014)
J. Vignollet, S. May, R.D. Borst, C. Verhoosel, Phase–field models for brittle and cohesive fracture. Meccanica **49**, 2587–2601 (2014)

B. Wcislo, L. Pamin, K. Kowalczyk-Gajewska, Gradient-enhanced damage model for large deformations of elastic-plastic materials. Arch. Mech. **65**, 407–428 (2013)

S. Wulfinghoff, E. Bayerschen, T. Böhlke, Conceptual difficulties in plasticity including the gradient of one scalar plastic field variable. PAMM. Proc. Appl. Math. Mech. **14**, 317–318 (2014)

H. Xiao, O.T. Bruhns, A. Meyers, Existence and uniqueness of the integrable–exactly hypoelastic equation and its significance to finite elasticity. Acta Mech. **138**, 31–50 (1999)

Higher Order Thermo-mechanical Gradient Plasticity Model: Nonproportional Loading with Energetic and Dissipative Components

16

George Z. Voyiadjis and Yooseob Song

Contents

Abstract

In this chapter, two cases of thermodynamic-based higher order gradient plasticity theories are presented and applied to the stretch-surface passivation problem for investigating the material behavior under the nonproportional loading condition. This chapter incorporates the thermal and mechanical responses of microsystems. It addresses phenomena such as size and boundary effects and in particular microscale heat transfer in fast-transient processes. The stored energy of cold work is considered in the development of the recoverable

G. Z. Voyiadjis (✉) · Y. Song
Department of Civil and Environmental Engineering, Louisiana State University, Baton Rouge, LA, USA
e-mail: voyiadjis@eng.lsu.edu; ysong17@lsu.edu

G. Z. Voyiadjis (ed.), *Handbook of Nonlocal Continuum Mechanics for Materials and Structures*, https://doi.org/10.1007/978-3-319-58729-5_14

counterpart of the free energy. The main distinction between the two cases is the presence of the dissipative higher order microstress quantities $\mathcal{S}_{ijk}^{\mathrm{dis}}$. Fleck et al. (Soc. A-Math. Phys. **470**:2170, 2014, ASME **82**:7, 2015) noted that $\mathcal{S}_{ijk}^{\mathrm{dis}}$ always gives rise to the stress jump phenomenon, which causes a controversial dispute in the field of strain gradient plasticity theory with respect to whether it is physically acceptable or not, under the nonproportional loading condition. The finite element solution for the stretch-surface passivation problem is also presented by using the commercial finite element package ABAQUS/standard (*User's Manual (Version 6.12)*. Dassault Systemes Simulia Corp., Providence, 2012) via the user-subroutine UEL. The model is validated by comparing with three sets of small-scale experiments. The numerical simulation part, which is largely composed of four subparts, is followed. In the first part, the occurrence of the stress jump phenomenon under the stretch-surface passivation condition is introduced in conjunction with the aforementioned three experiments. The second part is carried out in order to clearly show the results to be contrary to each other from the two classes of strain gradient plasticity models. An extensive parametric study is presented in the third part in terms of the effects of the various material parameters on the stress-strain response for the two cases of strain gradient plasticity models, respectively. The evolution of the free energy and dissipation potentials are also investigated at elevated temperatures. In the last part, the two-dimensional simulation is given to examine the gradient and grain boundary effect, the mesh sensitivity of the two-dimensional model, and the stress jump phenomenon. Finally, some significant conclusions are presented.

Keywords
Higher order gradient plasticity · Energetic · Dissipative · Stress jump · Non-proportional loading

Introduction

It is well known that the classical continuum plasticity theory cannot capture the size effect of the microstructure during the course of plastic deformation. Aifantis (1984) incorporated a material length scale into the conventional continuum plasticity model to capture the size effect and proposed a modified flow rule by including the gradient term $\beta \nabla^2 \varepsilon^p$ into the conventional flow rule as follows:

$$\tau_{eff} = R\left(\varepsilon^p\right) - \beta \nabla^2 \varepsilon^p \tag{1}$$

where τ_{eff} is the effective stress and is calculated by $\tau_{eff} = \sqrt{3\tau_{ij}\tau_{ij}/2}$ with the deviatoric stress tensor τ_{ij}, $R(\varepsilon^p) > 0$ is the conventional flow resistance, ε^p is the accumulated plastic strain, $\beta > 0$ is a material coefficient, and $\nabla^2 = \mathrm{Div}\nabla$ is the Laplacian operator. Hereafter, a number of mechanisms associated with geometrically necessary dislocations (GNDs) have been proposed within the framework of the strain gradient plasticity (SGP) theory. Generally, there are two different kinds of viewpoints in SGP theory in terms of the origin of the strengthening effect. Firstly, there is a specific argument that the strain gradient strengthening is purely energetic in the sense that GNDs originate in the blockage by grain boundaries and the pile-ups of dislocations has a backstress associated with the energetic strengthening (Fleck and Willis 2015). For example, Kuroda and Tvergaard (2010) argued that the term $\tau_{\text{eff}} + \beta\nabla^2\varepsilon^p$ calculated in Eq. 1 represents a total stress at the material point that activates the plastic straining, i.e., the generation and movement of dislocations. They also pointed out that the thermodynamic requirement on the plastic dissipation $\mathcal{D}$ is evaluated by $\mathcal{D} = \left(\tau_{eff} + \beta\nabla^2\varepsilon^p\right)\dot{\varepsilon}^p > 0$. This demonstration shows that the nonlocal term $\beta\nabla^2\varepsilon^p$ is naturally interpreted as an energetic quantity, which is consistent with the interpretation in Gurtin and Anand (2009) that the nonlocal term in Aifantis' formulation should be energetic. Secondly, there is another point of view that GNDs combine with the statistically stored dislocations (SSDs) to provide the forest hardening, which in turn, lead to the dissipative strengthening. For example, in Fleck and Hutchinson (2001), gradient term is implicitly considered as a dissipative quantity that causes the theory to violate the thermodynamic requirement on plastic dissipation. Fleck and Willis (2009a) developed a mathematical basis for phenomenological gradient plasticity theory corresponding to both rate dependent/independent behavior with the scalar plastic multiplier. The plastic work in Fleck and Willis (2009a) is taken to be purely dissipative in nature, and the thermodynamic microstresses are assumed to be dissipative. In their incremental form of plasticity theory, an associated plastic flow rule is assumed by means of the convex yield function, consequently, the positive plastic work is ensured. Fleck and Willis (2009b) developed a phenomenological flow theory version of SGP theory by extending their theory in Fleck and Willis (2009a) to isotropic and anisotropic solids with tensorial plastic multiplier. Fleck and Willis (2009b) argued that the microstress quantities should include a dissipative part; thus, it has been proposed that the term $\beta\nabla^2\varepsilon^p$ is additively decomposed into an energetic $(\cdot)_{\text{en}}$ and dissipative $(\cdot)_{\text{dis}}$ in order to develop a kinematic hardening theory. The dissipative stresses satisfy a yield condition with an associated flow plastic rule, while the free energy provides the standard kinematic hardening.

There has been a debate between Fleck, Willis, and Hutchinson (Fleck et al. 2014, 2015; Hutchinson 2012) and Gudmundson et al. (Gudmundson 2004; Gurtin and Anand 2009) for the last 15 years or so. Fleck and Hutchinson (2001) developed a phenomenological SGP theory using higher order tensors with a similar framework to that proposed by Aifantis (1984) and Muhlhaus and Aifantis (1991). Higher order stresses and additional boundary conditions have been involved in the theory to develop a generalization of the classical rate-independent J_2 flow theory of gradient plasticity. However, they do not discuss the compatibility of their theory with thermodynamic requirements on the plastic dissipation. Gudmundson (2004) and Gurtin and Anand (2009) pointed out that the formulation of Fleck and Hutchinson (2001) violates thermodynamic requirements on the plastic dissipation. Gurtin and Anand (2009) discussed the physical nature of nonlocal terms in the flow rules developed by Fleck and Hutchinson (2001) under isothermal condition and concluded that the flow rule of Fleck and Hutchinson (2001) does not always satisfy the thermodynamic requirement on plastic dissipation unless the nonlocal term is dropped. A formulation of Fleck and Hutchinson (2001) has been modified to meet this thermodynamic requirement by partitioning the higher order microstresses into energetic and dissipative components (Hutchinson 2012). In addition, Hutchinson (2012) classified the strain gradient version of J_2 flow theories into two classes: incremental theory developed by Fleck and Hutchinson and nonincremental theory developed by Gudmundson et al. (c.f. see Fleck et al. (2014, 2015), Gudmundson (2004), Gurtin and Anand (2005, 2009), and Hutchinson (2012) for details). The specific phenomenon in the nonincremental theory that exhibits a finite stress jump due to infinitesimal changes in plastic strain that may occur under the nonproportional loading is noted and its physical acceptance is also discussed in the work of Fleck et al. (2014, 2015). Hutchinson (2012) concluded that discontinuous changes with infinitesimal changes in boundary loads are physically suspect. (Despite the argument of J.W. Hutchinson, Acta Mech. Sin. 28, 4 (2012), there is another viewpoint to look at the stress jump phenomenon. The perspective taken in N.A. Fleck, J.R. Willis, ibid. 31, (2015) is that it is premature at this moment in time to judge whether a formulation associated with the stress jump is physically acceptable or not, therefore, an in-depth study of dislocation mechanism and microscale experiments with non-proportional loading history is needed.) Fleck et al. (2014, 2015) have shown this phenomenon with two plane strain problems, stretch-passivation problem, and stretch-bending problem, for nonproportional loading condition. In their work, it is noted that the dissipative higher order microstress quantities $\mathcal{S}_{ijk}^{\mathrm{dis}}$ always generate the stress jump for nonproportional loading problems.

In this chapter, two different cases of the high order SGP model with and without the dissipative higher order microstress quantities $\mathcal{S}_{ijk}^{\mathrm{dis}}$ are presented based on the new forms of the free energy and the dissipation potentials for eliminating an elastic loading gap. The presented model is applied to the stretch-surface passivation problem in order to compare the behavior of each case under the nonproportional loading condition. For this, the finite element solution for the stretch-surface passivation problem is presented by using the commercial finite element package ABAQUS/standard (2012) via the user-subroutine UEL and validated by comparing with three sets of small-scale experiments, which have been conducted by Han et al. (2008), Haque and Saif (2003), and Xiang and Vlassak (2006). An extensive numerical work is also carried out based on the one-dimensional and two-dimensional codes in order to compare the results from the two cases of the SGP model and to analyze the characteristics of the stress jump phenomenon.

Principle of Virtual Power

The principle of virtual power is used to derive the local equation of motion and the nonlocal microforce balance for volume V as well as the equations for local traction force and nonlocal microtraction condition for the external surface S, respectively. In the presence of varying temperature fields at the microstructure level, the formulation should incorporate the effects of the temperature gradient on the thermo-mechanical behavior of the material due to the microheterogeneous nature of material (Forest and Amestoy 2008). In this sense, it is assumed here that the plastic strain, plastic strain gradient, temperature, and temperature gradient contribute to the power per unit volume.

Moreover, as it is mentioned in section "Introduction," the effect of the interface plays a crucial role for the plastic behavior of the material at the microscale. An interface (grain boundary) separating grains $\mathcal{G}_1$ and $\mathcal{G}_2$ is taken into account here, and it is assumed that the displacement field is continuous, i.e. $u_i^{\mathcal{G}_1} = u_i^{\mathcal{G}_2}$, across the grain boundary (Fig. 1). As shown in this figure, a dislocation moving toward the grain boundary in grain $\mathcal{G}_1$ cannot pass through the grain boundary, but it is trapped and accumulated at the grain boundary due to the misalignment of the grains $\mathcal{G}_1$ and $\mathcal{G}_2$ that are contiguous to each other. In this sense, the grain boundary acts as an obstacle to block the dislocation movement; therefore, the yield strength of the material increases as the number of grain boundaries increases. By assuming that the interface surface energy depends on the plastic strain rate at the interface of the plastically deforming phase, the internal part of the principle of virtual power for the bulk P_{int} and for the interface P_{int}^I are expressed in terms of the energy contributions in the arbitrary subregion of the volume V and the arbitrary subsurface of the interface S^I, respectively, as follows:

$$P_{\text{int}} = \int_V \left(\sigma_{ij} \dot{\varepsilon}_{ij}^e + \mathcal{X}_{ij} \dot{\varepsilon}_{ij}^p + \mathcal{S}_{ijk} \dot{\varepsilon}_{ij,k}^p + \mathcal{A} \dot{T} + \mathcal{B}_i \dot{T}_{,i} \right) dV \tag{2}$$

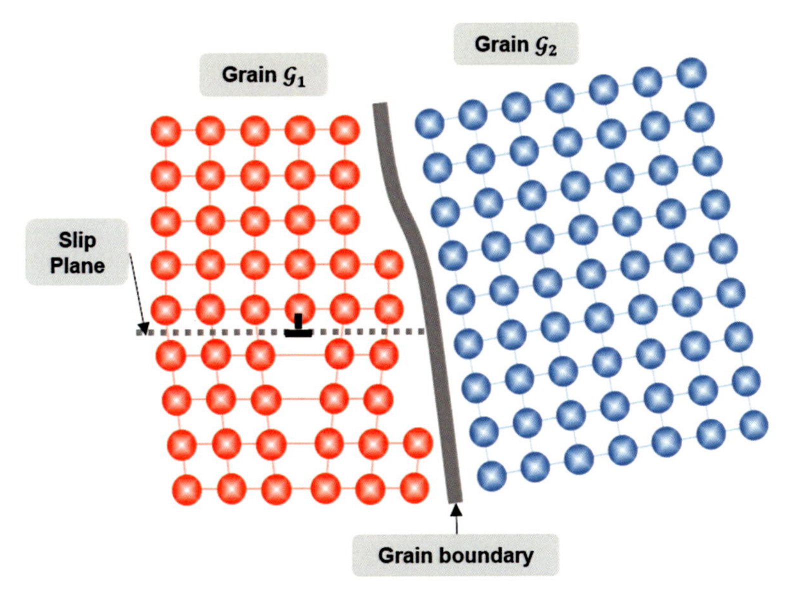

Fig. 1 The schematic illustration of the spatial lattice of two contiguous grains, $\mathcal{G}_1$ and $\mathcal{G}_2$, along with a single slip in grain $\mathcal{G}_1$ (Reprinted with permission from Voyiadjis et al. 2017)

$$P_{\text{int}}^{I} = \int_{S^I} \left(\mathbb{M}_{ij}^{I\mathcal{G}_1} \dot{\varepsilon}_{ij}^{pI\mathcal{G}_1} + \mathbb{M}_{ij}^{I\mathcal{G}_2} \dot{\varepsilon}_{ij}^{pI\mathcal{G}_2} \right) dS^I \tag{3}$$

where the superscripts "e," p," and "I" are used to express the elastic state, the plastic state, and the interface, respectively. The internal power for the bulk in the form of Eq. 2 is defined using the Cauchy stress tensor σ_{ij}, the backstress $\mathcal{X}_{ij}$ conjugate to the plastic strain rate $\dot{\varepsilon}_{ij}^{p}$, the higher order microstress S_{ijk} conjugate to the gradients of the plastic strain rate $\dot{\varepsilon}_{ij,k}^{p}$, and the generalized stresses $\mathcal{A}$ and $\mathcal{B}_i$ conjugate to the temperature rate $\dot{T}$ and the gradient of the temperature rate $\dot{T}_{,i}$, respectively. The internal power for the interface in the form of Eq. 3 is defined using the interfacial microscopic moment tractions $\mathbb{M}_{ij}^{I\mathcal{G}_1}$ and $\mathbb{M}_{ij}^{I\mathcal{G}_2}$ expending power over the interfacial plastic strain rates $\dot{\varepsilon}_{ij}^{pI\mathcal{G}_1}$ at $S^{I\mathcal{G}_1}$ and $\dot{\varepsilon}_{ij}^{pI\mathcal{G}_2}$ at $S^{I\mathcal{G}_2}$, respectively.

Moreover, since the plastic deformation, which is accommodated by the generation and motion of the dislocation, is influenced by the interfaces, Eq. 3 results in higher order boundary conditions generally consistent with the framework of a gradient type theory. These extra boundary conditions should be imposed at internal and external boundary surfaces or interfaces between neighboring grains (Aifantis and Willis 2005; Gurtin 2008). The internal power for the bulk is balanced with the external power for the bulk expended by the tractions on the external surfaces S and the body forces acting within the volume V as shown below:

$$P_{\text{ext}} = \int_V b_i v_i \, dV + \int_S \left(t_i v_i + m_{ij} \dot{\varepsilon}_{ij}^{p} + a\dot{T} \right) dS \tag{4}$$

where t_i and b_i are traction and the external body force conjugate to the macroscopic velocity v_i, respectively. It is further assumed here that the external power has terms with the microtraction m_{ij} and a, conjugate to the plastic strain rate $\dot{\varepsilon}_{ij}^{p}$ and the temperature rate $\dot{T}$, respectively, since the internal power in Eq. 2 contains the terms of the gradients of the plastic strain rate $\dot{\varepsilon}_{ij,k}^{p}$ and the gradients of the temperature rate $\dot{T}_{,i}$, respectively.

Making use of the principle of virtual power that the external power is equal to the internal power ($P_{\text{int}} = P_{\text{ext}}$) along with the integration by parts on some terms in Eq. 2, and utilizing the divergence theorem, the equations for balance of linear momentum and nonlocal microforce balance can be represented, respectively, for volume V, as follows:

$$\sigma_{ij,j} + b_i = 0 \tag{5}$$

$$\mathcal{X}_{ij} - \tau_{ij} - \mathcal{S}_{ijk,k} = 0 \tag{6}$$

$$\operatorname{div} \mathcal{B}_i - \mathcal{A} = 0 \tag{7}$$

where $\tau_{\text{ij}} = \sigma_{\text{ij}} - \sigma_{\text{kk}}\delta_{\text{ij}}/3$ is the deviatoric part of the Cauchy stress tensor and δ_{ij} is the Kronecker delta.

On the external surface S, the equations for local surface traction conditions and nonlocal microtraction conditions can be written, respectively, with the outward unit normal vector to S, n_k, as follows:

$$t_j = \sigma_{ij} n_i \tag{8}$$

$$m_{ij} = \mathcal{S}_{ijk} n_k \tag{9}$$

$$a = \mathcal{B}_i n_i \tag{10}$$

In addition, the interfacial external power P^I_{ext}, which is balanced with the interfacial internal power P^I_{int}, is expended by the macrotractions $\sigma^{\mathcal{G}_1}_{ij}(-n^I_j)$ and $\sigma^{\mathcal{G}_2}_{ij}(n^I_j)$ conjugate to the macroscopic velocity v_i, and the microtractions $\mathbb{S}^{I\mathcal{G}_1}_{ijk}\left(-n^I_k\right)$ and $\mathbb{S}^{I\mathcal{G}_2}_{ijk}\left(n^I_k\right)$ that are conjugate to $\dot{\varepsilon}^{pI\mathcal{G}_1}_{ij}$ and $\dot{\varepsilon}^{pI\mathcal{G}_2}_{ij}$, respectively, as follows:

$$P^I_{\text{ext}} = \int_{S^I} \left\{ \left(\sigma^{\mathcal{G}_2}_{ij} n^I_j - \sigma^{\mathcal{G}_1}_{ij} n^I_j \right) v_i + \mathbb{S}^{I\mathcal{G}_2}_{ijk} n^I_k \dot{\varepsilon}^{pI\mathcal{G}_2}_{ij} - \mathbb{S}^{I\mathcal{G}_1}_{ijk} n^I_k \dot{\varepsilon}_{ij}\, p^{I\mathcal{G}_1} \right\} dS^I \tag{11}$$

By equating $P^I_{\text{int}} = P^I_{\text{ext}}$ with considering the arbitrary variation of the plastic strain at the interface, the interfacial macro- and microforce balances can be obtained as follows:

$$\left(\sigma^{\mathcal{G}_1}_{ij} - \sigma^{\mathcal{G}_2}_{ij} \right) n^I_j = 0 \tag{12}$$

$$\mathbb{M}^{I\mathcal{G}_1}_{ij} + \mathbb{S}^{I\mathcal{G}_1}_{ijk} n^I_k = 0 \tag{13}$$

$$\mathbb{M}^{I\mathcal{G}_2}_{ij} - \mathbb{S}^{I\mathcal{G}_2}_{ijk} n^I_k = 0 \tag{14}$$

The microforce balance conditions in Eqs. 13 and 14 represent the coupling behavior in the grain interior at the interface to the behavior of the interface, since the microtractions $\mathbb{S}^{I\mathcal{G}_1}_{ijk} n^I_k$ and $\mathbb{S}^{I\mathcal{G}_2}_{ijk} n^I_k$ are the special cases of Eq. 8 for the internal surface of the interface.

Thermodynamic Formulation with Higher Order Plastic Strain Gradients

The first law of thermodynamics, which encompasses several principles including the law of conservation of energy, is taken into account in this chapter in order to develop a thermodynamically consistent formulation accounting for the thermo-mechanical behavior of small-scale metallic volumes during the fast transient process. In order to consider micromechanical evolution in the first law of thermodynamics, the enhanced SGP theory with the plastic strain gradient is employed for the mechanical part of the formulation, whereas the micromorphic model is employed for the thermal counterpart as follows (see the work of Forest and co-workers (Forest and Amestoy 2008; Forest and Sievert 2003)):

$$\rho \dot{e} = \sigma_{ij} \dot{\varepsilon}^{e}_{ij} + \mathcal{X}_{ij} \dot{\varepsilon}^{p}_{ij} + \mathcal{S}_{ijk} \dot{\varepsilon}^{p}_{ij,k} + \mathcal{A}\dot{T} + \mathcal{B}_i \dot{T}_{,i} - div q_i \tag{15}$$

$$\dot{e}^{I} = \mathbb{M}^{I}_{ij} \dot{\varepsilon}^{pI}_{ij} + q^{I}_{i} n^{I}_{i} \tag{16}$$

where ρ is the mass density, e is the specific internal energy, e^I is the internal surface energy density at the contacting surface, and q_i and q^I_i are the heat flux vectors of the bulk and the interface, respectively.

The second law of thermodynamics, or entropy production inequality as it is often called, yields a physical basis that accounts for the distribution of GNDs within the body along with the energy carrier scattering and requires that the free energy increases at a rate not greater than the rate at which work is performed. Based on this requirement, entropy production inequalities for the bulk and the interface can be expressed, respectively, as follows:

$$\rho \dot{s} T - \rho \dot{e} + \sigma_{ij} \dot{\varepsilon}^{e}_{ij} + \mathcal{X}_{ij} \dot{\varepsilon}^{p}_{ij} + \mathcal{S}_{ijk} \dot{\varepsilon}^{p}_{ij,k} - q_i \frac{T_{,i}}{T} + \mathcal{A}\dot{T} + \mathcal{B}_i \dot{T}_{,i} \geq 0 \tag{17}$$

$$\dot{s}^{I} T^{I} - q^{I}_{i} n^{I}_{i} \geq 0 \tag{18}$$

where s is the specific entropy and s^I is the surface density of the entropy of the interface.

The Energetic and Dissipative Components of the Thermodynamic Microstresses

Internal energy e, temperature T, and entropy s describing the current state of the material can be attributed to the Helmholtz free energy Ψ (per unit volume) such as

$$\rho \dot{s} T - \rho \dot{e} + \sigma_{ij} \dot{\varepsilon}^{e}_{ij} + \mathcal{X}_{ij} \dot{\varepsilon}^{p}_{ij} + \mathcal{S}_{ijk} \dot{\varepsilon}^{p}_{ij,k} - q_i \frac{T_{,i}}{T} + \mathcal{A}\dot{T} + \mathcal{B}_i \dot{T}_{,i} \geq 0. \tag{19}$$

By taking the time derivative of Eq. 19 for the bulk and the interface and substituting each into Eqs. 17 and 18, respectively, the nonlocal free energy (i.e., Clausius-Duhem) inequality for the bulk and the interface can be obtained as follows:

$$\sigma_{ij} \dot{\varepsilon}^{e}_{ij} + \mathcal{X}_{ij} \dot{\varepsilon}^{p}_{ij} + \mathcal{S}_{ijk} \dot{\varepsilon}^{p}_{ij,k} + \mathcal{A}\dot{T} + \mathcal{B}_i \dot{T}_{,i} - \rho \dot{\Psi} - \rho s \dot{T} - q_i \frac{T_{,i}}{T} \geq 0 \tag{20}$$

$$\mathbb{M}^{I}_{ij} \dot{\varepsilon}^{pI}_{ij} - \dot{\Psi}^{I} - s^{I} \dot{T}^{I} \geq 0 \tag{21}$$

In order to derive the constitutive equations within a small-scale framework, an attempt to account for the effect of nonuniform distribution of the microdefection with temperature on the homogenized response of the material is carried out in this chapter with the functional forms of the Helmholtz free energy in terms of its state variables. By assuming the isothermal condition for the interface (i.e., $\dot{T}^I = 0$), the Helmholtz free energy for the bulk and the interface are given, respectively, by

$$\Psi = \Psi\left(\varepsilon_{ij}^e, \varepsilon_{ij}^p, \varepsilon_{ij,k}^p, T, T_{,i}\right) \tag{22}$$

$$\Psi^I = \Psi^I\left(\varepsilon_{ij}^{pI}\right) \tag{23}$$

where the function Ψ is assumed to be smooth and the function Ψ^1 is assumed to be convex with respect to a plastic strain at the interface ε_{ij}^{pI}.

Taking time derivative of the Helmholtz free energy for the bulk $\dot{\Psi}$ and the interface $\dot{\Psi}^I$ give the following expressions, respectively

$$\dot{\Psi} = \frac{\partial \Psi}{\partial \varepsilon_{ij}^e}\dot{\varepsilon}_{ij}^e + \frac{\partial \Psi}{\partial \varepsilon_{ij}^p}\dot{\varepsilon}_{ij}^p + \frac{\partial \Psi}{\partial \varepsilon_{ij,k}^p}\dot{\varepsilon}_{ij,k}^p + \frac{\partial \Psi}{\partial T}\dot{T} + \frac{\partial \Psi}{\partial T_{,i}}\dot{T}_{,i} \tag{24}$$

$$\dot{\Psi}^I = \frac{\partial \Psi^I}{\partial \varepsilon_{ij}^{pI}}\dot{\varepsilon}_{ij}^{pI} \tag{25}$$

By substituting Eq. 24 into Eq. 20 for the bulk and Eq. 25 into Eq. 21 for the interface, and factoring out the common terms, one obtains the following inequalities:

$$\begin{aligned}\left(\sigma_{ij} - \rho\frac{\partial \Psi}{\partial \varepsilon_{ij}^e}\right)\dot{\varepsilon}_{ij}^e + \left(\mathcal{X}_{ij} - \rho\frac{\partial \Psi}{\partial \varepsilon_{ij}^p}\right)\dot{\varepsilon}_{ij}^p + \left(\mathcal{S}_{ijk} - \rho\frac{\partial \Psi}{\partial \varepsilon_{ij,k}^p}\right)\dot{\varepsilon}_{ij,k}^p \\ + \left(\mathcal{A} - \rho s - \rho\frac{\partial \Psi}{\partial T}\right)\dot{T} + \left(\mathcal{B}_i - \rho\frac{\partial \Psi}{\partial T_{,i}}\right)T_{,i} - \frac{q_i}{T}T_{,i} \geq 0\end{aligned} \tag{26}$$

$$\mathbb{M}_{ij}^I\dot{\varepsilon}_{ij}^{pI} - \rho\frac{\partial \Psi^I}{\partial \varepsilon_{ij}^{pI}}\dot{\varepsilon}_{ij}^{pI} \geq 0 \tag{27}$$

Guided by Eqs. 26 and 27, it is further assumed that the thermodynamic microstress quantities $\mathcal{X}_{ij}$, $\mathcal{S}_{ijk}$, $\mathcal{A}$, and $\mathbb{M}_{ij}^I$ are decomposed into the energetic and the dissipative components such as (Voyiadjis and Deliktas 2009; Voyiadjis and Faghihi 2012; Voyiadjis et al. 2014):

$$\mathcal{X}_{ij} = \mathcal{X}_{ij}^{\text{en}} + \mathcal{X}_{ij}^{\text{dis}} \tag{28}$$

$$\mathcal{S}_{ijk} = \mathcal{S}_{ijk}^{\text{en}} + \mathcal{S}_{ijk}^{\text{dis}} \tag{29}$$

$$\mathcal{A} = \mathcal{A}^{\text{en}} + \mathcal{A}^{\text{dis}} \tag{30}$$

$$\mathbb{M}_{ij}^{I} = \mathbb{M}_{ij}^{I,\text{en}} + \mathbb{M}_{ij}^{I,\text{dis}} \tag{31}$$

The components $\mathbb{M}_{ij}^{I,\text{en}}$ and $\mathbb{M}_{ij}^{I,\text{dis}}$ represent the mechanisms for the pre- and postslip transfer and thus involve the plastic strain at the interface prior to the slip transfer $\varepsilon_{ij}^{p^{I(\text{pre})}}$ and the one after the slip transfer $\varepsilon_{ij}^{p^{I(\text{post})}}$, respectively. The overall plastic strain at the interface can be obtained by the summation of both plastic strains such as:

$$\varepsilon_{ij}^{p^{I}} = \varepsilon_{ij}^{p^{I(\text{pre})}} + \varepsilon_{ij}^{p^{I(\text{post})}} \tag{32}$$

Substituting Eqs. 28, 29, and 30 into Eq. 26 for the bulk and Eq. 31 into Eq. 27 for the interface and rearranging them in accordance with the energetic and the dissipative parts give the following expressions:

$$\begin{aligned} &\left(\sigma_{ij} - \rho\frac{\partial\Psi}{\partial\varepsilon_{ij}^{e}}\right)\dot{\varepsilon}_{ij}^{e} + \left(\mathcal{X}_{ij}^{\text{en}} - \rho\frac{\partial\Psi}{\partial\varepsilon_{ij}^{p}}\right)\dot{\varepsilon}_{ij}^{p} + \left(\mathcal{S}_{ijk}^{\text{en}} - \rho\frac{\partial\Psi}{\partial\varepsilon_{ij,k}^{p}}\right)\dot{\varepsilon}_{ij,k}^{p} \\ &+ \left(\mathcal{A}^{\text{en}} - \rho s - \rho\frac{\partial\Psi}{\partial T}\right)\dot{T} + \left(\mathcal{B}_{i} - \rho\frac{\partial\Psi}{\partial T_{,i}}\right)\dot{T}_{,i} + \mathcal{X}_{ij}^{\text{dis}}\dot{\varepsilon}_{ij}^{p} \\ &+ \mathcal{S}_{ijk}^{\text{dis}}\dot{\varepsilon}_{ij,k}^{p} + \mathcal{A}^{\text{dis}}\dot{T} - \frac{q_{i}}{T}T_{,i} \geq 0 \end{aligned} \tag{33}$$

$$\left(\mathbb{M}_{ij}^{I,\text{en}} - \rho\frac{\partial\Psi^{I}}{\partial\varepsilon_{ij}^{pI}}\right)\dot{\varepsilon}_{ij}^{pI} + \mathbb{M}_{ij}^{I,\text{dis}}\dot{\varepsilon}_{ij}^{pI} \geq 0 \tag{34}$$

From the above equations with the assumption that the fifth term in Eq. 33 is strictly energetic, one can retrieve the definition of the energetic part of the thermodynamic microstresses as follows:

$$\begin{aligned} &\sigma_{ij} = \rho\frac{\partial\Psi}{\partial\varepsilon_{ij}^{e}};\ \mathcal{X}_{ij}^{\text{en}} = \rho\frac{\partial\Psi}{\partial\varepsilon_{ij}^{p}};\ \mathcal{S}_{ijk}^{\text{en}} = \rho\frac{\partial\Psi}{\partial\varepsilon_{ij,k}^{p}}; \\ &\mathcal{A}^{\text{en}} = \rho\left(s + \frac{\partial\Psi}{\partial T}\right);\ \mathcal{B}_{i} = \rho\frac{\partial\Psi}{\partial T_{,i}} \end{aligned} \tag{35}$$

$$\mathbb{M}_{ij}^{I,\text{en}} = \rho\frac{\partial\Psi^{I}}{\partial\varepsilon_{ij}^{pI}} \tag{36}$$

Hence, the residual respective dissipation is then obtained as:

$$\mathcal{D} = \mathcal{X}_{ij}^{\text{dis}}\dot{\varepsilon}_{ij}^{p} + \mathcal{S}_{ijk}^{\text{dis}}\dot{\varepsilon}_{ij,k}^{p} + \mathcal{A}^{\text{dis}}\dot{T} - \frac{q_{i}}{T}T_{,i} \geq 0 \tag{37}$$

$$\mathcal{D}^{I} = \mathbb{M}_{ij}^{I,\text{dis}}\dot{\varepsilon}_{ij}^{pI} \geq 0 \tag{38}$$

where $\mathcal{D}$ and $\mathcal{D}^{I}$ are the dissipation densities per unit time for the bulk and the interface, respectively. The definition of the dissipative thermodynamic microstresses can be obtained from the complementary part of the dissipation potentials $\mathcal{D}\left(\dot{\varepsilon}_{ij}^{p}, \dot{\varepsilon}_{ij,k}^{p}, \dot{T}, T_{,i}\right)$ and $\mathcal{D}^{I}\left(\dot{\varepsilon}_{ij}^{pI}\right)$ such as

$$\mathcal{X}_{ij}^{\mathrm{dis}} = \frac{\partial \mathcal{D}}{\partial \dot{\varepsilon}_{ij}^{p}}; \mathcal{S}_{ijk}^{\mathrm{dis}} = \frac{\partial \mathcal{D}}{\partial \dot{\varepsilon}_{ij,k}^{p}}; \mathcal{A}^{\mathrm{dis}} = \frac{\partial \mathcal{D}}{\partial \dot{T}}; -\frac{q_i}{T} = \frac{\partial \mathcal{D}}{\partial T_{,i}} \tag{39}$$

$$\mathbb{M}_{ij}^{I,\mathrm{dis}} = \frac{\partial \mathcal{D}^{I}}{\partial \dot{\varepsilon}_{ij}^{pI}} \tag{40}$$

One now proceeds to present the constitutive laws for both the energetic and the dissipative parts which are achieved by employing the free energy and the dissipation potentials, which relate the stresses to their work-conjugate generalized stresses. The functional forms of the Helmholtz free energy and dissipation potential and the corresponding energetic and dissipative thermodynamic microstresses for the aforementioned two different cases of the model, i.e., the case with the dissipative higher order microstress quantities $\mathcal{S}_{ijk}^{\mathrm{dis}}$ and the one without $\mathcal{S}_{ijk}^{\mathrm{dis}}$, are presented in the following sections.

Helmholtz Free Energy and Energetic Thermodynamic Microstresses

Defining a specific form of the Helmholtz free energy function Ψ is tremendously important since it constitutes the bases in deriving the constitutive equations. In this chapter, the Helmholtz free energy function is put forward with three main counterparts, i.e., elastic, defect, and thermal energy, as follows (Voyiadjis and Song 2017):

$$\Psi = \underbrace{\left\{\frac{1}{2\rho}\varepsilon_{ij}^{e} E_{ijkl}\varepsilon_{kl}^{e} - \frac{\alpha_t}{\rho}(T - T_r)\,\varepsilon_{ij}^{e}\delta_{ij}\right\}}_{\Psi^{e}(elastic)} + \underbrace{\left\{\frac{1}{\rho}\left(\underbrace{\frac{h}{r+1}\left(1-\left(\frac{T}{T_y}\right)^{n}\right)\varepsilon_{p}^{r+1}}_{\Psi_1^{d}} + \underbrace{\frac{G}{2}\ell_{en}^{2}\varepsilon_{lm,n}^{p}\varepsilon_{lm,n}^{p}}_{\Psi_2^{d}}\right)\right\}}_{\Psi^{d}(defect)} + \underbrace{\left\{-\frac{1}{2}\frac{c_{\varepsilon}}{T_r}(T - T_r)^2 - \frac{1}{2\rho}aT_{,i}T_{,i}\right\}}_{\Psi^{t}(thermal)} \tag{41}$$

where E_{ijkl} is the elastic modulus tensor, α_t is the coefficient of linear thermal expansion, T_r is the reference temperature, h is the hardening material constant corresponding to linear kinematic hardening, r $(0 < r < 1)$ is the isotropic hardening material constant, $\varepsilon_p = \sqrt{\varepsilon_{ij}^p \varepsilon_{ij}^p}$ is the accumulated plastic strain, T_y and n are the thermal material constants that need to be calibrated by comparing with the experimental data, ℓ_{en} is the energetic length scale that describes the feature of the short-range interaction of GNDs, G is the shear modulus for isotropic linear elasticity, c_ε is the specific heat capacity at constant stress, and a is a material constant accounting for the interaction between energy carriers such as phonon-electron. The term $(1 - (T/T_y)^n)$ in Eq. 41 accounts for the thermal activation mechanism for overcoming the local obstacles to dislocation motion.

The first term of the defect energy Ψ_1^d characterizes the interaction between slip systems, i.e., the forest dislocations leading to isotropic hardening. This term is further assumed to be decomposed into the recoverable counterpart $\Psi_1^{d,R}$ and nonrecoverable counterpart $\Psi_1^{d,\mathrm{NR}}$. The establishment of the plastic strain gradient independent stored energy of cold work with no additional material parameters is achievable with this decomposition.

The recoverable counterpart, $\Psi_1^{d,R}$, accounts for the stored energy of cold work. When the elasto-plastic solid is cold-worked, most of the mechanical energy is converted into heat, but the remaining contributes to the stored energy of cold work through the creation and rearrangement of crystal defects such as dislocations, point defects, line defects, and stacking faults (Rosakis et al. 2000). In this chapter, the plastic strain-dependent free energy, $\Psi_1^{d,R}$, accounting for the stored energy of cold work is derived by assuming that the stored energy is related to the energy carried by dislocations. Mollica et al. (2001) investigated the inelastic behavior of the metals subject to loading reversal by linking the hardening behavior of the material to thermo-dynamical quantities such as the stored energy due to cold work and the rate of dissipation. In the work of Mollica et al. (2001), it is assumed that the stored energy depends on the density of the dislocation network that increases with monotonic plastic deformation until it is saturated at some point. This points out that the material stores this energy for a certain range of the accumulated plastic strain, after which the material will mainly dissipate the external work supply.

For the derivation of the stored energy of cold work, one assumes that the energetic microstress $\mathcal{X}_{ij}^{\mathrm{en}}$, given later by Eq. 52, can be expressed by separation of variables as follows:

$$\mathcal{X}_{ij}^{\mathrm{en}}\left(\varepsilon_{ij}^p, T\right) = \sum\left(\varepsilon_{ij}^p\right)\mathcal{T}(T) \tag{42}$$

with $\sum\left(\varepsilon_{ij}^p\right) = h\varepsilon_{ij}^p \varepsilon_p^{r-1}$ and $\mathcal{T}(T) = \left(1 - \left(T/T_y\right)^n\right)$.

On the other hand, instead of using the plastic strain at the macroscale level to describe the plastic deformation, Σ can be defined at the microscale level using the Taylor law, which gives a simple relation between the shear strength and the dislocation density, as follows:

$$\sum = \varsigma \mathrm{Gb}\sqrt{\rho_t} \tag{43}$$

where ς is the statistical coefficient accounting for the deviation from regular spatial arrangements of the dislocation populations, b is the magnitude of the Burgers vector, and ρ_t is the equivalent total dislocation density and can be obtained by $(\Sigma/\varsigma \mathrm{Gb})^2$ from Eq. 43.

Here, it is assumed that the stored energy of cold work, the result of energy carried by each dislocation, results in an extra latent hardening which is recoverable and temperature independent. Thus, the recoverable energy of cold work can be put forward as follows:

$$\Psi_1^{d,R} = \mathbb{U}\rho_t \tag{44}$$

where $\mathbb{U}$ is the elastic deformation energy of a dislocation and can be approximately given by

$$\mathbb{U} = \frac{\mathrm{Gb}^2}{4\pi} \ln\left(\frac{R}{R_0}\right) \tag{45}$$

where R is the cut-off radius ($R \approx 10^3 b$) and R_0 is the internal radius ($b < R_0 < 10b$) (Meyers and Chawla 2009). By substituting $\rho_t = (\Sigma/\varsigma \mathrm{Gb})^2$ into Eq. 45 along with Eq. 42, one can obtain the stored energy of cold work as follows:

$$\Psi_1^{d,R} = \vartheta h^2 \varepsilon_p^{2r} \tag{46}$$

where by comparing the aforementioned ranges for R, R_0, and ς to the shear modulus G, ϑ can be expressed by

$$\vartheta = \frac{1}{4\pi\varsigma^2 G} \ln\left(\frac{R}{R_0}\right) \approx \frac{1}{G} \tag{47}$$

The nonrecoverable counterpart $\Psi_1^{d,\mathrm{NR}}$ accounting for the energetically based hardening rule that mimics the dissipative behavior by describing irreversible loading processes can then be derived as follows (Gurtin and Reddy 2009):

$$\Psi_1^{d,\mathrm{NR}} = \frac{h}{r+1}\left(1 - \left(\frac{T}{T_y}\right)^n\right)\varepsilon_p^{r+1} - \vartheta h^2 \varepsilon_p^{2r} \tag{48}$$

where ϑ is a constant that depends on the material microstructure.

The second term of the defect energy Ψ_2^d characterizes the short-range interactions between coupling dislocations moving on close slip planes and leads to the kinematic hardening. This defect energy Ψ_2^d is recoverable in the sense that by starting at any value of the accumulated plastic strain gradients, Ψ_2^d returns to its original value as the accumulated plastic strain gradients return to their original value.

One can now obtain the energetic thermodynamic forces by using the definitions in Eq. 35 along with the Helmholtz free energy given by Eq. 41 as follows:

$$\sigma_{ij} = E_{ij\text{kl}}\varepsilon^{e}_{\text{kl}} - \alpha_t \left(T - T_r\right)\delta_{ij} \tag{49}$$

$$\mathcal{A}^{\text{en}} = \rho s - \alpha_t \left(T - T_r\right)\varepsilon^{e}_{ij}\delta_{ij} - \frac{c_\varepsilon}{T_r}\left(T - T_r\right) - \frac{h\varepsilon_p^{r+1}}{r+1}\frac{n}{T_y}\left(\frac{T}{T_y}\right)^{n-1} \tag{50}$$

$$\mathcal{B}_i = -aT_{,i} \tag{51}$$

$$\mathcal{X}^{\text{en}}_{ij} = h\varepsilon^{p}_{ij}\left(1 - \left(\frac{T}{T_y}\right)^{n}\right)\varepsilon_p^{r-1} \tag{52}$$

$$\mathcal{S}^{\text{en}}_{ijk} = G\ell^2_{\text{en}}\varepsilon^{p}_{ij,k} \tag{53}$$

Here, according to the aforementioned decomposition of the first term of the defect energy into the recoverable and nonrecoverable counterparts, $\mathcal{X}^{\text{en}}_{ij}$ can be further decomposed into recoverable $\left(\mathcal{X}^{\text{en},R}_{ij}\right)$ and nonrecoverable $\left(\mathcal{X}^{\text{en,NR}}_{ij}\right)$ counterparts as follows:

$$\mathcal{X}^{\text{en},R}_{ij} = 2r\vartheta h^2\varepsilon^{p}_{ij}\varepsilon_p^{2r-2} \tag{54}$$

$$\mathcal{X}^{\text{en,NR}}_{ij} = \text{h}\varepsilon^{p}_{ij}\left(1 - \left(\frac{T}{T_y}\right)^{n}\right)\varepsilon_p^{r-1} - 2r\vartheta h^2\varepsilon^{p}_{ij}\varepsilon_p^{2r-2} \tag{55}$$

From the aforementioned physical interpretations of $\Psi_1^{d,R}$ and $\Psi_1^{d,\text{NR}}$, as well as $\mathcal{X}^{\text{en},R}_{ij}$ and $\mathcal{X}^{\text{en,NR}}_{ij}$ can be defined as the terms describing the reversible loading due to the energy carried by dislocations and the energetically based hardening rule that mimics dissipative behavior by describing irreversible loading processes, respectively.

Meanwhile, it is well known that the interface plays a role as the barrier to plastic slip in the early stages of plastically deforming phase, while it acts as a source of the dislocation nucleation in the later stages. The energetic condition in the area around the interface is affected by the long-range internal stress fields associated with constrained plastic flow which leads to the accumulated and pile-up of dislocations near the interface. Thus, the condition at the interface is determined by a surface energy that depends on the plastic strain state at the interface (Fredriksson and Gudmundson 2005).

The interfacial Helmholtz free energy per unit surface area of the interface is put forward under the guidance of Fredriksson and Gudmundson (2005) work such as (It should be noted that it is possible to introduce another form of the surface energy if it is convex in, ε^{pI}_{ij}.):

$$\Psi^I = \frac{1}{2} G \ell_{\text{en}}^I \varepsilon_{ij}^{p^I(\text{pre})} \varepsilon_{ij}^{p^I(\text{pre})} \tag{56}$$

where ℓ_{en}^I is the interfacial recoverable length scale.

By substituting the interfacial Helmholtz free energy per unit surface given by Eq. 56 into Eq. 36, the interfacial recoverable microstresses $\mathbb{M}_{ij}^{I,\text{en}}$ can be obtained as follows:

$$\mathbb{M}_{ij}^{I,\text{en}} = G \ell_{\text{en}}^I \varepsilon_{ij}^{p^I(\text{pre})} \tag{57}$$

As can be seen in Eq. 57, $\mathbb{M}_{ij}^{I,\text{en}}$ does not involve the plastic strain rate, which is related to the dislocation slip, and the temperature since the interfacial recoverable microstresses are activated by the recoverable stored energy.

Dissipation Potential and Dissipative Thermodynamic Microstresses

In this section, the dissipation potential functions for the aforementioned two cases of the SGP model are postulated, respectively. The first case is derived from the dissipation potential dependent on the plastic strain gradient, which leads to the nonzero dissipative thermodynamic microstress $\mathcal{S}_{ijk}^{\text{dis}} \neq 0$, while the other case is derived from the dissipation potential that is independent on the plastic strain gradient, which leads to $\mathcal{S}_{ijk}^{\text{dis}} = 0$.

Coleman and Gurtin (1967) pointed out that the dissipation potential function is composed of two parts, the mechanical part which is dependent on the plastic strain and its gradient and the thermal counterpart which shows the purely thermal effect such as the heat conduction. In this sense, and in the context of Eq. 37, the functional form of the dissipation potential, which is dependent on $\dot{\varepsilon}_{ij,k}^p$, for the former class can be put forward as (Voyiadjis and Song 2017):

$$\begin{aligned}\mathcal{D} = \frac{\sigma_y}{2\dot{p}} \left\{ \left(\frac{\dot{p}}{\dot{p}_0}\right)^m \left(1 - \left(\frac{T}{T_y}\right)^n\right) \left(\dot{\varepsilon}_{ij}^p \dot{\varepsilon}_{ij}^p\right) + \ell_{\text{dis}}^2 \left(\frac{\dot{p}}{\dot{p}_0}\right)^m \left(1 - \left(\frac{T}{T_y}\right)^n\right) \left(\dot{\varepsilon}_{\text{lm},k}^p \dot{\varepsilon}_{\text{lm},k}^p\right) \right\} \\ - \frac{b}{2} \dot{T}^2 - \frac{1}{2} \frac{k(T)}{T} T_{,i} T_{,i} \geq 0\end{aligned} \tag{58}$$

where σ_y is a material constant accounting for the yield strength, m is a nonnegative material constant for the rate sensitivity parameter, in which the limit $m \to 0$ corresponds to rate-independent material behavior, $\dot{p}_0$ is a constant for the reference flow rate, ℓ_{dis} is the dissipative length scale that corresponds to the dissipative effects in terms of the gradient of the plastic strain rate, b is the material constant accounting for the energy exchange between phonon and electron, and $k(T)$ is the thermal conductivity coefficient. The generalized dissipative effective plastic strain measure $\dot{p}$ is defined as a function of the plastic strain rate, the gradient of the plastic strain rate, and the dissipative length scale as follows:

$$\dot{\mathcal{p}} = \sqrt{\dot{\varepsilon}_{ij}^{p}\dot{\varepsilon}_{ij}^{p} + \ell_{\text{dis}}^{2}\dot{\varepsilon}_{\text{lm},k}^{p}\dot{\varepsilon}_{\text{lm},k}^{p}} \tag{59}$$

By using the dissipation potential given in Eq. 58 along with Eq. 39 and the assumption $k(T)/T = k_0 = \text{constant}$, the dissipative thermodynamic forces for the former case (Case I) can be obtained as follows:

$$\mathcal{X}_{ij}^{\text{dis}} = \sigma_y \left(\frac{\dot{\mathcal{p}}}{\dot{\mathcal{p}}_0}\right)^m \left(1 - \left(\frac{T}{T_y}\right)^n\right) \frac{\dot{\varepsilon}_{ij}^{p}}{\dot{\mathcal{p}}} \tag{60}$$

$$\mathcal{S}_{ijk}^{\text{dis}} = \sigma_y \ell_{\text{dis}}^2 \left(\frac{\dot{\mathcal{p}}}{\dot{\mathcal{p}}_0}\right)^m \left(1 - \left(\frac{T}{T_y}\right)^n\right) \frac{\dot{\varepsilon}_{ij,k}^{p}}{\dot{\mathcal{p}}} \tag{61}$$

$$\mathcal{A}^{\text{dis}} = -\mathrm{b}\dot{\mathrm{T}} \tag{62}$$

$$q_i = -k_0 T_{,i} \tag{63}$$

On the other hand, the functional form of the dissipation potential, which is independent of $\dot{\varepsilon}_{ij,k}^{p}$, for the latter case (Case II) can be postulated by setting $\ell_{\text{dis}} = 0$ in Eq. 58 as follows:

$$\mathcal{D} = \frac{\sigma_y}{2}\left(\frac{\dot{\mathcal{p}}_{\ell_{\text{dis}}=0}}{\dot{\mathcal{p}}_0}\right)^m \left(1 - \left(\frac{T}{T_y}\right)^n\right) \dot{\mathcal{p}}_{\ell_{\text{dis}}=0} - \frac{b}{2}\dot{T}^2 - \frac{1}{2}\frac{k(T)}{T} T_{,i} T_{,i} \geq 0 \tag{64}$$

where $\dot{\mathcal{p}}_{\ell_{\text{dis}}=0}$ is given by $\sqrt{\dot{\varepsilon}_{ij}^{p}\dot{\varepsilon}_{ij}^{p}}$ by setting $\ell_{\text{dis}} = 0$ in Eq. 59. By substituting Eq. 64 into Eq. 39, the dissipative thermodynamic forces for the latter case (Case II) can be obtained as follows:

$$\mathcal{X}_{ij}^{\text{dis}} = \sigma_y \left(\frac{\dot{\mathcal{p}}_{\ell_{\text{dis}}=0}}{\dot{\mathcal{p}}_0}\right)^m \left(1 - \left(\frac{T}{T_y}\right)^n\right) \frac{\dot{\varepsilon}_{ij}^{p}}{\dot{\mathcal{p}}_{\ell_{\text{dis}}=0}} \tag{65}$$

$$\mathcal{S}_{ijk}^{\text{dis}} = 0 \tag{66}$$

$$\mathcal{A}^{\text{dis}} = -\mathrm{b}\dot{\mathrm{T}} \tag{67}$$

$$q_i = -k_0 T_{,i} \tag{68}$$

Meanwhile, Gurtin and Reddy (2009) pointed out that the classical isotropic hardening rule, which is dissipative in nature, may equally well be characterized via a defect energy since this energetically based hardening rule mimics the dissipative behavior by describing loading processes that are irreversible. In this sense, as it was mentioned previously in this chapter, the energetic microstress $\mathcal{X}_{ij}^{\mathrm{en}}$ is further decomposed into $\mathcal{X}_{ij}^{\mathrm{en,NR}}$ that describes an irreversible loading process and $\mathcal{X}_{ij}^{\mathrm{en},R}$ that describes a reversible loading process due to the energy carried by the dislocations. The present framework follows the theorem of Gurtin and Reddy (2009) in that the theory without a defect energy is equivalent to the theory with a defect energy by replacing the dissipation $\mathcal{D}$ by an effective dissipation $\mathcal{D}_{\mathrm{eff}}$, which is defined as follows:

$$\mathcal{D}_{\mathrm{eff}} = \left(\mathcal{X}_{ij}^{\mathrm{dis}} + \mathcal{X}_{ij}^{\mathrm{en,NR}}\right)\dot{\varepsilon}_{ij}^{p} + \mathcal{S}_{ijk}^{\mathrm{dis}}\dot{\varepsilon}_{ij,k}^{p} + \mathcal{A}^{\mathrm{dis}}\dot{T} - \frac{q_i}{T}T_{,i} \tag{69}$$

where $\mathcal{X}_{ij}^{en,NR}$ may be viewed as the effectively dissipative microforce since it satisfies an effective dissipation inequality.

There are two main mechanisms affecting the energy dissipation during the dislocation movement in the grain boundary area. The first mechanism is related to an energy change in the grain boundary region. The macroscopic accumulated plastic strain at the grain boundary can be connected to the microscopic deformation of the grain boundary through the quadratic mean of the deformation gradient. Thus, the energy change after the onset of slip transmission to the adjacent grain is able to be approximately determined by a quadratic function of the deformation gradient at the microscale and hence the interfacial plastic strain at the macroscale. The other mechanism introduces the energy involved in the deformation of the grain boundary. This energy is mainly due to the energy dissipation during the dislocation movement and can be taken as a linear function of the interfacial plastic strain.

The interfacial dissipation potential $\mathcal{D}^I$ in the current study is postulated by combing the above-mentioned mechanisms as follows:

$$\mathcal{D}^I = \frac{\ell_{\mathrm{dis}}^I}{m^I + 1}\left(\sigma_y^I + h^I\varepsilon_p^{I(\mathrm{post})}\right)\left(1 - \frac{T^I}{T_y^I}\right)^{n^I}\left(\frac{\dot{\varepsilon}_p^{I(\mathrm{post})}}{\dot{p}_0^I}\right)^{m^I}\dot{\varepsilon}_p^{I(\mathrm{post})} \geq 0 \tag{70}$$

where ℓ_{dis}^I is the interfacial dissipative length scale, m^I and $\dot{p}_0^I$ are the viscous related material parameters, σ_y^I is a constant accounting for the interfacial yield stress at which the interface starts to deform plastically, h^I is an interfacial hardening parameter representing the slip transmission through the interface, T_y^I is the scale-independent interfacial thermal parameter at the onset of yield, n^I is the interfacial thermal parameter, and $\varepsilon_p^{I(\mathrm{post})} = \sqrt{\varepsilon_{ij}^{p^{I(\mathrm{post})}}\varepsilon_{ij}^{p^{I(\mathrm{post})}}}$ and $\dot{\varepsilon}_p^{I(\mathrm{post})} = \sqrt{\dot{\varepsilon}_{ij}^{p^{I(\mathrm{post})}}\dot{\varepsilon}_{ij}^{p^{I(\mathrm{post})}}}$ are defined, respectively, with the plastic strain at the interface after the slip transfer $\varepsilon_{ij}^{p^{I(\mathrm{post})}}$ and its rate $\dot{\varepsilon}_{ij}^{p^{I(\mathrm{post})}}$. The rate-dependency and temperature-dependency of the interfacial dissipation energy are clearly shown in Eq. 70 through the terms $\left(\dot{\varepsilon}_p^{I(\mathrm{post})}/\dot{p}_0^I\right)^{m^I}$ and $\left(1 - T^I/T_y^I\right)^{n^I}$, respectively.

The interfacial dissipative microstresses $\mathbb{M}_{ij}^{I,\text{dis}}$ can be obtained by substituting Eq. 70 into Eq. 40 as follows:

$$\mathbb{M}_{ij}^{I,\text{dis}} = \frac{\ell_{\text{dis}}^{I}}{m^{I}+1}\left(\sigma_y^I + h^I \varepsilon_p^{I(\text{post})}\right)\left(1-\frac{T^I}{T_y^I}\right)^{n^I}\left(\frac{\dot{\varepsilon}_p^{I(\text{post})}}{\dot{p}_0^I}\right)^{m^I}\frac{\dot{\varepsilon}_{ij}^{p^I(\text{post})}}{\dot{\varepsilon}_p^{I(\text{post})}} \tag{71}$$

By substituting Eqs. 71 and 57 into Eq. 31, one can obtain the interfacial microtraction $\mathbb{M}_{ij}^{I}$ as follows:

$$\begin{aligned}\mathbb{M}_{ij}^{I} &= G\ell_{\text{en}}^{I}\varepsilon_{ij}^{p^I(\text{pre})} \\ &\quad + \frac{\ell_{\text{dis}}^{I}}{m^{I}+1}\left(\sigma_y^I + h^I \varepsilon_p^{I(\text{post})}\right)\left(1-\frac{T^I}{T_y^I}\right)^{n^I}\left(\frac{\dot{\varepsilon}_p^{I(\text{post})}}{\dot{p}_0^I}\right)^{m^I}\frac{\dot{\varepsilon}_{ij}^{p^I(\text{post})}}{\dot{\varepsilon}_p^{I(\text{post})}}\end{aligned} \tag{72}$$

As can be seen in Eq. 72, a free surface, i.e., microfree boundary condition, at the grain boundary can be described by setting $\ell_{en}^{I} = \ell_{dis}^{I} = 0$ and it is also possible to describe a surface passivation, i.e., microclamped boundary condition, by setting $\ell_{en}^{I} \to \infty$ and $\ell_{dis}^{I} \to \infty$.

Flow Rule

The flow rule in the present framework is established based on the nonlocal microforce balance, Eq. 6, augmented by thermodynamically consistent constitutive relations for both energetic and dissipative microstresses. By substituting Eqs. 51, 53, 60, and 61 into Eq. 6, one can obtain a second-order partial differential form of the flow rule as follows (The flow rule, Eq. 73, corresponds to the SGP model (Case I), in which the functional form of the dissipation potential is dependent on $\dot{\varepsilon}_{ij,k}^{p}\left(\mathcal{S}_{ijk}^{dis} \neq 0\right)$. One can easily obtain the flow rule for the other case (Case II) of the SGP model, i.e. with $\mathcal{S}_{ijk}^{dis} = 0$, by setting $\ell_{\text{dis}} = 0$.) (Voyiadjis and Song 2017):

$$\begin{aligned}\tau_{ij} - \underbrace{\left(-G\ell_{en}^{2}\varepsilon_{ij,kk}^{p}\right)}_{-\mathcal{S}_{ijk,k}^{en}} &= h\varepsilon_{ij}^{p}\left(1-\left(\frac{T}{T_y}\right)^{n}\right)\varepsilon_p^{r-1} + \sigma_y\left(\frac{\dot{p}}{\dot{p}_0}\right)^{m}\left(1-\left(\frac{T}{T_y}\right)^{n}\right)\frac{\dot{\varepsilon}_{ij}^{p}}{\dot{p}} \\ &\quad - \left(\sigma_y\ell_{dis}^{2}\left(\frac{\dot{p}}{\dot{p}_0}\right)^{m}\left(1-\left(\frac{T}{T_y}\right)^{n}\right)\frac{\dot{\varepsilon}_{ij,kk}^{p}}{\dot{p}}\right)\end{aligned} \tag{73}$$

where the under-braced term $\mathcal{S}_{ijk,k}^{en}$ represents a backstress due to the energy stored in dislocations and results in the Bauschinger effect observed in the experiments

(Liu et al. 2015; Nicola et al. 2006; Xiang et al. 2005; Xiang and Vlassak 2006) and discrete dislocation plasticity (Nicola et al. 2006; Shishvan et al. 2010, 2011).

Thermo-mechanical Coupled Heat Equation

The evolution of the temperature field is governed by the law of conservation of energy (the first law of thermodynamics). The terms addressing heating as a result of the inelastic dissipation and thermo-mechanical coupling are involved for describing the evolution of the temperature field. The equation for the conservation of energy in this chapter is put forward as follows:

$$\sigma_{ij}\dot{\varepsilon}_{ij}^{e} + \mathcal{X}_{ij}\dot{\varepsilon}_{ij}^{p} + \mathcal{S}_{ijk}\dot{\varepsilon}_{ij,k}^{p} + \mathcal{A}\dot{T} + \mathcal{B}_i\dot{T}_{,i} - \text{div } q_i + \rho\mathcal{H}_{\text{EXT}} - \rho\dot{e} = 0 \quad (74)$$

where $\mathcal{H}_{\text{EXT}}$ is the specific heat from the external source.

By considering the effective dissipation potential given in Eq. 69 along with the equations for the entropy production (the second law of thermodynamics) previously described in Eq. 17, the relationship for the evolution of the entropy, which describes the irreversible process, can be derived as follows:

$$\rho\dot{s}T = \mathcal{D}_{eff} + \rho\mathcal{H}_{\text{EXT}} \quad (75)$$

By using Eq. 50 for solving the rate of the entropy $\dot{s}$, the evolution of the temperature field can be obtained as follows:

$$\begin{aligned}\rho c_0\dot{T} = {} & \underbrace{\mathcal{X}_{ij}^{\text{dis}}\dot{\varepsilon}_{ij}^{p} + \mathcal{X}_{ij}^{\text{en,IR}}\dot{\varepsilon}_{ij}^{p} + \mathcal{S}_{ijk}^{\text{dis}}\dot{\varepsilon}_{ij,k}^{p}}_{①} + \underbrace{\frac{k(T)}{2T}T_{,i}T_{,i} + \mathfrak{a}T_r\dot{T}_{,\text{ii}} - \text{b}T_r\ddot{T}}_{②} \\ & \underbrace{-\alpha_t\dot{\varepsilon}_{ij}^{e}\delta_{ij}T}_{③} \\ & \underbrace{-\left\{h\varepsilon_p^{r-1}\left(\frac{n}{T_y}\left(\frac{T}{T_y}\right)^{n-1}\right)\dot{\varepsilon}_{ij}^{p} + \frac{h\varepsilon_p^{r+1}}{r+1}\left(\frac{n(n-1)}{T_y^2}\left(\frac{T}{T_y}\right)^{n-2}\right)\dot{T}\right\}T}_{④} \\ & + \underbrace{\rho\mathcal{H}_{\text{EXT}}}_{⑤}\end{aligned} \quad (76)$$

where c_0 is the specific heat capacity at constant volume and is given by $c_0 = \text{constant} \cong c_\varepsilon T/T_r$. As shown in Eq. 76, the following terms are depicted, ① irreversible mechanical process, ② generalized heat conduction, ③ thermo-elastic coupling, ④ thermo-plastic coupling, and ⑤ heat source, which are involved in the evolution of the temperature. The third-order mixed derivation term $\mathfrak{a}T_r T_{,\text{ii}}$ in part ② introduces the microstructural interaction effect into the classical heat equation, in addition, the second-order time derivative term $\text{b}T_r\ddot{T}$ gives the thermal wave behavior effect in heat propagation.

By substituting the constitutive equations of the energetic microstresses given by Eqs. 49, 50, 51, 52, and 53 and the dissipative microstresses given by Eqs. 60, 61, 62, and 63 into Eq. 76 and defining three additional terms $\mathcal{t}_{\text{eff}} = k_0/\rho c_v$, $\mathcal{t}^{\text{en}} = \mathfrak{a}T_r/k_0$, and $\mathcal{t}^{\text{dis}} = \text{b}T_r/\rho c_v$, the evolution of temperature for the model, in which the functional form of the dissipation potential dependent on $\dot{\varepsilon}^p_{ij,k}$, i.e., $\mathcal{S}^{\text{dis}}_{ijk} \neq 0$, can be obtained as follows (Voyiadjis and Song 2017):

$$\begin{aligned}
&\left[1 + \frac{h\varepsilon_p^{r+1}}{\rho c_v (r+1)}\left(\frac{n(n-1)}{T_y}\left(\frac{T}{T_y}\right)^n\right)\right]\dot{T} \\
&= \frac{\dot{\varepsilon}^p_{ij}}{\rho c_v}\left[\left\{h\varepsilon_p^{r-1}\left(1-\left(\frac{T}{T_y}\right)^n\right) - 2r\vartheta h^2\varepsilon_p^{2r-1} - \text{nh}\varepsilon_p^{r-1}\left(\frac{T}{T_y}\right)^n\right\}\varepsilon^p_{ij}\right. \\
&\left. +\sigma_y\left(\frac{\dot{\mathcal{p}}}{\dot{\mathcal{p}}_0}\right)^m\left(1-\left(\frac{T}{T_y}\right)^n\right)\frac{\dot{\varepsilon}^p_{ij}}{\dot{\mathcal{p}}}\right] \\
&+ \frac{1}{\rho c_v}\left[\sigma_y \ell^2_{\text{dis}}\left(\frac{\dot{\mathcal{p}}}{\dot{\mathcal{p}}_0}\right)^m\left(1-\left(\frac{T}{T_y}\right)^n\right)\frac{\dot{\varepsilon}^p_{ij,k}\dot{\varepsilon}^p_{ij,k}}{\dot{\mathcal{p}}} - \alpha_t\dot{\varepsilon}^e_{ij}\delta_{ij}T + \rho\mathcal{H}_{\text{EXT}}\right] \\
&+ \left[\mathcal{t}_{\text{eff}}T_{,\text{ii}} + \mathcal{t}_{\text{eff}}\mathcal{t}^{\text{en}}\dot{T}_{,\text{ii}} - \mathcal{t}^{\text{dis}}\ddot{T}\right]
\end{aligned} \tag{77}$$

In the absence of the mechanical terms, Eq. 77 turns to the generalized heat equation including $\mathcal{t}^{\text{en}}$ and $\mathcal{t}^{\text{dis}}$. The evolution of temperature for the other case (Case II) of SGP model, i.e., with $\mathcal{S}^{\text{dis}}_{ijk} = 0$ can be obtained by setting $\ell_{\text{dis}} = 0$ in Eq. 77.

Finite Element Implementation of the Strain Gradient Plasticity Model

In this section, first a one-dimensional finite element model for the SGP model by Voyiadjis and Song (2017) is presented to investigate the size dependent behavior in the microscopic structures under macroscopically uniform uniaxial tensile stress. In a one-dimensional finite element implementation, the macroscopic partial differential equations for balance of linear momentum Eq. 5 with the macroscopic boundary conditions, $u_{x=0} = 0$ and $u_{x=L} = u^{\dagger}$ (prescribed), and the microscopic partial differential equations for nonlocal force balance Eq. 6 with the microscopic boundary conditions, $\left(\mathbb{M}^I - \mathcal{S}\right)_{x=0} = 0$ and $\left(\mathbb{M}^I - \mathcal{S}\right)_{x=L} = 0$, yield the following expressions in a global weak form, respectively (The finite element solutions, in this section, depend on the x-direction. A single crystal with the size of L bounded by two grain boundaries is analyzed (see Fig. 7 for details). Also a two dimensional case is presented for solving the problem of the square plate in the latter part of this chapter:

$$\int_0^L (\sigma \tilde{u}_{,x})\, \mathrm{dx} = 0 \tag{78}$$

$$\int_0^L \left[(\mathcal{X} - \tau)\, \tilde{\varepsilon}^p + \mathcal{S} \tilde{\varepsilon}^p_{,x} \right] \mathrm{dx} + \mathbb{M}^I \tilde{\varepsilon}^p_{x=L} - \mathbb{M}^I \tilde{\varepsilon}^p_{x=0} = 0 \tag{79}$$

where the arbitrary virtual fields $\tilde{u}$ and $\tilde{\varepsilon}^p$ are assumed to be kinematically admissible weighting functions in the sense that $\tilde{u}_{x=0} = \tilde{u}_{x=L} = 0$. (In the case of micro-clamped boundary condition, $\tilde{\varepsilon}^p_{x=0} = \tilde{\varepsilon}^p_{x=L} = 0$ is imposed at the grain boundaries to enforce the complete blockage of dislocations at the interface. In the case of micro-free boundary condition, on the other hand, the dislocations are free to move across the interface, which in turn, the present grain boundary flow rule is imposed.)

The user-element subroutine UEL in the commercial finite element package ABAQUS/standard (2012) is presented in this chapter in order to numerically solve the weak forms of the macroscopic and microscopic force balances, Eqs. 78 and 79, respectively. In this finite element solution, the displacement field u and the plastic strain field ε^p are discretized independently and both of the fields are taken as fundamental unknown nodal degrees of freedom. In this regard, the displacement field and corresponding strain field ε, and the plastic strain field and corresponding plastic strain gradient field $\varepsilon^p_{,x}$ can be obtained by using the interpolation as follows:

$$u(x) = \sum_{\xi=1}^{n_u} \mathrm{N}_u^\xi \mathcal{U}_u^\xi \quad \varepsilon(x) = \frac{\partial u(x)}{\partial x} = \sum_{\xi=1}^{n_u} \mathrm{N}_{u,x}^\xi \mathcal{U}_u^\xi \tag{80}$$

$$\varepsilon^p(x) = \sum_{\xi=1}^{n_{\varepsilon^p}} \mathrm{N}_{\varepsilon^p}^\xi \mathcal{E}_{\varepsilon^p}^\xi \quad \varepsilon^p_{,x}(x) = \frac{\partial \varepsilon^p(x)}{\partial x} = \sum_{\xi=1}^{n_{\varepsilon^p}} \mathrm{N}_{\varepsilon^p,x}^\xi \mathcal{E}_{\varepsilon^p}^\xi \tag{81}$$

where N_u^ξ and $\mathrm{N}_{\varepsilon^p}^\xi$ are the shape functions, and $\mathcal{U}_u^\xi$ and $\mathcal{E}_{\varepsilon^p}^\xi$ are the nodal values of the displacements and the plastic strains at node ξ, respectively. The terms n_u and n_{ε^p} represent the number of nodes per a single element for the displacement and the plastic strain, respectively. (If a one-dimensional three-noded quadratic element is employed, n_u and n_{ε^p} are set up as three. On the other hand, these parameters are set up as two in the case that a one-dimensional two-noded linear element is used. It should be noted that n_u and n_{ε^p} do not necessarily have to be same as each other in the present finite element implementation, even though both the displacement field and the plastic strain fields are calculated by using the standard isoparametric shape functions.)

Substituting Eqs. 80 and 81 into Eqs. 78 and 79 give the nodal residuals for the displacement r_u and the plastic strain r_{ε^p} for each finite element el as follows:

$$(r_u)_\xi = \int_{\text{el}} \left(\sigma \mathbb{N}^\xi_{u,x}\right) \mathrm{dx} \tag{82}$$

$$(r_{\varepsilon^p})_\xi = \int_{\text{el}} \left[(\mathcal{X} - \tau)\, \mathbb{N}^\xi_{\varepsilon^p} + \mathcal{S}\mathbb{N}^\xi_{\varepsilon^p,x}\right] \mathrm{dx} + \mathbb{M}^I \mathbb{N}^\xi_{\varepsilon^p} \tag{83}$$

where the term $\mathbb{M}^I \mathbb{N}^\xi_{\varepsilon^p}$ is applied only for the nodes on the interface which is at $x = 0$ and $x = L$.

The global coupled system of equations, $(r_u)_\xi = 0$ and $(r_{\varepsilon^p})_\xi = 0$, are solved using ABAQUS/standard (2012) based on the Newton-Raphson iterative scheme. Occasionally, the modified Newton-Raphson method, referred to as quasi Newton-Raphson method, is employed in the case that the numerical solution suffers a divergence during the initial increment immediately after an abrupt change in loading. The Taylor expansion of the residuals with regard to the current nodal values can be expressed by assuming the nodal displacement and the plastic strain in iteration ζ as $\mathcal{U}^\zeta_u$ and $\mathcal{E}^\zeta_{\varepsilon^p}$, respectively, as follows:

$$\begin{aligned}\left(r_u\big|_{\mathcal{U}^{\zeta+1}_u,\mathcal{E}^{\zeta+1}_{\varepsilon^p}}\right)_\xi &= \left(r_u\big|_{\mathcal{U}^{\zeta}_u,\mathcal{E}^{\zeta}_{\varepsilon^p}}\right)_\xi + \left(\frac{\partial r_u}{\partial \mathcal{U}^\xi_u}\bigg|_{\mathcal{U}^\zeta_u}\right)\Delta\mathcal{U}^\xi_u + \left(\frac{\partial r_u}{\partial \mathcal{E}^\xi_{\varepsilon^p}}\bigg|_{\mathcal{E}^\zeta_{\varepsilon^p}}\right)\Delta\mathcal{E}^\xi_{\varepsilon^p} \\ &\quad + \mathcal{O}\left(\left(\Delta\mathcal{U}^\xi_u\right)^2, \left(\Delta\mathcal{E}^\xi_{\varepsilon^p}\right)^2\right)\end{aligned} \tag{84}$$

$$\begin{aligned}&\left(r_{\varepsilon^p}\big|_{\mathcal{U}^{\zeta+1}_u,\mathcal{E}^{\zeta+1}_{\varepsilon^p}}\right)_\xi \\ &\quad = \left(r_{\varepsilon^p}\big|_{\mathcal{U}^{\zeta}_u,\mathcal{E}^{\zeta}_{\varepsilon^p}}\right)_\xi + \left(\frac{\partial r_{\varepsilon^p}}{\partial \mathcal{U}^\xi_u}\bigg|_{\mathcal{U}^\zeta_u}\right)\Delta\mathcal{U}^\xi_u + \left(\frac{\partial r_{\varepsilon^p}}{\partial \mathcal{E}^\xi_{\varepsilon^p}}\bigg|_{\mathcal{E}^\zeta_{\varepsilon^p}}\right)\Delta\mathcal{E}^\xi_{\varepsilon^p} \\ &\quad + \mathcal{O}\left(\left(\Delta\mathcal{U}^\xi_u\right)^2, \left(\Delta\mathcal{E}^\xi_{\varepsilon^p}\right)^2\right)\end{aligned} \tag{85}$$

where $\Delta\mathcal{U}^\xi_u = \left(\mathcal{U}^{\zeta+1}_u\right)_\xi - \left(\mathcal{U}^{\zeta}_u\right)_\xi$, $\Delta\mathcal{E}^\xi_{\varepsilon^p} = \left(\mathcal{E}^{\zeta+1}_{\varepsilon^p}\right)_\xi - \left(\mathcal{E}^{\zeta}_{\varepsilon^p}\right)_\xi$, and $\mathcal{O}\left(\left(\Delta\mathcal{U}^\xi_u\right)^2, \left(\Delta\mathcal{E}^\xi_{\varepsilon^p}\right)^2\right)$ are the big $\mathcal{O}$ notation to represent the terms of higher order than the second degree. The residual is ordinarily calculated at the end of each time step, and the values of the nodal displacements and the plastic strains are updated during the iterations. The increments in nodal displacements and the plastic strains can be computed by solving the system of linear equations shown in Eq. 86 with the Newton-Raphson iterative method:

$$\underbrace{\begin{bmatrix} K^{\text{el}}_{\text{uu}} & K^{\text{el}}_{u\varepsilon^p} \\ K^{\text{el}}_{\varepsilon^p{}_u} & K^{\text{el}}_{\varepsilon^p\varepsilon^p} \end{bmatrix}}_{\boldsymbol{K}^{\text{el}}} \begin{Bmatrix} \Delta\mathcal{U}^\xi_u \\ \Delta\mathcal{E}^\xi_{\varepsilon^p} \end{Bmatrix} = \begin{Bmatrix} \left(r_u\big|_{\mathcal{U}^\zeta_u,\mathcal{E}^\zeta_{\varepsilon^p}}\right)_\xi \\ \left(r_{\varepsilon^p}\big|_{\mathcal{U}^\zeta_u,\mathcal{E}^\zeta_{\varepsilon^p}}\right)_\xi \end{Bmatrix} \tag{86}$$

where K^{el} is the Jacobian (stiffness) matrix that needs to be defined in the user-subroutine for each element.

From Eqs. 84 and 85 along with the discretization for the displacements given by Eq. 80 and the plastic strains given by Eq. 81 at the end of a time step, and the functional forms of the energetic and dissipative higher order stresses defined in the previous sections, the Jacobian matrix for the case of the SGP model with $\mathcal{S}_{ijk}^{\mathrm{dis}} \neq 0$ can be obtained as follows:

$$K_{\mathrm{uu}}^{\mathrm{el}} = -\left.\frac{\partial r_u}{\partial \mathcal{U}_u^{\xi}}\right|_{\mathcal{U}_u^{\zeta}} = -\int_{\mathrm{el}} \left(E \mathbb{N}_{u,x}^{\xi} \mathbb{N}_{u,x}^{\xi}\right) \mathrm{dx} \tag{87}$$

$$K_{u\varepsilon^p}^{\mathrm{el}} = -\left.\frac{\partial r_u}{\partial \mathcal{E}_{\varepsilon^p}^{\xi}}\right|_{\mathcal{E}_{\varepsilon^p}^{\zeta}} = \int_{\mathrm{el}} \left(E \mathbb{N}_{\varepsilon^p}^{\xi} \mathbb{N}_{u,x}^{\xi}\right) \mathrm{dx} \tag{88}$$

$$K_{\varepsilon^p u}^{\mathrm{el}} = -\left.\frac{\partial r_{\varepsilon^p}}{\partial \mathcal{U}_u^{\xi}}\right|_{\mathcal{U}_u^{\zeta}} = \int_{\mathrm{el}} \left(E \mathbb{N}_{u,x}^{\xi} \mathbb{N}_{\varepsilon^p}^{\xi}\right) \mathrm{dx} \tag{89}$$

$$\begin{aligned}
\left(K_{\varepsilon^p\varepsilon^p}^{\mathrm{el}}\right)_{\mathcal{S}_{ijk}^{\mathrm{dis}}\neq 0} &= \left(-\left.\frac{\partial r_{\varepsilon^p}}{\partial \mathcal{E}_{\varepsilon^p}^{\xi}}\right|_{\mathcal{E}_{\varepsilon^p}^{\zeta}}\right)_{\mathcal{S}_{ijk}^{\mathrm{dis}}\neq 0} \\
&= \underbrace{\int_{\mathrm{el}} \left[\left\{ E + \left(\mathrm{rh}(\varepsilon^p)^r + \frac{\sigma_y (\dot{p})^{m-1}}{\Delta t (\dot{p}_0)^m}\right)\left(1 - \left(\frac{T}{T_y}\right)^n\right)\right\} \mathbb{N}_{\varepsilon^p}^{\xi}\mathbb{N}_{\varepsilon^p}^{\xi}\right.}_{\text{for bulk}} \\
&\quad + \underbrace{\left.\left\{\left(G\ell_{\mathrm{en}}^2 + \frac{\sigma_y \ell_{\mathrm{dis}}^2 (\dot{p})^{m-1}}{\Delta t (\dot{p}_0)^m}\right)\left(1 - \left(\frac{T}{T_y}\right)^n\right)\right\} \mathbb{N}_{\varepsilon^p,x}^{\xi}\mathbb{N}_{\varepsilon^p,x}^{\xi}\right] \mathrm{dx}}_{\text{for bulk}} \\
&\quad + \underbrace{\left[G\left(\ell_{\mathrm{en}}^I\right)^2 + \frac{\sigma_y^I \left(\ell_{\mathrm{dis}}^I\right)^2 \left(\dot{p}^I\right)^{m^I-1}}{\Delta t \left(\dot{p}_0^I\right)^{m^I}}\left(1 - \left(\frac{T}{T_y^I}\right)^{n^I}\right)\right] \mathbb{N}_{\varepsilon^p}^{\xi}\mathbb{N}_{\varepsilon^p}^{\xi}}_{\text{for interface}}
\end{aligned} \tag{90}$$

where Δt is a time step. The Jacobian matrix for the other case, i.e. $\left(K_{\varepsilon^p\varepsilon^p}^{\mathrm{el}}\right)_{\mathcal{S}_{ijk}^{\mathrm{dis}}=0}$, can be obtained by setting $\ell_{\mathrm{dis}} = 0$ in Eq. 90. The interfacial terms in $\left(K_{\varepsilon^p\varepsilon^p}^{\mathrm{el}}\right)_{\mathcal{S}_{ijk}^{\mathrm{dis}}\neq 0}$ and $\left(K_{\varepsilon^p\varepsilon^p}^{\mathrm{el}}\right)_{\mathcal{S}_{ijk}^{\mathrm{dis}}=0}$ are applied only for the nodes on the interface which is at $x = 0$ and $x = L$ in this chapter.

Experimental Validation of the Strain Gradient Plasticity Model

In this section, the present SGP model and corresponding finite element code by Voyiadjis and Song (2017) are validated by comparing with the experimental results from three sets of size effect experiments. In addition, the comparison between the

SGP model by Voyiadjis and Song (2017) and Voyiadjis and Faghihi (2014) is carried out to show the increase in accuracy of the model by Voyiadjis and Song (2017). To examine the applicability of the present finite element implementation to the various kinds of materials, it is considered that each set of three experiments involves the three different materials, viz. aluminum (Al), copper (Cu), and nickel (Ni).

Haque and Saif (2003) developed Micro-Electro Mechanical Systems (MEMS)-based testing techniques for uniaxial tensile testing of nanoscale freestanding Al thin films to explore the effect of strain gradient in 100 nm, 150 nm, and 485 nm thick specimens with average grain size of 50 nm, 65 nm, and 212 nm, respectively. The specimens with 99.99% pure sputter-deposited freestanding Al thin films are 10 μm wide and 275 μm long. All experiments are carried out in situ in SEM and the stress and strain resolutions for the tests are set 5 Mpa and 0.03%, respectively. In particular, the comparison between the present SGP model by Voyiadjis and Song (2017) and Voyiadjis and Faghihi (2014) is carried out to show the increase in accuracy of the present model. The calibrated material parameters as well as the general material parameters for the Al are presented in Table 1, and the numerical results from both the present model by Voyiadjis and Song (2017) and Voyiadjis and Faghihi (2014) are shown in Fig. 2 in conjunction with the experimental data of Haque and Saif (2003). As it is clearly shown in this figure, the size effect: Smaller is Stronger is observed on the stress-strain curves of the Al thin films. Furthermore, the calculated results of the present SGP model by Voyiadjis and Song (2017) display a

Table 1 The general and calibrated material parameters used for the validation of the proposed strain gradient plasticity model (Reprinted with permission from Voyiadjis and Song 2017)

General		**Aluminum**	**Copper**	**Nickel**
E (GPa)	Elastic modulus for isotropic linear elasticity	110	70	115
G (GPa)	Shear modulus for isotropic linear elasticity	48	27	44
ρ (g cm^{-3})	Density	8.960	2.702	8.902
c_3(J/g K)	Specific heat capacity at constant stress	0.385	0.910	0.540
Calibrated		**Aluminum**	**Copper**	**Nickel**
σ_y (MPa)	Yield stress	195	700	950
h (MPa)	Hardening material parameter	600	1,700	3,500
$\dot{p}_0$ (s^{-1})	Reference effective plastic strain rate	0.04	0.04	0.04
r	Nonlinear hardening material parameter	0.6	0.2	0.2
m	Non-negative rate sensitivity parameter	0.05	0.05	0.05
T_y ($\mathring{K}$)	Thermal material parameter	1,358	933	890
n	Temperature sensitivity parameter	0.3	0.3	0.3
ℓ_{en} (μm)	Energetic length scale	1.5 (1.0 μm)	0.9 (100 nm)	1.0
ℓ_{dis} (μm)	Dissipative length scale	2.5(1.0 μm)	8.0 (100 nm)	0.1

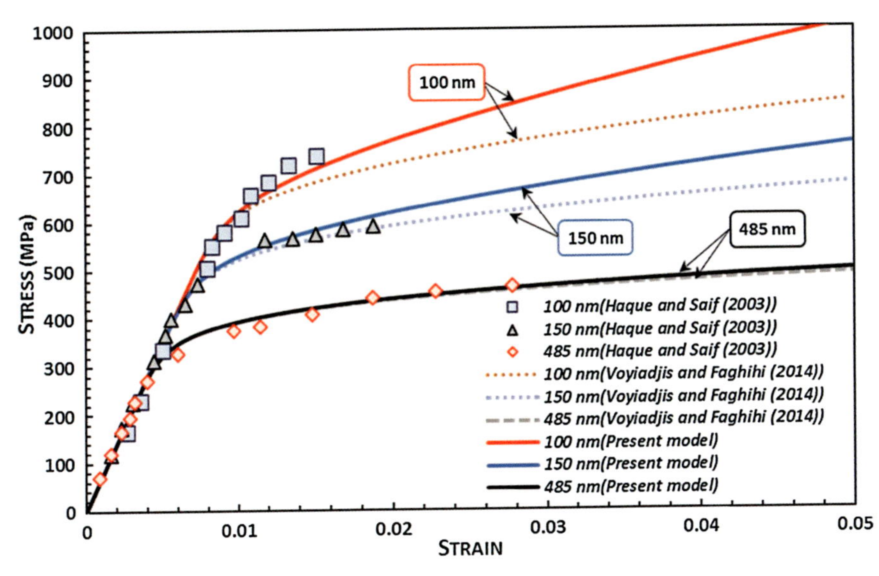

Fig. 2 The validation of the present strain gradient plasticity model by comparing the numerical results from the present model by Voyiadjis and Song (2017) with those from Voyiadjis and Faghihi (2014) and the experimental measurements from Haque and Saif (2003) on the stress-strain response of the sputter-deposited Al thin films (Reprinted with permission from Voyiadjis and Song 2017)

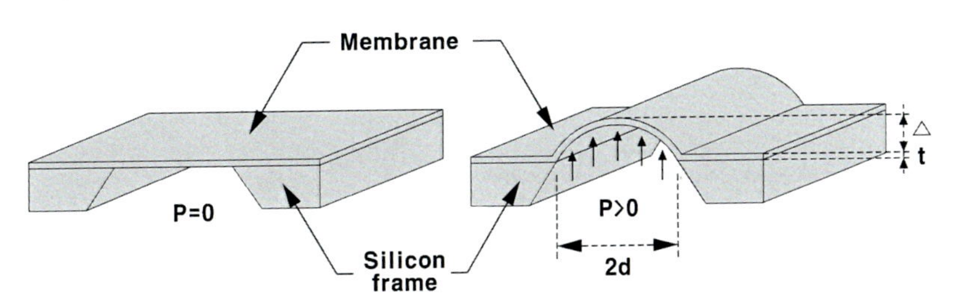

Fig. 3 Schematic representation of the plane strain bulge test technique presented by Xiang and Vlassak (2006). The stress-strain response of either a single material membrane or a stack of multiple material membranes adhered on a rigid Si frame is able to be obtained by using the following equations: $\sigma = P(d^2 + \Delta^2)/2t\Delta$ and$\varepsilon = \varepsilon_r + \{(d^2 + \Delta^2)/2d\Delta\} \arcsin(2d\Delta/d^2 + \Delta^2) - 1$ where ε_r is a residual strain in the membrane (Reprinted with permission from Voyiadjis and Song 2017)

tendency to be more coincident to the experimental data than those of Voyiadjis and Faghihi (2014).

Xiang and Vlassak (2006) investigated the size effects with a variety of film thicknesses on the plastic behavior of the freestanding electroplated Cu thin films by performing the plane strain bulge test. In this plane strain bulge test, the rectangular freestanding membranes surrounded by a rigid silicon (Si) frame are deformed in plane strain by applying a uniform pressure to one side of the membrane as shown in Fig. 3. The displacement and pressure resolutions for this bulge tests system are 0.3 μm and 0.1 kpa, respectively.

As can be seen in the work of Xiang and Vlassak (2006), the stress-strain curves of the Cu thin films with a passivation layer on both surfaces clearly show the size effects due to the presence of a boundary layer with high dislocation density near the film-passivation layer interfaces. In this sense, the bulge test of electroplated Cu thin films with both surfaces passivated by 20 nm of titanium (Ti) is considered here for the experimental validation of the present SGP model. In order to describe the passivation effect, the microclamped condition, which causes the dislocations to be completely blocked at the grain boundary, is imposed at both surfaces of the Cu thin films. Meanwhile, the experiments are performed with the various thicknesses of the Cu thin films of 1.0 μm, 1.9 μm, and 4.2 μm. The average grain sizes in all cases are given by 1.5 ± 0.05 μm, 1.51 ± 0.04 μm, and 1.5 ± 0.05 μm, respectively, which mean almost equal to each other.

The calibrated and general material parameters for the copper are presented in Table 1, and the comparison between the experimental measurements from the bulge tests and the calculated results from the present SGP model by Voyiadjis and Song (2017) is shown in Fig. 4. As it is clearly shown in this figure, the size effects according to the variation of the Cu thin film thicknesses is well observed in both the present SGP model and the experimental work of Xiang and Vlassak (2006). Moreover, the numerical results of the present model by Voyiadjis and Song (2017) are in good agreement with the experimental measurements.

Han et al. (2008) developed the microscale tensile testing system, which is composed of a high temperature furnace, a micro motor actuator and the Digital

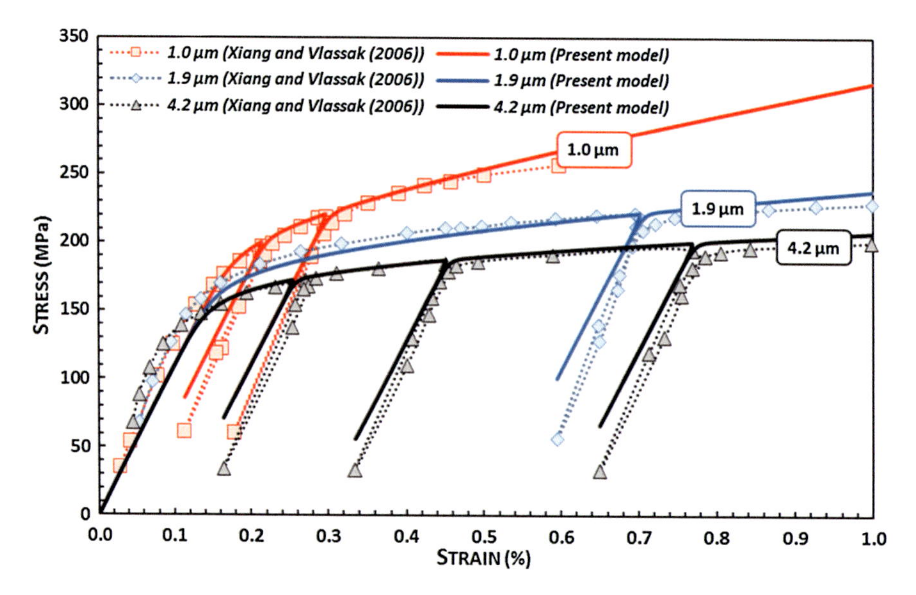

Fig. 4 The validation of the present strain gradient plasticity model by comparing to the experimental measurements from Xiang and Vlassak (2006) on the stress-strain response of the electroplated Cu thin films with the passivated layers on both sides (Reprinted with permission from Voyiadjis and Song 2017)

Image Correlation (DIC) system, for evaluating the mechanical properties of the Ni thin films at high temperatures. Dogbone-shaped specimens used in their experiments were made by Micro-electro mechanical system (MEMS) processes and the primary dimensions of the specimen are shown in Fig. 5.

The calibrated material parameters as well as the general parameters for Ni are presented in Table 1. The results of the microscale tensile tests at four different temperatures, i.e., 25 °C, 75 °C, 145 °C, and 218 °C, and corresponding numerical results from the present model (Voyiadjis and Song 2017) are shown in Fig. 6. As shown in this figure, it is clear from both the experimental and numerical results that

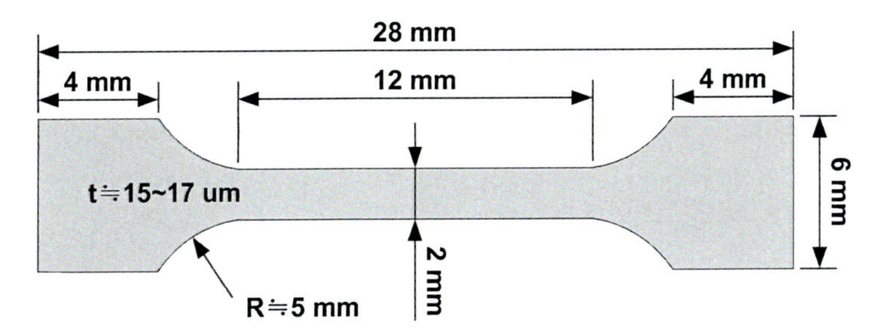

Fig. 5 The specimen dimensions for the experimental validation (Reprinted with permission from Voyiadjis and Song 2017)

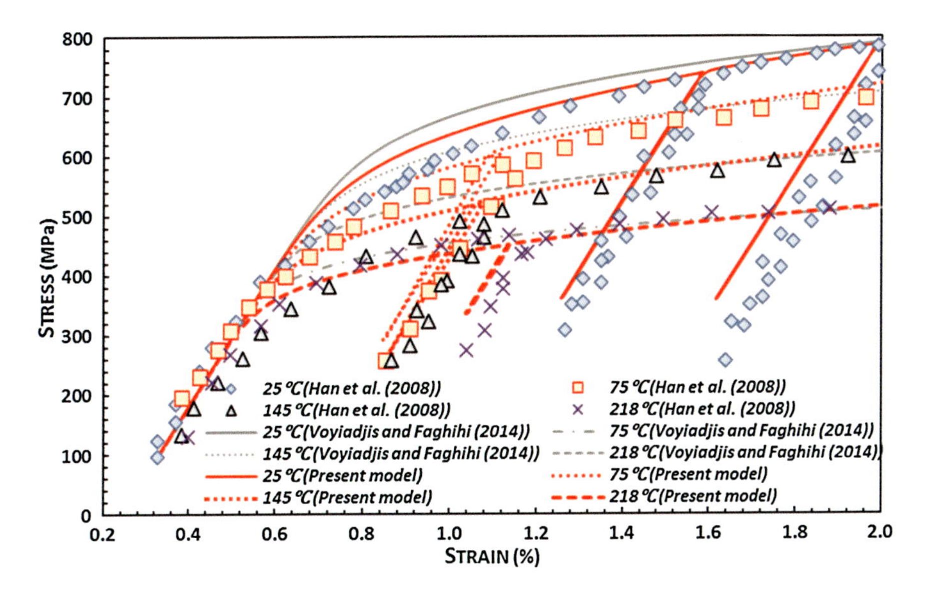

Fig. 6 The validation of the present strain gradient plasticity model (Voyiadjis and Song 2017) by comparing the numerical results with the model by Voyiadjis and Faghihi (2014) and the experimental measurements from Han et al. (2008) on the stress-strain response of Ni thin films (Reprinted with permission from Voyiadjis and Song 2017)

the Young's modulus is not affected by variations in temperature, while the yield and tensile strength decrease as the specimen temperature increases. In addition, Fig. 6 clearly shows that the Bauschinger effect is not affected very much by variations in the specimen temperature. Meanwhile, the calculated results of the present model (Voyiadjis and Song 2017) compare better to the experimental data than those of Voyiadjis and Faghihi (2014) (Fig. 6).

Stretch-Passivation Problem

The numerical solutions for the stretch-passivation problem with the two cases of the SGP model are presented in this section. The frameworks presented in this chapter represent the nonlocal flow rules in the form of partial differential equations when the microscopic force balances are integrated with the thermodynamically consistent constitutive equations. To interpret and analyze the physical phenomena characterized by the current frameworks is very complicated; in this sense, a one-dimensional numerical solution is presented first and extended to the two-dimensional one later in this chapter.

An initially uniform single grain with the size of L is used with two grain boundaries as shown in Fig. 7. The grain is assumed to be infinitely long along the x-direction and initially homogeneous; therefore, the solution depends only on the x-direction. In the one-dimensional stretch-passivation problem, the grain is deformed into the plastic regime by uniaxial tensile stretch with no constraint on plastic flow at the grain boundaries. At a certain point, the plastic flow is then constrained by blocking off the dislocations from passing out of the grain boundary, which leads the further plastic strain not to occur at the grain boundary.

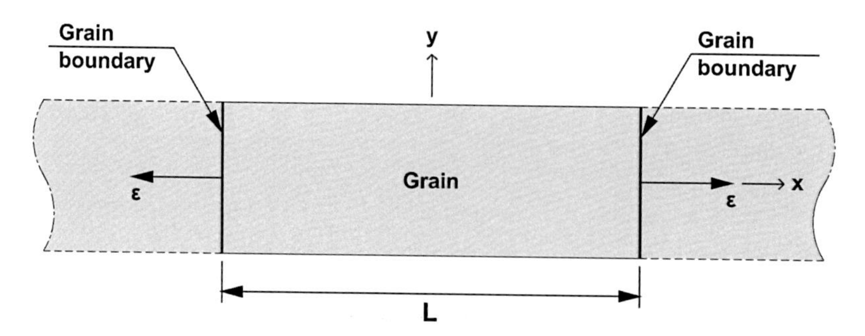

Fig. 7 One-dimensional model for a single grain with two grain boundaries (Reprinted with permission from Voyiadjis et al. 2017)

Numerical Results

In this section, an extensive numerical work is carried out based on the validated code in order to compare the results from the aforementioned two cases of the SGP model and to analyze the characteristics of the stress jump phenomenon. This section is largely composed of four subparts. In the first part, the occurrence of the stress jump phenomenon under the stretch-surface passivation condition is introduced in conjunction with three experiments used in section "Experimental Validation of the Strain Gradient Plasticity Model." The second part is focused on indicating that the results are contrary to each other for the two cases of the SGP model. An extensive parametric study is also conducted in terms of the various material parameters, and the evolution of the free energy involving the stored energy of cold work and the dissipation potentials during the plastic deformation are discussed in the third part. In the last part, the two-dimensional simulation is also given to examine the gradient and grain boundary effect, the mesh sensitivity of the two-dimensional model, and the stress jump phenomenon.

The numerical results on the stress-strain behaviors of Al and Cu thin films for the SGP model with the corresponding dissipative microstress quantities are presented in Figs. 8 and 9, respectively. In both simulations, a significant stress jump

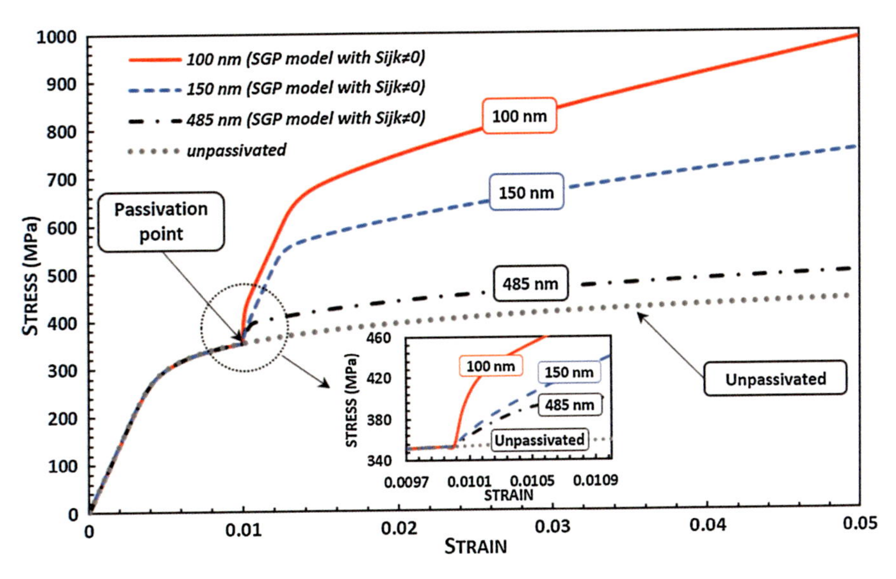

Fig. 8 A finite stress jump due to infinitesimal changes in the plastic strain. The numerical implementation of the SGP model with the dissipative potential dependent on $\dot{\varepsilon}^{p}_{ij,k}$, i.e., $\mathcal{S}^{\text{dis}}_{ijk} \neq 0$, is carried out based on the experiments of Haque and Saif (2003) with the various thicknesses of the Al thin films of 100 nm, 150 nm, and 485 nm (Reprinted with permission from Voyiadjis and Song 2017)

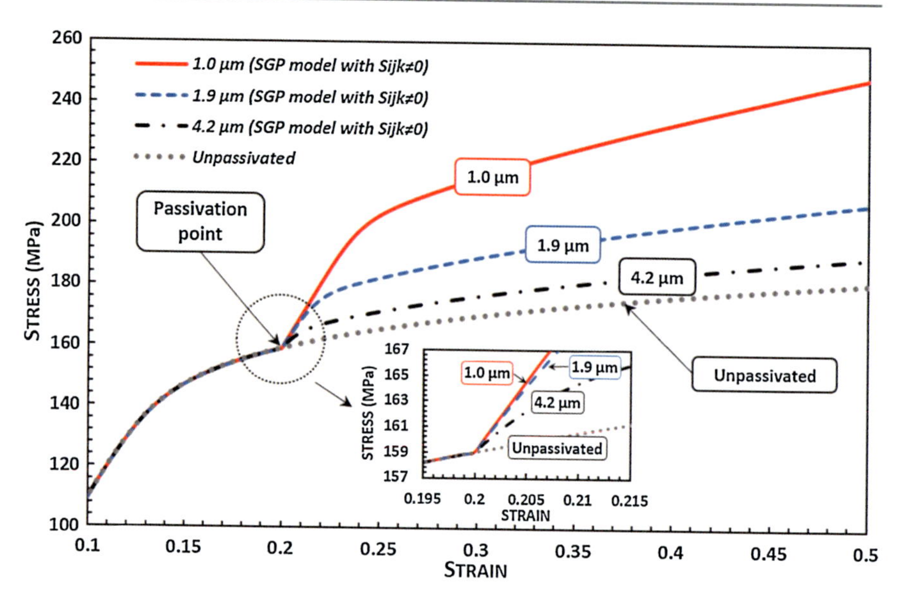

Fig. 9 A finite stress jump due to infinitesimal changes in the plastic strain. The numerical implementation of the SGP model with the dissipative potential dependent on $\dot{\varepsilon}^{p}_{ij,k}$, i.e., $\mathcal{S}^{\text{dis}}_{ijk} \neq 0$, is carried out based on the experiments of Xiang and Vlassak (2006) with the various thicknesses of the Cu thin films of 1.0 μm, 1.9 μm, and 4.2 μm (Reprinted with permission from Voyiadjis and Song 2017)

is observed at the onset of passivation. In particular, it is shown that the very first slopes immediately after the passivation increase as the film thicknesses decrease, viz. the dissipative length scales increase, in both simulations. Thus, the stress jump phenomenon is revealed to be highly correlated with the dissipative higher order microstress quantities $\mathcal{S}^{\text{dis}}_{ijk}$.

Figure 10 shows the numerical results on the stress-strain behavior of Ni thin films for the SGP model with the dissipative microstress quantities. As shown in this figure, the magnitudes of the stress jump are less than expected in all cases since the dissipative length scale ℓ_{dis} is set 0.1 which is much smaller than the energetic length scale $\ell_{\text{en}} = 1.0$. Nevertheless, the very first slopes immediately after the passivation are calculated as $E_{25^\circ\text{C}} = 58.0\,\text{GPa}$, $E_{75^\circ\text{C}} = 59.2\,\text{GPa}$, $E_{145^\circ\text{C}} = 72.6\,\text{GPa}$, and $E_{218^\circ\text{C}} = 105.0\,\text{GPa}$, respectively, and this shows the responses immediately after the passivation gets gradually closer to the elastic response $E = 115\,\text{GPa}$ as the temperature increases.

The numerical implementations to specify whether or not the stress jump phenomenon occurs under the stretch-surface passivation have been hitherto conducted within the framework of the SGP model with $\mathcal{S}^{\text{dis}}_{ijk} \neq 0$. Hereafter, the numerical simulations are given more focus on the direct comparison of the material response on the stress-strain curves between the two cases of the SGP models. The material parameters used for these implementations are presented in Table 2.

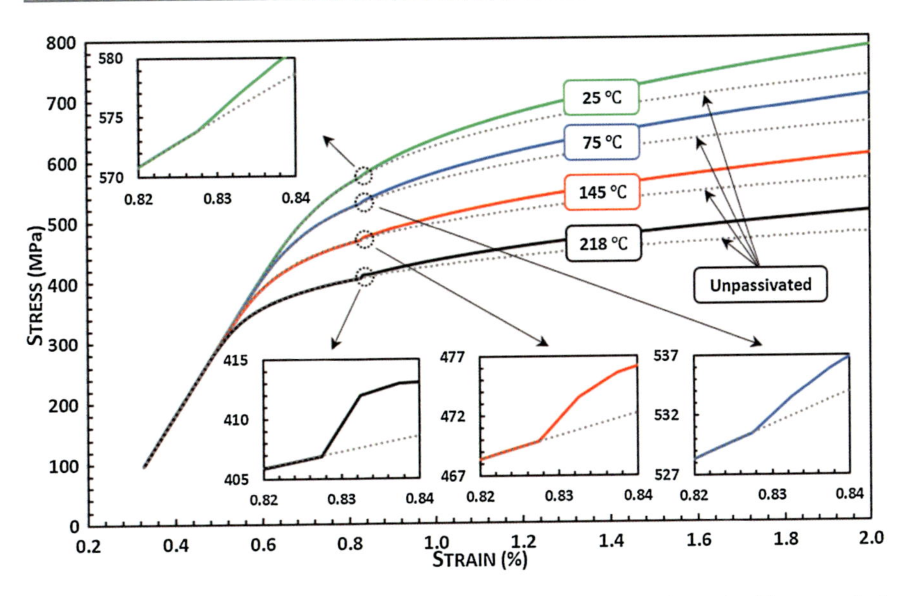

Fig. 10 A finite stress jump due to infinitesimal changes in the plastic strain. The numerical implementation of the SGP model with the dissipative potential dependent on $\dot{\varepsilon}^{p}_{ij,k}$, i.e., $S^{\text{dis}}_{ijk} \neq 0$, is carried out based on the experiments of Han et al. (2008) at the various temperatures of the Ni thin films of 25 °C, 75 °C, 145 °C, and 218 °C. The *circles* on the *curve* indicate the passivation point (Reprinted with permission from Voyiadjis and Song 2017)

Table 2 Material parameters for the numerical simulation (Reprinted with permission from Voyiadjis and Song (2017))

For grain (bulk)		
E	Elastic modulus for isotropic linear elasticity	100 (GPa)
σ_y	Yield stress	100 (MPa)
h	Hardening material parameter	200 (MPa)
$\dot{p}_0$	Reference effective plastic strain rate	0.04
r	Nonlinear hardening material parameter	0.1
m	Non-negative rate sensitivity parameter	0.3
T_0	Initial temperature	77 (K)
T_y	Thermal material parameter	1,000 (K)
n	Temperature sensitivity parameter	0.6
ρ	Density	$0.8570\ \text{g} \bullet \text{cm}^{-3}$
c_ε	Specific heat capacity at constant stress	0.265 (J/g K)
For grain boundary (interface)		
$\sigma^I{}_y$	Interfacial Yield stress	150 (MPa)
h^I	Interfacial hardening material parameter	300 (MPa)
$\dot{p}^I_0$	Interfacial reference effective plastic strain rate	0.04
n^I	Interfacial temperature sensitivity parameter	0.1
m^I	Interfacial rate sensitivity parameter	1.0
$T^I{}_y$	Interfacial thermal material parameter	700 (K)

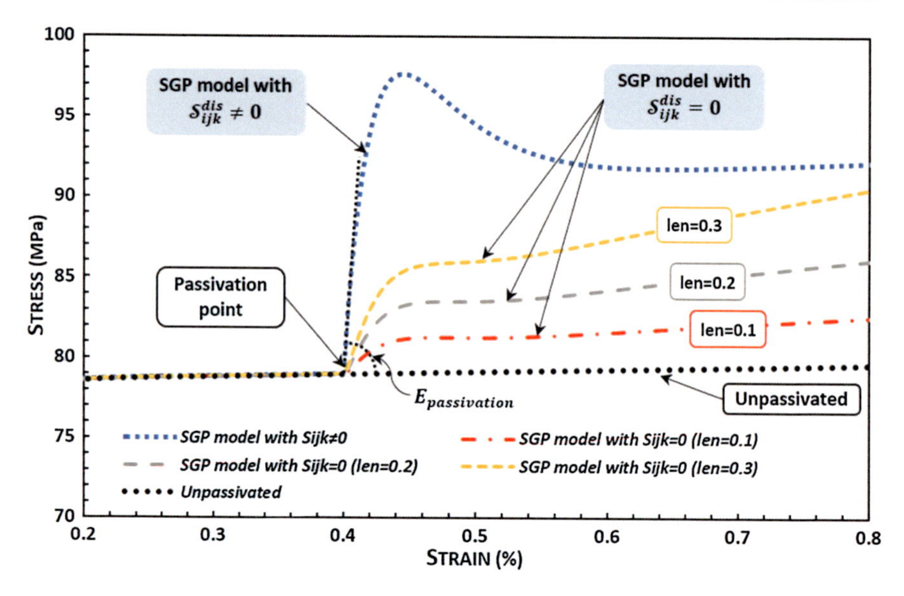

Fig. 11 Comparison of the results from two different cases of the present SGP model with the dissipation potential dependent on $\dot{\varepsilon}^{p}_{ij,k}$, i.e., $\mathcal{S}^{\text{dis}}_{ijk} \neq 0$, and with the dissipative potential independent on $\dot{\varepsilon}^{p}_{ij,k}$, i.e., $\mathcal{S}^{\text{dis}}_{ijk} = 0$. The results for the latter case are computed with three different values of energetic length scales $\ell_{\text{en}} = 0.1\,, 0.2$ and 0.3 (Reprinted with permission from Voyiadjis and Song 2017)

Figure 11 clearly shows this point by comparing the results from the two cases. The behavior of the SGP model with $\mathcal{S}^{\text{dis}}_{ijk} \neq 0$ is in stark contrast with that of the SGP case with $\mathcal{S}^{\text{dis}}_{ijk} = 0$ after the passivation point. A significant stress jump with the slope $E_{\text{passivation}}$ similar to the modulus of elasticity E is shown in the SGP case with $\mathcal{S}^{\text{dis}}_{ijk} \neq 0$. On the other hand, no elastic stress jump is observed in the SGP case with $\mathcal{S}^{\text{dis}}_{ijk} = 0$. This result is exactly in agreement with the prediction in Fleck et al. (2014, 2015). In the case that the dissipative potential is independent of $\dot{\varepsilon}^{p}_{ij,k}$, the contribution from the plastic strain gradients is entirely energetic as can be seen in section "Dissipation Potential and Dissipative Thermodynamic Microstresses." Both the increase in the yield strength in the early stages of passivation and subsequent hardening due to the effects of the plastic strain gradient are observed along with the increase of the energetic length scale as shown in Fig. 11.

The comparison of the results from the case of the present SGP model by Voyiadjis and Song (2017) with the dissipative potential dependent on $\dot{\varepsilon}^{p}_{ij,k}$ is shown in Fig. 12a with various dissipative length scales, i.e., $\ell_{\text{dis}} = 0.1\,,\ 0.3\,,\ 0.5\,,\ 1.0\,,\ 1.5\,,$ and 2.0. As can be seen in this figure, the magnitude of the stress jump significantly increases as the dissipative length scale increases, on the other hand, the stress jump phenomenon disappears as the dissipative length scale tends to zero. This is because the dissipative higher order microstress quantities $\mathcal{S}^{\text{dis}}_{ijk}$, which is the main cause of the stress jump, vanishes when the dissipative length scale ℓ_{dis} is

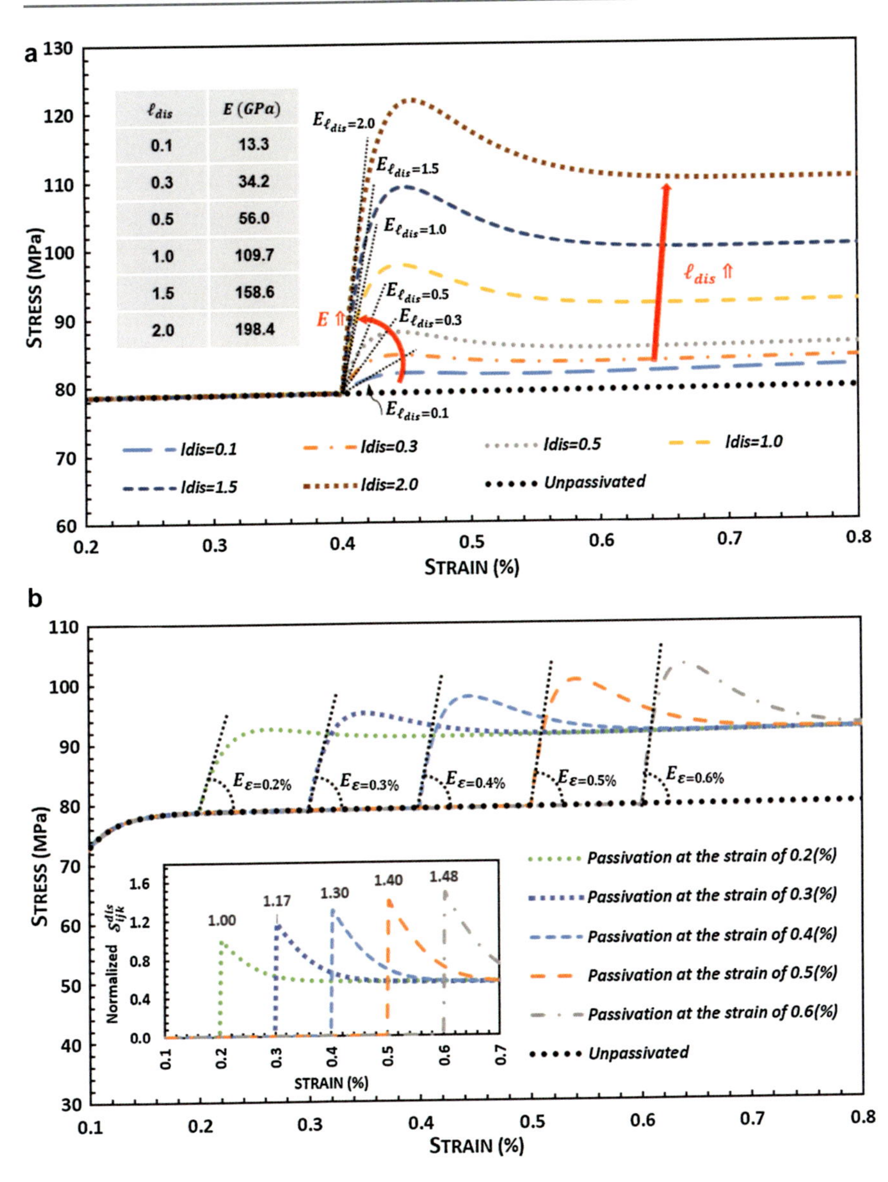

Fig. 12 Comparison of the results from the case of the present SGP model with the dissipative potential dependent on $\dot{\varepsilon}^{p}_{ij,k}$, i.e., $S^{\text{dis}}_{ijk} \neq 0$: (**a**) with various dissipative length scales ($\ell_{\text{dis}} = 0.1$, 0.3, 0.5, 1.0, 1.5, and 2.0) and (**b**) for various passivation points (ε = 20.2%, 0.3%, 0.4%, 0.5%, and 0.6% with identical energetic and dissipative length scales) (Reprinted with permission from Voyiadjis and Song 2017)

equal to zero. It is worth noticing that the very first behavior immediately after the passivation indicates a substantial difference with varying dissipative length scales. As the dissipative length scales increase from 0.1 to 2.0, the corresponding slopes of the very first response E also increase from 13.3 to 198.4 GPa. The main reason of this phenomenon in terms of the dissipative length scale is also because the dissipative higher order microstress quantities $\mathcal{S}_{ijk}^{\text{dis}}$ sharply increases with increasing ℓ_{dis}.

The comparison of the results from the case of the SGP model with the dissipative potential dependent on $\dot{\varepsilon}_{ij,k}^{p}$ is shown in Fig. 12b for various passivation points. The energetic and dissipative length scales are set identical in all cases. The magnitudes of the stress jump with the magnitudes of $\varepsilon = 20.2\%$, 0.3%, 0.4%, 0.5%, and 0.6% are obtained as 13.8 MPa, 16.2 MPa, 18.6 MPa, 21.1 MPa, and 23.5 MPa, respectively, and the values normalized by the value of the magnitude $\varepsilon = 20.2\%$ are calculated as 1.00, 1.17, 1.35, 1.52, and 1.70, respectively. The normalized higher order microstress quantities $\mathcal{S}_{ijk}^{\text{dis}}$ with the magnitudes of $\varepsilon = 20.2\%$, 0.3%, 0.4%, 0.5%, and 0.6% are obtained as 1.00, 1.17, 1.30, 1.40, and 1.48, respectively, as shown in Fig. 12b. Thus, it is worth noticing that the stress jump phenomenon is highly correlated with the dissipative higher order microstress quantities $\mathcal{S}_{ijk}^{\text{dis}}$. In addition, the very first responses immediately after the passivation also make a substantial difference with varying passivation points. The slopes of the very first responses for the magnitudes of $\varepsilon = 0.2\%$, 0.3%, 0.4%, 0.5%, and 0.6% are calculated as 64.8 GPa, 87.1 GPa, 109.7 GPa, 132.5 GPa, and 155.5 GPa, respectively.

The effects of various parameters on the mechanical behavior of the stretch-surface passivation problem are investigated by using a one-dimensional finite element. The numerical results reported in this parametric study are obtained by using the values of the material parameters in Table 2 unless it is differently mentioned.

The stress-strain graphs for various values of the hardening material parameter h are shown in Fig. 13a with two different cases of the SGP model. For the case with $\mathcal{S}_{ijk}^{\text{dis}} \neq 0$, the slopes of the very first response immediately after the passivation are obtained as $E_h = 109.7$ GPa in all simulations, and the corresponding magnitudes of the stress jump for each simulation are, respectively, obtained as $\sigma'_{h=100\text{MPa}} = 18.6$ MPa, $\sigma'_{h=200\text{MPa}} = 18.6$ MPa, $\sigma'_{h=300\text{MPa}} = 18.7$ MPa, $\sigma'_{h=400\text{MPa}} = 18.7$ MPa, and $\sigma'_{h=500\text{MPa}} = 18.7$ MPa. There is little difference between all the simulations. For the case with $\mathcal{S}_{ijk}^{\text{dis}} = 0$, no stress jump phenomena are observed in all simulations.

The effects of the nonnegative rate sensitivity parameter m on the stress-strain behavior for the two cases of the SGP model are represented in Fig. 13b. It is clearly shown in this figure that by increasing the rate sensitivity parameter, the stress jump phenomena are significantly manifested in the case of the SGP model with $\mathcal{S}_{ijk}^{\text{dis}} \neq 0$ in terms of both the slope of the very first response immediately after the passivation and the corresponding magnitude of the stress jump. In the case of the SGP model with $\mathcal{S}_{ijk}^{\text{dis}} = 0$, on the other hand, the material behavior is not affected a lot by the rate sensitivity parameter m. In the SGP model with $\mathcal{S}_{ijk}^{\text{dis}} \neq 0$, the slopes of the very first response immediately after the passivation (E_m) and the corresponding magnitudes of the stress jump (σ'_m) are obtained as $E_{m=0.2} = 96.7$ GPa, $E_{m=0.25} = 103.0$ GPa, $E_{m=0.3} = 109.7$ GPa, $E_{m=0.35} = 116.5$ GPa and $\sigma'_{m=0.2} = 14.8$ MPa, $\sigma'_{m=0.25} = 16.7$ MPa, $\sigma'_{m=0.3} = 18.6$ MPa, $\sigma'_{m=0.35} = 20.6$ MPa, respectively. Thus, it is clearly shown that both the slope of the very first response immediately after the passivation and the corresponding magnitude of the stress jump increase as the nonnegative rate sensitivity parameter m increases.

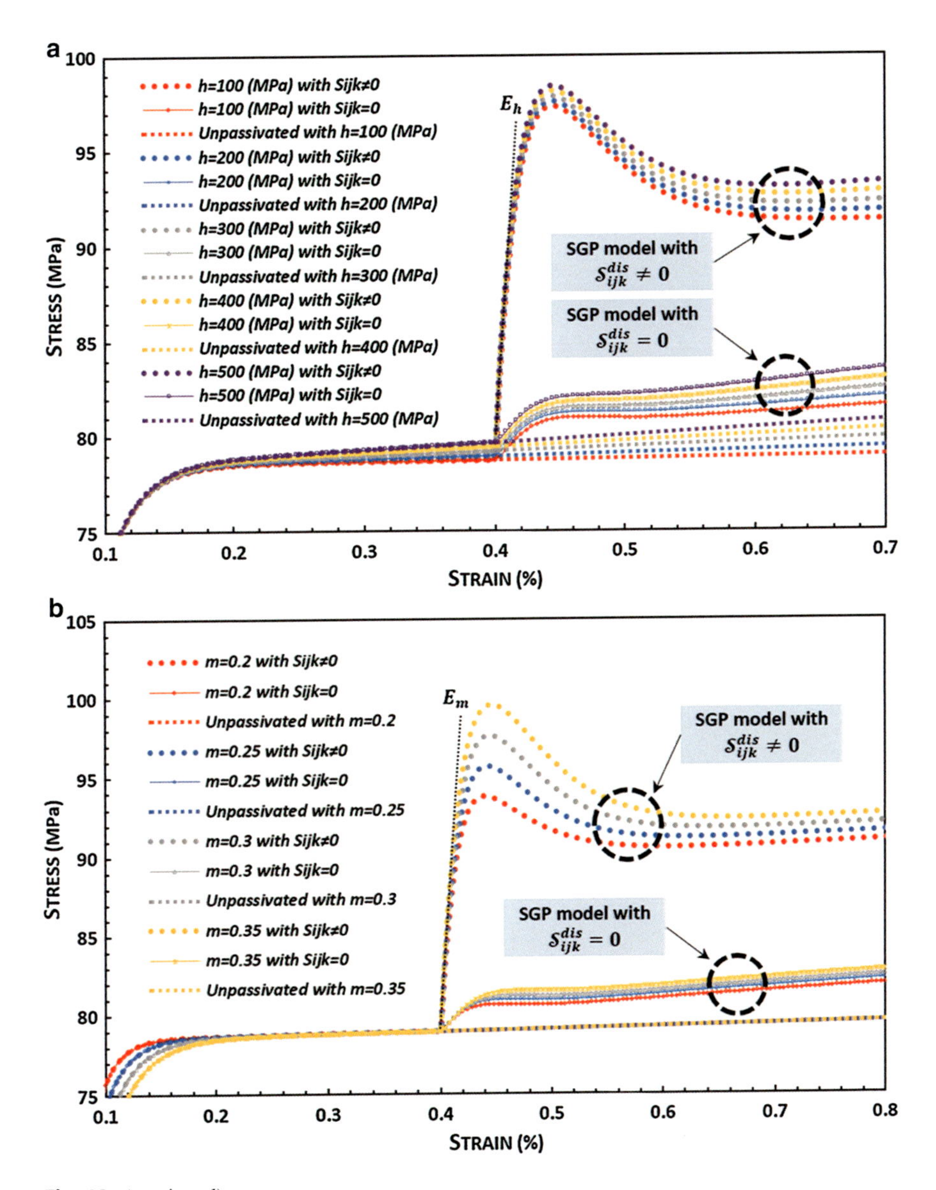

Fig. 13 (continued)

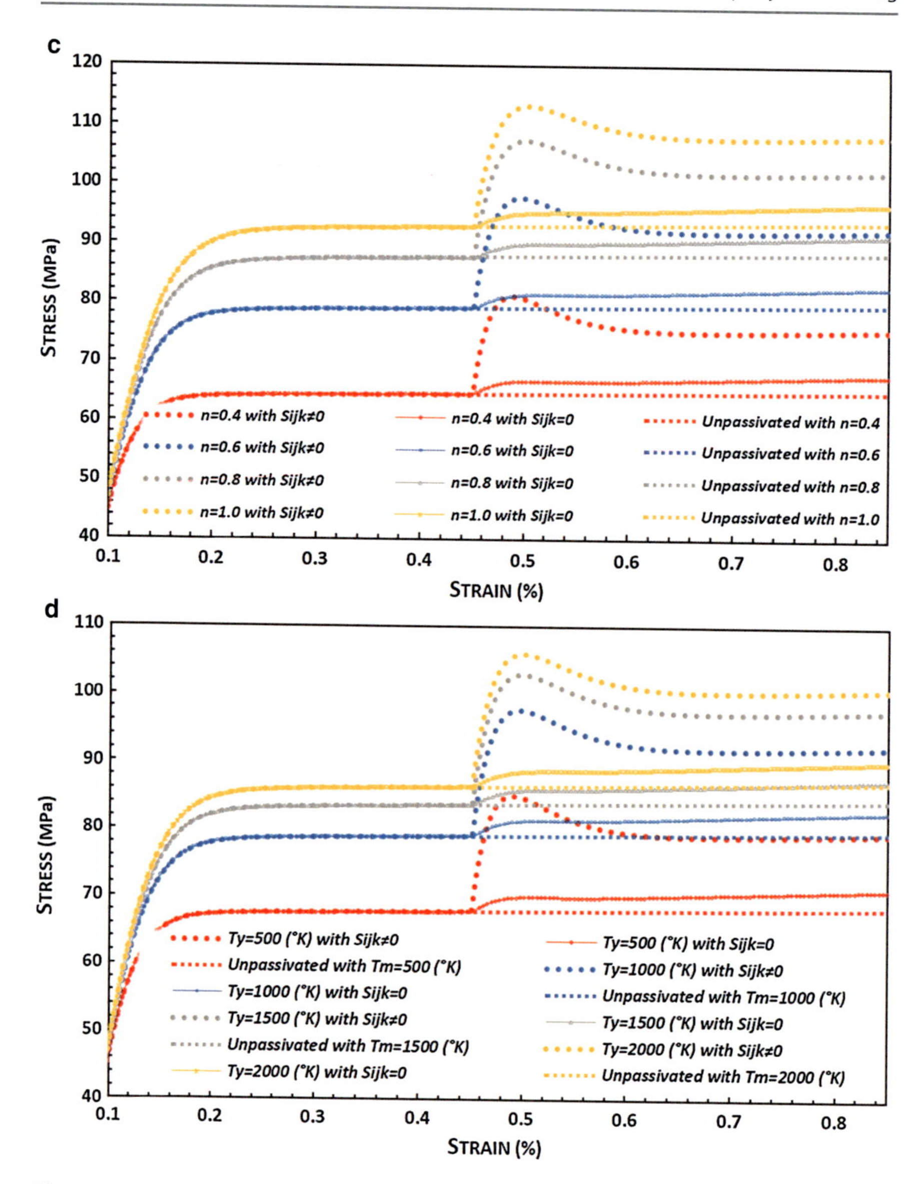

Fig. 13 (continued)

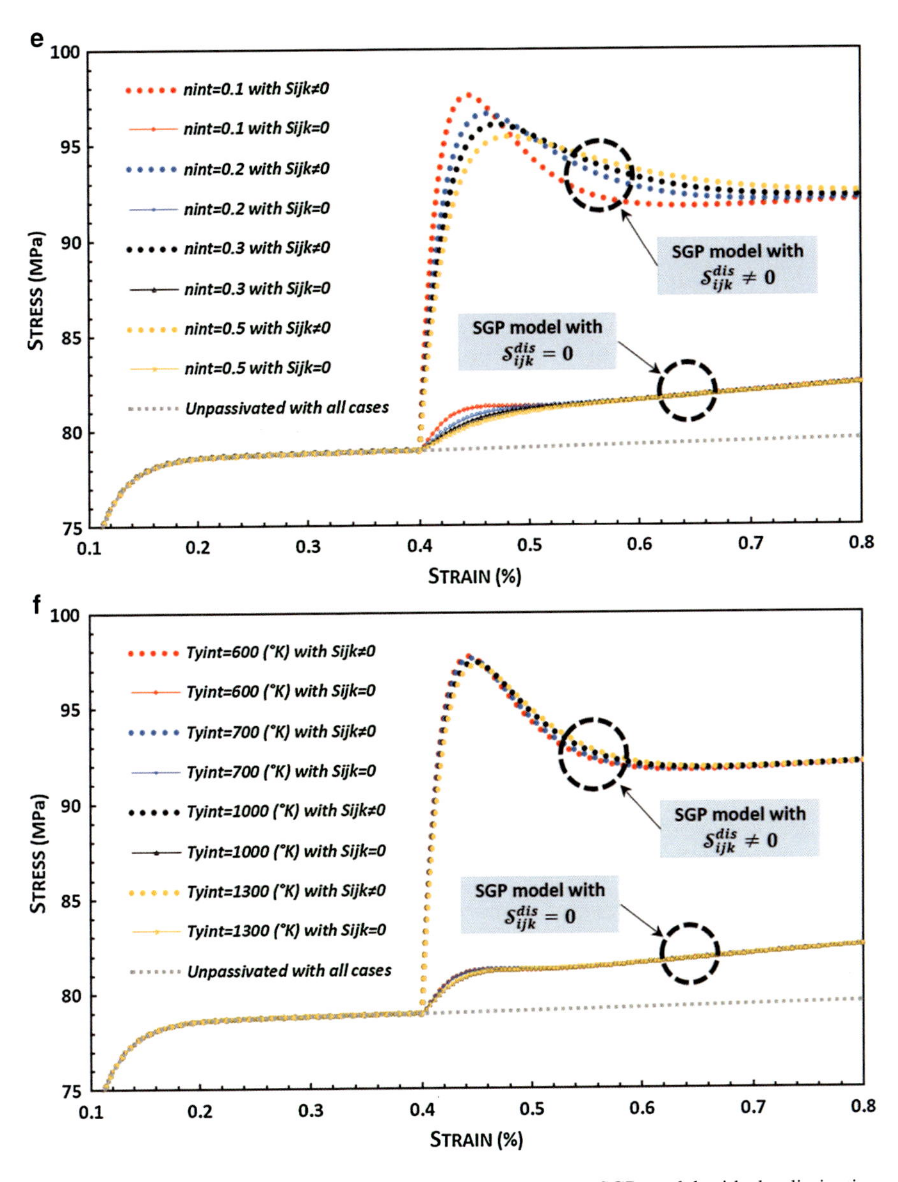

Fig. 13 Comparison of the results from the case of the present SGP model with the dissipative potential dependent on $\dot{\varepsilon}_{ij,k}^{p}$, i.e., $\mathcal{S}_{ijk}^{\text{dis}} \neq 0$ with the effects of: **a** the hardening material parameter h (100 MPa, 200 MPa, 300 MPa, 400 MPa, and 500 MPa), **b** the nonnegative rate sensitivity parameter m (0.2, 0.25, 0.3, and 0.35), **c** the temperature sensitivity parameter n (0.4, 0.6, 0.8, and 1.0), **d** the thermal material parameter T_y (500°K, 1,000°K, 1,500°K, and 2,000°K), **e** the interfacial temperature sensitivity parameter n^I (0.1, 0.2, 0.3 and 0.5), **f** the interfacial thermal material parameter T_y^I (600°K, 700°K, 1,000°K, and 11,300°K) (Reprinted with permission from Voyiadjis and Song 2017)

The effects of the temperature sensitivity parameter n on the stress-strain behavior for the two cases of the SGP model are represented in Fig. 13c. It is clearly shown in this figure that the yield stress significantly increases as the temperature sensitivity parameter n increases, while the strain hardening is not influenced a lot by this parameter. This is because the temperature affects the strain hardening mechanism through the dislocation forest barriers, while the backstress, i.e., energetic gradient hardening, is almost independent of the temperature. Meanwhile, the temperature sensitivity parameter n significantly affects the stress-strain response in the case of the SGP model with $\mathcal{S}_{ijk}^{\text{dis}} \neq 0$. In this case, the slopes of the very first response immediately after the passivation are obtained as $E_{n=0.4} = 112.5$ GPa, $E_{n=0.6} = 109.7$ GPa, $E_{n=0.8} = 107.8$ GPa, and $E_{n=1.0} = 106.6$ GPa, and the corresponding magnitudes of the stress jump are obtained as $\sigma'_{n=0.4} = 16.6$ MPa, $\sigma'_{n=0.6} = 18.6$ MPa, $\sigma'_{n=0.8} = 19.7$ MPa, and $\sigma'_{n=1.0} = 20.4$ MPa, respectively. Thus, in contrast with the rate sensitivity parameter m, the slope of the very first response immediately after the passivation decreases, while the corresponding magnitude of the stress jump increases as the temperature sensitivity parameter n increases.

The effects of the thermal material parameter T_y on the stress-strain behavior with the two cases of the SGP model are represented in Fig. 13d. Similar to the temperature sensitivity parameter n, the yield stress significantly increases as the thermal material parameter T_y increases, while the strain hardening is not influenced a lot by this parameter. In the case of the SGP model with $\mathcal{S}_{ijk}^{\text{dis}} \neq 0$, the slopes of the very first response immediately after the passivation (E_{T_y}) and the corresponding magnitudes of the stress jump (σ'_{T_y}) are obtained as $E_{T_y=500^\circ K} = 111.9$ GPa, $E_{T_y=1000^\circ K} = 109.7$ GPa, $E_{T_y=1500^\circ K} = 108.7$ GPa, $E_{T_y=2000^\circ K} = 108.1$ GPa and $\sigma'_{T_y=500^\circ K} = 17.1$ MPa, $\sigma'_{T_y=1000^\circ K} = 18.6$ MPa, $\sigma'_{T_y=1500^\circ K} = 19.2$ MPa, $\sigma'_{T_y=2000^\circ K} = 19.6$ MPa, respectively. These results from the simulations with various thermal material parameter T_y are very similar to those with the temperature sensitivity parameter n, in the sense that the slope of the very first response immediately after the passivation decreases, while the corresponding magnitude of the stress jump increases as the temperature sensitivity parameter n increases.

The effects of the interfacial temperature sensitivity parameter n^I on the stress-strain behavior for the two cases of the SGP model are presented in Fig. 13e. It is clearly shown in this figure that increasing interfacial temperature sensitivity parameter makes the grain boundary (interface) harder and results in less variation of the stress jump in both cases of the SGP model with $\mathcal{S}_{ijk}^{\text{dis}} \neq 0$ and $\mathcal{S}_{ijk}^{\text{dis}} = 0$. In the case of the SGP model with $\mathcal{S}_{ijk}^{\text{dis}} \neq 0$, the slopes of the very first response immediately after the passivation are obtained as $E_{n^I=0.1} = 109.7$ GPa, $E_{n^I=0.2} = 80.4$ GPa, $E_{n^I=0.3} = 70.3$ GPa, and $E_{n^I=0.5} = 62.1$ GPa, and the corresponding magnitudes of the stress jump are obtained as $\sigma'_{n^I=0.1} = 18.6$ MPa, $\sigma'_{n^I=0.2} = 17.6$ MPa, $\sigma'_{n^I=0.3} = 17.1$ MPa, and $\sigma'_{n^I=0.5} = 16.4$ MPa, respectively. From these results, it is easily observed that the slope of the very first response after the passivation and the corresponding magnitude of the stress jump increase as the interfacial temperature sensitivity parameter n^I decreases. In addition, the variations along with the different cases are shown to be more drastic by decreasing the interfacial temperature sensitivity parameter n^I. Similarly, for the case of the SGP model with $\mathcal{S}_{ijk} = 0$, decreasing the interfacial temperature sensitivity parameter makes the variation more radically as shown in Fig. 13e.

The effects of the interfacial thermal material parameter T_y^I on the stress-strain behavior for the two cases of the SGP model are represented in Fig. 13f. As can be seen in this figure, the overall characteristic from the simulation results with various T_y^I is similar to those with various n^I in the sense that increasing T_y^I makes the grain boundary (interface) harder and results in less variation of the stress jump in both cases of the SGP model with $\mathcal{S}_{ijk}^{\text{dis}} \neq 0$ and $\mathcal{S}_{ijk}^{\text{dis}} = 0$. In the case of the SGP model with $\mathcal{S}_{ijk}^{\text{dis}} \neq 0$, the slopes of the very first response immediately after the passivation are obtained as $E_{T_y^I=600^\circ K} = 114.0$ GPa, $E_{T_y^I=700^\circ K} = 109.7$ GPa, $E_{T_y^I=1000^\circ K} = 101.7$ GPa, and $E_{T_y^I=1300^\circ K} = 97.1$ GPa, and the corresponding magnitudes of the stress jump are obtained as $\sigma'_{T_y^I=600^\circ K} = 18.7$ MPa, $\sigma'_{T_y^I=700^\circ K} = 18.6$ MPa, $\sigma'_{T_y^I=1000^\circ K} = 18.4$ MPa, and $\sigma'_{T_y^I=1300^\circ K} = 18.3$ MPa, respectively.

Based on the calibrated model parameters of Ni (Table 1), the evolution of the various potentials during the plastic deformation are investigated with four different temperatures, i.e., 25 °C, 75 °C, 145 °C, and 218 °C, in this section. Figure 14 shows (a) the variation of the plastic strain dependent free energies (Ψ_1^d, $\Psi_1^{d,R}$and$\Psi_1^{d,\text{NR}}$), (b) plastic strain dependent dissipation rate $\mathcal{D}_1$ (i.e., plastic strain dependent term in Eq. 58), (c) plastic strain gradient-dependent free energy (Ψ_2^d) and dissipation rate $\mathcal{D}_2$ (i.e., plastic strain gradient dependent term in Eq. 58), and (d) amount of stored energy ($\Psi_1^d + \Psi_2^d$) and dissipated energy ($\mathbb{D} = \int(\mathcal{D}_1 + \mathcal{D}_2)\text{dt}$).

As can be seen in Fig. 14a, both free energies Ψ_1^d and $\Psi_1^{d,\text{NR}}$ decrease as the temperature increases. The stored energy of cold work $\Psi_1^{d,R}$ is also presented in this figure. One can observe that Ψ_1^d and $\Psi_1^{d,\text{NR}}$ have a strong temperature dependency, while the stored energy of cold work has no variation with varying temperatures. In addition, the stored energy of cold work tends to saturate after some critical point since the stored energy of cold work is proportional to the dislocation density that remains constant after the aforementioned critical point. Meanwhile, it is shown in Fig. 14b,c that rates of dissipation, $\mathcal{D}_1$ and $\mathcal{D}_2$, are dependent on the temperature such that both $\mathcal{D}_1$ and $\mathcal{D}_2$ increase with decreasing temperatures while Ψ_2^d shows no temperature dependency. The amount of stored and dissipated energies is shown in Fig. 14d. As it is shown in this figure, the amount of stored energy is larger than the dissipated energy and both decrease as the temperature increases.

Lastly, the one-dimensional finite element implementation is extended to the two-dimensional one. The simple tension problem of the square plate is solved to study the strain gradient effects, the mesh sensitivity of the model with the three cases according to the number of elements (100, 400, and 1,600 elements), and the stress jump phenomena under the abrupt surface passivation. Each edge of the plate has a length of w and the material parameters in Table 2 are used again for these simulations.

Figure 15a, b shows the stress-strain behavior of the plate with the various energetic length scales ($\ell_{\text{en}}/w = 0.0$, 0.1, 0.3, 0.5, 0.7, and 1.0 with $\ell_{\text{dis}}/w = 0.0$) and dissipative length scales ($\ell_{\text{dis}}/w = 0.0$, 0.1, 0.2, 0.3, 0.4,

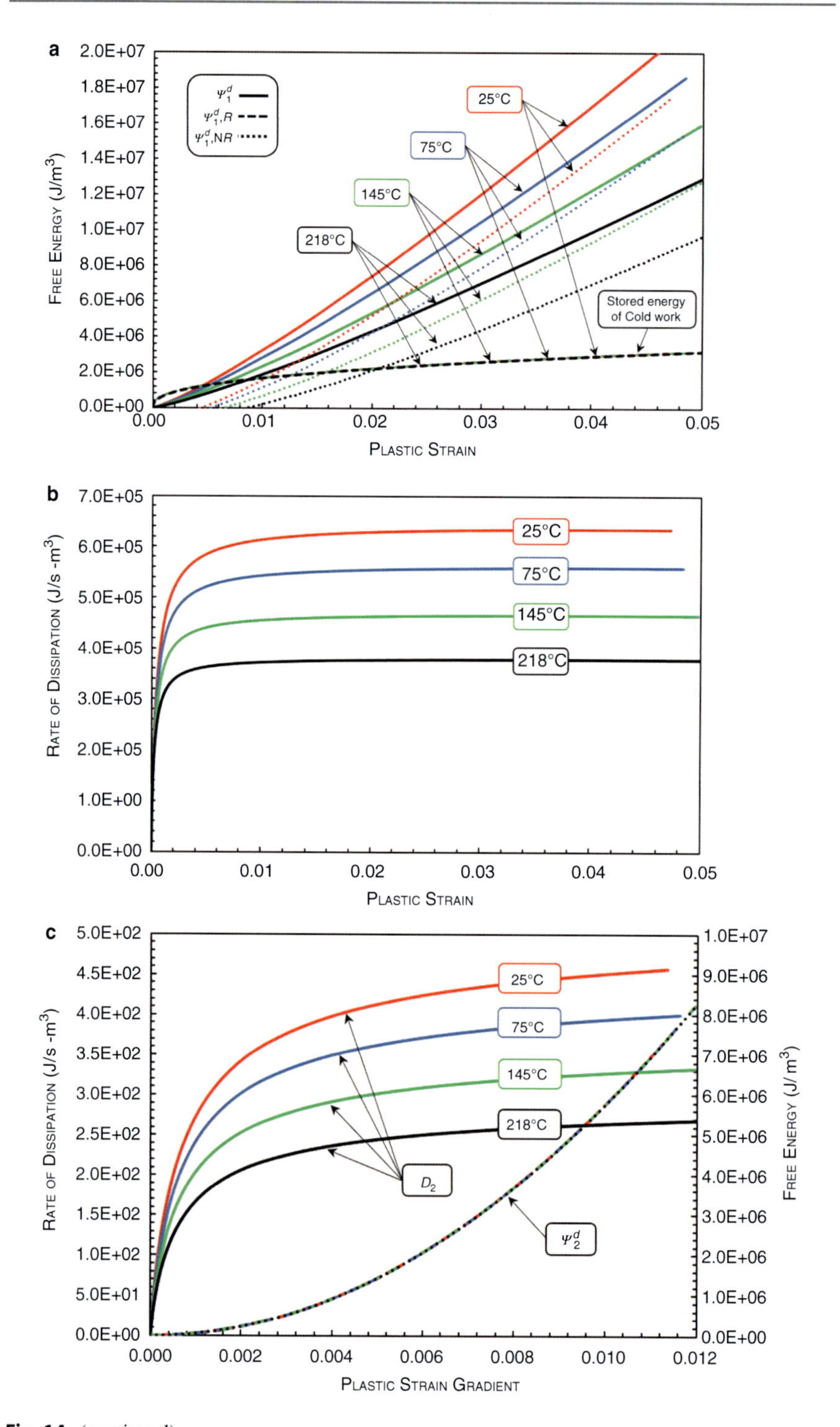

Fig. 14 (continued)

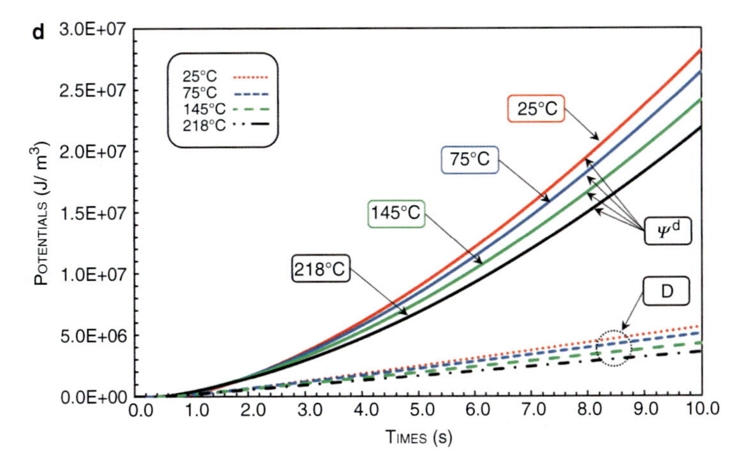

Fig. 14 The evolution of free energy and dissipation potentials based on the calibrated model parameters (Han et al. 2008): (**a**) plastic strain-dependent free energy, (**b**) plastic strain-dependent dissipation rate, (**c**) plastic strain gradient-dependent free energy and dissipation rate (The primary *y*-axis on the LHS: $\mathcal{D}_2$, the secondary *y*-axis on the RHS: Ψ_2^d), and (**d**) amount of stored and dissipated energies (Reprinted with permission from Voyiadjis and Song 2017)

and 0.5 with $\ell_{en}/w = 0.0$), respectively. In common with the one-dimensional simulations, the numerical results show the energetic hardening as the energetic length scale increases as well as the dissipative strengthening as the dissipative length scale increases.

Figure 15c shows the grain boundary effect of the square plate. It is well known that the grain boundary blocks the dislocation movement, which, in turn, leads to the strengthening of the material. The energetic and dissipative length scales reported in these simulations are set zero and the microclamped condition is imposed at the grain boundary, which is indicated by the bold line in the figure. The strengthening caused by increasing the grain boundary area is well observed as expected. For the simulations presented in Fig. 15a–c, 20 × 20 elements are used.

The mesh sensitivity of the two-dimensional numerical model is examined in terms of the energetic and dissipative length scales with 10 × 10, 20 × 20 and 40 × 40 mesh elements in Fig. 15d, e, respectively. Figure 15d shows the stress-strain behavior of the plate with the various energetic length scales ($\ell_{en}/w = 0.1,\ 0.5$ and 1.0 with $\ell_{dis}/w = 0$) compared to those in the absence of the gradient effects ($\ell_{en}/w = \ell_{dis}/w = 0$). The numerical results without the gradient terms significantly show the mesh sensitivity as expected, while for all nonzero values of ℓ_{en}/w, the numerical solutions show the mesh-independent behavior. In addition, Fig. 15d also shows the energetic hardening as the energetic length scale increases. The mesh-independent behavior is also observed with varying dissipative length scales $\ell_{dis}/w = 0.1,\ 0.2$ and 0.3 with $\ell_{en}/w = 0$ in Fig. 15e as with the case for the energetic length scale. The dissipative strengthening is also observed in this figure.

The stress jump phenomenon hitherto extensively studied in the one-dimensional finite element implementation is also examined for the two-dimensional simulation;

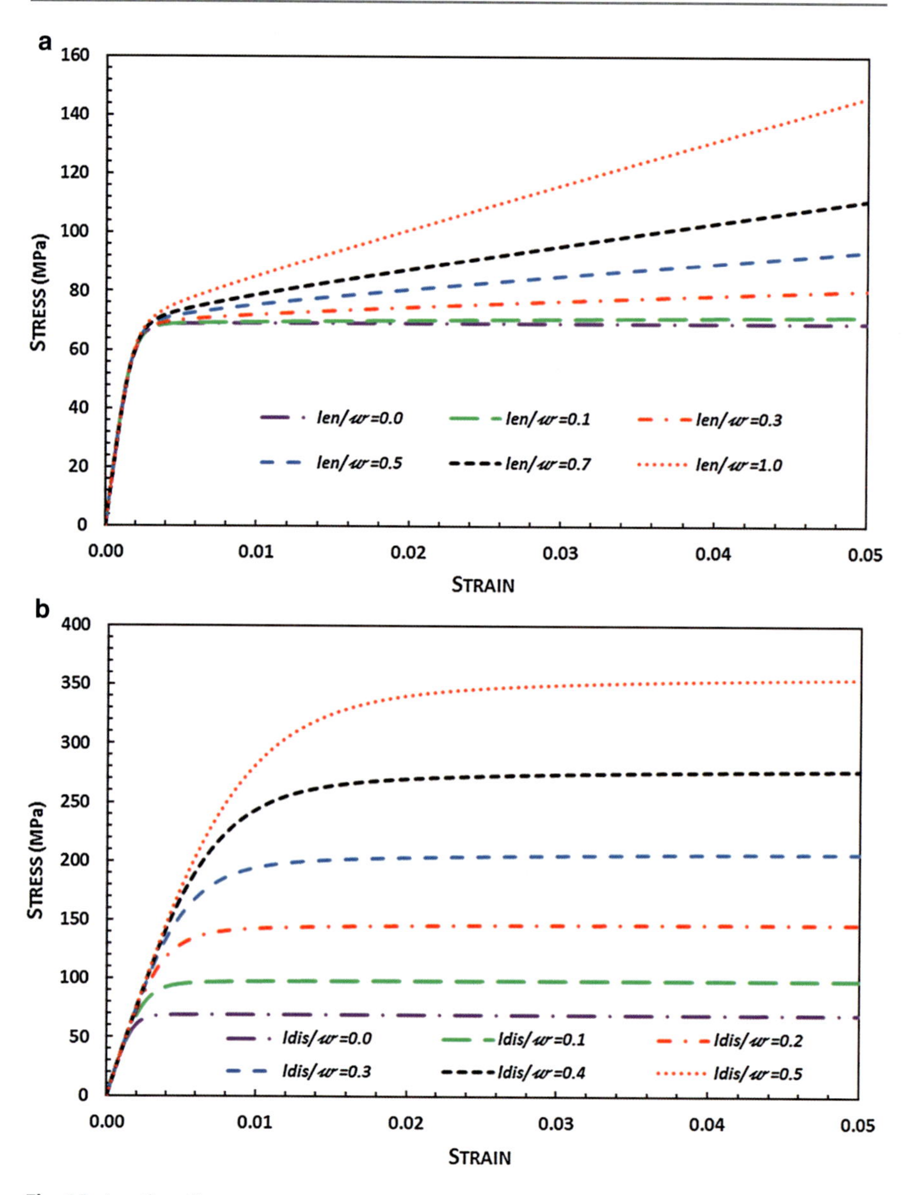

Fig. 15 (continued)

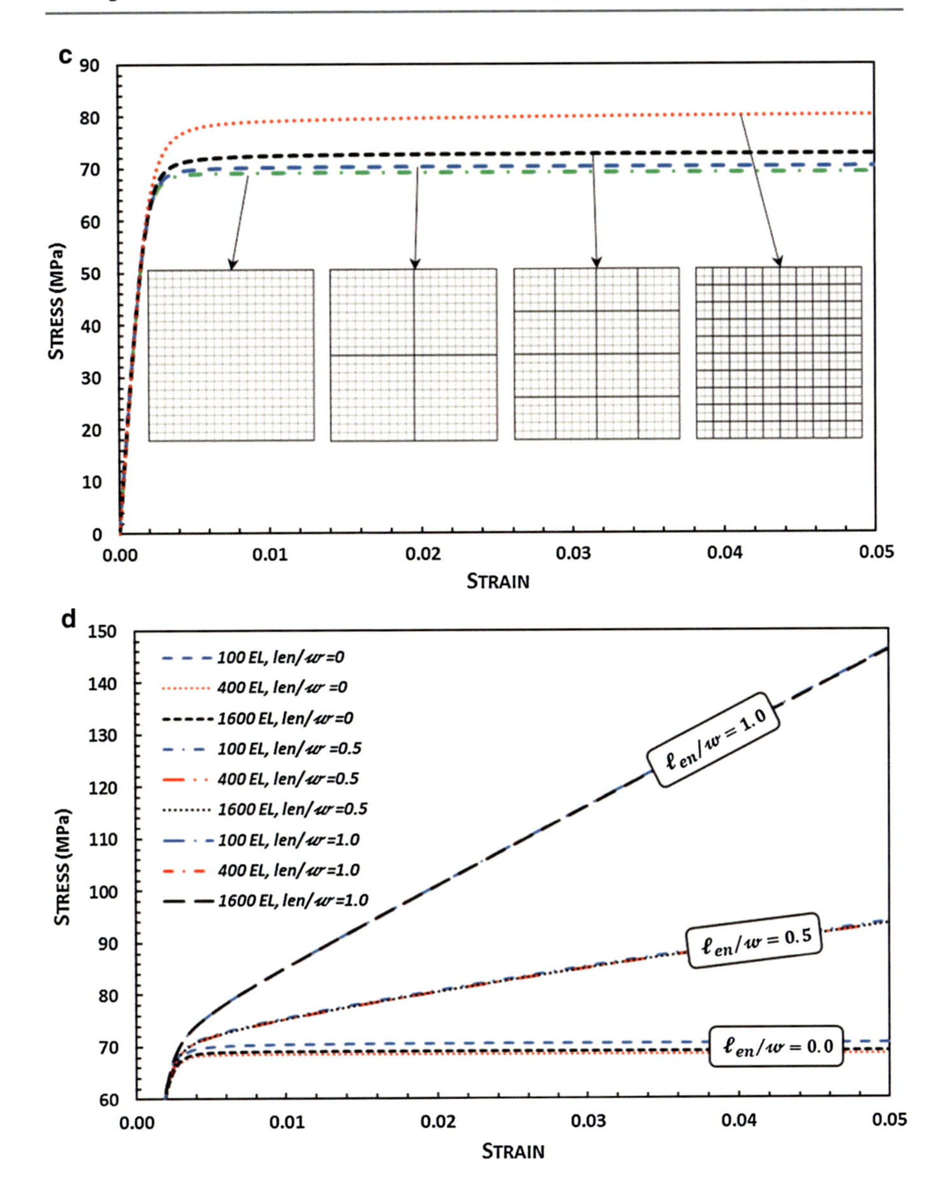

Fig. 15 (continued)

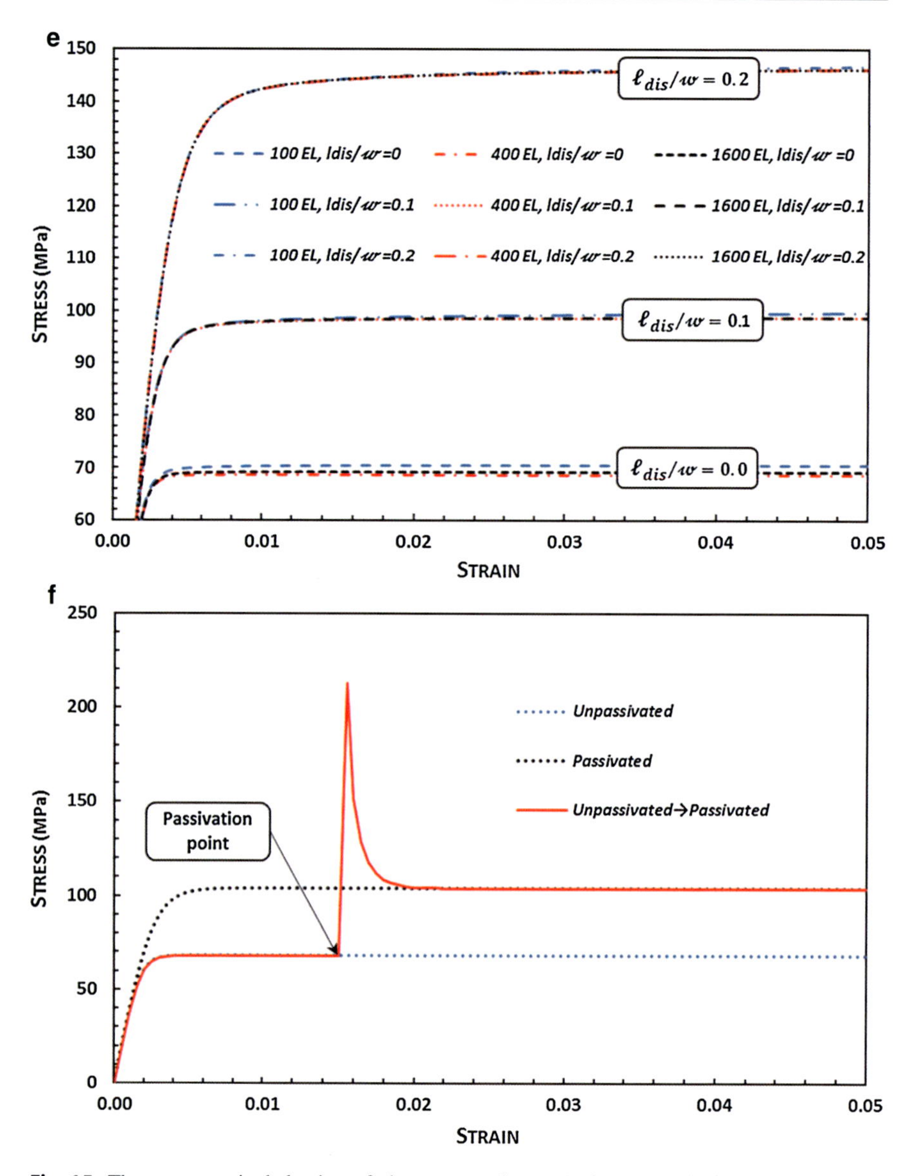

Fig. 15 The stress-strain behavior of the square plate: (**a**) the energetic hardening, (**b**) the dissipative strengthening, (**c**) the grain boundary effect, (**d**) the mesh sensitivity with varying ℓ_{en}, (**e**) the mesh sensitivity with varying ℓ_{dis}, and (**f**) the stress jump phenomenon (Fig. 15c, f: Reprinted with permission from Voyiadjis and Song (2017))

20×20 elements with $\ell_{\rm en}/w = 0.0$ and $\ell_{\rm dis}/w = 0.2$ are used for this simulation. Figure 15f shows the material behaviors of the unpassivated plate, the passivated plate, and the plate abruptly passivated at some point. As can be seen in this figure, the stress jump is significantly observed by the numerical results, which are identical to the one-dimensional case.

Conclusions

In this chapter, a phenomenological thermodynamic-based higher order gradient plasticity theory is presented and applied to the stretch-surface passivation problem for investigating the material behavior under the nonproportional loading condition. The thermodynamic potentials such as the Helmholtz free energy and the dissipation potential are established based on the concepts of the dislocation interaction mechanism and the thermal activation energy. The microstructural interface effect between two grains is also incorporated into the formulation, such that the present interfacial flow rule is able to account for the energy storage at the interface caused by the dislocation pile up as well as the energy dissipation through the interface caused by the dislocation transfer. The formulation is tested for two cases in the presence of the dissipative higher order microstress quantities $\mathcal{S}_{ijk}^{\rm dis}$. In the first case, the dissipation potential is dependent on the gradients of the plastic strain rate $\dot{\varepsilon}_{ij,k}^{p}$; as a result, $\mathcal{S}_{ijk}^{\rm dis}$ does not have a value of zero in this formulation. In the second case the dissipation potential is independent of $\dot{\varepsilon}_{ij,k}^{p}$, which in turn, $\mathcal{S}_{ijk}^{\rm dis}$ does not exist. It is noticed by Fleck et al. (2014, 2015) that $\mathcal{S}_{ijk}^{\rm dis}$ always gives rise to the stress jump phenomenon, which causes a controversial dispute in the field of SGP theory with respect to whether it is physically acceptable or not, under the nonproportional loading condition.

Prior to exploring the effect of the dissipative higher order microstress quantities $\mathcal{S}_{ijk}^{\rm dis}$ on the stress-strain behavior for the two cases of the SGP model with and without $\mathcal{S}_{ijk}^{\rm dis}$, the present model and corresponding finite element code by Voyiadjis and Song (2017) are validated by comparing with three sets of small-scale experiments. Particularly, each set of three experiments involving Al, Cu, and Ni are selected, respectively, to examine the applicability of the present finite element implementation to the various kinds of materials. The first experiment, which was performed by Haque and Saif (2003), is the uniaxial tensile testing of nanoscale freestanding Al thin films to explore the effect of strain gradient in 100 nm, 150 nm, and 485 nm thick specimens with average grain size of 50 nm, 65 nm, and 212 nm, respectively. The results clearly show the size effect on the stress-strain curves of the Al thin films; in addition, the calculated results of the present SGP model display a tendency to be more coincident to the experimental data than those of Voyiadjis and Faghihi (2014). For the second experimental validation, the experimental work of Xiang and Vlassak (2006) on the size effect in electroplated Cu thin films with various microscale thicknesses is selected since the effect of passivation on the stress-strain behavior of the Cu thin film is also considered in their work. The stress-strain curves from the numerical results of the present SGP (Voyiadjis and Song 2017) model are in good agreement with the experimental measurements. The size effect according to the variation of the film thicknesses is also well observed from the results. For the third experimental validation, the microtensile test on the temperature effect on Ni thin films by Han et al. (2008) is employed since, in the nano- or microsystems, the effect of the higher order gradient on temperature also needs to be considered for the fast transient behavior. From both the experimental and numerical results, it is shown that the Young's modulus is not affected by the variations in temperature, while the yield and tensile strength decrease as the specimen temperature increases. The calculated results of the present model (Voyiadjis and Song 2017) compare better to the experimental data than those of Voyiadjis and Faghihi (2014).

The numerical simulation part is largely composed of four subparts. The main purpose of the first part is to examine the occurrence of the stress jump phenomenon under the stretch-surface passivation condition in conjunction with the aforementioned three experiments. In all simulations, a stress jump is clearly observed at the onset of passivation. The second part is carried out in order to clearly show the results to be contrary to each other from the two cases of the SGP model. The conclusion in this part is drawn such that a significant stress jump with the slope $E_{\text{passivation}}$ similar to the modulus of elasticity E is shown in the case of the SGP model with $\mathcal{S}_{ijk}^{\text{dis}} \neq 0$, on the other hand, no elastic stress jump is observed in the case of the SGP model with $\mathcal{S}_{ijk}^{\text{dis}} = 0$. This result is exactly in agreement with the predictions in Fleck et al. (2014, 2015).

In the third part, an extensive parametric study is presented in terms of the effects of the dissipative length scale ℓ_{dis}, the onset point of passivation, the hardening material parameter h, the nonnegative rate sensitivity parameter m, the temperature sensitivity parameter n, the thermal material parameter T_y, the interfacial temperature sensitivity parameter n^I, and the interfacial thermal material parameter T_y^I on the stress-strain response for the two SGP cases, respectively. There are a number of conclusions worth mentioning here, namely: (1) the magnitude of the stress jump significantly increases as the dissipative length scale increases, on the other hand, the stress jump phenomenon disappears as the dissipative length scale comes closer to zero, (2) the slopes of the very first response E immediately after the passivation also increase as the dissipative length scales increase, (3) the stress jump phenomenon is highly correlated with the dissipative higher order microstress quantities $\mathcal{S}_{ijk}^{\text{dis}}$, in addition, the very first responses immediately after the passivation also make a substantial difference with varying passivation points, (4) the hardening material parameter h does not affect the stress jump significantly in the case of the SGP model with $\mathcal{S}_{ijk}^{\text{dis}} \neq 0$, (5) both the slope of the very first response immediately after the passivation and the corresponding magnitude of the stress jump substantially increase as the nonnegative rate sensitivity parameter m increases in the case of the SGP model with $\mathcal{S}_{ijk}^{\text{dis}} \neq 0$, (6) as the temperature-related parameters for the bulk such as the temperature sensitivity parameter n and the thermal material parameter T_y increase, the slope of the very first response immediately after the passivation decreases and the corresponding magnitude of the stress jump increases in the case of the SGP model with $\mathcal{S}_{ijk}^{\text{dis}} \neq 0$, (7) the slope of the very first response after the passivation, the corresponding magnitude of the stress jump increases and the variations along with the cases are shown more drastically by decreasing the temperature-related parameters for the interface, such as the interfacial temperature sensitivity parameter n^I and the interfacial thermal material parameter T_y^I in the case of the SGP model with $\mathcal{S}_{ijk}^{\text{dis}} \neq 0$, and finally (8) no stress jump is observed in all cases with $\mathcal{S}_{ijk}^{\text{dis}} = 0$.

Meanwhile, the plastic strain-dependent free energy accounting for the stored energy of cold work is derived in this chapter by assuming that the stored energy is related to the energy carried by dislocations. Accordingly, the variation of free energies and dissipation potentials during the plastic deformation are investigated with four different temperatures, i.e. 25 °C, 75 °C, 145 °C, and 218 °C. From the numerical results, it is shown that the stored energy of cold work has no temperature dependency; in addition, the stored energy of cold work tends to saturate after some critical point since the stored energy of cold work is proportional to the dislocation density that remains constant after the aforementioned critical point.

Lastly, the two-dimensional tension problem of the square plate (Voyiadjis and Song 2017) is solved to examine the mesh sensitivity of the model. The effects of the strain gradient and grain boundary are also studied. As expected, a strong mesh-dependence stress-strain behavior is observed in the case of no gradient effects, while the numerical results with the gradient effects show the mesh-independent behavior. The energetic hardening, the dissipative strengthening, the grain boundary strengthening, and the stress jump phenomena are well observed in common with the results from the one-dimensional simulation.

References

Abaqus, *User's Manual (Version 6.12)* (Dassault Systemes Simulia Corp., Providence, 2012)
E.C. Aifantis, J. Eng. Mater-T. ASME **106**, 4 (1984)
K.E. Aifantis, J.R. Willis, J. Mech. Phys. Solids **53**, 5 (2005)
B.D. Coleman, M.E. Gurtin, J. Chem. Phys. **47**, 2 (1967)
N.A. Fleck, J.W. Hutchinson, J. Mech. Phys. Solids **49**, 10 (2001)
N.A. Fleck, J.R. Willis, J. Mech. Phys. Solids **57**, 1 (2009a)
N.A. Fleck, J.R. Willis, J. Mech. Phys. Solids **57**, 7 (2009b)
N.A. Fleck, J.R. Willis, Acta Mech. Sinica **31**, 4 (2015)
N.A. Fleck, J.W. Hutchinson, J.R. Willis, P. Roy, Soc. A Math. Phys. **470**, 2170 (2014)
N.A. Fleck, J.W. Hutchinson, J.R. Willis, J. Appl. Mech-T. ASME **82**, 7 (2015)
S. Forest, M. Amestoy, C. R. Mecanique **336**, 4 (2008)
S. Forest, R. Sievert, Acta Mech. **160**, 1–2 (2003)
P. Fredriksson, P. Gudmundson, Mat. Sci. Eng. A Struct. **400** (2005)
P. Gudmundson, J. Mech. Phys. Solids **52**, 6 (2004)
M.E. Gurtin, J. Mech. Phys. Solids **56**, 2 (2008)
M.E. Gurtin, L. Anand, J. Mech. Phys. Solids **53**, 7 (2005)
M.E. Gurtin, L. Anand, J. Mech. Phys. Solids **57**, 3 (2009)
M.E. Gurtin, B.D. Reddy, Contin. Mech. Thermodyn. **21**, 3 (2009)

S. Han, T. Kim, H. Lee, H. Lee, Electronics System-Integration Technology Conference, 2008. ESTC 2008, 2nd (2008)
M.A. Haque, M.T.A. Saif, Acta Mater. **51**, 11 (2003)
J.W. Hutchinson, Acta Mech. Sinica **28**, 4 (2012)
M. Kuroda, V. Tvergaard, Int. J. Plasticity **26**, 4 (2010)
D.B. Liu, Y.M. He, L. Shen, J. Lei, S. Guo, K. Peng, Mat. Sci. Eng. A-Struct. **647** (2015)
M.A. Meyers, K.K. Chawla, *Mechanical Behavior of Materials* (Cambridge University Press, 2009), Cambridge, United Kingdom
F. Mollica, K.R. Rajagopal, A.R. Srinivasa, Int. J. Plast. **17**, 8 (2001)
H.B. Muhlhaus, E.C. Aifantis, Int. J. Solids Struct. **28**, 7 (1991)
L. Nicola, Y. Xiang, J.J. Vlassak, E. Van der Giessen, A. Needleman, J. Mech. Phys. Solids **54**, 10 (2006)
P. Rosakis, A.J. Rosakis, G. Ravichandran, J. Hodowany, J. Mech. Phys. Solids **48**, 3 (2000)
S.S. Shishvan, L. Nicola, E. Van der Giessen, J. Appl. Phys. **107**, 9 (2010)
S.S. Shishvan, S. Mohammadi, M. Rahimian, E. Van der Giessen, Int. J. Solids Struct. **48**, 2 (2011)
G.Z. Voyiadjis, B. Deliktas, Int. J. Eng. Sci. **47**, 11–12 (2009)
G.Z. Voyiadjis, D. Faghihi, Int. J. Plast. 30–31 (2012)
G.Z. Voyiadjis, D. Faghihi, J. Eng. Mater-T. ASME **136**, 4 (2014)
G.Z. Voyiadjis, Y. Song, Philos. Mag. **97**, 5 (2017)
G.Z. Voyiadjis, D. Faghihi, Y.D. Zhang, Int. J. Solids Struct. **51**, 10 (2014)
G.Z. Voyiadjis, Y. Song, T. Park, J. Eng. Mater-T. ASME **139**, 2 (2017)
Y. Xiang, J.J. Vlassak, Acta Mater. **54**, 20 (2006)
Y. Xiang, X. Chen, J.J. Vlassak, J. Mater. Res. **20**, 9 (2005)

Micropolar Crystal Plasticity

17

J. R. Mayeur, D. L. McDowell, and Samuel Forest

Contents

J. R. Mayeur
Theoretical Division, Los Alamos National Laboratory, Los Alamos, NM, USA
e-mail: jason.mayeur@gmail.com

D. L. McDowell
Woodruff School of Mechanical Engineering, School of Materials Science and Engineering, Georgia Institute of Technology, Atlanta, GA, USA
e-mail: david.mcdowell@me.gatech.edu

S. Forest (✉)
Centre des Materiaux, Mines ParisTech CNRS, PSL Research University, Paris, Evry Cedex, France
e-mail: samuel.forest@mines-paristech.fr

G. Z. Voyiadjis (ed.), *Handbook of Nonlocal Continuum Mechanics for Materials and Structures*, https://doi.org/10.1007/978-3-319-58729-5_48

Abstract

This chapter considers advances over the past 15 years achieved by the authors and coworkers on generalized crystal plasticity to address size and configuration effects in dislocation plasticity at the micron scale. The specific approaches addressed here focus on micropolar and micromorphic theories rather than adopting strain gradient theory as the starting point, as motivated by the pioneering ideas of Eringen (Eringen and Suhubi 1964; Eringen and Claus Jr 1969; Eringen 1999). It is demonstrated with examples that for isotropic elasticity and specific sets of slip systems, a dislocation-based formulation of micropolar or micromorphic type provides results comparable to discrete dislocation dynamics and has much in common with the structure of Gurtin's slip gradient theory (Gurtin 2002; Gurtin et al. 2007).

Keywords

Micropolar · Strain gradient · GNDs · Crystal plasticity · Finite elements

Introduction

The collective behavior of dislocations in a single crystal can be described by means of the continuum theory of dislocations. The material volume element is assumed to contain a suitable density of dislocations for the continuum theory of dislocations to be applicable. Nonhomogeneous plastic deformations induce material and lattice incompatibilities that are resolved by a suitable distribution of the dislocation density tensor field which can be interpreted as a second rank statistical mean for a population of arbitrary dislocations inside a material volume element (Kröner 1969; Cermelli and Gurtin 2001). Nye's fundamental relation linearly connects the dislocation density tensor to the lattice curvature field of the crystal. This fact has prompted many authors to treat a continuously dislocated crystal as a Cosserat continuum (Günther 1958; Kröner 1963; Schäfer 1969; Forest et al. 2000). The Cosserat approach records only the lattice curvature of the crystal but neglects the effect of the rotational part of the elastic strain tensor, which is a part of the total dislocation density tensor (Cordero et al. 2010). Full account of plastic incompatibility is taken in strain gradient plasticity theories, starting from the original work by Aifantis (1984) up to the work of Gurtin (2002). Formulation of crystal plasticity within the micromorphic framework is more recent and was suggested by Clayton et al. (2005) for a large spectrum of crystal defects, including point defects and disclinations. Limiting the discussion to dislocation density tensor effects, also called geometrically necessary dislocation (GND) effects, Cordero et al. (2010) showed, within a small deformation setting, how the micromorphic model can be used to predict grain and precipitate size effects in laminate crystalline materials. In particular, the micromorphic model is shown to deliver more general scaling laws than conventional strain gradient plasticity. These models represent extensions of the conventional crystal plasticity theory (cf. Teodosiu and Sidoroff

1976) that accounts for single crystal hardening and lattice rotation but does not incorporate the effect of the dislocation density tensor.

The layout of the chapter is as follows. Classical single crystal plasticity theory is first recalled in section "Finite Deformation Kinematics" including the thermodynamical framework and the definition of the continuum dislocation density tensor. The Cosserat generalization of crystal plasticity is presented in section "Micropolar Single Crystal Plasticity" with its relation to strain gradient plasticity. The section "Applications: Comparison to 2D Discrete Dislocation Dynamics Simulations" provides applications including a comparison between Cosserat constitutive laws and results from discrete dislocation dynamics in the case of constrained thin films in shear and bending.

Classical Single Crystal Plasticity

Finite Deformation Kinematics

The classical theory of finite deformation single crystal plasticity is based on a multiplicative decomposition of the deformation gradient into elastic and plastic parts,

$$\mathbf{F} = \mathbf{F}^e \mathbf{F}^p \tag{1}$$

where $\mathbf{F}^p$ describes the plastic deformation of the continuum that leaves the underlying lattice vectors unaltered and $\mathbf{F}^e$ describes the elastic stretching and rotation of the lattice relative to this intermediate, isoclinic configuration. The deformation gradient maps infinitesimal vectors from the reference to current configuration and can be expressed in terms of the referential gradient of the displacement field as

$$d\mathbf{x} = \mathbf{F} d\mathbf{X}, \quad \mathbf{F} = \mathbf{1} + \mathbf{H}, \quad \mathbf{H} = \mathbf{u}\nabla_0 \tag{2}$$

$\mathbf{H}$ is the distortion (or displacement gradient) tensor and has been introduced for use in subsequent sections. The theory is completed by supplying constitutive prescriptions for the relationships between $\mathbf{F}^e$ and the Cauchy stress and to provide an evolution equation for $\mathbf{F}^p$ (and any associated internal state variables) consistent with thermodynamics.

Thermodynamics

The standard nonpolar mechanical balance laws (neglecting inertial terms) are given in the current configuration as

$$\boldsymbol{\sigma} \cdot \nabla + \mathbf{f} = \mathbf{0}, \quad \boldsymbol{\sigma} = \boldsymbol{\sigma}^T \tag{3}$$

with the associated traction boundary condition, $\mathbf{t} = \boldsymbol{\sigma}\mathbf{n}$, where $\boldsymbol{\sigma}$ is the Cauchy stress tensor, $\mathbf{f}$ is a body force vector, $\mathbf{t}$ is the traction vector, and $\mathbf{n}$ is the unit normal to the external part of the boundary where tractions are specified. The total energy balance can be expressed as

$$\rho\dot{U} = \boldsymbol{\sigma} : \mathbf{D} - \nabla \cdot \mathbf{q} \tag{4}$$

where ρ is the current density, U is the specific internal energy, $\mathbf{D}$ is the rate of deformation tensor, and $\mathbf{q}$ is the heat flux vector. The entropy inequality which will be used to derive the state equations and guide the construction of plastic evolution equations is given as

$$\rho\dot{\eta} + \nabla \cdot \left(\frac{\mathbf{q}}{T}\right) \geq 0 \tag{5}$$

where η is the specific entropy, and T is temperature. Using the state relation $\psi = U - T\eta$, Eqs. (4) and (5) can be combined to obtain the Clausius-Duhem inequality

$$\boldsymbol{\sigma} : \mathbf{D} - \rho\left(\dot{\psi} + \dot{T}\eta\right) - \frac{1}{T}\mathbf{q} \cdot \nabla T \geq 0 \tag{6}$$

where the intrinsic (Δ^{intr}) and thermal (Δ^{th}) dissipation are defined as

$$\Delta^{intr} = \boldsymbol{\sigma} : \mathbf{D} - \rho\left(\dot{\psi} + \dot{T}\eta\right), \quad \Delta^{th} - \frac{1}{T}\mathbf{q} \cdot \nabla T \tag{7}$$

Next, we derive an expression for the stress power in the intermediate configuration by using the relationship between the Cauchy stress and the second Piola-Kirchhoff stress with respect to the intermediate configuration $\tilde{\mathbf{S}} = J^e \mathbf{F}^{e-1}\boldsymbol{\sigma}\mathbf{F}^{e-T}$ and by writing the velocity gradient in terms of the multiplicative decomposition, i.e.,

$$\mathbf{L} = \dot{\mathbf{F}}\mathbf{F}^{-1} = \dot{\mathbf{F}}^e\mathbf{F}^{e-1} + \mathbf{F}^e\dot{\mathbf{F}}^p\mathbf{F}^{p-1}\mathbf{F}^{e-1} \tag{8}$$

such that

$$\frac{1}{\rho}\boldsymbol{\sigma} : \mathbf{L} = \frac{1}{\tilde{\rho}}\left(\tilde{\mathbf{S}} : \dot{\tilde{\mathbf{E}}}^e + \boldsymbol{\Pi}^M : \tilde{\mathbf{L}}^p\right) \tag{9}$$

where $\tilde{\rho}$ is the density in the intermediate configuration, $\tilde{\mathbf{E}}^e$ is the elastic Green-Lagrange strain, $\boldsymbol{\Pi}^M$ is the Mandel stress tensor, and $\tilde{\mathbf{L}}^p$ is the plastic velocity gradient in the intermediate configuration which have the following definitions

$$\tilde{\mathbf{E}}^e = \frac{1}{2}\left(\mathbf{F}^{eT}\mathbf{F}^e - \mathbf{1}\right), \quad \boldsymbol{\Pi}^M = \mathbf{F}^{eT}\mathbf{F}^e\tilde{\mathbf{S}}, \quad \tilde{\mathbf{L}}^p = \dot{\mathbf{F}}^p\mathbf{F}^{p-1}. \tag{10}$$

Equation (9) identifies the appropriate power-conjugate variables in the intermediate configuration, from which it can be seen that $\tilde{\mathbf{E}}^e$ is the strain measure power-conjugate to $\tilde{\mathbf{\Pi}}^e$ and that $\mathbf{\Pi}^M$ is power-conjugate to $\tilde{\mathbf{L}}^p$. If we now introduce the Helmholtz free energy, $\psi = \hat{\psi}\left(\tilde{\mathbf{E}}^e, T, \zeta^\alpha\right)$, the following expression is obtained by inserting the right-hand side of Eq. (9) along with the chain rule expression of the time derivative of ψ into the Clausius-Duhem inequality

$$\left(\mathbf{\Pi}^e - \tilde{\rho}\frac{\partial \psi}{\partial \tilde{\mathbf{E}}^e}\right) : \dot{\tilde{\mathbf{E}}}^e - \tilde{\rho}\left(\eta + \frac{\partial \psi}{\partial T}\right)\dot{T} + \mathbf{\Pi}^M : \tilde{\mathbf{L}}^p - \tilde{\rho}\sum_\alpha \frac{\partial \psi}{\partial \zeta^\alpha} * \dot{\zeta}^\alpha \geq 0. \quad (11)$$

where ζ^α is a set of internal state variables. Here, "*" is an appropriate scalar product operator for the rank of tensor ζ^α. The state laws are then deduced as

$$\mathbf{\Pi}^e = \tilde{\rho}\frac{\partial \psi}{\partial \tilde{\mathbf{E}}^e}, \quad \eta = -\frac{\partial \psi}{\partial T}. \quad (12)$$

and the residual intrinsic dissipation is then expressed as

$$\Delta^{intr} = \mathbf{\Pi}^M : \tilde{\mathbf{L}}^p - \sum_\alpha r^\alpha * \dot{\zeta}^\alpha \geq 0, \quad r^\alpha = \tilde{\rho}\frac{\partial \psi}{\partial \zeta^\alpha}. \quad (13)$$

Therefore, thermodynamically consistent evolution equations for $\tilde{\mathbf{L}}^p$ and $\dot{\zeta}^\alpha$ may be derived by introducing a convex dissipation potential $\Omega = \hat{\Omega}\left(\mathbf{\Pi}^M, r^\alpha\right)$ such that

$$\tilde{\mathbf{L}}^p = \frac{\partial \Omega}{\partial \mathbf{\Pi}^M}, \quad \dot{\zeta}^\alpha = -\frac{\partial \Omega}{\partial r^\alpha} \quad (14)$$

Representative functional forms for $\hat{\psi}$ and $\hat{\Omega}$ are given for completeness below. It is typical to use a free energy that is quadratic with respect to its arguments, i.e.,

$$\tilde{\rho}\psi\left(\tilde{\mathbf{E}}^e, \zeta^\alpha\right) = \frac{1}{2}\tilde{\mathbf{E}}^e : \tilde{\mathbb{C}} : \tilde{\mathbf{E}}^e + \frac{1}{2}\sum_{\alpha,\beta} a^{\alpha\beta}\zeta^\alpha\zeta^\beta \quad (15)$$

where $\tilde{\mathbb{C}}$ is the fourth-order elasticity tensor in the intermediate configuration and $a^{\alpha\beta}$ is positive definite interaction matrix that describes the coupling between the ζ^α. Likewise, a typical power law potential for the inelastic evolution equations is given as

$$\Omega\left(\mathbf{\Pi}^M, r^\alpha\right) = \frac{\dot{\gamma}_0}{n+1}\sum_\alpha g^\alpha \left\langle \frac{\mathcal{F}^\alpha}{g^\alpha} \right\rangle^{n+1} \quad (16)$$

where n is the power law exponent, g^α is viscous stress, $\dot{\gamma}_0$ a typical strain rate parameter, and $\mathcal{F}^\alpha$ is a yield function. The brackets $\langle \bullet \rangle =$ Max $(\bullet, 0)$ have been introduced. The yield function is defined in terms of the resolved shear stress, τ^{α}, and the energetic flow resistance, r^{α}, as

$$\mathcal{F}^{\alpha} = \tau^{\alpha} - (r_0 + r^{\alpha}), \; \tau^{\alpha} = \mathbf{\Pi}^{M} : (\tilde{\mathbf{s}}^{\alpha} \otimes \tilde{\mathbf{n}}^{\alpha}) \tag{17}$$

The slip direction vector and the normal to the slip plane for slip system α are respectively denoted by $\tilde{\mathbf{s}}^{\alpha}$ and $\tilde{\mathbf{n}}^{\alpha}$ in the undistorted lattice configuration.

Deformation Incompatibility and the GND Density Tensor

Finite Deformation Kinematics

For a classical Cauchy continuum, compatibility of the displacement field requires that the curl of the deformation gradient vanishes, i.e.,

$$\oint_{c} d\mathbf{x} = \oint_{C} \mathbf{F} d\mathbf{X} = \mathbf{0} \Rightarrow \mathbf{F} \times \nabla_0 = \mathbf{0} \tag{18}$$

During an inhomogeneous elastic-plastic deformation, the elastic and plastic deformation maps are not compatible and can be used to quantify the heterogeneity of the deformation field in terms of the net Burgers on the intermediate configuration, i.e.,

$$\tilde{\mathbf{b}} = \oint_{\tilde{c}} d\tilde{\mathbf{x}} = \oint_{c} \mathbf{F}^{e-1} d\mathbf{x} = \oint_{C} \mathbf{F}^{p} d\mathbf{X} \tag{19}$$

Making use of Stokes' theorem, the last two expressions in Eq. (19) can be expressed in terms of surface integrals as

$$\tilde{\mathbf{b}} = \int_{s} \mathbf{A}^{e} \mathbf{n} da = \oint_{S} \mathbf{A}^{p} \mathbf{N} dA \tag{20}$$

where $\mathbf{A}^{e} = \mathbf{F}^{e-1} \times \nabla$ and $\mathbf{A}^{p} = \mathbf{F}^{p} \times \nabla_0$ are the corresponding two-point geometrically necessary dislocation density tensors that map from current to intermediate and reference to intermediate configurations, respectively.

There have been many works in the last two decades focused on incorporating the effects of GNDs into crystal plasticity modeling frameworks. It is beyond the scope of this chapter to attempt to review the myriad ways in which these extensions are carried out. The vast majority of generalized crystal plasticity models that incorporate the effects of GNDs do so by computing them from the plastic slip gradients (so-called slip gradient theories) rather than via gradients of $\mathbf{F}^{e-1}$. While the connection between slip gradients and GND densities can be established in a finite deformation context (see Kuroda and Tvergaard 2008), in this section, the presentation is limited to the linearized kinematic setting for ease of presentation and direct connection with the pioneering work of Nye (1953). The subsequent developments closely follow the presentation of Arsenlis and Parks (1999).

Linearized Kinematics

In the case of linearized kinematics, the multiplicative decomposition of the deformation gradient is replaced with an additive decomposition of the distortion tensor, i.e., $\mathbf{H} = \mathbf{H}^e + \mathbf{H}^p$ and Eq. (18) may conveniently be rewritten as

$$\mathbf{H} \times \nabla = 0 \tag{21}$$

The continuum GND density tensor, $\mathbf{A} \approx \mathbf{A}^e \approx \mathbf{A}^p$, can be equivalently expressed in terms of either the elastic or plastic distortion as

$$\mathbf{A} = -\mathbf{H}^e \times \nabla = \mathbf{H}^p \times \nabla \tag{22}$$

The lattice torsion-curvature is defined as the gradient of the lattice rotation vector (Nye 1953), i.e.,

$$\boldsymbol{\kappa} = \boldsymbol{\phi}\nabla \tag{23}$$

where $\boldsymbol{\phi}$ is given by

$$\boldsymbol{\phi} = -\frac{1}{2}\boldsymbol{\epsilon} : \ \text{skw}\,(\mathbf{H}^e) \tag{24}$$

where "skw" is the skew operator providing the antisymmetric part of the tensor. The notation $\boldsymbol{\epsilon}$ is used for the permutation tensor. Combining Eqs. (22), (23), and (24) the lattice torsion-curvature may be expressed as

$$\varkappa = \underbrace{-\mathbf{A}^{\mathrm{T}} + \frac{1}{2}\text{tr}\,(\mathbf{A})\,\mathbf{1}}_{\text{Nye curvature},\boldsymbol{\kappa}} - (\varepsilon^{\mathrm{e}} \times \nabla)^{\mathrm{T}} \tag{25}$$

where ε^{e} is the elastic strain tensor which is symmetric. As indicated in Eq. (25), the first two terms represent Nye's original torsion-curvature tensor since it was assumed that $\mathbf{H}^e \approx \text{skw}(\mathbf{H}^e)$.

Next, we seek an expression for $\mathbf{A}$ in terms of slip gradients which can be obtained from taking the curl of the plastic distortion. First, note that the discrete version of Nye's tensor, $\mathbf{A}^d$, for a population of straight edge and screw dislocations can be written as

$$\mathbf{A}^d = b\sum_{\alpha}\underbrace{\left(\varrho^{\alpha}_{\perp} - \varrho^{\alpha}_{\top}\right)}_{\varrho^{\alpha}_{G\perp}}\mathbf{s}^{\alpha} \otimes \mathbf{t}^{\alpha} + \underbrace{\left(\varrho^{\alpha}_{\odot} - \varrho^{\alpha}_{\otimes}\right)}_{\varrho^{\alpha}_{G\odot}}\mathbf{s}^{\alpha} \otimes \mathbf{s}^{\alpha} \tag{26}$$

where $\varrho_{\perp}^{\alpha}$ is the positive edge dislocation density, $\varrho_{\top}^{\alpha}$ is the negative edge dislocation density, $\varrho_{\odot}^{\alpha}$ is the positive screw dislocation density, $\varrho_{\otimes}^{\alpha}$ is the negative screw dislocation density, $\varrho_{G\perp}^{\alpha}$ is the edge GND density, and $\varrho_{G\odot}^{\alpha}$ is the screw GND density. The continuous expression of Nye's GND density tensor is then obtained as

$$\begin{aligned}\mathbf{A} &= \mathbf{H}^{p} \times \nabla \\ &= \sum_{\alpha} \gamma^{\alpha} \mathbf{s}^{\alpha} \otimes \mathbf{n}^{\alpha} \times \nabla \\ &= \sum_{\alpha} \left(-\nabla \gamma^{\alpha} \cdot \mathbf{s}^{\alpha}\right) \mathbf{s}^{\alpha} \otimes \mathbf{t}^{\alpha} + \left(\nabla \gamma^{\alpha} \cdot \mathbf{t}^{\alpha}\right) \mathbf{s}^{\alpha} \otimes \mathbf{s}^{\alpha}\end{aligned} \tag{27}$$

Comparing these two expressions for Nye's tensor, it may be shown that the continuum GND densities are given by slip gradients projected in the glide directions for the respective dislocation populations, i.e.,

$$\varrho_{G\perp}^{\alpha} = -b^{-1} \nabla \gamma^{\alpha} \cdot \mathbf{s}^{\alpha}, \quad \varrho_{G\odot}^{\alpha} = b^{-1} \nabla \gamma^{\alpha} \cdot \mathbf{t}^{\alpha} \tag{28}$$

An expression relating Nye's torsion-curvature tensor, $\boldsymbol{\kappa}$, to the GND densities is obtained by inserting Eq. (27) into the first two terms in Eq. (25)

$$\boldsymbol{\kappa} = -b \sum_{\alpha} \left[\varrho_{G\perp}^{\alpha} \mathbf{t}^{\alpha} \otimes \mathbf{s}^{\alpha} + \varrho_{G\odot}^{\alpha} \left(\mathbf{s}^{\alpha} \otimes \mathbf{s}^{\alpha} - \frac{1}{2} \mathbf{1} \right) \right] \tag{29}$$

We will revisit the expression given in Eq. (29) when motivating constitutive equations in subsequent sections.

Micropolar Single Crystal Plasticity

Finite Deformation Theory

The presentation is given here within the finite deformation framework before returning to the linearized case.

Kinematics

A micropolar continuum is a generalized continua with extra rotational degrees of freedom. Considering two sets of vector triads attached to each material point in the reference configuration, $\mathbf{X}$, there is an independent mapping of the two sets of vectors to the current configuration such that

$$d\mathbf{x}_i = \mathbf{F} d\mathbf{X}_i, \quad \mathbf{d}_i = \overline{\mathbf{R}} \mathbf{D}_i \quad \forall i = 1, 3 \tag{30}$$

where $\mathbf{F}$ is the usual deformation gradient (see Eq. (2)) and $\overline{\mathbf{R}}$ is the two-point tensor that maps the microstructure triad in the reference configuration, $\mathbf{D}_i$ to its image in the current configuration, $\mathbf{d}_i$. $\overline{\mathbf{R}}$ is the micropolar rotation tensor; the overbar is used to distinguish it from the rotational part of the deformation gradient obtained via the polar decomposition, i.e., $\mathbf{F} = \mathbf{RU}$. The micropolar rotation tensor is a proper orthogonal tensor such that the following relations hold:

$$\overline{\mathbf{R}}\,\overline{\mathbf{R}}^T = \mathbf{1}, \quad \overline{\mathbf{R}}\,(\mathbf{X}, t = 0) = \mathbf{1}, \quad \det\left(\overline{\mathbf{R}}\right) = 1 \tag{31}$$

The micropolar rotation field can also be expressed in terms of the axial vector field, $\overline{\phi}$, via

$$\overline{\mathbf{R}} = \exp\left(\overline{\mathbf{\Phi}}\right), \; \overline{\Phi}_{ij} = -\varepsilon_{ijk}\overline{\phi}_k \tag{32}$$

where $\overline{\mathbf{\Phi}}$ is the skew symmetric tensor associated with the axial vector $\overline{\phi}$. The three components of $\overline{\phi}$ along with the displacement field, $\mathbf{u}$, comprise the six independent degrees of freedom for the micropolar continuum. It has been shown by Eringen and Suhubi (1964) that a suitable set of invariant Lagrangian strain measures for the micropolar continuum may be defined as

$$\overline{\mathbf{U}} = \overline{\mathbf{R}}^T\mathbf{F}, \quad \mathbf{\Gamma} = \overline{\mathbf{R}}^T\left(\overline{\mathbf{R}}\nabla_0\right) \tag{33}$$

where $\overline{\mathbf{U}}$ is called the relative deformation tensor and $\mathbf{\Gamma}$ is the third-rank wryness (or torsion-curvature tensor. The Lagrangian micropolar strain is defined in terms of the relative distortion tensor as $\overline{\mathbf{E}} = \overline{\mathbf{U}} - \mathbf{1}$. Due to the antisymmetry with respect to its first two indices, it is convenient to express the torsion-curvature as a second-order tensor, i.e.,

$$\overline{\boldsymbol{K}} = -\frac{1}{2}\boldsymbol{\epsilon} : \mathbf{\Gamma} \tag{34}$$

The rates of the micropolar strain and torsion-curvature tensors are related to the velocity, $\mathbf{v} = \dot{\mathbf{u}}$, and the microstructural angular velocity, $\overline{\mathbf{\Omega}} = \dot{\overline{\mathbf{R}}}\,\overline{\mathbf{R}}^T$, as

$$\dot{\overline{\mathbf{U}}}\,\overline{\mathbf{U}}^{-1} = \overline{\mathbf{R}}^T\left(\mathbf{L} - \overline{\mathbf{\Omega}}\right)\overline{\mathbf{R}} \tag{35}$$

$$\dot{\overline{\boldsymbol{K}}}\,\overline{\mathbf{U}}^{-1} = \overline{\mathbf{R}}^T\left(\overline{\boldsymbol{\omega}}\nabla\right)\overline{\mathbf{R}} \tag{36}$$

where $\mathbf{L} = \dot{\mathbf{F}}\mathbf{F}^{-1}$ is the velocity gradient and $\overline{\boldsymbol{\omega}} = -1/2\boldsymbol{\epsilon} : \overline{\mathbf{\Omega}}$ is the axial vector associated with $\overline{\mathbf{\Omega}}$. From Eqs. (35) and (36), it is clear that the rate expressions on the LHS are pull-backs from the current configuration via the microrotation.

Thermodynamics

The mechanical balance laws under static equilibrium for the micropolar continuum are

$$\nabla \cdot \boldsymbol{\sigma}^T + \mathbf{f} = 0, \quad \nabla \cdot \mathbf{m}^T - \boldsymbol{\epsilon} : \boldsymbol{\sigma} + \mathbf{c} = 0 \tag{37}$$

where $\boldsymbol{\sigma}$ is the unsymmetric Cauchy force stress tensor, $\mathbf{f}$ is a body force vector, $\mathbf{m}$ is the couple stress tensor, and $\mathbf{c}$ is a body couple vector. These balance laws can be deduced using the principle of virtual work (Germain 1973; Forest and Sievert 2003). The energy balance and Clausius-Duhem inequality for the micropolar continuum are given, respectively, as

$$\rho \dot{U} = \boldsymbol{\sigma} : \left(\mathbf{L} - \overline{\boldsymbol{\Omega}}\right) + \mathbf{m} : \overline{\boldsymbol{\omega}} \nabla - \nabla \cdot \mathbf{q} \tag{38}$$

$$\hat{\boldsymbol{\sigma}} : \left(\dot{\overline{\mathbf{U}}}\, \overline{\mathbf{U}}^{-1}\right) + \hat{\mathbf{m}} : \left(\dot{\overline{\boldsymbol{K}}}\, \overline{\mathbf{U}}^{-1}\right) - \rho\left(\dot{\psi} + \eta \dot{T}\right) - \frac{1}{T}\mathbf{q} \cdot \nabla T \geq 0 \tag{39}$$

where $\hat{\boldsymbol{\sigma}}$ and $\hat{\mathbf{m}}$ are the Cauchy and couple stress tensors pulled back to the reference configuration via $\overline{\mathbf{R}}$, i.e.,

$$\hat{\boldsymbol{\sigma}} = \overline{\mathbf{R}}^T \boldsymbol{\sigma} \overline{\mathbf{R}}, \quad \hat{\mathbf{m}} = \overline{\mathbf{R}}^T \mathbf{m} \overline{\mathbf{R}} \tag{40}$$

The specific form of Eq. (39) was obtained by using the kinematic relations given in Eqs. (35) and (36). Let us first consider the case of finite micropolar thermoelasticity and assume that $\psi = \hat{\psi}\left(\overline{\mathbf{U}}, \overline{\boldsymbol{K}}, T\right)$. Taking the time derivative of ψ and inserting into Eq. (39) leads to

$$\left(\hat{\boldsymbol{\sigma}} \overline{\mathbf{U}}^{-T} - \rho \frac{\partial \psi}{\partial \overline{\mathbf{U}}}\right) : \dot{\overline{\mathbf{U}}} + \left(\hat{\mathbf{m}} \overline{\mathbf{U}}^{-T} - \rho \frac{\partial \psi}{\partial \overline{\boldsymbol{K}}}\right) : \dot{\overline{\boldsymbol{K}}} - \rho\left(\eta + \frac{\partial \psi}{\partial T}\right) \dot{T} \geq 0 \tag{41}$$

Therefore, the state laws for the micropolar material are

$$\hat{\boldsymbol{\sigma}} = \rho \frac{\partial \psi}{\partial \overline{\mathbf{U}}} \overline{\mathbf{U}}^T, \quad \hat{\mathbf{m}} = \rho \frac{\partial \psi}{\partial \overline{\boldsymbol{K}}} \overline{\mathbf{U}}^T, \quad \eta = -\frac{\partial \psi}{\partial T} \tag{42}$$

Next we must introduce elastic-plastic decompositions for $\overline{\mathbf{U}}$ and $\overline{\boldsymbol{K}}$. We start with the natural assumption that the hyperelastic relations will have the same form as Eq. (42) with respect to the elastic deformation measures, i.e.,

$$\hat{\boldsymbol{\sigma}} = \rho \frac{\partial \psi}{\partial \overline{\mathbf{U}}^e} \overline{\mathbf{U}}^{eT}, \quad \hat{\mathbf{m}} = \rho \frac{\partial \psi}{\partial \overline{\boldsymbol{K}}^e} \overline{\mathbf{U}}^{eT} \tag{43}$$

Consistent with this assumption, it can be shown (Sievert et al. 1998) that the appropriate decompositions are given as

$$\overline{\mathbf{U}} = \overline{\mathbf{U}}^e \overline{\mathbf{U}}^p, \quad \overline{\boldsymbol{K}} = \overline{\boldsymbol{K}}^e \overline{\mathbf{U}}^p + \overline{\boldsymbol{K}}^p \tag{44}$$

Substituting the rate forms of these decompositions back into the Clausius-Duhem inequality with $\psi = \hat{\psi}\left(\overline{\mathbf{U}}^e, \overline{\boldsymbol{K}}^e, T, \zeta^\alpha\right)$ yields

$$\Sigma : \left(\dot{\overline{\mathbf{U}}}^p \overline{\mathbf{U}}^{p-1}\right) + \boldsymbol{M} : \left(\dot{\overline{\boldsymbol{K}}}^p \overline{\mathbf{U}}^{p-1}\right) + \rho \sum_\alpha \frac{\partial \psi}{\partial \zeta^\alpha} * \dot{\zeta}^\alpha \geq 0 \tag{45}$$

where the driving forces for the plastic evolution equations are identified as

$$\Sigma = \overline{\mathbf{U}}^{eT} \hat{\boldsymbol{\sigma}} \overline{\mathbf{U}}^{e-T} + \overline{\boldsymbol{K}}^{eT} \hat{\mathbf{m}} \overline{\mathbf{U}}^{e-T}, \quad \boldsymbol{M} = \hat{\mathbf{m}} \overline{\mathbf{U}}^{e-T} \tag{46}$$

Specific constitutive equations must be provided for the free energy and plastic evolution equations to complete the formulation. The procedure for doing so can be undertaken in analogous fashion to what is done in the classical theory. However, there is some additional flexibility that is afforded in constructing the plastic evolution equations in this case compared to the classical theory (see Forest and Sievert (2003) for an in-depth discussion of single vs. multi-criterion flow rules). Rather than narrowing to specific constitutive choices in the current finite deformation context, we will discuss these issues within the small deformation framework in the subsequent sections. The principles guiding constitutive equation development are the same for both finite and infinitesimal deformations and we choose to discuss these aspects with respect to the theories used in the numerical simulations appearing later in the chapter. A more complete exposition of constitutive equation development in the finite deformation context has been given elsewhere (Forest et al. 1997; Forest 2012).

Linear Deformation Theory

Linearized Kinematics

In the case of small deformations and rotations, the following notation is introduced for the micropolar strain, $\overline{\varepsilon}$, and torsion-curvature, $\overline{\boldsymbol{\kappa}}$:

$$\overline{\mathbf{E}} = \overline{\mathbf{U}} - \mathbf{1} \approx \mathbf{u}\nabla - \overline{\boldsymbol{\Phi}} =: \overline{\varepsilon} = \overline{\varepsilon}^e + \overline{\varepsilon}^p \tag{47}$$

$$\overline{\boldsymbol{K}} = -\frac{1}{2}\boldsymbol{\epsilon} : \boldsymbol{\Gamma} \approx \overline{\phi}\nabla =: \overline{\boldsymbol{\kappa}} = \overline{\boldsymbol{\kappa}}^e + \overline{\boldsymbol{\kappa}}^p \tag{48}$$

As indicated in Eqs. (47) and (48), we assume an additive elastic-plastic decomposition of the strain and torsion-curvature tensors. Note that the symmetric part of $\overline{\varepsilon}$ is the classical small strain tensor: $\text{sym}(\overline{\epsilon}) = \epsilon = \text{sym}(\mathbf{H})$, and the skew-symmetric part is a measure of the difference between the continuum rotation,

$\boldsymbol{w} = \text{skw}\ (\mathbf{H})$, and microrotation: $\text{skw}\,(\overline{\varepsilon}) = \boldsymbol{w} - \overline{\boldsymbol{\Phi}}$. The additive decomposition of the distortion tensor, $\mathbf{H}$, into elastic and plastic parts is assumed (see section "Linearized Kinematics") and the micropolar plastic strain is defined to be equal to the plastic distortion such that the evolution equation of $\overline{\varepsilon}^p$ has the form

$$\dot{\overline{\varepsilon}}^p = \sum_{\alpha} \dot{\gamma}^{\alpha} \mathbf{s}^{\alpha} \otimes \mathbf{n}^{\alpha} \tag{49}$$

Therefore, elastic micropolar strain is defined as

$$\overline{\varepsilon}^e = \mathbf{H}^e - \overline{\boldsymbol{\Phi}} = \underbrace{\varepsilon^e}_{\text{sym}(\overline{\varepsilon}^e)} + \underbrace{\left(\boldsymbol{w}^e - \overline{\boldsymbol{\Phi}}\right)}_{\text{skw}(\overline{\varepsilon}^e)} \tag{50}$$

Equation (50) shows that the skew-symmetric part of the micropolar elastic strain is just the difference between the lattice rotation embodied by $\boldsymbol{w}^e$ and the microrotation. Using the relationship between the lattice torsion-curvature and GNDs presented in section "Linearized Kinematics" as motivation, Forest et al. (1997) proposed a micropolar plastic torsion-curvature evolution equation of the form (note there is a sign convention difference between the screw GND term presented here and what was originally proposed in that work):

$$\dot{\overline{\boldsymbol{\kappa}}}^{\mathrm{p}} = \sum_{\alpha} \left[\frac{\dot{\varphi}^{\alpha}_{\perp}}{L_{\perp}} \mathbf{t}^{\alpha} \otimes \mathbf{s}^{\alpha} + \frac{\dot{\varphi}^{\alpha}_{\odot}}{L_{\odot}} \left(\mathbf{s}^{\alpha} \otimes \mathbf{s}^{\alpha} - \frac{1}{2}\mathbf{1} \right) \right]. \tag{51}$$

Here, $\dot{\varphi}^{\alpha}_{\perp}$ and $L^{\alpha}_{\perp}$ are the plastic rotation rate and plastic length scale associated with edge GNDs and $\dot{\varphi}^{\alpha}_{\odot}$ and $L^{\alpha}_{\odot}$ are the analogous quantities for screw GNDs. Comparing Eqs. (29) and (51) reveals the relationship between the GND densities and the micropolar plastic torsion-curvature parameters, namely

$$\dot{\varrho}^{\alpha}_{G\perp} = -\frac{\dot{\varphi}^{\alpha}_{\perp}}{bL_{\perp}}, \quad \dot{\varrho}^{\alpha}_{G\odot} = -\frac{\dot{\varphi}^{\alpha}_{\odot}}{bL_{\odot}} \tag{52}$$

The introduced length scales are expected to be in the range of micron and submicron sizes as illustrated in the examples provided in this chapter.

Thermodynamics at Small Strains

Let us now revisit the Clausius-Duhem inequality for the micropolar material expressed in terms of the rates of linearized kinematic variables as a guide to constitutive equation development, i.e.,

$$\boldsymbol{\sigma} : \dot{\overline{\varepsilon}} + \mathbf{m} : \dot{\overline{\boldsymbol{\kappa}}} - \rho \left(\dot{\psi} + \eta \dot{T} \right) - \frac{1}{T} \mathbf{q} \cdot \nabla T \geq 0 \tag{53}$$

We assume a general form of the free energy that depends on the elastic strain, the elastic torsion-curvature, temperature, and a set of internal state variables, i.e., $\psi = \hat{\psi}\,(\overline{\varepsilon}^e, \overline{\boldsymbol{\kappa}}^e, T, \zeta^\alpha)$. Taking the chain rule expression for the time derivative of $\hat{\psi}$ along with the elastic-plastic decompositions for the strain and torsion-curvature and inserting into Eq. (53) and following the Coleman-Noll procedure, the following state laws are obtained:

$$\boldsymbol{\sigma} = \rho \frac{\partial \psi}{\partial \overline{\varepsilon}^e} \quad , \quad \mathbf{m} = \rho \frac{\partial \psi}{\partial \overline{\boldsymbol{\kappa}}^e} \quad , \quad \eta = -\frac{\partial \psi}{\partial T} \tag{54}$$

along with the expression for the intrinsic dissipation, i.e.,

$$\Delta^{intr} = \boldsymbol{\sigma} : \dot{\overline{\varepsilon}}^p + \mathbf{m} : \dot{\overline{\boldsymbol{\kappa}}}^p - \sum_\alpha r^\alpha * \dot{\zeta}^\alpha \geq 0, \qquad r^\alpha = \rho \frac{\partial \psi}{\partial \zeta^\alpha} \tag{55}$$

Elastic Free Energy Function

The most general form of the elastic strain energy for a linearized micropolar continuum is given by the quadratic form viz (Eringen 1999):

$$\rho \psi_e = \frac{1}{2} \overline{\varepsilon}^e : \mathbb{C} : \overline{\varepsilon}^e + \frac{1}{2} \overline{\boldsymbol{\kappa}}^e : \mathbb{D} : \overline{\boldsymbol{\kappa}}^e + \overline{\varepsilon}^e : \mathbb{E} : \overline{\boldsymbol{\kappa}}^e \tag{56}$$

where $\mathbb{C}$, $\mathbb{D}$, and $\mathbb{E}$ are fourth-order tensors of elastic moduli. However, the coupling moduli $\mathbb{E}$ are equal to zero for materials exhibiting point symmetry (Forest et al. 1997). Therefore, the stress and couple-stress constitutive equations for single crystals may be expressed as

$$\boldsymbol{\sigma} = \mathbb{C} : \overline{\varepsilon}^e \quad , \quad \mathbf{m} = \mathbb{D} : \overline{\boldsymbol{\kappa}}^e \tag{57}$$

For an elastically isotropic material, these expressions have the form

$$\begin{aligned} \boldsymbol{\sigma} &= \lambda \ \operatorname{tr}(\varepsilon^e)\,\mathbf{1} + 2\mu\, \varepsilon^e + 2\mu_c \left(\boldsymbol{\omega}^e - \overline{\boldsymbol{\Phi}}\right) \\ \mathbf{m} &= \alpha \ \operatorname{tr}(\overline{\boldsymbol{\kappa}}^e)\,\mathbf{1} + 2\beta \operatorname{sym}(\overline{\boldsymbol{\kappa}}^e) + 2\gamma \operatorname{skw}(\overline{\boldsymbol{\kappa}}^e) \end{aligned} \tag{58}$$

where λ and μ are the usual Lamé's constants and μ_c, α, β, and γ are nonstandard and/or higher-order elastic moduli. The coupling modulus, μ_c, gives rise to the skew symmetric part of the Cauchy stress and the couple-stress moduli, α, β, and γ, can be interpreted as elastic length scales; when they are normalized, for example, with respect to the shear modulus the resulting quantities have units of length, e.g., $\ell_e = \sqrt{\beta/\mu}$.

Flow Rules

The development of dissipative constitutive equations follows the standard approach utilizing a flow potential and associative flow rules. As discussed in Forest and Sievert (2003), one can formulate the rules either in terms of a single or multiple criteria, i.e., either a single flow potential for both strain and torsion-curvature or independent flow potentials for the respective deformation variables. Mutli-criterion theories are more general and require the specification of additional material parameters and should only be used if there is compelling need for this additional flexibility/complexity of the model. Representative multi-criterion micropolar single crystal plasticity models were presented by Forest et al. (2000) and Mayeur and McDowell (2013). The subsequent treatment of this section will focus on a single criterion framework since the numerical results presented later in the chapter were obtained using this type of model, and we have found it to suffice for most purposes. The plastic strain and torsion-curvature evolution equations are derived from a unified slip system flow potential, $\mathcal{F}^\alpha$, i.e.,

$$\dot{\bar{\varepsilon}}^p = \sum_\alpha \dot{\lambda}^\alpha \frac{\partial \mathcal{F}^\alpha}{\partial \boldsymbol{\sigma}}, \quad \dot{\bar{\boldsymbol{\kappa}}}^p = \sum_\alpha \dot{\lambda}^\alpha \frac{\partial \mathcal{F}^\alpha}{\partial \mathbf{m}} \tag{59}$$

Given the suggested kinematic forms for the plastic strain and torsion-curvature rates (Eqs. (49) and (51)), we propose a yield function, $\mathcal{F}^\alpha$, of the form

$$\mathcal{F}^\alpha = \hat{\tau}^\alpha - (r_0 + r^\alpha) \leq 0 \tag{60}$$

Here, $\hat{\tau}^\alpha$ is an effective resolved shear stress and r_0 is the initial yield strength. The effective resolved shear stress is defined with respect to the projections of the force and couple stress tensors as

$$\hat{\tau}^\alpha = \sqrt{\left|\tau^\alpha_{eff}\right|^2 + \left|\pi^\alpha_\perp / L_\perp\right|^2 + \left|\pi^\alpha_\odot / L_\odot\right|^2}. \tag{61}$$

where τ^α_{eff} is the resolved shear stress, $\pi_\perp$ is the resolved couple stress acting on edge GNDs, $\pi_\odot$ is the resolved couple stress acting on screw GNDs, and $L_\perp$, $L_\odot$ are normalizing length scales for edge and screw GNDs, respectively. The resolved shear and couple stresses are defined as

$$\tau^\alpha_{eff} = \boldsymbol{\sigma} : \ (\mathbf{s}^\alpha \otimes \mathbf{n}^\alpha) \tag{62}$$

$$\pi^\alpha_\perp = \mathbf{m} : (\mathbf{t}^\alpha \otimes \mathbf{s}^\alpha) \tag{63}$$

$$\pi^\alpha_\odot = \mathbf{m} : \left(\mathbf{s}^\alpha \otimes \mathbf{s}^\alpha - \frac{1}{2}\mathbf{1}\right) \tag{64}$$

The "*eff*" subscript has been applied to the resolved shear stress to emphasize that the driving force for slip has a contribution from the skew-symmetric part of the stress tensor and also to distinguish it from the classical resolved shear stress, which is computed using only the symmetric part of the Cauchy stress. The contribution of the skew-symmetric part of the Cauchy stress to τ^α_{eff} gives rise to gradient-dependent kinematic hardening which is elaborated upon further in section "Relationship to Slip Gradient Theory." The flow directions then follow as

$$\frac{\partial \mathcal{F}^\alpha}{\partial \boldsymbol{\sigma}} = \frac{\tau^\alpha_{eff}}{\hat{\tau}^\alpha} \mathbf{s}^\alpha \otimes \mathbf{n}^\alpha \tag{65}$$

$$\frac{\partial \mathcal{F}^\alpha}{\partial \mathbf{m}} = \frac{1}{L_\perp} \mathbf{t}^\alpha \otimes \mathbf{s}^\alpha \frac{\pi^\alpha_\perp / L_\perp}{\hat{\tau}^\alpha} + \frac{1}{L_\odot} \left(\mathbf{s}^\alpha \otimes \mathbf{s}^\alpha - \frac{1}{2} \mathbf{1} \right) \frac{\pi^\alpha_\odot / L_\odot}{\hat{\tau}^\alpha}. \tag{66}$$

Herein, we work within an elastic-viscoplastic setting and propose a power law expression for $\dot{\lambda}^\alpha$, i.e.,

$$\dot{\lambda}^\alpha = \dot{\lambda}_0 \left\langle \frac{\hat{\tau}^\alpha - (r_0 + r^\alpha)}{g^\alpha} \right\rangle^m \tag{67}$$

where $\dot{\lambda}_0$ is a reference effective deformation rate, g^α is a viscous drag stress, and m is an inverse rate sensitivity exponent. Inserting Eqs. (65), (66), and (67) into Eq. (59) and comparing with the expressions given in Eqs. (49) and (51), the expressions for the slip system deformation rates are obtained as

$$\dot{\gamma}^\alpha = \dot{\lambda}_0 \left\langle \frac{\hat{\tau}^\alpha - (r_0 + r^\alpha)}{g^\alpha} \right\rangle^m \frac{\tau^\alpha_{eff}}{\hat{\tau}^\alpha} \tag{68}$$

$$\dot{\varphi}^\alpha_\perp = \dot{\lambda}_0 \left\langle \frac{\hat{\tau}^\alpha - (r_0 + r^\alpha)}{g^\alpha} \right\rangle^m \frac{\pi^\alpha_\perp / L_\perp}{\hat{\tau}^\alpha} \tag{69}$$

$$\dot{\varphi}^\alpha_\odot = \dot{\lambda}_0 \left\langle \frac{\hat{\tau}^\alpha - (r_0 + r^\alpha)}{g^\alpha} \right\rangle^m \frac{\pi^\alpha_\odot / L_\odot}{\hat{\tau}^\alpha} \tag{70}$$

It is easily shown from Eqs. (68), (69), and (70) that the slip system plastic multiplier is related to the slip and curvature rates as

$$\dot{\lambda}^\alpha = \sqrt{|\dot{\gamma}^\alpha|^2 + \left|\dot{\varphi}^\alpha_\perp\right|^2 + \left|\dot{\varphi}^\alpha_\odot\right|^2} \tag{71}$$

Equation (71) reveals that $\dot{\lambda}^\alpha$ is an effective slip system deformation rate accounting for both slip and torsion-curvature deformation modes.

Internal State Variable Evolution

Consider the case where the evolution of the internal state variable associated with energetic isotropic hardening with the effective deformation rate, i.e.,

$$\dot{\varsigma}^\alpha = \dot{\lambda}^\alpha \tag{72}$$

Further, a quadratic dependence of the free energy on ς^α is assumed

$$\rho \psi_{in} = \frac{1}{2} \sum_\alpha \sum_\beta \mathrm{H}^{\alpha\beta} \varsigma^\alpha \varsigma^\beta, \quad \mathrm{H}^{\alpha\beta} = H_0 \tag{73}$$

where $\mathrm{H}^{\alpha\beta}$ is the hardening matrix and $H_0 > 0$ is the hardening modulus. The energetic isotropic hardening stress is then given as

$$r^{\alpha} = \sum_{\beta} \mathrm{H}^{\alpha\beta} \varsigma^{\beta} = H_0 \sum_{\beta} \varsigma^{\beta} \tag{74}$$

Given the preceding evolution equations, the intrinsic dissipation inequality may now be expressed in terms of slip system variables as

$$\Delta^{intr} = \sum_{\alpha} \left(\tau^{\alpha}_{eff} \dot{\gamma}^{\alpha} + \pi^{\alpha}_{\perp} \frac{\dot{\varphi}^{\alpha}_{\perp}}{L_{\perp}} + \pi^{\alpha}_{\odot} \frac{\dot{\varphi}^{\alpha}_{\odot}}{L_{\odot}} - r^{\alpha} \dot{\varsigma}^{\alpha} \right) \geq 0 \tag{75}$$

which may be further simplified and expressed as

$$\Delta = \sum_{\alpha} \left(\hat{\tau}^{\alpha} - r^{\alpha} \right) \dot{\lambda}^{\alpha} \geq 0 \tag{76}$$

For nonzero values of $\dot{\lambda}^{\alpha}$, Eq. (67) can be inverted to obtain the expression for $\hat{\tau}^{\alpha}$ as

$$\hat{\tau}^{\alpha} = g^{\alpha} \left(\frac{\dot{\lambda}^{\alpha}}{\dot{\lambda}_0} \right)^{\frac{1}{m}} + r_0 + r^{\alpha} \tag{77}$$

Inserting this expression into Eq. (76) yields

$$\Delta^{intr} = \sum_{\alpha} \left[g^{\alpha} \left(\frac{\dot{\lambda}^{\alpha}}{\dot{\lambda}_0} \right)^{\frac{1}{m}} + r_0 \right] \dot{\lambda}^{\alpha} \geq 0 \tag{78}$$

Therefore, the dissipation inequality is unconditionally satisfied for this set of constitutive equations.

While the illustrative example given in this particular section yields a simple linear isotropic hardening behavior, it is straightforward and often preferable in practice to use nonlinear hardening laws; general examples of such extensions were previously given by Forest et al. (1997) and Mayeur et al. (2011). In the sequel, we present a dislocation density-based hardening framework (Mayeur and McDowell 2013) that has been compared to a number discrete dislocation dynamics boundary value problems; some of which will be presented as representative applications of the theory.

Dislocation Density-Based Strength Model

Previously, Mayeur et al. (2011) and Mayeur and McDowell (2013) have employed both single and mutli-criterion flow rules with a variety of hardening laws to simulate size-dependent behavior observed in 2D discrete dislocation dynamics

(DDD) simulations to varying degrees of success. The hardening model presented is the simplest version capable of reproducing the observed DDD behavior.

We take the statistically stored dislocation density on each slip system, ϱ_S^α, as our primary internal state variable to describe the hardening behavior. The slip system yield stress is defined in terms of a Taylor relation that is assumed to depend only on the SSD density, i.e.,

$$r^\alpha = r_0 + c_1 \mu b \sqrt{\sum_\beta h^{\alpha\beta} \varrho_S^\beta} \tag{79}$$

where c_1 is a constant, b is the Burgers vector, and $h^{\alpha\beta}$ is an interaction matrix. We do not use a generalized Taylor relation that directly includes a dependence on the GND density, which is a commonly employed assumption in other classes of generalized single crystal models. It was shown (Mayeur and McDowell 2013) that a hardening model based on the generalized Taylor relation leads to excessive and unrealistic hardening as compared to the DDD simulations. The SSD density evolves according to a Kocks-Mecking (Mecking and Kocks 1981) relation which represents a competition between storage and recovery mechanisms until the steady-state value of dislocation density is reached:

$$\dot{\varrho}_S^\alpha = \frac{1}{b}\left(\frac{1}{\Lambda^\alpha} - 2y_c \varrho_S^\alpha\right) \dot{\lambda}^\alpha \tag{80}$$

Here, Λ^α is the mean free path between dislocations, and y_c is the capture radius for dislocation annihilation. The mean free path is defined in terms of the SSD density, an average dislocation junction strength, K, and an interaction matrix, $a^{\alpha\beta}$ as

$$\Lambda^\alpha = \frac{K}{\sqrt{\sum_\beta a^{\alpha\beta} \varrho_S^\beta}} \tag{81}$$

Since the SSD evolution equation is defined with respect to $\dot{\lambda}^\alpha$ rather than $\dot{\gamma}^\alpha$, it naturally includes scale-dependence by virtue of $\dot{\lambda}^\alpha$ being an effective slip system deformation measure (see Eq. (71)). We found this type of scale-dependent isotropic hardening described the DDD results better than either a direct dependence of the Taylor stress and/or the mean free path on the GND density. We note that size effects of initial yield stress, r_0, are not addressed by the hardening relations and may depend on the initial dislocation source distribution as well as obstacles and interfaces that impede slip.

Relationship to Slip Gradient Theory

In this section, we briefly discuss the relationship of micropolar crystal plasticity to slip gradient crystal plasticity theory in the form developed by Gurtin (2002, 2007).

A more in-depth comparison of the two model frameworks and simulation results has been given by Mayeur and McDowell (2014). Gurtin's theory is based on taking the slip system shears as continuum degrees of freedom and contains an additional balance equation – the so-called microforce balance – in addition to the classical force and angular momentum balances. The microforce balance is given as

$$\tau^{\alpha} + \nabla \cdot \boldsymbol{\xi}^{\alpha} - q^{\alpha} = 0 \tag{82}$$

where τ^{α} is the classical resolved shear stress, $\boldsymbol{\xi}^{\alpha}$ is a higher-order stress that is power-conjugate to the slip rate gradient, $\nabla\dot{\gamma}^{\alpha}$, and q^{α} is the stress power-conjugate to the slip rate, $\dot{\gamma}^{\alpha}$. In addition to Dirichlet micro-boundary conditions on the slip system shears, complementary micro-traction boundary conditions may also be prescribed in terms of the micro-traction Ξ^{α}, i.e.,

$$\Xi^{\alpha} = \boldsymbol{\xi}^{\alpha} \cdot \mathbf{n} \tag{83}$$

The macroscopic and microscopic scales are coupled by the presence of the resolved shear stress, $\tau^{\alpha} = \mathbf{n}^{\alpha} \cdot \boldsymbol{\sigma} \cdot \mathbf{s}^{\alpha}$, in the microforce balance. The microforce balance is a partial differential equation that governs the evolution of the slip system shears and can be interpreted as a nonlocal yield condition.

Thermodynamics

Thermodynamically consistent constitutive equations have been developed by Gurtin using a purely mechanical form of the 2nd law, i.e.,

$$\Delta^{intr} = \boldsymbol{\sigma} : \dot{\varepsilon}^{\mathrm{e}} + \sum_{\alpha} \left(q^{\alpha}\dot{\gamma}^{\alpha} + \boldsymbol{\xi}^{\alpha} \cdot \nabla\dot{\gamma}^{\alpha}\right) - \dot{\psi} \geq 0 \tag{84}$$

The free energy is assumed to depend on the elastic strain and the set of slip system shear gradients, $\nabla\vec{\gamma} = \left\{\nabla\gamma^{1}, \ldots, \nabla\gamma^{N}\right\}$, i.e.,

$$\psi = \hat{\psi}\left(\varepsilon^{e}, \nabla\vec{\gamma}\right) = \hat{\psi}_{e}\left(\varepsilon^{e}\right) + \hat{\psi}_{in}\left(\nabla\vec{\gamma}\right) \tag{85}$$

Expressing the time derivative of the free energy via the chain rule and inserting into Eq. (84) yields

$$\Delta^{intr} = \left(\boldsymbol{\sigma} - \frac{\partial\psi}{\partial\varepsilon^{e}}\right) : \dot{\varepsilon}^{e} + \sum_{\alpha}\left[q^{\alpha}\dot{\gamma}^{\alpha} + \left(\boldsymbol{\xi}^{\alpha} - \frac{\partial\psi}{\partial\nabla\gamma^{\alpha}}\right) \cdot \nabla\dot{\gamma}^{\alpha}\right] \geq 0 \tag{86}$$

It is assumed that $\boldsymbol{\xi}^{\alpha}$ can be decomposed additively into energetic and dissipative components such that

$$\boldsymbol{\xi}^{\alpha} = \boldsymbol{\xi}^{\alpha}_{en} + \boldsymbol{\xi}^{\alpha}_{d}, \quad \boldsymbol{\xi}^{\alpha}_{en} \doteq \frac{\partial\psi}{\partial\nabla\gamma^{\alpha}} \tag{87}$$

Following the Coleman-Gurtin (Coleman and Gurtin 1967) thermodynamic procedure, the state law for the Cauchy stress obtained is identical to that of Eq. (54)$_1$. Making use of Eq. (87), the reduced dissipation inequality is given as

$$\Delta^{intr} = \sum_{\alpha} \left(q^{\alpha} \dot{\gamma}^{\alpha} + \boldsymbol{\xi}_d^{\alpha} \cdot \nabla \dot{\gamma}^{\alpha} \right) \geq 0 \tag{88}$$

Classical quadratic elastic free energy potentials are employed within the Gurtin-type theory since the treatment of the macroscopic forces is unaltered. Representative constitutive equations for the defect energy and dissipative stresses are presented below.

Energetic Constitutive Equations (Defect Energy)

Several variations of the defect energy have been proposed in the literature and can be classified as either recoverable or nonrecoverable. A recoverable defect energy depends on the current values of the slip gradients and vanishes when there are no slip gradients regardless of prior loading history, whereas a nonrecoverable energy does not vanish upon unloading. A typical recoverable defect energy is given as (Reddy 2011)

$$\psi_d = \frac{1}{2} g_0 \ell_{en}^2 \sum_{\alpha} \left[c_{\vdash} \left(\tilde{\varrho}_{\vdash}^{\alpha} \right)^2 + c_{\odot} \left(\tilde{\varrho}_{\odot}^{\alpha} \right)^2 \right] \tag{89}$$

where g_0 is the initial flow stress, ℓ_{en} is an energetic length scale, and $c_{\vdash}/c_{\odot}$ are dimensionless constants defining the relative contributions of edge and screw GNDs, respectively. The edge and screw GND measures used in Eq. (89) are defined as

$$\tilde{\varrho}_{\vdash}^{\alpha} = -\nabla \gamma^{\alpha} \cdot \mathbf{s}^{\alpha}, \qquad \tilde{\varrho}_{\odot}^{\alpha} = -\nabla \gamma^{\alpha} \cdot \mathbf{t}^{\alpha} \tag{90}$$

Note the dimensional and sign difference of the screw GND term with respect to the GND density definitions used in section "Linearized Kinematics." The energetic microcouple stress vector corresponding to ψ_d is given as

$$\boldsymbol{\xi}_{en}^{\alpha} = -g_0 \ell_{en}^2 \left(c_{\vdash} \tilde{\varrho}_{\vdash}^{\alpha} \mathbf{s}^{\alpha} + c_{\odot} \tilde{\varrho}_{\odot}^{\alpha} \mathbf{t}^{\alpha} \right) \tag{91}$$

The microcouple stress vector lies in the slip plane and decomposes naturally into edge and screw components. Gurtin et al. (2007) view $\boldsymbol{\xi}_{en}^{\alpha}$ as reflective of a net distributed Peach-Koehler force acting on the GNDs since its components are perpendicular to the respective dislocation line directions and have units of force per unit length.

Dissipative Constitutive Equations

The dissipative constitutive equations are introduced with the aid of an effective slip system deformation rate, d^{α}, defined as

$$d^{\alpha} = \sqrt{|\dot{\gamma}^{\alpha}|^2 + L_d^2 \|\nabla_{\tan}^{\alpha} \dot{\gamma}^{\alpha}\|^2} \tag{92}$$

where L_d is a dissipative length scale and the tangential gradient operator is defined as $\nabla_{\tan}^{\alpha} f = (\nabla f \cdot \mathbf{s}^{\alpha}) \mathbf{s}^{\alpha} + (\nabla f \cdot \mathbf{t}^{\alpha}) \mathbf{t}^{\alpha}$. The scalar dissipative microstress is posited to have the functional form

$$q^{\alpha} = g^{\alpha} \left(\frac{d^{\alpha}}{d_0} \right)^{\frac{1}{m}} \frac{\dot{\gamma}^{\alpha}}{d^{\alpha}} \tag{93}$$

where g^{α} is a viscous drag stress, d_0 is the reference effective deformation rate, and m is the inverse rate-sensitivity exponent. The dissipative microcouple stress is introduced in an analogous manner, i.e.,

$$\boldsymbol{\xi}_d^{\alpha} = g^{\alpha} L_d^2 \left(\frac{d^{\alpha}}{d_0} \right)^{\frac{1}{m}} \frac{\nabla_{\tan}^{\alpha} \dot{\gamma}^{\alpha}}{d^{\alpha}} \tag{94}$$

Inserting Eqs. (93) and (94) into Eq. (88) and making use of Eq. (92) yields

$$\Delta^{intr} = \sum_{\alpha} \tilde{\tau}^{\alpha} d^{\alpha} \geq 0, \quad \tilde{\tau}^{\alpha} := g^{\alpha} \left(\frac{d^{\alpha}}{d_0} \right)^{\frac{1}{m}}. \tag{95}$$

Here, we introduce the implicitly defined effective dissipative stress, $\tilde{\tau}^{\alpha}$, that is power-conjugate to the effective slip system deformation rate d^{α}. It is straightforward to show that $\tilde{\tau}^{\alpha}$ can also be expressed in terms of the dissipative microstresses, i.e.,

$$\tilde{\tau}^{\alpha} = \left[(q^{\alpha})^2 + \|\boldsymbol{\xi}_d^{\alpha} / L_d\|^2 \right]^{1/2}. \tag{96}$$

Inverted Flow Rule Versus Microforce Balance

An examination of the micropolar and slip gradient theories reveals many similarities between the two sets of governing equations. Central to both theories is the notion that the presence of GNDs gives rise to both energetic and dissipative contributions to scale-dependent mechanical behavior. In general, nonlocal strengthening effects are manifested in both the isotropic and kinematic hardening responses of the material. A key component of the slip gradient theory is the microforce balance, which couples the macroscopic and microscopic responses and represents a nonlocal flow rule governing the evolution of the slip system shears. Typically, the $\nabla \cdot \boldsymbol{\xi}^{\alpha}$ term in the microforce balance induces kinematic hardening and the q^{α} term leads to isotropic hardening, although the exact nature of the hardening contributions depends on the specific constitutive forms employed. The noteworthy feature of second term in the microforce balance is that it is a function of second gradients of slip since $\boldsymbol{\xi}^{\alpha} = f\left(\nabla_{\tan}^{\alpha} \gamma^{\alpha}\right)$. The relationship between second gradients of slip

(GND gradients) and kinematic hardening is in accordance with other dislocation-based gradient single crystal plasticity theories (Yefimov et al. 2004a; Evers et al. 2004; Forest 2008).

A similar result is obtained for the micropolar theory by considering the inverted flow rule, which is obtained from the coaxiality relations $\tau^{\alpha}_{eff}/\hat{\tau}^{\alpha} = \dot{\gamma}^{\alpha}/\dot{\lambda}^{\alpha}$. Neglecting the isotropic energetic hardening terms in Eq. (77), the coaxiality relations may be rearranged and expressed as

$$\tau^{\alpha} + \tau^{\alpha}_{b} - g^{\alpha}\left(\frac{\dot{\lambda}^{\alpha}}{\dot{\lambda}_{0}}\right)^{\frac{1}{m}}\frac{\dot{\gamma}^{\alpha}}{\dot{\lambda}^{\alpha}} = 0. \tag{97}$$

Here, the resolved Cauchy stress, τ^{α}_{eff}, is explicitly written as two terms where the second term is a back stress, $\tau^{\alpha}_{b} := \frac{1}{2}\mathbf{t}^{\alpha}\cdot\left(\nabla\cdot\mathbf{m}^{\mathrm{T}}\right)$, related to the projection of the skew symmetric part of the stress tensor. The back stress may be expressed as a function of the couple stress since $2\ \mathrm{skw}(\boldsymbol{\sigma}) = \boldsymbol{\epsilon}\cdot(\nabla\cdot\mathbf{m}^{\mathrm{T}})$. Thus, the back stress in the micropolar theory is a function of lattice torsion-curvature gradients. This is analogous to the dependence exhibited by the slip gradient theory since second gradients of slip (GND gradients) are directly connected to lattice torsion-curvature gradients via Nye's relations (see section "Linearized Kinematics"). In both theories, the higher-order balance law leads to the coupling of the micro and macro scales and to the natural inclusion of gradient-dependent kinematic hardening. Compare Eq. (97) to the microforce balance augmented with the constitutive equation for q^{α} given in Eq. (93)

$$\tau^{\alpha} + \nabla\cdot\boldsymbol{\xi}^{\alpha} - \tilde{g}^{\alpha}\left(\frac{d^{\alpha}}{d_{0}}\right)^{\frac{1}{m}}\frac{\dot{\gamma}^{\alpha}}{d^{\alpha}} = 0. \tag{98}$$

The similarities between the two expressions given in Eqs. (97) and (98) are obvious, although some key differences emerge when the constitutive equations for $\mathbf{m}$ and $\boldsymbol{\xi}^{\alpha}$ are considered. To illustrate these differences, we compare the microforce balance and inverted micropolar flow rule augmented with simple constitutive equations for the 2D case (e.g., edge dislocations only). Inserting Eqs. (91) and (94) into Eq. (98), the following expression is obtained:

$$\begin{aligned}\tau^{\alpha} &+ g_{0}\ell^{2}_{en}c_{\vdash}\nabla\cdot\left[\left(\nabla\gamma^{\alpha}\cdot\mathbf{s}^{\alpha}\right)\mathbf{s}^{\alpha}\right] + L^{2}_{d}\nabla\cdot\left[g^{\alpha}\left(\frac{d^{\alpha}}{d_{0}}\right)^{\frac{1}{m}}\frac{\left(\nabla\dot{\gamma}^{\alpha}\cdot\mathbf{s}^{\alpha}\right)\mathbf{s}^{\alpha}}{d^{\alpha}}\right] \\ &- g^{\alpha}\left(\frac{d^{\alpha}}{d_{0}}\right)^{\frac{1}{m}}\frac{\dot{\gamma}^{\alpha}}{d^{\alpha}} = 0.\end{aligned} \tag{99}$$

An analogous expression for the inverted micropolar flow rule is obtained by using a simplified version of Eq. $(58)_2$ where $\mathbf{m} = 2\mu\ell^{2}_{e}\overline{\boldsymbol{\kappa}}^{e}$ in Eq. (97), i.e.,

$$\tau^{\alpha} + \mu\ell_e^2 \mathbf{t}^{\alpha} \cdot (\nabla \cdot \overline{\boldsymbol{\kappa}}^{\mathrm{T}}) - \frac{\mu\ell_e^2}{L_{\perp}} \mathbf{t}^{\alpha} \cdot \left(\nabla \cdot \sum_{\beta} \varphi_{\perp}^{\beta} \mathbf{s}^{\beta} \otimes \mathbf{t}^{\beta} \right) - g^{\alpha} \left(\frac{\dot{\lambda}^{\alpha}}{\dot{\lambda}_0} \right)^{\frac{1}{m}} \frac{\dot{\gamma}^{\alpha}}{\dot{\lambda}^{\alpha}} = 0. \tag{100}$$

Note that the energetic and dissipative length scales appear isolated in separate terms in Eq. (99), whereas the third term in Eq. (100) contains both the elastic and plastic length scales. Therefore, both energetic and dissipative gradient effects are eliminated when $\ell_e = 0$ in the micropolar theory; however, the dissipative gradient term is suppressed for $L_{\perp} \gg \ell_e^2$ irrespective of the value of ℓ_e. The separation of energetic and dissipative gradient hardening terms in the slip gradient theory is a direct consequence of the assumed additive decomposition of $\boldsymbol{\xi}$. Another, perhaps more subtle, difference between the two expressions is that the dissipative gradient term in the microforce balance is a function of slip rate gradients, while it is a function of total plastic curvature in the inverted micropolar flow rule. We remark that if the micropolar theory was cast in terms of an additive energetic/dissipative decomposition of the couple stress tensor rather than an additive elastic-plastic decomposition of the torsion-curvature tensor, the micropolar model would mirror the phenomenology of the slip gradient model in terms of a strict separation of energetic and dissipative gradient hardening effects, and the dissipative gradient hardening would be proportional to the rate of lattice torsion curvature.

Curvatures, Couple Stresses, and Gradient Terms

With the connection between the vectorial microcouple stress, $\boldsymbol{\xi}^{\alpha}$, and the second rank couple stress tensor, $\mathbf{m}$, established, additional analogies between the two models are explored. A comparison of Eqs. (90) and (52) reveals a relationship between the micropolar plastic torsion-curvature flow rates and the directional slip gradients, i.e.,

$$\dot{\varphi}_{\perp}^{\alpha} \leftrightarrow L_d \nabla \dot{\gamma}^{\alpha} \cdot \mathbf{s}^{\alpha} \quad , \quad \dot{\varphi}_{\odot}^{\alpha} \leftrightarrow L_d \nabla \dot{\gamma}^{\alpha} \cdot \mathbf{t}^{\alpha} \tag{101}$$

where $\leftrightarrow$ is used to signify that the two terms have analogous roles in the respective theories. This association is further evident when the effective deformation rate in Eq. (92) is expressed in component form as

$$d^{\alpha} = \sqrt{|\dot{\gamma}^{\alpha}|^2 + |L_d \nabla \dot{\gamma}^{\alpha} \cdot \mathbf{s}^{\alpha}|^2 + |L_d \nabla \dot{\gamma}^{\alpha} \cdot \mathbf{t}^{\alpha}|^2}. \tag{102}$$

which establishes $d^{\alpha} \leftrightarrow \dot{\lambda}^{\alpha}$ (see Eq. (71)). If the energetic isotropic hardening term in Eq. (76) is neglected and compared to the analogous expression in Eq. (95), we see that the two dissipation inequalities have identical forms. Therefore, it is straightforward to show that the resolved couple stresses in the micropolar theory can be related to the dissipative microcouple stresses of the slip gradient theory by expressing the dissipation inequality in a form that parallels Eq. (75). First, the dissipative constitutive equation for $\boldsymbol{\xi}_d^\alpha$ given in Eq. (94) is written in component form as

$$\boldsymbol{\xi}_d^\alpha = g^\alpha L_d^2 \left(\frac{d^\alpha}{d_0} \right)^{\frac{1}{m}} \frac{(\nabla \dot{\gamma}^\alpha \cdot \mathbf{s}^\alpha)\,\mathbf{s}^\alpha + (\nabla \dot{\gamma}^\alpha \cdot \mathbf{t}^\alpha)\,\mathbf{t}^\alpha}{d^\alpha} = \xi_{d\vdash}^\alpha \mathbf{s}^\alpha + \xi_{d\odot}^\alpha \mathbf{t}^\alpha \tag{103}$$

where $\xi_{d\vdash}^\alpha$ and $\xi_{d\odot}^\alpha$ are the edge and screw components of $\boldsymbol{\xi}_d^\alpha$. Inserting Eq. (103) into Eq. (88) and making use of the microforce balance, the dissipation inequality may be expressed as

$$\Delta^{intr} = \sum_\alpha \left[(\tau^\alpha + \nabla \cdot \boldsymbol{\xi}^\alpha)\,\dot{\gamma}^\alpha + \frac{\xi_{d\vdash}^\alpha}{L_d} (L_d \nabla \dot{\gamma}^\alpha \cdot \mathbf{s}^\alpha) + \frac{\xi_{d\odot}^\alpha}{L_d} (L_d \nabla \dot{\gamma}^\alpha \cdot \mathbf{t}^\alpha) \right] \geq 0. \tag{104}$$

A comparison of Eqs. (75) and (104) in conjunction with the relationships identified in Eq. (101) reveals that

$$\tau_{eff}^\alpha \leftrightarrow \tau^\alpha + \nabla \cdot \boldsymbol{\xi}^\alpha \tag{105}$$

$$\pi_\perp^\alpha / L_\perp \leftrightarrow \xi_{d\vdash}^\alpha / L_d \tag{106}$$

$$\pi_\odot^\alpha / L_\odot \leftrightarrow \xi_{d\odot}^\alpha / L_d \tag{107}$$

The association established in Eq. (105) was already apparent from a comparison of the inverted micropolar flow rule and the microforce balance, whereas the relationships established by Eqs. (106) and (107) were revealed by inserting the component form of $\boldsymbol{\xi}_d^\alpha$ into the reduced dissipation inequality. Furthermore, we can rewrite the effective dissipative stress in the slip gradient theory using the component form of $\boldsymbol{\xi}_d^\alpha$ as

$$\tilde{\tau}^\alpha = \sqrt{|\tau^\alpha + \nabla \cdot \boldsymbol{\xi}^\alpha|^2 + \left|\xi_{d\vdash}^\alpha / L_d\right|^2 + \left|\xi_{d\odot}^\alpha / L_d\right|^2}. \tag{108}$$

This expression is consistent with Eq. (61) of the micropolar theory given the relationships established by Eqs. (105), (106), and (107).

As a final remark, we remind the reader that the micropolar theory differs from slip gradient theory by its adoption of the rotation of the microstructure director vectors in Eq. (30) as an independent constitutive prescription. This also simplifies the numerical implementation since slip gradients need not be computed or estimated. The remainder of section "Micropolar Single Crystal Plasticity" explores applications of this single crystal micropolar framework. The chapter dedicated to micromorphic crystal plasticity considers further extension of the micropolar theory to a micromorphic theory that includes distortion of the director vectors.

Applications: Comparison to 2D Discrete Dislocation Dynamics Simulations

This section presents a comparison of two-dimensional micropolar crystal plasticity and discrete dislocation dynamics simulations of single crystal thin films subjected to a variety of loading conditions. Specifically, we examine three widely studied boundary value problems: constrained simple shear, pure bending (single and double slip configurations), and simple shear of a particle reinforced composite. These problems are routinely used as benchmarks to evaluate the ability of generalized single crystal plasticity models to accurately capture size effects. A micropolar single crystal model with the hardening described by Eqs. (79), (80), and (81) was implemented in an Abaqus/Standard Version 6.7.1 (2007) user element subroutine (UEL) for the simulations presented below. The element is a four node quadrilateral and the integration is performed using a B-bar technique to prevent volumetric locking. Both the displacement and rotation fields are interpolated using standard bilinear shape functions.

The micropolar model has been independently calibrated for each boundary value problem by fitting to both average (e.g., stress-strain curves) and microscopic (e.g., dislocation density distributions) deformation behavior. It was previously demonstrated that fitting to multiple aspects of the deformation behavior is necessary to obtain a unique set of micropolar constitutive parameters (Mayeur et al. 2011). As a result of the 2D nature of the problems considered, the calibration procedure is simplified in the sense that there are only a few parameters to be determined. Essentially, there are four fitting parameters that are summarized by the set: $\mathrm{M} = \{r_0, K, \ell_e, L_\perp\}$. All of the other material parameters are either known as inputs (e.g., classical elastic parameters, Burgers vector magnitude) or can be assigned standard values based on experience from classical local crystal plasticity models. For example, the viscoplastic parameters $\left(\dot{\lambda}^\alpha, g^\alpha, m\right)$ cannot be explicitly determined due to lack of DDD results at multiple strain rates, so reasonable values for these parameters are prescribed and held fixed throughout the calibration process. Note that only a single elastic and plastic length must be determined. For plane strain problems, the couple stress constitutive relation given in Eq. $(58)_2$ reduces to

$$m_{3i} = 2\beta\overline{\kappa}^e_{3i} = 2\mu\ell_e^2\overline{\kappa}^e_{3i}, \quad i = 1, 2, \tag{109}$$

where it has been assumed, without loss of generality, that $\beta = \gamma$. Recall, the couple stress constitutive parameter, β, is related to an elastic length scale parameter, $\ell_e = \sqrt{\beta/\mu}$. As discussed by Forest (2008), the nonclassical elastic constant, μ_c, Eq. $(58)_1$, is not a free fitting parameter in micropolar single crystal plasticity. Rather, it serves as an internal penalty constraint forcing the lattice rotations to coincide with the rotational part of the elastic distortion. Because of this constraint, the micropolar torsion-curvature is identified as the lattice torsion-curvature thereby making the connection between GNDs and the scale effects predicted by the model.

Since screw GNDs do not contribute to plastic torsion-curvature evolution in the 2D boundary value problems, there is a single plastic length scale parameter, $L_{\perp}$, to be determined.

Here, we briefly outline how the length scale parameters affect material response a general strategy for calibrating them to the DDD simulations. The elastic length scale parameter, ℓ_e, is related to the initial scale-dependent kinematic hardening modulus and the ratio $L_{\perp}/\ell_e$ dictates the saturation rate of gradient hardening. Larger $L_{\perp}/\ell_e$ ratios are associated with slower transients such that when $L_{\perp} \gg \ell_e$, the strengthening effect reduces to that of linear gradient kinematic hardening (Mayeur et al. 2011). The elastic length scale should also be small enough so as not to induce size-dependent effects in the elastic deformation regime. The plastic length scale, $L_{\perp}$, determines the magnitude of the GND density distributions; however, its influence is not independent of the prescribed value of ℓ_e. As shown in Mayeur et al. (2011), it is the ratio $L_{\perp}/\ell_e$ that dictates the maximum value of the local GND density fields. If ℓ_e is varied while holding $L_{\perp}/\ell_e$ and all other model parameters constant (including specimen dimensions), the resulting GND density field will be essentially unchanged. To illustrate how the length scale parameters influence the average stress strain response, we show two plots in Fig. 1 where the elastic length scale is varied while keeping the ratio $L_{\perp}/\ell_e$ fixed (here $L_p = L_{\perp}$) and another with fixed elastic length scale and varying $L_{\perp}/\ell_e$ for the problem of constrained simple shear (Mayeur and McDowell 2014).

The parameter r_0 is used to fit to the initial yield strength rather than a prescribed initial dislocation density since the DDD simulations are assumed to be initially dislocation free. However, in the planar double slip simulations, a negligibly small value of initial SSD density (10^{-6} μm^{-2}) is specified for each slip system so that SSD evolution is nonzero. This approach is taken in lieu of introducing a nucleation term in the SSD density evolution equation. It is worth mentioning that r_0 should not necessarily be treated as a fixed material constant due to the statistical variations in the source strength and spatial distribution in the discrete dislocation simulations, as well as the mean free path of initial obstacles or impenetrable interfaces. The initial yield point in the discrete simulations will depend significantly on the availability of weak sources in highly stressed regions. Given a fixed value for the initial slip system yield strength, approximate upper bounds can be established for ℓ_e and K by isolating the effects of the two distinct material strengthening mechanisms. For example, given a target stress-strain curve, the maximum value of ℓ_e is determined by assuming that all of the material hardening is due to the gradient-induced back stresses ($1/K = 0$), whereas the maximum value of K is determined by assuming that all of the material hardening is due to slip threshold hardening ($\ell_e = 0$). In general, both mechanisms will contribute to the material strengthening and the actual values will fall below the upper bounds. As discussed in Mayeur and McDowell (2013), either the unloading behavior or some other attribute of the local deformation field (e.g., shear strain distributions) must be used in order to differentiate between the relative hardening contributions. A list of micropolar constitutive parameters used in all of the simulations is given in Table 1.

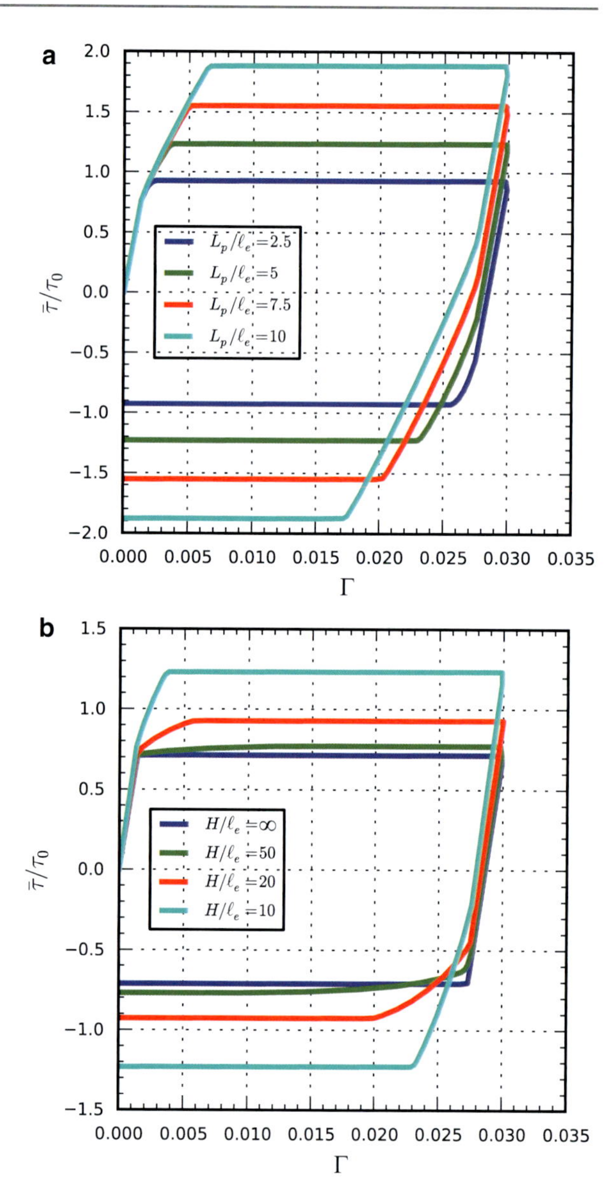

Fig. 1 Average shear stress-strain response for constrained simple shear with (**a**) different values of L_p/ℓ_e with $\ell_e = 100$ nm and (**b**) different elastic length scale parameters with $L_p/\ell_e = 5$. Slip threshold hardening is suppressed ($r^\alpha = r_0$)

Constrained Shear of Thin Films

Here, we compare results of the micropolar model to DDD results for a constrained thin film subjected to simple shear originally presented by Shu et al. (2001). Related studies were also conducted by Yefimov and van der Giessen (2005b) and Limkumnerd and van der Giessen (2008). A more detailed exposition of the micropolar results are contained in Mayeur and McDowell (2013). The specimen

Table 1 Summary of constitutive model parameters used in the micropolar single crystal simulations

Parameter	Symbol		Magnitude			Unit
–	–	Constrained shear	Bending SS	Bending DS	Composite	–
Shear modulus	μ	26.3	26.3	26.3	26.3	GPa
Poisson's ratio	ν	0.33	0.33	0.33	0.33	–
Couple modulus	μ_c	263	263	263	263	GPa
Elastic length scale	ℓ_e	10, 15	125, 125	300, 600	125	nm
Plastic length scale	$L_\perp$	45	562.5, 250	750, 700	125	nm
Reference threshold stress	r_0	12.78	10	10	13, 21, 30	MPa
Threshold stress coefficient	c_1	0.5	N/A	0.5	N/A	–
Burgers vector magnitude	b	0.25	0.25	0.25	nm	
Hardening matrix coefficients	$h^{\alpha\beta}$	$\delta^{\alpha\beta}$	N/A	1.0	N/A	–
Initial SSD density	ϱ_{S0}	10^{-6}	N/A	10^{-6}	N/A	μm^{-2}
Dislocation interaction coefficients	$a^{\alpha\beta}$	$\delta^{\alpha\beta}$	N/A	1.0	N/A	–
Dislocation segment length constant	K	16.67,18.18	N/A	160, 26	N/A	–
Dislocation capture radius	y_c	0	N/A	1.5	N/A	nm
Reference deformation rate	$\dot{\lambda}_0$	10^{-3}	10^{-3}	10^{-3}	10^{-3}	s^{-1}
Drag stress	g	5	5	5	5	MPa
Inverse rate sensitivity exponent	m	20	20	20	20	–

geometry is film oriented for symmetric slip with thickness, H, in the x_2-direction as shown in Fig. 2. The slip systems are oriented at $\pm 30°$ with respect to the x_2-direction. The top and bottom surfaces are modeled as rigid dislocation barriers (impenetrable), the bottom surface is fully constrained against displacement, and a uniform horizontal displacement is applied to the top surface while the vertical displacement is constrained. The load is applied under displacement control up to an

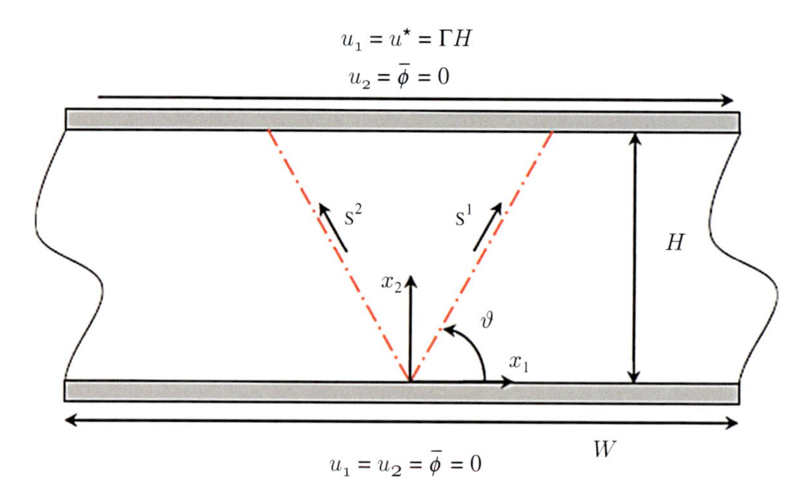

Fig. 2 Geometry schematic and boundary conditions for the constrained shear initial-boundary value problem

average strain of $\Gamma = 0.03$. The discrete dislocation problem was modeled as a unit cell of width, W, and thickness, H, and was spatially discretized with uniformly sized quadrilateral finite elements with an element size of $h_e = W/30$. The material was modeled as having zero initial dislocation density and the sources were distributed randomly throughout the spatial domain. Individual dislocation source strengths were determined by randomly sampling from a Gaussian distribution with a mean nucleation strength, $\overline{\tau}_{nuc} = 50$ MPa, and a standard deviation of $0.2\overline{\tau}_{nuc}$.

The displacement boundary conditions were applied at the top surface consistent with a constant strain rate of 10^3 s^{-1}. The constitutive parameters used in the simulations are representative of an aluminum single crystal.

$$\begin{array}{ll} u_1(x_1, 0, t) = u_2(x_1, 0, t) = \overline{\phi}(x_1, 0, t) = 0 & \\ u_1(x_1, H, t) = \Gamma(x_1, H, t)\, H, & u_2(x_1, H, t) = \overline{\phi}(x_1, H, t) = 0 \\ u_1\left(\frac{W}{2}, x_2, t\right) = u_1\left(-\frac{W}{2}, x_2, t\right) \quad , & u_2\left(\frac{W}{2}, x_2, t\right) = u_2\left(-\frac{W}{2}, x_2, t\right) \\ \overline{\phi}\left(\frac{W}{2}, x_2, t\right) = \overline{\phi}\left(-\frac{W}{2}, x_2, t\right) & \end{array} \tag{110}$$

We only used the data for the 1 μm thick film in the calibration process since this is the only thickness for which all relevant deformation fields have been reported. Consistent with earlier nonlocal single crystal plasticity simulations of this problem, the isotropic hardening response is assumed to be linear (Shu et al. 2001; Bittencourt et al. 2003). Due to the symmetry of the problem and since it is impossible to differentiate between self and latent hardening effects from the available DDD results, we assume that $h^{\alpha\beta} = a^{\alpha\beta} = \delta^{\alpha\beta}$. Results are presented for two sets of constitutive parameters. As shown in Fig. 3, the stress-strain curves are nearly identical for both fits and are in good agreement with the discrete dislocation results. Fit 1 uses slightly

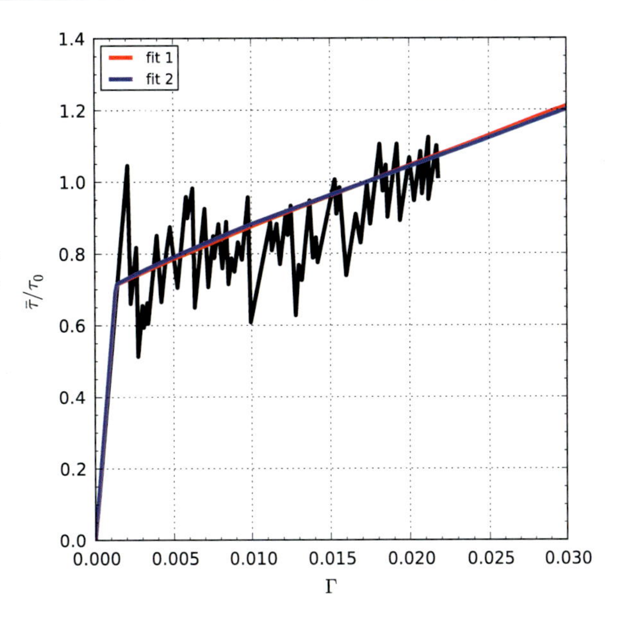

Fig. 3 Comparison of the average shear stress-strain response for $H = 1\ \mu$m. Results for two parameter sets are shown to illustrate how differences in the local fields can vary for an identical average response. The discrete dislocation results are from Shu et al. (2001) and are given by the solid black line

more threshold hardening ($K_1 = 16.67$ vs. $K_2 = 18.18$), whereas fit 2 has a larger contribution from gradient kinematic hardening ($\ell_{e2} = 15$ nm vs. $\ell_{e1} = 10$ nm). The differences in the relative proportions of isotropic versus gradient kinematic hardening are evident in the shear strain distributions shown in Fig. 4. The profiles for fit 1 have a blunted shape as compared to the rounded morphology observed for fit 2, which is consistent with the general observation that larger elastic length scales result in parabolic shear strain distributions. The signed GND density distributions for both fits are shown in Fig. 5a. The maximum GND densities at the boundary are marginally overpredicted and display steeper gradients in the near boundary regions than the discrete dislocation results, but compare favorably overall.

Pure Bending of Thin Films: Single Slip

The discrete dislocation results of the single slip bending configuration presented in this section were obtained by Yefimov et al. (2004b) and the micropolar results were given by Mayeur and McDowell (2011). The initial-boundary value problem is a thin film of width, W, and thickness, H, subjected to pure bending in a state of plane strain as sketched in Fig. 6. Considering a coordinate system attached to the midpoint of the film, the deformation is defined by the edge rotation angle, Θ, and is prescribed through a linear variation of the x_1 displacement component as a function of distance from the neutral axis:

$$u_1\left(\pm\frac{W}{2}, x_2, t\right) = \pm\Theta(t)x_2. \tag{111}$$

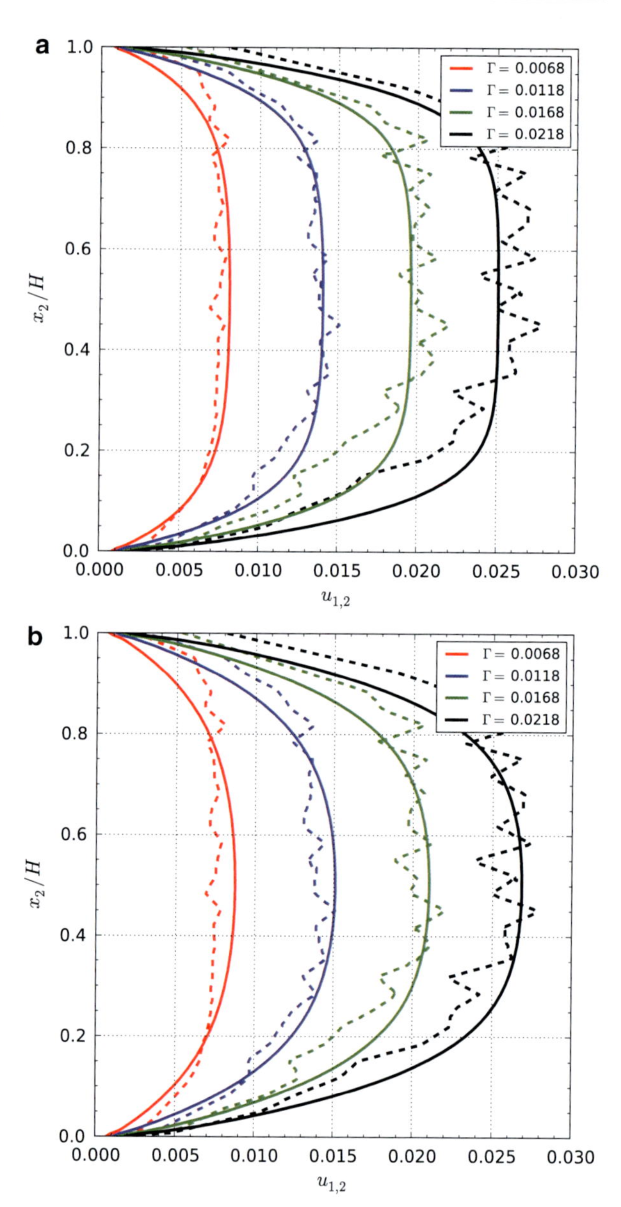

Fig. 4 Comparison of shear strain distributions at different levels of applied strain for $H = 1\ \mu$m: (**a**) fit 1 and (**b**) fit 2. The discrete dislocation results are from Shu et al. (2001) and are given by the dashed lines

The displacements are applied consistent with a constant average rotation rate, $10^{-3}\ \text{s}^{-1}$, until a final rotation angle of $\Theta = 0.02$ is reached. The top and bottom surfaces of the beam are traction-free. As shown in Fig. 6, a single slip system is oriented at an angle ϑ with respect to the x_1-axis, and slip is constrained to occur within region demarcated by the internal solid black lines. This restriction has been imposed in the discrete dislocation simulations in order to avoid the complication of

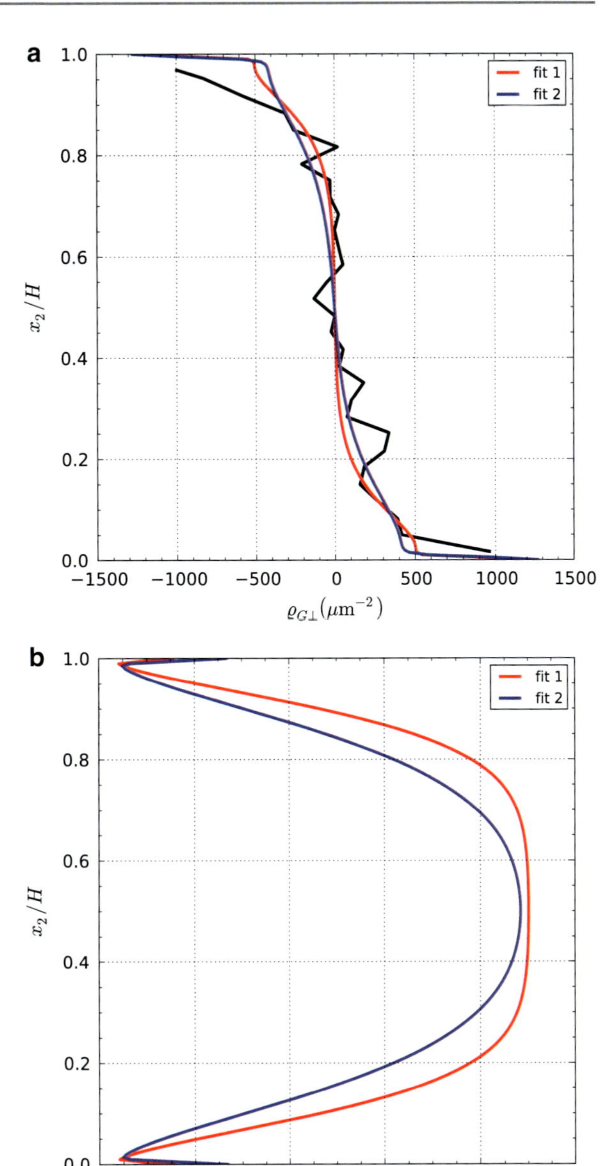

Fig. 5 Comparison of discrete dislocation and micropolar crystal plasticity dislocation density distributions at $\Gamma = 0.0168$ for $H = 1\ \mu$m: (**a**) signed GND density and (**b**) SSD density. The discrete dislocation result is from Shu et al. (2001) and are given by the solid black line

having dislocations exit the crystal through the lateral faces where the displacement boundary conditions are prescribed. Two different film thicknesses are considered for the single slip configuration, $H = 2$ and $4\ \mu$m, respectively, with a fixed width-to-thickness ratio of $W/H = 3$. Slip system orientations of 30° and 60° are studied and will be referred to using the shorthand notation ϑ_{30} and ϑ_{60} in the following.

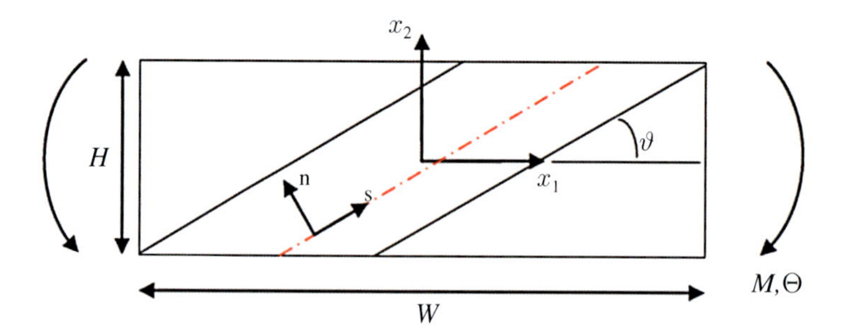

Fig. 6 Schematic of the geometry and slip system configuration for the single slip bending initial-boundary value problem

The average loading response is quantified by the bending moment, M, work-conjugate to Θ which is given by

$$M = \int_{-H/2}^{H/2} \sigma_{11}\left(\pm\frac{W}{2}, x_2\right) \; x_2 \; dx_2. \tag{112}$$

A thickness-independent measure of the average loading response is given by the normalized bending moment, M/M_{ref}, with the reference bending moment defined as: $M_{ref} = \frac{2}{3}\overline{\tau}_{nuc}\left(\frac{H}{2}\right)^2$. M_{ref} corresponds to the moment calculated from an assumed linear stress distribution over height of the beam with peak values of $\pm\overline{\tau}_{nuc}$ at the free surfaces. The mean critical nucleation stress, $\overline{\tau}_{nuc}$, is taken as 50 MPa in the discrete dislocation simulations. The micropolar finite element meshes, consistent with the DDD simulations, employ a uniform grid of bilinear quadrilaterals: 66×38 for ϑ_{30} and 155×30 for ϑ_{60}.

The normalized moment-rotation plots for both slip system orientations and thicknesses are plotted in Fig. 7, and they are in reasonably good agreement with the discrete dislocation results with respect to both the orientation and scale-dependence. Except for the 2 μm thick film for ϑ_{30}, the results are in good quantitative agreement with the initial yield strengths and nominal hardening rates predicted by the discrete dislocation model. The DDD results show that the yield strength for the 2 μm thick film for ϑ_{30} is lower than that of the 4 μm film in contradiction to an expected "smaller is stronger" trend, which underscores the stochastic nature of the initial flow stress obtained from DDD simulations. The discrete simulations display an approximately linear hardening rate that increases with decreasing film thickness, and the micropolar model shows similar trends although the rate is somewhat underestimated for the 2 μm thick film with ϑ_{60}. There is a substantially higher hardening rate for ϑ_{30} as compared to ϑ_{60} as shown in Fig. 7; this difference is primarily due to the way the boundary value problem is constructed with distinct elastic and plastic zones and is not a consequence of the local hardening behavior. Recall that the films are modeled as composite elastic-plastic materials with plastic deformation restricted to the interior region

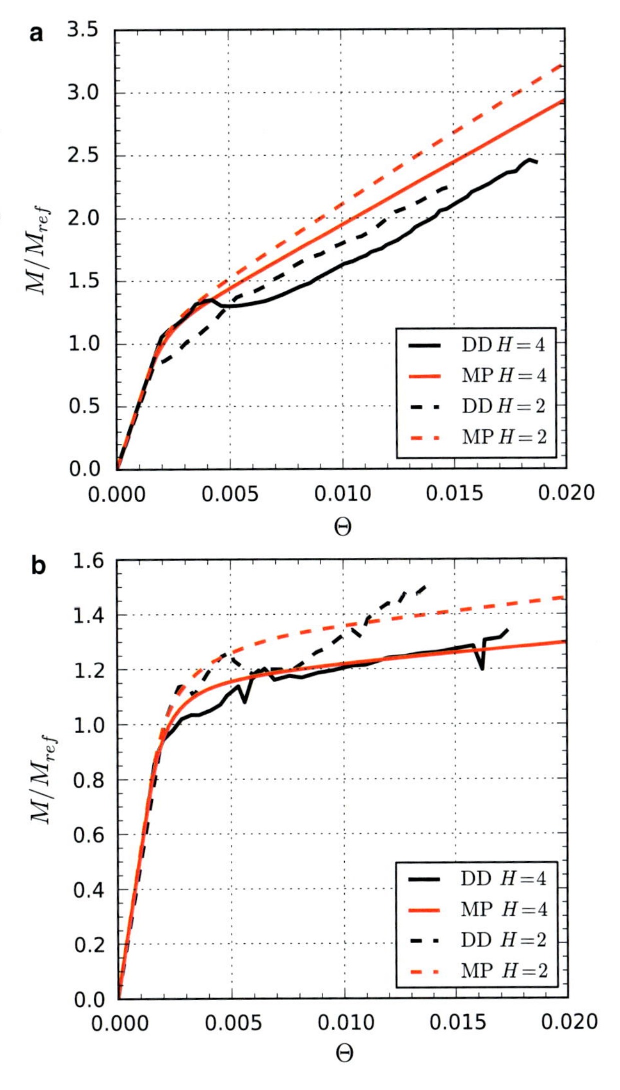

Fig. 7 Comparison of the discrete dislocation (DD) and micropolar crystal plasticity (MP) normalized moment-rotation response for the single slip bending configuration (**a**) $\vartheta = 30^{\circ}$ and (**b**) $\vartheta = 60^{\circ}$. The discrete dislocation results are from Yefimov et al. (2004b)

bounded by the solid lines parallel to the slip direction (see Fig. 6) for the sake of convenience in the DDD simulations. Therefore, the plastic zone size for ϑ_{30} is much smaller. The significantly higher apparent hardening rates for ϑ_{30} as compared to ϑ_{60} are essentially due to an increased volume fraction of the elastic phase and not dislocation hardening. In fact, a local crystal plasticity model with an elastic-perfectly plastic slip system level response would yield an apparent hardening rate comparable to, albeit lower than that shown in Fig. 7a. Of course, there is a component of the apparent hardening rate differences for the two orientations due to the relative misalignment of the slip and axial strain directions, and this is the portion associated with variations in GND distributions between the two cases.

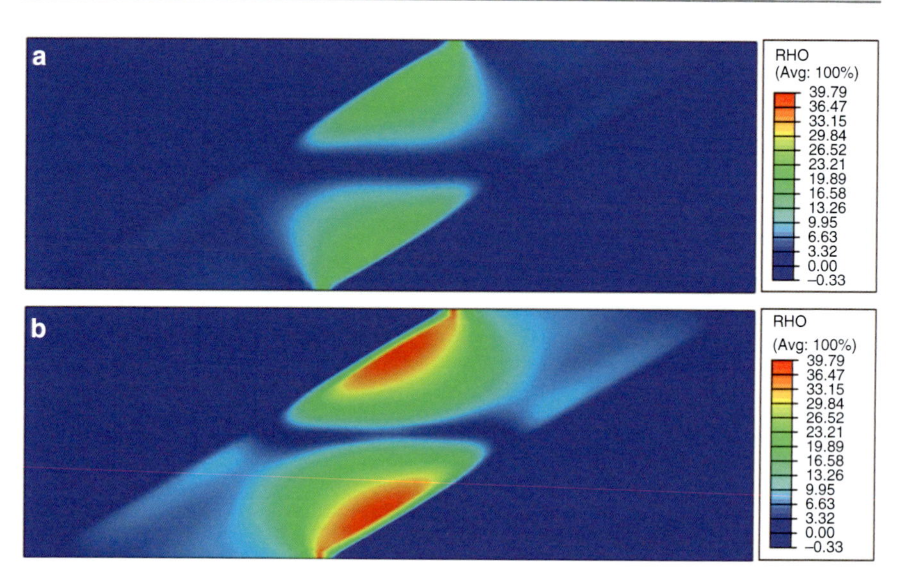

Fig. 8 Dislocation density distributions predicted by the micropolar crystal plasticity simulations at $\Theta = 0.015$ for $\vartheta = 30^{\circ}$: (**a**) $H = 4\ \mu\text{m}$ (**b**) $H = 2\ \mu\text{m}$

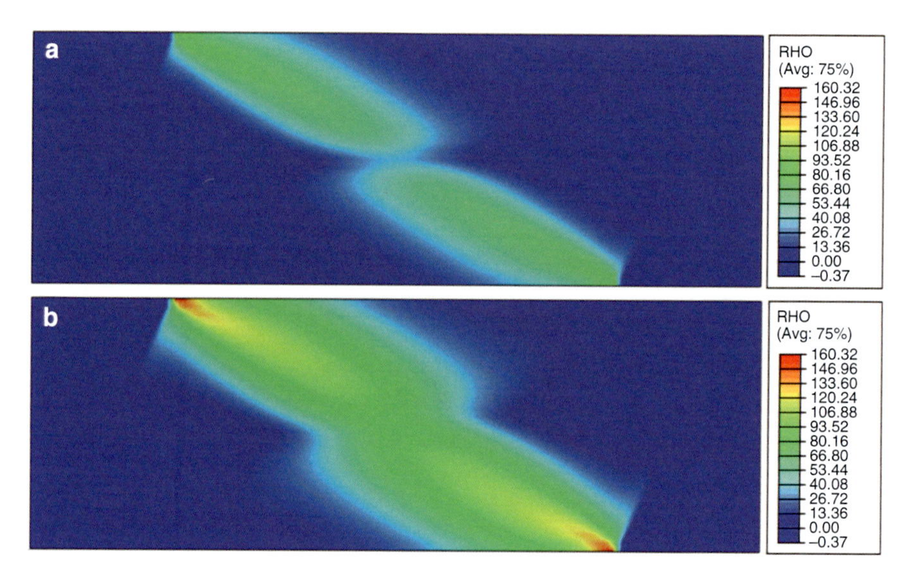

Fig. 9 Dislocation density distributions predicted by the micropolar crystal plasticity simulations at $\Theta = 0.015$ for $\vartheta = 60^{\circ}$: (**a**) $H = 4\ \mu\text{m}$ (**b**) $H = 2\ \mu\text{m}$

Dislocation density contour plots for both film thicknesses are shown in Figs. 8 and 9 for ϑ_{30} and ϑ_{60}, respectively, with $\Theta = 0.015$. In the ϑ_{30} film, the maximum dislocation densities are 39.8 μm^{-2} and 22.6 μm^{-2}, respectively, for 2 and 4 μm thick films. Dislocation-free zones are clearly observed along the neutral

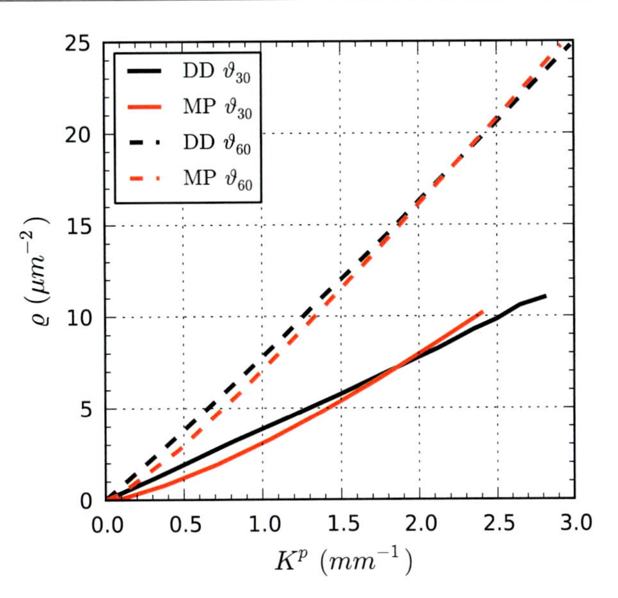

Fig. 10 Comparison of the discrete dislocation (DD) and micropolar crystal plasticity (MP) dislocation density evolution as a function of average plastic curvature for both orientations of the single slip configuration. The dashed curves are for ϑ_{60}. The discrete dislocation results are from Yefimov et al. (2004b)

axis for both film thicknesses, where the thickness of the dislocation-free region is approximately 3–4 times larger for the 4 μm thick film. As compared to the ϑ_{30} film, the dislocation density distributions for ϑ_{60} are markedly different. The morphology of the distribution can be characterized as having ellipsoidal-shaped lobes originating at the free surface near the corner of the elastic-plastic interface and extending perpendicular to elastic-plastic interface toward the neutral axis. The maximum local dislocation densities, 160 μm^{-2} ($H = 2\ \mu$m) and 98.3 μm^{-2} ($H = 4\ \mu$m), are significantly higher than in the ν_{30} film. The contour plots exhibit rather high dislocation density at the free surfaces that are generated in response to the strong rotational gradients that arise due to the compliance mismatch at the elastic-plastic interface. This is in contrast to the DDD simulations which exhibit dislocation-free zones at the free surfaces that are thought to be the result of an image force effect. It is possible that an image force effect might be imposed within the micropolar framework through an appropriately specified higher-order traction along the free surfaces; however, this avenue has yet to be pursued.

In Fig. 10, the total dislocation density computed over the entire volume is plotted as a function of the imposed deformation for both orientations with $H = 4\ \mu$m. It is shown that the micropolar model accurately captures the evolution as predicted by the DDD simulations. The dislocation density is computed by volume averaging the centroidal element values over the FE mesh, and the average plastic curvature, K^p, is calculated according to

$$K^p = \frac{2\Theta}{W} - \frac{M}{EI}. \tag{113}$$

Here, EI is the in-plane bending stiffness defining the elastic curvature. The total GND density required to accommodate an imposed bending angle, Θ, can be calculated according to Ashby (1970) in terms of K^p as

$$\hat{\rho} = \frac{K^p}{b_1}, \tag{114}$$

where b_1 is the magnitude of the x_1-component of the Burgers vector. The plot shows that the dislocation density increases in an approximately linear fashion with respect to the plastic curvature for both sets of simulations. Also, as expected from Eq. (114), the dislocation density at a given level of plastic curvature is higher for the ϑ_{60} film. Since the x_1-component of the Burgers vector is smaller for this orientation, more dislocations are needed to accommodate the imposed strain gradient.

Pure Bending of Thin Films: Double Slip

Next we study pure bending of thin films with a double slip system configuration as shown in Fig. 11. The DDD simulation results presented were originally reported by Yefimov and van der Giessen (2005a) and an analysis using micropolar crystal plasticity was presented by Mayeur (2010). As in the single slip case, two orientations are considered, ϑ_{30} and ϑ_{60} where the orientation angle is also defined as the angle between the x_1-axis and the slip direction for slip system 1. For each orientation, the second slip system is symmetrically aligned with respect to the x_2-axis. The films are partitioned into elastic and plastic phases as before and in accordance with the discrete dislocation simulations. The elastic zones are the triangular regions located at the top, bottom, left, and right ends of the film as shown in Fig. 11, and each slip system is only active in the slice of material parallel to the slip direction. Thus, the plastic zone can be divided into five regions: four single slip regions (two for each slip system) that are the outermost diagonal regions and a diamond-shaped double slip region located at the center of the film and is demarcated by the blue dotted lines in the schematic. The boundary conditions and

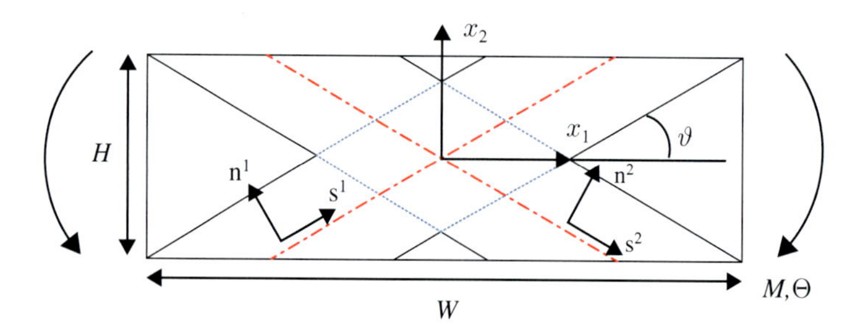

Fig. 11 Schematic of the geometry and slip system configuration for the double slip bending initial-boundary value problem

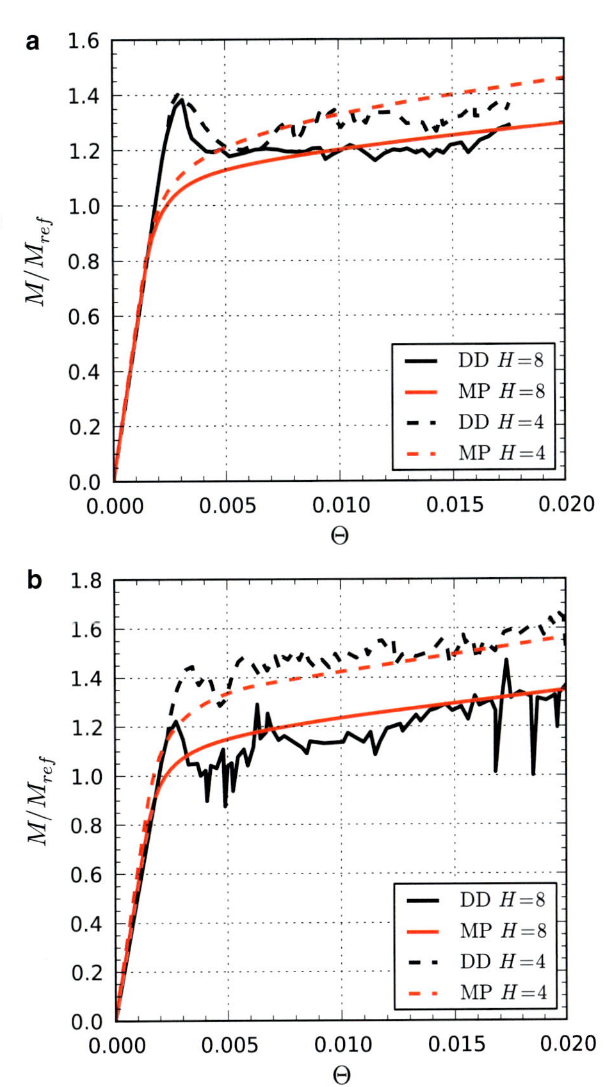

Fig. 12 Comparison of the discrete dislocation (DD) and micropolar crystal plasticity (MP) normalized moment-rotation response for the double slip bending configuration (**a**) $\vartheta = 30^\circ$ and (**b**) $\vartheta = 60^\circ$. The discrete dislocation results are from Yefimov and van der Giessen (2005a)

FE discretizations are the same as for the single slip configuration. Simulations are carried out for film thicknesses of 4 μm and 8 μm with fixed aspect ratio $W/H = 3$.

The normalized moment-rotation responses for both film thicknesses and orientations are plotted in Fig. 12 against the discrete dislocation results, and the results compare favorably. The response is similar for both orientations unlike the single slip configuration, where the hardening rate was much higher for the ϑ_{30} films due to larger effective film thickness resulting from the dominant influence of the elastic regions. However, the behavior for the ϑ_{60} films has a stronger scale-dependence as would be expected given that more GNDs are necessary to accommodate the

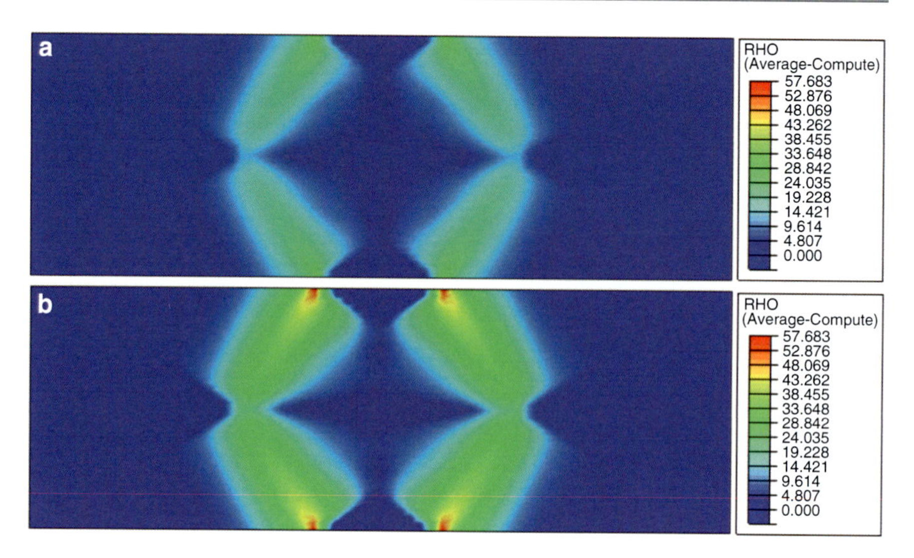

Fig. 13 Dislocation density distributions predicted by the micropolar crystal plasticity simulations at $\Theta = 0.02$ for $\vartheta = 30^{\circ}$: (**a**) $H = 8\ \mu\text{m}$ (**b**) $H = 4\ \mu\text{m}$

strain gradients for the ϑ_{60} film. In contrast to the single slip simulations, we found that different elastic length scales and dislocation multiplication constants, K, are required for each orientation to obtain a good match with the DDD results.

Figure 13 shows the total dislocation density contour plots for the ϑ_{30} films at $\Theta = 0.02$, and we note that the magnitude of the total dislocation density field is approximately three orders of magnitude larger than that of the SSD density field (not shown), which is characteristic of confined micro-scale dislocation plasticity. The maximum dislocation density is located at the free surfaces near the elastic-plastic interfaces for both film thicknesses, with peak values of 57.7 μm^{-2} and 32.3 μm^{-2} for the 4 μm and 8 μm thick films, respectively. In the case of the 8 μm thick film, there is dislocation-limited region (not dislocation-free) adjacent to the neutral axis separating the regions of higher dislocation density, whereas the dislocation distributions are continuous across the neutral axis for the 4 μm thick film. This is in contrast to the dislocation density fields for the single slip configuration (see Fig. 8) which exhibit a clear dislocation starved zone adjacent to the neutral axis. It is interesting that the maximum dislocation density values occur in the single slip regions and that the double slip region has a significantly lower density.

Figure 14 shows the total dislocation density contour plots for the ϑ_{60} films at $\Theta = 0.02$. The geometrical configuration of the elastic and plastic phases for the ϑ_{60} oriented crystal is such that there is no centrally located elastic zone and the majority of the plastic phase is a double slip region. Therefore, the dislocation density fields are continuous and smooth, in contrast to the ϑ_{30} orientation where the dislocation density field has a checkered type of pattern. The local maximum in the dislocation

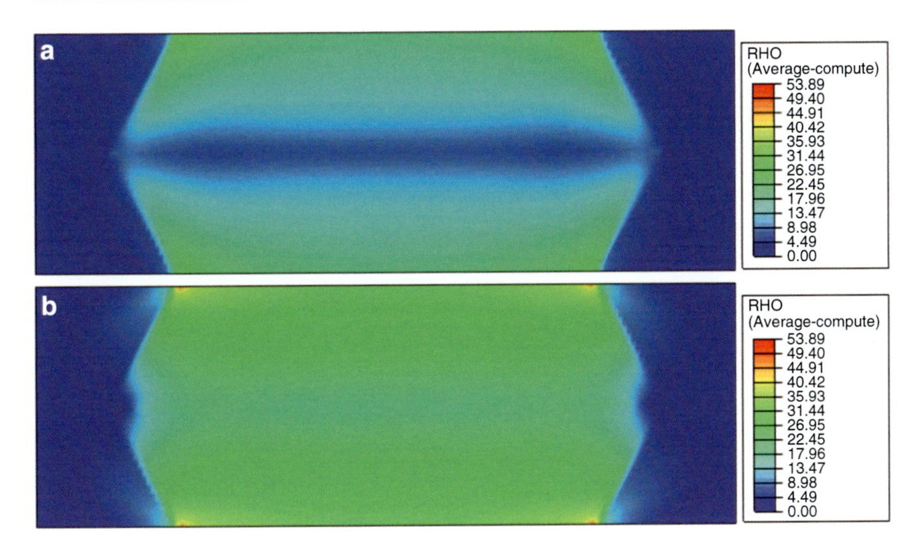

Fig. 14 Dislocation density distributions predicted by the micropolar crystal plasticity simulations at $\Theta = 0.02$ for $\vartheta = 60^{\circ}$: (**a**) $H = 8\ \mu\text{m}$ (**b**) $H = 4\ \mu\text{m}$

density fields are comparable for both orientations, thus the total (over the entire plastic region) dislocation density is much higher for the ϑ_{60} film.

The total film dislocation density is plotted for both slip orientations and thicknesses versus the average plastic curvature in Fig. 15. In general, the micropolar results compare favorably to the discrete dislocation results. The micropolar model captures the change in slope of the dislocation density-plastic curvature plot with the change in thickness for the ϑ_{30} oriented films, but not for the ϑ_{60} oriented films. An increase in the slope is predicted for the ϑ_{30} films but the micropolar results are nearly identical for the ϑ_{60} films. The DDD results for the 8 μm thick film show an increase in slope with increasing average plastic curvature, while the slope for the 4 μm remains essentially constant.

Simple Shear of a Metal Matrix Composite

In this application we study the size-dependent hardening of a metal matrix composite subjected to simple shear. The idealized particle reinforced system shown in Fig. 16 was previously analyzed using DDD simulations by Cleveringa et al. (1997, 1999) and Yefimov et al. (2004a) and micropolar crystal plasticity by Mayeur and McDowell (2015). The periodic unit cell consists of an elastic-viscoplastic matrix phase (white) with a single slip system parallel to the x_1 direction, reinforced by elastic particles (gray). The size of the unit cell is $2W \times 2H$ $\left(W = H\sqrt{3}\right)$ and the particles have dimensions $2W_f \times 2H_f$. Two distinct cases, denoted Material I and Material II, with different particle aspect ratios but the same area fraction, $A_f = (H_f W_f)/(HW) = 0.2$, are studied: the unit cell for Material I is reinforced

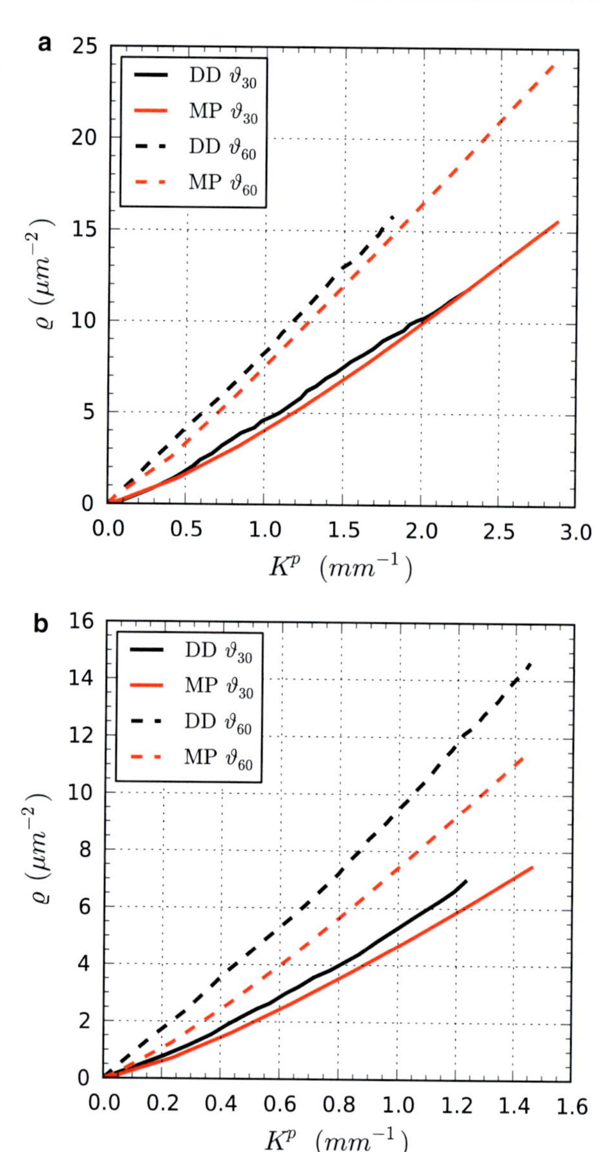

Fig. 15 Comparison of the discrete dislocation (DD) and micropolar crystal plasticity (MP) dislocation density evolution as a function of average plastic curvature for the double slip configuration (**a**) $H = 4\ \mu$m and (**b**) $H = 8\ \mu$m. The discrete dislocation results are from Yefimov and van der Giessen (2005a)

by square particles with $W_f = H_f = 0.416H$, while the unit cell for Material II is reinforced by rectangular particles with $H_f = 2W_f = 0.588H$. The two cases are differentiated such that Material I contains an unobstructed vein of matrix material that spans the unit cell, whereas the particles overlap in Material II and block slip. Since the area fraction of the elastic particles is the same for both morphologies, any observed differences in material response are due to the dislocation-particle interactions and not the phase volume fraction.

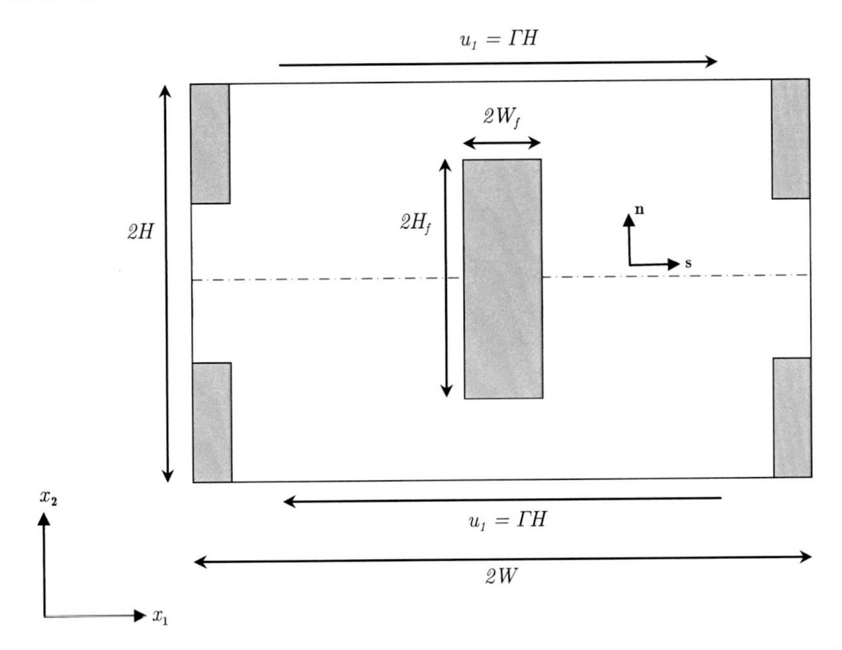

Fig. 16 Schematic of the geometry and slip system configuration for the metal matrix composite initial-boundary value problem

The composite is subjected to simple shear through displacement boundary conditions applied to top and bottom surfaces and these surfaces are assumed to be couple stress traction free. Periodic boundary conditions are enforced on displacements and microrotation at the left and right surfaces. The deformation is imposed at a shearing rate of $\dot{\Gamma} = 10^{-3}\text{s}^{-1}$ up to a unit cell shear strain of $\Gamma = 0.01$, at which point the material is unloaded back to zero strain. The boundary conditions are stated as:

$$\begin{aligned} u_1(x_1, \pm H, t) = \pm\Gamma(t)H, \quad u_2(x_1, \pm H, t) = 0 \\ u_1(-W, x_2, t) = u_1(W, x_2, t), \quad u_2(-W, x_2, t) = u_2(W, x_2, t) \\ \overline{\phi}(-W, x_2, t) = \overline{\phi}(W, x_2, t). \end{aligned} \tag{115}$$

Further, microrotation is assumed to be coupled at the matrix-particle interface. Simulations are performed for cell sizes of $H = \{0.5C, C, 2C\}$ where $C = 4000b$ with $b = 0.25$ nm. The FE mesh consists of 106×61 bilinear quadrilateral elements. The classical elastic properties of the particles are $\mu = 192.3$ GPa and $\nu = 0.17$, and the matrix constitutive parameters are listed in Table 1. These parameters are representative of silicon carbide particles embedded in an aluminum matrix. With regard to specifying the nonclassical elastic constants for inclusion, μ_c^I and β^I, we assume that $\mu_I^c = \mu_M^c$ and $\beta^I = \beta^M$ where the superscripts I and M refer to the inclusion and matrix, respectively. As discussed by Cordero et al. (2010), other choices are possible and perhaps should be considered in future work.

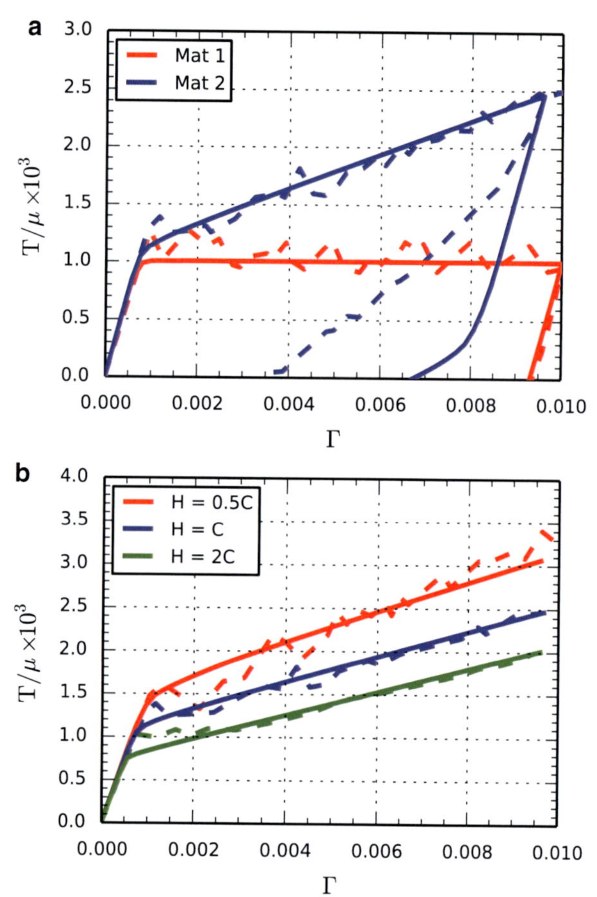

Fig. 17 Average stress-strain response for (**a**) Material I and II for $H = C$ and (**b**) Material II with variable slip threshold for different particle spacings. Dashed lines are discrete dislocation results (Yefimov 2004a) and solid lines are micropolar results

As shown in Fig. 17a, the unit-cell average stress-strain response predicted by the discrete dislocation model for Material I is nearly elastic-perfectly plastic, whereas Material II displays an approximately linear hardening rate. Material I does not harden because there are no obstacles to dislocation motion. In contrast, mobile dislocations in Material II are obstructed by the particles and form pileups and tilt walls at the matrix-particle interface. The unloading curve suggests the material strengthening in Material II is governed by the development of a strong back stress, as evidenced by the pronounced Bauschinger effect, while unloading is essentially elastic for Material I. The average stress-strain curves predicted by the micropolar model (solid lines in Fig. 17) are in good agreement with the discrete dislocation results during forward loading; however, the Bauschinger effect for Material II is significantly underestimated. Figure 17b shows the stress-strain curves for Material II for the three unit cell sizes. We found that it was necessary to use different r_0 values to obtain good agreement with the discrete dislocation results. Interestingly, the calibrated values, $r_0 = \{13, 21, 30\}$ MPa, show strong correlation with the Hall-

Petch relation $r_0 \propto \Lambda^{-1/2}$, where $\Lambda = \left(2\sqrt{3} - 0.588\right) H$ is the mean free path for Material II. The slope of the Hall-Petch relation is 40.51 MPa $\sqrt{\mu\text{m}}$. Note that the Hall-Petch relation was not assumed a priori, but rather is consistent with the result of parameter estimation.

The cumulative plastic slip distributions for both cases with $H = C$ and $\Gamma = 0.006$ as predicted by the micropolar model are shown in Fig. 18. The slip field morphology for Material I is characterized by intensely localized plasticity in the unreinforced veins of matrix material, whereas the slip morphology for Material II is characterized by highly localized bands that form along the top and bottom faces of the particles, but do not extend across the full width of the unit cell due the particle overlap. The cumulative slip distributions predicted by the micropolar model are consistent with the discrete dislocation simulations and are noticeably different than the predictions of local and low-order gradient theories, which show much higher levels of slip accumulation along the vertical matrix-particle interfaces.

The total dislocation density fields are plotted for all three cases of Material II in Fig. 19 at $\Gamma = 0.006$. There is a significant dislocation density accumulation

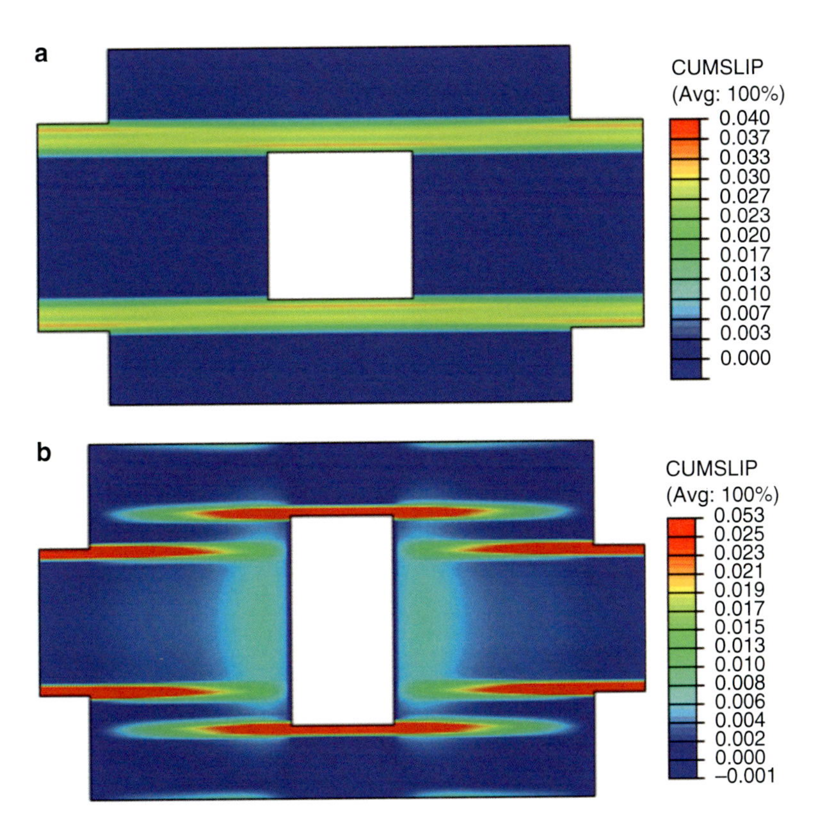

Fig. 18 Contours of cumulative slip at $\Gamma = 0.006$ for (**a**) Material I (**b**) Material II ($H = C$)

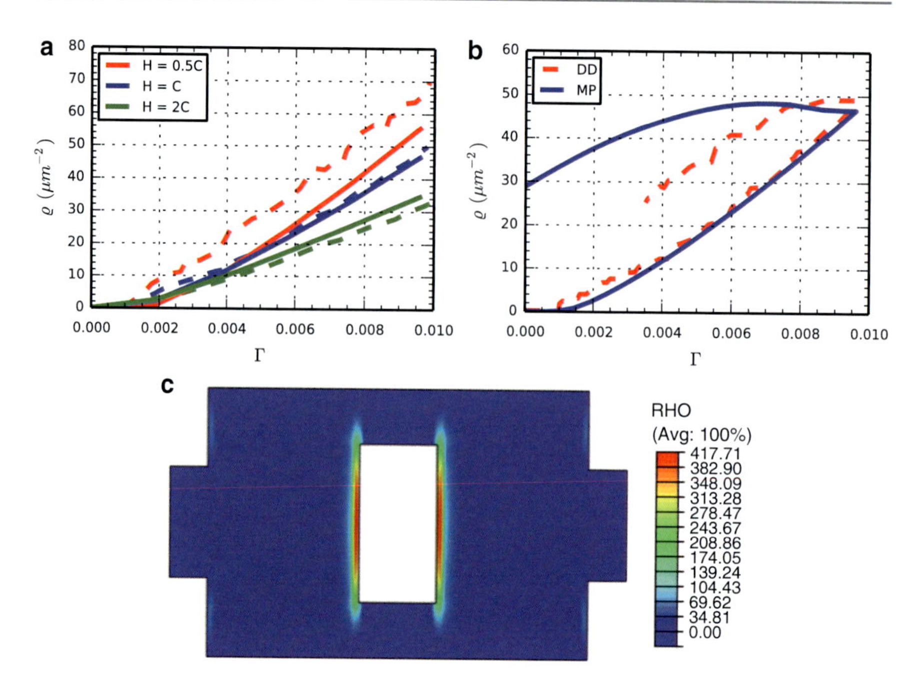

Fig. 19 Evolution of total matrix dislocation density versus applied strain as predicted by the micropolar model (solid lines) and discrete dislocation dynamics (dashed lines) (Yefimov et al. 2004a) (**a**) for various unit cell sizes during loading and (**b**) unloading for $H = C$. Contour of total dislocation density for Material II at $\Gamma = 0.006$ for the micropolar model $H = 0.5C$

along the vertical faces of the matrix-particle interface and sparse dislocation density distributed throughout the matrix. These GNDs are generated to accommodate the rotational gradients that develop at the interface (Ashby 1970). The total matrix dislocation density is plotted during loading as a function of applied shear strain for both the micropolar and discrete dislocation models for the three unit cell sizes in Fig. 19a and during loading/unloading for $H = C$ in Fig. 19b. The micropolar results are in good agreement with the discrete dislocation simulations for the two largest unit cells during loading, but there is a modest departure in the model predictions for the $H = 0.5C$ case. Interestingly, the discrete dislocation model predicts much higher rate of dislocation recovery upon unloading as compared to the micropolar model.

Overall, the predictions of the micropolar simulations are in reasonably good agreement with the discrete dislocation simulation results. The most significant discrepancy in the simulated material response is the underprediction of the Bauschinger effect and the rate of dislocation recovery upon unloading. We believe these discrepancies are largely related to the higher-order boundary conditions enforced at the matrix-particle interface. It is assumed that the lattice rotations at the matrix-particle interface are equal in these simulations, i.e., the finite element nodes along the matrix-particle interface are shared between the two materials. This represents a different boundary condition than what is enforced at the matrix-particle interface in the discrete dislocation simulation. The interface boundary condition enforced in the DDD models is one of equal displacements and zero slip at the vertical matrix-particle faces. Thus, the micropolar model enforces an additional constraint at the matrix-particle interface, which may overconstrain the material response upon unloading. It may be possible to achieve an improved unloading response by either modifying boundary conditions at the matrix-particle interface or the nonclassical elastic constants of the inclusion, μ_c^I and β^I, which will alter the way higher-order tractions are transmitted between the two phases.

Conclusions

Micropolar crystal plasticity is a specialized subset of the more general micromorphic crystal plasticity theory, discussed in a subsequent chapter, that accounts for size effects due to gradients of lattice rotation. The connection between the micropolar lattice torsion-curvature and Nye's GND tensor was established and related to concepts from slip gradient-based crystal plasticity. The advantage of the micropolar theory in comparison to the micromorphic and slip gradient-based theories is the reduced complexity in that it requires only three additional continuum degrees of freedom for the fully three-dimensional case.

A full treatment of the finite deformation kinematics and thermodynamic-based constitutive equations have been developed and placed in context as an extension of concepts of local crystal plasticity theory. A model employing linearized kinematics is then presented with an explicit set of constitutive equations that were used in finite element simulations of initial-boundary value problems previously solved using discrete dislocation dynamics (DDD). The simulation results demonstrate the ability of the micropolar theory to capture many of the salient features exhibited by the DDD simulations for a wide range of boundary value problems, including both the size-dependence of the stress-strain response and the evolution of the dislocation density.

An extended comparison to Gurtin-type slip gradient-based theories of higher-order single crystal plasticity was presented. The analysis highlights many striking theoretical similarities, which suggests that they will also share many of the same predictive capabilities. A few subtle, but key differences, with respect to the construction of dissipative constitutive equations are also discussed. For example, it is possible in the slip gradient-based theories to isolate gradient energetic and dissipative length scale effects, whereas they are coupled in the micropolar theory, i.e., one cannot have gradient dissipative effects in the absence of energetic gradient effects since there will be no driving force for plastic torsion-curvature evolution. However, it is noted that there is no principal restriction preventing one from casting the micropolar theory in terms of energetic-dissipative decomposition of the couple stress tensor rather than of the lattice torsion-curvature, which would then accommodate a true separation of energetic and dissipative gradient effects.

There is still much exciting work to be done in further developing the micropolar theory. Two areas of particular interest are the development of proper intermediate higher-order boundary conditions between fully constrained (micro-hard) and traction free (micro-free) and the proper description of interface boundary conditions and/or constitutive equations for grain and/or phase boundaries.

Acknowledgments JRM acknowledges the support of Los Alamos National Laboratory, operated by Los Alamos National Security LLC under DOE Contract DE-AC52-06NA25936. This work benefited from the support of the Laboratory Directed Research and Development Early Career award 20150696ECR. DLM would like to acknowledge the support of the Carter N. Paden, Jr. Distinguished Chair in Metals Processing.

References

Abaqus/Standard Version 6.7.1. Dassault Systèmes Simulia Corp. (2007)

E. Aifantis, On the microstructural origin of certain inelastic models. J. Eng. Mater. Technol. **106**, 326–330 (1984)

A. Arsenlis, D.M. Parks, Crystallographic aspects of geometrically-necessary and statistically-stored dislocation density. Acta Mater. **47**, 1597–1611 (1999)

M.F. Ashby, Deformation of plastically non-homogeneous materials. Philos. Mag. **21**(170), 399–424 (1970)

E. Bittencourt, A. Needleman, M.E. Gurtin, E. van der Giessen, A comparison of nonlocal continuum and discrete dislocation plasticity predictions. J. Mech. Phys. Solids **51**, 281–310 (2003)

P. Cermelli, M. Gurtin, On the characterization of geometrically necessary dislocations in finite plasticity. J. Mech. Phys. Solids **49**, 1539–1568 (2001)

J. Clayton, D. Bamman, D. McDowell, A geometric framework for the kinematics of crystals with defects. Philos. Mag. **85**, 3983–4010 (2005)

H.H.M. Cleveringa, E. van der Giessen, A. Needleman, Comparison of discrete dislocation and continuum plasticity predictions for a composite material. Acta Mater. **45**, 3163–3179 (1997)

H.H.M. Cleveringa, E. van der Giessen, A. Needleman, A discrete dislocation analysis of residual stresses in a composite material. Philos. Mag. A Phys. Condens. Matter Struct. Defects Mech Prop **79**, 893–920 (1999)

B.D. Coleman, M.E. Gurtin, Thermodynamics with internal state variables. J. Chem. Phys. **47**(2), 597–613 (1967)

N.M. Cordero, A. Gaubert, S. Forest, E.P. Busso, F. Gallerneau, S. Kruch, Size effects in generalised continuum crystal plasticity for two–phase laminates. J. Mech. Phys. Solids **58**, 1963–1994 (2010)

A. Eringen, *Microcontinuum Field Theories I: Foundations and Solids* (Springer, New York, 1999)

A.C. Eringen, W.D. Claus Jr., A micromorphic approach to dislocation theory and its relation to several existing theories. Technical Report TR-6, Princeton University Department of Aerospace and Mechanical Sciences (1969)

A.C. Eringen, E.S. Suhubi, Nonlinear theory of simple micro-elastic solids-I. Int. J. Eng. Sci. **2**, 189–203 (1964)

L.P. Evers, W.A.M. Brekelmans, M.G.D. Geers, Non-local crystal plasticity model with intrinsic SSD and GND effects. J. Mech. Phys. Solids **52**, 2379–2401 (2004)

S. Forest, Some links between Cosserat, strain gradient crystal plasticity and the statistical theory of dislocations. Philos. Mag. **88**, 3549–3563 (2008)

S. Forest, Generalized continuum modelling of crystal plasticity, in *Generalized Continua and Dislocation Theory* (Springer, Vienna, 2012), pp. 181–287

S. Forest, R. Sievert, Elastoviscoplastic constitutive frameworks for generalized continua. Acta Mech. **160**, 71–111 (2003)
S. Forest, G. Cailletaud, R.W. Sievert, A Cosserat theory for elastoviscoplastic single crystals at finite deformation. Arch. Mech. **49**, 705–736 (1997)
S. Forest, F. Barbe, G. Cailletaud, Cosserat modelling of size effects in the mechanical behaviour of polycrystals and multi-phase materials. Int. J. Solids Struct. **37**(46), 7105–7126 (2000)
P. Germain, The method of virtual power in continuum mechanics. Part 2: microstructure. SIAM J. Appl. Math. **25**, 556–575 (1973)
W. Günther, Zur Statik und Kinematik des Cosseratschen Kontinuums. Abh. Braunschw. Wiss. Ges. **10**, 195–213 (1958)
M.E. Gurtin, A gradient theory of single–crystal viscoplasticity that accounts for geometrically necessary dislocations. J. Mech. Phys. Solids **50**, 5–32 (2002)
M.E. Gurtin, L. Anand, S.P. Lele, Gradient single-crystal plasticity with free energy dependent on dislocation densities. J. Mech. Phys. Solids **55**, 1853–1878 (2007)
E. Kröner, On the physical reality of torque stresses in continuum mechanics. Int. J. Eng. Sci. **1**, 261–278 (1963)
E. Kröner, Initial studies of a plasticity theory based upon statistical mechanics, in *Inelastic Behaviour of Solids*, ed. by M. Kanninen, W. Adler, A. Rosenfield, R. Jaffee (McGraw-Hill, New York, 1969), pp. 137–147
M. Kuroda, V. Tvergaard, A finite deformation theory of higher-order gradient crystal plasticity. J. Mech. Phys. Solids **56**(8), 2573–2584 (2008)
S. Limkumnerd, E. van der Giessen, Study of size effects in thin films by means of a crystal plasticity theory based on DiFT. J. Mech. Phys. Solids **56**, 3304–3314 (2008)
J.R. Mayeur, Generalized continuum modeling of scale-dependent crystalline plasticity. Ph.D. thesis, Georgia Institute of Technology, 2010
J.R. Mayeur, D.L. McDowell, Bending of single crystal thin films modeled with micropolar crystal plasticity. Int. J. Eng. Sci. **49**, 1357–1366 (2011)
J.R. Mayeur, D.L. McDowell, An evaluation of higher-order single crystal strength models for constrained thin films subjected to simple shear. J. Mech. Phys. Solids **61**, 1935–1954 (2013)
J.R. Mayeur, D.L. McDowell, A comparison of Gurtin type and micropolar theories of generalized single crystal plasticity. Int. J. Plast. **57**, 29–51 (2014)
J.R. Mayeur, D.L. McDowell, Micropolar crystal plasticity simulation of particle strengthening. Model. Simul. Mater. Sci. Eng. **23**, 065007 (2015)
J.R. Mayeur, D.L. McDowell, D.J. Bammann, Dislocation-based micropolar single crystal plasticity: comparison of multi- and single criterion theories. J. Mech. Phys. Solids **59**, 398–422 (2011)
H. Mecking, U.F. Kocks, Kinetics of flow and strain-hardening. Acta Metall. **29**, 1865–1875 (1981)
J.F. Nye, Some geometrical relations in dislocated crystals. Acta Metall. **1**, 153–162 (1953)
B.D. Reddy, The role of dissipation and defect energy in variational formulations of problems in strain-gradient plasticity. Part 2: single-crystal plasticity. Contin. Mech. Thermodyn. **23**, 551–572 (2011)
H. Schäfer, Eine Feldtheorie der Versetzungen im Cosserat-Kontinuum. Z. Angew. Math. Phys. **20**, 891–899 (1969)
J.Y. Shu, N.A. Fleck, E. van der Giessen, A. Needleman, Boundary layers in constrained plastic flow: comparison of nonlocal and discrete dislocation plasticity. J. Mech. Phys. Solids **49**, 1361–1395 (2001)
R. Sievert, S. Forest, R. Trostel, Finite deformation Cosserat-type modelling of dissipative solids and its application to crystal plasticity. J. Phys. IV **8**, 357–364 (1998)
C. Teodosiu, F. Sidoroff, A theory of finite elastoviscoplasticity of single crystals. Int. J. Eng. Sci. **14**, 165–176 (1976)
S. Yefimov, E. van der Giessen, Multiple slip in a strain-gradient plasticity model motivated by a statistical-mechanics description of dislocations. Int. J. Solids Struct. **42**, 3375–3394 (2005a)
S. Yefimov, E. van der Giessen, Size effects in single crystal thin films: nonlocal crystal plasticity simulations. Eur. J. Mech. A Solids **24**, 183–193 (2005b)

S. Yefimov, I. Groma, E. van der Giessen, A comparison of a statistical-mechanics based plasticity model with discrete dislocation plasticity calculations. J. Mech. Phys. Solids **52**, 279–300 (2004a)

S. Yefimov, E. van der Giessen, I. Groma, Bending of a single crystal: discrete dislocation and nonlocal crystal plasticity simulations. Model. Simul. Mater. Sci. Eng. **12**, 1069 (2004b)

Micromorphic Crystal Plasticity

18

Samuel Forest, J. R. Mayeur, and D. L. McDowell

Contents

S. Forest (✉)
Centre des Materiaux, Mines ParisTech CNRS, PSL Research University, Paris, Evry Cedex, France
e-mail: samuel.forest@mines-paristech.fr

J. R. Mayeur
Theoretical Division, Los Alamos National Laboratory, Los Alamos, NM, USA
e-mail: jason.mayeur@gmail.com

D. L. McDowell
Woodruff School of Mechanical Engineering, School of Materials Science and Engineering, Georgia Institute of Technology, Atlanta, GA, USA
e-mail: david.mcdowell@me.gatech.edu

G. Z. Voyiadjis (ed.), *Handbook of Nonlocal Continuum Mechanics for Materials and Structures*, https://doi.org/10.1007/978-3-319-58729-5_49

Abstract

The micromorphic approach to crystal plasticity represents an extension of the micropolar (Cosserat) framework, which is presented in a separate chapter. Cosserat crystal plasticity is contained as a special constrained case in the same way as the Cosserat theory is a special restricted case of Eringen's micromorphic model, as explained also in a separate chapter. The micromorphic theory is presented along the lines of Aslan et al. (Int J Eng Sci 49:1311–1325, 2011) and Forest et al. (Micromorphic approach to crystal plasticity and phase transformation. In: Schroeder J, Hackl K (eds) Plasticity and beyond. CISM international centre for mechanical sciences, courses and lectures, vol 550, Springer, pp 131–198, 2014) and compared to the micropolar model in some applications. These extensions of conventional crystal plasticity aim at incorporating the dislocation density tensor introduced by Kröner (Initial studies of a plasticity theory based upon statistical mechanics. In: Kanninen M, Adler W, Rosenfield A, Jaffee R (eds) Inelastic behaviour of solids. McGraw-Hill, pp 137–147, 1969). and Cermelli and Gurtin (J Mech Phys Solids 49:1539–1568, 2001) into the constitutive framework. The concept of dislocation density tensor is equivalent to that of the so-called geometrically necessary dislocations (GND) introduced by Ashby (The deformation of plastically non-homogeneous alloys. In: Kelly A, Nicholson R (eds) Strengthening methods in crystals. Applied Science Publishers, London, pp 137–192, 1971). The applications presented in this chapter deal with pile-up formation in laminate microstructures and strain localization phenomena in polycrystals.

Keywords

Micromorphic medium · Crystal plasticity · Dislocation density tensor · Geometrically necessary dislocations · Strain gradient plasticity · Size effect

Introduction

The micromorphic approach to crystal plasticity represents an extension of the micropolar (Cosserat) framework which is presented in a separate chapter. Cosserat crystal plasticity is contained as a special constrained case in the same way as the Cosserat theory is a special restricted case of Eringen's micromorphic model, as explained also in a separate chapter. The micromorphic theory is presented along the lines of Aslan et al. (2011) and Forest et al. (2014) and compared to the micropolar model in some applications. These extensions of conventional crystal plasticity aim at incorporating the dislocation density tensor introduced by Kröner (1969) and Cermelli and Gurtin (2001) into the constitutive framework. The concept of dislocation density tensor is equivalent to that of so-called geometrically necessary dislocations (GND) introduced by Ashby (1971).

The links between the micromorphic continuum and the plasticity of crystalline materials have been recognized very early by Claus and Eringen (1969) and Eringen and Claus (1970). Lattice directions in a single crystal can be regarded as directors that rotate and deform as they do in a micromorphic continuum. The fact that lattice directions can be rotated and stretched in a different way than material lines connecting individual atoms, especially in the presence of static or moving dislocations, illustrates the independence between directors and material lines in a micromorphic continuum, even though their deformation can be related at the constitutive level.

The objective of the present chapter is to formulate a finite deformation micromorphic extension of conventional crystal plasticity to account for GND effects in single crystals. It also provides analytical predictions of size effects on the yield strength and kinematic hardening of laminate microstructures made of an elastic layer and an elastic–plastic single-crystal layer undergoing single slip. The theory is called the *microcurl* model because the evaluation of the curl of the microdeformation, instead of its full gradient, is sufficient to account for the effect of the dislocation density tensor.

The models proposed in this section for single crystals fall in the class of anisotropic elastoviscoplastic micromorphic media for which constitutive frameworks at finite deformations have been proposed in Forest and Sievert (2003), Lee and Chen (2003), Grammenoudis and Tsakmakis (2009), Sansour et al. (2010), and Regueiro 2010; see the corresponding chapter in this handbook. In fact, the micromorphic approach can be applied not only to the total deformation by introducing the microdeformation field but can also be restricted to plastic deformation, for specific application to size effects in plasticity, or to damage variables for application to regularized simulation of crack propagation, as proposed in Forest (2009, 2016) and Hirschberger and Steinmann (2009).

The outline of this chapter is as follows. The crystal plasticity model formulated within Eringen's micromorphic framework is presented at finite deformation in section "The Microcurl Model at Finite Deformation," together with its linearization. Size effects predicted by the model are illustrated in section "Size Effects in a Two-Phase Single-Crystal Laminate." Some constitutive laws involving the dislocation density tensor are discussed in section "Free Energy Potentials for Micromorphic Crystal Plasticity" with an application to cyclic plasticity in single crystals. Finally, the model is used to predict the response of polycrystalline metals and alloys in section "Grain Size Effects in Polycrystals."

The Microcurl Model at Finite Deformation

Model Formulation

Balance Equations

The degrees of freedom of the proposed theory are the displacement vector $\boldsymbol{u}$ and the microdeformation variable $\widehat{\chi}^p$, a generally nonsymmetric second-rank tensor. The field $\widehat{\chi}^p\,(\boldsymbol{X})$ is generally not compatible, meaning that it does not derive from a vector field. The exponent ^{p} indicates, in advance, that this variable will eventually be constitutively related to plastic deformation occurring at the material point. In particular, the microdeformation $\hat{\chi}^p$ is treated as an invariant quantity with respect to rigid body motion. The polar decomposition of the microdeformation contains the polar rotation $\overline{\mathbf{R}}$ used in the micropolar crystal plasticity theory and a symmetric microstretch tensor. As a result, when this microstretch tensor is close to the identity tensor, the micromorphic model reduces to the micropolar one.

A first gradient theory is considered with respect to the degrees of freedom. However, the influence of the microdeformation gradient is limited to its curl part because of the intended relation to the dislocation density tensor associated with the curl of plastic distortion. The following sets of degrees of freedom and of their gradients are therefore defined:

$$DOF = \{\boldsymbol{u}, \quad \hat{\chi}^p\}, \; GRAD = \{\mathbf{F} := 1 + \boldsymbol{u} \otimes \nabla_0, \; \mathbf{K} := \text{Curl } \hat{\chi}^p\} \tag{1}$$

The following definition of the curl operator is adopted:

$$\text{Curl } \hat{\chi}^p := \frac{\partial \hat{\chi}^p}{\partial X_k} \times \boldsymbol{e}_k, \; K_{ij} \;\; := \;\; \epsilon_{jkl} \frac{\partial \hat{\chi}^p_{ik}}{\partial X_l} \tag{2}$$

where ϵ_{ijk} is the permutation tensor.

The method of virtual power is used to derive the balance and boundary conditions, following Germain (1973). For that purpose, the power density of internal forces is defined as a linear form with respect to the velocity fields and their Eulerian gradients:

$$p^{(i)} = \boldsymbol{\sigma} \;\; : (\dot{\boldsymbol{u}} \otimes \nabla) + \mathbf{s} : \dot{\hat{\chi}}^p + \mathbf{M} : \text{curl } \dot{\hat{\chi}}^p, \; \forall \boldsymbol{x} \;\; \in \;\; V \tag{3}$$

Here, the conjugate quantities are the Cauchy stress tensor $\boldsymbol{\sigma}$, which is symmetric for objectivity reasons; the microstress tensor, $\mathbf{s}$; and the generalized couple-stress tensor $\mathbf{M}$. The curl of the microdeformation rate is defined as:

$$\text{curl } \dot{\hat{\chi}}^p := \;\; \epsilon_{jkl} \frac{\partial \dot{\hat{\chi}}^p_{ik}}{\partial x_l} \;\; \boldsymbol{e}_i \otimes \boldsymbol{e}_j = \dot{\mathbf{K}} \;\; \mathbf{F}^{-1} \tag{4}$$

The form of the power density of internal forces dictates the form of the power density of contact forces:

$$p^{(c)} = \boldsymbol{t} \cdot \dot{\boldsymbol{u}} + \mathbf{m} : \dot{\hat{\chi}}^p, \; \forall \boldsymbol{x} \quad \in \quad \partial V \tag{5}$$

where $\boldsymbol{t}$ is the usual simple traction vector and $\mathbf{m}$ is the double-traction tensor. The principle of virtual power is stated in the static case and in the absence of volume forces for the sake of brevity:

$$-\int_D p^{(i)} \; dV + \int_{\partial D} p^{(c)} \; dS = 0 \tag{6}$$

for all virtual fields $\dot{\boldsymbol{u}}, \dot{\widehat{\boldsymbol{\chi}}}^p$ and any subdomain $D \subset V$. By application of the Gauss divergence theorem, assuming sufficient regularity of the fields, this statement expands into:

$$\int_V \frac{\partial \sigma_{ij}}{\partial x_j} \dot{u}_i \, dV + \int_V \left(\in_{kjl} \frac{\partial M_{ik}}{\partial x_l} - s_{ij} \right) \dot{\widehat{\chi}}^p_{ij} \, dV$$
$$+ \int_{\partial V} \left(t_i - \sigma_{ij} n_j \right) \dot{u}_i \, dS + \int_{\partial V} \left(m_{ik} - \in_{jkl} M_{ij} n_l \right) \dot{\widehat{\chi}}^p_{ik} \, dS = 0, \quad \forall \dot{u}_i, \forall \dot{\widehat{\chi}}^p_{ij}$$

which leads to the two-field equations of balance of momentum and generalized balance of moment of momentum:

$$\text{div } \boldsymbol{\sigma} = 0, \text{ curl } \mathbf{M} + \mathbf{s} = 0, \forall \boldsymbol{x} \in V \tag{7}$$

and two boundary conditions:

$$\boldsymbol{t} = \boldsymbol{\sigma} \cdot \boldsymbol{n}, \ \mathbf{m} = \mathbf{M} \cdot \in \cdot \boldsymbol{n}, \ \forall \boldsymbol{x} \in \partial V \tag{8}$$

the index representation of the latter relation being $m_{ij} = M_{ik} \in_{kjl} n_l$. These balance equations can be compared to the corresponding ones in the chapter dedicated to the micropolar theory.

Constitutive Equations

The deformation gradient is decomposed into elastic and plastic parts in the form,

$$\mathbf{F} = \mathbf{F}^e \ \mathbf{F}^p \tag{9}$$

The isoclinic intermediate configuration is defined in a unique way by keeping the crystal orientation unchanged from the initial to the intermediate configuration following Mandel (1973). The plastic distortion $\mathbf{F}^p$ is invariant with respect to rigid body motions that are carried by $\mathbf{F}^e$. The current mass density is ρ, whereas the mass density of the material element in the intermediate configuration is $\tilde{\rho}$, such that $\tilde{\rho}/\rho = J_e := \det(\mathbf{F}^e)$. The elastic strain is defined as:

$$\tilde{\mathbf{E}}^e := \frac{1}{2} \left(\mathbf{F}^{eT} \ \mathbf{F}^e - 1 \right) \tag{10}$$

The microdeformation is linked to the plastic deformation via the introduction of a relative deformation measure, defined as:

$$\mathbf{e}^p := \mathbf{F}^{p-1} \ \widehat{\boldsymbol{\chi}}^p - 1 \tag{11}$$

This tensor $\mathbf{e}^p$ measures the departure of the microdeformation from the plastic deformation. The state variables are assumed to be the elastic strain, the relative deformation, the curl of microdeformation, and some internal variables, α:

$$STATE := \left\{\tilde{\mathbf{E}}^e, \mathbf{e}^p, \mathbf{K}, \alpha\right\} \tag{12}$$

The specific Helmholtz free energy density, ψ, is assumed to be a function of this set of state variables. In particular, in this simple version of the model, the curl of microdeformation is assumed to contribute entirely to the stored energy. In more sophisticated models, as proposed in Forest and Sievert (2003, 2006), Forest (2009), and Gurtin and Anand (2009), the relative deformation, the microdeformation, and its gradient can be split into elastic and plastic parts.

When the internal constraint $\mathbf{e}^p \equiv 0$ is enforced, the plastic microdeformation coincides with the plastic deformation so that the curl of the plastic microdeformation is directly related to the dislocation density tensor previously defined by:

$$\mathbf{K} := \text{Curl } \hat{\chi}^p \equiv \text{Curl } \mathbf{F}^p = J\mathbf{A}\mathbf{F}^{-T} \tag{13}$$

where $\mathbf{A}$ is the dislocation density tensor defined as the curl of the inverse elastic deformation.

The micromorphic model then reduces to strain gradient plasticity according to Gurtin (2002).

The dissipation rate density is the difference:

$$\Delta^{intr} := p^{(i)} - \rho\dot{\psi} \geq 0 \tag{14}$$

which must be positive according to the second principle of thermodynamics. When the previous strain measures are introduced, the power density of internal forces takes the following form:

$$\begin{aligned} p^{(i)} &= \boldsymbol{\sigma} : \dot{\mathbf{F}}^e\mathbf{F}^{e-1} + \boldsymbol{\sigma} : \left(\mathbf{F}^e\ \dot{\mathbf{F}}^p\ \mathbf{F}^{p-1}\ \mathbf{F}^{e-1}\right) + \mathbf{s} : \left(\mathbf{F}^p\ \dot{\mathbf{e}}^p + \dot{\mathbf{F}}^p\ \mathbf{e}^p\right) + \mathbf{M} : \dot{\mathbf{K}}\ \mathbf{F}^{-1} \\ &= \frac{\rho}{\tilde{\rho}}\boldsymbol{\Pi}^e : \dot{\tilde{\mathbf{E}}}^e + \frac{\rho}{\tilde{\rho}}\boldsymbol{\Pi}^M : \dot{\mathbf{F}}^p\ \mathbf{F}^{p-1} + \mathbf{s} : \left(\mathbf{F}^p\ \dot{\mathbf{e}}^p + \dot{\mathbf{F}}^p\ \mathbf{e}^p\right) + \mathbf{M} : \dot{\mathbf{K}}\ \mathbf{F}^{-1} \end{aligned} \tag{15}$$

where $\boldsymbol{\Pi}^e$ is the second Piola–Kirchhoff stress tensor with respect to the intermediate configuration and $\boldsymbol{\Pi}^M$ is the Mandel stress tensor:

$$\boldsymbol{\Pi}^e := J_e\ \mathbf{F}^{e-1}\ \boldsymbol{\sigma}\ \mathbf{F}^{e-T},\ \boldsymbol{\Pi}^M := J_e\ \mathbf{F}^{eT}\ \boldsymbol{\sigma}\ \mathbf{F}^{e-T} = \mathbf{F}^{eT}\ \mathbf{F}^e\ \boldsymbol{\Pi}^e \tag{16}$$

On the other hand,

$$\rho\dot{\psi} = \rho\frac{\partial\psi}{\partial\tilde{\mathbf{E}}^e} : \dot{\tilde{\mathbf{E}}}^e + \rho\frac{\partial\psi}{\partial\mathbf{e}^p} : \dot{\mathbf{e}}^p + \rho\frac{\partial\psi}{\partial\mathbf{K}} : \dot{\mathbf{K}} + \rho\frac{\partial\psi}{\partial\alpha}\dot{\alpha} \tag{17}$$

We compute:

$$\begin{aligned} J_e D = &\left(\boldsymbol{\Pi}^e - \tilde{\rho}\frac{\partial\psi}{\partial\tilde{\mathbf{E}}^e}\right) : \dot{\tilde{\mathbf{E}}}^e + \left(J_e\ \mathbf{F}^{pT}\mathbf{s} - \tilde{\rho}\frac{\partial\psi}{\partial\mathbf{e}^p}\right) : \dot{\mathbf{e}}^p \\ &+ \left(J_e\mathbf{M}\ \mathbf{F}^{-T} - \tilde{\rho}\frac{\partial\psi}{\partial\mathbf{K}}\right) : \dot{\mathbf{K}} \\ &+ \left(\boldsymbol{\Pi}^M + J_e\mathbf{s}\ \widehat{\boldsymbol{\chi}}^{pT}\right) : \dot{\mathbf{F}}^p\ \mathbf{F}^{p-1} - \tilde{\rho}\frac{\partial\psi}{\partial\alpha}\dot{\alpha} \geq 0 \end{aligned} \tag{18}$$

Assuming that the processes associated with $\dot{\tilde{\mathbf{E}}}^e$, $\dot{\mathbf{e}}^p$ and $\dot{\mathbf{K}}$ are nondissipative, the state laws are obtained:

$$\boldsymbol{\Pi}^e = \tilde{\rho}\frac{\partial\psi}{\partial\tilde{\mathbf{E}}^e},\ \mathbf{s} = J_e^{-1}\ \mathbf{F}^{p-T}\ \tilde{\rho}\frac{\partial\psi}{\partial\mathbf{e}^p},\ \mathbf{M} = J_e^{-1}\ \tilde{\rho}\frac{\partial\psi}{\partial\mathbf{K}}\ \mathbf{F}^T \tag{19}$$

The residual dissipation rate is:

$$J_e D = \left(\boldsymbol{\Pi}^M + J_e\mathbf{s}\ \widehat{\boldsymbol{\chi}}^{pT}\right) : \dot{\mathbf{F}}^p\ \mathbf{F}^{p-1} - R\dot{\alpha} \geq 0, \text{ with } R := \tilde{\rho}\frac{\partial\psi}{\partial\alpha} \tag{20}$$

At this stage, a dissipation potential that depends on stress measures, $\Omega(S, R)$, is introduced in order to formulate the evolution equations for plastic flow and internal variables:

$$\dot{\mathbf{F}}^p\ \mathbf{F}^{p-1} = \frac{\partial\Omega}{\partial\mathcal{S}},\ \text{with } \mathcal{S} := \boldsymbol{\Pi}^M + J_e\mathbf{s}\ \widehat{\boldsymbol{\chi}}^{pT} \tag{21}$$

$$\dot{\alpha} = -\frac{\partial\Omega}{\partial R} \tag{22}$$

where R is the thermodynamic force associated with the internal variable α and S is the effective stress conjugate to plastic strain rate, the driving force for plastic flow.

In the case of crystal plasticity, a generalized Schmid law is adopted for each slip system s in the form:

$$f^s\left(\mathcal{S}, \tau_c^s\right) = |\mathcal{S} : \mathbf{P}^s| - \tau_c^s \geq 0,\ \text{with } \mathbf{P}^s = \boldsymbol{l}^s \otimes \boldsymbol{n}^s \tag{23}$$

for activation of slip system s with slip direction, $\boldsymbol{l}^s$, and normal to the slip plane, $\boldsymbol{n}^s$. We call $\mathbf{P}^s$ the orientation tensor. The critical resolved shear stress is τ_c^s which may be a function of R in the presence of isotropic hardening. The kinematics of plastic slip follows from the choice of a dissipation potential, $\Omega(f^s)$, that depends on the stress variables through the yield function itself, f^s:

$$\dot{\mathbf{F}}^p\ \mathbf{F}^{p-1} = \sum_{s=1}^{N}\frac{\partial\Omega}{\partial f^s}\frac{\partial f^s}{\partial\mathcal{S}} = \sum_{s=1}^{N}\dot{\gamma}^s\ \mathbf{P}^s,\ \text{with } \dot{\gamma}^s = \frac{\partial\Omega}{\partial f^s}\,\text{sign}\left(\mathcal{S} : \mathbf{P}^s\right) \tag{24}$$

A possible viscoplastic potential is then:

$$\Omega\left(f^{s}\right)=\frac{K}{n+1}<\frac{f^{s}}{K}>^{n+1} \tag{25}$$

where K and n are viscosity parameters associated with viscoplastic slip, and the brackets stand for $<\cdot>=\mathrm{Max}(0,\cdot)$. The generalized resolved shear stress can be decomposed into two contributions:

$$\mathcal{S}:\mathbf{P}^{s}=\tau^{s}-x^{s},\ \text{with}\ \tau^{s}=\mathbf{\Pi}^{M}:\mathbf{P}^{s}\ \text{and}\ x^{s}=-\mathbf{s}\,\widehat{\boldsymbol{\chi}}^{pT}:\mathbf{P}^{s} \tag{26}$$

The usual resolved shear stress is τ^{s}, whereas x^{s} can be interpreted as an internal stress or back stress leading to kinematic hardening. The fact that the introduction of the effect of the dislocation density tensor or, more generally, of gradient of plastic strain tensor leads to the existence of internal stresses induced by higher-order stresses has already been noticed by Steinmann (1996); see also Forest (2008). The back stress component is induced by the microstress $\mathbf{s}$ or, equivalently, by the curl of the generalized couple-stress tensor, $\mathbf{M}$, via the balance Eq. (7).

Geometrically Linearized Model

When deformations and rotations remain sufficiently small, the previous equations can be linearized as follows:

$$\mathbf{F}=1+\mathbf{H}\simeq 1+\mathbf{H}^{e}+\mathbf{H}^{p},\ \mathbf{H}^{e}=\boldsymbol{\varepsilon}^{e}+\boldsymbol{\omega}^{e},\ \mathbf{H}^{p}=\boldsymbol{\varepsilon}^{p}+\boldsymbol{\omega}^{p} \tag{27}$$

where $\boldsymbol{\varepsilon}^{e}$ and $\boldsymbol{\omega}^{e}$ (resp. $\boldsymbol{\varepsilon}^{p}$, $\boldsymbol{\omega}^{p}$) are the symmetric and skew-symmetric parts of $\mathbf{F}^{e}-\mathbf{1}$ (resp. $\mathbf{F}^{p}-\mathbf{1}$). When microdeformation is small, the relative deformation is linearized as:

$$\mathbf{e}^{p}=\left(\mathbf{1}+\mathbf{H}^{p}\right)^{-1}\ \left(\mathbf{1}+\boldsymbol{\chi}^{p}\right)-\mathbf{1}\simeq\boldsymbol{\chi}^{p}-\mathbf{H}^{p},\ \text{with}\ \boldsymbol{\chi}^{p}=\widehat{\boldsymbol{\chi}}^{p}-1 \tag{28}$$

When linearized, the state laws (19) become:

$$\boldsymbol{\sigma}=\rho\frac{\partial\psi}{\partial\boldsymbol{\varepsilon}^{e}},\ \mathbf{s}=\rho\frac{\partial\psi}{\partial\mathbf{e}^{p}},\ \mathbf{M}=\rho\frac{\partial\psi}{\partial\mathbf{K}} \tag{29}$$

The evolution equations read then:

$$\dot{\boldsymbol{\varepsilon}}^{p}=\frac{\partial\Omega}{\partial\left(\boldsymbol{\sigma}+\mathbf{s}\right)},\ \dot{\alpha}=-\frac{\partial\Omega}{\partial R} \tag{30}$$

The most simple case of a quadratic free energy potential is first considered:

$$\rho\psi\left(\boldsymbol{\varepsilon}^{e},\mathbf{e}^{p},\mathbf{K}\right)=\frac{1}{2}\boldsymbol{\varepsilon}^{e}:\mathbb{C}:\boldsymbol{\varepsilon}^{e}+\frac{1}{2}H_{\chi}\mathbf{e}^{p}:\mathbf{e}^{p}+\frac{1}{2}A\mathbf{K}:\mathbf{K} \tag{31}$$

The usual four-rank tensor of elastic moduli is denoted by $\mathbb{C}$. The higher-order moduli have been limited to only two additional parameters: H_χ (unit MPa) and A (unit MPa.mm^2). Their essential impact on the prediction of size effects will be analyzed in the next section. It follows that:

$$\boldsymbol{\sigma} = \boldsymbol{C} : \boldsymbol{\varepsilon}^e, \; \mathbf{s} = H_\chi \mathbf{e}^p, \; \mathbf{M} = A\boldsymbol{K} \tag{32}$$

Large values of H_χ ensure that $\mathbf{e}^p$ remains small so that $\widehat{\chi}^p$ remains close to $\mathbf{H}^p$ and $\mathbf{K}$ is close to the dislocation density tensor. The yield condition for each slip system becomes:

$$f^s = |\tau^s - x^s| - \tau_c^s \tag{33}$$

with

$$x^s = -\mathbf{s} : \mathbf{P}^s = (\text{curl} \;\; \mathbf{M}) : \mathbf{P}^s = A\,(\text{curl} \;\; \text{curl} \;\; \boldsymbol{\chi}^p) : \mathbf{P}^s \tag{34}$$

Comparison Between Micropolar and Micromorphic Crystal Plasticity

Experimental techniques like Electron Back-Scatter Diffraction (EBSD) provide the field of lattice orientation and, consequently, of lattice rotation $\mathbf{R}^e$ during deformation. The rotation $\mathbf{R}^e$ appears in the polar decomposition of the elastic deformation $\mathbf{F}^e = \mathbf{R}^e\mathbf{U}^e$, where $\mathbf{U}^e$ is the lattice stretch tensor. Since

$$\boldsymbol{\alpha} = -\,\text{curl} \;\; \mathbf{F}^{e-1} = -\,\text{curl} \;\; \left(\mathbf{U}^{e-1} \cdot \mathbf{R}^{eT}\right) \tag{35}$$

the hypothesis of small elastic strain implies

$$\boldsymbol{\alpha} \simeq -\,\text{curl} \;\; \mathbf{R}^{eT} \tag{36}$$

This approximation also requires that the gradient of elastic strain is also small, which is not ensured even if the elastic strain is small. If, in addition, elastic rotations are small, we have:

$$\boldsymbol{\alpha} \simeq -\,\text{curl} \;\; (\mathbf{1} - \boldsymbol{\omega}^e) = \text{curl} \;\; \boldsymbol{\omega}^e \tag{37}$$

The small rotation axial vector is defined as:

$$\overset{\times_e}{\boldsymbol{\omega}} = -\frac{1}{2}\,\underset{\simeq}{\boldsymbol{\epsilon}} : \boldsymbol{\omega}^e, \; \boldsymbol{\omega}^e = -\underset{\sim}{\underline{\boldsymbol{\epsilon}}} \cdot \overset{\times_e}{\boldsymbol{\omega}} \tag{38}$$

or, in matrix notations:

$$[\omega^e] = \begin{bmatrix} 0 & \omega^e_{12} & -\omega^e_{31} \\ -\,\omega^e_{12} & 0 & \omega^e_{23} \\ \omega^e_{31} & -\omega^e_{23} & 0 \end{bmatrix} = \begin{bmatrix} 0 & -\overset{\times_e}{\omega}_3 & \overset{\times_e}{\omega}_2 \\ \overset{\times_e}{\omega}_3 & 0 & -\overset{\times_e}{\omega}_1 \\ -\,\overset{\times_e}{\omega}_2 & \overset{\times_e}{\omega}_1 & 0 \end{bmatrix} \tag{39}$$

The gradient of the lattice rotation field delivers the lattice curvature tensor. In the small deformation context, the gradient of the rotation tensor is represented by the gradient of the axial vector:

$$\boldsymbol{\kappa} := \overset{\times_e}{\boldsymbol{\omega}} \tag{40}$$

One can establish a direct link between curl $\boldsymbol{\omega}^e$ and the gradient of the axial vector associated with $\boldsymbol{\omega}$. For that purpose, the matrix form of curl $\boldsymbol{\omega}^e$ is derived according to:

$$[\text{curl} \quad \omega^e] = \begin{bmatrix} \omega^e_{12,3} + \omega^e_{31,2} & -\omega^e_{31,1} & -\omega^e_{12,1} \\ -\,\omega^e_{23,2} & \omega^e_{12,3} + \omega^e_{23,1} & -\omega^e_{12,2} \\ -\,\omega^e_{23,3} & -\omega^e_{31,3} & \omega^e_{23,1} + \omega^e_{31,2} \end{bmatrix} \tag{41}$$

or equivalently:

$$[\text{curl} \quad \omega^e] = \begin{bmatrix} -\overset{\times_e}{\omega}_{3,3} - \overset{\times_e}{\omega}_{2,2} & \overset{\times_e}{\omega}_{2,1} & \overset{\times}{\omega}_{3,1} \\ \overset{\times_e}{\omega}_{1,2} & -\overset{\times_e}{\omega}_{3,2} - \overset{\times_e}{\omega}_{1,1} & \overset{\times^e}{\omega}_{3,2} \\ \overset{\times_e}{\omega}_{1,3} & \overset{\times_e}{\omega}_{2,3} & -\overset{\times_e}{\omega}_{1,1} - \overset{\times_e}{\omega}_{2,2} \end{bmatrix} \tag{42}$$

from which it becomes apparent that:

$$\boldsymbol{\alpha} = \boldsymbol{\kappa}^T - (\text{trace } \boldsymbol{\kappa})\ \mathbf{1}, \quad \boldsymbol{\kappa} = \boldsymbol{\alpha}^T - \frac{1}{2}(\text{trace } \boldsymbol{\alpha})\ \mathbf{1} \tag{43}$$

This is a remarkable relation linking, with the context of small elastic strains (and in fact of small gradients of elastic strain) and rotations, the dislocation density tensor to lattice curvature. It is known as Nye's formula (Nye 1953).

As a conclusion, it appears that the Cosserat crystal plasticity model only considers the lattice curvature part contained in the full dislocation tensor. This seems to be a reasonable assumption. However, some significant differences can be found in the predictions of Cosserat vs. full micromorphic theory, as discussed in the reference Cordero et al. (2010a).

Size Effects in a Two-Phase Single-Crystal Laminate

A periodic two-phase single-crystal laminate under simple shear, whose unit cell is shown in Fig. 1, is considered, following Forest and Sedláček (2003a), Forest (2008), and Cordero et al. (2010a). This microstructure is composed of a hard elastic phase (h) and a soft elasto-plastic phase (s) where one slip system with slip direction normal to the interface between (h) and (s) is chosen. A mean simple glide $\overline{\gamma}$ is applied in the crystal slip direction $\boldsymbol{l}$ of the phase (s). The displacement and microdeformation fields take the form:

$$u_1 = \overline{\gamma} x_2,\ u_2(x_1),\ u_3 = 0,\ \chi^p_{12}(x_1),\ \chi^p_{21}(x_1) \tag{44}$$

within the context of small deformation theory. It follows that:

$$[\mathbf{H}] = \begin{bmatrix} 0 & \overline{\gamma} & 0 \\ u_{2,1} & 0 & 0 \\ 0 & 0 & 0 \end{bmatrix}$$

$$[\mathbf{H}^p] = \begin{bmatrix} 0 & \gamma & 0 \\ 0 & 0 & 0 \\ 0 & 0 & 0 \end{bmatrix} \quad [\mathbf{H}^e] = \begin{bmatrix} 0 & \overline{\gamma} - \gamma & 0 \\ u_{2,1} & 0 & 0 \\ 0 & 0 & 0 \end{bmatrix}$$

$$[\chi^p] = \begin{bmatrix} 0 & \chi^p_{12}(x_1) & 0 \\ \chi^p_{21}(x_1) & 0 & 0 \\ 0 & 0 & 0 \end{bmatrix} \quad [\mathrm{curl}\ \chi^p] = \begin{bmatrix} 0 & 0 & -\chi^p_{12,1} \\ 0 & 0 & 0 \\ 0 & 0 & 0 \end{bmatrix}$$

The resulting stress tensors are:

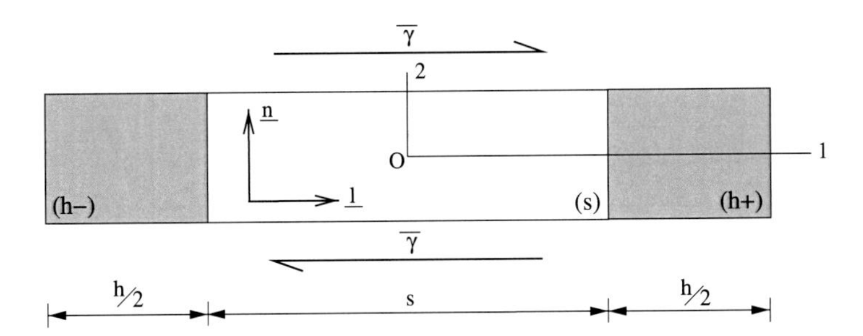

Fig. 1 Single slip in a periodic two-phase single-crystal laminate under simple shear: the gray phase (h) displays a purely linear elastic behavior, whereas the inelastic deformation of the white elasto-plastic phase (s) is controlled by a single-slip system $(\boldsymbol{n}, \boldsymbol{l})$

$$[\sigma] = \mu \begin{bmatrix} 0 & \overline{\gamma} - \gamma + u_{2,1} & 0 \\ \overline{\gamma} - \gamma + u_{2,1} & 0 & 0 \\ 0 & 0 & 0 \end{bmatrix}$$

$$[\mathrm{s}] = -H_\chi \begin{bmatrix} 0 & \gamma - \chi^p_{12} & 0 \\ -\chi^p_{21} & 0 & 0 \\ 0 & 0 & 0 \end{bmatrix}$$

$$[\mathrm{M}] = \begin{bmatrix} 0 & 0 & -A\chi^p_{12,1} \\ 0 & 0 & 0 \\ 0 & 0 & 0 \end{bmatrix} \quad [\mathrm{curl} \ \ \mathrm{M}] = \begin{bmatrix} 0 & -A\chi^p_{12,11} & 0 \\ 0 & 0 & 0 \\ 0 & 0 & 0 \end{bmatrix}$$

These forms of matrices are valid for both phases, except that $\gamma \equiv 0$ in the hard elastic phase. Each phase possesses its own material parameters, H_χ and A, the shear modulus, μ, being assumed for simplicity to be identical in both phases. The balance equation, $\mathbf{s} = -\mathrm{curl}\,\mathbf{M}$, gives $\chi^p_{21} = 0$ and the plastic slip:

$$\gamma = \chi^p_{12} - \frac{A}{H_\chi}\chi^p_{12,11}. \tag{45}$$

In the soft phase, the plasticity criterion stipulates that:

$$\sigma_{12} + s_{12} = \tau_c + H\gamma_{\mathrm{cum}}, \tag{46}$$

where H is a linear hardening modulus considered in this phase and γ_{cum} is the accumulated plastic slip as $\dot{\gamma}_{\mathrm{cum}} = |\dot{\gamma}|$. The following analytical resolution is done for the first loading branch, under monotonic loading. The slip direction, $\boldsymbol{l}$, has been chosen such that $\gamma > 0$ for this first loading branch so that we have: $\gamma_{\mathrm{cum}} = \gamma$. Considering Eqs. (45) and (46), we obtain the second-order differential equation for the microdeformation variable in the soft phase, χ^{ps}_{12},

$$\frac{1}{\omega^{s2}}\chi^{ps}_{12,11} - \chi^{ps}_{12} = \frac{\tau_c - \sigma_{12}}{H}, \text{ with } \omega^s = \sqrt{\frac{H^s_\chi H}{A^s\left(H^s_\chi + H\right)}}. \tag{47}$$

where $1/\omega^s$ is the characteristic length of the soft phase for this boundary value problem. The force stress balance equation requires σ_{12} to be uniform. It follows that the nonhomogeneous part of the differential equation is constant and then the hyperbolic profile of χ^{ps}_{12} takes the form:

$$\chi^{ps}_{12} = C^s \cosh(\omega^s x) + D, \tag{48}$$

where C^s and D are constants to be determined. Symmetry conditions $\left(\chi^{ps}_{12}(-s/2) = \chi^{ps}_{12}(s/2)\right)$ have been taken into account.

In the elastic phase, where the plastic slip vanishes, a hyperbolic profile of the microdeformation variable, χ_{12}^{ph}, is also obtained:

$$\chi_{12}^{ph} = C^h \cosh\left(\omega^h\left(x \pm \frac{s+h}{2}\right)\right), \text{ with } \omega^h = \sqrt{\frac{H_\chi^h}{A^h}}, \tag{49}$$

where, again, C^h is a constant to be determined, and symmetry conditions have been taken into account. It is remarkable that the plastic microvariable, χ_{12}^{ph}, does not vanish in the elastic phase, close to the interfaces, although no plastic deformation takes place. This is due to the transmission of double traction. Such a transmission has been shown in Cordero et al. (2010a) to be essential for size effects to occur. This point will be discussed in section "Size Effects in a Two-Phase Single-Crystal Laminate." The linear constitutive equation for the double-stress tensor in (32) can be interpreted, for the elastic phase, as nonlocal elasticity. That is why the corresponding characteristic length, $1/\omega^h$, will be kept of the order of nanometers in the presented simulation.

The coefficients C^s, D, and C^h can be identified using the interface and periodicity conditions:

- Continuity of χ_{12}^{p} at $x = \pm s/2$:

$$C^s \cosh\left(\omega^s \frac{s}{2}\right) + D = C^h \cosh\left(\omega^h \frac{h}{2}\right). \tag{50}$$

- Continuity of the double traction, as given in Eq. (8), $m_{12} = -M_{13}$ at $x = \pm s/2$:

$$A^s \omega^s C^s \sinh\left(\omega^s \frac{s}{2}\right) = -A^h \omega^h C^h \sinh\left(\omega^h \frac{h}{2}\right). \tag{51}$$

- Periodicity of displacement component u_2. We have the constant stress component:

$$\sigma_{12} = \mu\left(\overline{\gamma} - \gamma + u_{2,1}\right) \tag{52}$$

whose value is obtained from the plasticity criterion in the soft phase (Eq. 46):

$$\sigma_{12} = \tau_c + H\gamma_{cum} - A^s \chi_{12,11}^{ps}. \tag{53}$$

Still considering the first loading branch for which $\gamma_{cum} = \gamma$, it follows that:

$$u_{2,1}^s = \frac{\sigma_{12}}{\mu} - \overline{\gamma} + \gamma = \frac{\tau_c}{\mu} - \overline{\gamma} + \frac{A^s \omega^{s2} C^s}{H} \cosh\left(\omega^s x\right) + \frac{H+\mu}{\mu} D \tag{54}$$

in the soft phase and:

$$u^h_{2,1} = \frac{\sigma_{12}}{\mu} - \overline{\gamma} = \frac{\tau_c}{\mu} - \overline{\gamma} + \frac{H}{\mu} D \tag{55}$$

in the hard phase. The average on the whole structure,

$$\int_{-(s+h)/2}^{(s+h)/2} u_{2,1} dx = 0, \tag{56}$$

must vanish for periodicity reasons and gives

$$\left(\frac{\tau_c}{\mu} - \overline{\gamma}\right)(s+h) + \frac{2A^s\omega^s C^s}{H} \sinh\left(\omega^s \frac{s}{2}\right) + \frac{H\,(s+h) + \mu s}{\mu} D = 0 \tag{57}$$

The resolution of Eqs. (50), (51), and (57) gives:

$$C^s = \left(\frac{\tau_c}{\mu} - \overline{\gamma}\right)$$
$$\left[\frac{A^s\omega^s \sinh\left(\omega^s \frac{s}{2}\right)}{s+h}\left(\frac{H\,(s+h) + \mu s}{\mu}\left(\frac{\coth\left(\omega^s \frac{s}{2}\right)}{A^s\omega^s} + \frac{\coth\left(\omega^h \frac{h}{2}\right)}{A^h\omega^h}\right) - \frac{2}{H}\right)\right]^{-1} \tag{58}$$

$$D = -A^s\omega^s C^s \sinh\left(\omega^s \frac{s}{2}\right)\left(\frac{\coth\left(\omega^s \frac{s}{2}\right)}{A^s\omega^s} + \frac{\coth\left(\omega^h \frac{h}{2}\right)}{A^h\omega^h}\right) \tag{59}$$

$$C^h = -C^s \frac{A^s\omega^s \sinh\left(\omega^s \frac{s}{2}\right)}{A^h\omega^h \sinh\left(\omega^h \frac{h}{2}\right)}. \tag{60}$$

Figure 2 shows the profiles of plastic microdeformation and double traction in the two-phase laminate for different sets of material parameters and for a fraction of soft phase $(s), f_s = 0.7$. These profiles clearly show the continuity of χ^p_{12} and m_{12} at the interfaces. The different shapes presented are obtained for various values of the modulus A^s, the other material parameters being fixed and given in Table 1. Varying A^s modifies the mismatch with respect to the modulus A^h of the phase (h). Without mismatch the profile of χ^p_{12} is smooth at interfaces, while stronger mismatches lead to sharper transitions between the phases. Varying A^s also changes the intrinsic length scale $1/\omega^s$ of the phase (s). When the intrinsic length scale is small compared to the size of the microstructure, the microdeformation gradient can develop inside the phase (s) which leads to a rounded profile of the plastic microdeformation χ^p_{12} and to a double traction m_{12} localized at the interfaces. When the intrinsic length scale increases, the value of the double traction also increases at the interfaces (or equivalently, when decreasing the microstructure length scale, $l = s + h$, for a fixed

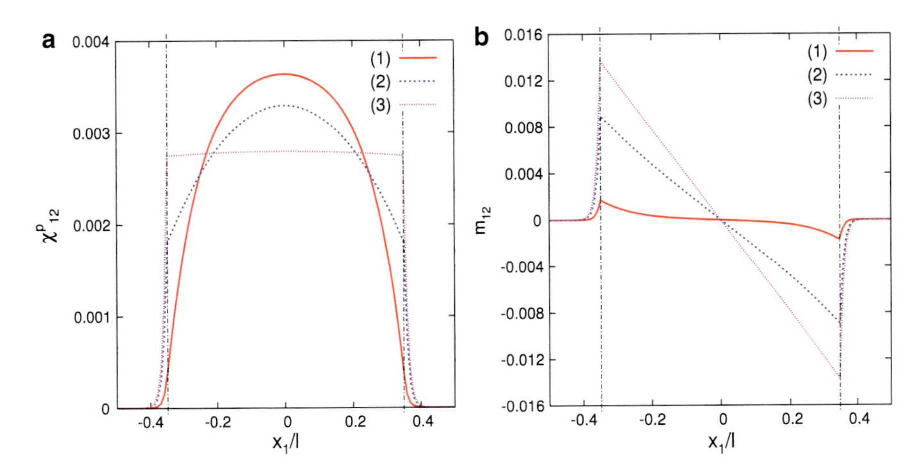

Fig. 2 Profiles of (**a**) plastic microdeformation χ^p_{12} and (**b**) double traction m_{12} in the two-phase microstructure with the *microcurl* model at 0.2% overall plastic strain obtained with the set of material parameters given in Table 1 and (1) with no mismatch between the moduli of the two phases, $A^h = A^s = 5.10^{-5}$ MPa.mm^2; (2) with a stronger mismatch, $A^h = 5.10^{-5}$ MPa.mm^2 and $A^s = 1.10^{-3}$ MPa.mm^2; and (3) $A^h = 5.10^{-5}$ MPa.mm^2 and $A^s = 5.10^{-2}$ MPa.mm^2. The associated intrinsic length scales, $1/\omega^s$, are, respectively, 100 nm, 449 nm, and 3.2 μm. In all three cases, the fraction of soft phase $f_s = 0.7$ and the microstructure size is fixed, $l = 1$ μm. The *vertical lines* indicate the position of interfaces

intrinsic length scale). When the intrinsic length scale becomes of the order of the size of the microstructure or even larger, the model starts to saturate so that χ^p_{12} becomes quasihomogeneous (flat profile) and the double traction is not localized anymore (linear profile). From Eq. (53) we derive the expression of the macroscopic stress tensor component, Σ_{12}, defined as the mean value of the stress component σ_{12} over the microstructure size, $l = (s + h)$:

$$\sum_{12} = \langle\sigma_{12}\rangle = \frac{1}{l}\int_{-\frac{1}{2}}^{\frac{1}{2}} \sigma_{12}dx = \tau_c + \frac{H}{f_s}\langle\gamma^{cum}\rangle - \frac{A^s}{f_s}\langle\chi^{ps}_{12,11}\rangle, \tag{61}$$

where brackets < > denote the average values over the microstructure unit cell. We obtain the mean plastic slip for the first loading branch from Eq. (45):

$$\langle\gamma\rangle = \left\langle \chi^{ps}_{12} - \frac{A^s}{H^s_\chi}\chi^{ps}_{12,11} \right\rangle = \frac{2A^s\omega^s C^s \sinh\left(\omega^s \frac{f_s l}{2}\right)}{Hl} + f_s D \tag{62}$$

where f_s is the fraction of soft phase. From this we obtain alternative expressions of C^s and D as functions of $\langle\gamma\rangle$,

Table 1 Set of material parameters used in the simulations. The intrinsic length scales, defined as $1/\omega^{h,s}$, induced by these parameters is of the order of 10 nm for the elastic phase (h) and 500 nm for the plastic phase (s)

	μ[MPa]	τ_c[MPa]	H[MPa]	H_χ[MPa]	A[MPa.mm^2]
Phase (s)	35,000	40	5000	500,000	1.10^{-3}
Phase (h)	35,000	–	–	500,000	5.10^{-5}

$$C^s = -\langle\gamma\rangle \left[A^s\omega^s \sinh\left(\omega^s \frac{f_s l}{2}\right)\left(f_s\left(\frac{\coth\left(\omega^s \frac{f_s l}{2}\right)}{A^s\omega^s} + \frac{\coth\left(\omega^h \frac{(1-f_s)l}{2}\right)}{A^h\omega^h}\right) - \frac{2}{Hl}\right)\right]^{-1} \tag{63}$$

$$D = \langle\gamma\rangle \left[f_s - \frac{2}{Hl}\left(\frac{\coth\left(\omega^s \frac{f_s l}{2}\right)}{A^s\omega^s} + \frac{\coth\left(\omega^h \frac{(1-f_s)l}{2}\right)}{A^h\omega^h}\right)^{-1}\right]^{-1} \tag{64}$$

which contain contributions from both the back stress and the isotropic hardening. The macroscopic stress takes the form:

$$\sum\nolimits_{12} = \tau_c + HD. \tag{65}$$

The hardening produced by the model is a combination of the kinematic hardening arising from the higher-order back stress component and the linear isotropic hardening introduced in (46). Its modulus, H^{tot}, is size-dependent and is obtained using Eqs. (64) and (65):

$$H^{tot} = H \left[f_s - \frac{2}{Hl}\left(\frac{\coth\left(\omega^s \frac{f_s l}{2}\right)}{A^s\omega^s} + \frac{\coth\left(\omega^h \frac{(1-f_s)l}{2}\right)}{A^h\omega^h}\right)^{-1}\right]^{-1} \tag{66}$$

One cycle of deformation $\overline{\gamma}$ has been considered to illustrate the kinematic hardening effects. In the absence of gradient effects, only isotropic hardening is visible. The *microcurl* model leads to an additional kinematic hardening component. When the size of the elasto-plastic phase (s) becomes large compared to the intrinsic length scale $1/\omega_s$, strain gradient effect is small, and the kinematic hardening arising from the *microcurl* model tends to vanish. Then the model reduces to conventional crystal plasticity theory, and the limit of the 0.2% macroscopic flow stress is:

$$\lim_{l \to \infty} \sum_{12|0.2} = \tau_c + \frac{H}{f_s} \langle \gamma^{cum} \rangle . \tag{67}$$

In contrast, the maximum extra stress, $\Delta\Sigma$, predicted by the model at small microstructure sizes can be computed as:

$$\Delta\Sigma = \lim_{l \to 0} \sum_{12} (\langle \gamma \rangle) - \lim_{l \to \infty} \sum_{12|0.2} = \frac{1 - f_s}{f_s} H_\chi \langle \gamma \rangle . \tag{68}$$

Figure 3 presents the predicted evolution of the macroscopic flow stress $\Sigma_{12|0.2}$ at 0.2% plastic strain (obtained by setting $\langle \gamma \rangle = 0.002$) as a function of the microstructure length scale l in a log–log diagram. This evolution is plotted using the material parameters given in Table 1 and for various values of the coupling modulus, $H_\chi^s = H_\chi^h = H_\chi$. The four lower curves are obtained for finite values of the modulus H_χ; they exhibit a tanh shape with saturation for large ($l > 10^{-2}$ mm) and small ($l < 10^{-5}$ mm) values of l. These saturations can be characterized by the limit given in Eq. (67) and the maximum extra stress, $\Delta\Sigma$, given in Eq. (68), respectively. A transition domain with strong size dependence is observed between these two plateaus. The limits and the maximum extra stress, the position of the transition zone, and the scaling law exponent in the size-dependent domain (slope in the log–log diagram) are directly related to the material parameters used in the model. In fact, the position of the size-dependent domain is controlled by the moduli $A^{h,s}$ (not illustrated here), while the maximum extra stress and the scaling law exponent are both controlled by the modulus H_χ, both increasing for higher values of H_χ as suggested by Fig. 3.

When H_χ is very small, we can deduce from Eq. (68) that $\Delta\Sigma$ vanishes, and consequently the scaling law exponent will tend to 0. The upper curve is obtained for $H_\chi \to \infty$; it no longer exhibits a tanh shape as no saturation occurs for small values of l, the limit $\Delta\Sigma \to \infty$ follows. This limit case will be described in the next subsection; it will be shown that in that case, a scaling law exponent of −2 is reached. Finally, the *microcurl* model can produce scaling law exponents ranging from 0 to −2.

Strain Gradient Plasticity as a Limit Case

In the proposed *microcurl* model, the modulus H_χ introduces a coupling between micro and macro variables. A high value of H_χ forces the plastic microdeformation χ^p to remain as close as possible to the macroplastic deformation $\mathbf{H}^p$. Consequently, it enforces the condition that $\mathbf{K}$ coincides with the dislocation density tensor. In this case, the *microcurl* model degenerates into the strain gradient plasticity model by Gurtin (2002). When applied to the laminate microstructure, the strain gradient plasticity model leads to the indeterminacy of the double-traction vector at the interfaces, due to the fact that no strain gradient effect occurs in the elastic phase; see Cordero et al. (2010a). The *microcurl* model can then be used to derive the missing

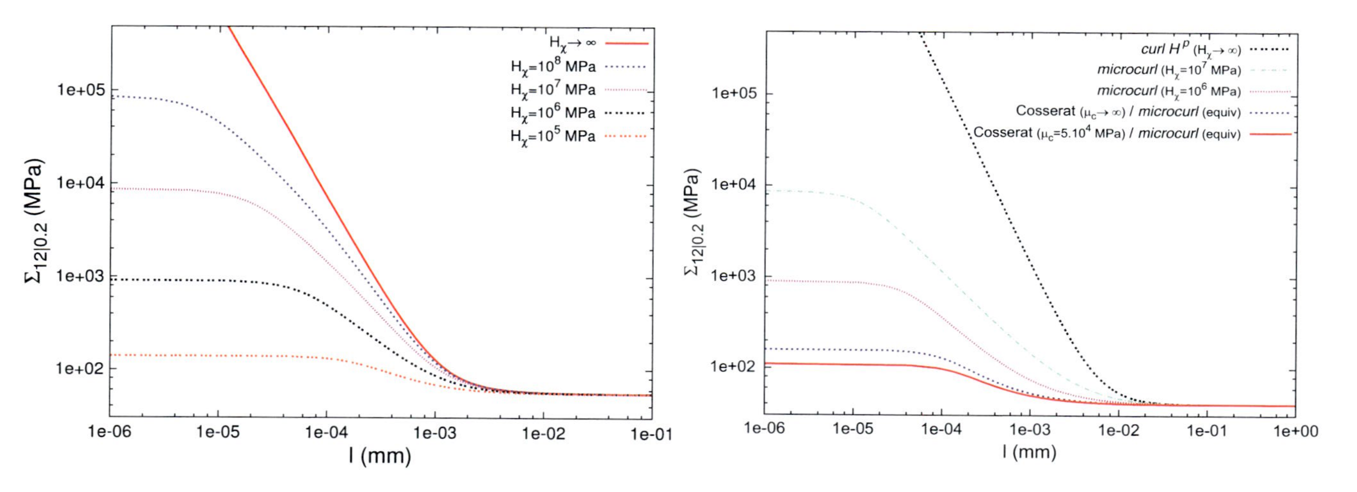

Fig. 3 Evolution of the macroscopic flow stress $\Sigma_{12|0.2}$ at 0.2% plastic strain as a function of the microstructure length scale l, plotted for different coupling moduli, $H_\chi \left(= H_\chi^s = H_\chi^h\right)$, on the left. The other material parameters are given in Table 1 and $f_s = 0.7$. On the right, comparison between the micropolar and micromorphic models

interface condition to be applied at the interface, by means of a limit process in the previous solution of the boundary value problem.

The limit $H_\chi \to \infty$ of the *microcurl* model can be used to determine the value of the double traction to be imposed at the interface, as follows:

$$\begin{aligned}\lim_{H_\chi\to\infty} m_{12}(s/2) &= \lim_{H_\chi\to\infty} A^s \chi^{ps}_{12,1}(s/2)\\ &= \lim_{H_\chi\to\infty} A^s\omega^s \sinh\left(\omega^s \frac{f_s l}{2}\right) C^s\\ &= \lim_{H_\chi\to\infty} \langle\gamma\rangle \left[\frac{2}{Hl} - f_s\left(\frac{\coth\left(\omega^s \frac{f_s l}{2}\right)}{A^s\omega^s} + \frac{\coth\left(\omega^h \frac{(1-f_s)l}{2}\right)}{A^h\omega^h}\right)\right]^{-1}\end{aligned}$$

Since $H_\chi \to \infty$, $1/\omega^h \to 0$ and $\coth(\omega^h\, h/2) \to 1$. Moreover, $\omega^s_\infty := \omega^s \to \sqrt{H/A^s}$.

Consequently,

$$\lim_{H_\chi\to\infty} m_{12}(s/2) = \langle\gamma\rangle \left[\frac{2}{Hl} - f_s \frac{\coth\left(\omega^s_\infty \frac{f_s l}{2}\right)}{A^s\omega^s_\infty}\right]^{-1} \tag{69}$$

Accordingly, the double traction is found to depend on the mean plastic slip. The characteristic length in the soft phase for the strain gradient plasticity model is found to be related to the ratio between the hardening modulus and the higher-order modulus, A^s. The limiting process can also be used to predict the response of the strain gradient plasticity model in the size effect zone. For that purpose, let us consider the limit of $\Sigma_{12|0\cdot 2}$, when H_χ goes to infinity. Indeed, when H_χ tends to infinity, the expression of D in Eq. (64) can be simplified. We consider sizes of the microstructures in the size effect zone, i.e., intermediate values of l. Since H_χ is very high, the term $\tanh(\omega^h(1-f_s)l/2)$ tends to 1. Considering that l is small enough, the term $l\,(\tanh(\omega_s f_s\, l/2))$ can be approximated by its Taylor expansion at the order 2, which leads to D of the form:

$$D \approx \frac{al + b}{cl^2 + dl + e} \tag{70}$$

where

$$a = \frac{\langle\gamma\rangle f_s}{2\sqrt{H_\chi}},\; b = \langle\gamma\rangle f_s A^h \left(1 + \frac{H}{H_\chi}\right) \tag{71}$$

$$c = -\frac{f_s^3 H\sqrt{A^h}}{12},\; d = \frac{f_s^2 H}{2\sqrt{H_\chi}},\; e = -\frac{f_s\sqrt{A^h}H}{H_\chi} \tag{72}$$

The terms a, d, and e tend to 0 when $H_\chi \to \infty$, so that:

$$D \approx \frac{12A_s \langle \gamma \rangle}{f_s^3 H l^2} \tag{73}$$

and for the macroscopic stress:

$$\Sigma_{12} \approx \tau_c + \frac{12A_s \langle \gamma \rangle}{f_s^3 l^2} \tag{74}$$

This expression indicates a l^{-2} scaling law for the strain gradient plasticity model. This scaling law differs from the Hall–Petch relation, $l^{-1/2}$, typical for grain size effects, and from the Orowan relation, l^{-1}, valid for precipitate size effects.

On the right Fig. 3 shows a comparison between the micropolar and micromorphic responses in the case of the two-phase laminate under shear. A saturation of the stress level is found for increasing values of the penalty modulus H_χ in the micropolar model at small scales. In contrast, the micromorphic response converges toward that of the strain gradient plasticity model and displays no limit at small scales. This is a fundamental difference between the lattice curvature-based and the dislocation density tensor-based models; see Cordero et al. (2010a).

Free Energy Potentials for Micromorphic Crystal Plasticity

The previous example showing that a simple quadratic potential with respect to the dislocation density tensor does not provide the satisfactory scaling law for the plastic behavior of the channel is an incentive for developing more physical constitutive laws for strain gradient plasticity. Such an attempt is presented in this section along the lines of Wulfinghoff et al. (2015).

Physically, the introduction of additional energy density terms may be motivated by the incomplete nature of the continuum theory. Clearly, the continuum description does not contain the full information on the discrete dislocation microstructure. In particular, single dislocations are not resolved. Instead, the continuum representation may be interpreted as a smoothed version of the real system, where information is lost deliberately. There is no reason to assume that the total elastic energy of the continuum representation coincides with the elastic energy of the real system including discrete dislocations. This is due to the loss of information as a result of the smoothing procedure (Mesarovic et al. 2010). Additional energy terms in gradient plasticity may therefore be interpreted as an attempt to partially compensate the error in the continuum elastic energy. This is done by taking into account available kinematical information on the dislocation microstructure as additional argument of the energy.

The optimal form of the energy is subject of current research. Most applications are based on a pragmatic quadratic approach (e.g., Cordero et al. 2012; Reddy et al.

2012; Wulfinghoff and Böhlke 2012; Miehe et al. 2014; Wulfinghoff et al. 2013a, b; Mesarovic et al. 2015).

Instead, a more reasonable approach seems to be based on a variable internal length scale as a function of the dislocation state (Groma et al. 2003; Mesarovic et al. 2010). The quadratic form was recently shown to provide physically unrealistic scaling in the size-dependent response of laminate microstructures under shear (Cordero et al. 2010b; Forest and Guéninchault 2013). Since quadratic forms are unusual in the classical dislocation theory, alternative free energy potentials were proposed in the past 10 years. Rank-one energies that are linear with respect to the GND densities have been shown to lead to a size-dependent yield stress in certain situations. Additionally motivated by line tension (and more elaborate) arguments, they are used by several authors (Ortiz and Repetto 1999; Conti and Ortiz 2005; Ohno and Okumura 2007; Kametani et al. 2012; Hurtado and Ortiz 2013).

Asymptotic methods can be used to derive alternative effective potential for distributions of edge dislocations. The asymptotic derivation of a logarithmic potential by De Luca et al. (2012) accounts for line tension effects at the macroscopic scale. Systematic derivations of back stress distributions were derived in Geers et al. (2013) by means of asymptotic methods.

The choice of a logarithmic energy is inspired by the statistical theory of dislocations of Groma et al. (2003, 2007) and Berdichevsky (2006a,b). Here, the internal length scale of the back stress is determined by the dislocation microstructure (see also Svendsen and Bargmann 2010; Forest and Guéninchault 2013). In the latter reference, the rank-one and logarithmic formulations were applied to strain gradient plasticity theories involving the full dislocation density tensor instead of the individual GND densities.

Formulation of Two Free Energy Potentials

It is assumed that the volumetric stored energy density has the form:

$$\rho\psi = W = W_{\mathrm{e}} + W_{\mathrm{g}} + W_{\mathrm{h}}, \tag{75}$$

with $W_e = (\varepsilon - \varepsilon^{\mathrm{p}}) \; : \mathbb{C} : (\varepsilon - \varepsilon^{\mathrm{p}})/2$. The expressions W_{h} and W_{g} are assumed to be functions of internal (hardening) variables α and the dislocation density tensor $\mathbf{A}$, respectively.

Size-independent isotropic hardening is accounted for by W_{h}, while W_{g} models size effects.

The following defect energies are investigated:

$$W_{\mathrm{g}}^{1} = cGb\,\|\mathbf{A}\|\,,\; W_{\mathrm{g}}^{\mathrm{ln}} = c_0\,\|\mathbf{A}\|\ln\frac{\|\mathbf{A}\|}{A_0}, \tag{76}$$

where c is a constant of order unity, G is the macroscopic shear modulus, b is the Burgers vector, A_0 is a constant, and c_0 is given by:

$$c_0 = \frac{Gb\beta}{2\pi\,(1-\nu)}, \tag{77}$$

where ν is Poisson's ratio and β is of order unity. The Euclidean norm of the dislocation density tensor is defined as: $\|\mathbf{A}\| = \sqrt{\mathbf{A}\cdot\mathbf{A}}$.

The rank-one energy W_g^1 can be motivated by simple line tension arguments; see Ortiz and Repetto (1999) and Hurtado and Ortiz (2012, 2013).

The logarithmic energy $W_g^{\ln}$ (Eq. (76)) is motivated by the form of the associated back stress (Forest and Guéninchault 2013). It turns out that the approach $W_g^{\ln}$ leads to a back stress which is formally close to the one derived in the statistical theory of Groma et al. (2003), given in 1D by

$$\frac{Gc_1}{2\pi\,(1-\upsilon)\,\rho}\partial^2_{x_1}\gamma \tag{78}$$

for a single-slip situation with slip direction e_1. Here, ρ denotes the total dislocation density. In the two-dimensional single-slip regime, the back stress involves the Laplacian of the plastic slip, as postulated by Aifantis (1987). However, the internal length scale is not interpreted as a material constant but determined by the dislocation microstructure, if $W_g^{\ln}$ is applied.

The stresses σ and $\mathbf{M}$ are assumed to be energetic, i.e.,

$$\boldsymbol{\sigma} = \frac{\partial W}{\partial \boldsymbol{\varepsilon}^e},\ \boldsymbol{M} = \frac{\partial W}{\partial \mathbf{A}} \tag{79}$$

If the stored energy is not differentiable at $\mathbf{A} = \mathbf{0}$, the symbol ∂ in Eq. $(79)_2$ is interpreted as a subdifferential operator (see, e.g., Han and Reddy 2013), i.e.,

$$\boldsymbol{M}\,|_{\alpha=0} \in \left\{\boldsymbol{M} : W_g\,(\mathbf{A}) - \boldsymbol{M}\cdot\mathbf{A} \geq 0\ \ \forall\mathbf{A}\right\}. \tag{80}$$

This can be interpreted as follows. If the stress $\boldsymbol{M}$ is applied at a material point, $\mathbf{A}$ will take a value which minimizes the expression $W_g(\mathbf{A}) - \mathrm{M}\cdot\mathbf{A}$. For small values of $\boldsymbol{M}$, the minimum is given by $\mathbf{A} = 0$. However, for sufficiently large values of $\boldsymbol{M}$, the value of $\mathbf{A}$ can be determined from the stationarity condition $\boldsymbol{M} = \partial_\alpha W_g$.

For example, if W_g is given by $W_g^1 = cGb\,\|\mathrm{A}\|$, it follows that $\mathbf{A} = 0$ if

$$\boldsymbol{M} : \mathbf{A} \leq W_g^1\,(\mathbf{A}) = cGb\,\|\mathbf{A}\|\ \forall\mathbf{A} \iff \boldsymbol{M} : \mathbf{A} \leq \|\boldsymbol{M}\|\,\|\mathbf{A}\| \leq cGb\,\|\mathbf{A}\|\ \forall\mathbf{A}. \tag{81}$$

Hence, it is found that:

$$\begin{cases} \boldsymbol{M} \in \{\boldsymbol{M} : \varphi\,(\boldsymbol{M}) \leq 0\}\,, \text{if } \mathbf{A} = 0 \\ \boldsymbol{M} = cGb\frac{\mathbf{A}}{\|\mathbf{A}\|}, \text{else.} \end{cases} \tag{82}$$

with $\varphi(\boldsymbol{M}) = \|\boldsymbol{M}\| - cGb$.

Note that the generalized stress $\boldsymbol{M}$ can be computed uniquely from $\mathbf{A}$ only if $\mathbf{A} \neq 0$. This makes analytical solutions as well as the numerical implementation difficult. Possible numerical strategies concerning this problem are discussed in Kametani et al. (2012) as well as Hurtado and Ortiz (2013).

Application to the Shearing of the Periodic Laminate

In this section, the two new potentials are applied to the elasto-plastic laminate microstructure already considered in section "Size Effects in a Two-Phase Single-Crystal Laminate"; see Fig. 1. The two promising candidates of the defect energy function W_g are investigated concerning their effect on the overall size effects as well as the dislocation pileup structures building up at impenetrable boundaries. The dislocation density tensor can be expressed in terms of the edge density $\rho_\perp = -\partial_{x1}\gamma$

$$\mathbf{A} = -\rho_\perp \boldsymbol{e}_1 \otimes \boldsymbol{e}_3. \tag{83}$$

The quantity $\rho_\perp$ represents the total Burgers vector amount per unit area of edge dislocations. Note that its unit (μm^{-1}) differs from the unit of the total line length per unit volume ρ, given by μm^{-2}.

Assuming the defect energy W_g to be a function of $||\mathbf{A}||$, the generalized stress $\boldsymbol{M}$ reads

$$\boldsymbol{M} = \partial_{\boldsymbol{A}} W_g = \partial_{\|\boldsymbol{A}\|} W_g \frac{\mathbf{A}}{\|\mathbf{A}\|} = M(x_1)\, \boldsymbol{e}_1 \otimes \boldsymbol{e}_3. \tag{84}$$

From the balance Eq. $(7)_2$, it follows that

$$s_{12} - M' = 0. \tag{85}$$

Throughout this section, the isotropic hardening contribution will be neglected, i.e., $W_h = 0$.

Rank-One Defect Energy

For the laminate, the following energy is adopted:

$$W_g^1 = cGb\,\|\mathbf{A}\| = cGb\,|\rho_\perp|, \tag{86}$$

where c is of order unity (Ortiz and Repetto 1999). According to Eq. (84), the generalized stress M reads:

$$\begin{aligned} M &= -\tfrac{\rho_\perp}{|\rho_\perp|} cGb = -\operatorname{sign} \rho_\perp cGb, \text{ if } |\rho_\perp| > 0 \\ &|M| \leq cGb, \text{ if } |\rho_\perp| = 0 \end{aligned} \tag{87}$$

where the second line follows from Eq. (82).

Subsequently, a monotonic shear deformation in the positive direction is prescribed such that the following relations hold in the soft phase: $\tau^{\text{eff}} \geq \tau^C, \quad \dot{\gamma} \geq 0$.

The principle of virtual power is written for the real field on the laminate unit cell V:

$$\int_V \boldsymbol{\sigma} : \dot{\boldsymbol{\varepsilon}} + \mathrm{s} : \dot{\mathbf{H}}^p + \mathrm{M} : \mathrm{curl}\ \dot{\mathbf{H}}^p dV = \int_{\partial V} \boldsymbol{t} \cdot \dot{\boldsymbol{u}}\, dS + \mathbf{m} : \dot{\mathbf{H}}^p dS \qquad (88)$$

The last term of the right-hand side vanishes due to the fact that $\mathbf{H}^p$ is periodic, whereas $\mathbf{m}$ is antiperiodic. The first term of the right-hand side coincides with the first term in the left-hand side, as can be shown by means of the Gauss theorem. As a result we obtain,

$$\int_{\partial V} \mathrm{s} : \dot{\mathbf{H}}^p + \mathrm{M} : \mathrm{curl}\ \dot{\mathbf{H}}^p\ dV = 0 \qquad (89)$$

For the laminate under single slip, this gives:

$$\int_{-s/2}^{s/2+h} s_{12}\ \dot{\gamma} + (-c\,Gb\ \mathrm{sign}\ \gamma_{,1})\,(-\dot{\gamma}_{,1})\ dx_1 = 0 \qquad (90)$$

Under monotonic loading, sign $\gamma_{,1}$ = sign $\dot{\gamma}_{,1}$, so that

$$\int_{-s/2}^{s/2+h} s_{12}\ \dot{\gamma} + c\,Gb\,|\dot{\gamma}_{,1}|\ dx_1 = 0 \qquad (91)$$

According to the Schmid law, $\sigma_{12} + s_{12} = \tau^{rmC}$ where the fields σ_{12}, s_{12} are uniform. The solution is such that the plastic slip field $\gamma(x_1)$ is uniform in $]-s/2, s/2[$ at each instant. So does $\dot{\gamma}(x_1) = \dot{\gamma}(0)$ in $]-s/2, s/2[$. It vanishes in $]s/2, s/2+h[$. It is therefore discontinuous at $\pm s/2$. As a result, the derivative of the plastic slip rate is the sum of two Dirac functions:

$$\dot{\gamma}_{,1}(x_1) = \dot{\gamma}(0)\left(\delta\left(x_1 + \frac{s}{2}\right) - \delta\left(x_1 - \frac{s}{2}\right)\right) \qquad (92)$$

The integration of these Dirac functions (in fact the absolute values due to (91)) on the interval $[-s/2, s/2 + h]$ finally gives

$$s\left(\tau^{rmC}c - \sigma_{12}\right)\dot{\gamma}(0) + 2c\,Gb\dot{\gamma}(0) = 0 \qquad (93)$$

The scaling law follows:

$$\sigma_{12} = \tau^{\mathrm{C}} + \frac{2c\,Gb}{s}. \qquad (94)$$

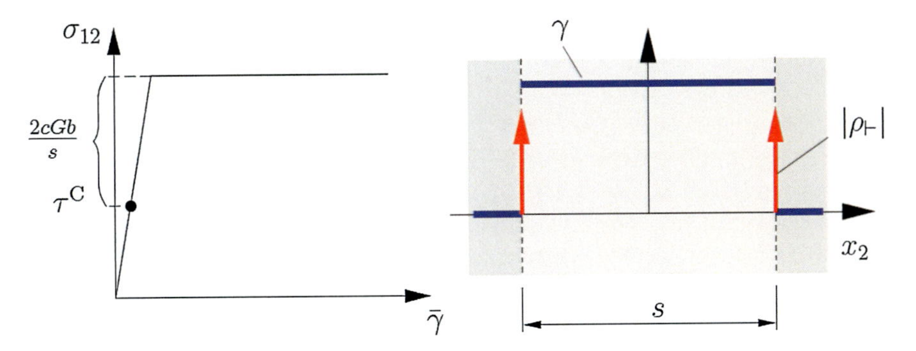

Fig. 4 Macroscopic shear stress–strain curve for the rank-one energy. The increase of the overall yield point scales inversely with the size of the soft phase, after Wulfinghoff et al. (2015)

This equation holds in the plastic regime. Clearly, the application of the rank-one energy increases the macroscopic yield stress by $2cGb/s$, i.e., the increase scales inversely with the size of the soft phase (see Fig. 4). The same scaling behavior has been found by Ohno and Okumura (2007) for a spherical grain, also using a rank-one energy. The authors concentrated on the overall mechanical response without having to compute the fields inside of the grains. As illustrated in Fig. 4, the dislocations localize in dislocation walls at the elasto-plastic interface.

For the material parameters of aluminum ($G = 26.12$ GPa and $b = 0.286$ nm) and $c = 1$, the size effect becomes important when the system size is below 10 μm. The plastic shear strain is constant in the bulk, i.e., the dislocations form singularities (walls) at the boundaries. The back stress is constant (w.r.t. space) in the bulk. During the first period, it increases and thereby impedes any plastic deformation. Therefore, the overall deformation is purely elastic during this period. At a certain point, the plastic deformation starts, and the back stress remains constant afterward. Its value is given by $2cGb/s$.

Logarithmic Energy

This section investigates the following defect energy:

$$W_{\mathrm{g}}^{\ln} = c_0 \, \|\mathbf{A}\| \ln \frac{\|\mathbf{A}\|}{A_0}, \tag{95}$$

with the constant c_0 as defined in Eq. (77). The energy is motivated by the statistical theory of dislocations by Groma et al. (2003). The authors derive a back stress term which involves the second gradient of slip as postulated by Aifantis (1987). However, their theory involves an internal length scale which is given by $1/\sqrt{\rho}$, where ρ denotes the total dislocation density.

In pure metals, the geometrical characteristics of the microstructure are essentially determined by the dislocation arrangement. This is a strong argument for a (variable) internal length scale, which is determined by the available dislocation field variables (instead of a constant length scale parameter; see also Forest and Sedláček (2003b) where this dependency is derived from a dislocation line tension model).

It is demonstrated subsequently that the approach (95) leads to a back stress which is similar to that of Groma et al. (2003). However, it should be mentioned that this energy is neither convex nor smooth with respect to the dislocation density tensor (a regularization will be discussed at a later stage).

For the laminate problem, the generalized stress M reads (see Eq. (84)):

$$M = -\operatorname{sign} \rho_\perp c_0 \left(\ln \frac{|\rho_\perp|}{A_0} + 1 \right). \tag{96}$$

In this section, rate-independent plasticity will be considered based on the yield criterion:

$$f = \left|\tau^{\text{eff}}\right| - \tau^{\text{C}} \leq 0. \tag{97}$$

Here, the effective stress reads:

$$\tau^{\text{eff}} = (\boldsymbol{\sigma} + \boldsymbol{s}) \cdot (\boldsymbol{l} \otimes \boldsymbol{n}) = \sigma_{12} + s_{12} \overset{(85)}{=} \sigma_{12} + M'. \tag{98}$$

With Eq. (96) and $M' = (\partial_{\rho\perp} M)\ (\partial_{x1}\rho_\perp)$, it follows that

$$\tau^{\text{eff}} = \tau - \frac{c_0}{|\rho_\perp|} \partial x_1 \rho_\perp = \tau + \frac{G\beta}{2\pi\,(1-\upsilon)} \frac{b}{|\rho_\perp|} \partial_{x_1}^2 \gamma. \tag{99}$$

Here, the second term can be interpreted as a back stress. Note that the back stress involves no internal length scale parameter. Instead, the internal length scale, $\sqrt{b/\rho_\perp}$, is determined by the dislocation microstructure. In contrast to the back stress of Groma et al. (2003), the internal length scale is determined by the GND density $\rho_\perp$ instead of the total density ρ. Hence, the influence of statistically stored dislocations (SSDs) is ignored for brevity. Therefore, a homogeneous initial GND density $|\rho_\perp| = A_0$ will be assumed to be given. In addition, it is assumed that the SSD density is equal or less than A_0. The soft phase is assumed to be under plastic loading, with $\tau^{\text{eff}} = \tau^{\text{C}}$ in the soft phase. In this case,

$$M' \overset{(85)}{=} s_{12} = -\left(\sigma_{12} - \tau^{\text{C}}\right) = \text{const.} \Rightarrow M = -\left(\sigma_{12} - \tau^{\text{C}}\right) x_1, \tag{100}$$

where, again, the constant of integration vanishes due to the symmetry requirement $|M(-s/2)| = |M(s/2)|$. The plastic slip γ can be derived from the equality of Eqns. (96) and (100), which yields a differential equation for γ. The solution reads:

$$\gamma = \frac{A_0 L}{e} \left(\exp\left(\frac{s}{2L}\right) - \exp\left(-a \frac{x_1}{L}\right) \right) \text{ with } L = \frac{c_0}{\sigma_{12} - \tau^{\text{C}}}, \tag{101}$$

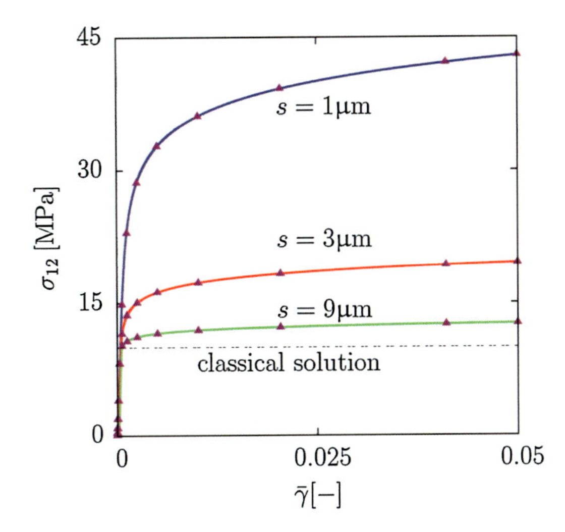

Fig. 5 Macroscopic stress–strain diagram for three different sizes. Analytical (*lines*) and regularized, numerical (*triangles*) solution for the logarithmic potential, after Wulfinghoff et al. (2015)

where the matching conditions $\gamma(-s/2) = \gamma(s/2) = 0$ have been exploited and where $e = \exp(1)$. The variable a is defined by $a = \text{sign}\,\gamma'$ and is assumed positive in $(-s/2, 0)$ and negative in $(0, s/2)$.

The macroscopic stress–strain relation follows:

$$\overline{\gamma} = \frac{A_0 L}{e\,(s+h)} \left(\exp\left(\frac{s}{2L}\right)(s-2L) + 2L \right) + \frac{\sigma_{12}}{G}. \tag{102}$$

The solution is evaluated for the following material parameters:

E [GPa]	ν	τ^{C} [MPa]	b [nm]	β	A_0/b [μm^{-2}]
70	0.34	10	0.286	1	1

A very thin hard phase with negligible width h is considered ($h/s = 10^{-6}$ for the analytical solution).

The macroscopic stress–strain curve (102) is illustrated in Fig. 5. A clear size effect is visible. Apparently, mainly the overall yield stress is affected. The hardening shows less size dependence. It is remarkable that the model provides a size-dependent yield stress and nonlinear kinematic hardening.

Since there is no distinct yield stress, the evaluation of the scaling behavior is based on the offset yield stress at 0.2% plastic strain. The offset yield stress as a function of the inverse of the size $1/s$ exhibits the same behavior as in the results obtained from the rank-one energy. It scales inversely with channel size; see Wulfinghoff et al. (2015).

Regularization of the Logarithmic Energy

The following regularization is introduced:

$$W_g = \begin{cases} \frac{1}{2}\frac{c_0}{b} l^2 \|\mathrm{A}\|^2, & \|\mathrm{A}\| < A_\mathrm{L} \\ c_0 \|\mathrm{A}\| \ln \frac{\|A\|}{A_0} + W_0, & \text{else.} \end{cases} \tag{103}$$

In the region of small GND densities, the energy is replaced by a quadratic potential. The internal length scale l, the transition density α_L, and the offset energy W_0 are chosen such that W_g, $\partial_{\|\alpha\|} W_g$ and $\partial^2_{\|\alpha\|} W_g$ are continuous at the transition point $\|\mathrm{A}\| = A_\mathrm{L}$. As a result

$$A_\mathrm{L} = A_0, \; l^2 = \frac{b}{A_0}, \; W_0 = \frac{c_0 A_0}{2}. \tag{104}$$

The regularized energy (103) is convex, normalized, and twice differentiable. The back stress for the laminate problem reads:

$$x = \begin{cases} -\frac{c_0}{A_0} \partial^2_{x_1} \gamma, & |\rho_\perp| < A_0, \\ -\frac{c_0}{|\rho_\perp|} \partial^2_{x_1} \gamma, & \text{else.} \end{cases} \tag{105}$$

Cyclic Behavior of the Laminate

The laminate is now submitted to one full cycle $\overline{\gamma} = \pm 0.05$. The hysteresis loops σ_{12} vs. $\overline{\gamma}$ for both rank-one and logarithmic potentials are represented in Fig. 6 for $s = 3$ μm. In the absence of isotropic hardening, the loops are stabilized after one full cycle. They are characterized by pure kinematic hardening. The influence of the back stress is clearly observable. The curves in Fig. 6 have been obtained numerically. One striking feature of the results is the nonconvexity of the obtained loops. According to the rank-one model, the first unloading stage is characterized by reverse plasticity at a constant negative shear stress. When $\overline{\gamma}$ goes through zero again, the overall shear stress experiences a jump of the same magnitude as computed analytically for monotonic loading in sections "Rank-One Defect Energy" and "Logarithmic Energy." The nonconvex loop obtained for the logarithmic potential is similar but smoother and displays smooth nonlinear kinematic hardening. A similar nonconvex hysteresis loop was obtained by Ohno and Okumura (2008) for the rank-one model.

The type of nonlinear kinematic hardening observed for both models can be identified with Asaro's type KIII model, corresponding to a *first-in/last-out* sequence of dislocation motion (Asaro (1975)). It is considered by Asaro as the most perfect form of recovery of plastic memory. Such stress–strain loops display inflection points that are observed in some materials, see Asaro (1975) for a Nimonic alloy, but such observations have also been made in several Nickel based superalloys. It is usually attributed to substructural recovery on the microscale,

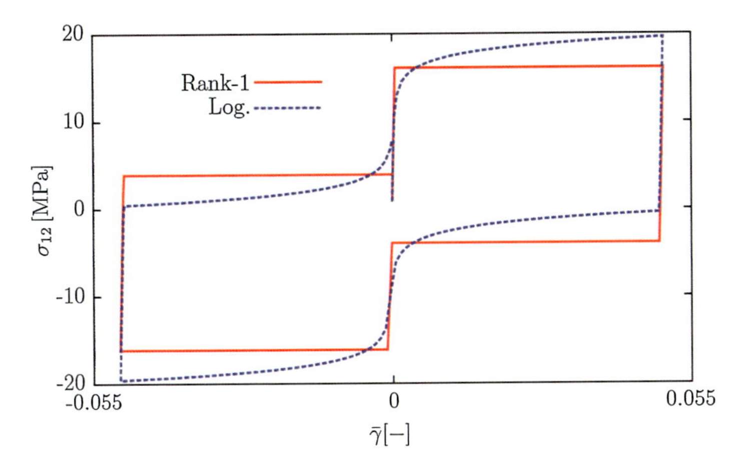

Fig. 6 Cyclic loading for s = 3μm

for instance pileup formation and destruction at γ' precipitates. In the present simple single-crystal model, it is the single active hardening mechanism induced by strain gradient plasticity and the presence of the hard phase in the laminate. It represents an accurate continuum description of dislocation piling-up and unpiling-up phenomena.

The experimental evidence of such nonconvex loops is illustrated in Figs. 7 and 8 in the case of polycrystalline Fe–Cr and Al–Cu–Mg alloys, respectively. The first loop in Fig. 7 (left) exhibits two inflection points, but the convexity is restored after a few cycles, and the usual shape with still a strong Bauschinger effect is retrieved in Fig. 7 (right). Figure 8 shows that the amount of plastic recoverability is controlled by the annealing degree of the dislocation microstructure. Further evidence of nonconvex loops in the cyclic behavior of FCC alloys can be found in the recent contribution by Proudhon et al. (2008) dealing with aluminum alloys. The common characteristics of these FCC alloys are that they all contain a population of nonshearable intragranular precipitates. This distribution of particles represents the first series of obstacles to be overcome by dislocations for the plasticity to start. The distance between precipitates presents a small scatter, and the average value is the characteristic length responsible for the size-dependent yield limit. This distance is comparable to the width s in our ideal laminate model. As illustrated by the TEM observations by Stoltz and Pelloux (1974, 1976), Taillard and Pineau (1982), and Proudhon et al. (2008), dislocation loops multiply around precipitates and can be destroyed after reverse loading unless the material is annealed before reversing the load, see Fig. 8, or unless the multiplication of forest dislocations or cross-slip effects limit the recoverability of cyclic plasticity. The effect has also been observed in nickel-base single-crystal superalloys for tension along <111>; see Fig. 9. The simulations based on the logarithmic potential provide smooth loops that are closer to the experimental shapes. Our simulations deal with ideal single-

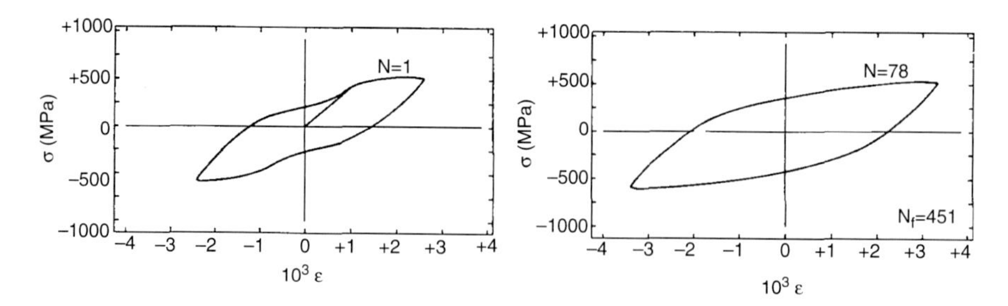

Fig. 7 Shape of the stress–strain hysteresis loop as a function of the number N of cycles for a Fe-19wt.%Cr alloy aged at 923 K for 72 h and mechanically tested at room temperature: N = 1 (left), N = 48 (right, N_f indicates the number of cycles to failure), after Taillard and Pineau (1982)

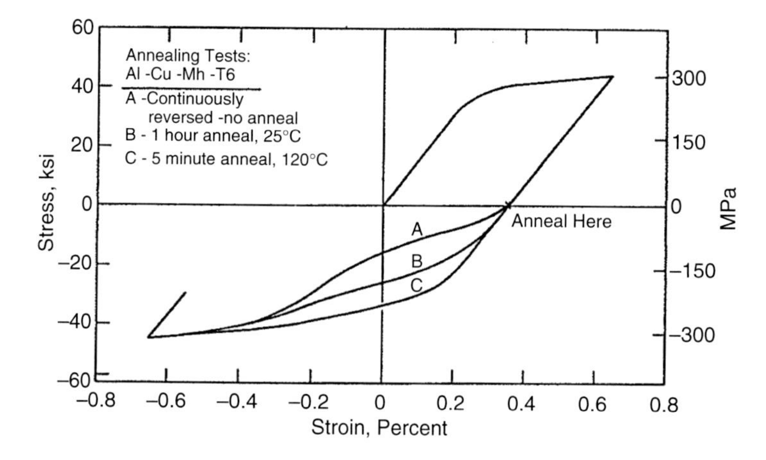

Fig. 8 Interrupted annealing hysteresis curves for an Al–Cu–Mg–T6 alloy tested at room temperature: A, continuously reverse load (no anneal); B, annealed 1 h/25 °C; and C, annealed 5 min/120 °C, after Stoltz and Pelloux (1976)

crystal laminates and simulations for polycrystals remain to be done. However, as shown by the two-dimensional strain gradient plasticity simulations performed by Ohno and Okumura (2008), based on the rank-one potential, the effect pertains for polycrystals. However, these authors did not recognize the physical reality of the simulated phenomena. Instead, they further developed the model to replace the rank-one energy potential by a dissipative formulation which restores the convexity of fatigue loops.

The two nonquadratic energy potentials represent continuum models of a discrete phenomenon which can be illustrated for a single dislocation source, as shown in Fig. 10. The cyclic response of a Frank–Read source, simulated by discrete dislocation dynamics (DDD) (Déprés et al. 2004; Chang et al. 2016), provides a nonconvex loop which is identical to the one predicted by the rank-one continuum model. This is related to the fact that an instability of the loop behavior is observed

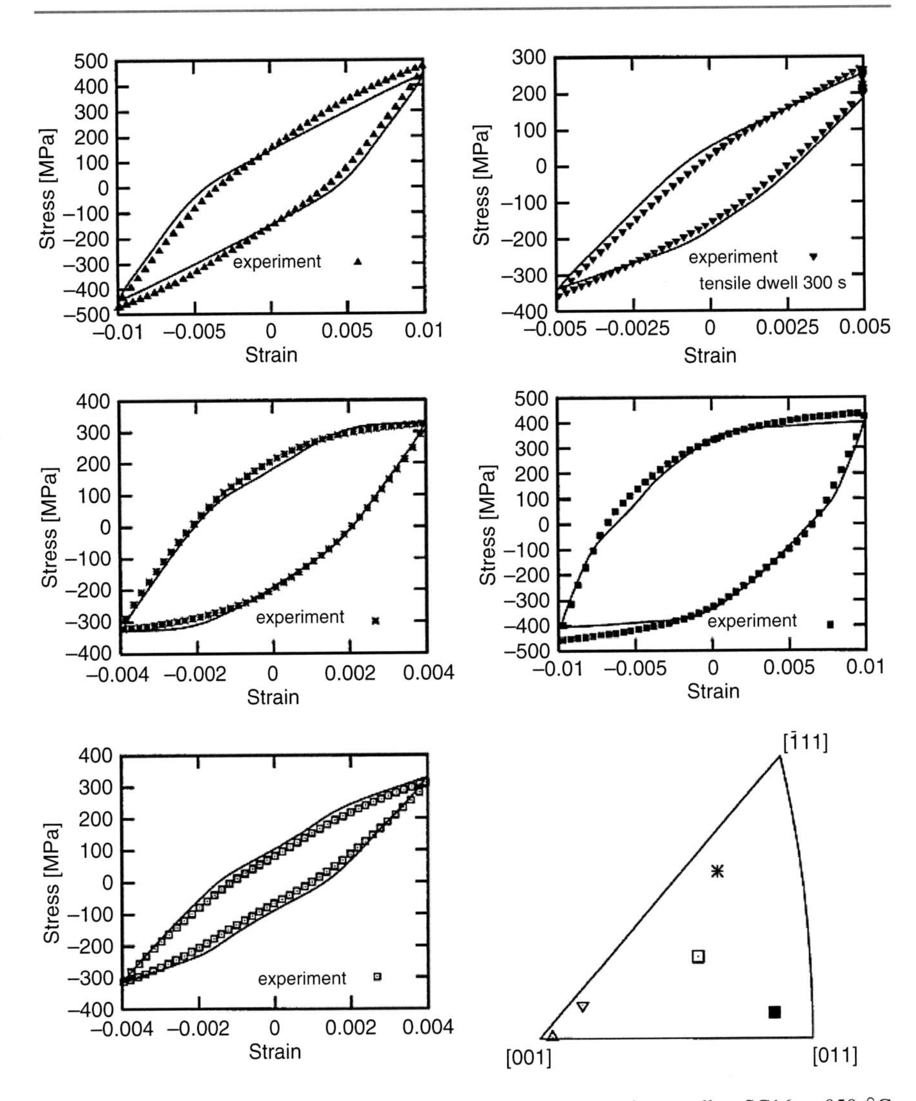

Fig. 9 Stabilized stress–strain loops for nickel-base single-crystal superalloy SC16 at 950 °C, experiment vs. simulation after Fedelich (2002)

for a critical stress that is inversely proportional to the length of the source. The scenario of dislocation source bowing and sudden propagation and multiplication can be reversed entirely in the absence of strong interaction with the dislocation forest and in the absence of cross-slip. This explains the concave shape of the stress–strain loop predicted by the DDD, which is accurately translated by the continuum model; see Fig. 6. Statistical effects of the collective behavior of dislocations finally destroy the recoverability of plastic deformation and the associated transmission of the single source behavior to the macroscopic response.

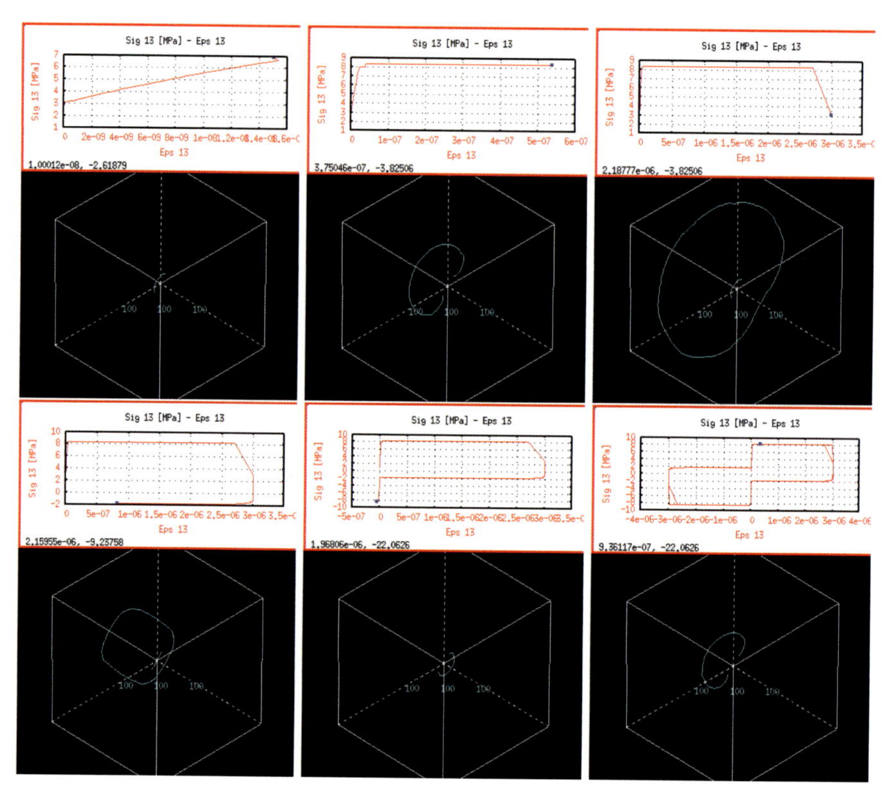

Fig. 10 Cyclic behavior of a single Frank–Read dislocation source simulated by discrete dislocation density. (Courtesy of Dr. M. Fivel)

Grain Size Effects in Polycrystals

The model is now applied to simulate the response of polycrystals and the effects of grain size.

The interface conditions at grain boundaries play a major role in the simulated size effects in the polycrystal behavior. No special interface law is considered in this work, although such physically motivated interface conditions exist in the literature; see Gurtin and Anand (2008). Instead, we consider the canonical interface conditions that arise from the formulation of the balance equations of the *microcurl* continuum model. These conditions are the continuity of displacement, $\boldsymbol{u}$, and the continuity of plastic microdeformation, $\boldsymbol{\chi}^p$. These conditions also include the continuity of the simple and double tractions, $\boldsymbol{t}$ and $\mathbf{M}$, described in Eq. (8). Continuity of displacement excludes grain boundary cracking and sliding. Continuity of plastic microdeformation is reminiscent of the fact that dislocations generally do not cross grain boundaries, especially for such random grain boundaries. Note that in the *microcurl* model, only the kinematic degrees of freedom $\boldsymbol{\chi}^p$ are continuous.

This is not the case of the plastic deformation, $\mathbf{H}^p$, which is treated here as an internal variable. However, due to the internal constraint discussed in section "Model Formulation," $\mathbf{H}^p$ closely follows the plastic microdeformation so that it is quasi-continuous at grain boundaries when the penalty coefficient, H_χ, is high enough. Conversely, lower values of H_χ may allow slightly discontinuous plastic deformation, which may be tentatively interpreted as dislocation sinking inside grain boundaries. The continuity of the associated tractions expresses the transmission of classical and generalized internal forces from one grain to another through grain boundaries. Such continuum models are then able to mimic in that way the development of dislocation pileups at grain boundaries (Forest and Sedláček 2003a).

More elaborate grain boundary behavior laws are necessary to go beyond the three possible interface conditions readily available according to the *microcurl* model: vanishing microdeformation, continuous microdeformation, or vanishing microtractions at grain boundaries. They require proper account of transmission and absorption rules for dislocations at grain boundaries. A simple and efficient strategy was proposed for the formulation and finite element implementation of such interface constitutive laws by Wulfinghoff et al. (2013a). The reader is referred to the references quoted therein for more advanced grain boundary behavior laws.

Boundary Value Problem for Polycrystals

The size effects exhibited by the solution of the boundary value problem are linked to an intrinsic length scale, l_s, introduced through the generalized moduli H_χ and A of Eq. (32) and defined as:

$$l_s = \sqrt{\frac{A}{H_\chi}}. \tag{106}$$

This intrinsic length scale has to be consistent with the fact that plasticity effects occur at scales ranging from hundreds of nanometers to a few microns. In addition, as stated in section "Model Formulation," the coupling modulus, H_χ, has to be chosen high enough to ensure that $\boldsymbol{\chi}^p$ and $\mathbf{H}^p$ are close. These requirements are guidelines for the choice of relevant generalized moduli H_χ and A. The sets of material parameters used in this section are chosen in that way.

The finite element simulations have been made on periodic 2D meshes of periodic polycrystalline aggregates generated by a method based on Voronoi tessellations (Fig. 11a, b). Quadratic isoparametric finite elements with reduced integration are used. The random distribution of the grain centers has been controlled so that their sizes are sensibly the same, around the mean grain size, d. A random orientation is assigned to each grain, and two slip systems are taken into account for simplicity. In 2D, the plastic behavior of FCC crystals can be simulated with 2D planar double slip by considering two effective slip systems separated by an angle of 2ϕ (Asaro 1983; Bennett and McDowell 2003). Figure 11c describes the geometry.

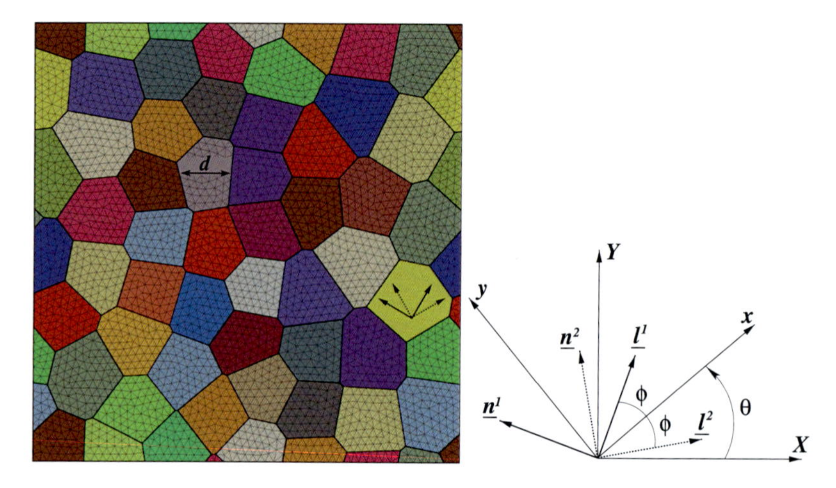

Fig. 11 Periodic meshes of the 2D periodic aggregates used in the finite element simulations including 52 grains. Two slip systems are taken into account in each randomly oriented grain. Various mean grain sizes, d, ranging from tens of nanometers to hundreds of microns, are investigated. On the right, description of the two effective slip systems for 2D planar double slip

The slip system pair is oriented by the angle θ, which is the grain orientation randomly fixed for each grain. For a FCC crystal $\phi = 35.1°$, it corresponds to the orientation of the close-packed planes in the crystal lattice of the grain.

Periodic homogenization for generalized continua is used to predict the effective response of the polycrystal. The displacement field is assumed to be of the form

$$\boldsymbol{u}(x) = \boldsymbol{E}.\boldsymbol{x} + \boldsymbol{v}(x), \tag{107}$$

with the fluctuation $\boldsymbol{v}$ periodic, meaning that it takes identical values at homologous points of the unit cell (Forest et al. 2001). The plastic microdeformation field, χ^p, is assumed to be periodic, meaning that no rotational macroscopic plastic deformation is imposed to the unit cell. Its components are equal at homologous opposite nodes. According to periodic homogenization, the simple and double tractions $\boldsymbol{t}$ and $\mathbf{m}$ are antiperiodic at homologous points of the unit cell.

Polycrystals are random materials so that the periodicity constraint may lead to a bias in the estimation of the effective properties. This boundary effect can be alleviated by considering several realizations of the microstructure and performing ensemble averaging (Zeghadi et al. 2007).

Overall Cyclic Response of a Polycrystalline Aggregate

The finite element simulations of the boundary value problem presented previously have been conducted under generalized plane strain conditions on aggregates with a relatively small number of grains. The aim here is not to obtain a representative

Table 2 Sets of material parameters used in the 24-grain aggregate case (Fig. 11a). The intrinsic length scale, $l_s = \sqrt{A/H_\chi}$, is given for each set

Set	μ [MPa]	τ_c [MPa]	H_χ [MPa]	A [MPa mm^2]	l_s [μm]
a	35,000	40	3.0 10^6	1.0 10^{-2}	5.8 10^{-2}
b	35,000	40	1.0 10^6	1.0 10^{-2}	1.0 10^{-1}
c	35,000	40	3.5 10^5	1.0 10^{-2}	1.7 10^{-1}
d	35,000	40	8.8 10^4	1.0 10^{-3}	1.1 10^{-1}

response but to catch the grain size effects and to explore qualitatively the impact of different sets of material parameters. In this section, a virtual material is considered with various intrinsic length scales. The macroscopic stress–strain curve is obtained by applying a cyclic simple shear loading controlled by the average strain component E_{12} on the aggregate of 52 grains with $d = 0.2$ μm and the set of material parameters labeled (c) in Table 2. The mean stress component Σ_{12} is then computed:

$$\Sigma_{12} = \frac{1}{V}\int_V \sigma_{12}\, dV,\ E_{12} = \frac{1}{V}\int_V \varepsilon_{12}\, dV, \tag{108}$$

where V denotes each polycrystal unit cell. The simulated response displays the kinematic hardening effect produced by the *microcurl* model. The stress–strain curves shown in Fig. 12 prove that this kinematic hardening is size dependent: it increases for smaller grains. Note that the observed overall kinematic hardening has two distinct sources: the intragranular back stress induced by plastic strain gradients and the intergranular internal stress that originates from the grain to grain plastic strain incompatibilities. The latter contribution is also predicted by classical crystal plasticity models.

Figure 13 presents the effect of the mean grain size, d, on the macroscopic flow stress at 1% plastic strain in the 52-grain aggregate in a log–log diagram for different intrinsic length scales, l_S, introduced through the sets of material parameters (labeled a, b, c, and d) given in Table 2. The considered loading conditions are still a simple shear test with periodic boundary conditions. The curves exhibit two plateaus for large (d > 20μm) and small ($d < 0.1$ μm) mean grain sizes with a transition domain in between. This tanh shape indicates that when d is large compared to the intrinsic length scale, l_s, strain gradient effects are small, and the kinematic hardening arising from the *microcurl* model vanishes. The model saturates when d is of the order of l_s or smaller. The transition domain exhibits a strong size dependence, the polycrystalline aggregate becoming harder for decreasing grain sizes. The position of the transition zone, the maximum extra stress (the distance between the two plateaus), and the scaling law exponent, m, in the size-dependent domain are controlled by the material parameters used in the model. The two latter effects are controlled by the coupling modulus, H_χ; they both increase for higher values of H_χ as shown in Fig. 13. The scaling exponent is defined as the slope in the log–log diagram in the inflection domain, reflecting the scaling law:

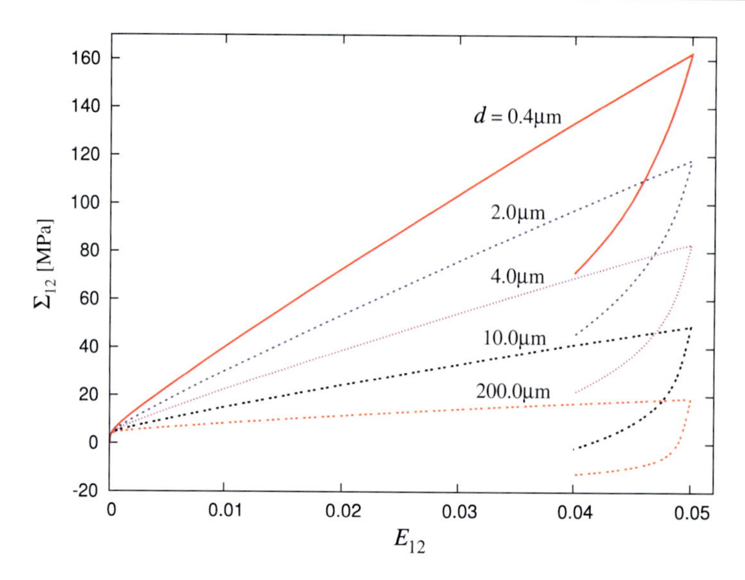

Fig. 12 Macroscopic stress–strain response of the 52-grain aggregate under simple shear for various mean grain sizes, d. The set of material parameters used is labeled (g) in Table 3

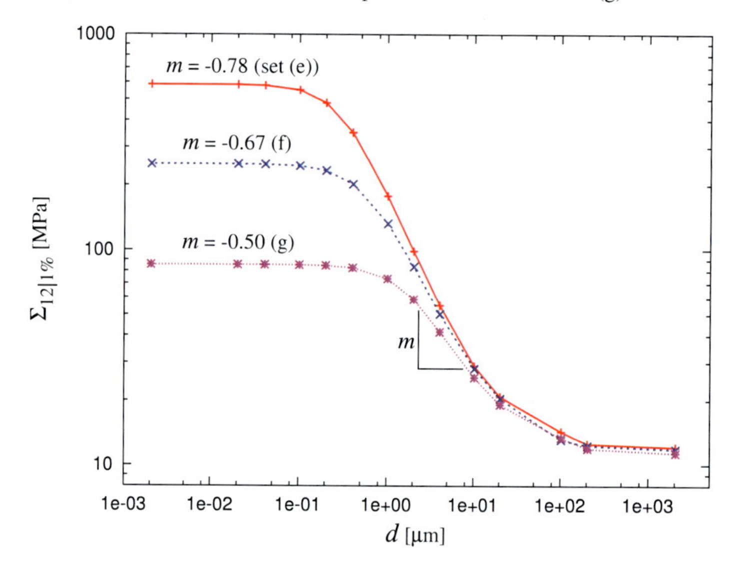

Fig. 13 Effect of the mean grain size, d, on the macroscopic flow stress, $\Sigma_{12|1\%}$, at 1% plastic strain. The results are obtained for the 52-grain aggregate using the different sets of material parameters given in Table 3. The scaling law exponent, m, is identified in each case

$$\Sigma_{12} \propto d^m. \tag{109}$$

It is obtained with the sets of material parameters given in Table 2. The computed values range from −0.26 to −0.64 including the well-known Hall–Petch exponent

Table 3 Sets of material parameters used in the 52-grain aggregate case (Fig. 11b)

Set	μ [MPa]	τ_c [MPa]	Q [MPa]	b	$h^{\alpha\alpha}$	$h^{\alpha\beta,\ \alpha \neq \beta}$
e	27,000	0.75	7.9	10.2	1	4.4
f	27,000	0.75	7.9	10.2	1	4.4
g	27,000	0.75	7.9	10.2	1	4.4

Set	H_χ [MPa]	A [MPa mm^2]	l_s [μm]
e	1.0 10^6	1.0 10^{-2}	1.0 10^{-1}
f	3.5 10^5	1.0 10^{-2}	1.7 10^{-1}
g	5.0 10^4	1.0 10^{-2}	4.5 10^{-1}

$m = -0.5$. In fact, it was shown in Cordero et al. (2010a) that values of m ranging from 0 to -2 can be simulated with the *microcurl* model in the case of two-phase microstructures. In each case, these values are obtained without classical isotropic hardening, meaning that the linear kinematic hardening produced by the model is able to reproduce a wide range of scaling laws. Note that conventional strain gradient plasticity models do not lead to *tanh*-shape curves but rather to unbounded stress increase for vanishingly small microstructures (Cordero et al. 2010a).

Grain Size Effects in Idealized Aluminum Polycrystals

Similar finite element simulations have been performed on the idealized aluminum aggregate of 52 grains of Fig. 11. An additional isotropic hardening component is addedas in (Méric et al. 1991) to obtain a more realistic response of large aluminum grains. The size-independent hardening law reads:

$$R^\alpha = \tau_c + Q \sum_\beta^n h^{\alpha\beta} \left(1 - \exp\left(-b\gamma_{\text{cum}}^\beta\right)\right), \tag{110}$$

where n is the number of slip systems (here $n = 2$), Q and b are material coefficients defining nonlinear isotropic hardening, $H^{\alpha\beta}$ is the interaction matrix, and $\gamma_{\text{cum}}^\beta$ is the accumulated micro-plastic slip on the slip system β. Cumulative plastic slip results from the integration of the differential equation $\dot{\gamma}_{\text{cum}}^\beta = \left|\dot{\gamma}^\beta\right|$. The material parameters used in these simulations are given in Table 3. The macroscopic stress–strain curves presented in Fig. 12 are obtained by applying a simple shear loading controlled by the average strain component E_{12} on the 52-grain aggregate with various mean grain sizes, d, taken in the size-dependent domain. The chosen set of material parameters has the label (g) in Table 3. These parameters are such that an acceptable description of aluminum polycrystals is obtained for large grains and that a Hall–Petch-like behavior is found in a plausible range of grain sizes. However, we did not attempt to calibrate the amplitude of the extra hardening so that simulation

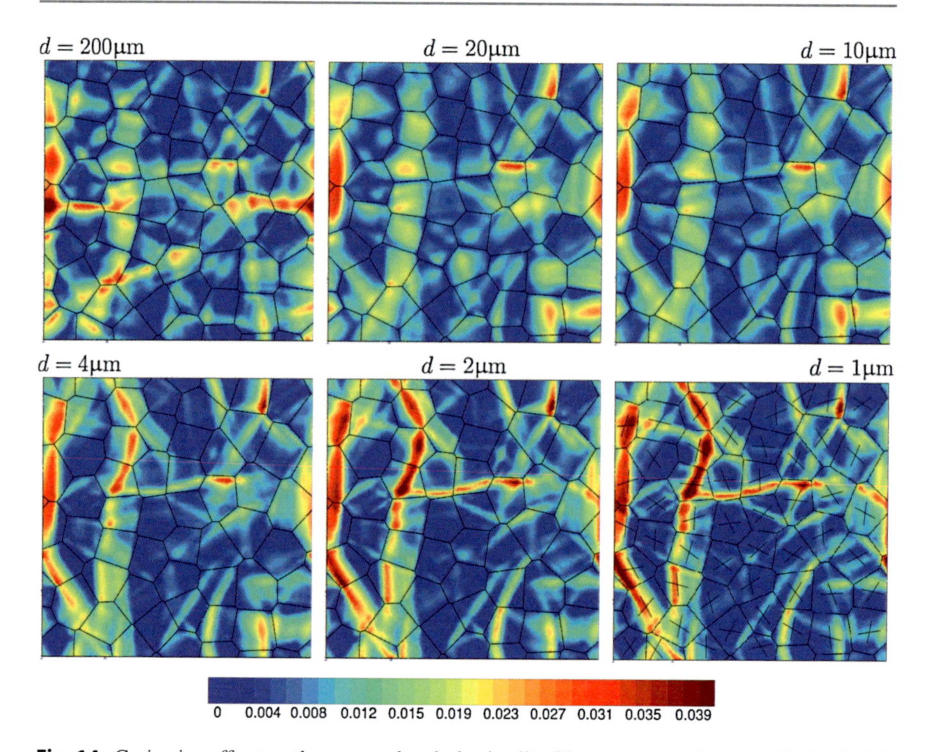

Fig. 14 Grain size effect on the accumulated plastic slip. These contour plots are obtained with the 52-grain aggregate for the same mean value of cumulative plastic strain $p = 0.01$. The set of material parameters (g) of Table 3 is used. The pairs of slip plane directions are represented for each grain on the 1 μm contour plot

predictions remain qualitative. The curves of Fig. 12 show again that the kinematic hardening produced by the model is strongly size dependent. The set of material parameters (g) of Table 3 gives the ideal Hall–Petch scaling law exponent $m = -0.5$.

An important output of the simulations is the dependence of the stress and strain fields in the grains of the polycrystal on grain size. Figures 14 and 15 show the contour plots of the field of accumulated plastic slip, computed as:

$$\dot{p} = \sqrt{\frac{2}{3}\dot{\boldsymbol{\varepsilon}}^p : \dot{\boldsymbol{\varepsilon}}^p}, \tag{111}$$

where ε^p is the symmetric part of the plastic deformation, $\mathbf{H}^p$, and the contour plots of the norm $\mathbf{A}$ of the dislocation density tensor,

$$\|\mathbf{A}\| = \sqrt{\mathbf{A} : \mathbf{A}}, \tag{112}$$

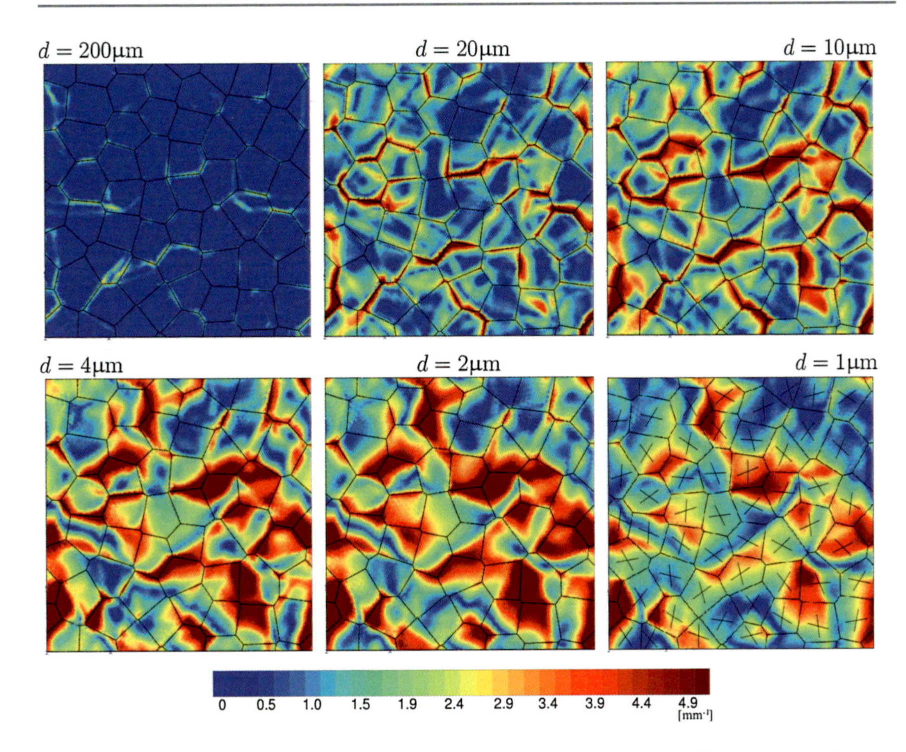

Fig. 15 Grain size effect on the norm of the dislocation density tensor. These contour plots are obtained with the 52-grain aggregate for the same mean value of macroscopic accumulated plastic strain $p = 0.01$. The set of material parameters (g) of Table 3 is used. The pairs of slip plane directions are represented for each grain on the 1 μm contour plot

respectively. The considered grain sizes are taken in the size-dependent domain where the evolution of the fields is assumed to be physically relevant. The chosen set of material parameters has the label (g) in Table 3; it corresponds to an intrinsic length scale $l_s = 0.45\mu$m and gives a scaling law exponent $m = -0.5$. The mean value of the accumulated plastic slip is the same for all cases; only its distribution varies with the size of the microstructure as shown in Fig. 14.

The first contour plot of each figure is obtained for $d = 200\ \mu\text{m} \gg l_s = 0.45\mu$m, at the very beginning of the size-dependent behavior domain according to Fig. 13. At this size, the simulated fields show that p is quite inhomogeneous and that some deformation bands appear; $||\mathbf{A}||$ is localized at the grain boundaries and almost vanishes in the grain cores. The contour plots obtained for $2\mu\text{m} < d < 20\ \mu$m show a significant evolution of both fields. One observes the formation of a network of strain localization bands with decreasing grain size. These bands are slip bands as they are parallel to the slip plane directions represented on the 1 μm contour plot of Fig. 14. They compensate the larger blue zones where plastic strain cannot develop due to the higher energy cost associated with its gradient. Plastic strain becomes stronger inside the localization bands. This is due to the fact that the contour plots are given for fixed macroscopic cumulative plastic strain mean value of p, which implies that the applied total strain is higher for small grain sizes as suggested by Fig. 12. The field of the norm of the dislocation density tensor is still high close to grain boundaries and spreads over the grain cores. The last contour plot of each figure is obtained for $d = 1$ μm, a size close to l_s. Here the model starts to saturate, which can be seen from the simulated fields. The field of p does not evolve anymore and $||\mathbf{A}||$ decreases. In fact, as l_s controls the strain gradient effects, strong strain gradients cannot develop because they become energetically too expensive when the grain size is too small.

Conclusions

The micromorphic crystal plasticity theory introduces independent plastic microdeformation degrees of freedom. It represents a relaxation of the strain gradient plasticity model. It contains as a special case the micropolar crystal plasticity model. The advantage of the micromorphic framework is that it provides a wider range of modeling possibilities regarding constitutive laws and boundary conditions, as it was illustrated for the description of Hall–Petch effects in polycrystals. It has also merits regarding computational mechanical aspects since its implementation is rather straightforward and can be used for strain gradient plasticity computations based on proper internal constraints. The advantage of the micropolar theory compared to the micromorphic one is its reduced number of degrees of freedom, 3 instead of 8/9 in 3D (8 dof if plastic incompressibility is enforced). The micropolar model incorporates the effect of the lattice curvature tensor, which represents an essential part of the dislocation density tensor.

A full set of constitutive equations has been formulated for micromorphic crystal plasticity by extending the framework of generalized standard materials based on the introduction of free energy and dissipation potentials.

Simple examples of the plasticity of sheared single-crystal layers show the ability of the continuum models to reproduce the essential features of the collective behavior of dislocations piling up in thin layers, channels, and laminate microstructures, as predicted by discrete dislocation dynamics. The scaling laws predicted by the continuum models strongly depend on the choice of the constitutive equations. For example, standard quadratic free energy potential was shown to be inadequate to reproduce Orowan-like size effects. Strongly nonlinear potentials, including the logarithmic free energy density, were shown to be more closely related to the physics of dislocations.

Size effects also strongly influence the strain localization behavior of polycrystals for which the formation of intense slip bands is predicted in micron-size grains by the micromorphic models. This feature is to be related to the high energy cost of dislocation pileup formation in small grains.

Special attention was given to the cyclic response of crystals, which is of the utmost importance for the prediction of fatigue lifetime of materials. The question of recovery of plastic strain has been shown to be a central issue. In the absence of strong forest interaction and cross-slip, total recovery of plastic strain gradients is possible, leading to nonconvex cyclic stress–strain loops as observed in some two-phase alloys. Note that results from discrete dislocation dynamics and the micropolar model were provided in the chapter dedicated to micropolar crystal plasticity. Similar comparisons can be found in the case of micromorphic model in the Reference Chang et al. (2016) where the continuum and discrete descriptions of dislocation pileups at interfaces are discussed.

Much work remains to be done in the development of predictive constitutive equations in micropolar and micromorphic crystal plasticity due to the complex underlying dislocation mechanisms. Especially, suitable modeling of grain boundary behavior remains a major issue for higher-order modeling of polycrystal plasticity.

Acknowledgments The first author is indebted to Dr. N.M. Cordero, Dr. S. Wulfinghoff, and Prof. E.B. Busso for their contribution to the presented micromorphic crystal plasticity theory. These contributions are duly cited in the references quoted in the text and listed below.

References

E. Aifantis, The physics of plastic deformation. Int. J. Plast. **3**, 211–248 (1987)

R.J. Asaro, Elastic–plastic memory and kinematic hardening. Acta Metall. **23**, 1255–1265 (1975)

R. Asaro, Crystal plasticity. J. Appl. Mech. **50**, 921–934 (1983)

Ashby, M., 1971. The deformation of plastically non-homogeneous alloys, in *Strengthening Methods in Crystals*, ed. by A. Kelly, R. Nicholson (Applied Science Publishers, London), pp. 137–192

O. Aslan, N.M. Cordero, A. Gaubert, S. Forest, Micromorphic approach to single crystal plasticity and damage. Int. J. Eng. Sci. **49**, 1311–1325 (2011)

V. Bennett, D. McDowell, Crack tip displacements of microstructurally small surface cracks in single phase ductile polycrystals. Eng. Fract. Mech. **70**(2), 185–207 (2003)

V. Berdichevsky, On thermodynamics of crystal plasticity. Scripta Mat. **54**, 711–716 (2006a)

V. Berdichevsky, On thermodynamics of crystal plasticity. Scr. Mater. **54**, 711–716 (2006b)

P. Cermelli, M. Gurtin, On the characterization of geometrically necessary dislocations in finite plasticity. J. Mech. Phys. Solids **49**, 1539–1568 (2001)

H.J. Chang, N.M. Cordero, C. Déprés, M. Fivel, S. Forest, Micromorphic crystal plasticity versus discrete dislocation dynamics analysis of multilayer pile-up hardening in a narrow channel. Arch. Appl. Mech. **86**, 21–38 (2016)

W. Claus, A. Eringen, Three dislocation concepts and micromorphic mechanics, in *Developments in Mechanics. Proceedings of the 12th Midwestern Mechanics Conference*, vol. 6, (1969), pp. 349–358

S. Conti, M. Ortiz, Dislocation microstructures and the effective behavior of single crystals. Arch. Ration. Mech. Anal. **176**, 103–147 (2005)

N. Cordero, A. Gaubert, S. Forest, E. Busso, F. Gallerneau, S. Kruch, Size effects in generalised continuum crystal plasticity for two–phase laminates. J. Mech. Phys. Solids **58**, 1963–1994 (2010a)

N.M. Cordero, A. Gaubert, S. Forest, E. Busso, F. Gallerneau, S. Kruch, Size effects in generalised continuum crystal plasticity for two-phase laminates. J. Mech. Phys. Solids **58**, 1963–1994 (2010b)

N.M. Cordero, S. Forest, E. Busso, S. Berbenni, M. Cherkaoui, Grain size effects on plastic strain and dislocation density tensor fields in metal polycrystals. Comput. Mater. Sci. **52**, 7–13 (2012)

L. De Luca, A. Garroni, M. Ponsiglione, Gamma-convergence analysis of Systems of Edge Dislocations: the self energy regime. Arch. Ration. Mech. Anal. **206**, 885–910 (2012)
C. Déprés, C.F. Robertson, M.C. Fivel, Low-strain fatigue in aisi 316l steel surface grains: a three-dimensional discrete dislocation dynamics modelling of the early cycles i. Dislocation microstructures and mechanical behaviour. Philos. Mag. **84**(22), 2257–2275 (2004)
Eringen, A., Claus, W., 1970. A micromorphic approach to dislocation theory and its relation to several existing theories, in *Fundamental Aspects of Dislocation Theory*, ed. by J. Simmons, R. de Wit, R. Bullough. National Bureau of Standards (US) Special Publication 317, vol. II (U.S. Government Printing Office, Washington, DC), pp. 1023–1062
B. Fedelich, A microstructural model for the monotonic and the cyclic mechanical behavior of single crystals of superalloys at high temperatures. Int. J. Mech. Sci. **18**, 1–49 (2002)
S. Forest, Some links between cosserat, strain gradient crystal plasticity and the statistical theory of dislocations. Philos. Mag. **88**, 3549–3563 (2008)
S. Forest, The micromorphic approach for gradient elasticity, viscoplasticity and damage. ASCE J. Eng. Mech. **135**, 117–131 (2009)
S. Forest, Nonlinear regularisation operators as derived from the micromorphic approach to gradient elasticity, viscoplasticity and damage. Proc. R. Soc. A **472**, 20150755 (2016)
S. Forest, N. Guéninchault, Inspection of free energy functions in gradient crystal plasticity. Acta. Mech. Sinica. **29**, 763–772 (2013) https://doi.org/10.1007/s10409-013-0088-0
S. Forest, R. Sedláček, Plastic slip distribution in two–phase laminate microstructures: Dislocation–based vs. generalized–continuum approaches. Philos. Mag. A **83**, 245–276 (2003a)
S. Forest, R. Sedláček, Plastic slip distribution in two–phase laminate microstructures: Dislocation–based vs. generalized–continuum approaches. Philos. Mag. A **83**, 245–276 (2003b)
S. Forest, R. Sievert, Elastoviscoplastic constitutive frameworks for generalized continua. Acta Mech. **160**, 71–111 (2003)
S. Forest, R. Sievert, Nonlinear microstrain theories. Int. J. Solids Struct. **43**, 7224–7245 (2006)
S. Forest, F. Pradel, K. Sab, Asymptotic analysis of heterogeneous Cosserat media. Int. J. Solids Struct. **38**, 4585–4608 (2001)
Forest, S., Ammar, K., Appolaire, B., Cordero, N., Gaubert, A., 2014. Micromorphic approach to crystal plasticity and phase transformation, in *Plasticity and Beyond*, ed. by J. Schroeder, K. Hackl. CISM International Centre for Mechanical Sciences, Courses and Lectures, no. 550 (Springer, Vienna), pp. 131–198
M. Geers, R. Peerlings, M. Peletier, L. Scardia, Asymptotic behaviour of a pile–up of infinite walls of edge dislocations. Arch. Ration. Mech. Anal. **209**, 495–539 (2013)
P. Germain, The method of virtual power in continuum mechanics. Part 2: microstructure. SIAM J. Appl. Math. **25**, 556–575 (1973)
P. Grammenoudis, C. Tsakmakis, Micromorphic continuum part I: strain and stress tensors and their associated rates. Int. J. Non–Linear Mech. **44**, 943–956 (2009)
I. Groma, F. Csikor, M. Zaiser, Spatial correlations and higher–order gradient terms in a continuum description of dislocation dynamics. Acta Mater. **51**, 1271–1281 (2003)
I. Groma, G. Györgyi, B. Kocsis, Dynamics of coarse grain grained dislocation densities from an effective free energy. Philos. Mag. **87**, 1185–1199 (2007)
M. Gurtin, A gradient theory of single–crystal viscoplasticity that accounts for geometrically necessary dislocations. J. Mech. Phys. Solids **50**, 5–32 (2002)
M. Gurtin, L. Anand, Nanocrystalline grain boundaries that slip and separate: a gradient theory that accounts for grain-boundary stress and conditions at a triple-junction. J. Mech. Phys. Solids **56**, 184–199 (2008)
M. Gurtin, L. Anand, Thermodynamics applied to gradient theories involving the accumulated plastic strain: the theories of Aifantis and Fleck and Hutchinson and their generalization. J. Mech. Phys. Solids **57**, 405–421 (2009)
W. Han, B. Reddy, *Plasticity: Mathematical Theory and Numerical Analysis* (Springer, New York, 2013)
C. Hirschberger, P. Steinmann, Classification of concepts in thermodynamically consistent generalized plasticity. ASCE J. Eng.Mech. **135**, 156–170 (2009)

D.E. Hurtado, M. Ortiz, Surface effects and the size-dependent hardening and strengthening of nickel micropillars. J. Mech. Phys. Solids **60**(8), 1432–1446 (2012)

D.E. Hurtado, M. Ortiz, Finite element analysis of geometrically necessary dislocations in crystal plasticity. Int. J. Numer. Methods Eng. **93**(1), 66–79 (2013)

R. Kametani, K. Kodera, D. Okumura, N. Ohno, Implicit iterative finite element scheme for a strain gradient crystal plasticity model based on self-energy of geometrically necessary dislocations. Comput. Mater. Sci. **53**(1), 53–59 (2012)

Kröner, E., 1969. Initial studies of a plasticity theory based upon statistical mechanics, in *Inelastic Behaviour of Solids*, ed. by M. Kanninen, W. Adler, A. Rosenfield, R. Jaffee (McGraw-Hill, New York/London), pp. 137–147

J. Lee, Y. Chen, Constitutive relations of micromorphic thermoplasticity. Int. J. Eng. Sci. **41**, 387–399 (2003)

J. Mandel, Equations constitutives et directeurs dans les milieux plastiques et viscoplastiques. Int. J. Solids Struct. **9**, 725–740 (1973)

L. Méric, P. Poubanne, G. Cailletaud, Single crystal modeling for structural calculations. Part 1: Model presentation. J. Eng. Mat. Technol. **113**, 162–170 (1991)

S.D. Mesarovic, R. Baskaran, A. Panchenko, Thermodynamic coarsening of dislocation mechanics and the size-dependent continuum crystal plasticity. J. Mech. Phys. Solids **58**(3), 311–329 (2010)

S. Mesarovic, S. Forest, J. Jaric, Size-dependent energy in crystal plasticity and continuum dislocation models. Proc. R. Soc. A **471**, 20140868 (2015)

C. Miehe, S. Mauthe, F.E. Hildebrand, Variational gradient plasticity at finite strains. Part III: local-global updates and regularization techniques in multiplicative plasticity for single crystals. Comput. Methods Appl. Mech. Eng. **268**, 735–762 (2014)

J. Nye, Some geometrical relations in dislocated crystals. Acta Metall. **1**, 153–162 (1953)

N. Ohno, D. Okumura, Higher-order stress and grain size effects due to self-energy of geometrically necessary dislocations. J. Mech. Phys. Solids **55**, 1879–1898 (2007)

N. Ohno, D. Okumura, Grain–size dependent yield behavior under loading, unloading and reverse loading. Int. J. Mod. Phys. B **22**, 5937–5942 (2008)

M. Ortiz, E. Repetto, Nonconvex energy minimization and dislocation structures in ductile single crystals. J. Mech. Phys. Solids **47**(2), 397–462 (1999)

H. Proudhon, W. Poole, X. Wang, Y. Bréchet, The role of internal stresses on the plastic deformation of the Al–Mg–Si–Cu alloy AA611. Philos. Mag. **88**, 621–640 (2008)

B.D. Reddy, C. Wieners, B. Wohlmuth, Finite element analysis and algorithms for single-crystal strain-gradient plasticity. Int. J. Numer. Methods Eng. **90**(6), 784–804 (2012)

R. Regueiro, On finite strain micromorphic elastoplasticity. Int. J. Solids Struct. **47**, 786–800 (2010)

C. Sansour, S. Skatulla, H. Zbib, A formulation for the micromorphic continuum at finite inelastic strains. Int. J. Solids Struct. **47**, 1546–1554 (2010)

P. Steinmann, Views on multiplicative elastoplasticity and the continuum theory of dislocations. Int. J. Eng. Sci. **34**, 1717–1735 (1996)

R. Stoltz, R. Pelloux, Cyclic deformation and Bauschinger effect in Al–Cu–Mg alloys. Scr. Metall. **8**, 269–276 (1974)

R. Stoltz, R. Pelloux, The Bauschinger effect in precipitation strengthened aluminum alloys. Metallurgical. Transactions **7A**, 1295–1306 (1976)

B. Svendsen, S. Bargmann, On the continuum thermodynamic rate variational formulation of models for extended crystal plasticity at large deformation. J. Mech. Phys. Solids **58**(9), 1253–1271 (2010)

R. Taillard, A. Pineau, Room temperature tensile properties of Fe-19wt.% Cr alloys precipitation hardened by the intermetallic compound NiAl. Mater. Sci. Eng. **56**, 219–231 (1982)

S. Wulfinghoff, T. Böhlke, Equivalent plastic strain gradient enhancement of single crystal plasticity: theory and numerics. Proc. R. Soc. A: Math. Phys. Eng. Sci. **468**(2145), 2682–2703 (2012)

S. Wulfinghoff, E. Bayerschen, T. Böhlke, A gradient plasticity grain boundary yield theory. Int. J. Plast. **51**, 33–46 (2013a)

S. Wulfinghoff, E. Bayerschen, T. Böhlke, Micromechanical simulation of the hall-petch effect with a crystal gradient theory including a grain boundary yield criterion. PAMM **13**, 15–18 (2013b)

S. Wulfinghoff, S. Forest, T. Böhlke, Strain gradient plasticity modeling of the cyclic behavior of laminate microstructures. J. Mech. Phys. Solids **79**, 1–20 (2015)

A. Zeghadi, S. Forest, A.-F. Gourgues, O. Bouaziz, Ensemble averaging stress–strain fields in polycrystalline aggregates with a constrained surface microstructure–part 2: crystal plasticity. Philos. Mag. **87**, 1425–1446 (2007)

Cosserat Approach to Localization in Geomaterials

19

Ioannis Stefanou, Jean Sulem, and Hadrien Rattez

Contents

Abstract

A renewed interest toward Cosserat or micropolar continuum has driven researchers to the development of specific models for upscaling discrete media such as masonry, granular assemblies, fault gouges, porous media, and biomaterials. Cosserat continuum is a special case of what is called micromorphic, generalized or higher-order continua. Due to the presence of

I. Stefanou · J. Sulem (✉) · H. Rattez
Navier (CERMES), UMR 8205, Ecole des Ponts, IFSTTAR, CNRS, Champs-sur-Marne, France
e-mail: ioannis.stefanou@enpc.fr; jean.sulem@enpc.fr

G. Z. Voyiadjis (ed.), *Handbook of Nonlocal Continuum Mechanics for Materials and Structures*, https://doi.org/10.1007/978-3-319-58729-5_10

internal lengths in its formulation, Cosserat continuum is quite attractive for addressing problems involving strain localization. It enables modeling the shear band thickness evolution, tracking the postlocalization regime, and correctly dissipating the energy when using numerical schemes. In this chapter, we summarize the fundamental governing equations of a Cosserat continuum under multiphysical couplings. Several examples of the numerical advantages of Cosserat continuum are also presented regarding softening behavior, strain localization, finite element formulation, reduced integration, and hourglass control. The classically used constitutive models in Cosserat elastoplasticity are presented and some common approaches for upscaling and homogenization in Cosserat continuum are discussed. Finally, a simple illustrative example of the adiabatic shearing of a rock layer under constant shear stress is presented in order to juxtapose a rate-independent Cosserat with a rate-dependent Cauchy formulation as far as it concerns strain localization.

Keywords
Cosserat continuum · Plasticity · Bifurcation theory · Multiphysical couplings · Strain localization · Shear locking · Upscaling · Faults

Introduction

Cosserat or micropolar continuum is a special case of what is called micromorphic, generalized, or higher-order continua. The general theory of micromorphic continua is general enough to describe various heterogeneous systems with microstructure. In Fig. 1, we briefly outline the various higher-order continuum theories, according to Germain (1973), and their special cases. Besides the classical continuum (also called Cauchy or Boltzmann continuum) and the Cosserat continuum, special cases of micromorphic continua are also the second gradient and the couple stress continuum theories.

In the last three decades, a renewed interest toward the Cosserat continuum has driven researchers to the development of specific continuous models for upscaling discrete media such as masonry, granular assemblies, fault gouges, porous media, and biomaterials (see, for instance, Besdo 1974; Anderson and Lakes 1994; Forest et al. 1999, 2001; Forest and Sab 1998; Stefanou et al. 2010; Sulem et al. 2011; Trovalusci et al. 2015; Masiani and Trovalusci 1996; Mühlhaus and Vardoulakis 1988; Bardet and Vardoulakis 2001; Pasternak and Mühlhaus 2006; Pasternak et al. 2006; de Buhan and Sudret 2000; among others). Cosserat continuum has seemed motivated by the advantages enclosed in its enhanced kinematics and the nonsymmetry of the stress tensor. When used for the formulation of continuum equivalent models for discrete media, Cosserat continuum allows to efficiently take into account high deformation gradients, relative particles' rotation, and scale effects. Moreover, Cosserat continuum enables the investigation of the phenomenon of wave dispersion, which characterizes the dynamic response of discrete media when the wavelength is comparable to the size of the microstructure (internal length).

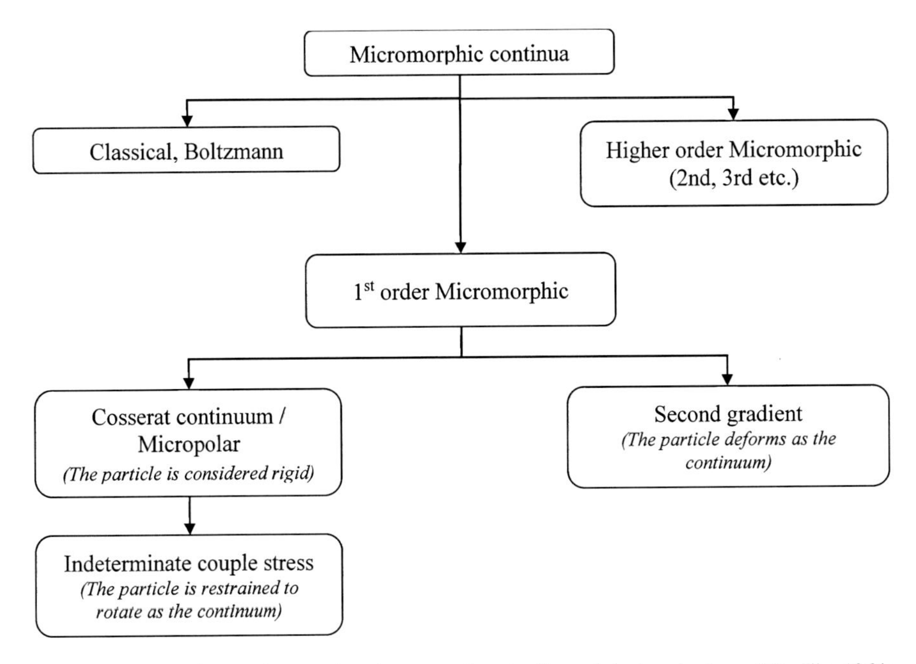

Fig. 1 Higher-order continuum theories according to Germain's terminology (Mindlin 1964; Eringen 1999)

Due to the presence of internal lengths in its formulation, Cosserat continuum is quite attractive for addressing problems involving strain localization. After the pioneering work of Mühlhaus and Vardoulakis (1988), who explained mathematically the shear band thickness and its evolution under shearing, Cosserat continuum became the first and nowadays one of the common modeling approaches for regularization of the ill-posed Cauchy continuum (Zervos et al. 2001; Papanastasiou and Zervos 2004; Papanastasiou and Vardoulakis 1992; De Borst and Sluys 1991; Godio et al. 2016a). Moreover, the correct simulation of the shear band thickness evolution (e.g., the thickness of the principal slip zone in a fault gouge) and of the postpeak behavior of a system, which Cosserat theory enables, guarantees correct energy dissipation.

This chapter is structured as follows. In section "The Cosserat Continuum," we give the general equations of a Cosserat continuum. In particular, the linear and angular momentum balance as well as the mass and energy balance equations and the second law of thermodynamics are extended for the Cosserat continuum. Under multiphysical couplings, involving chemistry and phase transformations, the above equations have to be adapted for momentum, mass, and energy exchange between the considered phases. Section "The Cosserat Continuum" is completed with a small paragraph discussing the additional terms due to this exchange. In section "Constitutive Laws for Cosserat Continua," we present some constitutive laws that are commonly used in Cosserat elasticity and elastoplasticity, including softening and multiphysical couplings. In section "Numerical Advantages of Cosserat Continuum," we show the numerical advantages of Cosserat continuum theory for solving problems when softening behavior and strain localization are encountered. The commonly used approaches for upscaling and homogenization of heterogeneous structures in the frame of Cosserat continuum are discussed in section "Upscaling and Homogenization." Finally, in section "Rock Shear Layer of Cosserat Continuum: Fault Mechanics," a simple example of the adiabatic shearing of a rock layer under constant shear stress is presented in order to highlight the similarities of a rate-independent Cosserat and a rate-dependent Cauchy continuum as far it concerns the conditions for which shear band localization takes place. It is worth mentioning that Cosserat continuum is a promising framework in fault mechanics as it enables to compute the thickness of the principal slip zone inside a fault core, its dependence upon the microstructure of the fault gouge (Sulem et al. 2011; Rattez et al. 2016a, b; Veveakis et al. 2013; Sulem and Stefanou 2016) and its impact on the seismic energy budget.

The Cosserat Continuum

Cosserat continuum equations can be derived in two different, but equivalent, ways, each one based on a different physical ansatz. The first one is based on the linear and angular momentum balance and the fundamental assumption of the existence of the so-called couple stresses at the material point (Fig. 2). The physical meaning

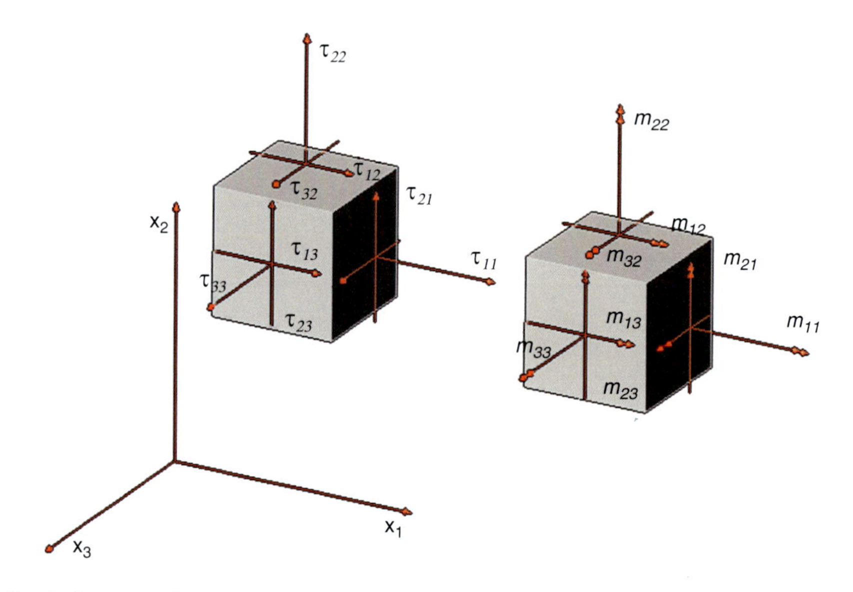

Fig. 2 Stresses and couple-stresses (moments) of Cosserat continuum (Stefanou et al. 2008)

of couple stresses, as well of their conjugate in energy kinematic measures, i.e., the gradient of the rotations (i.e., the curvatures), is a long-standing point of discussions in the scientific community and it is application-wise. For instance, in the Timoshenko theory of beams, which coincides with a one-dimensional Cosserat continuum, the Cosserat moment is simply the resultant section moment.

The second, alternative approach for deriving the Cosserat continuum equations is by postulating the form of the energy density function and by enriching directly the kinematics of the material point. Both ways are equivalent, but here we prefer to present the first one as only the momentum balance equations are needed. The energy of a Cosserat continuum can be then derived directly from the variational form of the Cosserat equilibrium equations.

Linear and Angular Momentum Balance

Let us consider a solid of volume V and its boundary ∂V. The linear momentum is given as follows:

$$P_i = \int_V \rho v_i \mathrm{dV} \tag{1}$$

where v_i is the velocity of a material point, $i = 1,2,3$ (in three dimensions), and ρ is the density. The resultant force is written as:

$$F_i = \int_V f_i \mathrm{dV} + \int_{\partial V} t_i \mathrm{dS} \tag{2}$$

where f_i is a body force per unit volume and t_i is the traction vector, with $t_i = \tau_{ij} n_j$ and n_j the normal unit vector at the boundary. Repeated indices denote summation over the repeated index. Applying the Newton's law $\frac{\mathrm{D}P_i}{\mathrm{Dt}} = F_i$, we obtain:

$$\tau_{ij,j} + f_i = \rho \frac{\mathrm{D}v_i}{\mathrm{Dt}} \tag{3}$$

In the case of a multiphase material, the velocity of each constituent phase has to be considered separately. Moreover, the momentum equations have to be written separately for each phase and then combined to give again Eq. 3, but with $\rho \frac{\mathrm{D}v_i}{\mathrm{Dt}} = (1-n)\rho^s \frac{D^s v_i^s}{\mathrm{Dt}} + n\rho^f \frac{D^f v_i^f}{\mathrm{Dt}}$, where, here, f stands for the fluid phase and s for the solid one, n is the Eulerian porosity (pore volume per unit volume of porous material in actual state), and $\frac{D^x(.)}{\mathrm{Dt}} = \frac{\partial(.)}{\partial t} + (.)_{,i} v_i^x$. The above equation is valid as far as no exchange or transformation takes place between the phases.

The moment of momentum about a point is:

$$H_i = \int_V e_{ijk} x_j \rho v_k \mathrm{dV} + \int_V I_{ij} \omega_j \mathrm{dV} \tag{4}$$

where e_{ijk} is the Levi-Civita perturbation symbol, I_{ij} the microinertia tensor, and ω_j the Cosserat rotational velocity of the material point. In the special case of an isotropic microinertia tensor, $I_{ij} = I_s \delta_{ij}$, where δ_{ij} is the Kronecker delta.

The resultant moment about the same point is written as:

$$L_i = \int_V \left(e_{\mathrm{ijk}} x_j f_k + \mu_i \right) \mathrm{dV} + \int_S e_{\mathrm{ijk}} x_j t_k \mathrm{dS} + \int_V m_i \mathrm{dS} \tag{5}$$

where μ_i is a body couple stress per unit volume and $m_i = m_{ij} n_j$ is the couple stress tractor, with m_{ij} the couple stress tensor. Euler's law $\frac{\mathrm{DH}_i}{\mathrm{Dt}} = L_i$, yields:

$$m_{ij,j} + e_{ijk} \tau_{\mathrm{kj}} = I_{ij} \frac{D\omega_j}{\mathrm{Dt}} \tag{6}$$

From the above equation, we observe that the stress tensor τ_{ij} is not symmetric due to the presence of the couple stress gradient and the microinertia terms. τ_{ij} can be decomposed into a symmetric σ_{ij} and an antisymmetric part $s_{\mathrm{ij}} : \tau_{\mathrm{ij}} = \sigma_{\mathrm{ij}} + s_{\mathrm{ij}}$. In this case, the symmetric part plays the role of the Cauchy stress tensor.

Mass Balance

The derivation of the pore pressure diffusion-generation equation from the mass balance follows the classical derivation procedure as in the classical Cauchy continuum. In particular, it becomes:

$$\frac{\partial p^f}{\partial t} = \Lambda \frac{\partial T}{\partial t} - \frac{1}{\beta^*} \frac{\partial \varepsilon}{\partial t} + \frac{1}{\beta^*} w_{i,i} \tag{7}$$

where $w_i = n \left(v_i^f - v_i^s \right)$ is the Darcy's fluid (relative) velocity, $\frac{\partial \varepsilon}{\partial t} = v_{i,i}$ is the volumetric strain rate, $\Lambda = \frac{\lambda^*}{\beta^*}$ is the undrained pore fluid pressurization coefficient, with $\lambda^* = n\lambda^f + (1-n)\lambda^s$, where λ^f is the fluid thermal expansion coefficient and λ^s is the solid thermal expansion coefficient, $\beta^* = n\beta^f + (1-n)\beta^s$ is the storage coefficient, where β^f is the fluid compressibility and β^s is the compressibility of the solid phase.

Assuming Darcy's law for the fluid flow, the above equation becomes:

$$\frac{\partial p^f}{\partial t} = \Lambda \frac{\partial T}{\partial t} - \frac{1}{\beta^*} \frac{\partial \varepsilon}{\partial t} + c_{\mathrm{hy}} p^f_{,\mathrm{ii}} \tag{8}$$

where $c_{hy} = \frac{k}{\eta_f \beta^*}$ is the hydraulic diffusivity with k the hydraulic permeability, which is assumed here isotropic and constant and, η_f the fluid viscosity.

Energy Balance

Similar to the mass balance equation, the derivation of the energy balance equation and in a second step of the heat equation follows the same steps as in the case of the Cauchy continuum. However, the work of the external, generalized forces and the kinetic energy have some additional terms due to the Cosserat rotations and moments. More specifically, assuming a micropolar solid phase and a nonmicropolar fluid, the kinetic energy rate is:

$$\begin{aligned}\frac{\mathrm{DK}}{\mathrm{Dt}} &= \frac{\mathrm{DK}^s}{\mathrm{Dt}} + \frac{\mathrm{DK}^f}{\mathrm{Dt}} \\ &= \frac{D}{\mathrm{Dt}}\left\{\int_V \left[\frac{1}{2}(1-n)\,\rho^s v_i^s v_i^s + \frac{1}{2}(1-n)\,I^s \omega_i \omega_i\right]\mathrm{dV} \right. \\ &\quad \left. + \int_V \frac{1}{2} n \rho^f v_i^f v_i^f\,\mathrm{dV}\right\}\end{aligned} \tag{9}$$

The internal energy is written as follows:

$$\begin{aligned}\frac{\mathrm{DE}}{\mathrm{Dt}} &= \frac{\mathrm{DE}^s}{\mathrm{Dt}} + \frac{\mathrm{DE}^f}{\mathrm{Dt}} \\ &= \frac{D}{\mathrm{Dt}}\int_V \left[(1-n)\,\rho^s \mathcal{E}^s + n\rho^f \mathcal{E}^f\right]\mathrm{dV}\end{aligned} \tag{10}$$

here ε^s and ε^f are the internal energy of the solid and fluid phase, respectively, per unit mass.

The work rate of external forces for a Cosserat continuum is given by:

$$\frac{\mathrm{DW}}{\mathrm{Dt}} = \int_{\partial V} \left(t_i^s v_i^s + m_i \omega_i + t_i^f v_i^f\right)\mathrm{dS} + \int_V \left(f_i^s v_i^s + \mu_i \omega_i + f_i^f v_i^f\right)\mathrm{dV} \tag{11}$$

Notice that due to the presence of two phases in the medium, the volumetric as well as the external generalized forces and tractions are split over the two phases. For a chemically inert system and in absence of any exchange between the solid and the fluid phase, the equations remain the same with Eqs. 3 and 6. According to the first law of thermodynamics, we have:

$$\frac{\mathrm{DK}}{\mathrm{Dt}} + \frac{\mathrm{DE}}{\mathrm{Dt}} = \frac{\mathrm{DQ}}{\mathrm{Dt}} + \frac{\mathrm{DW}}{\mathrm{Dt}} \tag{12}$$

where $\frac{\mathrm{DQ}}{\mathrm{Dt}} = -\int_{\partial V} h_i n_i \mathrm{dS} = -\int_V h_{i,i}\mathrm{dV}$ is the rate of heat input (given to V through its boundary ∂V) and h_i is the heat flux vector components. After some algebraic manipulations, the energy balance equation becomes:

$$\begin{aligned}&(1-n)\,\rho^{s}\frac{D^{s}\varepsilon^{s}}{Dt}+n\rho^{f}\frac{D^{f}\varepsilon^{f}}{Dt}+\rho^{f}\varepsilon^{f}_{,i}w_{i}\\&-\tau_{ij}\dot{\gamma}_{ij}-m_{ij}\dot{\kappa}_{ij}+h_{i,i}\\&+\left(p^{f}w_{i}\right)_{,i}-\rho^{f}\left(f_{i}-\frac{D^{f}v^{f}_{i}}{\mathrm{Dt}}\right)w_{i}=0\end{aligned} \tag{13}$$

where $\dot{\gamma}_{ij} = v_{i,j} + e_{ijk}\omega_k$ and $\dot{\kappa}_{ij} = \omega_{i,j}$. Neglecting the Cosserat effects and adapting the notation, the above equation coincides with Houlsby and Puzrin (2000) and Coussy (2004).

Assuming an elastoplastic constitutive law (see section "Constitutive Laws for Cosserat Continua"), the energy balance equation results to the heat equation for Cosserat continua:

$$\frac{\partial T}{\partial t}=c_{\mathrm{th}}\frac{\partial^{2}T}{\partial x_{i}\partial x_{i}}+\frac{1}{\rho C}\tau_{ij}\dot{\gamma}^{p}_{ij}+\frac{1}{\rho C}m_{ij}\dot{\kappa}^{p}_{ij} \tag{14}$$

where T is the temperature, $\dot{\gamma}^{p}_{ij}$ the plastic deformation rate tensor, and $\dot{\kappa}^{p}_{ij}$ the plastic curvature rate tensor and c_{th} the thermal diffusivity coefficient. For deriving the above equation, small deformations and an isotropic Fourier's law were considered.

Second Law of Thermodynamics

Following the same derivations, the second law of thermodynamics yields:

$$\int_{V}\left[\frac{D^{s}S}{\mathrm{Dt}}-\left(S-S^{f}\right)\left(\rho^{f}w_{i}\right)_{,i}+S^{f}_{,i}\rho^{f}w_{i}+\left(\frac{h_{i}}{T}\right)_{,i}\right]\mathrm{dV}\geq 0 \tag{15}$$

where S denotes the specific entropy of the mixture per unit mass and S^x the specific entropy of the specie x. The above equation coincides with the one of (Houlsby and Puzrin 2000) after proper substitutions and algebra. Under specific assumptions for the constitutive behavior of the solid and the fluid phases, the above equation can be further developed.

Phase Transitions Between the Species

Phase transitions from the solid to the fluid phase and vice versa, due to, for instance, dissolution/precipitation mechanisms or dehydration, has as a result the addition of supplementary terms in the above equations. In particular, if $\dot{r}_{s\to f} = -\dot{r}_{f\to s}$ expresses the rate of transition from the solid to the fluid phase, Eqs. 3, 6, 8, 13, and 15 become:

$$\tau_{ij,j} + f_i + \left(v_i^f - v_i^s\right) \dot{r}_{s\to f} = \rho \frac{\mathrm{Dv}_i}{\mathrm{Dt}} \tag{16}$$

$$m_{ij,j} + e_{ijk}\tau_{kj} + \alpha\omega_i \dot{r}_{s\to f} = I_{ij}\frac{D\omega_j}{\mathrm{Dt}} \tag{17}$$

$$\frac{\partial p^f}{\partial t} = \Lambda\frac{\partial T}{\partial t} - \frac{1}{\beta*}\frac{\partial \varepsilon}{\partial t} + \frac{1}{\beta*}w_{i,i} - c_{\mathrm{ch}}\dot{r}_{s\to f} \tag{18}$$

$$\begin{aligned}
&(1-n)\,\rho^s\tfrac{D^s\mathcal{E}^s}{\mathrm{Dt}} + n\rho^f\tfrac{D^f\mathcal{E}^f}{\mathrm{Dt}} + \rho^f\mathcal{E}^f_{,i}w_i \\
&-\tau_{ij}\dot{\gamma}_{ij} - m_{ij}\dot{\kappa}_{ij} + h_{i,i} \\
&+\left(p^f w_i\right)_{,i} - \rho^f\left(f_i - \tfrac{D^f v_i^f}{\mathrm{Dt}}\right)w_i \\
&+\dot{r}_{s\to f}\tfrac{1}{2}\left[\left(v_i^f - v_i^s\right)\left(v_i^f - v_i^s\right) + \alpha\omega_i\omega_i + 2\left(\mathcal{E}^f - \mathcal{E}^s\right)\right] = 0
\end{aligned} \tag{19}$$

$$\begin{aligned}
&\int_V\left[\tfrac{D^s S}{\mathrm{Dt}} - \left(S - S^f\right)\left(\rho^f w_i\right)_{,i} + S^f_{,i}\rho^f w_i + \left(\tfrac{h_i}{T}\right)_{,i}\right]\mathrm{dV} \\
&+\int_V \dot{r}_{s\to f}\left(S^f - S^s\right)\mathrm{dV} \geq 0
\end{aligned} \tag{20}$$

where $c_{\mathrm{ch}} = \frac{1}{\beta*}\frac{\rho^f - \rho^s}{\rho^f\rho^s}$ is the chemical pressurization coefficient and $\alpha = \frac{I_s}{\rho_s}$. The chemical pressurization coefficient may be insignificant in most of the cases of dissolution/precipitation reactions in rocks (Sulem and Stefanou 2016). However, this is not the case for dehydration or decarbonation reactions (Sulem and Famin 2009; Brantut et al. 2011).

Constitutive Laws for Cosserat Continua

Deriving constitutive relations for Cosserat continuum is not a trivial task. This is owed to the fact that Cosserat theory has intrinsic internal lengths and consequently the constitutive law is not scale invariant. In the simplest case, i.e., under small deformations, in a centrosymmetric, linear, elastic, isotropic Cosserat medium, the stresses are related to the generalized elastic deformation measures according to the following constitutive relations (Vardoulakis 2009):

$$\begin{aligned}
\tau_{ij} &= K\gamma^{\mathrm{el}}_{\mathrm{kk}}\delta_{ij} + 2G\left(\gamma^{\mathrm{el}}_{(ij)} - \tfrac{1}{3}\gamma^{\mathrm{el}}_{\mathrm{kk}}\delta_{ij}\right) + 2\eta_1 G\gamma^{\mathrm{el}}_{[ij]} \\
m_{ij} &= 4\mathrm{GR}^2\left(\kappa^{\mathrm{el}}_{(ij)} + \eta_2\delta_{ij}\kappa^{\mathrm{el}}_{\mathrm{kk}}\right) + 4\mathrm{GR}^2\eta_3\kappa^{\mathrm{el}}_{[ij]}
\end{aligned} \tag{21}$$

where K is the bulk modulus, G is the shear modulus, η_1, η_2, η_3 are positive material constants, and R is an internal length parameter, which for granular materials is often identified to the mean radius of the grains in the representative volume element (RVE). $\gamma_{(\mathrm{ij})}$ and $\gamma_{[\mathrm{ij}]}$ denote, respectively, the symmetric and antisymmetric parts of $\gamma_{\mathrm{ij}} = u_{i,j} + e_{\mathrm{ijk}}\theta_k$, where u_i are the components of

Table 1 Coefficients used for the generalized J2 invariants for Cosserat continuum (Sulem and Vardoulakis 1990)

	2D model	3D model
Static consideration	$h_i = \left\{\frac{3}{4}, \frac{-1}{4}, 1, 0\right\} g_i = \left\{\frac{3}{2}, \frac{1}{2}, 1, 0\right\}$	$h_i = \left\{\frac{2}{3}, \frac{-1}{6}, \frac{2}{3}, \frac{-1}{6}\right\} g_i = \left\{\frac{8}{5}, \frac{2}{5}, \frac{8}{5}, \frac{2}{5}\right\}$
Kinematical consideration	$h_i = \left\{\frac{3}{8}, \frac{1}{8}, \frac{1}{4}, 0\right\} g_i = \{3, -1, 4, 0\}$	$h_i = \left\{\frac{2}{5}, \frac{1}{10}, \frac{2}{5}, \frac{1}{10}\right\} g_i = \left\{\frac{8}{3}, \frac{-2}{3}, \frac{8}{3}, \frac{-2}{3}\right\}$

the displacement vector and θ_k the components of the Cosserat rotation vector (infinitesimal rotations). The Cosserat shear modulus, which expresses the stiffness related to the relative rotation of the particle (e.g. of a grain) with respect to the macrorotation of the continuum (e.g., assemblage of grains) is defined as $G_c = \eta_1 G$.

In a small deformations framework, the classical J2 plasticity criteria can be generalized for Cosserat continua. This approach is based on the generalization of the stress and strain invariants in order to account for the Cosserat couple stresses, curvatures, and rotations. Based on micromechanical considerations and averaging, the deviatoric stress and strain invariants take the following form (Rattez et al. 2016b; Sulem and Vardoulakis 1990; Vardoulakis and Sulem 1995):

$$\begin{aligned} q &= \sqrt{h_1\, \tau_{ij}^d \tau_{ij}^d + h_2 \tau_{ij}^d \tau_{\mathrm{ji}}^d + \frac{1}{R^2}\left(h_3\, m_{ij}^d m_{ij}^d + h_4\, m_{ij}^d m_{\mathrm{ji}}^d\right)} \\ \gamma &= \sqrt{g_1\, \gamma_{ij}^d \gamma_{ij}^d + g_2\, \gamma_{ij}^d \gamma_{\mathrm{ji}}^d + R^2\left(g_3\, \kappa_{ij}^d \kappa_{ij}^d + g_3\, \kappa_{ij}^d \kappa_{\mathrm{ji}}^d\right)} \end{aligned} \tag{22}$$

where τ_{ij}^d, m_{ij}^d, γ_{ij}^d, and κ_{ij}^d are the deviatoric parts of the stress, couple-stress, strain, and curvature tensors respectively. $h_{p.}$ and g_p ($p = 1,..,4$) are coefficients that can be determined by micromechanical considerations on the kinematics or the statics of an assemblage of grains in contact (Sulem and Vardoulakis 1990; Mühlhaus and Vardoulakis 1987) (see Table 1).

Notice that the internal length R appears directly in the above expressions. Once the above invariants are derived, various plasticity criteria can be used. We refer, for example, to the von Mises, the Drucker-Prager, or to the Cam-clay criterion, which are commonly used in metal, soil, and rock mechanics and they are expressed in terms of the q, p invariants. Volumetric and shear hardening parameters can be also defined allowing not only to reproduce the stress-strain response of a geomaterial, but also the shear band thickness evolution (Mühlhaus and Vardoulakis 1987), its thickness, and the postbifurcation regime, which is crucial in many applications as it is related to stress redistribution, collapse, and energy dissipation. In a small strains framework, the generalized deformation rate measures are decomposed in elastic and plastic parts as follows:

$$\begin{aligned} \dot{\gamma}_{ij} &= \dot{\gamma}_{ij}^{\mathrm{el}} + \dot{\gamma}_{ij}^{\mathrm{pl}} \\ \dot{\kappa}_{ij} &= \dot{\kappa}_{ij}^{\mathrm{el}} + \dot{\kappa}_{ij}^{\mathrm{pl}} \end{aligned} \tag{23}$$

Assuming a Drucker-Prager yield surface and plastic potential:

$$F = q + \mu p \leq 0 \tag{24}$$

$$Q = q + \beta p \tag{25}$$

where $p = \tau_{\mathrm{kk}}$, μ is the friction coefficient and β the dilatancy coefficient; the elastoplastic incremental generalized stress-strain relationships become:

$$\begin{aligned} \dot{\tau}_{ij} &= C^{\mathrm{ep}}_{ij\mathrm{kl}} \ \dot{\gamma}_{\mathrm{kl}} + D^{\mathrm{ep}}_{ij\mathrm{kl}} \ \dot{\kappa}_{\mathrm{kl}} + E^{\mathrm{ep}}_{ij\mathrm{kl}} \ \dot{T} \ \delta_{\mathrm{kl}} \\ \dot{\mu}_{ij} &= M^{\mathrm{ep}}_{ij\mathrm{kl}} \ \dot{\kappa}_{\mathrm{kl}} + L^{\mathrm{ep}}_{ij\mathrm{kl}} \ \dot{\gamma}_{\mathrm{kl}} + N^{\mathrm{ep}}_{ij\mathrm{kl}} \ \dot{T} \ \delta_{\mathrm{kl}} \end{aligned} \tag{26}$$

where

$$\begin{aligned} C^{\mathrm{ep}}_{ij\mathrm{kl}} &= C^{e}_{ij\mathrm{kl}} - \frac{<1>}{H_p} b^{Q}_{ij} b^{F}_{\mathrm{kl}} & D^{\mathrm{ep}}_{ij\mathrm{kl}} &= -\frac{<1>}{H_p} b^{Q}_{ij} b^{\mathrm{FM}}_{\mathrm{kl}} \\ E^{\mathrm{ep}}_{ij\mathrm{kl}} &= -\left(C^{e}_{ij\mathrm{kl}} - \frac{<1>}{H_p} b^{Q}_{ij} b^{F}_{\mathrm{kl}} \right) & L^{\mathrm{ep}}_{ij\mathrm{kl}} &= -\frac{<1>}{H_p} b^{Q}_{ij} b^{F}_{\mathrm{kl}} \\ M^{\mathrm{ep}}_{ij\mathrm{kl}} &= M^{e}_{ij\mathrm{kl}} - \frac{<1>}{H_p} b^{Q}_{ij} b^{\mathrm{FM}}_{\mathrm{kl}} & N^{\mathrm{ep}}_{ij\mathrm{kl}} &= \frac{<1>}{H_p} b^{Q}_{ij} b^{F}_{\mathrm{kl}} \end{aligned} \tag{27}$$

C^{e}_{ijkl} and M^{e}_{ijkl} are the elastic constitutive tensors for the stress and the couple stress, respectively, derived from Eq. 21, and

$$\begin{aligned} b^{F}_{\mathrm{kl}} &= \frac{\partial F}{\partial \tau_{ij}} C^{e}_{ij\mathrm{kl}}; \quad b^{Q}_{ij} = C^{e}_{ij\mathrm{kl}} \frac{\partial F}{\partial \tau_{\mathrm{kl}}}; \quad b^{\mathrm{FM}}_{\mathrm{kl}} = \frac{\partial F}{\partial \mu_{ij}} M^{e}_{ij\mathrm{kl}} \\ H_p &= \frac{\partial F}{\partial \tau_{ij}} C^{e}_{ij\mathrm{kl}} \frac{\partial Q}{\partial \tau_{\mathrm{kl}}} + \frac{\partial F}{\partial \mu_{ij}} M^{e}_{ij\mathrm{kl}} \frac{\partial Q}{\partial \mu_{\mathrm{kl}}} + H_s \end{aligned} \tag{28}$$

H_S is the hardening modulus due to shear plastic deformations, defined by:

$$\frac{\partial F}{\partial \gamma^{p}} = -H_s \tag{29}$$

Once the constitutive parameters of the material are calibrated, a bifurcation analysis can provide important information regarding strain localization evolution under shearing (Mühlhaus and Vardoulakis 1987). Various multiphysical phenomena (Sulem et al. 2011; Veveakis et al. 2012, 2013) as well as grain cataclasis (Rattez et al. 2016b) can also be considered. Predicting strain localization is of paramount importance in geotechnical applications involving strongly non-associated materials (common case) and in fault mechanics where the mutiphysical couplings, energy dissipation, softening behavior, and grain size evolution play the central role for earthquake triggering and coseismic slip (see next sections for more details).

It is worth emphasizing that the internal length(s) that are present in the Cosserat continuum theory have somehow to be determined or identified. This can be done either by using bottom-up, upscaling, and homogenization techniques that consider the microstructure in detail (see section "Upscaling and Homogenization") or through classical geotechnical tests. However, in practice of classical geotechnical testing, the calibration of these parameters is ignored.

Numerical Advantages of Cosserat Continuum

Mesh Dependency and Its Remedy

It is well known that when it comes to strain localization, the governing equations based on a classical Boltzmann continuum become ill-posed and when numerical approaches are used for integration of these equations, the result is mesh dependent (e.g., De Borst et al. 1993). With the term mesh dependency, we mean that the numerical results depend on the size and the type of the finite elements used. More specifically, the global behavior depends on the choice of the mesh and by refining it no convergence of the numerical results is observed. In Fig. 3 we

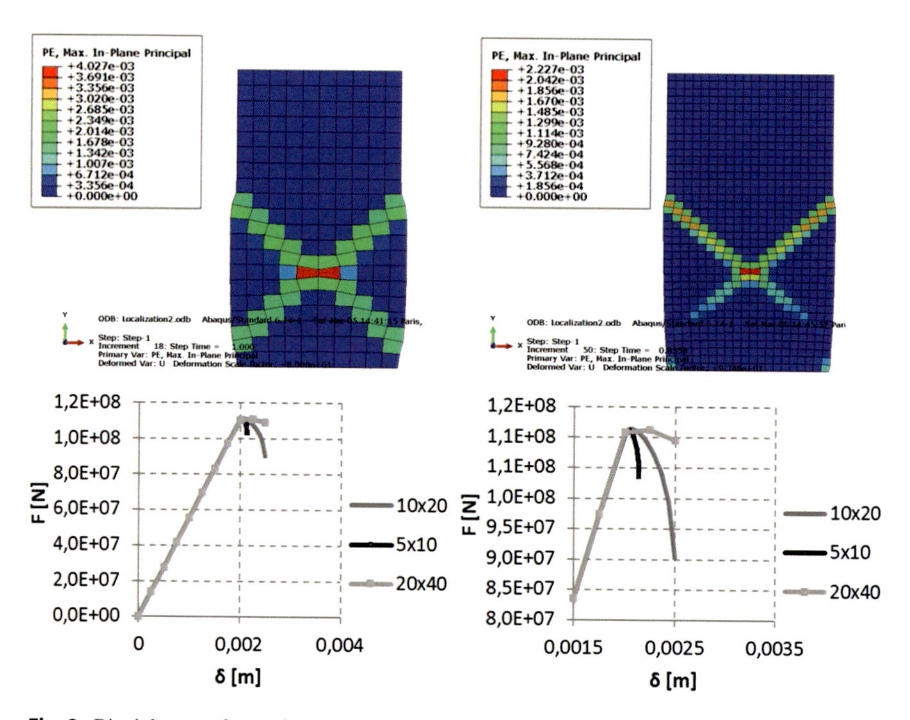

Fig. 3 Biaxial test of a softening material for different mesh sizes. The developed plastic deformations, the shear band thickness, and the total applied force – displacement depends on the mesh size

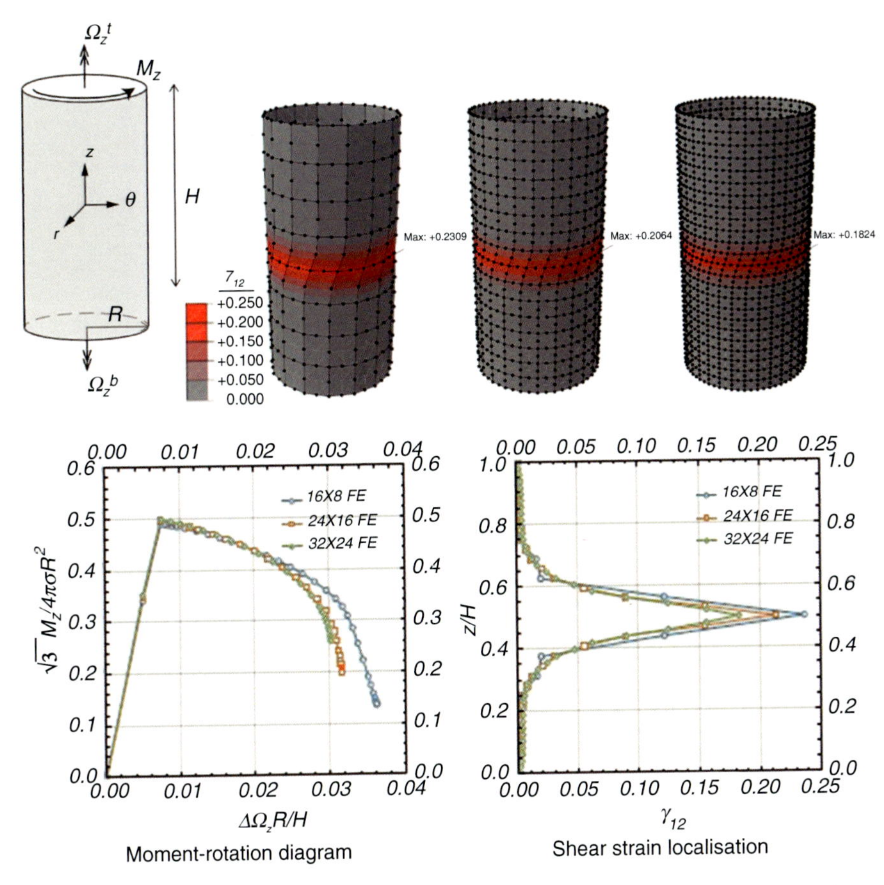

Fig. 4 *Above*: Deformation pattern and isocontour of in-plane shear strain distribution for different discretizations of the hollow cylinder. *Below*: Macroscopic behavior of the hollow cylinder (*left*) and localization of the in-plane shear strain (*right*) for different discretizations with the COSS8R finite element (FE) (Godio et al. 2015)

present this mathematical and numerical artifact, which is owed to the absence of internal lengths in the classical continuum. Moreover, the developed shear bands (contracting or dilating) are always one to two finite elements thick.

Cosserat continuum is one of the available techniques for regularizing mesh dependency during shear banding (De Borst and Sluys 1991; De Borst et al. 1993; De Borst 1984). In Fig. 4, we present the simulations that were carried out using COSS8R finite element, a homemade User Element for Abaqus, which is currently extended for use to any classical finite element code (Godio et al. 2016a). The problem treated is a thin hollow cylinder made of a softening material under shearing. Three different finite element discretizations were tested. In the same

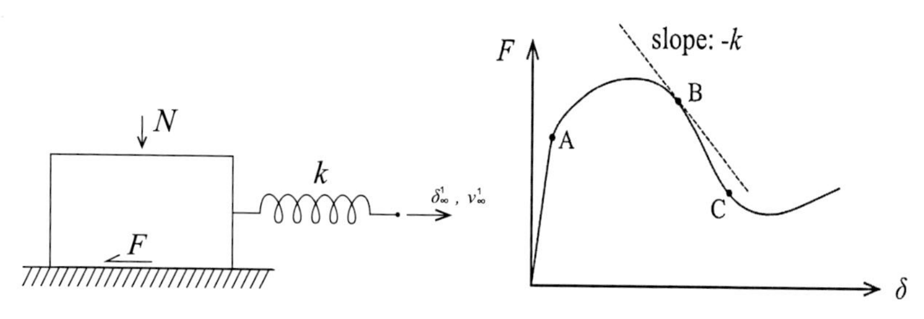

Fig. 5 *Left*: The classical spring-slider model used for understanding earthquake nucleation and energy partition (Scholz 2002; Kanamori and Brodsky 2004). *Right*: Schematic representation of the mechanical behavior of a rock interface (fault). From A to B, the slip is aseismic. Instability takes place at B (coseismic slip). The role of the softening branch is apparent

figure, we present the resulting macroscopic behavior of the cylinder in terms of the normalized reaction moment and the differential axial rotation. The onset of plastic deformations rapidly culminates with a softening branch, which is accompanied by strain localization.

Strain localization occurs exactly as in the shear layers studied by De Borst (1991). The curves show important mesh-independency, i.e., the thickness (width) of the localization zone does not depend on the number of elements falling inside this zone and the total force-axial rotation curve is independent of the mesh size. This is not only a fundamental feature of finite elements based on the Cosserat continuum, but it demonstrates also the fast convergence of the chosen Finite Element interpolation scheme (Godio et al. 2015).

Besides failure analysis in geotechnical engineering (Goodman 1989), the post-peak behavior of the geomaterial plays a crucial role in earthquake nucleation. In the idealistic, basic case of considering a fault zone as an interface with no multiphysical couplings and assuming a rate-independent behavior for the rock material, i.e., $F = F(\delta)$, where F is the interface friction and δ the slip (Fig. 5 left), bifurcation theory leads (see Mam and Quennehen 2016; Stefanou and Alevizos 2016) to the classical relationship for earthquake nucleation: $dF(\delta)/d\delta \leq -k$ (Scholz 2002). This condition is presented graphically in Fig. 5 (right). Instabilities are triggered (or induced) not at the peak, but in the softening branch, i.e., at B–C. Before that, there is slip but it is *aseismic*. This shows the importance of correctly capturing the postpeak behavior of the rock material (numerically and analytically). Moreover, correct simulation of the shear band thickness evolution (e.g., the thickness of the principal slip zone in a fault) guarantees that the dissipation of energy through the various multiphysical mechanisms that take place (e.g., temperature and pore pressure increase, shear heating, thermal pressurization, thermally induced chemical reactions, grain size evolution, etc.) is correctly calculated before and during the seismic slip.

Shear Locking, Reduced Integration and Physical Hourglass Control

In the above paragraph, we described the advantages of Cosserat continuum for inelastic behavior and strain localization compared to the classical continuum. Yet, Cosserat continuum numerical advantages do not stop there.

Shear locking is a common numerical problem when lower-order interpolation functions are used in finite element analysis. Shear locking is the phenomenon of extremely slow convergence of the numerical solution to the exact one, even when extremely fine discretizations are used. Due to the chosen interpolation functions, a large ratio of the energy input is erroneously transformed to shear strain energy instead of flexural (Zienkiewicz et al. 1971) resulting to a stiffer response. A numerical trick that it is commonly used in finite elements is reduced integration of the stiffness matrix. This alleviates shear locking without losing accuracy (in an incremental formulation, the right-hand side is completely/correctly integrated). However, this numerical trick comes with a price. Reduced integration of common Cauchy continuum finite elements (e.g., linear or quadratic polynomial interpolations) leads to a deficiency of the stiffness matrix. In other words, the stiffness matrix has some eigenvalues that are zero and consequently some deformation modes are associated to zero strain energy (spurious modes). For linear and quadratic Cauchy finite elements, these zero-energy modes have the geometrical shape of an hourglass. In traditional finite element technology, a small artificial/numerical stiffness is added to these deformation modes (represented by the corresponding to the zero eigenvalues eigenvectors). This technique is called hourglass control. However, when important stress gradients take place, hourglass control can fail.

Cosserat continuum offers a natural way to avoid hourglass control. It can be shown (Godio et al. 2015) that after reduced integration of the stiffness matrix, no zero eigenvalues appear in the stiffness matrix. In Fig. 6, we consider the well-discussed plane foundation problem (Zienkiewicz et al. 2013). Due to reduced integration, an hourglass mode is activated in the rigid element on top as a result of the applied force. This hourglass mode is able to propagate within a certain area

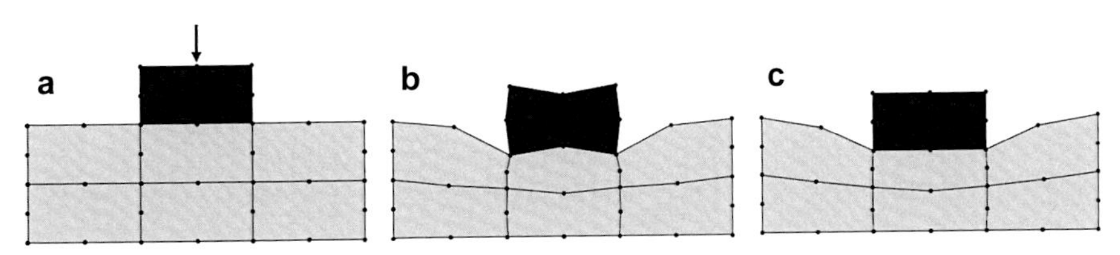

Fig. 6 Foundation of a rigid footing: (**a**) a highly rigid element resting upon a layer of flexible elements with reduced integration; (**b**) propagating zero-energy mode in an assembly of Cauchy (S8R) elements; (**c**) accurate results from a COSS8R subdivision (Godio et al. 2015)

in the elements below. On the contrary, the COSS8R element does not have this deficiency, as the rotational DOF perform an intrinsic action of hourglass control and no artificial hourglass control is needed.

Upscaling and Homogenization

Homogenization or upscaling methods is a class of methods that aim at deriving an equivalent continuum theory that describes the macroscopic behavior of heterogeneous systems, i.e., of systems with microstructure. (Asymptotic) homogenization is a mathematically rigorous, well-established theory for performing this task (Sánchez-Palencia 1986; Bakhvalov and Panasenko 1989; Pinho-da-Cruz et al. 2009; Charalambakis 2010). This method is based on the asymptotic expansion of the various state fields (displacements, deformations, stresses) in terms of a small quantity ε, which represents the ratio of the characteristic size of the 'elementary volume' over the overall size of the structure, and provides an equivalent to the heterogeneous system homogeneous continuum for $\varepsilon \to 0$. Besides the rigorous mathematical formulation of this approach, its main advantage is the ability to determine error estimators of the derived continuum for finite values of ε. However, when it comes to generalized continua, such as the Cosserat continuum, that possess internal lengths, the asymptotic limit $\varepsilon \to 0$ loses interest as it cancels out these internal length (Pradel and Sab 1998). In other words, by this pass to the limit, asymptotic homogenization erases any internal length that are related to the material's microstructure, which higher-order continuum theories, such as Cosserat, are in principle able to capture.

To overcome this problem, several alternative schemes have been proposed in the literature for upscaling heterogeneous systems (see Anderson and Lakes 1994; Forest and Sab 1998; Stefanou et al. 2008, 2010; Trovalusci et al. 2015; Bardet and Vardoulakis 2001; Pradel and Sab 1998; Sab and Pradel 2009; Rezakhani and Cusatis 2016; Godio et al. 2016b; Bacigalupo and Gambarotta 2011, 2012; Baraldi et al. 2016, among others). The majority of these schemes is based upon the homogeneous equivalent continuum concept (see Charalambakis 2010), in the sense that the derived Cosserat continuum shares (a) the same energy (internal energy, dissipation) and (b) the same kinematics with the heterogeneous, discrete medium. The classical asymptotic homogenization expansion ansatz that leads to a Cauchy continuum as the ratio of the size of the unit cell over the overall structure tends to zero is not followed in these approaches. Therefore, these heuristic approaches remain applicable even when the size of the microstructure is not infinitesimal as compared to the overall size of the system, or, in other words, when scale separation is no more applicable.

A typical example for applying and testing these upscaling methods is masonry-like structures. Masonry can be seen as a geomaterial whose building blocks are often quasi-periodically arranged in space. Moreover, the building blocks are at the human scale, which makes them an ideal toy model, contrary to granular media whose microstructure is small, shows geometric complexity and has to be statis-

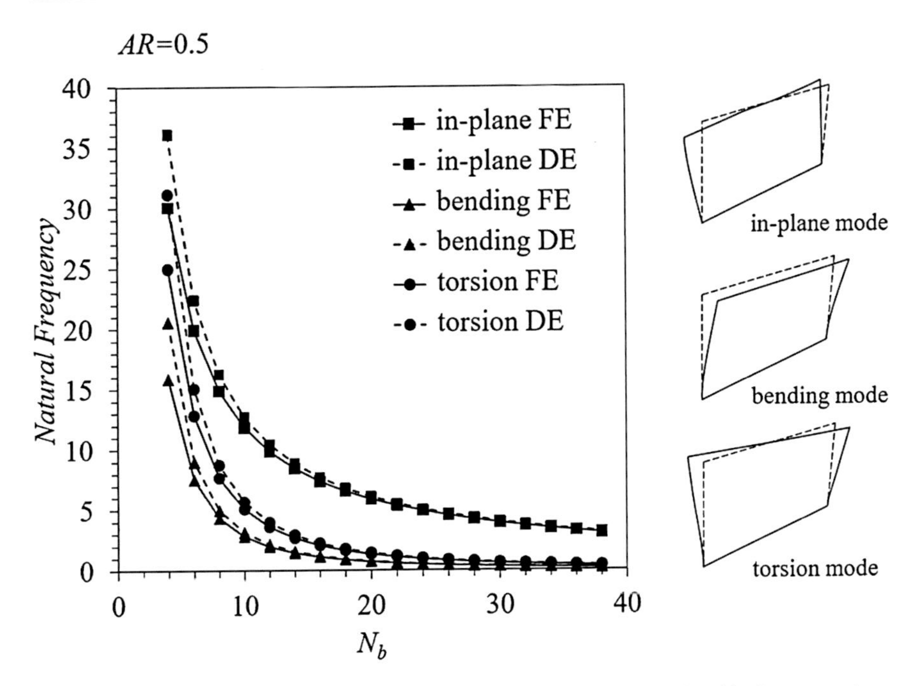

Fig. 7 Modal frequencies of a masonry panel versus the number of building blocks: comparison between the results extracted by discrete element method and by the use of the Cosserat homogenized plate model for masonry (Godio et al. 2016a; Stefanou et al. 2008)

tically described (Stefanou and Sulem 2014, 2016). When the upscaling scheme is correctly formulated, it is possible to capture the wave dispersion behavior of a heterogeneous system even when the wave length is comparable to the block size. Notice that in this case, the classical homogeneous Cauchy continuum approach fails as it is not a dispersive medium. In Fig. 7, we present the modal frequencies of a masonry panel which was up-scaled with Cosserat continuum (Stefanou et al. 2008; Godio et al. 2015) in function of the number of its building blocks. Even when the number of the building blocks is small, the Cosserat homogenized continuum model succeeds in representing the dynamics of the discrete heterogeneous structure. In Fig. 8, we present the out-of-plane-displacement contours of the first three flexural modal shapes of a homogenized masonry panel and a comparison with the flexural modal shapes provided by discrete elements (Godio et al. 2014).

In the case of granular media, the approach described in Bardet and Vardoulakis (2001, 2003) or in Rezakhani and Cusatis (2016) can be followed for upscaling and determining the effective parameters of an equivalent Cosserat continuum. It is worth pointing out that these upscaling techniques may provide valuable information on strain localization and energy dissipation in the absence of detailed experimental data as it is the case of the complex behavior of fault gouges. However, Cosserat continuum is effective only under shearing. In the case of pronounced extension or compaction, the Cosserat additional terms have no effect and at least a first-order micromorphic continuum has to be used (Fig. 1).

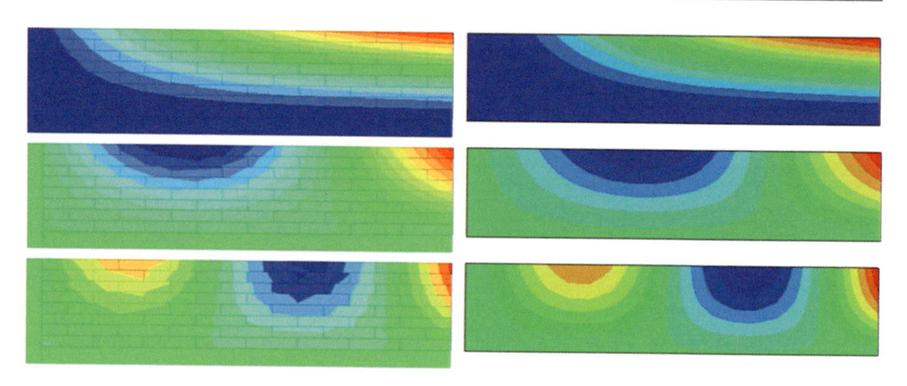

Fig. 8 Out-of-plane-displacement contours of the first three flexural modal shapes. *Left* Discrete elements solution. *Right* Homogenized Cosserat finite element solution (Godio et al. 2014)

Rock Shear Layer of Cosserat Continuum: Fault Mechanics

In this section, we give a simple example of Cosserat theory used for modeling the adiabatic shearing of a rock layer under constant shear stress. Two different modeling frameworks of at least, at first approximation, different physical assumptions are juxtaposed and compared as far as it concerns the conditions for which shear band localization takes place.

The first framework is the Cauchy continuum with rate-dependent constitutive law (viscoplasticity). Rate-dependent constitutive models and, in general, viscoplastic constitutive laws in the frame of Cauchy continuum are frequently used as, under some conditions, they can lead to finite thickness shear band formation. The relation between viscosity and shear band thickness (and consequently material length scale) has been discussed in several publications (e.g., Wang et al. 1996). The second modeling framework is Cosserat elastoplasticity.

Thermal softening is taken into account as a destabilizing mechanism that may lead to shear band localization. The complexity of the chosen constitutive laws and of the multiphysical couplings considered is kept to a minimum degree in order to reveal the salient futures of each framework and highlight their similarities or their differences regarding strain localization. For a more detailed modeling in the frame of Cosserat continuum involving thermo-poro-chemo-mechanical couplings and more elaborate constitutive laws for the rock material, the reader is referred to Sulem et al. (2011), Godio et al. (2016a), and Veveakis et al. (2012, 2013). The reader is referred to (Rice et al. 2014) for the Cauchy rate-dependent framework under thermo-poro-chemo-mechanical couplings.

The thickness of the rock layer is D and constant normal and shear stresses are applied at its boundaries as depicted in Fig. 9. Initially, the layer is considered to be in a state of homogeneous shear deformation.

In both models, it is assumed that all the plastic work is converted to heat and that Fourier's law is applicable. Under these assumptions, the heat equation is written in indicial notation as follows (see Eq. 14):

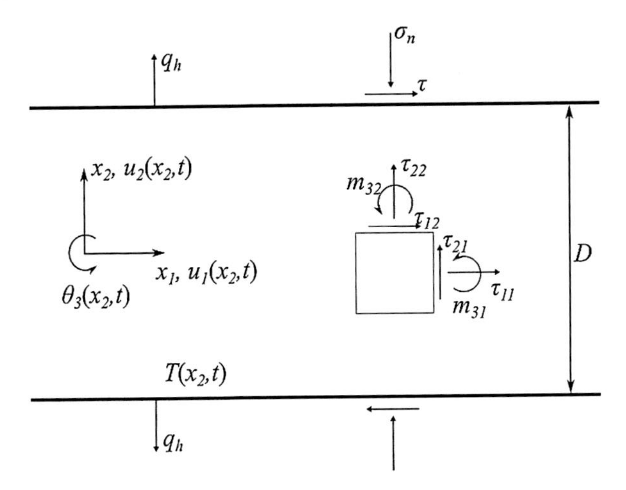

Fig. 9 Shearing of a rock layer: Cosserat rotational degree of freedom ω and couple stresses

$$\frac{\partial T}{\partial t} = c_{\text{th}}\frac{\partial^2 T}{\partial x_2^2} + \frac{1}{\rho C}\tau_{ij}\dot{\gamma}_{ij}^p + \frac{1}{\rho C}m_{ij}\dot{\kappa}_{ij}^p \tag{30}$$

where the infinite layer hypothesis was considered and consequently the derivatives in the x_1 and x_3 directions vanish (invariance). Small deformations are considered and the slip event is sufficiently rapid in order to justify adiabatic conditions at the boundaries of the layer. In the case of the Cauchy continuum, the third term in the right-hand side vanishes and the stress tensor is symmetric.

Cauchy Continuum with Rate-Dependent Constitutive Law

Let's assume a simple rate-dependent constitutive law for the shear stress at a point inside the shear layer:

$$\sigma_{12} = \tau_0 + H\dot{\gamma}_{12} + \xi\,(T - T_s) \tag{31}$$

where H is a mechanical hardening parameter (positive), ξ a thermal softening parameter (negative), T_s a reference temperature, and τ_0 the shear stress at steady state and reference temperature.

For a Cauchy continuum, the linear momentum balance is:

$$\sigma_{ij,j} = 0 \tag{32}$$

Inertia terms and body forces are neglected in this example. The angular momentum balance imposes the symmetry of the stress tensor, $\sigma_{\text{ij}} = \sigma_{\text{ji}}$. For a Cauchy continuum, $\gamma_{\text{ij}} = u_{i,j}$, where u_i is the displacement in the ith direction.

At steady state $T = T^* = T_s, \sigma_{12} = \sigma_{12}^* = \tau_0, \sigma_{22} = \sigma_{22}^* = \sigma_0, \dot{\gamma}_{12} = \dot{\gamma}_{12}^* = 0$, and $\dot{T}^* = 0$. This state will be stable as long as any perturbation does not grow in time. By perturbing the temperature and displacements fields at steady state, $T = T^* + \tilde{T}, u_i = u_i^* + \tilde{u}_i$) and by neglecting higher order terms, Eqs. 30, ,31, and 32 become:

$$\begin{aligned} &\tilde{\sigma}_{12} = H\ \dot{\tilde{\gamma}}_{12}^p + \xi\ \tilde{T}; \quad \frac{\partial \tilde{\sigma}_{12}}{\partial x_2} = 0; \quad \frac{\partial \tilde{\sigma}_{22}}{\partial x_2} = 0 \\ &\frac{\partial \tilde{T}}{\partial t} = c_{\mathrm{th}} \frac{\partial^2 \tilde{T}}{\partial x_2^2} + \frac{1}{\rho C} \sigma_{12}^* \dot{\tilde{\gamma}}_{12}^p \end{aligned} \tag{33}$$

The perturbations $\tilde{T}, \tilde{u}_i$, should fulfill the boundary conditions of the rock layer. Equation 33 together with the boundary conditions $\left.\frac{\partial \tilde{T}}{\partial z}\right|_{z=\pm\frac{D}{2}} = 0, \tilde{\sigma}_{12}\left(z = \pm\frac{D}{2}\right) = 0$ and $\tilde{\sigma}_{22}\left(z = \pm\frac{D}{2}\right) = 0$ form a linear system of partial differential equations, which admits solutions of the type $\tilde{u}_i = U_i e^{\mathrm{st}} \sin \frac{2\pi}{\lambda} z$, $\tilde{T} = \mathrm{T}e^{\mathrm{st}} \cos \frac{2\pi}{\lambda} z$, where s is the so-called growth coefficient and $\lambda = \frac{D}{N}, N = 1, 2, 3, \ldots$. By replacing into Eq. 33 we obtain: $s = -\frac{\xi \tau_0}{H \rho C} - \frac{4\pi^2 c_{\mathrm{th}}}{\lambda^2}$. The system is unstable when s>0 or, equivalently, when the wavelength of the perturbation is bigger than a critical wavelength $\lambda_{\mathrm{cr}}^{\mathrm{rd}}$: $\lambda > \lambda_{\mathrm{cr}}^{\mathrm{rd}} = 2\pi\sqrt{\frac{\mathrm{H}c_{\mathrm{th}}\rho C}{-\xi\tau_0}}$.

Cosserat Elastoplasticity

An elastic perfectly plastic constitutive behavior with thermal softening is assumed in this example. More advanced Cosserat constitutive models such as the Mühlhaus-Vardoulakis Cosserat plasticity model (Vardoulakis and Sulem 1995) might be used (see section "Constitutive Laws for Cosserat Continua"), but the advantage of this simple model is that analytical derivations are simple to perform, which permits a convenient comparison with the above rate-dependent model. By analogy with the Cauchy rate-dependent model presented in the previous paragraph, the yield surface is defined as:

$$F = \tau_{(12)} - \tau_0 - \xi\,(T - T_s) \ \leq 0 \tag{34}$$

where $\tau_{(\mathrm{ij})}$ denotes the symmetric part of the stress tensor. In this way, the same shear stress limit and thermal softening with the Cauchy model is retrieved if one neglects the rate-dependent term in Eq. 31. Nevertheless, because of the chosen yield surface (Eq. 34), the plastic curvatures are zero and therefore they do not contribute to the heat equation (Eq. 30). A centrosymmetric, linear elastic isotropic Cosserat medium is considered (see Eq. 21).

At steady state, we have a Cauchy continuum under homogeneous shear. In particular, $\dot{T}^* = 0, T = T^* = T_s, \tau_{(12)} = \tau_{(12)}^* = \tau_0, \tau_{[12]} = \tau_{[12]}^* = 0, m_{32} = m_{32}^* = 0$, and $\sigma_{22} = \sigma_{22}^* = \sigma_0$. This state will be stable as long as any perturbation does not grow in time. The temperature, the displacement, and the rotation fields at steady state are perturbed $\left(T = T^* + \tilde{T}, u_i = u_i^* + \tilde{u}_i, \theta_3 = \theta_3^* + \tilde{\theta}_3\right)$ as in the Cauchy case. The perturbations $\tilde{T}, \tilde{u}_i$, and $\tilde{\theta}_3$ have to fulfill the boundary conditions of the rock layer as in the Cauchy continuum case and additionally $\tilde{m}_{32}\left(z = \pm\frac{D}{2}\right) = 0$. A linear system is then formed which admits solutions of the form: $\tilde{u}_i = U_i e^{st} \sin\frac{2\pi}{\lambda}z, \tilde{\theta}_3 = \Theta_3 e^{st} \cos\frac{2\pi}{\lambda}z, \tilde{T} = \mathrm{T}e^{st} \cos\frac{2\pi}{\lambda}z$. The critical growth coefficient is then: $s = -\frac{16G\pi^4 R^2 \rho C c_{th}}{\lambda^2\left(4G\pi^2 R^2 \rho C + \left(8\pi^2 R^2 + \lambda^2\right)\xi\tau_0\right)}$ where we set $G_c = G$ for simplicity. The system is unstable when $s > 0$ or, equivalently when the wavelength of the perturbation is bigger than a critical wavelength λ_{cr}^{Cos}: $\lambda > \lambda_{cr}^{Cos} = 2\pi\sqrt{\frac{R^2(G\rho C + 2\xi\tau_0)}{-\xi\tau_0}} \approx 2\pi\sqrt{\frac{R^2 G\rho C}{-\xi\tau_0}}$. For typical values of the shear modulus, the applied shear stress at the boundary, the thermal softening parameter, and specific heat, it holds $G\rho C \gg \xi\tau_0$.

Discussion: Rate-Dependent Models Versus Cosserat Continuum

Even though both frameworks are based on different constitutive assumptions and micromechanisms, the resemblance of the expressions for the critical wavelength impels an analogy between the hardening parameter of the viscoplastic model H and the Cosserat internal length, which here is chosen equal to the mean grain radius: $Hc_{th} \sim R^2 G$. We observe that in the rate-dependent framework, a characteristic length appears from the combination of strain rate hardening and thermal diffusivity, whereas in the Cosserat framework, it is directly related to the material grain size. The hardening parameter H can be measured experimentally for a given rock and it generally decreases during shearing together with the size of the grains and the shear modulus, which also decrease due to important shearing and comminution. It is worth mentioning that the term R^2G represents the rolling stiffness of the grains, which, in comparison with the classical Cauchy continuum, rigidifies the system in the same way that the viscous term in the rate-dependent friction law does. If we take the example of a highly granulated fault gouge with a grain size of 10 μm and assuming a shear modulus $G = 300$ MPa, then for $c_{th} = 1\mathrm{mm}^2/\mathrm{s}$, the hardening parameter H is equal to $H = 0.03$ MPa s, which is in agreement with experimental measurements (Blanpied et al. 1995; Chester and Higgs 1992). Consistently, we observe the similar role of the diffusion length and of the Cosserat internal length in the control of the thickness of the localization zone.

Conclusion

In this chapter, we summarized the fundamental, balance equations of a Cosserat continuum under multiphysical couplings that involve a fluid and a solid phase. In the beginning, the fluid was considered not to react with the solid phase in order to present the basic thermo-hydro-mechanical framework. The additional terms due to fluid momentum, mass, and energy exchange were presented next in the case of chemically reactive fluids (thermo-hydro-chemo-mechanical couplings) resulting to phase transitions. Under appropriate, application-wise constitutive laws, the above equations can be used for modeling various physical systems by taking into account the size of the microstructure in a continuum mechanics framework.

Cosserat continuum has several advantages. First of all, it remedies the mesh dependency of the classical Cauchy continuum and leads to correct energy dissipation in inelasticity and softening. Notice that the latter is of particular importance in rock mechanics as it controls stress redistribution and failure, in fault mechanics as it determines the seismic energy budget, and in soil mechanics especially when nonassociative plastic flow behavior is involved. Several examples are presented in section "Numerical Advantages of Cosserat Continuum." Moreover, Cosserat continuum assures in a physical manner hourglass control in reduced integration finite element formulations.

Finally, a simple example of the adiabatic shearing of a rock layer under constant shear stress is presented in order to highlight the similarities of a rate-independent Cosserat and a rate-dependent Cauchy continuum as far it concerns strain localization conditions. Cosserat continuum is a promising framework for studying many problems where microstructure is of paramount importance. Faults is one of them (Sulem et al. 2011; Rattez et al. 2016a, b; Veveakis et al. 2013; Sulem and Stefanou 2016; Brantut et al. 2017).

References

W.B. Anderson, R.S. Lakes, Size effects due to Cosserat elasticity and surface damage iin closed-cell polymethacrylimide foam. J. Mater. Sci. **29**, 3–9 (1994)

A. Bacigalupo, L. Gambarotta, Computational two-scale homogenization of periodic masonry: characteristic lengths and dispersive waves. Comput. Methods Appl. Mech. Eng. **213–216**, 16–28 (2012)

A. Bacigalupo, L. Gambarotta, Multi-scale modelling of periodic masonry: characteristic lengths and dispersive waves, in *AIMETA 2011: 20th Conference of the Italian Association of Theoretical and Applied Mechanics*. University of Bologna. ISBN 978-88-906340-0-0 (2011)

N. Bakhvalov, G. Panasenko, *Homogenisation: Averaging Processes in Periodic Media: Mathematical Problems in the Mechanics of Composite Materials* (Springer, Berlin, 1989)

D. Baraldi, S. Bullo, A. Cecchi, Continuous and discrete strategies for the modal analysis of regular masonry. Int. J. Solids Struct. **84**, 82–98 (2016)

J.P. Bardet, I. Vardoulakis, The asymmetry of stress in granular media. Int. J. Solids Struct. **38**(2), 353–367 (2001)

J.P. Bardet, I. Vardoulakis, Reply to discussion by Dr. Katalin Bagi. Int. J. Solids Struct. **40**, 7683–7683 (2003)

D. Besdo, Ein Beitrag zur nichtlinearen Theorie des Cosserat-Kontinuums. Acta Mech. **20**, 105–131 (1974)

M.L. Blanpied, D.A. Lockner, J.D. Byerlee, Frictional slip of granite at hydrothermal conditions. J. Geophys. Res. **100**(B7), 13045 (1995)

N. Brantut, J. Sulem, A. Schubnel, Effect of dehydration reactions on earthquake nucleation: stable sliding, slow transients, and unstable slip. J. Geophys. Res. Solid Earth **116**, 1–16 (2011) July 2010

N. Brantut, I. Stefanou, J. Sulem, Dehydration-induced instabilities at intermediate depths in subduction zones. J. Geophys. Res. Solid Earth **122**(8), 6087–6107 (2017)
P. de Buhan, B. Sudret, Micropolar multiphase model for materials reinforced by linear inclusions. Eur. J. Mech. A/Solids **19**(4), 669–687 (2000)
N. Charalambakis, Homogenization techniques and micromechanics. A survey and perspectives. Appl. Mech. Rev. **63**(3), 30803 (2010)
F.M. Chester, N.G. Higgs, Multimechanism friction constitutive model for ultrafine quartz gouge at hypocentral conditions. J. Geophys. Res. **97**(B2), 1859 (1992)
O. Coussy, *Poromechanics* (Wiley, West Sussex, 2004)
R. de Borst, Possibilities and limitations of finite elements for limit analysis. Géotechnique (2), 199–210 (1984)
R. de Borst, Simulation of strain localization: a reappraisal of the Cosserat continuum. Eng. Comput. **8**(4), 317–332 (1991)
R. de Borst, L. J. Sluys, Localisation in a Cosserat continuum under static and dynamic loading conditions. Comput. Methods Appl. Mech. Eng. **90**(1–3), 805–827 (1991)
R. de Borst, L.J. Sluys, H.B. Mühlhaus, J. Pamin, Fundamental issues in finite element analyses of localization of deformation. Eng. Comput. **10**(2), 99–121 (1993)
A.-C. Eringen, *Microcontinuum Field Thoeries* (Springer Verlag, New York, 1999)
S. Forest, K. Sab, Cosserat overall modeling of heterogeneous materials. Mech. Res. Commun. **25**(4), 449–454 (1998)
S. Forest, R. Dendievel, G.R. Canova, Estimating the overall properties of heterogeneous Cosserat materials. Model. Simul. Mater. Sci. Eng. **7**(5), 829–840 (1999)
S. Forest, F. Pradel, K. Sab, Asymptotic analysis of heterogeneous Cosserat media. Int. J. Solids Struct. **38**(26–27), 4585–4608 (2001)
P. Germain, The method of virtual power in continuum mechanics. Part 2: microstructure. SIAM J. Appl. Math. **25**(3), 556–575 (1973)
M. Godio, I. Stefanou, K. Sab, J. Sulem, Cosserat elastoplastic finite elements for masonry structures. Key Eng. Mater. **624**, 131–138 (2014)
M. Godio, I. Stefanou, K. Sab, J. Sulem, Dynamic finite element formulation for Cosserat elastic plates. Int. J. Numer. Methods Eng. **101**(13), 992–1018 (2015)
M. Godio, I. Stefanou, K. Sab, J. Sulem, Multisurface plasticity for Cosserat materials: plate element implementation and validation. Int. J. Numer. Methods Eng. **108**(5), 456–484 (2016a)
M. Godio, I. Stefanou, K. Sab, J. Sulem, S. Sakji, A limit analysis approach based on Cosserat continuum for the evaluation of the in-plane strength of discrete media: application to masonry. Eur. J. Mech. A/Solids **66**, 168–192 (2016b)
R.E. Goodman, *Introduction to Rock Mechanics* (Wiley, New York, 1989), pp. 293–334
G.T. Houlsby, A.M. Puzrin, A thermomechanical framework for constitutive models for rate-independent dissipative materials. Int. J. Plast. **16**(9), 1017–1047 (2000)
H. Kanamori, E.E. Brodsky, The physics of earthquakes. Rep. Prog. Phys. **67**(8), 1429–1496 (2004)
K. Mam S. Quennehen, Réactivation de faille par géothermie profonde Implémentation d'une loi de frottement sous ABAQUS. Master Thesis, Ecole des Ponts ParisTech (2016)
R. Masiani, P. Trovalusci, Cosserat and Cauchy materials as continuum models of brick masonry. Meccanica **31**(4), 421–432 (1996)
R.D. Mindlin, Micro-structure in linear elasticity. Arch. Ration. Mech. Anal. **16**(1) (1964)
H.-B. Mühlhaus, I. Vardoulakis, The thickness of shear bands in granular materials. Géotechnique **37**(3), 271–283 (1987)
H.B. Mühlhaus, I. Vardoulakis, The thickness of shear hands in granular materials. Géotechnique **38**(3), 331–331 (1988)
P.C. Papanastasiou, I. Vardoulakis, Numerical treatment of progressive localization in relation to borehole stability. Int. J. Numer. Anal. Methods Geomech. **16**(6), 389–424 (1992)
P. Papanastasiou, A. Zervos, Wellbore stability analysis: from linear elasticity to postbifurcation modeling. Int. J. Geomech. **4**(1), 2–12 (2004)
E. Pasternak, A.V. Dyskin, H.B. Mühlhaus, Cracks of higher modes in Cosserat continua. Int. J. Fract. **140**(1), 189–199 (2006a)

E. Pasternak, A.V. Dyskin, Y. Estrin, Deformations in transform faults with rotating crustal blocks. Pure Appl. Geophys. **163**(9), 2011–2030 (2006b)

J. Pinho-da-Cruz, J.A. Oliveira, F. Teixeira-Dias, Asymptotic homogenisation in linear elasticity. Part I: mathematical formulation and finite element modelling. Comput. Mater. Sci. **45**(4), 1073–1080 (2009)

F. Pradel, K. Sab, Cosserat modelling of elastic periodic lattice structures. C. R. Acad. Sci. Ser. IIB Mech. **326**(11), 699–704 (1998)

H. Rattez, I. Stefanou, J. Sulem, E. Veveakis, T. Poulet, Insight into the different couplings involved in the weakening of faults and their effect on earthquake nucleation, in *American Geophysical Union Fall Meeting* (2016a)

H. Rattez, I. Stefanou, J. Sulem, Thermo-hydro mechanical couplings and strain localisation in 3D continua with microstructure. Part I : theory and linear stability analysis, under Prep. (2016b)

R. Rezakhani, G. Cusatis, Asymptotic expansion homogenization of discrete fine-scale models with rotational degrees of freedom for the simulation of quasi-brittle materials. J. Mech. Phys. Solids **88**, 320–345 (2016)

J.R. Rice, J.W. Rudnicki, J.D. Platt, Stability and localization of rapid shear in fluid-saturated fault gouge: 1. Linearized stability analysis. J. Geophys. Res. Solid Earth **119**(5), 4311–4333 (2014)

K. Sab, F. Pradel, Homogenisation of periodic Cosserat media. Int. J. Comput. Appl. Technol. **34**(1), 60 (2009)

E. Sánchez-Palencia, Homogenization in mechanics. A survey of solved and open problems. Rend. Sem. Mat. Univ **44**, 1–46 (1986)

C.H. Scholz, *The Mechanics of Earthquakes and Faulting*, 2nd edn. (Cambridge University Press, Cambridge, 2002)

I. Stefanou S. Alevizos, Fundamentals of bifurcation theory and stability analysis, in *Modelling of Instabilities and Bifurcation in Geomechanics, ALERT Geomaterials Doctoral School 2016*, ed. by J. Sulem, I. Stefanou, E. Papamichos, E. Veveakis (Aussois, 2016), http://alertgeomaterials.eu/data/school/2016/2016_ALERT_schoolbook.pdf

I. Stefanou, J. Sulem, Chemically induced compaction bands: triggering conditions and band thickness. J. Geophys. Res. Solid Earth **119**(2), 880–899 (2014)

I. Stefanou, J. Sulem, Existence of a threshold for brittle grains crushing strength: two-versus three-parameter Weibull distribution fitting. Granul. Matter **18**(2), 14 (2016)

I. Stefanou, J. Sulem, I. Vardoulakis, Three-dimensional Cosserat homogenization of masonry structures: elasticity. Acta Geotech. **3**(1), 71–83 (2008)

I. Stefanou, J. Sulem, I. Vardoulakis, Homogenization of interlocking masonry structures using a generalized differential expansion technique. Int. J. Solids Struct. **47**(11–12), 1522–1536 (2010)

J. Sulem, V. Famin, Thermal decomposition of carbonates in fault zones: slip-weakening and temperature-limiting effects. J. Geophys. Res. **114**(B3), B03309 (2009)

J. Sulem, I. Stefanou, Thermal and chemical effects in shear and compaction bands. Geomech. Energy Environ. **6**, 4–21 (2016)

J. Sulem, I. Vardoulakis, Bifurcation analysis of the triaxial test on rock specimens. A theoretical model for shape and size effect. Acta Mech. **83**(3–4), 195–212 (1990)

J. Sulem, I. Stefanou, E. Veveakis, Stability analysis of undrained adiabatic shearing of a rock layer with Cosserat microstructure. Granul. Matter **13**(3), 261–268 (2011)

P. Trovalusci, M. Ostoja-Starzewski, M.L. De Bellis, A. Murrali, Scale-dependent homogenization of random composites as micropolar continua. Eur. J. Mech. A/Solids **49**, 396–407 (2015)

I. Vardoulakis, *Lecture notes on Cosserat continuum mechanics with application to the mechanics of granular media* (NTU, Athens, 2009)

I. Vardoulakis, J. Sulem, *Bifurcation Analysis in Geomechanics* (Blackie, Glascow, 1995)

E. Veveakis, J. Sulem, I. Stefanou, Modeling of fault gouges with Cosserat continuum mechanics: influence of thermal pressurization and chemical decomposition as coseismic weakening mechanisms. J. Struct. Geol. **38**, 254–264 (2012)

E. Veveakis, I. Stefanou, J. Sulem, Failure in shear bands for granular materials: thermo-hydro-chemo-mechanical effects. Géotech. Lett. **3**, 31–36 (2013)

W.M. Wang, L.J. Sluys, R. De Borst, Interaction between material length scale and imperfection size for localisation phenomena in viscoplastic media. Eur. J. Mech. A/Solids **15**(3), 447–464 (1996)

A. Zervos, P. Papanastasiou, I. Vardoulakis, Modelling of localisation and scale effect in thick-walled cylinders with gradient elastoplasticity. Int. J. Solids Struct. **38**(30–31), 5081–5095 (2001)

O.C. Zienkiewicz, R.L. Taylor, J.M. Too, Reduced integration technique in general analysis of plates and shells. Int. J. Numer. Methods Eng. **3**(2), 275–290 (1971)

O.C. Zienkiewicz, R.L. Taylur, J.Z. Zhu, *The Finite Element Method* (Elsevier, Oxford, 2013)

Dispersion of Waves in Micromorphic Media and Metamaterials 20

Angela Madeo and Patrizio Neff

Contents

Abstract

In this contribution we discuss the interest of using enriched continuum models of the micromorphic type for the description of dispersive phenomena in metamaterials. Dispersion is defined as that phenomenon according to which the speed of propagation of elastic waves is not a constant, but depends on the

A. Madeo (✉)
SMS-ID, INSA-Lyon, Université de Lyon, Villeurbanne cedex, Lyon, France

Institut universitaire de France, Paris Cedex 05, Paris, France
e-mail: angela.madeo@insa-lyon.fr

P. Neff
Fakultät für Mathematik, Universität Duisburg-Essen, Essen, Germany
e-mail: patrizio.neff@uni-due.de

G. Z. Voyiadjis (ed.), *Handbook of Nonlocal Continuum Mechanics for Materials and Structures*, https://doi.org/10.1007/978-3-319-58729-5_12

wavelength of the traveling wave. In practice, all materials exhibit dispersion if one considers waves with sufficiently small wavelengths, since all materials have a discrete structure when going down at a suitably small scale. Given the discrete substructure of matter, it is easy to understand that the material properties vary when varying the scale at which the material itself is observed. It is hence not astonishing that the speed of propagation of waves changes as well when considering waves with smaller wavelengths.

In an effort directed toward the modeling of dispersion in materials with architectured microstructures (metamaterials), different linear-elastic, isotropic, micromorphic models are introduced, and their peculiar dispersive behaviors are discussed by means of the analysis of the associated dispersion curves. The role of different micro-inertias related to both independent and constrained motions of the microstructure is also analyzed. A special focus is given to those metamaterials which have the unusual characteristic of being able to stop the propagation of mechanical waves and which are usually called band-gap metamaterials. We show that, in the considered linear-elastic, isotropic case, the relaxed micromorphic model, recently introduced by the authors, is the only enriched model simultaneously allowing for the description of non-localities and multiple band-gaps in mechanical metamaterials.

Keywords

Dispersion · Microstructure · Metamaterials · Enriched continuum models · Relaxed micromorphic model · Multi-scale modeling · Gradient micro-inertia · Free micro-inertia · Complete band-gaps · Nonlocal effects

AMS 2010 subject classification: 74A10 (stress), 74A30 (non-simple materials), 74A60 (micro-mechanical theories), 74E15 (crystalline structure), 74M25 (micro-mechanics), 74Q15 (effective constitutive equations)

Introduction

The study of the dispersive behaviors of materials with respect to wave propagation is a central issue in modern mechanics. Dispersion is defined as that phenomenon for which the speed of propagation of waves in a given material changes when changing the wavelength (or, equivalently, the frequency) of the traveling wave (see, e.g., Achenbach 1973). This is a phenomenon which is observed in practically all materials as far as the wavelength of the traveling wave is small enough to interact with the heterogeneities of the material at smaller scales. Indeed, anyone knows that all materials are actually heterogeneous if considering sufficiently small scales: it suffices to go down to the scale of molecules or atoms to be aware of the discrete side of matter. It is hence not astonishing that the mechanical properties of materials are different when considering different scales and that such differences are reflected on the speed of propagation of waves.

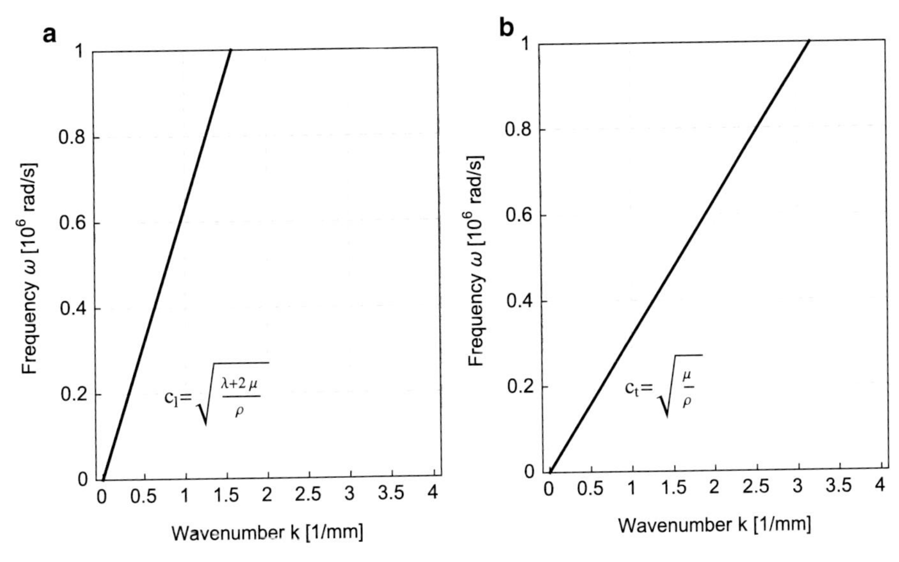

Fig. 1 Typical dispersion curves for longitudinal (**a**) and transverse (**b**) waves in (linear-elastic) Cauchy continua (λ and μ are the classical Lamé parameters and ρ is the average mass density). The slopes c_l and c_t are the phase velocities for longitudinal and transverse waves, respectively

The standard approach for the description of the mechanical behavior of materials at large scales is that of using classical Cauchy models which, by their intrinsic nature, are only capable to provide constant speeds of propagation (nondispersive behavior) depending on the elastic moduli of the material and on its mass density. Figure 1 shows the typical behavior of the dispersion curves for Cauchy continua (isotropic linear elasticity). The slope ω/k of the straight lines is called *phase velocity* and is a measure of the speed of propagation of waves in the considered continuum. As a matter of fact, this approach is very effective if one is not interested in the study of wave propagation at high frequencies, since dispersive behaviors are not activated when the wavelength is so large that the traveling wave cannot detect the presence of the underlying microstructure.

On the other hand, Cauchy continuum models are not adapted in all those situations in which dispersive behaviors cannot be neglected. This is the case, for example, when considering so-called *metamaterials*, i.e., materials with architectured microstructures which show dispersive behaviors at relatively low frequencies (large wavelengths). Metamaterials are man-made artifacts conceived arranging together small structural elements, usually in periodic or quasiperiodic patterns, in such a way that the resulting material possesses new unorthodox mechanical properties that are not shared by any classical material (see, e.g., Fig. 2). Since the characteristic sizes of microstructures which are usually encountered in mechanical metamaterials typically vary from microns to centimeters, it is not astonishing that such materials exhibit microstructure-related dispersive behaviors for wavelengths which are by far larger than those needed to unveil dispersion in more classical materials. In fact, it is not necessary to have waves with wavelengths that go down

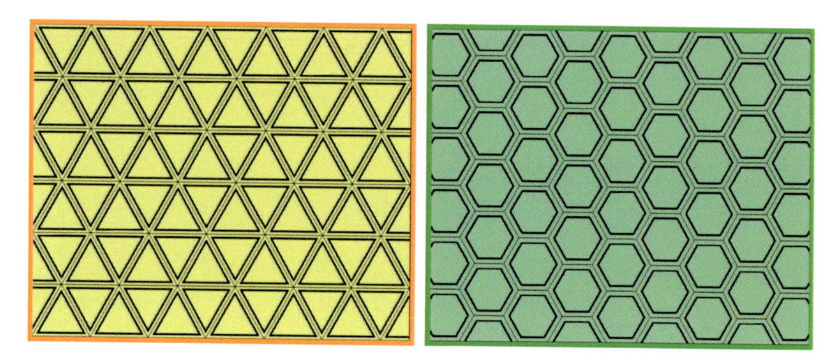

Fig. 2 Examples of periodic microstructures of metamaterials that are susceptible to unveil dispersive behaviors

at the scale of the crystals or of the molecules to start detecting dispersive behaviors. In conclusion, the phenomenon of dispersion in metamaterials cannot be neglected, even at relatively large scales, and models which are more adapted than the classical Cauchy theory need to be introduced.

In this optic, enriched continuum models (micromorphic, second gradient) have been proposed, already in the sixties by Mindlin (1963, 1964) and Eringen (1966, 1999), as suitable generalizations of Cauchy continuum mechanics to study dispersive behaviors. In particular, micromorphic models have shown their efficacy to model rather complex dynamical behaviors by accounting independently for global macroscopic vibrations of the unit cell, as well as for independent vibrations of the microstructures inside the unit cell itself. As we will show in detail, such models present dispersion patterns which are by far richer than the straight lines provided by classical linear elasticity.

As far as second gradient models are concerned, it must be said that they are able to account for some dispersion (as far as a suitable gradient micro-inertia is introduced), but they definitely do not allow for the description of all possible motions inside the unit cell: only those motions of the microstructure which are directly generated by a macro-deformation of the unit cell can be accounted for by such models. More particularly, the dispersion curves for a second gradient continuum are not so different from those of a Cauchy material presented in Fig. 1, except for the fact that they are not straight lines, thus allowing for the description of some dispersion. Nevertheless, due to the fact that the kinematics remains the same of Cauchy media (only the displacement field is introduced in second gradient theories), only two, slightly dispersive, acoustic curves can be found (see also Madeo et al. 2015, 2016d for more details) in such model. This is of course a strong limitation as far as modern metamaterials with extremely complex architectures are concerned.

Band-Gap Metamaterials

In the last years, advanced metamaterials have been conceived and manufactured which exhibit more and more unorthodox properties with respect to the propagation of mechanical and electromagnetic waves. Actually, materials which are able to "stop" or "bend" the propagation of waves of light or sound with no energetic cost have recently attracted the attention of scientists given the fact that they could suddenly disclose rapid and unimaginable technological advancements. In particular, metamaterials altering electromagnetic wave propagation are currently making the object of intense scientific research (see, e.g., Man et al. 2013; Armenise et al. 2010; Steurer and Sutter-Widmer 2007) for the potential they may have for immediate exotic applications. In fact, they might be used for rendering aircrafts or other vehicles undetectable to radar or for making objects invisible to the human eye. Many other incredible applications have been imagined for such metamaterials such as so-called super-lenses that would allow the human eye to see single viruses or nano-organisms. Notwithstanding the interest raised by such "electromagnetic metamaterials," they will not make the object of the present work which will be instead centered on the study of "mechanical metamaterials." As far as metamaterials interacting with mechanical waves are concerned, regrettably, the excitation directed toward their potential applications appears to be less intense than in the previous case, even if their possible fascinating effects on our life are only limited by our imagination. In fact, metamaterials altering elastic wave propagation could be used to build structures absorbing the higher frequencies of earthquake energy, or to conceive naval, automotive, and aeronautical vehicles that are able to absorb external solicitations and shocks thereby drastically improving their internal comfort. What's more, civil engineering structures which are built in the vicinity of sources of vibrations such as metro lines, tramways, train stations, and so forth would take advantage of the use of these metamaterials to ameliorate the enjoyment of internal and external environments. Based on the same principle, passive engineering devices perfectly able to insulate from noise could be easily conceived and produced at relatively low costs. The conception of waveguides used to optimize energy transfers by collecting waves in slabs or wires, as well as the design of wave screens employed to protect from any sort of mechanical wave, could also experience a new technological revolution. And many other unprecedented applications that, at this juncture, we cannot even envision could be abruptly disclosed once such metamaterials would become easily accessible.

In this contribution, we will focus our attention on those metamaterials which are able to "stop" wave propagation, i.e., metamaterials in which waves within precise frequency ranges cannot propagate. Such frequency intervals at which wave inhibition occurs are known as frequency band-gaps, and their intrinsic characteristics (characteristic values of the gap frequency, extension of the band-gap, etc.) strongly depend on the metamaterial microstructure. Such unorthodox dynamical behavior can be related to two main different phenomena occurring at the microlevel:

- Local resonance phenomena (Mie resonance): The microstructural components, excited at particular frequencies, start oscillating independently of the matrix so capturing the energy of the propagating wave which remains confined at the level of the microstructure. Macroscopic wave propagation thus results to be inhibited.

- Micro-diffusion phenomena (Bragg scattering): When the propagating wave has wavelengths which are small enough to start interacting with the microstructure of the material, reflection and transmission phenomena occur at the micro-level that globally result in an inhibited macroscopic wave propagation.

Such resonance and micro-diffusion mechanisms (usually a mix of the two) are at the basis of both electromagnetic and elastic band-gaps (see, e.g., Steurer and Sutter-Widmer 2007), and they are manifestly related to the particular microstructural topologies of the considered metamaterials. Indeed, it is well known (see, e.g., Steurer and Sutter-Widmer 2007; Armenise et al. 2010; Man et al. 2013) that the characteristics of the microstructures strongly influence the macroscopic band-gap behavior. As it has been said, we will be focused on mechanical waves, even if some of the used theoretical tools can be thought to be suitably generalized for the modeling of electromagnetic waves as well. Such generalizations could open new long-term research directions, for example, in view of the modeling of so-called phoxonic crystals which are simultaneously able to stop both electromagnetic and elastic wave propagation.

Band-Gap Mechanical Metamaterials and the Relaxed Micromorphic Model

As we have previously pointed out, enriched continuum models are well adapted to describe dispersion in metamaterials as well as specific motions of the microstructures inside the unit cells. Nevertheless, not all such enriched models are effective for the description of very complex metamaterials such as those exhibiting band-gap behaviors. As a matter of fact, it has been proved in a series of recent papers Madeo et al. (2014, 2015, 2016a,c,d) that the relaxed micromorphic model is the only enriched continuum model which is simultaneously able to account for:

- The description of the first (and sometimes the second) band-gap in mechanical metamaterials
- The presence of non-localities which are an intrinsic characteristic of microstructured materials, especially when high contrasts in the mechanical properties are present at the level of the microstructure.

Indeed, as we will show in the remainder of this contribution, the relaxed micromorphic model is the most suitable enriched continuum model to be used to simultaneously account for band-gaps and non-localities in mechanical metamaterials. A particular subclass of micromorphic models, the so-called internal variable models, allows for a simplified description of band-gaps, but are not able to account for non-localities related to the microscopic heterogeneities of metamaterials. Such models can be adapted for the description of a particular subclass of metamaterials that are conceived on the basis of the hypothesis of "separation of scales" which basically assumes that all the motions of the microstructure are confined inside the unit cell and do not have interactions with the motion of the adjacent cells (see, e.g., Pham et al. 2013; Sridhar et al. 2016).

It has to be underlined again that enriched continuum models of the type discussed in this paper are "macroscopic" models in the sense that they allow for the description of the averaged mechanical behavior of metamaterials with complex microstructures while remaining in the simplified framework of continuum mechanics. The main interest of using macroscopic theories for metamaterials can be found in the fact that they feature the introduction of only few parameters which are, in an averaged sense, reminiscent of the presence of the underlying microstructure. If, on the one hand, this fact provides a drastic modeling simplification which is optimal to proceed toward (meta-)structural design, some drawbacks can be reported which are mainly connected to the difficulty of directly relating the introduced macroscopic parameters to the specific characteristics of the microstructure (topology, microstructural mechanical properties, etc.). This difficulty is often seen as a limitation for the effective application of enriched continuum models. As a matter of fact, it is the authors' belief that such models are a necessary step if one wants to proceed toward the engineering design of *metastructures*, i.e., structures which are made up of metamaterials as building blocks. Of course, the proposed model will introduce a certain degree of simplification, but it is exactly this simplicity that makes possible to envision the following step which is that of proceeding toward the design of complex structures made of metamaterials.

To be more precise, as we will see, the relaxed micromorphic model is able to describe the onset of the first (and sometimes the second) band-gap in mechanical metamaterials. In order to catch more complex behaviors, the kinematics and the constitutive relations of the proposed model should be further enriched in a way that is not yet completely clear. Nevertheless, we do not see this fact as a true limitation since we intend to use the unorthodox dynamical behavior of some metamaterials exhibiting band-gaps to fit, by inverse approach, the parameters of the relaxed micromorphic model following what has been done, e.g., in Madeo et al. (2016a). This fitting, when successfully concluded for a certain number of specific metamaterials, will allow the design of metastructures by means of tools which are familiar to engineers, such as finite element codes.

Of course, as classical Cauchy models show their limits for the description of the dynamical behavior of metamaterials, even at low frequencies, the relaxed micromorphic model will show its limit for higher frequencies, yet remaining accurate enough to account for some macroscopic manifestations of microstructure.

The Fundamental Role of Micro-inertia in Enriched Continuum Models

As far as enriched continuum models are concerned, a central issue which is also an open scientific question is that of identifying the role of so-called micro-inertia terms on the dispersive behavior of such media. As a matter of fact, enriched continuum models usually provide a richer kinematics, with respect to the classical macroscopic displacement field alone, which is related to the possibility of describing the motions of the microstructure inside the unit cell. The adoption of such enriched kinematics (given by the displacement field u and the micro-distortion tensor P; see, e.g., Mindlin 1964; Eringen 1999; Ghiba et al. 2014; Madeo et al. 2014, 2015, 2016a,c,d; Neff et al. 2014b), as we will see, allows for the introduction of constitutive laws for the strain energy density that are able to describe the mechanical behavior of some metamaterials in the static regime. When the dynamical regime is considered, things become even more delicate since the choice of micro-inertia terms to be introduced in the kinetic energy density must be carefully based on:

- A compatibility with the chosen kinematics and constitutive laws used for the description of the static regime,
- The specific inertial characteristics of the metamaterial that one wants to describe (e.g., eventual coupling of the motion of the microstructure with the macro motions of the unit cell, specific resistance of the microstructure to independent motion, etc.).

It has to be explicitly mentioned that, in this contribution, we voluntarily limit ourselves to consider the linear-elastic, isotropic case with the aim of unveiling the most fundamental properties of enriched continuum models for the description of metamaterials in the dynamic regime. As a matter of fact, this simplification will allow us to understand the essence of each introduced term and its effect on the description of the mechanical behavior of particular metamaterials. Indeed, enriched continuum models of the micromorphic type featuring the description of band-gaps when introducing nonlinearities in the micro-inertia terms can also be found in the literature Stefano et al. (2011), but the interpretation of the introduced nonlinearities would be more complex to be undertaken on a phenomenological ground.

In the present contribution, for all the models that will be presented (relaxed micromorphic, Mindlin-Eringen micromorphic, internal variable, etc.), we will consider a kinetic energy of the type

$$J = \underbrace{\frac{1}{2}\rho \, \|u_{,t}\|^2}_{\text{Cauchy inertia}} + \underbrace{\frac{1}{2}\eta \, \|P_{,t}\|^2}_{\text{free micro-inertia}} + \underbrace{\frac{1}{2}\overline{\eta}_1 \, \| \operatorname{dev}\operatorname{sym} \nabla u_{,t}\|^2 + \frac{1}{2}\overline{\eta}_2 \, \| \operatorname{skew} \nabla u_{,t}\|^2 + \frac{1}{6}\overline{\eta}_3 \operatorname{tr}(\nabla u_{,t})^2}_{\text{gradient micro-inertia}}, \tag{1}$$

where ρ is the value of the average macroscopic mass density of the considered metamaterial, η is the free micro-inertia density, and the $\overline{\eta}_i, i = \{1, 2, 3\}$ are the gradient micro-inertia densities associated to the different terms of the Cartan-Lie decomposition of ∇u. The comprehension of the terms appearing in such expression of the kinetic energy density is fundamental for the correct understanding and exploitation of enriched continuum models for the description of the mechanical behavior of metamaterials in the dynamic regime.

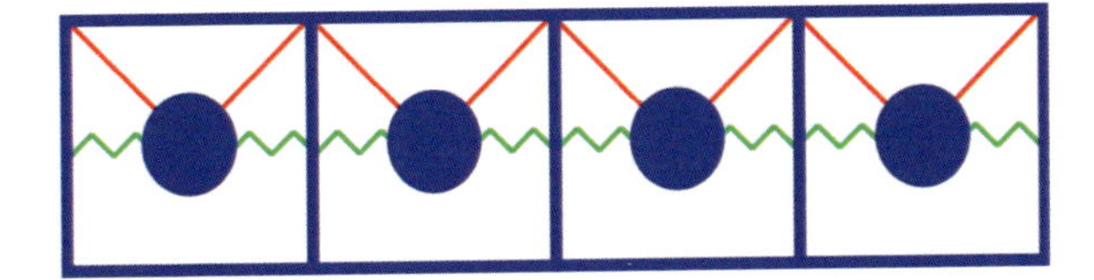

Fig. 3 Schematic representation of a system showing free and gradient micro-inertia

For this reason we discuss separately each term appearing in Eq. (1) by giving a general interpretation of its effect on the dynamics of metamaterials:

- The Cauchy inertia term $\frac{1}{2}\rho\,\|u_{,t}\|^2$ is the macroscopic inertia introduced in classical linear elasticity. It allows to describe the vibrations associated to the macroscopic displacement field. In an enriched continuum mechanical modeling framework, this means that such terms account for the inertia to vibration of the unit cells considered as material points (or representative volume elements) with apparent mass density ρ.
- The term $\frac{1}{2}\eta\,\|P_{,t}\|^2$ accounts for the inertia of the microstructure alone: with reference to Fig. 3, the free micro-inertia η can be thought to be due to the green springs connecting the blue masses inside the unit cells. The higher is the value of η, the stiffer we can imagine the green springs. We called η *free* micro-inertia (Madeo et al. 2016b) since it represents the inertia of the microstructure seen as a micro-system whose vibration can be independent of the vibration of the unit cells. An inertia term of this type is mandatory whenever one considers an enriched model of the micromorphic type, i.e., a model that features an enriched kinematics (u, P). Indeed, it would be senseless to introduce an enriched kinematics and an enriched constitutive form for the strain energy density and then avoid to introduce this free micro-inertia in the model. It would be like introducing a complex constitutive structure to describe in detail the mechanical behavior of microstructured materials while not giving to the model the possibility of activating the vibrations of such microstructures. The free micro-inertia allows us to account for the vibrations of the microstructures that typically appear for high frequencies (i.e., small wavelengths comparable with the characteristic size of the microstructure) in a huge variety of mechanical metamaterials. We explicitly mention the fact that the free micro-inertia term could take advantage of the Cartan-Lie decomposition of the micro-distortion tensor P in its dev sym (trace-free symmetric-), skew (skew-symmetric-) and tr (trace-) part as it is done for the following term. In this work we do not explicitly consider this possibility, even if we will evoke some of the effects of such a choice on the dispersion curves later on.
- The gradient micro-inertia term is of the type $\bar{\eta}\,\|\nabla u_{,t}\|^2$, and, when split using a Cartan-Lie decomposition, it takes the form shown in Eq. (1) (Madeo et al. 2016b). Such term allows to account for some specific vibrations of the microstructure which are directly related to the deformation of the unit cell at the higher scale. With reference to Fig. 3, the term $\bar{\eta}$ can be related to the effect that the red bars have on the motion of the blue masses. Increasing the value

of $\bar{\eta}$ is tantamount to make the red bars stiffer, so that we can finally say that a nonvanishing $\bar{\eta}$ constrains the motions of the microstructure to be somehow related to the macroscopic deformation of the unit cell. In fact, we can easily imagine that a deformation of the unit cell can induce a motion of the mass through the motion of the red bars. With respect to a similar metamaterial in which only the free micro-inertia η is present, we can imagine that a metamaterial with a nonvanishing gradient micro-inertia $\bar{\eta}$ is more "*local*" in the sense that the vibration of the microstructure is more confined to the unit cell than if only the green springs were present. This gradient micro-inertia term brings additional information with respect to the free micro-inertia term previously described, and this is translated in the behavior of some dispersion curves that, as we will see, can be flattened when increasing the value of $\bar{\eta}_1$, $\bar{\eta}_2$ or $\bar{\eta}_3$.

Notations

In this contribution, we denote by $\mathbb{R}^{3\times3}$ the set of real 3×3 second-order tensors, written with capital letters. We denote respectively by $\cdot$, $:$ and $\langle\cdot,\cdot\rangle$ a simple and double contraction and the scalar product between two tensors of any suitable order. (For example, $(A \cdot v)_i = A_{ij} v_j$, $(A \cdot B)_{ik} = A_{ij} B_{jk}$, $A : B = A_{ij} B_{ji}$, $(C \cdot B)_{ijk} = C_{ijp} B_{pk}$, $(C : B)_i = C_{ijp} B_{pj}$, $\langle v, w\rangle = v \cdot w = v_i w_i$, $\langle A, B\rangle = A_{ij} B_{ij}$ etc.) Everywhere we adopt the Einstein convention of sum over repeated indices if not differently specified. The standard Euclidean scalar product on $\mathbb{R}^{3\times3}$ is given by $\langle X, Y\rangle_{\mathbb{R}^{3\times3}} = \operatorname{tr}(X \cdot Y^T)$, and thus the Frobenius tensor norm is $\|X\|^2 = \langle X, X\rangle_{\mathbb{R}^{3\times3}}$. In the following we omit the index $\mathbb{R}^3, \mathbb{R}^{3\times3}$. The identity tensor on $\mathbb{R}^{3\times3}$ will be denoted by $\mathbb{1}$, so that $\operatorname{tr}(X) = \langle X, \mathbb{1}\rangle$.

We consider a body which occupies a bounded open set B_L of the three-dimensional Euclidian space $\mathbb{R}^3$ and assume that its boundary ∂B_L is a smooth surface of class C^2. An elastic material fills the domain $B_L \subset \mathbb{R}^3$, and we refer the motion of the body to rectangular axes Ox_i.

For vector fields v with components in $\mathrm{H}^1(B_L)$, i.e., $v = (v_1, v_2, v_3)^T$, $v_i \in \mathrm{H}^1(B_L)$, we define $\nabla v = ((\nabla v_1)^T, (\nabla v_2)^T, (\nabla v_3)^T)^T$, while for tensor fields P with rows in $\mathrm{H}(\operatorname{curl}; B_L)$, resp. $\mathrm{H}(\operatorname{div}; B_L)$, i.e., $P = \left(P_1^T, P_2^T, P_3^T\right)$, $P_i \in \mathrm{H}(\operatorname{curl}; B_L)$ resp. $P_i \in \mathrm{H}(\operatorname{div}; B_L)$ we define $\operatorname{Curl} P = \left((\operatorname{curl} P_1)^T, (\operatorname{curl} P_2)^T, (\operatorname{curl} P_3)^T\right)^T$, $\operatorname{Div} P = (\operatorname{div} P_1, \operatorname{div} P_2, \operatorname{div} P_3)^T$.

As for the kinematics of the considered micromorphic continua, we introduce the functions

$$\chi(X,t) : B_L \to \mathbb{R}^3, \qquad P(X,t) : B_L \to \mathbb{R}^{3\times3},$$

which are known as *placement* vector field and *micro-distortion* tensor, respectively. The physical meaning of the placement field is that of locating, at any instant t, the current position of the material particle $X \in B_L$, while the micro-distortion field describes deformations of the microstructure embedded in the material particle X

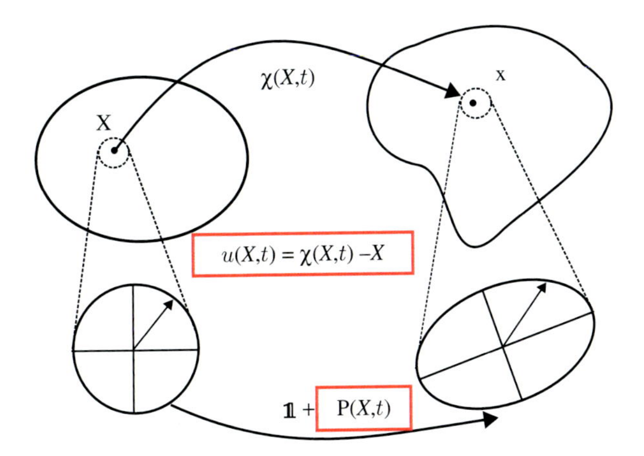

Fig. 4 Schematic representation of the enriched kinematics describing the motion of enriched models of the micromorphic type. A model of this type has 3 (components of u) + 9 (components of P) = 12 degrees of freedom (DOF)

(see Fig. 4). As it is usual in continuum mechanics, the displacement field can also be introduced as the function $u(X,t) : B_L \to \mathbb{R}^3$ defined as

$$u(X,t) = \chi(X,t) - X.$$

The Relaxed Micromorphic Model

The relaxed micromorphic model endows Mindlin-Eringen's representation with the second-order **dislocation density tensor** $\alpha = -\operatorname{Curl} P$ instead of the full gradient ∇P. (The dislocation tensor is defined as $\alpha_{ij} = -(\operatorname{Curl} P)_{ij} = -P_{ih,k}\epsilon_{jkh}$, where ϵ is the Levi-Civita tensor and Einstein notation of sum over repeated indices is used.) In the isotropic case, the energy of the relaxed micromorphic model reads (Ghiba et al. 2014; Madeo et al. 2014, 2015, 2016c; Neff et al. 2014a,b):

$$W = \underbrace{\mu_e \,\|\operatorname{sym}(\nabla u - P)\|^2 + \frac{\lambda_e}{2}(\operatorname{tr}(\nabla u - P))^2}_{\text{isotropic elastic-energy}} + \underbrace{\mu_c \,\|\operatorname{skew}(\nabla u - P)\|^2}_{\text{rotational elastic coupling}}$$

$$+ \underbrace{\mu_{\text{micro}} \,\|\operatorname{sym} P\|^2 + \frac{\lambda_{\text{micro}}}{2}(\operatorname{tr} P)^2}_{\text{micro-self-energy}} + \underbrace{\frac{\mu_e L_c^2}{2}\,\|\operatorname{Curl} P\|^2}_{\text{isotropic curvature}}, \tag{2}$$

where all the introduced constitutive parameters are assumed to be constant. The model is well-posed in the static and dynamical case including when $\mu_c = 0$; see Neff et al. (2014a) and Ghiba et al. (2014).

The relaxed micromorphic model counts 6 constitutive parameters in the isotropic case (μ_e, λ_e, μ_{micro}, λ_{micro}, μ_c, L_c). The characteristic length L_c is intrinsically related to nonlocal effects due to the fact that it weights a suitable combination of first-order space derivatives in the strain energy density (2). For a general presentation of the features of the relaxed micromorphic model in the anisotropic setting, we refer to Barbagallo et al. (2016).

The associated equations of motion in strong form, obtained by a classical least action principle, take the form (see Madeo et al. (2014, 2015, 2016d) and Neff et al. 2014a)

$$\rho\, u_{,tt} - \text{Div}[\mathcal{I}] = \text{Div}\,[\widetilde{\sigma}], \qquad \eta\, P_{,tt} = \widetilde{\sigma} - s - \text{Curl}\, m, \tag{3}$$

where

$$\begin{aligned}
\mathcal{I} &= \overline{\eta}_1 \,\text{dev}\,\text{sym}\, \nabla u_{,tt} + \overline{\eta}_2 \,\text{skew}\, \nabla u_{,tt} + \frac{1}{3}\overline{\eta}_3 \,\text{tr}\,(\nabla u_{,tt}), \\
\widetilde{\sigma} &= 2\,\mu_e \,\text{sym}\,(\nabla u - P) + \lambda_e \,\text{tr}\,(\nabla u - P)\,\mathbb{1} + 2\,\mu_c \,\text{skew}\,(\nabla u - P), \\
s &= 2\,\mu_{\text{micro}} \,\text{sym}\, P + \lambda_{\text{micro}} \,\text{tr}\,(P)\,\mathbb{1}, \\
m &= \mu_e L_c^2 \,\text{Curl}\, P.
\end{aligned} \tag{4}$$

The fact of adding a gradient micro-inertia in the kinetic energy (1) modifies the strong-form PDEs of the relaxed micromorphic model with the addition of the new term $\mathcal{I}$ (Madeo et al. 2016b). Of course, boundary conditions would also be modified with respect to the ones presented in Madeo et al. (2016a,d), but they will not be the object of study of the present contribution.

Dispersion Analysis

We rapidly recall in this subsection how, starting from the equations of motion in strong form for the relaxed micromorphic medium, it is possible to obtain the associated dispersion curves by following standard techniques. We start by making a plane-wave ansatz which means that we assume that all the 12 scalar components of the unknown fields (In what follows, we will not differentiate anymore the Lagrangian space variable X and the Eulerian one x. In general, such undifferentiated space variable will be denoted as $x = (x_1, x_2, x_3)^T$) $u(x,t)$ and $P(x,t)$ only depend on the component x_1 of the space variable x which is also assumed to be the direction of the traveling wave. With this unique assumption, together with the introduction of the new variables

$$\begin{aligned}
P^S &:= \frac{1}{3}\,\text{tr}\,(P), & P_{[ij]} &:= (\text{skew} P)_{ij} = \frac{1}{2}\left(P_{ij} - P_{ji}\right), \\
P^D &:= P_{11} - P^S, & P_{(ij)} &:= (\text{sym}\, P)_{ij} = \frac{1}{2}\left(P_{ij} + P_{ji}\right), \\
P^V &:= P_{22} - P_{33}, & i, j &= \{1, 2, 3\},
\end{aligned} \tag{5}$$

the equations of motions (3) can be simplified and rewritten, after suitable manipulation, as (see d'Agostino et al. (2016) and Madeo et al. (2014, 2015, 2016c) for additional details):

- A set of three equations only involving longitudinal quantities:

$$\begin{aligned}
\rho\,\ddot{u}_1\,\frac{2\,\overline{\eta}_1+\overline{\eta}_3}{3}\,\ddot{u}_{1,11} &= (2\,\mu_e+\lambda_e)\,u_{1,11} - 2\mu_e\,P^D_{,1} - (2\mu_e+3\lambda_e)\,P^S_{,1}\,,\\
\eta\,\ddot{P}^D &= \frac{4}{3}\,\mu_e\,u_{1,1} + \frac{1}{3}\,\mu_e L_c^2\,P^D_{,11} - \frac{2}{3}\,\mu_e L_c^2 P^S_{,11} - 2\,(\mu_e+\mu_{\text{micro}})\;P^D\,,\\
\eta\,\ddot{P}^S &= \frac{2\,\mu_e+3\,\lambda_e}{3}\,u_{1,1} - \frac{1}{3}\,\mu_e L_c^2 P^D_{,11} + \frac{2}{3}\,\mu_e L_c^2 P^S_{,11}\\
&\quad - (2\,\mu_e+3\,\lambda_e+2\,\mu_{\text{micro}}+3\,\lambda_{\text{micro}})\;P^S\,,
\end{aligned} \tag{6}$$

- Two sets of three equations only involving transverse quantities in the ξ-th direction, with $\xi = 2, 3$:

$$\begin{aligned}
\rho\,\ddot{u}_\xi\,\frac{\overline{\eta}_1+\overline{\eta}_2}{2}\,\ddot{u}_{\xi,11} &= (\mu_e+\mu_c)\,u_{\xi,11} - 2\,\mu_e\,P_{(1\xi),1} + 2\,\mu_c\,P_{[1\xi],1},\\
\eta\,\ddot{P}_{(1\xi)} &= \mu_e\,u_{\xi,1} + \frac{1}{2}\,\mu_e L_c^2\,P_{(1\xi),11} + \frac{1}{2}\,\mu_e L_c^2\,P_{[1\xi],11}\\
&\quad - 2\,(\mu_e+\mu_{\text{micro}})\;P_{(1\xi)},\\
\eta\,\ddot{P}_{[1\xi]} &= -\mu_c\,u_{\xi,1} + \frac{1}{2}\,\mu_e L_c^2\,P_{(1\xi),11} + \frac{1}{2}\,\mu_e L_c^2 P_{[1\xi],11} - 2\,\mu_c\,P_{[1\xi]},
\end{aligned} \tag{7}$$

- One equation only involving the variable $P_{(23)}$:

$$\eta\,\ddot{P}_{(23)} = -2\,(\mu_e+\mu_{\text{micro}})\,P_{(23)} + \mu_e L_c^2 P_{(23),11}, \tag{8}$$

- One equation only involving the variable $P_{[23]}$:

$$\eta\,\ddot{P}_{[23]} = -2\,\mu_c\,P_{[23]} + \mu_e L_c^2 P_{[23],11},$$

- One equation only involving the variable P^V:

$$\eta\,\ddot{P}^V = -2\,(\mu_e+\mu_{\text{micro}})\,P^V + \mu_e L_c^2 P^V_{,11}. \tag{9}$$

Once that this simplified form of the equations of motion is obtained, we look for a wave-form solution of the type

$$v_1 = \beta_1 \, e^{i(kx_1-\omega t)}, \qquad v_2 = \beta_2 \, e^{i(kx_1-\omega t)}, \qquad v_3 = \beta_3 \, e^{i(kx_1-\omega t)},$$
$$v_4 = \beta_4 \, e^{i(kx_1-\omega t)}, \qquad v_5 = \beta_5 \, e^{i(kx_1-\omega t)}, \qquad v_6 = \beta_6 \, e^{i(kx_1-\omega t)}, \tag{10}$$

where $\beta_1, \beta_2, \beta_3 \in \mathbb{C}^3$ and $\beta_4, \beta_5, \beta_6 \in \mathbb{C}$ are the unknown amplitudes of the considered waves, ω is the frequency, and k is the wavenumber and where, for compactness, we set

$$v_1 := \left(u_1, P_{11}^D, P^S\right), \qquad v_2 := \left(u_2, P_{(12)}, P_{[12]}\right), \qquad v_3 := \left(u_3, P_{(13)}, P_{[13]}\right),$$
$$v_4 := P_{(23)}, \qquad v_5 := P_{[23]}, \qquad v_6 := P^V. \tag{11}$$

Replacing the wave-form (10) and (11) in the equations of motion (6), (7), (8) and (9) and simplifying, we end up with the following systems of algebraic equations:

$$A_1(\omega,k)\cdot\beta_1 = 0, \qquad A_\tau(\omega,k)\cdot\beta_\tau = 0, \quad \tau = 2,3, \qquad A_4(\omega,k)\cdot\alpha = 0, \tag{12}$$

where we set $\alpha = (\beta_4, \beta_5, \beta_6)$ and

$$A_1(\omega,k) = \begin{pmatrix} -\omega^2\left(1+k^2\,\frac{2\overline{\eta}_1+\overline{\eta}_3}{3\rho}\right)+c_p^2\,k^2 & i\,k\,2\,\mu_e/\rho & i\,k\,\left(2\,\mu_e+3\,\lambda_e\right)/\rho \\ -i\,k\,\frac{4}{3}\,\mu_e/\eta & -\omega^2+\frac{1}{3}k^2c_m^2+\omega_s^2 & -\frac{2}{3}\,k^2c_m^2 \\ -\frac{1}{3}\,i\,k\,\left(2\,\mu_e+3\,\lambda_e\right)/\eta & -\frac{1}{3}\,k^2\,c_m^2 & -\omega^2+\frac{2}{3}\,k^2\,c_m^2+\omega_p^2 \end{pmatrix},$$

$$A_2(\omega,k) = A_3(\omega,k)$$
$$= \begin{pmatrix} -\omega^2\left(1+k^2\,\frac{\overline{\eta}_1+\overline{\eta}_2}{2\rho}\right)+k^2c_s^2 & i\,k\,2\,\mu_e/\rho & -i\,\frac{k}{\rho}\,\eta\,\omega_r^2, \\ -i\,k\,\mu_e/\eta, & -\omega^2+\frac{c_m^2}{2}k^2+\omega_s^2 & \frac{c_m^2}{2}k^2 \\ \frac{i}{2}\,\omega_r^2\,k & \frac{c_m^2}{2}k^2 & -\omega^2+\frac{c_m^2}{2}k^2+\omega_r^2 \end{pmatrix},$$
$$\tag{13}$$
$$A_4(\omega,k) = \begin{pmatrix} -\omega^2+c_m^2\,k^2+\omega_s^2 & 0 & 0 \\ 0 & -\omega^2+c_m^2\,k^2+\omega_r^2 & 0 \\ 0 & 0 & -\omega^2+c_m^2\,k^2+\omega_s^2 \end{pmatrix}.$$

In the definition of the matrices A_i, $i = \{1,2,3,4\}$, the following characteristic quantities have also been introduced:

$$\omega_s = \sqrt{\frac{2\,(\mu_e + \mu_{\text{micro}})}{\eta}}, \quad \omega_r = \sqrt{\frac{2\,\mu_c}{\eta}}, \quad \omega_p = \sqrt{\frac{(3\,\lambda_e + 2\,\mu_e) + (3\,\lambda_{\text{micro}} + 2\,\mu_{\text{micro}})}{\eta}},$$

$$c_m = \sqrt{\frac{\mu_e\, L_c^2}{\eta}}, \qquad c_p = \sqrt{\frac{\lambda_e + 2\mu_e}{\rho}}, \qquad c_s = \sqrt{\frac{\mu_e + \mu_c}{\rho}}. \tag{14}$$

Dispersion Curves for the Relaxed Micromorphic Model

The dispersion curves for the relaxed micromorphic model are the solutions $\omega = \omega(k)$ of the algebraic equations

$$\underbrace{\det A_1(\omega, k) = 0,}_{\text{longitudinal}} \qquad \underbrace{\det A_2(\omega, k) = \det A_3(\omega, k) = 0,}_{\text{transverse}} \qquad \underbrace{\det A_4(\omega, k) = 0.}_{\text{uncoupled}} \tag{15}$$

In order to plot such curves, we chose the values of the constitutive parameters to be those in Table 1.

These numerical values will be also kept for the simulations relative to the other models in the subsequent sections if not differently specified. (We explicitly remark that the values of the elastic coefficients must be chosen in such a way to respect the conditions of positive definiteness of the strain energy density (Madeo et al. 2014, 2015), i.e., $\mu_e > 0$, $\mu_{\text{micro}} > 0$, $3\lambda_e + 2\mu_e > 0$, $3\lambda_{\text{micro}} + 2\mu_{\text{micro}} > 0$, $\mu_c \geq 0$). We explicitly remark that the chosen values correspond to a relatively soft material, but completely analogous results can be obtained with any other combination of the parameters with the only constraint of respecting positive definiteness of the strain energy density.

We start by presenting in Fig. 5 the dispersion curves for the relaxed micromorphic model obtained with vanishing gradient micro-inertia $\overline{\eta}$. In the figures presented thereafter, we adopt the following acronyms:

Table 1 Values of the parameters used in the numerical simulations (left) and corresponding values of the Lamé parameters and of the Young modulus and Poisson ratio (right). For the formulas needed to calculate the homogenized macroscopic parameters starting from the microscopic ones see Barbagallo et al. (2016).

Parameter	Value	Unit
μ_e	200	MPa
$\lambda_e = 2\mu_e$	400	MPa
$\mu_c = 5\mu_e$	1000	MPa
μ_{micro}	100	MPa
λ_{micro}	100	MPa
L_c	1	mm
ρ	2000	kg/m^3
η	10^{-2}	kg/m^3
$\overline{\eta}$	10^{-1}	kg/m^3

Parameter	Value	Unit
λ_{macro}	82.5	MPa
μ_{macro}	66.7	MPa
E_{macro}	170	MPa

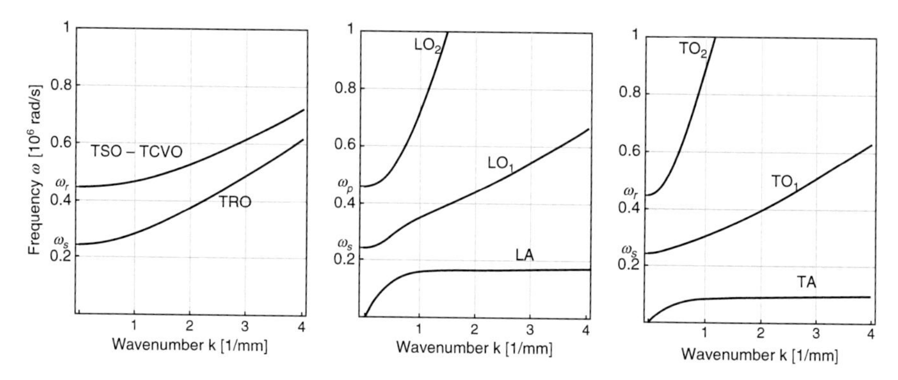

Fig. 5 Dispersion relations $\omega = \omega(k)$ of the **relaxed micromorphic model** for the uncoupled (*left*), longitudinal (*center*), and transverse (*right*) waves with nonvanishing free micro-inertia $\eta \neq 0$ and vanishing gradient micro-inertia $\overline{\eta} = 0$

- TRO: transverse rotational optic,
- TSO: transverse shear optic,
- TCVO: transverse constant-volume optic,
- LA: longitudinal acoustic,
- LO_1-LO_2: 1st and 2nd longitudinal optic,
- TA: transverse acoustic,
- TO_1-TO_2: 1st and 2nd transverse optic.

It can be seen from Fig.5 that the main characteristic feature of the relaxed micromorphic model is that of having a horizontal asymptote for the acoustic waves. It is exactly this feature that allows the relaxed model to account for the description of band-gaps in nonlocal band-gap metamaterials. Indeed, a complete band-gap can be identified in the interval of frequencies individuated by the horizontal asymptote of the longitudinal acoustic wave LA and the characteristic frequency ω_s. In such interval, the wavenumber becomes purely imaginary for all the considered waves (longitudinal, transverse, and uncoupled) which means that the wave can actually not propagate.

Now, we show in Fig. 6 the results obtained for nonvanishing gradient micro-inertia $\overline{\eta} \neq 0$ (here and in the sequel, we suppose in the expression (1) to have $\overline{\eta}_1 = 0$, $\overline{\eta}_2 = \overline{\eta}_3 = \overline{\eta}$, if not differently specified). The effect of the addition of such gradient micro-inertia term is that of flattening the optic curves so allowing for the possibility of the onset of a second longitudinal and transverse band-gap if the gradient inertia parameters are sufficiently high. The fact of having split the gradient micro-inertia in three parts allows to flatten the longitudinal and transverse waves separately. Moreover, we remark that the addition of gradient micro-inertiae $\overline{\eta}_1$, $\overline{\eta}_2$, and $\overline{\eta}_3$ has no effect on the cutoff frequencies ω_s, ω_p, and ω_r which only depend on the free micro-inertia η (and of course on the constitutive parameters). Indeed, if an equivalent Cartan-Lie decomposition of the free micro-inertia term would have been performed, the result would have been that of gaining the independence of the

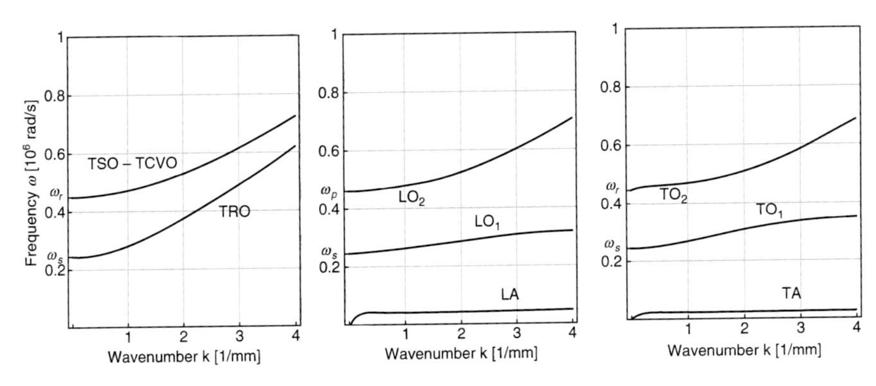

Fig. 6 Dispersion relations $\omega = \omega(k)$ of the **relaxed micromorphic model** for the uncoupled (*left*), longitudinal (*center*), and transverse (*right*) waves with nonvanishing free micro-inertia $\eta \neq 0$ and nonvanishing gradient micro-inertia $\bar{\eta} \neq 0$

three cutoff frequencies which could be then translated independently one from the other along the y-axis.

It is worth to explicitly remark that the observed flattening effect of the micro-inertia $\bar{\eta}$ can be interpreted as the fact that some modes propagate more slowly when $\bar{\eta}$ is nonvanishing. Indeed, it is well known (Achenbach 1973) that the slope of the dispersion curves in a point is a measure of the speed of propagation of the considered waves (phase velocity). With this definition in mind, it is easy to infer by direct comparison of Figs. 5 and 6 that the longitudinal and transverse waves are much slower in the second case, when $\bar{\eta} \neq 0$. On the other hand, we can also remark that the modes corresponding to the uncoupled waves are not affected by the introduction of the gradient micro-inertia $\bar{\eta}$.

With reference to previous works Madeo et al. (2014, 2015, 2016c,d), it is possible to observe that the oblique asymptotes of the waves LO_1, TO_1 as well as the asymptotes of the uncoupled waves are directly proportional to the value of the characteristic length L_c. This is equivalent to say that such waves propagate, thanks to the non-localities that are intrinsic of the proposed relaxed micromorphic model. When the gradient micro-inertia is introduced, the waves LO_1, TO_1 are flattened, so that the effect of non-localities results to be reduced. The non-local behavior on the uncoupled waves, instead, is not affected by the introduction of the gradient micro-inertia.

The Internal Variable Model

We recall (see Neff et al. 2014b) that the energy for the internal variable model does not include higher space derivatives of the micro-distortion tensor P and, in the isotropic case, takes the form

$$W = \underbrace{\mu_e \, \| \operatorname{sym} (\nabla u - P) \|^2 + \frac{\lambda_e}{2} (\operatorname{tr} (\nabla u - P))^2}_{\text{isotropic elastic-energy}} + \underbrace{\mu_c \, \| \operatorname{skew} (\nabla u - P) \|^2}_{\text{rotational elastic coupling}} \tag{16}$$

$$+ \underbrace{\mu_{\text{micro}} \, \| \operatorname{sym} P \|^2 + \frac{\lambda_{\text{micro}}}{2} (\operatorname{tr} P)^2}_{\text{micro-self-energy}},$$

Due to the absence of space derivatives of the micro-distortion tensor P in the strain energy density, such model intrinsically excludes the possibility of modeling non-localities. If, in a first instance, this can be a reasonable simplification for a certain class of metamaterials for which the "separation of scale" hypothesis can be verified (Pham et al. 2013; Sridhar et al. 2016), the fact of neglecting the effect of the deformation of the unit cell on the adjacent ones can be too restrictive, especially when metamaterials with high contrasts of the mechanical properties at the micro-level are considered.

The equations of motion for the internal variable model, obtained as the result of a least action principle based on the strain energy density (16) and the kinetic energy (1), are

$$\rho \, u_{,tt} - \operatorname{Div}[\mathcal{I}] = \operatorname{Div} [\widetilde{\sigma}], \qquad \eta \, P_{,tt} = \widetilde{\sigma} - s, \tag{17}$$

where

$$\begin{aligned} \mathcal{I} &= \overline{\eta}_1 \, \operatorname{dev} \operatorname{sym} \nabla u_{,tt} + \overline{\eta}_2 \, \operatorname{skew} \nabla u_{,tt} + \frac{1}{3} \overline{\eta}_3 \operatorname{tr} (\nabla u_{,tt}), \\ \widetilde{\sigma} &= 2 \, \mu_e \, \operatorname{sym} (\nabla u - P) + \lambda_e \, \operatorname{tr} (\nabla u - P) \, \mathbb{1} + 2 \, \mu_c \, \operatorname{skew} (\nabla u - P), \\ s &= 2 \, \mu_{\text{micro}} \, \operatorname{sym} P + \lambda_{\text{micro}} \, \operatorname{tr} (P) \, \mathbb{1}. \end{aligned} \tag{18}$$

We present the dispersion relations obtained for the internal variable model with a vanishing gradient inertia (Fig. 7) and for a nonvanishing gradient inertia (Fig. 8). We start by noticing in Fig. 7 that, suitably choosing the value of the cut-off frequencies, the internal variable model allows for the description of up to 2 band-gaps. The pattern of longitudinal and transverse waves, even if not including non-localities, still remains realistic for at least some particular metamaterials. On the other hand, this "absence of non-locality" drastically affects the uncoupled waves which become perfectly horizontal lines (local resonances of some modes inside the unit cell), which can be a too drastic simplification for three-dimensional metamaterials.

Figure 8 shows the effect of the addition of the gradient micro-inertia $\overline{\eta} \| \nabla u_{,t} \|^2$ on the internal variable model. By direct observation of Fig. 8, we can notice that suitably choosing the relative position of ω_r and ω_p, the internal variable model allows to account for 3 band-gaps. We thus have an extra band-gap with respect to

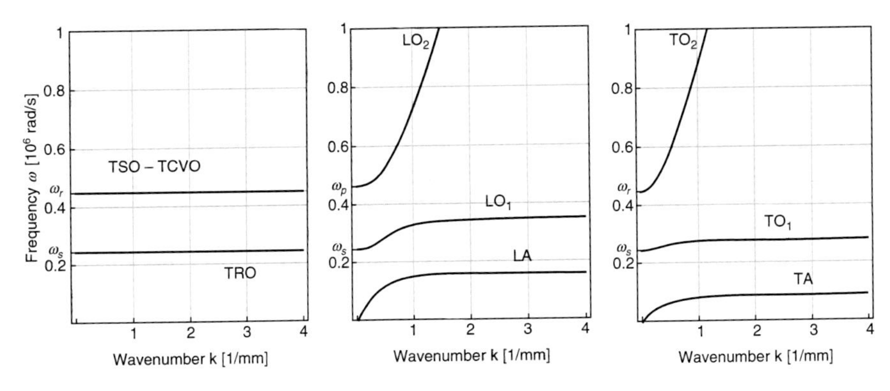

Fig. 7 Dispersion relations $\omega = \omega(k)$ of the **internal variable model** for the uncoupled (*left*), longitudinal (*center*) and transverse (*right*) waves with nonvanishing free micro-inertia $\eta \neq 0$ and vanishing gradient micro-inertia $\bar{\eta} = 0$

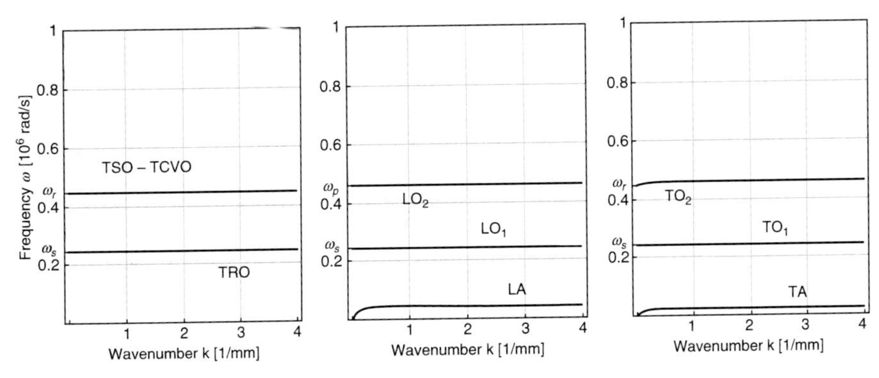

Fig. 8 Dispersion relations $\omega = \omega(k)$ of the **internal variable model** for the uncoupled (*left*), longitudinal (*center*), and transverse (*right*) waves with nonvanishing free micro-inertia $\eta \neq 0$ and nonvanishing gradient micro-inertia $\bar{\eta} \neq 0$

the case with vanishing gradient inertia (Fig. 7) and to the analogous case for the relaxed micromorphic model (see Fig. 6), but we are not able to consider non-local effects. The characteristic behavior of the dispersion curves becomes unrealistic in this case, given that almost only local resonances (dispersion curves which become horizontal straight lines) are provided. It turns out that the fact of excluding the possibility of describing non-local effects in metamaterials becomes too restrictive and unphysical for the description of the great majority of metamaterials.

The Standard Mindlin-Eringen Model

In this section we discuss the effect on the Mindlin-Eringen model of the addition of the gradient micro-inertia $\overline{\eta}\| \nabla u_{,t} \|^2$ to the classical terms $\rho\|u_{,t}\|^2 + \eta\|P_{,t}\|^2$. We recall that the strain energy density for this model in the isotropic case takes the form

$$W = \underbrace{\mu_e \,\|\, \mathrm{sym}\,(\nabla u - P)\|^2 + \frac{\lambda_e}{2}\,(\mathrm{tr}\,(\nabla u - P))^2}_{\text{isotropic elastic-energy}} + \underbrace{\mu_c \,\|\, \mathrm{skew}\,(\nabla u - P)\|^2}_{\text{rotational elastic coupling}}$$
$$+ \underbrace{\mu_{\text{micro}} \,\|\, \mathrm{sym}\, P\,\|^2 + \frac{\lambda_{\text{micro}}}{2}\,(\mathrm{tr} P)^2}_{\text{micro-self-energy}} + \underbrace{\frac{\mu_e L_c^2}{2}\,\|\nabla\, P\,\|^2}_{\text{isotropic curvature}}, \tag{19}$$

The dynamical equilibrium equations are

$$\rho\, u_{,tt} - \mathrm{Div}[\mathcal{I}\,] = \mathrm{Div}\,[\widetilde{\sigma}\,], \qquad \eta\, P_{,tt} = \widetilde{\sigma} - s - M, \tag{20}$$

where

$$\begin{aligned}
\mathcal{I} &= \overline{\eta}_1 \;\mathrm{dev}\,\mathrm{sym}\;\nabla u_{,tt} + \overline{\eta}_2 \;\mathrm{skew}\,\nabla u_{,tt} + \frac{1}{3}\overline{\eta}_3\,\mathrm{tr}\,(\nabla u_{,tt}),\\
\widetilde{\sigma} &= 2\,\mu_e\;\mathrm{sym}\;(\nabla u - P) + \lambda_e\;\mathrm{tr}\,(\nabla u - P)\;\mathbb{1} + 2\,\mu_c\;\mathrm{skew}\,(\nabla u - P),\\
s &= 2\,\mu_{\text{micro}}\;\mathrm{sym}\,P + \lambda_{\text{micro}}\;\mathrm{tr}\,(P)\;\mathbb{1},\\
M &= -\mu_e L_c^2\; \underbrace{\mathrm{Div}\nabla P}_{=\Delta P}.
\end{aligned} \tag{21}$$

Recalling the results of Madeo et al. (2014), we remark that when the gradient micro-inertia is vanishing ($\overline{\eta}_1 = \overline{\eta}_2 = \overline{\eta}_3 = 0$), the Mindlin-Eringen model does not allow the description of band-gaps (see Fig. 9), due to the presence of a straight acoustic wave. On the other hand, when switching on the parameters $\overline{\eta}_2$ and $\overline{\eta}_3$, some optic branches are flattened, so that a band-gap can be created suitably choosing the values of the characteristic cutoff frequencies (see Fig. 10). The analogous case for the relaxed micromorphic model (Fig. 5) allowed instead for the description of 2 band-gaps.

In conclusion, the classical Mindlin-Eringen model is suitable to account for dispersion, and for non-localities in metamaterials, it shows its limits for the description of band-gaps in metamaterials. In fact, the presence of a gradient micro-inertia is needed to create a single band-gap, while the relaxed micromorphic model is able to account for 2 band-gaps in the analogous case (see Fig. 6). This means that the relaxed micromorphic model is most suitable to model mechanical metamaterials exhibiting band-gaps also when they are manufactured in such a way that the gradient micro-inertia becomes non-negligible. Indeed, the description of the second band-gap occurring at higher frequencies becomes possible for the relaxed micromorphic model, but not for the Mindlin-Eringen one when considering suitable additional micro-inertia terms.

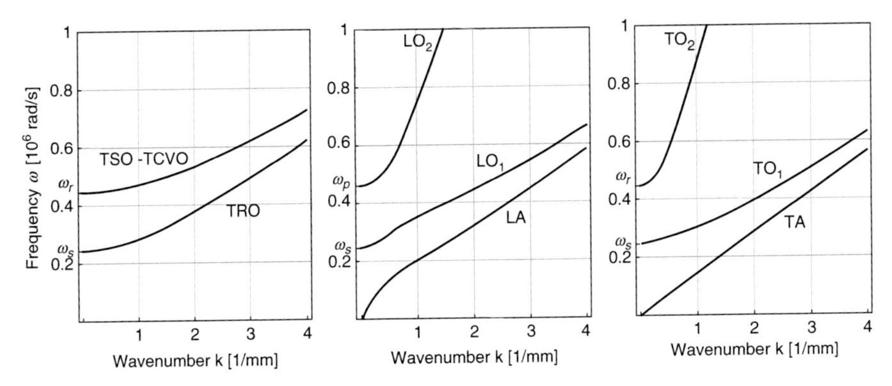

Fig. 9 Dispersion relations $\omega = \omega(k)$ of the **standard Mindlin-Eringen micromorphic model** for the uncoupled (*left*), longitudinal (*center*), and transverse (*right*) waves with nonvanishing free micro-inertia $\eta \neq 0$ and vanishing gradient micro-inertia $\bar{\eta} = 0$

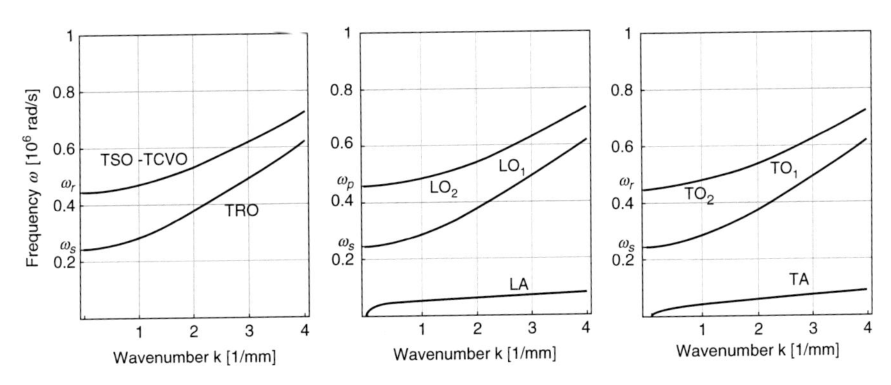

Fig. 10 Dispersion relations $\omega = \omega(k)$ of the **standard Mindlin-Eringen micromorphic model** for the uncoupled (*left*), longitudinal (*center*), and transverse (*right*) waves with nonvanishing free micro-inertia $\eta \neq 0$ and nonvanishing gradient micro-inertia $\bar{\eta} \neq 0$

The Micromorphic Model with Curvature $\|\mathrm{Div}P\|^2 + \|\mathrm{Curl}P\|^2$

The classical Mindlin-Eringen model can be reformulated by introducing the energy (see Madeo et al. 2016c):

$$W = \underbrace{\mu_e \,\| \operatorname{sym}(\nabla u - P)\|^2 + \frac{\lambda_e}{2}(\operatorname{tr}(\nabla u - P))^2}_{\text{isotropic elastic-energy}} + \underbrace{\mu_c \,\| \operatorname{skew}(\nabla u - P)\|^2}_{\text{rotational elastic coupling}}$$

$$+ \underbrace{\mu_{\text{micro}} \,\| \operatorname{sym} P \|^2 + \frac{\lambda_{\text{micro}}}{2}(\operatorname{tr} P)^2}_{\text{micro-self-energy}} + \underbrace{\frac{\mu_e L_c^2}{2}\left(\|\operatorname{Div} P\|^2 + \|\operatorname{Curl} P\|^2\right)}_{\text{augmented isotropic curvature}}, \tag{22}$$

in which the gradient of the micro-distortion tensor is decomposed in its Curl and Div part. The dynamical equilibrium equations are

$$\rho\, u_{,tt} - \underbrace{\mathrm{Div}[\mathcal{I}]}_{\text{new augmented term}} = \mathrm{Div}\,[\widetilde{\sigma}]\,, \qquad \eta\, P_{,tt} = \widetilde{\sigma} - s - M, \tag{23}$$

where

$$\begin{aligned}
\mathcal{I} &= \overline{\eta}_1\ \mathrm{dev\,sym}\ \nabla u_{,tt} + \overline{\eta}_2\ \mathrm{skew}\,\nabla u_{,tt} + \frac{1}{3}\overline{\eta}_3\,\mathrm{tr}\,(\nabla u_{,tt})\,,\\
\widetilde{\sigma} &= 2\,\mu_e\ \mathrm{sym}\,(\nabla u - P) + \lambda_e\ \mathrm{tr}\,(\nabla u - P)\,\mathbb{1} + 2\,\mu_c\ \mathrm{skew}\,(\nabla u - P)\,,\\
s &= 2\,\mu_{\mathrm{micro}}\ \mathrm{sym}\,P + \lambda_{\mathrm{micro}}\ \mathrm{tr}\,(P)\,\mathbb{1},\\
M &= -\mu_e L_c^2\ \underbrace{(\nabla\,(\mathrm{Div}\,P) - \mathrm{Curl}\,\mathrm{Curl}\,P)}_{\mathrm{Div}\nabla P = \Delta P}\,.
\end{aligned} \tag{24}$$

Simply rearranging the terms appearing in the definition of the tensor M, it can be remarked that the structure of the equation is equivalent to the one obtained in the standard micromorphic model with curvature $\frac{1}{2}\|\nabla P\|^2$; see Eq. (20) in section "The Standard Mindlin-Eringen Model." The results obtained for such case are thus completely superposable to those presented before for the classical Mindlin-Eringen micromorphic model (see Figs. 9 and 10).

The Micromorphic Model with Curvature $\|\mathrm{Div}P\|^2$

We have shown in the previous sections that the relaxed micromorphic model with curvature $\|\,\mathrm{Curl}\,P\,\|^2$ is an enriched model which allows to simultaneously account for non-localities and multiple band-gaps in mechanical metamaterials. We also showed that the full Mindlin-Eringen model with curvature $\|\nabla P\|^2$ can be equivalently reformulated with a curvature term of the type $\|\,\mathrm{Curl}\,P\,\|^2 + \|\mathrm{Div}\,P\,\|^2$. We concluded that the relaxed micromorphic model is well-adapted for the description of band-gap metamaterials, while the classical Mindlin-Eringen model, if equally suitable for describing dispersion and non-locality, is not performant for the modeling of band-gap metamaterials to the same extent. In this section we explore another possible micromorphic model with curvature term $\|\mathrm{Div}\,P\,\|^2$ (see Madeo et al. 2016c) which thus features an energy of the type

$$\begin{aligned}
W = &\underbrace{\mu_e\,\|\,\mathrm{sym}\,(\nabla u - P)\|^2 + \frac{\lambda_e}{2}\,(\mathrm{tr}\,(\nabla u - P))^2}_{\text{isotropic elastic-energy}} + \underbrace{\mu_c\,\|\,\mathrm{skew}\,(\nabla u - P)\|^2}_{\text{rotational elastic coupling}}\\
&+ \underbrace{\mu_{\mathrm{micro}}\,\|\,\mathrm{sym}\,P\,\|^2 + \frac{\lambda_{\mathrm{micro}}}{2}\,(\mathrm{tr}P)^2}_{\text{micro-self-energy}} + \underbrace{\frac{\mu_e L_d^2}{2}\,\|\mathrm{Div}\,P\,\|^2}_{\text{isotropic curvature}}\,.
\end{aligned} \tag{25}$$

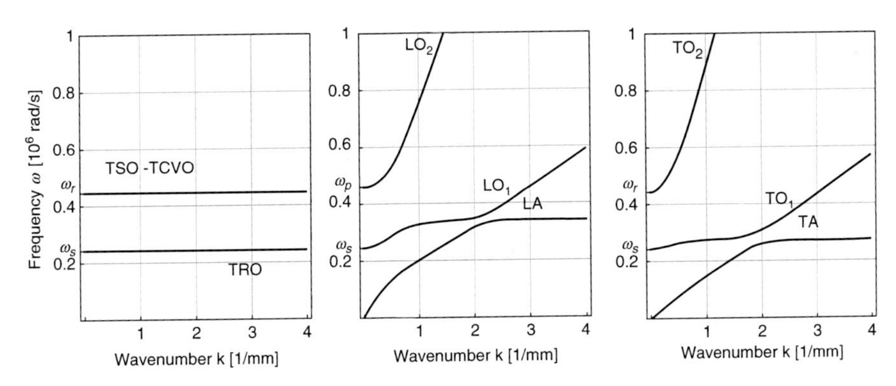

Fig. 11 Dispersion relations $\omega = \omega(k)$ of the **relaxed micromorphic model with curvature** $\|\text{Div} P\|^2$ for the uncoupled (*left*), longitudinal (*center*), and transverse (*right*) waves with nonvanishing free micro-inertia $\eta \neq 0$ and vanishing gradient micro-inertia $\overline{\eta} = 0$

The dynamical equilibrium equations are

$$\rho\, u_{,tt} - \text{Div}[\mathcal{I}] = \text{Div}\left[\widetilde{\sigma}\right], \qquad \eta\, P_{,tt} = \widetilde{\sigma} - s - M, \tag{26}$$

where

$$\begin{aligned}
\mathcal{I} &= \overline{\eta}_1 \text{ dev sym } \nabla u_{,tt} + \overline{\eta}_2 \text{ skew } \nabla u_{,tt} + \frac{1}{3}\overline{\eta}_3 \operatorname{tr}(\nabla u_{,tt}), \\
\widetilde{\sigma} &= 2\,\mu_e \text{ sym }(\nabla u - P) + \lambda_e \operatorname{tr}(\nabla u - P)\, \mathbb{1} + 2\,\mu_c \text{ skew }(\nabla u - P), \\
s &= 2\,\mu_{\text{micro}} \text{ sym } P + \lambda_{\text{micro}} \operatorname{tr}(P)\, \mathbb{1}, \\
M &= -\mu_e L_c^2\, \nabla\, (\text{Div} P)\,.
\end{aligned} \tag{27}$$

We start by presenting in Fig. 11 the dispersion relations obtained with a vanishing gradient micro-inertia (Fig. 11). Also in the case of the micromorphic model with only $\|\text{Div} P\|^2$, when considering a vanishing gradient micro-inertia, there always exist waves which propagate inside the considered medium independently of the value of the frequency, and the uncoupled waves assume a peculiar resonant behavior in which the frequency is independent of the wavenumber k (local resonances for some modes). The characteristic behavior observed for the longitudinal and transverse waves shows a significative coupling between the LO_1 and LA as well between as the TO_1 and TA waves, respectively. Such strong coupling does not allow for the presence of complete band-gaps. Moreover, since the non-localities introduced by the curvature $\|\text{Div} P\|^2$ are much weaker than those introduced by the term $\|\operatorname{Curl} P\|^2$, the behavior of the dispersion curves for the uncoupled waves is the same as that of a completely local model as the internal

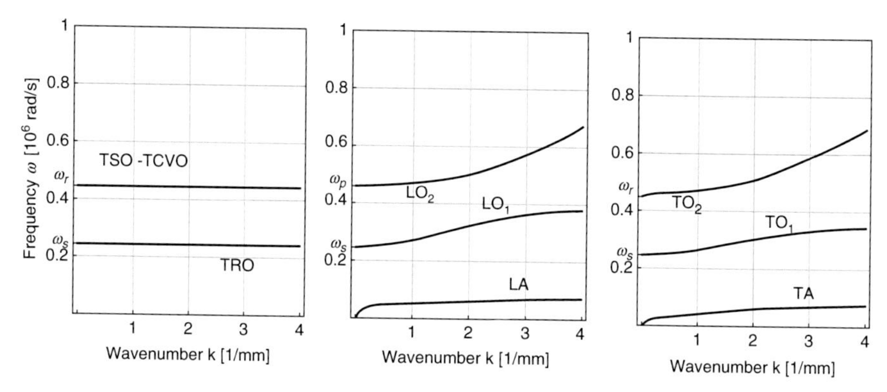

Fig. 12 Dispersion relations $\omega = \omega(k)$ of the **relaxed micromorphic model with curvature** $\|\mathrm{Div}\,P\|^2$ for the uncoupled (*left*), longitudinal (*center*), and transverse (*right*) waves with nonvanishing free micro-inertia $\eta \neq 0$ and nonvanishing gradient micro-inertia $\bar{\eta} \neq 0$

variable one (see the first of Fig. 7) and significantly deviates from the behavior observed for the relaxed micromorphic model (see the first of Fig. 5).

On the other hand, when switching on the gradient inertia (Fig. 12), a behavior analogous to the relaxed micromorphic model appears for the longitudinal and transverse waves, while the uncoupled waves definitely remain those of a local model. Thus, if the micromorphic model with curvature $\|\mathrm{Div}\,P\|^2$ can describe band-gaps when adding the gradient micro-inertia terms, it loses almost any capability of accounting for non-local effects. Indeed, the effect of the addition of the gradient micro-inertia is that of flattening the longitudinal and transverse waves which is equivalent to say that these waves become more "local." The main difference between Figs. 12 and 6 can be found in the uncoupled waves: if in the relaxed micromorphic model they are able to account for non-negligible non-local effects, this is not the case for the micromorphic model with curvature $\|\mathrm{Div}\,P\|^2$.

We can conclude that, independently of the type of micro-inertia which is considered, the micromorphic model with curvature $\|\mathrm{Div}\,P\|^2$ is not able to simultaneously account for band-gaps and non-localities, in strong contrast to what happens for the relaxed micromorphic case.

Conclusions

It has been shown in this contribution that the dispersive behaviors usually encountered in engineering metamaterials can be suitably accounted for by using enriched continuum models of the micromorphic type. The enriched kinematics that is peculiar of such enriched models allows for the introduction of extra degrees of freedom with respect to classical Cauchy continua, thus accounting for the description of independent motions of the microstructure inside the unit cells. Nevertheless, if all are suitable for modeling dispersive phenomena in microstructured materials, not all such models are equally adapted for describing the onset of band-gaps in mechanical metamaterials. The models proposed in this contribution differ one from the other basically for the expression of the curvature terms in the strain energy density, i.e., for the type of higher-order space derivatives of the micro-distortion tensor P which are accounted for.

After presenting in detail the dispersion curves associated to all the introduced isotropic, linear-elastic models, it is concluded that the relaxed micromorphic model is the most suitable enriched model which is simultaneously able to account for both band-gaps and non-localities in mechanical metamaterials.

The interest of introducing complex micro-inertia terms (gradient micro-inertia) accounting for a coupling of the motion of the microstructure with the deformation of the unit cell at the higher scale is also discussed. It is shown that the effect of adding such gradient micro-inertia is that of flattening the longitudinal and transverse dispersion curves, so that the final effect is that of lowering the speed of propagation of such waves at the macroscopic level. This can be equivalently interpreted as a "loss of non-locality" for the longitudinal and transverse waves which is substantially related to the presence of the gradient micro-inertia. As far as the uncoupled waves are concerned, their behavior is not affected by the addition of such new micro-inertia term, and the peculiar behavior of these waves for the relaxed micromorphic model makes evident the non-locality that such model is able to provide concomitantly to the presence of complete band-gaps.

The main interest of using continuum theories of the micromorphic type for the description of the behavior of materials with complex microstructures can be found in the fact that they feature the introduction of few parameters which are, in an averaged sense, reminiscent of the presence of the underlying microstructure. If, on the one hand, this fact provides a drastic modeling simplification which is optimal to proceed toward (meta-)structural design, some drawbacks can be reported which are mainly related to the difficulty of directly relating the introduced macroscopic parameters to the specific characteristics of the microstructure (topology, microstructural mechanical properties, etc.). The aforementioned difficulty of explicitly relating macro-parameters to micro-properties is often seen as a limitation for the effective application of enriched continuum models. As a matter of fact, it is the authors' belief that such models are a necessary step if one wants to proceed toward the engineering design of metastructures, i.e., structures which are made up of metamaterials as building blocks. Of course, the proposed model will introduce a certain degree of simplification, but it is exactly this simplicity that makes possible to envision the next step which is that of proceeding toward the design of complex (meta-) structures made of metamaterials.

To be more precise, the relaxed micromorphic model proposed here is able to describe the onset of the first (and sometimes the second) band-gap which occurs at lower frequencies. In order to catch more complex behaviors, the kinematics and the constitutive relations of the proposed model should be further enriched in a way that is not yet completely clear. Nevertheless, this fact has not to be seen as a limitation since it is possible to use the unorthodox dynamical behavior of some metamaterials exhibiting band-gaps to fit, by inverse approach, the parameters of the relaxed micromorphic model following what has been done, e.g., in Madeo et al. (2016a). This fitting, when successfully concluded for some specific metamaterials, will allow the setting up of the design of metastructures by means of tools which are familiar to engineers, such as finite element codes. Of course, as classical Cauchy models show their limits for the description of the dynamical behavior of metamaterials, even at low frequencies, the relaxed micromorphic model will show its limit for higher frequencies, yet remaining accurate enough for accounting for some important macroscopic manifestations of microstructure.

Acknowledgements Angela Madeo thanks the Institut Universitaire de France (IUF) for financial support, INSA-Lyon for the funding of the BQR 2016 "Caractérisation mécanique inverse des métamatériaux: modélisation, identification expérimentale des paramètres et évolutions possibles," as well as the CNRS-INSIS for the funding of the PEPS project.

References

J.D. Achenbach, *Wave Propagation in Elastic Solids* (North-Holland Publishing Company, Amsterdam, 1973)

M.N. Armenise, C.E. Campanella, C. Ciminelli, F. Dell'Olio, V.M.N. Passaro, Phononic and photonic band gap structures: modelling and applications. Phys. Procedia **3**(1), 357–364 (2010)

G. Barbagallo, M.V. d'Agostino, R. Abreu, I.-D. Ghiba, A. Madeo, P. Neff, Transparent anisotropy for the relaxed micromorphic model: macroscopic consistency conditions and long wave length asymptotics International Journal of Solids and Structures, https://doi.org/10.1016/j.ijsolstr.2017.01.030

M.V. d'Agostino, G. Barbagallo, I.-D. Ghiba, R. Abreu, A. Madeo, P. Neff, A panorama of dispersion curves for the isotropic weighted relaxed micromorphic model ZAMM, DOI: https://doi.org/10.1002/zamm.201600227

A.C. Eringen, Mechanics of micromorphic materials, in *Applied Mechanics* (Springer, Berlin/Heidelberg, 1966), pp. 131–138

A.C. Eringen, *Microcontinuum Field Theories* (Springer, New York, 1999)

I.-D. Ghiba, P. Neff, A. Madeo, L. Placidi, G. Rosi, The relaxed linear micromorphic continuum: existence, uniqueness and continuous dependence in dynamics. Math. Mech. Solids **20**(10), 1171–1197 (2014)

A. Madeo, P. Neff, I.-D. Ghiba, L. Placidi, G. Rosi, Band gaps in the relaxed linear micromorphic continuum. Zeitschrift für Angewandte Mathematik und Mechanik **95**(9), 880–887 (2014)

A. Madeo, P. Neff, I.-D. Ghiba, L. Placidi, G. Rosi, Wave propagation in relaxed micromorphic continua: modeling metamaterials with frequency band-gaps. Contin. Mech. Thermodyn. **27**(4–5), 551–570 (2015)

A. Madeo, G. Barbagallo, M.V. d'Agostino, L. Placidi, P. Neff, First evidence of non-locality in real band-gap metamaterials: determining parameters in the relaxed micromorphic model. Proc. R. Soc. A Math. Phys. Eng. Sci. https://doi.org/10.1098/rspa.2016.0169 **472**(2190)

A. Madeo, P. Neff, E.C. Aifantis, G. Barbagallo, M.V. d'Agostino, On the role of micro-inertia in enriched continuum mechanics Proceedings of the Royal Society A, https://doi.org/10.1098/rspa.2016.0722

A. Madeo, P. Neff, M.V. d'Agostino, G. Barbagallo, Complete band gaps including non-local effects occur only in the relaxed micromorphic model. Comptes Rendus Mécanique, **344**(11–12), 784–796

A. Madeo, P. Neff, I.-D. Ghiba, G. Rosi, Reflection and transmission of elastic waves in non-local band-gap metamaterials: a comprehensive study via the relaxed micromorphic model. J. Mech. Phys. Solids **95**, 441–479 (2016d)

W. Man, M. Florescu, K. Matsuyama, P. Yadak, G. Nahal, S. Hashemizad, E. Williamson, P. Steinhardt, S. Torquato, P. Chaikin, Photonic band gap in isotropic hyperuniform disordered solids with low dielectric contrast. Opt. Express **21**(17), 19972–19981 (2013)

R.D. Mindlin, Microstructure in linear elasticity. Technical report, Office of Naval Research, 1963

R.D. Mindlin, Micro-structure in linear elasticity. Arch. Ration. Mech. Anal. **16**(1), 51–78 (1964)

P. Neff, I.-D. Ghiba, M. Lazar, A. Madeo, The relaxed linear micromorphic continuum: well-posedness of the static problem and relations to the gauge theory of dislocations. Q. J. Mech. Appl. Math. **68**(1), 53–84 (2014a)

P. Neff, I.-D. Ghiba, A. Madeo, L. Placidi, G. Rosi, A unifying perspective: the relaxed linear micromorphic continuum. Contin. Mech. Thermodyn. **26**(5), 639–681 (2014b)

K. Pham, V.G. Kouznetsova, M.G.D. Geers, Transient computational homogenization for heterogeneous materials under dynamic excitation. J. Mech. Phys. Solids **61**(11), 2125–2146 (2013)

A. Sridhar, V.G. Kouznetsova, M.G.D. Geers, Homogenization of locally resonant acoustic metamaterials towards an emergent enriched continuum. Comput. Mech. **57**(3), 423–435 (2016)

G. Stefano, M.S. Greene, W.K. Liu, Characterization of heterogeneous solids via wave methods in computational microelasticity. J. Mech. Phys. Solids **58**(5), 959–974 (2011)

W. Steurer, D. Sutter-Widmer, Photonic and phononic quasicrystals. J. Phys. D Appl. Phys. **40**(13), 229–247 (2007)

PGSTL